ELECTRICAL MEASUREMENTS and MEASURING INSTRUMENTS

ELECTRICAL MEASUREMENTS AND MEASURING INSTRUMENTS

For Engineering Students preparing for B.E./B.Tech.; AMIE-Section B (India); Diploma and Competitive Examinations

Er. R.K. RAJPUT

M.E. (Hons.) Gold Medallist; Grad.-Mech. Engg. & Elect. Engg.; MIE (India); MSESI; MISTE; CE (India)

Recipient of :

"Best Teacher (Academic) Award"

"Distinguished Author Award"

"Jawahar Lal Nehru Memorial Gold Medal"

for an Outstanding Research Paper

(Institution of Engineers—India)

***Principal** (Formerly)*

- *Thapar Polytechnic College;*
- *Punjab College of Information Technology;*

PATIALA

S. CHAND PUBLISHING

S Chand And Company Limited

(ISO 9001 Certified Company)

RAM NAGAR, NEW DELHI - 110 055

S Chand And Company Limited
(ISO 9001 Certified Company)
Head Office: 7361, RAM NAGAR, QUTAB ROAD, NEW DELHI - 110 055
Phone: 23672080-81-82, 66672000 Fax: 91-11-23677446
www.**schandpublishing.com**; e-mail: **info@schandpublishing.com**

Branches:

Ahmedabad	:	Ph: 27541965, 27542369, ahmedabad@schandpublishing.com
Bengaluru	:	Ph: 22268048, 22354008, bangalore@schandpublishing.com
Bhopal	:	Ph: 4209587, bhopal@schandpublishing.com
Chandigarh	:	Ph: 2625356, 2625546, 4025418, chandigarh@schandpublishing.com
Chennai	:	Ph: 28410027, 28410058, chennai@schandpublishing.com
Cuttack	:	Ph: 2332580, 2332581, cuttack@schandpublishing.com
Dehradun	:	Ph: 2711101, 2710861, dehradun@schandpublishing.com
Guwahati	:	Ph: 2738811, 2735640, guwahati@schandpublishing.com
Hyderabad	:	Ph: 27550194, 27550195, hyderabad@schandpublishing.com
Jaipur	:	Ph: 2219175, 2219176, jaipur@schandpublishing.com
Jalandhar	:	Ph: 2401630, jalandhar@schandpublishing.com
Kochi	:	Ph: 2809208, 2808207, cochin@schandpublishing.com
Kolkata	:	Ph: 23353914, 23357458, kolkata@schandpublishing.com
Lucknow	:	Ph: 4065646, lucknow@schandpublishing.com
Mumbai	:	Ph: 22690881, 22610885, 22610886, mumbai@schandpublishing.com
Nagpur	:	Ph: 2720523, 2777666, nagpur@schandpublishing.com
Patna	:	Ph: 2300489, 2260011, patna@schandpublishing.com
Pune	:	Ph: 64017298, pune@schandpublishing.com
Raipur	:	Ph: 2443142, raipur@schandpublishing.com (Marketing Office)
Ranchi	:	Ph: 2361178, ranchi@schandpublishing.com
Sahibabad	:	Ph: 2771235, 2771238, delhibr-sahibabad@schandpublishing.com

First Edition 2008
Reprints 2009, 2012
Second Edition 2013, Reprint with Corrections 2015
Reprint 2019

ISBN : 978-81-219-2963-9 Code : 1010B 353

PRINTED IN INDIA

By Vikas Publishing House Pvt. Ltd., Plot 20/4, Site-IV, Industrial Area Sahibabad, Ghaziabad-201010 and Published by S Chand And Company Limited, 7361, Ram Nagar, New Delhi-110 055.

Other Books by the Same Author

1. Electronic Measurements and Instrumentation
2. Electrical and Electronic Measurements and Instrumentation
3. Mechatronics

PREFACE TO THE SECOND EDITION

I take pleasure in presenting the **"Second Edition"** of this book on **"Electrical Measurements and Measuring Instruments"**. The warm reception which the previous edition has received is a matter of much satisfaction to me.

In this edition, the book has been thoroughly revised and a new chapter (No. 12) namely **"Universities' Questions (latest) with Solutions"** has been added to make the book still more useful to the readers.

Any suggestions for improvement of the book are most welcome.

Er. R.K. RAJPUT

PREFACE TO THE FIRST EDITION

This treatise on the subject "**Electrical Measurements and Measuring Instruments**" contains comprehensive treatment of the subject matter in simple, lucid and direct language. It covers the syllabi of the various Indian Universities in this subject exhaustively.

The book contains eleven chapters in all, namely :

1. Units, dimensions and standards; 2. Measurements and measurement systems; 3. Measuring instruments; 4. Instrument transformers; 5. Potentiometers; 6. Measurement of resistance, inductance and capacitance; 7. Magnetic measurements; 8. High voltage measurements; 9. Illumination; 10. Digital measurements of electrical quantities; 11. Cathode ray oscilloscope (CRO).

All these chapters are saturated with much needed text, supported by simple and self-explanatory figures, and worked examples, wherever required. At the end of each chapter *"Highlights"*, *"Objective Type Questions"*, *"Theoretical Questions"* and *"Unsolved Examples"* have been added to make the book a comprehensive and complete unit in all respects.

The author's thanks are due to his wife Ramesh Rajput for extending all cooperation during preparation and proof reading of the manuscript.

As even before, I take this opportunity to express my gratitude to the Managing Director and all the concerned staff members of S. Chand & Company Ltd. who were involved so dedicatedly in the publication of this book.

Any suggestions for improvement of this book will be thankfully acknowledged and incorporated in the next edition.

Er. R.K. RAJPUT

CONTENTS

Measurement of Energy

CHAPTER

1 Units, Dimensions and Standards

1.1. Units – Definition of unit – Absolute units – Fundamental and derived units. **1.2. Dimensions; 1.3. Systems of electric units; 1.4. S.I. units** – Introduction to S.I. units – Definitions – Salient features of S.I. units – Conversion factors – Dimensions of quantities. **1.5. Dimensions in electrostatic and electromagnetic systems; 1.6. Determination of absolute units** – Absolute measurement of current – Absolute measurement of resistance. **1.7. Standards of measurement; 1.8. Standard of mass and length; 1.9. Atomic frequency and time standards; 1.10. E.m.f. standards** – Primary standard of e.m.f. – Saturated weston cell – Secondary standard of e.m.f. – Unsaturated weston cell – Zener diode reference standard – Josephon–Effect standard. **1.11. Resistance standards; 1.12. Inductance standards** – Standard and sub-standard of mutual inductance (M) – Standard and sub-standard of self inductance (L). **1.13. Capacitance standards** – Primary standard of capacitance – Secondary capacitance standard – *Highlights – Objective Type Questions – Theoretical Questions – Unsolved Examples.*

1.1 UNITS

1.1.1. Definition of Unit

The standard measure of each kind of physical quantity is called a **unit.**

Measurement implies *comparison with a standard value*. The magnitude of a physical quantity can be written as :

Magnitude of a physical quantity = (Numerical ratio) × (Unit)

The ***numerical ratio*** (or *numerical multiplier*) is the number of times the unit occurs in any given amount of the same quantity and, therefore, is called the *number of measures*. For example, when it is said that length is 8 metres, it indicates that the metre is the unit of length and that the number of units of length (*i.e.*, numerical ratio) is 8. So the physical quantity, length is defined by the unit metre. The number of measures has no physical meaning without unit.

1.1.2. Absolute Units

A system in which the various units are all expressed in terms of a small number of fundamental units is called an **absolute system of units.**

- The word "absolute" does not imply any special standard of accuracy; it has been associated with relating units to units of *length*, *mass* and *time*.
- Absolute measurements do not compare the measured quantity with arbitrary units of the same type but are made in terms of *fundamental units*.

1.1.3. Fundamental and Derived Units

In science and engineering, two kinds of units are used : *fundamental units* and *derived units.*

- Experience has shown that *three* basic ideas or units are adequate to describe quantitatively all the phenomena encountered in mechanical science. These are *"length"*, *"mass"* and *"time"* and are called ***fundamental units***. Since these units are fundamental to most other physical quantities besides those in mechanics, they are called the *"primary fundamental units"*.

Measures of certain physical quantities in the thermal, electrical, and illumination disciplines are also represented by fundamental units. These units are used only when these particular disciplines are involved and they may therefore be called as *"auxiliary fundamental units"*.

- *All other units which can be expressed in terms of the fundamental units are called* ***derived units.***

— Every derived unit *originates from some physical law or equation defining that unit.*
— A derived unit is always recognised by its *dimension*, which can be defined as the complete algebraic formula for the derived unit.

1.2 DIMENSIONS

The unique quality of every quantity which distinguishes it from all other quantities is called **dimension.**

The dimensional symbols for fundamental units of length, mass and time are L, M and T respectively. The various powers of the fundamental units represent the dimensions of any derived unit. For example, the dimensional symbols for the *derived units* of volume, speed and acceleration are $[L]^3$, $[LT^{-}]$ and $[LT^{-2}]$ respectively. These dimensional formulae of the derived units are very *useful for converting units from one system to another.*

Some derived units, for convenience, have been given new names such as the derived unit of force in the S.I. system is called the newton (N) instead of dimensionally correct name kg m/s^2.

All other mechanical quantities can be expressed in terms of three fundamental quantities, *i.e.*, length (L), mass (M) and time (T); a few of them are given below :

Force : mass × acceleration *i.e.* $[F] = [M][LT^{-2}] = [MLT^{-2}]$

Work : force × distance *i.e.* $[W] = [MLT^{-2}][L] = [ML^2T^{-2}]$

Power : $\dfrac{\text{Work}}{\text{Time}}$ *i.e.* $[P] = \dfrac{[ML^2T^{-2}]}{[T]} = [ML^2T^{-3}]$

Such expressions indicate the nature of derived quantities in terms of fundamental dimensions $[L]$, $[M]$ and $[T]$ and are known as ***dimensional equations.***

The dimensional equations help in serving the following *purposes* :

(i) Conversion from one system of units to another one.
(ii) Derivation of equations for physical quantities.
(iii) Checking the accuracy of an equation.

1.3 SYSTEMS OF ELECTRICAL UNITS

The British Association Committee on Practical Standards for Electrical Measurements adopted the centimetre, gram, and second as the fundamental units of length, mass and time respectively. It was called **CGS system** *of units.*

- *Two* systems of CGS units existed ; these are :
  - (*i*) One having only the *permittivity* K of the medium as well as the units of length, mass and time. This system was called *electrostatic* CGS system of units, normally abbreviated as ***ESU system** of units.*
  - (*ii*) The other system involves *permeability* μ as well as units of length, mass and time. This system was called *electromagnetic* CGS system of units, abbreviated as ***EMU system** of units.*
- The system which uses metre, kilogram and second as the units of length, mass and time is called **MKS system** *of units.*
- The concept of unit flux issuing from unit magnetic pole or from unit charge instead of a flux of 4π is called *rationalisation*. The system using this concept is called **rationalised system** *of units.*

— Rationalisation means the removal of quantity π from the places, where its appearance is unnatural, irrational and geometrically unjustified to places where it properly belongs.

1.4 S.I. UNITS

1.4.1. Introduction to S.I. units

S.I., the *International System of units*, are divided into three classes :

1. Base units
2. Derived units
3. Supplementary units.

From the scientific point of view, division of S.I. units into these classes is to a certain extent arbitrary, because it is not essential to the physics of the subject. Nevertheless, the General Conference, considering the advantages of a single, practical, world-wide system for international relations, for teaching and for scientific work, decided to base the international system on a choice of six well-defined units given in Table 1.1 below :

Table 1.1: S.I. Base Units

Quantity	*Name*	*Symbol*
length	metre	m
mass	kilogram	kg
time	second	s
electric current	ampere	A
thermodynamic temperature	kelvin	K
luminous intensity	candela	cd
amount of substance	mole	mol

The second class of S.I. units contains derived units, *i.e.*, units which can be formed by combining base units according to the algebraic relations linking the corresponding quantities. Several of these algebraic expressions in terms of base units can be replaced by special names and symbols can themselves be used to form other derived units.

Derived units may, therefore, be classified under three headings. Some of them are given in Tables 1.2, 1.3 and 1.4.

Table 1.2: Examples of S.I. Derived Units Expressed in Terms of Base Units

Quantity	*S.I. Units*	
	Name	*Symbol*
area	square metre	m^2
volume	cubic metre	m^3
speed, velocity	metre per second	m/s
acceleration	metre per second squared	m/s^2
density, mass density	kilogram per cubic metre	kg/m^3
concentration (of amount of substances)	mole per cubic metre	mol/m^3
activity (radioactive)	1 per second	s^{-1}
specific volume	cubic metre per kilogram	m^3/kg
luminance	candela per square metre	cd/m^2

Table 1.3: S.I. Derived Units with Special Names

Quantity	*S.I. Units*			
	Name	*Symbol*	*Expression in terms of other units*	*Expression in terms of S.I. base units*
frequency	hertz	Hz	–	s^{-1}
force	newton	N	–	$m.kg.s^{-2}$
pressure	pascal	Pa	N/m^2	$m^{-1}.kg.s^{-2}$
energy, work, quantity of heat power	joule	J	N.m	$m^2.kg.s^{-2}$
radiant flux, quantity of electricity	watt	W	J/s	$m^2.kg.s^{-3}$
electric charge	coulomb	C	A.s	s.A
electric tension, electric potential	volt	V	W/A	$m^2.kg.s^{-3}.A^{-1}$
capacitance	farad	F	C/V	$m^{-2}.kg^{-1}.s^4$
electric resistance	ohm	Ω	V/A	$m^{-2}.kg.s^{-3}.A^{-2}$
conductance	siemens	S	A/V	$m^{-2}.kg^{-1}.s^3.A^{-2}$
magnetic flux	weber	Wb	V.S.	$m^2.kg.s^{-2}.A^{-1}$
magnetic flux density	tesla	T	Wb/m^2	$kg.s^{-2}.A^{-1}$
inductance	henry	H	Wb/A	$m^2.kg.s^{-2}.A^{-2}$
luminous flux	lumen	lm	–	cd.sr
illuminance	lux	lx	–	$m^{-2}.cd.sr$

Table 1.4: Examples of S.I. Derived Units Expressed by Means of Special Names

Quantity	S.I. Units		
	Name	*Symbol*	*Expression in terms of S.I. base units*
dynamic viscosity	pascal second	Pa.s	$m^{-1}.kg.s^{-1}$
moment of force	metre newton	N.m	$m^2.kg.s^{-2}$
surface tension	newton per metre	N/m	$kg.s^{-2}$
heat flux density, irradiance	watt per square metre	W/m^2	$kg.s^{-3}$
heat capacity, entropy	joule per kelvin	J/K	$m^2.kg.s^{-2}.K^{-1}$
specific heat capacity, specific entropy	joule per kilogram kelvin	J/(kg.K)	$m^2.s^{-2}.K^{-1}$
specific energy	joule per kilogram	J/kg	$m^2.s^{-2}$
thermal conductivity	watt per metre kelvin	W/(m.K)	$m.kg.s^{-3}.K^{-1}$
energy density	joule per cubic metre	J/m^3	$m^{-1}.kg.s^{-2}$
electric field strength	volt per metre	V/m	$m.kg.s^{-3}.A^{-1}$
electric charge density	coulomb per cubic metre	C/m^3	$m^{-3}.s.A$
electric flux density	coulomb per square metre	C/m^2	$m^{-2}.s.A$
permittivity	farad per metre	F/m	$m^{-3}.kg^{-1}.s^4$
current density	ampere per square metre	A/m^2	–
magnetic field strength	ampere per metre	A/m	–
permeability	henry per metre	H/m	$m.kg.s^{-2}.A^{-2}$
molar energy	joule per mole	J/mol	$m^2.kg.s^{-2}mol^{-1}$
molar heat capacity	joule per mole kelvin	J/(mol.K)	$m^2.kg.s^{-2}.K^{-1}.mol^{-1}$

The S.I. units assigned to third class called "Supplementary units" may be regarded either as base units or as derived units. Refer to Tables 1.5 and 1.6.

Table 1.5: S.I. Supplementary Units

Quantity	S.I. Units	
	Name	*Symbol*
plane angle	radian	rad
solid angle	steradian	sr

Table 1.6: Examples of S.I. Derived Units Formed by Using Supplementary Units

Quantity	S.I. Units	
	Name	*Symbol*
angular velocity	radian per second	rad/s
angular acceleration	radian per second squared	1 rad/s^2
radiant intensity	watt per steradian	W/sr
radiance	watt per square metre steradian	$W\text{-}m^{-2}.sr^{-1}$

Table 1.7: S.I. Prefixes

Factor	Prefix	Symbol	Factor	Prefix	Symbol
10^{12}	tera	T	10^{-1}	deci	d
10^{9}	giga	G	10^{-2}	centi	c
10^{6}	mega	M	10^{-3}	milli	m
10^{3}	kilo	k	10^{-6}	micro	μ
10^{2}	hecto	h	10^{-9}	nano	n
10^{1}	deca	da	10^{-12}	pico	p
			10^{-15}	fasnto	f
			10^{-18}	atto	a

1.4.2. Definitions

The S.I. *seven* base units and two supplementary units are defined below :

(*i*) **Metre.** The metre is the length equal to 1,650, 763.73 wavelengths in vacuum of the radiation corresponding to the transition between the levels $2p_{10}$ and $5d_5$ of the krypton-86 atom.

(*ii*) **Kilogram.** One kilogram is equal to the mass of the international prototype of the kilogram.

(*iii*) **Second.** The second is the duration of 9,192,631,770 periods of the radiation corresponding to the transition between the two hyperfine levels of the ground state of the cesium-133 atom.

(*iv*) **Ampere.** One ampere is that constant current which, if maintained in two straight parallel conductors of infinite length, of negligible circular cross-section and placed one metre apart in vacuum, would produce between these conductors a force equal to 2×10^{-7} newton per metre of length.

(*v*) **Kelvin.** The kelvin is the fraction $\frac{1}{273.16}$ of thermodynamic temperature of the triple point of water.

(*vi*) **Candela.** The candela is the luminous intensity, in the perpendicular direction, of a surface of a $\frac{1}{600,000}$ square metre of a black body at a temperature of freezing platinum under a pressure of 101, 325 newton per square metre.

(*vii*) **Mole.** The mole is the amount of substance of a system which contains as many elementary entities as there are atoms in 0.012 kg, carbon 12. When the mole is used, the elementary entities must be specified and may be atoms, molecules, ions, electrons, other particles or specified groups of such particles.

(*viii*) **Radian.** The radian is the plane angle between two radii of a circle that cut off on the circle an arc equal in length to the radius.

(*ix*) **Steradian.** The steradian is the solid angle which, having its vertex in the centre of a sphere, cuts off an area of the surface of the sphere equal to that of a square with sides of length equal to the radius of the sphere.

Some Other Definitions :

Newton. The newton (N) is a derived unit of force and is defined as the unit of force per second, which when acting on a mass of 1 kilogram gives it an acceleration of one metre per second per second. Since acceleration due to gravity equals 9.81 m/s^2, one kilogram force equals 9.81 newtons.

Joule. The joule (J) is a derived unit of energy, work or quantity of heat and is defined as the work done when a force of one newton acts so as to cause a displacement of one metre. Energy is defined as the capacity to do work. A unit of energy in nuclear physics is the electron volt (eV) which is defined as the energy gained by an electron in rising through a potential difference of one volt.

$$1 \text{ eV} = 1.6021 \times 10^{-19} \text{ J.}$$

Watt. The watt (W) is a unit of power (*i.e.*, rate of doing work)

$$\text{Power in watts} = \frac{\text{work (or energy) in joules}}{\text{time in seconds}}$$

Thus 1 watt equals 1 joule/sec.

1 kilowatt-hour (kWh) = 1000 watt-hours = 3600000 joules.

Coulomb. The coulomb (C) is the derived unit of charge. It is defined as *the quantity of electricity passing a given point in a circuit when a current of I A is maintained for 1 second.*

$$Q = I.t$$

where, Q = Charge in coulombs,

I = Current in amperes, and

t = Time in seconds.

1 coulomb represents 6.24×10^{18} electrons.

Ohm. The ohm (Ω) is the unit of electric resistance and is defined as *the resistance in which a constant current of I A generates heat at the rate of 1 watt.*

Siemen. The siemen is a unit of electric conductance (*i.e.*, reciprocal of resistance). If a circuit has a resistance of 5 ohms, its conductance is 0.2 siemen. A more commonly used name for siemen is *mho* (℧).

Volt. The volt is a unit of potential difference and electromotive force. It is defined as *the difference of potential across a resistance of 1 ohm carrying a current of 1 ampere.*

Hertz. The hertz (Hz) is a unit of frequency. 1 Hz = 1 cycle per second.

Horse-power. It is a practical unit of mechanical output. BHP (British horse power or brake horse power) equals 746 watts. The metric horse power equals 735.5 watts. To avoid confusion between BHP and metric horse power, the mechanical output of machines in S.I. units is expressed in watts or kilowatts.

1.4.3. Salient Features of S.I. Units

The salient features of S.I. units are as follows :

1. It is a coherent system of units, *i.e.*, product or quotient of any two base quantities results in a unit resultant quantity. For example, unit length divided by unit time gives unit velocity.
2. It is a rationalised system of units, applicable to both magnetism and electricity.
3. It is a non-gravitational system of units. It clearly distinguishes between the units of mass and weight (force) which are kilogram and newton respectively.
4. All the units of the system can be derived from the base and supplementary units.
5. The decimal relationship between units of same quantity makes possible to express any small or large quantity as a power of 10.
6. For any quantity there is one and only one S.I. unit. For example, joule is the unit of energy of all forms such as mechanical, heat, chemical, electrical and nuclear. However, kWh will also continue to be used as unit of electrical energy.

Advantages of S.I. Units :

1. Units for many different quantities are related through a series of simple and basic relationship.
2. Being an absolute system, it avoids the use of factor '*g*', *i.e.*, acceleration due to gravity in several expressions in physics and engineering which had been a nuisance in all numericals in physics and engineering.
3. Being a rationalised system, it ensures all the advantages of rationalised MKSA system in the fields of electricity, magnetism, electrical engineering and electronics.
4. Joule is the only sole unit of energy of all forms and watt is the sole unit of power hence a lot of labour is saved in calculations.
5. It is a coherent system of units and involves only decimal coefficients. Hence it is a very convenient and quick system for calculations.
6. In electricity, all the practical units like volt, ohm, ampere, henry, farad, coulomb, joule and watt accepted in industry and laboratories all over the world for well over a century have become absolute in their own right in the SI system, without the need for any more practical units.

Disadvantages :

1. The non-S.I. time units 'minute' and 'hour' will still continue to be used until the clocks and watches are all changed to kilo seconds and mega seconds etc.
2. The base unit kilogram (kg) includes a prefix, which creates an ambiguity in the use of multipliers with gram.
3. S.I. units for energy, power and pressure (*i.e.*, joule, watt and pascal) are too small to be expressed in science and technology, and, therefore, in such cases the use of larger units, such as MJ, kW, kPa, will have to be made.
4. There are difficulties with regard to developing new S.I. units for apparent and reactive energy while joule is the accepted unit for active energy in S.I. systems.

1.4.4. Conversion Factors

1. Force :

$$1 \text{ newton} = \text{kg.m/sec}^2 = 0.012 \text{ kgf}$$
$$1 \text{ kgf} = 9.81 \text{ N}$$

2. Pressure :

$$1 \text{ bar} = 750.05 \text{ mm Hg} = 0.9869 \text{ atm} = 10^5 \text{ N/m}^2 = 103 \text{ kg/m-sec}^2$$
$$1 \text{ N/m}^2 = 1 \text{ pascal} = 10^{-5} \text{ bar} = 10^{-2} \text{ kg/m-sec}^2$$
$$1 \text{ atm} = 760 \text{ mm Hg} = 1.03 \text{ kgf/cm}^2 = 1.01325 \text{ bar} = 1.01325 \times 10^5 \text{ N/m}^2$$

3. Work, Energy or Heat :

$$1 \text{ joule} = 1 \text{ newton metre} = 1 \text{ watt-sec} = 2.7778 \times 10^{-7} \text{ kWh} = 0.239 \text{ cal} = 0.239 \times 10^{-3} \text{ kcal}$$
$$1 \text{ cal} = 4.184 \text{ joule} = 1.1622 \times 10^{-6} \text{ kWh}$$
$$1 \text{ kcal} = 4.184 \times 10^3 \text{ joule} = 427 \text{ kgf m} = 1.1622 \times 10^{-3} \text{ kWh}$$
$$1 \text{ kWh} = 8.6042 \times 10^5 \text{ cal} = 860.42 \text{ kcal} = 3.6 \times 10^6 \text{ joules}$$
$$1 \text{ kgf-m} = \left(\frac{1}{427}\right) \text{ kcal} = 9.81 \text{ joules}$$

4. Power :

$$1 \text{ watt} = 1 \text{ joule/sec} = 0.86 \text{ kcal/h}$$
$$1 \text{ h.p.} = 75 \text{ m kgf/sec} = 0.1757 \text{ kcal/sec} = 735.3 \text{ watts}$$
$$1 \text{ kW} = 1000 \text{ watts} = 860 \text{ kcal/h}$$

5. Specific heat :

$$1 \text{ kcal/kg-°K} = 4184 \text{ joules/kg-K}$$

6. Thermal conductivity :

$$1 \text{ watt/m-K} = 0.8598 \text{ kcal/h-m-°C}$$
$$1 \text{ kcal/h-m-°C} = 1.16123 \text{ watt/m-K} = 1.16123 \text{ joules/s-m-K.}$$

7. Heat transfer coefficient :

$$1 \text{ watt/m}^2\text{-K} = 0.86 \text{ kcal/m}^2\text{-h-°C}$$
$$1 \text{ kcal/m}^2\text{-h-°C} = 1.163 \text{ watt/m}^2\text{-K.}$$

The following conversion factors may be used to convert the quantities in non-S.I. units into S.I. units.

To convert	*To*	*Multiply by*
angstroms	m	10^{-10}
atmospheres	kg/m^2	10332
bars	kg/m^2	1.02×10^4
Btu	joules	1054.8
Btu	kWh	2.928×10^{-4}
circular mils	m^2	5.067×10^{-10}
cubic feet	m^3	0.02831
dynes	newtons	10^{-5}
ergs	joules	10^{-7}
ergs	kWh	0.2778×10^{-13}
feet	m	0.3048
foot-pounds	joules	1.356
foot-pounds	kg-m	0.1383
gauss	tesla	10^{-4}
grams (force)	newton	9.807×10^{-3}
horse power (metric)	watts	735.5
lines/sq. inch	tesla	1.55×10^{-5}
Maxwell	webers	10^{-8}
mho	siemens	1
micron	metre	10^{-6}
miles	km	1.609
mils	cm	2.54×10^{-3}
poundals	newton	0.1383
pounds	kilogram	0.454
pounds (force)	newtons	0.448
pounds/sq. ft.	N/m^2	47.878
pounds/sq. inch	N/m^2	6894.43

1.4.5. Dimensions of Quantities

Different units can be represented dimensionally in terms of units of length *L*, mass *M*, time *T* and current *I*. The dimensions can be derived as under :

1. Velocity = length/time = $L/T = LT^{-1}$
2. Acceleration = velocity/time = $LT^{-1}/T = LT^{-2}$
3. Force = mass × acceleration = MLT^{-2}
4. Charge (coulomb) = current × time = IT
5. Work or energy = force × distance = ML^2T^{-2}
6. EMF or potential = work/charge $= ML^2T^{-2}/IT = ML^2I^{-1}T^{-3}$
7. Power = work/time = $ML^2T^{-2}/T = ML^2T^{-3}$
8. Current density = current/area = $I/L^2 = IL^{-2}$
9. Resistance = emf/current = $ML^2I^{-1}T^{-3}/I = ML^2I^{-2}T^{-3}$
10. Electric flux density = electric flux or charge/area = $IT/L^2 = ITL^{-2}$
11. MMF = current × number of turns = I
12. Conductance = 1/resistance = $1/ML^2I^{-2}T^{-3} = I^2T^3M^{-1}L^{-2}$
13. Electric field intensity = volt/metre = $ML^2I^{-1}T^{-3}/L = MLI^{-1}T^{-3}$
14. Resistivity $= \dfrac{\text{resistant} \times \text{area}}{\text{length}} = (ML^2I^{-2}T^{-3})(L^2)/L = ML^3I^{-2}T^{-3}$
15. Magnetic field intensity (*H*) = MMF/length $= I/L = IL^{-1}$
16. Magnetic flux = emf × time = $(ML^2I^{-1}T^{-3})(T) = ML^2I^{-1}T^{-2}$
17. Magnetic flux intensity= magnetic flux/area $= (ML^2I^{-1}T^{-2})/L^2 = MI^{-1}T^{-2}$
18. Impedance = emf/current = $ML^2I^{-2}T^{-3}$
19. Admittance = I/impedance = $I^2T^3M^{-1}L^{-2}$
20. Inductance = magnetic flux/current $= ML^2T^{-2}I^{-1}/I = ML^2T^{-2}I^{-2}$
21. Capacitance = electric charge/potential $= IT/ML^2T^{-3}I^{-1} = M^{-1}L^{-2}T^4I^2$

1.5 DIMENSIONS IN ELECTROSTATIC AND ELECTROMAGNETIC SYSTEMS

1.5.1. Dimensions in Electrostatic System

In an electrostatic system the dimensions of permittivity $\in$ is taken as the *fourth* fundamental dimension.

1. **Charge.** The force exerted between two charges Q_1 and Q_2 according to Coulomb's law is given as:

$$F = \frac{Q_1Q_2}{\in d^2}$$

where, d = Distance between charges Q_1 and Q_2.
Inserting the dimensions of force and distance, we have

$$[MLT^{-2}] = \frac{[Q^2]}{[\in][L^2]}$$

$\therefore$ Dimensions of charge $[Q] = [\in^{1/2}\, M^{1/2}\, L^{3/2}\, T^{-1}]$ ...(1.1)

2. **Current.** Current is *charge per unit time.* Let (I) represent the dimensions of current.

$$[I] = \frac{[Q]}{[T]} = \frac{[\in^{1/2}\, M^{1/2} L^{3/2} T^{-1}]}{[T]} = [\in^{1/2}\, M^{1/2} L^{3/2} T^{-2}] \quad ...(1.2)$$

3. **Potential difference or e.m.f.** Potential difference is *work done per unit charge*

$$[E] = \frac{[W]}{[Q]} = \frac{[ML^2T^{-2}]}{[\in^{1/2}\, M^{1/2} L^{3/2} T^{-1}]} = [\in^{-1/2}\, M^{1/2} L^{1/2} T^{-1}] \quad ...(1.3)$$

4. **Capacitance.** Capacitance is the *ratio of charge to e.m.f. i.e.* $C = \frac{Q}{E}$

$\therefore$ The dimensions of capacitance are :

$$[C] = \frac{[Q]}{[E]} = \frac{[\in^{1/2}\, M^{1/2} L^{3/2} T^{-1}]}{[\in^{-1/2}\, M^{1/2} L^{1/2} T^{-1}]} = [\in L] \quad ...(1.4)$$

5. **Resistance.** Resistance is the *ratio of e.m.f. to current, i.e.,* $R = \frac{E}{I}$

$\therefore$ Dimensions of resistance are :

$$[R] = \frac{[E]}{[I]} = \frac{[\in^{-1/2}\, M^{1/2} L^{1/2} T^{-1}]}{[\in^{1/2}\, M^{1/2} L^{3/2} T^{-2}]} = [\in^{-1}\, L^{-1} T] \quad ...(1.5)$$

6. **Inductance.** Inductance, $L = \frac{\text{emf}}{\text{rate of change of current}} = \frac{E}{dI/dt}$

$\therefore$ Dimensions of inductance are :

$$[L] = \frac{[E]}{[I]/[T]} = \frac{[E][T]}{[I]} = \frac{[\in^{-1/2}\, M^{1/2} L^{1/2} T^{-1}][T]}{[\in^{1/2}\, M^{1/2} L^{3/2} T^{-2}]}$$

$$= [\in^{-1}\, L^{-1} T^2] \quad ...(1.6)$$

1.5.2. Dimensions in Electromagnetic System

In an electromagnetic system the *fourth* dimension is the dimension of permeability μ.

1. **Pole strength.** Force, $F = \frac{m_1 m_2}{\mu d^2}$

where, d = Distance between poles of strengths m_1 and m_2.
Writing the dimensions, we have :

$$[MLT^{-2}] = \frac{[m^2]}{[\mu][L^2]}$$

Thus, dimensions of pole strength are, $[m] = [\mu^{1/2}M^{1/2}L^{3/2}T^{-1}]$ ...(1.7)

2. **Magnetising force.** This force H is measured by force exerted on a unit pole.

∴ The dimensions of magnetising force are :

$$[H] = \frac{[F]}{[m]} = \frac{[MLT^{-2}]}{[\mu^{1/2}M^{1/2}L^{3/2}T^{-1}]} = [\mu^{-1/2}M^{1/2}L^{-1/2}T^{-1}] \quad ...(1.8)$$

3. **Current.** The magnetising force at the centre of a loop of radius r is given as :

$$H = \frac{2\pi I}{r} \text{ if } I \text{ is in e.m. units}$$

or,

$$[H] = \frac{[I]}{[L]}$$

∴ Dimensions of current are, $[I] = [H][L] = [\mu^{-1/2}M^{1/2}L^{1/2}T^{-1}]$ ...(1.9)

4. **Charge.** Charge is the *product of current and time, i.e.,* $Q = I \times t$

∴ The dimensions of charge are :

$$[Q] = [I][T] = [\mu^{-1/2}M^{1/2}L^{1/2}T^{-1}][T] = [\mu^{-1/2}M^{1/2}L^{1/2}] \quad ...(1.10)$$

5. **Potential difference.** It is the *work done per unit charge.*

The dimensions of potential difference are :

$$[E] = \frac{[W]}{[Q]} = \frac{[ML^2T^{-2}]}{[\mu^{-1}M^{1/2}L^{1/2}]} = [\mu^{1/2}M^{1/2}L^{3/2}T^{-2}] \quad ...(1.11)$$

6. **Capacitance.** The dimensions of capacitance are :

$$[C] = \frac{[Q]}{[E]} = \frac{[\mu^{-1/2}M^{1/2}L^{1/2}]}{[\mu^{1/2}M^{1/2}L^{3/2}T^{-2}]} = [\mu^{-1}L^{-1}T^2] \quad ...(1.12)$$

7. **Resistance.** The dimensions of resistance are :

$$[R] = \frac{[E]}{[I]} = \frac{[\mu^{1/2}M^{1/2}L^{3/2}T^{-2}]}{[\mu^{-1/2}M^{1/2}L^{1/2}T^{-1}]} = [\mu LT^{-1}] \quad ...(1.13)$$

8. **Inductance.** Inductance $L = \dfrac{E}{dI/dt}$

∴ The dimensions of inductance are :

$$[L] = \frac{[E]}{[I]/[T]} = \frac{[E][T]}{[I]} = \frac{[\mu^{1/2}M^{1/2}L^{3/2}T^{-2}][T]}{[\mu^{-1/2}M^{1/2}L^{1/2}T^{-1}]} = [\mu L] \quad ...(1.14)$$

Similarly the dimensions of other quantities can be found out.

Relationship Between Electrostatic and Electromagnetic Systems of Units :

The dimensions of charge in the above systems of units is given as :

$$[Q] = [\epsilon^{1/2}M^{1/2}L^{3/2}T^{-1}] \quad \text{e.s.u. system}$$

$$[Q] = [\mu^{-1/2}M^{1/2}L^{1/2}] \quad \text{e.m.u. system}$$

Equating the two dimensions, we get

$$\in^{1/2} M^{1/2}L^{3/2}T^{-1} = \mu^{-1/2}M^{1/2}L^{1/2}$$

or, $$[LT^{-1}] = [\mu^{-1/2} \in^{-1/2}]$$

But LT^{-1} are the dimensions of velocity,

$$\therefore \quad \frac{1}{\sqrt{\mu \in}} = \text{a velocity} \quad ...(1.15)$$

If we equate the dimensional equations of other quantities, we shall again get the same result, *i.e.*, some power of $\sqrt{\mu \in}$ is always involved. It should be understood that it is not possible to obtain separate expressions for μ and $\in$.

The permeability of free space μ_0 and permittivity of free space $\in_0$, in any system of units, are related by relation :

$$\mu_0 \in_0 = \frac{1}{v^2} \quad ...(1.16)$$

Where v is the velocity of light in system of units considered.

Eqn. (1.16) can be used to convert the dimensions of any electrical quantity from those of e.s.u. system to e.m.u. system, and vice-versa.

Example 1.1. *Derive the dimensions of the following quantities in L, M, T, I :*
(i) E.m.f. ; (ii) Magnetic flux density ; (iii) Electric flux density ; (iv) Current density ; (v) Permeability ; (vi) Permittivity ; (vii) Resistivity ; (viii) Conductivity.

Solution. (*i*) **E.m.f. (E) :**

$$\text{E.m.f.} = \frac{\text{Work done}}{\text{Charge}}$$

$$[E] = \frac{[ML^2T^{-2}]}{[IT]} = [ML^2T^{-3}I^{-1}] \quad \textbf{(Ans.)}$$

(*ii*) **Magnetic flux density (B) :**

$$\text{Magnetic flux density,} \quad B = \frac{\phi\,[\text{Flux}]}{A\,(\text{Area})}$$

$$\text{E.m.f.} = \text{Rate of change of flux linkages} = N\frac{d\phi}{dt}$$

or, $$\text{Flux } (\phi) = [\text{e.m.f.}][\text{time}]$$

$$\therefore \quad [B] = \frac{[\phi]}{[A]} = \frac{[\text{e.m.f.}][\text{time}]}{[\text{area}]} = \frac{[ML^2T^{-3}I^{-1}][T]}{L^2} = [MT^{-2}I^{-1}] \quad \textbf{(Ans.)}$$

(*iii*) **Electric flux density (D) :**

$$\text{Electric flux density,} \quad D = \frac{\text{Electric flux}}{\text{Area}} = \frac{\text{Charge}}{\text{Area}}$$

$$\therefore \quad [D] = \frac{[IT]}{[L^2]} = [L^{-2}TI] \quad \textbf{(Ans.)}$$

(iv) **Current density (J) :**

Current density, $J = \dfrac{\text{Current}}{\text{Area}}$

$\therefore \quad [J] = \dfrac{[I]}{[L^2]} = \mathbf{[L^{-2}I]}$ **(Ans.)**

(v) **Permeability (μ) :**

$$\text{Flux} = \text{Flux density} \times \text{area}$$

or, $[\phi] = [MT^{-2}I^{-1}]\,[L^2] = [ML^2T^{-2}I^{-1}]$

$$\text{M.m.f.} = \text{Turns} \times \text{current} = [I]$$

$$\text{Reluctance (S)} = \frac{\text{M.m.f.}}{\text{Flux}} \quad \therefore \quad [S] = \frac{[I]}{ML^2T^{-2}I^{-1}} = M^{-1}L^{-2}T^2I^2$$

Also, $\text{Reluctance} = \dfrac{\text{Length}}{\text{Permeability} \times \text{area}}$

$\therefore$ Permeability, $\mu = \dfrac{\text{Length}}{\text{Reluctance} \times \text{area}}$

or, $[\mu] = \dfrac{[L]}{[M^{-1}L^{-2}T^2I^2][L^2]} = \mathbf{[MLT^{-2}I^{-2}]}$ **(Ans.)**

(vi) **Permittivity (∈) :**

$$\text{Force} = \frac{Q_1Q_2}{\in d^2}, \quad \text{where } \in = \text{permittivity}$$

$$\therefore \quad \in = \frac{[\text{Charge}]^2}{[\text{Force}][\text{Distance}]^2}$$

$$= \frac{[\text{Current}]^2[\text{Time}]^2}{[\text{Force}][\text{Distance}]^2} = \frac{I^2T^2}{[MLT^{-2}][L^2]} = \mathbf{[M^{-1}L^{-3}T^4I^2]} \text{ (Ans.)}$$

(vii) **Resistivity (ρ) :**

Resistance $R = \dfrac{\text{E.m.f.}}{\text{Current}}$

or, $[R] = \dfrac{[ML^2T^{-3}I^{-1}]}{[I]} = [ML^2T^{-3}I^{-2}]$

Also, $\text{Resistance} = \dfrac{\text{Resistivity} \times \text{length}}{\text{Area}}$

$\therefore \quad [\text{Resistivity}] = \dfrac{[\text{Resistivity}][\text{area}]}{[\text{Length}]}$

or, $[\rho] = \dfrac{[ML^2T^{-3}I^{-2}]}{[L]} = \mathbf{[ML^3T^{-3}I^{-2}]}$ **(Ans.)**

(viii) **Conductivity (σ) :**

$$\text{Conductivity} = \frac{I}{\text{Resistivity}}$$

or, $[\sigma] = [\rho^{-1}] = \mathbf{[M^{-1}L^{-3}T^3I^2]}$ **(Ans.)**

Example 1.2. *Show that $\omega^2 LC$ is non-dimensional, ω being the angular frequency of the applied voltage.*

Solution. We know that dimensions of ω, L (inductance) and C (capacitance) are as under :

$$[\omega] = [T^{-1}];$$
$$[L] = ML^2T^{-2}I^{-2}]$$
$$(C) = [M^{-1}L^{-2}T^4I^2]$$

Now, $$[\omega^2 LC] = [\omega^2][L][C]$$
$$= [T^{-2}][ML^2T^{-2}I^{-2}][M^{-1}L^{-2}T^4I^2] = M^0L^0T^0I^0$$

Hence, $\omega^2 LC$ *is a non-dimensional quantity* - - - - -**Proved.**

Example 1.3. *The energy stored in a parallel plate capacitor per unit volume (energy density) is given by :*

$$w = K \in^a V^b d^c$$

where, K = *A constant,*

$\in$ = *Permittivity of medium,*

V = *Voltage between plates, and*

d = *Distance between plates.*

Determine the values of a, b, and c. **(Roorkee University)**

Solution. Dimensional equation $w = K\in^a V^b d^c$ *(Given)* ...(*i*)

Writing the dimensions of various quantities in LMTI system, we have :

Energy density, $$w = \frac{\text{Energy}}{\text{Volume}}$$

or, $$[w] = \frac{[ML^2T^{-2}]}{[L^3]} = [ML^{-1}T^{-2}]$$

Also, Force, $$F = \frac{Q_1Q_2}{\in d^2}$$

$\therefore$ Permittivity, $$\in = \frac{Q_1Q_2}{Fd^2}$$

or, $$[\in] = \frac{[I^2T^2]}{[MLT^{-2}][L^2]} = [M^{-1}L^{-3}T^4I^2]$$

and $$[\text{Voltage}] = [V] = [ML^2T^{-3}I^{-1}], \qquad d = [L]$$

Substituting the dimensions of various quantities in expression (*i*), we get :

$$[ML^{-1}T^{-2}] = [M^{-1}L^{-3}T^4I^{-2}]^a[ML^2T^{-3}I^{-1}]^b[L]^c$$

or $$[ML^{-1}T^{-2}] = [M]^{-a+b}\,[L]^{-3a+2b+c}\,[T]^{4a-3b}\,[I]^{2a-b}$$

Equating the components of M, L, T and I respectively, we have :

For M : $1 = -a + b$ ...(*ii*)

For L : $-1 = -3a + 2b + c$ ...(*iii*)

For T : $-2 = 4a - 3b$ ...(*iv*)

For I : $0 = 2a - b$ ...(*v*)

Solving expressions (*ii*), (*iii*), (*iv*) and (*v*), we get

$$a = 1, b = 2, c = -2$$

Hence, energy stored per unit volume,

$$w = K\epsilon V^2 d^{-2} = \frac{K\epsilon V^2}{d^2} \quad \textbf{(Ans.)}$$

Example 1.4. *The expression for mean torque T of an electrodynamic wattmeter is written in the form*

$$T \propto M^a E^b Z^c$$

where M is mutual inductance between fixed and moving coils, E-applied voltage, Z-impedance of coil.

From first principles determine the values of constants a, b and c.

(Punjab University)

Solution. $T \propto M^a E^b Z^c$ ...(*Given*)

or, $T = KM^a E^b Z^c$ ...(*i*)

where *K* is a dimensionless number.

Writing the dimensions of various quantities in *LMTI* system, we have :

$$[T] = [ML^2T^{-2}]$$
$$[M] = [ML^2T^{-2}I^{-2}]$$
$$[E] = [ML^2T^{-3}I^{-1}]$$
$$[Z] = [ML^2T^{-3}I^{-2}]$$

Substituting the dimensions of quantities involved in expression (*i*), we get :

$$[ML^2T^{-2}] = [ML^2T^{-2}I^{-2}]^a[ML^2T^{-3}I^{-1}]^b[ML^2T^{-3}I^{-2}]^c$$

or, $$[ML^2T^{-2}] = [M]^{a+b+c}\ [L]^{2a+2b+2c}\ [T]^{-2a-3b-3c}\ [I]^{-2a-b-2c}$$

Equating the components of *M*, *L*, *T* and *I* respectively, we get :

For M : $1 = a + b + c$ ...(*ii*)

For L : $2 = 2a + 2b + 2c$ ...(*iii*)

For T : $-2 = -2a - 3b - 3c$ ...(*iv*)

For I : $0 = -2a - b - 2c$ ...(*v*)

Solving above expressions, we get :

$$\boldsymbol{a = 1, b = 2,} \text{ and } \boldsymbol{c = -2} \quad \textbf{(Ans.)}$$

Example 1.5. *The following expression was found for eddy currents produced in a metallic former moving in the field of a permanent magnet :*

$$I_e = \frac{KBlbA}{(2b + l)\rho}$$

where, K = *A constant,*
B = *Flux density,*
l = *Length of former,*
b = *Width of former,*
A = *Area of former, and*
ρ = *Resistivity of conducting former.*

It is suspected that the term angular frequency is missing. Use LMTI system for dimensional analysis to detect the error.

Solution. The dimensions of L.H.S. = $[I]$

The dimensions of various quantities of RHS are :

$$[B] = [MT^{-2}I^{-1}]$$
$$[l] = [L]$$
$$[b] = [L]$$
$$[A] = [L^2]$$
$$[\rho] = [ML^3T^{-3}I^{-2}]$$

After substituting the dimensions of various quantities, we have :

$$\text{R.H.S.} = \frac{[MT^{-2}I^{-1}][L][L][L^2]}{[L][ML^3T^{-3}I^{-2}]} = [TI]$$

For making the two sides *dimensionally correct, the R.H.S. must be multiplied by a term having a dimension* T^{-1}, *i.e.,* ω.

∴ The correct expression is,

$$I_e = \frac{BlbA\omega}{(2b + l)\rho} \quad \textbf{(Ans.)}$$

Example 1.6. *Determine the indices in the following expression for eddy current loss W per cm of wire having circular cross-section :*

$$W = f^a B_{max}^b d^c \rho^g$$

where, f = *Frequency of the supply in Hz,*
B_{max} = *Maximum value of flux density,*
d = *Diameter of the wire,*
ρ = *Resistivity of wire material.* **(AMIE)**

Solution.

Dimensionally we can write : $W = f^a B_{\text{max}}^b d^c \rho^g$ ...(*Given*)

$$[V] = \frac{[F][d]}{[Q]} = \frac{[MLT^{-2}][L]}{[IT]} = [ML^2T^{-3}I^{-1}]$$

$$[W] = \frac{[V][I]}{[L]} = \frac{[ML^2T^{-3}I^{-1}][I]}{[L]} = [MLT^{-3}]$$

$$[f] = [T^{-1}]$$

$$[B_{\text{max}}] = \frac{[\phi]}{[A]} = \frac{[V][T]}{[A]} = \frac{[ML^2T^{-3}I^{-1}][T]}{[L^2]} = [MT^{-2}I^{-1}]$$

$$[d] = [L]$$

$$[\rho] = \frac{[R][A]}{[L]} = \frac{[V/I][A]}{[L]} = \frac{[ML^2T^{-3}I^{-1}][L^2]}{[I][L]} = [ML^3T^{-3}I^{-2}]$$

Substituting the dimensions of the quantities involved in the given expression, we get :

$$[MLT^{-3}] = [T^{-1}]^a\ [MT^{-2}I^{-1}]^b\ [L]^c\ [ML^3T^{-3}I^{-2}]^g$$

or, $$[MLT^{-3}] = [M]^{b+g}\ [L]^{c+3g}\ [T]^{-a-2b-3g}\ [I]^{-b-2g}$$

Equating the indices of M, L, T and I respectively, we get :

For M : $1 = b + g$ ...(*i*)

For L : $1 = c + 3g$ ...(*ii*)

For T : $-3 = -a - 2b - 3g$ ...(*iii*)

For I : $0 = -b - 2g$ ...(*iv*)

Solving the above expressions, we have :

$$a = 2\ ;\ b = 2\ ;\ c = 4\ ;\ g = -1 \quad \textbf{(Ans.)}$$

Hence the expression is : $W = f^2 B_{max}^2 d^4 \rho^{-1}$ **(Ans.)**

Example 1.7. *The following expression is obtained in deriving expression for resistance in Hay's bridge :*

$$R = \frac{\omega^2 R_1 R_2 R_3 C^2}{1+\omega^2 C R_2^2}$$

where R, R_1, R_2, R_3 are resistances; C, a capacitance and $\omega = 2\pi f$, where f is frequency, Hz

Determine whether the expression is dimensionally correct or incorrect. In case there is an error, point out the term which needs correction. **[Delhi University]**

Solution. $$R = \frac{\omega^2 R_1 R_2 R_3 C^2}{1+\omega^2 C R_2^2} \quad \text{...(Given)}$$

The dimensions of various terms in e.m. system are :

$$[R] = [\mu L T^{-1}]$$
$$[C] = [\mu^{-1} L^{-1} T^2]$$
$$[\omega] = [T^{-1}]$$

The dimensions of *L.H.S.* are :

$$[R] = [\mu L T^{-1}]$$

The dimensions of *R.H.S.* are :

$$\frac{[T^{-1}]^2[\mu LT^{-1}][\mu LT^{-1}][\mu LT^{-1}][\mu^{-1}L^{-1}T^2]^2}{1+[T^{-1}]^2\,[\mu^{-1}L^{-1}T^2][\mu LT^{-1}]^2} = \frac{[\mu LT^{-1}]}{1+[\mu LT^{-2}]}$$

From the above expressions we find that the dimensions of left hand and right hand terms do not balance and, therefore, the equation is dimensionally *incorrect*.

The dimensions of left hand terms are $[\mu LT^{-1}]$. The dimensions of the numerator of right hand terms are $[\mu LT^{-1}]$ and in the denominator we have two terms, *i.e.*, 1 and $\omega^2 C R_2^2$. Now 1 is a dimensionless quantity and, therefore, in order that the dimensions of left hand term balance those of the right hand term, the denominator of right hand term should also be *dimensionless* quantity. The dimensions of second term in the denominator are $[\mu LT^{-2}]$ and in order that it may be a dimensionless quantity it should be multiplied with a term having dimensions $[\mu^{-1}L^{-1}T^2]$ which are the *dimensions of capacitance*. Therefore, the second term in the denominator should contain another capacitance term or this term should be $(\omega^2 C R_2^2 \times C)$ or $\omega^2 C^2 R_2^2$. Therefore, the correct equation is :

$$R = \frac{\omega^2 R_1 R_2 R_3 C^2}{1+\omega^2 C^2 R_2^2} \quad \textbf{(Ans.)}$$

Example 1.8. *Check the following equation dimensionally :*

$$E = \frac{2Ir\omega^2 C\,(L + C^2 r^2)}{(1-\omega^2 LC)^2 - 4\omega^2 C^2 r^2}$$

where ω is the angular frequency of resonance. **[Punjab University]**

Solution.
$$E = \frac{2I r \omega^2 C\,(L + C^2 r^2)}{(1 - \omega^2 LC)^2 - 4\omega^2 C^2 r^2} \quad ...(Given)$$

The dimensions of various terms in M, L, T, Q system are :

$$[E] = [ML^2T^{-2}Q^{-1}]$$
$$[r] = [ML^2T^{-1}Q^{-2}]$$
$$[\omega] = [T^{-1}]$$
$$[L] = [ML^2Q^{-2}]$$
$$[C] = [M^{-1}L^{-2}T^2Q^2]$$
$$[I] = [T^{-1}Q]$$

The dimensions of L.H.S. are :

$$[E] = [ML^2T^{-2}Q^{-1}]$$

The dimensions of R.H.S. are :

$$\left[\frac{2I r \omega^2 C(L + C^2 r^2)}{(1 - \omega^2 LC)^2 - 4\omega^2 C^2 r^2}\right] =$$

$$\frac{\{[T^{-1}Q][ML^2T^{-1}Q^{-2}][T^{-2}][M^{-1}L^{-2}T^2Q^2]\} \times \{[ML^2Q^{-2}] + [M^{-2}L^{-4}T^4Q^4][M^2L^4T^{-2}Q^{-4}]\}}{\{[L^0M^0T^0Q^0] - [T^{-2}][ML^2Q^{-2}][M^{-1}L^{-2}T^2Q^2]\} - [T^{-1}]^2[M^{-2}L^{-4}T^4Q^4][M^2L^4T^{-2}Q^{-4}]}$$
$$= [T^{-2}Q][ML^2Q^{-2} + T^2]$$

This clearly shows that if the dimensions of left hand and right hand terms are to be dimensionally equal, then the terms in the second bracket of R.H.S. should be of dimensions ML^2Q^{-2}, which is possible if the expression is :

$$E = \frac{(2I r \omega^2 C)(L + C r^2)}{(1 - \omega^2 LC)^2 - 4\omega^2 C^2 r^2}$$

In this the dimensions of R.H.S. terms will be

$$= [T^{-2}Q]\ \{[ML^2Q^{-2}] + [M^{-1}L^{-2}T^2Q^2][M^2L^4T^{-2}Q^{-4}]$$
$$= [T^{-2}Q]\ \{[ML^2Q^{-2}] + [ML^2Q^{-2}]\}$$
$$= [T^{-2}Q][ML^2Q^{-2}]$$
$$= [ML^2T^{-2}Q^{-1}] = \text{Dimensions of L.H.S.}$$

Hence, *correct expression* must be,

$$E = \frac{2I r \omega^2 C\,(L + C r^2)}{(1 - \omega^2 LC)^2 - 4\omega^2 C^2 r^2} \quad \textbf{(Ans.)}$$

Example 1.9. *An expression of the following form was arrived at while working for a derivation:*

$$I = E\left\{\frac{1}{Z_1} + \frac{j\omega M}{Z_2}\left(\frac{1}{R} + \frac{C}{L}\right)\right\}$$

Show that there must be an algebraic error, and point out the term or terms which require correction.

Solution.
$$I = E\left\{\frac{1}{Z_1} + \frac{j\omega M}{Z_2}\left(\frac{1}{R} + \frac{C}{L}\right)\right\} \quad ...(Given)$$

The dimensions of various quantities in e.m. system are :

$$[I] = [\mu^{-1/2}\, L^{1/2}\, M^{1/2}\, T^{-1}]$$

$$[E] = [\mu^{1/2} L^{3/2} M^{1/2} T^{-2}]$$
$$[Z] = [\mu L T^{-1}]$$
$$[M] = [\mu L]$$
$$[C] = [\mu^{-1} L^{-1} T^2]$$
$$[R] = [\mu L T^{-1}]$$
$$[L] = [\mu L]$$
$$[\omega] = [T^{-1}]$$

The dimensions of L.H.S. are :

$$[I] = [\mu^{-1/2} L^{1/2} M^{1/2} T^{-1}]$$

The dimensions of R.H.S. are :

$$\left[E\left\{\frac{1}{Z_1} + \frac{j\omega M}{Z_2}\left(\frac{1}{R} + \frac{C}{L}\right)\right\}\right] = [\mu^{1/2} L^{3/2} M^{1/2} T^{-2}] \times \left\{\frac{1}{[\mu L T^{-1}]} + \frac{[T^{-1}][\mu L]}{[\mu L T^{-1}]}\left(\frac{1}{[\mu L T^{-1}]} + \frac{[\mu^{-1} L^{-1} T^2]}{[\mu L]}\right)\right\}$$
$$= [\mu^{1/2} L^{3/2} M^{1/2} T^{-2}] \times \{[\mu^{-1} L^{-1} T] + [\mu^{-1} L^{-1} T] + [\mu^{-2} L^{-2} T^2]\}$$
$$= \{[\mu^{-1/2} L^{1/2} M^{1/2} T^{-1}] + [\mu^{-1/2} L^{1/2} M^{1/2} T^{-1}] + [\mu^{-3/2} L^{-1/2} M^{1/2} T^0]\}$$

For the two sides to balance dimensionally, the dimensions of R.H.S. should be $[\mu^{-1/2} L^{1/2} M^{1/2} T^{-1}]$. We find that the dimensions of all the terms on the R.H.S. are not $[\mu^{-1/2} L^{1/2} M^{1/2} T^{-1}]$ and *hence there is an algebraic error.* This error is due to the last term having dimensions of $[\mu^{-3/2} L^{-1/2} M^{1/2} T^0]$. The error can be eliminated if it has the dimensions $[\mu^{-1/2} L^{-1/2} M^{1/2} T^{-1}]$. In order to achieve this we must multiply the last term by a quantity which has the dimensions

$$\frac{[\mu^{-1/2} L^{1/2} M^{1/2} T^{-1}]}{[\mu^{-3/2} L^{-1/2} M^{1/2} T^0]} = [\mathbf{\mu L T^{-1}}]$$

But $[\mu L T^{-1}]$ are the dimensions of resistance or impedance. Hence, the *last term of the R.H.S.,* $\frac{C}{L}$*, should be multiplied by a term having the dimensions of either resistance or impedance.* **(Ans.)**

Example 1.10. *The following expression was obtained while deriving the theory of Biff-circuit for testing of current transformer:*

$$\frac{I_p}{I_s} = \frac{R_2[(1+\omega^2 C^2 r_1^2)\{R_1^2(1+\omega^2 C^2 r_1^2) + 2R_1 r_1 + r_1^2\}]^{1/2}}{R_1(R_2 + r)(1+\omega^2 C^2 r_1^2)}$$

Examine the term dimensionally and suggest the correction, if any necessary.

[Bangalore University]

Solution. Writing the terms dimensionally in the S.I. units, we get :

$$[I_p] = [I_s] = [I]$$
$$[\omega] = [T^{-1}]$$
$$[R_1] = [R_2] = [r_1] = \frac{[V]}{[I]} = \frac{[ML^2T^{-3}I^{-1}]}{[I]} = [ML^2T^{-3}I^{-2}]$$
$$[C] = \frac{[I]}{[\omega][V]} = \frac{[I]}{[T^{-1}][ML^2T^{-3}I^{-1}]} = [M^{-1}L^{-2}T^4I^2]$$

The dimensions of L.H.S. are :

$$\frac{[I]}{[I]} = \text{Dimensionless quantity}$$

The dimensions of R.H.S. are :

$$\frac{R_2\,[(1+\omega^2C^2r_1^2)\{R_1^2\,(1+\omega^2C^2r_1^2)+2R_1r_1+r_1^2\}]^{1/2}}{R_1\,(R_2+r)\,(1+\omega^2C^2r_1^2)}$$

$$= \frac{\left[[ML^2T^{-3}I^{-2}][(M^0L^0T^0I^0+T^{-2}M^{-2}L^{-4}T^8I^4M^2L^4T^{-6}I^{-4})\left\{[M^2L^4T^{-6}I^{-4}](M^0L^0T^0I^0+T^{-2}M^{-2}L^{-4}T^8I^4M^2L^4T^{-1}I^{-4})+(ML^2T^{-3}I^{-2})(ML^2T^{-3}I^{-2})+M^2L^4T^{-6}I^{-4}\right\}\right]^{1/2}}{(ML^2T^{-3}I^{-2})(ML^2T^{-3}I^{-2}+ML^2T^{-3}I^{-2})(M^0L^0T^0I^0+T^{-2}M^{-2}L^{-4}T^8I^4M^2L^4T^{-6}I^{-4})}$$

Since the sum of terms which have the same dimensions as one another must have the same dimensions as its constituent terms, therefore, R.H.S. terms can be written dimensionally as :

$$\frac{[ML^2T^{-3}I^{-2}][M^2L^4T^{-6}I^{-4}+M^2L^4T^{-6}I^{-4}+M^2L^4T^{-6}I^{-4}]^{1/2}}{[ML^2T^{-3}I^{-2}][ML^2T^{-3}I^{-2}]}$$

$$= \frac{[ML^2T^{-3}I^{-2}][M^2L^4T^{-6}I^{-4}]^{1/2}}{[M^2L^4T^{-6}I^{-4}]} = \frac{M^2L^4T^{-6}I^{-4}}{M^2L^4T^{-6}I^{-4}}$$

= Dimensionless quantity.

Since the quantities on both sides are dimensionless, therefore, **no correction** *is required.* **(Ans.)**

Example 1.11. *Find the dimensions of each term in the following equation and make a dimensional check of equation :*

$$E = RI + j\omega LI + \frac{\mu_0 NAM}{\delta^2}\,v$$

N = Being a constant, it is dimensionless; $A = [L^2]$; $\delta = [L]$

where, E = e.m.f.; I = current; R = resistance; ωL = inductive reactance; μ_0 *= permeability of free space; v = the velocity; M = magneto-motive force; N = number of turns; A = area of pole pieces; δ = length of the air.*

Solution. $$E = RI + j\omega LI + \frac{\mu_0 NAM}{\delta^2}\,v \qquad \text{...(Given)}$$

The dimensional equations of the various terms are :

$$E = \frac{\text{Work done}}{\text{Charge}} = \frac{[ML^2T^{-2}]}{[IT]} = [ML^2T^{-3}I^{-1}]$$

$$R = \frac{\text{E.m.f.}}{\text{Current}} = \left[\frac{[ML^2T^{-3}I^{-1}]}{[I]}\right] = [ML^2T^{-3}I^{-2}]$$

$$\omega L = 2\pi fL = [T^{-1}\mu L]$$

μ_0 = Being a constant, it is dimensionless

$$v = [LT^{-1}]$$

Substituting these value in the given equation, we have :

$$[ML^2T^{-3}I^{-1}] = [ML^2T^{-3}I^{-2}][I] + [\mu LT^{-1}][I] + \frac{[L^2][I]}{[L^2]}[LT^{-1}]$$

$$= [ML^2T^{-3}I^{-1}] + [\mu LT^{-1}][I] + [LT^{-1}I]$$

But, $$[I] = [\mu^{-1/2} M^{1/2} L^{1/2} T^{-1}]$$

or, $$[\mu] = \frac{[MLT^{-2}]}{[I^2]}$$

$$\therefore \quad [ML^2T^{-3}I^{-1}] = [ML^2T^{-3}I^{-1}] + \frac{[MLT^{-2}][LT^{-1}I]}{[I^2]} + [LT^{-1}I]$$

$$= [ML^2T^{-3}I^{-1}] + [ML^2T^{-3}I^{-1}] + [LT^{-1}I]$$

Thus we observe that the *third term on the R.H.S. does not check dimensionally, hence the equation is inaccurate.* **(Ans.)**

1.6 DETERMINATION OF ABSOLUTE UNITS

The determination of the absolute values of electrical quantities is very essential for establishing a satisfactory system of measurements. *It involves the measurement of a unit in terms of physical quantities which are measured practically.* Thus if we can determine the units of, say, current and e.m.f. by practical measurements, standards for all other electrical quantities like resistance etc. can be set up in terms of current and e.m.f.

- There is not much difficulty in making absolute measurements provided that an accuracy of a few parts in a hundred is all that is required and in that case absolute measurements of current and resistance can be made in any reasonably well equipped laboratory by a number of methods.
- The determination of absolute unit of e.m.f. involves the measurement of energy which is difficult to measure and, therefore, the determination of absolute unit of e.m.f. is *not* usually attempted. Normal practice is to determine the units of *current* and *resistance* and to express the units of all other quantities *in terms of these quantities.*
- The unit of power is determined in terms of the units of current and potential difference.
- Mutual or self inductors are set up in forms for which the inductance can be calculated in terms of its dimensions. Similarly, the capacitance is also measured in terms of mutual inductance.

1.6.1. Absolute Measurement of Current

The dimensions of current are $[\mu^{-1/2} M^{1/2} L^{1/2} T^{-1}]$ or $[M^{1/2} L^{1/2} T^{-1}]$ if permeability μ is regarded as non-dimensional quantity. The dimensions of $(\text{current})^2$ are $[MLT^{-2}]$ which are the dimensions of the force; thus the absolute measurement of current involves the measurement of force. This force may be exerted in the following *two ways* :

1. By the current in a solenoid upon a suspended magnetic needle ... as in a *tangent or sine galvanometer.*

2. By the current in one part of the circuit upon other part of the circuit in series with it and carrying the same current ... as in an *electro-dynamometer* or a *current balance.*

- *"Galvanometer methods"* of determination of current suffer from the following *disadvantages* :

(i) There is always some uncertainty about the exact position of the poles of the magnetic needle used.

(ii) Horizontal component of the magnetic field of earth is required to be determined separately with great accuracy.

(*iii*) The following assumptions made in deriving the expression for deflection of galvanometer are not justifiable in practice :

— The needle is suspended exactly at the centre of coil.

— The axis of needle is horizontal.

— The needle is infinitesimally small.

— The plane of galvanometer coil lies exactly in the magnetic meridian, and is exactly vertical.

- In "*electro-dynamometer method*", the current is measured in terms of the torsion of suspension wire or of a bifilar suspension, therefore this method is also *not satisfactory*.
- The methods which make use of some form of current balance are the most satisfactory and, therefore, commonly used. The absolute measurement of current by *Rayleigh's current balance method* is described below.

Rayleigh current balance :

Fig. 1.1 shows a Rayleigh's current balance :

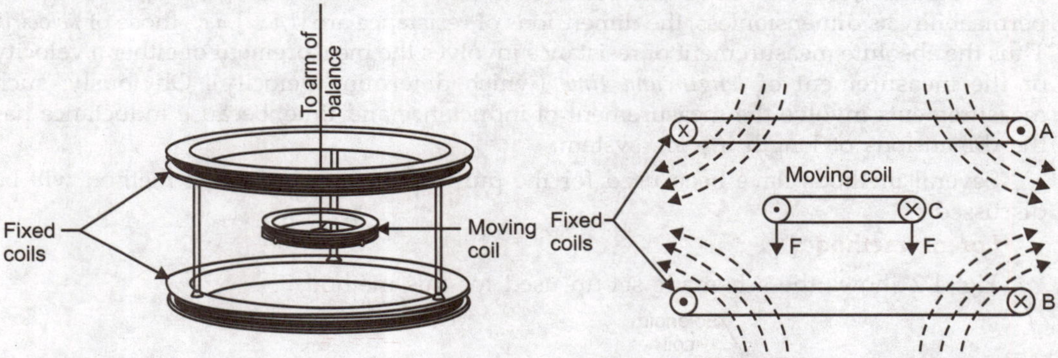

(a) Arrangement of coils in the Rayleigh current balance

(b) Arrangement of coils–working principle

Fig. 1.1. Rayleigh current balance — For absolute measurement of current.

— It consists of two fixed coils (A and B) and a small moving coil (C) which is suspended from the arm of a sensitive balance. The three coils are connected in *series*.

— If *I* is the value of current through them, then the force *F* exerted on the suspended coil is given as :

$$F = I^2 \frac{dM}{dx} \text{ newton} \qquad ...(1.17)$$

where, $\frac{dM}{dx}$ = The rate of change of mutual inductance between the fixed and moving coils. This is expressed in henry/metre.

— The value of $\frac{dM}{dx}$ can be obtained from the physical dimensions of the coils and *F* is purely a mechanical quantity which can be accurately measured by the balance. *Thus the value of I is dependent upon units of purely physical quantities and is independent of other electrical units.*

— For measuring the value *F*, it is usual to counterbalance the suspended coil when the force on it is upwards. The current through the suspended coil is then reversed thus reversing the force. The additional counterbalance weight gives the value 2*F*.

It should be noted that the expression for the force is in newton and since the weights used in counterbalancing the force are in kg, the value of '*g*' acceleration due to gravity must be known. The accuracy of the current measurement will depend on the accuracy with which the value of 'g' is known.

Precautions necessary for high accuracy :

1. Precision balance of special form should be used.
2. Much care is necessary in construction to ensure that the flexible leads, used for leading current into the moving coil or coils, must not exert any appreciable torque upon the moving system.
3. The cooling of the coils is very important. Water jackets are used for the fixed coils and water cooled chambers are provided for the moving coils.
4. In order to check the error due to distortion of magnetic field, bobbins made of non-magnetic materials should be used.

1.6.2. Absolute Measurement of Resistance

In e.m.u. system of units the resistance has the dimensions $[\mu LT^{-1}]$. Taking the permeability as dimensionless, the dimensions of resistance are $[LT^{-1}]$, *i.e.*, those of *velocity*. Thus the absolute measurement of resistance involves the measurement of either a velocity or the measurement of *length and time* (which determine velocity). Obviously, such measurements involve the measurement of inductance and time, because inductance has the dimensions of length in e.m. system.

Several methods have been used for the purpose, here only Lorenz method will be discussed :

Lorenz method :

Fig. 1.2 shows the schematic set-up used for this method :

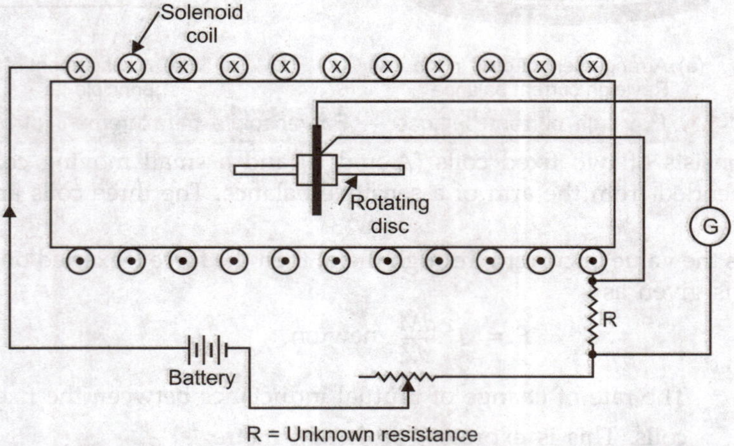

Fig. 1.2. Lorenz method – For absolute measurement of resistance.

— In this method/arrangement there is a circular metallic disc which is mounted concentrically inside a solenoid. The disc is rotated at a *uniform speed.*

— A steady current (*I*) is passed through the solenoid.

— *R* is a *low resistance whose absolute value is to be determined.* It is connected across the disc with the help of leads and two brushes. One brush presses the edge of the disc and the other brush makes a contact with the disc near its centre. A sensitive galvanometer *G* is included in one of the leads.

— The magnetising current (I) for the solenoid is supplied by a battery. This current passes the resistance R also.

— When current is passed through the solenoid, a magnetic field is produced. As the disc rotates an e.m.f. is induced in it. The current flowing through R produces a voltage drop across it. *The e.m.f. generated in the disc is balanced against the voltage drop* in resistance R, the balance being indicated by the galvanometer G. The *disc speed is adjusted till a balance is obtained.*

Let M be the mutual inductance between disc and solenoid in henry. Since inductance is defined as flux linkages per unit current, therefore, the value of *magnetic flux linkages* produced by the disc when a curent of I ampere passes through the solenoid is MI.

But, flux linkages of the disc = number of turns × flux linking the disc = flux linking the disc (since number of terms in a disc is one). Thus we conclude that the value of flux passing perpendicularly through the disc surface is MI weber when a current I ampere passes through the solenoid.

Let n be the speed of the disc in r.p.s. when the deflection of galvanometer is *zero*.

$$\therefore \quad \text{E.m.f. induced in the disc} = \text{Rate of change of flux linkages}$$

$$= MI \times n = MIn \text{ volt} \qquad \text{...(1.18)}$$

Also, voltage drop in resistance $R = IR$ volt

where, I and R are in ampere and ohm respectively.

Now, when the deflection of galvanometer is zero,

$$MIn = IR$$

or,

$$R = Mn \qquad \text{...(1.19)}$$

Knowing the value of mutual inductance M, which is calculated from the dimensions of the solenoid and disc, and from their relative positions and the value of rotational speed of disc n in r.p.s. which can be obtained by stroboscopic method, the value of unknown resistance (R) can be determined.

Precautions necessary for accuracy :

1. In order to determine the value of M accurately, both the disc and solenoid must be carefully constructed and their dimensions are accurately measured with the help of precision measuring instruments.

2. The dimensions of the solenoid relative to the disc diameter must be chosen such that the value of M may depend, as little as possible, upon the effective dimensions of the disc, which cannot be determined, when rotating, with great accuracy.

3. The effect of earth's magnetic field upon the e.m.f. induced should be made small. This is accomplished by arranging the plane of the disc to be in the magnetic meridian and making two measurements, one with current flowing in one direction and the other with the direction of current reversed.

4. The thermo-electric e.m.f. of the brush contacts should be reduced. This is done by using two phosphor-bronze discs of special construction and two solenoids.

5. The speed n should be as accurate as possible. It is determined by stroboscopic methods, or by a driven chronograph. It is advisable to incorporate flywheel to ensure uniform speed.

Example 1.12. *The following data relate to a simple form of Lorenz apparatus for absolute measurement of resistance :*

Diameter of solenoid = 60 mm

Length of solenoid = 600 mm

Number of turns of wire on the solenoid = 7500
Diameter of the rotating disc = 50 mm
Speed of the disc when the galvanometer indicates balance condition = 5845 r.p.m.
Calculate :
(*i*) *The mutual inductance between the disc and the coil in henry.*
(*ii*) *The value of resistance in ohm.*

Solution. Given: Radius of solenoid $r_s = \frac{d}{2} = \frac{60}{2}$, 30 mm or 0.03 m ; l_s = 600 mm or 0.6 m ; N = 7500 ; Radius of disc, $r_d = \frac{50}{2}$ = 25 mm = 0.025 m ; $n = \frac{5845}{60}$ = 97.417 r.p.s.

(*i*) **Mutual inductance, M :**

Intensity of field at the centre, for a long solenoid,

$$H = \frac{NI}{2\sqrt{(l/2)^2 + r_s^2}} \quad ...(i)$$

Field intensity at the centre, $$B = \mu_0 H = \frac{\mu_0 NI}{2\sqrt{(l_s/2)^2 + r_s^2}} \quad ...(ii)$$

Flux linked by the disc, $$\phi = B \times \text{area of the disc}$$

$$= \frac{\mu_0 NI}{2\sqrt{(l_s/2)^2 + r_s^2}} \times \pi r_d^2$$

$$= \frac{\pi \mu_0 r_d^2 NI}{2\sqrt{(l_s/2)^2 + r_s^2}} \quad ...(iii)$$

The disc is treated as a single turn winding.

∴ Flux linkages of disc, $$\psi = \frac{\pi \mu_0 r_d^2 NI}{2\sqrt{(l_s/2)^2 + r_s^2}} \quad ...(iv)$$

Mutual inductance between disc and solenoid,

$$M = \frac{\psi}{I} = \frac{\pi \mu_0 r_d^2 N}{2\sqrt{(l_s/2)^2 + r_s^2}}$$

$$= \frac{\pi \times 4\pi \times 10^{-7} \times (0.025)^2 \times 7500}{2\sqrt{(0.6/2)^2 + (0.03)^2}} \text{ H} = 3.07 \times 10^{-5} \text{ H}$$

$$= \mathbf{0.0307 \text{ mH}}$$ (Ans.)

(*ii*) **Value of resistance R :**

Resistance $R = Mn = 0.0307 \times 10^{-3} \times 97.417 = \mathbf{2.99\ m\Omega}$ (Ans.)

1.7 STANDARDS OF MEASUREMENT

The ***measurement standard*** *is the physical embodiment of the unit of measurement as well as that of its submultiple value.*

As the measurement is a process of comparison, therefore, regardless of our measurement method, we must employ a basis of comparison – *standardised units.* The standard must be precisely defined, and because different systems of units exist, the method of conversion from system to system must be mutually agreed upon.

— It is of significant importance that a relatonship between the standards and the read out scale of each measuring system must be established through a process known as *calibration.*

Classification of Standards :

The standards of measurement are often *classified* as follows :

1. International Standards.
2. Primary Standards.
3. Secondary Standards.
4. Working Standards.

1. **International standards :**

— These standards are defined on the basis of international agreement.

— They represent the units of measurements which are closest to the possible accuracy attainable with present-day technological and scientific methods.

— These are checked and evaluated regularly against *absolute measurements* in terms of the fundamental units.

— These standards are maintained at the International Bureau of Weights and Measures and are *not* available to the ordinary user of measuring instruments for the purposes of calibration and comparison.

- The improvements in the accuracy of absolute measurements have made the International units superfluous and they have been *replaced by absolute units.*

2. **Primary standards :**

The highest standard of either a base unit or a derived unit is called a **primary standard.**

- These standards essentially are copies of international prototypes and are kept throughout the world in national standard laboratories and institutions of similar standing.
- These standardized units are :
  — *Quite stable;*
  — *Independent;*
  — *Invariant;*
  — *Not relative but finite.*

While setting up the primary standards, the following points are given due considerations :

(*i*) Long time stability of the material.

(*ii*) Low temperature coefficient.

(*iii*) No deterioration in its characteristics with changes in humidity, temperature and other environmental conditions.

(*iv*) Rigidity of construction.

(*v*) Accuracy of machining.

(*vi*) Most sophisticated techniques to be used for measuring physical dimensions which determine the accuracy of the standard.

- The primary standards *constitute the ultimate basis of reference* and are used for the purpose of *"verification and calibration of secondary standards"*.
  — These standards have the *"highest possible accuracy"* but are *very expensive* to own and maintain.

— The primary standards are *not available for use outside the national laboratories.*

3. **Secondary Standards :**

The secondary standards are the *reference calibrated standards designed and calibrated from the primary standards.*

- These are sent periodically to the national standard laboratories for their calibration (the interval of calibration depends upon the accuracy and the type of standard being maintained).
- These standards are kept by the measurement laboratories and the industrial organisations to check and calibrate the general tools for their accuracy and precision.

4. **Working Standards :**

- These standards have an accuracy of *one order lower than that of the secondary standards.*
- These are the normal standards which are *used by the workers and technicians who actually carry out the measurements.*

1.8 STANDARDS OF MASS AND LENGTH

Standards of Mass :

Primary standards :

- *It is a* prototype kilogram kept at national physical laboratory of every country.
- These standards have an accuracy of 1 part in 10^8 and occasionally verified against the standard at the International Bureau.

Secondary standards :

- These are kept by the industrial laboratories, have an accuracy of 1 ppm (part per million) and may be checked against the primary standard kept at the national physical laboratories.

Working standards :

- These standards, which are used commercially, are available in a wide range of values to suit almost every application.
- These have an accuracy of the order of 5 ppm.
- These standards are verified against the secondary standards.

Standards of Length :

- The metre (the metric unit of length) was initially defined as the ten-millionth part of the meridianal quadrant through Paris. This was materially represented by the distance between two lines engraved on a platinum-irridium bar preserved at the International Bureau of Weights and Measures near Paris.
  - — Later on, metre was redefined more accurately in terms of an *optical standard,* namely, the orange-red radiation of *krypton atom*. The internationally specified krypton-86 discharge lamp, excited and observed under well-defined conditions, emits orange light whose wavelength now constitutes the basic standards of length to 1 ppm.
  - — The metre, the unit of length in S.I. system, is now defined as equal to 1,650,763.73 wavelengths in vacuum of the orange-red light radiation of the krypton-86 atom in its transition between level $^2p_{10}$ and $^5d_5$.

— In, 1983, the metre was redefined in terms of time which can be specified accurately using atomic standards. The metre is now the length travelled by light in vacuum in a time interval of $\frac{1}{299792458}$ second. The speed of the light therefore is now 299792458 m/s.

- The *working standards* of length, which are most widely used in industries, are precision gauge blocks, made of steel. These steel blocks have two parallel surfaces, at specified distance apart, with accuracy tolerances in the 0.5–0.25 micron range. Such precision gauges are cheaper in cost and of high accuracy.

1.9 ATOMIC FREQUENCY AND TIME STANDARDS

These days, due to technological advancements, it is very easy to construct clocks of great precision and accuracy by frequency conversion (because frequency $\propto \frac{1}{\text{time}}$).

The transition between two energy levels E_1 and E_2 of an atom is accompanied by the emission (or absorption) by radiation given by :

$$E_2 - E_1 = h\nu \quad \text{...(1.20)}$$

where, h = Planck's constant, and

ν = A physical constant (which is a frequency) provided that the energy states are not affected by external conditions such as magnetic fields. This constant depends only on the internal structure of the atom.

Since frequency is inversely proportional to time, atomic clocks can be constructed on the above mentioned principles.

— In a *passive* atomic resonator, transitions are induced by driving the atomic device with a signal derived from a quartz oscillator, which in turn is stabilised to produce the proper driving frequency.

— The *active* resonators are micro-wave masers such as atomic hydrogen masers, which produce stimulated emission in the microwave range.

The *main* "**atomic standards**" used are :

1. *Primary or absolute standards :*
   (*i*) Caesium beam standard.
   (*ii*) Hydrogen maser standard.
   - These standards do not need to be referenced to any other standard for checking or calibration.
2. *Secondary standards :*
   (*i*) Rubidium vapour standard.
   - This standard needs to be initially set up against a primary standard, and then periodically recalibrated.
   - Quartz oscillator is often used to provide feedback frequency to caesium or rubidium atomic standards, which are passive.

Advantages of atomic standards :

1. Results obtained with these standards are more uniform.
2. They form a better and convenient time base.
3. Their accuracy is much higher than that of standards based upon astronomical measurements.

4. They form the *primary frequency standard.*
   - Some of the atomic clocks are in operation which give a precision of 1 μs per day.

1.10 E.M.F. STANDARDS

The primary, secondary and working standards of e.m.f. are described below :

- Several types of voltaic cells were used as a voltage reference in the past. These cells were developed over a long period and to a high degree of excellence. Such cells *have a long life and serve as a standard in most satisfactory way.*

1.10.1. Primary Standard of E.m.f. – Saturated Weston Cell

The primary voltage standard is the *"Saturated normal Weston cell"*. It was developed by Edward Weston in 1892.

- This is a *voltaic cell.*

Construction :

Fig. 1.3 shows the construction of a saturated Weston cell :

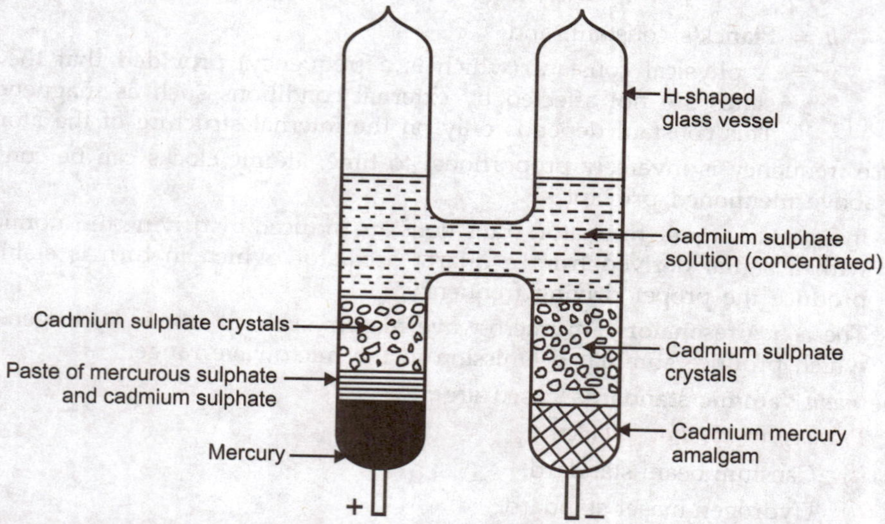

Fig. 1.3. Saturated Weston cell.

— It consists of a H-shaped glass vessel, each leg being 25 mm in diameter containing saturated solution of cadmium sulphate ($CdSO_4$) in the upper part, then crystals of cadmium sulphate, the paste of mercurous sulphate (Hg_2SO_4) and cadmium sulphate over the mercury at the bottom of one leg and crystals of cadmium sulphate and cadmium mercury amalgum at the bottom of second leg. The two legs are hermetically sealed.

— *"Cadmium sulphate solution"* acts as an *electrolyte, "mercury"* acts as a *+ve electrode,* cadmium mercury amalgam (a *"solution"* of 1 part of cadmium in 7 parts of mercury) acts as a *–ve electrode* and *"paste of mercurous sulphate and cadmium sulphate"* acts as a *depolariser.*

A simplified representation of the cell reaction is :

$$Cd + Hg_2SO_4 = CdSO_4 + 2Hg \qquad ...(1.21)$$

The e.m.f. of the cell remains absolutely constant so long as the amalgam has both a solid and liquid phase present.

— The connections of the cell to the external circuit are made by platinum wires sealed into the glass.

- Saturated cells are normally allowed to age for a year before they are considered to be of constant e.m.f. Following this period, their life span appears indefinite excluding the possibility of accidents.

Characteristics :

(i) The e.m.f. of the cell is *1.01864 volt.*

The e.m.f. of the cell changes with temperature according to the following equation :

$$E_t = E_{20} - 46 \times 10^{-6} (t - 20) - 0.95 \times 10^{-6} (t - 20)^2 + 10 \times 10^{-9} (t - 20)^3 \quad ...(1.22)$$

where, E_t = E.m.f. of cell at any temperature t°C, and

E_{20} = E.m.f. of cell at 20°C.

(ii) The cell voltage drops by 40 μV/°C. The cell is kept in an oil bath to control its temperature within 0.01°C.

(iii) It has a life span of about 10 to 20 years. The drift in the value of voltage is about 1 μV per year.

(iv) Its temperature coefficient is –40 μV/°C.

Uses :

- Since this cell is a standard for maintenance of the volt, therefore, it is *only used in standardising laboratories.*
- These cells are *not* used for ordinary work because of temperature requirements and also because they are not portable.

Example 1.13. *A standard cell has a voltage rating of 1.018500 V and internal resistance of 500 Ω. The insulation resistance between its terminals is 5.5 MΩ.*

Calculate :

(i) The current drain due to insulation resistance.

(ii) The terminal voltage.

Solution. *Given :* Voltage rating of the cell = 1.018500 V; Internal resistance = 500 Ω; Insulation resistance between the terminals = 5.5 MΩ.

(i) **The current drain :**

The current flowing through the insulation resistance

$$= \frac{1.018500}{500 + 5.5 \times 10^6} = 0.185\ \mu A$$

Thus there is a *constant drain of current of* **0.185 μA.** **(Ans.)**

(ii) **Terminal voltage :**

Voltage drop in internal resistance = $0.185 \times 10^{-6} \times 500$ V = 92.5 μV

∴ Terminal voltage = Voltage rating of the cell (e.m.f.) – voltage drop in internal resistance

$= 1.018500 - 92.5 \times 10^{-6} =$ **1.0184075 V** **(Ans.)**

1.10.2. Secondary Standard of e.m.f. – Unsaturated Weston Cell

The secondary standard of e.m.f. is unsaturated Weston cell (Fig. 1.4).

Construction :

— This cell is similar to a saturated cell except that it *does not have the cadmium sulphate crystals.*

— It has the same electrodes but the solution is at 4°C, and therefore, *unsaturated at room temperature.*

— In this cell, in addition, a 'sputum' or retaining member (porous plugs in this case) is used over each electrode to *hold the material in place* and hence these cells are considered *portable.*

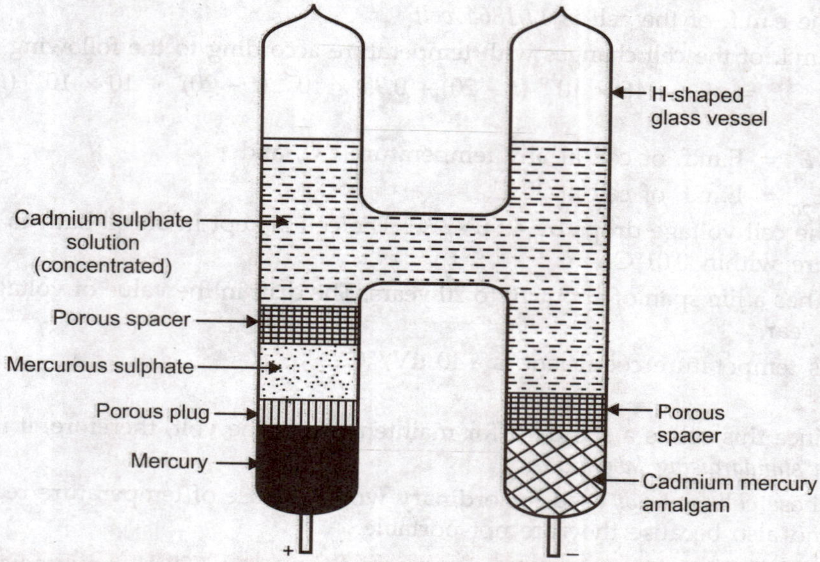

Fig. 1.4. Unsaturated Weston cell.

— The cells are enclosed in a bakelite case which is lined with 1.5 mm thick copper to help keep all parts of the cell at a *uniform temperature.*

— After being moved from one place to another, the cell should be allowed a few days to settle so that the results obtained from measurements involving cells are accurate.

Characteristics :

(i) The e.m.f. of new unsaturated cells varies; it lies between 1.0180 and 1.0194 volt. Therefore, each cell must be calibrated against the primary standard in order to know its exact e.m.f. The temperature at which this e.m.f. is measured should be specified.

(ii) Even though this type of cell becomes less stable with age, yet it is remarkably good.

(iii) The decrease in e.m.f. is about 30 to 50 μV per year.

(iv) Its life is shorter than a saturated type. These cells, if handled by qualified personnel in proper way, can last up to 20 years or even more. These cells should be checked every year or at an interval of two years against primary standard.

(v) Its temperature coefficient is –10 μV/°C. It has an accuracy of 0.005 per cent.

(vi) This cell has a positive temperature effect at one electrode and negative at the other, therefore, it is important to keep all parts of the cell at the same temperature.

Uses : Such cells are used as *secondary standard of volt.*

Precautions in the use of Standard Cells :

1. Mercurous sulphate is sensitive to light and, therefore, it should not be exposed to light.
2. Saturated cells should be kept in an upright position because if they are tilted there is always the danger of mixing of electrode materials. Therefore these cells cannot be shifted and are sent through special messengers.
3. Laboratory conditions, which would result in the condensation of moisture on the cell case, should be avoided.
4. The standard cells should be kept at dry places having uniform temperatures of about 15°C to 20°C for the purpose of storage, so that the hysteresis effect due to change in temperature and possibility of leakage currents owing to moisture on the insulating materials between the terminals of the cell are avoided.
5. The main limitation of the use of standard cells is the current drawn from the cell at any time; this should be negligibly small. Standard cell being a potential device gets injured if an appreciable current is drawn from it; a very minute current (not exceeding 10 μA) should be drawn from the cell and that also only for a short duration. Hence such cells should be used in measurements in *either with potentiometer or some other open-circuited device.* Protective resistance of high value is inserted in series with the cell to keep the current drawn to a minimum value while balancing the potentiometer.
6. Such a cell should never be short-circuited accidentally as voltage of the cell gets changed permanently on short-circuit. Of course with time, it recovers its voltage but not completely.

- A voltmeter should never be used for measuring the voltage of a standard cell due to the following reasons :

(*i*) Due to the current there is a possibility of damage to the cell;

(*ii*) This measurement is meaningless since the cells have a high internal resistance (about 600 to 800 ohms).

The voltage should be measured with a potentiometer only because at the balance the current drawn from the cell is zero. When using the cell in potentiometric work, a resistance of 20 kΩ should be inserted in series with the cell in order to avoid excessive drain of current when there is unbalance.

1.10.3. Zener Diode Reference Standard

Zener diode (a special junction diode) provides an alternative method of a reference standard. It is a *reverse-biased diode operating in the breakdown region.* It is *quite accurate for many measurement requirements.*

Fig. 1.5 shows the Zener diode in a typical application as a laboratory standard :

— The supply voltage E is usually much higher than the Zener voltage of diode assuring breakdown of the diode.

— Resistor R usually has a high value and is placed in series with the diode and battery. It serves to limit the

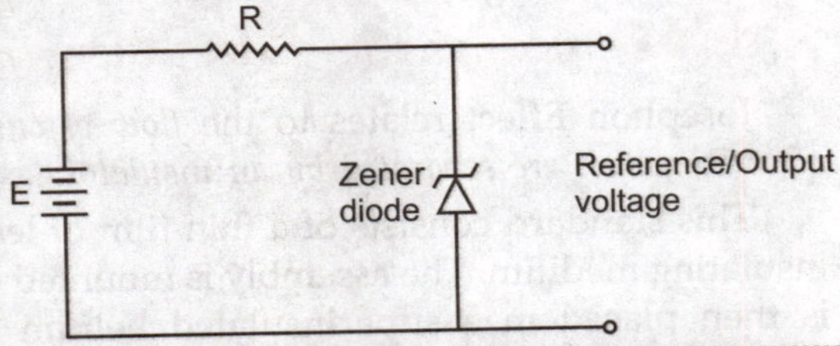

Fig. 1.5: Zener diode laboratory standard.

current through the diode during breakdown to a safe value. Also, when the supply voltage varies for any reason, most of this variation is taken up in the voltage drop across R with a very small change in diode current and as a result the *reference voltage* (*output voltage*) *remains practically the same.*

- Commercial voltage standards using zener diode essentially consist of :
  (*i*) A zener-controlled voltage source placed in a temperature-controlled environment to improve its long-term stability;
  (*ii*) A precision output voltage divider.
- If a standard voltage considerably greater than 5 V is to be provided, there is an advantage in using several low-voltage diodes in *series*, rather than one high-voltage diode as the temperature coefficient strongly depends upon the zener voltage for which the diode is made, being low for diodes of 4 to 5 volt zener voltage and higher for higher-voltage diodes.
- Versatile *laboratory working standards* have been developed with accuracies of the order of standard cell accuracy, in which zener-controlled voltage source is used in a temperature-controlled environment with a precision output voltage divider.

Note. The breakdown of Zener voltage depends upon the nature of semiconductor and the amount of doping and ranges from about *3V to several hundred volts.* If the diode is *heavily doped,* depletion layer will be *thin* and consequently the junction will breakdown at a *low reverse voltage.* On the other hand, a *lightly doped diode* will break down at *higher reverse voltage.* The breakdown phenomenon is reversible and harmless so long as the safe operating temperature is maintained.

Advantages :

1. The advantage of these standards over standard cell is that, unlike the standard cell, the *semiconductor device is immune to the effects of short circuit and overload conditions.*
2. For chemical cells, spring board or other applications where vibration is constantly present do not present ideal conditions. The Zener sources which are rugged may furnish the accuracy required under such conditions.
3. The Zener sources can be substituted for standard chemical cells in recording potentiometers which may be subjected to freezing temperatures. This is due to the fact that the standard cells cannot work below 4°C while Zener sources give good accuracies at all temperatures if properly compensated.
4. Whereas chemical cells can be built to voltage level of about 1 volt, the Zener sources can be built to any voltage level.
5. The constant voltage drop of Zener diode may be utilised in :
   — Voltage regulators;
   — Limiters;
   — Voltage reference units;
   — Compensators;
   — Clippers;
   — Voltage standards.

1.10.4. Josephon – Effect Standard

Josephon Effect relates to the *flow of current between two pieces of a superconducting material which are separated by an insulator.*

This standard consists of a thin film of lead separated by lead oxide which acts as an insulating medium. The assembly is mounted on a glase substrate. The probe, thus formed, is then placed in a superinsulated helium Dewar system. The tunnelling junction is excited by a microwave source having a frequency of nearly 9 GHz, and is biased by a supply current source. Under these conditions the Josephon junction produces a voltage as per the following equation :

$$E = \frac{Nhf}{2e} \qquad ...(1.23)$$

where, N = An integer (its value can be controlled by the bias current, thereby varying the voltage produced in discrete steps),

h = Planck's constant,

f = Frequency of microwave source, and

e = Electron charge.

- These voltage standards have an accuracy of about 0.05 ppm and, therefore, can be employed for calibration of other e.m.f. standards using potentiometers.

1.11 RESISTANCE STANDARDS

In S.I. system of units, the absolute value of the *ohm* is defined in terms of fundamental units of *length* and *mass*.

Absolute measurement of the ohm is done against a group of resistance standards, which are maintained at the International Bureau of Weights and Measures in Severes. This group is periodically checked against each other, and with similar groups of standards kept at other "standards laboratories" in the world.

Standard resistor. Fig. 1.6 shows the construction of a typical standard resistor :

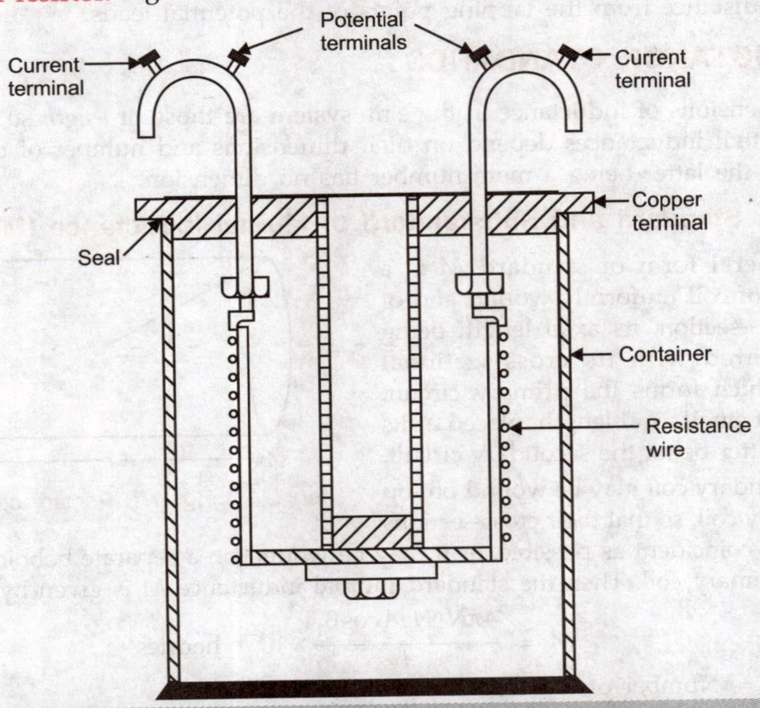

Fig. 1.6: Standard resistor.

— The resistance wire is made of an alloy like *manganium* having a composition : Nickel = 4%; Copper = 84%; Manganese = 12%. Manganin possesses the following properties :

(*i*) High resistivity (48×10^{-8} Ωm).

(*ii*) Low thermal e.m.f. with copper (Below 3 μV/°C).

(*iii*) Low temperature coefficient of resistance (40×10^{-6}/°C at 20°C).

— The resistance coil is so wound that it reduces stress.

— The coil is immersed in oil for temperature stability.

— The resistor is housed in a sealed moisture-free container.

— The connections to the resistance coil are silver soldered and the terminal hooks are made of nickel plated copper. The resistance of the standard at any temperature t°C is given by :

$$R_t = R_{25} + \alpha (t - 25) + \beta (t - 25)^2 \quad ...(1.24)$$

where, R_{25} = Normal value of resistance at 25°C in Ω, and

α, β = Thermal coefficients whose values are 10×10^{-6} and -0.5×10^{-6} respectively so that change in resistance for a change in temperature of 10°C is less than 50 ppm.

- The major difference between the *low resistance* standards and those used for *reference purpose* only (which are not required to carry large currents) is that the *former ones are fitted with current terminals as well as potential terminals* (See Fig. 1.6.)

— The *potential terminals* are the points, between which the normal resistance of the standard is measured. The *current terminals* are the points by means of which standard resistor is connected to supply circuit. For obtaining uniform distribution of current through the cross-section of the resistance material the current terminals should be at an appreciable distance from the tapping points of the potential leads.

1.12 INDUCTANCE STANDARDS

The dimensions of inductance in the e.m. system are those of *length* so standards of self and mutual inductances depend on their dimensions and number of turns of wire in them but the latter being a mere number has no dimension.

1.12.1. Standard and Substandard of Mutual Inductance (M)

The general form of standard M is a single layer of coil uniformly wound and of circular cross-section, its axial length being large compared with its cross-sectional diameter, which forms the primary circuit, with a coil of small axial length, placed at its centre, the latter being the secondary circuit.

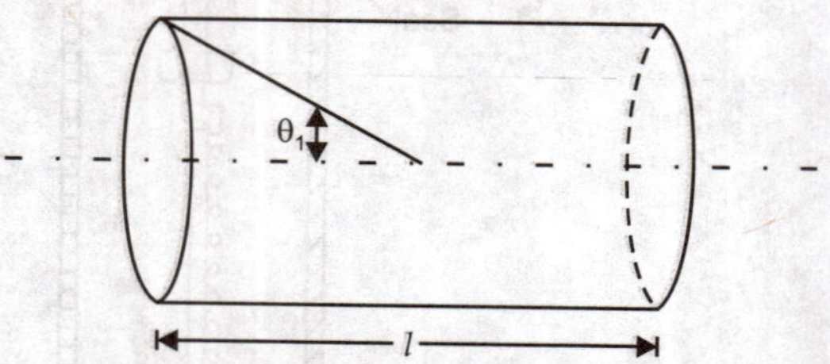

Fig. 1.7: Primary coil.

The secondary coil may be wound on top of the primary coil, so that their cross-sections are as nearly coincident as possible, or it may be wound on a separate bobbin and placed inside the primary coil. Then the standard mutual inductance M is given by :

$$M = \frac{4\pi N_p N_s A \cos\theta_1}{l} \times 10^{-9} \text{ henries}$$

where, N_p = Number of turns on the primary coil,

N_s = Number of turns on the secondary coil,

A = Cross-sectional area in sq. cm.,

l = Axial length of primary coil in cm, and

θ_1 = Angle between the axis of the coil and the line drawn from the centre point of the axis to a point on the circumference of an end turn of the coil as shown in Fig. 1.7.

Substandards of M are used as standard of mutual inductances for general laboratory purposes. Since the standards of *M* are not absolute standard, it is not essential that their dimensions shall be determined with great accuracy. The essential thing is that the substandard of *M* is as nearly as possible to the nominal value for which it is designed. When constructed they are compared with standard ones and their mutual inductance is adjusted.

Substandard *M* normally consists of two coils wound on a bobbin of *marble,* the end being separated by a flange. The wire is standard copper, with double silk coverings. After winding, the coils and bobbin are immersed in hot paraffin wax. When withdrawn and allowed to cool, the wax firmly fixes the wire in the coil in position.

Adjustment to the value of mutual inductance required is carried out by carrying the end of the coils through a further arc of circle in order to give the effect of a fraction of turn.

1.12.2. Standard and Substandard of Self Inductance (L)

The magnitudes of standard of *L* are calculated from formulae and performance is ensured by winding the coil of bare hard drawn copper wire under tension inside screw thread cut in a marble cylinder which should be very carefully ground so as to be as true cylindrical as possible.

As in the case of substandard of *M,* substandards *L* are constructed to have a nominal value which is usually a simple fraction of one henry. Such substandards are compared with standards L that are used as reference standards for general laboratory work.

1.13 CAPACITANCE STANDARDS

In capacitance standards, standard capacitors of fixed or variable values are used; the capacitance of such capacitors can be determined accurately by means of a formula and from dimensions.

Capacitance in electrostatic system of units has the dimensions of length and, therefore, the capacitance of absolute capacitors can be expressed in terms of their dimensions. Hence these dimensions should be known *very accurately* and should *not* change with the passage of time, once the capacitor has been made.

- The effect of time, frequency, temperature and other relevant factors on the capacitance of standard capacitors should be *as small as possible* and they should have *low dielectric losses and a high insulation resistance.*

1.13.1. Primary Standards of Capacitance

Primary standards of capacitance are of the following three types :

1. The concentric sphere type.
2. The concentric cylinder with guard ring.
3. The parallel plate type with guard plates.

The air is always preferred as dielectric because its *permittivity is definitely known and it is also free from absorption and dielectric loss.*

"Parallel plate capacitor" is used *rarely* as it requires very careful adjustments of plates to get the calculated value of capacitance accurately.

Standard air capacitors :

Usually these are *concentric cylinder type* capacitors in which *guard rings* of same diameter are used at both ends in order to shield the capacitor from external electrostatic influences and to define the effective area of electrodes accurately so that area to be used in calculation does not have any uncertainty.

— Air gap between the electrodes is kept large in order to withstand the applied voltage.

— Ends of the *inner cylinder* are made *bell shaped* and project beyond both ends of outer cylinder by a considerable length.

— Guard rings and *outer cylinder* are kept at same potential but are insulated from each other. Thus these capacitors are free from brush discharge and with efficient screening.

Fig. 1.8 shows a general sketch of an air capacitor.

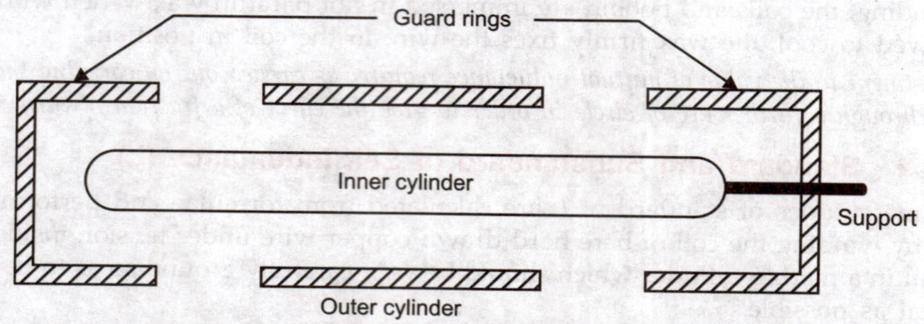

Fig. 1.8: Air capacitor.

- *These capacitors are used in measurement of dielectric loss and power factor of capacitors at high voltages.*

Compressed gas capacitors :

— In a compressed gas capacitor the gas used may be either air or nitrogen, but it should be *clean* and *dry*.

— In these capacitors gas filled at 10 bar increases the operating voltage as breakdown voltage of gas increases with the increase in pressure while the dielectric constant remains unchanged.

— These capacitors can be made for voltages up to 1500 kV.

— The capacitance of such capacitors is in the range of 10–200 pF and its loss angle is less than 0.00001 degree.

Precautions. Since air as dielectric has some disadvantages, therefore, following precautions need to be taken when using it as dielectric material in standard capacitors :

1. Owing to very low permittivity of air and long gap between electrodes, the air capacitor, even of small value, is *very bulky*.
2. Due to low dielectric strength of air long gap is required, so that air does not break down on the application of high voltage.
3. In air capacitors, dry air should be used otherwise dust particles settling in the gap between the plates may cause leakage troubles.
4. The electrodes of capacitors should be rigidly fixed in position by supports of some solid dielectrics such as fused quartz and amberite (which have no tendency to wrap and cause displacement of the plates from their original positions).

1.13.2. Secondary Capacitance Standards

The dimensions of the secondary capacitors need not to be known accurately since these are *calibrated capacitors*.

— In secondary standard capacitors a number of plates are used instead of only two as in the primary standards, so *greater value of capacitance can be obtained in small size.*

These have value up to about 0.02 μF as compared to 200 pF in case of primary standard capacitors.

— These capacitors have long period permeance of 1 part in 20,000 and temperature coefficient less than 1 part/10^5/°C.

— Their power factors are found to be less than 0.00001 at 1 kHz up to radio frequencies.

— These capacitors are properly screened to prevent the effect of stray electrostatic fields.

HIGHLIGHTS

1. The standard measure of each kind of physical quantity is called a *unit.*
2. A system in which the various units are all expressed in terms of a small number of fundamental units is called an *absolute system of units.*
3. The unique quality of every quantity which distinguishes it from all other quantities is called *dimension.*
4. S.I., the International system of units, has the following six base units : length; mass; time; electric current; thermodynamic temperature; luminous intensity.
5. *Rayleigh's current balance* method is employed for the absolute measurement of current.
6. *Lorenz method* is used for the absolute measurement of resistance.
7. The *measurement standard* is the physical embodiment of the unit of measurement as well as that of its submultiple value. The standards of measurement are often classified as :
   (*i*) International standards; (*ii*) Primary standards;
   (*iii*) Secondary standards; (*iv*) Working standards.
8. The primary voltage standard is *"Saturated Weston cell"*, while the secondary voltage standard is *"Unsaturated Weston cell"*.
9. Josephon-Effect relates to the flow of current between two pieces of a super-conducting material which are separated by an insulator.
10. Standard air capacitors are used in measurement of dielectric loss and power factor of capacitors at high voltages.

OBJECTIVE TYPE QUESTIONS

Choose the Correct Answer :

1. The maximum current that can be drawn from a saturated Weston cell without damaging it, is
   (*a*) 100 mA momentarily (*b*) 100 μA continuously
   (*c*) 100 μA momentarily (*d*) 1 mA momentarily.
2. The internal resistance of a saturated Weston cell is of the order of :
   (*a*) a few ohm (*b*) a few kilo ohm
   (*c*) a few hundreds of ohm (*d*) a fraction of ohm.
3. The decrease in the value of emf of an unsaturated Weston cell is :
   (*a*) about 1 μV per year (*b*) about 1 mV per year
   (*c*) about 5 – 10 μV per year (*d*) about 30-50 μV per year.

4. The symbol *nm* stands for :
   (*a*) newton metre (*b*) nano metre
   (*c*) nano milli (*d*) number of moles.
5. In measuring the absolute value of ampere using Rayleigh's balance, the value of gravitational constant g must be accurately known because
   (*a*) the expression for the force is in newton and weights used in counterbalancing the force are in kg.
   (*b*) value of g is different at different places.
   (*c*) corrections have to be made for the weight of moving coil.
   (*d*) none of the above.
6. The material of wires used for making resistance standards is usually
   (*a*) Manganin (*b*) Nichrome
   (*c*) Copper (*d*) Phosphor bronze.
7. In order to reduce skin effect in resistance standards when they are used on high frequency A.C. :
   (*a*) the diameter of the resistance wire should be small.
   (*b*) adjacent turns of the resistance coil should carry currents in the opposite direction.
   (*c*) they should be provided with four terminals.
   (*d*) all of the above.
8. The units whose sizes cannot be chosen independently are called
   (*a*) Derived units (*b*) Fundamental units
   (*c*) Absolute units (*d*) Auxiliary fundamental units.
9. The dimensions of inductance in CGS e.s. system are :
   (*a*) $[\in L^{-1} T^2]$ (*b*) $[\in^{-1} ML^{-1} T^2]$
   (*c*) $[\in^{-1} LT^{-2}]$ (*d*) $[\in^{-1} L^{-1} T^2]$
10. In Lorenz method for absolute measurement of ohm, two measurements are made one with current in one direction and the other with current in the opposite direction. This is done to
    (*a*) eliminate the effect of thermal emfs.
    (*b*) eliminate the effect of earth's magnetic field.
    (*c*) eliminate the effects of both thermal emfs and earth's magnetic field.
    (*d*) eliminate the effect of temperature on coil and disc.
11. Zener sources when used as voltage standards
    (*a*) are reverse biased.
    (*b*) can be designed to cover a wide range of voltages.
    (*c*) are immune to the effects of short circuits.
    (*d*) all of the above.
12. The output voltage Josephon effect standard
    (*a*) can be changed in a continuous manner.
    (*b*) can be changed only in discrete steps.
    (*c*) can be changed continuously as well as in discrete steps.
    (*d*) cannot be changed.
13. Caesium beam standard is normally used as primary atomic standard for frequency because
    (*a*) caesium beam standard has an accuracy 3 parts in 10^{12}, has short-term stability, low drift and small size.

(b) hydrogen maser standard, though it is the most stable frequency source, has a large size.

(c) quartz crystal's operating frequency changes with change in load capacitance and also the drift in frequency is asymptotic with time.

(d) all of the above.

14. The disadvantages of atomic standards are :

(a) Their accuracy is lower than that of standards based upon astronomical measurements.

(b) The time taken for making measurements is long.

(c) The results obtained with atomic standards are not uniform.

(d) None of the above.

15. The dimensions of resistance CGS e.m. system are :

(a) $[\mu^{-1} LT^{-1}]$ (b) $[\mu^{-1} MLT^{-1}]$

(c) $[\mu LT^{-1}]$ (d) $[\mu LT^{-2}]$.

16. The base units in SI system are :

(a) metre, kilogramme, second

(b) metre, kilogramme, second, ampere

(c) metre, kilogramme, second, ampere, kelvin, candela, mole

(d) metre, kilogramme, second, ampere, kelvin, candela.

17. The international standard of length is defined in terms of

(a) length of earth's meridian passing through Paris.

(b) distance between two lines engraved on a platinum-iridium bar.

(c) wavelength in vacuum, of radiation of krypton – 86 atom in its two specified transitions.

(d) none of the above.

18. The most stable primary atomic standard for frequency is :

(a) Caesium beam standard (b) Hydrogen maser standard

(c) Quartz standard (d) Rubidium vapour standard.

ANSWERS

1. (c)	**2.** (c)	**3.** (d)	**4.** (b)	**5.** (a)	**6.** (a)	**7.** (a)
8. (a)	**9.** (d)	**10.** (c)	**11.** (d)	**12.** (b)	**13.** (d)	**14.** (d)
15. (c)	**16.** (c)	**17.** (c)	**18.** (b).			

THEORETICAL QUESTIONS

1. Define the following terms :

(i) Units (ii) Absolute units

(iii) Fundamental units (iv) Derived units.

2. What is the difference between rationalised and unrationalised systems of units?

3. Discuss S.I. system of units.

4. How many base and supplementary units exist in S.I. system? Give their definitions.

5. Define base units of S.I. system.

6. What are the advantages of S.I. units?

7. Prove that $\frac{1}{\sqrt{\mu_0 \epsilon_0}}$ = velocity of light, where μ_0 and ϵ_0 are respectively the permeability and permittivity of free space.

8. What is meant by dimension of an electrical quantity? Obtain the dimensions of resistance and current.

9. Derive dimensional equations for current and e.m.f. in both e.s. and e.m. systems.
10. Derive the dimensions of the following quantities in L, M, T, I system :
    (*i*) Magnetic flux; (*ii*) Inductance; (*iii*) Capacitance; (*iv*) Magnetic flux density; (*v*) MMF; (*vi*) Reluctance; (*vii*) Magnetising force; (*viii*) Electric flux; (*ix*) Permittivity; (*x*) Resistance; (*xi*) Permeability.
11. Explain briefly an experiment for the absolute measurement of resistance. Give the precautions necessary to ensure accuracy of measurement.
12. Describe the method of absolute measurement of ohm.
13. Explain a method for the absolute measurement of current. Give the necessary precautions.
14. Discuss with the help of a diagram the principle and working of one form of current balance.
15. Explain the constructional details of a Weston standard cell and give the precautions in using standard cells.
16. Describe Josephon-effect standard for e.m.f. Explain how the output voltage can only be varied in discrete steps.
17. Explain briefly primary standard for mutual inductance.

UNSOLVED EXAMPLES

1. Prove that the following equation is dimensionally correct :

$$e = Blv$$

where, e = E.m.f. generated in a conductor,
B = Flux density,
l = Length of conductor, and
v = Peripheral velocity of conductor.

2. The pull of an electromagnet is given by the expression :

$$F \propto B^a A^b \mu^c$$

where, B = The flux density in the air gap,
A = Cross-sectional area of the gap, and
μ = Permeability of the air in the gap.

Find the values of a, b and c by analysing the equation dimensionally.

[**Ans.** $a = 2; b = 1; c = -1$]

3. Use dimensional equations to find out whether the following expression is dimensionally correct or not :

$$R_4 = \frac{R_2(1+\omega^2 R_3^2 C_3^2)}{\omega^2 R_1 R_3 C_3}$$

where R_1, R_2, R_3, R_4 are resistances, C_3 is a capacitance, and $\omega = 2\pi f$ (f being frequency in Hz).

$\left(\textbf{Ans.}\ \text{Incorrect, correct eqn.}: R_4 = \dfrac{R_2\,(1+\omega^2 R_3^2 C_3^2)}{\omega^2 R_1 R_3 C_3^2}\right)$

4. Using dimensional analysis check whether the following expression is correct or not.

$$R = \frac{1+\omega^2 C^2 R^2}{\omega^2 CR}$$

$\left(\textbf{Ans.}\ \text{No, correct expression}: R = \dfrac{1+\omega^2 C^2 R^2)}{\omega^2 C^2 R}\right)$

5. Prove that the following equations are dimensionally correct :

(i) $r = \dfrac{R\omega^2 M_1 M_2}{R^2 + \omega^2 L^2}$; (ii) $C = \dfrac{1}{\omega^2 M}$

where, r, R = Resistances,
M_1, M_2, M = Mutual inductances,
L = Self inductance,
C = Capacitance, and
ω = Angular frequency.

6. In one of the bridges, an expression for self inductance was derived as under :

$$L = \frac{R_2 R_3 C_4^2}{1 + \omega^2 C_4^2 R_4^2}$$

Find out if this expression is correct or not. If not correct, suggest a suitable correction.

$\left(\textbf{Ans.}\ \text{No, correct expression} : L = \dfrac{R_2 R_3 C_4}{1 + \omega^2 C_4^2 R_4^2}\right)$

7. The following expression was obtained in the course of a calculation :

$$I = \frac{V\,\omega\, M}{[(\omega M + R_1 R_2)^2 + \omega^2 L_1 L_2 R_1^2]^{1/2}}$$

Check whether the equation is dimensionally correct. If not, suggest any corrections.

$\left(\textbf{Ans.}\ \text{Incorrect; } I = \dfrac{V\omega M}{[(\omega^2 M^2 + R_1 R_2)^2 + \omega^2 L_1 L_2 R_1^2]^{1/2}}\right)$

8. In the course of calculation the following expression for current in a circuit was arrived at and it is suspected that an erorr has been made in it :

$$I = \frac{\omega ME}{[(\omega^2 M^2 + R_1 R_2)^2 + \omega^2 L_1 R_1^2]^{1/2}}$$

where, E = Voltage,
w = Angular velocity,
M = Mutual inductance,
L_1 = Self inductance, and
R_1, R_2 = Resistances.

Examine if the expression is dimensionally correct and if necessary, make a correction.

$\left(\textbf{Ans.}\ \text{No; Correct expression} : I = \dfrac{\omega ME}{[(\omega^2 M^2 + R_1 R_2)^2 + \omega^2 L_1 L_2 R_1^2]^{1/2}}\right)$

9. In a simple form of Lorenz apparatus for absolute measurement of resistance, the solenoid, is 50 mm in diameter and 500 mm long. It is uniformly wound with 8000 turns of wire. The rotating copper disc is 40 mm in diameter and the galvanometer indicates balance condition when the speed of the disc is 5952 r.p.m.

Calculate :

(i) The mutual inductance between the disc and the coil in henry.

(ii) The value of resistance in ohm. [**Ans.** (i) 0.0251 mH; (ii) 2.49 mΩ]

10. A standard cell has a voltage rating of 1.018500 V and an internal resistance of 500 Ω. The insulation resistance between its terminals is 5 MΩ. Find the current drain due to insulation resistance and the terminal voltage. [**Ans.** 0.204 μA; 1.018398 V]

CHAPTER

2 Concepts of Measurements and Measurement Systems

2.1. Introduction to measurement; 2.2. Significance of measurement; 2.3. Methods of measurement; 2.4. Modes of measurement; 2.5. Generalised measurement system and its functional elements; 2.6. Input-output configurations of measuring instruments and measuring systems; 2.7. Applications of measurement systems; 2.8. Static and dynamic characteristics of instruments; 2.9. Errors in measurements. *Highlights – Objective Type Questions – Theoretical Questions – Unsolved Examples.*

2.1 INTRODUCTION TO MEASUREMENT AND INSTRUMENTATION

2.1.1. Measurement

Measurement *is the act, or the result of a quantitative comparison between a predetermined standard and an unknown magnitude.* The procedure and apparatus employed for obtaining the comparison, however, must be *provable*; the procedure for this is called *calibration.*

Probably it has rightly been said: *"Whatever exists, exists in some amount"*. The determination of the amount is what measurement is all about.

Measurand. *The physical quantity or the characteristic condition which is the object of measurement in an instrumentation system is variously termed as "**measurand**", "measurement variable", "instrumentation variable" or "process variable".*

The measurand may be :

— Fundamental quantity, *e.g.,* length, mass and time;

— Derived quantity, *e.g.,* speed, velocity, acceleration, pressure.

The word *measurand* is used to designate the particular physical parameter being observed and quantified; that is, the *input quantity* to the measuring process. The act of measurement produces a *result.* Fig. 2.1 shows a *fundamental measuring process.*

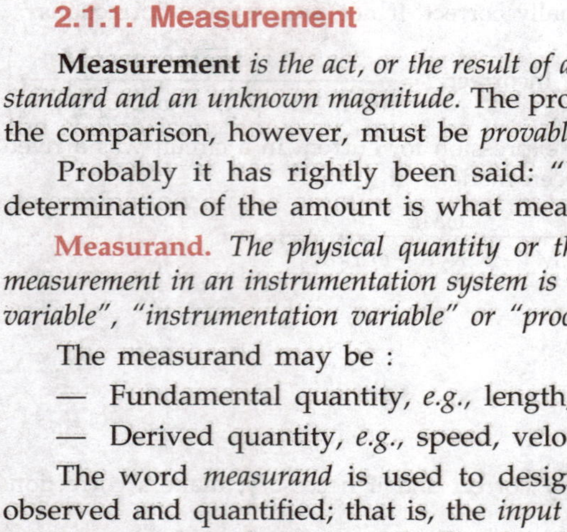

Fig. 2.1. Fundamental measuring process.

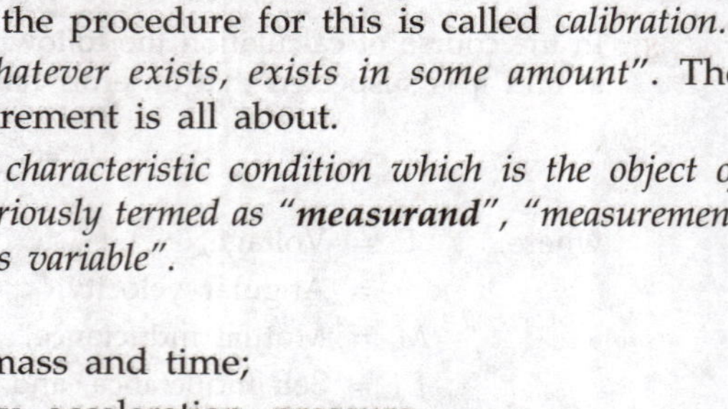

Fig. 2.2. Measurand and measurement.

— Whereas the input to a measuring system is known as *measurand*, the output is called *measurement* (See Fig. 2.2).

The *basic requirements* for getting meaningful result of measurement are:

(*i*) The standard employed for comparison purpose *must be accurately defined* and should be *commonly acceptable.*

(*ii*) The standard must be of the *same character* as the measurand and usually, but not always, is prescribed and defined by a legal or recognized agency or organisation, *e.g.,* Indian Bureau of Standards (IBS) or the International Organisation of Standards (ISO).

(*iii*) The apparatus used and method adopted for the comparison purposes must be *provable.*

- Several measurands of interest *vary with time in a continuous manner over a range of magnitude.*

For example, the speed of the automobile, as it starts from rest, has some magnitude at every instant during its motion. A sensor that responds to velocity will produce an output signal having a line variation analogous to the time change in automobile's speed. Such a signal is referred to as **analog signal** *because it is analogous to a continuous physical process.*

The analog signal, which has a value at every instant of time, becomes a **digital signal.**

— A digital signal is a *set of discrete numbers each corresponding to the value of the analog signal at a single specific instant of time.*

— The digital signal contains *no information* about the value of the analog signal at time *other than sample times.*

The mechanical quantities such as pressure, stress, strain, temperature, fluid-flow rates behave time-wise in an analog manner. However, by converting an analog signal to an equivalent digital signal for the purpose of signal conditioning and/or read out oftenly following distinct *advantages* are obtained:

(*i*) *Noise problems are reduced or sometimes eliminated completely.*

(*ii*) *Data transmission becomes simpler.*

(*iii*) *Computers are designed to process digital information, and direct numerical display or recording is more easily accomplished by manipulating digital quantities.*

2.1.2. Instrumentation

The technology of using instruments to measure and control the physical and chemical properties of materials is called **Instrumentation.**

- When the instruments are used for the *measurement and control of industrial manufacturing, conversion or treatment process,* the term **"process instrumentation"** is used.
- *When the measuring and controlling instruments are combined so that measurements provide impulses for remote automatic action, the result is called a* **Control system.**

2.2 SIGNIFICANCE OF MEASUREMENT

The significance of measurements is discussed below:

- Measurements play a very significant role in every branch of scientific research and engineering processes which include the following:

  — Control systems;

— Process instrumentation;

— Data reduction.

- The whole area of automation or automatic control is based on measurements. The very concept of control is based on the comparison of the actual condition and the desired performance. *The exactness of error depends on the precision and accuracy of measurement made.*
- The measurements confirm the validity of a hypothesis and also add to it its understanding. This eventually leads to new discoveries that require new and sophisticated measuring techniques.
- Through measurements a product can be designed or a process be operated with maximum efficiency, minimum cost, and with desired degree of reliability and maintainability.

2.3 METHODS OF MEASUREMENT

The broad classification of methods of meaurement is as follows:

1. Direct comparison methods.
2. Indirect comparison methods.

1. Direct comparison method:

In this method the parameter to be measured is *directly compared with either a primary or a secondary standard.*

Direct comparison is quite commonly used for measurements of length. However, for measurement of mass the problem becomes much more intricate since it is just not possible for human beings to distinguish between wide margins of mass.

- Generally, this method is not always the most accurate or the best; it is not sensitive enough also.

2. Indirect comparison method:

In this method the *comparison* is *done with a standard through the use of a calibrated system.*

These methods for measurement are used in those cases where the desired parameter to be measured is *difficult to be measured directly, but it has got some correlation with some other parameter which can be easily measured.*

Example: The elimination of bacteria in milk is directly dependent upon its temperature. Thus the bacteria elimination can be measured indirectly by *measuring the temperature of the milk.*

— In indirect measurements an *empirical relation is generally established between the measurement actually made and the results that are desired.*

- When the primary purpose of making a measurement is to determine quality of a product, then quality should be measured *directly*. However, in case direct measurement is not possible, then *indirect* measurement should be made.
- *The measurement systems used in engineering applications make use of indirect methods for measurement purposes.*

— A measurement system consists of a *transducing element* which converts the quantity to be measured with an *analogous signal*. This signal after being processed by some *intermediate means* is then fed to the end devices which provide the measurement results.

2.4 MODES OF MEASUREMENT

Following are the *three* modes of measurement :

1. Primary measurements.
2. Secondary measurements.
3. Tertiary measurements.

1. Primary measurements :

- In this case the *sought value of a parameter is determined by comparing it directly with "reference standards"*.
- There is *no conversion* of measurand in terms of length.

Examples :

(*i*) Measurement of time by counting the number of strokes of a clock.
(*ii*) Matching of two lengths when determining the length of an object with a ruler.
(*iii*) Matching of two colours when judging the temperature of a red hot steel.

2. Secondary measurements :

The indirect measurements involving 'one translation' are called ***secondary measurements.***

Examples :

(*i*) The pressure measurement by manometers.
(*ii*) The temperature measurement by mercury-in-glass thermometers.

3. Tertiary measurements:

The indirect measurements involving 'two conversions' are called ***tertiary measurements.***

Examples :

(*i*) The measurement of temperature of an object by thermocouple (Fig. 2.3). The primary signal (temperature of object) is transmitted to a thermocouple which generates a *voltage* which is a function of the temperature (first translation/ conversion). The voltage in turn is applied to a voltmeter through a pair of wires. The second conversion is then voltage into length. The tertiary signal is transmitted to brain of the observer.

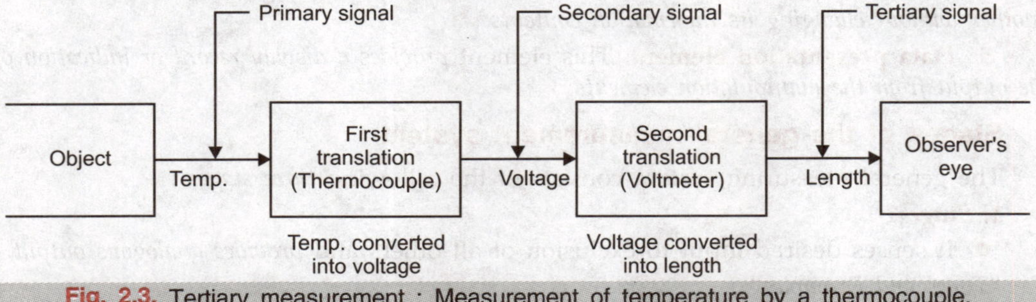

Fig. 2.3. Tertiary measurement : Measurement of temperature by a thermocouple.

(*ii*) The measurement of static pressure by a Bourdon tube pressure gauge.
(*iii*) The measurement of the speed of a rotating shaft by means of an electric tachometer.

The unit of a measuring system where translation of measurand takes place is called the *transducer or translator.*

- Majority of measurement systems are tertiary systems and they include the whole range of mechanical, electrical, pneumatic electro-mechanical and electro-pneumatic instruments.

Measurements may also be divided into following *two categories* :

1. **Contact type.** In this case the sensor of the measuring device *contacts* the controlled medium.
2. **Non-contact type.** Here the sensor *does not contact* the controlled medium.

- Non-contact measurements include *optical, radioactive and others.*

2.5 GENERALISED MEASUREMENT SYSTEM AND ITS FUNCTIONAL ELEMENTS

- Fig. 2.4 shows a measurement system (generalized) with different components (called *elements*).

The various elements are :

1. Primary sensing element. It is an element that is sensitive to the measured variable. The sensing elements *sense the condition, state or value of the process variable by extracting a small part of energy from the measurand, and then produces an output which reflects this condition, state or value of the measurand.*

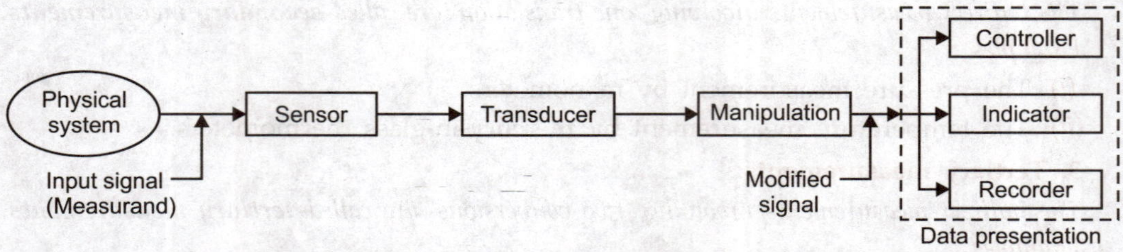

Fig. 2.4. Generalised measurement system.

2. Variable conversion of transducer element. This element *converts the signal from one physical form to another* without changing the information content of the signal.

3. Manipulation element. This element *operates* on the signal according to some *mathematical rule without changing the physical nature of the variable.*

4. Data transmission element. This element *transmits the signal from one location to another without changing its information contents.*

5. Data presentation element. This element *provides a display record or indication of the output from the manipulation elements.*

Stages of the general measurement system :

The general measuring system consists of the following *three stages* :

1. *Stage-1 :*

- It senses desired input to exclusion of all others and *provides analogous output.*

"Types and Examples" :

Mechanical : Contacting spindle, spring-mass, elastic devices (*e.g.*, Bourdon tube for pressure, proving ring for force), gyro.

Electrical : Contacts, resistance, capacitance, inductance, piezoelectric crystals and polymers, thermocouple, semiconductor junction, etc.

Optical : Photographic film, photoelectric diodes and transistors, photomultiplier tubes, holographic plates.

Hydro-pneumatic : Buoyant float, orifice, venturi, vane, propeller.

2. *Stage-2 :*

- *It modifies transduced signal into form usable by final stage.*
- It *usually increases amplitude and/or power* depending on requirement.
- It may also *selectively filter unwanted component or convert signal into pulsed form.*

"Types and Examples" :

Mechanical : Gearing, cranks, slides, connecting links, cams, etc.

Electrical : Amplifying or attenuating systems, bridges, filters, telemetering systems, various special purpose I.C. (Integrated circuit) devices.

Optical : Mirrors, lenses, optical filters, optical fibres, spatial filters (pin hole, slit).

Hydro-pneumatic: Piping, valving, dashpots, plenum chambers.

3. *Stage-3 :*

- It *provides an indication or recording* in form that can be evaluated by an unaided human sense or by a controller.
- It *records data* digitally on a computer.

"Types and Examples" :

Indicators Displace types : Moving pointer and scale, moving scale and index, light beam and scale, electron beam and scale (oscilloscope), liquid column.

Digital types : Direct alphanumeric readout.

Recorder : Digital printing, inked pen and chart, direct photography, magnetic recording (hard disk or type).

Processor and computers : Various types of computing systems, either special purpose or general, used to feed read out/recording devices and/or controlling systems.

Controllers : All types.

Illustration of measuring systems :

I. Bourdon tube pressure gauge:

Fig. 2.5 shows the simple Bourdon tube pressure gauge – an example of a simple measurement system.

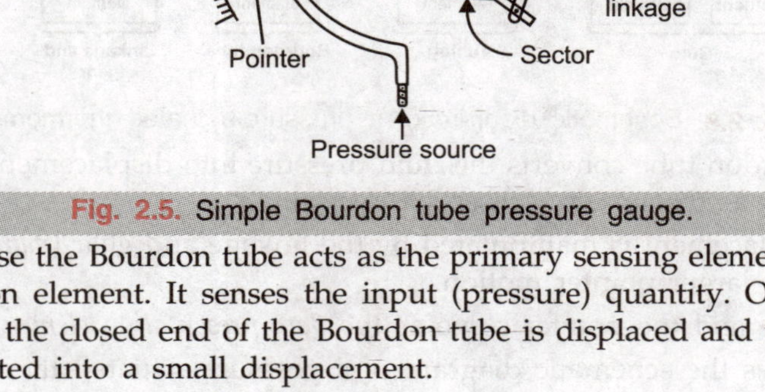

Fig. 2.5. Simple Bourdon tube pressure gauge.

— In this case the Bourdon tube acts as the primary sensing element and a variable conversion element. It senses the input (pressure) quantity. On account of the pressure, the closed end of the Bourdon tube is displaced and thus the pressure is converted into a small displacement.

— The closed end of the Bourdon tube is connected to a gearing arrangement through mechanical linkage. The gearing arrangement *amplifies* the small displacement and consequently the pointer rotates through a large angle. Thus, the mechanical linkage acts as a data transmission element while the gearing arrangement acts as a data manipulation element.

— The final data presentation stage consists of a pointer and dial arrangement which, when calibrated with known pressure inputs, gives an indication of the pressure signal applied to the Bourdon tube.

Fig. 2.6 shows the schematic diagram of this measurement system.

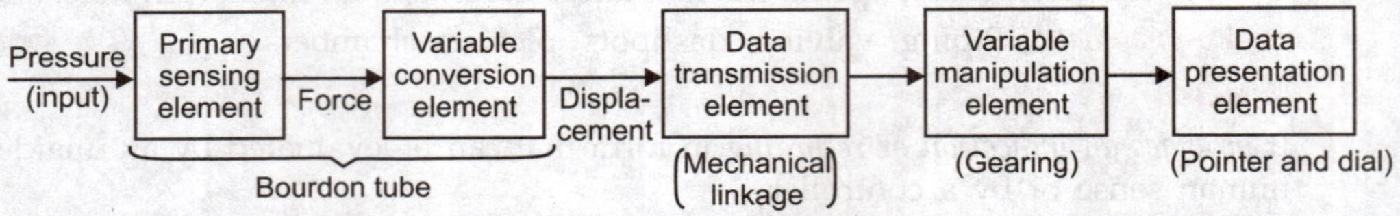

Fig. 2.6. Schematic diagram of a Bourdon tube pressure gauge.

II. Pressure-actuated thermometer :

Fig. 2.7 shows a pressure-actuated thermometer.

— The liquid-filled bulb acts as the *primary sensor* and *variable conversion element* since a temperature change results in pressure build-up within the bulb, due to the constrained thermal expansion of filling liquid.

— The pressure tubing is employed to transmit the pressure to the Bourdon tube and thus functions as the *data transmission element.*

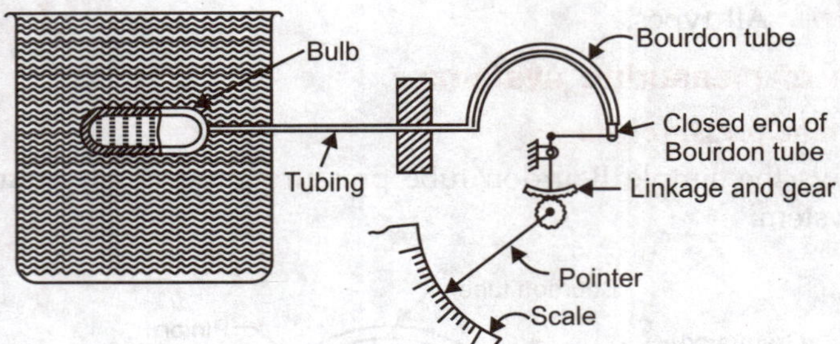

Fig. 2.7. Pressure-actuated thermometer.

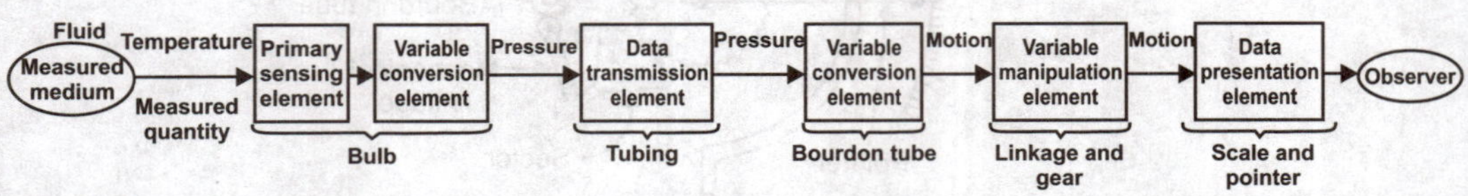

Fig. 2.8. Schematic diagram of a pressure-actuated thermometer.

— The Bourdon tube converts the fluid pressure into displacement of its tip, and as such acts as the *variable conversion element*.

— The displacement is manipulated by the linkage and gear (*manipulation elements*) to give a larger pointer motion.

— The scale and the pointer serve as the *data presentation element*.

Fig. 2.8 shows the schematic diagram of a pressure-actuated thermometer.

2.6 INPUT-OUTPUT CONFIGURATIONS OF MEASURING INSTRUMENTS AND MEASURING SYSTEMS

Fig. 2.9 shows the generalised input-output configuration of measurement systems. The various inputs to the measurement system are *classified* as follows :

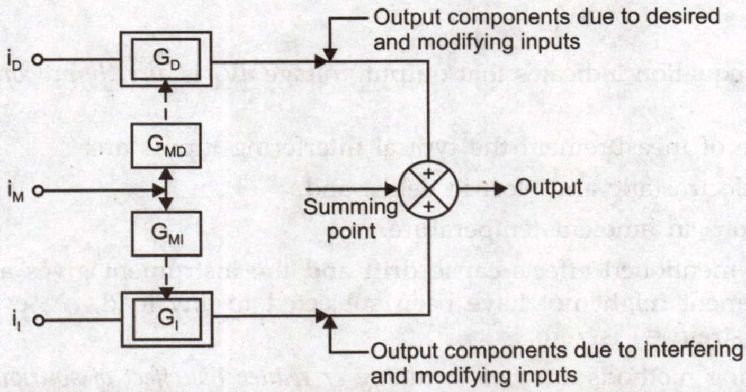

Fig. 2.9. Generalised input-output configuration of measurement systems.

1. Desired inputs.
2. Interfering inputs.
3. Modifying inputs.

1. Desired inputs :

The quantities for which the instrument or the measurement system is specifically designed to measure and respond are called ***desired inputs.***

The desired input i_D produces an output component according to an input-output relation symbolised by G_D; here G_D (*Transfer function*) represents the mathematical operation necessary to obtain the necessary output from the input.

- The transfer function G_D is a representative of a wide range of functions from a *constant in the case of linear systems to a statistical function used for statistical measurements.*

2. Interfering inputs :

The quantities to which an instrument or a measurement system becomes unintentionally sensitive are called ***interfering inputs.***

The interfering inputs i_I would produce an output component according to input-output relation symbolised by G_I.

3. Modifying inputs :

The inputs which cause a change in input-output relationships for either desired input or interfering input or for both are called ***modifying inputs.***

The modifying input i_M would cause a change in G_D and/or G_I. The specific manner in which i_M affects G_D and G_I is represented by G_{MD} and G_{MI} respectively.

For illustration, let us consider *strain measurement by a strain gauge.* When the gauge is strained due to mechanical loading, the change in output voltage is given by,

$$dV_0 = \frac{V_s}{4} \cdot \frac{dR}{R}$$

where, dV_0 = Output voltage,

V_s = Supply voltage (battery), and

dR = Change in resistance of the strain gauge which constitutes one arm of the Wheatstone bridge circuit (assuming, $R_1 = R_2 = R_3 = R_4 = R$)

In terms of strain $\in$ and strain gauge factor, F,

$$dV_0 = \frac{V_s}{4} \cdot F \in$$

The above equation indicates that output voltage dV_0 is *directly proportional to applied strain.*

In this case of measurement the typical interfering inputs are:

- Stray electrostatic and electric fields, and
- Variations in ambient temperature.

The above mentioned effects cause drift and the instrument gives an output even though the element might not have been subjected to any load, *i.e.*, even though the desired input (strain $\in$) is zero.

The following methods are *used to nullify or reduce the effect of spurious inputs on the indicated output of an instrument:*

1. Signal filtering.
2. Compensation by opposing inputs.
3. Output correction.
4. Use of high amplification and stable/accurate feedback device.
5. Use of instrument elements which are inherently sensitive only to the desired input (*Example :* The strain gauge may be made of a material with a very low temperature coefficient of resistance and temperature independent gauge factor).

2.7 APPLICATIONS OF MEASUREMENT SYSTEMS

The instruments and measurement systems are employed for the following *applications :*

1. Monitoring of processes and operations.
2. Control of processes and operations.
3. Experimental engineering analysis.

1. Monitoring of processes and operations :

In these types of applications the measuring instruments *simply indicate the value or condition of parameter under study and do not serve any control function.*

Examples :

(*i*) Water and electric energy meters installed in homes.

(*ii*) An ammeter or a voltmeter indicates the value of current or voltage being monitored at a particular instant.

2. Control of processes and operations :

There is a strong association between measurements and control. The instruments find a very useful application in automatic control systems.

- A common example is the typical refrigeration system which employs a thermostatic control. A temperature measuring device (often a bimetallic element) senses the room temperature, thus providing the information necessary for proper functioning of the control system.

3. Experimental engineering analysis :

The engineering problems can be solved by theoretical as well as experimental methods; several applications require the use of both the methods.

Following are the *uses* of experimental engineering analysis :

(*i*) To determine system parameters, variable and performance indices.

(*ii*) To formulate the generalised empirical relationship in cases where there is no proper theoretical backing.

(*iii*) To test the validity of theoretical predictions.

(*iv*) To solve mathematical relationships with the help of analogies.

2.8 STATIC AND DYNAMIC CHARACTERISTICS OF INSTRUMENTS

2.8.1. Introduction

The performance of an instrument is described by means of quantitative qualities, termed as **characteristics**; these characteristics having been broken down into the sub areas of *static characteristics* and *dynamic characteristics.*

- *The **static characteristics** pertain to a system where the quantities to be measured are constant or vary slowly with time. Performance criteria based on dynamic relations (involving rapidly varying quantities) constitute **dynamic characteristics**.*

— The static characteristics, in a real sense, also influence the quality of measurement under dynamic conditions, but these characteristics (static) *show up as non-linear or statistical effects* in otherwise linear differential equations giving the dynamic characteristics. *These effects would make the differential equations analytically unmanageable and so the conventional approach is to treat the two aspects of the problem separately*. Thus, even though these effects influence the dynamic behaviour, the differential equations of dynamic performance generally *neglect* the effects of dry friction, backlash, hysteresis, statistical scatter etc.

- *The overall performance of an instrument is judged by a semiquantitative superimposition of the static and dynamic characteristics.*

2.8.2. Definitions Relating to Measuring Instruments

Some important terms used in connection with measuring instruments are discussed below :

1. **True or actual value.** *The actual magnitude of a signal input to a measuring system which can only be approached and never evaluated is termed as "**true or actual value**".*
2. **Indicated value.** *It is the magnitude of a variable indicated by a measuring instrument.*
3. **Correction.** *The revision applied to the critical value so that the final result obtained improves the worth of the result is called "**correction**".*
4. **Overall error.** *It is the difference of the scale reading and the true value.*

   — When the instrument is properly designed and correctly adjusted, the consistent bias in error is very rare.
5. **Range.** *The region between the limits within which an instrument is designed to operate for measuring, indicating or recording a physical quantity is called the "**range of the instrument**".*
6. **Sensitivity.** *The ratio of output response to a specified change in the output is called "**sensitivity**".*

- The minimum change in the measured variable which produces an effective response of the instrument is called *"Resolution sensitivity"*. It is also called *"discrimination"*.
- The lowest level of measured variable which produces effective response of the instrument is called *"Threshold sensitivity"*.

7. **Scale sensitivity.** *It is defined as the ratio of a change in scale reading to the corresponding change in pointer deflection.*
8. **Scale readability.** The scale readability (in analog instruments) *indicates the closeness with which the scale can be read.*
9. **Repeatability.** It is defined as the *variation of scale reading*; it is random in nature.
   - *It is a measure of closeness with which a given input can be measured over and over again.*
10. **Accuracy.** It may be defined as *conformity with or closeness to an accepted standard value* (*true value*).
    - Accuracy of an instrument is influenced by factors like static error, dynamic error, reproducibililty, dead zone.
11. **Uncertainty.** Uncertainty denotes the *range of error, i.e.*, the region in which one guesses the error to be.
12. **Precision.** It refers to the *degree of agreement* within a group of measurements.
    - It is usually expressed in *terms of the deviation in measurement.*
13. **Drift.** *An undesired gradual departure of the instrument output over a period of time that is unrelated to changes in input, operating conditions or lead is called "**drift**".*
14. **Linearity or non-linearity.** *Deviation of transducer output curve from a specified straight line.* The *"non-linearity"* may be: (*i*) *Terminal linearity* (deviation from a straight line through the end points; (*ii*) *Best-fit linearity* (deviation from the straight line which gives minimum errors, both plus and minus.
15. **Dead zone.** *It is the range within which variable can vary without being detected.*
16. **Dead time.** *It is the time before the instrument begins to respond after the measured quantity has been changed.*
17. **Speed of response.** *The quickness of an instrument to read the measured variable is called "**speed of response**".*
18. **Reproducibility.** *The degree of closeness with which the same value of a variable may be measured at different times is called "**reproducibility**".*
19. **Tolerance.** *It is the range of inaccuracy which can be tolerated in measurements.*
20. **Backlash.** It is defined as the *maximum distance or angle through which any part of a mechanical system may be moved in one direction without applying appreciable force or motion to the next part in a mechanical system.*
21. **Stiction (static friction).** *It is the force or torque that is necessary just to initiate motion from rest.*
22. **Noise.** It may be defined as *extraneous disturbance generated in a measuring system which conveys no meaningful information w.r.t. desired signal.*

2.8.3. Static Characteristics

Measurements of applications in which parameter of interest is more or less constant or varies very slowly with time are called **static measurements.** *A set of criteria (e.g., "accuracy", "error", "reproducibility", "drift", "sensitivity", "dead zone") that provide meaningful description of measurements under static conditions are called* **static characteristics.**

The main *static characteristics* may be summed up as follows :

(i)	Accuracy	*(ii)*	Sensitivity
(iii)	Reproducibility	*(iv)*	Drift
(v)	Static error	*(vi)*	Dead zone.

2.8.3.1. Accuracy, errors and correction

Accuracy :

It is the degree of correctness with which a measuring means yields the "true value" with reference to accepted engineering standards.

Accuracy is determined as the maximum amount by which the result differs from the true value (the average of an infinite number of measured values when the average deviation due to the various contributing factors tends to zero). It is *almost impossible to determine experimentally the true value. The true value is not indicated by any measurement system due to the loading effects, lags and mechanical problems (e.g., wear, hysteresis, noise pick-up etc.).*

Accuracy of the measured signal depends upon the following *factors* :

- Variation of the signal being measured;
- Intrinsic accuracy of the instrument itself;
- Accuracy of the observer;
- Whether or not the quantity is being truly impressed upon the instrument.

Errors :

The accuracy of any measurement system is measured in terms of its *errors.* Since it is impossible to measure the true value of a quantity, as such, an approximation of the *"true value"* obtained by sufficiently extended series of measurements and also taking into parameters and conditions to which corrections may be applied, we obtain what is called *"best-measured value"* of the quantity. *The difference between the best measured value and the true value of the quantity is known as* **static error.**

The absolute value of error does not indicate precisely the accuracy of measurements. It is the *relative static error* which provides the quality of measurement. The **relative static error** *is defined as the ratio of absolute static error to the true value of the quantity under measurement.*

Let, V_m = Measured value of quantity,

V_t = True value of quantity,

E_s = Static (or absolute) error of measurement, and

E_r = Relative static error.

Then, $$E_s = V_m - V_t \quad ...(2.1)$$

and, $$E_r = \frac{\text{Absolute error}}{\text{True value}} = \frac{E_s}{V_t} \quad ...(2.2)$$

or, %age relative error, $$\% E_r = E_r \times 100 \quad ...(2.3)$$

We have, $$V_t = V_m - E_s \quad ...[\text{From eqn. (2.1)}]$$

$$= V_m - E_r \cdot V_t \quad ...[\text{From eqn. (2.2)}]$$

or, $$V_t + E_r \cdot V_t = V_m$$

or, $$V_t = \frac{V_m}{1 + E_r} \quad ...(2.4)$$

However, when E_r is small, $E_r << 1$,

Eqn. (2.4) may be written as :

$$V_t = V_m (1 - E_r) \quad ...(2.5)$$

- The error may be either +*ve* or –*ve*. For +*ve* static errors the instrument reads *high* whereas it reads *low* for –*ve* static errors.

Correction.

The difference between the true value and the measured value of a quantity is called **static correction** (C_s).

i.e.,

$$C_s = V_t - V_m \ (= -E_s) \quad ...(2.6)$$

Eqn. (2.6) indicates that the static correction of the instrument reading is of the same magnitude as the error, but *opposite in sign* (*i.e.,* $C_s = -E_s$).

"Accuracy" may be *expressed as* :

(*i*) Point accuracy, or

(*ii*) Percent of scale range, or

(*iii*) Percent of true value.

(*i*) *Point accuracy.* In this case, the accuracy of an instrument is stated for one or more points in a range.

Examples :

(*a*) A given thermometer may be stated to read ± 0.2°C between 50°C and 100°C;

(*b*) A scale of length may be read within ± 0.2 mm.

(*ii*) *Percent of scale range.* Here, the error is calculated on the basis of maximum value of the scale.

$$\% \text{ Error} = \left(\frac{V_m - V_t}{\text{Max. scale value}}\right) \times 100 \quad ...(2.7)$$

Eqn. (2.7) refers to the intrinsic accuracy of the instrument itself and does not include procedural or personal performance.

(*iii*) *Percent of true value.* In this case the absolute error of measurement is expressed as %age of true value of the unknown quantity.

i.e.,

$$\%\text{Error} = \left(\frac{V_m - V_t}{V_t}\right) \times 100 \quad ...(2.8)$$

— Here, with a drop in true value, the size of error decreases.

— In this type of expression of accuracy, the percentage error is the maximum for any point in the instrument range.

The percent accuracies on the basis of actual scale or full-scale reading are normally used to express *equipment accuracies* and do not include procedural or personnel performance which is actually included in the total error.

Example 2.1. *A voltmeter reads 112.68 V. If the true value of the voltage is 112.6 V, determine the following :*

(i) The static error.

(ii) The static correction for the voltmeter.

Solution. *Given :* $V_m = 112.68$ volts; $V_t = 112.6$ volts.

(*i*) **Static error, E_s :**

$$E_s = V_m - V_t = 112.68 - 112.6 = +0.08 \text{ V} \quad \textbf{(Ans.)}$$

(ii) Static correction, C_s :

$$C_s = -E_s = -\mathbf{0.08\ V} \quad \textbf{(Ans.)}$$

Example 2.2. *A thermometer reads 92.35°C and the static correction given in the correction curves is – 0.07°C. Determine the true value of the temperature.*

Solution. *Given :* $V_m = 92.35°C; \quad C_s = -0.07°C.$

True value, V_t :

We know that, $E_s = V_m - V_t$

or, $-C_s = V_m - V_t$ $(\because C_s = -E_s)$

or, $V_t = V_m + C_s$ ...[Eqn (2.6)]

$$= 92.35 - 0.07 = \mathbf{92.28°C} \quad \textbf{(Ans.)}$$

Example 2.3. *An analog indicating instrument with a scale range of 0–5.0V shows a voltage of 2.65 V. The true value of a voltage is 2.70 volts.*

(i) What are the values of absolute error and correction ?

(ii) Express the error as a function of the true value and full scale deflection.

Solution. *Given :* $V_m = 2.65\ V; \quad V_t = 2.70\ V$

(i) Absolute error, E_s :

$$E_s = V_m - V_t = 2.65 - 2.70 = \mathbf{-0.05\ V} \quad \textbf{(Ans.)}$$

Absolute correction, C_s :

$$C_s = -E_s = -(-0.05) = \mathbf{0.05\ V} \quad \textbf{(Ans.)}$$

(ii) Relative error, E_r :

As a function of the *true value* :

$$E_r = \frac{E_s}{V_t} = \frac{-0.05}{2.70} = -0.0185, \text{ or, } \mathbf{-1.85\%} \quad \textbf{(Ans.)}$$

As a function of *full scale deflection :*

$$E_s = \frac{E_s}{(5.0-0)} = \frac{-0.05}{5} = -0.01, \text{ or, } \mathbf{-1\%} \quad \textbf{(Ans.)}$$

Example 2.4. *A pressure indicator showed a reading as 42 bar on a scale range of 0–50 bar. If the true value was 41.4 bar, determine :*

(i) Static error.

(ii) Static correction.

(iii) Relative static error.

Solution. *Given :* $V_m = 42$ bar ; $V_t = 41.4$ bar ; Scale range = 0–50 bar

(i) Static error, E_s :

$$E_s = V_m - V_t = 42 - 41.4 \text{ bar} = \mathbf{+\ 0.6\ bar} \quad \textbf{(Ans.)}$$

(ii) Static correction, C_s :

$$C_s = -E_s = -(+\ 0.6) = \mathbf{-0.6\ bar} \quad \textbf{(Ans.)}$$

(iii) Relative error, E_r :

$$E_r = \frac{E_s}{V_t} = \frac{0.6}{41.4} = 0.0145 \text{ or } \mathbf{1.45\%} \quad \textbf{(Ans.)}$$

Example 2.5. *A pressure gauge of range 50 bar is stated to have an error of ± 0.15 bar when calibrated by the manufacturer. Determine :*

(i) The percentage error on the basis of maximum scale value.

(ii) The possible error as a percentage of the indicated value when a reading of 10 bar is obtained in a test.

Solution. *Given* : Pressure range = 50 bar ; Error = ± 0.15 bar

(*i*) *Percentage error on the basis of maximum scale value*

$$= \pm \frac{0.15}{50} \times 100 = \mathbf{\pm 0.3\%} \quad \textbf{(Ans.)}$$

(*ii*) *Percentage error on the basis of indicated value of 10 bar pressure*

$$= \pm \frac{0.15}{10} \times 100 = \mathbf{\pm 1.5\%} \quad \textbf{(Ans.)}$$

The above results indicate that the pressure gauge is *more unreliable at the lower end of its gauge.*

Example 2.6. *A system comprising a transmitter, a relay and a receiver element has been used to measure the pressure at a remote point. If the specified accuracy limits for transmitter, relay and receiver are within ± 0.3%, ±1.4% and ± 0.9% estimate the maximum possible error and root-square accuracy of the measurement system.*

Solution. Maximum possible error (*least accuracy*) = ± (0.3 + 1.4 + 0.9) = **± 2.6%** **(Ans.)**

Least root-square accuracy or root-sum square error

$$= \pm\sqrt{(0.3)^2 + (1.4)^2 + (0.9)^2} = \mathbf{\pm 1.69\%} \quad \textbf{(Ans.)}$$

The above results indicate that the error is possibly as large as ± 2.6% but probably not larger than ± 1.69%.

- In an instrument system, *root-square accuracy is more justifiable because it is not probable that all the elements of the system will have the greatest static error at the same point and at the same time.*

2.8.3.2. Static calibration

Static calibration *is a process by which all the static performance characteristics are obtained in one form or another.* In general, static calibration refers to a situation in which all inputs, whether desirable, interfering or modifying except one, are kept at some constant values. Then the one input under study is varied over some range of constant values, which causes the output (s) to vary over some range of constant values. Thus an *output-input relationship is developed which comprise a static calibration valid under the stated constant conditions of all the other inputs.*

- The procedure of calibration involves a comparison of the particular instrument with either :
  (*i*) a *primary standard.*
  (*ii*) a *secondary standard with a higher accuracy than the instrument to be calibrated,* or
  (*iii*) an *instrument of known accuracy.*

— All working instruments must be calibrated against some reference instruments *which have a higher accuracy.* The reference instruments in turn must be calibrated against instruments of still higher grade of accuracy or against primary standards or against other standards of known accuracy.

- The following **steps** are necessary in *performing a calibration :*
  1. *Examine the construction of the instrument,* and identify and list all the *possible inputs.*

2. Decide as best as possible which of the *inputs will be significant* in the application for which instrument is to be calibrated.
3. Procure apparatus that *will allow to vary all the significant inputs over the ranges considered necessary.* Procure standards to measure each input.
4. By holding some inputs constant, varying others, and recording the output(s), *develop the desired static input-output relations.*

2.8.3.3. Range and span

Range. *The difference between the largest and the smallest reading of the instrument is called the "Range of an instrument"*. The range is expressed by stating the lower and upper values.

Span *represents the algebraic difference between the upper and lower range values of the instrument.*

If the highest point of calibration is S_{max} units while the lowest is S_{min} units and that the calibration is continuous between the points, then we say that the *instrument range is between* S_{min} *and* S_{max}.

The *instrument* ***span*** is given by, $S_{max} - S_{min}$.

The above definitions apply both to analog as well as digital instruments.

Error calibration curve : Refer to Fig. 2.10.

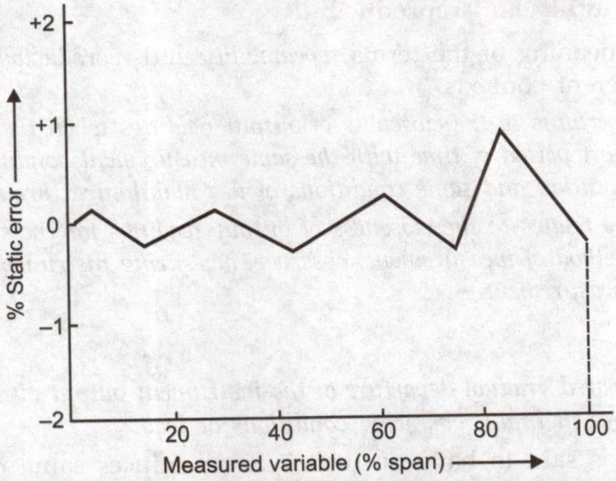

Fig. 2.10. Error calibration curve.

- *It is an error curve which can be used for correcting instrument readings.*
- The curve is plotted by calibrating the instrument against a suitable standard at a number of points on the scale.

  Examples :

  (i) *Range :* 2 kN/m^2 to 50 kN/m^2;
  Span : 50 – 2 = 48 kN/m^2

  (ii) *Range :* –5°C to 90°C ;
  Span : 90 – (–5) = 95°C.

- While determining the range of the instrument the frequency range must be taken into consideration. *"Frequency range"* is defined as *frequency over which measurements can be performed with a specified degree of accuracy.*

  Example : A moving voltmeter may have a range of 0–120 V range and 0–125 Hz frequency range with an accuracy of 0.5 or 1 per cent of full scale reading.

Example 2.7. *A pressure gauge is calibrated 5 bar to 60 bar. The accuracy is specified with ± 0.20 per cent of instrument span. What is the maximum static error ?*

Solution. Span of pressure gauge = 60 – 5 = 55 bar

$$\therefore \quad \text{Maximum static error} = \frac{\pm 0.20 \times 55}{100} = \pm \mathbf{0.11\ bar} \quad \textbf{(Ans.)}$$

2.8.3.4. Scale readability

The term scale readability is frequently used in analog type of instruments. *It indicates the closeness with which the scale of an analog type of instrument can be read.*

The readability depends upon many *factors* such as :

(*i*) Number of graduations;
(*ii*) Spacing of graduations;
(*iii*) Size of pointer;
(*iv*) Parallax effects;
(*v*) Discriminating power of the observer.

The readability, thus, is logically implied by the number of significant figures; the higher the number of significant figures, the better the readability of the instrument scale.

2.8.3.5. Repeatability and reproducibility

Although the meaning of the terms *repeatability* and *reproducibility* is same but they are applied in different contexts.

Repeatability *pertains to the closeness of output readings when the same input is applied repetitively over a short period of time with the same measurement conditions, same instrument and observer, same location and same conditions of use maintained throughout.*

Reproducibility *relates to the closeness of output readings for the same input when there are changes in the method of measurement, observer, measuring instrument, location, conditions of use and time of measurement.*

2.8.3.6. Drift

Drift *is an undesired gradual departure of the instrument output over a period of time that is unrelated to changes in input, operating conditions or load.*

An instrument is said to have no drift if it reproduces same readings at different times for same variation in measured variable.

The drift may be caused by the following *factors* :

(*i*) High mechanical stresses developed in some parts of instruments and systems.
(*ii*) Wear and tear.
(*iii*) Mechanical vibrations.
(*iv*) Temperature changes.
(*v*) Stray electric and magnetic fields.
(*vi*) Thermal e.m.fs.

Examples of drift :

- Drift occurs in *"resistance thermometers and thermocouples" due to the contamination of the metal and a change in its atomic or metallurgical structure.*
- Drift may occur in *"flow meters" due to wear and erosion of the orifice plate, nozzle or venturimeter.*

Classification of drift:

Drift may be *classified* as follows :

1. Zero drift; 2. Span drift; 3. Zonal drift.

1. Zero drift :

Drift is called *zero drift if the whole of instrument calibration gradually shifts over by the same amount.*

— It may be due to *permanent set or slippage* and can be corrected by *shifting pointer position.*

2. Span drift :

If the calibration from zero upwards changes proportionally, it is called "span (or sensitivity) drift".

— It may be due to change in spring gradient etc.

3. Zonal drift :

When the drift occurs only over a portion of span of an instrument, it is called "zonal drift".

— In industrial instruments, drift is an undesirable quantity since it is rarely apparent and cannot be easily compensated for.

- *Drift occurs very slowly and can be checked only by periodic inspection and maintenance.*

2.8.3.7. Accuracy and precision

Accuracy. *The closeness with which an instrument reading approaches the true value of the quantity being measured is called* **accuracy**. Accuracy is determined as the maximum amount by which the result differs from the true value. Accuracy of an instrument is influenced by factors like static error, dynamic error, reproducibility, dead zone.

Precision. The term 'precise' means clearly or sharply defined. *"Precision" is a measure of reproducibility of measurement.* It is usually expressed in terms of the deviation in measurement.

— ***Significant figure*** *is an indication of the precision of measurement.* Significant figures convey actual information regarding the magnitude of precision.

Following **examples** will illustrate the difference in meaning of the two terms *accuracy* and *precision* :

(*i*) Let us consider the case of an ammeter having high precision which can take readings to 1/150 of an ampere, but has a wrong zero adjustment. Now we can take readings to 1/150 of an ampere which are consistent and clearly defined since the ammeter is as *precise* as ever. However, the readings taken with this ammeter are *not accurate,* since they do not conform to truth because of faulty adjustment of zero.

(*ii*) Take the case of a micrometer which is normal in every respect but its anvil is displaced from its true position. When the readings are taken by this instrument, these would be clearly defined and consistent and we would say that the micrometer is as *precise* as ever. However, the readings do not conform to the truth as anvil is displaced from its true position. Thus, the readings taken by this micrometer are *not accurate.*

- It is worth noting that *precision is a necessary prerequisite to accuracy but it does not guarantee accuracy.*

2.8.3.8. Sensitivity

The ratio of the magnitude of output signal to the input signal or response of measuring system to the quantity being measured is called **sensitivity.**

It is represented by the slope of the calibration curve if the ordinates are expressed in actual units.

— When the calibration curve is linear [Fig. 2.11 (*a*)], the sensitivity is constant. However, if the calibration curve is non-linear [Fig. 2.11 (*b*)], the sensitivity is different at different points, being the slope of curve in various points.

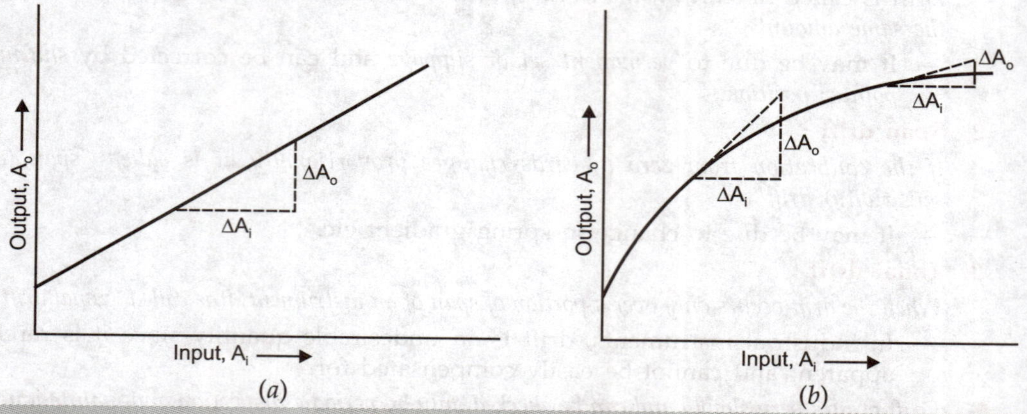

Fig. 2.11. Static sensitivity.

Static sensitivity, in general, is defined as :

$$\text{Static sensitivity } (S) = \frac{\text{Infinitesimal change in output}}{\text{Infinitesimal change in input}} = \frac{\Delta A_o}{\Delta A_i} \quad ...(2.9)$$

Similarly, *Inverse sensitivity* or *deflection factor* or *scale factor* = $\frac{\Delta A_o}{\Delta A_i}$ ...(2.10)

- *The sensitivity of an instrument should be high and thus the range should not be high in comparison to the value being measured.*
- Sensitivity has a wide range of units, and these depend upon the instrument or measurement system being investigated.

The *following points are worth noting :*

— In case the form of input to and output from the measurement system used with electrical/electronic equipment is same, the term **gain** is used instead of *sensitivity.*

— The *increase* in displacement with the optical and mechanical instruments is described by the term **amplification.**

— When the input or output signal is *changing with time,* the term **transfer function** or **transfer operator** is used (in place of the terms *sensitivity, gain or amplification*— which apparently mean the same and describe a relationship between output and input).

Example 2.8. *A pressure gauge, which has linear calibration curve, has radius of scale line as 90 mm and pressure of zero to 60 pascals is displayed over an arc of 300°. Determine the sensitivity of the gauge as a ratio of scale length to pressure.*

Solution.

$$\text{Full scale deflection} = 300° = 300 \times \frac{\pi}{180} = \frac{5\pi}{3} \text{ rad.}$$

$$\text{Length of scale} = \frac{5\pi}{3} \times 90 = 150\,\pi \text{ mm}$$

$$\therefore \quad \textbf{Sensitivity} = \frac{\text{Scale length}}{\text{Calibration pressure}}$$

$$= \frac{150\pi}{60} = \mathbf{2.5\pi\ mm/Pa} \quad \text{(Ans.)}$$

Example 2.9. *In a Wheatstone bridge, a change of 6Ω in the unknown arm of the bridge is required to produce a change in deflection of 2.4 mm of the galvanometer. Determine :*
(i) Sensitivity; *(ii) Deflection factor.*

Solution. *Given* : Magnitude of output response = 2.4 mm;
Magnitude of input = 6 Ω

$$\therefore \quad (i)\ \textbf{Sensitivity} = \frac{2.4}{6} = \mathbf{0.4\ mm/\Omega} \quad \text{(Ans.)}$$

and,
$$(ii)\ \textbf{Deflection factor} = \frac{6}{2.4} = \mathbf{2.5\ \Omega/mm} \quad \text{(Ans.)}$$

Example 2.10. *A pressure measuring system consists of a piezoelectric transducer, a charge amplifier and an ultra violet charge recorder; their sensitivities are 6.8 pC/bar, 0.0032 V/pC and 16 mm/V respectively. Determine the deflection on the chart for a pressure change of 20 bar.*

Solution. With the elements arranged in series, the *overall sensitivity* of the measuring system (S)

= Sensitivity of piezoelectric transducer (S_1) × sensitivity of the charge amplifier (S_2) × sensitivity of ultra violet charge recorder (S_3).

$$= 6.8\frac{pC}{\text{bar}} \times 0.0032\frac{V}{pC} \times 16\frac{\text{mm}}{V} = 0.348\frac{\text{mm}}{\text{bar}}$$

Also,
$$\textit{Sensitivity} = \frac{\text{Change of output signal}}{\text{Change of input signal}}$$

or,
$$0.348 = \frac{\text{Output}}{20\ \text{bar}}$$

$$\therefore \quad \textit{Deflection on the chart} = 0.348 \times 20 = \mathbf{6.96\ mm} \quad \text{(Ans.)}$$

2.8.3.9. Linearity

The ability to reproduce the input characteristics symmetrically is called **linearity**. It can be expressed by the straight line equation.

The linearity is simply a measure of maximum deviation of any of the calibration points from the straight line (drawn by using the method of least square from the given calibration data). Fig. 2.12 shows the actual calibration curve and a straight line drawn from the origin using method of least squares.

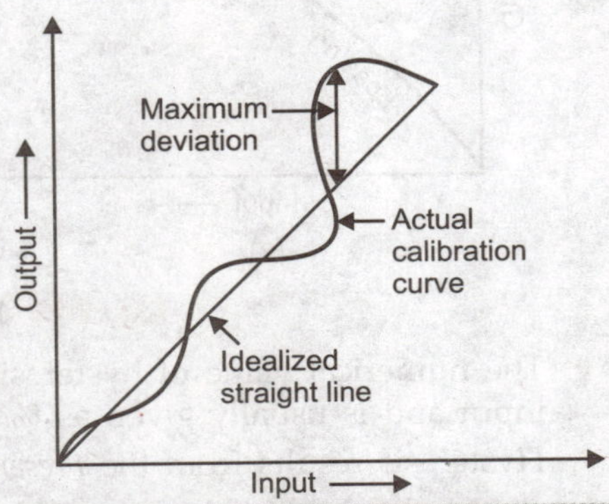

Fig. 2.12. Linearity exhibited as deviation between actual calibration curve and idealized straight line relationship.

Any departure from straight line relationship is non-linearity. The non-linearity may be due to the following *factors :*

(i) Viscous flow or creep;
(ii) Non-linear elements in the measurement device;
(iii) Mechanical hysteresis;
(iv) The elastic after-effects in the mechanical system.

The two common types of non-linearity are :

1. **Terminal linearity.** It is the deviation from a straight line through the end points.
2. **Best fit linearity.** It is the deviation from the straight line which gives minimum errors, both plus and minus.
   - **Linearity** (*i.e.,* output is linearly proportional to input) is considered to be one of the best characteristics of an instrument or measurement system. This is so, because the conversion from a scale reading to the corresponding value of input quantity is very convenient. However, it may be understood that a non-linear behaviour does not essentially lead to inacuracy; an instrument having non-linear calibration curve may be highly accurate as ever.
   - Since accuracy and linearity are related closely to each other, it is better to keep the non-linearity as small as possible so as to result in small errors.

2.8.3.10. Hysteresis

The maximum differences in output at any measured value within the specified range when approaching the point first with increasing and then with decreasing input may be termed as **hysteresis.**

- It is a phenomenon which shows different output effects when loading and unloading. It is *non-coincidence of loading and unloading curves.*

Figure 2.13(*a*) shows output and input curves (loading and unloading) for an instrument which has no friction due to sliding parts. The non-coincidence of loading and unloading curves is on account of *internal friction* or *hysteresis damping.*

Figure 2.13(*b*) shows the input-output relations of instruments which do not have internal friction but have external sliding friction.

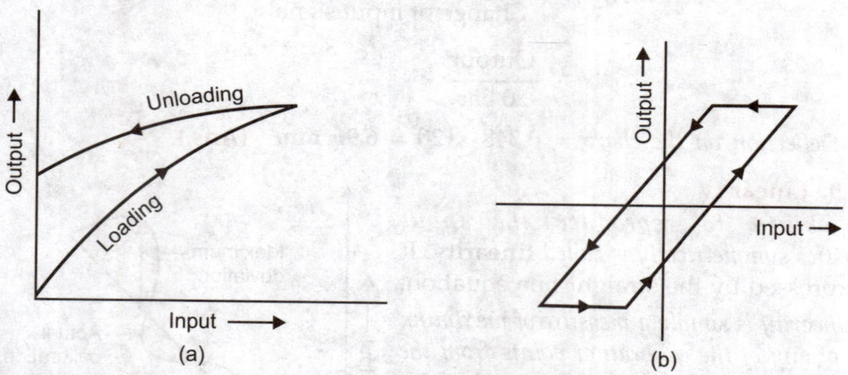

Fig. 2.13. Hysteresis effects.

- The numerical value of hysteresis can be specified in terms of either output or input and is usually given as *%age of full scale.*
- Hysteresis results from the *presence of irreversible phenomenon such as :*
  — Mechanical friction;
  — Slack motion in bearings;
  — Magnetic and thermal effects.

2.8.3.11. Threshold and resolution

Threshold :

The minimum value below which no output change can be detected when the input of an instrument is increased gradually from zero is called the **threshold** *of the instrument.* Thus threshold defines minimum value of input which is necessary to cause a detectable change from zero output.

— In a *digital system,* threshold is the input signal necessary to cause one least significant digit of the output reading to change.

- Threshold may be caused by *backlash or internal noise.*

Resolution or discrimination :

When the input is slowly increased from some arbitrary (non-zero) input value, it is observed that the output does not change at all until a certain increment is exceeded; this increment is called ***Resolution or discrimination*** *of the instrument.* Thus resolution defines the smallest change of input for which there will be a change of output.

— In case of *analog instruments,* the resolution is determined by the observer's ability to judge the position of a pointer on a scale. Resolution is usually reckoned to be no better than ± 0.2 of the smallest division of the scale.

— In case of digital instruments, resolution is determined by the number of neon tubes taken to show the measured value.

- **Threshold** *defines the smallest measurable input* while the **resolution** *defines the smallest measurable input change.*
- *"Threshold"* and *"resolution"* may be *expressed as an actual value or as a fraction or percentage of full scale value.*

Example 2.11. *A transducer measures a range of 0–200 N force with a resolution of 0.15 per cent of full scale. What is the smallest change which can be measured by this transducer ?*

Solution. *Given :* Range of force : 0 – 200 N

Resolution : 0.15% of full scale

Smallest change which can be measured $= \frac{0.15}{100} \times 200 = \mathbf{0.3\ N}$ **(Ans.)**

Example 2.12. *Determine the resolution of a moving coil voltmeter having a uniform scale with 50 divisions; the full scale reading is 50 V and $\frac{1}{10}$ of a scale division can be estimated with a fair degree of certainty.*

Solution. One scale division $= \frac{50}{50} = 1$ V

∴ *Resolution* $= \frac{1}{10} \times 1 = \mathbf{0.1\ V}$ **(Ans.)**

Example 2.13. *Determine the resolution of a digital voltmeter which has read-out range 0 to 9999 counts and its full scale reading is 9.999 V.*

Solution. *Resolution of the digital voltmeter*

$$= \frac{1}{9999} \times 9.999$$

$$= 0.001 \text{ V or } \mathbf{1\ mV}$$ **(Ans.)**

2.8.3.12. Dead zone and dead time

Dead zone :

Refer to Fig. 2.14. *The largest change of input quantity for which there is no output of the instrument is termed as* ***dead zone.*** It may occur due to friction in the instrument which does not allow pointer to move till sufficient driving force is developed to overcome the friction loss.

- *"Dead zone" is caused by backlash and hysteresis in the instrument.*

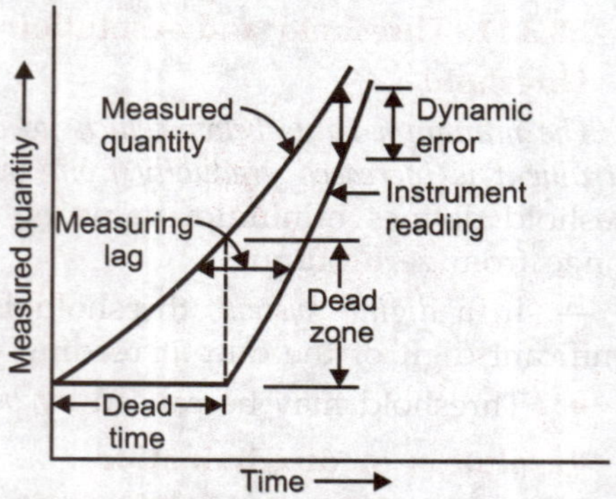

Fig. 2.14. Dead zone and dead time.

Dead time :

Refer to Fig. 2.14. *The time required by a measurement system to begin to respond to a change in the measurand is termed as* ***dead time.*** It represents the time before the instrument begins to respond after the measured quantity has been changed.

Example 2.14. *The calibration range of a certain pyrometer is 300°C to 800°C. If the dead zone in it is 0.11 per cent of span, determine the temperature change which might occur before it is detected.*

Solution. Span of the pyrometer = 800 – 300 = 500°C

$$\therefore \quad \text{Dead zone} = \frac{0.11}{100} \times 500 = 0.55°\text{C}$$

Hence, a change of 0.55°C must occur before it is detected. **(Ans.)**

2.8.3.13. Loading effects

Under ideal conditions an element used for signal sensing, conditioning, transmission and detection should not change/distort the original signal. The sensing element should not use any energy or take least energy from the process so as not to change the parameter being measured. However, under practical condition, it has been observed that the introduction of any element in a system results invariably *in extraction of the energy* from the system thereby *distorting the original signal.* This distortion may take the form of *attenuation, waveform distortion, phase shift etc.* Consequently the ideal measurements become impossible.

The incapability of the system to faithfully measure the input signal in undistorted form is called the ***loading effect.*** This results in *loading errors.*

- The loading effects, in a measurement system, not only occur in *detector-transducer stage* but also occur in *signal conditioning and signal presentation stages* as well. The loading problem is carried right down to the basic elements themselves.

The loading effects may occur on account of both electrical and mechanical elements; these are *due to impedances of the various elements connected in a system.* The mechanical impedances may be treated similar to electrical impedances.

- *Loading effects occur due to extraction of energy from the measurand* (as earlier stated).

The transfer of energy requires the specification of two variables. The physical variables which determine the flow of energy in all dynamical systems can be classified as :

(*i*) Through variables; (*ii*) Across variables.

— *"Through variables"*, also called flow variables, are those which can be specified and measured at *one point in a space.* These are extensive variables in the sense, that their magnitude depends upon the extent of the system taking part in the energy transfer.

— *"Across variables"*, also called effort variables and transvariables, are those which can be specified by *two points in a space;* usually one point is a reference point. These are intensive variables whose magnitude is independent of the medium being considered.

All objects in a measurement system or any dynamical system involve a measurand or defined relationship between an "across variable" and a "through variable".

Static stiffness and static compliance :

In some instruments when input signal is applied to them under *static conditions,* the output signal is a faithful representation of the input signal. But when these instruments are subjected to *dynamic conditions* (*i.e.,* going from one steady state to another one), they produce loading effects since energy is required to perform these operations. Under such conditions concepts of static stiffness and static compliance (instead of the concepts of impedance and admittance) need to be introduced.

It may be understood that *while impedance and admittance are concerned with* ***power drain,*** *stiffness and compliance are concerned with* ***energy drain.***

Mechanical Stiffness, in *general,* is defined as :

$$K_g = \frac{\text{Across variable}}{\int (\text{Through variable})\, dt} \quad ...(2.11)$$

Compliance is defined as *inverse of stiffness* (as admittance is inverse of impedance) and is given by

$$C_g = \frac{1}{K_g} = \frac{\text{Through variable}}{\int (\text{Across variable})\, dt} \quad ...(2.12)$$

Example 2.15. *(a) Discuss the concept of impedance loading and impedance matching as applied to a measurement system.*

(b) An unknown resistance is connected in series with a milliammeter of very small internal resistance. A voltmeter with internal resistance 125 kΩ is connected across the unknown resistance.

Determine the loading error when :

(i) the voltmeter and milliammeter read 180 V and 6 mA respectively.

(ii) the same voltmeter and milliammeter with another resistance read 60 V and 1.2 A respectively. Comment upon the results.

Solution. ***(a)* Impedance loading and impedance matching :**

In a generalised measurement system, the sensing element senses the condition, state or value of the process variable by *extracting a small part of energy from the measurand,* and then processes an output which reflects this condition, state or value of the measurand. Due to this energy extraction the measurand is always disturbed by the measurement act and consequently a perfect measurement becomes *theoretically impossible.*

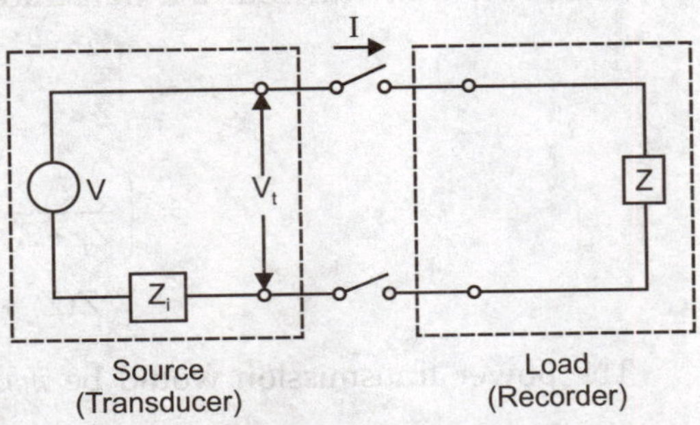

Fig. 2.15. Loading effect of recorder on the transducer element.

Examples : (*i*) The reading shown by the hand tachometer would vary with the pressure with which it is pressed against the shaft; (*ii*) A milliammeter would introduce additional resistance in the circuit and thereby change the flow of current by a significant amount.

The *interaction of system* (mentioned above) is referred to as *"loading effect"* and the *term loading when applied to a general electrical system is called* **impedance loading.** While designing an instrument efforts should be made to ensure that the *alteration due to system interaction is minimal.*

Refer to Fig. 2.15. Consider a transducer element with internal impedance, Z_i (= $R_i + jX_i$), R and X being the resistance and reactance respectively.

Let, V = Voltage developed in the transducer, and

V_t = Terminal voltage.

Then, when output terminals are open (*i.e.*, no-load condition),

$$V_t = V \quad ...(i)$$

When the transducer is connected in series with a recorder of input impedance Z (= $R + jX$), a current flows through the circuit, the value of which is given as,

$$I = \frac{V}{Z_i + Z}$$

and,

$$V_t = V - I Z_i = V - \left(\frac{V}{Z_i + Z}\right) Z_i$$

$$= V\left[1 - \frac{Z_i}{Z_i + Z}\right] = \frac{Z}{Z_i + Z} \cdot V$$

i.e.,

$$V_t = \frac{Z}{Z_i + Z} \cdot V \quad ...(ii)$$

[In ideal cases, when impedance loading is zero (*i.e.*, $Z_i = 0$), eqn. (*ii*) transforms to eqn. (*i*)]

From eqn. (*ii*), it may be observed that for the *loading effect to be minimal, the following conditions should be satisfied :*

(*i*) Z_i (transducer impedance) should be *small.*

(*ii*) Z (impedance of the measuring device) must be *high.* (It is due to this reason that vacuum tube voltmeters having an internal resistance to the tune of 10 MΩ are used in the measurement of the voltage).

The *power delivered* from the transducer to the external load,

$$P = \frac{V_t^2}{Z}$$

$$= \left(\frac{Z}{Z_i + Z} \cdot V\right)^2 \cdot \frac{1}{Z} \quad ...[\text{Substituting for } V_t \text{ from eqn. } (ii)]$$

$$= V^2 Z (Z_i + Z)^{-2}$$

The power transmission would be *maximum* when, $\frac{dP}{dZ} = 0$

i.e.,

$$\frac{dP}{dZ} = V^2\left[(Z_i + Z)^{-2} \times 1 + Z \times (-2)(Z_i + Z)^{-3}\right] = 0$$

or, $$(Z_i + Z)^{-2} - 2Z(Z_i + Z)^{-3} = 0$$

or, $$1 - 2Z(Z_i + Z)^{-1} = 0$$

or, $$(Z_i + Z) - 2Z = 0$$

or, $$Z = Z_i \qquad ...(iii)$$

Eqn. (*iii*) indicates that maximum power will be delivered from the transducer when the impedance of the external load just *matches* its internal impedance. This aspect is referred to as ***impedance matching*** *in electrical circuits.*

If $Z_i = R_i$ and $Z = R$

Then, $V_t = \frac{R}{R_i + R}$, and $R = R_i$.

(*b*) Refer to Fig. 2.16. *Given* : Internal resistance of the voltmeter, $R_v = 125$ kΩ;

Readings of voltmeter and ammeter in two cases :

(*i*) 180 V, 6 mA; (*ii*) 60 V, 1.2 A

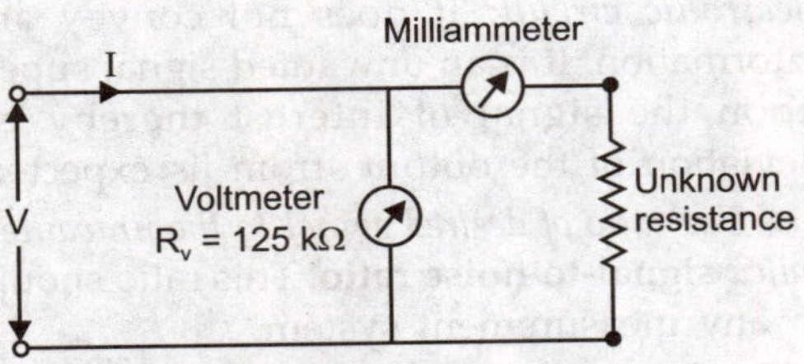

Fig. 2.16.

(*i*) Total resistance of the circuit,

$$R_T = \frac{V}{I} = \frac{180}{6 \times 10^{-3}} = 30 \text{ k}\Omega$$

Since the milliammeter is stated to have negligible resistance, the total resistance as calculated above equals the apparent resistance ($R_{appt.}$) of the resistor.

i.e., $R_{appt.} = 30$ kΩ

The actual value of resistance of the resistor is obtained from the condition that voltmeter connected is parallel with the resistor, *i.e.*,

$$\frac{1}{R_T} = \frac{1}{R_{act.}} + \frac{1}{R_v}$$

or, $$\frac{1}{R_{act.}} = \frac{1}{R_T} - \frac{1}{R_v} = \frac{R_v - R_T}{R_T \times R_v}$$

or $$R_{act.} = \frac{R_T \times R_v}{R_v - R_T} = \frac{30 \times 125}{125 - 30} = 39.474 \text{ k}\Omega$$

$\therefore$ $$\textit{Percentage loading error} = \frac{R_{act.} - R_{appt.}}{R_{act.}}$$

$$= \frac{39.474 - 30}{39.474} \times 100 = \mathbf{24\%} \quad \textbf{(Ans.)}$$

(*ii*) Total resistance, $R_T = \frac{60}{1.2} = 50\ \Omega$

i.e., $R_{appt.} = 50\ \Omega$

$$R_{act.} = \frac{R_T \times R_v}{R_v - R_T} = \frac{50 \times 125}{125 - (50/1000)} = 50.02\ \Omega$$

$$\therefore \quad \textit{Percentage loading error} = \frac{50.02 - 50}{50.02} \times 100 = \mathbf{0.04\%} \quad \textbf{(Ans.)}$$

From the above results it may be concluded that :

- When a voltmeter with high internal resistance is connected across two points in a *high resistance circuit*, the loading effect is appreciable and, therefore, the voltmeter gives *misleading voltage reading*.
- The same voltmeter has a *negligible loading error* when connected across a *low resistance circuit*, and it gives more *reliable voltage reading*.

2.8.3.14. Noise

Noise *is a spurious current or voltage extraneous to the current or voltage of interest in an electrical or electronic circuit.* It does not convey any useful information. It is an unwanted signal superimposed upon the signal of interest thereby causing a deviation of the output from its expected value.

The ratio of desired signal to the unwanted signal is called **signal-to-noise ratio**. This ratio should be *large* in any measurement system.

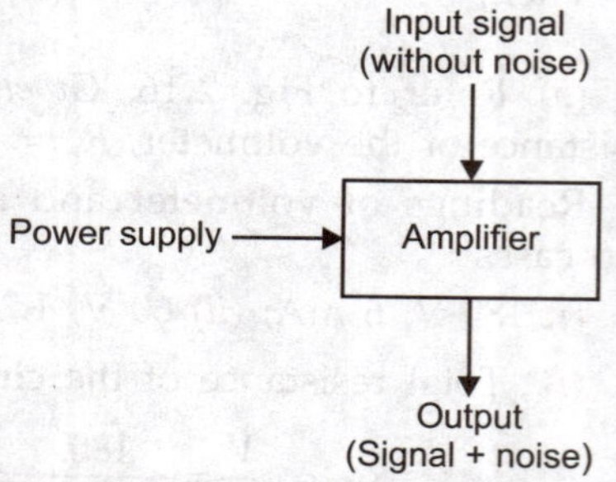

Fig. 2.17. Block diagram of an amplifier for illustration of different types of noise signals.

The noise may be broadly *classified* as follows :

(*i*) Generated noise;

(*ii*) Conducted noise;

(*iii*) Radiated noise.

Refer to Fig. 2.17. It shows a block diagram of an amplifier for illustration of different types of noise signals.

(*i*) Generated noise :

- The internally generated noise is on account of components like resistors, capacitors, transistors, etc.
- The temperature dependent noise in a resistor which increases with internal heating (I^2R loss) or with an increase in the ambient temperature is called *"Johnson noise"*.
- The wide band noise, produced within a resistor, due to vibrations produced by the thermal effects, covering a wide frequency range is called *"White noise"*.

  The internally generated noise in resistor can be *reduced* by :

  (*a*) lowering the internal temperature;

  (*b*) using special film and glass substrates.

  The internally generated noise in semiconductor and vacuum tube devices on account of random movement of charges is called *"Shot noise"*. It is difficult to reduce this type of noise; however, it can be reduced to some extent by the use of selective filtering.
- The internally generated noise may also be caused by the changing electric fields in the region between plates of a capacitor and changing magnetic fields around in an inductor. In this case noise signal has a fixed frequency (since fields change in orderly manner) and as such its magnitude can be reduced by using a filter tuned to this frequency.

***(ii)* Conducted noise :**

- This type of noise may be caused by the power supply (to the amplifier) which may have ripples or random deviations.
- 50 Hz power supply and the harmonics contained in it are one of the common sources for conducted noise.
- It can be reduced by using filters in the leads to trap out the noise.

***(iii)* Radiated noise :** Radiated noise consists of unwanted signals (relating to electric or magnetic fields or disturbances in the environments around the amplifier) which are radiated into the interior of the amplifier.

- The electromagnetic impulses radiated from ignition wiring of spark plugs are one of the common sources of radiated noise.
- It can be *reduced by proper shielding.*

2.8.4. Dynamic Characteristics

When an instrument is required to measure a time-varying process variable, one has to be concerned with "*dynamic characteristics*"; these *characteristics quantify the dynamic relation between the input and output.*

2.8.4.1. Dynamic response

The evaluation of the ability of a system to faithfully transmit and present all the pertinent information included in the input signal and to exclude all else, is called the **response.**

The behaviour of the system when inputs vary with time (i.e., inputs are dynamic in nature) and so does the output, is called **dynamic response.** Such a response in measurement systems is found to occur in industrial, aerospace and biological applications.

The dynamic inputs are of two types :

(i) Steady state periodic; *(ii)* Transient.

The "*steady state periodic*" quantity is one whose magnitude has a definite repeating time, whereas the time variation of a ***"transient"*** magnitude does not repeat.

- The response of a measurement system subjected to a time varying input can be divided into the following two parts :

  (i) Steady state response; *(ii)* Transient response.

  "Steady state response" is simply the response when time reaches infinity.

  "Transient response" in measurement systems is defined as the part of response which goes to zero as time becomes large.

- **Periodically varying inputs may be :**

  — A *purely sinusoidal input* like the A.C. voltage with a frequency of 50 Hz.

  or

  — A complex time varying repetitive input which is a *combination of the fundamental* and *harmonics.*

When the systems are subjected to periodically varying inputs, they exhibit in their response a magnitude and phase relationship which is *different* from that of the input signal because of the *energy storage elements. The output is not a faithful representation of the input.* The harmonic contents of the input are not reproduced in the output both in magnitude and phase relationship as in the input. *The output is distorted both in terms of magnitude as well as phase relationships.*

2.8.4.2. Dynamic characteristics of a measurement system

The *dynamic characteristics* of a measurement system are :

1. Speed of response ... *Desirable*
2. Measuring lag ... *Undesirable*
3. Fidelity ... *Desirable*
4. Dynamic error ... *Undesirable*

1. Speed of response. The *"speed of response"* or *"responsiveness"* is defined as the *rapidity with which a measurement system responds to changes in the measured quantity.*

2. Measuring lag. It refers to *retardation or delay in the response of a measurement system to changes in measured quantity.*

The lag *is caused by conditions such as capacitance, inertia, or resistance.*

The measuring lags are of the following two types :

(i) Retardation type lag. In this type of measuring lag the response of the measurement system *begins immediately after a change in measured quantity has occurred.*

(ii) Time delay type lag. In this case the response of the measurement system *begins after a dead time after the application of the input.* If the measured quantity varies at a fast rate, the dead time has a severe adverse effect on the performance of the system. The measurement lags of this type are *very small* and are of the order of a fraction of a second and hence can be *ignored.*

3. Fidelity. It is defined as the *degree to which a measurement system indicates changes in the measured quantity without any dynamic error.* It refers to the ability of the system to reproduce the output in the same form as the input.

4. Dynamic error. The dynamic error, also called **"measurement error"**, is the *difference between the true value of the quantity changing with time and the value indicated by the measurement system if no static error is assumed* (see Fig. 2.14.)

- *The maximum amount by which the pointer moves beyond the steady state is called "overshoot".*

2.8.4.3. 'Dynamic' analysis of measurement systems

The dynamic behaviour of measurement systems is studied in the following two domains :

1. Time domain analysis.
2. Frequency domain analysis.

1. Time domain analysis :

In this the input signal is applied to the measurement system and the behaviour of the system is studied as a *function of time.* The dynamic response of the system to different types of inputs, which are a function of time, is analysed at different intervals of time after the application of the input signal. In most cases, the actual input signals vary in random fashion with respect to time and therefore cannot be mathematically defined. Consequently the performance of a system can be analysed (in the time domain analysis) by using the following standard *test signals/inputs :*

(i) Step input;

(ii) Ramp input;

(iii) Parabolic input;

(iv) Impulse input.

2. Frequency domain analysis :

This type of analysis of a system *pertains to the steady state response of the system to a sinusoidal input*. Here, the system is subjected to a sinusoidal input and the system response is studied with *frequency as the independent variable.*

- *Frequency response.* It is the maximum frequency of the measured variable that an instrument is capable of following without error. The usual requirement is that the frequency of measurand should not exceed 60 per cent of the natural frequency of the measuring instrument.

Standard test signals/inputs :

The most common standard inputs used for dynamic analysis are discussed below :

1. ***Step function :*** Refer to Fig. 2.18(*a*). It is a *sudden change from one steady value to another.*

It is mathematically represented by the relationship :

$$x = 0 \quad \text{at} \quad t < 0$$
$$x = x_c \quad \text{at} \quad t \geq 0$$

where x_c is a constant value of the input signal x_i.

- The "*transient response*" *indicates the capacity of the system to cope with change in the input signal.*

2. ***Ramp or linear function :*** Refer to Fig. 2.18(*b*). In this case, the *input varies linearly with time.*

This input is mathematically represented as :

$$x = 0 \quad \text{at} \quad t < 0$$
$$x = \psi \quad \text{at} \quad t \geq 0$$

where, ψ is the slope of the input versus time relationship.

- *The ramp-response becomes indicative of the steady state error in following the changes in the input signal.*

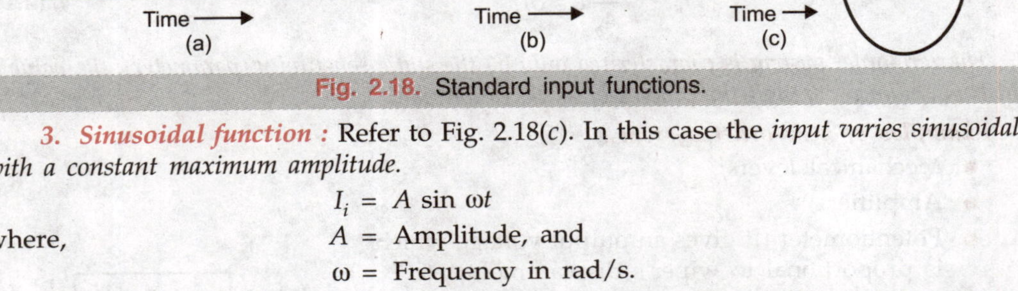

Fig. 2.18. Standard input functions.

3. ***Sinusoidal function :*** Refer to Fig. 2.18(*c*). In this case the *input varies sinusoidally with a constant maximum amplitude.*

$$I_i = A \sin \omega t$$

where, A = Amplitude, and

ω = Frequency in rad/s.

- *The frequency or harmonic response is a measure of the capability of the system to respond to inputs of cyclic nature.*

A general measurement system can be mathematically described by the following differential equation:

$$(A_nD^n + A_{n-1}D^{n-1} + ... + A_1D + A_0)I_o = (B_mD^m + B_{m-1}D^{m-1} + ... + B_1D + B_0)I_i \quad ...(2.13)$$

where, A's and B's = Constants, depending upon the physical parameters of the system,

D^k = Operative derivative of the order k,

I_o = The information out of the measurement system, and

I_i = The input information.

The order of the measurement system is generally classified by the value of the power of n.

- *Zero-order system* : $n = 0$ and $A_1, A_2, A_3 .... A_n = 0$
- *First-order system* : $n = 1$ and $A_2, A_3, A_4 .... A_n = 0$
- *Second-order system* : $n = 2$ and $A_3, A_4, A_5 .... A_n = 0$

The above method of classification is used for most of the instruments and systems.

Although general equation can be solved by various methods, we shall be using methods of D-operator for getting its solution.

2.8.4.4. Zero, first and second order systems

2.8.4.4.1. Zero-order system

Figure 2.19 shows the block diagram of a 'Zero-order system'. In this case the output of the measuring system (ideal) is directly proportional to input, no matter how the input varies. The *output is faithful reproduction of input without any distortion or time lag*.

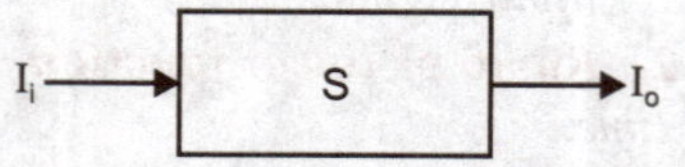

Fig. 2.19. Block diagram for Zero-order system.

The behaviour of the zero-order system is represented by the following mathematical solution :

$$I_o = S\,I_i \quad ...(2.14)$$

where, I_o = Information out of the measuring system,

S = Sensitivity of the system, and

I_i = Input information.

This equation is obtained by putting $n = 0$ in the general equation (2.13),

i.e., $$A_oI_o = B_oI_i$$

or $$I_o = \frac{B_o}{A_o}I_i = SI_i \quad ...(2.15)$$

The zero-order system is characterised only by the static sensitivity (parameter), the value of which is obtained through the process of static calibration.

Examples of zero-order system :

- Mechanical levers;
- Amplifiers;
- Potentiometer (It gives an output voltage which is proportional to wiper's displacement) etc.

2.8.4.4.2. First-order system

Figure 2.20 shows the block diagram of a 'First-order system'.

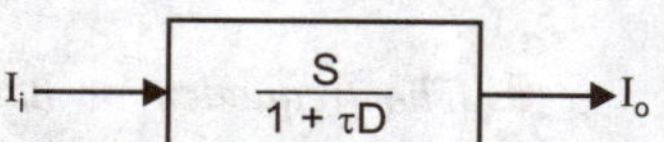

Fig. 2.20. Block diagram for First-order system.

The behaviour of a first-order system is given by following first-order differential equation :

$$A_1 \frac{dI_o}{dt} + A_o I_o = B_o I_i \quad ...(2.16)$$

(This equation is obtained by inserting $n = 1$ in the general equation).

Eqn. (2.16) may be written in standard form as follows:

$$\frac{A_1}{A_o}\frac{dI_o}{dt} + I_o = \frac{B_o}{A_o} I_i \quad ...(2.17)$$

or,

$$\tau \frac{dI_o}{dt} + I_o = SI_i \quad ...(2.18)$$

where,

$$\tau = \frac{A_1}{A_0} = \text{Time constant, and}$$

$$S = \frac{B_o}{A_o} = \text{Sensitivity.}$$

Using D-operator, we get

$$\left[\text{where,} \quad D = \frac{d}{dt}, \quad \text{and} \quad D^2 = \frac{d^2}{dt^2}\right]$$

$$\tau DI_o + I_o = SI_i$$

or,

$$I_o(\tau D + 1) = SI_i$$

or,

$$\frac{I_o}{I_i} = \frac{S}{I + \tau D} \quad ...(2.19)$$

Equation (2.19) gives the standard form of *transfer operator* for first-order system.

Examples of first-order system :

- Velocity of a true falling mass;
- Air pressure build-up in bellow;
- Measurement of temperature by mercury-in-glass thermometers;
- Thermistors and thermocouples;
- Resistance-capacitance network.

2.8.4..4.3. Second-order system

Fig. 2.21 shows the block diagram of 'Second-order system':

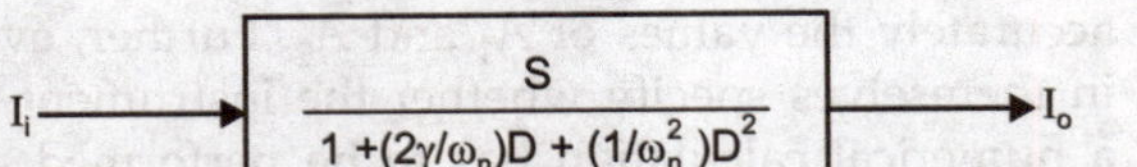

Fig. 2.21. Block diagram for Second-order system.

The behaviour of a second-order system is given by the following differential equation (obtained by putting $n = 2$ in the general equation);

$$A_2 \frac{d^2 I_o}{dt^2} + A_1 \frac{dI_o}{dt} + A_o I_o = B_o I_i \quad ...(2.20)$$

Dividing the above equation by A_o, we have

$$\frac{A_2}{A_o}\frac{d^2 I_o}{dt^2} + \frac{A_1}{A_0}\frac{dI_o}{dt} + I_o = \frac{B_o}{A_o} I_i \quad ...(2.21)$$

Let, $\omega_n = \sqrt{\frac{A_o}{A_2}}$ = Undamped natural frequency, rad/s,

$\gamma = \frac{A_1}{2\sqrt{A_0 A_2}}$ = Damping ratio, dimensionless, and

$S = \frac{B_o}{A_o}$ = Static sensitivity or steady-state gain.

Then, by substituting these values in eqn. 2.21, we get

$$\frac{I}{\omega_n^2} \cdot \frac{d^2 I_o}{dt^2} + \frac{2\gamma}{\omega_n} \cdot \frac{dI_o}{dt} + I_o = SI_i \quad ...(2.22)$$

or, in terms of D-operator, we have

$$\left(\frac{D^2}{\omega_n^2} + \frac{2\gamma}{\omega_n} D + 1\right) I_o = SI_i$$

or, $$\frac{I_o}{I_i} = \frac{S}{\frac{1}{\omega_n^2} D^2 + \frac{2\gamma}{\omega_n} D + 1} \quad ...(2.23)$$

Examples of second-order system :

- Piezoelectric pickup;
- Spring-mass system (used for acceleration and force measurements);
- Pen control system on X-Y plotters;
- U.V. galvanometer, etc.

Damping ratio :

In the design of instruments a term which is very frequently used is the "*damping ratio*" (γ) defined as the *ratio of the actual value of coefficient of viscous friction in movement and the value required to produce critical damping.*

i.e., $$\gamma = \frac{A_1}{2\sqrt{A_o A_2}}$$

This dimensionless term is very useful because to determine its value, it is not necessary that the values of A_1, A_0 and A_2 may be known. In practice it is not easy to determine accurately the values of A_1 and A_2. Further, even if these values are known, they do not in themselves specify whether the instrument is under, over or critically damped, since a numerical calculation has to be performed with them first. Therefore, designers find "*damping ratio*" as a very convenient measure of the amount of damping present in the movement.

The terms *damping ratio* (γ) and *underdamped natural frequency* (ω_n) immediately conjure up a physical picture of the response of an instrument and both of the quantities are very easy to measure. Thus γ and ω_n easily do away with quantities A_2, A_1 and A_0.

2.8.4.5. First-order system responses

The complete solution of an equation which describes the dynamical behaviour of a system consists of the following *two parts :*

(*i*) **Complementary function.** *It corresponds to the short time or transient response.*

(*ii*) **Particular integral.** *It refers to the long time steady state response.*

The transfer operator form of the first-order system is given by :

$$\frac{I_o}{I_i} = \frac{S}{I + \tau D}$$

When S (static sensitivity or steady state gain) equals *unity*, we get

$$(I + \tau D)I_o = I_i \quad ...(2.24)$$

Now we shall obtain the solution of this equation for different standard inputs (The solutions are not mathematically rigorous, but are practical).

Transient response (Complementary function):

The transient response from the auxiliary equation is obtained by putting input I_i equal to zero :

i.e., $$(1 + \tau D)I_{o,t} = 0 \quad ...(2.25)$$

(Subscript t refers to the transient value)

Let the solution be of the form :

$$I_{o,\, t} = A\, e^{mt}$$

where, m is an algebraic variable.

or, $$(1 + \tau D)A\, e^{mt} = 0$$

or, $$A\, e^{mt} + \tau \cdot \frac{d}{dt}(A\, e^{mt}) = 0$$

or, $$A\, e^{mt} + \tau \cdot A\, m\, e^{mt} = 0$$

$$A\, e^{mt}\, (1 + \tau \cdot m) = 0$$

$\therefore$ $$m = -\frac{1}{\tau}$$

Then, $$I_{o,\, t} = A\, e^{mt} = Ae^{-t/\tau} \quad ...(2.26)$$

The transient response of a first-order system is *same for different standard inputs.*

Steady state response (Particular integral) :

The steady state response is given by

$$(1 + \tau D)I_{o,s} = I_i \quad ...(2.27)$$

(Subscript s refers to the steady state value)

or, $$I_{o,s} = (1 + \tau D)^{-1}\, I_i$$

$$= (1 - \tau D + \text{terms in } D^2 \text{ and higher})\, I_i \quad ...(2.28)$$

1. Step input :

Since the input I_i is a step of constant magnitude; its differential equals zero, and subsequently, we get

$$I_{o,s} = (1 - \tau D)I_i = I_i \quad ...(2.29)$$

Total response = Transient response + steady state response

or $$I_o = Ae^{-t/\tau} + SI_i \quad ...(2.30)$$

The constant A is evaluated from the initial conditions as follows :

At $t = 0,\quad I_o = 0$

$\therefore$ $0 = A + SI_i$ or, $A = -SI_i$

$\therefore$ $$I_o = \underbrace{-I_i\, e^{-t/\tau}}_{\text{Transient}} + \underbrace{I_i}_{\text{Steady state}}$$

or, $$I_o = I_i(1 - e^{-t/\tau}) \quad ...(2.31)$$

or, $$\frac{I_o}{I_i} = (1 - e^{-t/\tau}) \quad ...(2.32)$$

... in *non-dimensional form.*

Salient features (with step input) :

Following are the salient features of first-order system with step input :

(*i*) The transient response of the first-order system is *time dependent;* as the time passes, its value decreases (Refer to eqn. 2.32) and after a very long time the value becomes zero approximately. Thus magnitude of output (I_o) will be same as input (I_i) when the time is very large.

(*ii*) The speed of response relates to the time constant τ. A large τ indicates that response of the system is slow, whereas a small τ represents a fast system response. Thus in order to get good fidelity (*i.e.,* for accurate dynamic measurements) efforts should be made to minimise the value of τ.

(*iii*) Refer to Fig. 2.22, which shows the time response of a first-order system to a step-input; when $t = \tau$, $\frac{I_o}{I_i} = (1 - e^{-1}) = 0.632$. Thus, the *time constant* (τ), *for a rising exponential function, is defined as the time to reach* 63.2% *of its steady state value*. The time constant for a *decaying function* would correspond to the time taken to fall to *36.8% of its initial value.*

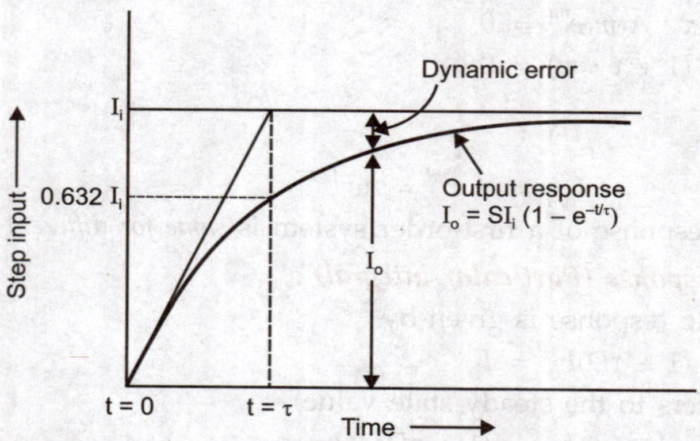

Fig. 2.22. Time response of a first-order system to step input.

(*iv*) Dynamic error (*i.e.,* vertical difference between the input and output response curve),

$$\begin{aligned} E_{dy.} &= I_i - I_o \\ &= I_i - I_i(1 - e^{-t/\tau}) \\ &= -I_i\, e^{-t/\tau} \end{aligned} \quad ...(2.33)$$

or, $$\frac{E_{dy.}}{I_i} = -e^{-t/\tau} \quad ...(2.34)$$

... (*in dimensionless form*)

(*v*) In case the measurand has an initial value of I_{initial} at $t = 0$, then the output I_o at any instant t is given by

$$I_o = I_i[1 - e^{-t/\tau}] + I_{\text{initial}}\, e^{-t/\tau}$$

or, $$I_o = I_i + (I_{\text{initial}} - I_i)\, e^{-t/\tau} \qquad ...(2.35)$$

(vi) The speed response of a system is defined in terms of *settling time* (it is time taken by the system to reach and remain within a certain percentage tolerance band of the final state value). Smaller the settling time, faster is the response. Typical values of tolerance band are 2% and 5% settling times.

For unit step input, the output reaches the values given in Table 2.1 at various intervals of time. This table also shows the per unit dynamic error at different intervals of time.

Table 2.1: Per Unit Output and Dynamic Error for a Step Input for a First-order System

t/τ	*Per unit output* (I_o/I_i)	*Per unit dynamic error* (E_{dy}/I_i)
0	0	1.000
1	0.632	0.368
2	0.865	0.135
3	0.950	0.050
4	0.982	0.018
5	0.993	0.007
∞	1.000	0

Thus, 5% settling time means that the system has reached its specified value after a time which is thrice the time constant (Table 2.1).

2. Ramp Input :

Consider that the input signal varies linearly with time so that ramp input I_i is prescribed by the relation $I_i + \psi t$, where ψ is constant. Then the governing differential equation is given by

$$(1 + \tau D) I_o = \psi t \qquad ...(2.36)$$

Now, *Transient response* (complimentary function) is given as:

$$I_{o,t} = A\, e^{-t/\tau} \qquad \text{... as before}$$

and, *steady state response* (particular integral) will be;

$$I_{o,s} = (1 - \tau D + \text{terms in } D^2 \text{ and higher})\, \psi t = \psi t - \tau \frac{d}{dt}(\psi \tau)$$

or, $$I_{o,s} = \psi t - \psi \tau \qquad ...(2.37)$$

∴ Complete response = Transient response + steady state response

$$I_o = A\, e^{-t/\tau} + (\psi t - \psi \tau) \qquad ...(2.38)$$

The value of constant A can be evaluated by applying the initial condition,

At, $t = 0 \quad I_0 = 0$

or, $0 = A - \psi\tau \quad \therefore \quad A = \psi\tau$

∴ $$I_o = \psi t - \psi\tau + \psi\tau \times e^{-t/\tau} = \psi(t - \tau) + \psi\tau\, e^{-t/\tau} \qquad ...(2.39)$$

or, $$I_o = \psi[t - \tau(1 - e^{-t/\tau})] \qquad ...(2.39a)$$

Fig. 2.23 shows the time response of a first-order system to a ramp input.

The dynamic error,

$$E_{dy.} = I_i - I_o$$
$$= \psi t - [\psi t - \psi\tau + \psi\tau\, e^{-t/\tau}]$$

$$= \underbrace{\psi\tau}_{\text{steady}} - \underbrace{\psi\tau\, e^{-t/\tau}}_{\text{Transient}} \quad ...(2.40)$$

or,
$$\frac{E_{dy.}}{\psi\tau} = 1 - e^{-t/\tau} \quad ...(2.41)$$

... (in dimensionless form)

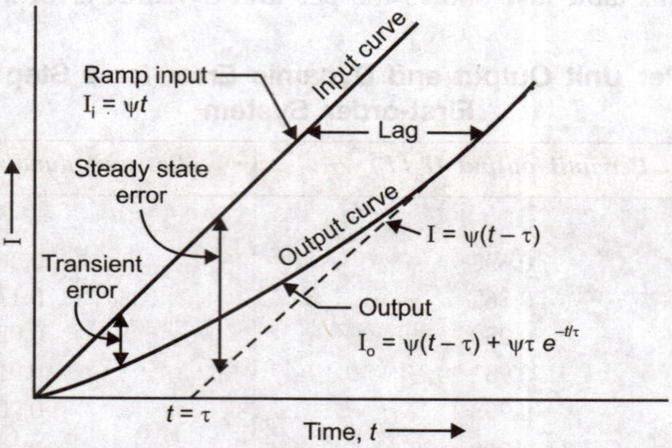

Fig. 2.23. Time response of a first-order system to a ramp input.

Salient features (with ramp input) :

(*i*) The term $\psi\tau$ being independent of time continues to exist and so it is called the steady state error. The term $\psi\tau\, e^{-t/\tau}$ gradually decreases with time and hence is called the *transient error*.

— Since the steady state error is directly proportional to τ (time constant), therefore, the larger the value of τ, the larger will be the magnitude of the error.

— When τ is made small the transient error decreases rapidly; this implies that the system attains the steady state at a faster pace.

(*ii*) The output response curve always lags behind the input curve by a constant amount known as ***lag***.

3. Sinusoidal (Harmonic) input :

The frequency analysis of a system pertains to the steady state response of the system to a sinusoidal input. In this analysis, the system is subjected to a sinusoidal input and the system response is studied with frequency as the independent variable. The sinusoid is a unique input signal, and the resulting output signal for a linear system is sinusoidal in the steady state. However, the output signal differs from the input waveform in amplitude and phase.

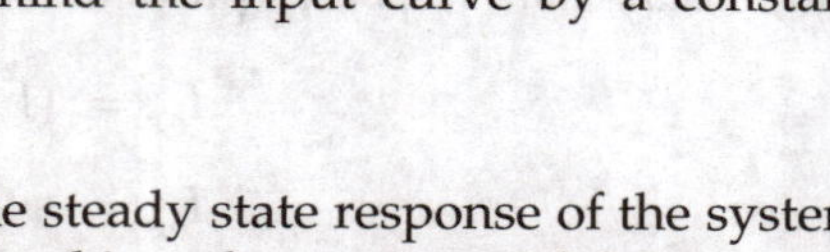

Fig. 2.24.

In order to determine the frequency response of sinusoidal input to a first-order system, let us replace the transfer operator D by a factor $j\omega$ in the input/output relationship; then we get,

$$\frac{I_o}{I_i} = \frac{1}{1 + D\tau} = \frac{1}{1 + j\omega\tau} \quad ...(2.42)$$

where, ω = Input frequency, rad/s, and

$j = \sqrt{(-1)}$

In a frequency response the following two quantities are of interest : Refer to Fig. 2.24.

(i) *Amplitude ratio or modulus* $\left(\frac{I_o}{I_i}\right)$. It prescribes the size of the output amplitude relative to the input amplitude.

(ii) *Phase shift* of output relative to input.

For the first-order system represented by equation (2.42),

$$\text{Modulus} = \sqrt{1+(\omega\tau)^2}$$

$$\text{Argument/Phase angle} = \tan^{-1}(\omega\tau) \qquad ...(2.43)$$

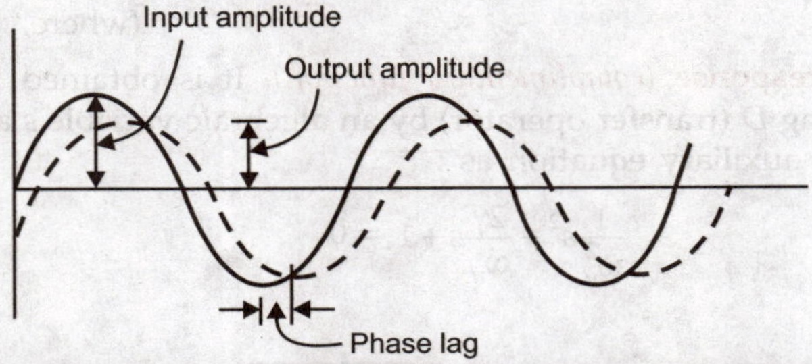

Fig. 2.25. Relationship between an input frequency and corresponding output frequency.

***Salient features (with sinusoidal input)*:** Refer to Fig. 2.25.

(i) When a system is subjected to a sinusoidal input with frequency ω, its output will also be sinusoidal, but the magnitude of the output amplitude necessarily may not be the same (as the input one). *The ratio of the amplitude* (often called *attenuation*) is given as :

$$\frac{I_o}{I_i} = \frac{1}{\sqrt{1+(\omega\tau)^2}} \qquad ...(2.44)$$

Thus, with the increase in input frequency, the amplitude ratio decreases.

(ii) The output from the system may not necessarily be in phase with the input; and the phase difference is given by

$$\phi \text{ (phase angle)} = -\tan^{-1} (\omega\tau) \qquad ...(2.45)$$

–ve indicates that output *lags* behind the input. When $\omega = \frac{1}{\tau}$, the phase lag is $\frac{\pi}{4}$ or 45°.

As the accuracy of an instrument measuring dynamic input depends upon the time constant, therefore, smaller the time constant, greater the accuracy; for phase shift to be small, the time period τ should be small.

(iii) When the input and output signals are given by the relations :

$I_i = A \sin \omega t$, and $I_o = B \sin (\omega t + \phi) = zA \sin (\omega t + \phi)$,

Then the amplitude ratio may be represented as follows :

$$K = \left|\frac{I_o}{I_i}\right| = \frac{z}{\sqrt{1+(\omega\tau)^2}} \qquad ...(2.46)$$

In order to produce amplitude of sinewave without any attenuation ($K = 1$) we must use an instrument whose time constant, $\tau = \frac{\sqrt{z^2-1}}{\omega}$.

2.8.4.6. Second-order system responses

In case of typical second-order system having unit static sensitivity, the homogeneous equation is given by :

$$\frac{I_o}{I_i} = \frac{1}{\left(\frac{1}{\omega_n^2}\right)D^2 + \left(\frac{2\gamma}{\omega_n}\right)D + 1}$$

or,
$$\left[\left(\frac{1}{\omega_n^2}\right)D^2 + \left(\frac{2\gamma}{\omega_n}\right)D + 1\right] I_o = I_i \quad ...(2.47)$$

(where, γ = damping ratio)

(a)* Transient response *(complimentary function). It is obtained from the auxiliary equation by replacing D (transfer operator) by an algebraic variable s and putting I_i equal to zero; we get the auxiliary equation as :

$$\frac{1}{\omega_n^2}s^2 + \frac{2\gamma}{\omega_n}s + 1 = 0$$

The roots are :

$$s_1, s_2 = \frac{\frac{-2\gamma}{\omega_n} \pm \sqrt{\left(\frac{2\gamma}{\omega_n}\right)^2 - \frac{4}{\omega_n^2}}}{\frac{2}{\omega_n^2}}$$

$$= -\gamma\omega_n \pm \sqrt{\gamma^2\omega_n^2 - \omega_n^2}$$

or,
$$s_1, s_2 = -\gamma\omega_n \pm \omega_n\sqrt{\gamma^2 - 1} \quad ...(2.48)$$

The *transient solution* has the accepted form,

$$I_{o,t} = Ae^{s_1 t} + Be^{s_2 t}$$

where, A and B = Arbitrary constants to be determined from initial conditions, and

s_1, s_2 = Roots of the auxiliary equation (The roots may be real and different, real and equal or imaginary and *that determines* the *nature of transient response of the system*).

The response of the system is of the following three types depending upon the roots of the characteristic equations :

(i) Overdamped systems.
(ii) Critically damped systems.
(iii) Under-damped systems.

***(i)* Over-damped systems.** In this case $\gamma > 1$ and the roots are *real and unequal.*

- There is *heavy damping* and the system responds to the final steady-state value without any oscillations but in a *sluggish manner.*
- *No overshoot* in step response and no *"resonance"* (resonance refers to the output signal *greater* in magnitude than the ideal output) in the *frequency response.*

- The overdamped systems, owing to their sluggish response, are usually unsuitable for several control applications.

(ii) *Critically-damped systems.* In this case $\gamma = 1$ and the roots are *real and equal.*

- The system has a *quick and smooth response to the final steady state value without any oscillations.*
- *No overshoot in the step response and no resonance in the frequency response.*

(iii) *Under-damped systems.* For an undamped system $\gamma < 1$ and roots of the characteristic equation are a complex conjugate pair; these are given as

$$s_1, s_2 = -\gamma\omega_n \pm j\omega_n\sqrt{(1-\gamma^2)}$$
$$= -\gamma\omega_n \pm j\omega_d \quad \text{...(2.49)}$$

where, $\omega_d = \omega_n\sqrt{(1-\gamma^2)}$; this quantity is called *"damped natural frequency"* of the system (This is the frequency at which the damped system freely oscillates when disturbed).

- Such systems *take a long time to reach steady state, but have quick initial response.*
- In these systems, there are oscillations in the step response and resonance effects in the frequency response for values of $\gamma > 0.707$.
- Majority of instruments and control are generally underdamped (*light damping*).

(b) The steady state response (*particular integral*). It is given by :

$$\left(\frac{1}{\omega_n^2}D^2 + \frac{2\gamma}{\omega_n}D + 1\right)I_{o,s} = I_i$$

or,
$$I_{o,s} = \left(1 + \frac{2\gamma}{\omega_n}D + \frac{1}{\omega_n^2}D^2\right)^{-1} I_i$$
$$= \left(1 - \frac{2\gamma}{\omega_n}D + \text{terms in } D^2 \text{and higher}\right)I_i \quad \text{...(2.50)}$$

1. Step input : Fig. 2.26 shows the time response of a second-order system to a step input.

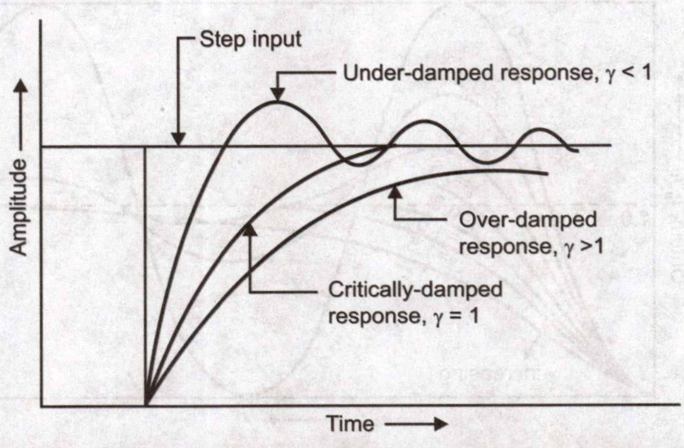

Fig. 2.26. Time response of a second-order system to a step input.

Since the input I_i is a step of constant magnitude, its differential equals zero and we have,

$$I_{o,\,s} = \left(1-\frac{2\gamma}{\omega_n}D\right)I_i = I_i$$

$\therefore$ The complete response

$$I_o = I_{o,s} + I_{o,t}$$

or, $$I_o = I_i A\, e^{s_1 t} + B e^{s_2 t} \quad ...(2.51)$$

For the under-damped system, the complex conjugate pair of roots are given by

$$s_1,\, s_2 = -\gamma\omega_n \pm j\omega_d$$

$\therefore$ $$I_o = I_i + A e^{-(\gamma\omega_n + j\omega_d)t} + B e^{-(\gamma\omega_n - j\omega_d)t} \quad ...(2.52)$$

Replacing the complex exponentials by sines and cosines, we get

$$I_o = I_i + e^{-\gamma\omega_n t}(A\cos w_d t + B\sin w_d t) \quad ...(2.53)$$

By applying the initial conditions,

At $t = 0$, $I_o = 0$ and $\frac{dI_o}{dt} = 0$,

we get the values of the constants A and B as :

$$A = -I_i, \text{ and}$$

$$B = \frac{-I_i\gamma}{\sqrt{1-\gamma^2}}$$

Inserting these values in equation (2.53), we have

$$I_o = I_i\left[1 - e^{-\gamma\omega_n t}\left\{\cos\omega_d t + \frac{\gamma}{\sqrt{1-\gamma^2}}\sin\omega_d t\right\}\right] \quad ...(2.54)$$

Fig. 2.27 shows the transient response of a second-order system to a unit step function for different values of damping factor γ; *the curves indicate the overshoot and oscillations increase with a reduced damping in the system.*

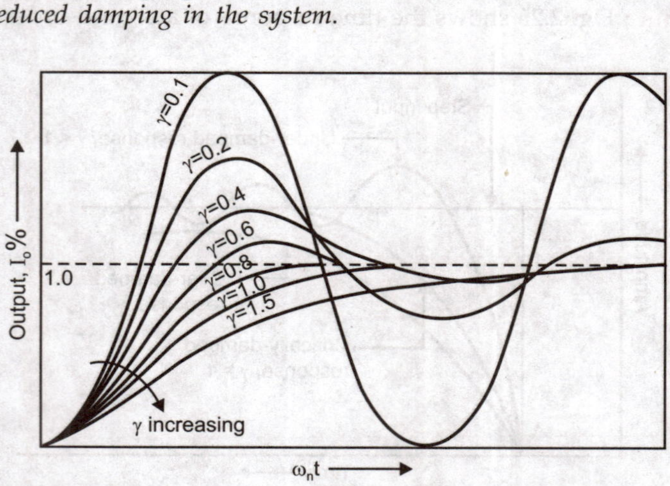

Fig. 2.27. Transient response of a second-order system to a unit step function input for different values of damping factor γ.

2. Sinusoidal (Harmonic) input :

When a sinusoidal input is given to the system, its steady state response is determined by replacing *the operator D by jω in the input/output relationship, as given below :*

$$\frac{I_o}{I_i} = \frac{1}{\frac{1}{\omega_n^2}D^2 + \frac{2\gamma}{\omega_n}D + 1} = \frac{\omega_n^2}{D^2 + 2\gamma\omega_n D + \omega_n^2}$$

$$= \frac{\omega_n^2}{(j\omega)^2 + 2\gamma\omega_n(j\omega) + \omega_n^2} = \frac{\omega_n^2}{(\omega_n^2 - \omega^2) + j(2\gamma\omega_n\omega)} \quad ...(2.55)$$

where, ω = Input frequency in rad/s, and

$j = \sqrt{-1}$

The denominator is a complex number having :

$$\text{Modulus} = \sqrt{\left[\left(\omega_n^2 - \omega^2\right)^2 + (2\gamma\omega_n\omega)^2\right]}$$

$$\text{Argument} = \tan^{-1}\left\{\frac{2\gamma\omega_n\omega}{\omega_n^2 - \omega^2}\right\} \quad ...(2.56)$$

∴ The amplitude ratio,

$$\left|\frac{I_o}{I_i}\right| = \frac{\omega_n^2}{\sqrt{\left[\left(\omega_n^2 - \omega^2\right)^2 + (2\gamma\omega_n\omega)^2\right]}} \quad ...(2.57)$$

and, the system phase lag,

$$\phi = \tan^{-1}\left\{\frac{2\gamma\omega_n\omega}{\omega_n^2 - \omega^2}\right\} \quad ...(2.58)$$

...[using eqn. (2.56)

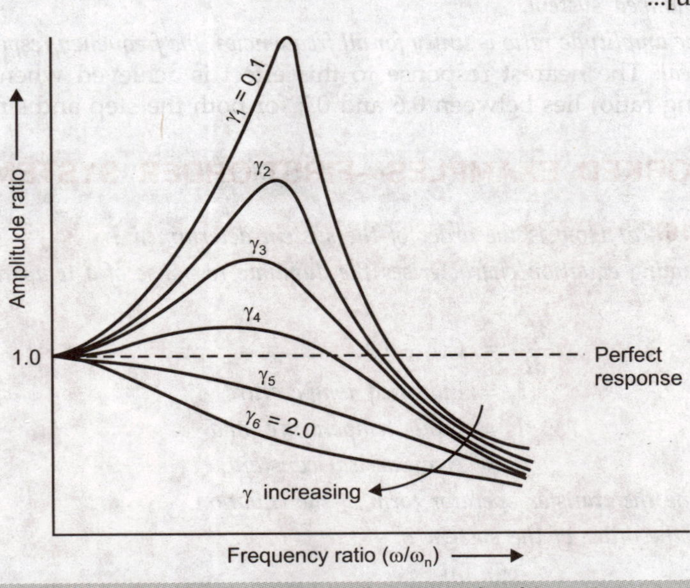

Fig. 2.28. Step response of a second-order system.

Fig. 2.28 and Fig. 2.29 show the variation of amplitude ratio and phase lag versus frequency ratio (ω/ω_n) for various values of damping ratio (γ). From these graphs we observe that the *salient features* of the *steady state response* of a second-order system are :

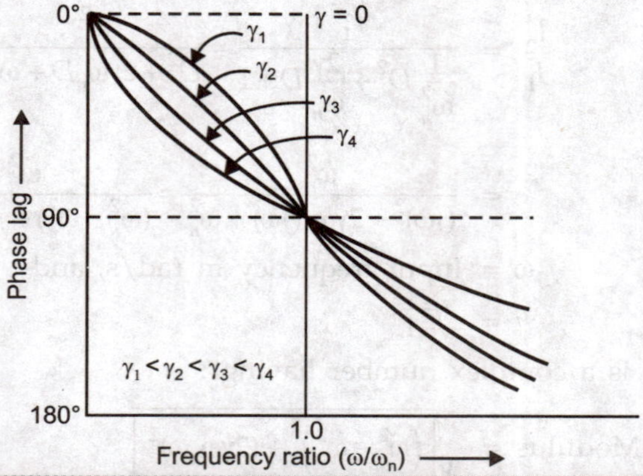

Fig. 2.29. Plot between phase lag and frequency ratio for a second-order system.

(*i*) As *frequency ratio* $\to 0$:
Amplitude ratio $\to 1$;
Phase lag $\to 0°$

(*ii*) As frequency ratio $\to \infty$:
Amplitude ratio $\to 0$;
Phase lag $\to 180°$

(*iii*) When frequency ratio = 1 :
Amplitude ratio $\to \infty$ in undamped systems ($\gamma = 0$)
Phase lag $\to -90°$ in all the systems.
This condition is known as "*resonance*" and can *result in destructive oscillations in lightly damped system.*

(*iv*) *When the amplitude ratio is unity for all frequencies, the frequency response is considered to be ideal.* The nearest response to this effect is achieved when the value of γ (damping ratio) lies between 0.6 and 0.7 for both the step and sinusoidal inputs.

WORKED EXAMPLES—FIRST-ORDER SYSTEMS

Example 2.16. *(a) How is the order of the system determined ?*

(b) The following equation characterises the dynamic response of a temperature measuring instrument :

$$\frac{dI_o}{dt} = C(I_i - I_o)$$

where, I_o = *Indicated temperature,*
I_i = *Input temperature, and*
C = *A numerical constant.*

(i) Determine the transfer operator form of the equation.
(ii) What is the order of the system ?

Solution. (*b*) *Given equation :* $\frac{dI_o}{dt} = C(I_i - I_o)$

(i) Transfer operator form :

The given equation can be rewritten as :

$$\frac{1}{C} \cdot \frac{dI_o}{dt} = I_i - I_o$$

or, $$\frac{1}{C} \cdot \frac{dI_o}{dt} + I_o = I_i$$

or, $$I \cdot \frac{dI_o}{dt} + I_o = I_i \qquad \text{(where, } \tau = \text{time constant} = \frac{1}{C}\text{)}$$

or, $$(\tau D + 1)I_o = I_i$$

∴ The transfer operator form of the equation is given by :

$$\frac{I_o}{I_i} = \frac{1}{(\tau D + 1)} \quad \textbf{(Ans.)}$$

(ii) Order of the system :

Since the highest differential in the denominator of the transfer operator is *unity*, therefore, the temperature measuring instrument has a *first-order system.*

The grouping of the measurement and control systems is done according to the *order of the highest differential in the denominator* of the system transfer operation.

Examples :

- First-order system : $$\frac{I_o}{I_i} = \frac{2}{1+3D}$$
- Second-order system : $$\frac{I_o}{I_i} = \frac{6}{D^2+3D+4}, \text{ or}$$

$$\frac{I_o}{I_i} = \frac{14}{(1+0.3D)(1+0.2D)}.$$

Example 2.17. *Formulate the governing equation for a first-order system-temperature measurement by a thermal measuring element (say a thermometer or thermocouple).*

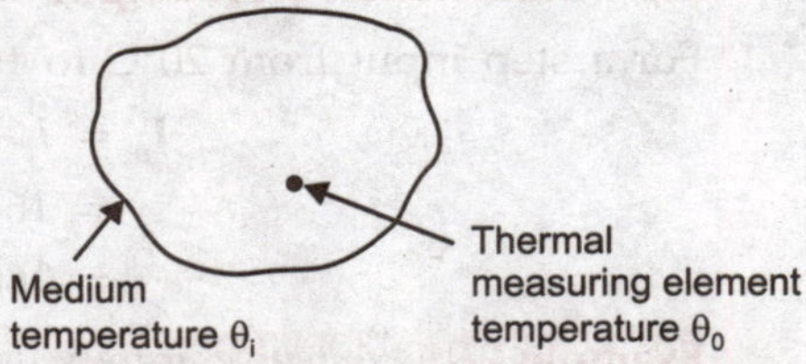

Fig. 2.30. Thermal element.

Solution. Refer to Fig. 2.30.

Let I_i = Temperature of the medium,

I_o = Temperature indicated by the thermal measuring instrument (say a thermometer or thermocouple),

A = Exposed area of the thermal measuring element,

h = Convective heat transfer coefficient,

m = Mass of thermal element, and

c = Specific heat of the element.

Then, the rate of heat flux into the element is,

$$Q = hA(\theta_i - \theta_0) \qquad ...(i)$$

The rate of enthalpy gain by the element

$$= mc\frac{d\theta_0}{dt} \qquad ...(ii)$$

Since the rate of heat flow equals the rate of enthalpy gain by the element, therefore, equating (*i*) and (*ii*) we get :

$$mc\frac{d\theta_0}{dt} = hA(\theta_i - \theta_o)$$

or,
$$\frac{mc}{hA}\cdot\frac{d\theta_o}{dt} + \theta_o = \theta_i$$

$$\tau\frac{d\theta_o}{dt} + \theta_o = \theta_i$$

where, $\tau = \frac{mc}{hA}$ is known as *time constant* of the system.

In terms of *D*-operator (where $D = \frac{d}{dt}$), we have

$$(\tau D + 1)\theta_0 = \theta_i$$

or,
$$\frac{\theta_o}{\theta_i} = \frac{1}{\tau D + 1} \quad \text{...Required equation.}$$

which is an equation of *first-order*. **(Ans.)**

Example 2.18. *A thermometer, idealised as a first-order system with a time constant of 2.2 seconds, is suddenly given an input of 160°C from 0°C.*

(i) What will be the reading of the thermometer after 1.2 seconds ?

(ii) Determine its reading if it is initially held at 20°C.

Solution. *Given :* I_i = 160°C; t = 1.2s; t = 2.2s; I_{initial} = 20°C.

(*i*) Thermometer's reading after 1.2s :

We know that,
$$I_o = I_i(1 - e^{-t/\tau}) \quad \text{...[Eqn. (2.31)]}$$
$$= 160[1 - e^{-(1.2/2.2)}] = \mathbf{67.27°C} \quad \textbf{(Ans.)}$$

(*ii*) Thermometer's reading if it was initially held at 20°C :

For a step input from 20°C to 160°C, we have

$$I_o = I_i + (I_{\text{initial}} - I_i)\, e^{-t/\tau} \quad \text{...[Eqn. (2.35)]}$$
$$= 160 + (20 - 160)\, e^{-(1.2/2.2)}$$
$$= 160 + (20 - 160) \times 0.5796 = \mathbf{78.86°C} \quad \textbf{(Ans.)}$$

Example 2.19. *A temperature sensing device can be modelled as a first-order system with a time constant of 5 seconds. It is suddenly subjected to a step input of 30°C – 160°C. Calculate the temperature indicated by the device after 10 seconds after the start of the process.*

Solution. *Given :* τ = 5s; I_{initial} = 30°C, I_i = 160°C; t = 10s.

Temperature after 10 seconds is calculated as follows :

$$I_o = I_i + (I_{\text{initial}} - I_i)\, e^{-t/\tau} \quad \text{...[Eqn. (2.35)]}$$
$$= 160 + (30 - 160)\, e^{-10/5}$$
$$= 160 - 130 \times 0.1353 = \mathbf{142.4°C} \quad \textbf{(Ans.)}$$

Example 2.20. *A temperature sensitive transducer when subjected to sudden temperature change takes 9 seconds to reach equilibrium conditions (Three time constants). Calculate the time taken by the transducer to read half of the temperature difference.*

Solution. Time taken to reach equilibrium condition = $3\tau = 9s$ (*Given*)

$\therefore$ Time constant, $\tau = \frac{9}{3} = 3s$

Time taken by the transducer to read half of the temperature difference is calculated as follows :

$$I_o = I_i(1 - e^{t/\tau}) \qquad \text{...[Eqn. (2.31)]}$$

or, $$\frac{I_o}{I_i} = 1 - e^{-t/\tau}$$

or, $$0.5 = 1 - e^{-t/3}$$

or, $$e^{-t/3} = 0.5 \quad \text{or,} \quad e^{t/3} = 2$$

$\therefore$ $$t = \mathbf{2.08s} \quad \textbf{(Ans.)}$$

Example 2.21. *A resistance temperature detector has a steady state gain of 2.96 Ω/°C and a time constant of 4.8 seconds. It is subjected to a step change of 80°C in temperature. If its stable resistance before the step change is 90Ω, write down the time-domain equation for resistance and find its value after 12 seconds of application of step input.*

Solution. Step input in terms of resistance

= Steady stage gain × change of temperature

= 0.296 × 80 = 23.68 Ω step input in terms of resistance.

In order to determine the time domain equation for resistance, the value of initial resistance must be added to it.

i.e., $$R_t = 23.68(1 - e^{-t/4.8}) + 90 \quad \textbf{(Ans.)}$$

...*Time domain equation*

$\therefore$ The value of resistance at $t = 12s$.

$$R_{12} = 23.68(1 - e^{-12/4.8}) + 90 = \mathbf{111.74\ \Omega} \quad \textbf{(Ans.)}$$

Example 2.22. *A thermometer initially at a temperature of 15°C is suddenly plunged into a liquid bath maintained at 140°C. After a time interval of 4 seconds, the thermometer indicated a reading of 75°C. Determine :*

(i) The time constant for the thermometer.

(ii) The indicated temperature after five times constant; comment upon this result.

Solution. *Given :* $I_{initial} = 15°C$; $I_i = 140°C$; $t = 4s$; $I_o = 75°C$.

(*i*) The time constant for the thermometer, τ :

When a first-order system is subjected to a step input from $I_{initial}$ to I_i, the indicated temperature is given by the relation :

$$I_o = I_i + (I_{initial} - I_i)e^{-t/\tau} \qquad \text{...[Eqn. (2.35)]}$$

$$75 = 140 + (15 - 140)e^{-4/\tau}$$

or, $$e^{-4/\tau} = \frac{75-140}{15-140} = 0.52$$

or, $$e^{4/\tau} = 1.923$$

Taking logs on both sides, we get

$$\frac{4}{\tau} = 0.6539$$

$\therefore$ $$\tau = \mathbf{6.12\ s} \quad \textbf{(Ans.)}$$

(*ii*) Indicated temperature after five times constant, I_o :

Substituting the value in the above equation, we get

$$I_o = 140 + (15 - 140)\, e^{-5\tau/\tau}$$
$$= 140 + (15 - 140)e^{-5} = \mathbf{139.16°C} \quad \text{(Ans.)}$$

Comments : After a time interval equivalent to five times constants, the thermometer reaches the *equilibrium condition* (*i.e.*, it indicates a temperature which is approximately equal to the maximum step input).

Example 2.23. *A temperature sensitive transducer used to measure the temperature of a furnace has been idealised as a first-order system subjected to ramp input. If the maximum permissible error in temperature measurement is limited to 3.9°C, calculate the time constant of the transducer when the furnace temperature increases at the rate of 0.2°C/s.*

Solution. *Given :* Dynamic error, $E_{dy.}$ = 3.9°C; slope, ψ = 0.2°C/s.

For the ramp input, the input signal varies linearly with time ($I_i - \psi t$) and the dynamic or measurement error,

$$E_{dy.} = \underbrace{\psi\tau}_{\text{Steady}} - \underbrace{\psi\tau\, e^{-t/\tau}}_{\text{Transient}} \qquad \text{...[Eqn. (2.40)]}$$

Since the transient error gradually vanishes with time, we have

$$E_{dy.} = \psi\tau$$

$$\therefore \quad \text{Time constant,} \quad \tau = \frac{E_{dy.}}{\psi} = \frac{3.9}{0.2} = \mathbf{19.5\ s} \quad \text{(Ans.)}$$

Example 2.24. *A balloon (weather) carrying a temperature sensing device (of first-order, time constant 8 seconds) rises through the atmosphere at 5 m/s. It supplies information about temperature and altitude through radio signals. At 2500 m height, a temperature indication of 30°C has been received. If the temperature varies with altitude at a uniform rate of 0.011°C/m, determine the time altitude at which 30°C temperature occurs.*

Solution. Time constant of temperature sensing device, τ = 8s

Rate of rise of the balloon = 5 m/s

Temperature indicated at an altitude of 2500 m = 30°C

Rate of temperature variation with altitude = 0.011°C/m

True altitude at 30°C :

Rate of change of temperature with time $y = 0.011 \times 5 = 0.055$°C/s

For a ramp signal applied to a first-order system, the error, when transient conditions die away,

$$E_{dy.} = \psi\tau = 0.055 \times 8 = 0.44°C$$

$$\therefore \quad \text{Error in amplitude} = \frac{0.44}{0.011} = 40\ \text{m}$$

$$\therefore \quad \textit{Actual altitude} = 2500 - 40 = \mathbf{2460\ m} \quad \text{(Ans.)}$$

Example 2.25. *A thermometer with time constant of 50 seconds is used for measuring temperature cycling with 500 s time period. Calculate :*

(i) The ratio of output to input.

(ii) The time lag.

Solution. Time constant of Thermometer, τ = 50 s

Time period, t = 500 s

Now, frequency of temperature variation,

$$\omega = \frac{2\pi}{t} = \frac{2\pi}{500} = 0.01257\ \text{rad/s}$$

(i) The ratio of output to input, $\frac{I_o}{I_i}$:

$$\frac{I_o}{I_i} = \frac{1}{\sqrt{1+(\omega\tau)^2}}$$

$$= \frac{1}{\sqrt{1+(0.01257\times 50)^2}} = \mathbf{0.8467} \quad \textbf{(Ans.)}$$

(ii) Time lag :

Phase shift = $\tan^{-1}(\omega\tau) = \tan^{-1}(0.01257 \times 50) = 32.15°$ or 0.561 rad.

$\therefore$ Time lag = $\frac{1}{\omega} \times$ phase shift in radian

$$= \frac{1}{0.01257} \times 0.561 = \mathbf{44.63\ s} \quad \textbf{(Ans.)}$$

Example 2.26. *During a particular process, the air temperature cycle at the rate of 1 cycle every 4 minutes. The time constant of the temperature measuring device is 20 seconds. Determine :*

(i) The variation in the indicated temperature if true (input) temperature has a sinusoidal variation of ± 25°C.

(ii) The time by which the maximum reading of the thermometer lags the true maximum value.

Solution. *Given :* Air temperature variation = 1 cycle every 4 minutes

Time constant of the device, τ = 20 s

Sinusoidal variation of the input, I_i = ± 25°C

Now, t = 4 minutes

$$\therefore \quad f = \frac{1}{t} = \frac{1}{4\times 60} = \frac{1}{240} \text{ cycles/s}$$

and, $$\omega = 2\pi f = 2\pi \times \frac{1}{240} = 0.0262 \text{ rad./s}$$

(i) The variation in the indicated temperature (I_o) :

The steady state (when transient errors are omitted) solution for the sinusoidal input is given as :

$$\frac{I_o}{I_i} = \frac{1}{\sqrt{(1+\omega\tau)^2}} \sin(\omega t - \phi) \quad \text{...[Eqn. (2.44)]}$$

where, I_o = Output temperature as indicated by the thermometer,

I_i = True temperature, and

ϕ = Phase lag = $\tan^{-1}(\omega\tau)$

Substituting the value of ω and τ (calculated above) in the above equation, we get

$$\frac{I_o}{I_i} = \frac{I}{\sqrt{1+(0.0262\times 20)^2}} = 0.886$$

$$I_o = \pm 25 \times 0.886 = \mathbf{\pm 22.15°C} \quad \textbf{(Ans.)}$$

(ii) The lag (in seconds) :

Phase lag = $\tan^{-1}(\omega\tau)$

or, $\phi = \tan^{-1}(0.0262 \times 20) = 27.65°$ or 0.4826 rad.

The lag in seconds

$$= \frac{1}{\omega} \times \text{phase lag in radian}$$

$$= \frac{1}{0.0262} \times 0.4826 = \textbf{18.42 seconds} \quad \textbf{(Ans.)}$$

Example 2.27. *A first-order instrument having an input of sine-wave pattern is required to measure signals with frequency response of 120 cycles per second. If an amplitude inaccuracy of 4 per cent is permissible, calculate the maximum time constant and time lag at 90 cycles per second.*

Solution. *Given :* Input frequency = 120 cycles per second ;

Amplitude inaccuracy permissible = 4%.

Now, $\omega = 2\pi f = 2\pi \times 120 = 754$ rad/s

Maximum time constant τ :

Corresponding to 4% amplitude inaccuracy, the amplitude ratio of output and input signals is $\left(1 - \frac{4}{100}\right) = 0.96.$

For a first-order instrument, the amplitude ratio is given as :

$$\frac{I_o}{I_i} = \frac{1}{\sqrt{1 + (\omega\tau)^2}} \qquad \text{...[Eqn. (2.44)]}$$

or, $$0.96 = \frac{1}{\sqrt{(1 + 754\tau)^2}}$$

or, $$(1 + 754\tau)^2 = \left(\frac{1}{0.96}\right)^2 = 1.085$$

or, $$1 + 754\tau = 1.0416$$

∴ $$\tau = \mathbf{5.517 \times 10^{-5}\ s} \quad \textbf{(Ans.)}$$

Time lag at 90 cycles per second :

We know that for a sinusoidal input,

$$\phi = \tan^{-1}(\omega\tau) = \tan^{-1}(2\pi \times 90 \times 5.517 \times 10^{-5})$$
$$= 1.787° \quad \text{or} \quad 0.03119 \text{ rad}$$

∴ $$\text{Time lag} = \frac{1}{\omega} \times \text{phase lag in radian}$$

$$= \frac{1}{2\pi \times 90} \times 0.03119 = \mathbf{5.515 \times 10^{-5}\ s} \quad \textbf{(Ans.)}$$

Example 2.28. *The temperature of a furnace is found to vary sinusoidally between 520°C and 580°C with a periodic time of 50 seconds. A thermocouple system with a time constant of 10 seconds is employed to measure the furnace temperature. Determine :*

(i) The maximum and minimum values that will be indicated by the thermocouple.

(ii) The phase shift and the corresponding time lag between the temperature signals and the thermocouple input signals.

Solution. *Given :* Range of variation of temperature = 520° to 580°C

Periodic time, t = 50 seconds

Time constant of the thermocouple system, τ = 10 seconds.

(i) The maximum and minimum values of the temperature :

The temperature would oscillate about the mean value of $\frac{580+520}{2}$ = 550°C, with an amplitude of 30°C.

Angular frequency of oscillation,

$$\omega = 2\pi f = \frac{2\pi}{t} = \frac{2\pi}{50} = 0.1256 \text{ rad/s}$$

The amplitude ratio, after the transient effects die away, is given by :

$$\frac{I_o}{I_i} = \frac{1}{\sqrt{1+(\omega\tau)^2}} = \frac{1}{\sqrt{1+(0.1256\times 10)^2}} = 0.6229$$

$$\therefore \quad I_o = 0.6229 \times 30 = 18.7°\text{C}$$

$\therefore$ *Maximum temperature* = 550 + 18.7 = **568.7°C** **(Ans.)**

Minimum temperature = 550 − 18.7 = **531.3°C** **(Ans.)**

(ii) Phase shift and time lag :

$$\text{Phase shift} = \tan^{-1}(\omega\tau)$$

$$= \tan^{-1}(0.1256 \times 10) = \mathbf{51.47°}, \text{ or, } \mathbf{0.898 \text{ rad.}} \quad \textbf{(Ans.)}$$

$$\text{Time lag} = \frac{1}{\omega} \times \text{phase shift in radian}$$

$$= \frac{1}{0.1256} \times 0.898 = \mathbf{7.15 \text{ s}} \quad \textbf{(Ans.)}$$

Fig. 2.31 shows the variation of actual and indicated temperatures of furnace.

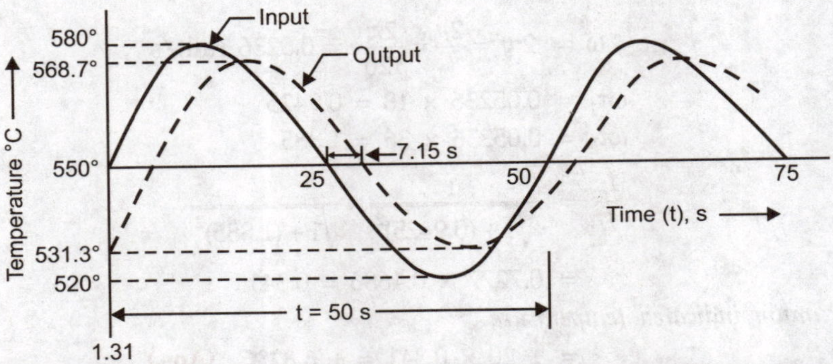

Fig. 2.31. Variation of actual and indicated temperatures of furnace.

Example 2.29. *A first-order instrument is subjected to a sinusoidal input given by relation : I_i = 0.35 sin 25t. If the instrument has a time constant of 0.3 second, develop an expression for the corresponding output.*

Solution. *Given :* Sinusoidal input relation, I_i = 0.35 sin 25 t

Time constant of the instrument, τ = 0.3 s.

Expression for the output :

The amplitude ratio, after the transient effects die away, is given as :

$$\frac{I_o}{I_i} = \frac{1}{\sqrt{1+(\omega\tau)^2}} = \frac{1}{\sqrt{1+(25\times 0.3)^2}} = 0.132$$

(Here, ω = 25 ..... from the given relation)

∴ Magnitude of output,

$$I_o = 0.132 \times \text{magnitude of input}$$
$$= 0.132 \times {}^*0.35 = 0.0462 \qquad \text{(* from the given relation)}$$

Phase shift,
$$\phi = \tan^{-1}(\omega\tau)$$
$$= \tan^{-1}(25 \times 0.3) = 82.4°$$

∴ $$I_o = \mathbf{0.0462 \sin (25t - 82.4°)}$$ ...*(Required expression)* **(Ans.)**

Example 2.30. *A thermometer (its glass bulb protected by a well) is subjected to a temperature which is changing ± 20°C every 120 seconds. The measurement system may be idealised by a double capacity system with time constants of 18 seconds for the bulb and 36 seconds for the well.*

Determine the maximum value of temperature indicated by the thermometer. After how much time of the actual occurrence of temperature, it will be indicated by the instrument (i.e., time lag) ?

Solution. *Given :* Rate of change of temperature : ± 20°C every 120 seconds

Time constant for the bulb = 18 seconds

Time constant for the well = 36 seconds.

Maximum value of the temperature indicated :

The amplitude ratio (I_o/I_i) for a double capacity system is given as :

$$\frac{I_o}{I_i} = \frac{1}{\sqrt{1+(\omega\tau_1)^2}} \times \frac{1}{\sqrt{1+(\omega\tau_2)^2}}$$

and, the angle of lag, $\phi = \tan^{-1}(\omega\tau_1) + \tan^{-1}(\omega\tau_2)$

From the given data :

$$\omega = 2\pi f = \frac{2\pi}{t} = \frac{2\pi}{120} = 0.5236 \text{ rad./s}$$

$$\omega\tau_1 = 0.05236 \times 18 = 0.9425$$

$$\omega\tau_2 = 0.05236 \times 36 = 1.885$$

$$\frac{I_o}{I_i} = \frac{1}{\sqrt{1+(0.9425)^2}} \times \frac{1}{\sqrt{1+(1.885)^2}}$$

$$= 0.7277 \times 0.4686 = 0.341$$

∴ *Maximum indicated temperature*

$$= \pm 20 \times 0.341 = \mathbf{\pm 6.82°C} \quad \textbf{(Ans.)}$$

Time lag :

$$\text{Angle of lag} = \tan^{-1}(0.9425) + \tan^{-1}(1.885)$$
$$= 43.30° + 62.05° = 105.35° \text{ or } 1.839 \text{ rad.}$$

∴ $$\text{Time lag (delay)} = \frac{1}{\omega} \times \text{angle of lag (in rad.)}$$

$$= \frac{1}{0.05236} \times 1.839 = \mathbf{35.12 \text{ seconds}} \quad \textbf{(Ans.)}$$

Example 2.31. *An input of 2 sin 2t + 0.5 sin10t is applied to a first-order instrument having a time constant of 0.3 second. Find the output.*

Solution. *Given :* Input equation :

$$I_i = 2 \sin 2t + 0.5 \sin 10t$$

Time constant, $\tau = 0.3$ s

Output :

The output of the instrument to an input 2 sin 2*t* + 0.5 sin 10*t* when time constant is 0.3s, is given as:

$$I_o = \frac{1}{\sqrt{1+(2\times0.3)^2}}\sin[2t-\tan^{-1}(2\times0.3)]+\frac{1}{\sqrt{1+(10\times0.3)^2}}\sin[10t-\tan^{-1}(10\times0.3)]$$

$$= 0.857 \sin (2t - 30.96°) + 0.316 \sin(10t - 71.56°) \quad \textbf{(Ans.)}$$

Example 2.32. *A first-order system having a time constant of 0.15 second is used to measure a signal prescribed by the relation: I_i = 2 sin 2t + 0.2 cos 8t. Develop an expression for the corresponding output.*

Solution. *Given* : Relation followed by the signal : $I_i = 2 \sin 2t + 0.2 \cos 8t$

Time constant of the system, $\tau = 0.15$ s

Expression for the output :

The concept of sinusoidal transfer function can be applied for the frequency response only if input, I_i, is either a sinewave or sum of sinewaves of different frequencies and amplitudes. In order to satisfy this condition, the given input need to be rewritten in the following form :

$$I_i = 2 \sin 2t - 0.2 \sin [8t + \pi \text{ (or } 180°)]$$

Further, the first-order measuring systems are *linear one*, and therefore the output can be obtained by using the *superposition principle*.

- For the sinusoidal input : $I_i = 2 \sin 2t$

$$\frac{I_o}{I_i} = \frac{1}{\sqrt{1+(\omega\tau)^2}} = \frac{1}{\sqrt{1+(2\times0.15)^2}} = 0.9578$$

Magnitude of output = 0.9578 × 2 = 1.9156

Phase shift, $\phi = \tan^{-1}(\omega\tau) = \tan^{-1}(2 \times 0.15) = 16.7°$

∴ Output, $I_o = 1.9156 \sin (2t - 16.7°)$

- For the sinusoidal input : $I_i = -0.2 \sin (8t + \pi)$

$$\frac{I_o}{I_i} = \frac{1}{\sqrt{1+(\omega\tau)^2}} = \frac{1}{\sqrt{1+(8\times0.15)^2}} = 0.6402$$

Magnitude of output = 0.2 × 0.6402 = 0.128

Phase shift, $\phi = \tan^{-1}(\omega\tau) = \tan^{-1}(8 \times 0.15) = 50.19°$

∴ Output, $I_o = -0.128 \sin(8t + \pi - 50.19°)$

Hence, *overall output*, $\boldsymbol{I_o} = 1.9156 \sin(2t - 16.7°) - 0.128 \sin(8t + \pi - 50.19°)$ **(Ans.)**

SECOND-ORDER SYSTEMS

Example 2.33. *Formulate the governing equation for a second-order system–spring mass system with damping.*

Solution. Refer to Fig. 2.32.

Let, x_i = Input displacement,

x_o = Output displacement,

k = Stiffness of the spring,

C_d = Viscous damping coefficient, and

γ = Damping ratio.

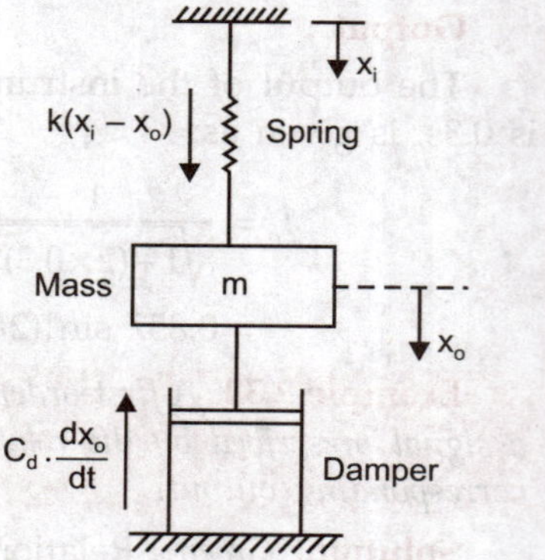

Fig. 2.32. Spring-mass system.

The *forces* acting on the mass are :

(i) As both ends of the springs are free to move, therefore

Spring force = Spring stiffness × displacement of one end of the spring relative to other

$= k(x_i - x_o)$, acting *downward*.

(ii) One end of the dashpot fixed; there is a reaction force acting in the *upward* direction.

Damping force = Damping coefficient × velocity

$$= C_d \times \frac{dx_o}{dt}$$

For translational systems, Newton's law states that,

$$\Sigma \text{ Force} = \text{Mass} \times \text{acceleration} = m\frac{d^2x_o}{dt^2}$$

or,
$$m \cdot \frac{d^2x_o}{dt^2} = k(x_i - x_o) - C_d\frac{dx_o}{dt}$$

or,
$$mD^2x_o + C_dDx_o + kx_o = kx_i \qquad \left(\text{where } D = \frac{d}{dt}, \text{ and } D^2 = \frac{d^2}{dt^2}\right)$$

or,
$$x_i = \left(\frac{m}{k}D^2 + \frac{C_d}{k}D + 1\right)x_o \qquad \textit{...Required equation}$$

which is an equation of *second-order* type. **(Ans.)**

Comparing the above expression with the standard second-order form, we have :

Undamped natural frequency,

$$\omega_n = \sqrt{\frac{k}{m}}$$

and
$$\frac{2\gamma}{\omega_n} = \frac{C_d}{k}$$

$$\therefore \quad \text{Damping ratio, } \gamma = \frac{C_d\omega_n}{2k} = \frac{C_d}{2k}\sqrt{\frac{k}{m}}$$

or,
$$\gamma = \frac{C_d}{2\sqrt{km}} \qquad \textit{...Required equation.}$$

Example 2.34. *(a) Write down the expression describing the motion of linear and rotational displacement systems of the second-order.*

(b) The pen arrangement of U.V. recorder (second-order system) has a mass of 4.5g. Calculate the percentage reduction in mass if it is desired to have 15 per cent increase in natural frequency of the recorder.

Solution. *(a)* • The expression for *linear displacement* (spring-mass-damper) *system* is given by :

$$m\frac{d^2I_o}{dt^2}+C_d\frac{dI_o}{dt}+k\,I_o = k\,I_i \quad ...(i)$$

where, m = Mass (kg),
C_d = Viscous damping force (Ns/m),
k = Spring stiffness (N/m),
I_o = Output reading, and
I_i = Input reading.

• For the *rotational system*, the expression may be written as :

$$J\frac{d^2I_o}{dt^2}+C_d\frac{dI_o}{dt}+q\,I_o = q\,I_i$$

where, J = Inertia (kg m^2), and
q = Torsional stiffness.

Comparing these expressions with differential equation in the standard form,

$$\frac{1}{\omega_n^2}\cdot\frac{d^2I_o}{dt^2}+\frac{2\gamma}{\omega_n}\cdot\frac{d\,I_o}{dt}+I_o = I_i, \text{ we get}$$

Natural frequency, $\omega_n = \sqrt{\frac{k}{m}} = \sqrt{\frac{q}{J}}$, as the case may be. **(Ans.)**

(b) Given : m = 4.5 g; Percentage increase required = 15%.

Using subscripts 1 and 2 for initial and final values respectively, we have :

$$\omega_{n1}^2 = \frac{k}{m_1}, \text{ and, } \omega_{n2}^2 = \frac{k}{m_2}$$

$$\therefore \quad m_2 = m_1\times\left(\frac{\omega_{n1}}{\omega_{n2}}\right)^2$$

But, $\omega_{n2} = 1.15\,\omega_{n1}$ ...(Given)

$$\therefore \quad m_2 = m_1\times\left(\frac{\omega_{n1}}{1.15\,\omega_{n1}}\right)^2 = 0.756\,m_1$$

∴ *Percentage reduction in mass*

$$= \left(\frac{m_1-m_2}{m_1}\right) = \left(\frac{1-0.756m_1}{m_1}\right)\times 100 = \mathbf{24.4\%} \quad \textbf{(Ans.)}$$

Example 2.35. *A second-order system follows the differential equation given below :*

$$\frac{d^2I_o}{dt^2}+3\frac{dI_o}{dt}+30\,I_o = 30\,I_i$$

where, I_o and I_i are output and input quantities respectively. Determine the following :

(i) Damping ratio.

(ii) Damped natural frequency.

(iii) Static sensitivity.

(iv) Time constant.

Solution. The standard form of the differential equation of a second-order system is given as :

$$\frac{1}{\omega_n^2}\cdot\frac{d^2I_o}{dt^2}+\frac{2\gamma}{\omega_n}\cdot\frac{dI_o}{dt}+I_o = k\,I_i \qquad ...(i)$$

Since the term I_o in eqn. (*i*) has a unit coefficient, therefore to recast the given equation in the standard form, let us divide it throughout by 30; we get

$$\frac{1}{30}\cdot\frac{d^2I_o}{dt^2}+\frac{1}{10}\frac{dI_o}{dt}+I_o = I_i \qquad ...(ii)$$

Comparing eqns. (*i*) and (*ii*), we get

$$\omega_n^2 = 30;\ \frac{2\gamma}{\omega_n}=\frac{1}{10}=0.1;\ k = 1$$

∴ Natural frequency, $\omega_n = \sqrt{30} = 5.477$ rad./s

(*i*) Damping ratio, γ :

$$\frac{2\gamma}{\omega_n} = 0.1, \text{ or } \gamma = \frac{\omega_n}{2}\times 0.1$$

or,
$$\gamma = \frac{5.477}{2}\times 0.1 = \mathbf{0.274} \quad \textbf{(Ans.)}$$

(*ii*) Damped natural frequency, ω_d :

$$\omega_d = \omega_n\sqrt{1-\gamma^2} = 5.477\sqrt{1-0.274^2} = \mathbf{5.267\ rad./s} \quad \textbf{(Ans.)}$$

(*iii*) Static sensitivity :

Static sensitivity, k = **1** **(Ans.)**

(*iv*) Time constant, τ :

$$\tau = \frac{1}{\omega_n}=\frac{1}{5.477} = \mathbf{0.1826\ s} \quad \textbf{(Ans.)}$$

Example 2.36. *The following data pertain to a torque sensing transducer connected to the shaft of an electric motor which drives a load :*

Input torque = Sinusoidal with frequency varying at 1.95 rad./s; Maximum permissible error = ± 8 per cent; Moment of inertia of load, J = 0.14 kg m²; Torsional constant of the shaft, q = 1.22 Nm/rad. Calculate the following :

(i) Natural frequency.
(ii) Damping ratio.
(iii) Damped natural frequency.
(iv) Time constant of the system.

Solution. *Given* : ω = 1.95 rad/s; Max. permissible error = ± 8%; J = 0.14 kg m²; q = 1.22 Nm/rad.

(*i*) Natural frequency, ω_n :

$$\omega_n = \sqrt{\frac{q}{J}}=\sqrt{\frac{1.22}{0.14}} = \mathbf{2.95\ rad./s} \quad \textbf{(Ans.)}$$

(*ii*) Damping ratio, γ :

Also, normalised frequency ratio, $r = \frac{\omega}{\omega_n}=\frac{1.95}{2.95}=0.661$

Corresponding to maximum permissible error of ± 8%, the amplitude ratio is 1.08. Further the amplitude ratio (M) is prescribed by the relation :

$$M = \frac{1}{\sqrt{(1-r^2)^2 + (2\gamma r)^2}}$$

or,

$$1.08 = \frac{1}{\sqrt{(1-0.661^2)^2 + (2\gamma \times 0.661)^2}}$$

$$= \frac{1}{\sqrt{0.317 + 1.747\gamma^2}}$$

or,

$$0.317 + 1.747\gamma^2 = \left(\frac{1}{1.08}\right)^2 = 0.857$$

$\therefore$ γ = **0.555** **(Ans.)**

***(iii)* Damped natural frequency, ω_d :**

$$\omega_d = \omega_n\sqrt{1-\gamma^2}$$

$$= 2.95\sqrt{1-(0.555)^2} = \mathbf{2.454\ rad./s} \quad \textbf{(Ans.)}$$

***(iv)* Time constant, τ :**

$$\tau = \frac{1}{\omega_n} = \frac{1}{2.95} = \mathbf{0.339\ s} \quad \textbf{(Ans.)}$$

Example 2.37. *(a) Discuss the time domain parameters for the response of step input to second-order system.*

(b) When a step input was given to a second-order system, the measurements revealed that the system had an overshoot of 12 per cent in a rise time of 0.22 second. Determine :

(i) Effective damping ratio.

(ii) Undamped natural frequency of the system.

Solution. (*a*) In an under-damped system, the time domain parameters of interest are (Refer to Fig. 2.33) :

(*i*) How long does the system take to practically reach its final steady value ?

(*ii*) How fast the system moves to follow the applied input ?

(*iii*) How oscillatory is the system ?

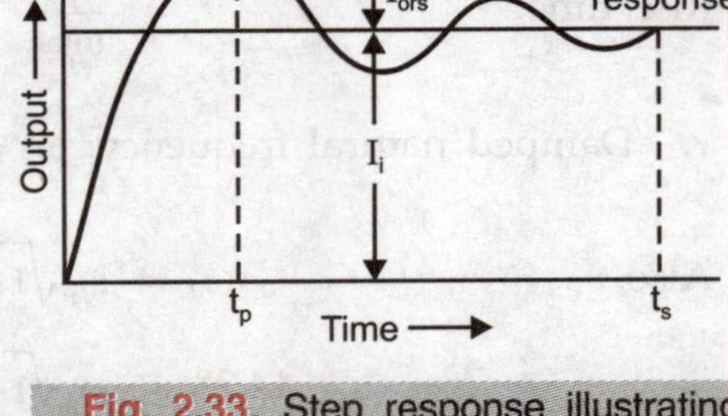

Fig. 2.33. Step response illustrating response, rise and settling times.

- **Settling time (t_s)** is defined as the *time required for the system to reach a steady state.*
- **Rise time (t_r)** is the *time required for the system to rise from 0 to 100% of its final value.*
- **Peak time (t_p)** is the *time taken to reach the first peak (overshoot).*

The percentage overshoot $= \frac{I_{ors}}{I_i} \times 100$.

Mathematical analysis of the under-damped system would show that :

$$\textit{Rise time } (t_r) = \frac{\pi}{\omega_d}$$

$$\textit{Peak time } (t_p) = \frac{\pi}{\omega_n\sqrt{1-\gamma^2}}$$

$$\textit{Overshoot} = \exp\left[\frac{-\pi\gamma}{\sqrt{(1-\gamma^2)}}\right]$$

(b) *Given* : Percentage overshoot = 12%; Rise time = 0.22 s.

(i) Effective damping ratio, γ :

$$\text{Percentage overshoot} = \exp\left[\frac{-\pi\gamma}{\sqrt{1-\gamma^2}}\right]\times 100$$

or,
$$12 = \exp\left[\frac{-\pi\gamma}{\sqrt{1-\gamma^2}}\right]\times 100$$

Rearranging and taking logs on both sides, we get

$$\log_e\left(\frac{12}{100}\right) = \frac{-\pi\gamma}{\sqrt{1-\gamma^2}}$$

$$-2.12 = \frac{-\pi\gamma}{\sqrt{1-\gamma^2}}$$

or,
$$0.675\sqrt{1-\gamma^2} = \gamma$$

$$0.456 - 0.456\,\gamma^2 = \gamma^2 \quad \text{...(Squaring both sides)}$$

$$\therefore \quad \gamma = \sqrt{\frac{0.456}{1.456}} = \mathbf{0.56} \quad \textbf{(Ans.)}$$

(ii) Undamped natural frequency, f_n :

Rise time,
$$t_r = \frac{\pi}{\omega_d}$$

$$\therefore \text{ Damped natural frequency, } \omega_d = \frac{\pi}{t_r} = \frac{\pi}{0.22} = 14.28 \text{ rad./s}$$

Also,
$$\omega_d = \omega_n\sqrt{1-\gamma^2} \quad (\text{where, } \omega_n = \text{undamped natural frequency})$$

$$14.28 = \omega_n\sqrt{1-0.56^2}$$

or,
$$\omega_n = \frac{14.28}{\sqrt{1-0.56^2}} = 17.24 \text{ rad./s}$$

$$\therefore \quad f_n = \frac{\omega_n}{2\pi} = \frac{17.24}{2\pi} = \mathbf{2.74\ Hz} \quad \textbf{(Ans.)}$$

Example 2.38. *A second-order instrument has a natural frequency of 5 Hz and a damping ratio of 0.62. If the excitation frequency of the system is 7 Hz, determine the error due to the proximity of excitation frequency with the natural frequency of the system.*

Solution. *Given :* Natural frequency of the instrument, $f_n = 5$ Hz.

Damping ratio, $\gamma = 0.62$

Excitation frequency of the system, $f = 7$ Hz

Now, the ratio of excitation frequency to natural frequency

$$r = \frac{\omega}{\omega_n} = \frac{f}{f_n} = \frac{7}{5} = 1.4$$

$$\text{Amplitude ratio, } M = \frac{1}{\sqrt{(1-r^2)^2 + (2\gamma r)^2}} = \frac{1}{\sqrt{(1-1.4^2)^2 + (2\times 0.62\times 1.4)^2}} = 0.504$$

$\therefore$ The *"error"* due to proximity of excitation frequency with the natural frequency of the system

$$= (1 - 0.504) \times 100 = \mathbf{49.6\%} \quad \textbf{(Ans.)}$$

Example 2.39. *While measuring force by a diaphragm type pressure transducer (second-order system), the system is found to have a natural frequency of 800 cycles per second (cps) and a damping ratio of 62 per cent.*

Determine the frequency range over which the amplitude ratio corresponding to sinusoidal input deviates by a maximum amount of 12 per cent.

Solution. *Given :* Natural frequency of the system = 800 cps

Damping ratio, $\gamma = 62$ per cent

Maximum amount of deviation of amplitude ratio = 12 per cent.

Corresponding to a maximum deviation of 12%, the amplitude ratio (M) is either 0.88 or 1.12.

Now,
$$M = \frac{1}{\sqrt{(1-r^2)^2 + (2\gamma r)^2}}$$

where, r = Frequency ratio $= \dfrac{f}{f_n} = \dfrac{\omega}{\omega_n}$

(*i*) *When M = 1.12 :*

$$1.12 = \frac{1}{\sqrt{(1-r^2)^2 + (2\times 0.62\times r)^2}}$$

or,
$$(1 - r^2)^2 + (2 \times 0.62\, r)^2 = \left(\frac{1}{1.12}\right)^2 = 0.797$$

or,
$$1 + r^4 - 2r^2 + 1.538\, r^2 = 0.797$$

or,
$$r^4 - 0.462\, r^2 + 0.203 = 0$$

The solution has imaginary roots and this implies that with $\gamma = 0.62$, the *output would never be 1.12 times the output for any frequency.*

(*ii*) *When M = 0.88 :*

$$0.88 = \frac{1}{\sqrt{(1-r^2)^2 + (2\times 0.62\times r)^2}}$$

or, $(1 - r^2)^2 + (2 \times 0.62\, r)^2 = \left(\frac{1}{0.88}\right)^2 = 1.29$

or, $1 + r^4 - 2r^2 + 1.538\, r^2 = 1.29$

or, $r^4 - 0.462\, r^2 - 0.29 = 0$

or, $$r^2 = \frac{0.462 \pm \sqrt{(0.462)^2 + 4 \times 0.29}}{2} = 0.817$$

or, $r = 0.904$

$\therefore$ $f = f_n \times r = 800 \times 0.904 = 723$ cps

Thus, the deviation remains within 12 per cent of output for the *frequency range* **0 – 723 cps** **(Ans.)**

Example 2.40. *A unit amplitude sinusoidal input signal having a frequency of 0.6 Hz is given to a second-order control system of transfer function :* $\frac{I_o}{I_i} = \frac{8}{D^2 + 4D + 20}$.

Develop an expression for the steady state response of the system and determine the corresponding output amplitude, output frequency and phase lag.

Solution. Frequency, $\omega = 2\pi f = 2\pi \times 0.6 = 3.77$ rad/s

As the sinusoidal input signal has a *unit* amplitude, therefore,

$$I_i = 1 \sin (3.77\, t)$$

Putting $D = j\omega$ in the given relation, we get

$$\frac{I_o}{I_i} = \frac{8}{(j\omega)^2 + 4j\omega + 20} = \frac{8}{(3.77j)^2 + 4 \times 3.77j + 20}$$

$$= \frac{8}{-14.21 + 15.08j + 20} = \frac{8}{5.79 + 15.08j} = \frac{1}{0.724 + 1.885j}$$

The denominator has a magnitude of

$$\sqrt{(0.724)^2 + (1.885)^2} = 2.02$$

$$\text{Phase lag, } \phi = \tan^{-1}\frac{1.885}{0.724} = 69° = 1.204 \text{ rad}$$

$$\text{Magnitude of output} = \frac{1}{2.02} \times \text{magnitude of input}$$

$$= \frac{1}{2.02} \times 1 = 0.495$$

$\therefore$ The required expression for the output in the sinusoidal form is :

$$I_o = 0.495 \sin (3.77\, t - 69°)$$ **(Ans.)**

Hence, *corresponding output amplitude* = **0.495** (*as compared with unit amplitude of input*) **(Ans.)**

Output frequency = **3.77** rad/s (*same as that of input signal*) **(Ans.)**

Phase lag = **69°** (*between input and output*) **(Ans.)**

2.9 ERRORS IN MEASUREMENTS

2.9.1. Introduction

Errors and uncertainties are *inherent* in the process of making any measurement and in the instrument with which the measurements are made. The study of error is important as a step in finding ways of reducing them, and also as means of estimating the reliability of final results.

The accuracy and precision of a measuring instrument depends upon its design, proper selection of material for each and every part, and skill of manufacturing.

2.9.2. Limiting (or Guarantee) Errors

In most instruments the accuracy is generated to be *within a certain percentage of full scale reading.* The manufacturer has to specify the deviations from the *nominal value* of a particular quantity. *The limits of these deviations from the specified value are defined as* **Limiting errors** *or* **Guarantee errors.**

The **relative error** is *defined as the ratio of the error to the specified (nominal) magnitude of a quantity :*

If, Q_s = The specified quantity or true quantity of a component, and

Q_a = The actual or erroneous quantity ;

Then, error = $Q_a - Q_s$ (= δQ)

and, Relative limiting error (also called accuracy),

$$E_r = \frac{Q_a - Q_s}{Q_s} = \frac{\delta Q}{Q_s}$$

Percentage limiting error,

$$\% \, E_r = E_r \times 100$$

If a component has a relative limiting error $\pm E_r$, then limiting values of the component, $Q_a = Q_s (1 \pm E_r)$. Thus if the specified magnitude of a resistor is 100 Ω with a limiting error of ± 4 ohms, then the magnitude of the resistor will lie between the limits 100 ± 4, *i.e.,* 96 ohms and 104 ohms. In other words, the manufacturer of the resistor guarantees its value to lie between 96 and 104 ohms.

Example 2.41. *A wattmeter having a range of 500 W has an error of ± 1.5 per cent of full scale deflection. If the true power is 50 W, what would be the range of the readings ?*

If the error is specified as percentage of true value, what would be the range of the readings ?

Solution. *Given :* Range of the wattmeter = 500 W

Error (% age of full scale deflection) = ± 1.5%

True or specified power, Q_s = 50 W

- When the error is specified as a *percentage of full scale deflection*, the magnitude of limiting error at full scale

$$= \pm\left(\frac{1.5}{100}\right) \times 500 = \pm 7.5 \text{ W}$$

Thus the wattmeter reading when the true reading is 50 W may be 50 ± 7.5, *i.e.,* between **42.5 and 57.5 W** **(Ans.)**

- Relative error, $E_r = \frac{7.5}{50} \times 100 = 15\%$

Now suppose the error is specified as percentage of true value, then,

$$\text{Magnitude of error} = \pm \frac{1.5}{100} \times 50 = \pm 0.75 \text{ W, and}$$

The meter may read between 50 ± 0.75 or **49.25 and 50.75 W.** **(Ans.)**

Example 2.42. *The guaranteed accuracy of a flow meter working on thermal principles is ± 3 per cent of full scale reading of 2.5 × 10^{-6} m^3/s. If the flow measured by this meter is 1.25 × 10^{-6} m^3/s, calculate the limiting error in per cent. Comment upon the results.*

Solution. *Given :* Guaranteed accuracy of flow meter,

$$E_r = \pm 3\% \text{ (of full scale reading)}$$

Full scale reading, $Q_s = 2.5 \times 10^{-6} \text{ m}^3/\text{s}$

Flow measured by the meter = $1.25 \times 10^{-3} \text{ m}^3/\text{s}$.

Limiting error (Per cent) :

The magnitude of limiting error at full scale of the meter is,

$$\delta Q = E_r \times Q_s$$
$$= \pm 0.03 \times 2.5 \times 10^{-6}$$
$$= \pm 0.075 \times 10^{-6} \text{ m}^3/\text{s}$$

The magnitude of the flow being measured is $1.25 \times 10^{-6} \text{ m}^3/\text{s}$;

∴ The relative error at this flow,

$$E_r' = \frac{\delta Q}{Q_s} = \frac{0.075 \times 10^{-6}}{1.25 \times 10^{-6}} = 0.06$$

∴ The value of flow being measured is,

$$Q = Q_s(1 \pm E_r') = 1.25 \times 10^{-6} (1 \pm 0.06)$$
$$= 1.25 \times 10^{-6} \pm 0.075 \times 10^{-6} \text{ m}^3/\text{s}$$

Hence, *Percentage limiting error,*

$$\% E_r = \pm \frac{0.075 \times 10^{-6}}{1.25 \times 10^{-6}} \times 100 = \pm 6\% \quad \textbf{(Ans.)}$$

Comments. From above results we observe that when the meter reading is full scale value of 2.5 × 10^{-6} m^3/s, the flow meter is guaranteed to have a limiting error of ± 3%, but when the meter is reading its half scale value of 1.25 × 10^{-6} m^3/s, the limiting error is ±6%. Thus when reading is halved, the limiting error is doubled. *This indicates that the per cent limiting error will be greater if smaller values are read by the meter.* Therefore, while selecting instruments, special care should be taken as regards the range; the values to be measured *should not lie in the lower third of the range.* Hence as far as possible, meter which will read up their scale should be used.

Resultant error of a device composed of several components :

A measuring instrument/device/system may be formed by combining several components, each having its limiting error. The limiting error of the combination can be computed in the following ways depending on how the various components are connected in the total system:

> When two or more quantities, each having a limiting error, are combined, it is advantageous to be able to compute the limiting error of the combination. The limiting error can be easily found by *considering the relative increment of the function if the final result is in the form of an algebraic equation.*

(i) When the final result is either sum or difference of outputs of various components :

Let, $q_1, q_2, q_3, \ldots\ldots\ldots\ldots q_n$ = Outputs of each device, and

Q = The final result ;

Then, Q (sum or difference) $= q_1 \pm q_2 \pm q_3 \pm \ldots\ldots\ldots\ldots\ldots\ldots \pm q_n,$

and,
$$\frac{\delta Q}{Q} = \pm\left[\frac{q_1}{Q}\cdot\frac{\delta q_1}{q_1} + \frac{q_2}{Q}\cdot\frac{\delta q_2}{q_2} + \frac{q_3}{Q}\cdot\frac{\delta q_3}{q_3} + \ldots\right] \qquad \ldots(2.59)$$

Equation (2.59) shows that the *resultant limiting error in this case is equal to the sum of the products formed by multiplying the individual relative errors by the ratio of each term to the function.*

(ii) When the result is either product or quotient of output of various components :

In this case, $Q = q_1 \times q_2 \times q_3 \times \ldots q_n$

or, $Q = q_1 \div q_2 \div q_3 \div \ldots q_n$

or, $Q = q_1 \times q_2 \div q_3 \times q_4 \div \ldots q_n$

Then,
$$\frac{\delta Q}{Q} = \pm\left(\frac{\delta q_1}{q_1} + \frac{\delta q_2}{q_2} + \frac{\delta q_3}{q_3} + \ldots\ldots\frac{\delta q_n}{q_n}\right) \qquad \ldots(2.60)$$

In other words, the *relative limiting error of the system is the sum of relative errors of all devices/components.*

(iii) When the final output is composite factor of output of various components :

In this case,
$$Q = q_1^{\pm n_1} \times q_2^{\pm n_2} \times q_3^{\pm n_3} \ldots\times q_n^{\pm n_n}$$

Then,
$$\frac{\delta Q}{Q} = \pm\left[n_1\frac{\delta q_1}{q_1} + n_2\frac{\delta q_2}{q_2} + n_3\frac{\delta q_3}{q_3} + \ldots\right] \qquad \ldots(2.61)$$

Example 2.43. *The three resistors R_1, R_2 and R_3 have the following ratings :*

$$R_1 = 25\ \Omega \pm 4\%,\ R_2 = 65\ \Omega \pm 4\%;\ R_3 = 45\ \Omega \pm 4\%$$

Determine the following :

(i) Limiting value of the resultant resistance.

(ii) Per cent limiting error of series combination of resistance.

Solution. *Given :*
$$R_1 = 25\ \Omega \pm 4\%$$
$$R_2 = 65\ \Omega \pm 4\%$$
$$R_3 = 45\ \Omega \pm 4\%.$$

(i) Limiting value of the resultant resistance :

The values of the resistances are :

$$R_1 = 25 \pm \frac{4}{100}\times 25 = 25 \pm 1.0\ \Omega$$

$$R_2 = 65 \pm \frac{4}{100}\times 65 = 65 \pm 2.6\ \Omega$$

$$R_3 = 45 \pm \frac{4}{100}\times 45 = 45 \pm 1.8\ \Omega$$

∴ The limiting value of resultant resistance,

$$R = (25 + 65 + 45) \pm (1.0 + 2.6 + 1.8) = (135 \pm 5.4)\ \Omega$$

Magnitude of resistance = **135 Ω** and, *error* = **± 5.4 Ω** **(Ans.)**

(ii) Per cent limiting error of the series combination of resistance :

Per cent limiting error of the combination

$$= \pm\frac{5.4}{135}\times 100 = \pm\ \mathbf{4\%} \quad \textbf{(Ans.)}$$

Example 2.44. *A power meter and an ammeter are used to measure resistance of an unknown resistance. Find the limiting error in the measurement of resistance when the limiting errors in the measurement of power and current are ± 1.2% and 0.8% respectively.*

Solution. *Given :* Limiting error in the measurement of power (*P*) = $\frac{\delta P}{P} = 1.2\%$

Limiting error in the measurement of current (*I*) = $\frac{\delta I}{I} = 0.8\%$

We know that, $R = \frac{P}{I^2}$

∴ Limiting error in the measurement of resistance,

$$\frac{\delta R}{R} = \pm\left(\frac{\delta P}{P} + 2\frac{\delta I}{I}\right)$$

$$= \pm\ (1.2 + 2 \times 0.8) = \pm\ \mathbf{2.8\%} \quad \textbf{(Ans.)}$$

Example 2.45. *The unknown resistance in a Wheatstone bridge is measured by using three known resistances such that :*

$$R_4 = \frac{R_2 R_3}{R_1}$$

where, $R_1 = 50 \pm 0.5\%\ \Omega;\ R_2 = 500 \pm 0.5\%\ \Omega;\ R_3 = 440 \pm 0.5\%\ \Omega$

Determine the magnitude of the unknown resistance (R_4) and the limiting error in per cent and in ohms (for the unknown resistance).

Solution. *Given :* $R_1 = 50 \pm 0.5\%\ \Omega;\ R_2 = 500 \pm 0.5\%\ \Omega;\ R_3 = 440 \pm 0.5\%\ \Omega$

The unknown resistance is given by

$$R_4 = \frac{R_2 R_3}{R_1} = \frac{500\times 440}{50} = \mathbf{4400\ \Omega} \quad \textbf{(Ans.)}$$

The relative limiting error of the unknown resistance is,

$$\frac{\delta R_4}{R_4} = \pm\left(\frac{\delta R_2}{R_2} + \frac{\delta R_3}{R_3} + \frac{\delta R_1}{R_1}\right)$$

$$= \pm\ (0.5 + 0.5 + 0.5) = \pm\ \mathbf{1.5\%} \quad \textbf{(Ans.)}$$

Limiting error in ohms $= \frac{1.5}{100}\times 4400 = \pm\ \mathbf{66\ \Omega}$ **(Ans.)**

i.e., The guaranteed values of resistance are between **4334** (*i.e.,* 4400 – 66) **Ω** and **4466** (*i.e.,* 4400 + 66) **Ω** **(Ans.)**

Example 2.46. *A cantilever beam of length l, width b and depth d is used to measure the force F. The deflection is measured using an LVDT of accuracy ± 0.1 mm. The length, width and depth are measured to an accuracy of ± 0.01 mm and their values are 45 mm, 4.5 mm and 0.9 mm respectively. The modulus of rigidity is known accurate to 0.2 %. If the deflection noted was 1.8 mm, determine the error in force measured.*

Given force, $F = \frac{3Ebd^3}{12l^3}\cdot y$

where, E = *Modulus of elasticity;* b = *width,* d = *depth, and*

l = *length of the beam ;* y = *deflection.*

Solution. The *limiting error in force F is,*

$$\frac{\delta F}{F} = \pm\left(\frac{\delta E}{E} + \frac{\delta b}{b} + 3\frac{\delta d}{d} + 3\frac{\delta l}{l} + \frac{\delta y}{y}\right)$$

$$= \pm\left(0.2 + \frac{0.01}{4.5} + 3\times\frac{0.01}{0.9} + 3\times\frac{0.01}{45} + \frac{0.1}{1.8}\right)$$

$$= \pm\ (0.2 + 0.00222 + 0.03333 + 0.00067 + 0.05555)$$

$$= \pm\ \mathbf{0.2917\%}$$ **(Ans.)**

Example 2.47. *A dynamometer is used to measure the output power of a rotating shaft. The output power is given by the following relation :*

$$P = \frac{2\pi\times 9.81\ FLR}{t\times 10^6}\ kW$$

where, F = *Force at the end of torque arm (kg),*

L = *Length of torque arm (mm),*

R = *Number of revolutions during time t, and*

t = *Time for test run (s).*

The given test data are

$F = 4.26 \pm 0.02$ *kg;* $L = 382 \pm 1.2$ *mm;*

$R = 1192 \pm 1.0$ *revolutions;* $t = 60 + 0.50$ *s*

The errors are limiting (absolute) errors.

(i) The magnitude of power.

(ii) The magnitude of the limiting error in the computed power.

Solution. *Given :* $F = 4.26 \pm 0.02$ kg; $L = 382 \pm 1.2$ mm; $R = 1192 \pm 1.0$ revolutions; $t = 60 + 0.50$ s

(i) The magnitude of power *P* :

$$P = \frac{2\pi\times 9.81\times FLR}{t\times 10^6}\ \text{kW} \qquad \text{...(Given)}$$

Inserting the various values in the above expression, we get

$$P = \frac{2\pi\times 9.81\times 4.26\times 382\times 1192}{60\times 10^6} = \mathbf{1.993\ kW}$$ **(Ans.)**

(*ii*) Limiting errors in the computed power :

Total per unit error in computed power,

$$\frac{\delta P}{P} = \pm\left(\frac{\delta F}{F} + \frac{\delta L}{L} + \frac{\delta R}{R} + \frac{\delta t}{t}\right)$$

$$= \pm\left(\frac{0.02}{4.26} + \frac{1.2}{382} + \frac{1.0}{1192} + \frac{0.50}{60}\right)$$

$$= \pm\ (0.00469 + 0.00314 + 0.00084 + 0.00833) = \pm\ 0.017$$

Limiting error in computer power

$$\delta P = \pm 0.017 \times P = \pm\ 0.017 \times 1.993 = \mathbf{0.0339\ kW}$$

The result is written as: P = **1.993 ± 0.0339 kW** **(Ans.)**

Example 2.48. *(a) How are "known errors" treated ?*

(b) During a test the current was measured as 26.5 A flowing in a resistor of 0.12 Ω. Later on it was discovered that the ammeter reading was low by 1.1% and the marked resistance was high by 0.25%.

Determine the true power as a percentage of the power that was originally calculated.

Solution. (a) Known errors. When the error of quantity of an instrument is known, the effect of this error, when combined with errors, can be computed in a manner similar to the combinations of the limiting errors. However, the difference is that in case of known errors the *signs of relative errors are given and must be presented in the calculations.*

(b) True value of current $I = 26.5(1 + 0.011) = 26.791$ A

True value of resistance $R = 0.12(1 - 0.0025) = 0.1197$ Ω

True value of power $= I^2R = (26.791)^2 \times 0.1197 = 85.916$ W

Measured value of power $= (26.5)^2 \times 0.12 = 84.27$ W

$$\therefore \quad \frac{\text{True power}}{\text{Measured power}} \times 100 = \frac{85.916}{84.27} \times 100 = \mathbf{101.9\%} \quad \textbf{(Ans.)}$$

Alternatively :

$$\text{Power, } P = I^2R$$

$$\text{Total relative error} = \frac{\delta P}{P} = 2 \cdot \frac{\delta I}{I} + \frac{\delta R}{R}$$

$$= 2(-0.011) + 0.0025 = -0.0195$$

$$\frac{\text{True power}}{\text{Measured power}} = \frac{1}{1 + \text{relative error}}$$

$$= \frac{1}{1 - 0.0195} = \mathbf{101.9\%} \quad \textbf{(Ans.)}$$

2.9.3. Types of Errors

In measurement systems, the errors could originate from several sources. Broadly these may be *classified* as follows :

1. *Gross errors*
2. *Systematic errors :*
   - (i) Instrumental errors;
   - (ii) Environmental errors;
   - (iii) Observational errors.
3. *Random errors.*

2.9.3.1. Gross errors

These errors occur due to *human mistakes* in reading instruments and recording and calculating results of measurement. Although it is probably impossible to eliminate the gross error completely, yet one should try to anticipate and correct them.

The mathematical analysis of gross errors is impossible since these may occur in different amounts. While some gross errors may be easily detected, others may go unnoticed.

These errors can be *avoided* by adopting *two means :*

(i) Immense care should be taken while taking the reading and recording the data.

(ii) Two, three, or even more readings should be taken for the quantity being measured.

• Designers of instruments to some extent can take care to avoid gross errors but much would depend upon the interest and competence of the observer.

2.9.3.2. Systematic errors

The systematic errors are repeated consistently with the repetition of the experiment and are caused by *such effects as* :

— Sensitivity shift;

— Zero off-set;

— Known non-linearity.

• These errors *cannot* be determined by direct and repetitive observations of the measurand made each time with same technique.

These errors can be located only by having repeated measurements under different conditions or with different equipment and where possible by an entirely different method.

These errors may be *instrumental, environmental or observational errors.*

1. Instrumental errors :

These errors arise *due to :*

— Inherent shortcomings in the instruments.

— Misuse of the instruments.

— Loading effects of instruments.

Instrumental errors are inherent in instruments due to construction, calibration or operation of the instruments. *These errors may cause the instrument to read too low or too high.*

Some *"examples"* of such errors are :

(*i*) Mechanical friction and wear, backlash, yielding of supports, pen or pointer drag and hysteresis of elastic members due to aging.

(*ii*) Constructional faults resulting from finite width of knife edges, excessive friction at the mating parts, lost motion due to necessary clearance in gear teeth etc.

(*iii*) Improper selection and poor maintenance of the instrument.

(*iv*) Unavoidable physical phenomenon due to friction, capillary attraction and imperfect rarefaction.

Assembly errors. These errors may be due to the following *reasons* :

(*i*) Displaced scale *i.e.,* incorrect fitting of the scale zero with respect to the actual zero position of the movement.

(*ii*) Non-uniform division of the scale.

(*iii*) Bent or distorted pointer.

The assembly errors do not change with time, so these can be easily discovered and rectified.

Misuse of instruments. A good instrument used in an unintelligent way may give erroneous results.

• *Loading effects.* In several measurement techniques, the detector/sensor utilizes the energy of the medium, thus altering slightly the parameter to be measured, resulting in errors.

2. Environmental errors :

The environmental errors are due to conditions external to the measuring device (*e.g.,* effects of pressure, temperature, humidity, dust, vibrations, external magnetic or electrostatic fields). These errors are much more troublesome than assembly errors.

These errors can be *eliminated or reduced* by taking the following measures :

(*i*) Using the instrument in controlled conditions of pressure, temperature, humidity in which it was originally assembled and calibrated.

(*ii*) Measure deviations in the local conditions from the calibrated ones and then apply a suitable correction to the instrument readings.

(*iii*) Automatic compensation for departures from the calibrated conditions by using sophisticated devices.

(*iv*) Make a complete new calibration under the local conditions.

(*v*) In case it is suspected that external magnetic or electrostatic fields can affect the readings of the instruments, magnetic or electrostatic shields may be provided.

(*vi*) Apply computed corrections. Normally, efforts are made to avoid the use of applications of computed corrections, but where these corrections are needed and are necessary, they are incorporated for the computation of the results.

3. Observational errors :

These errors occur due to *carelessness of operators.* Even when an instrument has been properly selected, carefully installed and faithfully calibrated, shortcoming in the measurement occurs due to certain failings on the part of the observer.

The observational errors may be due to the following *reasons :*

(*i*) *Parallax*—Since the parallax errors arise on account of pointer and the scale not being in the same plane, we can eliminate this error by having the pointer and the scale in the same plane.

(*ii*) Wrong scale reading, and wrong recording of data.

(*iii*) Inaccurate estimates of average reading, lack of ability to interpolate properly between graduations.

(*iv*) Incorrect conversion of units in between consecutive readings, and non-simultaneous observation of interdependent quantities.

(*v*) Tendency to read high or low, anticipate a signal and read too soon—Personal bias.

- Modern electrical instruments have *digital display of output* which completely eliminates the errors on account of human observational or sensing power as the output is in the *form of digits.*

2.9.2.3. Random errors

- The random errors are *accidental, small and independent.*
- *They vary in an unpredictable manner.*
- The magnitude and direction of these errors *cannot* be predicted from a knowledge of measurement system; however these errors are *assumed to follow the law of probabilities.*

The most common causes of these errors are :

(*i*) Friction in instrument movement.

(*ii*) Backlash in the movement.

(*iii*) Parallax errors between pointer and scale.

(*iv*) Finite dimensions of the pointer and scale divisions.

(*v*) Hysteresis in elastic members.

(*vi*) Mechanical vibrations.

2.9.4. Sources of Errors

Besides the errors so far enumerated (of the physical and mechanical nature) and the errors which necessarily result from the faulty calibration of a measuring system, the errors arise from the *following sources also* :

(*i*) **Noise.** *It is defined as any signal that does not convey useful information.*

— Noise signals/errors can be reduced to a minimum level through filtering, careful selection of components, shielding and isolation of the entire measuring system.

(*ii*) **Response time.** *It is defined as the time taken by the instrument to show 63.2% change in a reading to a step input*. This factor contributes to the uncertainty of the measurement.

(*iii*) **Design limitations.** In the design of the instrument, certain inevitable factors such as friction and resolving power lead to uncertainty of measurements.

(*iv*) **Energy exchanged by interaction.** As earlier pointed out, whenever the energy required for operating the measuring system is extracted from the measurand, the value of the latter is altered to a greater or lesser extent. This alternation is dependent upon the capacity of the system.

(*v*) **Transmission.** During the transmission of information from the primary sensing element to the indicator, the signal may be *attenuated* due to any of the following *reasons* :

— It may suffer loss through leakage ;

— It may be absorbed or otherwise consumed in the communication channel;

— It may be distorted by resonance, attenuation or delay phenomenon whose actions are selective on various signal components.

(*vi*) **Deterioration of measuring system.** The examples of some types of deterioration which occur in a measuring system and result in a source of error in measurement are :

— Change in resistance of a circuit element through strain relief ;

— Alteration of thermocouple characteristics through oxidizing or reducing atmosphere;

(*vii*) **Ambient influences on measuring systems.**

(*viii*) **Errors of observation and interpretation.**

2.9.5. Statistical Analysis of Test Data

The *systematic and random errors* are evaluated and studied by *statistical procedures* which make it possible to state from a limited group of data the most probable value of a quantity, the probable uncertainty of a single observation, and the probable limits of uncertainty of the best value that can be derived from the data.

— The object of the statistical methods based on laws of chance, which operate only on random errors and not on systematic errors, is to achieve consistency (precision) of value and not their accuracy.

The experimental data is obtained in the following *two forms of tests* :

(*i*) Single-sample test;

(*ii*) Multi-sample test.

(*i*) **Single-sample test.** *A single measurement (or succession of measurements) done under identical conditions excepting for time is known as "single-sample test".*

(ii) **Multi-sample test.** In a multi-sample test, *repeated measurements of a given quantity are taken under different test conditions such as using different instruments, different ways of measurement and by employing different observers.* This type of test does not materialise result by simply making measurements with the same equipment, procedure, technique and same observer.

- For getting the exact value of the quantity under measurement, tests should be conducted using as many different procedures, techniques or experiments as practicable. It should be understood that the *statistical means which help us to arrive at correct results are only valid for multi-sample tests.*

2.9.5.1. Statistical averages

When several multi-sample observations are taken experimentally there is a scatter of the data about some central value, one of the methods, presenting the result in the form of a histogram (also called a *frequency distribution curve*). Thus, histogram (Fig. 2.34) is a pictorial representation of the test data where the ordinates indicate the number of observed readings of a particular value. With more and more data taken at smaller and smaller increments, the histogram finally changes into a *smooth curve.* The smooth curve is symmetrical with respect to the central value.

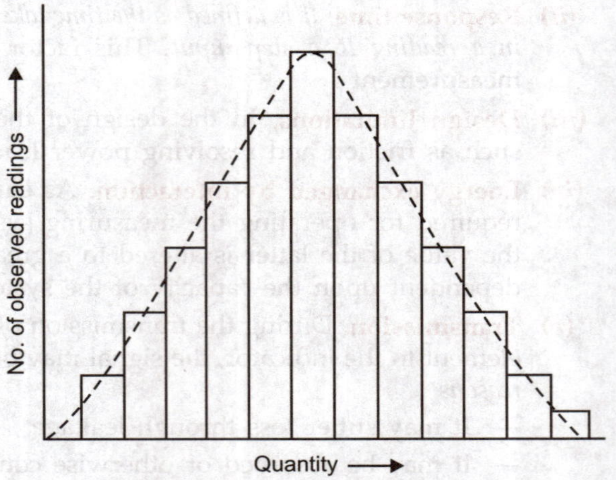

Fig. 2.34. Histogram.

(i) **Arithmetic mean (AM).** *The most probable value of measured variable (variate) is the arithmetic mean of the number of readings taken.* We arrive at the best approximation if the number of readings of the same quantity are very large. The mean is computed by summing all the values and dividing by the number of measurements. If $q_1, q_2, q_3 \dots q_n$ be the set of measured values of a *quantity,* then the arithmetic mean $\bar{q}$ will be given by,

$$\bar{q} = \frac{q_1 + q_2 + q_3 + \dots q_n}{n} = \frac{1}{n}\sum_{i=1}^{n} q_i \quad \dots(2.62)$$

(ii) **Geometric mean (G.M.).** It is defined as the nth root of the product of n terms.

$$\text{Geometric mean (G.M)} = \sqrt[n]{q_1 \times q_2 \times q_3 \dots \times q_n} \quad \dots(2.63)$$

(iii) **Median.** *The middle value of a set of an "odd" number of readings, if variables are arranged in numerical order, is called the "median".* For an *"even"* number of readings, the median is the *arithmetic average of the two central readings.*

(iv) **Mode.** *The value of the variable that occurs most frequently is termed as "mode".*

- For a symmetrical distribution curve, the three measures of central value, *viz., 'mean', 'median' and 'mode', are equal.*

2.9.5.2. Dispersion from mean

The property which denotes the extent to which the values are dispersed about the central value is termed as **dispersion** *or* **stored** *or* **scatter.**

A large dispersion indicates that some factors involved in the measurement process are not under close control and therefore it becomes difficult to estimate the measured quantity with confidence and definiteness.

The simplest possible measure of dispersion is the **Range** *which is the difference between greatest and least values of data.*

(i) Deviation. *The departure of the observed reading from the arithmetic mean of the group of readings is termed as* ***deviation***. Let the deviation of reading q_1 be d_1 and that of reading q_2 be d_2 etc.

Then,

$$d_1 = q_1 - \bar{q}$$
$$d_2 = q_2 - \bar{q}$$
$$\cdots \quad \cdots$$
$$d_n = q_n - \bar{q} \qquad \text{...(2.64)}$$

and,

$$\bar{q} = \frac{\Sigma(q_n - d_n)}{n} \qquad \text{...(2.65)}$$

$$\text{Algebraic sum of deviations} = d_1 + d_2 + d_3 + \ldots\ldots d_n$$
$$= (q_1 - \bar{q}) + (q_2 - \bar{q}) + (q_3 - \bar{q}) + \ldots\ldots(q_n - \bar{q})$$
$$= (q_1 + q_2 + q_3 + \ldots\ldots + q_n) - n\bar{q} = 0$$
$$(\because q_1 + q_2 + q_3 + \ldots + q_n) = n\bar{q}$$

Thus, the *algebraic sum of deviations is zero.*

The above mathematical result implies that the deviations can be *positive as well as negative.*

(ii) Average deviation. *It is defined as the sum of the absolute values of deviations divided by the number of readings.*

Average deviation may be expressed as

$$\bar{d} = \frac{|d_1| + |d_2| + |d_3| + \ldots\ldots + |d_n|}{n}$$

$$= \frac{\Sigma|d|}{n}, \text{ or, } \sum_{i=1}^{n} \frac{|q_i - \bar{q}|}{n} \qquad \text{...(2.66)}$$

The average deviation is taken as a measure of accuracy of measurement.

- The average deviation is an *indication of the precision of the instruments* used in making the measurements. *Highly precise instruments yield a low average deviation between readings.*

(iii) Standard deviation. *"Standard deviation"* (or *"mean square deviation"*) of an infinite number of data is defined *as the square root of the sum of individual deviations squared divided by the number of readings.*

Thus, standard deviation

$$\sigma = \sqrt{\frac{d_1^2 + d_2^2 + d_3^2 + \ldots\ldots + d_n^2}{n}}$$

$$= \sqrt{\frac{\Sigma d^2}{n}} = \sqrt{\sum_{i=1}^{n} \frac{(q_i - \bar{q})^2}{n}} \qquad \text{...(2.67)}$$

- In practice, however, the number of observations is *finite*. When the number of observations is *greater than 20*, standard deviation is denoted by the symbol σ while if it is *less than* 20, the symbol used is *s*. The standard deviation of a finite number of data is given by :

$$s = \sqrt{\frac{d_1^2 + d_2^2 + d_3^2 + ... + d_n^2}{n-1}} = \sqrt{\frac{\Sigma d^2}{n-1}} \quad ...(2.68)$$

Variance. *The square of the standard deviations is called* ***variance.***

Variance, $V = \text{(Standard deviation)}^2$

$$= \sigma^2 = \frac{d_1^2 + d_2^2 + d_3^2 ...... d_n^2}{n} = \frac{\Sigma d^2}{n} \quad ...(2.69)$$

But when the number of observation is less than 20,

$$\text{Variance,} \quad V = s^2 = \frac{\Sigma d^2}{n-1} \quad ...(2.70)$$

2.9.5.3. Best value from a sample of readings

Let, q_b = The best value (approximately the true value) from a set of readings/data.

Then, the deviation or individual component error,

$$d_i = q_i - q_b \quad \text{where, } i = 1, 2, 3, ...n$$

This deviation may be *+ve, zero* or *–ve.*

(*i*) Sum of deviations :

$$\sum_{i=1}^{n} d_i = \sum_{i=1}^{n} (q_i - q_b) \quad ...(2.71)$$

In case of a *best value,* the sum of deviations is equal to *zero.*

$$\text{i.e.,} \quad \sum_{i=1}^{n} (q_i - q_b) = 0 \quad ...(2.72)$$

$$\text{or,} \quad q_b = \frac{1}{n}\sum_{i=1}^{n} q_i \text{, which is the arithmetic mean.}$$

Thus, the arithmetic mean may be interpreted as the best value around which the chances of error being *+ve* or *–ve* likely to be equal.

(*ii*) $d_i = q_i - q_b$

Squaring both sides, we have

$$d_i^2 = q_i^2 - 2q_i q_b + q_b^2$$

Sum of the square of deviations,

$$\sum_{i=1}^{n} d_i^2 = \sum_{i=1}^{n} q_i^2 - 2q_b \sum_{i=1}^{n} q_i + nq_b^2 \quad ...(2.73)$$

The sum of the square of deviations, corresponding to the *best,* would be *minimum.* This condition can be obtained by setting the deviation w.r.t. q_b equal to zero;

$$\text{i.e.,} \quad \frac{d}{dq_b}\left[\sum_{i=1}^{n} d_i^2\right] = 0 \quad ...(2.74)$$

or,
$$\frac{d}{dq_b}\left[\sum_{i=1}^{n} q_i^2 - 2q_b \sum_{i=1}^{n} q_i + nq_b^2\right] = 0$$

As all the $q_i^2 s$ are constant, the differentiation gives,

$$-2\sum_{i=1}^{n} q_i + 2nq_b = 0$$

or,
$$q_b = \frac{1}{n}\sum_{i=1}^{n} q_i \qquad ...(2.75)$$

which is again the arithmetic mean. Obviously, *the arithmetic mean is the most probable value or the best value which one can obtain from a set of scattered measurement data.*

Example 2.49. *By using a micrometer screw the following readings were taken of a certain physical length :*

1.34, 1.38, 1.56, 1.47, 1.42, 1.44, 1.53, 1.48, 1.40, 1.59 mm

Assuming that only random errors are present, calculate the following :

(i) Arithmetic mean,

(ii) Average deviation,

(iii) Standard deviation, and

(iv) Variance.

Solution. *Given* : Readings of the micrometer screw : 1.34, 1.38, 1.56, 1.47, 1.42, 1.44, 1.53, 1.48, 1.40, 1.59 mm.

(i) Arithmetic mean :

We know that arithmetic mean,

$$\bar{q} = \frac{1}{n}\sum_{i=1}^{n} q_i \qquad ...(\text{Eqn. } 2.62)$$

$$= \frac{q_1 + q_2 + q_3 + q_4 + q_5 + q_6 + q_7 + q_8 + q_9 + q_{10}}{10}$$

$$= \frac{1.34 + 1.38 + 1.56 + 1.47 + 1.42 + 1.44 + 1.53 + 1.48 + 1.40 + 1.59}{10}$$

= **1.461 mm** **(Ans.)**

(ii) Average deviation :

From eqn. (2.64), the *deviations* are :

$$d_1 = q_1 - \bar{q} = 1.34 - 1.461 = -0.121$$
$$d_2 = q_2 - \bar{q} = 1.38 - 1.461 = -0.081$$
$$d_3 = q_3 - \bar{q} = 1.56 - 1.461 = +0.099$$
$$d_4 = q_4 - \bar{q} = 1.47 - 1.461 = +0.009$$
$$d_5 = q_5 - \bar{q} = 1.42 - 1.461 = -0.041$$
$$d_6 = q_6 - \bar{q} = 1.44 - 1.461 = -0.021$$
$$d_7 = q_7 - \bar{q} = 1.53 - 1.461 = +0.069$$

$$d_8 = q_8 - \bar{q} = 1.48 - 1.461 = +0.019$$

$$d_9 = q_9 - \bar{q} = 1.40 - 1.461 = -\ 0.061$$

$$d_{10} = q_{10} - \bar{q} = 1.59 - 1.461 = 0.129$$

$\therefore$ *Average deviation,*

$$\bar{d} = \frac{\Sigma|d|}{n} \left(\text{or } \sum_{i=1}^{n} \frac{|q_i - \bar{q}|}{n} \right)$$

$$= \frac{0.121 + 0.081 + 0.099 + 0.009 + 0.041 + 0.021 + 0.069 + 0.019 + 0.061 + 0.129}{10}$$

= **0.065** **(Ans.)**

It may be noted that for calculation of average deviations, the *signs of deviations are not considered.*

(iii) **Standard deviation :**

Since the number of readings is 10, which is *less than 20,* therefore equation (2.68) is used for calculation of standard deviation.

$$\text{Standard deviation, } s = \sqrt{\frac{\Sigma d^2}{n-1}} \qquad \text{...[Eqn. (2.68)]}$$

$$\text{or, } s = \sqrt{\frac{(-0.121)^2 + (-0.081)^2 + (+0.099)^2 + (0.009)^2 + (-0.041)^2 + (-0.021)^2 + (0.069)^2 + (0.019)^2 + (-0.061)^2 + (0.129)^2}{(10-1)}}$$

$$= \sqrt{\frac{0.05869}{9}} = \mathbf{0.0807\ mm} \quad \textbf{(Ans.)}$$

(iv) **Variance :**

$$\textit{Variance, } V = s^2 = (0.0807)^2 = \mathbf{0.00651\ mm^2} \quad \textbf{(Ans.)}$$

Example 2.50. *Eight different students tuned in the circuit for resonance and the values of resonant frequency in kHz were recorded as :*

412, 428, 423, 415, 426, 411, 423, 416

Calculate the following :

(i) Arithmetic mean,

(ii) Average deviation,

(iii) Standard deviation, and

(iv) Variance.

Solution. *Given* : Values of resonant frequency : 412, 428, 423, 415, 426, 411, 423, 416

(i) **Arithmetic mean :**

Arithmetic mean,

$$\bar{q} = \frac{\Sigma q}{n} = \frac{412 + 428 + 423 + 415 + 426 + 411 + 423 + 416}{8}$$

= **419.25 kHz** **(Ans.)**

(ii) **Average deviation :**

The deviations are :

$$d_1 = q_1 - \bar{q} = 412 - 419.25 = -7.25 \text{ kHz}$$

$$d_2 = q_2 - \bar{q} = 428 - 419.25 = +8.75 \text{ kHz}$$

$$d_3 = q_3 - \bar{q} = 423 - 419.25 = 3.75 \text{ kHz}$$

$$d_4 = q_4 - \bar{q} = 415 - 419.25 = -4.25 \text{ kHz}$$

$$d_5 = q_5 - \bar{q} = 426 - 419.25 = +6.75 \text{ kHz}$$

$$d_6 = q_6 - \bar{q} = 411 - 419.25 = -8.25 \text{ kHz}$$

$$d_7 = q_7 - \bar{q} = 423 - 419.25 = +3.75 \text{ kHz}$$

$$d_8 = q_8 - \bar{q} = 416 - 419.25 = -3.25 \text{ kHz}$$

∴ *Average deviation,*

$$\bar{d} = \frac{\Sigma|d|}{n}$$

$$= \frac{7.25 + 8.75 + 3.75 + 4.25 + 6.75 + 8.25 + 3.75 + 3.25}{8}$$

$$= \mathbf{5.75 \text{ kHz}}$$ **(Ans.)**

(For the calculation of average deviation, the *signs of deviations are not considered*)

***(iii)* Standard deviation :**

Since the number of readings is 8, which is less than 20, therefore eqn. (2.68) is used for calculation of standard deviation.

$$\text{Standard deviation, } s = \sqrt{\frac{\Sigma d^2}{n-1}} \qquad \text{...[Eqn (2.68)]}$$

$$\text{or, } s = \sqrt{\frac{(-7.25)^2 + (+8.75)^2 + (+3.75)^2 + (-4.25)^2 + (+6.75)^2 + (-8.25)^2 + (+3.75)^2 + (-3.25)^2}{(8-1)}}$$

$$= \mathbf{6.54 \text{ kHz}}$$ **(Ans.)**

***(iv)* Variance :**

$$\textit{Variance, } V = s^2 = (6.54)^2 = \mathbf{42.77 \ (kHz)^2}$$ **(Ans.)**

2.9.5.4. Normal or Gaussian curve of errors

In experimental work the random effects are normally studied with reference to the Gaussian or normal error distribution. Here, the following *assumptions* are made :

(i) All observations include a large number of small random disturbing effects, and these small random errors may be both *+ve* and *–ve*. Thus, there is a strong central tendency, *i.e., there is a greater probability of small errors and large errors of both +ve and –ve values.*

(ii) There is equal probability of *+ve* and *–ve* disturbing effects. As such the probability curve of errors plotted against the magnitude of error will be *symmetrical about the zero value.*

The law of probability states that the normal occurrence of deviations from average value of an infinite number of measurements or observations can be expressed as :

$$P = \frac{1}{\sigma\sqrt{2\pi}} e^{-x^2/2\sigma^2} \qquad \text{...(2.76)}$$

where, P = Number of readings at any deviation 'd', *i.e.,* probability of occurrence of deviation d,

σ = Standard deviation, and

x = Magnitude of deviation from mean.

Fig. 2.35 shows a Gaussian distribution curve in which *P* has been plotted as a function of *x*. The curve is symmetrical about the mean value and may be regarded as the limiting form of a histogram. The two values of standard deviation σ_1 and σ_2 ($\sigma_1 > \sigma_2$) have been illustrated in the given figure. *Larger the value of σ, the flatter the curve and hence larger the expected error of all measurements.*

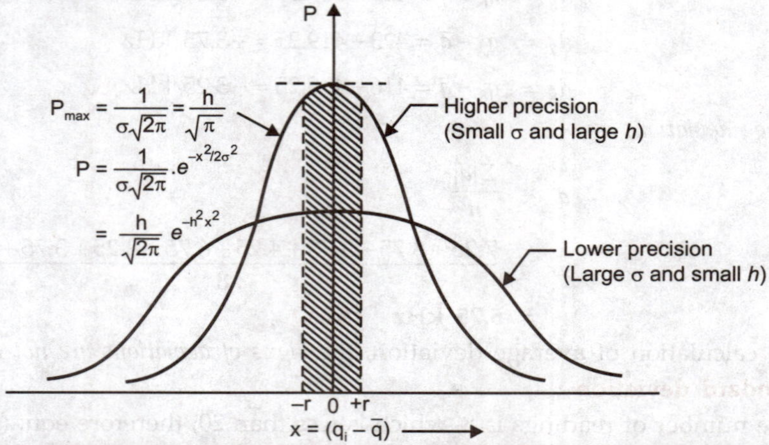

Fig. 2.35. Gaussian distribution curve.

If we integrate *P* between the limits $x = \pm x$, we shall obtain the area under the curve between these two values of *x*. Because the curve is related to the probability that the measured lies within same range, we have

$$\int_{-x}^{+x} P dx = 1.0 \qquad ...(2.77)$$

i.e., area under the probability curve is *unity*.

If the total number of readings are represented by unity, then the fraction of the total number of readings lying between x_1 and x_2 will be represented by the area under the curve between these two values of deviation ;

i.e., $$n_{1-2} = \int_{-x}^{+x} P dx$$

If the area between x_1 and x_2 is 0.6, then 60 per cent of the deviations fall between x_1 and x_2.

Precision index. It may be observed from the normal distribution function that the maximum probability occurs when deviation is *zero, i.e.,* $q = \bar{q}$ and the corresponding value of probability is,

$$P_{max} = \frac{1}{\sigma\sqrt{2\pi}} = \frac{h}{\sqrt{\pi}} \qquad ...(2.78)$$

where, $h = \frac{1}{\sigma\sqrt{2}}$ is called the **precision index**.

A large value of *h* represents high precision of the data because the probability of occurrence of variates in a given range falls off *rapidly* as the deviation increases because the variates tend to cluster (become closer) into a narrow range. On the other hand, a small value of h represents low precision because the probability of occurrences of variates in a given range falls off *gradually* as the deviation increases; this is because the variates are spread over a wide range.

Probable error : Let us consider the two points $-r$ and $+r$ marked in Fig. 2.35. These points are so located that the area bounded by the curve, the X-axis and the ordinates erected at $x = -r$ and $x = +r$ is equal to half of the total area under the curve. That is half the deviations lie between $x = \pm r$. The location of the points r can be worked out as follows :

$$\int_{-r}^{+r} P.dx = \frac{1}{2}$$

or,
$$\frac{h}{\sqrt{\pi}}\int_{-r}^{r} e^{-h^2x^2} dx = \frac{1}{2}$$

This gives *probable error,*

$$r = \frac{0.4769}{h} \qquad ...(2.79)$$

Average deviation for the normal curve :

In case of normal curve,

Average deviation,

$$\bar{d} = \int_{-\infty}^{+\infty} |x| P\,dx$$

($|x|$ represents the numerical value of deviation)

$$= \frac{h}{\sqrt{\pi}}\int_{-\infty}^{+\infty} e^{-h^2x^2} x.dx = \frac{2h}{\sqrt{\pi}}\int_{0}^{+\infty} e^{-h^2x^2} x.dx$$

$$= \frac{2h}{\sqrt{\pi}}\int_{0}^{+\infty} e^{-h^2z} dz \qquad \left[\begin{array}{l}\text{where, } z = x^2\\ dz = 2x.dx\\ x.dx = \dfrac{dz}{2}, \text{and}\\ \int e^{ax}dx = \dfrac{e^{ax}}{a}\end{array}\right]$$

$$= \frac{2h}{\sqrt{\pi}} \times \frac{1}{2}\left[\frac{e^{-h^2z}}{-h^2}\right]_0^{+\infty}$$

$$= \frac{h}{\sqrt{\pi}}\left[\frac{0-1}{-h^2}\right] = \frac{h}{\sqrt{\pi}} \times \frac{1}{h^2} = \frac{1}{h\sqrt{\pi}}$$

$$= \frac{r}{0.4769\sqrt{\pi}} = \frac{r}{0.8453}$$

i.e.,
$$\bar{d} = \frac{r}{0.8453} \qquad ...(2.80)$$

Standard deviation for the normal curve :

We know that standard deviation σ is given by :

$$\sigma = \sqrt{\frac{\Sigma d^2}{n}} \qquad ...[\text{Eqn. } (2.67)]$$

Following a method similar to that followed above, we have

$$\sigma^2 = \frac{2h}{\sqrt{\pi}} \int_0^{+\infty} e^{(-h^2x^2)} x^2 . dx = \frac{1}{2h^2}$$

or, $$\sigma = \frac{1}{\sqrt{2}h} \quad ...(2.81)$$

$$= \frac{r}{0.4769 \times \sqrt{2}}$$

$$= \frac{r}{0.6745} \quad ...(2.82)$$

From eqns. (2.80) and (2.82), we have *probable error,*

$$r = 0.8453 \bar{d} \quad ...(2.83)$$

and, $$r = 0.6745\ \sigma \quad ...(2.84)$$

Also, *Standard deviation of mean,*

$$\sigma_m = \frac{\sigma}{\sqrt{n}} \quad ...(2.85)$$

and, *Standard deviation of standard deviation,*

$$\sigma_\sigma = \frac{\sigma}{\sqrt{2n}} = \frac{\sigma_m}{\sqrt{2}} \quad ...(2.86)$$

2.9.5.5. Probability tables

The probability of an individual measurement to fall within the deviations x_1 and x_2 from the mean is given by,

$$P(x_1, x_2) = \frac{h}{\sqrt{\pi}} \int_{x_1}^{x_2} e^{(-h^2x^2)} dx$$

for a normal Gaussian curve.

The parameter $h = \frac{1}{\sigma\sqrt{2}}$ and the mean value $\bar{q}$ may be known from the raw data. Since the evaluation of the integral is a cumbersome and a tedious task, hence, **probability tables** have been prepared from which the value of probability can be directly read off.

Let $$y = \sqrt{2}hx = \frac{x}{\sigma},$$

or, $$x = \sigma \cdot y$$

Now, $$P(x_1, x_2) = P(y_1, y_2)$$

$$= \frac{1}{\sqrt{2\pi}} \int_{y_1}^{y_2} e^{(-y^2/2)} dy \quad ...(2.87)$$

where, $$y_1 = \sqrt{2}hx_1$$

and, $$y_2 = \sqrt{2}hx_2$$

Eqn. (2.87) may be written as :

$$P(y_1, y_2) = \frac{1}{\sqrt{2\pi}}\int_0^{y_2}(e)^{(-y^2/2)}dy - \frac{1}{\sqrt{2\pi}}\int_0^{y_1}e^{(-y^2/2)}.dy \qquad ...(2.88)$$

Each of the two integrals in equation (2.88) is a function of a single parameter y_1 or y_2. Therefore, probability,

$$P(0, y) = \frac{1}{\sqrt{2\pi}}\int_0^{y}(e)^{(-y^2/2)}dy \qquad ...(2.89)$$

The results of this integral are given in Table 2.2 and can be used to compute the probability between the limits x_1 and x_2. Tabulated values of the integral represent the area under the normalised Gaussian error curve for $y = 0$ and $y = y$ as shown in Fig. 2.35.

As the curve is *symmetrical* the same table may be used for *–ve* deviations replacing y by $-y$ in equation (2.89) since the value of probability is not affected by this substitution.

— Suppose we are required to find the probability that the deviation lies between $\pm\,\sigma$. For the *+ve* deviation σ,

$$y = \frac{q-\bar{q}}{\sigma} = \frac{\sigma}{\sigma} = 1$$

Table 2.2 indicates the probability value $P = 0.3413$ for $y = 1.0$. Due to symmetrical nature of the error curve, an equal probability would exist for the deviations to lie between $-\,\sigma$ and 0. Hence the probability that deviation lies between $\pm\,\sigma$ works out to be $2 \times 0.3413 = 0.6826$.

Table 2.2: Probability Tables

y	0	0.01	0.02	0.03	0.04	0.05	0.06	0.07	0.08	0.09
0.0	0.0000	0.0040	0.0080	0.0120	0.0160	0.0199	0.0239	0.0279	0.0319	0.0359
0.1	0.0398	0.0438	0.0478	0.0517	0.0557	0.0596	0.0636	0.0675	0.0714	0.0754
0.2	0.0793	0.0832	0.0871	0.0910	0.0948	0.0987	0.1026	0.1064	0.1103	0.1141
0.3	0.1179	0.1217	0.1255	0.1293	0.1331	0.1368	0.1406	0.1443	0.1480	0.1517
0.4	0.1554	0.1591	0.1628	0.1664	0.1700	0.1736	0.1772	0.1808	0.1844	0.1879
0.5	0.1915	0.1950	0.1985	0.2019	0.2054	0.2088	0.2123	0.2157	0.2190	0.2224
0.6	0.2258	0.2291	0.2324	0.2357	0.2389	0.2422	0.2454	0.2486	0.2518	0.2549
0.7	0.2580	0.2612	0.2642	0.2673	0.2704	0.2734	0.2764	0.2794	0.2823	0.2852
0.8	0.2881	0.2910	0.2939	0.2967	0.2996	0.3023	0.3051	0.3079	0.3106	0.3133
0.9	0.3159	0.3186	0.3212	0.3238	0.3264	0.3289	0.3315	0.3340	0.3365	0.3389
1.0	0.3413	0.3438	0.3461	0.3485	0.3508	0.3531	0.3554	0.3577	0.3599	0.3621
1.1	0.3643	0.3665	0.3686	0.3708	0.3729	0.3749	0.3770	0.3790	0.3810	0.3830
1.2	0.3849	0.3869	0.3888	0.3907	0.3925	0.3944	0.3962	0.3980	0.3997	0.4015
1.3	0.4032	0.4049	0.4066	0.4082	0.4099	0.4115	0.4131	0.4147	0.4162	0.4177
1.4	0.4192	0.4207	0.4222	0.4236	0.4251	0.4265	0.4279	0.4292	0.4306	0.4319
1.5	0.4332	0.4345	0.4357	0.4370	0.4382	0.4394	0.4406	0.4418	0.4430	0.4441
1.6	0.4452	0.4463	0.4474	0.4485	0.4495	0.4505	0.4515	0.4525	0.4535	0.4545
1.7	0.4554	0.4564	0.4573	0.4582	0.4591	0.4599	0.4608	0.4616	0.4625	0.4633
1.8	0.4641	0.4649	0.4656	0.4664	0.4671	0.4678	0.4686	0.4693	0.4700	0.4706

1.9	0.4713	0.4719	0.4726	0.4732	0.4738	0.4744	0.4750	0.4756	0.4762	0.4767
2.0	0.4773	0.4778	0.4783	0.4788	0.4793	0.4798	0.4803	0.4808	0.4812	0.4817
2.1	0.4821	0.4826	0.4830	0.4834	0.4838	0.4842	0.4846	0.4850	0.4854	0.4857
2.2	0.4861	0.4865	0.4868	0.4871	0.4875	0.4878	0.4881	0.4884	0.4887	0.4890
2.3	0.4893	0.4896	0.4898	0.4901	0.4904	0.4906	0.4909	0.4911	0.4913	0.4916
2.4	0.4918	0.4920	0.4922	0.4925	0.4927	0.4929	0.4931	0.4932	0.4934	0.4936
2.5	0.4938	0.4940	0.4941	0.4943	0.4945	0.4946	0.4948	0.4949	0.4951	0.4952
2.6	0.4953	0.4955	0.4956	0.4957	0.4959	0.4960	0.4961	0.4962	0.4963	0.4964
2.7	0.4965	0.4966	0.4967	0.4968	0.4969	0.4970	0.4971	0.4972	0.4973	0.4974
2.8	0.4974	0.4975	0.4976	0.4977	0.4977	0.4978	0.4979	0.4980	0.4980	0.4981
2.9	0.4981	0.4982	0.4983	0.4983	0.4984	0.4984	0.4985	0.4985	0.4986	0.4986
3.0	0.4987	0.4987	0.4987	0.4988	0.4988	0.4989	0.4989	0.4989	0.4990	0.4990
3.1	0.4990	0.4991	0.4991	0.4991	0.4992	0.4992	0.4992	0.4992	0.4993	0.4993
3.2	0.4993	0.4993	0.4994	0.4994	0.4994	0.4994	0.4994	0.4995	0.4995	0.4995
3.3	0.4995	0.4995	0.4996	0.4996	0.4996	0.4996	0.4996	0.4996	0.4996	0.4997
3.4	0.4997	0.4997	0.4997	0.4997	0.4997	0.4997	0.4997	0.4997	0.4998	0.4998
3.5	0.4998	0.4998	0.4998	0.4998	0.4998	0.4998	0.4998	0.4998	0.4998	0.4998
3.6	0.4998	0.4999	0.4999	0.4999	0.4999	0.4999	0.4999	0.4999	0.4999	0.4999
3.7	0.4999	0.4999	0.4999	0.4999	0.4999	0.4999	0.4999	0.4999	0.4999	0.4999
3.8	0.4999	0.4999	0.4999	0.4999	0.4999	0.4999	0.4999	0.5000	0.5000	0.5000

– Further, let it be required to find the probable error which represents the interval about the mean in which lies half of the data. This corresponds to symmetrical area about the mean which has a value of 0.5. Only half of this area lies to the right $y = 0$ and accordingly we look for the value of y in the probability table that corresponds to 0.25. The corresponding value of $y =$ **0.6745** and accordingly the probable error is given by : *Probable error* $= \pm\ 0.6745\ \sigma$.

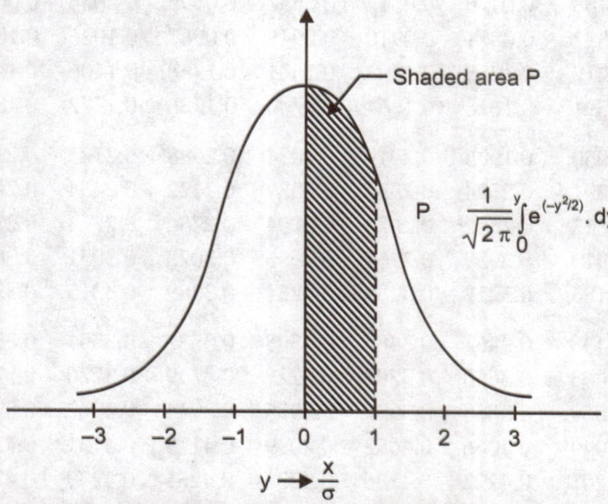

Fig. 2.36. Graphical representation of probability.

Example 2.51. *While measuring a temperature the following ten readings were recorded : 39.6, 39.9, 39.7, 39.9, 40.0, 39.8, 39.9, 39.8, 40.4 and 39.7°C.*

Calculate the following :

(i) The mean,

(ii) The standard deviation,

(iii) The probable error of one reading,

(iv) The probable error of mean, and

(v) The range.

Solution. The statistical values of the given date are computed with the help of the following table :

Observation number	*Temperature reading, °C (q)*	*Deviation,* $d = q - \bar{q}$	d^2
1	39.6	– 0.27	0.0729
2	39.9	+ 0.03	0.0009
3	39.7	– 0.17	0.0289
4	39.9	+ 0.03	0.0009
5	40.0	+ 0.13	0.0169
6	39.8	– 0.07	0.0049
7	39.9	+ 0.03	0.0009
8	39.8	– 0.07	0.0049
9	40.4	+ 0.53	0.2809
10	39.7	– 0.17	0.0289
$n = 10$	$\Sigma q = 398.7$ $\bar{q} = \frac{\Sigma q}{n} = \frac{398.7}{10} = 39.87$		$\Sigma d^2 = 0.441$

(i) The mean value of the reading :

$$\bar{q} = \frac{\Sigma q}{n} = \frac{398.7}{10} = \mathbf{39.87°C} \quad \textbf{(Ans.)}$$

(ii) The standard deviation :

$$s = \sqrt{\frac{\Sigma d^2}{(n-1)}}$$ ...since the number of observations is 10 (*i.e.*, less than 20).

$$= \sqrt{\frac{0.441}{10-1}} = \mathbf{0.22°C} \quad \textbf{(Ans.)}$$

(iii) Probable error of one reading :

$$r_1 = 0.6745s = 0.6745 \times 0.22 = \mathbf{0.15°C} \quad \textbf{(Ans.)}$$

(iv) Probable error of mean :

$$r_m = \frac{r_1}{\sqrt{n-1}} = \frac{0.15}{\sqrt{10-1}} = \mathbf{0.05°C} \quad \textbf{(Ans.)}$$

(v) Range :

$$Range = 40.4 - 39.6 = \mathbf{0.8°C} \quad \textbf{(Ans.)}$$

Example 2.52. *In a test temperature, measurements of temperature were made 100 times with variations in apparatus and procedures. After applying the corrections the following results were obtained :*

Temperature °C	*197*	*198*	*199*	*200*	*201*	*202*	*203*	*204*	*205*
Frequency of Occurrence	*2*	*4*	*10*	*24*	*36*	*14*	*5*	*3*	*2*

Calculate the following :

(i) Arithmetic mean,

(ii) Average deviation,

(iii) Standard deviation,

(iv) Variance,

(v) Probable error of one reading,

(vi) Probable error of the mean, and

(vii) Standard deviation of the standard deviation.

Solution. The computations in tabular form are given below :

Temperature °C *T*	*Frequency of occurrence, f*	$T \times f$	*Deviation* $d\left(=T-\frac{\Sigma T\times f}{n}\right)$	$f \times d$	d^2	$f \times d^2$
197	2	394	− 3.77	− 7.54	14.213	28.426
198	4	792	− 2.77	− 11.08	7.673	30.692
199	10	1990	− 1.77	− 17.7	3.133	31.330
200	24	4800	− 0.77	− 18.48	0.593	14.232
201	36	7236	+ 0.23	+ 8.28	0.053	1.908
202	14	2828	+ 1.23	+ 17.22	1.513	21.182
203	5	1015	+ 2.23	+ 11.15	4.973	24.865
204	3	612	+ 3.23	+ 9.69	10.433	31.299
205	2	410	+ 4.23	+ 8.46	17.893	35.786
Total	$n = 100$	$\Sigma T \times f =$ 20077		$\Sigma\lvert f \times d\rvert$ = 109.6		$\Sigma f \times d^2$ = 219.72

***(i)* Arithmetic mean :**

$$\text{Arithmetic mean} = \frac{\Sigma T\times f}{n} = \frac{20077}{100} = \mathbf{200.77°C} \quad \textbf{(Ans.)}$$

***(ii)* Average deviation :**

$$\sigma = \frac{\Sigma\lvert f\times d\rvert}{n} = \frac{109.6}{100} = \mathbf{1.096°C} \quad \textbf{(Ans.)}$$

***(iii)* Standard deviation :**

$$\textit{Standard deviation, } \sigma = \sqrt{\frac{\Sigma f\times d^2}{n}} = \sqrt{\frac{219.72}{100}} = \mathbf{1.482°C} \quad \textbf{(Ans.)}$$

***(iv)* Variance :**

$$\text{Variance, } V = \sigma^2 = (1.482)^2 = \mathbf{2.196°C^2} \quad \textbf{(Ans.)}$$

***(v)* Probable error of one reading :**

Probable error of one reading,

$$r_1 = 0.6745\ \sigma = 0.6745 \times 1.482 \simeq \mathbf{1°C} \quad \textbf{(Ans.)}$$

***(vi)* Probable error of the mean :**

$$r_m = \frac{r_1}{\sqrt{n}} = \frac{1(°C)}{\sqrt{100}} = \mathbf{0.1°C} \quad \textbf{(Ans.)}$$

(vii) Standard deviation of the standard deviation :

Standard deviation of the mean,

$$\sigma_m = \frac{\sigma}{\sqrt{n}} = \frac{1.482}{\sqrt{100}} = 0.148°C$$

∴ Standard deviation of the standard deviation,

$$\sigma_\sigma = \frac{\sigma_m}{\sqrt{2}} = \frac{0.148}{\sqrt{2}} = \mathbf{0.1046°C} \quad \textbf{(Ans.)}$$

Example 2.53. *A known current of 80A is measured by an ammeter. If 40% of the readings are within 0.8A of true value, determine the following :*

(i) The standard deviation for the meter.

(ii) The probability of an error of 1.2 A.

Solution. (i) Standard deviation for the meter :

Considering a normal distribution curve, half the readings, *i.e.*, 20% of the total readings have a *+ve* error and the remaining 20% have a *–ve* error.

From probability tables, it is found that corresponding to a probability of $P(y) = 0.2$,

$$y = \frac{x}{\sigma} = 0.5248$$

∴ Standard deviation,

$$\sigma = \frac{x}{y} = \frac{0.8}{0.5248} = \mathbf{1.524} \quad \textbf{(Ans.)}$$

(ii) Probability of an error of 1.2 A :

Now,
$$y = \frac{x}{\sigma} = \frac{1.2}{1.524} = 0.787$$

Corresponding to $y = 0.787$, probability is 0.2842

∴ Probability of an error of $1.2A = 2 \times 0.2842 = 0.5684$

Thus about 57% of the readings are *within ± 1.2A of the true value.* **(Ans.)**

Example 2.54. *An underdamped galvanometer was energized 100 times under the same carefully controlled experimental conditions and maximum deflection was read in each case. The readings were distributed about a mean value of 21.9 mm and had a probable error of 2.1 mm. How many of 100 readings would you estimate exceeded 25 mm ?*

Solution. Deviation, $x = 25 - 21.9 = 3.1$ mm

Probable error, $r = 2.1$ mm

From eqn. 2.82, standard deviation,

$$\sigma = \frac{r}{0.6745} = \frac{2.1}{0.6745} = 3.1$$

$$\text{Ratio, } y = \frac{x}{\sigma} = \frac{3.1}{3.1} = 1.0$$

Corresponding to ratio $y = \frac{x}{\sigma} = 1.0$, the area under the curve is 0.3413 (from table 2.2).

∴ Number of readings having a deviation within ± 3.1 mm

$$= 2 \times 0.3413 \times 100 = 68$$

Number of readings exceeding a deviation of ± 3.1 mm

$$= 100 - 68 = 32$$

Hence, the number of readings having a deviation of + 3.1 mm

$$= \frac{32}{2} = 16.$$

Thus the *number of readings exceeding a maximum deflection of 25 mm, i.e., (21.9 + 3.1) mm is 16.* **(Ans.)**

Example 2.55. *A machine shop was assigned the task to manufacture 10,000 aluminium rods of a nominal length of 20 mm; the rod length was stipulated neither to exceed 20.25 mm and nor to be smaller than 19.5 mm. When inspected for quality control, it was found that 1000 of the rods were too long to fit into a gauge set at 20.25 mm. Determine the number of remaining 9000 rods which will conform to the specifications. Assume that the measurement data conforms to Gaussian normal distribution curve.*

Solution. In case of normal distribution, there is equal probability of the *+ve* and *–ve* errors about the mean. Therefore 5000 rods have lengths greater than 20 mm and 5000 rods have lengths smaller than 20 mm. Since 1000 rods have length greater than 20.25 mm, therefore,

The number of rods where length lies between 20 mm and 20.25 mm

$$= 5000 - 1000 = 4000$$

∴ Probability that 4000 rods have a value greater than 20 mm and less than 20.25 mm,

$$P(y) = \frac{4000}{10,000} = 0.4$$

From probability tables, it is found that corresponding to a probability of $P(y) = 0.4$,

$$y = \frac{x}{\sigma} \simeq 1.3 \qquad \text{...(From table 2.2)}$$

∴ Standard deviation,

$$\sigma = \frac{x}{1.3} = \frac{20.25 - 20.00}{1.3} = \frac{0.25}{1.3} = 0.1923$$

Now, for rods with nominal length of 19.5 mm and 20 mm,

$$y = \frac{x}{\sigma} = \frac{(20 - 19.5)}{0.1923} = 2.60$$

Corresponding probability, $P(y) = 0.4953$ ...(From table 2.2)

∴ Number of rods that have lengths between 19.5 mm and 20 mm

$$= 10000 \times 0.4953 = 4953$$

Hence, *total number of rods whose lengths lie between specified limits of 19.5 mm to 20.25 mm,*

$$= 4000 + 4953 = \mathbf{8953} \quad \textbf{(Ans.)}$$

Example 2.56. *The speed of a hydraulic turbine model, which is running at 1500 rpm, is measured by a tachometer. The hydraulic turbine is subject to variations in speed. For a sample of 20 readings at this speed, how many readings would you expect between 1485 and 1515. Assume that the tachometer gives a normal set of deviations with precision index of 0.04 at 1500 rpm.*

Solution. *Given* : Deviation, $d = q - \bar{q} = \pm 15$ r.p.m.

and, precision index, $h = 0.04$

Standard deviation, $\sigma = \frac{1}{\sqrt{2}h}$ ...[Eqn. (2.81)]

$$= \frac{1}{\sqrt{2} \times 0.04} = 17.68$$

$$y = \frac{x}{\sigma} = \frac{q - \bar{q}}{\sigma} = \frac{15}{17.68} = 0.8484$$

Corresponding to $y = 0.8484$, as read from the probability tables

$$P(y) = 0.3015$$ ...(From table 2.2)

Probability of an error ± 15 r.p.m. = 2 × 0.3015 = **0.603**

Thus we expect 60% of 0.6 × 20 = **12 readings** *to lie between 1485 and 1515 r.p.m.* **Ans.**

Example 2.57. *The depth of water, which had a nominal value of 15 cm, was measured with a hook. Measurements were taken 40 times and 10 readings were found to lie outside a particular range. Assuming that the measurements conform to normal distribution with precision index 9 cm^{-1}, determine the prescribed range.*

Solution. *Given* : Nominal value of depth of water = 15 cm

Total number of times the measurements were taken = 40

No. of measurement readings found to lie outside a particular range = 10

Precision index, $h = 9$ cm^{-1}

Prescribed range :

Probability of falling within a particular range

$$= \frac{(40 - 10)}{40} = 0.75$$

Half of these measurements have a *+ve* error and half have a *–ve* error.

$$P(y) = \frac{0.75}{2} = 0.375$$

From probability tables, it is found that corresponding to a probability

$$P(y) = 0.375,$$

$$y = \frac{x}{\sigma} = \frac{q - \bar{q}}{\sigma} = 1.15$$ ...(From table 2.2)

Now, standard deviation, $\sigma = \frac{1}{\sqrt{2}h}$ ...[Eqn. (2.81)]

$$= \frac{1}{\sqrt{2} \times 9} = 0.0786$$

$\therefore$ $x = (q - \bar{q}) = 1.15 \times 0.0786 = 0.0904$ cm

Thus 75 per cent of depth measurements lie within the **range (15 ± 0.0904) cm. (Ans.)**

Example 2.58. *(a) A mass flow meter, using electromagnetic principle, is set to record a large number of readings. The fixed mass flow rate is 100 kg/s; half of the readings lie between 104.8 and 95.2 kg/s. The sample measurements give a straight line between the (0, 50%) points on the probability coordinates. Calculate the precision index of the instrument.*

(b) An alarm rings when the flow meter gives a reading of 88 kg/s.

(i) How many false alarms are expected in the month of November if the flow is checked automatically 4 times a day ?

(ii) What will be the value of precision index if one wants to reduce the false alarm rate by a factor of 2 ?

Solution. (*a*) The overall probability is 0.5. Considering normal distribution, 0.25 of the measurements have a +*ve* error and the rest 0.25 of the measurements have a –*ve* error.

Corresponding to a probability of $P(y) = 0.25$, from probability tables, we have

$$y = \frac{x}{\sigma} = 0.675 \qquad \text{...(From table 2.2)}$$

$$\therefore \text{ Standard deviation, } \quad \sigma = \frac{x}{0.675} = \frac{4.8}{0.675} = 7.11$$

$$\textbf{Precision index,} \quad h = \frac{1}{\sqrt{2}\sigma} = \frac{1}{\sqrt{2} \times 7.11} = \mathbf{0.099} \quad \textbf{(Ans.)}$$

(*b*) (*i*) For a mass flow rate of 88 kg/s.

$$y = \frac{x}{\sigma} = \frac{100 - 88}{7.11} = 1.69$$

Corresponding probability $P(y) = 0.45$ ...(From table 2.2)

Thus it is expected that (0.5 – 0.45) = 0.05 or 5% of the flow measurements will fall in the range of false alarms.

Number of measurements in the month of November = 30 × 4 = 120

∴ *Expected false alarms* = 120 × 0.05 = **6** **(Ans.)**

(*ii*) Reduced number of false alarms $= \frac{1}{2} \times 6 = 3$

Probability of false alarms $= \frac{3}{120} = 0.025$ (or 2.5%)

Probability of data to lie in the tolerant band,

$$P(y) = 0.5 - 0.025 = 0.475$$

Corresponding to $P(y) = 0.475$

$$y = \left(\frac{x}{\sigma}\right) = 1.96 \qquad \text{...(From table 2.2)}$$

$$\text{or, Standard deviation, } \quad \sigma = \frac{x}{1.96} = \frac{100 - 88}{1.96} = 6.12$$

$$\therefore \textit{ Precision index,} \quad h = \frac{1}{\sqrt{2}\sigma} = \frac{1}{\sqrt{2} \times 6.12} = \mathbf{0.1155} \quad \textbf{(Ans.)}$$

2.9.5.6. Odds

Odds *is the number of chances that a particular reading will occur when error limit is specified.* Thus the probability of occurrence can be stated in terms of *odds*.

Example. When the error limits are specified as ± 0.6745σ, the chances are that 50 per cent of observations will lie between the above limits or in other words we can say odds are 1 to 1.

The odds can be calculated as under :

$$\text{Probability of occurrence} = \frac{\text{Odds}}{\text{Odds} + 1} \qquad \text{...(2.90)}$$

Certain deviations, their probability of occurrence and the corresponding odds are given in Table 2.3.

Table 2.3: Deviations, Probability and Odds

Deviation	*Probability*	*Odds*
$\pm$ 0.6745 σ	0.5000	1 to 1
$\pm$ σ	0.6826	2.15 to 1
$\pm$ 2σ	0.9546	21 to 1
$\pm$ 3σ	0.9974	369 to 1

2.9.5.7. Specification of measurement data

The results of statistical analysis of multi-sample data are normally expressed as deviations about a mean value; the deviations are expressed as follows :

(*i*) **Probable error.** The result is expressed as $\overline{q} \pm 0.6745\sigma$, which implies that 0.50 (50%) of the readings are within this limit and odds are 1 to 1. Obviously there is an even probability that a new reading will lie within these limits.

(*ii*) **Standard deviation.** The result is expressed as $\overline{q} \pm \sigma$. In this case, the error limit is the standard deviation. It implies that 0.6828 (about 68%) of the readings are within the limits $\sigma = \pm 1$ and the odds are 2.15 to 1. Thus there is approximately a 2 to 1 possibility that a new reading will fall within this limit.

(*iii*) **± 2σ limit.** The result is expressed as $\overline{q} \pm 2\sigma$, which implies that 0.9546 (about 95%) of the readings are within the given limits and the odds are 21 to 1.

(*iv*) **± 3σ limit.** The result is expressed as $\overline{q} \pm 3\sigma$, which implies that 0.9974 (about 99%) of the readings are within the given limits. The odds of any observation falling within this limit are 369 to 1.

2.9.5.8. Confidence interval and confidence level

The range of deviation from the mean value within which a certain fraction of all values are expected to lie is called **confidence interval.**

The probability that the value of a randomly selected observation will lie in the range of confidence interval is termed as **confidence level.**

When the number of observations is large and their errors are random and follow Gaussian distribution, the various confidence intervals about the mean value $\overline{q}$ are given in Table 2.4.

Table 2.4: Confidence Intervals

Confidence level	*Confidence interval*	*Values lying outside confidence interval*
0.500	$\overline{q} \pm 0.674\sigma$	1 in 2
0.800	$\overline{q} \pm 1.282\sigma$	1 in 5
0.900	$\overline{q} \pm 1.645\sigma$	1 in 10
0.950	$\overline{q} \pm 1.960\sigma$	1 in 20
0.990	$\overline{q} \pm 2.576\sigma$	1 in 100
0.999	$\overline{q} \pm 3.291\sigma$	1 in 1000

If the number of observations is small (say *less* than 20) and the standard deviation is not accurately known, the confidence interval must be *broadened*. The standard deviation, in this case, is computed as :

$s = \sqrt{\frac{\Sigma d^2}{n-1}}$; this standard is multiplied by a suitable factor to establish the confidence interval.

2.9.5.9. Probability graph paper

Prior to undertaking statical analysis of the measurement data, the experimenter needs to check that the data is in conformity with the normal distribution. To accomplish this, the data is plotted on a probability graph paper wherein the abscissa is the value of a particular reading (on a linear scale) and the ordinate has a percentage of readings at or below the value of abscissa (on a logarithmic scale). A straight line is then drawn through the data points; however, while fitting the curve much weightage is not given to the points at low and high ends of the plot. *If the 50 percent reading at the ordinate exactly corresponds to averages of all the data points, the data is considered to follow the Gaussian distribution.*

2.9.5.10. Rejection of test data–Chauvenet's criterion

During experimentation, it is observed that some points are noticeably different from the majority of the data. Such points can be determined and subsequently eliminated by applying Chauvenet's criterion.

Chauvenet's criterion states as follows :

"An observed reading may be rejected if the probability of obtaining the particular deviation from the mean is less than $\frac{1}{2n}$". The quantity *n* refers to the number of observations which are large enough to presume that the data follows the Gaussian error distribution.

Let us consider that during a measurement act ten readings have been made. Then,

$$\frac{1}{2n} = \frac{1}{2 \times 10} = 0.05 \text{ or } 5\%$$

Thus readings with 95 percent of probability are to be accepted, and less than that can be rejected. Corresponding to probability

$$P(y) = \frac{0.95}{2} = 0.475; \; y = 1.96$$

Results based on such calculations for different number of readings are given in table 2.5.

Table 2.5: Chauvenet's Criterion for Rejecting Data

Number of readings	*Ratio of maximum acceptable deviation to standard deviation, d_{max}/σ*
2	1.15
3	1.38
4	1.54
5	1.65
6	1.73
7	1.80
10	1.96
15	2.13

25	2.33
50	2.57
100	2.81
300	3.14
500	3.29
1000	3.48

When applying Chauvenet's criterion, in order to eliminate any dubious data, the following *procedure* is adopted :

(i) Using all data points calculate the standard deviation.

(ii) Compare the deviation of the individual readings with standard deviation.

(iii) If the ratio of deviation of a reading to the standard deviation exceeds the limits given in Table 2.5, reject the readings.

(iv) Calculate again the mean value and the standard deviations by excluding the rejected readings from the data.

- A criterion (*based upon "confidence intervals"*) used for discarding a data point is that its deviation from the mean exceeds four times the probable error of a single reading. This results in discarding a data outside a confidence interval for a single reading at a confidence level of 0.993.

— A better criterion which does not involve the evaluation of probable error when set of data points is small and standard deviation is not accurately known, *is to discard a reading that lies outside the interval corresponding to confidence level of 0.99 for a single observation* (Refer to Tables 2.4 and 2.5). On this basis not more than one reading in 100 would lie outside their range.

— A still better method is to use confidence interval corresponding to a confidence level of 0.95 in order to scrutinize the measurement procedure adopted.

Example 2.59. *Ten copper rods, selected at random, were found to a have the following lengths in metres : 5.30, 5.73, 6.77, 5.26, 4.33, 5.45, 6.09, 5.64, 5.81, 5.75.*

Determine any reading that can be rejected by applying Chauvenet's criterion. The ratio of maximum deviation to standard deviation should not exceed 1.96.

Solution. The computations in a tabular form are given below :

S.No.	Length q, m	Deviation $d = q - \bar{q}$	d^2	$\frac{\|d\|}{s}$
1	5.30	– 0.313	0.09797	0.5
2	5.73	+ 0.117	0.01369	0.19
3	6.77	+ 1.157	1.33865	1.85
4	5.26	– 0.353	0.12461	0.56
5	4.33	– 1.283	1.64609	2.05
6	5.45	– 0.163	0.02657	0.26
7	6.09	+ 0.477	0.22753	0.76
8	5.64	+ 0.027	0.00073	0.04
9	5.81	+ 0.197	0.03881	0.31
10	5.75	+ 0.137	0.01877	0.22
$n = 10$	$\Sigma q = 56.13$ $\bar{q} = \frac{56.13}{10} = 5.613$		$\Sigma d^2 = 3.53342$	

Arithmetic mean, $\bar{q} = \frac{\Sigma q}{n} = \frac{56.13}{10} = 5.613$

Standard deviation, $s = \sqrt{\frac{\Sigma d^2}{n-1}}$ ... since the number of readings is less than 20.

$$= \sqrt{\frac{3.53342}{10}} = 0.626$$

It is given that for 10 readings the ratio of deviation to standard deviation is not to exceed 1.96 and therefore **reading number 5, *i.e.*, 4.33 m,** *should be rejected.* **(Ans.)**

2.9.5.11. Method of least squares

- Let us consider a set of n readings $q_1, q_2, q_3, \ldots q_n$. The sum of squares of their deviations for some mean value Q_m is,

$$S = \sum_{i=1}^{n} (q_1 - Q_m)^2$$

For minimizing S with respect to mean value Q_m, we have

$$\frac{\partial S}{\partial Q_m} = 2\left(\sum_{i=1}^{n} q_1 - nQ_m\right) = 0$$

or, $$Q_m = \frac{1}{n}\sum_{i=1}^{n} q_i$$

But, $$\frac{1}{n}\sum_{i=1}^{n} q_i = \bar{q}$$

$\therefore$ $$Q_m = \bar{q} \quad \ldots(2.91)$$

Thus the *mean value which minimizes the sum of the squares of the deviation is the arithmetic mean.*

- Now consider two variables u and v measured over a range of values and let us find the linear function connecting y and x.

Let the linear function be :

$$v = au + b$$

Now let us minimize the quantity,

$$S = \sum_{i=1}^{n} [v_i - (au_i + b)]^2$$

This is accomplished by taking derivatives of S w.r.t. a and b and setting them equal to zero. This gives

$$nb + a\Sigma u_i = \Sigma v_i \quad \ldots(i)$$

and, $$b\Sigma u_i + a\Sigma u_i^2 = \Sigma u_i\, v_i \quad \ldots(ii)$$

Solving, the above eqns. (*i*) and (*ii*), we get

$$a = \frac{n\Sigma u_i v_i - (\Sigma u_i)(\Sigma v_i)}{n\Sigma u_i^2 - (\Sigma u_i)^2} \quad \ldots(2.92)$$

$$b = \frac{(\Sigma v_i)(\Sigma u_i^2) - (\Sigma u_i v_i)(\Sigma u_i)}{n\Sigma u_i^2 - (\Sigma u_i)^2} \quad ...(2.93)$$

The standard deviations may be found as :

$$s_a = \sqrt{\frac{n}{n\Sigma u_i^2 - (\Sigma u_i)^2}} s_v \quad ...(2.94)$$

and

$$s_b = \sqrt{\frac{\Sigma u_i^2}{n\Sigma u_i^2 - (\Sigma u_i)^2}} s_v \quad ...(2.95)$$

where, $s_v = \sqrt{\frac{1}{n}\sum (au_i + b - v_i)^2}$

Standard deviation of u is,

$$s_u = \sqrt{\frac{1}{n}\sum \left(\frac{v_i - b}{a} - u_i\right)^2} = \frac{s_v}{a} \quad ...(2.96)$$

Example 2.60. *The data given below are expected to follow a linear relationship :*

$$v = au + b$$

u	1.8	4.6	6.6	9.0	11.4	13.4
v	2.2	3.2	5.2	6.4	8.0	10.0

(i) Obtain the best linear relation, and

(ii) Calculate the standard deviation.

Solution. *Given* : $v = au + b$ .... Linear relationship ;

Number of readings, $n = 6$

The computations in a tabular form are given below :

u_i	v_i	$u_i v_i$	u_i^2
1.8	2.2	3.96	3.24
4.6	3.2	14.72	21.16
6.6	5.2	34.32	43.56
9.0	6.4	57.60	81.0
11.4	8.0	91.20	129.96
13.4	10.0	134.00	179.56
$\Sigma u_i = 46.8$	$\Sigma v_i = 35$	$\Sigma u_i v_i = 335.8$	$\Sigma u_i^2 = 458.48$

(*i*) Best linear relation :

Using equations (2.92) and (2.93), we have

$$a = \frac{(6) \times (335.8) - (46.8) \times (35)}{(6) \times (458.48) - (46.8)^2} = \frac{376.8}{560.64} = 0.672$$

$$b = \frac{(35) \times (458.48) - (335.8)(46.8)}{(6)(458.48) - (46.8)^2} = \frac{331.36}{560.64} = 0.591$$

The best linear relation is : $\boldsymbol{v = 0.672\, u + 0.591}$ **(Ans.)**

(ii) **Standard deviation :**

The computations for finding standard deviation, in a tabular form, are given below :

u_i	v_i	$au_i + b$	$au_i + b - v_i$	$(au_i + b - v_i)^2$
1.8	2.2	1.8006	– 0.3994	0.1595
4.6	3.2	3.6822	+ 0.4822	0.2325
6.6	5.2	5.0262	– 0.1738	0.0302
9.0	6.4	6.6390	+ 0.239	-0.0571
11.4	8.0	8.2518	+ 0.2518	0.0634
13.4	10.0	9.5958	– 0.4074	0.1660
				$\Sigma(au_i + b - v_i)^2 = 0.7087$

∴ Standard deviation of v,

$$s_v = \sqrt{\frac{1}{n}\Sigma(au_i + b - v_i)^2} = \sqrt{\frac{0.7087}{6}} = \pm\mathbf{0.3436} \quad \textbf{(Ans.)}$$

From eqn. (2.96), standard deviation of u,

$$s_u = \frac{s_v}{a} = \frac{\pm 0.3436}{0.672} = \pm\mathbf{0.511} \quad \textbf{(Ans.)}$$

From eqns. (2.94 and 2.95) the standard deviations in a and b are :

$$s_a = \sqrt{\frac{n}{n\Sigma u_i^2 - (\Sigma u_i)^2}}\, s_v$$

$$= \sqrt{\frac{6}{6 \times 458.48 - (46.8)^2}} \times (\pm 0.3436) = \pm\mathbf{0.0355} \quad \textbf{(Ans.)}$$

$$s_b = \sqrt{\frac{\Sigma u_i^2}{n\Sigma(u_i^2) - (\Sigma u_i)^2}}\, s_v$$

$$= \sqrt{\frac{458.48}{6 \times 458.48 - (46.8)^2}} \times (\pm 0.3436) = \pm\mathbf{0.3107} \quad \textbf{(Ans.)}$$

Example 2.61. *(a) Enumerate the factors which should be considered while selecting an instrument for a specific measurement application.*

(b) The iron losses in a ferromagnetic material used in a transformer vary with frequency f of the supply given to the transformer. For a particular transformer these iron losses were determined at various frequencies with a constant flux density in the ferromagnetic material; the results obtained are given below :

Frequency, f (Hz) :	550	700	850	1000
Iron losses, P (mW) :	23	31	37	61

Assuming the iron losses to have general form $P = af^2 + bf$, *determine the constants a and b to achieve best fit for the four measured values, using method of least squares.*

Solution. *(a)* The following *factors* need to be considered for the selection of an instrument to be used for a specific measurement application :

1. Easy calibration.
2. Convenience and ease in reading the instrument.
3. The maximum and minimum values the input variables are expected to assume.

4. Nature of the input signal.
5. Non-interference with measuring system.
6. Accuracy expected from the instrument.
7. Cost criterion.
8. Specific application of the data.
9. Required form of data.
10. Negligible or well-known environmental errors.

(*b*) *Given* : General form $P = af^2 + bf$

or, $$\frac{P}{f} = af + b$$

The above relation may be expressed as

$$v = au + b$$

which is a straight line relationship between, $v = \frac{P}{f}$ and $u = f$,

i.e., $u = f$: 550 700 850 1000

$v = \frac{P}{f}$: 0.04182 0.04429 0.05529 0.061

The computations are tabulated below :

u_i	v_i	$u_i v_i$	u_i^2
550	0.04182	23	0.302×10^6
700	0.04429	31	0.49×10^6
850	0.05529	47	0.722×10^6
1000	0.0610	61	1×10^6
$\Sigma u_i = 3100$	$\Sigma v_i = 0.2024$	$\Sigma u_i v_i = 162$	$\Sigma u_i^2 = 2.514 \times 10^6$

In order to find the values of *a* and *b* for the best straight line fit using the method of least squares, let us use the equations (2.92) and (2.93) as follows :

$$a = \frac{n\Sigma u_i v_i - (\Sigma u_i)(\Sigma v_i)}{n\Sigma u_i^2 - (\Sigma u_i)^2} \quad \text{...[Eqn (2.92)]}$$

$$= \frac{4 \times 162 - (3100)(0.2024)}{4 \times 2.514 \times 10^6 - (3100)^2} = \frac{20.56}{446000} = 46.1 \times 10^{-6}$$

$$b = \frac{(\Sigma v_i)(\Sigma u_i^2) - (\Sigma u_i v_i)(\Sigma u_i)}{n\Sigma u_i^2 - (\Sigma u_i)^2} \quad \text{...[Eqn. (2.93)]}$$

$$= \frac{0.2024 \times 2.514 \times 10^6 - 162 \times 3100}{4 \times 2.514 \times 10^6 - (3100)^2}$$

$$= \frac{6633.6}{446000} = 14.87 \times 10^{-3}$$

Hence, *the best fit relationship is* :

$$P = 46.1 \times 10^{-6} f^2 + 14.87 \times 10^{-3} f \; mW \quad \textbf{(Ans.)}$$

Example 2.62. *The final result of a measurement depends on two measurements in series, each having values of 50 and 100 units with errors of ± 2% and ± 1% such that final result is sum of both. Determine the error of final result if errors are considered as :*

(i) Limiting errors, and

(ii) Standard deviations.

Solution. Error of final result :

(i) Limiting errors :

When individual errors are limiting errors, then fractional errors of combination

$$= \pm\left(\frac{q_1}{q_1+q_2}\times\frac{\delta q_1}{q_1}+\frac{q_2}{q_1+q_2}\times\frac{\delta q_2}{q_2}\right)$$

$$= \pm\left(\frac{50}{50+100}\times 0.02+\frac{100}{50+100}\times 0.01\right)$$

$$= 0.0133 \text{ or } \mathbf{1.33\%}$$ **(Ans.)**

(ii) Standard deviations :

When individual errors are standard deviations, then errors in individual measurements are 2% of 50 and 1% of 100, *i.e.*, 1 and 1

$$\text{Thus, resultant error} = \sqrt{1^2+1^2} = \sqrt{2}$$

$$= \frac{\sqrt{2}}{(50+100)}\times 100\% = \mathbf{0.943\%}$$ **(Ans.)**

HIGHLIGHTS

1. *Measurement* is the act, or the result, of a quantitative comparison between a predetermined standard and an unknown magnitude. Measurements provide us with a means of describing various phenomena in quantitative terms.
2. The physical quantity or the characteristic condition which is the object of measurement in an instrumentation system is variously termed as "*measurand*", "*measurement variable*", "*instrumentation variable*" or "*process variable*".
3. The technology of using instruments to measure and control the physical and chemical properties of materials is called "*instrumentation*".
4. Methods of measurements are : (*i*) Direct method, (*ii*) Indirect method.
   — In *direct comparison method* the parameter to be measured is directly compared with either a primary or a secondary standard.
   — In *indirect comparison method* the comparison is done with a standard through the use of a calibrated system.
5. The three modes of measurement are :
   (*i*) Primary measurements, (*ii*) Secondary measurements, and (*iii*) Tertiary measurements.
6. A generalised measurement system consists of the following components/ elements:
   (*i*) Primary sensing element;

(ii) Variable conversion or transducer element;
(iii) Manipulation element;
(iv) Data transmission element;
(v) Data presentation element.

7. The *static characteristics* pertain to a system where the quantities to be measured are constant or vary slowly with time. The main static characteristics are : *Accuracy, sensitivity, reproducibility, drift, static error and dead zone.*
8. The *relative static error* is defined as the ratio of absolute static error to the true value of the quantity under measurement.
9. The difference between the true value and the measured value of a quantity is called *static correction*.
10. *Static calibration* is a process by which all the static performance characteristics are obtained in one form or another.
11. The difference between the largest and the smallest reading of the instrument is called the *range* of the instrument.
12. *Span* represents the algebraic difference between the upper and lower range values of the instrument.
13. An error curve which can be used for correcting instrument readings is called *"Error calibration curve"*.
14. *Scale readability* indicates the closeness with which the scale of an analog type of instrument can be read.
15. *Drift* is an undesired gradual departure of the instrument output over a period of time that is unrelated to the changes in input, operating conditions or load.
16. The closeness with which an instrument reading approaches the true value of the quantity being measured is called *accuracy*.
17. *Precision* is a measure of reproducibility of measurements.
18. The ratio of the magnitude of the output signal to the input signal or response of measuring system to the quantity being measured is called *sensitivity*.
19. The ability to reproduce the input characteristics symmetrically is called *linearity*.
20. The maximum differences in output at any measured value within the specified range when approaching the point first increasing and then with decreasing input may be termed as *hysteresis*.
21. The minimum value below which no output change can be detected when the input of an instrument is increased gradually from zero is called the *threshold* of an instrument.
22. *Threshold* defines the smallest measurable input while the *resolution* defines the smallest measurable input change.
23. The largest change of input quantity for which there is no output of the instrument is termed as *dead zone*.
24. The time required by a measurement system to begin to respond to a change in the measurement is called *dead time*.
25. The incapability of the system to faithfully measure the input signal in undistorted form is called the *loading effect*.
26. *Noise* is a spurious current or voltage extraneous to the current or voltage of interest in an electrical or electronic circuit.

27. *"Dynamic characteristics"* quantify the dynamic relation between the input and output.
28. The evaluation of the ability of a system to faithfully transmit and present all the pertinent information included in the input signal and to exclude all else, is called *response.*
29. The behaviour of the system when inputs vary with time (*i.e.,* inputs are dynamic in nature) and so does the output is called *dynamic response.*
30. Response of a measurement system may be :
    (*i*) Transient response; (*ii*) Steady state response.
31. The rapidity with which a measurement system responds to changes in the measured quantity is called the *speed of response* or *"responsiveness".*
32. *Measuring lag* refers to retardation or delay in the response of a measurement system to changes in measured quantity.
33. *Fidelity* is defined as the degree to which the measurement system indicates changes in the measured quantity without any dynamic error.
34. The *dynamic error* or *measurement error* is the difference between the true value of the quantity changing with time and the value indicated by the measurement system if no static error is assumed.
35. "Examples" of the systems :
    (*i*) *Zero-order systems* : Mechanical levers, amplifiers etc.
    (*ii*) *First-order systems* : Velocity of a free falling mass; air pressure buildup in bellows, thermistors and thermocouples etc.
    (*iii*) *Second-order systems* : Piezoelectric pick-ups, spring-mass system, pen control system on X-Y plotters etc.
36. *Damping ratio* (γ) is defined as the ratio of the actual value of coefficient of viscous friction in movement and the value required to produce critical damping.
37. The limits of the deviations from the specified value are defined as *limiting errors* or *guarantee errors.*
38. The *relative error* is defined as the ratio of the error to the specified (nominal) magnitude of a quantity.
39. The errors are classified as :
    1. Gross errors
    2. Systematic errors :
        (*i*) Instrumental errors.
        (*ii*) Environmental errors.
        (*iii*) Observational errors.
    3. Random errors.
40. *Sources of errors are :*
    Noise, Response time; Design limitations; Energy exchanged by interaction; Transmission; Determination of measuring system; Ambient influences on measuring systems, Errors of observation and interpretation.
41. *Arithmetic mean* of the number of readings taken is the most probable value of measured variable.
42. *Geometric mean* is defined as the *n*th root of the product of *n* terms.

43. *Median* is the middle value of a set of an "odd" number of readings, if variables are arranged in numerical order. For an "even" number of readings, the median is the arithmetic average of the two central readings.
44. *Mode* is the value of the variable that occurs most frequently.
45. *Dispersion* (or *spread* or *scatter*) is the property which denotes the extent to which the values are dispersed about the central value.
46. The departure of the observed readings from the arithmetic mean of the group of readings is termed as *deviation*.
47. *Standard deviation* of an infinite number of data is defined as the square root of the sum of individual deviations squared divided by the number of readings.
48. The square of the standard deviation is called *variance*.
49. The "*law of probability*" states that the normal occurrence of deviations from the average value of an infinite number of measurements or observations can be expressed as :

$$P = \frac{1}{\sigma\sqrt{2\pi}} e^{-x^2/2\sigma^2}$$

where P = Number of readings at any deviation 'd', *i.e.*, probability of occurrence of deviations 'd',

σ = Standard deviation, and

x = Magnitude of deviation from mean.

50. Chauvenet's criterion states as follows :

"An observed reading may be rejected if the probability of obtaining the particular deviation from the mean is less than $\frac{1}{2n}$". The quantity n refers to the number of observations which are large enough to presume that the data follows the Gaussian error distribution.

OBJECTIVE TYPE QUESTIONS

Choose the Correct Answer :

1. The purpose of instruments is to
   (*a*) allow measurements to be made (*b*) transmit the information
   (*c*) change signals (*d*) any of the above.
2. The measurement refers to which of the following ?
   (*a*) Primary signal (*b*) Measured variable
   (*c*) Output (*d*) All of the above.
3. The measurement of a quantity
   (*a*) is an act of comparison of an unknown quantity with another quantity.
   (*b*) is an act of comparison of an unknown quantity with a known quantity whose accuracy may be known or may not be known.
   (*c*) is an act of comparison of an unknown quantity with a predefined acceptable standard which is accurately known.
   (*d*) none of the above.
4. In present-day measurement systems
   (*a*) direct methods are commonly used.

(*b*) one of the direct methods is limited but indirect methods are commonly used.
(*c*) both direct and indirect methods are commonly used.
(*d*) all of the above.

5. The measurand is
(*a*) output (*b*) measured variable
(*c*) secondary signal (*d*) any of the above.

6. The output stage of a generalised measurement system may comprise
(*a*) manipulator (*b*) detector-transducer
(*c*) indicating or recording unit (*d*) all of the above.

7. A null type of instrument as compared to a deflection type instrument has
(*a*) a higher accuracy (*b*) a lower sensitivity
(*c*) a faster response (*d*) all of the above.

8. In an intelligent information system used for the measurement of a physical quantity
(*a*) the digital computer must perform all or nearly all the signal and information processing.
(*b*) the observer must perform all or nearly all the signal and information processing.
(*c*) the digital computer is not required.
(*d*) none of the above.

9. A high grade of slip gauges preserved in a factory and not put into general use would be a
(*a*) primary standard (*b*) secondary standard
(*c*) tertiary standard (*d*) any of the above.

10. The temperature measurement by a thermocouple is
(*a*) primary measurement (*b*) secondary measurement
(*c*) tertiary measurement (*d*) any of the above.

11. Purely mechanical instruments cannot be used for dynamic measurements because they have
(*a*) high inertia (*b*) large time constant
(*c*) higher response time (*d*) all of the above.

12. The usage of electronic instruments is becoming more extensive because they have
(*a*) a high sensitivity and reliability.
(*b*) a fast response and compatibility with digital computers.
(*c*) the capability to respond to signals from remote places.
(*d*) all of the above.

13. The intermediate stage of an instrument is used to
(*a*) sense the primary (input) signal.
(*b*) indicate or record the measurements.
(*c*) amplify and transmit the secondary signal (signal after it has been transduced).
(*d*) any of the above.

14. The function of the signal manipulating element, in a generalised measurement system, is to
(*a*) change the quantity under measurement to an analogous signal.
(*b*) change the magnitude of the input signal while retaining its identity.
(*c*) to perform non-linear operations like filtering, chopping and clipping and clamping.
(*d*) none of the above.

15. A modifying input to a measurement system can be defined as an input,
(*a*) which changes the input-output relationship for desired inputs only.
(*b*) which changes the input-output relationship for desired as well as interfering inputs.
(*c*) which changes the input-output relationship for interfering inputs only.
(*d*) none of the above.

16. The ratio of the output to input change for a given measuring system is referred to as
(*a*) sensitivity (*b*) linearity
(*c*) stability (*d*) none of these.

17. The largest change in the measured variable which produces no instrument response is known as
(*a*) threshold (*b*) dynamic error
(*c*) dead zone (*d*) none of these.

18. The conformity of the output to the true value of the measurement is called
(*a*) precision (*b*) accuracy
(*c*) sensitivity (*d*) none of these.

19. The resolution of a system refers to
(*a*) smallest change in the measurand that can be measured
(*b*) true value of the input
(*c*) retardation of the response (*d*) none of these.

20. The least change of the measured variable which can be detected at the output of the measuring system is
(*a*) least count (*b*) sensitivity
(*c*) discrimination (*d*) any of these.

21. State which of the following describes the linearity of instrument :
(*a*) Range of inaccuracy which can be tolerated.
(*b*) Largest change in the measured value which produces no instrument response.
(*c*) Relationship between the output and the input.
(*d*) None of these.

22. In measurement systems, which of the following are undesirable static characteristics?
(*a*) Sensitivity and accuracy (*b*) Drift, static error and dead zone
(*c*) Reproducibility and non-linearity
(*d*) Drift, static error, dead zone and non-linearity.

23. The values of static stiffness and compliance in a measurement system determine the amount of
(*a*) power drain from a system (*b*) current drain from a system
(*c*) energy drain from a system (*d*) none of the above.

24. In the centre zero analog ammeter having a range of –10A to +10A, there is a detectable change of the pointer from its zero position on either side of the scale only if the current reaches a value of 1 A. The ammeter has a
(*a*) resolution of 1 A (*b*) dead zone of 1 A
(*c*) dead zone of 2 A (*d*) none of the above.

25. The input resistance of a cathode ray oscilloscope is of the order of
(*a*) tens of ohm (*b*) megaohms
(*c*) kilohms (*d*) fraction of an ohm.

26. Johnson noise is
(*a*) caused by thermal agitation of free electrons carrying currents thereby modulating the current.

(*b*) noise carried into a circuit through conductors.
(*c*) noise of an electromagnetic origin that is radiated into a circuit.
(*d*) shot noise which results from random emission of electrons across a *p-n* junction.

27. A pressure measurement instrument is calibrated between 10 bar and 250 bar. The scale span of the instrument is

(*a*) 10 bar (*b*) 250 bar
(*c*) 240 bar (*d*) 260 bar.

while the scale range is

(*a*) 10 bar (*b*) 250 bar
(*c*) 240 bar (*d*) 260 bar.

28. A pressure gauge is calibrated from 0-50 kN/m^2. It has a uniform scale with 100 scale divisions. One-fifth of a scale division can be read with certainty. The gauge has a

(*a*) resolution of 0.1 kN/m^2 (*b*) threshold of 0.1 kN/m^2
(*c*) dead zone of 0.2 kN/m^2 (*d*) a resolution of 0.5 kN/m^2.

29. A D.C. circuit can be represented by an internal voltage source of 50 V with an output resistance of 100 kΩ. In order to achieve 99 per cent accuracy for voltage measurement across its terminals, the voltage measuring device should have

(*a*) a resistance of at least 10 MΩ (*b*) a resistance of 100 kΩ
(*c*) a resistance of at least 10 Ω (*d*) none of the above.

30. In A.C. circuits, the connection of measuring instruments causes loading errors which may affect

(*a*) only the magnitude of quantity being measured
(*b*) only the phase of the quantity being measured
(*c*) both the magnitude and phase of the quantity being measured
(*d*) magnitude, phase and also the waveform of the quantity being measured.

31. Backlash is

(*a*) gradual departure of instrument output from its calibrated value.
(*b*) the ability of the device to give identical output when repeated or measurements are made with the same input signal.
(*c*) maximum angle or distance through which any part of mechanical system may be moved in one direction without causing motion of the next part.
(*d*) unwanted signal tending to obscure the transducer signal.

32. In measurement systems, which of the following static characteristics are desirable ?

(*a*) Accuracy (*b*) Sensitivity
(*c*) Reproducibility (*d*) All of the above.

33. In force-current analogy stiffness constant K is analogous to :

(*a*) inductance (*b*) reciprocal of inductance
(*c*) capacitance (*d*) reciprocal of capacitance.

34. In force-voltage analogy viscous friction constant B is analogous to

(*a*) resistance (*b*) reciprocal of resistance
(*c*) inductance (*d*) capacitance.

35. Which of the following are the desirable dynamic characteristics of a measurement system?

(*a*) Fast response, fidelity, measuring lag and dynamic error
(*b*) Fast response and lag
(*c*) Fidelity and measuring lag
(*d*) Fast response and fidelity.

36. Which of the following errors can arise, as a result of mistakes in reading, parallax, improper instrument location and inadequate lighting ?

(*a*) Construction error (*b*) Transmission error
(*c*) Observation error (*d*) Translation error.

37. Gearing, backlash, friction between moving parts, and scale accuracies are generally known as

(*a*) instrument errors (*b*) interference errors
(*c*) calibration errors (*d*) interaction errors.

38. Errors which may be variable both in magnitude and nature (positive or negative) are classified as

(*a*) hysteresis errors (*b*) random errors
(*c*) systematic errors (*d*) interaction errors.

39. The characteristic of an observer to read either systematically higher or systematically lower than other observers is referred to as

(*a*) insensitivity (*b*) lack of precision
(*c*) lack of discrimination (*d*) operator bias.

40. The maximum probability has a large value for small values of

(*a*) standard deviation (*b*) precision index
(*c*) uncertainty (*d*) average deviation.

41. What refers to the "number of changes that a particular reading will occur when the error limit is specified" ?

(*a*) Probability (*b*) Probable error
(*c*) Odds (*d*) Confidence interval.

42. Systematic errors in a Bourdon tube pressure gauge may be caused by

(*a*) friction in the pins and gears of the amplifying mechanism
(*b*) incorrect zero setting of the pointer
(*c*) variation of atmospheric pressure
(*d*) incorrect readings of the scale due to parallax.

43. If the confidence level is 0.95, then the values laying outside the confidence interval are

(*a*) 1 in 5 (*b*) 1 in 20
(*c*) 1 in 100 (*d*) 1 in 1000.

44. Uncertainty distribution is used for

(*a*) analysis of multi-sample data
(*b*) analysis of single-sample data
(*c*) analysis of both single and multi-sample data
(*d*) none of the above.

45. The mean deviation $\overline{d}$ in terms of deviations from the mean value of *n* reading is

(*a*) $\dfrac{\Sigma|d|}{n}$ (*b*) $\dfrac{\Sigma d}{n}$

(*c*) $\dfrac{\sqrt{\Sigma d^2}}{n}$ (*d*) $\sqrt{\dfrac{\Sigma d^2}{n}}$.

ANSWERS

1. (*a*)	**2.** (*c*)	**3.** (*c*)	**4.** (*b*)	**5.** (*b*)	**6.** (*c*)	**7.** (*a*)
8. (*a*)	**9.** (*b*)	**10.** (*c*)	**11.** (*d*)	**12.** (*d*)	**13.** (*c*)	**14.** (*b*)
15. (*b*)	**16.** (*a*)	**17.** (*c*)	**18.** (*b*)	**19.** (*a*)	**20.** (*c*)	**21.** (*c*)
22. (*d*)	**23.** (*c*)	**24.** (*c*)	**25.** (*b*)	**26.** (*a*)	**27.** (*c*)	**28.** (*a*)
29. (*a*)	**30.** (*d*)	**31.** (*c*)	**32.** (*d*)	**33.** (*b*)	**34.** (*a*)	**35.** (*d*)
36. (*c*)	**37.** (*a*)	**38.** (*b*)	**39.** (*d*)	**40.** (*b*)	**41.** (*c*)	**42.** (*b*)
43. (*b*)	**44.** (*b*)	**45.** (*a*).				

THEORETICAL QUESTIONS

1. What do you understand by the term "measurement" ?
2. Describe the significance of measurements.
3. Explain briefly the term "measurand".
4. State the basic requirements for getting the meaningful result of the measurement.
5. Explain the difference between an analog signal and a digital signal.
6. Define the term "instrumentation".
7. How are standards of measurement classified ?
8. Explain clearly the difference between primary, secondary and working standards.
9. Explain briefly the following methods of complete solutions to complex problem in mechanical design :
   (*i*) The empirical method.
   (*ii*) The rational method.
   (*iii*) The experimental method.
10. Distinguish between direct and indirect methods of measurement.
11. Describe the mechanics and power types of measurements. Explain with suitable examples.
12. What are primary, secondary and tertiary measurements ? Explain with examples.
13. Explain the difference between the following :
    (*i*) Measurand and measurement.
    (*ii*) Measurement and instrument.
    (*iii*) Input and output.
14. Draw a block diagram representation of a generalised measurement system. Identify the various elements and point out the functions performed by each element/component.
15. Explain briefly various stages of the general measurement system.
16. Describe briefly the following examples of measurement systems :
    (*i*) Bourdon tube pressure gauge.
    (*ii*) Pressure-actuated thermometer.
17. What do you mean by characteristics of a measuring instrument ?
18. What are static characteristics ? How do they differ from dynamic characteristics ?
19. Define the following terms as applied to the measuring instrument :
    (*i*) True value (*ii*) Correction
    (*iii*) Range (*iv*) Sensitivity
    (*v*) Accuracy (*vi*) Linearity
    (*vii*) Dead zone (*viii*) Backlash.

Static characteristics :

20. Enumerate the main static characteristics of measuring instruments.
21. Explain briefly the terms : Accuracy, errors and correction.
22. Explain the following terms :
    (*i*) Static error (*ii*) Static correction
    (*iii*) Relative error (*iv*) Percentage relative error.
23. How can "accuracy" be expressed ? Explain briefly.
24. What do you mean by "Static calibration" ? Give the steps which are necessary in performing a calibration.
25. Differentiate between "Scale range" and "Scale span" giving suitable examples.
26. What is an "Error calibration curve" ?
27. What do you mean by "Scale readability" ?
28. Differentiate between the terms "Repeatability" and "Reproducibility".
29. Explain the term "Drift" using suitable examples.
30. How is "drift" classified ?
31. Explain briefly the following :
    (*i*) Zero drift (*ii*) Span drift (*iii*) Zonal drift.
32. Differentiate between the terms "Accuracy" and "Precision" with suitable examples.
33. What is "sensitivity" ?
34. Define the term "Deflection factor".
35. What is "linearity" ? Explain briefly.
36. What do you mean by the term "Non-linearity" ? Explain briefly "Terminal linearity" and "Best fit linearity".
37. Describe briefly the term "Hysteresis".
38. Explain clearly the difference between "Threshold" and "Resolution".
39. Explain briefly the following :
    (*i*) Dead zone; (*ii*) Dead time.
40. What do you mean by the term "loading effect" as applied to measuring instruments ? Explain briefly.
41. Explain briefly the terms "Static stiffness" and "Static compliance".
42. What is "noise" ?
43. What is signal to noise ratio ?
44. Explain briefly the following :
    (*i*) Generated noise; (*ii*) Conducted noise; (*iii*) Radiated noise.

Dynamic characteristics :

45. Define the following terms :
    (*i*) Response; (*ii*) Dynamic response of a measuring system
46. Explain briefly the following types of dynamic inputs :
    (*i*) Transient response; (*ii*) Steady state periodic.
47. Define the following terms :
    (*i*) Transient response; (*ii*) Steady state response.
48. Define the dynamic response of a system and distinguish between steady state response and transient response.
49. Define the following terms :
    (*i*) Speed of response; (*ii*) Measuring lag;
    (*iii*) Fidelity; (*iv*) Dynamic error;
    (*v*) Overshoot.

50. What are the different standard inputs for studying the dynamic response of a system ? Explain them briefly.
51. Derive the equations for time response of a first-order system when subject to :
    (*i*) Unit step input; (*ii*) Unit ramp input.
52. Derive the expression for time response of a second-order underdamped system when subjected to unit ramp input. Show that the nature of the response is same as that for a unit step input.
53. Derive the expression for the magnitude and phase of a first-order system when subjected to a sinusoidal input signal. Draw the magnitude and phase versus frequency plots.
54. Derive the expressions for the frequency response of a second-order system. Draw the magnitude and phase versus frequency plots.
55. Sketch and explain the response of a first-order system when subjected to step input signal.
56. Sketch and explain the response of a first-order system subjected to ramp input signal. Indicate thereon the transient error, steady error and the time lag.
57. Describe the dynamics for sinusoidal input to a first-order system.
58. Develop an expression for the amplitude ratio and phase difference of indicated reading and the input signal which is of sinusoidal form.
59. What is the importance of time constant in appropriate measuring instruments ?

Errors in measurements :

60. What are limiting errors ? Explain briefly.
61. Explain briefly the term *"relative error"*.
62. How are errors classified ?
63. What are "Gross errors " ? How can these be avoided ?
64. What are "Systematic errors" ? Explain briefly the following systematic errors with examples :
    (*i*) Instrumental errors.
    (*ii*) Environmental errors.
    (*iii*) Observational errors.
65. How can the systematic errors be located ?
66. What are assembly errors ? Can they be easily discovered and rectified ? If yes, why ?
67. State the factors which give rise to observational errors.
68. What are random errors ? State the common causes of their occurrence.
69. Enumerate the various sources of errors.
70. Explain briefly the following sources of errors :
    (*i*) Noise (*ii*) Response time
    (*iii*) Design limitations (*iv*) Energy exchanged by interaction
    (*v*) Transmission (*vi*) Deterioration of measuring system.
71. What are single-sample and multi-sample tests ? Explain briefly.
72. Explain briefly the following terms :
    (*i*) Arithmetic mean (*ii*) Geometric mean
    (*iii*) Median (*iv*) Mode.
73. Describe briefly the following terms :
    (*i*) Deviation (*ii*) Average deviation
    (*iii*) Standard deviation (*iv*) Variance.
74. What is the law of probabilities ? Explain briefly.
75. Explain briefly the term "Precision index".
76. Write a short not on "Probability tables".

77. While specifying measurement data, how are the deviations expressed ?

78. Explain briefly the following terms :

(i) Confidence interval. (ii) Confidence level.

79. What is "Chauvenet's criterion" ? Expain briefly.

UNSOLVED EXAMPLES

1. A meter reads 136.6 V and the true value of the voltage is 136.52 V. Determine :

(i) The static error; (ii) The static correction for this instrument

[**Ans.** (i) + 0.08 V (ii) – 0.08 V]

2. An analog indicating instrument with a scale range of 0 – 10.0 V shows a voltage of 5.30 V. The true value of a voltage is 5.40 volts.

(i) What are the values of absolute error and correction ?

(ii) Express the error as a function of the true value and full scale deflection.

[**Ans.** (i) – 0.1 V, 0.1 V, (ii) 1.85 %, – 1 %]

3. A pressure indicator showed reading as 89 bar on a scale range of 0 – 100 bar, but the true value was 88.5 bar. Determine :

(i) Static error, (ii) Static correction, and

(iii) Relative static error. [**Ans.** (i) + 0.5 bar, (ii) – 0.5 bar, (iii) 0.565%]

4. A thermometer is calibrated 100°C to 150°C. The accuracy is specified within ± 0.25 per cent of instrument span. What is the maximum static error ? [**Ans.** ± 0.125°C]

5. In a Wheatstone bridge a change of 9 Ω in the unknown arm of the bridge is required to produce a change in the deflection of 3 mm of the galvanometer. Determine (i) Sensitivity, and (ii) Deflection factor. [**Ans.** (i) 0.333 mm/Ω; (ii) 3Ω/mm]

6. A pressure measuring system consists of a piezoelectric transducer, a charge amplifier and an ultra violet charge recorder; their sensitivities are 8.5 pC/bar, 0.004 V/pC and 20 mm/V respectively. Determine the deflection on the chart for a pressure change of 25 bar. [**Ans.** 17 mm]

7. Determine the resolution of a moving coil voltmeter having a uniform scale with 100 divisions; the full scale reading is 100 V and $\frac{1}{10}$ of a scale division can be estimated with a fair degree of certainty. [**Ans.** 0.1 V]

8. The calibration range of a certain pyrometer is 200°C to 900°C. If the dead zone in it is 0.12 per cent of the span, determine the temperature change which might occur before it is detected. [**Ans.** 0.84°C]

9. On the application of a step input of 100 bar to a pressure gauge, the pointer swings to pressure of 102.5 bar and finally comes to rest at 101.3 bar. Determine :

(i) The overshoot of the gauge readings and express it as a percentage of the final reading.

(ii) The percentage error of the gauge. [**Ans.** (i) 1.2 bar, 1.18%, (ii) 1.3%]

10. A thermometer has a time constant of 2.5 seconds. It is quickly taken from a temperature 0°C to liquid bath having a temperature 200°C. Calculate the temperature indicated by the thermometer after a time of 1.5 seconds.

Determine its reading when it was initially held at 25°C. [**Ans.** 90.24°C, 103.96°C]

11. A temperature sensitive device can be modelled as first-order system with a time constant of 6 seconds. It is suddenly subjected to a step input of 25°C – 150°C. Calculate the temperature indicated by the device after 10 seconds after the start of the process.

[**Ans.** 126.4°C]

12. A temperature sensitive transducer when subjected to sudden temperature change takes 10 seconds to reach the equilibrium conditions (five times constant). Calculate the time taken by the transducer to read half of the temperature difference. [**Ans.** 1.39s]

13. A resistance temperature detector has a steady state gain of 0.3925 Ω/°C and a time constant of 5.5 seconds. It is subjected to a step change of 75°C in temperature. If its stable resistance before the step change is 100 Ω, write down the time-domain equation for resistance and find its value after 15 seconds of applications of step input.

[**Ans.** $R_t = 29.44\ (1 - e^{-t/5.5}) + 100$, 127.5 Ω]

14. A thermometer initially at a temperature of 20°C is suddenly plunged into a liquid bath maintained at 150°C. After a time interval of 3 seconds, the thermometer indicated a temperature of 95°C. Determine.

(*i*) The time constant for the thermometer.

(*ii*) The indicated temperature after five times constant; comment upon this result.

[**Ans.** (*i*) 3.487 s, (*ii*) 149.12°C]

15. A temperature sensitive transducer used to measure the temperature of a furnace has been idealised as a first-order system subjected to ramp input. If the maximum permissible error in temperature measurement is limited to 4.2 °C, calculate the time constant of the transducer when the furnace temperature increases at the rate of 0.14/°C. [**Ans.** 30°C]

16. A weather balloon carrying a temperature sensing device (of first-order, time constant 10 seconds) rises through the atmosphere at 6m/s. It supplies information about temperature and altitude through radio signals. At 2800 m height a temperature indication of 35°C has been received. If the temperature varies with altitude at a uniform rate of 0.01°C/m, determine the true altitude at which 35°C temperature occurs. [**Ans.** 2740 m]

17. The temperature of a furnace is found to vary sinusoidally between 400°C and 450°C with a periodic time of 60 seconds. A thermocouple system with a time constant of 8 seconds is employed to measure the furnace temperature. Determine :

(*i*) The maximum and minimum values that will be indicated by the thermocouple.

(*ii*) The phase shift and corresponding time lag between the temperature signals and the thermocouple input signals. [**Ans.** (*i*) 444.2°C, 405.8 °C, (*ii*) 39.76°, 6.669s]

18. A first-order system having a time constant of 0.1 second is used to measure a signal prescribed by the relation; $I_i = 3\ \sin 2t + 0.4\ \cos 10t$. Develop an expression for the corresponding output. [**Ans.** $I_o = 2.94\ \sin\ (2t - 11.31°) - 0.283\ \sin\ (10t + \pi - 45°$]

19. A liquid thermometer has its glass bulb protected by a well. The system can be represented by a double capacity system with time constants of 40s for the well and 20s for the bulb. The thermometer is subjected to a cyclic change of ± 10°C which occurs every 120 seconds. Find the maximum value of the indicated temperature and the delay time.

[**Ans.** ± 2.97°C; 36.8s]

20. An input of $1\sin 2t + 0.3\sin 20t$ is applied to a first-order instrument having a time constant of 0.2 second. Find the output. [**Ans.** $0.93\ \sin\ (2t - 21.8°) + 0.073\ \sin\ (20t - 76°$]

21. The pen arrangement of a U.V. recorder has a mass of 4.8 g. What should be the percentage reduction in mass if it is desired to have 18 per cent increase in the natural frequency of the recorder ? Consider the pen motion to conform to a second-order system.

[**Ans.** 28.18%]

22. Determine the damping ratio, damped natural frequency, static sensitivity and time constant for a second-order system which follows the differential equation given as : $\frac{d^2I_o}{dt^2} + 2\frac{dI_o}{dt} + 25I_o = 25I_i$ where I_o and I_i are the output and input quantities respectively.

[**Ans.** 0.2; 4.899 rad/s; 1.0; 0.2s]

23. While measuring force by a diagram type pressure transducer (second-order system), the system is found to have a natural frequency of 900 cycles per second (cps) and damping ratio of 60 per cent.

Determine the frequency range over which the amplitude ratio corresponding to sinusoidal input deviates by a maximum amount of 10 per cent. [**Ans.** 0 – 824 cps]

24. A unit amplitude sinusoidal input signal having a frequency of 0.5 Hz is given to second-order control system of transfer function : $(I_o/I_i) = 8/(D^2 + 4D + 20)$.

Develop an expression for the steady state response of the system and determine the corresponding output, amplitude, output frequency and phase lag.

[**Ans.** I_o = 0.4587 (sin πt – 46.04°); 0.4587; π; 46.04°]

25. A torque sensing transducer is connected to the shaft of an electric motor which drives a load that has a moment of interia of 0.20 kg-m^2. If the torsional constant is 1.5 Nm/rad, what should be the damping ratio of the system when a simusoidal torque varying at frequency of 2 rad/s its to be reproduced with a maximum error of + 10 per cent ?

[**Ans.** 0.534]

26. A second-order instrument has a natural frequency of 4 Hz and a damping ratio of 0.66. If the excitation frequency of the system is 6Hz, determine the error due to the proximity of excitation frequency with the natural frequency of the system. [**Ans.** 57.3%]

Errors in measurements :

27. A wattmeter having a range of 2000 W has an error of ± 1 per cent of full scale deflection. If the true value is 200 W, what would be the range of the readings ?

If the error is specified as percentage of the true value, what would be the range of the readings ? [**Ans.** 180 to 220 W; 198 W and 202 W]

28. The guaranteed accuracy of a flow meter working on thermal principles is ± 4 per cent of full scale reading of 10.0 × 10^{-6} m^3/s. If the flow measured by this meter is 5 × 10^{-6} m^3/s, calculate the limiting error in per cent. [**Ans.** ± 8 percent]

29. A – 300 Pa manometer has a guaranteed accuracy of 2 per cent of full scale reading. The pressure measured by this instrument is 150 Pa. Calculate the limiting error in per cent.

[**Ans.** 4 per cent]

30. Three resistors have the following ratings :

R_1 = 74 Ω ± 5%, R_2 = 150 Ω ± 5%; R_3 = 100 Ω + 5%.

Determine the magnitude and limiting error in ohms and in per cent of the resistance of these resistances connected in series. [**Ans.** 324 ± 16.20; ± 5%]

31. The resistance of a circuit is found by measuring current flowing and the power fed into the circuit. Find the limiting error in the measurement of resistance when the limiting errors in the measurement of power and current are 0.9% and ± 0.6% respectively.

[**Ans.** ± 2.1%]

32. The unknown resistance in a Wheatstone bridge is measured by using three known resistances such that $R_4 = \dfrac{R_2 R_3}{R_1}$;

where, R_1 = 100 ± 0.5% Ω ; R_2 = 500 ± 5% Ω ; R_3 = 280 ± 0.5% Ω.

Determine the magnitude of the unknown resistance (R_4) and the limiting error in per cent and in ohms (for the unknown resistance). [**Ans.** 1400 Ω ; 1.5% ± 21 Ω].

33. A cantilever beam of length l, width b and depth d is used to measure the force F. The deflection is measured using an LVDT of accuracy ± 0.1 mm. The length, width and depth are measured to an accuracy of ± 0.01 mm and their values are 50 mm and 1.0 mm respectively. The modulus of rigidity is known accurate to 0.2%. If the deflection noted was 2 mm, determine the error in force measured.

Given force $F = \dfrac{3Ebd^3}{12l^3} \cdot y$

where, E = modulus of elasticity; l, b and d = length, width and depth of the beam respectively; y = deflection. [**Ans.** ± 0.2820%]

34. A dynamometer is used to measure the output power of a rotating shaft. The output power is given by the following relation : $P = \dfrac{2\pi \times 9.81\, FLR}{t \times 10^6}\, kW$

where F = force at the end of torque arm (kg); L = length of torque arm (mm); R = number of revolutions during time t; t = time for test run (s).

The given test data are : $F = 4.58 + 0.02$ kg ; $L = 397 \pm 7.3$ mm ; $R = 1202 \pm 1.0$ revolutions; $t = 60 + 0.50$ s.

The errors are limiting (absolute) errors.

Determine the magnitude of power and magnitude of the limiting error in the computed power. **[Ans.** 2.245 kW ; ± 0.038 kW]

35. During a test the current was measured as 30.4A, flowing in a resistor of 0.105 Ω. Later on its was discovered that the ammeter reading was low by 1.2% and the marked resistance was high by 0.3%. Determine the true power as a percentage of the power that was originally calculated. **[Ans.** 102.1%]

36. A circuit was tuned for resonance by eight different students, and the values of resonant frequency in KHz was recorded as 532, 548, 543, 535, 546, 531, 543 and 536. Calculate the following :

(*i*) Arithmetic mean. (*ii*) Average deviation, and
(*iii*) Variance. **[Ans.** (*i*) 539.25 kHz, (*ii*) 5.75 kHz; (*iii*) 6.54 kHz, (*iv*) 42.77 $(kHz)^2$]

37. During a test run, measurements of temperature were made 100 times with variation in apparatus and procedure. After applying corrections for known systematic errors, the following data were obtained :

Temperature °C :	397	398	399	400	401	402	403	404	405
Frequency of Occurrence :	1	3	12	23	37	16	4	2	2

Calculate the following :

(*i*) Arithmetic mean, (*ii*) Mean deviation,
(*iii*) Standard deviation, (*iv*) Probable error of one reading,
(*v*) Probable error of the mean, and
(*vi*) Standard deviation of the standard deviation.

[Ans. (*i*) 400.78°C, (*ii*) 1.028°C, (*iii*) 1.380°C, (*iv*) 0.93°C, (*v*) 0.093°C, (*vi*) 0.0796°C]

38. The speed of a hydraulic turbine mode, which is running at 1000 r.p.m., is measured by a tachometer. The hydraulic turbine is subjected to variations in speed. For a sample of 20 readings at this speed, how many readings would you expect between 990 and 1010 r.p.m.? Assume that the tachometer gives a normal set of deviation with precision index of 0.04 at 1000 r.p.m. **[Ans.** 9 readings]

39. The depth of water, which had a nominal value of 18 cm, was measured with a hook. Measurements were taken 24 times and 6 readings were found to lie outside a particular range. Assume that the measurements conform to normal distribution with precision index 8 cm^{-1}. Determine the prescribed range. **[Ans.** (18 ± 0.1016) cm]

40. (*a*) A mass flow meter using electromagnetic principle is set to record a large number of readings. The fixed mass flow rate is 50 kg/s; half of the readings lie between 52.4 and 47.6 kg/s. The sample measurements give a straight line between the (0, 0.50%) points on the probability coordinates. Calculate the precision index of the instrument.

(*b*) An alarm rings when the flow meter gives a reading of 44 kg/s.

(*i*) How many false alarms are expected in the month of June if the flow is checked automatically eight times a day ?

(*ii*) What will be the value of precision index if one wants to reduce the false alarm rate by a factor of two ? **[Ans.** (*a*) 0.199; (*b*) (*i*) 12, (*ii*) 0.231]

41. The data given below are expected to follow a linear relationship : $v = u + ab$

u :	0.9	2.3	3.3	4.5	5.7	6.7
v :	1.1	1.6	2.6	3.2	4.0	5.0

Using least square analysis : (*i*) obtain the best linear relation, and (*ii*) calculate the standard deviation.

[Ans. $v = 0.672u + 0.296$, (*ii*) $s_v = \pm 0.1716$; $s_u = \pm 0.255$; $s_a = \pm 0.0355$; $s_b = \pm 0.155$]

42. The final result of a measurement depends on two measurements in series, each having values of 25 and 50 units with errors of ±2% and ±1%, such that result is sum of both. Determine the error of final result if errors are considered as (*i*) Limiting errors, (*ii*) Standard deviations. **[Ans.** (*i*) 1.34%; (*ii*) 0.943%]

CHAPTER

3 Measuring Instruments

3.1 Introduction to instruments; 3.2. Classification of instruments; 3.3. Selection of insruments; 3.4. Functions of instruments; 3.5. Comparison of analog and digital instruments; 3.6. Electrical instruments; 3.7. Galvanometers – D' Arsonval galvanometer – Ballistic galvanometer – Grassot fluxmeter – Vibration galvanometer – Duddall's oscillograph; **3.8. Ammeters, Voltmeters** and **Ohmmeter** – Introduction – Types of instruments mainly used as ammeters and voltmeters – Errors in ammeters and voltmeters – *Moving iron instruments* – *Moving-coil instruments* – Permanent magnet moving-coil instruments – ohmmeters – *Electrodynamic or dynamometer type instruments* – *Hot-wire instruments* – *Thermocouple instruments* – *Rectifier instruments* – *Electrostatic instruments*; **3.9. Wattmeters and measurements of power** – Types of wattmeters – Dynamometer wattmeter – Induction wattmeter – Electrostatic wattmeter – Thermal wattmeter – Measurement of power in single-phase circuits – Measurement of power in 3-phase circuits – Measurement of power at high frequencies (AF/RF) – Bolometer method of power measurement – calorimetric method for measurement of large amount of RF power – Output power meter; **3.10. Energy meters and energy measurement** – General aspects – Essential characteristics of energy meters – Types of energy meters – Motor meters – Motor-driven meter – Watt-hour meter – Induction type of watt-hour meter – Induction type single-phase energy meters – Maximum demand indicators/meters – Measurement of VAh and VARh – Testing of energy meters; **3.11. Miscellaneous measuring instruments** – Power factor meters – Frequency meters – Rotational speed measurements – Synchromoscopes – Phase sequence indicators. *Highlights – Objective Type Questions – Theoretical Questions – Unsolved Examples.*

3.1 INTRODUCTION TO INSTRUMENTS

The **instrument** *serves as an extension of human faculties and enables the man to determine the value of unknown quantity or variable which his unaided human faculties cannot measure.* A measuring instrument provides information about the physical value of some variable being measured.

— The instrument would sense a physical parameter (e.g., velocity, pressure, temperature, etc.), process and translate it into a format and range which can be interpreted by the observer.

— The instrument *must also provide the controls* by which the operator can obtain, manipulate and respond to the information.

- An instrument, in *simple cases,* consists of a *single unit* which gives an output reading or signal according to the unknown variable (measurand) applied to it. Under *more complex measurement situations,* a measuring instrument may consist of *transducing elements* which convert the measurand to analogous form. The *analogous signal* is then processed by some intermediate means and then fed to the end devices for presenting the measurement results for the purpose of display, record and control.

- The instruments made by man not only retain their characteristics for extended periods of time but are also accurate and sensitive in their response.
- *Measurements* involve the use of *instruments* as a physical means of determining quantities and variables.

3.2 CLASSIFICATION OF INSTRUMENTS

The instruments may be *classified* as follows:

1. Absolute and secondary instruments.
2. Analog and digital instruments.
3. Mechanical, electrical and electronic instruments.
4. Manual and automatic instruments.
5. Self-contained and remote indicating instruments.
6. Self-operated and power-operated instruments.
7. Deflection and null output instruments.

1. Absolute and secondary instruments:

(i) Absolute instruments :

These intruments give the magnitude of the quantity under measurement in terms of physical constants of the instrument.

Examples : Tangent galvanometer and Rayleigh's current balance.

- These instruments are *seldom used except in standard institutions.*

(ii) Secondary instruments :

These instruments are so constructed that the quantity being measured can only be measured observing the output indicated by the instrument.

These instruments are *calibrated* by comparison with an absolute instrument or another secondary instrument which has already been calibrated against an absolute instrument.

Examples : Voltmeter, glass thermometer, pressure gauge etc.

- The secondary instruments find *wide use in every sphere of measurement.*

2. Analog and digital instruments:

(i) Analog instruments :

The signals of an analog unit vary in a continuous fashion and can take on infinite number of values in a given range.

Examples : Fuel gauge, ammeters and voltmeters, wrist watch, speedometer of an automobile, etc.

(ii) Digital instruments :

Signals varying in *discrete steps* and taking on a finite number of different values in a given range are *digital signals* and the corresponding instruments are of *digital type.*

Examples : Odometer of an automobile, calibrated balance of a platform scale, timer on a scoreboard, etc.

- The digital devices have the following *advantages* :
  - *(i)* High accuracy.
  - *(ii)* High speed.
  - *(iii)* Elimination of human operational errors.

- Owing to the application of digital computers for data handling, reduction and in automatic controls, the importance of *digital instrumentation* is increasing very fast. Thus, it becomes necessary to have both analog-to-digital (A/D) converters at input to the computers and digital-to-analog (D/A) converters at the output to the computers.

3. Mechanical, electrical and electronic instruments:

(i) Mechanical instruments :

- These instruments are very reliable for static and stable conditions.
- They are *unable to respond rapidly* to the measurements of dynamic and transient conditions due to the fact that they have moving parts that are rigid, heavy and bulky and consequently have a large mass. Mass presents *inertia problems* and hence these instruments cannot faithfully follow the rapid changes which are involved in dynamic instruments.
- Most of the mechanical instruments *cause noise pollution*.

(ii) Electrical instruments :

- The electrical methods of indicating the output of detectors *are more rapid than that of mechanical methods.*
- Unfortunately, an electrical system normally depends upon a mechanical meter as an indicating device. This mechanical movement has some inertia due to which the *frequency response* of these instruments is *poor*.

(iii) Electronic instruments :

- Most of the scientific and industrial instruments require *very fast responses.* Such requirements cannot be met with by mechanical and electrical instruments.
- These instruments use *semiconductor devices. In electronic devices, since the only movement involved is that of electrons, the response time is extremely small owing to very small inertia of electrons.*
- With the use of electronic devices, *very weak signals can be detected by using pre-amplifiers and amplifiers.*

Comparison between mechanical and electrical/electronic instruments:

Advantages and disadvantages of mechanical instruments:

Advantages :

(*i*) Simple in design and use.
(*ii*) More durable due to rugged construction.
(*iii*) Relatively cheaper in cost.
(*iv*) For operation of the instruments, no external power supplies are required.
(*v*) Quite reliable and accurate for the measurement of parameters which are stable and non-variant with time.
(*vi*) Long history of development and successful use.

Disadvantages :

(*i*) Poor frequency response to dynamic and transient measurements.
(*ii*) Large forces are required to overcome mechanical friction.
(*iii*) Incompatibility when remote indication or control is needed.
(*iv*) Cause noise pollution.

Advantages of electrical/electronic techniques of measurements:

(*i*) Less power consumption and less load on the system being measured.
(*ii*) Non-contact measurements are possible.

(*iii*) Light, compact and more reliable instrumentation.

(*iv*) Remote indication and recording feasible.

(*v*) The amplification produced is greater than that produced by a mechanical contrivance.

(*vi*) The mathematical processing of signals like summation, differentiating and integrating is possible.

(*vii*) The frequency and transient response is good.

(*viii*) Greater flexibility.

(*ix*) Lighter in construction.

- The electronic instruments are *widely used in detection of electro-magnetically produced signals such as radio, video and microwaves.*

- The use of these instruments has made it possible to *build analog and digital computers* which are quite instrumental in the modern developments in the fields of science and technology.

4. Manual and automatic instruments :

- In case of *manual* instruments services of an operator are required.

Example : Measurement of temperature by a resistance thermometer incorporating a Wheatstone bridge in its circuit (an operator is required to indicate the temperature being measured).

- In an *automatic* type of instrument, an operator is *not* required.

Example : Measurement of temperature by mercury-in-glass thermometer.

5. Self-contained and remote indicating instruments :

- A self-contained instrument *has all its different elements in one physical assembly.*

- In a remote indicating instruments, the primary sensing element *may be located at an adequate long distance* from the secondary indicating element. Such types of instruments are *finding wide use in the modern instrumentation technology.*

6. Self-operated and power-operated instruments :

- A self-operated instrument does not require any outside power for its operation; the output energy is supplied wholly or almost wholly by the input signal.

Examples : Mercury-in-glass thermometer, Dial indicator etc.

The power operated instruments are those instruments which require some auxiliary source of power such as electricity, compressed air, hydraulic supply etc. for their operation. In such cases, the input signal supplies only an insignificant portion of the output power.

Example : Electro-mechanical measurement system shown in Fig. 3.1.

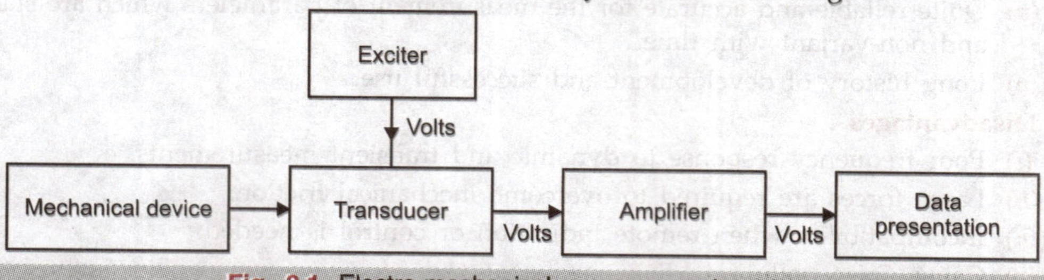

Fig. 3.1. Electro-mechanical measurement system.

7. Deflection and null output instruments :

- *In a "**deflection type**" instrument, the deflection of the instrument provides a basis for determining the quantity under measurement.* The measured quantity produces some physical effect which deflects or produces a mechanical displacement of the moving system of the instrument. An opposing effect is built in the instrument which tries to oppose the deflection or the mechanical displacement of the moving system. The balance is achieved when opposing effect *equals* the cause, producing the deflection or mechanical movement. The deflection or mechanical displacement at the point of balance then gives the value of measured quantity (which depends upon the calibration of the instrument).

Examples : Permanent magnet moving coil (PMMC) ammeter etc.

— These instruments are more suited for measurements under *dynamic conditions* than null type of instruments whose intrinsic response in slower.

• In a *"**null type**" of instrument, a zero or null indication leads to determination of the magnitude of measured quantity.* The null condition is dependent upon some other known conditions.

Example : D.C. potentiometer.

Advantages :

(*i*) The "null type" instruments are *more accurate* than deflection type instruments.

(*ii*) These instruments can be *highly sensitive* as compared with deflection type instruments.

3.3 SELECTION OF INSTRUMENTS

The following are some important factors/points which are very essential to be considered for the proper selection of a correct instrument for any application :

1. The *accuracy* expected from the instrument.

   The following most significant and fundamental rule, which is perhaps the most important rule in measurement, must always be kept in mind :

   "Never demand an accuracy of measurement higher than that which is really needed, and never forget that each degree of accuracy, if demanded, is likely to have a disproportionate effect on the complexity and cost of measuring apparatus".

2. *When are the final data required* (*i.e.* at the time of taking the measurement, or at later one)?
3. The *cost criterion* (*i.e.,* the cost of the measuring process).
4. *In what form the data should be displayed* (*i.e.,* indicating, recording, integrating etc.)?
5. Whether quantity to be measured has *constant value or is it a time variant*?

   — In the measurement of time variant quantities the manner in which the quantity to be measured changes has a significant effect on the type of measuring instruments to be used. *The physical characteristics of the measuring device must always be taken into account since these characteristics determine the amount of interaction which takes place between the measurand quantity and the instrument.*

3.4 FUNCTIONS OF INSTRUMENTS

Following are the *three* main functions of instruments :

1. Indicating function.
2. Recording function.
3. Controlling function.

1. Indicating function :

— The value of the quantity can be read by the movement of the needle on a calibrated scale provided on the instrument. Readings can be taken to any fraction within the limitations of the instrument and human eye.

Examples : (*i*) The deflection of a pointer of a *'speedometer'* indicates the speed of the automobile at that moment.

(*ii*) The use of a *'pressure gauge'* for indicating pressure.

— These days digital display (*i.e.*, the reading in numericals) is also commonly used. In fact, where highly accurate and exact readings are desired, use of digital readouts is recommended.

2. Recording function :

— In several cases the instrument continuously records, with pen and ink, the value of measured quantity against some other variable or against time. A graph paper is provided whose speed can be adjusted. A recording instrument can be used to record more than one number of quantities also, in which case, either the recording of different parameters may be done by different pens or same pen may record one parameter after other.

Example : A 'potentiometric type of recorder' used for monitoring temperature records the instantaneous values of temperatures on a strip chart recorder.

3. Controlling function :

— The controlling function is perhaps one of the most important functions particularly in the field of industrial control processes. Here, the information is used by the instrument/system to control the original measured quantity.

Examples : (*i*) *Floats* for liquid level control.

(*ii*) *Thermostats* for temperature control.

Essential requirements of a measuring instrument :

The essential requirements of a measuring instrument are :

1. Its introduction into the circuit where measurements are to be made, should not change the circuit configuration.
2. Its power consumption should be small.

3.5 COMPARISON OF ANALOG AND DIGITAL INSTRUMENTS

Qualitatively, the *"electrical indicating instruments"* are widely used for measurement of current, voltage, resistance and power. These instruments can be *classified* as follows :

1. Analog instruments.
2. Digital instruments. (For details of digital instruments refer to *chapter* 10)

• *The deflection type instruments with a scale and movable pointer are called* ***analog instruments.***

— The measured parameter is converted into electrical signal which further actuates some electro-mechanical device to move the pointer continuously as the parameter changes.

• *The instruments which use logic circuits and techniques to obtain a measurement and then display it in numerical reading (digital) form are called* ***digital instruments.***

The digital instruments claim the following *advantages over analog instruments* :

(*i*) Greater accuracy.

(*ii*) Better resolution.

(*iii*) Automatic polarity and zeroing.

(*iv*) Easy readability.

Table 3.1 shows the comparison between analog and digital type instruments.

Table 3.1. Comparison between Analog and Digital type Instruments

S.No.	Aspects	Analog type instruments	Digital type instruments
1.	*Information form*	As the position of a pointer against a calibrated scale or dial.	As a number.
2.	*Possibility of human error*	Exists	Does not exist.
3.	*Best possible accuracy*	+ 0.25%	+ 0.005% or better.
4.	*Resolution*	One in several hundreds.	One part in several hundred thousands.
5.	*Presence of moving parts*	Moving parts involved.	These instruments can be made without moving parts.
6.	*Construction*	Simple in construction and direct reading type; can perform under unfavourable conditions.	Since these instruments involve electronics, proper environmental conditions are essential.
7.	*Rate of change of parameter*	These instruments enable the operator to judge the rate of change of parameter by seeing the needle movement.	Change of digital reading does not give any knowledge of rate of change of parameter.
8.	*Time required to observe the reading*	If exact reading is required operator takes more time as he has to guess the approximate tenths of small division.	Reading of digital meters is very fast.
9.	*Auxiliary power requirement*	These instruments require no auxiliary source of power for actuation but derive driving power for indicating system from the process.	These instruments require auxiliary source of power.
10.	*Mobility*	Can be portable also.	Usually stationary type.
	Examples :	Examples of pointer-dial output devices are : • Micrometer and platform scales; • Manometers and Bourdon-tube pressure gauge; • Mercury in glass and filled system thermometers; • Speedometer of an automobile; • Common voltmeters and ammeters etc.	Examples of Digital output devices are : • Digital ammeters and voltmeters; • Electronic and mechanical counters; • Odometers; • Yes-No light (On or Off); • Time on a scoreboard etc.

Pointer-scale analog indicators :

In analog instruments, the value of the measured parameter is indicated by positioning of the indicating pointer against a calibrated scale. This purpose can be achieved either by moving the pointer with relation to a stationary scale (fixed-scale moving-pointer indicators) or the scale may be moved with relation to a fixed reference (moving-scale fixed-pointer indicators).

Multi-pointer indicator. This type of indicator contains more than one number of pointers and above each point the identification number of the medium being measured is marked. *Usually this arrangement is used in recorders and not in indicators.*

Multi-range indicators. An instrument with multi-range indicators has *different scales for different ranges;* The choice of a particular scale is made by a selector switch.

1. ***Single-point indicators:***

The *fixed-scale and movable-pointer indicators,* available in a variety of forms, are shown in Fig. 3.2(*a*).

Fig. 3.2(*b*) shows the fixed-pointer and movable-scale indicators.

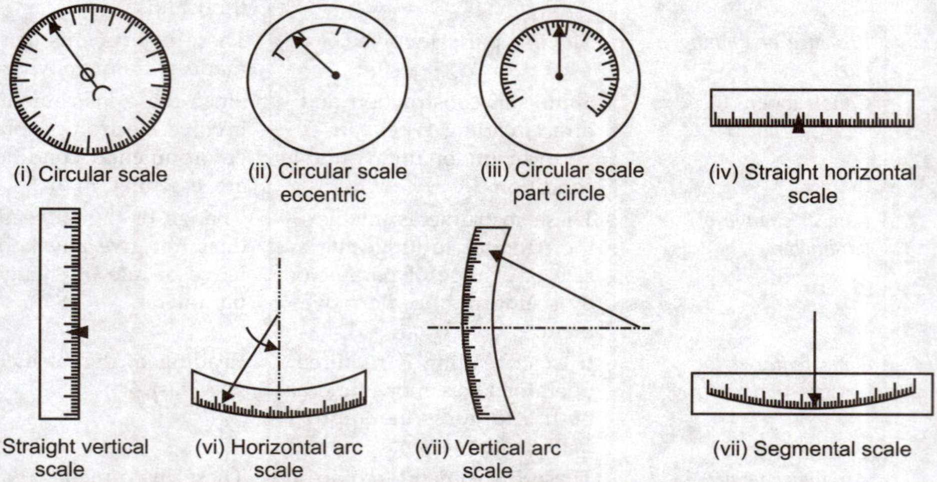

Fig. 3.2 (a). Fixed-scale and movable-pointer indicators.

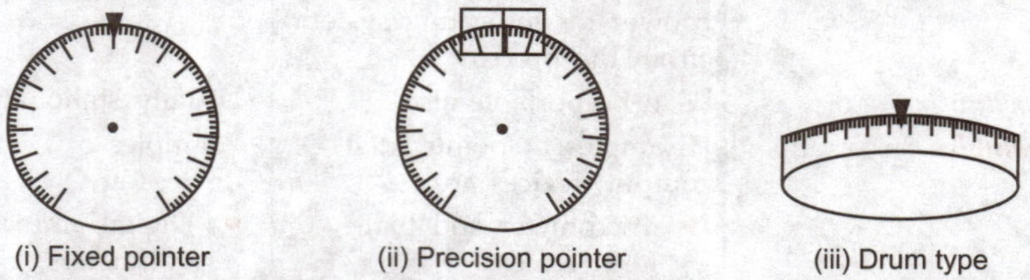

Fig. 3.2 (b). Fixed-pointer and movable-scale indicators.

The readability of graduated dials is influenced by the following factors :

- The shape and length of the pointer.
- The number, spacing, length and thickness of scale marking.
- The system of numbering of the scale marks.
- The size and design of the numerals.

2. ***Multi-point, multi-pointer and multi-range indicators:***

Multi-point indicator. In this system, the indicator pointer can be connected to a number of inputs, one at a time with the help of a selector switch.

The selector switch may be operated either manually or automatically after a pre-determined time. The observed reading is multiplied by a factor corresponding to the particular measurand.

Generally, such systems are confined where measurable variables are of electrical signals as the selection is accomplished by switching electrical circuits. However, gas selector switches also exist which connect one gas pipe at a time to the measuring instrument and are well-designed to avoid leakage of gas.

3.6 ELECTRICAL INSTRUMENTS

3.6.1. Introduction and Classification

The instruments used for all electrical measurements are called *measuring instruments.* They include *ammeters, voltmeters, wattmeters, energy meters* etc. The various electrical instruments may broadly be divided into two groups :

1. **Absolute instruments.** Absolute instruments are those instruments which indicate the quantity to be measured in terms of the *constants of the instrument* (dimensions, turns etc.) and in order to find out the quantity in the practical units it is necessary to *multiply such deflections with an instrument constant.*

- *No previous calibration or comparison is necessary in this case.*
- The most common absolute instrument is *tangent* galvanometer which gives the measured current in terms of tangent of the deflected angle, the radius and the number of turns of the galvanometer.
- Such insruments are rarely used (the use being merely confined within laboratories as standardising instruments).

2. **Secondary instruments.** Secondary instruments are those in which the value of electrical quantity to be measured can be determined from the *deflection of instrument only* when they have been *pre-calibrated* by comparison with an absolute instrument.

- The deflection of the instrument gives directly the quantity to be measured.
- These instruments are most generally used in everyday work.

Secondary instruments may also be *classified* as follows :

1. **Indicating instruments.** Indicating instruments are those which indicate the instantaneous value of the electrical quantity being measured at the time at which it is being measured. Their indications are given by pointers moving over calibrated scales.

Example. *Ammeters, voltmeters* and *wattmeters.*

2. **Recording instruments.** Recording instruments are those which give a *continuous record* of the variations of an electrical quantity over a selected period of time. The pointer in these types of instruments is an infed pen which leaves a trace on a paper put over a moving drum.

3. **Integrating instruments.** Integrating instruments are those which measure the total quantity of electricity delivered in a particular time.

Example. *Ampere-hour and watt-hour meters.*

Electrical measuring instruments may also be *classified* as follows :

1. **According to the quantity being measured :**

Ammeters. ...... for measuring the magnitude of current.

Voltmeters. ...... for measuring voltages.

Ohmmeters and resistance bridges. ...... for measuring resistances.

Wattmeters. ...... for power measurements.

Watt-hour meters. ...... for energy measurements.

Frequency meters. ...... for frequency measurements.

Power factor meters. ...... for power-factor measurements.

2. **According to the kind of current :**

Instruments are classified into *D.C., A.C.* and *A.C./D.C. instruments.*

3. **According to accuracy limits :**

4. **According to the principle of operation :**

Instruments are grouped into :

— Moving coil
— Moving iron
— Electrodynamic
— Induction
— Hot-wire
— Thermoelectric
— Rectifier types.

5. **According to the type of indication :**

Instruments may be :

— Indicating type
— Recording type.

6. **According to application :**

— Switch board
— Portable.

3.6.2. Electrical Principles of Operation

All electrical measuring instruments depend for their action on any of many physical effects of electric current or potential. The following are the effects generally used in their manufacture :

(*i*) **Magnetic effect.** ...... Voltmeters, ammeters, wattmetes, power factor meters etc.

(*ii*) **Thermal effect.** ...... Ammeters, voltmeters, maximum demand meters etc.

(*iii*) **Chemical effect.** ...... D.C. ampere hour meters (integrating meters).

(*iv*) **Electrostatic effect.** ...... Voltmeters which can indirectly be used as ammeters and wattmeters.

(*v*) **Electro-magnetic induction effect.** .... Voltmeters, ammeters, wattmeters and integrating meters used in A.C. only.

• **Difference between an ammeter and a voltmeter :**

An ammeter and a voltmeter work on the same principle. The ammeter has a *low resistance* so that when it is connected in **series** with any circuit, it does not change the current. The voltmeter has a *high resistance* and it is so designed that when connected in **parallel** to the circuit for measuring voltages it does not take appreciable current.

An ammeter of low range can be used as a voltmeter by connecting an external resistance in series with it.

3.6.3. Essential Features of Indicating Instruments

Indicating instruments possess *three essential features* :

1. **Deflecting device.** ...... whereby a mechanical force is produced by the electric current, voltage or power.
2. **Controlling device.** ...... whereby the value of deflection is dependent upon the magnitude of the quantity being measured.

3. **Damping device.** ...... to prevent oscillation of the the moving system and enable the latter to reach its final position quickly.

3.6.3.1. Deflecting device

A deflecting device produces a deflecting torque which is caused by anyone of these effects: thermal effect, chemical effect, electrostatic effect etc.; with the help of this deflecting torque the needle or the pointer moves from zero position to the final position. The arrangement of the deflecting device with each type of instrument will be discussed individually.

3.6.3.2. Controlling devices

These are two types of controlling devices :

(*i*) Spring control. (*ii*) Gravity control.

(*i*) ***Spring control.*** Fig. 3.3 shows a commonly used spring control arrangement. It utilises two spiral hair springs, 1 and 2, the inner ends of which are attached to the spindle S. The outer end of spring 2 is fixed while that of 1 is attached to a lever, the adjustment of which gives zero adjustments. The two springs 1 and 2 are wound in *opposite directions* so that when the moving system is deflected, one spring winds up while the other unwinds, and the controlling torque is due to the *combined torsions* of the springs.

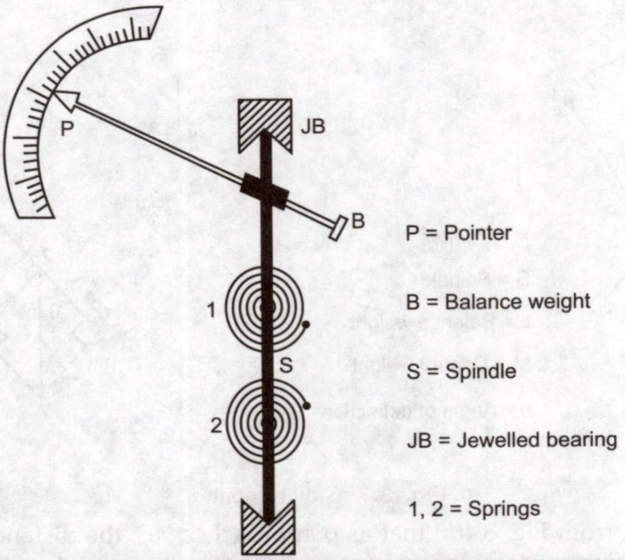

Fig. 3.3. Spring control.

Since the torsional torque of a spiral spring is proportional to the angle of twist, the controlling torque (T_c) is directly proportional to the angular deflection of the pointer (θ).

i.e., $T_c \propto \theta$

The spring material should have the following *properties* :

(*i*) It should be non-magnetic.

(*ii*) It must be of low temperature co-efficient.

(*iii*) It should have low specific resistance.

(*iv*) It should not be subjected to fatigue.

- The exact expression for controlling the torque is :

$$T_c = k\,\theta \qquad ...(3.1)$$

where k is spring constant (or control constant, or torsion constant or restoring constant). Its value is given by :

$$k = \frac{Ebt^3}{12l} \text{ N-m/rad.; the angle } \theta \text{ is in radians}$$

where, E = Young's modulus of spring material,

b = Width of spring (m),
t = Thickness of spring (m), and
l = Length of spring (m).
} *Flat spiral spring*

(*ii*) **Gravity control.** With gravity control, weights L and M are attached to the spindle S [Fig. 3.4 (*a*)], the function of L being to *balance* the weight of the pointer P. Weight M therefore provides the controlling torque. When the pointer is at zero, M hangs vertically downwards. When P is deflected through angle θ, the controlling torque is equal to (weight of M × distance d) and is therefore proportional to the *sine* of the angular deflection [Fig. 3.4 (*b*)]

i.e., $T_c \propto \sin \theta$

The degree of control is adjusted by screwing the weight up or down the carrying system.

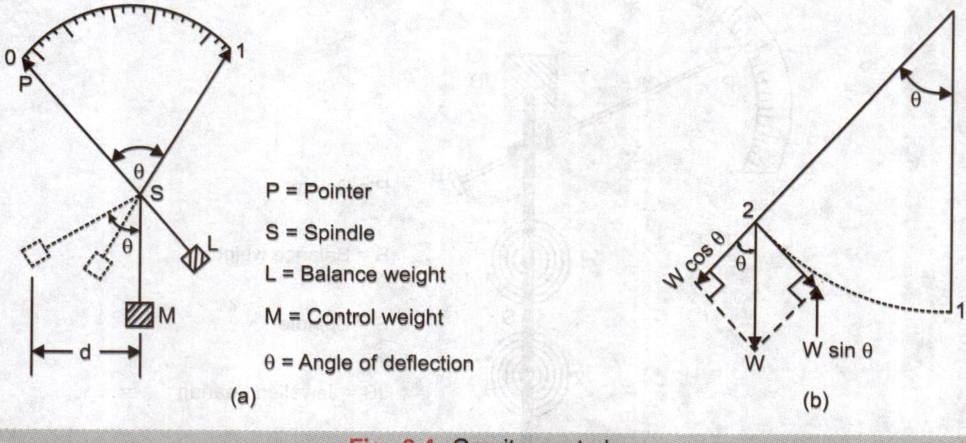

Fig. 3.4. Gravity control.

It may be seen from Fig. 3.4(*b*) that as θ approaches 90°, the distance 1-2 increases by a relatively small amount for a given change in the angle that when θ is just increasing from its zero value. Hence, *gravity-controlled instruments have scales which are not uniform but are cramped or crowded at their lower ends.*

Advantages :

1. The gravity controlled instrument is cheaper than corresponding spring-controlled instrument.
2. It is not subjected to fatigue.
3. It is unaffected by temperature.

Disdvantages:

1. Gravity control gives a cramped scale.
2. The instrument must be levelled before use.

Example 3.1. *The length, width and thickness of the control spring of an instrument are 390 mm, 0.52 mm and 0.075 mm. The Young's modulus of the spring material is 110 GN/m². Determine the torque exerted by the spring when it is turned through 80°.*

Solution. *Given :* l = 390 mm; b = 0.52 mm; t = 0.075 mm; T_c = 110 GN/m²;

$\theta = 80° = 80 \times \frac{\pi}{180}$ rad. or 1.396 rad.

We know that, controlling torque,

$$T_c = \frac{Ebt^3}{12l} \cdot \theta \qquad \text{[Eqn. (3.1)]}$$

$$= \frac{(110 \times 10^9) \times (0.52 \times 10^{-3}) \times (0.075 \times 10^{-3})^3}{12 \times (390 \times 10^{-3})} \times 1.396$$

$= \mathbf{5.156 \times 10^{-6}}$ **Nm** **(Ans.)**

Example 3.2. *The toque of an ammeter varies as the square of the current through it. If a current of 5A produces a deflection of 90°, what deflection will occur for a current of 3A when the instrument is :*

(i) Spring-controlled; *(ii) Gravity controlled.* **(Jadavpur University)**

Solution. Since deflecting torque varies as (current)², we have $T_d \propto I^2$

For spring control : $T_c \propto \theta$ $\quad\therefore\quad$ $\theta \propto I^2$

For gravity control : $T_c \propto \sin\theta$ $\quad\therefore\quad$ $\sin\theta \propto I^2$

(i) ***For spring control :*** $90° \propto 5^2$ and $\theta° \propto 3^2$

$$\theta = 90° \times \frac{3^2}{5^2} = \mathbf{32.4°} \quad \textbf{(Ans.)}$$

(ii) ***For gravity control :*** $\sin 90° \propto 5^2$, and $\sin\theta° \propto 3^2$

$$\therefore \quad \frac{\sin\theta}{\sin 90°} = \frac{3^2}{5^2} = \frac{9}{25}$$

or,
$$\sin\theta = \frac{9}{25}$$

$$\therefore \quad \theta = \sin^{-1}(9/25) = \mathbf{21.2°} \quad \textbf{(Ans.)}$$

Example 3.3. *In a spring controlled indicating instrument the control is exerted by two springs. The allowable maximum stress is 30 MN/m² and Young's modulus of spring material is 120 GN/m².*

Determine the suitable dimensions for spring if the deflecting torque corresponding to a full scale deflection of 90° is 1.2 × 10⁻³ Nm. Assume the width of spring strip as 0.6 mm.

(Roorkee University)

Solution. *Given :* Allowable maximum stress, σ_{max} = 30 MN/m²; E = 120 GN/m²;

$T_d = 1.2 \times 10^{-3}$ Nm; $\theta = 90°$ or $\frac{\pi}{2}$ rad.; b = 0.6 mm

Dimensions of spring, *l, t* :

As there are two controlling springs, the controlling torque (T_c) caused by each spring is equal to half of the deflecting torque (T_d) *i.e.*, controlling torque of each spring,

$$T_c = \frac{1.2 \times 10^{-3}}{2} = 0.6 \times 10^{-3} \text{ Nm} \qquad ...(i)$$

We know that, $T_c = \dfrac{Ebt^3}{12l}\theta$ [Eqn. (3.1)]

or, $$\frac{t^3}{l} = \frac{12T_c}{Eb\theta} = \frac{12\times 0.6\times 10^{-3}}{(120\times 10^9)\times(0.6\times 10^{-3})\times(\pi/2)} = 0.6366 \times 10^{-10} \quad ...(i)$$

The maximum stress (σ_{max}) developed in the spring must be well below the elastic limit of the spring material at the maximum deflection of the moving system in order to avoid failure and to preserve long period stability; it is given by the expression :

$$\frac{l}{t} = \frac{E\theta}{2\sigma_{max}}$$

or, $$\frac{l}{t} = \frac{120\times 10^9 \times(\pi/2)}{2\times(30\times 10^6)} = 3142 \quad ...(ii)$$

Multiplying (*i*) and (*ii*), we get

$$t^2 = 0.6366 \times 10^{-10} \times 3142 = 0.2 \times 10^{-6}$$

or, *Thickness of each strip*, $t = \sqrt{0.2\times 10^{-6}} \simeq 4.5 \times 10^{-4}$ m or **0.45 mm** **(Ans.)**

Substituting this value of t in eqn. (*ii*) we get

$$l = 4.5 \times 10^{-4} \times 3142 = \mathbf{1.414\ m}$$ **(Ans.)**

3.6.3.3. Damping devices

Owing to the inertia of the moving system, when subjected to the deflecting and restoring torques, number of vibrations will be produced before finally coming to rest. To-avoid this, a *damping torque* is required which opposes the motion and ceases when the pointer comes to rest. The degree of damping should be adjusted to a value which is sufficient to enable the pointer to rise quickly to its deflected position without overshooting. In that case, the instrument is said to be *dead-beat*. If the instrument is over-damped the movement is very slow (and the instrument becomes lethargic) as shown in Fig. 3.5.

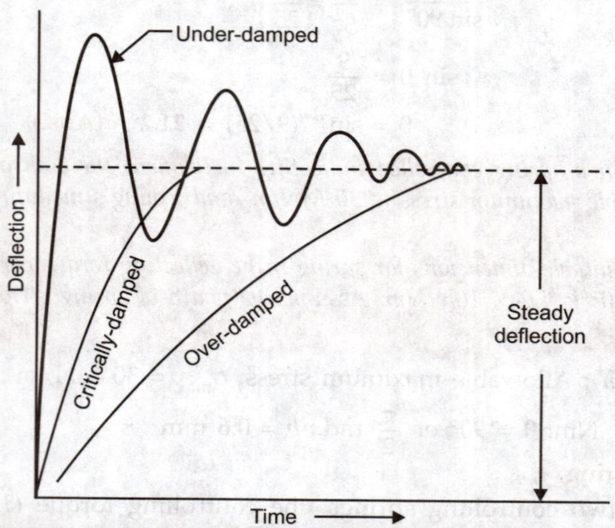

Fig. 3.5. Damping curves.

Damping can be provided by the following *methods* :

1. Air damping.

2. Eddy current damping.
3. Fluid friction damping.

1. ***Air damping.*** Fig. 3.6 shows an arrangement for obtaining air damping. It consists of a thin metal vane *MV* attached to the spindle *S*; the vane moves in a sector-shaped box *B*. *Any tendency of the moving system to oscillate is damped by the action of the air on the vane.*

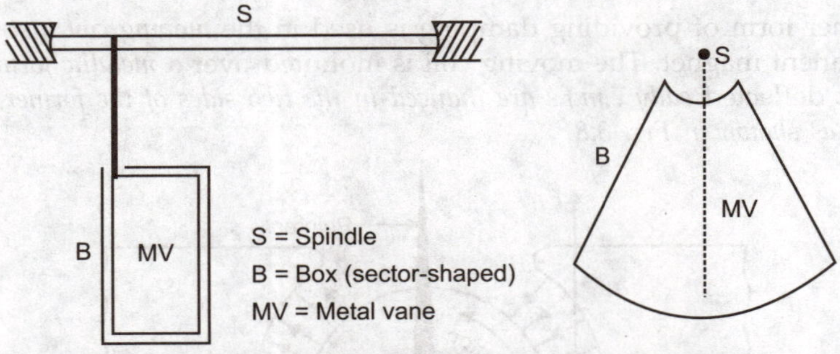

Fig. 3.6. Eddy current damping.

2. ***Eddy current damping.*** This method of damping is based on the *principle* that when a conducting, non-magnetic material is moved in a magnetic field, an e.m.f. is induced in it which causes currents called the *eddy-currents*. Due to these eddy currents, a force exists between them and the field. Due to Lenz's law this force is always in opposition to the force causing rotation of the conducting material, thus, it provides the necessary damping.

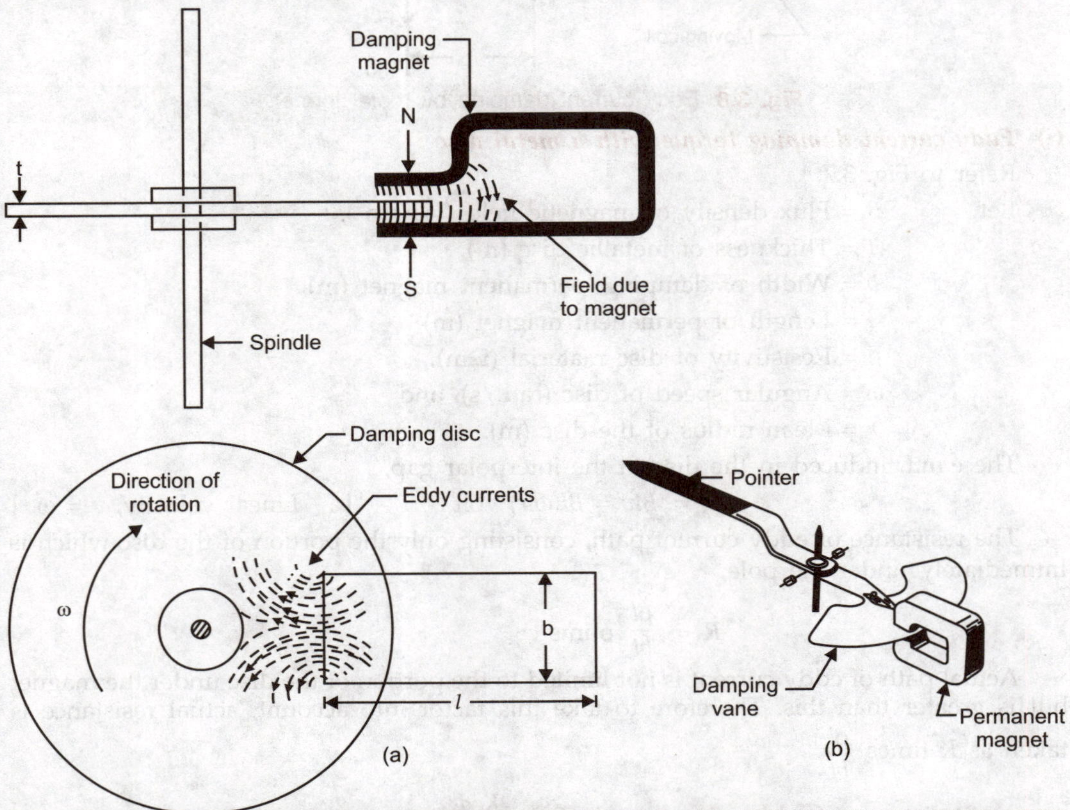

Fig. 3.7. Eddy current damping by metallic disc.

- One form of eddy-current damping is shown in Fig. 3.7. Here a copper or aluminium disc, carried by a spindle, can move between the poles of permanent magnet. If the disc moves clockwise, the e.m.f.s induced in the disc circulate eddy currents as shown dotted. It follows from Lenz's law that these currents exert a force *opposing* the motion producing them, namely the clockwise movement of the disc.
- Another form of providing damping is used in the *moving coil instruments* using permanent magnet. The moving coil is mounted over a *metallic former*. When the coil is deflected *eddy e.m.f.s are induced in the two sides of the former, causing eddy forces as shown in Fig. 3.8.*

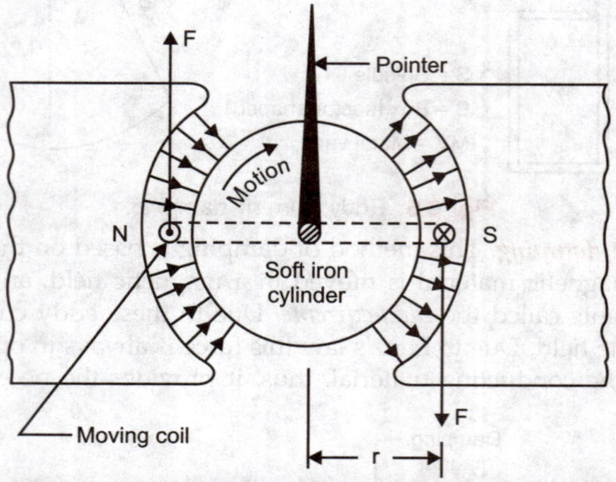

Fig. 3.8. Eddy current damping by metal former.

(i) Eddy current damping torque with a metal disc :

Refer to Fig. 3.7.

Let, B = Flux density of magnetic field (Wb/m^2),

t = Thickness of metallic disc (m),

b = Width of damping/permanent magnet (m),

l = Length of permanent magnet (m),

ρ = Resistivity of disc material (Ωm),

ω = Angular speed of disc (rad./s), and

r = Mean radius of the disc (m).

The e.m.f. induced in the disc in the interpolar gap,

$$E_e = Blv = Bl(\omega.r) \text{ volt} \qquad [\because \text{ Linear velocity, } v = \omega.r]$$

The resistance of eddy current path, consisting only the portion of the disc which is immediately under the pole,

$$R = \frac{\rho l}{bt} \text{ ohms}$$

Actual path of eddy current is not limited to the portion of the disc under the magnet but is greater than this. Therefore to take this factor into account, actual resistance is taken as K times $\frac{\rho l}{bt}$.

$\therefore$ Resistance of eddy current path, $R_e = \frac{K\rho d}{bt}$

where, K = a constant which depends upon radial position of poles, it is *always greater than unity.*

$$\text{Eddy current, } I_e = \frac{E_e}{R_e} = \frac{Bl\omega r}{K\rho l/(bt)} = \frac{Brbt\omega}{K\rho}$$

$$\textit{Damping force, } F_D = BI_e l = B \times \frac{Brbt\omega}{K\rho} \times l = \frac{B^2 rlbt\omega}{K\rho} \quad ...(3.2)$$

$$\textit{Damping torque, } T_d = F_D \times r = \frac{B^2 rlbt\omega}{K\rho} \times r = \frac{B^2 r^2 lbt\omega}{K\rho} \quad ...(3.3)$$

$$\textit{Damping constant, } k_D = \frac{T_D}{\omega} = \frac{B^2 r^2 lbt}{K\rho} \text{ Nm/rad. s}^{-1} \quad ...(3.4)$$

— The magnitude of damping torque can be varied by adjusting the radial position of the poles; damping torque (T_D) decreases with the movement of the magnet towards the edge of the disc and becomes zero when the centres of the poles are at the edge of the disc.

— The above phenomenon is employed in the induction type instruments by placing a damping magnet (permanent type) at the opposite side of the disc from the electromagnet, so that the disc *serves for damping as well as for operating purposes.*

(ii) Eddy current damping with a metal former :

Fig. 3.8, shows a metallic former moving in the field of a permanent magnet.

Let, B = Strength of magnetic field (Wb/m^2),
l = Length of the former (m),
d = Diameter of the former (m),
b = Width of the former (m),
t = Thickness of the former (m), and
ω = Angular speed of former (rad./s).

$$\text{Linear velocity of former, } v = \omega \times \frac{d}{2}$$

$$\text{E.m.f. generated in the former, } E_e = 2Blv = 2Bl\left(\omega \times \frac{d}{2}\right) = Bld\omega$$

Resistance of path of eddy currents,

$$R_e = \frac{\text{Resistivity} \times \text{Length}}{\text{Area}} = \frac{\rho \times 2(l+d)}{b \times t} = \frac{2\rho(l+d)}{bt}$$

$$\text{Eddy current, } I_e = \frac{E_e}{R_e} = \frac{Bld\omega}{2\rho(l+d)/(bt)} = \frac{Blbtd\omega}{2\rho(l+d)}$$

$$\therefore \textit{ Damping force, } F_d = BI_e l = B \times \frac{Blbtd\omega}{2\rho(l+d)} = \frac{B^2 l^2 btd\omega}{2\rho(l+d)} \quad ...(3.5)$$

$$\textit{Damping torque, } T_D = F_D \times d = \frac{B^2 l^2 btd\omega}{2\rho(l+d)} \times d = \frac{B^2 l^2 d^2 bt\omega}{2\rho(l+d)} \quad ...(3.6)$$

$$\textit{Damping constant, } \mathbf{K_D} = \frac{T_D}{\omega} = \frac{B^2 l^2 d^2 bt}{2\rho(l+d)} \text{ Nm/rad. s}^{-1} \quad ...(3.7)$$

— It is obvious from eqn. (3.6) that damping torque (T_D) can be varied by varying the thickness (t) of the metal former.

3. ***Fluid friction damping.*** Fig. 3.9 shows the method of fluid friction damping. Here light vanes are attached to the spindle of the moving system. The vanes are dipped into a pot of damping oil and are completely submerged by the oil. The motion of the moving system is always opposed by the friction of the damping oil on the vanes. The damping force thus created always increases with the increase in velocity of vanes. There is no damping force when the vanes are stationary.

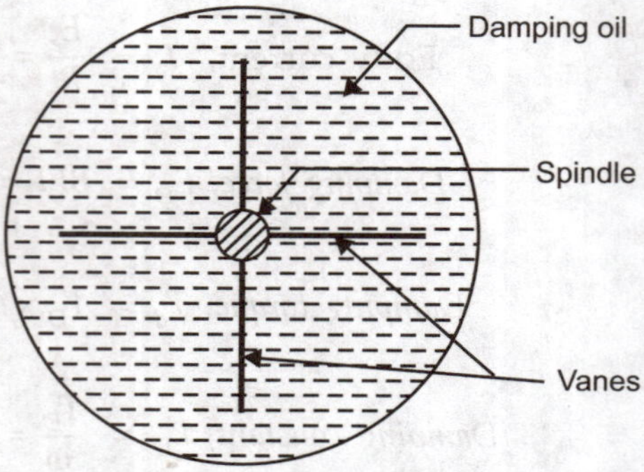

Fig. 3.9. Fluid friction damping.

The damping oil used must have the following *properties* :

(*i*) Must be a good insulator.
(*ii*) Should be non-evaporating.
(*iii*) Should not have corrosive action upon the metal of the vane.
(*iv*) The viscosity of the oil should not change with the temperature.

Though in this method of damping, no case is required as in the air friction damping but it is not much used due to the following *disadvantages* :

(*i*) Objectionable creeping of oil.
(*ii*) Using the instrument always in vertical position and its obvious unsuitability for use in portable instruments.

☞ The various type of indicating instruments used for measurement of current, voltage, power and energy along with suitability for the *type of measurement, type of control, type of damping, special features and applications are given below in tabular form :*

S.No.	*Type of instrument*	*Suitability for type of measurement*	*Type of control*	*Type of damping*	*Remarks*
1.	***Moving iron***	D.C. or A.C. measurements (current and voltage)	Spring or gravity control	Air friction	• Cheapter to manufacture. • Very accurate on A.C. or D.C., if properly designed. • *Mostly used as an indicating instrument.*
2.	***Moving coil*** (*i*) *Permanent magnet type*	D.C. or A.C. measurements (current and voltage only)	Spring control	Eddy current	• Most accurate type for D.C. measurements. • *Most widely used for measurement of D.C. voltage, current and resistance.*
	(*ii*) *Dynamometer type*	D.C. or A.C. measurements	Spring control	Air friction	• Mainly used as *wattmeter*. • Also used as standard meter for calibration and as transfer instrument.

3.	***Thermal*** (*i*) *Hot wire*	D.C. or A.C. measurements (current, voltage, and power)	Spring control	Eddy current	• Same calibration for both D.C. and A.C. • Free from errors due to frequency, wave form and external field when used on A.C., therefore, these are particularly used for A.C. requirements. Since its disadvantages outweigh its advantages, hence it is *now little used.*
	(*ii*) *Thermo-couple*	D.C. or A.C. measurements (current and voltage)	Spring control	Eddy current	• Free from errors due to frequency, wave form and external field when used on A.C. • *Used for measurement of current and voltage at power frequencies upto 100 MHz.*
4.	***Electrostatic***	D.C. or A.C. (voltage only)	Spring or gravity control	Air friction	• Very low power consumption and can be made to cover a large range of volts (above 500 V).
5.	***Induction***	A.C. measurements (current, voltage, power and energy) only	Spring control	Eddy current	• Expensive and degree of accuracy not high. • Mainly used for *measurement of power and energy in A.C. circuits.*
6.	***Rectifiers***	D.C. or A.C.	Spring control	Eddy current	• These are PMMC instruments used in conjunction with rectifying devices for A.C. measurements (current and voltage) from about 20 Hz to 20 kHz

3.7 GALVANOMETERS

A **galvanometer** *is an instrument used for detecting presence of small currents or voltages in a closed circuit or for measuring their magnitudes.*

However, measurement of magnitude is avoided as much as possible, for accuracy in reading a deflection is seriously limited.

Several types of galvanometers have been devised, including *moving coil* and *magnet types.* Special instruments are needed for some purposes, but the great majority of the galvanometers in use now are the *permanent-magnet moving coil type, normally referred as* **"D' Arsonval type"**. This is *D.C. galvanometer.*

Other types of galvanometers are :

Ballistic galvanometer. *It is used to measure the quantity of electricity (charge) passed through it.* In "*magnetic measurements*", this quantity of electricity is due to an instantaneous e.m.f. induced in a search coil connected across the ballistic galvanometer.

The main advantage of *moving coil instrument lies in its freedom from the effects of stray fields and the ease with which a long and uniform scale may be attained.* These are the features that have caused it to be adopted, in a modified form, for D.C. ammeters and voltmeters.

Vibration galvanometer. These galvanometers are most widely used as "*tuned detectors*". These galvanometers are of the following types :

(*i*) Moving magnet type.

(*ii*) Moving coil type.

— The moving coil type galvanometer is more generally used, the moving magnet having the disadvantage of being seriously affected by magnetic fields of the resonance frequency, unless adequately screened.

Galvanometers find their principal applications in *bridge* and *potentiometer measurements.* Therefore, a galvanometer in addition to being sensitive, should have :

— A stable zero;

— A short periodic time;

— Nearly critical damping.

3.7.1. D'Arsonval Galvanometer

3.7.1.1. Construction and Working

Fig. 3.10 shows the essential features of a D'Arsonval galvanometer :

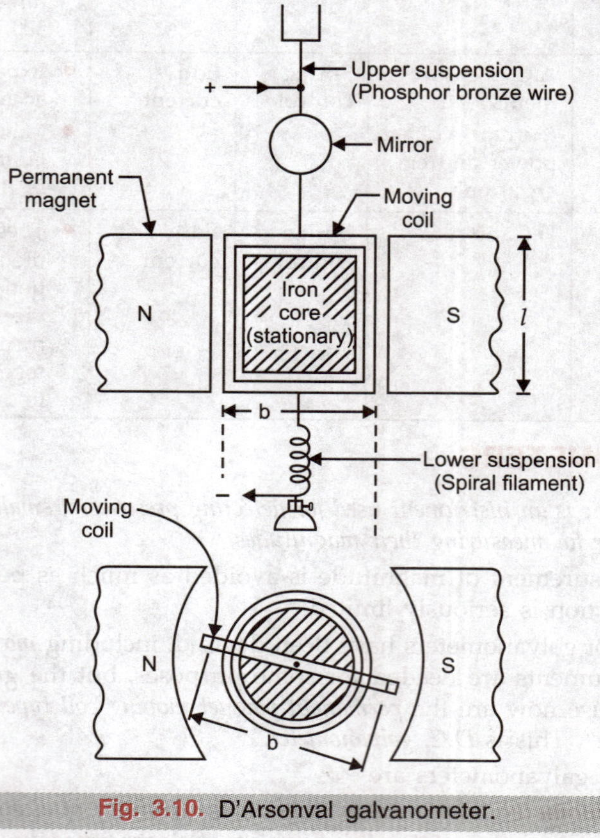

Fig. 3.10. D'Arsonval galvanometer.

Construction :

— It consists of a circular or rectangular coil of very fine insulated wire with many turns usually wound on an aluminium frame or bobbin suspended between the poles of a permanent magnet.

— The coil is suspended by a phosphor bronze filament which acts as one lead-in for the coil. The other lead-in wire is a very flexible spiral wire at the bottom of the coil. *The elasticity of the filament sets up a moderate torque in opposition to the rotation of the coil.*

— The poles of the magnet are usually cylindrical in shape, and a cylindrical soft iron core is placed between the poles and coaxial, with them.

The cylindrical form of poles faces and core possesses the following *advantages* :

(*i*) The length of air gap is reduced so that the amount of flux linking the coil is increased, thus increasing the sensitivity of the instrument;

(*ii*) The flux between the pole faces and core is practically radial. In radial field the deflection of coil is directly proportional to the current in the coil and, therefore, a uniform scale is obtained.

Working/Operation :

— When current flows through the coil, a deflecting force proportional to the flux density, the coil current, and dimensions of the coil makes the coil to rotate on its vertical axis.

— The "*deflecting force is opposed by the restraining force*" of the suspension filament so that coil does not continue to rotate as in a motor but turns until the deflecting torque is *balanced* by the restraining force.

— Since the deflecting torque is directly proportional to the coil current, therefore, the amount of deflection of the coil indicates the magnitude of current flowing in the coil. The amount of deflection and hence the magnitude of curent flowing through the coil may be indicated by a pointer attached to the moving element and moving over a calibrated scale.

For the adjustment of coil and also for *zero adjustment*, a "*torsion head*" is provided.

— In a *sensitive galvanometer*, a small "*mirror*" is attached to the moving coil to indicate the position of the coil. This arrangement may be used in *two* ways : (*i*) To mount a telescope and scale in front of the galvanometer so that the observer reads the scale reflected by the mirror; (*ii*) Another arrangement (more common and less tiring to the user) consists of a lamp and scale device in which a spot of light is projected on the mirror and reflected to the scale. The scale is usually mounted 1 m away from the mirror although 0.5 m is used sometimes for compactness.

• There is a "*damping torque*" present owing to production of eddy currents in the metal former on which the coil is mounted. Damping is also obtained by connecting a low resistance across the galvanometer terminals. Damping torque depends upon the resistance and we *can obtain critical damping by adjusting the value of resistance.*

3.7.1.2. Torque equation

Refer to Fig. 3.10 :

Let, N = Number of turns in coil,

B = Flux density, Wb/m^2,

I = Current through the coil,

l = Length of coil side (vertical), m

b = Width of coil, m,

θ = Deflection of pointer, rad., and

C = Spring (or controlling or restoring) constant, Nm/rad.

Force on each side of coil = BIl newtons [∵ Field is radial]

Since the coil has N turns and each turn has two sides, therefore,

$$\text{Total force} = 2NBIl \text{ newtons}$$

This force acts at a radius $\frac{b}{2}$.

∴ *Deflecting torque,* T_d = Force × radius

$$\text{or,} \quad T_d = 2NBIl \times \frac{b}{2} = NBIlb = NBIA \text{ Nm} \quad ...(3.8)$$

where, A = Area of the coil ($= l \times b$), m^2

N, B, A are constants for a galvanometer.

$$\therefore \text{ Deflecting torque, } T_d = GI \quad ...(3.9)$$

G is called the *displacement constant* (= NBA) of galvanometer. *Controlling torque* exerted by the suspension at deflection θ,

$$T_c = C\theta \quad ...(3.10)$$

For steady deflection :

$$T_d = T_c \quad ...(3.11)$$

$$GI = C\theta$$

$$\text{or,} \quad \theta = \frac{GI}{C} = \left(= \frac{NBIA}{C} \right) \quad ...[3.11(a)]$$

Several times the deflection is measured on a scale kept at a distance r metres from a mirror. A beam of light is reflected on the scale by a small mirror which is mounted on the moving system (Fig. 3.11). The distance r is usually kept as 1 metre. If the moving system rotates through angle θ, the light beam is turned through angle 2θ. Then,

Deflection *oa i.e.*, $d = (r) \times (2\theta)$ metres.

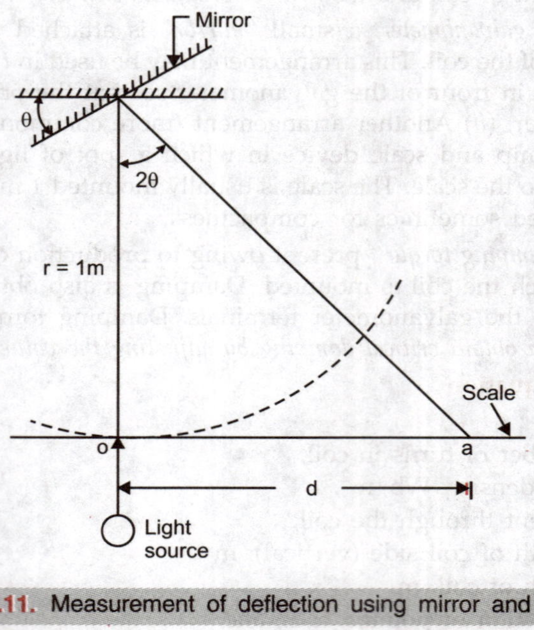

Fig. 3.11. Measurement of deflection using mirror and scale.

3.7.1.3. Galvanometer Constants

The galvanometer has *four* constants known as *"Intrinsic constants."* These are :

1. Displacement constant.
2. Inertia constant.
3. Damping constant.
4. Control or restoration constant.

1. *Displacement constant :*

- The deflection torque is given by $T_d = GI$ (eqn. 3.9) where G is the *displacement constant* of the galvanometer and is equal to $NBlb$.
- The units of G are Nm/A.

2. *Inertia constant :*

- Owing to inertia of moving system, a *retarding torque* is produced.
- This torque (T_j) is dependent upon the moment of inertia of moving system and the angular acceleration.

$$T_j = J \cdot \frac{d^2\theta}{dt^2} \quad \text{...(3.12)}$$

where, J = Moment of inertia of moving system about the axis of rotation, kg-m^2, and

θ = Deflection at any time t, rad.

3. *Damping constant :*

- *Damping* is provided by friction due to motion of the coil air and also by induced electrical effects if a closed circuit is provided.
- *Damping torque* (T_D) is usually assumed to be proportional to the angular velocity of the system and may be expressed as :

$$T_D = D\frac{d\theta}{dt} \quad \text{...(3.13)}$$

where, D = Damping constant.

4. *Control or restoration constant :*

- A *controlling torque* (T_c) is produced due to *elasticity* of the system which tries to restore the moving system back to its original position, and may be expressed as :

$$T_c = C\theta$$

where, C = Control constant, Nm/rad.

3.7.1.4. Dynamic response

The deflecting torque T_d (*GI*) *causes* the motion, while the inertia torque $\left(T_j = J\frac{d^2\theta}{dt^2}\right)$, damping torque $\left(T_D = D\frac{d\theta}{dt}\right)$ and controlling torque ($T_c = C\theta$) opposite the motion. The *"equation of motion"* is given as :

$$T_j + T_D + T_C = T_d$$

or,

$$\underbrace{J\frac{d^2\theta}{dt^2}}_{\text{Inertia torque}} + \underbrace{D\frac{d\theta}{dt}}_{\text{Damping torque}} + \underbrace{C\theta}_{\text{Controlling torque}} = \underbrace{GI}_{\text{Deflecting torque}} \quad \text{...(3.14)}$$

Eqn. (3.14) is well known *second order differential equation.* Use of Laplace transform provides an easy solution. Taking Laplace transforms of both sides and writing Laplace transform of response θ as θ(*s*), we get

$$\theta(s) = \frac{GI}{s(Js^2 + Ds + C)} = \frac{GI/C}{s\left[\frac{s^2}{\omega_n^2} + \frac{2\gamma s}{\omega_n} + 1\right]} \quad ...(3.15)$$

where, $\omega_n = \sqrt{\frac{C}{J}}$

and, $\gamma = \frac{D}{2J\omega_n}\left(= \frac{D}{\sqrt{2JC}}\right)$ ...[3.15(*a*)]

where, γ is known as *"damping ratio"*;

ω_n = *Natural* frequency of oscillation (rad./sec.), and

$$f_n = \frac{\omega_n}{2\pi}, \quad \text{and} \quad T = \frac{1}{f_n}$$

The characteristic equation is :

$$s^2 + 2\gamma\omega_n s + \omega_n^2 = 0 \quad ...[3.16(a)]$$

or, $$s^2 + \frac{D}{J}s + \frac{C}{J} = 0 \quad ...[3.16(b)]$$

or, $$s = \frac{-\frac{D}{J} \pm \sqrt{\frac{D^2}{J^2} - \frac{4C}{J}}}{2}$$

∴ The roots of characteristic equation are :

$$s_{1,2} = \frac{-D}{2J} \pm \sqrt{\frac{D^2}{4J^2} - \frac{C}{J}} \quad ...(3.17)$$

The actual response depends on the expression under the radical in eqn. (3.17)

1. If $\frac{D^2}{4J^2} < \frac{C}{J}$ or $D^2 < 4CJ$, the response is *underdamped;*
2. If $\frac{D^2}{4J^2} = \frac{C}{J}$ or $D^2 = 4CJ$, the response is *critically damped;*
3. If $\frac{D^2}{4J^2} > \frac{C}{J}$ or $D^2 > 4CJ$, the response is *overdamped.*

— When damping is *critical"*, the pointer reaches its final position $\theta_d \left(= \frac{GI}{C}\right)$ *without any oscillation.*

The ***damping ratio*** *γ is the ratio of actual value of D to the value of D for critical damping.*

i.e., $$\gamma = \frac{D_{actual} \text{ (or D)}}{D_{critical}} = \frac{D}{2\sqrt{CJ}} \quad ...(3.18)$$

or,
$$\frac{D}{2J} = \frac{\gamma(2\sqrt{CJ})}{2J} = \gamma\sqrt{\frac{C}{J}} = \gamma\omega_n \qquad ...(3.19)$$

and,
$$\sqrt{\frac{C}{J} - \left(\frac{D}{2J}\right)^2} = \sqrt{\omega_n^2 - \gamma^2\omega_n^2} = \omega_n\sqrt{1-\gamma^2} \qquad ...(3.20)$$

Therefore, the roots, s_1, s_2 in eqn. (3.17) can be written as :

$$s_{1,\,2} = -\gamma\omega_n \pm j\omega_n\sqrt{1-\gamma^2} \qquad ...(3.21)$$

The damped frequency of oscillation ω_d is :

$$\omega_d = \omega_n\sqrt{1-\gamma^2} \qquad ...(3.22)$$

or,
$$f_d = f_n\sqrt{1-\gamma^2} \qquad ...[3.22(a)]$$

Eqn. (3.21) can be written as :

$$s_{1,\,2} = -\gamma\omega_n \pm j\omega_d \qquad ...(3.23)$$

The *response of the three conditions of damping* is given by :

1. $$\theta = \frac{GI}{C}\left[1 - \frac{e^{-\gamma\omega_n t}}{\sqrt{1-\gamma^2}}\sin\left\{\omega_n\sqrt{(1-\gamma^2)}\,t + \sin^{-1}\sqrt{(1-\gamma^2)}\right\}\right] \qquad ...(3.24)$$

for $\gamma < 1$ *or underdamped motion.*

2. $$\theta = \frac{GI}{C}\left[1 - (1+\omega_n t)\,e^{-\omega_n t}\right] \qquad ...(3.25)$$

for $\gamma = 1$ *or critically damped motion.*

3. $$\theta = \frac{GI}{C}\left[1 - \frac{e^{-\gamma\omega_n t}}{\sqrt{\gamma^2-1}}\sinh\left\{\omega_n\sqrt{(\gamma^2-1)}\,t + \sinh^{-1}\sqrt{\gamma^2-1}\right\}\right] \qquad ...(3.26)$$

for $\gamma > 1$ *or overdamped motion.*

The system response for the underdamped, critically damped and overdamped cases is shown in Fig. 3.12.

☞ *Following observations are made :*

- When the system is *underdamped* the pointer makes a number of oscillations about the final position θ_d before setting down. Thus an underdamped response *leads to waste of time.*
- *Overdamped* response also leads to *wastage of time* because the system shows a sluggish behaviour.
- In *critical damping,* the time required for setting down to the final value θ_d is the *minimum.*

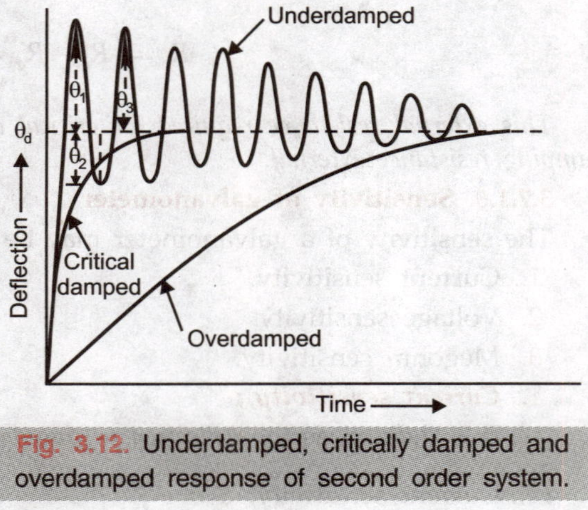

Fig. 3.12. Underdamped, critically damped and overdamped response of second order system.

3.7.1.5. Effect of galvanometer resistance on damping

Even when the galvanometer circuit is *open*, the *air friction and eddy current damping are present.*

When the galvanometer circuit is closed, a current I flows in this circuit. I is given by the equation :

$$IR = E - G\frac{d\theta}{dt} \qquad ...(3.27)$$

where, R = Resistance of galvanometer circuit,

E = Applied voltage, and

$G\frac{d\theta}{dt}$ = Back e.m.f. induced in the galvanometer coil due to its motion in the magnetic field. The constant, as defined earlier, equals NBA.

Now the equation of motion is :

$$J\frac{d^2\theta}{dt^2} + D'\frac{d\theta}{dt} + C\theta = \frac{G}{R}\left(E - G\frac{d\theta}{dt}\right) \qquad ...(3.28)$$

or, $$J\frac{d^2\theta}{dt^2} + \left(\frac{G^2}{R} + D'\right)\frac{d\theta}{dt} + C\theta = \frac{GE}{R} = GI \qquad ...(3.29)$$

The term $\left(\frac{G^2}{R} + D'\right)$ in eqn. (3.29) replaces the term D in eqn. (3.14). If *damping constant D' is small and negligible,* the galvanometer will be *critically damped* when :

$$\frac{G^2}{R} = \sqrt{4CJ} = 2\sqrt{CJ} \qquad ...(3.30)$$

∴ Resistance R for *critical damping* is given by,

$$R_c = \frac{G^2}{2\sqrt{CJ}} \qquad ...(3.31)$$

If R_g is the resistance of galvanometer, the external series resistance R_e required for *critical damping* is given as :

$$R_e = R_c - R_g = \frac{G^2}{2\sqrt{CJ}} - R_g \qquad ...(3.32)$$

This external resistance required for critical damping is called **CDRX** *or* **ECDR**, *critical damping resistance external.*

3.7.1.6. Sensitivity of galvanometer

The sensitivity of a galvanometer may be defined as follows :

1. Current sensitivity.
2. Voltage sensitivity.
3. Megohm sensitivity.

1. ***Current sensitivity :***

— It is defined as the *deflection produced by unit current.*

i.e., Current sensitivity, $S_I = \frac{\theta}{I}$

But, $$\theta = \frac{GI}{C} = \frac{G}{C} \text{ rad./A} \quad ...(3.33)$$

— In case of galvanometer attached with a *mirror* it may be defined as *deflection in mm* produced on a scale *at a distance of one metre from the galvanometer mirror by unit current.*

i.e. *Current sensitivity,* $$S_I = \frac{d}{I \times 10^6} \text{ mm per microampere} \quad ...(3.34)$$

where, d = The deflection on scale in mm.

— Sometimes the current density is defined as the *current required to cause a deflection of one scale division.*

2. *Voltage sensitivity :*

— It is defined in terms of *deflection in scale divisions caused by unit voltage applied across the galvanometer.*

— In case of a galvanometer attached with a *mirror*, the voltage sensitivity is defined *as the deflection produced in mm on a scale at a distance of 1 m from the mirror by a unit voltage impressed on the galvanometer.*

i.e., *Voltage sensitivity,* $$S_V = \frac{d}{V \times 10^6} \text{ mm}/\mu\text{V} \quad ...(3.35)$$

3. *Megohm sensitivity :*

— It is defined as the *resistance required in MΩ in the circuit to cause a deflection of one scale division on impressing 1 V.*

— In case of galvanometer attached with a *mirror* the megohm sensitivity is defined as the *resistance in MΩ placed in series with the galvanometer through which one volt of impressed voltage will cause a deflection of 1 mm on a scale at a distance of 1m from the mirror.* This is *reicprocal of current density.*

- *A sensitive galvanometer is one which produces a large deflection for a small current.* Hence, *for a sensitive galvanometer the displacement constant G should be large and control constant C should be small.*

— As displacement constant $G = NBA$, therefore, the sensitivity can be increased by increasing the number of turns, using high flux density and by increasing the dimensions of the coil. For a given coil area A (= lb) and a constant value of flux density B, the value of G can be increased by *increasing the number of turns.* If the winding area is fixed, a large number of turns can only be wound if we use *thin wire.* Hence *for a sensitive galvanometer* the value of resistance of coil R_g is large.

— The sensitivity can also be *increased* by making the control constant C *small.*

3.7.1.7. Types of D.C. galvanometer

D'Arsonval type galvanometers, which are D.C. galvanometers, are made in three general types :

1. Portable point type.
2. Laboratory reflecting type.
3. Box type.

1. *Portable point type :*

— These are small instruments, similar in construction to other types of portable instruments. They may be pivoted or may have taut suspension.

— This type of galvanometer has sensitivity from 0.5 to 4.0 mm/μA and has a period from 2.5 to 3 seconds.

— They have CDRX from 20 to 1800 ohms and coil resistance from 12 to 250 ohms.

● *Light spot and pointer D.C. galvanometer* has sensitivities from 50.4 to 252.0 mm/μA with period of 2 seconds.

It has CDRX from 100 to 3500 ohms and coil resistance from 100 to 1000 ohms.

2. ***Laboratory reflecting type :***

— These are highly sensitive instruments.

— Reflecting lamp and scale type have sensitivities from 250 to 100,000 mm/μA and period of 5 to 40 seconds.

— They have CRDX from 900 to 1000 kΩ and a coil resistance of 80 to 800 ohms.

3. ***Box type :***

— They have taut suspension and do not require exact levelling.

— The box includes the complete optical system and, therefore, it is not necessary to make separate mounting for galvanometer, lamp and scale etc. They are *self contained.*

— Their sensitivity ranges from 200 to 2000 mm/μA having a period of 2.5 to 3 seconds.

— Their CDRX is 400 to 25000 ohms having coil resistance of 25 to 550 ohms. The characteristics of different types of galvanometers are given in Table 3.1.

Table 3.1. Characteristic of Moving Coil Galvanometers

Type of galvanometer	*Sensitivity in mm/μA*	*Period in seconds*	*Resistance in ohms*	
			CDRX	*Coil*
Pointer	0.5 1.0 4.0	2.5 3 3	20 110 1800	12 25 250
Light spot and pointer	50.4 2520	2 2	100 3500	10 1000
Reflecting lamp and scale	250 2000 10,000 100,000	5 6 14 40	900 10,000 22,000 100,000	80 650 500 800
Box type self contained	200 2000	2.5 3	400 2500	25 550

3.7.1.8. Galvanometer shunt

In bridge and potentiometer measurements, during preliminary balancing work, the potential across the galvanometer circuit may be high due to unbalance. If galvanometer with sufficient sensitivity for final balance is used; the large unbalance that probably exists at the initial stage is likely to *damage the coil of the suspension.* In addition, excessive currents that may flow, send the light spot so far off the scale that the *deflection is not readable* and experimenter does not know what to do. Therefore, it is *necessary to use protective parallel resistances.* The simplex procedure is to use :

(*i*) A *high resistance in series with the galvanometer* to limit the current and therefore the deflection produced by the *available e.m.f. under conditions of unbalances;*

(ii) A *low resistance connected across the terminals of the galvanometer* thus by-passing most of the currents. This reduces the sensitivity of galvanometer circuit so that the response of the instrument to a given current is reduced.

— A *low resistance shunt will heavily overdamp the* instrument so that its motion will be *sluggish.*

— For both series and shunt resistances the *time required* to attain the final deflection is *excessive.*

— As a *general rule, a combination of series and a shunt resistance is preferable which will give the desired reduction in sensitivity and at the same time the proper damping characteristics.*

Ayrton universal shunt :

— **Fig. 3.13** shows an arrangement in which a combination of series and parallel resistances is incorporated. The lower part of the resistance (*i.e. BC* below the sliding contact) is in *parallel*, the upper part of resistance (*i.e. AB* above the sliding contact) is in *series* with the galvanometer.

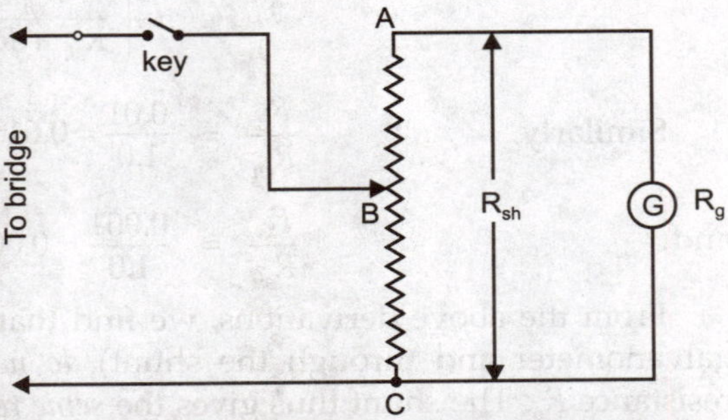

Fig. 3.13. Galvanometer shunt.

When the slider is near the lower end, a small portion of the resistance is in parallel with the galvanometer and the major part is in series with the galvanometer. The current divides itself in two parts, the major going through the shunt and only a small part through the galvanometer, thus protecting it and giving it small sensitivity.

When the sliding contact is moved up, the shunt resistance increases and the series resistance decreases. This increases the current through the galvanometer and thus the sensitivity is increased. But, there is a loss of sensitivity (as compared with galvanometer alone) with the sliding contact even at the top point, this loss, however, is not serious if R_{sh} is very high in comparison to galvanometer resistance R_g.

When key is *open* (see Fig. 3.13), there is closed circuit through the galvanometer coil and R_{sh} and this provides a damping circuit. R_{sh} should, accordingly be selected *to have a value approximately equal to that of critical damping resistance of the galvanometer* (somewhat higher is satisfactory) as the galvanometer response is excellent; if considerably too high, a fixed shunt can be added across R_{sh}. As CDRX is usually several times the coil resistance, the loss of sensitivity due to use of shunt of this value is *not serious.*

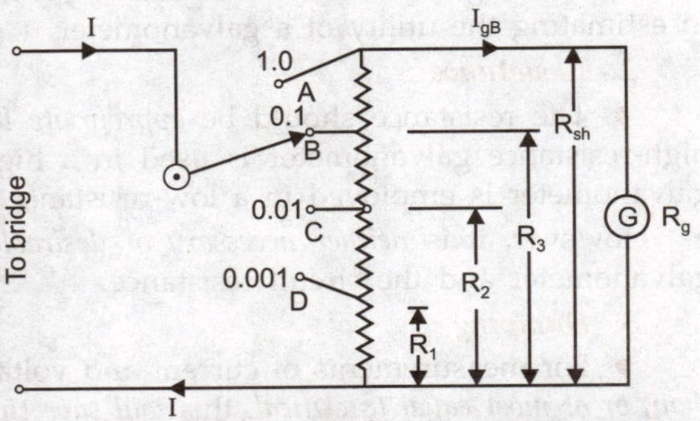

Fig. 3.14. Ayrton universal shunt.

— **Fig. 3.14** shows the construction of *"Ayrton universal shunt"*, using a dial type switch. The resistance of each shunt can be *proportioned* to give the desired result, as explained below :

When the contact is on stud *A* (or 1.0),

$$I_{gA} = I.\frac{R_{sh}}{R_{sh}+R_g}, \text{ where } I \text{ is the total current}$$

When the contact is on stud B (or 0.1),

$$I_{gB} = I.\frac{R_3}{R_3+(R_{sh}-R_3+R_g)} = I.\frac{R_3}{R_{sh}+R_g}$$

Thus, we have,
$$\frac{I_{gB}}{I_{gA}} = \frac{I.\left(\frac{R_3}{R_{sh}+R_g}\right)}{I.\left(\frac{R_{sh}}{R_{sh}+R_g}\right)} = \frac{R_3}{R_{sh}} = \frac{0.1}{1.0} = 0.1$$

Similarly,
$$\frac{R_2}{R_{sh}} = \frac{0.01}{1.0} = 0.01$$

and,
$$\frac{R_3}{R_{sh}} = \frac{0.001}{1.0} = 0.001$$

From the above derivations, we find that the relative values of current (through the galvanometer and through the shunt) *do not depend* upon the value of galvanometer resistance R_g. The shunt thus gives the *same relative current values* for the various steps for all galvanometers although *not the same fraction of the total current*; therefore, it is called a *Universal shunt* :

- The *"Ayrton shunt" is very convenient in bridge measurements as the galvanometer sensitivity can be increased as bridge balance is approached.*

3.7.1.9. Selection of galvanometer

The following *features* should be considered while making choice of a galvanometer :

1. ***Sensitivity :***

- The sensitivity should be *sufficient* for the work in hand so that measurements may be made without undue fatigue and loss of time.
- Although sensitivity is important, yet it should not be the only feature considered in estimating the utility of a galvanometer.

2. ***Resistance :***

- The resistance should be *appropriate* for the measurement in hand. Usually a high-resistance galvanometer is used in a high-resistance circuit and a low-resistance galvanometer is employed in a low-resistance circuit.

However, it is *neither necessary or desirable* to have an exact matching between a galvanometer and the circuit resistance.

3. ***Damping :***

- For measurements of current and voltage, the damping should be *somewhat less than, or at most equal to critical,* this *will save time.*

In general, a galvanometer should be selected whose *critical damping resistance matches reasonably well with the resistance of the circuit* in which it is to be used.

4. ***Period :***

- In order that the instrument will respond quickly to the current, the time of vibration of the moving system should be *as short as possible.*

5. ***Promptness of response :***

- The promptness of response implies a *short period* and *critical damping.*

6. ***Freedom from mechanical disturbances :***

- The symmetric making and mounting the system about the axis of rotation contributes much to the system stability when it is subjected to mechanical disturbances.
- When the taut suspension galvanometers are properly balanced, they are, in general, *less expensive* than slack-suspension instruments to mechanical disturbances.

7. ***Freedom from stray-field effects :***

It is of significant importance that the indications of the instrument be not influenced by the unavoidable variations of the local field.

- The D'Arsonval galvanometer is usually relatively free from such trouble as its coil operates in a *strong field.* However, in case of instruments with weak operating field the use of an *astatic system* or of *magnetic shielding* in necessary.
- Moving magnet galvanometers are sensitive to external field and must be very carefully shielded.

8. ***Definiteness of zero reading :***

- The *zero reading* should be *definite.*
- The deflections should come promptly to their final values with no viscous actions of the suspension.

9. ***Law of deflection :***

- The deflection, throughout its useful range, should be proportional to the current.

10. ***Coil damping :***

- A suitable device should be provided for damping the coil and holding it rigidly in place during transport.
- A zero adjustment accessible from outside the case is essential.

11. *Visibility of suspended parts.*
12. *Accessibility for repairs.*
13. The *mirror* should be of the *smallest practical dimensions.*

Example 3.4. *Determine the angle through which coil turns when a deflection of 48 mm is observed on a scale of a galvanometer placed at a distance of 650 mm from the mirror.*

Solution. Refer to Fig. 3.11

Given: d = 48 mm or 0.048 m; r = 650 mm or 0.65 m.

The angle turned, θ :

The angle through which the mirror or coil turns may be found from the following expression :

$$d = r \times 2\theta$$

or,
$$\theta = \frac{d}{2r} = \frac{0.048}{2\times0.65} = 0.0369 \text{ rad. or } \mathbf{2.11°} \quad \textbf{(Ans.)}$$

Example 3.5. *Determine the displacement constant of a suspended coil in a galvanometer of a moving coil vibrating type having 30 turns, 4.5 mm × 2.5 mm mean area and situated in a magnetic field of 0.8 tesla.*

Solution. *Given:* N = 30; A = 4.5 mm × 2.5 mm = 11.25 mm^2 or 11.25 × 10^{-6} m^2; B = 0.8 Wb/m^2

Displacement constant, G :

Displacement constant is given by :

$$G = NBA$$
$$= 30 \times 0.8 \times 11.25 \times 10^{-6} = \mathbf{270 \times 10^{-6}\ Nm/A} \quad \textbf{(Ans.)}$$

Example 3.6. *The suspended coil of a galvanometer has a 5 mm × 2.5 mm mean area and is situated in a magnetic field of 1.1 tesla. The moment of inertia of moving parts is 0.25 × 10^{-6} kg-m and the control string constant is 35 × 10^{-6} Nm/radian. If a current of 12 mA produces a deflection of 110°, calculate the number of turns of the suspended coil.*

Solution. *Given* : $A = 5 \text{ mm} \times 2.5 \text{ mm} = 5 \times 2.5 \times 10^{-6} \text{ m}^2$ or $12.5 \times 10^{-6} \text{ m}^2$; $B = 1.1 \text{ Wb/m}^2$; $J = 0.25 \times 10^{-6}$ kg-m; $C = 35 \times 10^{-6}$ Nm/rad.; $\theta = 110°$ or 1.92 radians.

Number of turns of the suspended coil, N :

We know that for steady deflection,

$$GI = C\theta \quad \text{[Eqn. (3.11)]}$$

But, $$G = NBA$$

$\therefore$ $$NBAI = C\theta$$

or, $$N = \frac{C\theta}{BAI} = \frac{35 \times 10^{-6} \times 1.92}{1.1 \times 12.5 \times 10^{-6} \times 0.012} = \mathbf{407} \quad \textbf{(Ans.)}$$

Example 3.7. *Fig. 3.15 shows a galvanometer system in which a wire of 0.12 mm in diameter is used as a torsion member. The wire is made of phosphor bronze for which modulus of rigidity is 40 GN/m². The polar moment of inertia for a round wire is $\frac{\pi}{32} d^4$, where d is the diameter of the wire.*

A collimated ray of light is reflected by the galvanometer mirror shown. The radius from the mirror to the scale is 180 mm.

Calculate the following :

(i) Torsional constant;

(ii) Maximum allowable torque if the shear stress is limited to 100 MN/m²;

(iii) Sensitivity of the instrument in m/Nm;

(iv) Maximum possible scale deflection.

(Roorkee University)

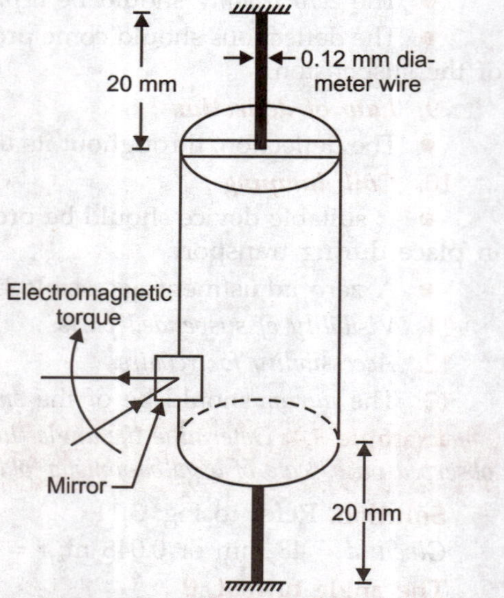

Fig. 3.15. A galvanometer moving system.

Solution. *Given* : $d = 0.12$ mm or 0.12×10^{-3} m; $l = 20$ mm = 0.02 m

G (modulus of rigidity) = 40 GN/m²; r = 180 mm or 0.18 m; τ (shear stress) = 100 MN/m²

***(i)* Torsional constant, C :**

For a round wire : $$\frac{T}{I_p} = \frac{\tau}{r} = \frac{G\theta}{l} \quad ...(i)$$

[where, I_p (Polar moment of inertia) = $\frac{\pi}{32} d^4$ ... *given*]

$\therefore$ Deflection per unit torque, $$\frac{\theta}{T} = \frac{l}{GI_p}$$

Since there are two wires, one at the top and the other at the bottom, the deflection is half of that for single wire.

$$\therefore \quad \frac{\theta}{T} = \frac{l}{2GI_p} = \frac{l}{24\times(\pi/32)\,d^4} = \frac{0.02}{2\times(40\times10^9)\times(\pi/32)\times(0.12\times10^{-3})^4}$$

$$= 12280.5 \text{ rad./Nm}$$

Hence, torsional constant, $C = \frac{T}{\theta} = \frac{1}{12280.5} = \mathbf{81.43 \times 10^{-6}}$ **Nm/rad.** (Ans.)

(*ii*) **Maximum allowable torque, T :**

$$T = \frac{\tau I_p}{r} = \frac{\tau I_p}{(d/2)} = \frac{I\times(\pi/32)\,d^4}{(d/2)} = \frac{\tau\times\pi d^3}{16}$$

Since there are two wires, the maximum allowable torque is :

$$T = \frac{2\times\tau\times\pi d^3}{16} = \frac{\pi}{8}\tau d^3$$

$$= \frac{\pi}{8}\times(100\times10^6)\times(0.12\times10^{-3})^3$$

$$= \mathbf{67.86 \times 10^{-6}\ Nm} \quad \text{(Ans.)}$$

(*iii*) **Sensitivity of the instrument in m/Nm :**

The scale is kept at a radius of 0.18 m, and therefore the sensitivity of the scale is :

$$S = 2 \times (12280.5) \times 0.18 = \mathbf{4421\ m/Nm} \quad \text{(Ans.)}$$

A factor of 2 has been introduced because if mirror turns through an angle θ, the collimated light turns through 2θ.

(*iv*) **Maximum possible scale deflection :**

The deflection for maximum torque = $4421 \times 67.86 \times 10^{-6}$ m

= 0.3 m or **300 mm** (Ans.)

Example 3.8. *The coil of a moving coil galvanometer has 300 turns and is suspended in a uniform magnetic field of 0.1 Wb/m². The control constant is 0.2 × 10⁻⁶ Nm per radian. The coil is 2 cm wide and 2.5 cm high with a moment of inertia of 0.15 × 10⁻⁶ kg m². If the galvanometer resistance in 250 Ω calculate the value of resistance which when connected across the galvanometer terminals will give critical damping, assuming damping to be entirely electromagnetic.*

(Punjab University)

Solution. *Given* : $N = 300$; $B = 0.1$ Wb/m²; $C = 0.2 \times 10^{-6}$ Nm/rad; $b = 2$ cm or 0.02 m; $l = 2.5$ cm or 0.025 m; $J = 0.15 \times 10^{-6}$ kg m²; $R_g = 250\ \Omega$.

External resistance for critical damping, CDRX :

Total resistance required for critical damping,

$$R_c = \frac{G^2}{2\sqrt{CJ}} \qquad \text{[Eqn. (3.31)]}$$

or, $$R_c = \frac{(NBA)^2}{2\sqrt{CJ}} \qquad (\because\ G = NBA)$$

$$= \frac{[300\times0.1\times(0.025\times0.02)]^2}{2\sqrt{0.2\times10^{-6}\times0.15\times10^{-6}}} = 650\ \Omega$$

∴ External resistance required for critical damping,

$$CDRX = R_c - R_g = 650 - 250 = \mathbf{400\ \Omega} \quad \text{(Ans.)}$$

Example 3.9. *The coil of a moving coil galvanometer is wound on a non-magnetic former whose height and width are both 25 mm. It moves in a constant field of 0.1 Wb/m². The moment of inertia of its moving parts is 0.3 × 10^{-6} kg-m² and the control spring constant is 32 × 10^{-6} Nm/rad.*

Calculate :

(i) The number of turns that must be wound on the former to produce a deflection of 140° with a current of 12 mA.

(ii) The resistance of coil to produce critical damping, all damping being assumed as electromagnetic. **(Bangalore University)**

Solution. *Given* : b = 25 mm or 0.025 m; B = 0.1 Wb/m²; $J = 0.3 \times 10^{-6}$ kg-m²;

$C = 32 \times 10^{-6}$ Nm/rad.; $\theta = 140° = 140 \times \frac{\pi}{180} = 2.44$ rad.; I = 12 mA or 0.012 A.

(i) Number of turns, N :

For steady deflection, $GI = C\theta$ [Eqn. (3.11)]

$\therefore$ Displacement constant, $$G = \frac{C\theta}{I}$$

$$= \frac{32\times10^{-6}\times2.44}{0.012} = 6.5 \times 10^{-3} \text{ Nm/A}$$

Also, $$G = NBlb$$

$\therefore$ $$N = \frac{G}{Blb} = \frac{6.5\times10^{-3}}{0.1\times0.025\times0.025} = 104 \text{ (Ans.)}$$

(ii) Resistance of galvanometer for critical damping, R_c :

We know that, $$R_c = \frac{G^2}{2\sqrt{CJ}} = \frac{(6.5\times10^{-3})^2}{2\sqrt{32\times10^{-6}\times0.3\times10^{-6}}} = 6.82\ \Omega \text{ (Ans.)}$$

Example 3.10. *In a moving coil galvanometer with free time period of 2.5 seconds, a current of 2.4 mA gives a deflection of 250 mm on a scale at a distance of 1.4 m. The moving system has a moment of inertia of 3.8 × 10^{-6} kg-m².*

Determine the total resistance in the galvanometer circuit for critical damping.

Solution. *Given* : T_0 (Free time period) = 2.5 s; I = 2.4 μA; d = 250 mm or 0.25 m; r = 1.4 m; $J = 3.8 \times 10^{-6}$ kg-m².

Total resistance in the galvanometer for critical damping, R_c :

We know that, $$R_c = \frac{G^2}{2\sqrt{CJ}} \quad ...(i)$$

For steady deflection : $$GI = C\theta$$

or, $$G = \frac{C\theta}{I} = \frac{C\times(d/2r)}{I} \quad \left(\because \theta = \frac{d}{2r}\right)$$

or, $$G = \frac{C\times[0.25/(2\times1.4)]}{2.4\times10^{-6}} = 3.72 \times 10^4 C$$

Also, free (or natural) time period

$$T_n = 2\pi\sqrt{\frac{J}{C}}$$

or, $$C = \left(\frac{2\pi}{T_n}\right)^2 J = \left(\frac{2\pi}{2.5}\right)^2 \times 3.8 \times 10^{-6} = 24 \times 10^{-6} \text{ Nm/rad.}$$

Inserting the values in eqn. (*i*), we get

$$R_c = \frac{(3.72 \times 10^4 \times 24 \times 10^{-6})^2}{2\sqrt{(24 \times 10^{-6}) \times (3.8 \times 10^{-6})}} = \mathbf{41733\ \Omega} \quad \textbf{(Ans.)}$$

Example 3.11. *A moving coil galvanometer deflects 220 mm on a scale at a distance of 1.2 m from the mirror when a current of 2.2 μA passes through it. The free time period of the galvanometer is 3.8 seconds. The galvanometer is dead beat when the total resistance in the galvanometer is 19500 Ω.*

Determine the moment of inertia of the galvanometer moving system.

Solution. *Given :* d = 220 mm or 0.22 m; r = 1.2 m; I = 2.2 μA; T_0 = 3.8 s; R_c = 19500 Ω

Moment of inertia of the system, *J* :

For steady deflection : $$GI = C\theta$$

or, $$G = \frac{C\theta}{I} = \frac{C \times (d/2r)}{I}$$

or, $$G = \frac{C \times [0.22/(2 \times 1.2)]}{2.2 \times 10^{-6}} = 4.167 \times 10^4\ C$$

Also, free time period, $$T_n = 2\pi\sqrt{\frac{J}{C}}$$

or, $$3.8 = 2\pi\sqrt{\frac{J}{C}}$$

or, $$J = \left(\frac{3.8}{2\pi}\right)^2 \times C = 0.3657\ C$$

Critical damping resistance, $$R_c = \frac{G^2}{2\sqrt{CJ}}$$

or, $$19500 = \frac{(4.167 \times 10^4 C)^2}{2\sqrt{C \times (0.3657C)}} = \frac{(4.167 \times 10^4)^2\ C^2}{2\sqrt{0.3657} \times C}$$

or, $$C = \frac{19500 \times 2\sqrt{0.3657}}{(4.167 \times 10^4)}$$

$$= 0.1358 \times 10^{-4} \text{ Nm/rad.}$$

and, $$J = 0.3657\ C = 0.3657 \times 0.1358 \times 10^{-4}$$

$$= \mathbf{4.966 \times 10^{-6}\ kg\text{-}m^2} \quad \textbf{(Ans.)}$$

Example 3.12. *The following data relate to a moving coil galvanometer which has a former of a non-conducting material :*

The current sensitivity of the instrument = 0.012 μA/mm at 1 m

The period of undamped oscillation = 5.5 seconds

The displacement constant of the instrument = 4.8 × 10^{-3} Nm/A

Calculate :

(*i*) *Control constant.*

(*ii*) *Inertia constant.*

(*iii*) *Total circuit resistance for instrument to be dead beat.*

Solution. *Given :* Current sensitivity, S_I = 0.0012 μA/mm at 1 m; T_n = 5.5 s; G = 4.8 × 10^{-3} Nm/A.

(*i*) Control constant, C :

Suppose that the instrument is reflecting type. Therefore, if the final steady deflection is θ, the spot moves through an angle 2θ. The spot moves a distance 1 mm at 1 m with a current of 0.0012 μA or in other words a current of 0.0012 μA deflects the instrument through,

$$\theta = \frac{d}{2r} = \frac{(1/1000)}{2 \times 1} = 0.5 \times 10^{-3} \text{ rad.}$$

∴ Control constant,
$$C = \frac{GI}{\theta} = \frac{4.8 \times 10^{-3} \times 0.0012 \times 10^{-6}}{0.5 \times 10^{-3}}$$

$$= \mathbf{11.52 \times 10^{-9} \text{ Nm/rad.}}$$ **(Ans.)**

(*ii*) Inertia constant, *J* :

Undamped (free or natural time) period,

$$T_n = 2\pi\sqrt{\frac{J}{C}}$$

or, Inertia constant,
$$J = \frac{CT_n^2}{4\pi^2} = \frac{11.52 \times 10^{-9} \times (5.5)^2}{4\pi^2}$$

$$= \mathbf{8.827 \times 10^{-9} \text{ kg-m}^2}$$ **(Ans.)**

(*iii*) Total circuit resistance for critical damping, R_c :

Total circuit resistance for critical damping or dead beat,

$$R_c = \frac{G^2}{2\sqrt{CJ}} = \frac{(4.8 \times 10^{-3})^2}{2\sqrt{11.52 \times 10^{-9} \times 8.827 \times 10^{-9}}}$$

$$= \mathbf{1142.4\ \Omega}$$ **(Ans.)**

Example 3.13. *A d'Arsonval galvanometer has a rectangular coil wound on an aluminium former of resistivity 30 × 10^{-9} Ωm. The sides of the former, each of effective length 35 mm, move in a radial field of uniform flux density 0.12 Wb/m^2 against a control torque of 20 × 10^{-6} Nm/rad. The width of the former is 25 mm, the ends being effectively outside the magnetic field, and the moment of inertia of the moving system is 90 × 10^{-9} kg-m^2. Calculate the cross-sectional area of the aluminium former if it is to provide critical damping, other sources of damping being neglected.* **(Nagpur University)**

Solution. *Given : r* = 30 × 10^{-9} Ωm; *l* = 35 mm or 0.035 m; *b* = 0.025 m; *B* = 0.12 Wb/m^2; *J* = 90 × 10^{-9} kg-m^2; control torque, C = 20 × 10^{-6} Nm/rad.

Cross-sectional area of aluminium former, *a* :

Total damping factor (D_f) in the absence of all other damping except the damping produced by *former* is given by :

$$D_f = \frac{G^2}{N^2 R_f}$$

$$= \frac{N^2 B^2 l^2 b^2}{N^2 R_f} = \frac{B^2 l^2 b^2}{R_f}$$

For critical damping, $D = D_c = 2\sqrt{CJ}$ [Eqn. (3.18)]

$$\therefore \quad \frac{B^2 l^2 b^2}{R_f} = 2\sqrt{CJ}$$

Hence, resistance of former, $R_f = \dfrac{B^2 l^2 b^2}{2\sqrt{CJ}}$

$$= \frac{(0.12)^2 \times (0.035)^2 \times (0.025)^2}{2\sqrt{20 \times 10^{-6} \times 90 \times 10^{-9}}} = 4108.8 \times 10^{-6}\ \Omega$$

Length of eddy currents, $L = 2(l + b)$

$$= 2(0.035 + 0.025) = 0.12 \text{ m}$$

Now, $R_f = \dfrac{\rho L}{a} = \dfrac{30 \times 10^{-9} \times 0.12}{a} = 4108.8 \times 10^{-6}\ \Omega$

$$\therefore \quad a = \frac{30 \times 10^{-9} \times 0.12}{4108.8 \times 10^{-6}} = 0.876 \times 10^{-6} \text{ m}^2 \text{ or } \mathbf{0.876\ mm^2} \quad \textbf{(Ans.)}$$

Example 3.14. *A moving coil galvanometer has the following data :*

Number of turns = 320

Flux density = 0.12 Wb/m^2

Control constant = 0.12×10^{-6} *Nm/rad.*

Coil size = 21 *mm* × 22 *mm*

Moment of inertia = 0.2×10^{-6} $kg\text{-}m^2$

Galvanometer resistance (including external resistance) = 2100 Ω

Determine :

(i) Total resistance for critical damping.

(ii) Damping ratio.

(iii) Frequency of undamped oscillations.

(iv) Frequency of damped oscillations.

Solution. *Given:* $N = 320$; $B = 0.12\ Wb/m^2$; $C = 0.12 \times 10^{-6}$ Nm/rad.; $A = 21 \times 22 \times 10^{-6} = 4.62 \times 10^{-4}\ m^2$; $J = 0.2 \times 10^{-6}$ kg-m²; $R_{actual} = 2100\ \Omega$.

(i) Total resistance for critical damping, R_c :

$$R_c = \frac{G^2}{2\sqrt{CJ}} = \frac{(NBA)^2}{2\sqrt{CJ}}$$

$$= \frac{[320 \times 0.12 \times (4.62 \times 10^{-4})]^2}{2\sqrt{0.12 \times 10^{-6} \times 0.2 \times 10^{-6}}} = \mathbf{1015.8\ \Omega} \quad \textbf{(Ans.)}$$

(*ii*) **Damping ratio, γ :**

$$\gamma = \frac{D}{D_{critical}}$$

$$= \frac{G^2/R_{actual}}{G^2/R_{critical}} = \frac{R_{critical}}{R_{actual}} = \frac{1015.8}{2100} = \mathbf{0.4837} \quad \textbf{(Ans.)}$$

(*iii*) **Frequency of undamped oscillations, f_n :**

Frequency of undamped oscillations, f_n is given by

$$\omega_n = \sqrt{\frac{C}{J}} = \sqrt{\frac{0.12\times10^{-6}}{0.2\times10^{-6}}} = 0.7746$$

$$\therefore \quad f_n = \frac{0.7746}{2\pi} = \mathbf{0.1233\ Hz} \quad \textbf{(Ans.)}$$

(*iv*) **Frequency of damped oscillations, f_d :**

$$f_d = f_n\sqrt{1-\gamma^2} \qquad \text{[Eqn. 3.22(a)]}$$

$$= 0.1233\sqrt{1-(0.4837)^2}$$

$$= \mathbf{0.1079\ Hz} \quad \textbf{(Ans.)}$$

Example 3.15. *The following data relate to a moving coil galvanometer :*

Number of turns of the coil	*= 340*
Flux density in the magnetic field	*= 0.12 Wb/m*^{2}
Control constant	*= 0.22 × 10*$^{-6}$ *Nm/rad.*
Dimensions of the coil	*= 25 mm (wide) × 30 mm (high)*
Moment of inertia	*= 0.2 × 10*$^{-6}$ *kg-m*^{2}
Galvanometer resistance	*= 350 Ω*
Resistance connected in series with the galvanometer	*= 3150 Ω*

Determine :

(*i*) *The resistance to be connected to galvanometer for critical damping.*
(*ii*) *Relative damping.*
(*iii*) *Logarithmic decrement.*
(*iv*) *Frequency of free oscillations.*
(*v*) *First maximum deflection.*
(*vi*) *Periods of undamped and damped oscillations.* **(Roorkee University)**

Solution. *Given:* $N = 340$; $B = 0.12$ Wb/m^2; $C = 0.22 \times 10^{-6}$ Nm/rad.; $A = 25$ mm × 30 mm = 750 mm^2 or 0.00075 m^2; $J = 0.2 \times 10^{-6}$ kg-m^2; $R_g = 350$ Ω; Resistance connected in series with galvanometer = 3150 Ω.

(*i*) **External resistance required for critical damping, CDRX :**

Total resistance required for critical damping

$$R_c = \frac{G^2}{2\sqrt{CI}} = \frac{(NBA)^2}{2\sqrt{CJ}} = \frac{(340\times0.12\times0.00075)^2}{2\sqrt{0.22\times10^{-6}\times0.2\times10^{-6}}} = 2232\ \Omega$$

$$\therefore \quad CDRX = R_c - R_g = 2232 - 350 = \mathbf{1882\ \Omega} \quad \textbf{(Ans.)}$$

(*ii*) **Relative damping, γ :**

$$\text{Damping, } D = \frac{G^2}{R_{actual}} = \frac{G^2}{(350+3150)} = \frac{G^2}{3500}$$

Critical damping, $D_c = \frac{G^2}{R_c} = \frac{G^2}{2232}$

$$\therefore \quad \gamma = \frac{D}{D_c} = \frac{(G^2/3500)}{(G^2/2232)} = \frac{2232}{3500} = \mathbf{0.638} \quad \textbf{(Ans.)}$$

(*iii*) Logarithmic decrement, λ :

Logarithmic decrement λ is given by :

$$\lambda = \frac{\pi\gamma}{\sqrt{1-\gamma^2}} = \frac{\pi \times 0.638}{\sqrt{1-(0.638)^2}} = \mathbf{2.6} \quad \textbf{(Ans.)}$$

(*iv*) Frequency of free oscillations, f_n :

$$f_n = \frac{1}{2\pi}\sqrt{\frac{C}{J}} = \frac{1}{2\pi}\sqrt{\frac{0.22\times 10^{-6}}{0.2\times 10^{-6}}} = \mathbf{0.167\ Hz} \quad \textbf{(Ans.)}$$

(*v*) First maximum deflection, θ_1 :

First maximum deflection θ_1, in terms of steady deflection θ_f, is given by :

$$\theta_1 = \theta_f\left[1+\exp.\left(-\frac{\pi\gamma}{\sqrt{1-\gamma^2}}\right)\right]$$

$$= \theta_f\left[1+\exp.\left(-\frac{\pi\times 0.638}{\sqrt{1-0.638^2}}\right)\right] = \mathbf{1.074\ \theta_f} \quad \textbf{(Ans.)}$$

(*vi*) Periods of undamped and damped oscillations, T_n, T_d :

Frequency of damped oscillations, f_d :

$$f_d = f_n\sqrt{1-\gamma^2} \qquad \text{[Eqn. 3.22(a)]}$$

$$= 0.167\sqrt{1-(0.638)^2} = \mathbf{0.128\ Hz} \quad \textbf{(Ans.)}$$

$\therefore$ Period of undamped oscillations, $T_n = \frac{1}{f_n} = \frac{1}{0.167} = \mathbf{5.98\ s}$ **(Ans.)**

Period of damped oscillations, $T_d = \frac{1}{f_l} = \frac{1}{0.128} = \mathbf{7.81\ s}$ **(Ans.)**

Example 3.16. *The following data relate to a moving coil galvanometer :*

Number of turns of the coil	*= 300*
Flux density in the uniform magnetic field	*= 0.06 Wb/m²*
Torsion constant	*= 1.6 × 10⁻⁷ Nm/radian*
Dimensions of the coil	*= 20 mm (width) × 25 mm (height)*
Moment of inertia	*= 1.7 × 10⁻⁷ kg-m²*
Galvanometer resistance	*= 250 Ω*
Total damping	*= 4.5 × 10⁻⁸ Nm/radian/second.*

Determine :

(*i*) *The deflection of galvanometer in radian, and in mm when a current of 3 μA is flowing through the galvanometer, the scale is at a distance of 1.6 m from the mirror.*

(ii) *The current sensitivity.*
(iii) *The voltage sensitivity.*
(iv) *The megohm sensitivity.*
(v) *Frequency of oscillations.*
(vi) *Relative damping.*
(vii) *Logarithmic decrement.* **(Delhi University)**

Solution. *Given* : $N = 300$; $B = 0.06$ Wb/m^2; $C = 1.6 \times 10^{-7}$ Nm/radian; $A = 20 \times 10^{-3} \times 25 \times 10^{-3} = 0.0005$ m^2; $J = 1.7 \times 10^{-7}$ kg-m^2; $R_g = 250\ \Omega$; Total damping, $D = 4.5 \times 10^{-8}$ Nm/rad./s; $I = 3\ \mu$A; $r = 1.6$ m.

(i) **The deflection of galvanometer :**

The deflection of galvanometer in radian :

$$\theta = \frac{GI}{C} = \frac{(NBA)\,I}{C}$$

$$= \frac{300 \times 0.06 \times 0.0005 \times 3 \times 10^{-6}}{1.6 \times 10^{-7}} = \mathbf{0.1687\ rad.}\quad \textbf{(Ans.)}$$

The deflection of galvanometer in *mm*:

$$d = r \times 2\theta = 1.6 \times 2 \times 0.1687 \times 1000 \text{ mm} = \mathbf{539.8\ mm}\quad \textbf{(Ans.)}$$

(ii) **The current sensitivity, S_I :**

$$S_I = \frac{d \text{ in mm}}{I \text{ in } \mu\text{A}} = \frac{539.8}{3} = \mathbf{179.9\ mm/\mu A}\quad \textbf{(Ans.)}$$

(iii) **Voltage sensitivity, S_V :**

$$S_V = \frac{d \text{ in mm}}{V \text{ in microvolts}} = \frac{d \text{ in mm}}{R \times I \text{ in } \mu\text{A}}$$

$$= \frac{539.8}{250 \times 3} = \mathbf{0.7197\ mm/\mu V}\quad \textbf{(Ans.)}$$

(iv) **The megohm sensitivity :**

Current required for a deflection of 1 mm $= \frac{1}{179.9}\ \mu$A

$$\therefore \quad \text{Megohm sensitivity} = \frac{1}{\frac{1}{179.9} \times 10^{-6}} = \mathbf{179.9\ M\Omega/mm}\quad \textbf{(Ans.)}$$

(v) **Frequency of oscillations :**

$$\text{Frequency of oscillations} = \frac{1}{2\pi}\sqrt{\frac{C}{J} - \frac{D^2}{4J^2}}$$

$$= \frac{1}{2\pi}\sqrt{\frac{1.6 \times 10^{-7}}{1.7 \times 10^{-7}} - \frac{(4.5 \times 10^{-8})^2}{4(1.7 \times 10^{-7})^2}}$$

$$= \frac{1}{2\pi}\sqrt{0.9412 - 0.0175} = \mathbf{0.153\ Hz}\quad \textbf{(Ans.)}$$

(vi) **Relative damping, γ :**

$$\gamma = \frac{D}{\sqrt{2JC}} \qquad \text{...[Eqn. 3.15(a)]}$$

$$= \frac{4.5 \times 10^{-8}}{\sqrt{2 \times 1.7 \times 10^{-7} \times 1.6 \times 10^{-7}}} = \mathbf{0.193} \quad \textbf{(Ans.)}$$

(*viii*) **Logarithmic increment, λ :**

Logarithmic increment is given by :

$$\lambda = \frac{\pi\gamma}{\sqrt{1-\gamma^2}} = \frac{\pi \times 0.193}{\sqrt{1-(0.193)^2}} = \mathbf{0.618} \quad \textbf{(Ans.)}$$

Example 3.17. *The following data relate to a panel type PMMC instrument which has a full scale deflection 90° = 100 scale divisions, with a current of 1.2 mA.*

The period of oscillations = 0.6 s

In order to measure the spring constant, a small weight having a gravitational force of 98.1 × 10^6 N is placed at a distance of 110 mm from the horizontal axis of rotation with the horizontal pointer acting as lever arm; the resultant deflection = 40 divisions.

The first maximum deflection, when a current of 1.2 mA is suddenly passed through the meter = 115 divisions

The value of flux density in the air gap measured with a Hall probe = 0.26 Wb/m²

The length and average diameter of the coil = 18 mm and 16 mm respectively.

Determine the following :

(*i*) *Spring constant.*
(*ii*) *Moment of inertia.*
(*iii*) *Number of turns.*
(*iv*) *Damping ratio.*
(*v*) *Equivalent resistance of the short-circuited winding represented by the damping action of the former of the coil.*

Solution. *Given* : Full scale deflection = 90° = 100 scale divisions, or, 1 scale division = 0.9° = 0.0157 rad.; I = 1.2 mA or 0.0012 A; T_n = 0.6 s; B = 0.26 Wb/m²; l = 18 mm or 0.018 m; d (or b) = 16 mm or 0.016 m.

(*i*) **Spring constant, *C* :**

Restoring torque, $T_c = 98.1 \times 10^{-6} \times (110 \times 10^{-3}) = 10.79 \times 10^{-6}$ Nm

Deflection, θ = 40 div. = 40 × 0.0157 rad. = 0.628 rad.

$$\therefore \quad \text{Spring constant, } C = \frac{T_c}{\theta} = \frac{10.79 \times 10^{-6}}{0.628} = \mathbf{17.18 \times 10^{-6}\ Nm/rad.} \quad \textbf{(Ans.)}$$

(*ii*) **Moment of inertia, *J* :**

$$\text{Period of free oscillations } T_0 = 2\pi\sqrt{\frac{J}{C}}$$

$$\therefore \quad J = \frac{CT_n^2}{4\pi^2} = \frac{17.18 \times 10^{-6} \times (0.6)^2}{4\pi^2}$$

$$= \mathbf{0.1567 \times 10^{-6}\ kg\text{-}m^2} \quad \textbf{(Ans.)}$$

(*iii*) **Number of turns, *N* :**

Control torque for 40 scale divisions = 10.79 × 10^{-6} Nm (calculated above)

∴ Control torque for full scale deflection (100 div.),

$$T_c = 10.79 \times 10^{-6} \times \frac{100}{40} = 26.97 \times 10^{-6}\ \text{Nm}$$

∴ Deflecting torque at full scale,

$$T_d = NBlbI = N \times 0.26 \times 0.018 \times 0.016 \times 0.0012 = 89.86\ N \times 10^{-9}\ \text{Nm}$$

At equilibrium : $T_d = T_c$

or, $89.86\ N \times 10^{-9} = 26.97 \times 10^{-6}$

∴ Number of turns, $N = \dfrac{26.97 \times 10^{-6}}{89.86 \times 10^{-9}} = \mathbf{300}$ **(Ans.)**

(iv) **Damping ratio, γ :**

First maximum deflection, θ_1 = 115 division = 1.15 θ_f

(where θ_f = final steady deflection)

But, the value of θ_1 in terms of θ_f can be expressed as :

$$\theta_1 = \theta_f \left[1 + \exp.\left(-\frac{\pi\gamma}{\sqrt{1-\gamma^2}}\right)\right]$$

or, $$\frac{\theta_1}{\theta_f} = 1 + \exp.\left(-\frac{\pi\gamma}{\sqrt{1-\gamma^2}}\right)$$

or, $$1.15 = 1 + \exp.\left(-\frac{\pi\gamma}{\sqrt{1-\gamma^2}}\right)$$

or, $$\exp.\left(-\frac{\pi\gamma}{\sqrt{1-\gamma^2}}\right) = 0.15$$

or, $$\frac{\pi\gamma}{\sqrt{1-\gamma^2}} = 1.897$$

or, $$\frac{\gamma}{\sqrt{1-\gamma^2}} = \frac{1.897}{\pi} = 0.604$$

Squaring both sides, we get

$$\gamma^2 = 0.604^2\ (1 - \gamma^2)$$

or, $$\gamma^2\ (1 + 0.604^2) = 0.604^2$$

or, $$\gamma = \left(\frac{0.604^2}{1 + 0.604^2}\right)^{1/2} = \mathbf{0.517}$$ **(Ans.)**

(v) **Damping constant, D :**

We know that, $$\gamma = \frac{D}{2\sqrt{CJ}}$$ [Eqn. (3.18)]

or, $$D = \gamma \times 2\sqrt{CJ} = 0.517 \times 2 \times \sqrt{17.18 \times 10^{-6} \times 0.1567 \times 10^{-6}}$$

$$= \mathbf{1.696 \times 10^{-6}\ Nm/rad.\ s^{-1}}$$ **(Ans.)**

(*vi*) **Resistance of former, R_f :**

Considering the former to be a single-turn coil, the damping constant is,

$$D_{\text{former}} = \frac{(Blb)^2}{R_f}$$

$$\therefore \quad R_f = \frac{(Blb)^2}{D_{\text{former}}} = \frac{(0.26 \times 0.018 \times 0.016)^2}{1.696 \times 10^{-6}} = \mathbf{3.3 \times 10^{-3}\ \Omega} \quad \textbf{(Ans.)}$$

Example 3.18. *The deflections of a galvanometer with no current and a steady state current of 6 μA are zero and 80 mm respectively. Its first maximum deflection, after a step function of voltage which produces a steady state current is applied, is 115 mm. The maximum deflection in the next cycle is 90 mm.*

Determine : (*i*) *Current sensitivity.*
(*ii*) *Logarithmic decrement.*
(*iii*) *Relative damping.*

Solution. *Given* : $I = 6\ \mu A$; $\theta_f = 80$ mm ; $\theta_1 = 115$ mm ; $\theta_3 = 90$ mm

(*i*) **Current sensitivity, S_I :**

$$S_I = \frac{\theta_f}{I} = \frac{80}{6} = \mathbf{13.33\ mm/\mu A} \quad \textbf{(Ans.)}$$

(*ii*) **Logarithmic decrement, λ :**

First maximum deflection θ_1 is given by :

$$\theta_1 = \theta_f \left[1 + \exp\left(-\frac{\pi\gamma}{\sqrt{1-\gamma^2}} \right) \right] \quad ...(i)$$

Similarly, *second maximum* deflection θ_3 is given by :

$$\theta_3 = \theta_f \left[1 + \exp\left(-\frac{3\pi\gamma}{\sqrt{1-\gamma^2}} \right) \right] \quad ...(ii)$$

Hence, from (*i*) and (*ii*), we have :

$$\frac{\theta_1 - \theta_f}{\theta_3 - \theta_f} = \exp\left(\frac{2\pi\gamma}{\sqrt{1-\gamma^2}} \right)$$

But logarithmic decrement, λ is given by :

$$\lambda = \frac{\pi\gamma}{\sqrt{1-\gamma^2}}$$

$$\therefore \quad \frac{\theta_1 - \theta_f}{\theta_3 - \theta_f} = \exp\ (2\lambda)$$

or, $$\frac{115 - 80}{90 - 80} = \exp\ (2\lambda) \quad \text{or,} \quad 3.5 = \exp\ (2\gamma)$$

or $$\lambda = \mathbf{0.626} \quad \textbf{(Ans.)}$$

(*iii*) **Relative damping, γ :**

$$0.626 = \frac{\pi\gamma}{\sqrt{1-\gamma^2}}, \text{ or }, \quad \frac{\gamma}{\sqrt{1-\gamma^2}} = \frac{0.626}{\pi} = 0.1993$$

or, $\gamma^2 = (1 - \gamma^2) \times 0.1993^2$ ...squaring both sides

or, $\gamma^2(1 + 0.1993^2) = 0.1993^2$

or, $\gamma = \left(\dfrac{0.1993^2}{1 + 0.1993^2}\right) = \mathbf{0.195}$ **(Ans.)**

Example 3.19. *The Ayrton universal shunt has a total resistance of 9000 Ω and galvanometer has a resistance 3000 Ω. Determine the multiplying power of shunt for 1000 Ω, 3000 Ω, 6000 Ω tapping.*

Solution. Total shunt resistance, $R_{sh} = 9000\ \Omega$

Galvanometer resistance, $R_g = 3000\ \Omega$

$$\text{Multiplying power of shunt, } N = \frac{I}{I_g} = \frac{R_{sh} + R_g}{R_1}$$

∴ (*i*) $N = \dfrac{R_{sh} + R_g}{R_1} = \dfrac{9000 + 3000}{1000} = \mathbf{12}$ **(Ans.)**

...when $R_1 = 1000\ \Omega$

(*ii*) $N = \dfrac{9000 + 3000}{3000} = \mathbf{4}$ **(Ans.)**

...when $R_1 = 3000\ \Omega$

(*iii*) $N = \dfrac{9000 + 3000}{6000} = \mathbf{2}$ **(Ans.)**

...when $R_1 = 6000\ \Omega$

3.7.2. Ballistic Galvanometer

3.7.2.1. Construction of ballistic galvanometer

A ballistic galvanometer is also a *permanent magnet moving coil instrument and has D'Arsonval type construction.*

- *It is used to measure quantity of electricity, i.e., charge passing through it in a given time. Thus, it is an integrating instrument and is used extensively in* ***magnetic measurement.***

The quantity of electricity is the result of an e.m.f. induced instantaneously in a *search coil* connected to galvanometer terminals when the magnetic flux linking the coil is changed. *The 'charge' is proportional to the e.m.f. induced in the search coil.* By noting the maximum value of the deflection, the amount of charge that has passed through the galvanometer is measured. The *proportionality* between the deflection and amount of charge is *valid only* if the whole of charge has passed through the galvanometer before any appreciable deflection of the moving system occurs. Therefore, this instrument has a *large inertia.* Very often, *additional weights* are added to the moving system so that *inertia is high as compared to the restoring moment due to the suspension.* Therefore, natural period of vibration is *large,* about 10 to 15 seconds.

The damping is *small so that first deflection* (which indicates the total charge) *is large.*

- It is desirable that the instrument is *dead beat.* The galvanometer is short circuited by a key so that the moving system is brought to rest immediately.
  - — The *damping is only electromagnetic.*
  - — Air friction damping is *avoided because it is indeterminate.* Electromagnetic can be determined from the *constants of the instrument.*
- It is necessary that the moving coil is free from magnetic material.

- The terminals, coil and internal connections are of *copper* so as to *avoid thermoelectric effects at the junctions.*
- In the best ballistic galvanometer, the suspension is *non-conducting*, the *current being led into the coil by delicate spiral of very thin copper strip.*

3.7.2.2. Theory of ballistic galvanometer

In a ballistic galvanometer, as discussed earlier, the coil moves after the charge to be measured has passed through it. Obviously, during the movement of the coil, there is no current flowing through it. Hence the equation of motion is :

$$J\frac{d^2\theta}{dt^2} + D\frac{d\theta}{dt} + C\theta = 0 \qquad \text{...(3.36)}$$

where, J = Moment of inertia,
D = Damping constant, and
C = Controlling or restoring constant.

Since *damping is extremely small*, the approximate solution of the eqn. (3.36) is :

$$\theta = U_{\text{exp.}}\ [-(D/2J)t]\ \sin(\omega_0 t + \phi) \qquad \text{...(3.37)}$$

(where ω_0 (= ω_n) is the natural frequency of oscillations)

At the start of motion, when $t = 0$, $\theta = 0$ and hence $\phi = 0$.

$$\therefore \quad \theta = U_{\text{exp.}}\ [-(D/2J)t]\ \sin \omega_0 t \qquad \text{...(3.38)}$$

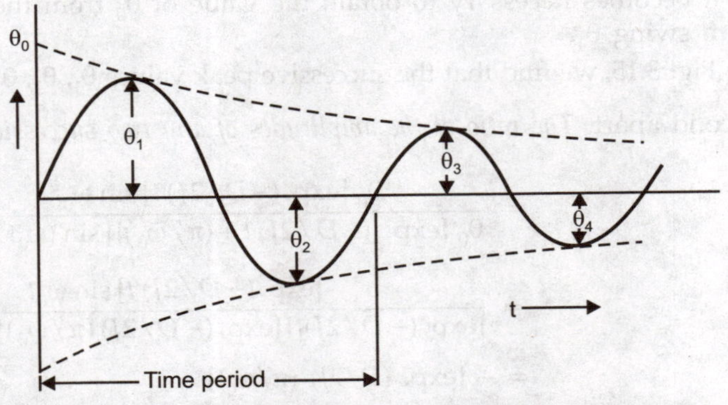

Fig. 3.16

During the passage of charge, at any instant, there will be a deflecting torque of GI acting on the coil. If t is the time taken by the whole charge to pass through, the torque impulse due to this charge is given as : $\int_0^t GI\,dt$

But, $$\int_0^t I\,dt = Q$$

$\therefore$ Torque impulse = GQ

This must be equal to the change of angular momentum produced *i.e.* $J.\ \alpha$, where α is the angular velocity of the coil at the end of impuse period.

$$\therefore \quad GQ = J\alpha \qquad \text{...(3.39)}$$

or, $$\alpha = \frac{GQ}{J} \qquad \text{...[3.39(a)]}$$

Differentiating eqn. (3.38), we get

$$\frac{d\theta}{dt} = U\left[\exp.\{-(D/2J)\,t\}\cos\omega_0 t\,.\,\omega_0 + \sin\omega_0 t\,.\exp.\{-(D/2J)\,t\}\left(-\frac{D}{2J}\right)\right]$$

Since duration of the passage of charge is very small, at the end of the passage, $t \simeq 0$, so that from above,

$$\frac{d\theta}{dt} = U\omega_0 \qquad \text{...(3.40)}$$

However, at this time, $\frac{d\theta}{dt} = \alpha = \frac{GQ}{J}$, $\therefore \quad \frac{GQ}{J} = U\omega_0$

or, $$Q = \frac{J\omega_0}{G}U \qquad \text{...(3.41)}$$

Now, *U being the amplitude which the oscillations would* have if the damping were *zero*, it may be called *undamped swing* θ_0.

$\therefore$ $$Q = \frac{J\omega_0}{G}\theta_0 \qquad \text{...(3.42)}$$

$$(\text{or, } Q \propto \theta_0)$$

However, in practice, due to the presence of small amount of damping, the successive oscillations diminish exponentially (See Fig. 3.16). Even the first swing θ_1 is much less than θ_0. Hence, it becomes necessary to obtain the value of θ_0 from the observed value of first maximum swing θ_1.

Referring to Fig. 3.15, we find that the successive peak values θ_1, θ_2, θ_3 etc. are ϕ radian apart or $\frac{\phi}{\omega_0}$ second apart. *The ratio of the amplitudes of any two successive peaks is*

$$= \frac{\theta_0\,[\exp.(-D/2J)\,t]\sin\omega_0 t}{\theta_0\,[\exp.\{(-D/2J)\,t + (\pi/\omega_0)\}]\sin(\omega_0 t + \pi)}$$

$$= \frac{[\exp.(-D/2J)\,t]\sin\omega_0 t}{[\exp.(-D/2J)t][\exp.(-D/2J)(\pi/\omega_0)](-\sin\omega_0 t)}$$

$$= -\,[\exp.\,(D/2J)\,(\pi/\omega_0)]$$

$\therefore$ $$\frac{\theta_1}{\theta_2} = \frac{\theta_2}{\theta_3} = \frac{\theta_3}{\theta_4} = \ldots\ldots [\exp.\,(D/2J)\,(\pi/\omega_0)] \qquad \text{...(3.43)}$$

Let $\exp.\,[(D/2J)\,(\pi/\omega_0)] = \delta^2$, where δ is called the **damping factor*

*Now, $$\frac{\theta_1}{\theta_2} = \frac{\theta_2}{\theta_3} = \ldots\ldots \frac{\theta_{n-1}}{\theta_n} = \delta^2$$

$\therefore$ $$\frac{\theta_1}{\theta_2}\times\frac{\theta_2}{\theta_3}\times\frac{\theta_3}{\theta_4}\times\ldots\ldots\times\frac{\theta_{n-1}}{\theta_n} = (\delta^2)^{n-1}$$

or, $$\frac{\theta_1}{\theta_n} = (\delta^2)^{n-1} \quad \text{or} \quad \delta = \left(\frac{\theta_1}{\theta_n}\right)^{1/2(n-1)}$$

Hence, δ may be obtained by observing the first and *n*th swing.

The time period of oscillation, $T_0 = \frac{2\pi}{\omega_0}$. If damping is very small $\theta_0 = \theta_1$, $t = \frac{T_0}{4} = \frac{\pi}{2\omega_0}$ as a very close approximation.

Putting $t = \dfrac{\pi}{2\omega_0}$ in eqn. (3.38), we have :

$$\theta_1 = \theta_0 \text{ exp. } [-(D/2J)(\pi/2\omega_0)] \sin(\pi/2), \text{ or, } \theta_1 = \theta_0 . \delta^{-1}$$

or, $$\theta_0 = \delta\theta_1 \quad \text{...(3.44)}$$

or, Undamped swing = Damping factor × Ist swing

Suppose, a steady current of I_s flowing through the galvanometer produces a steady deflection θ_s, then

$$C\theta_s = GI_s \quad \text{...(3.45)}$$

or, $$G = \frac{C\theta_s}{I_s} \quad [3.45(a)]$$

Since damping is small, $\omega_0 = \sqrt{\dfrac{C}{J}}$

or, $$C = J\omega_0^2 = J \times \left(\frac{2\pi}{T_0}\right)^2 = 4\pi^2 \frac{J}{T_0^2}$$

Inserting this value of C in eqn. 3.45 (*a*), we get :

$$G = \frac{4\pi^2 J\theta_s}{T_0^2 I_s} \quad \text{...(3.46)}$$

Substituting this value of G in eqn. (3.42), we have :

$$Q = \frac{J\omega_0\theta_0}{4\pi^2 J\theta_s/(T_0^2 I_s)} = \frac{J\omega_0\theta_0 \times T_0^2 I_s}{4\pi^2 J\theta_s} = \frac{J \times (2\pi/T_0)\,\theta_0 T_0^2 I_s}{4\pi^2 J\theta_s}$$

$$\left(\because\ \omega_0 = \frac{2\pi}{T_0}\right)$$

or $$Q = \frac{T_0}{2\pi} \cdot \frac{I_s}{\theta_s} \cdot \delta\theta_1 \quad \text{...(3.47)}$$

Alternatively, let *quantity $(D/2J)(\pi/\omega_0)$ be called the "logarithmic decrement λ".*

Since, $$\delta^2 = e^{\lambda},$$

$$\delta = e^{\lambda/2} = 1 + (\lambda/2) + \frac{(\lambda/2)^2}{2!} + \ldots \simeq \left(1 + \frac{\lambda}{2}\right)$$

when λ is ***small.*

**Since $\delta^2 = e^{\lambda}$, taking logs we have : $2 \ln(\delta) = \lambda \ln(e) = \lambda$

$$\therefore \quad \frac{\lambda}{2} = \ln(\delta) = \ln\left(\frac{\theta_1}{\theta_n}\right)^{1/2(n-1)} = \frac{1}{2(n-1)} \ln\left(\frac{\theta_1}{\theta_n}\right)$$

$$\therefore \quad \lambda = \frac{1}{n-1} \ln\left(\frac{\theta_1}{\theta_n}\right) \quad \text{...[3.47(a)]}$$

Hence, from eqn. (3.47), we have :

$$Q = \frac{T_0}{2\pi} \cdot \frac{I_s}{\theta_s}\left(1+\frac{\lambda}{2}\right)\theta_1 \qquad ...(3.48)$$

In general, eqn. (3.48) may be expressed as :

$$Q = k\theta_1 \qquad ...[3.48(a)]$$

where, $k = \frac{T_0}{2\pi} \cdot \frac{I_s}{\theta_s}\left(1+\frac{\lambda}{2}\right)$ is a constant and is found by *calibration*. Since the value of k depends upon the damping and shunting of the galvanometer it is essential that the resistance of the galvanometer circuit during calibration shall be the *same* as it is when the galvanometer is being used for testing purposes.

3.7.2.3. Calibration of a ballistic galvanometer

*The experimental determination of the constants of a ballistic galvanometer is termed as its **"calibration"**.*

There are *three* methods of calibration of ballistic galvanometer :

1. Using a mutual inductance.
2. Using a capacitor.
3. Using Hibbert magnetic standard.

1. Using a mutual inductance :

This method is *most commonly used for the calibration purposes.*

Fig. 3.17(*a*) shows the schematic arrangement for the calibration of a ballistic galvanometer with mutual inductance. The mutual inductance is usually in the form of a standard solenoid with a secondary coil.

A *standard solenoid* (*primary coil*) consists of a long coil wound on a cylinder. The length of cylinder is very large as compared with its diameter. At the centre of the solenoid a *secondary coil* is placed whose length is *small*. The secondary coil is connected to the ballistic galvanometer (B.G.). When a current is passed through the solenoid, the field strength (H) at the centre is :

$$H = \frac{N_1 I}{l}\cos\theta$$

or,

$$H = \frac{N_1 I}{l} \times \frac{(l/2)}{\sqrt{(l/2)^2 + (r_1)^2}} = \frac{N_1 I}{2\sqrt{(l/2)^2 + (r_1)^2}} \qquad ...(3.49)$$

where, N_1 = Number of turns of the solenoid,

I = Current through the solenoid,

l = Length of the solenoid, and

r_1 = Mean radius of solenoid.

Flux density at the centre of solenoid, $B = \mu_0 H$

Flux linking with the secondary coil,

$$\phi = B \times \text{area of secondary coil}$$

i.e.

$$\phi = B \times \pi r_2^2 = \pi B r_2^2 = \pi\mu_0 H r_2^2 \qquad ...(3.50)$$

where, r_2 = Mean radius of the secondary coil.

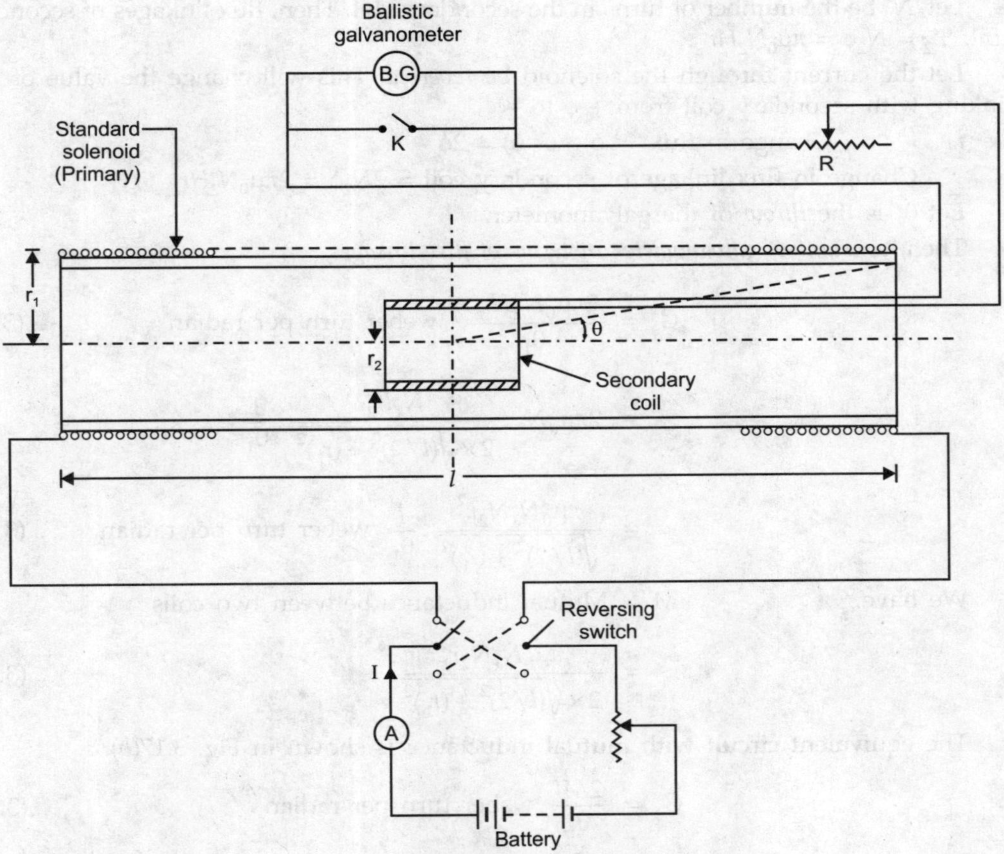

(a) Schematic arrangement

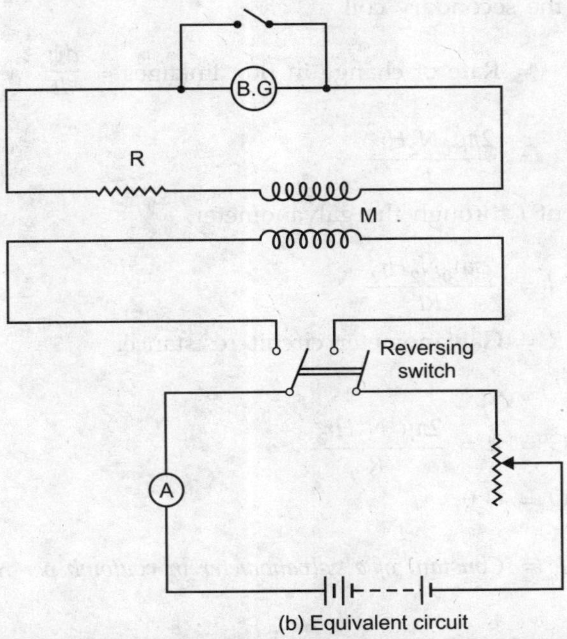

(b) Equivalent circuit

Fig. 3.17. Calibration of ballistic galvanometer with mutual inductance.

Let, N_2 be the number of turns in the secondary coil. Then, flux linkages of secondary coil, $\Psi_2 = N_2\phi = \pi\mu_0 N_2 H r_2^2$.

Let the current through the solenoid be *reversed.* This will change the value of flux linking with secondary coil from $+\phi$ to $-\phi$.

i.e., Change in flux $= \phi - (-\phi) = 2\phi$

$\therefore$ Change in flux linkage of secondary coil $= 2N_2\phi = 2\pi\mu_0 N_2 H r_2^2$

Let θ_1 is the *throw* of the galvanometer.

Then, *constant of galvanometer in terms of flux linkages,*

$$G_\phi = \frac{2\pi\mu_0 N_2 H r_2^2}{\theta_1} \text{ weber turn per radian} \quad ...(3.51)$$

$$= 2\pi\mu_0 N_2 \cdot \frac{N_1 I r_2^2}{2\times\sqrt{(l/2)^2 + (r_1)^2}} \cdot \frac{1}{\theta_1}$$

$$= \frac{\pi\mu_0 N_1 N_2 r_2^2}{\sqrt{(l/2)^2 + (r_1)^2}} \cdot \frac{I}{\theta_1} \text{ weber turn per radian.} \quad ...(3.52)$$

We have, M = Mutual inductance between two coils

$$= \frac{\pi\mu_0 N_1 N_2 r_2^2}{2\times\sqrt{(l/2)^2 + (r_1)^2}} \quad ...(3.53)$$

The equivalent circuit with mutual inductance is shown in Fig. 3.17(*b*).

$$\therefore \quad G_\phi = \frac{2MI}{\theta_1} \text{ weber turn per radian} \quad ...(3.54)$$

Let us assume that the change in flux takes place in time t.

Then, E.m.f. induced in the secondary coil

$$= \text{Rate of change in flux linkages} = \frac{d\psi}{dt}$$

$$= \frac{2\pi\mu_0 N_2 H r_2^2}{t}$$

This e.m.f. sends a current I through the galvanometer,

$$I = \frac{2\pi\mu_0 N_2 H r_2^2}{Rt}$$

where, R = Galvanometer circuit resistance.

Charge passing in time t,

$$Q = It = \frac{2\pi\mu_0 N_2 H r_2^2}{R} \quad ...(3.55)$$

But, $$Q = G_q\theta_1 \quad ...(3.56)$$

where, G_q = *Constant of a galvanometer in coulomb per radian.*

∴ Constant of galvanometer,

$$G_q = \frac{Q}{\theta_1} = \frac{2\pi\mu_0 N_2 H r_2^2}{R} \cdot \frac{1}{\theta_1} = \frac{\pi\mu_0 N_1 N_2 r_2^2 I}{R\sqrt{(l/2)^2 + (r_1)^2}} = \frac{2MI}{R\theta_1} \quad ...(3.57)$$

Thus the galvanometer constant can be obtained in terms of coulomb per radian.

- "*Ballistic sensitivity*" and "*Flux linkage sensitivity*" are sometimes used in place of galvanometer constants and are defined as :

Ballistic sensitivity, $$S_q = \frac{\theta_1}{Q} \text{ rad/C} = \frac{1}{G_q} \quad ...(3.58)$$

Flux linkage sensitivity, $$S_\phi = \frac{\theta_1}{\Delta\psi} \text{ rad./weber turn} = \frac{1}{G_\phi} \quad ...(3.59)$$

Also, $$S_q = RS_\phi \quad ...(3.60)$$

2. Using a capacitor :

In this method, a capacitor (C) which has been charged to a known voltage (V) is discharged through the galvanometer. The throw of the galvanometer is observed. Let this throw be θ.

∴ *Constant of galvanometer,* $$G_q = \frac{Q}{\theta} = \frac{CV}{\theta} \text{ coulomb/rad.}$$

- The *disadvantages* of this method are :

(i) The capacitance of condenser is difficult to be determined accurately under all the conditions.

(ii) The damping of the galvanometer during calibration is different from that during testing.

3. Using Hibbert magnetic standard :

This method of calibration of ballistic galvanometer is *very simple and convenient,* and so is used in the *laboratories.*

Fig. 3.18 shows Hibbert magnetic standard :

— It consists of a circular bar magnet and an iron yoke which have a narrow annular gap between them. A brass tube carrying a coil can slide through the gap.

— The brass tube is taken to its top position and is then released with the help of a trigger. It falls under gravity and it slides through the gap (See Fig. 3.18), its coil cuts through the magnetic field of a permanent magnet and therefore an e.m.f. is produced in the coil.

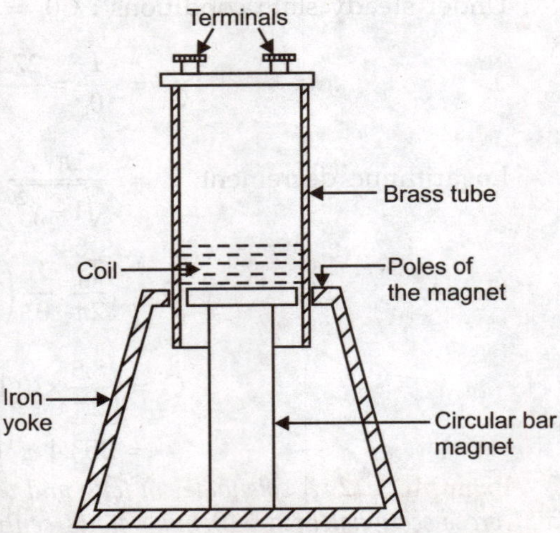

Fig. 3.18. Hibbert magnetic standard.

— The coil is connected across the terminals of the ballistic galvanometer, the *throw* of which can be observed. The rate at which coil cuts through the field is constant for a particular apparatus. The *e.m.f. per turn is constant.*

— By the use of the standard, the flux linkages which produce an observed throw of galvanometer can be known by knowing the air gap flux and number of turns in the coil. Hence the *galvanometer constant can be calculated.*

Example 3.20. *A ballistic galvanometer has a free period of 10 seconds and gives a steady deflection of 200 divisions with a steady current of 0.1 mA. A charge of 121 μC is instantaneously discharged through the galvanometer giving rise to a first maximum deflection of 100 divisions. Calculate the decrement of the resulting oscillations.* **(Bombay University)**

Solution. *Given* : T_0 = 10 s; θ_s = 200; I_s = 0.1 mA; Q = 121 μC ; θ_1 = 100

Decrement of the resulting oscillations, λ :

We know,
$$Q = \frac{T_0}{2\pi} \cdot \frac{I_s}{\theta_s}\left(1+\frac{\lambda}{2}\right)\theta_1 \qquad \text{[Eqn. (3.48)]}$$

Substituting the various values, we get

$$121 \times 10^{-6} = \frac{10}{2\pi} \times \frac{0.1\times10^{-3}}{200}\left(1+\frac{\lambda}{2}\right)\times 100$$

or, λ = **1.04** **(Ans.)**

Example 3.21. *A ballistic galvanometer has a resistance of 140 Ω and an undamped period of 8.5 seconds. A steady e.m.f. of 3.8 mV produces a deflection of 220 mm. If the deflection produced is 720 mm and relative damping is 0.78, determine the quantity of electricity discharged from the capacitor.*

Solution. R = 140 Ω; T_0 = 8.5 s; V = 3.8 mV; θ_s = 220 mm; θ_1 = 720 mm; γ = 0.78

Quantity of electricity discharged, Q :

Steady state current,
$$I_s = \frac{3.8\times10^{-3}}{140} = 27.14 \times 10^{-6}\ \text{A}$$

Under steady state conditions : $C\theta_s = GI_s$ [Eqn. (3.45)]

$$\therefore \quad \frac{C}{G} = \frac{I_s}{\theta_s} = \frac{27.14\times10^{-6}}{220} = 0.123 \times 10^{-6}\ \text{A/mm}$$

Logarithmic decrement,
$$\lambda = \frac{\pi\gamma}{\sqrt{1-\gamma^2}} = \frac{\pi\times0.78}{\sqrt{1-(0.78)^2}} = 3.9$$

Hence,
$$Q = \frac{T_0}{2\pi} \cdot \frac{I_s}{\theta_s}\left(1+\frac{\lambda}{2}\right)\theta_1 \qquad \text{[Eqn. (3.48)]}$$

$$Q = \frac{8.5}{2\pi}\times(0.123\times10^{-6})\left(1+\frac{3.9}{2}\right)\times720$$

$$= 353.4 \times 10^{-6}\ \text{C} = \mathbf{353.4\ \mu C}$$ **(Ans.)**

Example 3.22. *A solenoid 1 m long and wound with 960 turns has a search coil of 60 turns and cross-sectional area 340 mm² at its centre. Reversing a current of 3.5 A in the solenoid causes a deflection of 4 divisions in a ballistic galvanometer connected to the search coil.*

Calculate :

(i) The galvanometer constant in flux linkages per division.

(ii) The flux linkage sensitivity.

Solution. *Given* : l = 1 m; N_1 = 960; N_2 = 60; A = 340 mm²; θ_1 = 4 divisions; I = 3.5 A.

(*i*) **Galvanometer constant, G_ϕ :**

Flux density at the centre of solenoid $B = \dfrac{\mu_0 N_1 I}{l}$

Flux linking with search coil, $\phi = B \times A = \dfrac{\mu_0 N_1 I A}{l}$

Flux linkages of search coil, $\psi = N_2 BA = \dfrac{\mu_0 N_1 N_2 IA}{l}$

Change of flux linkages, $\Delta\psi = 2\psi = \dfrac{2\mu_0 N_1 N_2 IA}{l}$ (since current is *reversed*)

$$\text{Galvanometer constant, } G_\phi = \frac{2\psi}{\theta_1} = \frac{2\mu_0 N_1 N_2 IA}{l\,\theta_1}$$

$$= \frac{2 \times 4\pi \times 10^{-7} \times 960 \times 60 \times 3.5 \times (340 \times 10^{-6})}{1 \times 4}$$

$$= \mathbf{43.07 \times 10^{-6}} \textbf{ weber turn per division} \quad \textbf{(Ans.)}$$

(*ii*) **The flux linkage sensitivity, S_ϕ :**

$$S_\phi = \frac{1}{G_\phi} = \frac{1}{43.07 \times 10^{-6}}$$

$$= \mathbf{23.22 \times 10^3} \textbf{ divisions per weber turn} \quad \textbf{(Ans.)}$$

Example 3.23. *The discharge of a capacitor through a ballistic galvanometer produces a damped frequency of 0.13 Hz and successive swings of 130, 102 and 85.4 mm.*

Calculate :

(*i*) *Logarithmic decrement.*

(*ii*) *Damping ratio.*

(*iii*) *Period of undamped and damped oscillations.*

Solution. *Given :* $f_d = 0.13$ Hz; $\theta_1 = 130$ mm; $\theta_3 = 85.4$ mm

(*i*) **Logaritmic decrement, λ :**

We know that, $$\lambda = \frac{1}{n-1} \ln\left(\frac{\theta_1}{\theta_n}\right) \qquad [3.47(a)]$$

or, $$\lambda = \frac{1}{3-1} \ln\left(\frac{130}{85.4}\right) = \mathbf{0.21} \quad \textbf{(Ans.)}$$

(*ii*) **Damping ratio, γ :**

$$\text{Logarithmic decrement, } \lambda = \frac{\pi\gamma}{\sqrt{1-\gamma^2}}$$

or, $$0.21 = \frac{\pi\gamma}{\sqrt{1-\gamma^2}}$$

Squaring both sides, we get

$$(0.21)^2 (1 - \gamma^2) = \pi^2\gamma^2$$

or, $$\gamma^2(\pi^2 + 0.21^2) = 0.21^2$$

or, $$\gamma = \left(\frac{0.21^2}{\pi^2 + 0.21^2}\right)^{1/2} = \mathbf{0.067} \quad \textbf{(Ans.)}$$

(iii) **Period of damped and undamped oscillations :**

Frequency of undamped oscillations, $$f_n = \frac{f_d}{\sqrt{1-\gamma^2}} = \frac{0.13}{\sqrt{1-0.067^2}} = 0.1303 \text{ Hz}$$

$\therefore$ Period of damped oscillations, $$T_d = \frac{1}{f_d} = \frac{1}{0.13} = \mathbf{7.69s} \quad \textbf{(Ans.)}$$

and, Period of undamped oscillations, $$T_n \text{ (or } T_0) = \frac{1}{f_n} = \frac{1}{0.1303} = \mathbf{7.67s} \quad \textbf{(Ans.)}$$

Example 3.24. *A ballistic galvanometer having 200 Ω resistance gives a throw of 80 divisions when the flux through search coil to which it is connected is reversed. The galvanometer constant is 120 μC/division. The search coil has 1500 turns, a mean area 4500 mm^2 and a resistance 230 Ω.*

Determine :

(i) Flux density in which the reversal of the coil takes place.

(ii) Ballistic and flux linkages sensitivities of the galvanometer.

Solution. *Given :* R_g = 200 Ω; θ_1 = 80 divisions; G_q = 120 μC/div.; N_{sc} = 1500; A_{sc} = 4500 mm^2 or 4500 × 10^{-6} m^2; R_{sc} = 230 Ω

(i) **Flux density in which the reversal of the coil takes place, *B* :**

Resistance of galvanometer circuit, $R = R_g + R_{sc}$ = 200 + 30 = 230 Ω

Galvanometer constant, G_q = 120 μC/div.

Now, $$G_q = \frac{Q}{\theta_1} = \frac{It}{\theta_1} = \frac{Et}{R\theta_1}$$

But change in flux linkages $\Delta\psi = 2\psi$ since the current is reversed.

$\therefore$ $$E = \frac{2\psi}{t} = \frac{2\phi N}{t} = \frac{2BAN}{t}$$

Hence, $$G_q = \frac{2BAN}{t} \cdot \frac{t}{R\theta_1} = \frac{2BAN}{R\theta_1}$$

$\therefore$ Flux density, $$B = \frac{G_q R\theta_1}{2AN} = \frac{120\times10^{-6}\times230\times80}{2\times(4500\times10^{-6})\times1500} = \mathbf{0.16\ Wb/m^2} \quad \textbf{(Ans.)}$$

(ii) **The ballistic sensitivity and flux linkages sensitivity :**

$\therefore$ Ballistic sensitivity, $$S_q = \frac{1}{G_q} = \frac{1}{120} = 8.333 \times 10^{-3} \text{ div./μC}$$

$$= \mathbf{8333\ div./C} \quad \textbf{(Ans.)}$$

Flux linkages sensntivity, $$S_\phi = \frac{S_q}{R}$$ ...[Eqn. (3.60)]

or, $$S_\phi = \frac{8333}{230} = \mathbf{36.23\ weber\ turn/div.} \quad \textbf{(Ans.)}$$

Example 3.25. *A galvanometer with an undamped period of 8 seconds and a current sensitivity of 1.1 mA/div. is connected to a search coil in a circuit of 3500 Ω resistance. Determine the flux linkage change in search coil to produce a first swing of 80 divisions if the logarithmic decrement is 0.18.*

Solution. T_0 (or T_n) = 8 s; Current sensitivity, $\frac{I_s}{\theta_s}$ = 1.1 μA/div.; R = 3500 Ω;

θ_1 = 80 divisions; λ = 0.18.

Change in flux linkages, Δψ :

Charge, $$Q = \frac{T_0}{2\pi} \cdot \frac{I_s}{\theta_s}\left(1+\frac{\lambda}{2}\right)\theta_1 \quad \text{[Eqn. (3.48)]}$$

$$= \frac{8}{2\pi} \times 1.1 \times 10^{-6}\left(1+\frac{0.18}{2}\right) \times 80 = 122.13 \times 10^{-6}\ C$$

Let Δψ be the change in flux linkages in time t.

Then, $$\text{e.m.f., } E = \frac{\Delta\psi}{t}$$

Also, $$\text{charge } Q = It = \frac{E}{R} \cdot t = \frac{E}{R} \times \frac{\Delta\psi}{E} = \frac{\Delta\psi}{R}$$

∴ $$\Delta\psi = RQ = 3500 \times 122.13 \times 10^{-6} = \mathbf{427.5 \times 10^{-3}}\ \textbf{Weber turn}$$ **(Ans.)**

Example 3.26. *A ballistic galvanometer gives a first maximum deflection of 50° for a discharge of 900 μC. Find the quantity of electricity which when discharged through this galvanometer gives rise to a spot deflection of 8 divisions on a millimetre circular scale 1 metre away.*

Solution. *Given* : $\theta_1 = 50°$ or $50 \times \frac{\pi}{180}$ = 0.873 rad.; Q = 900 μC; d = 8 mm or 0.008 m; r = 1 m

Charge Q :

Now, $$Q = G_q\,\theta_1, \quad \therefore \quad G_q = \frac{Q}{\theta_1} = \frac{900}{0.873} = 1030.9\ \mu\text{C/rad.}$$

Angle turned through by reflected ray for a swing of 8 mm,

$$\theta = \frac{d}{2r} = \frac{0.008}{2 \times 1} = 0.004 \text{ rad.}$$

∴ Charge for a swing of 0.004 radian, Q = 1030.9 × 0.004 = **4.12 μC** **(Ans.)**

3.7.3. Grassot Fluxmeter

This instrument is a special type of ballistic galvanometer in which the *controlling torque is very small and electromagnetic damping is heavy. It is used for measurement of **flux**.*

Construction :

Fig. 3.19 shows the construction of a Grassot fluxmeter :

— It consists of a moving coil of small cross-section suspended by means a single silk thread from a spring support and hanging with its parallel sides in the narrow air gap of a permanent magnet (See Fig. 3.19). The coil is formerless and the air friction damping is negligible.

— The current is led into the coil by spirals of very thin, annealed silver strips. By this construction the *controlling torque is reduced to a minimum.*

— The instrument is usually fitted with a pointer (attached to the moving system) and a scale, although it may also be used as a reflecting instrument. The *scale is graduated in terms of flux-turns.*

Working :

— When the instrument terminals are connected to a *search coil* and the flux linking with the search coil is changed, the moving system of the flux meter is deflected and rotates through an angle which depends upon the change in flux-turns.

— The instrument coil rotates during the whole period of the flux change but stops as soon as the flux change ceases because of its high electromagnetic damping. The *high damping is obtained by reducing the resistance of the circuit and search coil to the minimum possible.* Though theoretically, the value of the resistance of the circuit comprising the instrument coil and search coil does not affect the sensitivity of the instrument but it should be kept fairly small (below 20 Ω), otherwise the damping will not be adequate.

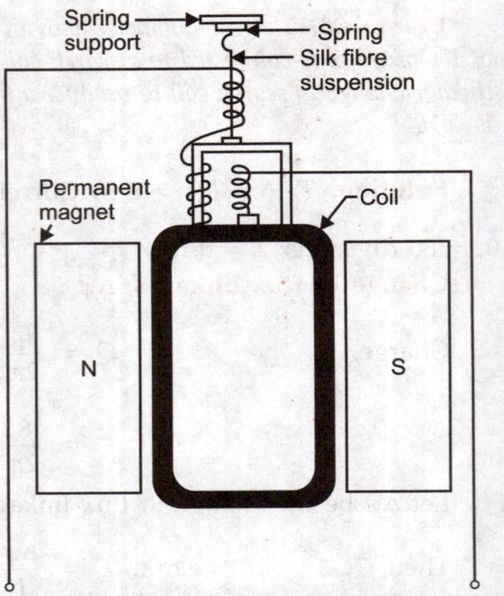

Fig. 3.19. Grassot fluxmeter.

— Theoretically, the time taken for the change in flux does not affect the deflection but in practice if the time flux change exceeds about 30 seconds serious errors are introduced due to *friction*, the *slight controlling torque due to the current leads and the effect of incorrect levelling combined with slight unbalance of the moving parts.*

Advantages :

The instrument possesses the following *advantages* :

1. It is *very portable* and has *scale calibrated directly in weber-turns.*

2. The length of time taken for the change in the flux producing the deflection need not be small. The deflection obtained, for a given change of flux interlinking with the search coil connected to the instrument, will be the *same* whether the time taken for the change be a fraction of a second or as much as one or two minutes.

Disadvantages :

The fluxmeter is *less sensitive and accurate than the ballistic galvanometer.*

> *Note :* • In the complete absence of controlling torque, the instrument would remain in its deflected position indefinitely. Actually the pointer returns very slowly to zero, but *readings may be taken by observing the difference in deflection at the beginning and end of the change in flux to be measured without waiting for the pointer to return to zero, the scale being uniform.*
>
> • In some modern instruments, the coil is fitted with pivots and mounted in jewelled bearings, the current being led into the coil by fine ligaments. This form of construction is rather *more robust* than the suspended instruments.

Theory of Fluxmeter :

Let, N_{sc} = Number of turns in search coil,

R_{sc}, L_{sc} = Resistance and inductance of search coil respectively,

R_{fm}, L_{fm} = Resistance and inductance of fluxmeter respectively,

ϕ = Flux linking with search coil,
i = Current in the circuit at any instant, and
θ = Deflection of the instrument at any instant.

The *equation of motion* is given as :

$$J\frac{d^2\theta}{dt^2} + D\frac{d\theta}{dt} + C\theta = Gi$$

Assuming control torque to be negligibly small and also the air friction damping is small, the equation of motion reduces to :

$$J\frac{d^2\theta}{dt^2} = Gi \qquad \text{...(3.61)}$$

as D and C are both zero. (The effect of electromagnetic damping is considered later in expression for current i)

The e.m.f. due to any change of flux linked with search coil,

$$e_{sc} = N_{sc} = \frac{d\phi}{dt}$$

Simultaneously, due to the movement of the fluxmeter coil in the field of the magnet, a rotational e.m.f. is induced in the meter coil. This e.m.f., $e_{fm} = G \cdot \frac{d\theta}{dt}$. Besides this there are voltage drops in the resistance and inductance of the circuit. Therefore, we can write :

$$e_{sc} = e_{fm} + \left(L_{fm} + L_{sc}\right)\frac{di}{dt} + (R_{fm} + R_{sc})\,i \qquad \text{...(3.62)}$$

or, $$N_{sc} \cdot \frac{d\phi}{dt} = G \cdot \frac{d\theta}{dt} + \left(L_{fm} + L_{sc}\right)\frac{di}{dt} + (R_{fm} + R_{sc})\,i$$

or, $$i = \frac{N_{sc} \cdot \frac{d\phi}{dt} - G \cdot \frac{d\theta}{dt} - (L_{fm} + L_{sc})\frac{di}{dt}}{(R_{fm} + R_{sc})}$$

$\therefore$ The equation of motion is :

$$J \cdot \frac{d^2\theta}{dt^2} = \frac{G}{(R_{fm} + R_{sc})}\left[N_{sc} \cdot \frac{d\phi}{dt} - G \cdot \frac{d\theta}{dt} - (L_{fm} + L_{sc})\frac{di}{dt}\right] \qquad \text{...(3.63)}$$

or, $$N_{sc} \cdot \frac{d\phi}{dt} = \frac{J(R_{fm} + R_{sc})}{G} \cdot \frac{d^2\theta}{dt^2} + G \cdot \frac{d\theta}{dt} + (L_{fm} + L_{sc})\frac{di}{dt}$$

or, $$N_{sc} \cdot \frac{d\phi}{dt} = \frac{J(R_{fm} + R_{sc})}{G} \cdot \frac{d\omega}{dt} + G \cdot \frac{d\theta}{dt} + (L_{fm} + L_{sc})\frac{di}{dt} \qquad \text{...(3.64)}$$

where, ω = angular velocity of moving coil at instant t.

If T is the time taken by the change in flux, then we have :

$$\int_0^T N_{sc} \cdot \frac{d\phi}{dt} \cdot dt = \int_0^T \frac{J(R_{fm} + R_{sc})}{G}\frac{d\omega}{dt} \cdot dt + \int_0^T G \cdot \frac{d\theta}{dt} \cdot dt + \int_0^T (L_{fm} + L_{sc})\frac{di}{dt} \cdot dt$$

or,
$$\int_{\phi_1}^{\phi_2} N_{sc}\,.\,d\phi = \int_{\omega_1}^{\omega_2} \frac{J(R_{fm}+R_{sc})}{G}\,d\omega + \int_{\theta_1}^{\theta_2} G\,.\,d\theta + \int_{i_1}^{i_2} (L_{fm}+L_{sc})\,di$$

$$\therefore \quad N_{sc}\,(\phi_2 - \phi_1) = \frac{J(R_{fm}+R_{sc})}{G}(\omega_2-\omega_1) + G(\theta_2-\theta_1) + (L_{fm}+L_{sc})(i_2-i_1) \quad ...(3.65)$$

where, ϕ_1, ϕ_2 = Inter linking fluxes,
ω_1, ω_2 = Angular velocities, and
i_1, i_2 = Currents.

The suffices 1 and 2 indicate respectively values at the *beginning* and at the *end* of the change.

Now, since the angular velocities and currents are zero at both the beginning and end of the change, therefore,

$$\omega_1 = \omega_2 = 0, \quad \text{and} \quad i_1 = i_2 = 0$$

Inserting these values in eqn. (3.65), we have :

$$N_{sc}\,(\phi_2 - \phi_1) = G_2\,(\theta_2 - \theta_1) \quad ...(3.66)$$

or, if ϕ is the change in 0 flux and θ the change in fluxmeter deflection, then,

$$\phi = \left(\frac{G}{N_{sc}}\right)\theta \quad ...(3.67)$$

When the permanent magnet field of fluxmeter is uniform for all positions of the moving coil, G is a constant. Thus the *change in the value of flux is directly proportional to the change in deflection and hence the instrument will have a uniform scale.*

Use of shunt with Fluxmeter :

When the values of flux are very high, the deflection of a fluxmeter, even when a single-turn search coil is used, may exceed the scale length. In these circumstances, the range may be increased by employing a low-resistance, non-inductive shunt.

Fig. 3.20 shows the circuit using a shunt with a fluxmeter.

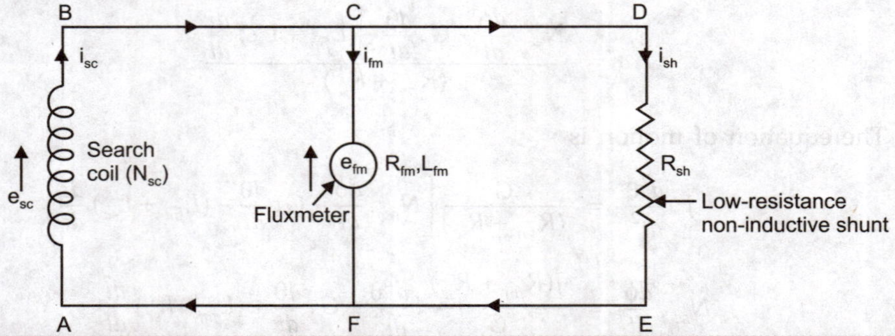

Fig. 3.20. Fluxmeter with shunt.

Let, R_{sc}, R_{fm}, R_{sh} = Resistances of search coil, fluxmeter and shunt respectively,
L_{sc}, L_{fm} = Inductances of search coil and fluxmeter respectively,
i_{sc}, i_{fm}, i_{sh} = Currents through search coil, fluxmeter and shunt respectively,
e_{sc} = e.m.f. induced in search coil because of change in flux change $= N_{sc}\dfrac{d\phi}{dt}$, and
e_{fm} = e.m.f. generated in fluxmeter coil because of rotation $= G.\dfrac{d\theta}{dt}$

- Applying Kirchhoff's law for the mesh ABCF, we get :

$$e_{sc} = i_{sc} R_{sc} + L_{sc} \frac{di_{sc}}{dt} + I_{fm} \cdot R_{fm} + L_{fm} \cdot \frac{di_{fm}}{dt} + e_{fm}$$

or, $$N_{sc} \cdot \frac{d\phi}{dt} - L_{sc} \frac{di_{sc}}{dt} - L_{fm} \cdot \frac{di_{fm}}{dt} - G \cdot \frac{d\theta}{dt} = i_{sc} R_{sc} + i_{fm} \cdot R_{fm}$$

Integrating as before, we have :

$$N_{sc} (\phi_2 - \phi_1) - L_{sc} (i_{sc2} - i_{sc1}) - L_{fm} (i_{fm2} - i_{fm1}) - G(\theta_2 - \theta_1)$$
$$= R_{sc} \int_0^T i_{sc}\, dt + R_{fm} \int_0^T i_{fm}\, dt \quad ...(3.68)$$

Since the currents are zero at both the beginning and the end of the flux change, therefore,

$$i_{sc2} = i_{sc1} = 0; \; i_{fm2} = i_{fm1} = 0$$

Thus, eqn. (3.68) reduces to :

$$N_{sc} (\phi_2 - \phi_1) - G(\theta_2 - \theta_1) = R_{sc} \int_0^T i_{sc}\, dt + R_{fm} \int_0^T i_{fm}\, dt \quad ...(3.69)$$

The equation of motion is :

$$J \frac{d^2\theta}{dt^2} = G i_{fm}$$

or, $$i_{fm} = \frac{J}{G} \cdot \frac{d^2\theta}{dt^2} = \frac{J}{G} \cdot \frac{d\omega}{dt} \quad ...(3.70)$$

where ω is the angular velocity at any instant.

Hence, $$\int_0^T i_{fm} dt = \frac{J}{G} \int_0^T \frac{d\omega}{dt} \cdot dt = \frac{J}{G} \int_{\omega_1}^{\omega_2} d\omega$$

or, $$\int_0^T i_{fm} dt = \frac{J}{G} (\omega_2 - \omega_1)$$

But, $$\omega_2 = \omega_1 = 0$$

∴ $$\int_0^T i_{fm} dt = 0$$

Hence eqn. (3.69) reduces to,

$$N_{sc} (\phi_2 - \phi_1) - G(\theta_2 - \theta_1) = R_{sc} \int_0^T i_{sc} dt \quad ...(3.71)$$

- Applying Kirchhoff's law to mesh *CDEF*, we have :

$$e_{fm} + L_{fm} \cdot \frac{di_{fm}}{dt} + i_{fm} R_{fm} = i_{sh} R_{sh}$$

or, $$G.\frac{d\theta}{dt} + L_{fm}.\frac{di_{fm}}{dt} = -i_{fm}R_{fm} + i_{sh}R_{sh} = -i_{fm}R_{fm} + (i_{sc} - i_{fm})R_{sh}$$

or, $$G.\frac{d\theta}{dt} + L_{fm}.\frac{di_{fm}}{dt} = -(R_{fm} + R_{sh})\,i_{fm} + i_{sc}R_{sh}$$

Integrating as before, we get :

$$G(\theta_2 - \theta_1) + L_{fm}(i_{fm2} - i_{fm1}) = -(R_{fm} + R_{sh})\int_0^T i_{fm}dt + R_{sh}\int_0^T i_{sc}dt$$

But $i_{fm2} - i_{fm1} = 0$ and $\int_0^T i_{fm}dt = 0$

Hence, $$G(\theta_2 - \theta_1) = R_{sh}\int_0^T i_{sc}dt \qquad ...(3.72)$$

or, $$\int_0^T i_{sc}dt = \frac{G(\theta_2 - \theta_1)}{R_{sh}}$$

Putting this value of $\int_0^T i_{sc}dt$ in eqn. (3.71), we have :

$$N_{sc}\,(\phi_2 - \phi_1) - G(\theta_2 - \theta_1) = \frac{R_{sc}}{R_{sh}}G(\theta_2 - \theta_1)$$

or, $$N_{sc}\,(\phi_2 - \phi_1) = G\frac{R_{sc} + R_{sh}}{R_{sh}}(\theta_2 - \theta_1) \qquad ...(3.73)$$

If $(\phi_2 - \phi_1) = \phi$, and, $(\theta_2 - \theta_1) = \theta$

we have, $$\phi = \frac{G}{N_{sc}}.\frac{R_{sc} + R_{sh}}{R_{sh}}.\theta \qquad ...(3.74)$$

Comparison of eqn. (3.74) with corresponding eqn. (3.67) of unshunted fluxmeter shows that the *multiplying factor* for the shunted fluxmeter is : $\frac{R_{sc} + R_{sh}}{R_{sh}}$

It may be noted that this factor involves the search coil resistance (R_{sc}), but *not* the fluxmeter resistance (R_{fm}).

Difference between fluxmeter and ballistic galvanometer :

Fluxmeter differs from ballistic galvanometer in following *respects* :

1. Fluxmeter has *very small controlling torque* in comparison to ballistic galvanometer.
2. Fluxmeter has *heavy electromagnetic damping* as compared to ballistic galvanometer.
3. The fluxmeter is *less sensitive and accurate* than the ballistic galvanometer.
4. The indication of fluxmeter is *independent of time taken by the flux changes.* This is of great importance in measurement of flux linking with highly inductive circuit where the flux changes may be relatively slow rendering the use of ballistic galvanometer impractical.

Example 3.27. *The following constants relate to a certain fluxmeter :*

Air gap flux density $= 0.06$ *Wb/m*^{2}

Turns on moving coil $= 50$

Area of moving coil $= 800$ *mm*^{2}

If the flux linking a 12 turn search coil of 220 mm^2 *area connected to the fluxmeter is reversed in a uniform field of 0.6 Wb/m*^{2}*, calculate the deflection of the fluxmeter.*

Solution. *Given* : $B_a = 0.06$ Wb/m^2; $N_{mc} = 50$; $A_{mc} = 800$ mm^2; $N_{sc} = 12$; $A_{sc} = 220$ mm^2; $B_f = 0.6$ Wb/m^2.

Deflection of the fluxmeter, θ :

Constant of fluxmeter, $G = N_{mc} \times B_a \times A_{mc} = 50 \times 0.06 \times (800 \times 10^{-6}) = 2400 \times 10^{-6}$

Flux linking with the search coil $= B_f \times A_{sc} = 0.6 \times (220 \times 10^{-6}) = 132 \times 10^{-6}$ Wb

Change in flux linking the coil, $\phi = 2 \times 132 \times 10^{-6} = 264 \times 10^{-6}$ Wb

Also, change in flux $\phi = \dfrac{G}{N_{sc}} . \theta$ [Eqn. (3.67)]

$$\therefore \qquad \theta = \frac{\phi \times N_{sc}}{G} = \frac{264 \times 10^{-6} \times 12}{2400 \times 10^{-6}} = \mathbf{1.32\ rad.\ or\ 75.6°} \quad \textbf{(Ans.)}$$

Example 3.28. *A fluxmeter is connected to a search coil having 600 turns and a mean area of 600 mm*^{2}*. The search coil is placed at a centre of a solenoid 1 metre long, wound with 900 turns. When a current of 6A is reversed, there is a deflection of 30 scale divisions.*

Calculate the calibration in flux linkages per scale division.

Solution. $N_{sc} = 600$; $A_{sc} = 600$ mm^2; $l = 1$m; $l_{sd} = 1$m; $N_{sd} = 900$; $I = 6$A; $\theta = 30$ scale divisions.

Flux linkages per scale division :

Field strength of solenoid, $H_{sd} = \dfrac{N_{sd} \times I}{l_{sd}} = \dfrac{900 \times 6}{1} = 5400$ A/m (assuming it to be very long)

Flux density, $B_{sd} = \mu_0 H_{sd} = 4\pi \times 10^{-7} \times 5400 = 6.78 \times 10^{-3}$ Wb/m^2

Flux linkages of fluxmeter search coil $= N_{sc} \times B_{sd} \times A_{sc}$

$= 600 \times 6.78 \times 10^{-3} \times (600 \times 10^{-6})$

$= 2.44 \times 10^{-3}$ Weber turn

Change in flux linkages $= 2 \times (2.44 \times 10^{-3}) = 4.88 \times 10^{-3}$ weber turn

Flux linkages per scale division $= \dfrac{4.88 \times 10^{-3}}{30} = \mathbf{0.1627 \times 10^{-3}}$ **weber turn** **(Ans.)**

Example 3.29. *A fluxmeter when used with a single turn search coil gives full scale deflection if a given uniform field linking with the search coil is reversed. It is proposed to extend the range of fluxmeter to measure flux densities six times the above value using the same search coil and method of reversals. If the resistance of search coil is 1.2 Ω, find the resistance of the shunt to be connected in parallel with the fluxmeter.*

Solution. *Given :* $R_{sc} = 1.2\ \Omega$

The range of instrument is extended $\dfrac{R_{sc} + R_{sh}}{R_{sh}}$ times by the use of a shunt

...[From eqn. 3.74]

where, R_{sc} = Resistance of search coil, and

R_{sh} = Resistance of shunt.

Now, we have : $\frac{R_{sc}+R_{sh}}{R_{sh}} = 6$

or, $\frac{1.2+R_{sh}}{R_{sh}} = 6$

$\therefore$ R_{sh} = **0.24 Ω** **(Ans.)**

Example 3.30. *A magnetic circuit excited by an m.m.f. of 8500 AT has a reluctance of 125,000 AT/Wb. A fluxmeter is used to measure the flux. If the fluxmeter has 125 divisions and flux linkages required for a deflection of one division are 0.16 × 10^{-3} weber turn, calculate the resistance of the shunt required for use with the search coil. Number of turns of search coil is 1 and its resistance is 0.03 Ω. The measurement is made by switching off the excitation.*

Solution. *Given* : m.m.f. = 8500 AT; Reluctance = 1,25,000 AT/Wb; N_{sc} = 1; R_{sc} = 0.03 Ω; Number of divisions on fluxmeter = 125; Flux linkages required for deflection of one division = 0.16 × 10^{-3} weber turn.

$$\text{Flux to be measured} = \frac{\text{m.m.f.}}{\text{reluctance}} = \frac{8500}{1,25,000} = 0.068 \text{ Wb}$$

Value of flux when the fluxmeter shows full scale deflection

$$= \frac{0.16\times10^{-3}\times125}{1} = 20 \times 10^{-3} \text{ Wb}$$

$$\text{Ratio } \frac{\text{flux to be measured}}{\text{flux for full scale deflection}} = \frac{0.068}{20\times10^{-3}} = 3.4$$

Thus, we have : $\frac{R_{sc}+R_{sh}}{R_{sh}} = \frac{0.03+R_{sh}}{R_{sh}} = 3.4$

$\therefore$ R_{sh} = **0.0125 Ω** **(Ans.)**

Example 3.31. *A fluxmeter is connected to a search coil of 100 turns and the mean area of the coil is 5 × 10^{-4} m^2. The search coil is placed at the centre of a standard solenoid, 1 metre long, uniformly wound with 800 turns. When a current of 5A is reversed, a deflection of 10 scale divisions is obtained with the fluxmeter. Calculate the calibration constant of the instrument in weber-turn per division.*

Solution. *Given* : N_{sc} = 100; A_{sc} = 5 × 10^{-4} m^2; l_{sd} = 1 m ; N_{sd} = 800; I = 5A; θ = 10 scale divisions.

Instrument constant :

Since the axial length is large as compared with the mean diameter of the solenoid, therefore, the field strength in the neighbourbood of the centre of the solenoid core is *assumed to be uniform.*

$$H_{sd} = \frac{N_{sd}I}{l_{sd}} = \frac{800\times5}{1} = 4000 \text{ AT/m}$$

The flux passing through the search coil,

$$\phi = \beta_{sd} \times A_{sc} = \mu_0\mu_r H_{sd} \times A_{sc}$$

or, $$\phi = 4\pi \times 10^{-7} \times 1 \times 4000 \times 5 \times 10^{-4} = 25.13 \times 10^{-7} \text{ Wb}$$

(μ_r = 1 for air medium)

Weber-turn = Flux × number of turns on search coil

$$= \phi \times N_{sc} = 25.13 \times 10^{-7} \times 100 = 25.13 \times 10^{-5}$$

Change in number of weber-turns on the reversal of the solenoid current

$$= 2 \times 25.13 \times 10^{-5} = 50.6 \times 10^{-5} \text{ weber-turn.}$$

Assuming uniform scale,

$$\textit{Instrument constant} = \frac{\text{Change in weber-turn on reversal of solenoid current}}{\text{Corresponding deflection}}$$

$$= \frac{50.6 \times 10^{-5}}{10}$$

$$= \mathbf{5.06 \times 10^{-5}} \textbf{ Weber-turn per scale division} \quad \textbf{(Ans.)}$$

3.7.4. Vibration Galvanometers

Vibration galvanometers are of two types : (*i*) *Moving magnet*, (*ii*) *Moving coil*. The latter type is more generally used because moving magnet type has the *disadvantage* of being seriously affected by magnet fields of the resonant frequency (unless adequately screened). The **moving-coil** *galvanometers are not appreciably affected by such fields*. Hence only moving coil type will be discussed here.

These galvanometers are of *"D'Arsonval type"* having a moving coil suspended between the poles of a permanent magnet. When an alternating current is passed through the moving coil, an *alternating deflecting torque is produced which makes the coil vibrate with a frequency equal to the frequency of the current passing*. On account of the moving parts, the amplitude of vibrations is *small*. However, if natural frequency of the moving system is made equal to the frequency of the current, *mechanical resonance* is obtained and the moving system vibrates with a *large amplitude*.

- These galvanometers are the most widely used *"**tuned detectors**" particularly as null point detectors in A.C. bridges*. They are manufactured for various frequencies from 5 Hz upto 1000 Hz; these are most commonly used below 1200 Hz above which range they are considerably more sensitive than telephone.

Construction :

Fig. 3.21(*a*) shows the construction of a Duddell's moving coil vibration galvanometer :

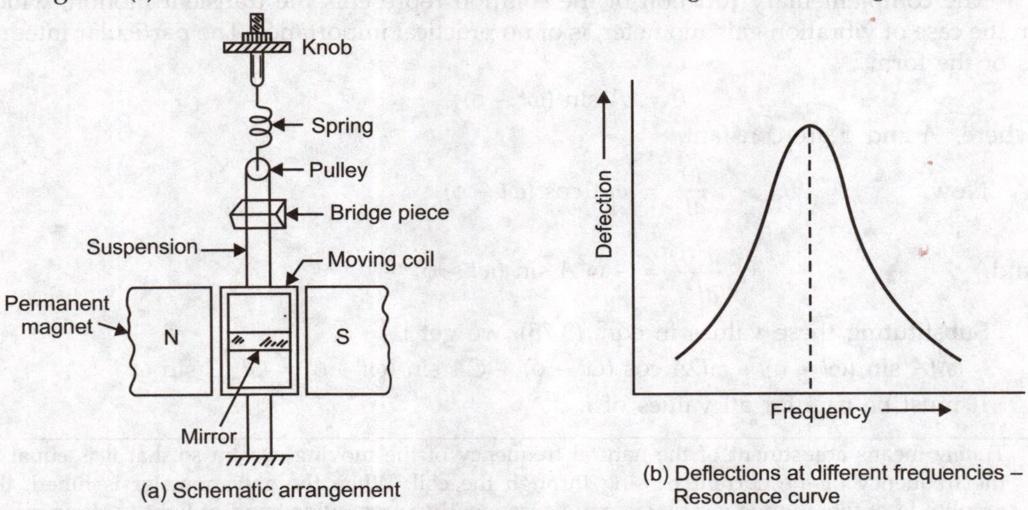

Fig. 3.21. Vibration galvanometer.

— The moving coil (consisting of a fine bronze or platinum silver sire) is suspended between poles of a strong permanent magnet. This wire passes over a small pulley at the top and is pulled tight by a spring attached to the pulley, the tension of the spring can be adjusted by turning a milled head attached to the spring.

Generally, the suspension is rigidly held by small movable clamps on bridges, and the active length of the suspension is varied by sliding the clamps along the suspension to change the natural response frequency.

— The moving system carries a small *mirror* upon which a light beam is thrown.

Working :

— When an alternating current is passed through the coil, an alternating torque is applied to it so that the reflected spot of light on the scale is drawn out in the form of band of light. The light of this band of light is *maximum* if the natural frequency of oscillation of coil *coincides* with the supply frequency due to *resonance.*

— The *tuning of the coil is done by either changing the length of suspension or by adjusting tension in the suspension. The former is achieved by raising or lowering bridge piece against which the bifilar loop presses while the latter is achieved by turning the knob [See Fig. 3.21(*a*)].

- By making the damping very small, the *resonance curve* of the galvanometer can be made sharply-peaked [Fig. 3.21(*b*)]. In that case, the instrument discriminates sharply against frequencies other than its own natural frequency. That is, its reflection becomes very small even when the frequency of the applied current differs by a very small amount from its resonance frequency.

Theory :

Let the value of current passing through the moving coil at an instant t be :

$$i = I_{max} \sin \omega t$$

Then, the *equation of motion* of the coil is :

$$J\frac{d^2\theta}{dt^2} + D\frac{d\theta}{dt} + C\theta = Gi = GI_{max} \sin \omega t \quad \text{...(3.75)}$$

where J, D and C have the usual meanings and G is the *deflection constant*.

The complementary function of the solution represents the transient motion, which in the case of vibration galvanometer, is of no practical importance. The particular integral is of the form :

$$\theta = A \sin (\omega t - \phi)$$

where, A and ϕ are constant.

Now, $$\frac{d\theta}{dt} = \omega A \cos (\omega t - \phi)$$

and, $$\frac{d^2\theta}{dt^2} = -\omega^2 A \sin (\omega t - \phi)$$

Substituting these values in eqn. (3.75), we get :

$$-\omega^2 JA \sin (\omega t - \phi) + \omega DA \cos (\omega t - \phi) + CA \sin (\omega t - \phi) = GI_{max} \sin \omega t$$

It must be true for all values of i.

---

* *Tuning* means adjustment of the natural frequency of the moving system so that it is equal to the frequency of the current passing through the coil. When the galvanometer is tuned, the amplitude of vibration is very large and consequently a very wide band of light is observed on the scale.

When, $\omega t = \phi$ $\omega DA = GI_{max} \sin \phi$ ...(i)

When, $(\omega t - \phi) = \frac{\pi}{2}$ $-\omega^2 JA + CA = GI_{max} \cos \phi$ ...(ii)

Since the phase angle ϕ of oscillations is of *no* practical significance, it may be eliminated by squaring and adding eqns. (*i*) and (*ii*).

$$\therefore \quad \omega^2 D^2 A^2 + A^2 (C - \omega^2 J)^2 = G^2 I_{max}^2$$

or,
$$A = \frac{GI_{max}}{\sqrt{[(\omega D)^2 + (C - \omega^2 J)^2]}} \quad ...(3.76)$$

This represents the amplitude A of the resulting oscillation for a sinusoidally alternating current of peak value I_{max} flowing through the moving coil of the galvanometer.

- The *current sensitivity* of a vibration galvanometer is about 50 mm/μA with a scale 1 m away. The *sensitivity is almost inversely proportional to the frequency.*

Comparison of ballistic galvanometer with vibration galvanometer :

1. Both are of D'Arsonval type.
2. They differ in details of usage rather in fundamental ideas.
3. Ballistic galvanometers have long period of the order of 20 to 30 seconds compared to 4 to 6 seconds for vibration type.
4. In vibration galvanometer, current through it is steady as long as the switch is pressed while in ballistic case it is transitory.
5. *"Vibration galvanometers" are more suitable for use at power and low audio frequencies while ballistic type can be used for higher frequencies.*

Example 3.32. *The following data relate to a vibration galvanometer tuned to a frequency of 50 Hz :*

Inertia constant = 0.012 kg-m²

Damping constant = 0.028 Nm/rad.s⁻¹

Calculate the ratio of sensitivity for fundamental to that for third harmonic of the galvanometer.

Solution. *Given :* $f = 50$ Hz; $J = 0.012$ kg-m^2; $D = 0.028$ N-m/rad.-s^{-1}

The amplitude of vibration of frequency ω is :

$$A = \frac{GI_{max}}{\sqrt{[\omega^2 D^2 + (C - \omega^2 J)^2]}} \quad \text{[Eqn. 3.76]}$$

∴ The ratio of amplitudes of fundamental and third harmonic is :

$$\frac{A_1}{A_3} = \frac{\sqrt{(3\omega_1 D)^2 + (C - 9\omega_1^2 J)^2}}{\sqrt{(\omega_1 D)^2 + (C - \omega_1^2 J)^2}}$$

But, for a *tuned galvanometer*, $\omega_1 = \sqrt{\frac{C}{J}}$ or $C = \omega_1^2 J$

$$\therefore \quad \frac{A_1}{A_3} = \frac{\sqrt{(3\omega_1 D)^2 + (\omega_1^2 J - 9\omega_1^2 J)^2}}{\omega_1 D}$$

$$= \frac{\sqrt{(3\omega_1 D)^2 + (8\omega_1^2 J)^2}}{\omega_1 D}$$

$$= \frac{\sqrt{(3\times\pi\times 50\times 0.028)^2 + \{8\times(2\pi\times 50)^2\times 0.012\}^2}}{2\pi\times 50\times 0.028}$$

$$= \mathbf{1077} \quad \textbf{(Ans.)}$$

3.7.5. Duddell's Oscillograph

This instrument is essentially a vibration galvanometer having a *low period of vibrations.* The natural frequency is about 10 kHz.

Construction :

Fig. 3.22 shows the schematic arrangement of a Duddell's oscillograph element :

— It consists of a single loop of thin phosphor-bronze strip which forms the moving system and is called a *vibrator loop*. This loop is situated in the field of a permanent magnet. Highly saturated electromagnets are used when high sensitivities are required.

— The vibrator loop is formed by passing the strip over an ivory pulley as shown in the figure.

— A spring tension adjusting device is attached to the ivory pulley

— A small mirror for reflecting light is cemented to the loop midway between the bridge pieces.

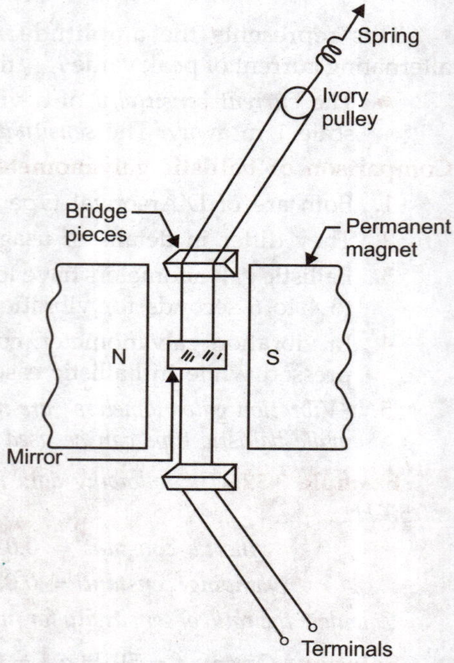

Fig. 3.22. Duddell's oscillograph element.

Such types of galvanometer can be constructed with diameter 3 mm to 6 mm and about 75 mm length and are called *pencil galvanometer*. They consist of moving element of the galvanometer and are set in a fixed frame of the permanent magnets. Two or more units per cm can be constructed and 50 or more galvanometers are included in a single oscillograph to record many waveforms simultaneously.

Working :

— When an alternating current is passed through the loop, an alternating torque acts upon the loop and it vibrates.

— When a beam of light is cast upon the mirror, it is reflected. As the loop is vibrating, the reflected beam will move backwards and forwards.

— If a wave shape is to be recorded, a photographic film is passed at a constant speed in a direction perpendicular to that of movement of the reflected beam of light. The speed of film depends upon the frequency of the wave being recorded. The wave shape is recorded on the film as a result of the two perpendicular motions *i.e.* of the film and the reflected light.

The oscillograph is calibrated by passing a known value of direct current in the loop and recording the deflection produced by the current.

— If, however, the waveform is *not to be recorded* but is *only to be observed,* the arrangement as shown in Fig 3.23 is used. The reflected beam of light from the vibrator mirror falls on the plain mirror and is further relected on to a screen. The plain mirror is rocked by a cam which is driven by a synchronous motor which in turn is supplied from the source whose waveform is to be observed. The plain mirror driven by the cam gives forward motion of the reflected beam on the screen. The forward motion is carried out for $1\frac{1}{2}$ cycles of the wave, and the plain mirror is returned to the original position during the next half cycle. During this cycle, the light is cut off by rotating shutter as shown in the figure. This is repeated continuously and the wave appears stationary on the screen on account of persistence of vision.

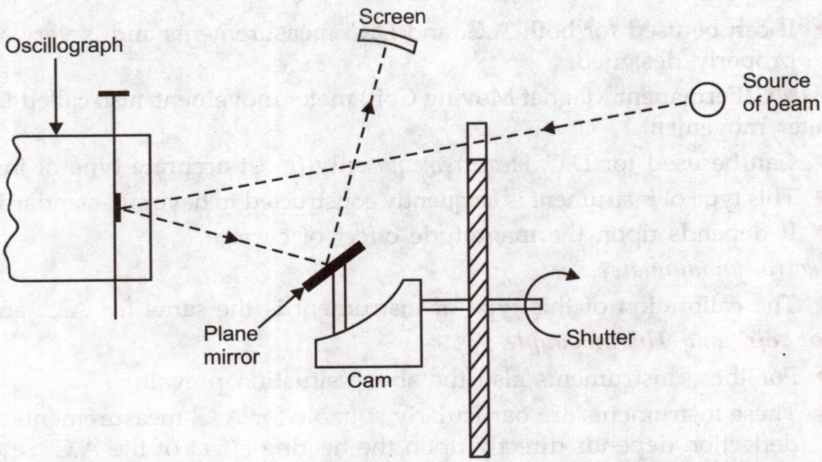

Fig. 3.23. Observation of waveform with Duddell's oscilloscope.

- Duddell's oscillographs are *very useful at low voltages and low frequencies*. They can be provided with a large member of elements and therefore a number of quantities may be *observed or recorded simultaneously*. They provide *convenient facilities for photographic purposes*.

However, their use is *limited to frequencies upto 500 Hz if an accurate record of the waveform is required.* The use of these instruments is quite *complicated.* Hence, these oscillographs are not used these days and are being replaced by cathode ray oscillographs (CROs).

3.8 ANALOG AMMETERS, VOLTMETERS AND OHMMETERS

3.8.1. Introduction

- *Analogue* ammeters and voltmeters are classified together, since there is no basic difference in their operating principles. The action of all ammeters and voltmeters, except those of the electrostatic variety, depends upon a deflecting torque principle by an electric current. In an ammeter, this torque is produced by the current to be measured, or by a definite fraction of it. In a voltmeter, it is produced by a current that is proportional to the voltage to be measured. Hence, both *voltmeters and ammeters are essential current measuring devices.*

— The ***"ammeter"*** has a *low resistance* so that when it is connected in ***series*** with any circuit, it does not change the current. Consequently, there is a small voltage drop and small power is absorbed.

The *"voltmeter"* has a *high resistance* and it is so designed that when connected in *parallel* to the circuit for measuring voltages it does *not* take appreciable current; consequently power consumed is small.

— *An ammeter of low range can be used as a voltmeter by connecting an external resistance in series with it.*

3.8.2. Types of Instruments mainly used as Ammeters and Voltmeters

The following types of instruments are mainly used as *ammeters* and *voltmeters.*

1. ***Moving iron :***
   - It is the *most commonly used* form of indicating instrument, as well as the *cheapest.*
   - It can be used for both A.C. and D.C. measurements and is very accurate, if properly designed.
2. *PMMC* (Permanent Magnet Moving Coil) meter movement; also called *D' Arsonval* meter movement :
   - Can be used for *D.C. measurements* only (most accurate type of instrument).
   - This type of instrument is frequently constructed to have sub-standard accuracy.
   - It depends upon the magnitude effect of current.
3. ***Electrodynamometer :***
   - The calibration of this type of instrument is the same for A.C. and D.C.
4. ***Hot wire and Thermocouple :***
   - For these instruments also the above situation prevails.
   - These instruments are particularly suitable for A.C. measurements since their deflection depends directly upon the heating effect of the A.C. *i.e.,* upon the r.m.s. value of the current. Their readings are therefore *independent* of the frequency.
5. ***Rectifier.***
6. ***Electrostatic :***
   - These instruments used as voltmeters have the advantage that *their power consumption is exceedingly small.*
   - They can be made to cover a large range of voltages and can be constructed to have sub-standard accuracy.
7. ***Induction type :***
   - The induction principle is most *generally used for watt-hour meters. This principle is not preferred for use in ammeters and voltmeters because of the comparatively high cost and inaccuracy of the instrument.*

3.8.3. Errors in Ammeters and Voltmeters

The common errors in ammeters and voltmeters are on account of the following *factors :*

(i) Friction.

(ii) Change in temperature.

(iii) Mechanical unbalance.

(iv) Variation of the electrical and mechanical properties of material employed with time.

1. Friction error :

— The *friction of pivot in the jewel* produces a frictional torque which opposes the deflection of the moving element and as a result, the pointer comes to rest a short distance from its correct position; this causes an error in readings of the instrument.

This error is more serious for sensitive instruments designed for low operating torque.

— In order to *reduce the effect of friction torque,* and consequently the error produced by it the weight of the moving system must be made as small as possible compared with the operating force, *i.e.,* the *ratio of torque to weight must be large.* In high quality instruments, the ratio of deflecting torque of full scale to weight of moving system is rarely much less than 0.01 Nm/kg.

From the point of view of a small friction torque a *vertical spindle is generally preferred to a horizontal one.*

2. Temperature error :

— The most serious error is produced by the heat generated in the instrument, or by changes in ambient (room) temperature, is *due to a change in the resistance of the working coil.* Such a change of resistance is of *little importance in ammeters* but in *voltmeters,* in which working current should be directly proportional to the applied voltage, it is *essential that the resistance of the instrument remains as nearly constant as possible.*

The power loss in the instrument should be small, and resistance coils which are likely to produce appreciable heating should be mounted, if possible, in such a position that they are *well-ventilated.*

— In order to *eliminate temperature errors, the working coil is wound with copper wire and is of comparatively low resistance.* A high *"swamping resistance"* of material whose resistance temperature coefficient is small, is connected in series with the coil, so that, although the resistance of the coil may change considerably, the *change in total resistance of circuit is small.*

Other errors resulting from heating may be caused by *expansion of control spring, or of other parts of the instrument,* although such errors are usually *small.*

3. Observational error :

— These errors are due to *mis-reading of the scale, parallex in readings* and *errors of estimation.* Precision of readings becomes difficult if the pointer has a broad tip. *Interpolation errors* may enter, particularly if the scale readings are too large or too small.

— *Parallex error* is caused by the observer *not* having his line of sight on the pointer exactly at right angles to the plane of the scale. Such an error is eliminated by providing *a mirror beneath the scale and a knife-edged pointer.*

3.8.4. Moving-Iron Instruments

Moving-iron instruments are commonly used in *laboratories and switch boards* at commercial frequencies because they are *very cheap and can be manufactured with required accuracy.*

3.8.4.1. Principle of working

The general principle of working of a moving-iron instrument (which is strictly relevant to *attraction type* of moving-iron instruments) is explained below :

— The moving element of the system consists of a plate or vane of soft iron or of high permeability steel. This iron vane is so situated that it can move in a magnetic field produced by a stationary coil.

— When the coil is excited, it becomes an electromagnet and iron vane moves in such a way so as to increase the flux of the electromagnet; this is because the vane tries to occupy a position of minimum reluctance. Thus, the force produced is always in such direction so as to increase the inductance of the coil (since inductance is *inversely proportional to reluctance of magnetic circuit of the coil*).

3.8.4.2. Theory of the moving-iron instruments

● An expression for the torque of a moving iron instrument may be derived by considering the energy relations when there is a small increment in current supplied to the instrument. When this happens there will be a *small deflection* $d\theta$ of the pointer and *work will be done.* This is given by $T_d \cdot d\theta$, where T_d is the deflecting torque of the instrument. There will also be a *change in the energy stored owing to inductance.*

Let, I = Initial current,

L = Inductance of the instrument, and

θ = Deflection.

Now, if the current increases by dI, then deflection changes by $d\theta$ and inductance by dL. In order to effect an increment dI in the current, there must be an increase in the applied voltage given by :

$$e = \frac{d}{dt}(LI) = I\frac{dL}{dt} + L\cdot\frac{dI}{dt} \quad \text{...(i)}$$

The *electrical energy supplied* is,

$$e\,I\,dt = I^2\,dL + IL\,dI \quad \text{...(ii)}$$

The stored energy changes from

$$\frac{1}{2}I^2L \text{ to } \frac{1}{2}(I+dI)^2(L+dL) \quad \text{...(iii)}$$

Hence the *change in stored energy* is

$$= \frac{1}{2}(I+dI)^2(L+dL) - \frac{1}{2}I^2L$$

or,

$$= \frac{1}{2}(I^2 + 2\,I\,dI + dI^2)(L+dL) - \frac{1}{2}I^2L$$

Neglecting second-and higher-order terms in small quantities, this reduces to :

$$IL\,dI + \frac{1}{2}I^2\,dL$$

Invoking the principle of conservation of energy, we have :

Electrical energy supplied = Increase in stored energy + mechanical work done

$$I^2dL + IL\,dI = IL\,dI + \frac{1}{2}I^2\,dL + T_d\cdot d\theta$$

or,

$$T_d \cdot d\theta = \frac{1}{2}I^2dL$$

or,

$$T_d = \frac{1}{2}I^2\frac{dL}{d\theta} \quad \text{...(3.77)}$$

where, T_d is in Nm, I in amps., L in henry, and θ in radians.

• The moving system is provided with control springs and in turn the deflecting torque T_d is balanced by the controlling torque, $T_c = C\theta$, where C is the control spring constant (Nm/rad.) and θ is the deflection (rad.).

At steady deflection state :

$$T_c = T_d$$

or,
$$C\theta = \frac{1}{2} I^2 \frac{dL}{d\theta}$$

or, Deflection,
$$\theta = \frac{1}{2} \frac{I^2}{C} \frac{dL}{d\theta} \quad ...(3.78)$$

It is evident from eqn. (3.78) that the *deflection is proportional to square of the r.m.s. value of the operating current.* The deflecting torque is, therefore, *unidirectional* whatever may be the polarity of the current.

3.8.4.3. Classification of moving-iron instruments

Moving-iron instruments can be divided into two types :

1. *Attraction type* ..... in which a sheet of soft iron is *attracted towards a solenoid.*
2. *Repulsion type* ..... in which two parallel rods or strips of soft iron, magnetised inside a solenoid, are regarded as *repelling each other.*

1. Attraction type :

Fig. 3.24, shows the sectional front and an end elevation of the attracted-iron type instrument.

It consists of a solenoid (or coil) C and oval shaped soft-iron disc D in such a way that it can move in or cut of the solenoid. To this iron, a pointer P is attached so that it may deflect along with the moving iron over a graduated scale. The soft-iron disc is made of sheet metal specially shaped to give a scale as nearly uniform as possible.

— When the current to be measured (or a definite fraction of the current to be measured or proportional to the voltage to be measured) is passed through the solenoid, a magnetic field is set up inside the solenoid, which in turn magnetises the iron. Thus the soft-iron disc is attracted into the solenoid/coil, causing the spindle and the pointer to rotate. *Damping* is provided by vane V attached to the spindle and moving in an *air chamber*, and *control* is by *hair spring.*

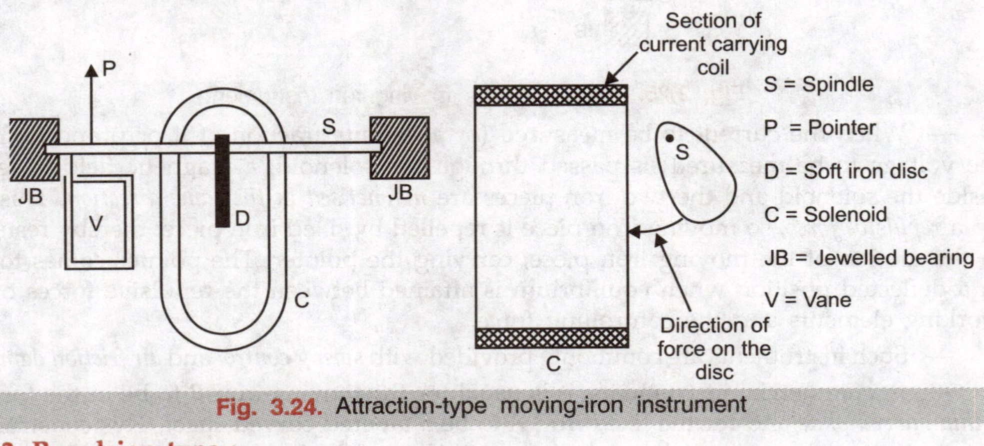

Fig. 3.24. Attraction-type moving-iron instrument

2. Repulsion type :

Repulsion-type moving-iron instrument is shown in Fig. 3.25 :

— Here, there are two irons, one fixed (*A*) and the other mounted on a short arm fixed (*B*) to the instrument spindle. The two irons lie in the magnetic field due to a solenoid/coil *C*. When there is no current in the coil the two iron pieces (moving one and fixed one) are almost touching each other and the pointer rests on zero position.

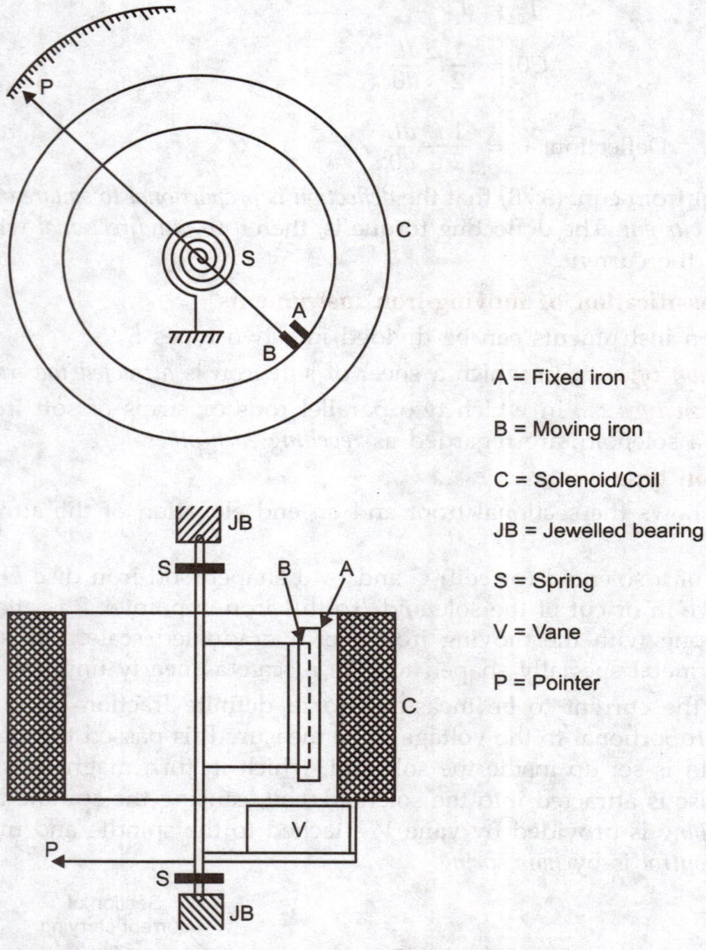

Fig. 3.25. Repulsion-type moving-iron instrument

— When the current to be measured (or a definite fraction of it or proportional to the voltage to be measured) is passed through the solenoid, a magnetic field is set up inside the solenoid and the two iron pieces are *magnetised in the same direction.* This sets up a *repulsive force*, so moving iron piece is repelled by fixed iron piece, thereby resulting in the motion of the moving iron piece, carrying the pointer. The pointer comes to rest in a deflected position when equilibrium is attained between the repulsive forces of the working elements and the controlling force.

— Such instruments are commonly provided with *spring control* and *air friction damping.*

— In commercial instruments, it is usual for the moving-iron *B* to be in the form of a *thin curved plate* and for the fixed iron *A* to be a tapered curved sheet. This construction can be arranged to give *a longer and more uniform scale* than is possible with the rods shown in Fig. 3.25.

Deflecting torque in moving-iron instruments :

In both the attraction and repulsion type moving-iron instruments it is found that for a given position of the moving system, the value of the *deflecting torque is proportional to the square of the current, so long as the iron is working below saturation.* Hence, if the current waveform is as shown in Fig. 3.26, the variation of the deflecting torque is represented by the dotted wave. If the supply frequency is 50 Hz, the torque varies between zero and a maximum 100 times a second, so that the moving system (due to its inertia) takes up a position corresponding to the mean torque, where

mean torque ∝ mean value of the square of the current = kI^2

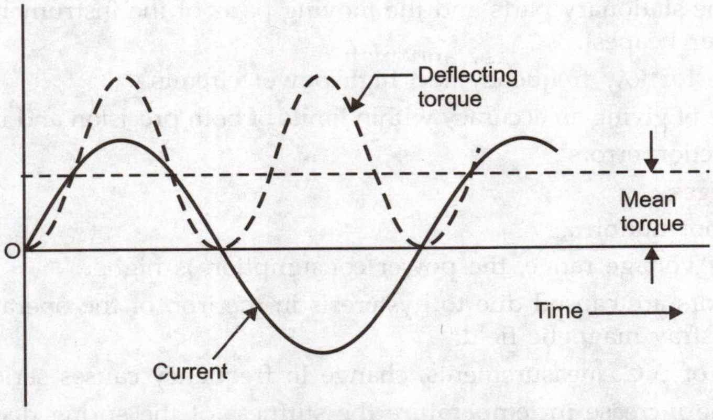

Fig. 3.26. Deflecting torque in a moving-iron instrument

where, k = a constant for a given instrument, and

I = r.m.s. value of the current.

Thus the moving-iron instrument can be used to *measure both direct current and alternating current,* and in the latter case the instrument gives the r.m.s. value of the current.

Owing to the deflecting torque being proportional to the square of the current, the scale divisions are not uniform, being cramped at the beginning and open at the upper end of the scale.

Note. For both types of instruments (attraction-type and repulsion-type), the necessary magnetic field is produced by the *ampere-turns of a current-carrying coil.*

— In case the instrument is to be used as an *ammeter,* the coil has *comparatively few turns of thick wire so* that the ammeter has *low resistance* because it is connected in *series* with the circuit.

— In case it is to be used as a *voltmeter,* the coil has *high impedance* so as to draw as small a current as possible since it is connected in *parallel with the circuit.* As current through the coil is small it has large *number of turns* in order to produce sufficient ampere-turns.

3.8.4.4. Comparison between attraction and repulsion types of instruments

Although, in general, it may said that both types of these instruments have the same advantages and are subject to same limitations, yet they differ as follows :

— An *attraction type instrument* will usually have a *lower inductance* that the corresponding repulsion type instrument, and voltmeters will therefore be *accurate over a wider range of frequency and there is a greater possibility of using shunts and* ammeters.

— Repulsion instruments, however, are *more suitable for economical production in manufacture, and a nearly uniform scale is more easily obtained.* These instruments are *much more common than the attraction type.*

3.8.4.5. Advantages, disadvantages and applications of moving-iron instruments

Advantages :

1. Can be used both in D.C. as well as in A.C. circuits.
2. Robust and simple in construction.
3. Possess high operating torque.
4. Can withstand overload momentarily.
5. Since the stationary parts and the moving parts of the instruments are simple so they are cheapest.
6. Suitable for low frequency and high power circuits.
7. Capable of giving an accuracy within limits of both precision and industrial grades.
8. Less friction errors.

Disadvantages :

1. Scales not uniform.
2. For low voltage range, the power consumption is higher.
3. The errors are caused due to hysteresis in the iron of the operating system and due to stray magnetic field.
4. In case of A.C. measurements, change in frequency causes serious error.
5. With the increase in temperature the stiffness of the spring decreases.

- There is a *difference* between D.C. and A.C. calibrations because of inductance effect of the instrument and eddy currents on A.C. The instruments should be calibrated for frequency at which these are to be tested.

Applications :

- The moving iron instruments are widely used as ammeters and voltmeters for D.C. and A.C.
- These instruments can be easily constructed for ranges of 0.1 A to 30 A without the use of a shunt and a wide range of voltages.

3.8.4.6. Errors in moving-iron instruments

The various types of errors which occur in moving-iron instruments are discussed as under :

I. Errors with both D.C. and A.C.

(i) Errors due to hysteresis :

— Because of hysteresis in the iron parts of the moving system, readings are higher for descending values but lower for ascending values of current (and voltage).

— The hysteresis error is almost completely *eliminated* by using Mumetal or Perm-alloy, which have negligible hysteresis loss.

(ii) Errors due to stray fields :

— External stray magnetic fields are liable to affect adversely the accurate functioning of the instrument.

— These errors can be *minimized* by using an iron case or a thin iron shield over the working parts.

(iii) *Error due to temperature changes :*

— The effect of temperature changes on moving-iron instruments arise chiefly from temperature coefficient of spring; the error may be 0.02 per cent per °C change in temperature.

— The errors in *voltmeter* are caused due to self-heating of coil and series resistance. Therefore, the series resistance should be made of material like *manganin* which has a small temperature coefficient. The value of series resistance should be very large as compared with the coil resistance in order to minimize errors due to self-heating. The series resistance is about ten times the coil resistance in case of switch board instruments.

II. Errors with A.C. only

(i) Errors due to change in frequency : The changes in frequency may cause errors due to *changes of reactance of the working coil* and also due to changes of magnitude of *eddy currents* set up in the metal parts of instrument.

(a) Reactance of instrument coil :

If the frequency of the supply changes, the reactance of the coil will change. This causes change in the current in coil of the voltmeter and hence an error in measurement. Since an ammeter is in series with the circuit, the current in its coil is the same as load current. Hence, *change in frequency causes an error in voltmeter reading only.*

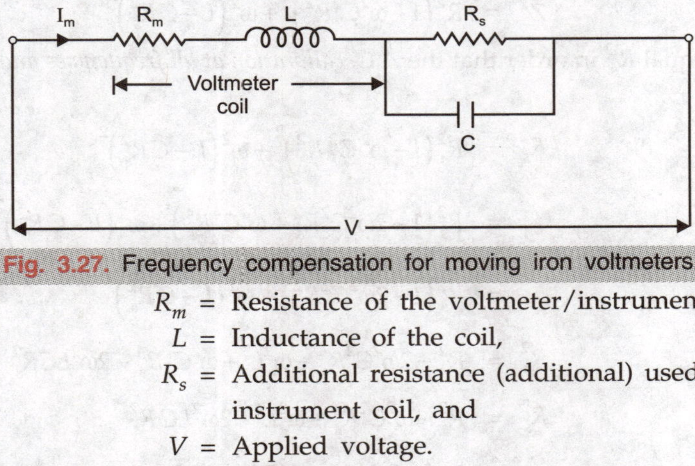

Fig. 3.27. Frequency compensation for moving iron voltmeters.

Let, R_m = Resistance of the voltmeter/instrument coil,

L = Inductance of the coil,

R_s = Additional resistance (additional) used in *series* with instrument coil, and

V = Applied voltage.

Then the current through the voltmeter coil,

$$I_m = \frac{V}{\sqrt{(R_m + R_s)^2 + \omega^2 L^2}} \qquad \text{...(3.79)}$$

If the frequency is *lower* than the rated value, I_m would be higher and the error will be *positive.* If the frequency is *higher* than rated value, I_m would be lower and error will be *negative.*

One of method of compensation for frequency error is to connect a capacitor C across the series resistance R_s as shown in Fig. 3.27. The impedance of the circuit,

$$\bar{Z} = R_m + j\omega L + \frac{R_s}{1 + j\omega C R_s}$$

$$= R_m + j\omega L + \frac{R_s}{1 + j\omega C R_s} \times \frac{1 - j\omega C R_s}{1 - j\omega C R_s}$$

or, $$\bar{Z} = R_m + j\omega L + \frac{R_s - j\omega C R_s^2}{1+\omega^2 C^2 R_s^2} \qquad ...(3.80)$$

The value of C can be found from eqn. (3.80) by equating Z to resistance $R_m + R_s$. Generally $R_m << R_s$. Then,

$$\bar{Z} = j\omega L + \frac{R_s - j\omega C R_s^2}{1+\omega^2 C^2 R_s^2}$$

Since $\omega C R_s << 1$, we can write

$$\bar{Z} = j\omega L + \left(R_s - j\omega C R_s^2\right)\left(1-\omega^2 C^2 R_s^2\right)$$

$$= j\omega L + R_s - j\omega C R_s^2 - \omega^2 C^2 R_s^3 + j\omega^3 C^3 R_s^4$$

$$= R_s - \omega^2 C^2 R_s^3 + j\left[\omega L - \omega C R_s^2 (1-\omega^2 C^2 R_s^2)\right]$$

$$\simeq R_s - \omega^2 C^2 R_s^3 + j\left[\omega L - \omega C R_s^2\right] \quad \text{..... as } \omega^2 C^2 R_s^2 << 1$$

$$= R_s\left(1-\omega^2 C^2 R_s^2\right) + j\omega\left(L - C R_s^2\right)$$

or, $$\bar{Z}^2 = R_s^2\left(1-\omega^2 C^2 R_s^2\right)^2 + \omega^2\left(L - C R_s^2\right)^2$$

This must equal R_s^2 in order that the *A.C. calibration at all frequencies and D.C. calibration is the same.*

$$R_s^2 = R_s^2\left(1-\omega^2 C^2 R_s^2\right)^2 + \omega^2\left(L - C R_s^2\right)^2$$

$$= R_s^2\left(1-2\omega^2 C^2 R_s^2 + \omega^4 C^4 R_s^4\right) + \omega^2\left(L - C R_s^2\right)^2$$

$$= R_s^2\left(1-2\omega^2 C^2 R_s^2\right) + \omega^2\left(L - C R_s^2\right)^2 \quad \text{...... as } \omega^4 C^4 R_s^4 << 1$$

or, $$R_s^2 = R_s^2 - 2\omega^2 C^2 R_s^4 + \omega^2 L^2 + \omega^2 C^2 R_s^4 - 2\omega^2 L C R_s^2$$

or, $$R_s^2 = R_s^2 - \omega^2 C^2 R_s^4 + \omega^2 L^2 - 2\omega^2 L C R_s^2$$

or, $$\omega^2 L^2 - 2\omega^2 L C R_s^2 - \omega^2 C^2 R_s^4 = 0$$

or, $$L^2 - 2LC R_s^2 - C^2 R_s^4 = 0$$

or, $$L = \frac{2C R_s^2 \pm \sqrt{4C^2 R_s^4 + 4C^2 R_s^4}}{2}$$

or, $$L = \frac{2C R_s^2 \pm 2.83\, C R_s^2}{2} = 2.41 C\, R_s^2$$

∴ $$C = \frac{L}{2.41\, R_s^2} = 0.41\frac{L}{R_s^2} \qquad ...(3.81)$$

Because of the approximations involved in deriving eqn. (3.81), the *above value of C will provide compensation only for a limited frequency range (upto about 100 Hz only).*

(*b*) *Eddy currents* : These errors are caused by eddy currents induced in the iron parts of the instruments.

Refer to Fig. 3.28.

Let, M = Mutual inductance between the instrument coil and the iron parts,

I = Current in the instrument coil,

I_e = Eddy current,

I_{e1}, I_{e2} = Component of eddy current I_e,

R_e, L_e = Resistance and inductance of eddy current path, and

$E_e = \omega MI$ = Induced voltage due to current I in the instrument coil; it lags I by 90°.

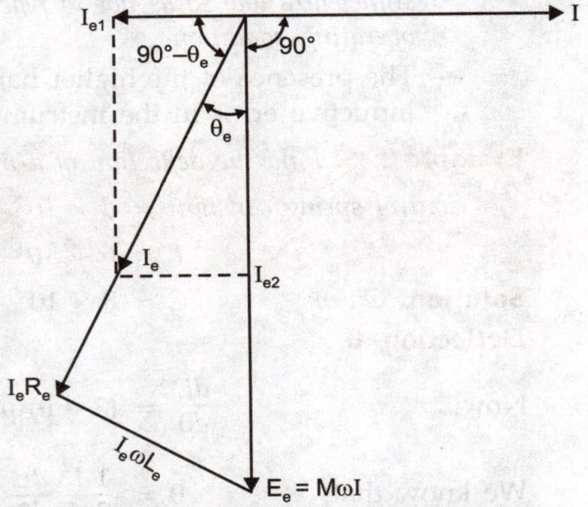

Fig. 3.28. Effect of eddy currents in moving-iron instruments.

Now,
$$I_e = \frac{E_e}{\sqrt{R_e^2 + \omega^2 L_e^2}} = \frac{\omega MI}{\sqrt{R_e^2 + \omega^2 L_e^2}} \quad ...(3.82)$$

$$I_e \text{ lags } E_e \text{ by an angle } \theta_e = \tan^{-1}\left(\frac{\omega L_e}{R_e}\right)$$

A component of this current, $I_{e1} = I_e \cos (90° - \theta_e) = I_e \sin \theta_e$ opposes the instrument current I and sets up an opposite field thus reducing the torque on the moving system. Referring to phase diagram shown in Fig. 3.28, we have :

$$I_{e1} = I_e \sin \theta_e = \frac{\omega MI}{\sqrt{R_s^2 + \omega^2 L_e^2}} \times \frac{\omega L_e}{\sqrt{R_e^2 + \omega^2 L_e^2}}$$

$$= \frac{\omega^2 ML_e I}{R_e^2 + \omega^2 L_e^2}$$

$$\therefore \quad I_{e1} \simeq \frac{\omega^2 ML_e I}{R_e^2} \quad \text{when } R_e >> \omega L_e \textit{ i.e., } \omega \textit{ is small}$$

$$\simeq \frac{MI}{L_e} \quad \text{when } \omega L_e >> R_e \textit{ i.e., } \text{when } \omega \textit{ is large}$$

Thus we find that *at low frequencies the eddy current error increases with square of the frequency while at high frequencies the error is practically constant.*

- *For reducing eddy current error the metal parts are placed as far out of the field as possible.*

(ii) Waveform error :

— The moving-iron instruments may be seriously affected by waveform, both on account of the change in the form of flux waveform and, in the case of voltmeters, the effect of harmonics upon the inductance.

— A moving-iron instrument with moderately saturated iron calibrated on a sine wave current and used on a peaky current wave gives readings *considerably lower than actual ones.* Contrary a flat top wave causes the instrument to read too high. This is the reason for *keeping the induction in iron sufficiently low so as not to reach the bend of the magnetisation curve under any operating condition.*

— The presence of the higher harmonics on any considerable scales causes the inductive error in the instrument and makes it to read too low.

Example 3.33. *Find the deflection of a moving-iron ammeter having the following data :*

Control spring constant = 8×10^{-6} Nm/rad.; Current = 6A,

$$L = 6 + 3\theta - 0.5\,\theta^2\ \mu H.$$

Solution. *Given:* $C = 8 \times 10^{-6}$ Nm/rad.; $I = 6$A; $L = 6 + 3\theta - 0.5\,\theta^2\ \mu$H

Deflection, θ :

Now, $$\frac{dL}{d\theta} = (3 - \theta)\ \mu\text{H/rad. or } (3 - \theta) \times 10^{-6}\ \text{H/rad.}$$

We know that, $$\theta = \frac{1}{2}\frac{I^2}{C}\frac{dL}{d\theta} \quad \text{...[Eqn. (3.78)]}$$

or, $$\theta = \frac{1}{2} \times \frac{6^2}{8 \times 10^{-6}} \times (3 - \theta) \times 10^{-6}$$

$$\theta = 2.25 \times 10^6 (3 - \theta) \times 10^{-6} = 2.25(3 - \theta)$$

or, $$\theta = \frac{2.25 \times 3}{3.25} = 2.077\ \text{rad.} = \mathbf{119°}$$ **(Ans.)**

Example 3.34. *The inductance of a moving-iron ammeter is given by the expression :*

$$L = (12 + 5\theta - 2\theta^2)\mu H,$$

where θ is the angular deflection in radians from zero position.

Determine : (i) The spring constant.

(ii) The angular deflection in radians for a current of 10 A if the deflection for a current of 5A is 30°.

(Delhi University)

Solution. *Given :* $L = (12 + 5\theta - 2\theta^2) \times 10^{-6}$ H

(*i*) Spring constant, *C* :

Now, $$\frac{dL}{d\theta} = (5 - 4\theta) \times 10^{-6}\ \text{H/rad.} \quad \text{...(i)}$$

Also, $$\theta = \frac{1}{2}\frac{I^2}{C}\frac{dL}{d\theta} \quad \text{...[Eqn. (3.78)]}$$

or, $$\frac{dL}{d\theta} = \frac{2C\theta}{I^2} \quad \text{...(ii)}$$

From (*i*) and (*ii*), we get

$$(5 - 4\theta) \times 10^{-6} = \frac{2C\theta}{I^2}$$

Inserting $\theta = 30°$ or $\frac{\pi}{6}$ rad. and $I = 5$A in the above expression, we get :

$$\left(5 - 4 \times \frac{\pi}{6}\right) \times 10^{-6} = \frac{2 \times C \times (\pi/6)}{5^2}$$

or, $$2.906 \times 10^{-6} = 0.0419\ C$$

∴ $$C = \frac{2.906 \times 10^{-6}}{0.0419} = \mathbf{69.36 \times 10^{-6}\ Nm/rad.}$$ **(Ans.)**

***(ii)* Angular deflection for a current of 10A, θ :**

Substituting I = 10A, in the equation we get :

$$\theta = \frac{1}{2}\times\frac{10^2}{69.36\times10^{-6}}\times(5-4\theta)\times10^{-6}$$

or, $$\theta = 0.72(5 - 4\theta) = 3.6 - 2.88\theta$$

∴ $$\theta = \frac{3.6}{3.88} \text{ rad.} = 0.928 \text{ rad. or } \mathbf{53.2°} \text{ (Ans.)}$$

Example 3.35. *The inductance of a moving iron ammeter with a full scale deflection of 90° at 1.5A is given by the expression :*

$$L = 180 + 400\theta - 4\theta^2 - \theta^3 \ \mu H$$

where, θ is the deflection in radians from the zero position, calculate :

(i) Spring constant.

(ii) The angular deflection of the pointer for a current of 1.0 A. **(Nagpur University)**

Solution: *Given :* $$L = (180 + 40\theta - 4\theta^2 - \theta^3) \times 10^{-6} \text{ H},$$

$$\theta = 90° \text{ or } \frac{\pi}{2} \text{ rad. at 1.5 A current}$$

***(i)* Spring constant, C :**

Now, $$\frac{dL}{d\theta} = (40 - 8\theta - 3\theta^2) \times 10^{-6} \text{ H rad.} \quad ...(i)$$

Also, $$\theta = \frac{1}{2}\frac{I^2}{C}\frac{dL}{d\theta} \quad ...[\text{Eqn. (3.78)}]$$

or, $$\frac{dL}{d\theta} = \frac{2C\theta}{I^2} \quad ...(ii)$$

From *(i)* and *(ii)*, we have :

$$(40 - 8\theta - 3\theta^2) \times 10^{-6} = \frac{2C\theta}{I^2}$$

Substituting θ = π rad., and I = 1.5 A in the above expression, we get,

$$\left[40 - 8\times\frac{\pi}{2} - 3\times\left(\frac{\pi}{2}\right)^2\right]\times10^{-6} = \frac{2C\times(\pi/2)}{(1.5)^2}$$

or, $$20.03 \times 10^{-6} = 1.396 \text{ C}$$

∴ $$C = \mathbf{14.35 \times 10^{-6} \text{ Nm/rad.}} \text{ (Ans.)}$$

***(ii)* The angular deflection for a current of 1.0 A, θ :**

Substituting I = 1.0 A in the above equation, we get

$$\theta = \frac{1}{2}\times\frac{(1)^2}{14.35\times10^{-6}}\times(40-8\theta-3\theta^2)\times10^{-6}$$

$$28.7\theta = 40 - 8\theta - 3\theta^2$$

or, $$3\theta^2 + 36.7\theta - 40 = 0$$

or, $$\theta = \frac{-36.7 \pm \sqrt{(36.7)^2 + 4\times3\times40}}{6} = 1.007 \text{ rad.} = \mathbf{57.7°} \text{ (Ans.)}$$

Example 3.36. *The inductance of a moving-iron instrument is given by the expression :*

$$L = (0.01 + K\theta)^2 \ \mu H,$$

where θ is the angular deflection in radians from zero position. The instrument angular deflections corresponding to currents of 2A and 5A are 45° and 90° respectively. Determine the value of K.

(Bangalore University)

Solution. *Given :* $L = (0.01 + K\theta)^2 \ \mu H$

Value of K :

Now, $$\frac{dL}{d\theta} = 2(0.01 + K\theta)K \ \mu H \quad ...(i)$$

Also, $$\theta = \frac{1}{2}\frac{I^2}{C}\frac{dL}{d\theta} \quad ...[\text{Eqn. (3.28)}]$$

or, $$\frac{dL}{d\theta} = \frac{2C\theta}{I^2} \quad ...(ii)$$

From (*i*) and (*ii*), we have :

$$\frac{2C\theta}{I^2} = 2(0.01 + K\theta)K \times 10^{-6}$$

Substituting $\theta = 45°$ or $\frac{\pi}{4}$ radian when $I = 2A$ and $= 90°$ or $\frac{\pi}{2}$ when $I = 5A$, we get:

$$\frac{2C\times(\pi/4)}{(2)^2} = 2\left(0.01 + K\times\frac{\pi}{4}\right)K\times10^{-6} \quad ...(i)$$

and, $$\frac{2C\times(\pi/2)}{(5)^2} = 2\left(0.01 + K\times\frac{\pi}{2}\right)K\times10^{-6} \quad ...(ii)$$

Dividing (*i*) by (*ii*), we get :

$$\frac{50}{16} = \frac{0.01+\frac{\pi}{4}K}{0.01+\frac{\pi}{2}K}$$

or, $$50\left(0.01+\frac{\pi}{2}K\right) = 16\left(0.01+\frac{\pi}{4}K\right)$$

or, $$0.5 + 25\pi K = 0.16 + 4\pi K$$

or, $$21\pi K = -0.34$$

or, $$K = -\frac{0.34}{21\pi} = \mathbf{-5.15 \times 10^{-3}} \ \textbf{(Ans.)}$$

Example 3.37. *The law of deflection of a moving iron ammeter is given by the expression :*

$$I = 4\theta^n$$

where θ is the deflection in radians and n is constant. The self-inductance when the meter current is zero is 10 mH. The spring constant is 0.16 Nm/rad.

(*i*) *Establish an expression for self-inductance of the meter as a function of θ and n.*

(*ii*) *Calculate the meter current and deflection that corresponds to a self inductance of 60 mH, when n = 0.75.*

(Gorakhpur University)

Solution. *Given :* $I = 4\theta^n$; $L = 10$ mH, when $I = 0$; $C = 0.16$ Nm/rad.

(i) Expression for self-inductance :

Deflection, $\theta = \frac{1}{2}\frac{I^2}{C}\frac{dL}{d\theta}$ radian ...[Eqn. (3.78)]

or, $\frac{dL}{d\theta} = \frac{2C\,\theta}{I^2}$

or, $dL = \frac{2C}{I^2}\theta \cdot d\theta$

Substituting $I = 4\theta^n$ in the above expression, we get :

$$dL = \frac{2C\,\theta}{(4\theta^n)^2} \cdot d\theta = \frac{1}{8} C\,\theta^{1-2n} \cdot d\theta$$

Integrating the above expression, we get

$$L = \frac{C\,\theta^{2-2n}}{8(2-2n)} + A = \frac{C}{16(1-n)}(\theta)^{2-2n} + A \quad ...(i)$$

where, A = constant of integration.

We have $I = 4\theta^n$; thus $\theta = 0$, when $I = 0$. It is given that when $I = 0$, $L = 10$ mH = 0.01 H

Inserting this in eqn. (i), we get :

$$10 \times 10^{-3} = 0 + A \quad \text{or} \quad A = 10 \times 10^{-3}$$

Hence the *expression for self-inductance as a function of θ and n* is given by :

$$L = \frac{0.16}{16(1-n)}(\theta)^{2-2n} + 0.01\,H$$

or $$L = \frac{1}{100(1-n)}(\theta)^{2-2n} + 0.01\,H \quad \textbf{(Ans.)} \quad ...(ii)$$

(ii) Deflection and meter current, θ, I :

When $n = 0.75$, $L = 60$ mH = 0.06 H (*Given*)

Therefore, by putting $L = 0.06$ H and $n = 0.75$ in (ii), we get

$$0.06 = \frac{1}{100(1-0.75)}(\theta)^{2-2\times0.75} + 0.01$$

or, $$0.05 = 0.04\,(\theta)^{0.5}$$

$$\therefore \quad \theta = \left(\frac{0.05}{0.04}\right)^2 = 1.5625 \text{ rad. or } \mathbf{89.5°} \quad \textbf{(Ans.)}$$

Current, $I = 4\theta^n = 4 \times (1.5625)^{0.75} = $ **5.59 A (Ans.)**

Example 3.38. *The full scale deflecting torque of a 10 A moving-iron ammeter is 4 × 10⁻⁵ Nm. Determine in μH/rad. the rate of change of self-inductance of the instrument at full scale.* **(Gorakhpur University)**

Solution. *Given :* $T_d = 4 \times 10^{-5}$ Nm; $I = 10$ A

We know that; $T_d = \frac{1}{2} I^2 \frac{dL}{d\theta}$ ...[Eqn. (3.77)]

$$\therefore \quad \frac{dL}{d\theta} = \frac{2T_d}{I^2} = \frac{2\times4\times10^{-5}}{(10)^2} = \mathbf{0.8\ \mu H/rad.} \quad \textbf{(Ans.)}$$

Example 3.39. *A moving-iron voltmeter reads correctly on 250 V D.C. If 250 V, 50 Hz A.C. is applied to it, what will be the reading of the voltmeter? The instrument coil has a resistance of 500 Ω and an inductance of 1H and series non-inductive resistance is 2 kΩ.*

With what value of capacitance must the series resistance be shunted to make the meter read correctly at 50 Hz ? **(UPTU)**

Solution. *Given :* $R = 500\ \Omega;\ L = 1\text{H};\ R_s = 2\ \text{k}\Omega$ or $2000\ \Omega$.

Voltmeter reading when connected to 250 V, 50 Hz supply :

Total resistance of the moving iron voltmeters $R + R_s = 50 + 2000 = 2500\ \Omega$. When the instrument is used on D.C. system the inductance of the coil will not effect the instrument readings, so if instrument reads correctly, then when the instrument is used on A.C. system the reading of the voltmeter will be *affected* by *the inductance of the coil* and the readings given by the instrument will be *incorrect*.

The impedance of the instrument circuit at 50 Hz,

$$Z = \sqrt{(R+R_s) + (2\pi/L)^2}$$

$$= \sqrt{(2500)^2 + (2\pi \times 50 \times 1)^2} = 2520\ \Omega$$

Current drawn by the instrument, when connected to 250 V, 50 Hz A.C. supply,

$$I = \frac{V}{Z} = \frac{250}{2520}\ \text{A}$$

Since voltmeter reads correctly on D.C. supply on 250 V, so its current corresponding to 250 V

$$= \frac{V}{R+R_s} = \frac{250}{500+2000} = 0.1\ \text{A}$$

∴ *Voltmeter reading when connected to 250 V, 50 Hz supply*

$$= 250 \times \frac{(250/2520)}{0.1} = \mathbf{248\ V}\quad \textbf{(Ans.)}$$

- From eqn. (3.81) value of capacitance required to make the circuit non-inductive,

$$C = 0.41\frac{L}{R_s^2} = 0.41 \times \frac{1}{(2000)^2} = \mathbf{0.1\ \mu F}\quad \textbf{(Ans.)}$$

Example 3.40. *The coil of a 250 V moving-iron voltmeter has a resistance of 500 Ω and inductance of 1 H. The current taken by the instrument when placed on 250 V, D.C. supply is 0.05 A. Determine the percentage error when the instrument is placed on 250 V, A.C. supply at 100 Hz.*

Solution. Total ohmic resistance, $R = \frac{250}{0.05} = 5000\,\Omega$

(Original calibration of the instrument is with direct current)

Reactance of the coil, $X_L = 2\pi fL = 2\pi \times 100 \times 1 = 628\ \Omega$

Coil impedance, $Z = \sqrt{R^2 + X_L^2} = \sqrt{(5000)^2 + (628)^2} = 5039\,\Omega$

∴ Voltage reading on A.C. $= \frac{250 \times 5000}{5039} = 248\ \text{V}$

∴ Error = 248 – 250 = – 2V

$$\textbf{Percentage error} = \frac{2}{250} \times 100 = \mathbf{0.8\%}\ (low) \quad \textbf{(Ans.)}$$

Example 3.41. *A 15-volt moving-iron voltmeter has a resistance of 300 Ω and an inductance of 0.12 H. Assuming that this instrument reads correctly on D.C., what will be its readings on A.C. at 15 volts when frequency is (i) 25 Hz and (ii) 100 Hz ?*

Solution. On *D.C., only ohmic resistance* is involved and the voltmeter reads correctly. But on *A.C.*, it is the *impedance* of the instrument which has to be taken into account.

(i) **When frequency is 25 Hz :**

Impedance at 25 Hz,

$$Z = \sqrt{R^2 + X_L^2} = \sqrt{R^2 + (2\pi fL)^2}$$

$$= \sqrt{(300)^2 + (2\pi \times 25 \times 0.12)^2} = 300.6\,\Omega$$

$$\therefore \quad \textbf{Voltmeter reading} = 15 \times \frac{300}{300.6} = \mathbf{14.97\ V} \quad \textbf{(Ans.)}$$

(ii) **When frequency is 100 Hz :**

Impedance at 100 Hz,

$$Z = \sqrt{(300)^2 + (2\pi \times 100 \times 0.12)^2} = 309.3\,\Omega$$

$$\therefore \quad \textbf{Voltmeter reading} = 15 \times \frac{300}{309.3} = \mathbf{14.55\ V} \quad \textbf{(Ans.)}$$

Incidently, it may be noted that as the frequency is increased, the impedance of the voltmeter is also increased. Hence, the current is decreased and, therefore, the voltmeter readings are lower.

Example 3.42. *An A.C. voltmeter with a maximum scale reading of 50 V has an inductance of 0.09 H and a total resistance of 500 Ω. The magnetising coil is wound with copper wire having a resistance of 50 Ω and the remainder of the circuit consists of a non-inductive resistance in series with it. Find the value of capacitance, that should be placed across the non-inductive series resistance, to make the instrument read correctly both on D.C. as well as on A.C. (of 50 Hz) circuits.* **(Karnataka University)**

Solution. *Given :* L = 0.09 H; $R + R_s$ = 500 Ω; R (Resistance of magnetising coil) = 50 Ω; f = 50 Hz.

The total circuit resistance, $R + R_s = 500\ \Omega$

∴ Non-inductive series resistance, $R_s = 500 - R = 500 - 50 = 450\ \Omega$

Let C be the capacitance to be placed across R_s to make the instrument read correctly on D.C. as well as on 50 Hz A.C. supply. This will be true only when,

$$C = 0.41 \frac{L}{R_s^2} \qquad \text{...[Eqn.(3.81)]}$$

or,

$$C = 0.41 \times \frac{0.09}{(450)^2} = 0.1822 \times 10^{-6} \text{ or } \mathbf{0.1822\ \mu F} \quad \textbf{(Ans.)}$$

Example 3.43. *The coil of a 150 V moving-iron voltmeter has a resistance of 400 Ω and an inductance of 0.75 H. The coil is made of copper which has a resistance temperature coefficient of 0.004 / °C. The current consumed by the instrument when placed on a 150 V. D.C. supply is 0.05 A. The series resistance of the voltmeter is of Manganin with a resistance temperature coefficient of 0.00015 / °C.*

Calculate :

(i) Resistance temperature coefficient of the instrument.

(ii) Alteration of the reading between D.C. and A.C. at 100 Hz.

(iii) Capacitance of the capacitor necessary to eliminate the frequency errors.

(Punjab University)

Solution. *Given :* $V = 150$ volts; $R = 400\ \Omega$; $L = 0.75$ H, $\alpha_{cu} = 0.004/°C$;
$I = 0.05$ A; $\alpha_{mang.} = 0.00015/°C$

(i) **Resistance temperature coefficient of instrument, $\alpha_{inst.}$:**

$$\text{Total resistance of instrument } (R + R_s) = \frac{V}{I} = \frac{50}{0.05} = 3000\ \Omega$$

∴ Series (swamping) resistance, $R_s = 3000 - R = 3000 - 400 = 2600\ \Omega$

Now, change in resistance of coil/°C = $0.004 \times 400 \times 1 = 1.6\ \Omega$

Change in swamping resistance/°C = $0.0015 \times 2600 \times 1 = 0.39\ \Omega$

Total change in resistance of instrument circuit/°C = $1.6 + 0.39 = 1.99\ \Omega$

$$\therefore \quad \alpha_{inst.} = \frac{\text{Change in resistance/°C}}{\text{Total resistance}} = \frac{1.99}{3000} = \mathbf{0.000633/°C} \text{ (Ans.)}$$

(ii) **Error in the reading at 100 Hz :**

Reactance of coil at 100 Hz, $X_L = 2\pi f L = 2\pi \times 100 \times 0.75 = 471.2\ \Omega$

$$\text{Impedance of instrument at 100 Hz} = \sqrt{(3000)^2 + (471.2)^2} = 3037\ \Omega$$

$$\text{Current drawn by the instrument at 100 Hz} = \frac{V}{Z} = \frac{150}{3037} = 0.0494 \text{ A}$$

$$\therefore \quad \text{Reading of the instrument at 100 Hz} = \frac{0.0494}{0.05} \times 150 = 148.2 \text{ V}$$

$$\%\ \textit{Error in the reading} = \frac{148.2 - 150}{150} \times 100 = \mathbf{1.2\%\ (\textit{low})} \text{ (Ans.)}$$

(iii) **Value of C :**

In order to eliminate frequency error, the value of capacitance to be connected across R_s,

$$C = 0.41 \frac{L}{R_s^2} \quad \text{...[Eqn. (3.81)]}$$

or,
$$C = 0.41 \times \frac{0.75}{(2600)^2} = \mathbf{0.0455\ \mu F} \text{ (Ans.)}$$

3.8.5. Moving-coil Instruments – Classification

The moving-coil instruments are of the following two types :

1. Permanent magnet type ...... can be used for D.C. only.
2. Dynamometer type .... can be used for both A.C. and D.C.

3.8.6. Permanent Magnet Moving-coil (PMMC) Instruments

The permanent magnet moving-coil instrument is the *most accurate type for "D.C. measurements"*.

The working principle of these instruments is the same as that of the D'Arsonval type of galvanometers, the difference being that a direct reading instrument is provided with a pointer and a scale.

3.8.6.1. Principle, construction and working

A permanent-magnet moving coil-type instrument works on the principle that *"when a current-carrying conductor is placed in a magnetic field, it is acted upon by a force which tends to move it to one side and out of the field"*.

Construction : Refer to Fig. 3.29.

- The instrument consists of a permanent magnet *M* and a rectangular coil *C* which consists of insulated copper wire wound on light aluminium frame fitted with polished steel pivots resting in jewel bearings. The magnet is made of Alnico and has soft-iron pole-pieces *PP* which are bored out cylindrically.
- The rectangular coil *C* is free to move in air gaps between the soft-iron pole pieces and a soft-iron cylinder *A* (central core), supported by a brass plate (not shown).

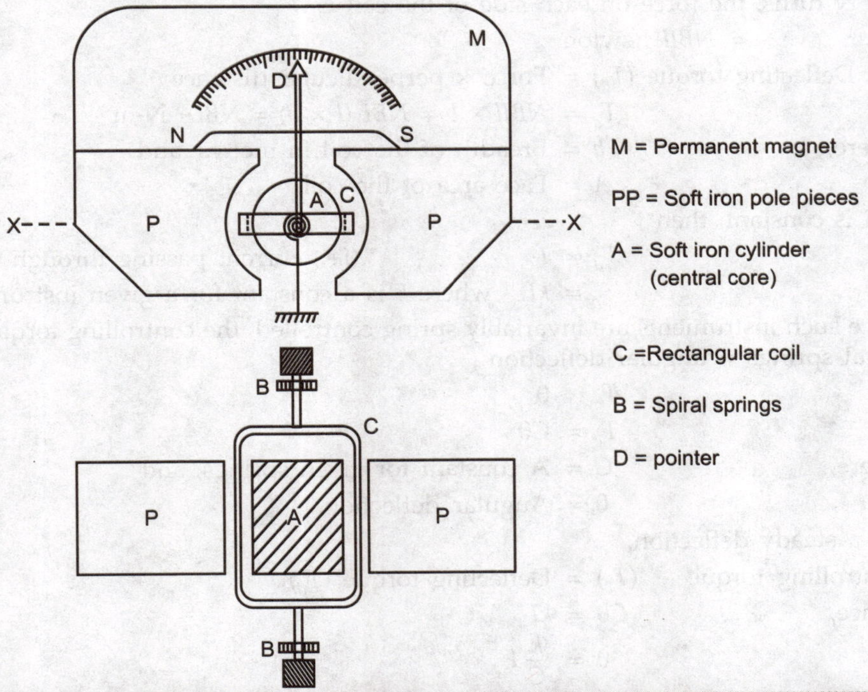

Fig. 3.29. Permanent-magnet moving-coil instrument.

The *functions of the central core A* are :

(*i*) To intensify the magnet field by reducing the length of air gap across which the magnetic flux has to pass.

(*ii*) To give a radial magnetic flux of uniform density, thereby enabling the scale to be uniformly divided.

— The movement of the coil is controlled by two phosphor bronze hair springs *BB* (one above and one below), which additionally serve the purpose of leading the current in and out of the coil. The two springs are spiralled in *opposite directions* for neutralizing the effects of changes in temperature.

— The aluminium frame not only provides support for the coil but also *provides damping by eddy currents induced in it.*

Deflecting torque. Refer to Fig. 3.30. When current is passed through the coil, forces are set up on its both sides which produce *deflection torque.* If I amperes is the current passing through the coil, the magnitude of the force (F) experienced by each of its sides is given by :

$$F = BIl \text{ newton}$$

where, B = flux density in Wb/m^2, and

l = length or depth of coil in metres.

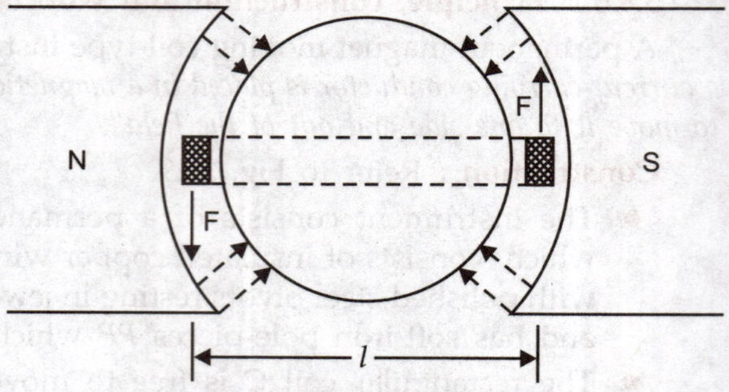

Fig. 3.30.

For N turns, the force on each side of the coil is

$$= NBIl \text{ newton}$$

$\therefore$ Deflecting torque (T_d) = Force × perpendicular distance

or, $$T_d = NBIl \times b = NBI\,(l \times b) = NBIA \text{ N-m} \qquad ...(3.83)$$

where, b = Breadth of the coil in metres, and

A = Face area of the coil.

If B is constant, then

$$T_d \propto I \qquad (i.e., \text{ current passing through the coil})$$

$$= kI \quad \text{where } k \text{ is a constant for a given instrument.}$$

Since such instruments are invariably spring controlled, the controlling torque (T_c) of the spiral springs ∝ angular deflection

i.e., $$T_c \propto \theta$$

or, $$T_c = C\theta$$

where, C = A constant for given springs, and

θ = Angular deflection.

For a steady deflection,

Controlling torque (T_c) = Deflecting torque (T_d)

Hence, $$C\theta = kI$$

$\therefore$ $$\theta = \frac{k}{C} I \qquad ...(3.84)$$

i.e., The *deflection is proportional to current* and the *scale is therefore uniformly divided.* PMMC instrument is *unsuitable for A.C. measurements unless the current is rectified.*

Ranges :

D.C. ammeters :

(*i*) *Without shunt.........0/5 micro-amperes up to 0/30 micro-amperes.*

(*ii*) *With internal shunts........upto 0/2000 amperes.*

(*iii*) *With external shunts.......upto 0/5000 amperes.*

D.C. voltmeters :

(*i*) *Without series resistance.......0/100 milli-volts.*

(*ii*) *With series resistance.......upto 20000 or 30000 volts.*

- The PMMC system can be converted into an instrument to measure D.C. as well as A.C. quantities like current, voltage and resistance etc. It can carry a maximum current of I_m, without any modification and can withstand a maximum D.C. voltage, $V = I_m R_m$.

The various modifications of PMMC meter movement may be summed up a follows :

1. ***D.C. instruments :***
   - (*i*) D.C. ammeter – By using a *shunt resistor.*
   - (*ii*) D.C. voltmeter – By using *series multiplier resistor.*
   - (*iii*) Ohmmeter – By using *battery and series resistor.*
2. ***A.C. instruments :***
   - (*i*) Audio-frequency (AF) A.C. ammeter or voltmeter – By using a *rectifier.*
   - (*ii*) Radio-frequency (RF) ammeter or voltmeter – By using a *thermocouple.*

3.8.6.2. Temperature compensation

The PMMC basic movement is not inherently insensitive to temperature, but it may be *temperature-compensated* by the appropriate use of series and shunt resistors.

- Complete temperature compensation may be accomplished by using swamping resistors as shown in Fig. 3.31. The swamping resistor is made of manganin (having temperature coefficient of practically zero) and combined with copper in the ratio of 20 : 1 to 30 : 1. The total resistance of the coil and swamping resistor increases slightly with rise in temperature but only just enough to counteract the change of springs and magnet, so that the overall temperature effect is zero.

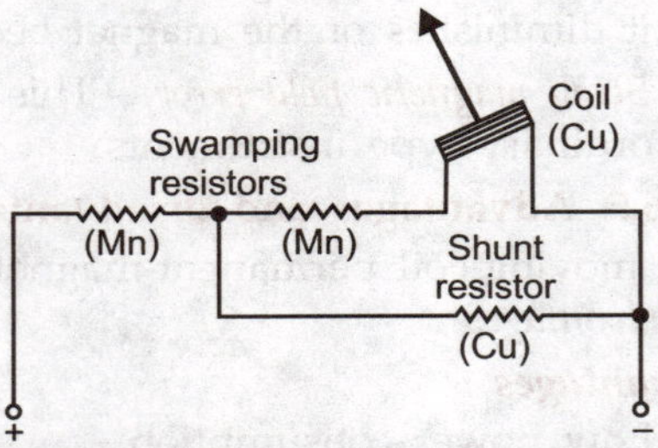

Fig. 3.31. Temperature compensatiousing series and shunt resistors.

— The main "*disadvantage*" of the use of swamping resistors is a reduction in the full-scale sensitivity of the movement; because a higher applied voltage is necessary to sustain the full-scale current.

3.8.6.3. Characteristics of PMMC meter movement :

The following *three* characteristics of PMMC meter movement will be discussed :

1. ***Full-scale deflection current (I_m) :***

- *It is the current needed to deflect the pointer all the way to the right to the last mark on the calibrated scale.*
- Typical valves of the I_m for D'Arsonval movement vary from 2 μA to 30 mA.
  — It may be remembered that for smaller currents, the number of turns in the moving coil should be *more* so that the magnetic field produced by the coil is strong enough to react with the field of permanent magnet for producing reasonable deflection of the pointer. *Fine wire* has to be used for reducing the weight of the moving coil but this increases its resistance.
  — For heavy currents, thick wire with lesser number of turns is needed so that resistance of the moving coil is comparatively less.

2. ***Internal resistance (R_m) :***

- *It is the D.C. ohmic resistance of the wire of the moving coil.*
- Typical values of R_m range from 1-2Ω for a 30 mA movement to 2 kΩ for a 50 μA movement.

3. ***Sensiivity (S) :***

- It is also known as *current-sensitivity* or *sensitivity factor.*
- It is given by the reciprocal of full scale deflection current I_m.

$$\therefore \qquad S = \frac{1}{I_m} \text{ ohm/volt} \qquad ...(3.85)$$

— The sensitivity of a meter movement depends on the *strength of the permanent magnet and number of turns in the coil.* Larger the number of turns, smaller the amount of current required to produce full-scale deflection and hence, higher the sensitivity.

3.8.6.4. Errors in PMMC instruments

The following *errors* usually occur in PMMC instruments :

1. *Frictional error* – This can be avoided by winding of coil carefully.
2. *Temperature error* – It can be avoided as discussed in Art. 3.8.3.
3. *Error due to weakening of permanent magnet* – It can be eliminated completely by weakening of mechanical control and attaching a symmetrical iron or steel needle to the coil. The magnetic needle is so arranged that magnetic control provided by it diminishes or the magnet becomes weaker.
4. *Stray magnetic field error* – This error can be avoided by using an iron case for ordinary type instruments.

3.8.6.5. Advantages and disadvantages

The moving-coil permanent-magnet type instrument have the following *advantages* and *disadvantages :*

Advantages :

(*i*) Low power consumption.
(*ii*) Their scales are uniform.
(*iii*) No hysteresis loss.
(*iv*) High torque/weight ratio.
(*v*) They have very effective and efficient eddy-current damping.
(*vi*) Range can be extended with shunts or multipliers.
(*vii*) No effect of stray magnetic field as intense polarised or unidirectional field is employed.

Disadvantages :

(*i*) Somewhat costlier as compared to moving-iron instruments.
(*ii*) Cannot be used for A.C. measurements.
(*iii*) Friction and temperature might introduce errors as in case of other instruments.
(*iv*) Some errors are set in due to the ageing of control springs and the permanent magnets.

☞ • Moving-coil permanent-magnet instrument can be *used as* :

(*i*) "*Ammeters*" — By using a *low resistance shunt.*
(*ii*) "*Voltmeters*" — By using a *high series resistance.*
(*iii*) "*Fluxmeters*" — By *eliminating the control spring.*
(*iv*) "*Ballistic galvanometers*" — By *making control springs of large moment of inertia.*

Example 3.44. *The coil of an instrument has 38 turns. The mean width and axial length of the coil are 25 mm and 20 mm respectively. If the flux density is 0.12 Wb/m², calculate the torque on the moving coil for a current of 12 mA through the coil.*

Solution. Given : $N = 38$; $b = 25$ mm or 0.025 m; $l = 20$ mm or 0.02 m;
$B = 0.12 \text{ Wb/m}^2$; $I = 12 \text{ mA} = 0.012 \text{ A}$

Torque on the moving coil,

$$T_d = NBIlb \qquad \text{[Eqn. 3.83]}$$
$$= 38 \times 0.12 \times 0.012 \times 0.02 \times 0.025$$
$$= \mathbf{2.736 \times 10^{-5} \text{ Nm}} \quad \text{(Ans.)}$$

Example 3.45. *The resistance of a moving voltmeter is 11 kΩ. The moving coil has 100 tuns and is 40 mm long and 30 mm wide. The flux density in the air gap is 0.05 Wb/m².*

Determine the deflection produced by 220 V if the spring control gives a deflection of 1° for a torque of 20 × 10⁻⁷ Nm. **(Allahabad University)**

Solution. *Given* : $R = 11000\ \Omega$; $N = 100$; $l = 40$ mm or 0.04 m; $b = 30$ mm or 0.03 m; $B = 0.05\ \text{Wb/m}^2$; $V = 220$ V; $T_d = 20 \times 10^{-7}$ Nm/degree deflection.

Value of deflection, θ :

$$I = \frac{V}{R} = \frac{220}{11000} = 0.02 \text{ A}$$

For steady deflection state,

Deflecting torque, T_d = Controlling torque, T_c

$\therefore \quad NBIlb = 20 \times 10^{-7} \times \theta$

or, $$\theta = \frac{NBIlb}{20 \times 10^{-7}} = \frac{100 \times 0.05 \times 0.02 \times 0.04 \times 0.03}{20 \times 10^{-7}} = \mathbf{60°} \quad \textbf{(Ans.)}$$

Example 3.46. *A moving coil millivoltmeter has a resistance of 20 Ω and full scale deflection of 120° is reached when a potential difference of 100 mV is applied across its terminals. The moving coil has the effective dimensions of 3.1 cm × 2.6 cm and is wound with 120 turns. The flux density in the gap is 0.15 Wb/m².*

Determine the control constant of the spring and suitable diameter of copper wire for coil winding if 55 per cent of total instrument resistance is due to coil winding. ρ for copper = 1.73 × 10⁻⁸ Ωm. **(M.S. University of Baroda)**

Solution. *Given* : $R = 20\ \Omega$; Full scale deflection, $\theta = 120°$; Voltage applied for full scale deflection = 100 mV = 0.1 V; $l = 3.1$ cm = 0.031 m; $b = 2.6$ cm or 0.026 m; $N = 120$; $B = 0.15\ \text{Wb/m}^2$; $\rho_{copper} = 1.73 \times 10^{-8}\ \Omega\text{m}$.

Spring constant, *C* :

Full scale deflection current, $I = \frac{V}{R} = \frac{0.1}{20} = 0.005$ A

Full scale deflection torque, $T_d = NBIlb$

$= 120 \times 0.15 \times 0.005 \times 0.031 \times 0.026 = 72.54 \times 10^{-6}$ Nm

For steady deflection state, $T_d = T_c$

or, $T_c = C\theta = 72.54 \times 10^{-6}$

or, $$C = \frac{72.54 \times 10^{-6}}{120} = \mathbf{6.045 \times 10^{-7}\ Nm/degree} \quad \textbf{(Ans.)}$$

Diameter of copper wire, *d* :

Resistance of coil winding, $R = \frac{55}{100} \times 20 = 11\ \Omega$

Length of copper wire = (0.031 + 0.026) × 2 × 120 = 13.68 m

Area of cross-section of wire, $a = \frac{\rho l}{R}$

or, $$a = \frac{1.73 \times 10^{-8} \times 13.68}{11} = 2.15 \times 10^{-8} \text{ m}^2$$

But, $a = \frac{\pi}{4} d^2$

$$\therefore \quad d = \sqrt{\frac{4a}{\pi}} = \sqrt{\frac{4 \times 2.15 \times 10^{-8}}{\pi}} = \mathbf{1.655 \times 10^{-4} \ m \ or \ 0.1655 \ mm} \quad \textbf{(Ans.)}$$

Example 3.47. *The following data relate a moving coil instrument:*

Number of turns (wound on non-inductive former) = 200
Width of the former = 20 mm
Height of the former = 25 mm
Flux density of the field = 0.1 Wb/m²
Moment of the inertia of the moving parts = 3 × 10⁻⁷ kg-m²
Torque produced by the control spring = 90 × 10⁻⁷ Nm/rad.
Calculate the current in the coil to produce a deflection 120°. **(AMIE)**

Solution. *Given* : $N = 200$; $b = 20$ mm or 0.02 m; $l = 25$ mm or 0.025 m; $B = 0.1$ Wb/m²;
$T_c = 90 \times 10^{-7}$ Nm/rad.; $\theta = 120°$ or $120 \times \frac{\pi}{180} = \frac{2\pi}{3}$ rad.

Current in the coil, *I* :

For steady deflection state,

$$T_d = T_c$$

$$NBIlb = 90 \times 10^{-7} \times \frac{2\pi}{3}$$

or, $200 \times 0.1 \times I \times 0.025 \times 0.02 = 90 \times 10^{-7} \times \frac{2\pi}{3}$

$$I = \mathbf{1.88 \ mA} \quad \textbf{(Ans.)}$$

Example 3.48. *The coil of a moving coil voltmeter is 40 mm long and 30 mm wide and has 100 turns on it. The control spring exerts a torque of 240 × 10⁻⁶ Nm when the deflection is 100 divisions on full scale. If the flux density of the magnetic field in the air gap is 1.0 Wb/m², calculate the resistance that must be put in series with the coil to give 1 V/division. The resistance of the voltmeter coil may be neglected.* **(Punjab University)**

Solution. *Given* : $l = 40$ mm or 0.04 m; $b = 30$ mm or 0.03 m;
$N = 100$; $T_c = 240 \times 10^{-6}$ Nm; $B = 1.0$ Wb/m²

Resistance, R :

Deflecting torque at full scale deflection,

$$T_d = NBIlb = 100 \times 1.0 \times I \times 0.04 \times 0.03 = 0.12 \ I \ \text{Nm}$$

At final steady state position,

$$T_d = T_c$$

or, $0.12 \ I = 240 \times 10^{-6}$

$\therefore$ Current at full scale deflection, $I = \frac{240 \times 10^{-6}}{0.12} = 0.002$ A

Voltage across the instrument $= IR = 0.002 \ R$

$$\therefore \quad \text{Volts per division} = \frac{0.002\,R}{100}$$

This value should be equal to 1 in order to get 1 volt per division.

$$\therefore \quad \frac{0.002\,R}{100} = 1$$

or, $$R = \frac{100}{0.002} = 50000\ \Omega \text{ , or , } \mathbf{50\ k\Omega} \quad \textbf{(Ans.)}$$

3.8.6.6. Extension of range

D.C. ammeters :

Shunt resistor : Since the coil winding of a basic movement of a D.C. ammeter (PMMC galvanometer) is small and light, it *can carry only very small currents.* When it is required to measure large current, it is necessary to bypass the major part of the current through a resistance, called a *shunt*, as shown in Fig. 3.32. The resistance of the shunt can be calculated as follows.

Let, R_m = Internal resistance of the meter,

R_{sh} = Resistance of the shunt,

I_m = Current through the meter,

I_{sh} = Shunt current, and

I = Current of the circuit to be measured.

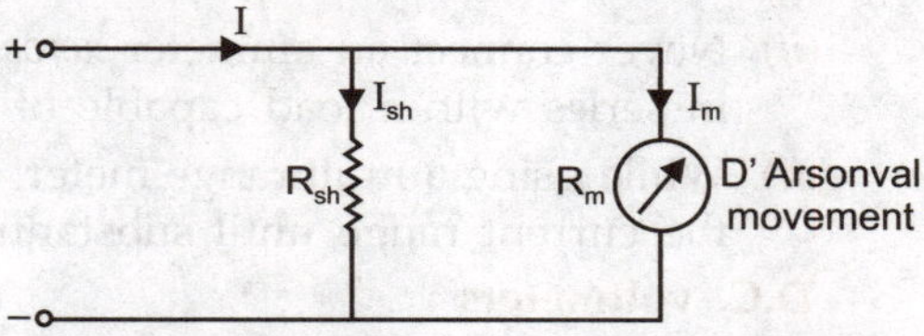

Fig. 3.32. Basic D.C. ammeter circuit.

Thus, $$I = I_m + I_{sh}$$

$$(\text{or } I_{sh} = I - I_m).$$

Also, since the voltage drop across the shunt and the meter/instrument is same,

$$I_{sh}R_{sh} = I_mR_m \quad \text{...(3.86)}$$

or, $$R_{sh} = \frac{I_mR_m}{I_{sh}}$$

or, $$R_{sh} = \frac{I_mR_m}{(I - I_m)} \quad \text{...(3.87)}$$

The shunt resistance with PMMC meter movement may consist of a length of constant-temperature resistance wire within the case of the instrument or it may be an external (manganin or constantan) shunt having a very low resistance.

— The *"external shunt"* consists of evenly spaced sheets of resistance material welded into a large block of heavy copper on each end of the sheets.

— The resistance material has a very low temperature coefficient, and a low thermoelectric effect exists between the resistance material and copper.

— External shunts of this type are normally used for measuring very large currents.

— **Fig. 3.33** shows a simple ***multirange ammeter*** in which the current range of the D.C. ammeter may be further

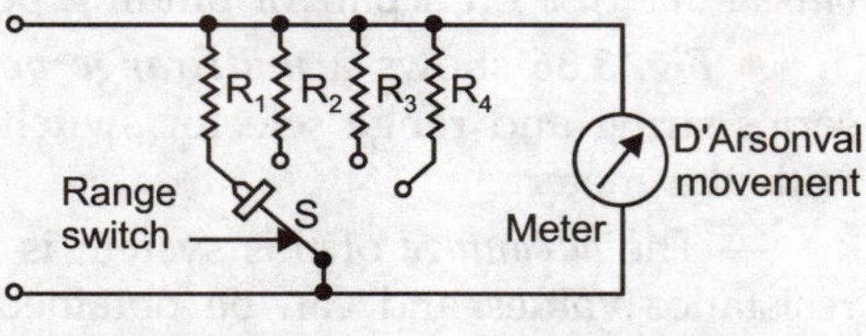

Fig. 3.33. Simple multirange ammeter.

extended by a number of shunts, selected by a *range switch.*

— *Fig. 3.34* shows a *universal or ayrton shunt,* the use of which eliminates the possibility of having the meter in the circuit without a shunt. This *advantage* is gained at the cost of a slightly higher overall meter resistance.

— The Ayrton shunt provides an excellent opportunity *to apply basic network theory to a practical circuit.*

- D.C. ammeters are commercially available in a large number of ranges, from 20 μA to 50A full-scale for a *self-contained* meter and to 500 A for a meter with *external shunt.*

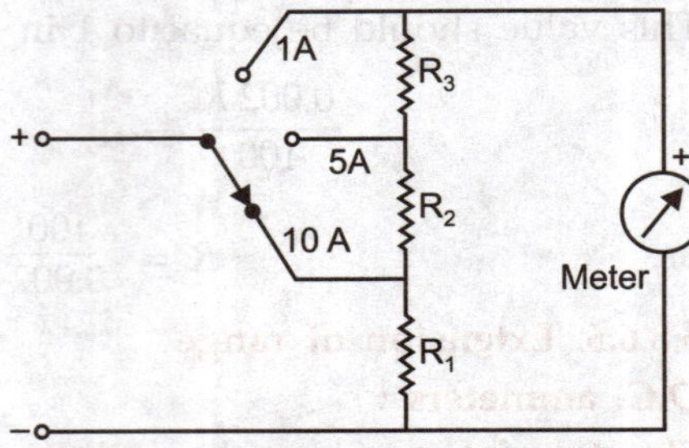

Fig. 3.34. Universal or Ayrton shunt.

While using an ammeter in measurement work, the following ***precautions*** should be observed :

(*i*) Observe the correct polarity.

(*ii*) Never connect an ammeter *across* an e.m.f. source. Always connect an ammeter in series with a load capable of limiting the current.

(*iii*) While using a multirange meter, first use the higher current range; then decrease the current range until substantial deflection is obtained.

D.C. voltmeters :

The basic D'Arsonval movement can be converted into a D.C. voltmeter by adding a series resistor or *multiplier,* as shown in Fig. 3.35. The multiplier *limits* the current through the meter so as not to *exceed* the value of the full-scale deflection current (I_{fsd}). The value of multiplier, required to extend the voltage range, is calculated from Fig. 3.35 as follows :

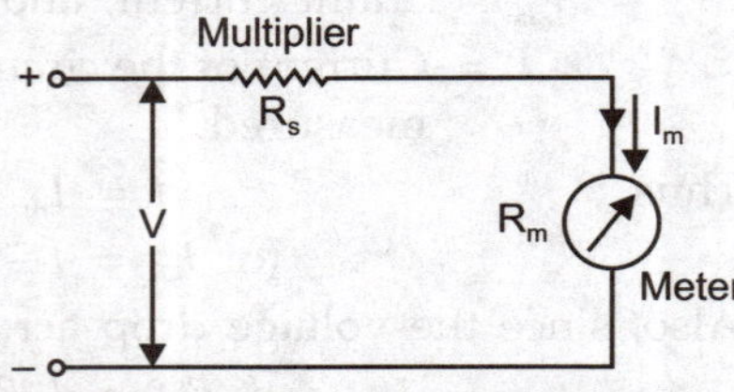

Fig. 3.35. Basic D.C. voltmeter circuit.

Let, I_m = Full scale deflection current of the meter (I_{fsd})

R_m = Internal resistance of the meter,

R_s = Multiplier resistance, and

V = Voltage of the circuit to be measured.

Now, $V = I_m\,(R_s + R_m)$

or,
$$R_s = \frac{V - I_m R_m}{I_m} = \frac{V}{I_m} - R_m \quad ...(3.88)$$

— For moderate ranges upto 500 V, the multiplier is usually mounted *inside* the case of the voltmeter. However, for *higher voltages,* the multiplier may be mounted separately *outside the case* on a pair of binding posts *to avoid excessive heating inside the case.*

- *Fig. 3.36* shows a ***multirange voltmeter*** where the multipliers are connected in a series string and range selector switches the appropriate amount of resistance in series with the meter.

— The *advantage* of this system is that all multipliers except the first have standard resistance values and can be obtained commercially in precision tolerances. The low-range multiplier, R_4, is the only special resistor that must be manufactured to meet the specific circuit requirements.

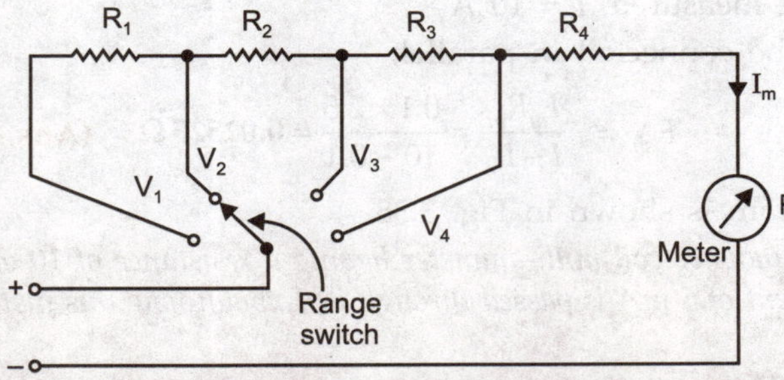

Fig. 3.36. Multirange voltmeter.

Loading effect of a voltmeter :

If the resistance of a voltmeter is *not high* as compared to the resistance of the circuit across which it is connected, the measured voltage will be *less* than the actual one. The decrease in voltage may be negligible or it may be appreciable depending on the sensitivity and input resistance of the voltmeter. *It is called **loading effect** because the voltmeter loads down the circuit across which it is connected.*

Loading effect can be minimised by using a voltmeter whose resistance is as high as possible as compared to that of the circuit across which it is connected.

— In case of an *electronic voltmeter "there is no loading problem because its input resistance is very high (10 MΩ or more).*

• It may be remembered that whereas *"accuracy"* is always required in instruments, *"sensitivity" is needed only in special applications where loading distorts which is being measured.*

Example 3.49. *A milliammeter of 2.5 ohms resistance reads upto 100 milliamperes. Calculate the resistance which is necessary to enable it to be used as :*

(i) A voltmeter reading upto 10 V.

(ii) An ammeter reading upto 10 A.

Draw the connection diagram in each case.

Solution. Resistance of the milliammeter, $R_m = 2.5\ \Omega$

Maximum current of the milliammeter, I_m = 100 mA = 0.1 A.

(i) Voltage to be measured, V = 10 volts

Resistance to be connected in series.

$$\mathbf{R_s} = \frac{V}{I_m} - R_m = \frac{10}{0.1} - 2.5 = \mathbf{97.5\ \Omega} \quad \textbf{(Ans.)}$$

Connection diagram is shown in Fig. 3.37.

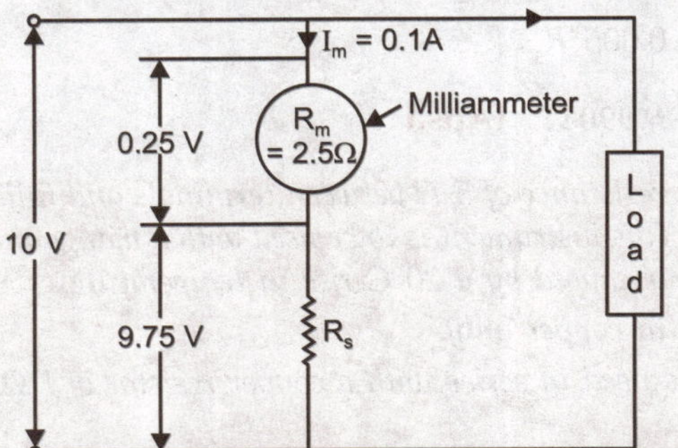

Fig. 3.37.

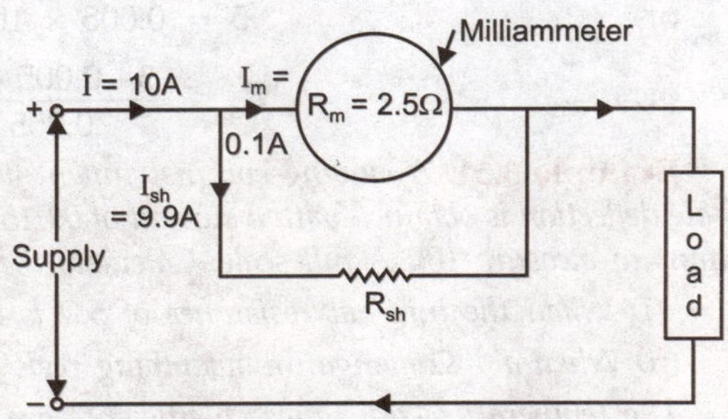

Fig. 3.38.

(*ii*) Current to be measured, I = 10 A

∴ Resistance to be connected in parallel,

$$R_{sh} = \frac{I_m R_m}{I - I_m} = \frac{0.1 \times 2.5}{10 - 0.1} = \mathbf{0.02525\ \Omega} \quad \textbf{(Ans.)}$$

Connection diagram is shown in Fig. 3.38.

Example 3.50. *A moving-coil milli-ammeter having a resistance of 10 ohms gives full scale deflection when a current of 5 mA is passed through it. Explain how this instrument can be used for measurement of :*

(i) Current upto 1 A.

(ii) Voltage upto 5 V.

Solution. Resistance of the milli-ammeter, R_m = 10 Ω

Full scale deflection current, I_m = 5 mA = 0.005 A.

(i) To measure current upto 1 A :

Resistance of the shunt, R_{sh} : Refer to Fig. 3.39.

Since voltage drop across the milliammeter and the shunt are equal

$$\therefore \quad I_m R_m = I_{sh} R_{sh} = (I - I_m)\, R_{sh}$$

$$\therefore \quad \mathbf{R_{sh}} = \frac{I_m R_m}{(I - I_m)} = \frac{0.005 \times 10}{(1 - 0.005)} = \mathbf{0.05025\ \Omega} \quad \textbf{(Ans.)}$$

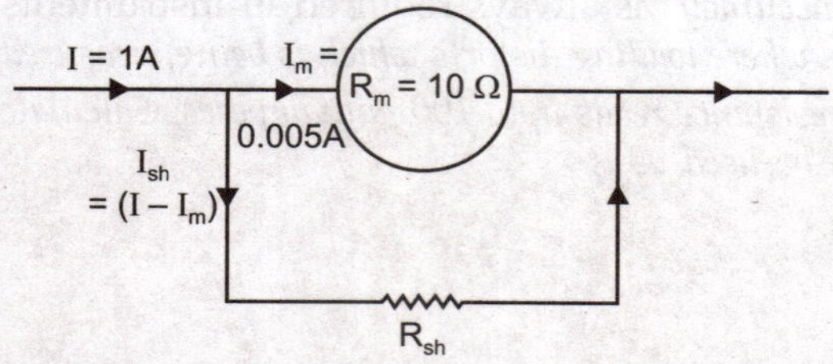

Fig. 3.39.

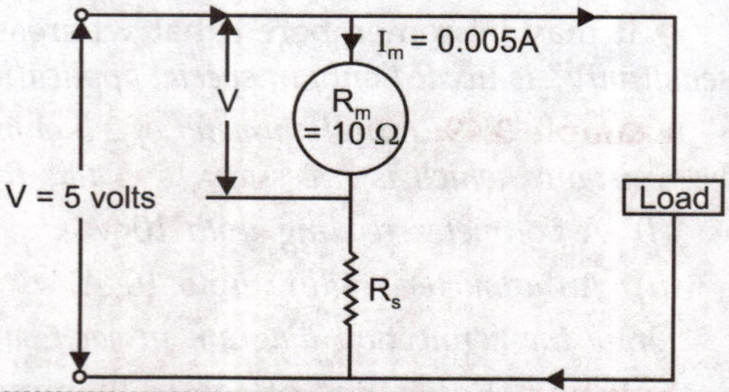

Fig. 3.40.

(ii) To measure voltage upto 5 V :

The value of external series resistance, R : Refer to Fig. 3.40.

Now, voltage across supply loads

= Voltage drop across the milliammeter + voltage drop across external series resistance R_s

$$\therefore \quad V = I_m R_m + I_m R_s$$

$$\text{or,} \quad 5 = 0.005 \times 10 + 0.005\, R_s$$

$$\text{or,} \quad R_s = \frac{5 - 0.005 \times 10}{0.005} = \mathbf{990\ \Omega} \quad \textbf{(Ans.)}$$

Example 3.51. *A moving-coil instrument has a resistance of 5 Ω between terminals and full-scale deflection is obtained with a current of 0.015 A. This instrument is to be used with a manganin shunt to measure 100 A full scale. Calculate the error caused by a 20°C rise in temperature.*

(i) When the internal resistance of 5Ω is due to copper only.

(ii) When a 4 Ω manganin swamping resistor is used in series with a copper resistor of 1 Ω.

The temperature-resistance co-efficients are :

Copper : α_c = 0.4% per °C, Manganin : α_m = 0.015% per °C.

Solution. Resistance of the instrument, $R_m = 5\Omega$

Current through the instrument, $I_m = 0.015$ A

Current to be measured, $I = 100$ A

Current through the shunt,

$$I_{sh} = I - I_m = 100 - 0.015 = 99.985 \text{ A}$$

Voltage across the shunt

$$I_m R_m = 5 \times 0.015 = 0.075 \text{ V}$$

∴ Shunt resistance, $R_{sh} = \dfrac{0.075}{99.985} = 0.00075\ \Omega$ $[\because I_m R_m = (I - I_m) R]$

Shunt resistance after a rise of 20°C

$$= 0.00075\ (1 + 20 \times 0.00015) = 0.000752\ \Omega.$$

(*i*) The instrument resistance (which is wholly copper) after a rise of 20°C

$$= 5\ (1 + 20 \times 0.004) = 5.4\ \Omega$$

Hence, current through the instrument corresponding to 100 A in the line

$$= \frac{0.000752}{(5.4 + 0.000752)} \times 100 = 0.01392 \text{ A}$$

Reading of the instrument = $0.01392 \times 100/0.015 = 92.8$ A

∴ **Percentage error** = 100 – 92.8 = **7.2% (*low*)** **(Ans.)**

(*ii*) The instrument resistance after a rise of 20°C

$$= 1\ (1 + 20 \times 0.004) + 4\ (1 + 20 \times 0.00015)$$

$$= 1.08 + 4.012 = 5.092\ \Omega$$

Instrument current with a line current of 100 A

$$= \frac{0.000752}{(5.092 + 0.000752)} \times 100 = 0.01476 \text{ A}$$

∴ Instrument reading $= 0.01476 \times \dfrac{100}{0.015} = 98.4$ A

Percentage error = 100 – 98.4 = **1.6% (*low*)** **(Ans.)**

3.8.7. Ohmmeters

The ohmmeters may be of the following two types :

1. Series-type ohmmeter.
2. Shunt-type ohmmeter.

3.8.7.1. Series-type ohmmeter

Fig. 3.41 shows the elements of a simple single-range series ohmmeter :

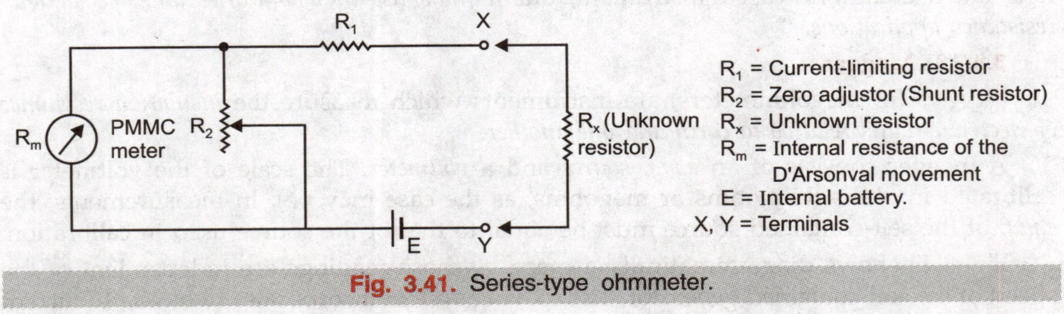

Fig. 3.41. Series-type ohmmeter.

— It consists of a D'Arsonval movement connected in series with a resistance and a battery to a pair of terminals (*X, Y*) to which the unknown resistance (R_x) is connected. The current through the meter then depends on the magnitude of unknown resistor, and the meter indication is proportional to the value of the unknown, provided that the calibration problems are taken into account,

— When $R_x = 0$ (*i.e.,* terminals *X* and *Y* shorted), *maximum* current flows in the circuit. Under this condition, shunt resistor R_2 is adjusted until the meter indicates full-scale current (I_{sd}); this position of the pointer is marked "$0\ \Omega$" on this scale.

— When $R_x = \infty$ (*i.e.,* terminals *X* and *Y* are open), the current in the circuit drops to *zero* and the meter indicates zero current, which is then marked "∞" on the scale.

Intermediate markings may be placed on the scale by connecting different known values of R_x to the instrument.

— The accuracy of these scale markings depends on the repeating accuracy of the meter and the tolerances of the calibrating resistors.

3.8.7.2. Shunt-type Ohmmeter :

Fig. 3.42. shows the circuit diagram of a shunt-type ohmmeter :

— It consists of a battery in series with an adjustable resistor R_1 and PMMC meter movement. The unknown resistor R_x is connected across terminals *X* and *Y*, in parallel with the meter. An *off-on switch* is provided to disconnect the battery from the circuit when the instrument is not used.

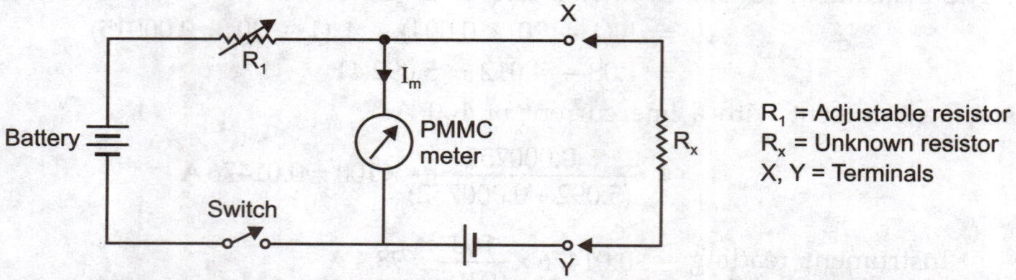

Fig. 3.42. Shunt-type ohmmeter.

— When $R_x = 0\ \Omega$ (*X* and *Y* terminals shorted), the meter current is zero.

— When $R_x = \infty$ (*X* and *Y* terminals open), the current finds a path only through the meter, and by the appropriate selection of the value R_1, the pointer can be made to read full scale.

The ohmmeter therefore has the 'zero' mark at the left hand side of the scale (no current) and the "infinite" mark on the right hand side of the scale (full-scale deflection).

- This type of ohmmeter is particularly suited to the measurement of *low-value resistors*. It is not a commonly used instrument, but it *finds use in laboratories and special low-resistance applications.*

3.8.7.3. Megger

Meggers (or megohmmeters) are instruments which measure the *insulation resistance of electric circuits relative to earth and one another.*

A megger consists of an *e.m.f. source* and a *voltmeter*. The scale of the voltmeter is calibrated in ohms (kilo-ohms or megohms, as the case may be). In measurements, the *e.m.f.* of the self-contained source must be equal to that of the source used in calibration.

Fig. 3.43 shows diagrammatically a megger whose readings are independent of the speed of the self-contained generator. The moving system incorporates two coils 1 (current coil) and 2 (pressure coil) mounted on the same shaft and *placed in the field of a permanent magnet* (now shown) 90° apart. The generator energizes the two coils over separate wires. Connected in series with one coil is a fixed resistance R_1 (or several different resistances in order to extend the range of the instrument). The unknown resistance R_x is connected in series with the other coil. *The currents in the coils interact with the magnetic field and produce opposing torques.*

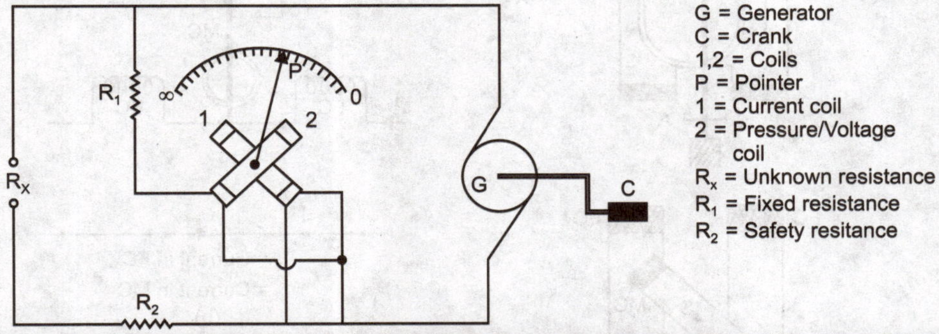

Fig. 3.43. Circuit diagram of megger.

The deflection of the moving system depends on the ratio of the currents in the coils and is independent of the applied voltage. The unknown resistance is read *directly* from the scale of the instrument. (The accuracy of measurement is unaffected by variations in the speed of the generator between 60 and 180 r.p.m.).

3.8.8. Electrodynamic or Dynamometer Type Moving Coil Instruments

3.8.8.1. Introduction

In this type of instrument, the *operating field is produced by another fixed coil* and *not by a permanent magnet*.

- These instruments are capable of service as *transfer instruments* (A *'**transfer instrument**' is one that may be calibrated with a D.C. source and then used without modification to measure A.C.*). Indeed, their principal use as ammeters and voltmetes in laboratory and measurement work is for the transfer calibration of working instruments and *as standards for calibration of other instruments a their accuracy is very high.*
- These instruments are employed as *A.C. voltmeters and ammeters both in the range of power frequencies and lower part of the audio frequency range.* They are used as *watt-meters, var-meters and with some modification as power factor meters and frequency meters.*

3.8.8.2. Construction and working

Fig. 3.44 shows a dynamometer type instrument :

The *main components* of the instrument are :

(*i*) Field system (Fixed coil).
(*ii*) Moving system (Moving coil).
(*iii*) Control system (Hair springs).
(*iv*) Damping system (Air friction damping).
(*v*) Shielding.
(*vi*) Cases and scales.

— These instruments essentially consist of *fine wire moving coil* placed in the magnetic field produced by *another fixed coil* when carrying currents.

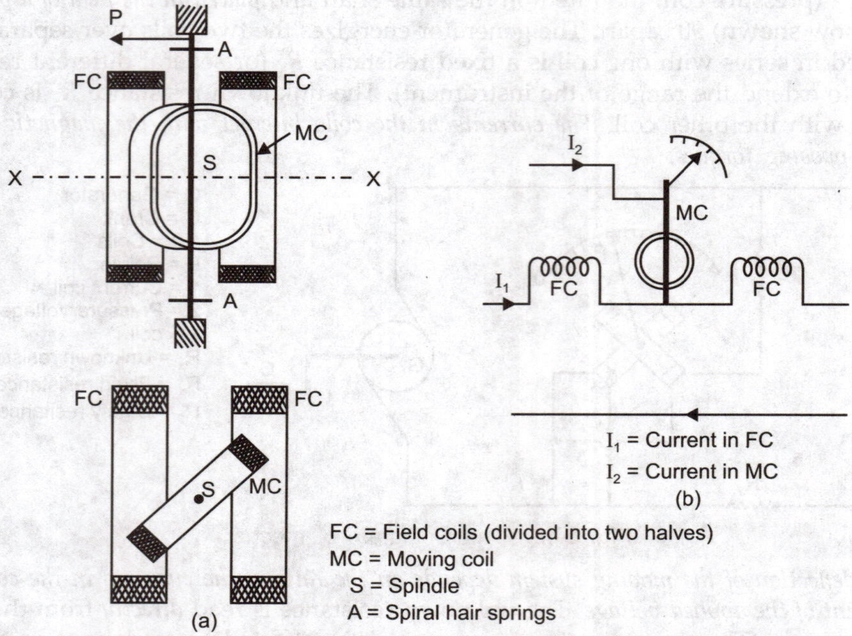

Fig. 3.44. Electrodynamic or dynamometer type instrument.

— The coils are usually *air cored to avoid hysteresis, eddy currents and other errors* when the instrument is used on A.C. The *fixed coil FC* is divided into *two halves* placed close together and *parallel* to each other in order to provide a *fairly uniform field,* within the range of the movement of the moving coil.

— The diagram in Fig. 3.44(*a*) shows a sectional elevation through fixed coil *FC* and the lower diagram represents a sectional plan on *XX*. The moving coil *MC* is carried by a spindle S and the *controlling torque is exerted by spiral hair spring A,* which may also serve to lead the current into and out of MC.

— In these instruments *air friction damping is used* and may be either piston type or vane type. Eddy current damping *cannot* be used as introduction of a permanent magnet for the purpose would *distort* the working magnetic field of the instrument.

— The complete assembly is surrounded by a *laminated steel shield to protect the instrument from external magnetic field* which may affect the operation of the instrument.

— Laboratory standard instruments are usually contained in *highly polished wooden* cases. The case is supported by adjustable levelling screws. A spirit level is also provided to ensure proper levelling. The *scales* are hand drawn, using machine sub-dividing equipment.

Deflecting torque. The deflecting torque is due to *interaction of the magnetic fields produced by currents in the fixed and moving coils.*

— Fig. 3.45(*a*) shows the magnetic field due to current flowing through *FC* (I_1) in the direction indicated by the dots and cross.

— Fig. 3.45(*b*) shows the magnetic field due to current (I_2) in *MC*.

— Fig. 3.45(*c*) shows the combined effect of the above magnetic fields. By combining these magnetic fields it will be seen that when currents (I_1 and I_2) flow simultaneously through *FC* and *MC*, the resultant magnetic field is *distorted* and effect is to exert a clockwise torque on *MC*.

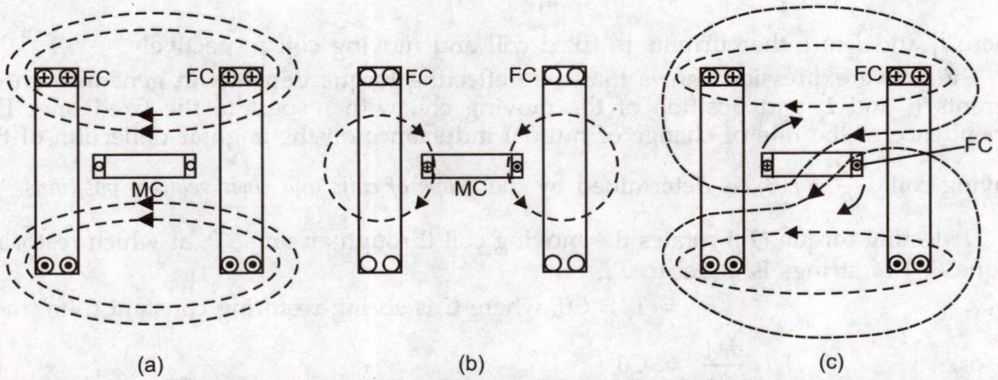

Fig. 3.45. Magnetic fields due to fixed and moving coils.

Since *MC* is carrying current (I_2) at right angles to the magnetic field produced by *FC*, *deflecting torque,*

$$T_d \propto I_1 \times I_2$$

or, $\quad T_d = KI_1I_2 \quad$ where, K is a constant.

Since the instrument is spring-controlled, the *restoring or control torque* (T_c) is proportional to the angular deflection θ.

$$T_c \propto \theta \quad \text{or} \quad T_c = K'\theta$$

The two torques (T_d and T_c) are equal and opposite in the final deflection position.

i.e., $\quad T_d = T_c$

or, $\quad KI_1I_2 = K'\theta$

or, $\quad \theta \propto I_1I_2$

or, $\quad \theta \propto I^2 \quad$ since $I_1 \propto I$ and $I_m \propto I$

Expression for torque in terms of mutual inductance M between fixed and moving coils :

The total energy stored in the magnetic field of the fixed coil,

$$E = \frac{1}{2}L_1I_1^2 + \frac{1}{2}L_2I_2^2 + I_1I_2M \quad ...(3.89)$$

where L_1 and L_2 are the self inductances of the fixed coil and moving coil respectively.

The interaction of these currents give rise to a deflecting torque, T_d that sets the moving coil in a position for which the energy of the coil magnetic field attains maximum value.

Deflecting torque, $\quad T_d = \dfrac{dE}{d\theta} = \dfrac{1}{2} \cdot \dfrac{dL_1}{d\theta} \cdot I_1^2 + \dfrac{1}{2}\dfrac{dL_2}{d\theta} I_2^2 + I_1I_2 \dfrac{dM}{d\theta}$

where $d\theta$ is the increase in angular deflection at which the field energy increases by dE.

Since coil inductances L_1 and L_2 are constant, therefore, dL_1 and dL_2 each is equal to zero and T_d is given by :

$$T_d = I_1I_2\frac{dM}{d\theta} \quad ...(3.90)$$

For direct current :

$$T_d = I_1 I_2 \frac{dM}{d\theta}$$

where I_1 and I_2 are the currents in fixed coil and moving coil respectively.

The above expression shows that the deflecting torque depends in general both on currents I_1 and I_2 and position of the moving coil with respect to the fixed one. The dependence of the rate of change of mutual inductance on the angular deflection of the moving coil $\frac{dM}{d\theta} = f(\theta)$ is determined by the *shape of coils and their relative positions.*

Deflecting torque (T_d) rotates the moving coil through an angle θ, at which restoring torque (T_c) of strings is equal to T_d.

$\therefore \quad T_d = T_c = C\theta$, where C is spring restoring constant (Nm/rad.)

or, $$I_1 I_2 \frac{dM}{d\theta} = C\theta$$

or, Deflection, $$\theta = \frac{I_1 I_2}{C} \cdot \frac{dM}{d\theta} \qquad \text{...(3.91)}$$

For alternating current :

When the coils carry alternating currents i_1 and i_2 the instantaneous deflecting torque,

$$(T_d)_i = i_1 i_2 \frac{dM}{d\theta} \qquad \text{...(3.92)}$$

The average deflecting torque over a complete cycle,

$$(T_d)_{av.} = \frac{1}{T}\int_0^T T_{di}\, dt$$

or, $$(T_d)_{av.} = \frac{dM}{d\theta}\frac{1}{T}\int_0^T i_1 i_2\, dt \qquad \text{...(3.93)}$$

where, T = Time period for one complete cycle.

If currents i_1 and i_2 are *sinusoidal and are displaced by a phase angle* ϕ, *i.e.,*

$$i_1 = I_{max.1} \sin \omega t, \quad \text{and} \quad i_2 = I_{max.2} \sin(\omega t - \phi),$$

then, the *average deflecting torque,*

$$(T_d)_{av.} = \frac{dM}{d\theta} \cdot \frac{1}{T}\int_0^T i_1 i_2\, dt$$

$$= \frac{dM}{d\theta} \cdot \frac{1}{2\pi}\int_0^{2\pi} I_{max.1} \sin \omega t \,.\, I_{max.2} \sin(\omega t - \phi)\, d(\omega t)$$

or, $$(T_d)_{av.} = \frac{I_{max.1} \cdot I_{max.2}}{2} \cos\phi \frac{dM}{d\theta} = I_1 I_2 \cos\phi \frac{dM}{d\theta} \qquad \text{...(3.94)}$$

where I_1 and I_2 are r.m.s. values of current flowing in the coils.

At steady deflection state,

$$T_d = T_c$$

$$\text{or,} \qquad I_1 I_2 \cos\phi \frac{dM}{d\theta} = C\theta$$

$$\therefore \qquad \text{Deflection, } \theta = \frac{I_1 I_2}{C} \cos\phi \cdot \frac{dM}{d\theta} \qquad \text{...(3.95)}$$

From the above discussion it is obvious that for a *sinusoidal alternating currents the deflecting torque and the deflection are determined by the product of r.m.s. values of coil currents and cosine of the phase angle between them.* It is clear that the *dynamometer type of instruments can be used on both D.C. and A.C.*

Use of the instrument as an ammeter. When the instrument is used as an ammeter then same current passes through both moving coil (*MC*) and fixed coils (*FC*) as shown in Fig. 3.46. In this case, $I_1 = I_2 = I$, hence $\theta \propto I^2$ or $I \propto \sqrt{\theta}$. The connections of Fig. 3.46 are used when *small currents* are to be measured.

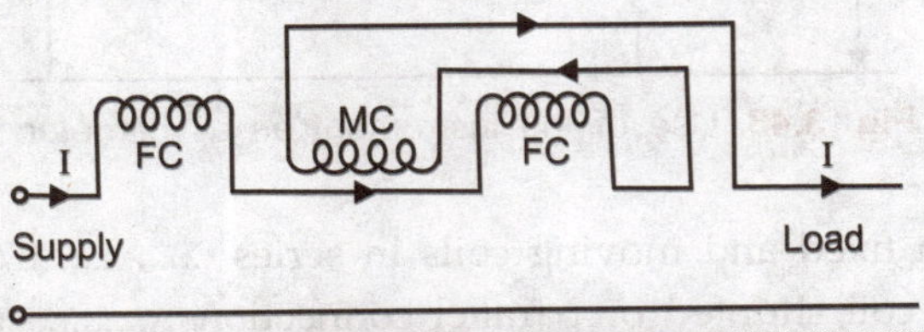

Fig. 3.46. Measurement of small currents.

In the case of heavy currents, a shunt is used to limit current through the moving coil as shown in Fig. 3.47.

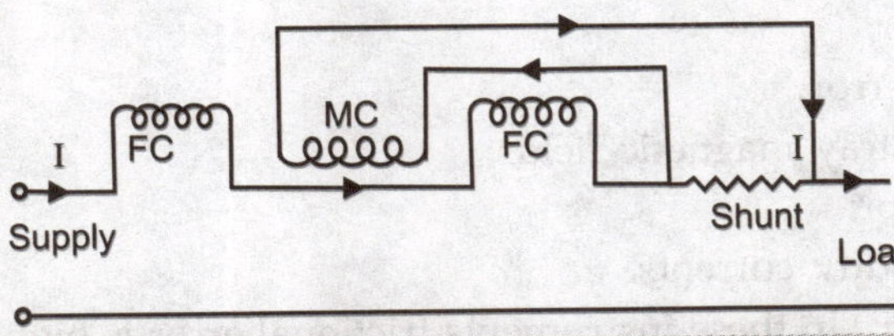

Fig. 3.47. Measurement of heavy currents.

Use of the instrument as voltmeter. When the instrument is used as a voltmeter, the fixed and moving coils are used in series along with high resistance as shown in Fig. 3.48.

Here again, $I_1 = I_2 = I$,

where, $\qquad I = \frac{V}{R}$ ......in D.C. circuits

and, $\qquad I = \frac{V}{Z}$ ......in A.C. circuits

$\therefore \qquad \theta \propto V \times V$ or $\theta \propto V^2$

or, $\qquad V = \sqrt{\theta}$

- Thus, whether the instrument is used as an ammeter or voltmeter, its scale is *uneven* through the whole of its range and is cramped or crowded near the zero in particular.

Note: When the dynamometer instrument is used to measure an alternating current or voltage, the moving coil due to its inertia takes up a position where the average deflecting torque over one cycle is balanced by the restoring torque of the spiral springs. For that position, the deflecting torque is proportional to the mean value of the square of current or voltage, and the instrument scale can therefore be *calibrated to read the r.m.s. value.*

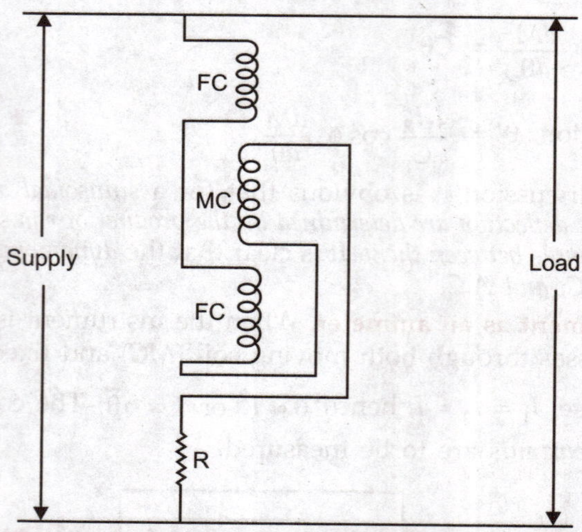

Fig. 3.48. Use of the instrument as a voltmeter.

Ranges :

Ammeter. (*i*) With fixed and moving coils in series .... 0/0.01 A – 0/0.05 A

(*ii*) With moving coil shunted or parallel connections .... upto 0/30 A.

Voltmeters. Upto 0–750 volts.

3.8.8.3. Errors in dynamometer type instruments

Following errors usually occur in these instruments :

(*i*) Frictional error;

(*ii*) Temperature error;

(*iii*) Error due to stray magnetic fields;

(*iv*) Frequency error;

(*v*) Error due to eddy currents.

1. ***Frictional error :*** In these instruments frictional error is high. So, in order to have high accuracy of measurement this error needs to be minimised to the lowest possible, this is achieved as follows :

(*i*) By a reasonable reduction in weight;

(*ii*) By proper selection of materials for bearings and spindle, and adequate polishing of these parts.

However, these measures increase the cost of the instrument and make it more sensitive to overloads and mechanical impacts.

2. ***Temperature error :***

- High currents are carried by coils which produce heat. *Self-heating of coils produces errors.*
- High accuracy instruments contain *temperature compensating resistors* which tend to *neutralise* the effects of changes in temperature.

3. ***Errors due to stray magnetic fields :*** In order to avoid the influence of stray fields a good amount of *screening is necessary*.

- In precision instruments the influence of external magnetic fields can be minimised by using *"Astatic system"*.

— *"Astatic"* dynamometer type instruments are constructed with two similar sets of fixed and moving coils mounted on the same shaft. The two pairs of fixed coils are so connected that their *magnetic fields are in opposition. The currents in the moving coils are also opposed* and, in consequence, the deflection torques applied to the moving coils act in the same direction. Since the two fields are in opposition and, therefore, any uniform external field reduces the field of one coil and increases by an equal amount the field of the other coil, the deflecting torque produced by one coil is increased while that produced by the other coil is reduced by an equal amount and, thus *net torque on account of external magnetic field is zero.*

4. ***Frequency error :*** This error in these instruments is largely due to variation of self-reactance of coils, with frequency.

- In order to reduce frequency error in dynamometer ammeters, the ratio of currents in fixed and moving coils *i.e.* $\frac{I_1}{I_2}$ should be independent of frequency. This requires that *time constant* $\left(\frac{L}{R}\right)$ *of two circuits should be same.*
- For reducing frequency errors in voltmeters the *coil winding is made a very small part of the circuit.* This makes inductive reactance ($X_L = 2\pi fL$), a *small fraction* of total impedance, and therefore $Z = \sqrt{R^2 + (2\pi fL)^2} \simeq R$. This means frequency does not affect the calibration of the instrument. In fact, dynamometer voltmeters may be used, in general, within their guaranteed accuracy from D.C. to about 125 Hz.

5. ***Error due to eddy currents :***

- The eddy currents are induced in metal parts of the instrument and develop a torque because of coupling between the moving coil and the neighbouring metal part.
- This error can be *reduced* by keeping the metal in the supports of the coil and other structural parts to the *minimum possible* and selecting the metal, to be used, of *high resistivity.*

3.8.8.4. Advantages and disadvantages

Advantages :

1. Can be used on both D.C. as well as A.C. systems (as the instrument has square law response).
2. They are free from hysteresis and eddy current errors, because of absence of iron in the operating parts of the instrument.
3. It is possible to construct ammeters upto 10 A and voltmeters upto 600 V with precision grade accuracy.
4. Very useful for accurate measurement of r.m.s. values of voltages irrespective of waveforms.
5. Owing to precision grade accuracy and same calibration for D.C. and A.C. measurements, these instruments are used as *transfer and calibration instruments.*

Disadvantages :

1. Since torque/weight ratio is small, such instruments have *low sensitivity.*
2. The *scale is not uniform* because $\theta \propto \sqrt{I}$.
3. Cost of these instruments is *higher* in comparison to those of moving-iron instruments. So, these are *only used as voltmeters and ammeters for precision measurements.*
4. *Higher frictional losses.*

5. These instruments are sensitive to overloads and mechanical impacts. Therefore, they *must be handled with care.*
6. In these instruments the operating current is large and consequently they have a *higher power consumption than PMMC instruments.*

Example 3.52. *The mutual inductance of a 25 A electrodynamometer ammeter changes uniformly at a rate of 0.0035 μH/degree. The torsion constant of the controlling spring is 10^{-6} Nm/degree. Determine the angular deflection for full scale.* **(Poona University)**

Solution. *Given* : $I_1 = I_2 = I = 25$ A; $\frac{dM}{d\theta} = 0.0035$ μH/degree; $C = 10^{-6}$ Nm/degree

Angular deflection for full scale, θ :

Now, $$T_d = I^2 \frac{dM}{d\theta} \quad \text{[Eqn. (3.90)]}$$

... instrument being ammeter ($I_1 = I_2$)

Since spring constant is given Nm/degree, therefore, we must express $\frac{dM}{d\theta}$ in H/rad. if deflection is to be found in degrees.

$$\therefore \quad \frac{dM}{d\theta} = \frac{0.0035 \times 10^{-6}}{(\pi/180)} = 0.2 \times 10^{-6} \text{ Nm/rad.}$$

Full scale deflection torque,

$$T_d = T_c = C\theta = 10^{-6}\theta \text{ Nm}$$

Substituting this value of T_d in the above eqn., we get :

$$10^{-6}\theta = (25)^2 \times 0.2 \times 10^{-6}$$

or $$\theta = \mathbf{125°} \quad \textbf{(Ans.)}$$

Example 3.53. *A 10 A electrodynamometer is controlled by a spring having a torsion constant of 0.1 × 10^{-6} Nm/degree. The full scale deflection is 110°. Determine the inductance of the instrument when measuring a current of 10 A. The mutual inductance at 0° deflection is 2 μH and the change in mutual inductance is linear as the deflection.* **(Nagpur University)**

Solution. *Given* : $C = 0.1 \times 10^{-6}$ Nm/degree; θ = 110°, I = 10 A; M at 0° deflection = 10A

Inductance of the instrument :

Full scale deflecting torque, $T_d = T_c = C\theta = 0.1 \times 10^{-6} \times 110 = 11.1 \times 10^{-6}$ Nm

Also, $$T_d = I^2 \frac{dM}{d\theta}$$

$$\therefore \quad \frac{dM}{d\theta} = \frac{T_d}{I^2} = \frac{11.1 \times 10^{-6}}{(10)^2} = 0.111 \text{ μH/rad.}$$

Change in inductance, $$dM = 0.111 \times 10^{-6} \times 110 \times \frac{\pi}{180} = 0.213 \text{ μH}$$

∴ *Total inductance of the instrument* = Initial inductance + change in inductance

= 2 + 0.213 = **2.213 μH** **(Ans.)**

Example 3.54. *The spring constant of a 10 A dynamometer wattmeter is 10.5 × 10^{-6} Nm/radian. The variation of inductance with angular position of moving system is practically linear over the operating range, the rate of change being 0.078 mH/radian. If the full-scale deflection of the instrument is 83 degrees, calculate the current required in the voltage coil at full scale on D.C. circuit.* **(Nagpur University)**

Solution. *Given* : $C = 10.5 \times 10^{-6}$ Nm/rad.; $10.5 \times 10^{-6} \times \frac{\pi}{180}$ Nm/degree; $\frac{dM}{d\theta} = 0.078$ mH/rad.; $\theta = 83°$.

Current required in voltage coil, I_2 :

$$T_d = T_c = C\theta = \left(10.5 \times 10^{-6} \times \frac{\pi}{180}\right) \times 83 = 15.21 \times 10^{-6} \text{ Nm}$$

Also, $$T_d = I_1 I_2 \frac{dM}{d\theta} \quad \text{[Eqn. (3.90)]}$$

$\therefore$ $$15.21 \times 10^{-6} = 10 \times I_2 \times 0.078$$

or, $$I_2 = 19.5 \times 10^{-6} \text{ A} \quad \text{or} \quad \mathbf{19.5\ \mu A}$$ **(Ans.)**

Example 3.55. *In a certain dynamometer ammeter the mutual inductance M varies with deflection θ (expressed in degees) as :*

$$M = -8 \cos(\theta + 30°) \text{ mH}$$

Find the deflecting torque produced by a direct current of 60 mA corresponding to a deflection of 60°.

Solution. *Given* : $M = -8 \cos(\theta + 30°)$ mH ; $I = 60$ mA or 0.06 A; $\theta = 60°$

Deflecting torque, T_d :

Rate of change of mutual inductance with deflection,

$$\frac{dM}{d\theta} = \frac{d}{d\theta}[-8\cos(\theta + 30°)] = 8 \sin(\theta + 30°) \text{ mH}$$

$\frac{dM}{d\theta}$ at a deflection of 60° *i.e.*, $\left(\frac{dM}{d\theta}\right)_{\theta = 60°} = 8 \sin(60° + 30°) \text{ mH} = 8 \times 10^{-3}$ H/degree

$\therefore$ Deflecting torque, $T_d = I^2 \frac{dM}{d\theta} = (0.06)^2 \times 8 \times 10^{-3} = \mathbf{28.8\ \mu Nm}$ **(Ans.)**

Example 3.56. *In a torsion-head type electrodynamometer a current of 36 A requires a deflection of 90° in the head to give balance.*

Determine :

(i) The range of the instrument (maximum angle 360°)

(ii) For what current will the deflection be 180° ?

(iii) What will be the deflection for a current of 30 A ?

Solution. In torsion-head electrodynamometer,

$$\text{Torque, } T \propto (\text{current})^2$$

Also, torque, $T \propto \theta$ (deflection)

$\therefore$ $$\theta \propto I^2$$

or, $$I \propto \sqrt{\theta}$$

(i) Current corresponding to deflection of 360°,

$$I_2 = I_1 \times \sqrt{\frac{\theta_2}{\theta_1}} = 36 \times \sqrt{\frac{360}{90}} = \mathbf{72\ A}$$ **(Ans.)**

(ii) Current corresponding to deflection of 180°,

$$I_3 = I_1 \times \sqrt{\frac{\theta_2}{\theta_1}} = 36 \times \sqrt{\frac{180}{90}} = \mathbf{50.9\ A}$$ **(Ans.)**

(*iii*) Deflection corresponding to current 30 A,

$$\theta_4 = \theta_1 \times \left(\frac{I_4}{I_1}\right)^2 = 90 \times \left(\frac{30}{36}\right)^2 = \mathbf{62.5°} \quad \textbf{(Ans.)}$$

Example 3.57. *A 50 V range spring controlled, electrodynamic voltmeter having a square law scale response takes 0.05 A on D.C. for full scale deflection of 90°. The control constant is 0.5×10^{-6} Nm/degree and the initial mutual inductance of the instrument is 0.25 H. Find the true potential difference across the instrument when it reads 50 V at 50 Hz.*

Solution. *Given* : $V = 50$ volts; $I = 0.05$ A at $\theta = 90°$; $C = 0.5 \times 10^{-6}$ Nm/degree; Initial inductance = 0.25 H; $f = 50$ Hz.

True p.d. across the instrument :

Full scale deflection controlling torque, $T_c = 0.5 \times 10^{-6} \times 90 = 45 \times 10^{-6}$ Nm

For steady deflection state, $T_d = T_c$

$$\therefore \quad I^2 \frac{dM}{d\theta} = 45 \times 10^{-6}$$

or,
$$\frac{dM}{d\theta} = \frac{45 \times 10^{-6}}{I^2} = \frac{48 \times 10^{-4}}{(0.05)^2} = 18 \text{ mH/rad.}$$

Total change in mutual inductance for deflection of 90° or $\frac{\pi}{2}$ rad.,

$$dM = 18 \times 10^{-3} \times \frac{\pi}{2} = 0.02827 \text{ H}$$

$\therefore$ Total mutual inductance (M) = Initial mutual inductance + change in mutual inductance

$= 0.25 + 0.02827 \simeq 0.2783$ H

Reactance of the instrument at 50 Hz $= 2\pi fM = 2\pi \times 50 \times 0.2783 = 87.43\ \Omega$

Resistance of the instrument, $R = \frac{V}{I} = \frac{500}{0.05} = 1000\ \Omega$

Impedance of the instrument at 50 Hz, $Z = \sqrt{(1000)^2 + (87.43)^2} = 1003.8\ \Omega$

$\therefore$ *True p.d. across the instrument* $= IZ = 0.05 \times 1003.8 =$ **50.19 V** **(Ans.)**

Example 3.58. *A 50 V range spring controlled electrodynamic voltmeter has an initial inductance of 0.25 H, the full scale deflection torque of 0.4×10^{-4} Nm and full scale deflection current of 50 mA.*

Determine the difference between D.C. and 50 Hz A.C. readings at (i) 50 V and (ii) 25 V if the voltmeter inductance increases uniformly over the full scale of 90°. **(Delhi University)**

Solution. *Given* : Initial inductance = 0.25 H; $T_d = 0.4 \times 10^{-4}$ Nm; $I = 50$ mA or 0.05A

Now,
$$T_d = I^2 \frac{dM}{d\theta}$$

or,
$$0.4 \times 10^{-4} = (0.05)^2 \times \frac{dM}{d\theta}$$

or,
$$\frac{dM}{d\theta} = \frac{0.4 \times 10^{-4}}{(0.05)^2} = 0.016 \text{ H/rad.}$$

Total change in inductance for full scale deflection,

$$dM = 0.016 \times \left(90 \times \frac{\pi}{180}\right) = 0.0251 \text{ H}$$

$\therefore$ Total mutual inductance, $M = 0.25 + 0.251 = 0.2751$ H

At 50 V :

The resistance of voltmeter, $R = \frac{V}{I} = \frac{50}{0.05} = 1000\ \Omega$

The impedance while measuring the voltage of 50 V at 50 Hz A.C.,

$$Z = \sqrt{(1000)^2 + (2\pi \times 50 \times 0.2751)^2} = 1003.7\ \Omega$$

$$\therefore \quad \text{Voltmeter reading} = \frac{50}{1003.7} \times 1000 = 49.8 \text{ V}$$

Hence, *difference in readings* = 50 – 49.8 = **0.2 V** **(Ans.)**

At 25 V :

Current through instrument, $I' = \frac{25}{1000} = 0.025$ A

$$\text{Deflection, } \theta' = \theta \times \frac{(I')^2}{(I)^2}, \text{ since } \frac{dM}{d\theta} \text{ is constant}$$

$$= \frac{\pi}{2} \times \left(\frac{0.025}{0.05}\right)^2 = 0.3927 \text{ rad.}$$

$$\text{Mutual inductance at 25 V} = \text{Initial inductance} + \frac{dM}{d\theta} \times \theta'$$

$$= 0.25 + 0.016 \times 0.3927 = 0.2563 \text{ H}$$

$$\text{Impedance at 25 V,} \quad Z' = \sqrt{(1000)^2 + (2\pi \times 50 \times 0.2563)^2} = 1003.24\ \Omega$$

$$\therefore \quad \text{Voltmeter reading} = \frac{25}{1003.24} \times 1000 = 24.92 \text{ V}$$

Hence, difference in reading = 25 – 24.92 = **0.08 V** **(Ans.)**

THERMAL INSTRUMENTS (HOT-WIRE AND THERMOCOUPLE)

3.8.9. Hot-wire Instruments

- *These instruments utilise the expansion of wire heated by current being measured.*
- These instruments are primarily meant for being used as ammeters but can be adopted as voltmeters by connecting a high resistance in series with them.
- The hot wire instruments are suited both for A.C. and D.C. work.

Construction. Fig. 3.49 shows the working parts of the instrument :

— It consists of platinum-iridium (It can *withstand oxidation at high temperatures*) wire *AB* stretched between a fixed end *B* and the tension-adjusting screw at *A*. The diameter of this wire is about 0.2 mm.

— A phosphor bronze wire is attached to the hot wire and in turn a fine silk thread *R* is attached to phosphor bronze wire as shown in the figure.

— The silk thread *R* passes around a pulley *E* and the other end of it is fixed to spring *S*.

— A light pointer P and thin aluminium disc L are carried by the spindle upon which the pulley is also mounted.

— M is the damping magnet; eddy current damping is used for these instruments.

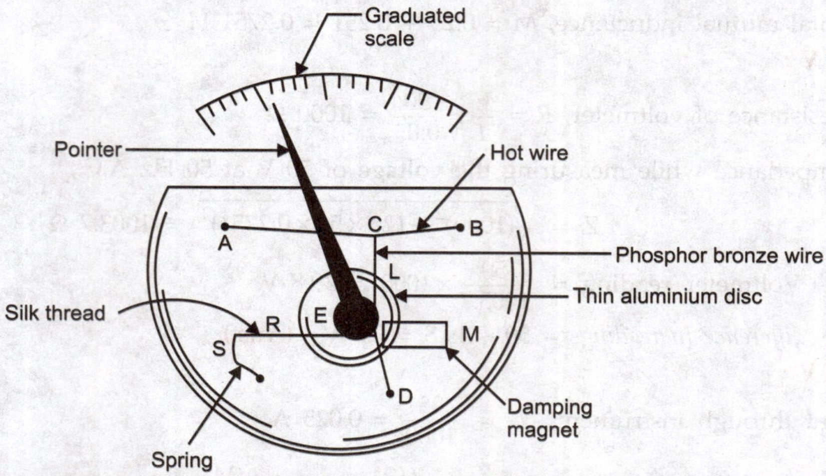

Fig. 3.49. Hot-wire instrument.

Working. When the current to be measured (or a definite fraction of it or proportional to the voltage to be measured) is passed through the wire AB, it expands due to heating effect of the current and slack in the phosphor bronze wire which is taken up by string S through the silk thread R and the moving system is rotated due to motion of the pulley E. Thus the pointer P moves over a calibrated scale.

It will be seen that the deflection of the pointer is proportional to the extension of AB which in itself proportional to I^2. Hence, *deflection is proportional to* I^2. If spring control is used then $T_c \propto \theta$.

Hence, $$\theta \propto I^2 \quad ...(3.96)$$

So, these instruments have a *square law type scale. They read the "r.m.s. value of current" and their readings are independent of its form and frequence.*

- The hot wire should be made *as thin as possible so that it may attain steady temperature as quick as possible when current flows through it*. At the same time the wire should be of *such size that it may be capable of withstanding the normal mechanical stresses* developed in the instrument due to handling, vibration, tension etc.
- In order to minimise the error due to uneven expansion between wire and its fixtures, the base of the instrument is made of material of coefficient of expansion equal to that of hot wire.

Magnification of the expansion :

- *Refer to Fig. 3.50 (a)* :

Let, l = Length of hot wire, and

δL = Increase in length of wire due to flow of current I through it.

Then, the sag in wire due to expansion,

$$S = \sqrt{\left(\frac{L+\delta L}{2}\right)^2 - \left(\frac{L}{2}\right)^2}$$

$$= \sqrt{\frac{2L\,\delta L + (\delta L)^2}{4}}$$

$$\text{or,} \qquad S = \sqrt{\frac{L\,.\,\delta L}{2}} \qquad \text{...(3.97)}$$

neglecting (δL) being very small

The *magnification factor,*

$$M = \frac{S}{\delta L} = \frac{\sqrt{L\,.\,\delta L/2}}{\delta L} = \sqrt{\frac{L}{2\,\delta L}} \qquad \text{...(3.98)}$$

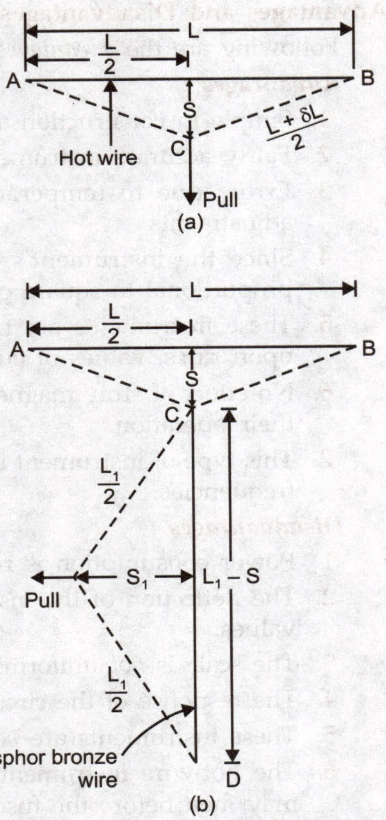

Fig. 3.50.

- *Refer to Fig. 3.50(b) :*

Further magnification is obtained by connecting a phosphor bronze wire of length L_1 to the centre of the wire *AB*.

Let the sag in this case be S_1.

$$\text{Now,} \qquad S_1 = \sqrt{\left(\frac{L_1}{2}\right)^2 - \left(\frac{L_1 - S}{2}\right)^2}$$

$$= \sqrt{\frac{L_1^2}{4} - \frac{L_1^2}{4} + \frac{2L_1S - S^2}{4}}$$

$$\text{or,} \qquad = \sqrt{\frac{L_1S}{2}}$$

...neglecting S^2, being very small

$$= \sqrt{\frac{L_1}{2}} \times \sqrt{\frac{L.\delta L}{2}} \qquad \text{....substituting the value of } S \text{ from eqn. (3.97)}$$

$$\text{or,} \qquad S_1 = \sqrt{\frac{L_1}{2\sqrt{2}}}\,(L.\delta L)^{1/4} \qquad \text{...(3.99)}$$

$$\text{or} \qquad S_1 \propto (\delta L)^{1/4} \propto \sqrt{\text{current}} \qquad \text{since } \delta L \propto (\text{current})^2$$

The *magnification factor,*

$$M = \frac{S_1}{\delta L} = \frac{\sqrt{\frac{L_1}{2\sqrt{2}}}\,.\,(L.\delta L)^{1/4}}{\delta L} = \sqrt{\frac{L_1}{2\sqrt{2}}}\left(\frac{L}{(\delta L)^3}\right)^{1/4} \qquad \text{...(3.100)}$$

Ranges :

Ammeter : (*i*) Without use of shunt : 0–1 A.

(*ii*) Without use of shunt by using device like dividing the wire paths into several parallel paths : 0–5 A.

Voltmeter : Upto 400 V by using high non-inductive resistance in series with the instrument.

Advantages and Disadvantages :

Following are the *advantages* and *disadvantages* of hot-wire instruments :

Advantages :

1. Simple in construction and cheap in cost.
2. Fairly accurate instruments.
3. Errors due to temperature variations can be made negligible with suitable adjustments.
4. Since the instrument's operation depends upon the heating effect which is proportional to square of current, it can be used on A.C. and D.C. both.
5. These instruments are free from waveform errors since their readings depend upon r.m.s. values of current.
6. No effect of stray magnetic fields because no magnetic effects are used to cause their operation.
7. This type of instrument is quite suitable for measurement of current at very high frequencies.

Disadvantages :

1. Power consumption is relatively high.
2. The deflection of the instrument is *not* the same for ascending and descending values.
3. The scale is *not* uniform.
4. The response of the circuit is *very slow* as the wire takes time to heat up.
5. These instruments are *very delicate.*
6. The hot-wire instruments cannot take overload because wire is very fine and may melt before the fuse can protect it.
7. Changes in room temperature causes errors which shifts the zero as a result of which it has to be reset frequently.

- In fact, although the accuracy of hot-wire instruments for measurement of current and voltage at higher frequency is higher than others, yet due its disadvantages they are *practically obsolete.*

Example 3.59. *The working wire of a single sag hot wire instrument is 15 cm long and is made-up of platinum-silver with a coefficient of linear expansion of 16×10^{-6}. The temperature rise of the wire is 85°C and the sag is taken up at the centre. Find the magnification.*

(*i*) *With no initial sag;*

(*ii*) *With an initial sag of 1 mm.* **(Calcutta University)**

Solution. *Given :* $L = 15$ cm or 0.15 m; $\alpha = 16 \times 10^{-6}$; Temperature rise, $\Delta t = 85°C$

Magnification, *M* :

Increase in length of wire when heated through 85°C,

$$\delta L = \alpha L.(\Delta t) = 16 \times 10^{-6} \times 0.15 \times 85 = 2.04 \times 10^{-4} \text{ m}$$

(*i*) *With no initial sag :*

$$M = \sqrt{\frac{L}{2.\delta L}} \qquad \text{[Eqn. (3.98)]}$$

$$= \sqrt{\frac{0.15}{2 \times 2.04 \times 10^{-4}}} = \mathbf{19.17} \quad \textbf{(Ans.)}$$

(ii) With an initial sag of 1 mm :

$$\text{Sag, } S = \sqrt{\frac{L.\delta L}{2}} \qquad \text{[Eqn. (3.97)]}$$

$$= \sqrt{\frac{0.15 \times 2.04 \times 10^{-4}}{2}} = 3.91 \times 10^{-3} \text{ m}$$

Net increase in sag, $S' = S - \text{initial sag} = 3.91 \times 10^{-3} - 1 \times 10^{-3} = 2.91 \times 10^{-3}$ m

$$\therefore \qquad M = \frac{S'}{\delta L} = \frac{2.91 \times 10^{-3}}{2.04 \times 10^{-4}} = \mathbf{14.26} \quad \textbf{(Ans.)}$$

3.8.10. Thermocouple Instruments

Working principle :

*When two wires of different materials are joined together to form two junctions and if these junctions are maintained at different temperatures, a current flows around a circuit. This is known as **Seebeck effect** and the system is known as a "thermocouple".* Seebeck arranged 25 metals in the order of their thermoelectric properties, these are : Bi-Ni-Co-Pd-Pt-U-Cu-Mu-Ti-Hg-Pb-Sn-Cr-Mo-Ph-Ir-Au-Ag-Zn-W-Cd-Fe-As-Sb-Te.

The current flows across the hot junction from the former to the latter metal of the above series, e.g., if copper and iron are used to form a thermocouple, the current flows from copper to iron at the hot junction and iron to copper at the cold junction (Fig. 3.51). If the copper wire is cut, an e.m.f. appears across the open circuit.

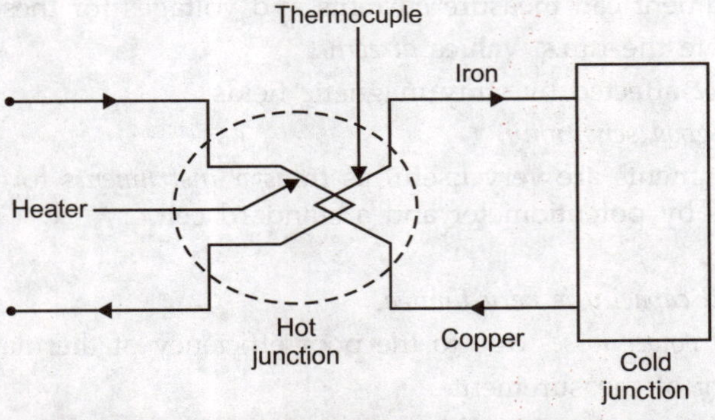

Fig. 3.51. Seebeck effect.

Construction and Working :

Fig. 3.52 shows the construction of a thermocouple instrument :

— It consists of a *heater element* (through which the current being measured is passed), *thermocouple* and *a PMMC instrument* to measure the e.m.f. produced by Seebeck effect.

— The heater wire and one junction of the thermocouple are in contact at the hot junction.

— The heater converts electrical energy to heat. A part of this heat is transferred to the thermocouple at the hot junction. Due to Seebeck effect an *e.m.f. is produced and it is measured by the PMMC instrument* (D.C. moving coil microvoltmeter). *The deflection is proportional to thermoelectric e.m.f. which in turn is proportional to the difference between the temperatures of hot and cold junctions*. This *temperature difference is proportional to* I^2.

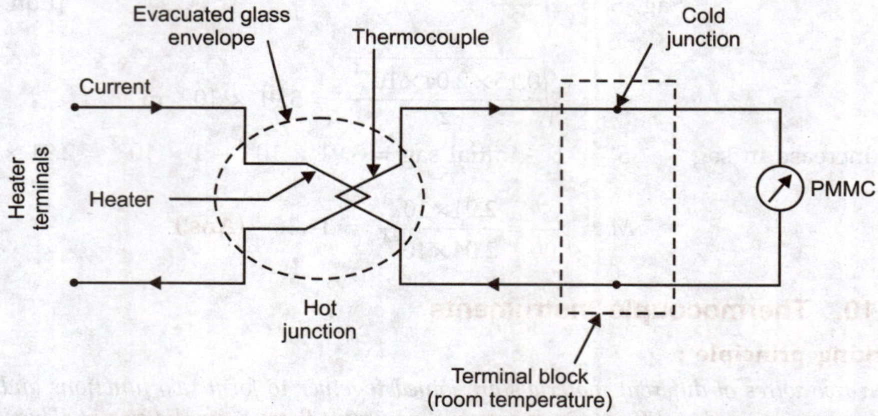

Fig. 3.52. Thermocouple instrument.

The scale of PMMC meter can be *calibrated to indicate the r.m.s. value*.

- The commonly used junctions for thermocouple ammeters are *copper-constantan* and *iron-constantan*.

Advantages and disadvantages :

Advantages :

1. These instruments are suitable for *very high frequencies* upto *about 50 MHz*. No other instrument can measure currents and voltages for these high frequencies.
2. They indicate the r.m.s. values *directly*.
3. They are *not* affected by stray magnetic fields.
4. They have *good sensitivity*.
5. These instruments are very useful as *transfer instruments* for calibration of D.C. instruments by potentiometer and a standard cell.

Disadvantages :

1. The *overload capacity is very limited*.
2. *Considerable power losses* due to the poor efficiency of thermal conversion.
3. *Low accuracy* of measurement.
4. The millivoltmeters used with thermo-elements must be necessarily more sensitive and delicate than those used with shunts, and therefore, *require careful handling*.

3.8.11. Rectifier Type Instruments

3.8.11.1 Introduction

These are not separate types of instruments, but rather a means of using a D'Arsonval movement, in conjunction with a *rectifier*, to change A.C. to D.C. Thus, a direct current movement can be adopted for use with alternating current as shown in Fig. 3.53. Rectifier type meters, using copper oxide rectifiers, are useful at low frequencies and will give good indication upto about 20 kHz.

- Rectifier instruments can operate well into the RF (radio-frequency) range with the proper use of silicon or germanium rectifiers.
- Measurement of current and voltage at several hundred MHz is possible.
- This type of meter is *more sensitive than any other type of A.C. meter*.

3.8.11.2. Rectifier type elements

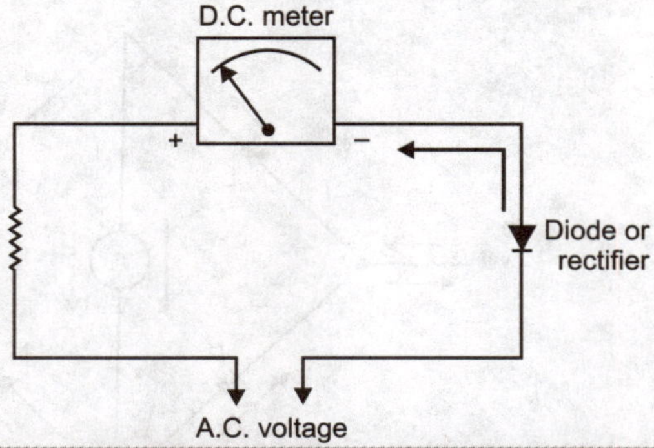

Fig. 3.53. A D.C. moving-coil meter can be used to measure A.C. voltage by putting a diode or rectifier in the meter circuit.

- A *rectifier element is used to convert A.C. to D.C. so that a unidirectional current flows through the PMMC movement.*
- The rectifier type of instruments use a copper oxide or a selenium cell, a germanium or silicon diode.
  - — *Copper oxide elements* have a peak inverse voltage (PIV) of about 2 V while *selenium elements* have a PIV of about 10 V and also they can handle only limited amount of current and hence are becoming *obsolete.*
  - — Germanium diodes have a PIV of the order of 300 V, and a current rating of 100 mA (appr.). *Low current silicon diode* rectifiers have a PIV of upto 1000 V, and a current rating of the order of 500 mA. For these reasons germanium and silicon diodes find extensive applications in rectifier instruments these days.

It may be mentioned here that though silicon diodes can carry currents as high as 85 A per rectifying elements, but these *high current silicon diodes are seldom used in indicating instruments and find wide applications in power supply circuits.*

3.8.11.3. Characteristics of rectifier instruments

The *characteristics* of rectifier instruments are given below :

1. They provide an economical and practical means of A.C. measurements in radio and communication circuits at audio frequencies (1 kHz to 10 kHz) and in other A.C. circuits where small available power makes it necessary to use sensitive low-energy instruments.
2. A rectifier instrument is, in general, approximately *fifty times* more sensitive than either an electrodynamometer or a moving-iron instrument.
3. In most instances rectifier instruments have *essentially linear scales.*
4. They commonly possess sensitivity of the order of 1000 to 2000 ohms per volt and more.
5. The power consumed by a rectifier instrument is generally several times as high as its permanent magnet moving coil (PMMC) mechanism because of the resistance of its rectifier.
6. Shunting of rectifier instruments is not practical because of the change in resistance of the rectifier with both temperature and the amount of the current.
7. Rectifier instruments are manufactured as micrometers and milliammeters in ratings from 100 μA to 15 mA for full scale deflection. The higher rating of 15 mA is more or less governed by the special size of the rectifier in comparison with the mechanism with which it is to be used.

3.8.11.4. Rectifier ammeters and voltmeters

Rectifier ammeters. The *rectifier ammeters usually* consists of four rectifier elements *L*, *M*, *N* and *P* arranged in the form of a bridge (as shown in Fig. 3.54) and A represents a moving-coil ammeter. The apex of the black triangle (shown in the bridge) indicates the direction in which the resistance is low.

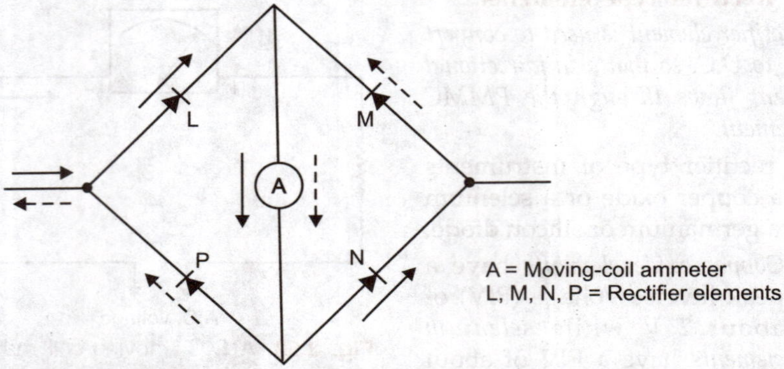

Fig. 3.54. Bridge circuit for full-wave rectification.

As shown in Fig. 3.54 during the half-cycles current flows through elements L and N from left to right as shown by the *full arrows.* During other half-cycles, the current flows through M and P, as shown by *dotted arrows.* The waveform of the current flowing through A is therefore as shown in Fig. 3.55. The deflection of moving-coil ammeter A, thus depends upon the *average value of the current;* and the scale of A can be calibrated to read the r.m.s. value of the current on the assumption that the waveform of the latter is sinusoidal with a form factor of 1.11.

Rectifier voltmeters. In a *rectifier voltmeter,* A is a milliammeter and the bridge circuit of Fig. 3.54 is connected in *series with a suitable resistor.*

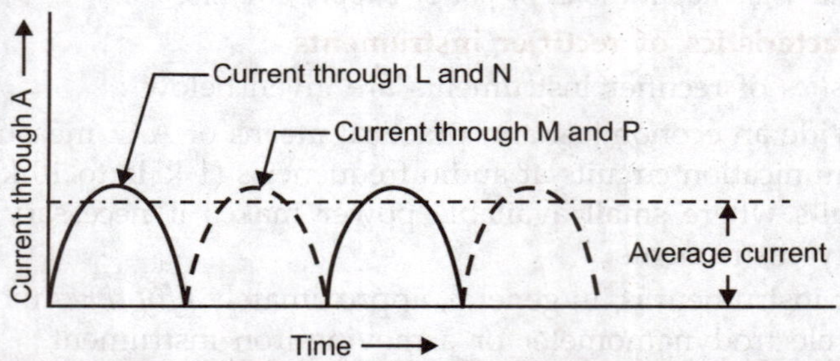

Fig. 3.55. Waveform of current through moving-coil ammeter.

Ranges :

(*i*) 0–100 μA; upto 0–100 mA

(*ii*) 0–1.0 V; upto 0–250 V without external resistance

(*iii*) Frequency range 20 Hz to 20 kHz.

☞ **Precaution :** The circuit to the instrument should *never be opened* while the rectifier is connected to the line, because this will result in *full voltage impressed across the rectifier plates, which may get damaged due to it.*

3.8.11.5. Sensitivity of rectifier type instruments

- The *D.C. sensitivity* of a rectifier type instrument is given by :

$$S_{D.C.} = \frac{1}{I_{fsd}} \ \Omega/V \qquad ...(3.101)$$

where, I_{fsd} = The current required to produce full scale deflection.

- For a half-wave rectifier with *sinusoidal input* the *A.C. sensitivity* is given as :

$$S_{A.C.} = 0.45\ S_{D.C.} \quad ...(3.102)$$

Similarly, for a full-wave rectifier type instrument, with *sinusoidal input*, the A.C. sensitivity is given by :

$$S_{A.C.} = 0.9\ S_{D.C.} \quad ...(3.103)$$

These relations are not valid for non-sinusoidal input waveforms such as square, triangular and sawtooth waveforms. Further these relations are *based on the assumption that the forward resistances of diodes are zero and the reverse resistance infinite.*

3.8.11.6. Extension of range of rectifier instruments as voltmeters :

Let us support we are to extend the range of a rectifier instrument which uses a *PMMC instrument having a D.C. sensitivity of* $S_{D.C.}$

Further, let v = Voltage drop across the PMMC instrument,

V = Applied voltage,

R_m = Resistance of the meter, and

R_d = Diode resistance.

Then, for D.C. operation the values of series resistance (multiplier) needed is,

$$R_s = \left(\frac{V}{I_{fsd}}\right) - R_m - R_d \quad ...(3.104)$$

or, $$R_s = S_{D.C.} V - R_m - R_d$$ for *half-wave rectification*

$$= \left(\frac{V}{I_{fsd}}\right) - R_m - 2R_d$$

or, $$R_s = S_{D.C.} V - R_m - 2Rd \quad ...(3.105)$$

...for *full-wave rectification*

For A.C. voltmeters :

$$R_s = \left(\frac{V}{I_{fsd}}\right) - R_m - R_d$$

$$= S_{A.C.} V - R_m - R_d = 0.45\ S_{D.C.} V - R_m - R_d \quad ...(3.106)$$

...for *half-wave rectification*

or, $$R_s = \left(\frac{V}{I_{fsd}}\right) - R_m - 2R_d = 0.9\ S_{D.C.}\ V - R_m - 2R_d \quad ...(3.107)$$

...for *full-wave rectification*

- *Following factors affect the performance of rectifier type instruments :*
  - (*i*) Waveform;
  - (*ii*) Resistance of rectifier;
  - (*iii*) Temperature changes;
  - (*iv*) Capacitance of rectifier;
  - (*v*) Sensitivity.
- *"Loading effects of rectifier instruments"* :

The sensitivity of rectifier type of voltmeters, using either half or full wave rectification, is *less* than the sensitivity of D.C. voltmeters. Therefore, the *loading effects of an A.C. rectifier voltmeter is greater than that of a corresponding D.C. voltmeter.*

3.8.11.7. Advantages of rectifier instruments

1. The primary advantage of the rectifier *voltmeter* is that it is *far more sensitive* as compared to other types of voltmeter suitable for measuring A.C. voltages.

2. Metal rectifiers can be incorporated in universal instruments, such as ammeter, thereby enabling a moving-coil milliammeter to be used in combination with shunt and series resistances to measure various ranges of D.C. and D.C. *voltage,* and in combination with a *bridge rectifier and suitable resistors* to measure various ranges of A.C. and A.C. *voltage.*
3. These instruments can be employed at frequencies well above the range of other A.C. instruments of the ordinary type (from about 20 Hz to 20 kHz).
4. They incorporate the high torque, efficient damping of the moving-coil instrument with *linear scale shape, except of low ranges.*
5. They have a very low power consumption.
6. Their accuracy is about ± 5% under normal operating conditions.
7. These instruments have a much lower operating current for voltmeters (or higher value of ohm per volt) than other A.C. instruments, with the exception of electrostatic and valve voltmeters.

Example 3.60. *An alternating current is measured by a rectifier and a hot-wire ammeters and the readings are found to be 31 A and 33 A respectively.*

Determine the form factor of the current wave.

Solution. R.m.s. value of current under measurement,

$$I_{\text{r.m.s.}} = \text{Hot-wire ammeter reading} = 33 \text{ A}$$

Average value of current under measurement,

$$I_{\text{av.}} = \text{Rectifier ammeter reading} = 31 \text{ A}$$

$$\therefore \quad \text{From factor, } K_f = \frac{I_{r.m.s.}}{I_{av.}} = \frac{33}{31} = \mathbf{1.064} \quad \textbf{(Ans.)}$$

Example 3.61. *A PMMC ammeter gives reading of 35 mA when connected across two opposite corners of bridge rectifier, the other two corners of which are connected in series with a capacitor to 100 kV, 50 Hz supply.*

Determine the capacitance.

Solution. Since PMMC ammeter gives average reading, therefore, average value of current (I_{av}) = 35 mA or 0.035 A.

Assuming sinusoidal waveform *i.e.* form factor 1.11,

$$\text{R.m.s. value of current } (I_{\text{r.m.s.}}) = 1.11 \times 0.035 = 0.0388 \text{ A}$$

Neglecting resistance of the instrument and assuming the capacitance connected is of C farads,

$$I_{\text{r.m.s.}} = \frac{V}{X_C} = \frac{V}{(1/2\pi fC)} = 2\pi fCV$$

$$\text{or,} \quad C = \frac{I_{r.m.s.}}{2\pi fV} = \frac{0.0388}{2\pi \times 50 \times 100 \times 10^3} = 1235 \times 10^{-12} \text{ or } \mathbf{1235 \text{ pF}} \quad \textbf{(Ans.)}$$

Example 3.62. *A PMMC ammeter is connected across appropriate points of a bridge having resistance of each arm of 40 Ω and connected across a supply of voltage $v = 10 \sin \theta + 0.4 \sin 3\theta$.*

Determine the reading of the ammeter if its resistance is 20 Ω.

Solution. *Given* : Instantaneous applied voltage, $v = 10 \sin \theta + 0.4 \sin 3\theta$; Resistance of each arm = 40 Ω; Ammeter resistance = 20 Ω.

Ammeter reading, $I_{av.}$:

Total resistance to the flow of current, $R = 2 \times 40 + 20 = 100\ \Omega$

Instantaneous current through the ammeter,

$$i = \frac{V}{R} = \frac{10 \sin \theta + 0.4 \sin 3\theta}{100} = 0.1 \sin \theta + 0.004 \sin 3\theta$$

Since PMMC reads average value, therefore, reading of PMMC ammeter,

$$I_{av.} = \frac{1}{\pi}\int_0^{\pi} (0.1 \sin \theta + 0.004 \sin 3\theta)\, d\theta$$

$$= \frac{1}{\pi}\left| -0.1 \cos \theta - \frac{0.004}{3} \cos 3\theta \right|_0^{\pi}$$

$$= \frac{1}{\pi}\left(0.2 + \frac{0.008}{3}\right) = 0.0645 \text{ A or } \mathbf{64.5\ mA} \quad \textbf{(Ans.)}$$

Example 3.63. *A moving coil ammeter, a hot wire ammeter and a resistance of 100 Ω are connected in series with a rectifying device across a sinusoidal alternating supply of 200 V. If the rectifying device has a resistance of 100 Ω to current in one direction and of 500 Ω in the opposite direction, calculate the readings on the two ammeters.* **(UPTU)**

Solution. Reading of moving coil ammeter :

R.m.s. current in one direction,

$$I_{r.m.s.1} = \frac{200}{100 + 100} = 1\text{A}$$

Average current in one direction, $I_{av.1} = \frac{I_{r.m.s.1}}{K_f} = \frac{1}{1.11} = 0.9$ A

Similarly r.m.s. value of current in other direction,

$$I_{r.m.s.2} = \frac{200}{500 + 100} = 0.333 \text{ A}$$

Average current in other direction,

$$I_{av.2} = \frac{I_{r.m.s.2}}{K_f} = \frac{0.333}{1.11} = 0.3 \text{ A}$$

Average current over the whole cycle,

$$I_{av.1} - I_{av.2} = \frac{0.9 - 0.3}{2} = 0.3 \text{ A}$$

Hence, the *reading of moving coil ammeter* = **0.3 A (Ans.)**

Reading of hot-wire ammeter :

Average heating effect during *positive* half cycle

$$= (I_{r.m.s.})^2 \times R = (1)^2 \times R = R,$$

where R is the resistance of hot-wire ammeter

Similarly, average heating effect during *negative* half cycle

$$= (0.333)^2 \times R = 0.11\ R$$

Average heating effect over the complete cycle

$$= \frac{1}{2}(R + 0.11R) = 0.555\ R$$

If I is the direct current which produces the same heating effect, then

$$I^2R = 0.555\ R$$

or, $$I = \sqrt{0.555} = 0.745\ \text{A}$$

Hence, *reading of hot-wire ammeter* = **0.745 A (Ans.)**

Example 3.64. *A sinusoidal alternating voltage of 250 V peak is applied to a circuit containing a rectifying device which entirely prevents current flowing in one direction, and offers a non-reactive resistance of 200 Ω to the flow of current in the other direction. Determine the readings on :*

(i) Hot-wire ammeter; *(ii) Moving coil in the circuit*

(Punjab University)

Solution. *Given* : V_{max} = 250 V; Resistance of rectifying device, R = 200 Ω

Maximum value of current flowing through the instruments,

$$I_{max} = \frac{V_{max}}{R} = \frac{250}{200} = 1.25\ \text{A}$$

Reading of hot-wire ammeter, $I_{r.m.s.}$:

The hot-wire ammeter *reads r.m.s. value*, thus,

$$I_{r.m.s.} = \sqrt{\frac{1}{2\pi}\int_0^{\pi} I_{max}^2 \sin^2\theta\, d\theta}$$

$$= \sqrt{\frac{I_{max}^2}{2\pi}\int_0^{\pi}\left(\frac{1-\cos 2\theta}{2}\right) d\theta}$$

$$= \sqrt{\frac{I_{max}^2}{2\pi}\times\frac{\pi}{2}} = \frac{I_{max}}{2} = \frac{1.25}{2} = \mathbf{0.625\ A}\ \textbf{(Ans.)}$$

Reading of moving coil ammeter, $I_{av.}$:

The moving coil ammeter *reads average value*, thus,

$$I_{av.} = \frac{1}{2\pi}\int_0^{\pi} I_{max} \sin\theta\, d\theta$$

$$= \frac{I_{max}}{2\pi}\left|-\cos\theta\right|_0^{\pi} = \frac{1.25}{\pi} = \mathbf{0.398\ A}\ \textbf{(Ans.)}$$

Example 3.65. *A permanent magnet moving coil type full-wave bridge rectifier ammeter is used to measure current in a load connected across a supply of voltage :*

$$v = 5 \sin\theta + 0.2 \sin 3\theta$$

Determine the reading of the ammeter, if its resistance is 30 Ω. Assume dynamic resistance of each diode as 35 Ω under forward biased condition. **(UPTU)**

Solution. Refer to Fig. 3.56

Given : Supply voltage, $v = 5 \sin\theta + 0.2 \sin 3\theta$; R_m = 30 Ω; Resistance of each diode (forward biased) = 35 Ω.

Reading of the ammeter :

Total resistance to the flow of current (through D_2 – PMMC – D_4 or through D_3 – PMMC – D_1),

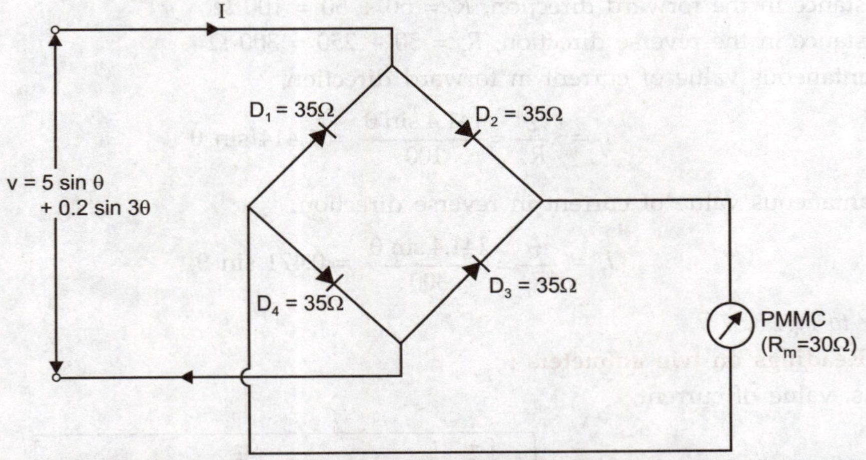

Fig. 3.56. Reading of the ammeter.

$$R = 35 + 30 + 35 = 100\ \Omega$$

Now, $$i = \frac{v}{R} = \frac{5 \sin \theta + 0.2 \sin 3\theta}{100} = 0.05 \sin \theta + 0.002 \sin 3\theta$$

The PMMC reads *average value,* so,

$$I_{av.} = \frac{1}{\pi}\int_0^{\pi} (0.05 \sin \theta + 0.002 \sin 3\theta)\, d\theta$$

$$= \frac{1}{\pi}\left[-0.05 \cos \theta - \frac{0.002}{3} \cos 3\theta\right]_0^{\pi}$$

$$= \frac{1}{\pi}[-0.05\, (\cos \pi - \cos 0) - 0.000667\, (\cos 3\pi - \cos 0)]$$

$$= \frac{1}{\pi}[-0.05\, (-1-1) - 0.000667\, (-1-1)]$$

$$= \frac{1}{\pi}(0.1 + 1.334 \times 10^{-3}) = 0.03225 \text{ A} \quad \text{or} \quad 32.25 \text{ mA}$$

Hence, *reading of the ammeter* = **32.25 mA** **(Ans.)**

Example 3.66. *A resistance of 50 Ω is connected in series with a rectifying device, a moving coil ammeter and a dynamometer type ammeter. The circuit is connected across a sinusoidal A.C. supply of 100 V. The resistance of the rectifying device is 50 Ω in one direction and 250 Ω in the opposite direction.*

Calculate :

(i) Reading on two ammeters.

(ii) Power supplied from mains.

(iii) Power dissipated in rectifying elements. **(Anna University)**

Solution. R.m.s. value of supply voltage, $V = 100$ V

Maximum value of the voltage, $V_{max} = \sqrt{2} \times 100 = 141.4$ V

Instantaneous value of the voltage, $v = 141.4 \sin \theta$ V

Resistance in the forward direction, $R_f = 50 + 50 = 100\ \Omega$
Resistance in the reverse direction, $R_r = 50 + 250 = 300\ \Omega$
Instantaneous value of current in forward direction,

$$i_f = \frac{v}{R_f} = \frac{141.4 \sin\theta}{100} = 1.414 \sin\theta$$

Instantaneous value of current in reverse direction,

$$i_r = \frac{v}{R_r} = \frac{141.4 \sin\theta}{300} = 0.471 \sin\theta$$

Refer to Fig. 3.57 :

(*i*) Readings on two ammeters :

R.m.s. value of current,

$$I_{r.m.s.} = \sqrt{\frac{1}{2\pi}\left[\int_0^{\pi}(1.414 \sin\theta)^2\, d\theta + \int_{\pi}^{2\pi}(0.471 \sin\theta)^2\, d\theta\right]}$$

$$= \sqrt{\frac{1}{2\pi}\left[\int_0^{\pi} 2 \sin^2\theta\, d\theta + \int_{\pi}^{2\pi} 0.222 \sin^2\theta\, d\theta\right]}$$

$$= \sqrt{\frac{1}{2\pi}\left[2\int_0^{\pi}\left(\frac{1-\cos 2\theta}{2}\right) d\theta + 0.222\int_{\pi}^{2\pi}\left(\frac{1-\cos 2\theta}{2}\right)\right]}$$

$$= \sqrt{\frac{1}{2\pi}\left\{\left[\theta - \frac{\sin 2\theta}{2}\right]_0^{\pi} + 0.111\left[\theta - \frac{\sin 2\theta}{2}\right]_{\pi}^{2\pi}\right\}}$$

or, $$I_{r.m.s.} = \sqrt{\frac{1}{2\pi}\{\pi + 0.111(2\pi - \pi)\}} = 0.745\ \text{A}$$

i.e., Reading of dynamometer ammeter = **0.745 A** **(Ans.)**

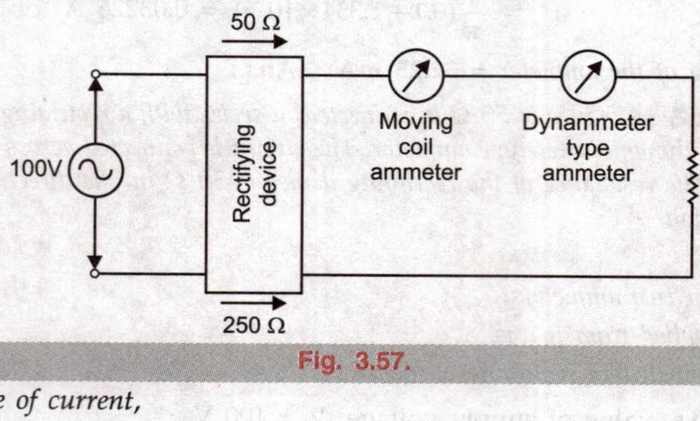

Fig. 3.57.

Average value of current,

$$I_{av.} = \frac{1}{2\pi}\left[\int_0^{\pi} 1.414 \sin\theta\, d\theta + \int_{\pi}^{2\pi} 0.417 \sin\theta\, d\theta\right]$$

$$= \frac{1}{2\pi}\left\{[-1.414 \cos\theta]_0^{\pi} + [-0.417 \cos\theta]_{\pi}^{2\pi}\right\}$$

$$= \frac{1}{2\pi}\left[\{-1.414 (\cos\pi - \cos 0)\} + \{-0.417 (\cos 2\pi - \cos\pi)\}\right]$$

$$= \frac{1}{2\pi}\left[\{-1.414 (-1-1)\} + \{-0.417 (1+1)\}\right]$$

$$= \frac{1}{2\pi}(2.828 - 0.834) = 0.317 \text{ A}$$

i.e., Reading of moving coil ammeter = **0.317 A (Ans.)**

(*ii*) **Power supplied from mains :**

Power supplied from mains = Power supplied in the forward half cycle + power supplied in the backward half cycle

$$= \frac{1}{2}\left(\frac{V^2}{R_f} + \frac{V^2}{R_r}\right) = \frac{1}{2}\left[\frac{(100)^2}{100} + \frac{(100)^2}{300}\right]$$

= **66.67 W (Ans.)**

(*iii*) **Power dissipated in rectifying element :**

Power consumed in 50 Ω resistor = I_2 R = $(0.745)^2 \times 50$ = 27.75 Ω

Power dissipated in rectifier = Total power supplied – power consumed in resistor

= 66.67 – 27.75 ≃ **39 W (Ans.)**

3.8.12. Electrostatic Instruments

3.8.12.1. Working principle

The underlying principle of operation of electrostatic instruments is the force of attraction between electric charges on neighbouring plates between which potential difference is maintained. This force gives rise to a deflecting torque. Unless the p.d. is adequately large, the force is small. Hence such instruments are *used for measurement of very high voltages.*

- The instruments are *almost always used as voltmeters and that too more as a laboratory rather than as industrial instruments.*

Such instruments *may be used with the help of external components, to measure current and power.*

3.8.12.2. Force and torque equations

Linear motion : Refer to Fig. 3.58. Let a voltage V be applied between two parallel plates, one fixed (A) and the other movable (B). The capacitance between the plates is C and the energy stored is $\frac{1}{2}CV^2$. If the voltage is increased by dV, the force of attraction (F) will increase and the movable plate will move by a distance dx.

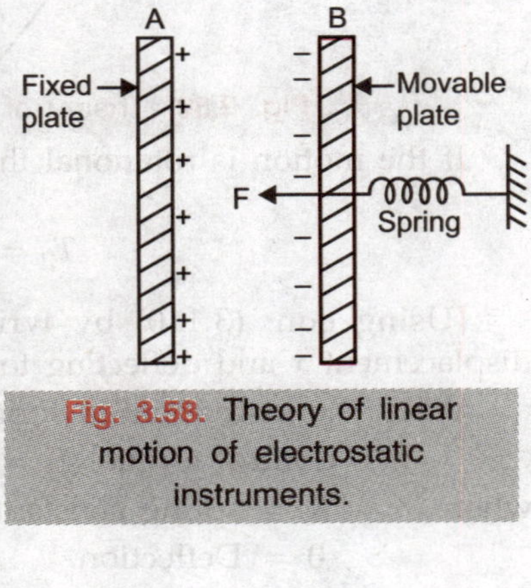

Fig. 3.58. Theory of linear motion of electrostatic instruments.

The capacitance current,

$$i = \frac{dQ}{dt} = \frac{d}{dt}(CV)$$

$$= C\frac{dV}{dt} + V\frac{dC}{dt} \quad \text{...(3.108)}$$

The energy input is,

$$Vi\ dt = V^2.dC + CV.dV \quad \text{...(3.109)}$$

The change in energy stored in electric field

$$= \frac{1}{2}(C + dC)(V + dV)^2 - \frac{1}{2}CV^2$$

$$= \frac{1}{2}(C + dC)(V^2 + 2VdV + dV^2) - \frac{1}{2}CV^2$$

$$= \frac{1}{2}\left(CV^2 + CVdV + CdV^2 + V^2dC + 2VdVdC + dC.V^2\right) - \frac{1}{2}CV^2$$

$$= \frac{1}{2}V^2dC + CV.dV$$

neglecting the higher order terms as they are small quantities.

Invoking law of conservation of energy, we have :

Energy input = Increase in energy stored + mechanical work done

or, $$V^2.dC + CV.dV = \frac{1}{2}V^2.dC + CV.dV + F.dx$$

or, $$\text{Force, } F = \frac{1}{2}V^2\frac{dC}{dx} \text{ newton} \quad \text{...(3.110)}$$

Rotational motion : Refer to Fig. 3.59.

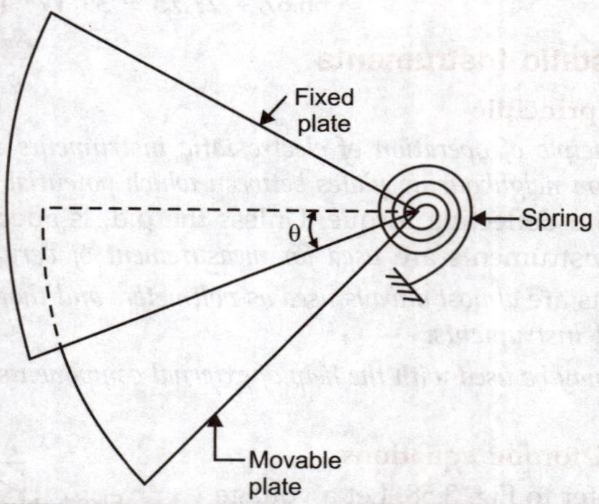

Fig. 3.59. Theory of rotational motion of electrostatic instruments.

If the motion is rotational the deflecting torque is given as :

$$T_d = \frac{1}{2}V^2\frac{dC}{d\theta} \quad \text{...(3.111)}$$

[Using eqn. (3.110) by writing an angular displacement θ in place of linear displacement x and deflecting torque T_d in place of force F].

When the instrument is spring controlled or has a suspension,

Control torque, $T_c = K\theta$

where, K = spring constant, and

θ = Deflection.

At steady deflection state, $T_d = T_c$

$$\therefore \quad \frac{1}{2}V^2\frac{dC}{d\theta} = K\theta$$

or, deflection, $$\theta = \frac{1}{2}\frac{V^2}{K}\frac{dC}{d\theta} \quad ...(3.112)$$

- Because the deflection is *proportional to the square of voltage* to be measured, the instrument *can be used on both A.C. and D.C.*
- Since the instrument exhibits *a square law response,* therefore, the *scale is non-uniform and is compressed at lower end.*

3.8.12.3. Classification of electrostic instruments

There are two general types of such instruments :

1. The *quadrant type* ...... Used upto 20 kV
2. The *attracted disc type* .... Used upto 500 kV.

3.8.12.4. Quadrant electrometer

Construction. Fig. 3.60(*a*) shows a *quadrant electrometer :*

— It consists of *four fixed metal double quadrants* arranged to form a *circular box.* The opposite pairs are connected together. There are *short air gaps between the quadrants.* Inside the box, a *double sector shaped needle* is suspended by a phosphor bronze or silver quartz thread. The needle is *equidistant* from top and bottom quadrants.

— The needle carries a mirror and the deflection read off by means of a *lamp and scale arrangement.*

An alternate arrangement may be mounting the needle on a spindle and the controlling troque is provided by a spring attached to the spindle. The deflecting torque moves a pointer and the deflection is read on a calibrated scale. For these instruments, *eddy current damping is used.*

— Usually, a high resistance is connected in series with an electrostatic voltmeter, particularly if the instrument is of the suspended-vane type. This resistance is used for limiting the current which can flow if there is accidental connection between a vane and a quadrant.

Principle of operation. The fixed quadrants are connected together and the voltage to be measured is applied between the fixed quadrants and the moving needle. *Due to charge accmulated, an electrostatic force is set up because of which the needle rotates. The suspension exerts a controlling torque and the needle takes up a position where the deflecting torque is equal to the controlling torque.*

Types of connections. The quadrant electrometer may be used in *two ways* :

(*i*) ***Heterostatic connection.*** This connection is shown in Fig. 3.60(*b*). A high tension (HT) battery is used for charging the needle to a potential considerably above that of the quadrants to which the negative of the voltage to be measured is connected.

(*ii*) ***Idiostatic connection.*** This connection is shown in Fig. 3.60(*c*). The needle is directly connected to one pair of quadrants.

Generally, *idiostatic connection is used.*

- It can be shown that with the four quadrants and a double sector shaped needle, as shown in Fig. 3.60, the deflecting torque (T_d) in Nm is given as :

$$T_d = \frac{\in r^2}{d}.V^2 \quad ...(3.113)$$

where, $\in = \in_0$ (since air is the dielectric) = Absolute permittivity = 8.854×10^{-12} F/m,

r = Radius of each quadrant m,

d = Distance of needle from the top or bottom plate, m, and

V = Voltage to be measured, volts.

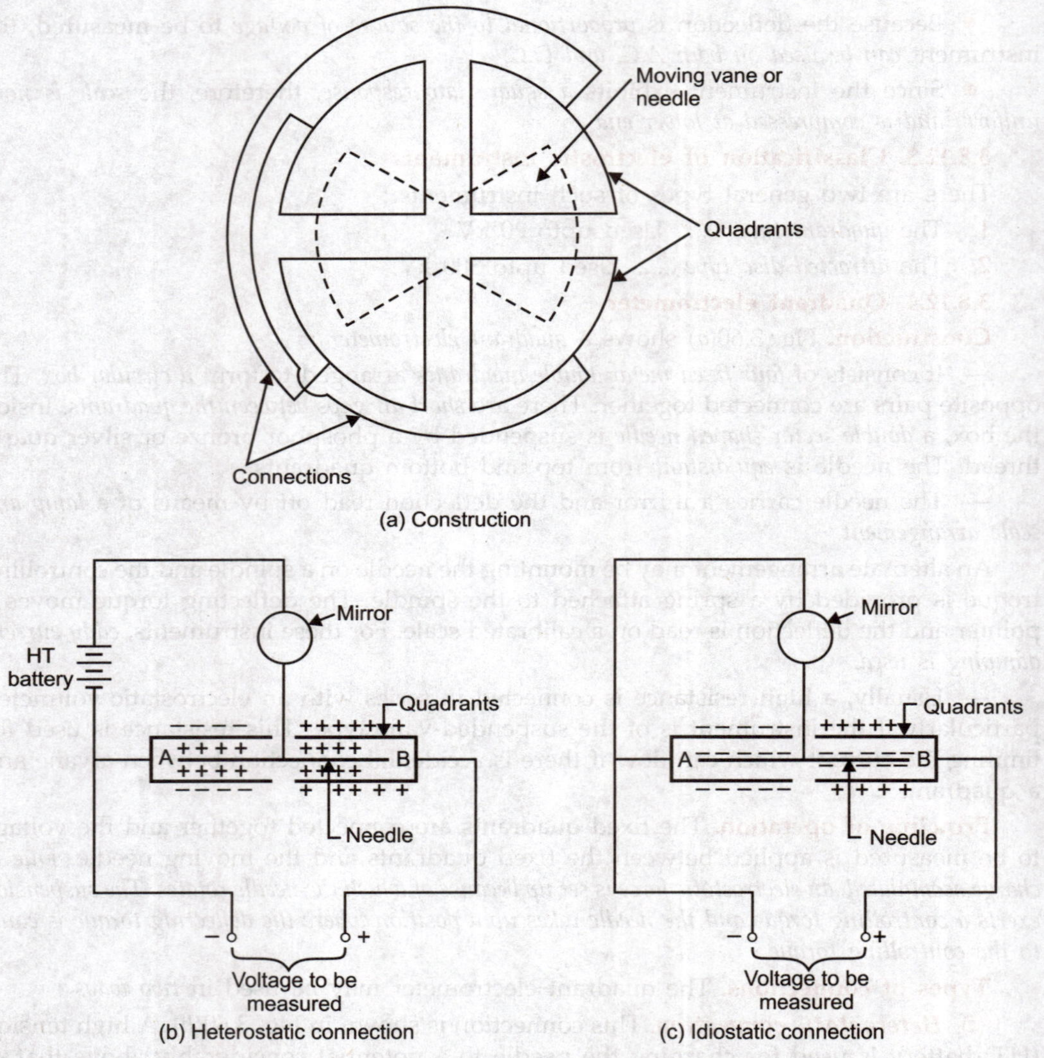

Fig. 3.60. Quadrant electrometer.

The controlling torque is provided by the spring. Thus,

$$T_c = K\theta$$

At steady deflection state,

$$T_d = T_c$$

i.e.,

$$\frac{\in_0 r^2}{d} . V^2 = K\theta$$

$$\text{or,} \qquad \theta = \frac{\epsilon_0 r^2}{Kd} . V^2 \qquad ...(3.114)$$

Hence, with *idiostatic connection* the instrument gives a *square law response.*

Kelvin multicellular voltmeter :

The construction shown in Fig. 3.60 has a rather small deflecting torque. To increase sensitivity, a multicellular electrostatic voltmeter devised by Lord Kelvin is used.

Fig. 3.61 shows the components of a *Kelvin multicellular voltmeter* :

— It has a large number of *fixed quadrants and needles*. The large number of quadrants is needed to *give a high deflecting torque even at low voltages.*

— The moving system is suspended by a *phosphor bronze suspension to reduce bearing friction.*

— The *coach spring* provides *protection* against breaking of suspension due to vibration.

— The *torsion head* is provided for *zero adjustment*.

— Damping is done by using an *oil dashpot*.

- This voltmeter is suitable for *even low voltages in the range of 100 to 1000 V*.

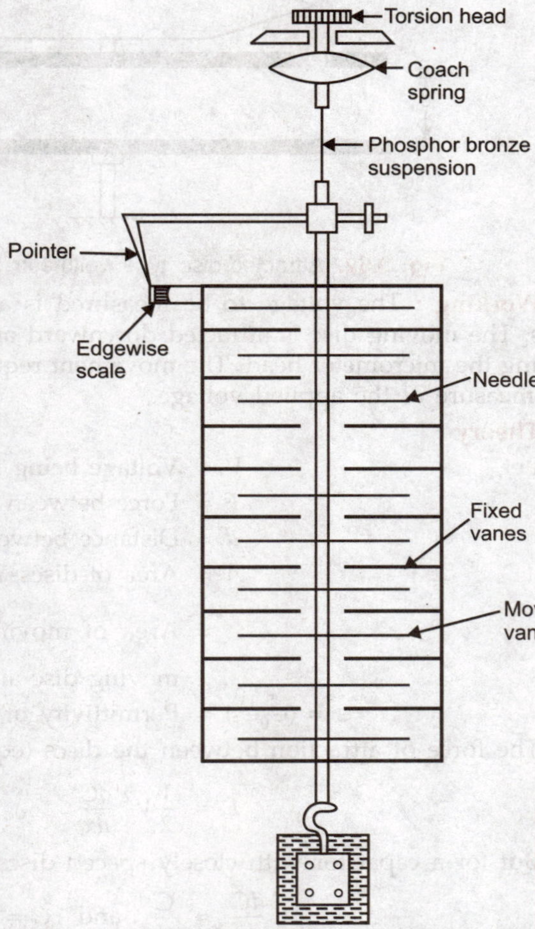

Fig. 3.61. Kelvin multicellular voltmeter.

3.8.12.5. Attracted-disc type voltmeter

Construction: This instrument is *suitable for only high voltages* (more than 10 kV).

Fig. 3.62 shows the main constructional features of an attracted-disc type voltmeter (also known as Kelvin absolute electrometer) :

— It consists of two discs, the fixed disc and the moving disc.

— The moving disc is supported by coach spring which is suspended from a micrometer head as shown.

— The moving disc is provided with a *guard ring to reduce the fringing effects.* The effective area of the moving disc is its actual area plus half the area of air gap.

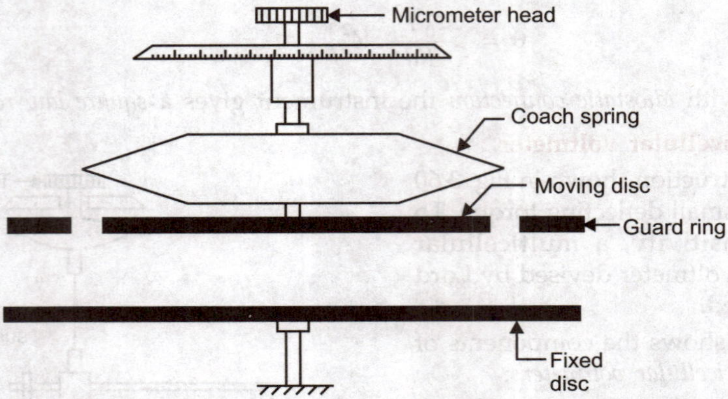

Fig. 3.62. Attracted-disc type voltmeter (Kelvin absolute electrometer)

Working : The voltage to be measured is applied between the fixed and moving discs. The moving disc is attracted downward and is brought back to zero position by turning the micrometer head. The movement required to return the disc to zero position is a measure of the applied voltage.

Theory :

Let, V = Voltage being measured, volts,

F = Force between discs, N,

d = Distance between discs, m,

A = Area of discs, m^2,

= Area of moving disc + $\frac{1}{2}$ area of air gap between moving disc and guard ring, and

$\in = (\in_0 \in_r)$ = Permittivity of dielectric, F/m.

The force of attraction between the discs (eqn. 3.110),

$$F = \frac{1}{2} V^2 \frac{dC}{dx}$$

But for a capacitor with closely spaced discs/plates,

$$\frac{dC}{dx} = \frac{C}{d} \quad \text{and} \quad C = \frac{\in A}{d}$$

$$\therefore \quad F = \frac{1}{2} \in A \cdot \frac{V^2}{d^2} \quad ...(3.115)$$

Hence,

$$V = \sqrt{\frac{2Fd^2}{\in A}} \quad ...(3.116)$$

The above theory shows that such an instrument will give an *absolute determination of potential difference*, as p.d. is given in terms of force and linear dimensions. The deflecting force is only adequate if the voltage to be measured is high..

— In order to ensure good insulation, screening, and freedom from corona effects, special construction may be necessary.

— A modern development is to enclose the working parts of the instruments within a *highly evacuated chamber*.

The superior dielectric strength of a high vacuum, in comparison with that of air at normal pressures, enables the clearance between the plates to be *reduced*, thus *giving an increased force for a given voltage. The use of a compressed-gas dielectric is an alternative means of obtaining higher dielectric strength and increased sensitivity.*

3.8.12.6. Advantages and disadvantages

Following are the *advantages and disadvantages of electrostatic instruments*.

Advantages :

1. Suitable for both A.C. and D.C.
2. Measure r.m.s. values.
3. Draw negligible power from the mains.
   — For operation on D.C., the initial current is the charging current which decreases exponentially and the final current is zero.
   — For operation on A.C., the current is inversely proportional to the impedance of the measuring circuit.
4. Free from hysteresis and eddy current errors (because there is no iron in the working system).
5. Free from frequency and waveform errors.
6. Especially suitable for high voltages.
7. Owing to negligible power loss there is no heating error.
8. Can be manufactured with first grade accuracy.
9. Cause little disturbance to circuits to which they are connected.

Disadvantages :

1. Low deflecting torque especially when the voltage required to be measured is less than 500 V.
   However, Kelvin multicellular voltmeter can be designed for use for measurements of voltages as low as 30 V.
2. Fragile and expensive and cannot be made robust.
3. Non-uniform scale.
4. Liable to friction errors.

3.8.12.7. Range extension of electrostatic voltmeters

The range of electrostatic voltmeters can be extended by the use of *multipliers* which are in form of :

1. Resistance potential divider ------ *Can be used both for direct and alternating voltages.*
2. Capacitance potential divider ------ Useful *only for alternating voltages.*

1. Resistance potential divider : This divider consists of a high non-inductive resistance across a small portion which is attached to the electrostatic voltmeter as shown in Fig. 3.63.

Let, R = The resistance of the whole of the potential divider across which the voltage V under measurement is applied, and

r = The resistance of the portion of the divider across which voltmeter is connected.

Then, the *multiplying factor* is given by :

$$\frac{V}{v} = \frac{R}{r} \qquad ...(3.117)$$

The above eqn. (3.117) is true for D.C. circuits but for A.C. circuits, the capacitance of the voltmeter (which is in parallel with r) has to be taken into account. Since this capacitance is variable, it is advisable to calibrate the voltmeter alongwith its multiplier.

2. **Capacitance potential divider :** In this method, the voltmeter may be connected in series with a single capacitor C and put across the voltage V which is to be measured [Fig. 3.64(*a*)] or a number of capacitor may be joined in series to form the potential divider and the voltmeter may be connected across one of the capacitors as shown in Fig. 3.64(*b*).

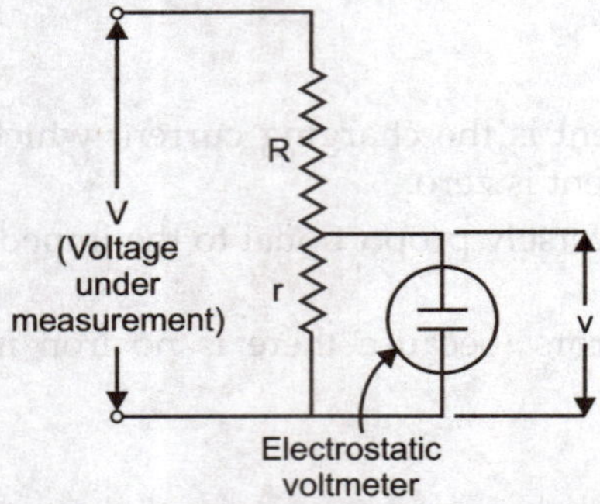

Fig. 3.63. Resistance potential divider.

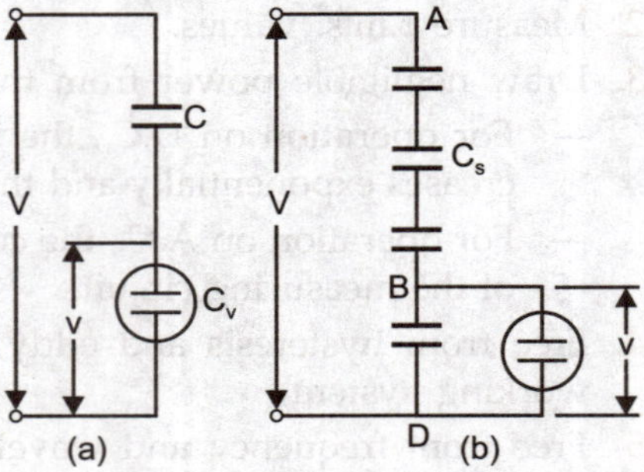

Fig. 3.64. Capacitance potential divider.

Referring to Fig. 3.64(*a*), the multiplying factor is given by :

$$\frac{V}{v} = \frac{\text{Reactance of total circuit}}{\text{Reactance of voltmeter}}$$

Now, capacitance of the total circuit $= \dfrac{CC_v}{C + C_v}$

and, its reactance $= \dfrac{1}{\omega \times \text{capacitance}} = \dfrac{C + C_v}{\omega CC_v}$

Reactance of the voltmeter $= \dfrac{1}{\omega C_v}$

$$\therefore \quad \frac{V}{v} = \frac{(C + C_v)/\omega CC_v}{1/\omega C_v} = \frac{C + C_v}{C}$$

$$\therefore \quad \text{Multiplying factor, } \frac{V}{v} = \frac{C + C_v}{C} = 1 + \frac{C_v}{C} \qquad ...(3.118)$$

Example 3.67. *An absolute electrometer uses a movable circular plate 80 mm in diameter. During a voltage measurement the distance between the plates is 3.2 mm and the force of attraction is 0.003 N. The medium is air. ϵ_0 = 8.85 × 10^{-12} F/m. Determine V.*

Solution. *Given* : $A = \frac{\pi}{4} \times (0.08)^2 = 5.0265 \times 10^{-3}$ m^2; d = 3.2 mm or 0.0032 m; F = 0.03 N; $\epsilon = \epsilon_0 = 8.85 \times 10^{-12}$ F/N

We know that, $$V = \sqrt{\frac{2Fd^2}{\epsilon A}} \qquad \text{[Eqn. (3.116)]}$$

or, $$V = \sqrt{\frac{2 \times 0.03 \times (0.0032)^2}{8.85 \times 10^{-12} \times 5.0265 \times 10^{-3}}} = \mathbf{1175\ V} \quad \text{(Ans.)}$$

Example 3.68. *In an electrometer, the movable plate is 110 mm in diameter. When 12 kV is applied between the movable plate and the fixed plate, the force is 0.006 N. Find the change in capacitance for 1.2 mm movement of the movable plate.*

Solution. *Given :* Diameter of movable plate = 110 mm or 0.11 m; V = 12 kV; F = 0.006 N; x = 1.2 mm.

We know that,
$$F = \frac{1}{2} V^2 \frac{dC}{dx} \quad \text{[Eqn. (3.110)]}$$

$$0.006 = \frac{1}{2} \times (12 \times 10^3)^2 \times \frac{dC}{dx}$$

or,
$$\frac{dC}{dx} = \frac{2 \times 0.006}{(12 \times 10^3)^2} = 83.33 \times 10^{-12} \text{ F/m}$$

∴ Change in capacitance, $dC = 83.33 \times 10^{-12} \times (1.2 \times 10^{-3})$

$\simeq 0.1 \times 10^{-12}$ F or **0.1 pF (Ans.)**

Example 3.69. *An electrostatic voltmeter has full scale deflection of 90°. The control constant of the spring is 8×10^{-6} Nm/rad. When reading is zero, the capacitance is 12 pF. What is the capacitance when the instrument reads 1500 V and the deflection is 90°.*

Solution. *Given :* $\theta = 90° = \frac{\pi}{2}$ rad.; $K = 8 \times 10^{-6}$ Nm/rad.; C = 12 pF, when reading is zero; V = 1500 volts.

At steady deflection state,
$$T_d = T_c$$

or,
$$\frac{1}{2} V^2 \frac{dC}{d\theta} = K\theta \quad \text{[Eqn. (3.111)]}$$

or,
$$\frac{dC}{d\theta} = \frac{2K\theta}{V^2} = \frac{2 \times 8 \times 10^{-6} \times (\pi/2)}{(1500)^2} = 0.1117 \times 10^{-10} \text{ F/rad.}$$

Change in capacitance, $dC = 0.1117 \times 10^{-10} \times \frac{\pi}{2} = 0.1755 \times 10^{-10}$ F

∴ New value of capacitance $= 12 \times 10^{-12} + 0.1755 \times 10^{-10} = 29.55 \times 10^{-12}$ F

$= 29.55 \times 10^{-12}$ F or **29.55 pF (Ans.)**

Example 3.70. *An electrostatic voltmeter is constructed with 6 parallel, semicircular fixed plates equally-spaced at 4 mm intervals and 5 interleaved semi-circular movable plates that move in planes midway between the fixed plates, in air. The movement of the movable plates is about an axis through the centre of the circles of the plates system, perpendicular to the planes of the plates. The instrument is spring-controlled. If the radius of the movable plates is 4 cm, calculate the spring constant if 10 kV corresponds to a full-scale deflection of 100°. Neglect fringing, edge-effects and plate thickness.* **(Bombay University)**

Solution. *Given :* No. of fixed plates = 6; No. of movable plates = 5; d (Distance between fixed and movable plates) $= \frac{4}{2}$ = 2 mm or 0.002 m ; r = 4 cm or 0.04 m ; V = 10 kV ; $\theta = 100°$ or $100 \times \frac{\pi}{180}$ = 1.745 rad.

Spring constant, *K* :

Total number of plates (both fixed and movable) is 11, hence there are *10 parallel plate capacitors* (See Fig. 3.65)

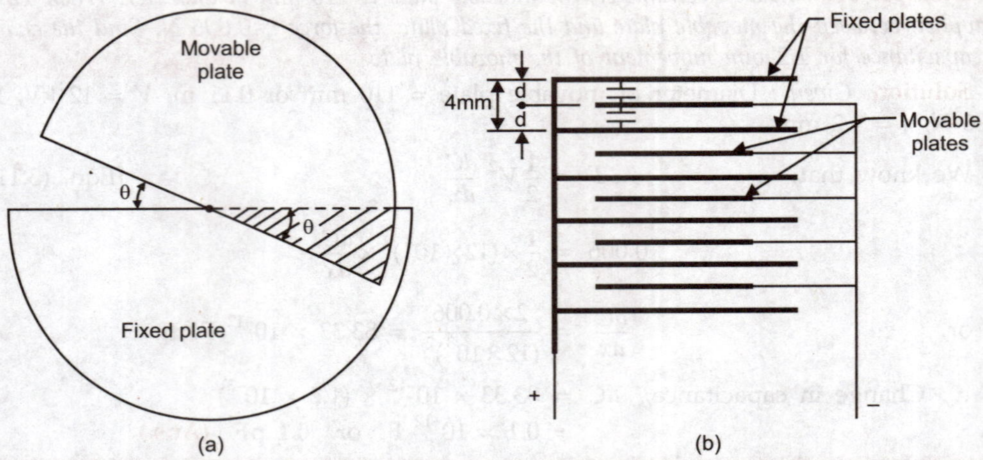

Fig. 3.65.

Suppose, the movable plates are rotated into the fixed plaes by an angle of θ radian. Then, overlap area between one fixed and one movable semi-circular plate is,

$$A = \frac{1}{2}r^2\theta = \frac{1}{2}\times 0.04^2 \times \theta = 8 \times 10^{-4}\,\theta$$

Capacitance of each of ten parallel-plate capacitors,

$$C_1 = \frac{\epsilon_0 A}{d} = \frac{8.854\times 10^{-12}\times 8\times 10^{-4}\,\theta}{2\times 10^{-3}} = 3.54 \times 10^{-12}\,\theta \text{ F}$$

Total capacitance, $C = 10 \times C_1 = 10 \times 3.54 \times 10^{-12}\,\theta = 35.4 \times 10^{-12}\,\theta$ F

$$\therefore \quad \frac{dC}{d\theta} = 35.4 \times 10^{-12} \text{ F/rad.}$$

Deflecting torque, $T_d = \frac{1}{2}V^2 . \frac{dC}{d\theta}$ Nm [Eqn. (3.111)]

$$= \frac{1}{2}\times(10\times 10^3)^2 \times 35.4\times 10^{-12} = 17.7 \times 10^{-4} \text{ Nm}$$

If K is the spring constant, then control torque is,

$$T_c = K\theta$$

At steady deflection state,

$$T_d = T_c$$

$$17.7 \times 10^{-4} = K \times 1.745$$

$$\therefore \quad K = \mathbf{10.14 \times 10^{-4}\ Nm/rad.}$$ **(Ans.)**

Example 3.71. *The reading '100' of a 120 V electrostatic voltmeter is to represent 10,000 volts when its range is extended by the use of a capacitor in series. If the capacitance of the voltmeter at the above reading is 70 pF, find the capacitance of the capacitor multiplier required.*

Solution. *Given :* V = 10,000 volts; v = 100 volts, C_v = 70 pF.

We know that, multiplying factor $= \frac{V}{v} = 1 + \frac{C_v}{C}$ [Eqn. 3.118)]

Inserting the various values in the above eqn., we get

$$\frac{10,000}{100} = 1+\frac{70}{C}$$

or, $$\frac{70}{C} = \frac{10,000}{100}-1=99$$

or, $$C = \frac{70}{99} \text{ pF} = \mathbf{0.707 \text{ pF}} \quad \textbf{(Ans.)}$$

Example 3.72. *A current of 0.5 + 0.3 sin ωt – 0.2 sin 2ωt is passed through the circuit shown in Fig. 3.66. Determine the reading of each instrument if ω = 10^6 rad./s.*

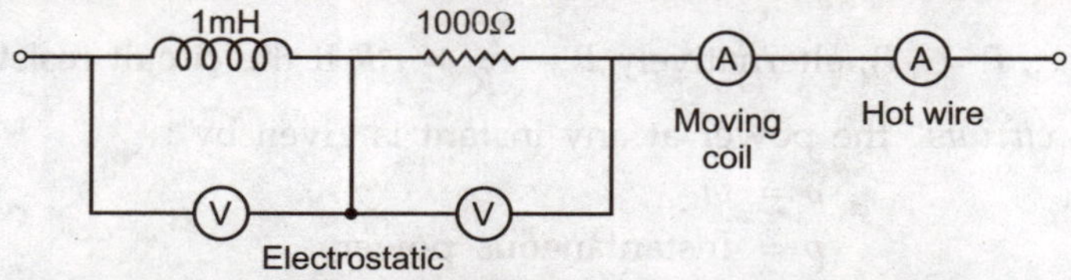

Fig. 3.66.

Solution. *Given :* $i = 0.5 + 0.3 \sin \omega t - 0.2 \sin 2\omega t$; $\omega = 10^6$ rad./s.

Reading of each instrument :

Let, the instantaneous value of current,

$$i = I_0 + I_{1max} \sin \omega t + I_{2max} \sin 2\omega t$$

or, $$i = 0.5 + 0.3 \sin \omega t - 0.2 \sin 2\omega t \quad \textit{(Given)}$$

Average value of a sinusoidal quantity over a number of cycles or over a time very much greater than the time period of a cycle is equal to zero.

$$\therefore \quad I_{av.} = i = 0.5 \text{ A}$$

Hence, the *reading of moving-coil instrument* = **0.5 A** **(Ans.)**

R.m.s. value of current, $$I = \sqrt{I_0^2 + \left(\frac{I_{1max}}{\sqrt{2}}\right)^2 + \left(\frac{I_{2max}}{\sqrt{2}}\right)^2}$$

$$= \sqrt{(0.5)^2 + \left(\frac{0.3}{\sqrt{2}}\right)^2 + \left(\frac{0.2}{\sqrt{2}}\right)^2} = \mathbf{0.561 \text{ A}}$$

Hence, the *reading of hot-wire instrument* = **0.561 A** **(Ans.)**

Reading of "electrostatic voltmeter across 1000 Ω resistance",

$$V_R = 0.561 \times 1000 = \mathbf{561 \text{ V}} \quad \textbf{(Ans.)}$$

Instantaneous value of voltage across 1 mH inductor,

$$v_L = L\frac{di}{dt} = (1\times 10^{-3})\times\frac{d}{dt}(0.5 + 0.3 \sin \omega t - 0.2 \sin 2\omega t)$$

$$= (1 \times 10^{-3}) \times \omega(0.3 \cos \omega t - 0.4 \cos 2\omega t)$$

$$= 300 \cos \omega t - 400 \cos 2\omega t \quad ...(\because \omega = 10^6 \text{ rad./s})$$

Hence, *"reading of electrostatic voltmeter across 1 mH inductor",*

$$V_L = \sqrt{\left(\frac{300}{\sqrt{2}}\right)^2 + \left(\frac{400}{\sqrt{2}}\right)^2} \simeq \mathbf{354\ V} \quad \textbf{(Ans.)}$$

3.9 WATTMETERS AND MEASUREMENT OF POWER

3.9.1. Introduction

3.9.1.1. Power in D.C. and A.C. circuits

Power *may be defined as the rate at which energy is transformed or made available.*

- In *"D.C. circuits"* the power may be measured either by a wattmeter or by an ammeter and voltmeter, the product of whose readings gives the power in the circuit (*i.e.*, $P = VI$); alternatively $P = \frac{V^2}{R} = I^2R$ if the circuit resistance is known.
- In *"A.C. circuits"* the power at any instant is given by :

$$p = vi \qquad \text{...(3.119)}$$

where, p = Instantaneous power,

v = Instantaneous voltage, and

i = Instantaneous current.

Thus, if both the current and voltage waves are sinusoidal, the current lagging in phase by an angle ϕ, and

$$v = V_{max} \sin \omega t,$$

and, $$i = I_{max} \sin (\omega t - \phi),$$

Then instantaneous power p is given by,

$$p = vi$$
$$= V_{max} I_{max} \sin \omega t \cdot \sin (\omega t - \phi)$$

or, writing θ for ωt,

$$p = V_{max} I_{max} \sin \theta \sin (\theta - \phi)$$

The mean power in given by :

$$P = \frac{1}{2\pi}\int_0^{2\pi} V_{max} I_{max} \sin\theta \sin(\theta - \phi)\, d\theta \qquad \text{...(3.120)}$$

$$= \frac{V_{max} I_{max}}{2\pi}\int_0^{2\pi} \frac{\cos\phi - \cos(2\theta - \phi)}{2} d\theta$$

$$= \frac{V_{max} I_{max}}{4\pi}\left[\theta\cos\phi - \frac{\sin(2\theta-\phi)}{2}\right]_0^{2\pi}$$

$$= \frac{V_{max} I_{max}}{2}\cos\phi = \frac{V_{max}}{\sqrt{2}} \cdot \frac{I_{max}}{\sqrt{2}} \cdot \cos\phi$$

or, $$P = VI \cos \phi \qquad \text{...(3.121)}$$

where V and I are r.m.s. values of voltage and current.

The fact that the power factor ($\cos \phi$) is involved in the expression for the power means that a *wattmeter must be used instead of the merely an ammeter and voltmeter, since the latter method takes no account of power factor.*

3.9.1.2. Wattmeter connections/measurements in single-phase A.C. circuits

Fig. 3.67 shows a wattmeter connected in a single-phase A.C. circuit.

— The *"current coil"* of the instrument carries the load current, while the *"pressure or voltage coil"* carries a current proportional to, and in phase with, the voltage. The deflection of the wattmeter depends upon the currents on these coils and upon the power factor.

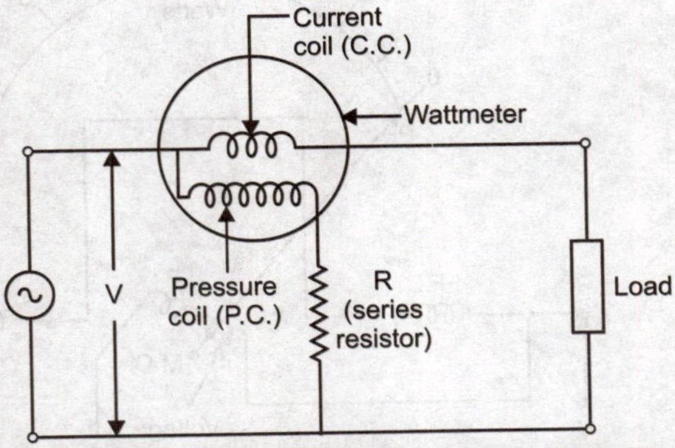

Fig. 3.67. Wattmeter connections.

— Inductance in the voltage-coil circuit should be avoided as far as possible, since it causes the voltage-coil current to lag behind the applied voltage. *A high non-inductance resistance is connected in series with the voltage coil in order that the reactance of the coil itself shall be small in comparison with the resistance of the whole voltage-coil circuit and also, of course, to ensure that the current taken by the voltage coil shall be small.*

WATTMETERS

3.9.2. Types of Wattmeters

A **wattmeter** is *a combination of an ammeter and a voltmeter* and, therefore, consists of two coils known as *current coil* and *pressure coil. The operating torque is produced due to interaction of fluxes on account of currents in current and pressure coils.*

There are following *three* types of wattmeters :

1. Dynamometer wattmeters. } Most commonly used
2. Induction wattmeters. }
3. Electrostatic wattmeters.

3.9.3. Dynamometer Wattmeter

3.9.3.1. Description

Fig. 3.68 shows the connection of dynamometer for measuring power. If the coils are connected so that a value of current *proportional to the load voltage* flows in one, and a value of current *proportional to the load current* flows in the other, the meter may be calibrated directly in watts. This is true because the indication depends upon the product of the two magnetic fields. The *strength of the magnetic fields depends upon the values of currents flowing* through the coils. If one current is proportional to load voltage and other current is the load current, then the meter can be calibrated in terms of watts or true power consumed by the load.

Let v = Supply voltage,

i = Load current, and

R = Resistance of the moving coil circuit.

Current through fixed coils, $i_f = i$,

Current through the moving coil, $i_m = \frac{v}{R}$

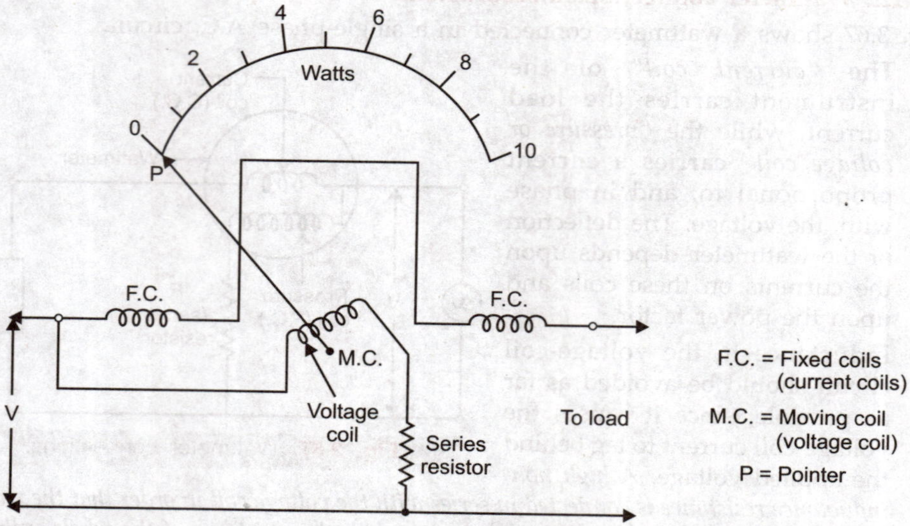

Fig. 3.68. Connection of dynamometer for measuring power

$$\text{Deflecting torque, } T_d \propto i_f \times i_m \propto \frac{iv}{R} \qquad ...(3.122)$$

— *For a D.C. circuit the deflecting torque is thus proportional to the power.*

— For any circuit with *fluctuating torque, the instantaneous torque is proportional to instantaneous power.* In this case due to inertia of moving parts, the deflection will be proportional to the *average torque i.e.,* the deflection will be proportional to the *average power.* For sinusoidal alternating quantities the average power is $VI \cos \phi$, where

V = r.m.s. value of voltage,

I = r.m.s. value of current, and

ϕ = phase angle between V and I.

Hence an electrodynamic instrument, when connected as shown in Fig. 3.68, indicates the power, irrespective of the fact it is connected in an A.C. or D.C. circuit.

Effect of mutual inductance on the deflecting torque :

The mutual inductance between the fixed and moving coils varies as the moving coil moves and the deflecting torque is affected due to change in mutual inductance. Hence it is necessary to study the effect of change of mutual inductance on the deflecting torque.

Refer to Fig. 3.69 :

Let, i_1, i_2 = Instantaneous values of currents in two coils, and

V, I = R.m.s. values of voltage and current being measured, and

M = Mutual inductance between the current coil and potential or pressure coil.

Then, the instantaneous torque of a dynamometer type of instrument,

$$T_i = i_1 i_2 \frac{dM}{d\theta} \qquad ...(3.123)$$

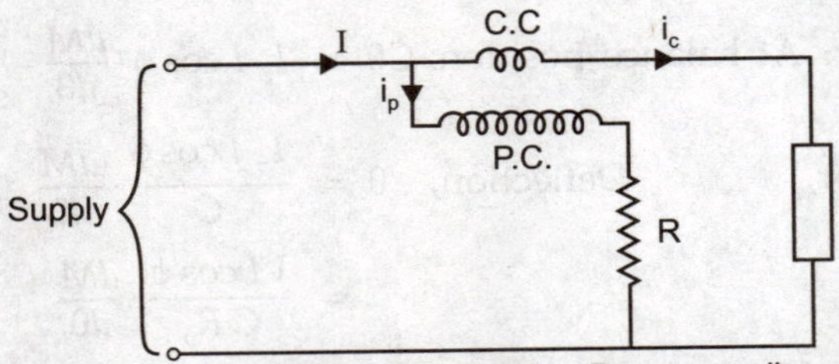

C.C = Current coil, P.C. = Pressure coil

Fig. 3.69. Circuit of dynamometer wattmeter.

Now, instantaneous value of voltage across the pressure coil circuit,

$$v = \sqrt{2}\, V \sin \omega t$$

If the pressure coil circuit has a very high resistance, it can be *treated as purely resistive.* Therefore, current i_p in the pressure coil is in phase with the voltage and its instantaneous value is,

$$i_p = \frac{v}{R_p} = \frac{\sqrt{2}V \sin \omega t}{R_p} = \sqrt{2}\, I_p \sin \omega t$$

where, $I_P = \frac{V_p}{R_p}$ = r.m.s. value of current in pressure coil circuit, and

R_p = Resistance of pressure-coil circuit.

If the current in the current coil *lags* the voltage in phase by an angle ϕ, instantaneous value of current through current coil is,

$$i_c = \sqrt{2}\, I \sin (\omega t - \phi)$$

$\therefore$ Instantaneous torque,

$$T_i = \sqrt{2}\, I_p \sin \omega t \times \sqrt{2}\, I \sin(\omega t - \phi) \frac{dM}{d\theta}$$

$$= 2\, I_p\, I \sin \omega t \sin(\omega t - \phi) \frac{dM}{d\theta}$$

or, $$T_i = I_p\, I[\cos\phi - \cos(2\omega t - \phi)] \frac{dM}{d\theta} \quad ...(3.124)$$

From above it is obvious that there is a component of power which varies as *twice the frequency of current and voltage* (refer to term containing 2 ωt).

Average deflecting torque $(T_d)_{av.}$ is given by :

$$(T_d)_{av.} = \frac{1}{T}\int_0^T T_i \cdot d(\omega t)$$

$$= \frac{1}{T}\int I_p I[\cos\phi - \cos(2\omega t - \phi)] \frac{dM}{d\theta} \cdot d(\omega t)$$

or, $$(T_d)_{av.} = I_p\, I \cos\phi \cdot \frac{dM}{d\theta} \quad ...(3.125)$$

or, $$(T_d)_{av.} = \frac{V}{R_p} \cdot I \cos\phi \frac{dM}{d\theta} \left(= \frac{P}{R_p} \cdot \frac{dM}{d\theta} \right) \quad ...(3.126)$$

Controlling torque exerted by springs, $T_c = C\theta$

where, C = Spring constant, and

θ = Final steady deflection.

Since the moving system of the instrument cannot follow the rapid variations in torque (the torque has a double frequency, component), it will take up a position at which the average deflection torque is equal to the restoring torque of the springs.

$$\therefore \text{ At balance position: } C\theta = I_p I \cos\phi \cdot \frac{dM}{d\theta} \quad \text{...(3.127)}$$

or, Deflection,

$$\theta = \frac{I_p I \cos\phi}{C} \cdot \frac{dM}{d\theta} \quad \text{...(3.128)}$$

$$= \frac{VI \cos\phi}{C R_p} \cdot \frac{dM}{d\theta} \quad \text{...(3.129)}$$

$$= kVI \cos\phi \cdot \frac{dM}{d\theta} \quad \text{...(3.130)}$$

$$= \left(k \cdot \frac{dM}{d\theta}\right) \cdot P \quad \text{...(3.131)}$$

where, P = Average power being measured = $VI \cos \phi$

and,

$$k = \frac{1}{CR_p} \quad \text{...(3.132)}$$

It is evident from eqn. (3.131) that the *deflection is directly proportional to the power being measured and the scale is essentially uniform over the range in which* $\frac{dM}{d\theta}$ *is constant.* By suitable design, the mutual inductance between fixed and moving coils can be made to vary linearly with angle over a range of 40° to 50° on either side of zero mutual inductance position. If the position of zero mutual inductance is at the midscale, the scale will be uniform over 80° to 100° which covers almost the entire range of scale.

Ranges :

(i) Current circuit : 0.25 A to 100 A with employing current transformers (CTs)

(ii) Potential circuit : 5 V to 750 V without employing potential transformers (PTs).

3.9.3.2. Types of dynamometer wattmeter

Dynamometer wattmeters may be divided into *two classes* :

1. Suspended-coil torsion instruments.
2. Pivoted-coil, direct indicating instruments.

1. ***Suspended-coil torsion wattmeters.*** These instruments are used largely as *standard wattmeters.*
   - — The moving, or voltage, coil is suspended from a torsion head by a metallic suspension which serves as a lead to the coil. This coil is situated entirely inside the current, or fixed coils and the winding in such that the system is astatic. Errors due to external magnetic fields are thus avoided.
   - — The torsions head carries a scale, and when in use, the moving coil is bought back to the zero position by turning this head, the *number of divisions turned through when multiplied by a constant for the instrument gives the power.*
   - — Eddy currents are eliminated as far as possible by winding the current coils of stranded wire and by using no metal parts within the region of the magnetic field of the instrument.
   - — The mutual inductance errors are completely eliminated by making zero position of the coil such that the angle between the planes of moving coil and fixed coil = 90°, *i.e.*, the mutual inductance between the fixed and moving coils is zero.
   - — The *elimination of pivot friction* makes possible the construction of *extremely sensitive and accurate* electrodynamic instruments of this pattern.

2. ***Pivoted-coil direct-indicating wattmeters.*** These instruments are commonly used as *switchboard or portable instruments.*
   - In these instruments, the *fixed coil* is wound in two halves, which are placed parallel to another at such a distance, that uniform uniform field is obtained. The *moving coil* is wound of such a size and pivoted centrally so that it does not project outside the field coils at its maximum deflection position.
   - The *springs* are provided for controlling the movement of the moving coil, which also serves as current leads to the moving coil.
   - The *damping* is provided by using a damping vane attached to the moving system and moving in a sector-shaped box.
   - The reading is indicated directly by the pointer attached to the moving system and moving over a calibrated scale.
   - The eddy current errors, within the region of the magnetic field of the instrument, are minimised by the use of non-metallic parts of high resistivity material.

3.9.3.3. Advantages and disadvantages

The advantages and disadvantages of dynamometer type wattmeters are as under :

Advantages :

1. The scale of the instrument is uniform (because deflecting torque is proportional to true power in both the cases *i.e.*, D.C. and A.C. and the instrument is spring controlled.)
2. High degree of accuracy can be obtained by careful design, hence these are used for calibration purposes.

Disadvantages :

1. The error due to the inductance of pressure coil at low power factor is very serious (unless special features are incorporated to reduce its effect).
2. Stray field may effect the reading of the instrument. To reduce it, magnetic shielding is provided by enclosing the instrument in an iron case.

Example 3.73. *A wattmeter is used to measure power in the circuit with the help of the following equation :*

$$P = E^2/R$$

where limiting values of voltage and resistance are, E = 200 V + 1% and R = 100 Ω ± 5%. Calculate :

(*i*) *The nominal power consumed.*

(*ii*) *The limiting error of power in per cent and watts.* **(UPTU)**

Solution. *Given :* Equation for power, $P = \frac{E^2}{R}$; Limiting values : $E = 200$ V + 1%;

$R = 100\ \Omega \pm 5\%$

(*i*) **The nominal power consumed, *P* :**

$$P = \frac{E^2}{R} = \frac{(200)^2}{1000} = 40 \text{ W} \quad \textbf{(Ans.)}$$

(*ii*) **The limiting errors of power in per cent and watts :**

$$P = \frac{E^2}{R}$$

Taking log on both sides, we get :

$$\log_e P = 2 \log_e E - \log_e R$$

Differentiating w.r.t. P, we have :

$$\frac{1}{P} = 2 \cdot \frac{1}{E} \cdot \frac{dE}{dP} - \frac{1}{R} \cdot \frac{dR}{dP}$$

or,
$$\frac{dP}{P} = 2\frac{dE}{E} - \frac{dR}{R}$$

Hence, the *relative limiting error of P in percent is,*

$$\frac{\delta P}{P} = \pm\left(2\frac{\delta E}{E} + \frac{\delta R}{R}\right)$$

$$= \pm (2 \times 1 + 5) = \pm 7\% \text{ (Ans.)}$$

Limiting error in watts $= \pm\frac{7}{100} \times 40 = 2.8$ **W (Ans.)**

3.9.3.4. Errors in dynamometer wattmeters

The various types of *errors* in dynamometer wattmeters are :

1. Error due to pressure-coil inductance.
2. Error due to pressure-coil capacitance.
3. Error due to eddy currents.
4. Error due to power lose in pressure coil or current coil.
5. Error due to friction.
6. Error due to temperature.
7. Error due to stray fields.

1. *Error due to pressure-coil inductance :*

Let, r_p = Resistance of the pressure-coil,

L_p = Inductance of the pressure coil,

R = Resistance connected in series with the pressure coil, and

I_p = Current through the pressure coil.

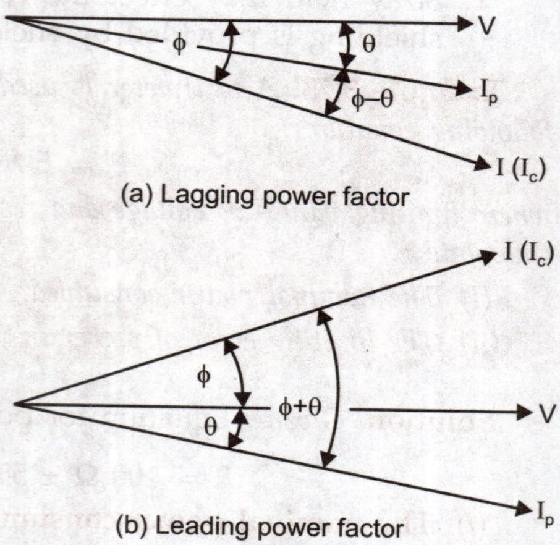

Fig. 3.70. Phasor diagram.

Then, $I_p = \dfrac{V}{\sqrt{(r_p + R)^2 + \omega^2 L_p^2}}$,

lagging behind the supply voltage V by,

$$\theta = \tan^{-1}\left(\frac{\omega L_p}{r_p + R}\right)$$

where, ω is the angular velocity of supply.

If ϕ is the phase angle (lagging) between load current I and supply voltage V, then phase angle between the currents of current coil (I_c) and pressure coil (I_p) will be $\phi - \theta$; then deflection of wattmeter $\alpha\ I_p\ I_c \cos(\phi - \theta)$.

or, Deflection $\alpha \ \dfrac{V}{Z_p} I \cos(\phi - \theta)$ ...since $I_p = \dfrac{V}{Z_p}$

where, Z_p is the impedance of pressure coil and $I_c = I$,

or, Deflection $\alpha \ \dfrac{V}{R + r_p} \ I \cos\theta \cos(\phi - \theta)$ ...since $Z_p = \dfrac{R + r_p}{\cos\theta}$

When $L_p = 0$, then $\theta = 0$, and the wattmeter deflection would be proportional to $\dfrac{V}{R + r_p} \ I \cos\phi$ at all frequencies and power factors.

Now, $$\frac{\text{True power}}{\text{Power indicated by the wattmeter}} = \frac{VI\cos\phi/(R + r_p)}{VI\cos\theta\cos(\phi - \theta)/(R + r_p)}$$

$$= \frac{\cos\phi}{\cos\theta \cdot \cos(\phi - \theta)} \quad ...(3.133)$$

or, $$\text{True power} = \frac{\cos\phi}{\cos\theta \cdot \cos(\phi - \theta)} \times \text{Reading of wattmeter} \quad ...[3.133(a)]$$

The term $\dfrac{\cos\phi}{\cos\theta \cdot \cos(\phi - \theta)}$ for *lagging p.f.* is known as *correction factor.*

Correction factor for leading p.f. $= \dfrac{\cos\phi}{\cos\theta \cdot \cos(\phi + \theta)}$

$$\text{Error} = \text{Actual reading} - \text{true reading}$$

$$= \frac{VI}{R + r_p}\cos\theta \cdot \cos(\phi - \theta) - \frac{VI}{R + r_p}\cos\phi$$

$$= \left[1 - \frac{\cos\phi}{\cos\theta\cos(\phi - \theta)}\right]\frac{VI}{R + r_p}\cos\theta\cos(\phi - \theta)$$

$$= \left[1 - \frac{\cos\phi}{\cos\theta\cos(\phi - \theta)}\right] \times \text{Actual reading}$$

$$= \left[1 - \frac{\cos\phi}{\cos\phi + \sin\phi\sin\theta}\right] \times \text{Actual reading}$$

... when cos ϕ is *assumed equal to unity*

$$= \left(\frac{\cos\phi + \sin\phi\sin\theta - \cos\phi}{\cos\phi + \sin\phi\sin\theta}\right) \times \text{Actual reading}$$

$$= \left(\frac{\sin\phi\sin\theta}{\cos\phi + \sin\phi\sin\theta}\right) \times \text{Actual reading}$$

Dividing numerator and denominator by sin ϕ, we have :

$$\textbf{Error} = \left(\frac{\sin\theta}{\cot\phi + \sin\theta}\right) \times \text{Actual reading} \quad ...(3.134)$$

Eqn. (3.134) indicates that *instrument gives high reading on lagging power factor and low reading on leading power factor.*

2. *Error due to pressure coil capacitance :*

— The pressure-coil circuit may have capacitance as well as inductance, this being largely due to inter-turn capacitance in the high-value series resistor. The effect produced is similar to that of inductance in the circuit, except that the pressure-coil current tends to *lead* the applied voltage instead of to lag behind it. This causes the wattmeter to *read low, on lagging power factors* of the load, by increasing the angle between the load and voltage-coil contents.

— The effect of frequency will be, of course, to vary the phase angle between V and I_p, the *angle increasing with increase of frequency.*

— If the capacitance reactance of pressure-coil circuit is *equal* to its inductive reactance, there will be *no error* due to these effects since the two individual errors will *neutralise* one another.

3. *Error due to eddy currents :*

— The eddy currents induced in the solid metal parts of the instrument, by the alternating magnetic field of the current coil, *alter* the magnitude and phase of this field, and so produce an error.

— The phase of the induced e.m.fs will be 90° behind the inducing flux, *i.e.*, rather more than 90° behind the main current in the current coil. The eddy current is practically in phase with its e.m.f. and this current sets up a magnetic field which, combined with that of the current coil, produces a resultant magnetic field which is less than that of the current coil alone and which also lags behind the current-coil field by a small angle.

— The eddy current error is *not easily calculable* and may be serious if care is not taken to ensure that any solid metal parts (which should be avoided as far as possible) are well removed from the current coil.

If current coil is designed for heavy currents, it should consist of *stranded conductors* in order to minimize the eddy currents flowing in the current coil itself.

4. *Error due to power loss in pressure coil (P.C.) or current coil (C.C.).* Fig. 3.71 shows two methods of connecting a wattmeter in circuit; *neither measures the power in the load directly, without corrections, even neglecting the errors discussed above.*

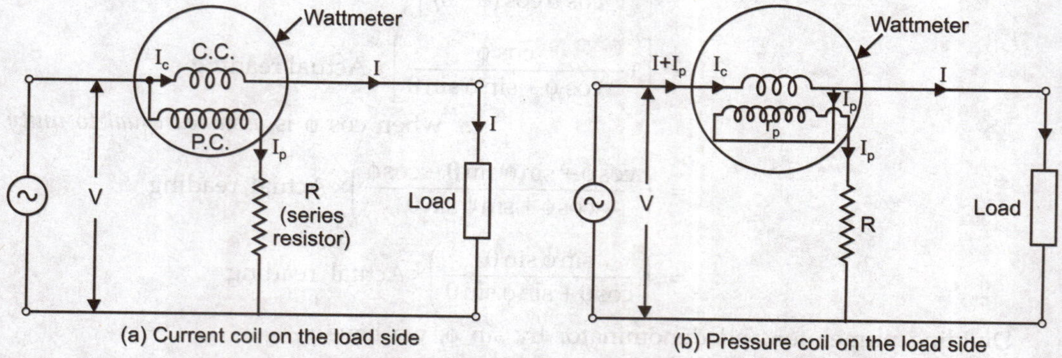

Fig. 3.71. Alternative wattmeter connections.

- In the method illustrated in Fig. 3.71(*a*), the pressure coil is connected on the "supply" side of the current coil, the voltage applied to the pressure coil is higher than that of the load on account of the voltage drop in the current coil.

Power indicated by the wattmeter

$$= \text{Power consumed by load + power loss in } \textit{current coil}$$

$$= \text{Power consumed by load} + I^2R_c \qquad ...(3.135)$$

where, R_c = The resistance of current coil.

- In the method illustrated in Fig. 3.71(*b*), the current coil carries the small current taken by the pressure coil, in addition to the load current.

∴ Power indicated by wattmeter

$$= \text{Power consumed by load + Power loss in } \textit{pressure coil}$$

$$= \text{Power consumed by load} + I_p^2(R + r_p) \qquad ...(3.136)$$

- If the load current is *small* the voltage drop in the current coil is small, so that the first method of connection introduces a *very small error*. On the other hand, *if the load current is large*, the power lost in the voltage coil will be small compared with the power in the load, and the second method of connection is *better*.

Compensation for power loss in pressure/voltage coil :

— In some wattmeters a *"compensating coil"* is used to eliminate error due to current coil carrying the pressure-coil current in addition to the load current when the connections are as in Fig. 3.71(*b*).

— This compensating coil is as *nearly as possible identical and coincident with the current coil*, so that if it were connected in series with the latter, and a current passed through the two coils-connected so that this magnitude effects were in opposition– the *resultant magnetic field would be zero*. Actually, the compensating coil is connected in series with the pressure coil, but in such a way that its magnetic effect opposes that of the current coil and *neutralises the pressure-coil component of the current in the current coil*. Thus, if no-load current flows in the instrument, the deflection should be zero, since the resultant current-coil field should be zero. The connections of this method of compensation are shown in Fig. 3.72.

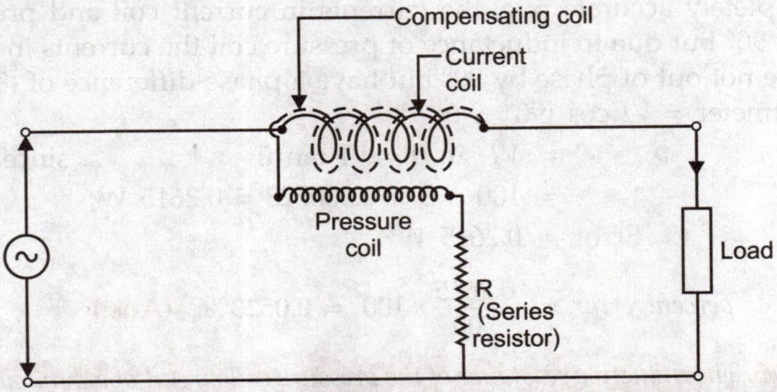

Fig. 3.72. Connections of compensating coil.

5. ***Error due to friction.*** Owing to availability of small deflecting torque, the frictional error becomes relatively important in such an instrument.

- In order *to reduce friction error* it becomes necessary that the *weight of moving system be reduced to the minimum possible on one hand and great care must be taken in the pivoting on the other hand.*

6. ***Error due to temperature.*** The indication of a wattmeter is affected by changes in room temperature. This is because any change in room temperature changes the resistance of the pressure coil and the stiffness of the springs. These effects are opposite in nature

and would nearly neutralize each other, if the pressure coil circuit were composed of copper and of a resistance alloy having a negligible resistance temperature coefficient in the ratio of 1 : 10.

7. *Errors due to stray fields.* The dynamometer wattmeter has a relatively weak operating field and therefore, it is particularly affected by stray magnetic fields resulting in the serious errors. Hence these instruments should be *shielded* against effects of stray magnetic fields.

- *Laminated iron shields are used in portable laboratory instruments while steel cases are sometimes provided to shield the switchboard instruments.*
- Precision type of wattmeters are *not* provided with shields in order to keep down errors caused by eddy currents and to avoid the small D.C. errors caused by permanent magnet of the shields; such wattmeters use *astatic system.*

Example 3.74. *A dynamometer type wattmeter is connected across 100 V, 50 Hz supply with a load of 5 A of zero power factor. The inductance and resistance of the pressure coil are 5 mH and 3000 Ω respectively. If the voltage drop in the current coil of the wattmeter is negligible, calculate the percent error in the reading of wattmeter for full scale of 500 W.* **(AMIE)**

Solution. *Given* : Supply voltage = 100 V, 50 Hz; I = 5A, zero p.f., L_p = 5 mH or 0.005 H; $r_p + R$ = 3000 Ω; wattmeter reading for full scale = 500 W.

Percent error in wattmeter reading :

The angle θ by which the current in pressure coil lags behind the voltage

$$= \tan^{-1} \frac{\omega L_p}{(r_p + R)}$$

$$= \tan^{-1}\left(\frac{2\pi \times 50 \times 0.005}{3000}\right) = \tan^{-1} 0.000523$$

Since the power factor of load is zero, the true power is zero. If the instrument would have been completely accurate one, the currents in current coil and pressure coil were out of phase by 90° but due to inductance of pressure coil the currents in current coil and pressure coil are not out of phase by 90° but have a phase difference of (90° ± θ), and the reading of wattmeter = $VI \cos (90° \pm \theta)$

or, $= VI \sin \theta = VI \tan \theta$ ... since θ is very small

$= 100 \times 5 \times 0.000523 = 0.2615$ W

or, Error = 0.2615 W

Hence, *percent error* $= \frac{0.2615}{500} \times 100 =$ **0.0523%** **(Ans.)**

Example 3.75. *The inductive reactance of the pressure coil circuit of a dynamometer wattmeter is 0.35 percent of its resistance at normal frequency and the capacitance is negligible. Calculate the correction factor and percentage error due to reactance for load having 0.5 p.f. lagging.* **(Nagpur University)**

Solution. *Given* : X_L = 0.0035 R; cos φ = 0.5 lagging.

Now, $\frac{X_L}{R} = \tan \theta = 0.0035$

∴ $\theta = \tan^{-1} (0.0035) = 0.2°$

$\sin \theta = \tan \theta = \theta = 0.2°$ ...since angle θ is very small

$$\cos\theta = \cos 0.2°$$
$$\cos\phi = 0.5 \quad \text{or} \quad \phi = \cos^{-1}(0.5) = 60°$$
$$\cot\phi = 0.5773$$

$$\therefore \quad \textit{Correction factor} = \frac{\cos\phi}{\cos\theta\cdot\cos(\phi-\theta)} \qquad \text{[Eqn. (3.133)]}$$

$$= \frac{\cos 60°}{\cos(0.2°)\cos(60°-0.2°)} = \mathbf{0.994} \ \textbf{(Ans.)}$$

$$\%\text{age error} = \frac{\sin\theta}{\cot\phi+\sin\theta}\times 100 \qquad \text{[Eqn. (3.134)]}$$

$$= \frac{0.0035}{0.5773+0.0035}\times 100 = \mathbf{0.6\%} \ \textbf{(Ans.)}$$

Example 3.76. *A wattmeter has a current coil of 0.1 Ω resistance and a pressure coil of 6500 Ω resistance. Calculate the percentage errors, due to resistance only, with each of the methods of connection, when reading the input to an apparatus which takes :*

(i) 12 A at 250 V with unity power factor, and

(ii) 12 A at 250 V with 0.4 power factor. **(Punjab University)**

Solution. *Given :* Resistance of coil, R_c = 0.1 Ω; Resistance of pressure coil, R_p = 6500 Ω; I = 12 A; V = 250 V; cos ϕ = unity; 0.4.

Percentage error :

(*i*) *When p.f. is unity, i.e., cos ϕ = 1 :*

Power input to the apparatus,

$$P = VI\cos\phi$$
$$= 250 \times 12 \times 1 = 3000 \text{ W}$$

There are only *two possible ways* in which the pressure coil can be connected as shown in Figs. 3.71(*a*) and 3.71(*b*) respectively.

- *For connection 3.71(a)*, the error is caused due to *power lost in the current coil.*

Power lost in current coil = I^2R_c = $(12)^2$ × 0.1 = 14.4 W

$$\therefore \quad \textit{Percentage error} = \frac{I^2R_c}{VI\cos\phi}\times 100 = \frac{14.4}{3000}\times 100 = \mathbf{0.48\%} \ \textbf{(Ans.)}$$

- *For connection 3.71(b)*, the error is due to *power lost in pressure coil.*

$$\text{Power lost in pressure coil} = \frac{V^2}{R_p} = \frac{(250)^2}{6500} = 9.6 \text{ W}$$

$$\therefore \quad \textit{Percentage error} = \frac{V^2/R_p}{VI\cos\phi} = \frac{9.6}{3000}\times 100 = \mathbf{0.32\%} \ \textbf{(Ans.)}$$

(*ii*) *When p.f. is 0.4, i.e., cos ϕ = 0.4 :*

$$P = VI\cos\phi = 250 \times 12 \times 0.4 = 1200 \text{ W}$$

Power lost in current coil and pressure coil is the *same* as worked out earlier in part (*i*), since *supply voltage and current are same.*

- *Percentage error* for connection shown in *Fig. 3.71(a)*

$$= \frac{I^2R_c}{VI\cos\phi}\times 100 = \frac{14.4}{1200}\times 100 = \mathbf{1.2\%} \ \textbf{(Ans.)}$$

- *Percentage error* for connection shown in *Fig. 3.71(b)*

$$= \frac{V^2/R_p}{VI\cos\phi} = \frac{9.6}{1200}\times 100 = \mathbf{0.8\%} \text{ (Ans.)}$$

Example 3.77. *A wattmeter has a current coil of 0.03 Ω resistance and a pressure coil of 6000 Ω resistance. Calculate the percentage error if the wattmeter is so connected that :*

(i) The current coil is on the load side;

(ii) The pressure coil is on the load side.

(a) If the load takes 20 A at a voltage of 220 V and 0.6 power factor in each case;

(b) What load current would give equal errors with the two connections ?

(Nagpur University)

Solution. *Given :* $R_c = 0.03\ \Omega$; $R_p = 6000\ \Omega$; $I = 20$ A; $V = 220$ V; $\cos\phi = 0.6$

(*a*) Power consumed by the load = $VI\cos\phi = 220 \times 20 \times 0.6 = 2640$ W

(*i*) *When the current coil is on the load side :*

Power loss in current coil = $I^2R_c = (20)^2 \times 0.03 = 12$ W

$$\therefore \quad \text{Percentage error} = \frac{12}{2640}\times 100 = \mathbf{0.45\%} \text{ (Ans.)}$$

(*ii*) *When the pressure coil is on the load side :*

For this connection, the wattmeter measures the *loss in pressure coil.*

Now, Power loss in pressure coil

$$= \frac{V^2}{R_p} = \frac{(220)^2}{6000} = 8.07 \text{ W}$$

$$\therefore \quad \text{Percentage error} = \frac{8.07}{2640}\times 100 = \mathbf{0.31\%} \text{ (Ans.)}$$

(*b*) *For equal error for the two connections; we have :*

$$I^2R_c = \frac{V^2}{R_p}$$

or,
$$I^2 = \frac{V^2}{R_c \times R_p} = \frac{(220)^2}{0.03\times 6000} = 268.89$$

$$\therefore \quad I = \mathbf{16.4\ A} \text{ (Ans.)}$$

Example 3.78. *An electrodynamic wattmeter is employed to measure power in a single-phase circuit. The load voltage is 200 V and the load current is 5 A at a lagging power factor of 0.1. The wattmeter potential coil has a resistance of 12000 Ω and inductance of 120 mH. Determine the percentage error in the wattmeter reading.* **(Kanpur University)**

Solution. *Given :* $V = 200$ volts; $I = 5$ A; $\cos\phi = 0.1$ (lagging); $R_p = 12000\ \Omega$; $L = 120$ mH $= 0.12$ H.

Percentage error :

Let the wattmeter be connected in the circuit as shown in Fig. 3.71(*b*).

$$\text{Load power} = VI\cos\phi = 200 \times 5 \times 0.1 = 100 \text{ W}$$

$$\phi = \cos^{-1}(0.1) = 84.26°$$

$$\text{Power lost in the coil} = \frac{V^2}{R_p} = \frac{(200)^2}{12000} = 3.33 \text{ W}$$

Neglecting the inductance of the voltage coil the reading of wattmeter would be 100 + 3.33 = 103.33 W.

Reactance of pressure coil circuit, $X_p = 2\pi \times 50 \times 0.12 = 37.7\ \Omega$

$$\theta = \tan^{-1}\left(\frac{X_p}{R_p}\right) = \tan^{-1}\left(\frac{37.7}{12000}\right) = 0.18°$$

$$\text{Wattmeter reading} = \text{True reading} \times \frac{\cos\theta \times \cos(\phi - \theta)}{\cos\phi} \quad \text{[Eqn. 3.133(a)]}$$

$$= 103.33 \times \frac{\cos(0.18°) \times \cos(84.26° - 0.18°)}{\cos(84.26°)} = 106.56 \text{ W}$$

$\therefore$ *Percentage error* $= \dfrac{106.56 - 100}{100} \times 100 =$ **6.56% (*high*)** **(Ans.)**

Example 3.79. *A dynamometer wattmeter is used to measure the power factor of a 20 μF capacitor. The pressure coil of the wattmeter having a resistance 1000 Ω and an inductive reactance of 15 Ω is connected across a 50 Hz supply. The current coil of the wattmeter, a variable resistor R and the capacitor are connected in series across the same supply. The wattmeter deflection is made zero by adjusting the value of R to 1.65 Ω. If the current coil resistance is 0.1 Ω and its inductance negligible, determine the power factor of the capacitor.* **(Bombay University)**

Solution. *Given:* $C = 20\ \mu F = 20 \times 10^{-6}$ F; $R_p = 1000\ \Omega$; $X_L = 15\ \Omega$; $f = 50$ Hz; $R = 1.65\ \Omega$; $R_c = 0.1\ \Omega$.

Power factor of capacitor :

Phase angle of pressure coil circuit, $\theta = \tan^{-1}\left(\dfrac{X_L}{R_p}\right) = \tan^{-1}\left(\dfrac{15}{1000}\right) = 0.86°$

Capacitance reactance, $X_C = \dfrac{1}{2\pi \times 50 \times (20 \times 10^{-6})} = 159.2\ \Omega.$

Let r be the series resistance of the capacitor. Total resistance of the load circuit including resistance of current coil

$$= 0.1 + 1.65 + r = (1.75 + r)\ \Omega$$

$\therefore$ ϕ = Phase angle of load circuit $= \tan^{-1}\left(\dfrac{159.2}{1.75 + r}\right)$

The load circuit has a *leading* p.f. as the reactance is *capacitance. For leading p.f.,*

Reading of wattmeter $\propto \cos\theta \cos(\phi + \theta)$

For a wattmeter reading to be zero,

$$\cos\theta \cos(\phi + \theta) = 0$$

or, $\cos(\phi + \theta) = 0$

or, $\phi + \theta = 90°$

or, $\phi = 90 - \theta = 90° - 0.86° = 89.14°$

or, $\tan^{-1}\left(\dfrac{159.2}{1.75 + r}\right) = 89.14°$

or, $\dfrac{159.2}{1.75 + r} = \tan(89.14°) = 66.62$

or, $$r = \frac{159.2}{66.62} - 1.75 = 0.64\ \Omega$$

Impedance of a capacitor, $$Z_C = \sqrt{r^2 + (X_C)^2} = \sqrt{(0.64)^2 + (159.2)^2} = 159.2\ \Omega$$

$\therefore$ $$\textit{Power factor of the capacitor} = \frac{r}{Z_C} = \frac{0.64}{159.2} = \mathbf{0.004}\ \textbf{(Ans.)}$$

Example 3.80. *A dynamometer wattmeter reading correctly on D.C. is used to measure power in circuit of resistance of 2 Ω and inductance of 0.25 H. The supply is 200 V at 50 Hz and the pressure coil circuit of wattmeter has a resistance of 1000 Ω and inductance of 5.6 mH. What is the actual reading of the wattmeter ?*

Neglect the impedance of the current coil circuit. Assume that pressure coil is connected on the load side of the instrument. **(Punjab University)**

Solution. *Given:* Load resistance, $R_L = 2\Omega$; Load resistance, $X_L = 2\pi fL = 2\pi \times 50 \times 0.25 = 78.5\ \Omega$; $V = 200$ volts; $R_p = 1000\ \Omega$; $L_p = 5.6$ mH $= 0.0056$ H.

Actual reading of wattmeter :

Load impedance, $$\bar{Z}_L = R_L + jX_L = (2 - j78.5)\ \Omega$$

Load current, $$\bar{I}_L = \frac{V}{\bar{Z}_L} = \frac{200}{2 + j78.5} = (0.0648 - j2.55)\text{A}$$

Pressure coil circuit impedance, $$\bar{Z}_p = R_p + jX_{Lp}$$
$$= 1000 + j(2\pi \times 50 \times 0.0056) = (1000 + j1.76)\ \Omega$$

Pressure coil current, $$\bar{I}_p = \frac{V}{\bar{Z}_p} = \frac{200}{1000 + j1.76} = (0.2 - \text{j}0.000352)\text{A}$$

$$\theta = \tan^{-1}\left(-\frac{0.000352}{0.2}\right) = -0.1°$$

Current coil current, $$\bar{I}_c = \bar{I}_L + \bar{I}_p$$
$$= (0.0648 - j2.55) + (0.2 - j0.000352)$$
$$= 0.2648 - j2.550352 = 2.56\ \text{A}$$

$$\phi = \tan^{-1}\left(\frac{-2.550352}{0.2648}\right) = -84.07°$$

$$\phi - \theta = -84.07 - (-0.1°) = -83.97°$$
$$\cos(\phi - \theta) = \cos(-83.97°) = 0.105$$
$$\cos\theta = \cos(-0.1°) \simeq 1$$

$\therefore$ Actual reading of wattmeter

$$= VI_c \cos\theta \cos(\phi - \theta)$$
$$= 200 \times 2.56 \times 1 \times 0.105 = \mathbf{53.76\ W}\ \textbf{(Ans.)}$$

Example 3.81. *The current coil of a wattmeter is connected in series with an ammeter and an inductive load. A voltmeter and the voltage coil are connected across a 100 Hz supply. The ammeter reading is 4.5 A and the voltmeter and wattmeter readings are respectively 240 V and 23 W. The inductance of voltage circuit is 10 mH and its resistance 2000 Ω. If the voltage drop across the ammeter and current coil are negligible, what is the percentage error in the wattmeter reading ?* **(Banglore University)**

Solution. *Given:* f = 100 Hz; I = 4.5 A; V = 240 V; Wattmeter reading = 23 W; L = 10 mH = 0.01 H; R_p = 2000 Ω.

Percentage error in wattmeter reading :

Inductive reactance of pressure coil = $2\pi fL = 2\pi \times 100 \times 0.01 = 6.28\ \Omega$

Phase angle of pressure coil circuit, $\theta = \tan^{-1}\left(\frac{X_L}{R_p}\right) = \tan^{-1}\left(\frac{6.28}{2000}\right) = 0.18°$

For an inductive load, reading of wattmeter $\propto \cos\theta \cos(\phi - \theta)$

True power $\propto \cos\phi$, where $\cos\phi$ = load power factor

$\therefore$ True power $= \frac{\cos\phi}{\cos\theta\cos(\phi-\theta)} \times$ reading of wattmeter ...(*i*)

But, True power $= I^2R = (4.5)^2 \times R = (4.5)^2 Z\cos\phi$

where, R = Resistance of load

Impedance of load, $Z = \frac{V}{I} = \frac{240}{4.5} = 53.33\ \Omega$

$\therefore$ True power $= (4.5)^2 \times 53.33 \cos\phi$

Substituting the various values in (*i*), we get

$$(4.5)^2 \times 53.33 \cos\phi = \frac{\cos\phi}{\cos\theta\cos(\phi-\theta)} \times 23$$

or, $$\cos\theta\cos(\phi-\theta) = \frac{23}{(4.5)^2 \times 53.33} = 0.0213$$

or, $\cos(0.18°)\cos(\phi - 0.18°) = 0.0213$

or, $$\cos(\phi - 0.18°) = \frac{0.0213}{\cos(0.18°)} = 0.0213$$

or, $$\phi = \cos^{-1}(0.0213) + 0.18° = 88.96°$$

Percentage in wattmeter reading

$$= \frac{\sin\theta}{\cot\phi + \sin\theta} \times 100 \quad \text{[Eqn. (3.134)]}$$

$$= \frac{\sin(0.18°)}{\cot(88.96°) + \sin(0.18°)} \times 100 = \mathbf{14.75\%}$$ **(Ans.)**

Example 3.82. *The current coil of a dynamometer is connected to a 24 V D.C. source in series with a 5Ω resistor. The potential circuit is connected through an ideal rectifier in series with a 50 Hz source of 100 V. The inductance of pressure circuit and current coil resistance are negligible. Compute the reading of the wattmeter.* **(AMIE)**

Solution. Current through the current coil $= \frac{24}{5} = 4.8$ A.

The pressure coil is energised by an ideal rectifier. Therefore, the pressure coil carries current during one half cycle and in the other half cycle there is no current in it. This means there is *deflecting torque on the meter during one half cycle.*

$\therefore$ Reading of wattmeter = Average power over a cycle

$$= \frac{1}{2\pi}\left[\int_0^{\pi} vi\, d\theta\right] = \frac{1}{2\pi}\left[\int_0^{\pi}(\sqrt{2}\times 100)\sin\theta \times 4.8\, d\theta\right]$$

$$= \frac{\sqrt{2}\times 100\times 4.8}{2\pi}\left[\int_0^{\pi}\sin\theta\cdot d\theta\right]$$

$$= 108.04[-\cos\theta]_0^{\pi} = \mathbf{216\ W}\ \textbf{(Ans.)}$$

Example 3.83. *A dynamometer wattmeter is used to measure the power in an inductive load having a phase angle of 87°. The potential coil has a resistance of 1000 Ω and inductance of 6 mH. The potential coil current is negligible. The impedance of current coil is also negligible. Find the % error in wattmeter reading.*

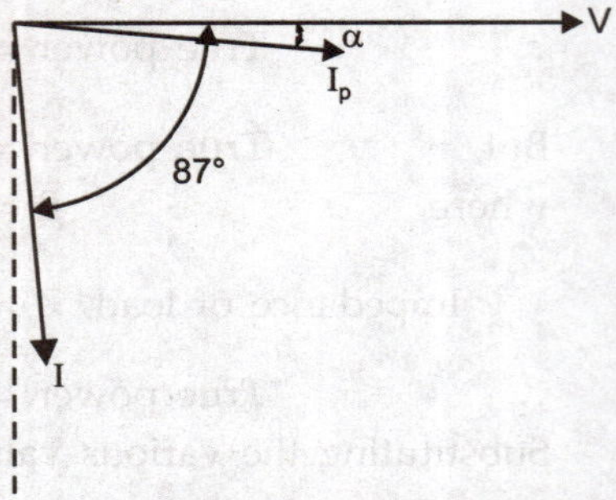

Fig. 3.73.

Solution. Fig. 3.73 shows the phasor diagram. V and I are the load voltage and load current. I_p is the potential coil current. The current I is lagging the voltage by 87°. If the potential coil were purely resistive, I_p would be in phase with V and the reading of the wattmeter would be correct. Because of inductance of potential coil, I_p is lagging by $\angle\alpha$. The power factor as seen by the wattmeter is $\cos(87 - \alpha)°$.

$$\alpha = \tan^{-1}\left(\frac{2\pi\times 50\times 6\times 10^{-3}}{1000}\right) = 0.108°$$

$$\%\ \text{error} = \frac{VI\cos(87 - 0.108)° - VI\cos 87°}{VI\cos 87°}\times 100 = \mathbf{3.6\%}\ \textbf{(Ans.)}$$

3.9.3.5. Low power factor electro-dynamometer type wattmeters

Ordinary electro-dynamometer wattmeter is not suitable for measuring power in circuits having low power factors due to the following *reasons:*

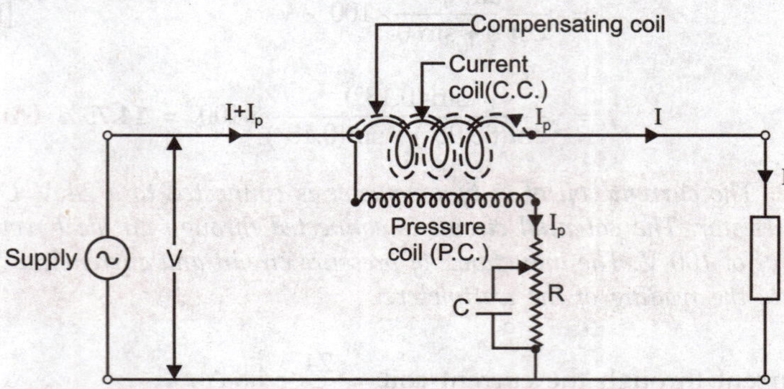

Fig. 3.74. Low power factor electro-dynamometer type wattmeter.

(*i*) Small deflecting torque on the moving system even when the current and pressure coils are fully excited.

(*ii*) Introduction of large error due to inductance of pressure coil at low power factors.

Thus, the following *special features are incorporated* in an electro-dynamometer type wattmeter to make it suitable for measurement of power in low p.f. circuits:

1. ***Pressure coil circuit :***
   - — The pressure coil circuit is made of *low resistance* so that the current flowing through it is increased to give an *increased operating torque.*
   - — The pressure coil current in low p.f. wattmeter may be as much as 10 times the value used for high p.f. meters.
2. ***Compensation for pressure coil current.*** This compensation is same as given in Fig. 3.72.
3. ***Compensation for inductance of pressure coil :***
   - — The error caused by pressure coil inductance is $\dfrac{\sin\theta}{\cot\phi + \sin\theta}$ times the actual reading of wattmeter. Now, when the *p.f. is low* the value of ϕ is large and, therefore, the *error is large.*
   - — The error caused by inductance of pressure coil is compensated by connecting a capacitor across a part of series resistance in the pressure coil circuit shown in Fig. 3.74.
4. ***Small control torque.*** Low p.f. wattmeters are *designed to have a small control torque so that they give full scale deflection for power factor as low as 0.1.*

3.9.3.6. Calibration of dynamometer wattmeters

The various methods of calibration of dynamometer wattmeters are enumerated and discussed as follows :

A. *D.C. calibration :*
   1. By comparison with standard wattmeter.
   2. By D.C. potentiometer.

B. *A.C. calibration :*
   1. By comparison with standard wattmeter.
   2. By A.C. potentiometer.
   3. By using resistive load.
   4. By electrostatic wattmeter.

A. D.C. calibration :

1. ***By comparison with standard wattmeter :***
   - — In this method, the current coils of test wattmeter (W_T) and standard wattmeter (W_S) are connected in *series,* and the two voltage circuits are connected in *parallel.*
   - — It is used to employ different sources of supply for voltage and current circuits, the *source for current being a low-voltage supply source.* This arrangement with separate supplies is known as *"fictitious load (phantom loading)"* and is almost invariably employed in wattmeter testing.

   It may be noted that an equalizing connection is required to maintain the wattmeter coils at the same voltage; this arrangement keeps the losses in the current regulating rheostat small and the total power consumed during the test is only a small fraction of the wattmeter indication.

   Fig. 3.75 shows the circuit diagram for such a test. The voltmeter and ammeter are used for checking the voltage and current, but the *calibration is a direct comparison between the readings of two wattmeters W_T and W_S.*

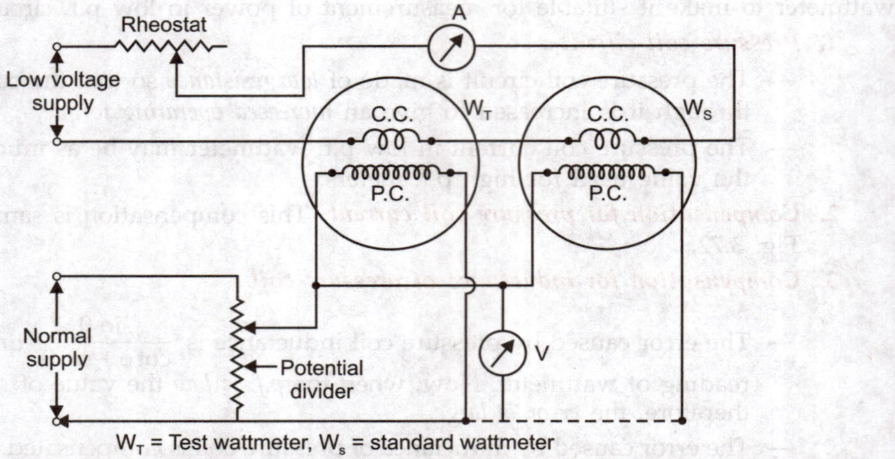

Fig. 3.75. D.C. calibration of dynamometer wattmeter with standard wattmeter.

- *Errors due to stray fields are eliminated by including reversing switch* in both supplies and taking the *mean readings* obtained with connections as shown and with both supplies *reversed.*

2. ***By D.C. potentiometer.*** Refer to Art. 5.2.

B. A.C. calibration :

1. ***By comparison with standard wattmeter.*** This method is similar to the corresponding D.C. test, the circuit being shown in Fig. 3.75.

— The low voltage supply may be obtained conveniently from a step-down transformer.

— The test may be carried out at various power factors by direct loading, using chokes or capacitors in addition to resistance loads.

In fictitious loading arrangement a phase-shifting transformer may be employed for shifting the phase of the voltage applied to the circuits with respect to current in the current coils.

- This method is *not very convenient and involves wastage of huge amount of power.*

2. ***By A.C. potentiometer.*** This method is similar to the D.C. potentiometer method (Art. 5.2) *and will normally include fictitious loading and arrangement for varying the p.f. The true power is given by VI cos ϕ; V, I and ϕ are measured by potentiometer.*

3. ***By using resistive load.*** In this method, a non-inductive resistance (standard) is used as a load and power consumed by this load is obtained from its resistance and the voltage across it. The voltage across the load may be measured with an electrostatic voltmeter.

- This is an *accurate method of testing low range wattmeters at unity power factor.*

4. ***By electrostatic wattmeter.*** The arrangement of calibrating a wattmeter by an electrostatic wattmeter is an *expensive arrangement* (two alternators of similar electrical design coupled together and driven by a variable speed motor) and as such it is justifiable where large number of tests at varying power factors are to be conducted.

3.9.4. Induction Wattmeter

3.9.4.1. Description

Induction wattmeters can be used in *A.C. circuit only* (in contrast with, dynamometer wattmeters can be used both in D.C. and A.C. circuits) and *are useful only when the frequency and supply voltage are constant.*

The *operation* of all induction instruments depends on the *production of torque due to reaction between a flux* ϕ_1 (whose magnitude depends on the current or voltage to be measured) *and eddy currents induced in a metal disc or drum by another flux* ϕ_2 (whose magnitude also depends on the current or voltage to be measured). Since the magnitude of eddy currents also depends on the flux producing them, the *instantaneous value of the deflecting torque is proportional to the square of the current or voltage under measurement and the value of mean deflecting torque is proportional to the mean square of the current or voltage.*

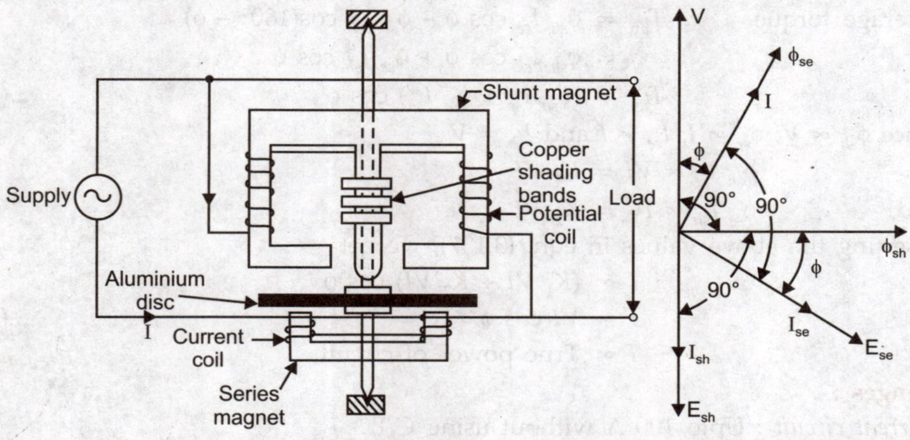

(a) Construction (b) Phasor diagram

Fig. 3.76. Induction wattmeter.

Construction. Fig. 3.76(*a*) shows the construction of an induction wattmeter :

— It has two laminated electromagnets. The upper electromagnet (called magnet shunt) has two coils in series which are wound such that the direction of fluxes set up by them are same. They are connected in series with a *high resistance across the main supply*. Thus these two coils serve as *pressure coils*.

— One or more copper shading bands are fitted on the central limb of the upper electromagnet. The induced currents in these bands set up their own fluxes. The total flux of the upper electromagnet is the resultant of the main flux and the fluxes of currents in the shading bands. The positions of these bands are adjusted so that this *resultant flux lags the applied voltage by 90°*.

— The two coils of the lower electromagnet (called series magnet) are also in series and carry the circuit current. They are also wound such that their *fluxes are in the same direction.*

— A thin aluminium disc supported by jewelled bearings is mounted on the spindle. The spindle also carries a hair spring (for providing torque) and a pointer.

Theory. Fig. 3.76(*b*) shows the phasor diagram.

Let the load current *I* lag the circuit voltage *V* by an angle ϕ as shown in the figure. The flux ϕ_{sh} of shunt magnet lags behind *V* by 90° (this is done by adjusting the position of shading bands).

Assuming that hysteresis and saturation effects in the iron are negligible the series magnet flux ϕ_{se} *is taken proportional to and in phase with load current I.*

E.m.fs E_{sh} and E_{se} are induced in the disc by shunt magnet flux ϕ_{sh} and series magnet flux ϕ_{se} respectively and lag behind their respective fluxes by 90°. Eddy currents I_{sh} and I_{se} are set up by the induced e.m.fs E_{sh} and E_{se} respectively and are in phase with their respective e.m.fs.

Now, the two opposite torques given by $\phi_{sh}I_{se}$ and $\phi_{se}I_{sh}$ will act on the disc. The instantaneous values of resultant torque is the difference of the two *i.e.,*

$$T_{res.} = \phi_{sh}\, I_{se} - \phi_{se}\, I_{sh}$$

where, ϕ_{sh}, ϕ_{se}, I_{sh} and I_{se} are the *instantaneous values.*

Since the phase angle between ϕ_{sh} and I_{se} is ϕ and phase angle between ϕ_{se} and I_{sh} is $(180° - \phi)$, therefore,

Average torque, $T_{av.} \propto \phi_{sh}\, I_{se} \cos\phi - \phi_{se}\, I_{sh} \cos(180° - \phi)$

$$\propto \phi_{sh}\, I_{se} \cos\phi + \phi_{se}\, I_{sh} \cos\phi$$

or, $$T_{av.} \propto (\phi_{sh}\, I_{se} + \phi_{se}\, I_{sh}) \cos\phi \quad ...(3.137)$$

Since $\phi_{sh} \propto V$; $\phi_{se} \propto I$; $I_{se} \propto I$ and $I_{sh} \propto V$

$\therefore$ $$\phi_{sh}\, I_{se} \propto VI = K_1\, VI$$

and, $$\phi_{se}\, I_{sh} \propto IV = K_2\, VI$$

Inserting the above values in eqn. (3.137), we get

$$T \propto (K_1\, VI + K_2\, VI) \cos\phi$$

or, $$T \propto VI \cos\phi \quad ...(3.138)$$

or, $T \propto$ True power of circuit.

Ranges :

Current circuit : Upto 100 A without using C.T.

Voltage circuit : Upto 750 V without using P.T.

3.9.4.2. Advantages and disadvantages

Advantages :

1. Fairly long scale (extending over 300°).
2. Free from the effects of stray fields.
3. Good damping.
4. Practically free from frequency errors.

Disadvantages :

Sometimes subjected to serious temperature error because the main effect of temperature is on the resistance of the eddy current paths.

3.9.4.3. Comparison between dynamometer and induction wattmeters

The comparison between dynamometer and induction wattmeters is given in tabular form below :

S. No.	Aspects	Dynamometer wattmeters	Induction wattmeters
1.	*Working torque*	Comparatively weaker	Comparatively stronger
2.	*Power consumption*	Comparatively low	Comparatively higher
3.	*Weight of moving system*	Comparatively low	Comparatively higher

4.	*Type of scale*	Uniform	Uniform and long
5.	*Degree of accuracy*	High degree of accuracy if carefully designed.	Less accurate; it is accurate only at stated frequency and temperature.
6.	*Use*	Both D.C. and A.C. systems.	Can be used on A.C. system only.

Example 3.84. *A voltage 200 sin ωt + 50 sin (3ωt – 30°) + 40 sin(5ωt + 220°)V is applied to the pressure coil of a wattmeter. The current through the current coil is 10 sin (ωt – 10°) + 2 cos(3ωt – 150°) + 2 cos(5ωt + 150°)A. Find :*

(i) The reading of the wattmeter.

(ii) The contribution of each harmonic to the total power.

Solution. *Given:*

$$v = 200 \sin \omega t + 50 \sin (3\omega t - 30°) + 40 \sin(5\omega t + 220°)\text{V}$$

$$i = 10 \sin (\omega t - 10°) + 2 \cos(3\omega t - 150°) + 2 \cos(5\omega t + 150°)\text{A}$$

$$= 10 \sin (\omega t - 10°) + 2 \sin(3\omega t - 60°) + 2 \sin(5\omega t + 240°)\text{A}$$

(*i*) The reading of the wattmeter :

$$\text{Average power, } P = \frac{1}{2\pi}\int_0^{2\pi} ei \, d(\omega t)$$

The average value of all product terms involving different frequencies is zero. Only products of current and voltage of the same frequency need be considered to calculate power.

For nth harmonic,

$$P_n = \frac{1}{2\pi}\int_0^{2\pi} V_{n\,\max.}\, I_{n\,\max.} \sin(n\,.\,\omega t)\sin(n\,.\,\omega t - \phi_n)\, d(\omega t)$$

$$= \left(\frac{V_{n\,\max.}}{\sqrt{2}}\right)\left(\frac{I_{n\,\max.}}{\sqrt{2}}\right)\cos\phi_n$$

$$\therefore \quad P = \frac{200\times10}{2}\cos 10° + \frac{50\times2}{2}\times\cos 30° + \frac{40\times2}{2}\cos 20°$$

or, $\quad P = 984.81 + 43.30 + 37.59 =$ **1065.7 (Ans.)**

(*ii*) The contribution of each harmonic to total power :

Contribution of fundamental component $= \frac{984.81}{1065.7}\times100 =$ **92.41% (Ans.)**

Contribution of third harmonic $= \frac{43.30}{1065.7}\times100 =$ **4.063% (Ans.)**

Contribution of fifth harmonic $= \frac{37.59}{1065.7}\times100 =$ **3.527% (Ans.)**

Example 3.85. *A voltage 50 + 25 sin ωt is applied to a series R-L circuit having a resistance of 10 Ω and inductance 0.1 H. A wattmeter is connected in the circuit to measure power. Find the reading of the wattmeter. ω = 2π × 50 rad./sec.*

Solution. The inductance offers zero reactance to D.C.

For first harmonic,

$$\bar{Z}_1 = 10 + j(2\pi \times 50 \times 0.1) = 10 + j31.46\ \Omega = 32.97\angle 72.34°$$

$$i = \frac{50}{10} + \frac{25}{32.97}\sin(\omega t - 72.34°)$$

$$= 5 + 0.76 \sin(\omega t - 72.34°)$$

$$P = 50 \times 5 + \frac{25 \times 0.76}{2}\cos 72.34° = \mathbf{252.9\ W}\ \textbf{(Ans.)}$$

3.9.5. Electrostatic Wattmeter

These wattmeters are used for the *measurement of small amounts of power, particularly when the voltage is high and power factor low.* This type of wattmeter is also used for *measurement of dielectric loss of cables on alternating voltage and for calibration of wattmeters and energy meters.*

Electrostatic wattmeter consists of a quadrant electrometer used with a non-inductive resistor R as shown in Fig. 3.77.

Let i, e, v, v_1 and v_2 be the instantaneous load current, instantaneous potentials of the load, needle and of the two pairs of quadrants respectively, as shown in the figure.

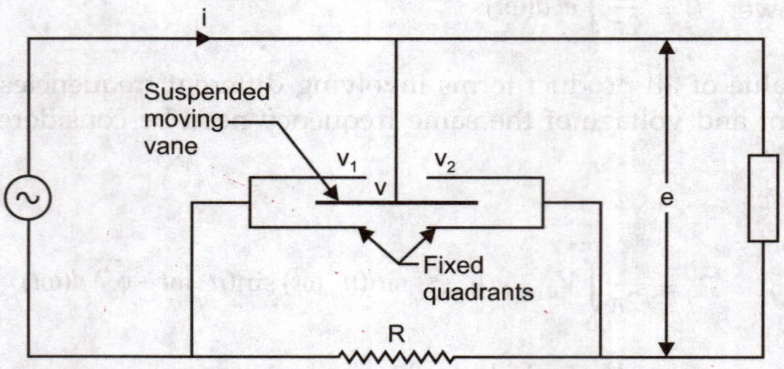

Fig. 3.77. Electrostatic wattmeter.

Instantaneous torque, $T \propto (v - v_1)^2 - (v - v_2)^2$

Since, $v_1 - v_2 = e$, the potential across load, and

$v - v_1$ = supply voltage

= voltage across load + voltage drop in non-inductive resistance

$= e + iR$

$\therefore$ Instantaneous torque, $T \propto (e + iR)^2 - e^2$

$$\propto e^2 + 2eiR + i^2R^2 - e^2$$

$$\propto 2eiR + i^2R^2$$

or, $$T \propto 2R\left(ei + \frac{i^2R}{2}\right) \quad ...(3.139)$$

i.e., Instantaneous torque is proportional to the instantaneous power in the load plus half of the power lost in non-inductive resistance.

Advantages and Disadvantages :

- Electrostatic wattmeter is a *precision instrument* and should be used as such.
- *It is free from errors on account of waveforms, frequency and eddy currents.*
- It has a *very small working torque.*

3.9.6. Thermal Wattmeter

Fig. 3.78 shows the arrangement for measuring power with thermocouples :

This method uses two similar thermocouples (**1** and **2**) whose outputs are connected in *opposition* with a galvanometer in between. R_h is the resistance of each thermocouple heating element. R is a high series resistance, and between C and D is a low resistance R_2 capable of carrying the load current i. The resistance R_2 develops a potential difference which depends upon the load current, together with the current of one heater, and the series resistance R carries the current of both heaters.

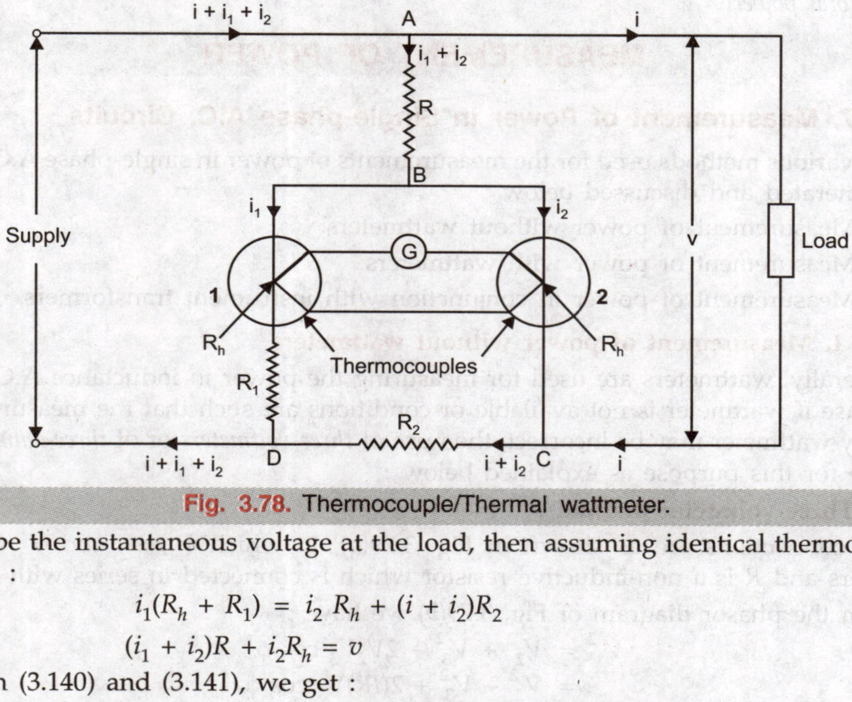

Fig. 3.78. Thermocouple/Thermal wattmeter.

If v be the instantaneous voltage at the load, then assuming identical thermocouples, we have :

$$i_1(R_h + R_1) = i_2 R_h + (i + i_2)R_2 \quad ...(3.140)$$

$$(i_1 + i_2)R + i_2 R_h = v \quad ...(3.141)$$

From (3.140) and (3.141), we get :

$$i_1 = \frac{v(R_h + R_2) + iR_2(R_h + R)}{(R + R_h)(R_h + R_1) + R(R_h + R_2)} \quad ...(3.142)$$

and,

$$i_2 = \frac{v(R_h + R_1) - iR_2 R}{(R_h + R)(R_h + R_1) + R(R_h + R_2)} \quad ...(3.143)$$

E.m.f. of *thermocouple-1*, $e_1 \propto$ heat produced by heating element, or,

$$e_1 \propto i_1^2 R_h$$

E.m.f. of *thermocouple-2*, $e_2 \propto i_2^2 R_h$

E.m.f. across galvanometer $G = e_1 - e_2$

Instantaneous deflecting torque of galvanometer,

$$T_{inst.} = K_1(e_1 - e_2) = K_2(i_1^2 - i_2^2) \quad ...(3.144)$$

where, K_1 and K_2 are constants.

Substituting the values of i_1 and i_2 in eqn. (3.144), we get

$$T_{inst.} = K(C_1\, vi + C_2 i^2)$$

where, C_1 and C_2 are constants involving the resistances.

If $R_h + R$ is not very different from R_h, then $C_2 i^2$ may be *neglected and* $T_{inst.} \propto vi$ *or instantaneous power.*

Thus, *galvanometer may be calibrated to read the power.*

— The commercial thermal wattmeters employ a number of thermocouples connected in the form of a chain in order to increase the output. C.Ts and P.Ts are also used with these instruments.

— For high frequency measurements careful shielding is required.

- The thermal wattmeters can be *used for measurement of power is several circuits and the sum of their outputs can be applied to a recording potentiometer which records the total power.*

MEASUREMENT OF POWER

3.9.7. Measurement of Power in Single-phase A.C. Circuits

The various methods used for the measurements of power in single-phase A.C. circuits are enumerated and discussed below :

1. Measurement of power without wattmeters.
2. Measurement of power with wattmeters.
3. Measurement of power in conjunction with instrument transformers.

3.9.7.1. Measurement of power without wattmeters

Generally, wattmeters are used for measuring the power in inductance A.C. circuits but in case if wattmeter is not available or conditions are such that the measurement of power by wattmeter may be incorrect, then use of *three-voltmeters* or of *three-ammeters* can be made for this purpose as explained below :

(i) Three-voltmeter method :

The connections are as shown in Fig. 3.79(*a*), in which V_1, V_2 and V_3 are three voltmeters and R is a non-inductive resistor which is connected in series with the load.

From the phasor diagram of Fig. 3.79(*b*) we have :

$$V_1^2 = V_2^2 + V_3^2 + 2V_2V_3 \cos \phi \qquad ...(3.145)$$

$$= V_2^2 + V_3^2 + 2(IR)V_3 \cos \phi$$

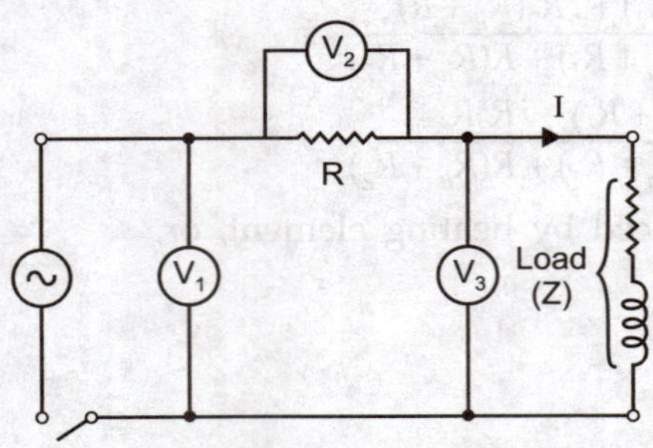

(a) Connection diagram

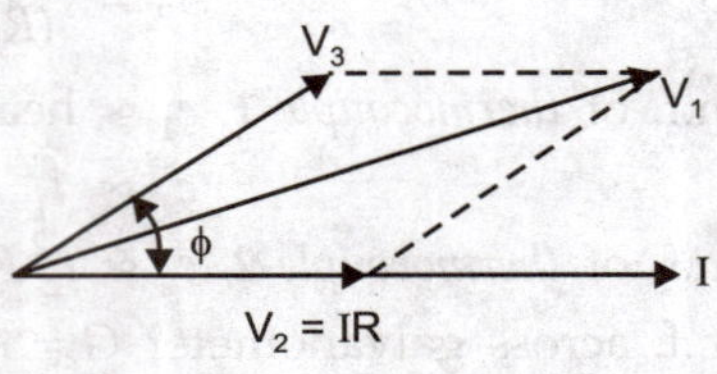

(b) Phasor diagram

Fig. 3.79. Three-voltmeter method of measuring single-phase power.

$$= V_2^2 + V_3^2 + 2R(V_3\ I \cos\phi)$$

$$= V_2^{\ 2} + V_3^{\ 2} + 2PR$$ ...since power in the inductive circuit, $P = V_3 I \cos\phi$

or, Power, $$P = \frac{V_1^2 - V_2^2 - V_3^2}{2R} \quad ...(3.146)$$

From eqn. (3.145) *power factor* of the circuit is given by :

$$\cos\phi = \frac{V_1^2 - V_2^2 - V_3^2}{2V_2V_3} \quad ...(3.147)$$

- *The assumptions are made that the current in the resistor R is the same as the load current.*

Disadvantages :

(*i*) Supply voltage higher than normal voltage is required because an additional resistance R is connected in series with the load Z (inductive circuit).

(*ii*) Even small errors in measurement of voltages may cause serious errors in the value of power determined by this method.

(*ii*) Three-ammeter method :

The connection diagram and phasor diagram respectively are shown in Fig. 3.80(*a*) and (*b*). The current measured by ammeter A_1, is the vector sum of the load current and that taken by the non-inductive resistor R, this latter being in phase with V.

From the phasor diagram :

$$I_1^2 = I_2^2 + I_3^2 + 2I_2I_3\cos\phi \quad ...(3.148)$$

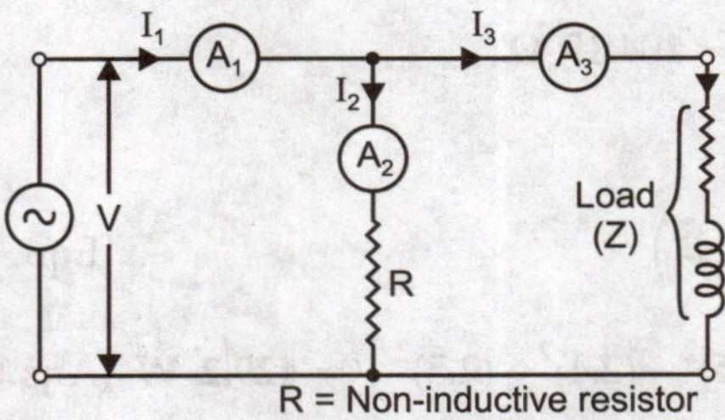

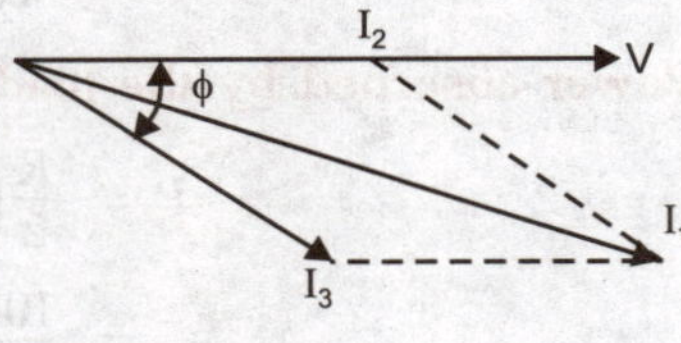

Fig. 3.80. Three-ammeter method of measuring single-phase power.

But, $$I_2 = \frac{V}{R}$$

$\therefore$ $$I_1^2 = I_2^2 + I_3^2 + 2\times\frac{V}{R}\times I_3\cos\phi$$

Hence, the power $P = VI_3\cos\phi$, is given by :

$$P = \frac{R}{2}\left(I_1^2 - I_2^2 - I_3^2\right) \quad ...(3.149)$$

Also, $$\cos\phi = \frac{I_1^2 - I_2^2 - I_3^2}{2I_2I_3} \quad ...(3.150)$$

... From eqn. (3.148)

- The *"advantage"* of this method is that the value of power determined is *independent of supply frequency and waveforms.* The disadvantages of measurement of power by three-voltmeter method are overcome in this method.

Example 3.86. *The following readings were obtained from three voltmeters used for a single-phase power measurement :*

V_2 = 180 volts across a non-inductive resistor; V_3 = 200 volts across an inductive load; V_1 = 300 volts across the two in series.

Calculate the power factor of the inductive load. **(Gorakhpur University)**

Solution. *Given :* $V = 180$ V; $V_3 = 200$ V; $V_1 = 300$ V

Power factor, cos φ :

We know that : $$\cos\phi = \frac{V_1^2 - V_2^2 - V_3^2}{2V_2V_3}$$ ...[Eqn. (3.147)]

or, $$\cos\phi = \frac{(300)^2 - (180)^2 - (200)^2}{2 \times 180 \times 200} = \mathbf{0.244} \text{ (Ans.)}$$

Example 3.87. *An inductive load takes a current of 2.5 A; a non-inductive resistor connected in parallel takes 2.4 A, when connected across 250 V supply. The total current taken from the supply is 4.5 A. Calculate :*

(i) Power absorbed by the load.

(ii) Load impedance.

(iii) Power factor of the load. **(Punjab University)**

Solution. *Given :* $I_3 = 2.5$ A; $I_2 = 2.4$ A; $I_1 = 4.5$ A; $V = 250$ V.

Non-inductive resistance, $R = \frac{V}{I_2} = \frac{250}{2.4} = 104.17\ \Omega.$

(*i*) Power absorbed by the load, P :

$$P = \frac{R}{2}\left(I_1^2 - I_2^2 - I_3^2\right)$$ ...[Eqn. (3.149)]

$$= \frac{104.17}{2}\left[(4.5)^2 - (2.4)^2 - (2.5)^2\right] = \mathbf{429.2\ W} \text{ (Ans.)}$$

(*ii*) Load impedance, Z :

$$Z = \frac{V}{I_3} = \frac{250}{2.5} = \mathbf{100\ \Omega} \text{ (Ans.)}$$

(*iii*) Power factor of the load, cos φ :

$$\cos\phi = \frac{I_1^2 - I_2^2 - I_3^2}{2I_2I_3}$$ ...[Eqn. (3.150)]

$$= \frac{(4.5)^2 - (2.4)^2 - (2.5)^2}{2 \times 2.4 \times 2.5} = \mathbf{0.687} \text{ (Ans.)}$$

3.9.7.2. Measurement of power with wattmeter

Refer to Art. 3.9.1.2.

3.9.7.3. Measurement of power in conjunction with instrument transformers

Power measurements are made in high voltage circuits connecting the wattmeter to the circuit through current and potential transformers as shown in Fig. 3.81:

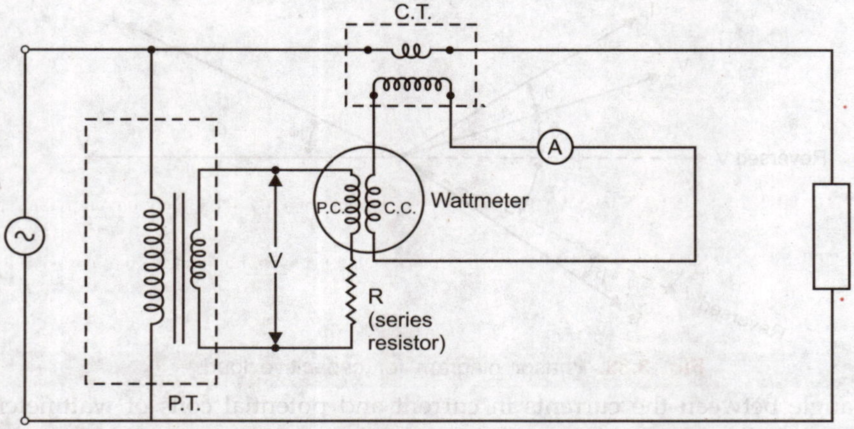

Fig. 3.81. Measurement of power with wattmeter in conjunction with instrument transformers in single-phase A.C. circuits.

— The primary winding of C.T. is connected in series with the load and the secondary winding is connected in series with an ammeter and the current coil (C.C.) of a wattmeter.

— The primary winding of the potential transformer (P.T.) is connected across the supply lines and a voltmeter and the pressure coil (P.C.) of the wattmeter are connected in parallel with the secondary winding of the transformer. One secondary terminal of each transformer and the casings are earthed.

— When the wattmeter is used in conjunction with instrument transformer, then *correction should be applied for ratio and phase angle errors of instrument transformers.*

Let, V, I and $\cos \phi$ be the load voltage, current and power factor respectively.
Further, let, current in C.C. of wattmeter = Current in secondary of C.T. = I_s
Voltage across pressure coil of wattmeter = Voltage across secondary of P.T. = V_s
Current in pressure coil of wattmeter = I_p lagging behind V_s by a small angle θ due to inductance of pressure coil.

Phase angle of P.T. = δ
Phase angle of C.T. = β

From phasor diagram shown in Figs. 3.82 and 3.83 we have :

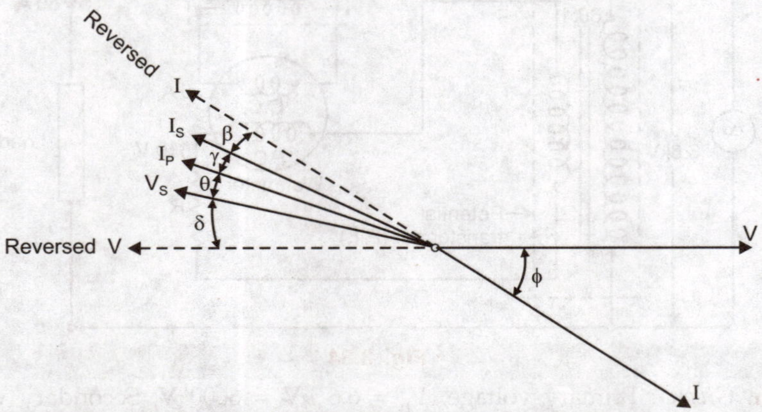

Fig. 3.82. Phase diagram for inductive load.

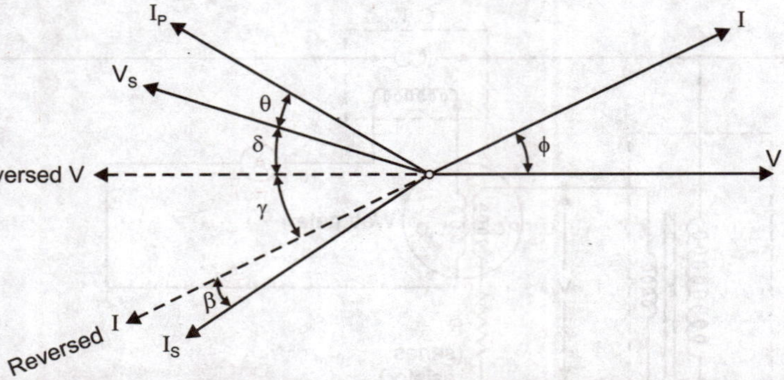

Fig. 3.83. Phasor diagram for capacitive load.

Phase angle between the currents in current and potential coils of wattmeter,

$$\gamma = \phi - \theta - \beta \pm \delta \quad \text{... for } \textit{inductive} \text{ loads} \quad ...(3.151)$$

and,

$$\gamma = \phi + \theta + \beta \pm \delta \quad \text{... for } \textit{capacitive} \text{ loads} \quad ...(3.152)$$

Since phase angles of P.T. may be lagging or leading.

Neglecting transformation ratio errors of instrument transformers,

$$\text{Correction factor, } K = \frac{\cos\phi}{\cos\theta\cos(\phi - \theta - \beta \pm \delta)} \quad \text{...for } \textit{inductive} \text{ loads} \quad ...(3.153)$$

and,

$$K = \frac{\cos\phi}{\cos\theta\cos(\phi + \theta + \beta \pm \delta)} \quad \text{...for } \textit{capacitive} \text{ loads} \quad ...(3.154)$$

True power = K × Actual ratio of C.T. × actual ratio of P.T. × wattmeter reading ...(3.155)

Example 3.88. *Power of a single phase 6.6 kV load drawing a current of about 50 A is required to be measured by means of a wattmeter having volt-terminals marked as 110 V and current terminals as 5 A. Draw the circuit diagram showing the connection of the wattmeter using the instrument transformers. Write suitable transformation ratio against each instrument transformer shown in the diagram.* **(UPTU)**

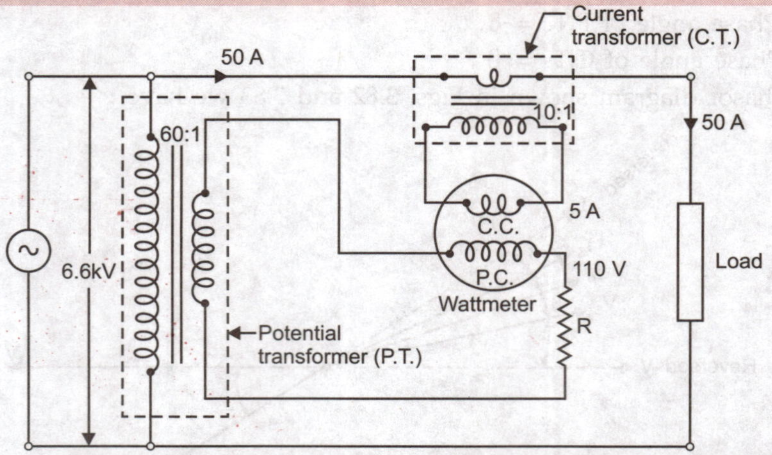

Fig. 3.84

Solution. *Given* : Primary voltage, V_p = 6.6 kV = 6600 V; Secondary voltage, V_s = 110 V; Primary current, I_p = 50 A; I_s = 5A.

$$\text{Transformation ratio of P.T.} = \frac{V_p}{V_s} = \frac{6600}{110} = \mathbf{60} \text{ (Ans.)}$$

$$\text{Transformation ratio of C.T.} = \frac{I_p}{I_s} = \frac{50}{5} = \mathbf{10} \text{ (Ans.)}$$

Fig. 3.84 shows the circuit diagram indicating the connection of the wattmeter using the instrument transformers. The transformation ratio against each instrument is also shown in the figure.

Example 3.89. *A 100 V, 5 A range wattmeter is used in conjunction with instrument transformers to measure power consumed by a 6 kV single-phase load taking 100 A at 0.5 p.f. (lagging). The potential and current transformers used have nominal ratios of 60 : 1 and 20 : 1, ratio errors of 1% and –0.5% and phase angle errors of –1° and +2° respectively. Assuming that there is no error in the wattmeter, find out the error in the measurement of power in the circuit due to the errors of instrument transformers.*

Solution. *Given:* V = 6 kV = 6000 volts; I = 100 A; $\cos \phi$ = 0.5 (lagging) Phase angle of P.T., $\delta = -1°$; Phase angle of C.T. = 2°.

Percentage error in measurement :

$$\text{Actual power consumed} = VI \cos \phi$$
$$= 6000 \times 100 \times 0.5 = 3{,}00{,}000 \text{ W} = 300 \text{ kW}$$

Since, $\cos \phi = 0.5$ (lagging),

$\therefore$ $\phi = \cos^{-1}(0.6) = 60°$

Also, $\theta = 0°$ since wattmeter reads correctly

Now, phase angle between the currents in C.C. and P.C. of wattmeter,

$$\gamma = \phi - \theta - \beta + \delta \quad \text{...[Eqn. (3.151)]}$$
$$= 60° - 0° - 2° - 1 = 57°$$

$$\text{Correction factor, } K = \frac{\cos \phi}{\cos \theta \ \cos \gamma} \quad \text{...[Eqn. (3.153)]}$$

$$= \frac{\cos 60°}{\cos 0° \ \cos 57°} = 0.918$$

Actual transformation ratio of C.T. = 20(1 – 0.005) = 19.9

Actual transformation ratio of P.T. = 60(1 + 0.01) = 60.6

$$\text{Power indicated by wattmeter} = \frac{\text{Actual power consumed}}{K \times \text{actual ratio of P.T.} \times \text{Actual ratio of C.T.}}$$

$$= \frac{300}{0.918 \times 60.6 \times 19.9} = 0.27 \text{ kW}$$

$$\text{True reading of the wattmeter} = \frac{\text{Actual power consumed}}{\text{Nominal ratio of P.T.} \times \text{Nominal ratio of C.T.}}$$

$$= \frac{300}{60 \times 20} = 0.25 \text{ kW}$$

$$\therefore \quad \text{Percentage error} = \frac{\text{Power indicated} - \text{true reading}}{\text{True reading}} \times 100$$

$$= \frac{0.27 - 0.25}{0.25} \times 100 = \mathbf{8\%} \ (\textit{high}) \ \text{(Ans.)}$$

3.9.8. Measurement of Power in 3-Phase Circuit

- The power in 3-phase load can be measured by using the following methods :
  1. Three-wattmeter method.
  2. Two-wattmeter method.
  3. One-wattmeter method.

3.9.8.1. Three-wattmeter method

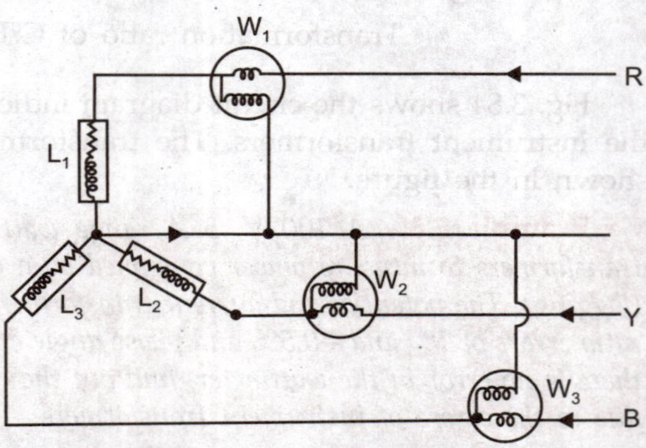

Fig. 3.85. Star-connected load.

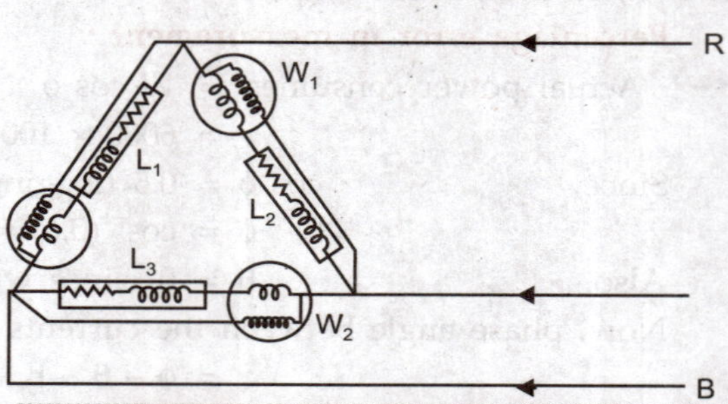

Fig. 3.86. Delta-connected load.

Figs. 3.85 and 3.86 show the connection diagram for star-connected and delta-connected loads respectively. As indicated in the figures, three wattmeters are connected in each of the three phases of the load whether star or delta connected. The current coil of each wattmeter carries the current of one phase only and the pressure coil measures the phase-voltage of the phase. Hence, *each wattmeter measures the power in a single phase. The total power in the load is given by the algebraic sum of the readings of the three wattmeters.*

While using this method, following *difficulty* is met with :

- In case of star-connected load it is not always possible to get a neutral point which is required for connections (Fig. 3.85).
- In case of delta-connected load, under ordinary conditions it is not generally feasible to break into the phases of the load.

To measure power it is not necessary to use three wattmeters, two wattmeters can be used for the purpose as explained in the Article 3.9.8.2.

3.9.8.2. Two-wattmeter method – balanced load

The total power consumed by a *balanced load* can be found by using two wattmeters (Figs. 3.87 and 3.88). When load is assumed inductive in Fig. 3.87 the vector diagram for such a balanced star-connected load is shown in Fig. 3.89.

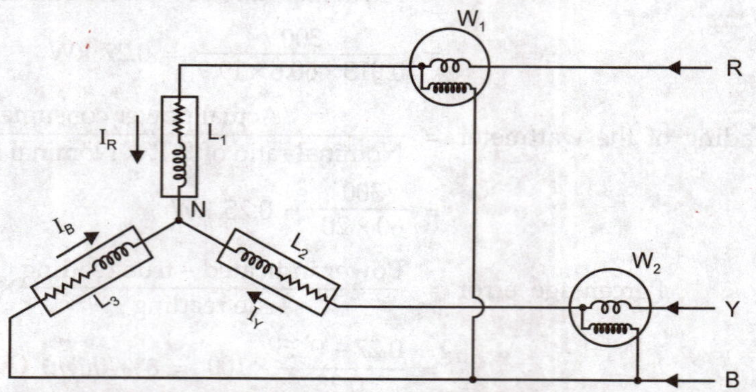

Fig. 3.87. Star-connected load

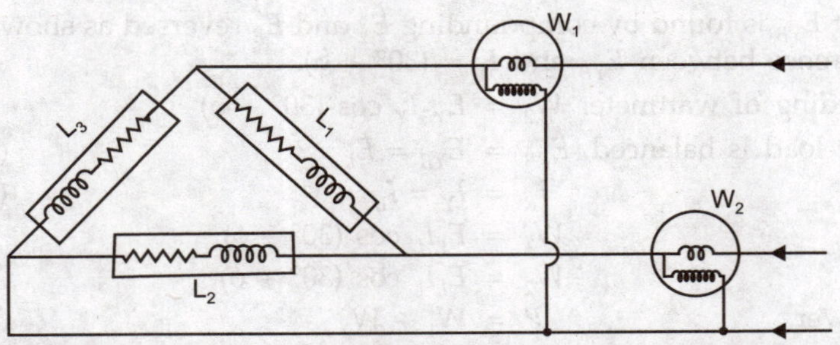

Fig. 3.88. Delta-connected load.

Let us consider the problem in terms of r.m.s. values (instead of instantaneous values).

Let, E_R, E_Y, E_B = r.m.s. values of the three phase voltages, and

I_R, I_Y, I_B = r.m.s. values of the currents.

Since these voltages and currents are assumed sinusoidal, they can be represented by vectors, the currents lagging behind their respective phase voltages by ϕ.

Refer to Fig. 3.89.

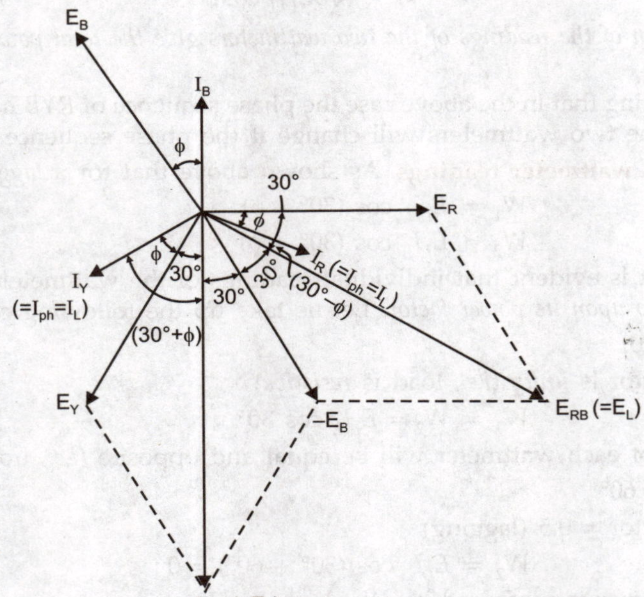

Fig. 3.89. Vector diagram – two-wattmeter method.

Current through wattmeter, $W_1 = I_R$

Potential difference across pressure coil of wattmeter W_1

$$= E_{RB} = E_R - E_B \qquad \text{(Vectorially)}$$

The value of E_{RB} is found by compounding E_R and E_B reversed as shown in Fig. 3.89. It may be observed that phase difference between E_{RB} and $I_R = (30° - \phi)$.

$\therefore$ Reading of wattmeter $W_1 = E_{RB} I_R \cos(30° - \phi)$ ...(i)

Similarly, current through wattmeter $W_2 = I_Y$

Potential difference across pressure coil of wattmeter W_2

$$= E_{YB} = E_Y - E_B \qquad \text{(Vectorially)}$$

The value of E_{YB} is found by compounding E_Y and E_B reversed as shown in Fig. 3.89. The phase difference between E_{YB} and $I_Y = (30° + \phi)$.

$\therefore$ Reading of wattmeter $W_2 = E_{YB} I_Y \cos (30° + \phi)$ ...(*ii*)

Since the load is balanced, $E_{RB} = E_{YB} = E_L$ (Line voltage)

and, $I_R = I_Y = I_L$ (Line current)

$$\therefore \quad W_1 = E_L I_L \cos (30° - \phi)$$

$$\text{and,} \quad W_2 = E_L I_L \cos (30° + \phi)$$

$\therefore$ Total power,

$$P = W_1 + W_2$$
$$= E_L I_L \cos (30° - \phi) + E_L I_L \cos (30° + \phi)$$
$$= E_L I_L [\cos (30° - \phi) + \cos (30° + \phi)]$$
$$= E_L I_L [\cos 30° \cos \phi + \sin 30° \sin \phi + \cos 30° \cos \phi - \sin 30° \sin \phi]$$
$$= E_L I_L (2 \cos 30° \cos \phi) = E_L I_L \times 2 \times \frac{\sqrt{3}}{2} \cos \phi$$
$$= \sqrt{3} E_L I_L \cos \phi$$

i.e.,

$$P = \sqrt{3} E_L I_L \cos \phi \quad \text{...(3.156)}$$

Hence the *sum of the readings of the two wattmeters give the total power consumption in the 3-phase load.*

It is worth noting that in the above case the phase sequence of *RYB* has been assumed, the readings of the two wattmeters will change if the phase sequence is reversed.

Variations in wattmeter readings. As shown above that for a *lagging power factor*

$$W_1 = E_L I_L \cos (30° - \phi)$$

$$\text{and,} \quad W_2 = E_L I_L \cos (30° + \phi)$$

From above it is evident that individual readings of the wattmeters *not only depend on the load but also upon its power factor.* Let us take up the following cases :

(*i*) When $\phi = 0$

i.e., power factor is *unity* (*i.e.,* load is *resistive*)

Then, $W_1 = W_2 = E_L I_L \cos 30°$

The reading of each wattmeter will be equal and opposite (*i.e.,* up-scale reading).

(*ii*) When $\phi = 60°$

i.e., power factor = 0.5 (lagging)

Then, $W_2 = E_L I_L \cos (30° + 60°) = 0$

Hence, the power is measured by W_1 above.

(*iii*) When $90° > \phi > 60°$

i.e., $0.5 > \text{p.f.} > 0$

Then W_1 is still *positive* but reading of W_2 is *reversed*. For a leading p.f., conditions are just the opposite of this. In that case, W_1 will read *negative* because the phase angle between the current and voltage is *more than* 90°. For getting the total power, the reading of W_2 is to be *subtracted* from that of W_1. Under this condition, W_2 will read *'down scale' i.e., backwards. Hence, to obtain a reading on W_2, it is necessary to reverse either its pressure coil or current coil, usually the former.*

All readings taken after reversal of pressure coil are to be taken as negative.

(*iv*) When $\phi = 90°$

(*i.e.,*p.f. = 0 *i.e.,* pure inductive or capacitive load)

Then, $W_1 = E_L I_L \cos(30° - 90°) = E_L I_L \sin 30°$

and, $W_2 = E_L I_L \cos(30° + 90°) = -E_L I_L \sin 30°$

These two readings are equal in magnitude but opposite in sign

$\therefore \quad W_1 + W_2 = 0$

So far we have considered *lagging angles* (taken as *positive*). Now let us discuss how the readings of wattmeters change when the power factor is *leading* one.

- For $\phi = +60°$ (lag) : $W_2 = 0$
- For $\phi = -60°$ (lead) : $W_1 = 0$

Thus we find that for angles of lead the readings of the two wattmeters are interchanged. Hence, when the power is *leading* :

$$W_1 = E_L I_L \cos(30° + \phi)$$

$$W_2 = E_L I_L \cos(30° - \phi).$$

Power factor – When the load is 'balanced'. When load is balanced with a *lagging power factor* and the voltage and currents are sinusoidal :

$$W_1 + W_2 = E_L I_L \cos(30° - \phi) + E_L I_L \cos(30° + \phi) = \sqrt{3} E_L I_L \cos\phi \qquad ...(i)$$

Similarly, $W_1 - W_2 = E_L I_L \cos(30° - \phi) - E_L I_L \cos(30° + \phi)$

$= E_L I_L \cos(2 \sin\phi \sin 30°) = E_L I_L \sin\phi \qquad ...(ii)$

Dividing (*ii*) by (*i*), we get

$$\tan\phi = \frac{\sqrt{3}(W_1 - W_2)}{(W_1 + W_2)} \qquad ...(3.157)$$

For a *leading power*, this expression becomes

$$\tan\phi = -\frac{\sqrt{3}(W_1 - W_2)}{(W_1 + W_2)} \qquad ...(3.158)$$

After finding tan ϕ, hence ϕ, the value of power factor cos ϕ can be found (from trigonometrical tables).

One important point which must be kept in mind is that if W_2 reading has been taken after reversing the pressure coil *i.e.*, if W_2 is *negative*, then the eqn., (3.158) becomes

$$\tan\phi = \sqrt{3}\frac{W_1 - (-W_2)}{W_1 + (-W_2)}$$

or, $$\tan\phi = \sqrt{3}\frac{W_1 + W_2}{W_1 - W_2} \qquad ...(3.159)$$

The power factor may also be expressed in terms of ratio of the readings of the two wattmeters.

Let, $$\frac{\text{Smaller reading}}{\text{Larger reading}} = \frac{W_2}{W_1} = \alpha$$

Then from eqn. (3.157) above, we have

$$\tan\phi = \frac{\sqrt{3}\left[1 - \left(\frac{W_2}{W_1}\right)\right]}{\left[1 + \left(\frac{W_2}{W_1}\right)\right]} = \frac{\sqrt{3}(1-\alpha)}{(1+\alpha)}$$

We know that, $\sec^2 \phi = 1 + \tan^2 \phi$

or, $\frac{1}{\cos^2 \phi} = 1 + \tan^2 \phi$, or, $\cos^2 \phi = \frac{1}{1+\tan^2 \phi}$

$$\cos \phi = \frac{1}{\sqrt{1+\tan^2 \phi}} = \frac{1}{\sqrt{1+\left[\frac{\sqrt{3}(1-\alpha)}{1+\alpha}\right]^2}}$$

$$= \frac{1}{\sqrt{1+3\left(\frac{1-\alpha}{1+\alpha}\right)^2}} = \frac{1+\alpha}{\sqrt{(1+\alpha)^2+3(1-\alpha)^2}}$$

$$= \frac{1+\alpha}{\sqrt{1+\alpha^2+2\alpha+3(1+\alpha^2-2\alpha)}} = \frac{1+\alpha}{\sqrt{4+4\alpha^2-4\alpha}}$$

$$= \frac{1+\alpha}{2\sqrt{1-\alpha+\alpha^2}}$$

i.e., $$\cos \phi = \frac{1+\alpha}{2\sqrt{1-\alpha+\alpha^2}} \quad ...(3.160)$$

If a curve is plotted between α and $\cos \phi$, then the curve obtained will be as shown in Fig. 3.90, this curve is called **watt-ratio curve.**

Reactive volt amperes *(with two wattmeters)* **:**

We know that, $\tan \phi = \frac{\sqrt{3}(W_1 - W_2)}{W_1 + W_2}$ [Eqn. (3.157)]

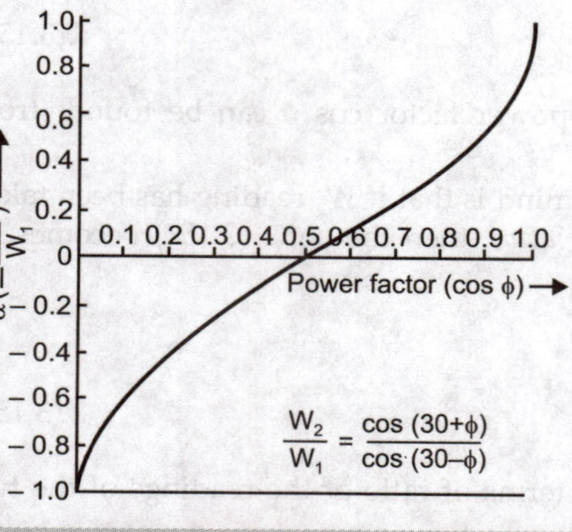

Fig. 3.90. Watt-ratio curve.

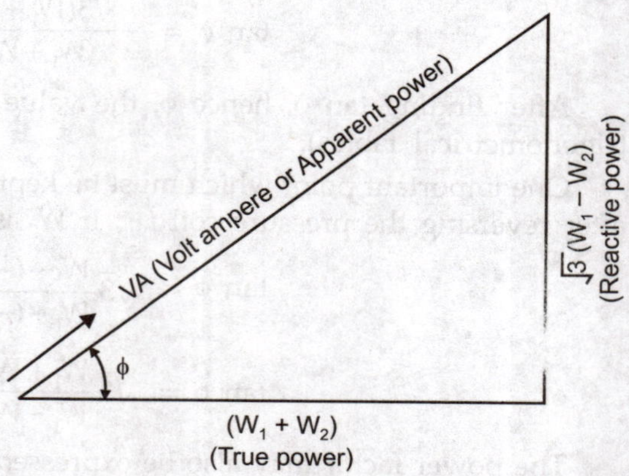

Fig. 3.91

As the tangent of the angle of lag between phase current and phase voltage of a circuit is always equal to the ratio of reactive power to the true power (Fig. 3.91) hence, in case of a balanced load, the reactive power is given by $\sqrt{3}$ *times the difference of the readings of the two wattmeters* used to measure the power of a 3-phase circuit by two-wattmeter method.

Mathematical proof is as follows :

$$\sqrt{3}\,(W_1 - W_2) = \sqrt{3}\,[(E_L I_L \cos (30° - \phi) - E_L I_L \cos (30° + \phi)]$$

$$= \sqrt{3}\, E_L I_L\ [(\cos 30° \cos \phi + \sin 30° \sin \phi) - (\cos 30° \cos \phi - \sin 30° \sin \phi)]$$

$$= \sqrt{3}\, E_L I_L\ [\cos 30° \cos \phi + \sin 30° \sin \phi - \cos 30° \cos \phi + \sin 30° \sin \phi]$$

$$= \sqrt{3}\, E_L I_L\ (2 \sin 30° \sin \phi) = \sqrt{3}\, E_L I_L \sin \phi$$

i.e., $$W_r = \sqrt{3}\,(W_1 - W_2) = \sqrt{3}\, E_L I_L \sin \phi. \quad ...(3.161)$$

3.9.8.3. Blondel's theorem

According to Blondel's theorem (for measurement of power in a *poly-phase system*) :

"If a network is supplied through N conductors, the total power is measured by summing the readings of N wattmeters so arranged that a current element of a wattmeter is in each line and the corresponding voltage element is connected between that line and a common point. If the common point is located on one of the lines, then the power may be measured by N-1 wattmeters."

- The most common applications of a general *theorem* known as ***"Blondel's theorem"*** *include the two-wattmeter method of measurement of power in 3-phase, 3-wire load circuits and three-wattmeter method of measurement of power in 3-phase, 4-wire load circuits.*

— The total power supplied in a N-wire circuit is given by the sum of the readings of (N – 1) wattmeters, so arranged that *a current coil of a wattmeter in each wire and the corresponding potential coil is connected between the wire and a common point on the system.*

— When the common point is located on one of the wires, the potential difference across the potential coil circuit of the wattmeter whose current coil is in that wire is *zero*, and the wattmeter has a zero reading. *Thus only (N – 1) wattmeters are required e.g., two wattmeters for a 3-wire circuit and three wattmeters for a 4-wire circuit.*

3.9.8.4. One-wattmeter method

In this method, the current coil is connected in any one line and the pressure coil is connected alternately between this and the other two lines as shown in Fig. 3.90. The two readings so obtained, for a balanced load, correspond to those obtained by normal two-wattmeter method.

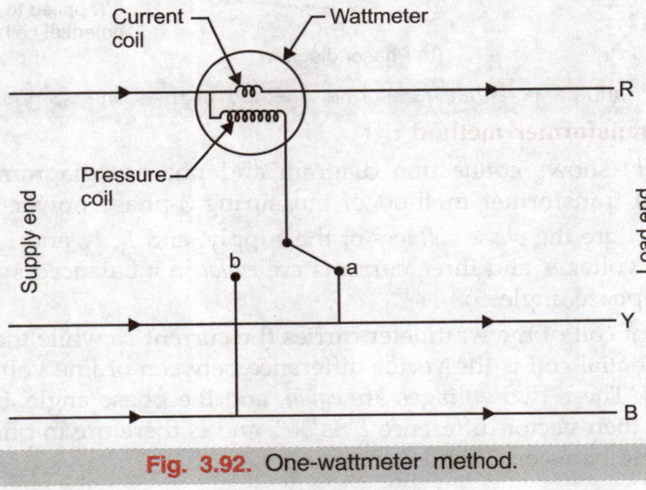

Fig. 3.92. One-wattmeter method.

This method is not of as much universal application as the two-wattmeter method because it is restricted to *fairly balanced loads only*. However, it may be conveniently applied, for instance, when it is desired to find the power input to a factory motor *in order to check the load upon the motor*.

3.9.8.5. "Barlow's method" for a balanced 3-phase circuit using instrument transformers with one wattmeter

Barlow's method is used for measuring the power in a balanced 3-phase circuit, with the *load delta-connected, which does not involve the provision of an artificial neutral point and in which the wattmeter gives the total power directly without any multiplying factor.*

There are two alternative methods of connection, one using a P.T. of 1 : 1 ratio, and the other using two currents transformers each of ratio 1 : 1.

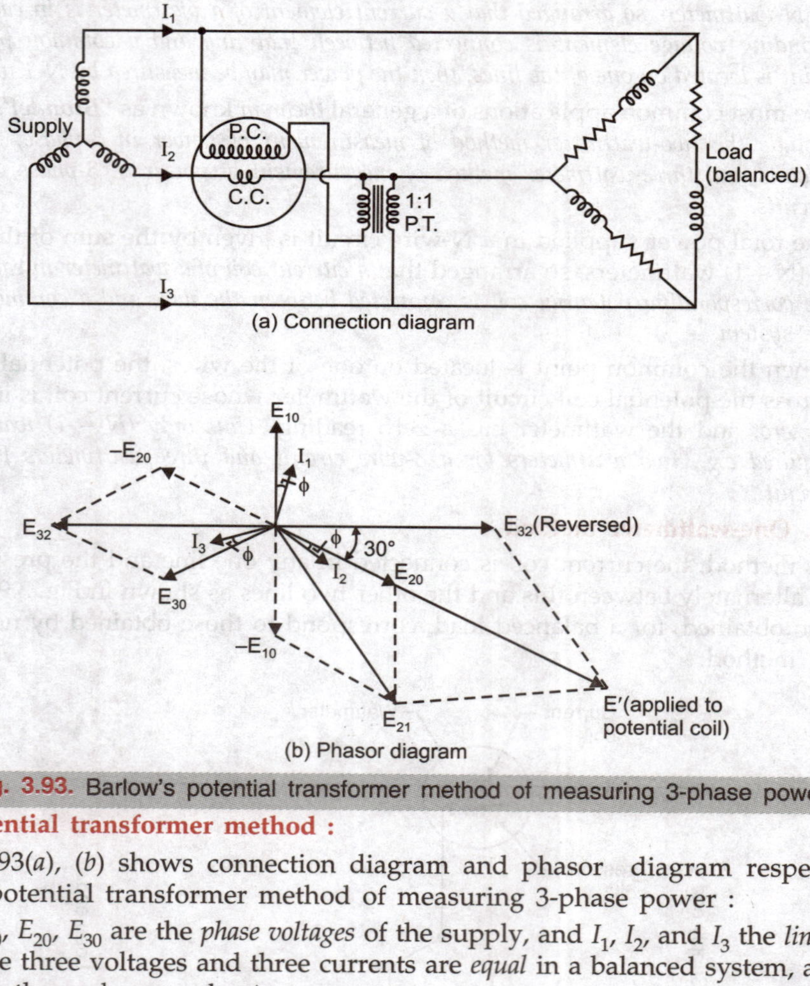

Fig. 3.93. Barlow's potential transformer method of measuring 3-phase power.

1. Potential transformer method :

Fig. 3.93(*a*), (*b*) shows connection diagram and phasor diagram respectively of Barlow's potential transformer method of measuring 3-phase power :

— E_{10}, E_{20}, E_{30} are the *phase voltages* of the supply, and I_1, I_2, and I_3 the *line currents.* The three voltages and three currents are *equal* in a balanced system, as are also the three phase angles ϕ.

— The current coil of the wattmeter carries the current I_2, while the voltage applied to the potential coil is the vector difference between of line voltages E_{21}, and E_{32} (reversed). These two voltages are *equal*, and the phase angle, between either of them and their vector difference E' is 30°, and is therefore in phase with E_{20}. The phase angle between E' and I_2 is thus ϕ.

— The wattmeter measures $E'I_2 \cos \phi$.

Now, if $E_{10} = E_{20} = E_{30} = E$, then,

$$E_{21} = E_{32} = \sqrt{3}E \text{ and, thus}$$

$$E' = \sqrt{3}E_{21} = \sqrt{3}\times\sqrt{3}E = 3E$$

The power measured by the wattmeter is 3EI cos φ (since $I_1 + I_2 + I_3 = I$), which is the mean value of the total power in the balanced circuit. The ratio error and phase angle of the P.T. are, of course, *neglected* in the above reasoning.

2. Current transformer method :

The connection diagram and phasor diagram of current transformer method of measuring 3-phase power are shown respectively in Fig. 3.94(*a*), (*b*) :

— Since the load is balanced,

$$I_1 = I_2 = I_3 = I$$

and, $E_{10} = E_{20} = E_{30} = E$

and, the phase angles ϕ are *all equal*.

— The vector E', which is the vector difference of the voltages E_{10} and E_{20}, is the voltage applied to the potential coil and equals $\sqrt{3}E$.

— The currents in the secondary circuits of current transformers 1 and 2 are I_1 reversed and I_2 reversed. The connection to current coil of the wattmeter are such that the current in it is the vector sum of I_1 and I_2 reversed and is represented by the vector I'. The value of this current is $\sqrt{3}I$, and the phase angle between E' and I' is ϕ. Then,

$$\text{Power measured by wattmeter} = E'I' \cos \phi$$

$$= \sqrt{3}E \times \sqrt{3}I \cos \phi = 3EI \cos \phi$$

$$= \text{Total power in load.}$$

Again, the current-transformer ratio and phase-angle errors are *neglected.*

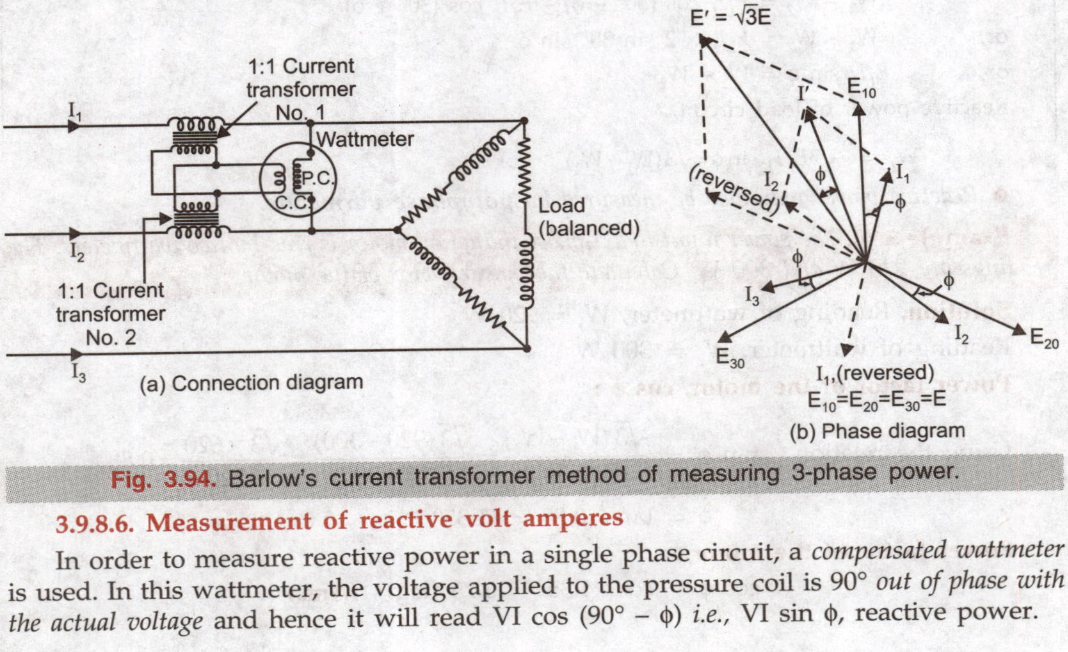

Fig. 3.94. Barlow's current transformer method of measuring 3-phase power.

3.9.8.6. Measurement of reactive volt amperes

In order to measure reactive power in a single phase circuit, a *compensated wattmeter* is used. In this wattmeter, the voltage applied to the pressure coil is 90° *out of phase with the actual voltage* and hence it will read VI cos (90° – φ) *i.e.*, VI sin φ, reactive power.

In balanced three-phase circuit the reactive power can be determined by using one wattmeter. The necessary connections are shown in Fig. 3.95. The current coil is inserted in one line and the pressure coil is connected across the other two lines.

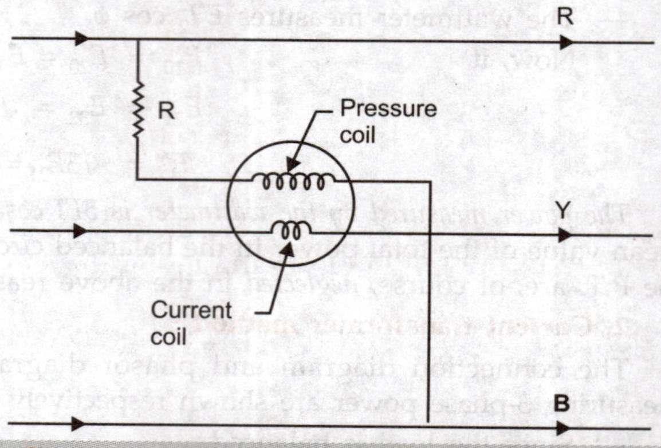

Fig. 3.95. Measurement of reactive volt-amperes.

The current following through the current coil of the wattmeter.

$$= I_Y = I_{ph} \text{ (say)}$$

The potential difference across the potential coil of wattmeter,

$$E = E_R - E_B$$

$= \sqrt{3}E_{ph}$ leading vector E_Y by 90°

$= \sqrt{3}E_{ph}$ leading the vector of current I_Y by (90° + ϕ)

∴ Reading of the wattmeter (Fig. 3.96)

$$= \sqrt{3}E_{ph}I_{ph}\cos(90° + \phi)$$

$$= -\sqrt{3}E_{ph}I_{ph}\sin\phi = -W_r$$

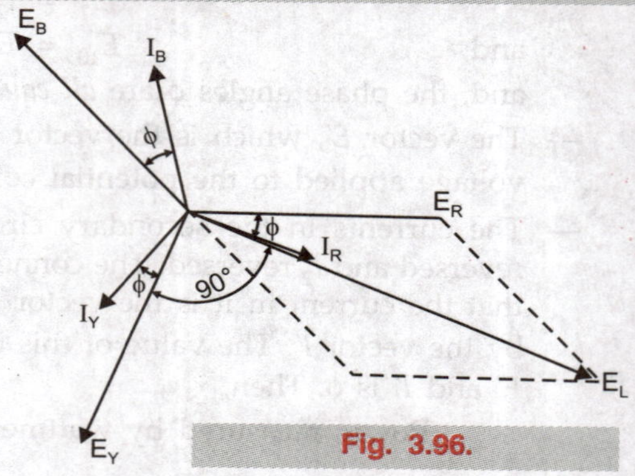

Fig. 3.96.

Total reactive power of the circuit,

$$= 3E_{ph}I_{ph}\sin\phi = -\sqrt{3}\,W_r$$

[Reactive power (as earlier stated) can also be determined from two wattmeter readings connected for measurement of power.

$$W_1 - W_2 = E_LI_L \cos(30° - \phi) - E_LI_L \cos(30° + \phi)$$

or, $$W_1 - W_2 = E_LI_L \times 2 \sin 30° \sin\phi$$

or, $$E_LI_L \sin\phi = W_1 - W_2$$

Reactive power of load circuit,

$$W_r = \sqrt{3}E_LI_L\sin\phi = \sqrt{3}(W_1 - W_2).]$$

- *Reactive power may also be measured by* ***polyphase varmeters.***

Example 3.90. *The power input to a 3-phase induction motor is read by two wattmeters. The readings are 920 W and 300 W. Calculate the power factor of the motor.*

Solution. Reading of wattmeter, $W_1 = 920$ W

Reading of wattmeter, $W_2 = 300$ W

Power factor of the motor, cos ϕ :

Using the relation, $$\tan\phi = \frac{\sqrt{3}(W_1 - W_2)}{(W_1 + W_2)} = \frac{\sqrt{3}(920 - 300)}{(920 + 300)} = \frac{\sqrt{3} \times 620}{1220} = 0.88$$

∴ $$\phi = \tan^{-1} 0.88 = 41.35°$$

Power factor of the motor,

$$\cos\phi = \cos 41.35° = \mathbf{0.75 \text{ (lag)}}$$ **(Ans.)**

Example 3.91. *While performing a load test on a 3-phase wound-rotor induction motor by two-wattmeter method, the readings obtained on two wattmeters were + 14.2 kW and – 6.1 kW and the line voltage was 440 V. Calculate :*

(i) True power drawn by the motor, (ii) Power factor, and (iii) Line current.

Solution. Reading of wattmeter, $W_1 = 14.2$ kW

Reading of wattmeter, $W_2 = -6.1$ kW

Line voltage, $E_L = 440$ V

(*i*) **True power drawn by the motor**

$$= 14.2 - 6.1 = \mathbf{8.1\ kW} \quad \textbf{(Ans.)}$$

(*ii*)
$$\tan \phi = \frac{\sqrt{3}\,(W_1 - W_2)}{(W_1 + W_2)} = \frac{\sqrt{3}\,[14.2-(-6.1)]}{[14.2+(-6.1)]} = \frac{\sqrt{3}\times 20.3}{8.1} = 4.34$$

$$\therefore \quad \phi = \tan^{-1} 4.34 = 77°$$

∴ **Power factor,** $\cos \phi = \cos 77° = \mathbf{0.2249\ (lag).}$ **(Ans.)**

(*iii*) **Line current, I_L :**

Using the relation, $P = \sqrt{3}\, E_L I_L \cos\phi$

$$8.1 \times 1000 = \sqrt{3} \times 440 \times I_L \times 0.2249$$

$$\therefore \quad I_L = \frac{8.1\times 1000}{\sqrt{3}\times 440\times 0.2249} = 47.26\ \text{A}$$

Hence, *line current* = **47.26 A (Ans.)**

Example 3.92. *A 3-phase, 440 V motor load has a power factor of 0.6. Two wattmeters connected to measure the power show the input to be 25 kW. Find the reading on each instrument.*

Solution. Input power, = 25 kW

Line voltage, $E_L = 440$ V

Power factor of the motor load, $\cos \phi = 0.6$

W_1 ; W_2 :

Using the relation : $\tan \phi = \dfrac{\sqrt{3}\,(W_1 - W_2)}{W_1 + W_2}$ ...(*i*)

Now, $\cos \phi = 0.6, \quad \therefore \phi = 53.13°$

$\therefore \quad \tan \phi = \tan 53.13° = 1.333$

Also $W_1 + W_2 = 25$ kW (*Given*) ...(*ii*)

Substituting these values in (*i*), we get

$$1.333 = \frac{\sqrt{3}\,(W_1 - W_2)}{25}$$

$$\therefore \quad W_1 - W_2 = \frac{1.333 \times 25}{\sqrt{3}} = 19.24\ \text{kW} \quad \text{...(iii)}$$

From (*ii*) and (*iii*), we get

$\mathbf{W_1 = 22.12\ kW;}$

and $\mathbf{W_2 = 2.88\ kW}$ **(Ans.)**

Example 3.93. *In a 3-phase circuit, two wattmeters used to measure power indicate 1200 W and 600 W respectively. Find the power factor of the circuit :*

(*i*) *When both wattmeter readings are positive.*

(*ii*) *When the latter is obtained by reversing the current coil connections.*

Solution. (*i*) *When both wattmeter readings are positive :*

Reading of wattmeter, W_1 = 1200 W

Reading of wattmeter, W_2 = 600 W

We know that,
$$\tan\phi = \frac{\sqrt{3}\,(W_1 - W_2)}{W_1 + W_2} = \frac{\sqrt{3}(1200 - 600)}{(1200 + 600)} = 0.577$$
$$\phi = \tan^{-1} 0.577 = 30°$$
$$\textbf{Power factor} = \cos\phi = \cos 30° = \mathbf{0.866\ (lag).}\ \textbf{(Ans.)}$$

(*ii*) *When the reading of wattmeter* W_2 *is obtained by reversing the coil connection :*

Reading of wattmeter, W_1 = 1200 W

Reading of wattmeter, W_2 = −600 W

We know that,
$$\tan\phi = \frac{\sqrt{3}\,(W_1 - W_2)}{W_1 + W_2} = \frac{\sqrt{3}(1200 - (-600))}{(1200 + (-600))} = \frac{\sqrt{3} \times 1800}{600}$$
$$= 5.196$$

or
$$\phi = \tan^{-1} 5.196 = 79.1°$$

Hence,
$$\textbf{power factor} = \cos\phi = \cos 79.1° = \mathbf{0.1889.}\ \textbf{(Ans.)}$$

Example 3.94. *In order to measure the power input and the power factor of an over-excited synchronous motor two wattmeters are used. If the meters indicate (– 3.5 kW) and (+ 8.0 kW) respectively, calculate :*

(*i*) *Power factor of the motor;* (*ii*) *Power input to the motor.*

Solution. (*i*) *Since an over-excited synchronous motor runs with a leading power factor,* we should use the relation,
$$\tan\phi = -\frac{\sqrt{3}(W_1 - W_2)}{(W_1 + W_2)}$$

Moreover it is W_1 that gives negative reading and not W_2. Hence W_1 = −3.5 kW.

$$\therefore \quad \tan\phi = -\frac{\sqrt{3}(-3.5 - 8)}{(-3.5 + 8)} = \frac{\sqrt{3} \times 11.5}{5.5} = 3.62$$
$$\therefore \quad \phi = \tan^{-1} 3.62 = 74.6° \text{ (lead)}$$
$$\therefore \quad \textbf{Power factor} = \cos\phi = \cos 74.6° = \mathbf{0.2655\ (lead).}\ \textbf{(Ans.)}$$

(*ii*)
$$\textbf{Power input} = W_1 + W_2 = -3.5 + 8 = \mathbf{4.5\ W}\ \textbf{(Ans.)}$$

Example 3.95. *Two wattmeters are used to measure power input to a 1.5 kV, 50 Hz, 3-phase motor running on full-load at an efficiency of 85 per cent. Their readings are 250 kW and 80 kW respectively. Calculate :*

(*i*) *Input,* (*ii*) *Power factor,*

(*iii*) *Line current, and* (*iv*) *Output.*

Solution. *Since the motor is running at full-load, its power factor must be greater than 0.5. Hence,* W_2 *reading is positive.*

$$\therefore \quad W_1 = +250 \text{ kW} \quad \text{and} \quad W_2 = +80 \text{ kW}$$

(*i*) **Input**
$$= W_1 + W_2 = 250 + 80 = \mathbf{330\ kW.}\ \textbf{(Ans.)}$$

(*ii*)
$$\tan\phi = \frac{\sqrt{3}\,(W_1 - W_2)}{W_1 + W_2} = \frac{\sqrt{3}(250 - 80)}{(250 + 80)} = 0.892$$
$$\therefore \quad \phi = \tan^{-1} 0.892 = 41.74°$$

and,
$$\textbf{power factor} = \cos\phi = \cos 41.74° = \mathbf{0.746\ (lag).}\ \textbf{(Ans.)}$$

(*iii*) Power, $P = \sqrt{3}\, E_L I_L \cos\phi$

$$\therefore \quad I_L = \frac{P}{\sqrt{3}\, E_L \cos\phi} = \frac{330\times1000}{\sqrt{3}\times1.5\times1000\times0.746} = 170.27$$

Hence, **line current = 170.27 A. (Ans.)**

(*iv*) **Output** = Input × efficiency = 330 × 0.85 = **280.5 kW. (Ans.)**

Example 3.96. *Two wattmeters are used to measure power input to a synchronous motor. Each of them indicates 60 kW. If the power factor be changed to 0.866 leading, determine the reading of the two wattmeters, the total input power remaining the same.*

Solution. Reading of wattmeter, $W_1 = 60$ kW

Reading of wattmeter, $W_2 = 60$ kW.

First case. In the first case, *both wattmeters read equal and positive. Hence, motor must be running at unity power factor.*

Second case. *Power factor is 0.866 leading.*

In this case : $W_1 = E_L I_L \cos(30° + \phi)$

$W_2 = E_L I_L \cos(30° - \phi)$

$$\therefore \quad W_1 + W_2 = \sqrt{3}\, E_L I_L \cos\phi$$

$$W_1 - W_2 = -E_L I_L \sin\phi$$

$$\therefore \quad \tan\phi = -\frac{\sqrt{3}(W_1 - W_2)}{(W_1 + W_2)}$$

Since, $\cos\phi = 0.866$ (*Given*)

$$\therefore \quad \phi = \cos^{-1} 0.866 = 30°$$

and, $\tan\phi = \tan 30° = \frac{1}{\sqrt{3}}$

$$\therefore \quad \frac{1}{\sqrt{3}} = -\frac{\sqrt{3}\,(W_1 - W_2)}{120} \qquad [\because W_1 + W_2 = 120 \text{ kW } (\textit{Given})]$$

$$\therefore \quad W_1 - W_2 = -\frac{120}{3} = -40 \qquad \text{...}(i)$$

But $W_1 + W_2 = 120$ ...(*ii*)

From (*i*) and (*ii*), we get **W_1 = 40 kW, W_2 = 80 kW. (Ans.)**

Example 3.97. *Two wattmeters connected to read the total power in a 3-phase system supplying a balanced load read 10.5 kW and –2.5 kW respectively. Calculate the total power and power factor.*

Also, explain the significance of (*i*) *equal wattmeter readings and* (*ii*) *a zero reading on one wattmeter.*

Solution. *Given :* $W_1 = 10.5$ kW; $W_2 = -2.5$ kW.

For two wattmeter method,

total power = $W_1 + W_2$ = 10.5 + (–2.5) = **8.0 kW. (Ans.)**

We know that,
$$\tan\phi = \frac{\sqrt{3}\,(W_1 - W_2)}{W_1 + W_2} = \frac{\sqrt{3}\,[10.5 - (-2.5)]}{[10.5 + (-2.5)]} = 2.8145$$

or, $\phi = 70.44°$

∴ **Power factor,** $\cos \phi = \cos 70.44° = $ **0.335. (Ans.)**

(*i*) For readings of the two wattmeters to be equal,

$$W_1 = W_2$$

or, $$\sqrt{3}E_L I_L \cos(30° - \phi) = \sqrt{3}\, E_L I_L \cos(30° + \phi)$$

or, $$\cos(30° - \phi) = \cos(30° + \phi)$$

∴ $$\phi = 0° \quad \text{or} \quad \cos \phi = 1$$

i.e., for *unity power factor, the reading of two wattmeters are equal.* **(Ans.)**

(*ii*) As reading of wattmeters are

$$W_1 = \sqrt{3}\, E_L I_L \cos(30° - \phi), \text{ and}$$

$$W_2 = \sqrt{3}\, E_L I_L \cos(30° + \phi),$$

for *reading of one of the wattmeters to be zero* ϕ *must be* 60°, *which makes* $W_2 = \sqrt{3}\, E_L I_L \cos 90° = 0$. **(Ans.)**

Example 3.98. *A star connected 3-phase load consists of three similar impedances. When load is connected to a 3-phase 400 V, 50 Hz supply it takes 30A line current at a power factor of 0.8 lagging.*

(*i*) *Calculate the total power taken by the load.*

(*ii*) *Calculate the resistance and reactance of each phase of the load.*

(*iii*) *If power is measured by two-wattmeter method, then calculate the reading of each wattmeter.*

Solution. *Given :* $E_L = 400$ V; $f = 50$ Hz; $I_L = (I_{ph}) = 30$ A, $\cos \phi = 0.8$ lagging.

(*i*) **Total power taken by the load, *P* :**

$$P = \sqrt{3}\, E_L I_L \cos \phi$$

$$= \sqrt{3} \times 400 \times 30 \times 0.8 = 16627 \text{ W} \quad \text{or} \quad \mathbf{16.627\ kW.}$$ **(Ans.)**

(*ii*) **Resistance and reactance of each phase of the load, *R, X* :**

$$E_{ph} = \frac{E_L}{\sqrt{3}} = \frac{400}{\sqrt{3}} = 230.95 \text{ V}$$

∴ Load impedance, $$Z = \frac{E_{ph}}{I_{ph}} = \frac{230.95}{30} = 7.7\ \Omega$$

∴ Load resistance, $R = Z \cos \phi = 7.7 \times 0.8 =$ **6.16 Ω. (Ans.)**

and, load resistance, $X = Z \sin \phi = 7.7 \times 0.6 =$ **4.62 Ω. (Ans.)**

(*iii*) **Reading of each wattmeter, W_1, W_2 :**

Reading of wattmeter 1, $W_1 = E_L I_L \cos(30° + \phi)$

Reading of wattmeter 2, $W_2 = E_L I_L \cos(30° - \phi)$

where, $\phi = \cos^{-1} 0.8 = 36.9°$

∴ $W_1 = 400 \times 30 \cos(30° + 36.9°) = 4708$ W or **4.708 kW. (Ans.)**

and, $W_2 = 400 \times 30 \cos(30° - 36.9°) = 11913$ W or **11.913 kW. (Ans.)**

Example 3.99. *A 3-phase, 3-wire, 415 V system supplies a balanced load of 20 A at a power factor 0.8 lag. The current coil of wattmeter 1 is in phase R and of wattmeter 2 in phase B. Calculate :*

(i) The reading on 1 when its voltage coil is across R and Y ;
(ii) The reading on 2 when its voltage coil is across B and Y ;
(iii) The reading on 1 when its voltage coil is across Y and B.

Solution. *Given* : E_L = 415 V, I_L = 20 A, cos ϕ = 0.8 (lag) or $\phi = \cos^{-1} 0.8 = 36.87°$ (lag)

(*i*) *The reading on 1 when its voltage coil is across R and Y and current coil is in R phase*

$$= E_L I_L \cos (30° - \phi) = 415 \times 20 \times \cos (30° - 36.87°)$$
$$= 8240 \text{ W} \quad \text{or} \quad \mathbf{8.24\ kW} \quad \textbf{(Ans.)}$$

(*ii*) *Reading on wattmeter 2 when its voltage coil is across B and Y and current coil is in phase B*

$$= E_L I_L \cos (30° + \phi) = 415 \times 20 \cos (30° + 36.87°)$$
$$= 3260 \text{ W} \quad \text{or} \quad \mathbf{3.26\ kW} \quad \textbf{(Ans.)}$$

(*iii*) *When the current coil is in one phase* (*phasor R*) *and the voltage coil is across other two phases* (*Y and R*), the wattmeter reads $\dfrac{\text{total } kVAR}{\sqrt{3}}$ and,

$$\text{Total } kVAR = \sqrt{3}\,(W_1 - W_2)$$

$$\therefore \quad \text{Wattmeter reading} = \frac{\text{Total } kVAR}{\sqrt{3}} = \frac{\sqrt{3}\,(W_1 - W_2)}{\sqrt{3}}$$
$$= W_1 - W_2 = 8240 - 3260 = \mathbf{4980\ VAR} \quad \textbf{(Ans.)}$$

Example 3.100. *The ratio of the readings of two wattmeters connected to measure power in a balanced 3–φ, 3 wire load is 5 : 3. The load is known to be inductive with a lagging power factor. Calculate the power factor of the load.* **(Punjab University)**

Solution. Let the ratio of wattmeter reading be *r i.e.*, $\dfrac{W_2}{W_1} = r$

Now,
$$\tan \phi = \sqrt{3}\left(\frac{W_1 - W_2}{W_1 + W_2}\right) = \sqrt{3}\left[\frac{1 - \dfrac{W_2}{W_1}}{1 + \dfrac{W_2}{W_1}}\right] = \sqrt{3}\left(\frac{1-r}{1+r}\right)$$

and, power factor,
$$\cos \phi = \frac{1}{\sec \phi} = \frac{1}{\sqrt{1 + \tan^2 \phi}} = \frac{1}{\sqrt{1 + 3\left(\dfrac{1-r}{1+r}\right)^2}}$$

Substituting $r = \dfrac{W_2}{W_1} = \dfrac{3}{5} = 0.6$ in the above expression, we get

$$\cos \phi = \frac{1}{\sqrt{1 + 3\left(\dfrac{1-0.6}{1+0.6}\right)^2}} = \mathbf{0.918\ (lag)} \quad \textbf{(Ans.)}$$

Example 3.101. *A symmetrical star-connected three-phase load is made of three identical coils having an internal resistance of 12 Ω and inductance of 142 mH. The power taken by this load when connected to a balanced 3-phase, three wire supply of 400 V, 50 Hz is measured by* *using the two-wattmeter method. The wattmeters used are similar and are of type of 400 V, 5 A, UPF, class 1 accuracy.*

(*i*) *Determine expected readings of W_1 and W_2.*

(*ii*) *Compute the total power.*

Solution. Line voltage, $E_L = 400$ V

Phase voltage, $E_{ph} = \frac{400}{\sqrt{3}} = 231$ V

Coil resistance, $R_{ph} = 12\ \Omega$

Coil inductive reactance, $X_{ph} = 2\pi f L_{ph} = 2\pi \times 50 \times 142 \times 10^{-3} = 44.61\ \Omega$

Coil impedance, $Z_{ph} = \sqrt{R_{ph}^2 + X_{ph}^2} = \sqrt{12^2 + 44.61^2} = 46.2\ \Omega$

Coil phase angle, $\phi = \cos^{-1} \frac{R_{ph}}{Z_{ph}} = \cos^{-1}\left(\frac{12}{46.2}\right) = 79.94°$ (lagging)

Phase current, $I_{ph} = \frac{E_{ph}}{Z_{ph}} = \frac{231}{46.2} = 5$ A

(*i*) **Readings of wattmeters W_1, W_2 :**

Readings of wattmeter, $W_1 = E_L I_L \cos(30° - \phi)$

$= 400 \times 5.0 \cos(30° - 74.94°) =$ **1415.7 W** **(Ans.)**

($\because I_L = I_{ph} = 5.0$ A)

Reading of wattmeter, $W_2 = E_L I_L \cos(30° + \phi)$

$= 400 \times 5.0 \times \cos(30° + 74.24°) =$ **– 515.6 W** **(Ans.)**

(*ii*) **Total power :**

Total power $= W_1 + W_2$

$= 1415.7 + (-515.6) =$ **900.1 W** **(Ans.)**

Example 3.102. *Two wattmeters are being used to measure power of a balanced load of 30 A at power factor 0.8 lagging being supplied by a 3-phase, 3-wire, 440 V supply. Calculate :*

(*i*) *Power consumed* (*ii*) *Reading of wattmeter No. 1*

(*iii*) *Reading of wattmeter No. 2.*

Solution. Line voltage, $E_L = 440$ V

Line current $I_L = 30$ A

Power factor, $\cos\phi = 0.8$

Phase angle, $\phi = \cos^{-1} 0.8 = 36.87°$

(*i*) **Power consumed,** $P = \sqrt{3}\, E_L I_L \cos\phi = \sqrt{3} \times 440 \times 30 \times 0.8$

$= 18290$ W or **18.29 kW** **(Ans.)**

(*ii*) **Reading of wattmeter No. 1**

$= E_L I_L \cos(30° - \phi) = 440 \times 30 \cos(30° - 36.87°)$

$= 13105$ W or **13.105 kW** **(Ans.)**

(*iii*) **Reading of wattmeter No. 2**

$= E_L I_L \cos(30° + \phi) = 440 \times 30 \times \cos(30° + 36.87°)$

$= 5185$ W or **5.185 kW (app.)** **(Ans.)**

3.9.8.7. Polyphase meters

*A **polyphase meter** is used to measure power in a polyphase (3-phase) system.*

It works on the *principle that two single-phase wattmeters are enough to measure power in 3-phase system (balanced or unbalanced)*.

It is a double element wattmeter consisting of two separate elements mounted one above the other in one and same case. The moving coils of each unit are rigidly coupled together axially by a rod of insulating material.

The instruments give the total power directly because the total deflecting torque action on the moving system is equal to the algebraic sum of torques produced by the two separate working systems.

— For avoiding mutual interference between the units, these are magnetically shielded from each other by placing a nickel iron plate between them.

— The electrical connections of this type of meter for measurement of power in a 3-phase system are identical with those of two-wattmeter method of measuring power in 3-phase system.

- The "*advantage*" of such an instrument is that the total power in circuit is given by *direct indication* instead of indirectly by addition of readings of the two wattmeters.

Its "*disadvantage*" is that it is *not* as accurate as a single-phase wattmeter.

3.9.9. Measurement of Power at High Frequencies (AF/RF)

Quite often we are required to measure the output power of electronic devices such as amplifiers and oscillators. In such devices the frequencies involved are generally *audio frequencies* (up to 15 kHz appr.) or *radio frequencies* (above 15 kHz). The dynamometer type of wattmeters are not suitable for measurement of power at high frequencies since they are designed to measure power at power frequencies.

Out of a few methods of power measurements at high frequencies, the simplest one is to connect a "*dummy load*" to the output terminals of the device under test. Then either the voltage across the dummy load or the current through the dummy load is measured using instruments like *rectifier or thermocouple types since these instruments can work at high frequencies*. If R is the resistance of dummy load, then power output can be calculated as follows :

Power, $$P = \frac{V^2}{R} \qquad \text{...(3.162)}$$

where, V = Voltage across the dummy load, and

I = Current through the dummy load.

Fig. 3.97. shows the basic circuit for power measurement at "*audio frequencies*".

— The resistance R (non-inductive type) which forms the *dummy load* is connected across the output terminals of the device under test. The value of R should be nearly equal to that of output impedance of the device.

— The A.C. voltmeter (a high resistance rectifier type of instrument) when used with a fixed value of R, its scale can be directly calibrated in terms of power output at audio frequencies. Such A.C.

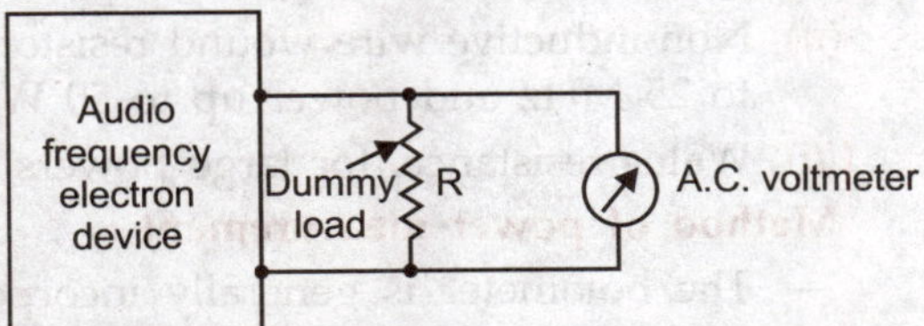

Fig. 3.97. Power measurement at Afrequencies.

voltmeters are provided with *decibel scales*.

Fig. 3.98 shows the basic circuit for power measurement at *radio frequencies* (RF)

— The output tank of amplifier dummy or oscillator (consisting of inductance L_1 and capacitance C) is transmitted to the dummy load consisting of inductance L_4, capacitance C_1 and resistance R in series with a thermocouple ammeter A, by means of an inductive link (L_2, L_3) that couples L_1 and L_2 and L_3 and L_4.

— Using capacitor C_1, the dummy load is tuned to resonance at the resonant frequency of the tank circuit. Thus at resonant frequency the dummy load is only R.

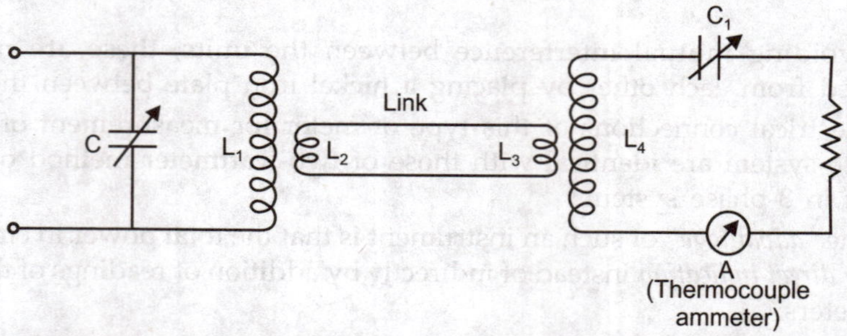

Fig. 3.98. Measurement of power at radio frequencies.

— A is directly calibrated to read the value of power.

— For the measurement of power at high frequencies, *effective shielding must be used.*

3.9.10. Bolometer method of power measurement

At *AF* and *RF*, it is easier to measure voltage, current and impedance, than to measure power. In this range direct measurement of power is not done. The power is calculated from the relation : $P = \dfrac{V^2}{R} = I^2R$ (as stated earlier) .

At *'microwave frequencies', voltage, current and impedance are difficult to determine*. Their values *may vary* at different points in the circuit, and they are affected appreciably by small changes in geometry. Therefore, in this range, *direct measurement of power is more accurate* and the actual load is replaced by a *dummy load.*

The *requirements* of a dummy load are :

(*i*) Low skin effect;
(*ii*) Low reactances;
(*iii*) Ability to dissipate the required amount of power; and
(*iv*) A suitable value equal to that of the actual load.

The dummy load may take any of the following *forms* :

(*i*) Bolometer element for small powers;
(*ii*) Non-inductive wire wound resistors, cooled by air, or water, for frequencies up to 25 MHz and power up to 50 W; and
(*iii*) Water resistance, for large powers up to 50 W etc.

Method of power measurement :

— The bolometer is generally incorporated into a bridge network, so that small changes in resistance can be readily detected and the power measurement performed by substitution of low frequency power.

— In one method of power measurement, the bridge is initially balanced with low frequency (or bias) power. When *RF* power is applied to bolometer, *low frequency power is withdrawn until the bridge is balanced again.*

In any case, whatever bridge technique is used, the *underlying assumption of measurement is that equal amount of low frequency and RF power produce the same heating effect and resistance changes in the bolometer.*

Bolometer :

— *Bolometric measurements are based on the dissipation of the RF power in a small temperature sensitive resistive element, called a "**Bolometer**".*

— The bolometer may be a *short ultra thin wire* having a positive temperature coefficient (PTC) of resistance, called *"Baretter"*, or a bead of semiconductor having a negative temperature coefficient (NTC) called *"Thermistor"*. Both Baretters and Thermistors can measure small powers, of the order of a *fraction of microwatts*. They can also indicate or monitor large amounts of power by inserting a directed compiler.

— The RF power to be measured heats the bolometer and causes a change in its electrical resistance, which serves as an indication of the magnitude of power.

Bolometer bridge :

One of the most commonly used *absorption type meters* is called *"bolometer bridge"*.

It is basically a bridge circuit where one of the arms contains a temperature-sensitive resistor (R_T), as shown in Fig. 3.99.

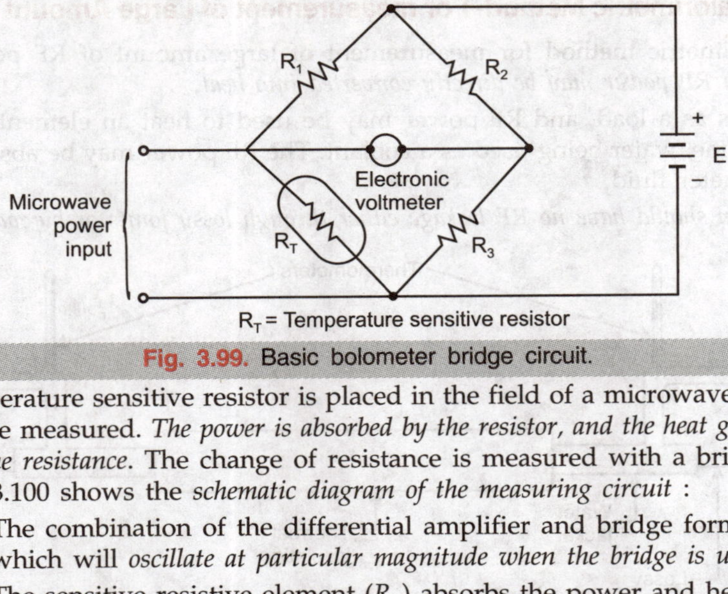

Fig. 3.99. Basic bolometer bridge circuit.

The temperature sensitive resistor is placed in the field of a microwave signal whose power is to be measured. *The power is absorbed by the resistor, and the heat generated causes a change in the resistance.* The change of resistance is measured with a bridge circuit.

Fig. 3.100 shows the *schematic diagram of the measuring circuit* :

— The combination of the differential amplifier and bridge form an oscillator which will *oscillate at particular magnitude when the bridge is unbalanced.*

— The sensitive resistive element (R_T) absorbs the power and heats itself until the bridge is very nearly balanced. The D.C. bias is then adjusted until *exact balance is achieved.*

— Now the element is placed in the microwave field. The element absorbs the power and is heated, which causes a change in its resistance, thereby unbalancing the bridge.

— *The imbalance is in a direction opposite to that of the initial cold resistance. This action causes the output from the oscillator to decrease so as to effect a balance.* The electronic voltmeter circuit measures the amount of this power decrease from the oscillator and displays the measurement as a *power increase which is that supplied by the microwave field.*

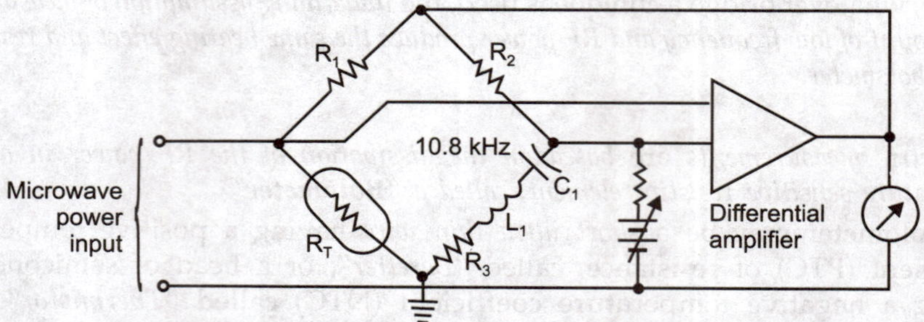

Fig. 3.100. Simplified diagram of the measurement circuit of a modern bolometer type power meter.

Uses of bolomoter :

(*i*) The bolometer is customarily used for power measurements in the range 0.01 – 10 mW.

(*ii*) A comparison-type bolometer is quite suitable in the range of 10 mW to 10 W.

(*iii*) Power measurements above 10 W can be made by *using attenuators and a low-power bolometer.*

3.9.11. Calorimetric Method–For measurement of Large Amount of RF Power

The calorimetric method for measurement of large amount of RF power uses the *principle that a RF power may be directly converted into heat.*

Water acts as a load, and RF power may be used to heat an element such as long transmission line, water being used as a coolant. The RF power may be absorbed directly in the calorimeter fluid.

The *system should have no RF leakage either through lossy joints or by radiation.*

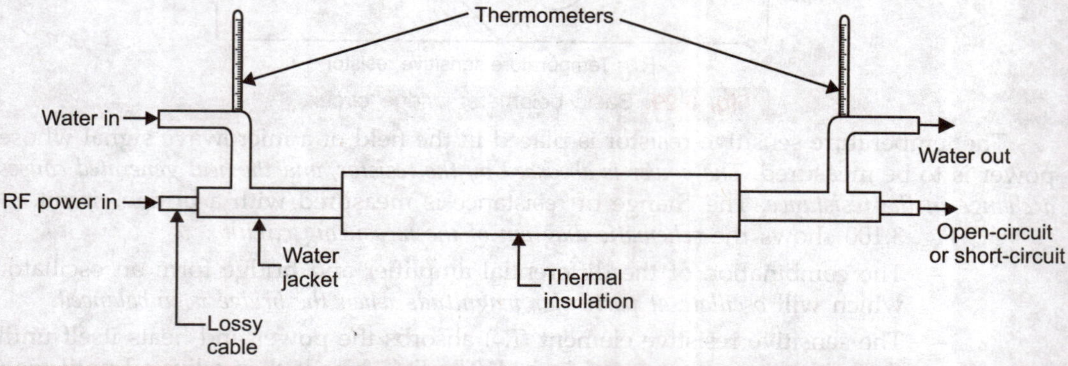

Fig. 3.101. Measurement of power in a lossy cable using the calorimetric method.

The power radiated (P) is given by the relation :

$$P = m \times c_p \times \Delta t \quad ...(3.163)$$

where, m = Mass flow rate of the fluid (water),

c_p = Specific heat of fluid, and

Δt = Temperature rise.

Alternatively, the relationship between temperature rise and power dissipated may be established by dissipating a known amount of A.C. (50Hz) or D.C. power in the calorimeter system.

- Fig. 3.101 shows an arrangement for the power measurement in a lossy cable using the calorimetric method. Here the difference in temperature is measured by thermometers.

3.9.12. Ouput power meter

An *"output power wattmeter"* is designed to directly measure the output in an arbitrary load. In this instrument *a set of resistive loads* are provided to be selected for *power measurements*. Besides this, the instrument *can be used to measure impedance and frequency response characteristics*.

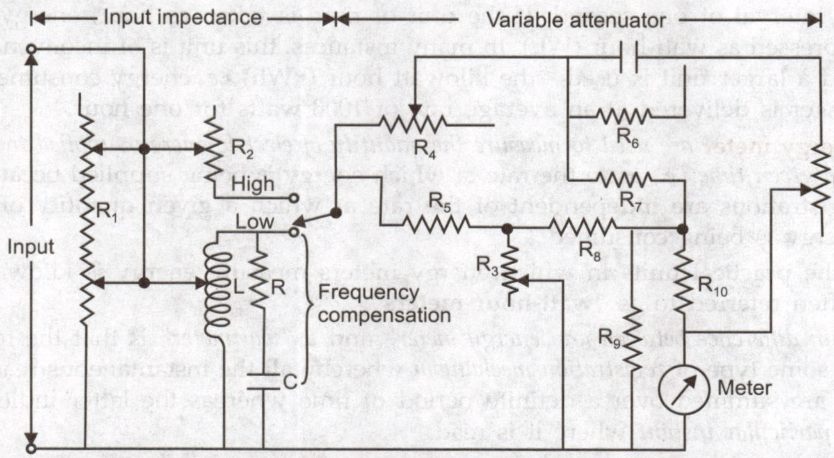

Fig. 3.102. Output-meter circuit.

Fig. 3.102 shows a simple circuit for an output-meter :

- The input impedance network consists of two tapped resistances and a coil. The input impedance can be varied in steps from 2.5 Ω to 40 kΩ.
  — *At low impedance values,* the coil shunts a portion of R_1 – an in resistance results in power turns of the coil. This arrangement keeps the meter reading *proportional to the energy dissipated* by the resistor.
  — At *high impedance values*, the coil is replaced by another tapped resistance R_2.
- The R-C frequency compensator is in parallel with the coil. At low frequency, capacitive reactance X_C is high; it decreases as frequency increases. This *compensates* for the losses of the coil at low frequencies.
  — The frequency range is generally between 20 Hz and 20 kHz.
- Resistance R_3 is the control in a T-network. This amounts to a variable meter shunt, which is used to *extend the range of the meter*.
  — The lowest meter range is normally 5 mW, but it can be extended in decade steps of 50 mW, 500 mW etc.
- The rest of the circuit is a *combination of a calibration and frequency compensation network.*
- This type of meter may have a midscale accuracy of ± 2% at 1 kHz. The accuracy, over the frequency range of 20Hz – 15kHz, may be within ± 2% of the 1kHz value.
- The input meter is subjected to a waveform error when the input is other than sinusoidal.

- *Practical instruments for measuring the power output of oscillators, amplifiers, transformers, transducers and low frequency lines use an input impedance of a tapped transformer with 48 impedance setting.*

This meter can also be used :

(*i*) To measure output impedance by adjusting the maximum power.

(*ii*) To check the frequency response characteristics of audio frequency devices.

3.10 ENERGY METERS AND MEASUREMENT OF ENERGY

3.10.1. General Aspects

- **Energy** *is the total power delivered or consumed after a time internal.* Thus energy is the *product of power and time.* Its unit is *joule* or *watt-second* which is 1 watt over an interval of one second. If the unit of time is taken as hour, energy is then expressed as watt-hour (Wh). In many instances, this unit is of inconvenient size and a larger unit is used - the kilowatt hour (kWh) *i.e.* energy consumed when power is delivered at an average rate of 1000 watts for one hour.
- **Energy meter** *are used to measure the quantity of electric energy supplied to a circuit in a given time i.e.,* as to the rate at which energy is being supplied because their registrations are independent of the rate at which a given quantity of electric energy is being consumed.

Since the practical units in which energy meters measure energy is kilowatt-hour, they are often referred to as **"watt-hour meters"**.

The main difference between an "energy meter" and a *"wattmeter"* is that the former is fitted with some type of *registration mechanism* whereby all the instantaneous readings of the power are summed over a definite period of time whereas the latter indicates the value at a *particular instant* where it is read.

3.10.2. Essential Characteristics of Energy Meters

The *essential characteristics of energy meters* are given below :

1. They must be simple in design and must not contain any parts which may rapidly deteriorate.
2. The readings may be given directly by the dials and must avoid any multiplying factors.
3. The casing of the meter should be dust, water and insect proof.
4. Permanency of calibration is a prime requisite and to attain it, the friction at the pivots etc., and retarding torque of the magnetic brakes must remain constant. The magnets should be so placed that they are not affected in their strength by the magnetic field of the current coil.
5. The friction losses should be minimum (being unavoidable) and must remain practically constant over long periods of time. This requires that the moving parts should be light, and the jewels and pivots should be of best quality and kept in good order.
6. There should be minimum possible friction in the counter device. The torque of the meter should be high so that the unavoidable irregularities in friction may not cause inaccuracies.
7. The energy meter should maintain its accuracy under reasonably varying conditions of voltage and load.
8. The energy loss in the meter itself must be small.

3.10.3. Types of Energy Meters

Energy meters are generally of the following three types :

1. *Electrolytic meters* ..... Their operation depends on electrolytic action.
2. *Motor meters* ..... They are really small electric motors.
3. *Clock meters* ...... They function as clock mechanisms.

Here only motor meters will be discussed.

3.10.4. Motor Meters

The motor meters can be used in *D.C. as well as in A.C. circuits.* In principle the motor meter is a small motor of D.C. or A.C. type whose *instantaneous speed of rotation is proportional to the circuit current in case of an ampere-hour meter and to the power of the circuit in case of a watt-hour meter.*

The following are the essential parts of the motor meters :

1. **An operating torque system.** It produces a torque and causes the moving system to rotate continuously.

2. **A braking device.** It is usually a permanent magnet, known as *brake magnet.* This brake magnet induces current in some part of the moving system which in turn produces the braking torque. Thus the braking torque is *proportional to the induced currents* whereas the induced currents are *proportional to the speed of the moving system* (and hence the *braking torque is proportional to the speed of the moving system* (*disc*). When the braking torque is equal to the driving torque the moving system attains a steady speed.

3. **Revolution registering device.** This device is obtained by having a worm cut on the spindle of the instrument. The worm engages with a pinion and thus drives the train of wheels and registers ampere-hours and watt-hours directly.

Types of motor meters. The various types of motor meters are :

(*i*) *Mercury motor meters* ------ Normally used on D.C. circuits.
(*ii*) *Commutator motor meters* ------ Can be used both for D.C. as well A.C. work
(*iii*) *Induction motor/energy meters* ------ Used only on A.C. circuit

3.10.5. Motor-driven Meter—Watt-hour Meter

The motor-driven meter shown in Fig. 3.103 can be used on direct or alternating current. It contains a small motor and an aluminium retarding disc. *The field winding is connected in series with the load, and the field strength is proportional to the load current.*

The armature is connected *across* the source, and the *current in the armature is proportional to the source or line voltage.* The torque produced in the armature is *proportional to the power* consumed by the load.

The armature shaft drives a series of counters that are calibrated in watt-hours. The power that is used can be read directly from the dials.

The aluminium disc attached to the armature is used to control the armature speed. The disc turns is used can be read direcly from the dials.

The aluminium disc attached to the armature is used to control the armature speed. The disc turns in a magnetic field produced by the permanent magnets, and the retarding force increases as the rotation increases and stops when the disc stops. The retarding force is produced by the aluminium conductor cutting through the lines of force of the permanent magnets. This is a form of magnetic damping.

The meter must overcome the friction of the bearing and indicators at *very light loads.* A portion of the field is produced by the armature current (coil in series with the armature

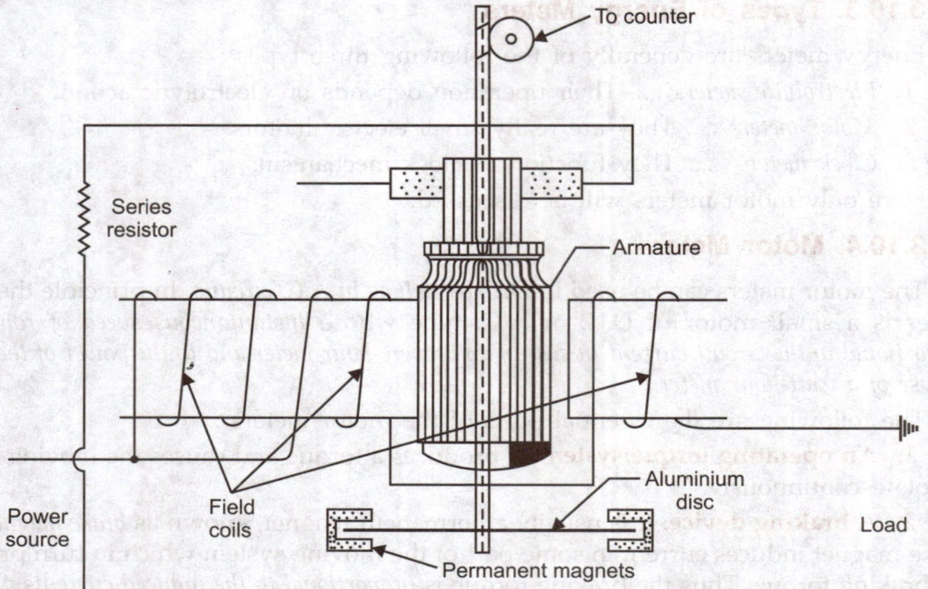

Fig. 3.103. Motor-driven meter (watt-hour meter) designed to operate on direct or alternating current.

winding). This coil is (called as compensating coil) wound to aid the field and is adjusted to the point where it just overcomes the meter friction.

3.10.6. Induction Type Watt-hour Meter

This is the *most commonly used meter on A.C. circuits for measurement of energy.*

Advantages :

(*i*) Simple in operation.

(*ii*) High torque/weight ratio.

(*iii*) Cheap in cost.

(*iv*) Correct registration even at very low power factor.

(*v*) Unaffected by temperature variations.

(*vi*) More accurate than commutator type energy meter on light loads (owing to absence of a commutator with its accompanying friction).

3.10.7. Induction type Single-phase Energy Meters

These are, by far, the most common form of A.C. meters met with in every-day *domestic* and *industrial installations.* These meters measure electric energy in kWh.

3.10.7.1. Construction and theory of operation

The principle of these meters is practically the same as that of induction watt meters. Instead of the control spring and the pointer of the wattmeter, the watt-hour meter, (energy meter) employs a *brake magnet and a counter attached to the spindle.* Just like other watt-hour meters, the eddy currents induced in the aluminium disc by the brake magnet due to the revolution of the disc, are utilised to control the continuously rotating disc.

Construction. The construction of a typical meter of this type is shown in Fig. 3.104(*a*); the brake-magnet and recording wheel-train being omitted for clearances. It consists of the following :

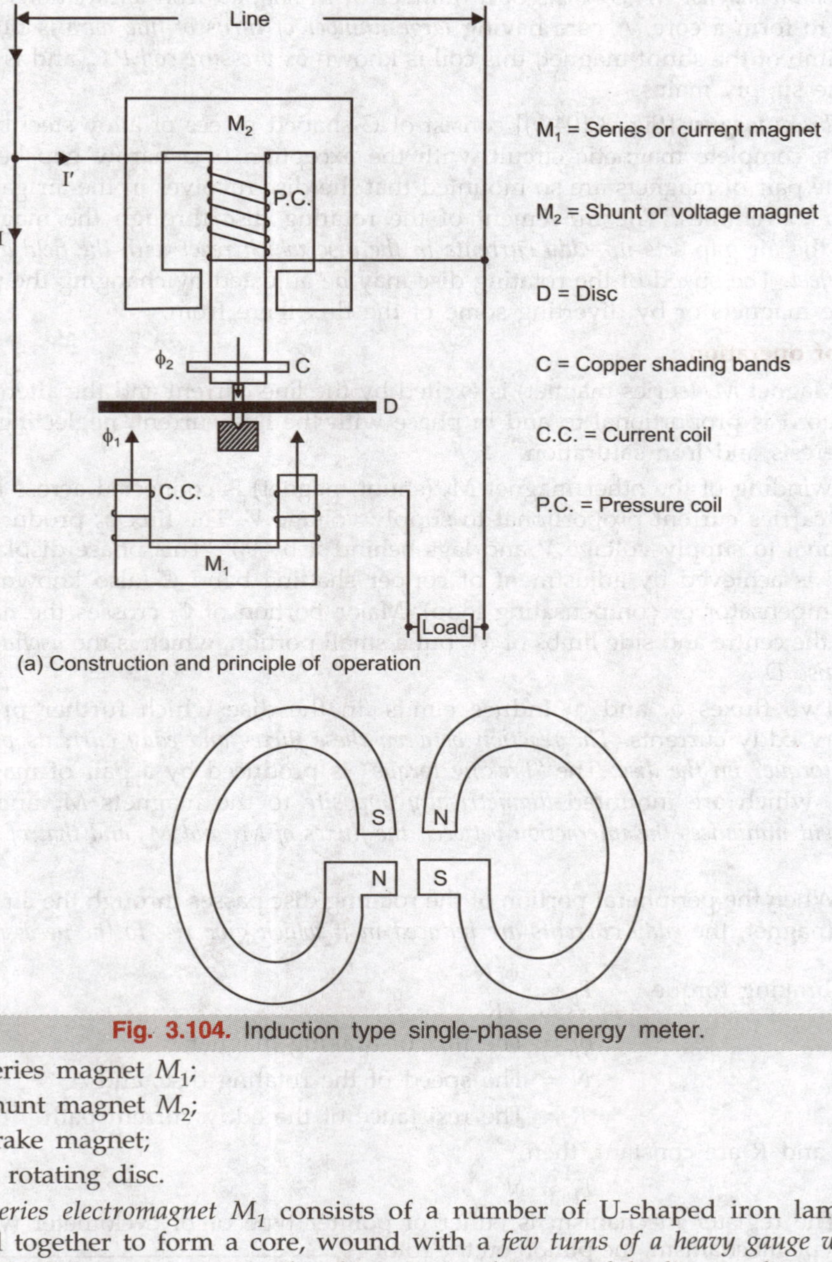

Fig. 3.104. Induction type single-phase energy meter.

(*i*) Series magnet M_1;

(*ii*) Shunt magnet M_2;

(*iii*) Brake magnet;

(*iv*) A rotating disc.

The *series electromagnet* M_1 consists of a number of U-shaped iron laminations assembled together to form a core, wound with a *few turns of a heavy gauge wire.* This wound coil is known as *current coil* and is connected in one of the lines and in series with the load to be metered. The series electromagnet is energized and sets up a magnetic field cutting through the rotating disc, when load current flows through the current coil C.C. The rotating disc is an aluminium disc mounted on a vertical spindle and supported on a sapphire cup contained in a bottom screw. The bottom pivot, which is usually removable, is of hardened steel, and the end, which is hemispherical in shape, rests in the sapphire cup. The top pivot (not shown) merely serves to maintain the spindle in a vertical position under working condition and does not support any weight or exert any appreciable thrust in any direction.

The *shunt magnet* M_2 consists of a number of M shaped iron laminations assembled together to form a core. A core having *large number of turns of fine wire* is fitted on the middle limb of the shunt magnet, this coil is known as *pressure coil P.C.* and is connected across the supply mains.

The *brake magnet* [Fig. 3.104(*b*)] consist of C shaped pieces of alloy steel bent round to form a complete magnetic circuit, with the exception of a narow gap between the poles. The pair of magnets are so mounted that the disc revolves in the air gap between the polar extremities. The movement of the rotating disc through the magnetic field crossing the air gap *sets up eddy currents in the disc which react with the field and exerts a braking effect.* The speed of the rotating disc may be adjusted by changing the position of the brake magnets or by diverting some of the flux there from.

Theory of operation :

— Magnet M_1 (series magnet) is excited by the line current and the alternating flux ϕ_1, produced is proportional to and in phase with the line current, neglecting of course the hysteresis and iron saturation.

The winding of the other magnet M_2 (shunt magnet) is connected across the supply line and carries current proportional to supply voltage V. The flux ϕ_2 produced by it is proportional to supply voltage V and lags behind it by 90°. This phase displacement of exact 90° is achieved by adjustment of copper shading band C (also known as power factor compensator or compensating loop). Major portion of ϕ_2 crosses the narrow gap between the centre and side limbs of M_2 but a small portion, which is the *useful flux passes through disc D.*

— Two fluxes ϕ_1 and ϕ_2 induce e.m.fs. in the disc which further produce the circulatory eddy currents. *The reaction between these fluxes and eddy currents produces the "driving torque" on the disc.* The "*braking torque*" is produced by a pair of magnets [Fig. 3.104(*b*)], which are mounted *diametrically opposite* to the magnets M_1 and M_2. *This arrangement minimises the interaction between the fluxes of* M_1 *and* M_2 *and that of the braking magnet.*

— When the peripheral portion of the rotating disc passes through the air gap of the braking magnet, the *eddy currents are induced in it which give rise to the necessary torque.*

The braking torque, $$T_b \propto \frac{\phi_b^2 N}{R} \qquad \text{...(3.164)}$$

where, ϕ_b = The flux of braking magnet,

N = The speed of the rotating disc, and

R = The resistance of the eddy current path.

If ϕ_b and R are constant, then,

$$T_b \propto N \qquad \text{...(3.165)}$$

— The register mechanism is either of pointer type or of cyclometer type. In the pointer type mechanism, the pinion on the rotor shaft drives, with the help of a suitable train of reduction gears, a series of five or six pointers rotating on dials marked with ten equal divisions. The gearing between different pointers is such that each pointer advances by $\frac{1}{10}$th of a revolution for a complete revolution of the adjacent pointer on the main rotor disc in the train of gearing as shown in Fig. 3.105.

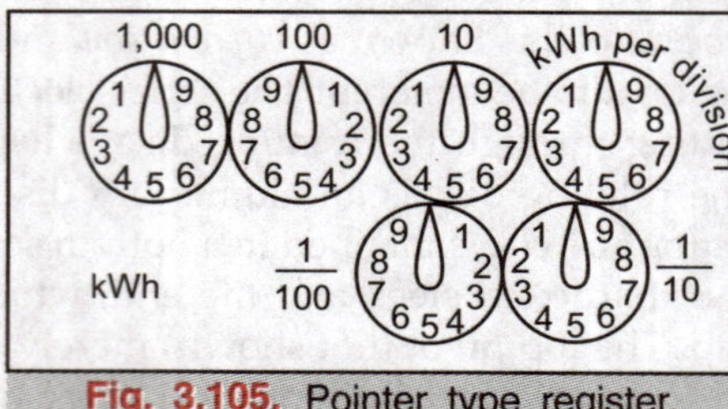

Fig. 3.105. Pointer type register.

- It can be derived that the driving torque, is given by :

$$T_d \propto \omega\, \phi_{1max}\, \phi_{2max} \sin \alpha \qquad ...(3.166)$$

where, ϕ_{1max} and ϕ_{2max} = The maximum fluxes produced by the magnets M_1 and M_2, and

α = The angle between the *fluxes* = $(90 - \phi)°$ ...(See Fig. 3.106)

Assuming that fluxes are proportional to the current, we have :

Current through the winding of $M_1 = I$ ....the line current

Current through the winding of $M_2 = \dfrac{V}{\omega L}$

where, L = Inductance of the pressure/voltage coil (The pressure coil is made highly inductive with negligible resistance; hence $\phi_2 \propto \dfrac{V}{\omega L}$)

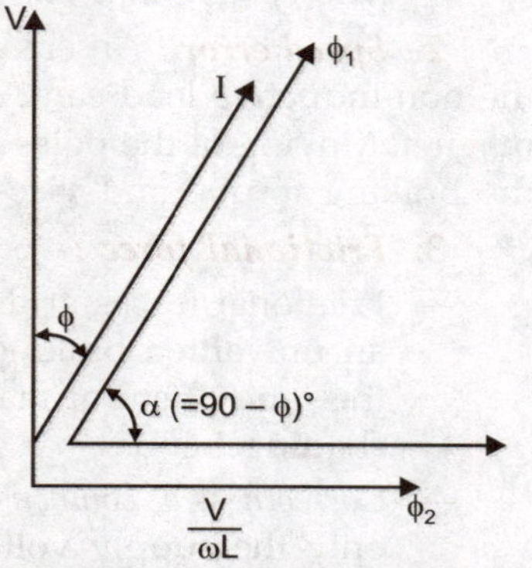

Fig. 3.106.

By inserting various values in eqn. (3.166), we get :

$$T_d \propto \omega \,.\, I \,.\, \frac{V}{\omega L} \sin (90 - \phi)°$$

or, $$T_d \propto VI \cos \phi \propto \text{Power} \qquad ...(3.167)$$

Also, $$T_b \propto N$$

The disc achieves a steady speed N when the two torques are equal *i.e.*, when,

$$T_d = T_b$$

$\therefore$ $$N \propto \text{power} \qquad ...(3.168)$$

Hence, in a given period of time, the number of revolutions $\int_0^t N dt$ *is proportional to* $\int_0^t W \,.\, dt$ *i.e. the electrical energy consumed.*

3.10.7.2. Errors, compensation and adjustments

I. Sources of errors :

The various sources of errors in an induction-type energy meter are given below :

(i) *Incorrect magnitude of the fluxes.* These may arise from abnormal voltages and load currents.

(ii) *Incorrect phase relation of fluxes.* These may arise from defective lagging, abnormal frequencies, changes in the iron losses etc.

(iii) *Unsymmetrical magnetic structure.* The disc may go on rotating while no current is being drawn but pressure coils alone are excited.

(iv) *Changes in the resistance of the disc.* It may occur due to changes in temperature.

(v) *Changes in the strength of the drag magnets.* It may be due to temperature or ageing.

(vi) Phase-angle errors due to lowering of power factor.

(vii) Abnormal friction of moving parts.

(viii) Badly distorted waveform.

(ix) Changes in the retarding torque due to the disc moving through the field of the current coils.

Errors and **compensations** are discussed below :

1. ***Phase error :***

— Normally the flux due to shunt magnet does *not* lag behind the supply voltage by exactly 90° due to the fact that the coil has some resistance. Consequently the *torque is not zero at zero power factor*. This is known as *"phase error"*.

— This error is *compensated by means of adjustable copper bands placed over the central limb of the shunt magnet.*

2. ***Speed error :*** An error in the speed of the motor (*i.e.* speed error) when tested on the non-inductive load can be eliminated by correctly adjusting the position of the brake magnet. Moving of the poles of the braking magnet towards the centre of the disc reduces the braking torque and vice-versa.

3. ***Frictional force :***

— Frictional forces at the rotor bearings and in the register mechanism give rise to an unwanted braking torque on the disc rotor. This can be reduced by making the ratio of shunt magnet flux and series magnet flux large with the help of two shading bands.

— *Correctness of compensation* is achieved when the rotor does not run on load with only the supply voltage connected.

4. ***Creeping :***

— The slow but continuous rotation of motor when only the voltage *i.e.,* the pressure coil excited but no current flowing in the circuit is called *"creeping"*. It may be *caused due to friction compensation, vibration, stray magnetic field or the voltage supply in excess of the normal.*

— To overcome this creeping on no load, *two holes are drilled in the disc on the diameter, i.e., on the opposite side of the spindle.* This causes sufficient distortion of the field to prevent rotation when one of the holes comes under one of the poles of shunt magnet.

5. ***Temperature error :*** The errors due to temperature variations of the various instruments are usually *small* because the various effects produced tend to neutralise one another.

6. ***Overload compensation :***

— Under load conditions, the disc revolves continuously in the field of the series magnet and, therefore, there is a dynamically induced e.m.f. in the disc because of this rotation. This e.m.f. causes eddy currents which interact with the field of the series magnet to produce a braking torque. This braking torque is obviously *proportional to the square of the load current* and it assumes increasing importance as the load rises. Thus at *high values of load current the registration tends to be lower than the actual.*

— To *minimise the self braking action, the full-load speed of the disc is kept as low as possible* and the current coil flux is made *small* in comparison to the potential coil flux so that the dynamically induced e.m.f. responsible for self braking action is *very small* as compared with statically induced e.m.f. produced by flux alteration at the rate of 50 Hz and which is responsible for the production of driving torque.

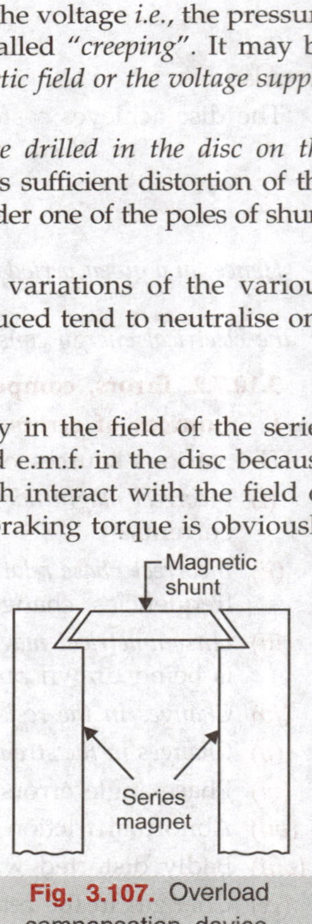

Fig. 3.107. Overload compensation device.

— It is also customary to add an *overload compensator*, which usually takes the form of a saturable magnetic shunt for the series magnet core as shown in Fig. 3.107. The magnetic shunt, at high values of load current, *diverts some of the series magnetic flux and so compensation for self-braking action.*

7. ***Voltage compensation :***

— Voltage variations may introduce errors due to non-linear magnetic characteristics of the shunt magnet core and also due to self braking torque which is approximately proportional to the square of the supply voltage.

Since a certain amount of variation is allowed in the declared voltage of supply, therefore, energy meters must be compensated for this variation.

— *Compensation for variations* in supply voltage is provided by making use of a *saturable magnetic shunt which diverts a greater proportion of the flux into the active path when the voltage rises.* The compensation can be conveniently provided by *increasing the reluctance of the side limbs of the shunt magnet; this is done by providing holes in the side limbs.*

II. Energy meter adjustments :

In energy meter, some adjustments are carried out so that they give correct reading and their errors are within allowable limits. Following is the sequence of these adjustments :

(*i*) *Preliminary light load adjustment :*

— The disc is so positioned that the holes are *not* underneath the electromagnets.

— Rated voltage is applied to the voltage coil with no current through the current coil.

— The light load device is adjusted until the disc just fails to start.

(*ii*) *Full load unity factor adjustment :*

— The pressure/voltage coil is connected across the rated supply voltage and rated full current at unity p.f. is passed through the current coils.

— The position of the brake magnet is adjusted to vary the braking torque so that the meter revolves at the correct speed with required limits of errors.

(*iii*) *Lag adjustment :*

— The pressure coil is connected across rated supply voltage and rated full load current is passed through the current coil at 0.5 p.f. lagging.

— The lag device is adjusted till the meter runs at correct speed.

(*iv*) With rated supply voltage, rated full current and unity p.f., the speed of the motor is checked and full load unity p.f. and low p.f. adjustments are repeated until the desired accuracy limits are reached for both the conditions.

(*v*) *Light load adjustment :*

— Rated supply voltage is applied across the pressure coil and a very low current (about 5% of full load) is passed through the meter at unity p.f.

— This adjustment is done so that the meter runs at correct speed.

(*vi*) Full load-unity p.f. and light load adjustments are again carried out until the speed is correct for both loads *i.e.*, full load as well as light load.

(*vii*) The performance is rechecked at 0.5 p.f. lagging.

(*viii*) *Creep adjustment :*

— As a final check on light load adjustment, the pressure coil is excited by 110% of rated voltage with zero load current.

— If the light load adjustment is correct, the meter should not creep under the above conditions.

3.10.7.3. Polyphase energy meters

In polyphase circuits, the energy can be measured by a group of single-phase energy meters connected as required by Blondel's theorem; the *total energy consumed being the sum of the readings of all energy meters (–ve sign being used for the reading of any meter which runs backward)*.

However, this is never done in practice as it would be both more expensive and more troublesome than the use of polyphase meters.

In *commercial measurement, polyphase meters are used.*

- As in the case of wattmeters for measuring power in polyphase circuits, polyphase energy meters are also constructed by combining two or more single-phase energy meters in one unit.

Polyphase meters may be *'multi-disc type'* or *'single disc type'*. In the multi-disc type, *each element drives a separate disc.* In the single disc type meter, *all the elements drive the same disc.* This disc is slotted or laminated in sector to prevent interaction between eddy currents produced by one element with the flux produced by another element.

Two element energy meter :

Fig. 3.108, shows a two element energy meter used for measurement of energy in 3-phase, 3-wire circuit :

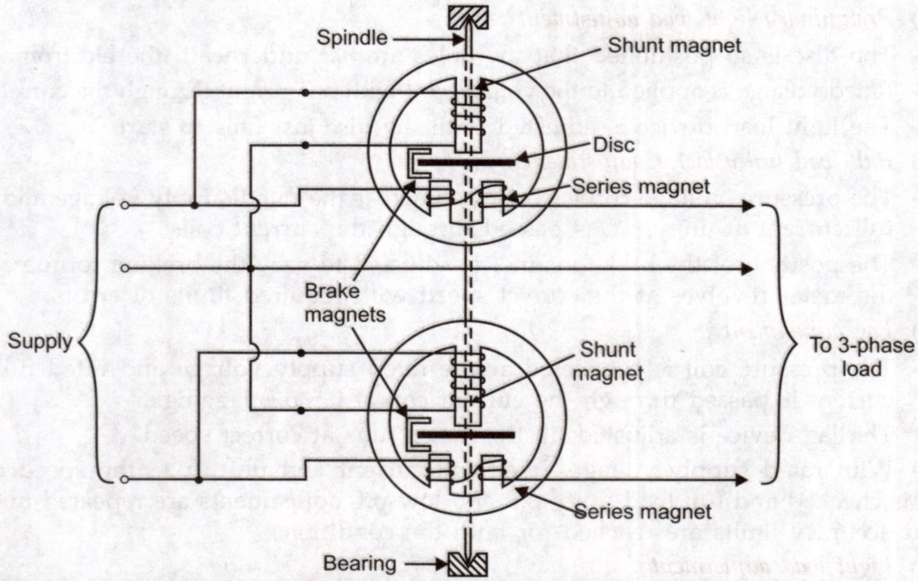

Fig. 3.108. Two element energy meter.

— It consists of two discs mouned on common spindle.

— The moving system drives a single gear train. Each unit is provided with its own shading ring, shading bands and brake magnet for adjustments.

— It is essential *that the driving torque of two elements be exactly the same for the same power passing through them.* This condition is checked by *energising the coils from a single circuit,* the two pressure coils being connected in parallel and the circuit coils in series, with one reversed to oppose the other. This arrangement causes the two torques to be in opposition and if the two torques are *equal,* the disc will not rotate. If there is slight motion, the magnetic shunt is adjusted till the speed of the discs become zero.

— The lag, light load and unity p.f. adjustments are made independently for each element.

— For measurement of energy in 3-phase, 4-wire unbalanced system three element energy meters are used.

- Polyphase energy meters are subjected to *errors due to interaction between the elements,* this may be reduced by suitable positioning and screening of the elements. They also have *relatively large friction error owing to the greater weight of the rotating parts.*

These meters are little *cheaper* than the corresponding number of single-phase meters and have the clear advantage of *giving the total power on a single register.*

- When the currents and voltages of the circuits to be measured are high then use of instrument transformers (C.T.s and P.T.s) is made with energy meters just as with wattmeters. If the line voltage exceeds 440 V CTs and PTs are almost invariably used irrespective of the magnitude of current in order to *insulate meter from the line.*

Example 3.103. *A 5 A, 230 V meter on full load unity power factor test makes 60 revolutions in 360 seconds. If the normal disc speed is 520 revolutions per kWh, what is the percentage error ?*

(Punjab University)

Solution. Energy consumed in 360 seconds

$$= \frac{VI \cos \phi \times t}{3600 \times 1000} \text{ kWh}$$

$$= \frac{230 \times 5 \times 1 \times 360}{3600 \times 1000} = 0.115 \text{ kWh}$$

where, t is in seconds.

Energy recorded by the meter

$$= \frac{60}{520} = 0.11538 \text{ kWh}$$

$$\therefore \quad \text{\%age error} = \frac{\text{Energy registered/recorded} - \text{energy actually consumed}}{\text{Energy actually consumed}} \times 100$$

$$= \frac{0.11538 - 0.115}{0.115} \times 100 = \mathbf{0.33\%} \ (\textit{fast}).$$ **(Ans.)**

Example 3.104. *The constant of a 230 V, 50 Hz, single-phase energy meter is 185 revolutions per kWh. The meter takes 190 seconds for 10 revolutions while supplying a non-inductive load of 4.5 A at normal voltage. What is the percentage error of the instrument ?*

Solution. Energy consumed in 190 seconds

$$= \frac{VI \cos \phi}{1000} \times t = \frac{230 \times 4.5 \times 1}{1000} \times \frac{190}{3600} = 0.0546 \text{ kWh}$$

[cos ϕ = 1, since load supplied is non-inductive]

$$\text{Energy registered by the meter} = \frac{10}{185} = 0.054 \text{ kWh}$$

$$\therefore \quad \text{\%age error} = \frac{0.054 - 0.0546}{(0.054)} = \mathbf{0.06\%} \ (\textit{slow}).$$ **(Ans.)**

Example 3.105. *The name plate of a meter reads "1 kWh = 15000 revolutions". In a check up, the meter completed 150 revolutions during 45 seconds. Calculate the power in the circuit.*

Solution. Power metered in 150 revolutions

$$= 1 \times 150/1500 = 0.01 \text{ kWh}$$

If *P* kilowatt is the power in the circuit, then energy consumed in 45 seconds

$$= \frac{P \times 45}{3600} \text{ kWh} = 0.0125\ P \text{ kWh}$$

Equating the two amounts of energy, we have

$$0.0125\ P = 0.01$$

$$\therefore \quad P = 0.8 \text{ kW} = \mathbf{800\ W} \quad \textbf{(Ans.)}$$

Example 3.106. *A 230 V ampere-hour type meter is connected to a 230 V D.C. supply. If the meter completes 225 revolutions in 10 minutes when carrying 14 A, calculate :*

(i) The kWh registered by the meter.

(ii) The percentage error of the meter above or below the original calibration.

The timing constant of the meter is 40 A-s/revolution.

Solution. During 225 revolutions the meter would register 40 × 225 A-s or coloumbs. Since time taken is 10 minutes or 600 seconds it corresponds to a current of $\frac{40 \times 225}{600} = 15$ A.

(*i*) Energy recorded by the meter $= \frac{VIt}{1000}$ kWh where, *t* is in hour

$$= \frac{230 \times 15}{1000} \times \frac{10}{60} = 0.575 \text{ kWh}$$

(*ii*) Actual energy consumed $= \frac{230 \times 14}{1000} \times \frac{10}{60} = 0.5367$ kWh

$$\textbf{Percentage error} = \frac{0.575 - 0.5367}{0.5367} \times 100 = \mathbf{7.14\%}\ (\textit{fast}) \quad \textbf{(Ans.)}$$

Example 3.107. *A single phase kWh meter makes 500 revolutions per kWh. It is found on testing as making 40 revolutions in 58.1 seconds at 5 kW full load. Find out the percentage error.* **(Nagpur University)**

Solution. Number of revolutions made by the meter in one hour testing

$$= \frac{40}{58.1} \times 3600 = 2478.5 \text{ rev.}$$

These revolutions correspond to an energy of 5 × 1 kWh

$$\therefore \quad \text{Number of revolution per kWh} = \frac{2478.5}{5} = 495.7 \text{ rev.}$$

Hence, Percentage error $= \frac{500 - 495.7}{495.7} \times 100 = \mathbf{0.867\%}$ **(Ans.)**

Example 3.108. *An energy meter is designed to make 100 revolutions of the disc for one unit of energy. Calculate the number of revolutions made by it when connected to a load carrying 40 A at 230 V and 0.4 power factor for an hour. If it actually makes 360 revolutions, find the percentage error.* **(Allahabad University)**

Solution. Energy consumed in one hour $= 230 \times 40 \times 0.4 \times \frac{1}{1000} = 3.68$ kWh

Number of revolutions the meter should make when it is correct

$$= 3.68 \times 100 = 368 \text{ rev.}$$

Number of revolutions actually made = 360 rev.

$$\therefore \quad \text{Percentage error} = \frac{360 - 368}{368} \times 100 = -2.17\% \textit{ i.e. } \mathbf{2.17\%} \text{ (slow)} \quad \textbf{(Ans.)}$$

Example 3.109. *A 50 A, 230 V energy meter in a full load test makes 61 revolutions in 37 seconds. If the normal speed of the disc is 520 revolutions/kWh, compute the percentage error.* **(Nagpur University)**

Solution. Assuming unity power factor :

$$\text{Energy consumed in kWh in 37 seconds} = \frac{230 \times 50 \times 37}{3600 \times 1000}$$

$$= 0.1182 \text{ kWh}$$

Number of revolutions corresponding to this energy

$$= 520 \times 0.1182 = 61.464 \text{ rev.}$$

The meter makes 61 revolutions,

$$\therefore \quad \% \textit{ Error} = \frac{61 - 61.464}{61.464} = \mathbf{0.755\%} \text{ (slow)} \quad \textbf{(Ans.)}$$

Example 3.110. *The constant of a 25 A, 220 V meter is 500 rev./kWh. During a test at full load of 4.4 kW, the disc makes 50 revolutions in 83 seconds. Calculate the meter error.* **(UPTU)**

Solution. Number of revolutions which the meter should make in one hour at full load

$$= \frac{(4.4 \times 1000) \times 1 \times 500}{1000} = 2200 \text{ rev.}$$

This corresponds to speed of $\frac{2200}{60} = 36.7$ r.p.m.

Correct time for 50 revolutions $= \frac{50}{(36.7/60)} = 81.7$ s

Hence, meter is *slow* by 83 – 81.7 = 1.3 s

$$\therefore \quad \% \textit{ Error} = \frac{1.3}{81.7} \times 100 = \mathbf{1.59\%} \text{ (Slow)} \quad \textbf{(Ans.)}$$

Example 3.111. *A 16 A ampere-hour meter with a dial marked in kWh, has an error of +2.5% when used on 250 V circuit. Find the percentage error in the registration of the meter if it is connected for an hour in series with a load taking 3.2 kW at 200 V.* **(Nagpur University)**

Solution. Number of revolutions which the meter should make in one hour at full load

$$= \frac{16 \times 250}{1000} = 4 \text{ kWh}$$

$$\text{Correct reading} = 4 + \frac{2.5}{100} \times 4 = 4.1 \text{ kWh}$$

Meter constant on a 3.2 kW load at 200 V = $\frac{3.2 \times 1000}{200}$ = 16 A

Since on the given load meter constant is same as the normal current of meter, hence in one hour it would give a correct reading of 4.1 kWh.

But actual load = 3.2 kWh

∴ Error = 4.1 – 3.2 = 0.9 kWh

Hence, % *Error* = $\frac{0.9}{3.2} \times 100$ = **28.12%** **(Ans.)**

Example 3.112. *The disc of an energy meter makes 600 revolutions per unit of energy. When a 1 kW load is connected the disc rotates at 10.2 r.p.m. If the load is on for 12 hours how many units are recorded as error ?* **(Allahabad University)**

Solution. Energy actually consumed = 1 (kW) × 12 (h) = 12 kWh

Total number of revolutions made by the disc during the period of 12 hours

= (10.2 × 60) × 12 = 7344 rev.

Since 600 revolutions record 1 kWh, therefore, energy recorded by the meter

$= \frac{7344}{600} = 12.24$ kWh

Hence, *extra* unit recorded as *error* = 12.24 – 12 = **0.24 kWh** **(Ans.)**

Example 3.113. A *D.C. ampere-hour meter is rated at 5A, 250 V. The declared constant is 5A- s/rev. Express the constant in revolutions per kWh. Also calculate the full load speed of the meter.*

Solution. Meter constant = 5A-s/rev.

= 5 × 250 W-s/rev. with rated voltage of 250 V

$= \frac{5 \times 250}{1000 \times 3600}$ kWh/rev. $= \frac{1}{2880}$ kWh/rev.

∴ *Meter constant in terms of revolution per kWh* = **2880** **(Ans.)**

Energy consumption in 1 min. at full load = $\frac{5 \times 250 \times (1/60)}{1000} = \frac{1}{48}$ kWh

∴ *Full load speed* = Meter constant (rev./kWh) × Energy consumption (kWh/min.)

$= 2880 \times \frac{1}{48}$ = **60 r.p.m.** **(Ans.)**

Example 3.114. *A meter whose constant is 750 revolutions per kWh makes 15 revolutions in 30 seconds. Determine the load in kW.* **(Allahabad University)**

Solution. Energy consumed in 30 seconds

$= \frac{\text{No. of revolutions made}}{\text{Meter constant in rev./kWh}} = \frac{15}{750} = 0.02$ kWh

∴ *Load in kW* $= \frac{\text{Energy consumed in kWh}}{\text{Time in hours}} = \frac{0.02}{(30/3600)}$ = **2.4 kW** **(Ans.)**

Example 3.115. *A correctly adjusted single phase 240 V watt-hour meter has a meter constant of 600 revolutions per kWh. Determine the speed of the disc for a current of 8A at a lagging power factor of 0.6.* **(Punjab University)**

Solution. *Given* : V = 240 volts ; I = 8A ; cos ϕ = 0.6 ; meter constant = 600 rev./kWh.

Speed of the disc, N :

Energy supplied per minute = $\frac{VI\cos\phi}{1000} \times t$ kWh

$$= \frac{240 \times 8 \times 0.6}{1000} \times \frac{1}{60} = 0.0192 \text{ kWh/min.}$$

$\therefore$ *Speed of the disc, N* = Energy supplied/min. × meter constant (rev./kWh)

= 0.0192 × 600 = **11.52 rev./min.** **(Ans.)**

Example 3.116. *A 230 V, 50 Hz single-phase energy meter has a constant of 200 revolutions per kWh. While supplying a non-inductive load of 4.4 A at normal voltage, the meter takes 3 minutes for 10 revolutions. Calculate the percentage error of the instrument.* **(Bombay University)**

Solution. Energy consumed in 3 minutes

$$= \frac{VI \cos\phi \times t}{1000} \text{ kWh}$$

$$= \frac{230 \times 4.4 \times 1 \times (3/60)}{1000} = 0.0506 \text{ kWh}$$

[$\because$ Power factor ($\cos \phi$) for non-inductive load is unity]

$$\text{Energy consumption registered} = \frac{\text{No. of revolutions made}}{\text{Meter constant in rev./kWh}}$$

$$= \frac{10}{200} = 0.05 \text{ kWh}$$

$$\therefore \quad \%\ Error = \frac{\text{Energy registered} - \text{Energy consumed}}{\text{Energy consumed}} \times 100$$

$$= \frac{0.05 - 0.0506}{0.0506} \times 100 = -\mathbf{1.186\%} \quad \textbf{(Ans.)}$$

Example 3.117. *A 230 V, single-phase domestic energy meter has a constant load of 4A passing through it for 6 hours at unity power factor. If the meter disc makes 2,208 revolutions during this period, what is the meter constant in revolutions/kWh ?*

Calculate the power factor of the load if the number of revolutions made by the meter are 1,472 when operating at 230 V and 5A for 4 hours. **(AMIE)**

Solution. Energy consumed in 6 hours

$$= \frac{VI \cos\phi \times t}{1000} = \frac{230 \times 4 \times 1 \times 6}{1000} = 5.52 \text{ kWh}$$

$$\therefore \quad \textit{Meter constant} = \frac{2208}{5.52} = \textbf{400 rev./kWh} \quad \textbf{(Ans.)}$$

In the *second case* :

$$\text{Energy consumed} = \frac{\text{Revolutions made}}{\text{Meter constant}}$$

$$= \frac{1472}{400} = 3.68 \text{ kWh}$$

Also, $\quad \text{Energy consumed} = \dfrac{VI \cos\phi \times t}{1000}$

$$\therefore \quad 3.68 = \frac{230 \times 5 \times \cos\phi \times 4}{1000}$$

or, $\quad \cos\phi = \dfrac{3.68 \times 1000}{230 \times 5 \times 4} = \mathbf{0.8} \quad$ **(Ans.)**

Example 3.118. *In a test run of 30 minutes duration with a constant current of 5A, the ampere-hour meter registers 525 Wh. The supply voltage is 220 V. Calculate the error and state whether the meter is fast or slow.* **(Punjab University)**

Solution. *Given* : t = 30 minutes or 0.5 h ; I = 5 A; Consumption registered by the meter = 525 Wh or 0.525 kWh; Supply voltage, V = 220 volts.

% Error :

$$\text{Energy supplied} = \frac{VIt}{1000} = \frac{220 \times 5 \times 0.5}{1000} = 0.55 \text{ kWh}$$

$$\therefore \quad \%\ Error = \frac{\text{Energy registered} - \text{Energy consumed}}{\text{Energy consumed}} \times 100$$

$$= \frac{0.525 - 0.55}{0.55} \times 100 = -4.54\% \textit{ i.e. } \mathbf{4.54\%}\ (\textit{slow}) \quad \textbf{(Ans.)}$$

Example 3.119. *The meter constant of a 230 V, 10 A wattmeter is 1800 revolutions per kWh. The meter is tested at half load and rated voltage and unity power factor. The meter is found to make 80 revolutions in 138 seconds. Determine the meter error at half load.*

(Bangalore Univeristy)

Solution. Actual energy consumed at half load during 69 seconds

$$\frac{VI \cos \phi \,.\, t}{1000} = \frac{230 \times (10/2) \times 1 \times (138/3600)}{1000} = 44.08 \times 10^{-3} \text{ kWh}$$

$$\text{Energy recorded} = \frac{\text{No. of revolutions made}}{\text{Rev./kWh}} = \frac{80}{1800} = 44.44 \times 10^{-3} \text{ kWh}$$

$$\therefore \quad \%\ Error = \frac{44.44 \times 10^{-3} - 44.08 \times 10^{-3}}{44.08 \times 10^{-3}} \times 100 = \mathbf{0.817\%}\ (\textit{fast}) \quad \textbf{(Ans.)}$$

Example 3.120. *An A.C. energy meter is tested for half hour run at a supply voltage of 230 V and a load current of 10 A at 0.8 lag power factor. The dial reading at the start was 57.35 and at the end of the test was 58.25. The meter constant is 1200 revolutions per kWh. The meter made 890 revolutions in 30 minutes run.*

Calculate : (i) Error in registration.

(ii) Error in r.p.m. of the meter. **(AMIE)**

Solution. *Given* : Supply voltage, V = 230 volts ; I = 10 A ; $\cos \phi$ = 0.8 lag p.f.

Energy consumption recorded by the meter = 58.25 – 57.35 = 0.9 kWh

***(i)* Error in registration :**

Actual energy consumed in half an hour

$$= \frac{VI \cos \phi \,.\, t}{1000} = \frac{230 \times 10 \times 0.8 \times (30/60)}{1000} = 0.92 \text{ kWh}$$

$\therefore$ *Error in registration* = Energy recorded – Energy actually consumed

= 0.90 – 0.92 = – 0.02 kWh *i.e.,* **0.02 kWh** (*less*) **(Ans.)**

***(ii)* Error in r.p.m. of the metre :**

Actual revolutions required to be made by the meter for an energy consumption of 0.90 kWh = 1200 × 0.90 = 1080 rev.

Revolutions made by the meter = 890

$$\therefore \quad \textit{Error in r.p.m.} = \frac{890 - 1080}{30} = \mathbf{-6.33\ r.p.m.} \quad \textbf{(Ans.)}$$

Example 3.121. *A single-phase energy meter of the induction type is rated 230 V, 10A, 50 Hz and has a meter constant of 600 rev./kWh when correctly adjusted. If 'quadrature' adjustment is slightly disturbed so that the lag is 85°, calculate the percentage error at full load, 0.8 power factor lag.* **(Bombay University)**

Solution. *Given* : Supply voltage, $V = 230$ volts, $I = 10$ A ; $f = 50$ Hz ; Meter constant $= 600$ rev./kWh ; $\cos \phi = 0.8$ (lag)

Percentage error at full load :

Deflecting torque, $T_d \propto \sin \alpha$, where α is the angle between the alternating fluxes due to series and shunt magnets.

When the voltage flux lagging adjustment is disturbed so that the phase angle between the voltage flux and the voltage is *less than 90°; the error is introduced.*

$$\phi = \cos^{-1}(0.8) = 36.87°$$

$$\alpha' = 85° - 36.87 = 48.13°, \text{ instead of } \alpha = 90° - 36.87° = 53.13°$$

$$\therefore \quad \% \ Error = \frac{\sin \alpha' - \sin \alpha}{\sin \alpha} \times 100$$

$$= \frac{\sin 48.13° - \sin 53.13°}{\sin 53.13°} \times 100 \simeq -\mathbf{6.92\%} \quad \textbf{(Ans.)}$$

Example 3.122. *A 3-phase, 2 element energy meter has a constant of 0.2 revolutions of disc per kWh. The meter is being used with a P.T. of ratio 22 kV/220 V and C.T. of ratio 100/5A.*

If the line voltage is 220 V, current is 10 A, time to complete 10 revolutions is 30 seconds on unity power factor, determine the error expressed as a percentage of the correct reading. **(Gorakhpur University)**

Solution. Actual energy consumed during the test period

$$= \sqrt{3} \times \text{ratio of P.T.} \times \text{ratio of C.T.} \times V_s I_s \cos \phi \times t \times 10^{-3} \text{ kWh}$$

$$= \sqrt{3} \times \frac{22000}{220} \times \frac{100}{5} \times 220 \times 10 \times 1 \times \frac{30}{3600} \times 10^{-3} = 63.5 \text{ kWh}$$

Energy recorded by the meter during the test period

$$= \frac{\text{No. of revolutions made}}{\text{Meter constant in rev./kWh}} = \frac{10}{0.2} = 50 \text{ kWh}$$

$$\% \ Error = \frac{50 - 63.5}{62.5} \times 100 = -21.3\% \quad i.e. \ \mathbf{21.3\%} \ \textbf{(\textit{low})} \quad \textbf{(Ans.)}$$

3.10.8. Maximum Demand Indicators/Meters

3.10.8.1. Introduction

Firstly, we shall discuss how to determine the cost of generating electric energy and hence the rates or tarrifs of charging for this energy. The cost consists of :

(*i*) Standing charges which are independent of the output.

(*ii*) Running or operating charges which are proportional to output.

— *The size or capacity of the generating plant and hence necessary capital investment is determined by the* ***maximum demand*** *of the system supplied by the generating plant.*

— The demand of a system is its load requirement, usually in kW or kVA, averaged over a suitable and specified interval of time of short duration. Thus the *"maximum demand" of an installation is defined as the greatest of all the demands which have occured during a given period, such as a day, a month or a year.*

— The instrument which measures the maximum demand is called maximum demand indicator or meter.

— The primary requirement of maximum demand meters is that they shall record maximum power taken by a consumer during a particular period. These meters should be so designed that any sudden momentary increase in loads such as due to short circuits on the system or due to high starting current of motors are not taken into account. Therefore, *these meters must be designed to record the average power over successive predetermined periods; these periods may be of 15 to 30 minutes duration.*

3.10.8.2. Types of maximum demand indicators

Maximum demand indicators are of the following *four* types :

1. Recording demand indicators.
2. Average demand indicators (Merz price maximum demand indicator)
3. Thermal type maximum demand indicator.
4. Digital maximum demand indicators.

Out of all these, only average demand indicators will be described here.

3.10.8.3. Average demand indicator (Merz Price maximum demand indicator)

This indicator is not in the *form of a separate instrument,* but is a fitting which can be used with any type of motor on clock meter to indicate the maximum consumption during a half-hour or other period throughout a quarter of year (or other period between consecutive readings of the instrument).

- *This indicator can be used to record either maximum current or maximum power.*

Construction and operation : Refer to Fig. 3.109.

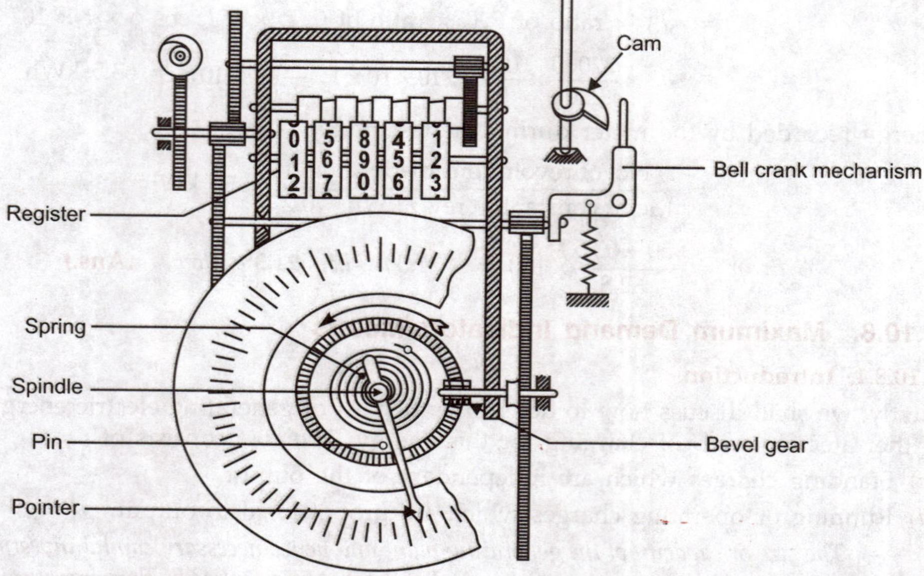

Fig. 3.109. Merz Price maximum demand indicator.

— It consists of a separate dial fitted inside the instrument, the pointer of which is driven by the spindle of the moving system of the meter through a train of gears and pin.

— Normally, the pointer is driven forward for half an hour period or other period known as integrating time period and the energy consumed during this period is indicated on the dial by the pointer. At the end of the time period, by means of a time switch and an operating mechanism the pointer is disconnected from the gearing, returned to zero and reconnected to the gearing.

— A maximum demand pointer is advanced by the driving pointer, but is *held by friction when not being driven*. This pointer will be advanced to a new position when one half-hour's energy consumption *exceeds* than that in any preceding half hour, but will not be affected during the half periods when the energy consumption is *lower* than that corresponding to its existing position.

— There are a number of variations in details of construction. For instance, the cam may be replaced by an electromagnetic relay and a clutch may be substituted for the bell crank releasing device.

The *average demand* can be calculated as follows :

$$\text{Average demand in kW} = \frac{\text{Maximum energy recorded over a time interval in kWh}}{\text{Time interval in hours}} \quad ...(3.169)$$

● *These meters can also be used to measure the maximum demand in terms of kVAh or kVARh by attaching to a suitable meter element which will measure such quantities.*

Advantages :

(*i*) Uniform scale.

(*ii*) More accurate than the thermal type.

Disadvantages :

(*i*) More expensive than thermal type, since it involves mechanisms of greater complexity.

(*ii*) The serious disadvantage is that if the maximum demand occurs sometime after the start of one time interval and continues over only a part of the next interval, the real maximum is *not* indicated since it is split up into two different time intervals (thermal lagged meters do not suffer from this disadvantage since their indication does not depend upon preselected times as their integration is continuous).

3.10.9. Measurement of VAh and VARh

If a circuit has a supply voltage V, a load current I, and cos ϕ as the p.f. then we can write :

$$(VI)^2 = (VI)^2 (\cos^2\phi + \sin^2\phi)$$

or,

$$(VI)^2 = (VI \cos\phi)^2 + (VI \sin\phi)^2$$

Multiplying both sides by h^2 where h is the time in hours,

$$(VIh)^2 = (VIh \cos\phi)^2 + (VIh \sin\phi)^2 \quad ...(3.170)$$

$$(VAh)^2 = (Wh)^2 + (VARh)^2 \quad ...(3.171)$$

where, VAh = Volt-ampere hours,

Wh = Watt-hours, and

$VARh$ = volt-ampere hours *reactive*

Expressing the eqn. (3.171) in kilo-units, use have :

$$(kVAh)^2 = (kWh)^2 + (kVARh)^2 \quad ...(3.172)$$

Eqns. (3.170), (3,171), and (3.172) hold *good only for the condition that* cos ϕ *is maintained constant;* in practice, this condition is not usually met. Thus measurement of *VAh* or *kVAh* is not a simple matter. *Meters required for these measurements are a combination of watt-hour meters and reactive volt-ampere hour meters.*

3.10.9.1. Measurement of volt-ampere hours reactive (VARh)

Sometimes the metering of volt-ampere hours *reactive* (*VARh*) of a circuit is necessary in connection with electricity tariffs. For this, a meter is required whose motoring action is proportional to $VI \sin \phi$ or $VI \cos (90° - \phi)$, where ϕ is the power factor angle. The measurement of *VARh* may, therefore, be affected by employing a watt-hour meter, in which either the voltage flux or current flux is given a phase displacement of 90°. Thus an induction watt-hour meter with voltage flux in phase with the voltage and current flux in phase with the current will register *VARh*.

- In order to measure VARh in single-phase circuits, use of specially compensated watt-hour meter is made. *Phase compensation is accomplished by suitable combination of resistance, capacitance and inductance.*

Normal induction type energy meter can be used to register *VARh* with the help of following schemes, these schemes apply to 3-phase, 3-wire system :

1. Single element method.
2. Crossed phase method.
3. Auto-transformer method.

1. Single element method :

Fig. 3.110 illustrate the single element method of measuring *VARh* in a 3-phase, 3-wire system.

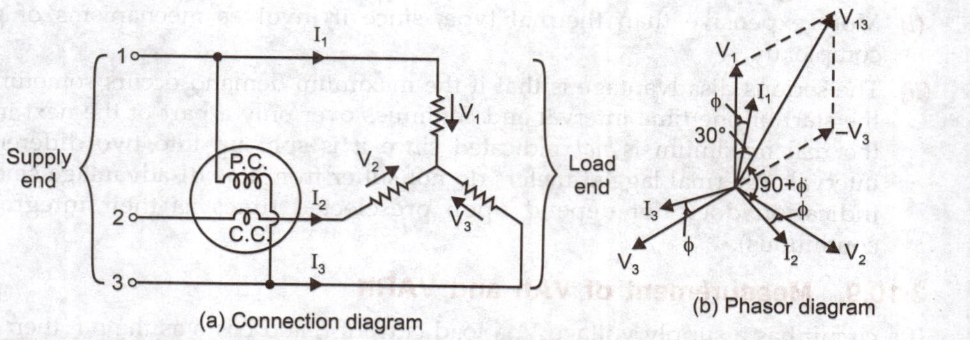

Fig. 3.110. Single element mehod of measuring VARh in a 3-phase, 3-wire system.

A normal single-phase induction type energy meter is used. The current coil is connected in series with one of the phase while the pressure coil is connected across the other two phases as shown in Fig. 3.110(*a*). This connection is same as used for measurement of reactive power.

$$\text{Total number of revolutions} = K \int V_{13} I_2 \cos (90 + \phi)° \, dt$$

$$= -\sqrt{3}K \int VI \sin \phi \, dt$$

$$= -\sqrt{3}K \times (\text{VARh of each phase}) \quad ...(3.173)$$

Thus, if a system is balanced the meter registers the total *VARh*.

- The *disadvantage* of this method is that it *can only be used for balanced loads.*

2. Crossed phase method :

This method can be employed for *unbalanced currents in a 3-phase, 3-wire system.*

Fig. 3.111, illustrates the cross phase method of measuring *VARh* in a 3-phase, 3-wire system :

— In this method a 2-element polyphase energy meter is used. A neutral point is obtained by a *choke* as shown in Fig. 3.111(*a*).

— Each current is associated with a voltage differing in phase from the voltage normally used for power or energy measurements [See Fig. 3.111(*b*)].

— In this method, we must ensure that torque acting on the two elements (Element-1 and Element-2) are *opposite* in direction.

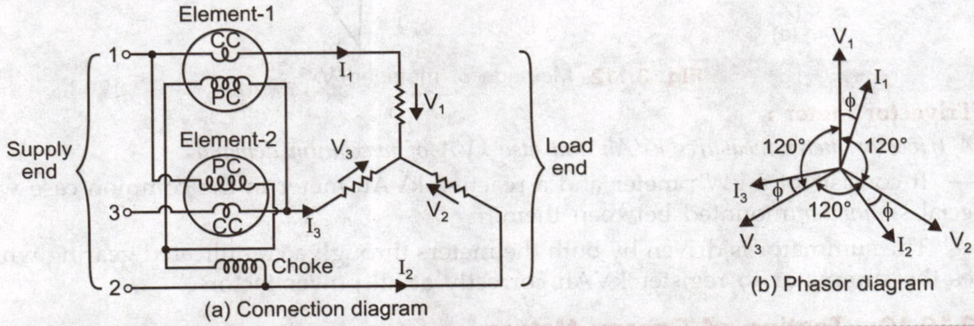

Fig. 3.111. Crossed phase method of measuring VARh in 3-phase, 3-wire system.

— Since the voltage across the pressure coil is $\frac{1}{\sqrt{3}}$ of that used for power or energy measurements, evidently the registration requires a multiplier of value $\sqrt{3}$, this is evident from the following :

Now, Total number of revolutions $= K\int [V_3 I_1 \cos(120° + \phi) - V_1 I_3 \cos(120° - \phi)]\, dt$

In use of balanced currents, total number of revolutions

$$= K\int VI\,[\cos(120° + \phi) - \cos(120° - \phi)]\, dt$$

$$= K\int (\sqrt{3}\, VI \sin\phi)\, dt \qquad ...(3.174)$$

3. Auto-transformer method :

This method uses a two element watt-hour meter with a phase shifting arramgement employing two open delta-connected auto-transformers.

3.10.9.2. Measurement of volt-ampere hours (VAh)

Fig. 3.112 illustrate the two alternative methods of metering *VAh.*

- In **Fig. 3.112(*a*)** meter *L* is the watt-hour meter while meter *M* registers volt-ampere hours. The meter *M* may be maximum demand indicator which gives *VA* demand.

  The reading of meter *B* is in *VAh*

- In **Fig. 3.112(*b*)** meter *L* records watt-hours (*Wh*) while meter *N* registers volt-ampere hours reactive (*VARh*).

Then, $\quad VAh = \sqrt{(Wh)^2 + (VARh)^2}$

$$= \sqrt{(\text{Reading of meter } L)^2 + (\text{Reading of meter } N)^2} \quad ...(3.175)$$

This principle of vector addition is utilised for the constructin of a meter which measures *VAh* directly.

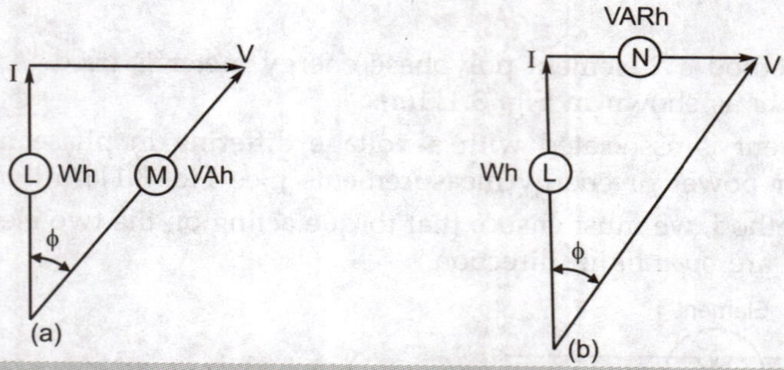

Fig. 3.112. Methods of metering VAh.

Trivector meter :

A trivector meter measures kVAh and also kVA of maximum demand.

— It consists of a kWh meter and a reactive kVAh meter in the common case with a special *summator* mounted between them.

— The summator is driven by both the meters through a complicated gearing which makes the summator to register kVAh correctly at all power factors.

3.10.10. Testing of Energy Meters

3.10.10.1. General aspects

Prior to the installation of energy meters in domestic or commercial premises, it is essential by law, to test the accuracy of the meter. The errors, under specified conditions, must lie within permissible limits.

A.C. energy meter must be tested for the following *conditions :*

(*i*) At 5% of marked current with unity p.f.

(*ii*) At 100% or 125% of marked current with unity p.f.

(*iii*) At one intermediate load with unity p.f.

(*iv*) At marked current and 0.5 lagging p.f.

Besides this, the meters should also be subjected to following *tests* :

(*i*) *Starting test.* The meter should start and run at 0.5% of marked value of current and with normal voltage (Accuracy not to be checked at this current).

(*ii*) *Creep test.* The meter should not *creep* (should not revolve through more than one revolution) with an applied voltage of 110% of its marked value and with the current circuit open.

3.10.10.2. Apparatus required for meter testing

The approved apparatus for testing of energy meters is listed below :

A. Standard apparatus :

(*i*) D.C. potentiometer, calibrated by National Physical Laboratory.

B. Substandard apparatus :

(*i*) Indicating wattmeter

(*ii*) Rotating meter

(*iii*) Ammeter

(*iv*) Voltmeter

(*v*) Electrolytic ampere-hour meter.

- *Time standard* – Ship's chronometer or pendulum-type clock.
- *Time substandard* – Stop-watch or other suitable tuning device.

3.10.10.3. Methods of testing motor meters

Three main methods of testing motor meters are described below :

Method A : *"Long-period dial tests," using substandard rotating meters.*

— In this method, a rotating substandard meter is used to measure the amount of energy passing through the meter being tested in a given time.

— The current circuits of two meters are connected in *series* and thus carry the same current while the voltage circuits are connected in *parallel* across the same supply.

— The meters are started and stopped *simultaneously.*

— The registrations of two meters are compared at the end of the test period (which is an hour or even more) and the error is calculated as under :

Let, K_x = Registration of test meter, in kWh

K_s = Registration of substandard meter, in kWh,

Then,

$$\% \text{ Error} = \frac{K_x - K_s}{K_s} \times 100 \quad \text{...(3.176)}$$

This is the *most rigorous* of the three methods of testing.

Advantages :

(*i*) Reliability only on dial indications of the test meter.

(*ii*) The complete elimination of errors due to starting or stopping of substandard stop-watches.

(*iii*) Avoidance of any necessity for holding the load absolutely constant.

Disadvantage. The main diadvantage is the length of time required for a single accurate test and thus for the complete set of tests.

- This method is *unsuitable if adjustments to the meters have to be made during testing.*

Method B : *"Short period test" using substandard or rotating meters.*

This test extends over a *short period* and therefore it is known as a *short period test.*

— A rotating substandard meter is used in this method also.

— The current coils of the substandard meter and the meter under test are connected in series, and their pressure coils are connected in *parallel.*

— Necessary provision is made to start and stop the two meters simultaneously.

- In this method either *"revolutions"* of the disc of meter under test are compared with those made by the disc of substandard meter during the same time or *"speeds"* of the test and the substandard meters are directly compared.

Let, R_x = Number of revolutions per kWh for the meter under test,

R_s = Number of revolutions per kWh for the substandard meter,

R_{xt} = Total number of revolutions made by the meter under test,

R_{st} = Total number of revolutions made by the substandard meter during the test interval,

N_x = Speed of disc of the meter under test,

N_s = Speed of disc of the substandard meter, and

t = Time interval in hours.

(i) When revolutions are compared :

Energy recorded by the meter under test during test interval = $\dfrac{R_{xt}}{R_x}$

Energy recorded by the substandard meter during test interval = $\dfrac{R_{st}}{R_s}$

$$\therefore \quad \% \text{ Error} = \frac{(R_{xt}/R_x)-(R_{st}/R_s)}{(R_{st}/R_s)} \times 100 \quad ...(3.177)$$

(ii) When speeds are compared :

Energy recorded by the meter under test = $\dfrac{N_x.t}{R_x}$

Energy recorded by the substandard meter = $\dfrac{N_s.t}{R_s}$

$$\therefore \quad \% \text{ Error} = \frac{(N_x.t/R_x)-(N_s.t/R_s)}{(N_s t/R_s)} = \left(\frac{N_x R_s}{N_s R_x} - 1\right) \times 100 \quad ...(3.178)$$

- An additional requirement of this method is one *"long period dial test" extending over an hour at full load. This is done to check the registering mechanism and also the effect of time.* The meter is then again subjected to *short time test at full load to check for any effect due to heating.*

Advantages :

(i) Less time required comparatively.

(ii) The complete elimination of error owing to starting or stopping of substandard or stop watches.

(iii) Avoidance of any necessity for holding the load absolutely constant.

Method C : *Precision test*

— In this method precision grade indicating instruments are used as reference standards.

— There indicating instruments are connected in the circuit of meter under test.

— The current and voltage are held constant during the test.

— The number of revolutions made by the meter disc during the test are recorded. The time taken is also measured.

Now, Energy recorded by the meter under test = $\dfrac{R_{xt}}{R_x}$ kWh

Energy computed from the readings of the indicating instrument = kW × t

where, R_{xt} = Total number of revolutions made by the disc of meter under test,

R_x = Number of revolutions per kWh for the meter under test, and

t = Time in hour

$$\therefore \qquad \% \text{ Error} = \frac{(R_{xt}/R_x)}{kW \times t} \times 100 \qquad ...(3.179)$$

- In addition to this test, a 'long period dial test' is required to check the registration mechanism.

Caution : Before starting any of the test mentioned above on a watt-hour meter, its potential circuit must be connected to the supply for one hour in order to enable the self-heating of the potential coil to stabilise.

3.10.10.4. Phantom or Fictitious loading

When the meter under test has high current rating, a test with *actual loading arrangements would involve a considerable waste of power. This can be avoided by using phantom or fictitious loading.*

In phantom loading arrangement, the pressure circuit is supplied from a circuit of required normal voltage and the current circuit from a separate low voltage supply (The rated current can be circulated through the current circuit since the impedance of this circuit is very low). The total power supplied for the test, in this arrangement, is the sum of power supplied due to the small pressure coil current at normal voltage and that due to the circuit current supplied at low voltage. Thus, the total power required for testing the meter with phantom loading is *comparatively very small.*

- For testing of meter of high capacity while in service, the pressure circuit of the meter is supplied from the line in the normal way while the current circuit of the meter is removed from the consumer load circuit and supplied from a source of low voltage.

3.10.10.5. Testing of A.C. meters

— When A.C. meters are to be tested, it is not only necessary to vary the voltage and load current, but, in addition, the *relative phases of the two quantities must be altered;* in other words the *p.f. of the load must be variable also.*

— Variation of the p.f. may be brought about by making up the load circuit with variable resistors, inductors and capacitors, and adjusting these as required. This method is, however, inconvenient in most cases, since it involves the waste of the power in the load and necessitates several fairly expensive pieces of apparatus, these objections being particularly apparent if the capacity of the meters is high.

  - Due to this reason the load employed for testing purposes is usually a *fictitious one,* as in the case of D.C. meter testing. Provision must, of course, be made for *varying the phase of the current in the current circuit relative to that of voltage circuit in addition to the variation of the magnitude of the current.*

— This phase of current is usually altered in one of the following *two* ways :

(i) Using two similar alternators coupled together and driven by a motor, with the *stator of one of them capable of rotation through any desired angle relative to the position of the stator of the other.*

(ii) Using a *phase-shifting transformer.*

- **Fig. 3.113** shows the connections for the testing of the a single-phase meter X, using two coupled alternators A_1 and A_2 for the supply to the voltage-and current-coil circuits respectively.

— The stator of A_1 is capable of rotation for phase variation.

— A_2 is connected first of all to a step-down transformer T whose secondary then supplies the current coils of the two meters *X* and *S*, the latter being either a substandard indicating wattmeter or a standard watt-hour meter, according to the method of test employed.

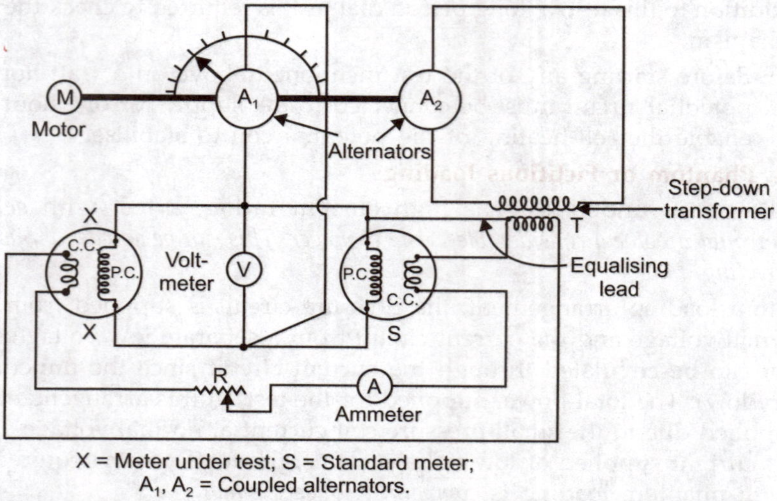

Fig. 3.113. Connections for the testing of single-phase meters.

— The variable resistor *R* is for adjustment of the current in the current-coil circuit.

— *V* and *A* are a voltmeter and ammeter for the measurement of the load voltage and current.

A phase-shifting transformer may, of course, be used instead of the two alternators.

— The *equilising lead* shown is necessary *when a fictitious load is used*, in order to ensure that p.d. between the current and voltage coils of the meters is zero, as it is when they are in service.

After making adjustment of power factor the load as desired for the test, the errors of the meters under test can be determined either by timing a certain number of revolutions and measuring the load by an indicating wattmeter or by comparing their rates of revolutions with that of a standard watt-hour meter connected in the circuit so that it may measure the same load.

It may noted that the *link between the current and voltage coils in each meter must be disconnected for this test.*

- Fig. 3.114 shows the connections for the testing of 3-phase, 3-wire meters :
  - — In this method, 3-phase supply is used for the following two purposes : (*i*) To supply the stator of a phase-shifting transformer; (*ii*) To supply three step-down transformers, two of whose secondaries supply the current coils of the meter.
  - — A regulating choke coil is connected in the primary circuits of these transformers.
  - — The motor of the phase-shifting transformer supplies two auto-transformers having variable secondary tappings. These supply the voltage coils of the meters, the latter consisting of a double standard wattmeter and one or more meters to be tested. These are connected as shown.
  - — A_1, A_2 and A_3 are ammeters and V_1 and V_2 voltmeters, for adjustment of the load.

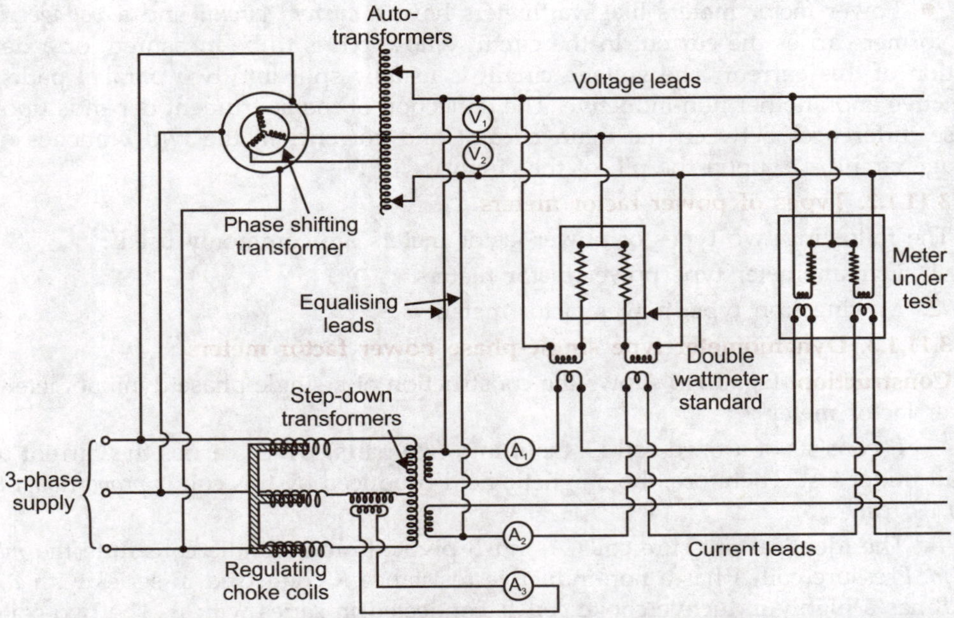

Fig. 3.114. Connections for testing of 3-phase, 3-wire energy meters.

3.10.10.6. Measurement of kVA

For charging the industrial consumer on power factor tariff, in which consumer is penalized for low p.f. of his load, *maximum demand kVA indicator is required for measurement of maximum demand in kVA.*

One type of this instrument is essentially an induction type A.C. ampere-hour meter consisting of a shaded pole, a continuously rotating disc and a braking magnet of permanent magnet type. It is used with a timing device of the Merz type and records the maximum ampere hours during successive time intervals of 15 minutes. The dial of the instrument is *graduated in terms of kVA* corresponding to the circuit rated voltage. *The readings of the instrument are independent of p.f. of the load* and frequency of supply within such variations of the latter might be expected in the normal supply system.

3.11. MISCELLANEOUS MEASURING INSTRUMENTS

Under this heading we shall discuss the following instruments :

1. Power factor meters.
2. Frequency meters.
3. Rotational speed measurements.
4. Synchronoscope.
5. Phase sequence indicators.

3.11.1. Power Factor Meters

3.11.1.1. Introduction

- *Power-factor meters or phase meters* indicate the power factor of a circuit *directly*, as distinct from instruments from whose readings the power factor may be obtained by *dividing the watts supplied by the volt-amperes in the circuit.*

The accuracy obtained with the use of power factor meters is sufficient for most purposes other than high precision testing.

- Power factor meters like wattmeters have a *current circuit* and a *voltage circuit*. The former carries the current in the circuit whose p.f. is to be measured, or a definite fraction of this current. The voltage circuit is usually split into two parallel paths, one inductive and another non-inductive. The deflection of the instrument depends upon the phase differences between the main current and currents in the two branches of the voltage circuit, i.e. upon the p.f. of the circuit.

3.11.1.2. Types of power factor meters

The following two types of power-factor meters are commonly used :

1. Dynamometer type power factor meters.
2. Moving iron type, power factor meter.

3.11.1.3. Dynamometer type single-phase power factor meters

Construction. Fig. 3.115 shows the construction of a single-phase dynamometer type power factor meter :

— It consists of a fixed coil *FF* (split into two parts) which carries the current of the circuit under test. Therefore, the magnetic field produced by this coil is *proportional to the main current.*

— The identical pressure coils *A* and *B* pivoted on a spindle constitute the *moving system.* Pressure coil *A* has a non-inductive resistance *R* connected in series with it, and coil *B* has a highly inductive choke coil *L*, connected in series with it. The two coils are *connected across the voltage of the circuit. The values of R and L are so adjusted that the two coils carry the same value of current at normal frequency, i.e.,* $R = \omega L$.

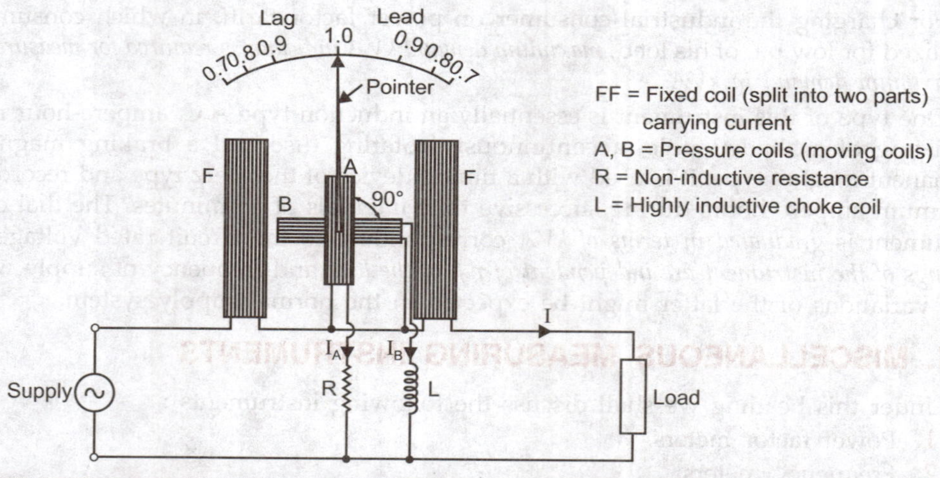

Fig. 3.115. Single-phase dynamometer type power factor meter.

— The current through the coil *A* is in phase with the circuit voltage while that through the coil *B* lags the voltage by an angle δ which is nearly equal to 90°. The angle between the planes of coils is also made equal to δ. Connections to moving coils are made through thin silver or gold ligaments which are *extremely flexible* and this gives a *minimum control effect on the moving system.*

Theory. Let us assume (to simplify the problem) that the current through the coil *B* lags the voltage by exactly 90°. Also that the angle between planes of coils is exactly 90° (*i.e.*, $\delta = 90°$).

Fig. 3.116 illustrates the *action of the instrument.*

Let us assume that the field of the two fixed coils is uniform and in the direction arrowed. The torque on each coil for a given coil current will be maximum when the coil is parallel to the field, *i.e.*, along XX. Suppose that, when the system power factor angle is ϕ, the coils take up a position of equilibrium displaced θ from the vertical. Then the torques due to the two coils must be *equal and opposite*. Now, since the current in coil A is in phase with the system voltage, and the field in which it moves is proportional to the system current, then coil A is essentially a wattmeter movement displaced $90° - \theta$ from the maximum torque position. Then the torque of A,

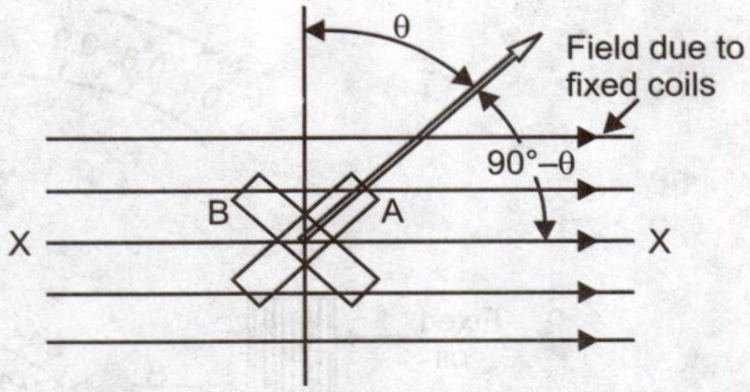

Fig. 3.116.

$$T_A = kVI \cos\phi \cos(90° - \theta) \qquad ...(3.180)$$

where, k is a constant.

Similarly, since the current in coil B lags 90° on the system voltage, coil B is in a *sine-meter movement* displaced θ from the maximum torque position. Then torque on B is,

$$T_B = kVI \sin\phi \cos\theta \qquad ...(3.181)$$

Hence, in the equilibrium position,

$$T_A = T_B$$

i.e., $kVI \cos\phi \cos[90° - \theta] = kVI \sin\phi \cos\theta$

or, $\cos\phi \sin\theta = \sin\phi \cos\theta$

or, $\tan\theta = \tan\phi$

or, $\theta = \phi$

Therefore, the *deflection of the instrument is a measure of phase angle of the circuit.*

- It may be noted that the instrument must be designed for, and calibrated at, the frequency of the supply on which it is to be used. In case the meter is *used for any other frequency or if supply contains harmonics it will give rise to serious errors in the indication due to change in the value of reactance of choke coil.*

3.11.1.4. Dynamometer-type power factor meter for balanced 3-phase load

Fig. 3.117, shows the connections of a 3-phase power-factor meter, the *readings of which are correct only when the load is balanced :*

— In this instrument the two moving coils are fixed with their planes 120° apart, and are connected across two different phases of the supply circuit, the fixed coils being connected in the third phase and carrying the current in the line. There is now no necessity for phase-splitting by artificial means, since the required phase displacement between the currents in the two moving coils can be obtained from the supply itself, as shown.

— *Provided the two moving coils are 120° apart, the angle through which the pointer is deflected from the unity power factor position is equal to the phase angle of the circuit.*

- The 3-phase instrument gives indications which are *independent of frequency and waveform,* since the currents in the two moving coils are both affected in the *same way* by any change of frequency.

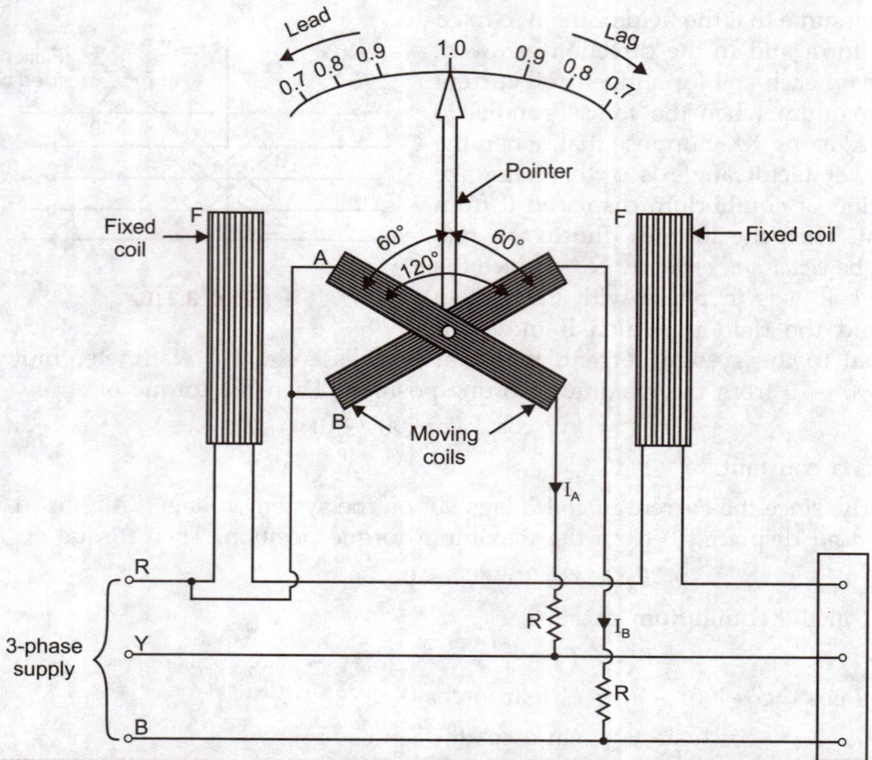

Fig. 3.117. Dynamometer-type power factor meter for balanced 3-phase load.

Advantage : The dynamometer type instrument is *more accurate* than moving iron type.

Disadvantages :

(*i*) Limited scale arc.

(*ii*) The connecting ligaments or threads to the moving coils are required, which subject the moving system of the instrument to a minimum of restraint, this is difficult to accomplish to a satisfactory degree due to low working torque obtainable with this instrument.

3.11.1.5. Moving-iron power factor meters

There are two types of moving iron power factor meters :

1. Rotating field type.
2. Alternating field type.

1. Rotating field type for 3-phase loads :

Fig. 3.118 illustrates the principle of the Westing house instrument of this type :

— A rotating field is set up by the three coils A_1, A_2 and A_3, which are supplied from the 3-phase mains through current transformers.

— The coil B is placed at the centre of this system of coils, and is connected in series with a resistance, across two lines of the supply. Inside coil B is a short pivoted iron rod, with sector-shaped iron pieces S_1 and S_2 at its ends. Damping vanes and a pointer are also carried on the same spindle. The moving system is shown in Fig. 3.118(*b*).

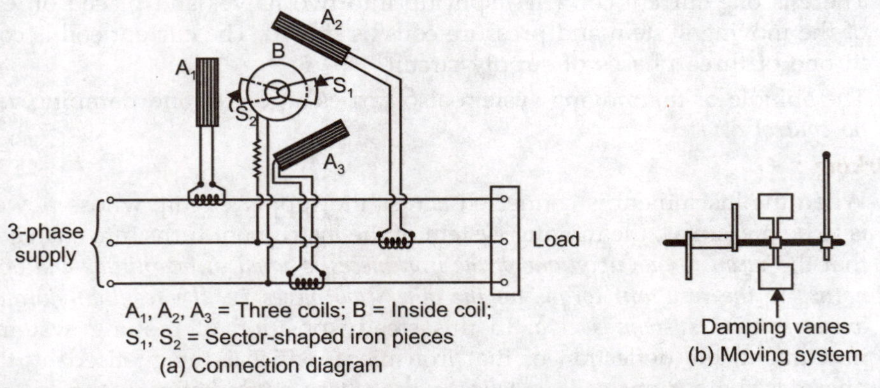

Fig. 3.118. Rotating field type moving iron power factor meter for a balanced 3-phase load.

There are no controlling forces.

— The moving-iron system carries an alternating flux. It does not tend to rotate continuously, but sets in a definite position which depends upon the relative phases of the current in coil B, which is practically in phase with the voltage, and of the currents in coil A.

— The deflection of the moving system is approximately equal to the angle of phase displacement between current and voltage in the 3-phase circuit.

— Owing to the presence of iron in the instrument and the dependence of the reactance of coil B upon frequency it becomes necessary to calibrate the instrument at the normal frequency.

2. Alternating field type :

Fig. 3.119 shows an alternating field type (Lipman) moving iron power factor meter :

Construction :

— It consists of three sets of moving irons (B_1, B_2, B_3) mounted on a single spindle one above the other with non-magnetic separating pieces (S) in between them and spaced at 120° in different parallel planes.

— Each iron piece is magnetised by a separate coil known as *pressure coil*. The three pressure coils (C_1, C_2, C_3) are star-connected to the 3-phases of supply.

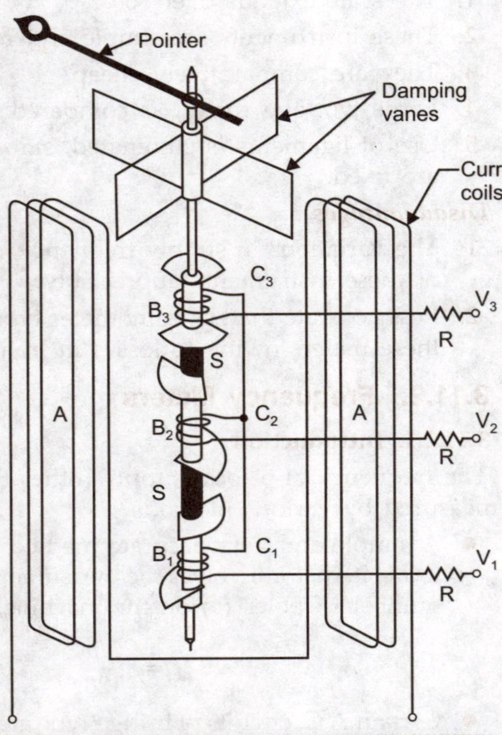

Fig. 3.119. Alternating field type (Lipman) moving iron power factor meter.

— There is one current coil (*AA*) split up into two halves and placed on each side of the moving system and pressure coils as shown. The current coil is connected in one of three phases of supply circuit.

— The spindle of the moving system also carries a pointer and damping vanes but *no control device.*

Working :

— When the instrument is connected across the supply circuit, whose power factor is to be measured, the moving system of the instrument turns into such a position that the *mean torque upon one of the iron pieces is equal in magnitude and opposite in nature to the resultant torque on the other two pieces i.e. the resultant torque on the whole moving system is zero.* In this steady position of moving system of the instrument, the deflection of that iron piece, which is magnetised by the same phase as the current coil, is *equal* to the phase angle between the voltages and currents of 3-phase supply circuit provided that the effects of iron losses and of inductance of pressure coils are negligible.

— As the three pressure coils are at different levels, no resultant rotating magnetic field is set up by them hence any tendency for the moving system to rotate continuously following the "*drag*" of such a field is *eliminated.*

- The arranement shown in Fig. 3.119 is *suitable for 3-phase balance system but it can be suitably modified for use on unbalanced 3-phase circuits and for two phases and single-phase circuits.*

Advantages :

1. The scale extends over 360°.
2. These instruments are simple and robust in construction.
3. They are comparatively cheap.
4. Large working forces, as compared with those in dynamometer type.
5. Use of ligaments is eliminated, since all the coils in a moving iron instruments are fixed.

Disadvantages :

1. The variations in supply frequency, voltage and waveform affect the calibration of these instruments appreciably.
2. Less accurate than dynamometer type instruments, since errors are introduced in these meters owing to losses in iron parts.

3.11.2. Frequency Meters

3.11.2.1. Introduction

The frequency of a local supply (other than that of the controlled grid system) can be measured by various methods.

- A simple and very accurate method of determining frequency of supply from a local alternator involves the measurement of speed (*N*) of the machine and knowing number of poles (*p*) on the machine, the frequency (*f*) supply is given by :

$$f = \frac{Np}{120} \quad ...(3.182)$$

- Certain A.C. circuits of bridge type are also used for the measurement of frequency over a wide range but most of them are used when the *source of alternating supply is not accessible or where the frequency is in the audible range.*

- Commercial frequency (*i.e.* frequency of the order of 50–100 Hz) is measured by portable commercial instruments, known as *frequency meters* or *indicators.* Although the accuracy of such instruments is less than that of above methods but sufficient for normal uses.

3.11.2.2. Types of frequency meters

The different types of frequency meters are :

1. Mechanical resonance type.
2. Electrical resonance type
3. Weston type.
4. Ratiometer type.
5. Saturable core type.
6. Digital frequency meters.

- Measurement of audio and radio frequencies will also be discussed.

3.11.2.3. Mechanical resonance type (Vibrating reed type) frequency meter

This meter works on the *principle of the mechanical resonance* of thin flat steel reeds arranged alongside and close to an electromagnet. This meter is also known as *vibrating reed frequency meter.*

Construction :

— It consists of a number of thin steel strips called *reeds.* These reeds are placed in a row alongside and close to an electromagnet as shown in Fig. 3.120(*a*).

— The electromagnet has a laminated iron core and its coil is connected in series with a resistance, across the supply whose frequency is to be measured.

— Each reed is approximately about 4 mm wide and 0.5 mm thick. All the reeds are not exactly similar to each other; they *have either slightly different dimensions* or *carry different weights or flags at their tops.*

The natural frequency of vibration of the reeds depends upon their weights and dimensions.

— The reeds are so designed and arranged that the natural frequency of one reed differs by one or half a cycle from that of reed next to it. So if a frequency meter has a range of 47–53 Hz then natural frequency of first reed will be 47 Hz, 47.5 Hz for second reed, 48 Hz for the third reed and so on.

The reeds are fixed at the bottom end and are free at the top end. Since the reeds on a frequency meter are arranged to be viewed end on, they have a portion bent over at free end to *serve as a flag* as shown in Fig. 3.120(*b*). The flags are painted *white* to provide *maximum contrast* against their *black* background.

Working :

— When the instrument is connected across the supply whose frequency is to be measured, an alternating flux is set up. Due to this alternating field an attractive force is experienced upon the reeds after every half cycle. Consequently the reeds tend to vibrate but only the reed whose actual frequency is double of supply frequency will be in resonance and vibrate most (with maximum amplitude). Normally the vibration of other reeds is so slight as to be unobservable.

— If two adjacent reeds vibrate with equal amplitude then the supply frequency will be half way between the frequencies of the two adjacent reeds. The usual range of frequency meters of this type is about 6 Hz (say from 47 Hz to 53 Hz).

Fig. 3.120(*c*) illustrates the *vibration of reeds at different frequencies.*

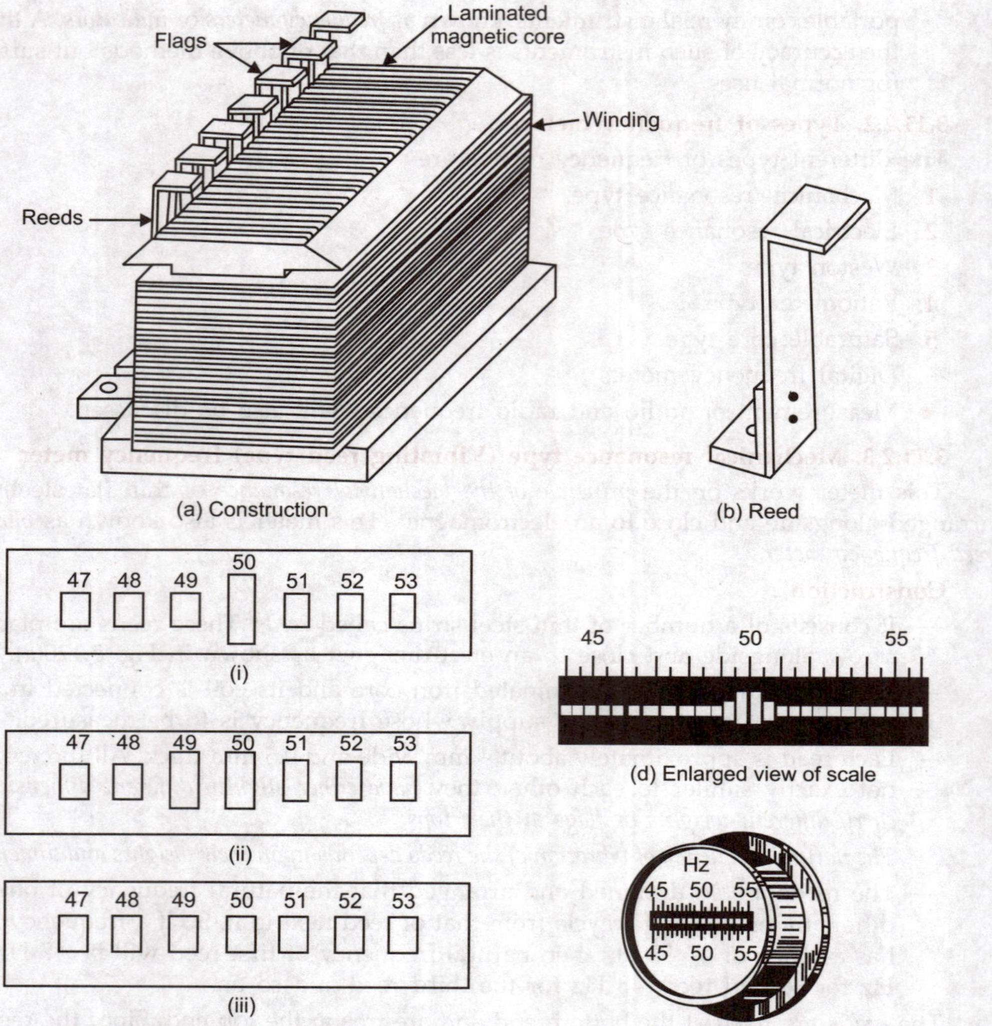

Fig. 3.120. Vibrating reed type frequency meter.

Fig. 3.120(c)–(i) illustrates that when supply frequency is 50 Hz, the reed having natural frequency of 50 Hz vibrates with maximum amplitude.

Fig. 3.120(c)–(ii) illustrates that when frequency is 49.5 Hz then reeds having natural frequencies of 49 Hz and 50 Hz will vibrate with equal and maximum amplitude.

Fig. 3.120(c)–(iii) illustrates that neither of the reeds has considerable amplitude, which indicates that either magnetic coil is *unexcited* or frequency under measurement is *out of range.*

- *The frequency range of a given set of reeds may be doubled by polarising the magnets.* The polarisation may be accomplished by using a D.C. winding in addition to the A.C. winding or by using a permanent magnet.

Advantage. *The indications of the meter are independent of the waveform of the applied voltage and of the magnitude of the voltage,* provided that the voltage should be high enough to provide sufficient amplitude for the reed vibration so as to make its reading reliable.

Disadvantage. These instruments cannot read much closer than half the frequency difference between adjacent reeds. Thus they *cannot be used for precision measurements.*

3.11.2.4. Electrical resonance type frequency meter

Fig. 3.121 shows a very simple form of indicating frequency meter, whose action is dependent upon *electrical resonance.*

Construction :

— It consists of a *laminated iron core of varying cross-section, a fixed magnetising coil* mounted at one end of the iron core and a *moving coil* with a pointer attached to it and pivoted so that it can move freely over the iron core as shown.

— The moving coil is connected across a capacitor C and the fixed coil is connected across the supply circuit, whose frequency is to be measured.

Working :

— When the magnetising coil is connected across the supply circuit, current *I* flows through it and a flux ϕ in phase with current is set up in the iron core and an e.m.f. *e* lagging behind the flux ϕ by 90° is induced in the moving coil.

The current in the moving coil be lagging or leading the induced e.m.f. *e*, depending upon whether the circuit of the moving coil is largely inductive or capacitive, as shown in Fig. 3.121(*b–i*) and (*b–ii*) respectively. The current in the moving coil will be *in phase* with *e* [Fig. 3.121 (*b–iii*)] when inductance reactance (X_L) is *equal* to capacitive reactance (X_C).

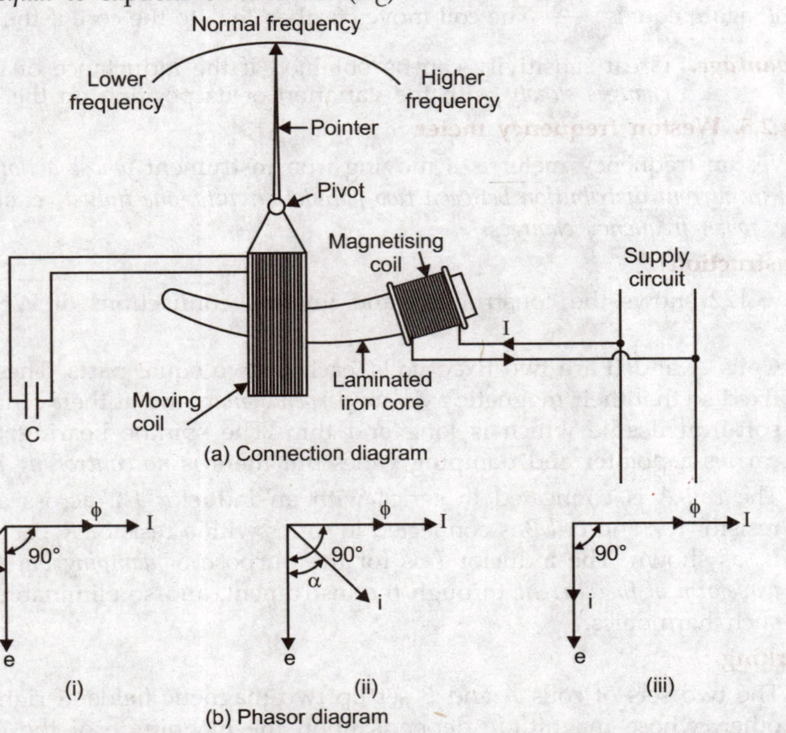

Fig. 3.121. Electrical resonance type frequency meter.

— The torque on the moving system (which has *no controlling device*) is given by :

$$T \propto Ii \cos (90° \pm \alpha) \qquad ...(3.183)$$

where, I = The magnetising coil current,
i = The moving coil current, and
α = The phase angle between current and e.m.f. induced in the moving coil.

From the eqn. (3.183) it is evident that the torque on the moving system will be *zero* when X_L (or ωL)= X_C $\left(\text{or } \frac{1}{\omega C}\right)$

- Referring now to the actual operation of the instrument, the capacitive reactance $\frac{1}{\omega C}$ $\left(\text{or } \frac{1}{2\pi fC}\right)$ is *constant for any given frequency*. The inductive reactance ωL ($2\pi fL$) depends upon the *position of the pivoted coil on the core. The nearer this coil approaches the magnetising coil the greater is its inductance.* The pivoted coil is pulled towards the magnetising coil, therefore until $\omega L = \frac{1}{\omega C}$ when, as already seen, the torque is *zero*. Obviously the *circuit of the moving coil is in resonance when the torque is zero.*

The value of the capacitor C is so chosen that the pivoted coil takes up a convenient mean position when the frequency is at its normal value. If the frequency increases $\frac{1}{\omega C}$ decreases, and the coil moves *"off"* the core, so that its inductance is reduced to the point where ωL again equals $\frac{1}{\omega C}$. The coil moves farther *"on"* to the core if the frequency falls.

Advantage. Great sensitivity can be obtained if the inductance of the moving coil *changes slowly* with the variation of its position on the core.

3.11.2.5. Weston frequency meter

A Weston frequency meter is a moving iron instrument *whose action depends on the variation in current distribution between two parallel circuits, one inductive and the other non-inductive, when frequency changes.*

Construction :

Fig. 3.122 shows the construction and internal connections of Weston frequency meter :

— Coils A and B are two fixed coils, each in two equal parts. These two coils are fixed so that their magnetic axis are *perpendicular*, and at their centres is a pivoted soft-iron needle which is long and thin. The spindle bearing the needle also carries a pointer and damping vanes but there is *no controlling device.*

— The *coil-A* is connected in series with an inductor L_A, across a non-inductive resistor R_A, and *coil B* is connected in series with a resistor R_B, across an inductor L_B as shown. The inductor L is for the purpose of *damping out harmonics in the waveform of the current* through the instrument, and so eliminating errors due to such harmonics.

Working :

— The two sets of coils A and B set up two magnetic fields at right angle to each other, whose magnitude depends upon the magnitude of the current flowing through the respective sets of coils. Both of these fields act upon the pivoted iron needle, so the position taken by the soft-iron needle depends upon the magnitude of two fields or of currents in coils A and B. The values of various resistances and inductances are so chosen that *for normal frequency of supply*, the voltage drop across non-inductive resistor R_A and inductance L_B sends equal currents through the two pairs of coils A and B so that the needle takes such a position that the *pointer is at the centre* as shown in the figure.

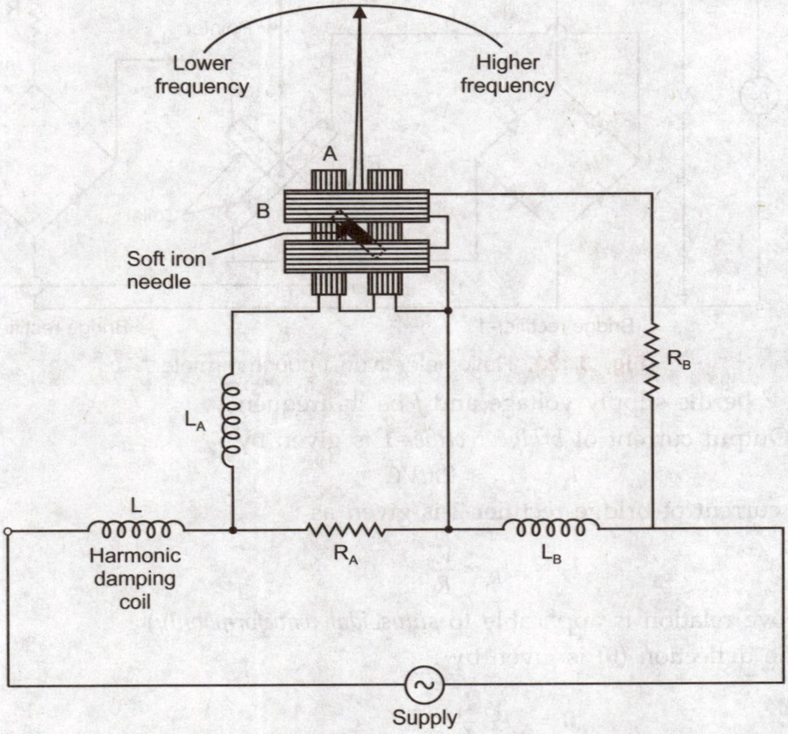

Fig. 3.122. Weston frequency meter.

— When the supply frequency *increases* above the normal one, the voltage across R_A remains more or less same and reactance of coil L_A increases due to increase in frequency, so the current through coil-A *decreases.* Simultaneously the current through the coil-B *increases* on account of increase of reactance of inductive coil L_B resulting in increase of voltage drop across the coil-B. Thus the magnitude of field of coil-A decreases and that of coil-B increases and the needle sets itself more in line with axis of coil-B; the pointer, therefore, swings over to the right indicating increased frequency.

Decrease of frequency obviously has the opposite effect, and these variations of frequency are followed by the pointer as shown.

3.11.2.6. Ratiometer type frequency meter

Fig. 3.123 shows the construction and connection diagram for a ratiometer type frequency meter :

— This type of frequency meter consists of a *ratiometer which gives a clear relationship between the current ratio and the deflection.*

— The two coils of this ratiometer are fed with rectified output currents of two separate bridge rectifiers as shown.

— The input sides of the two bridge rectifiers are connected to A.C. supply whose frequency is to be measured. Input sides of one of the bridges rectifiers has a series capacitance C and the other has series resistance R.

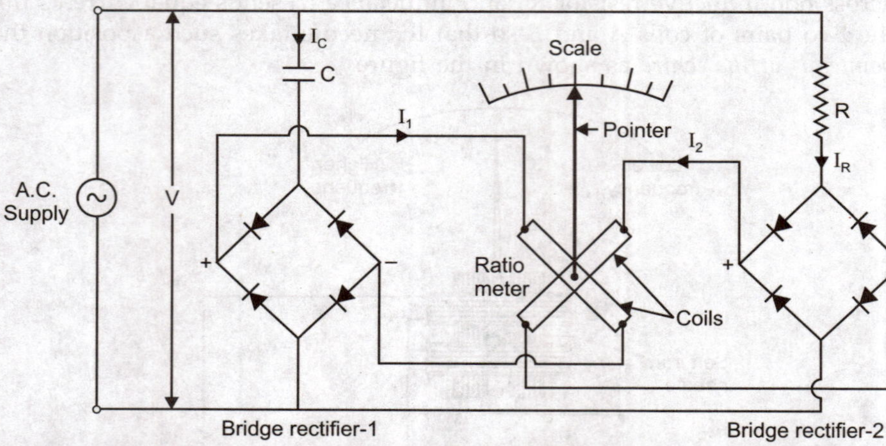

Fig. 3.123. Ratiometer type frequency meter.

- Let V be the supply voltage and f be its frequency.

Then, Output current of *bridge rectifier-1* is given by :

$$I_1 \propto I_C \propto 2\pi fVC \qquad \text{...(3.184)}$$

Output current of bridge rectifier-2 is given as :

$$I_2 \propto I_R \propto \frac{V}{R} \qquad \text{...(3.185)}$$

(The above relation is applicable to *sinusoidal waveform only*)

Then, the deflection (θ) is given by :

$$\theta = K.\frac{I_1}{I_2}$$

$$= K_1 \times \frac{2\pi fVC}{(V/R)} = 2\pi K_1 CRf$$

Since K_1, C and R are constants.

$$\therefore \qquad \theta = K_2 f \qquad \text{...(3.186)}$$

where K_2 = a constant $2\pi K_1 CR$

— Thus, the instrument has a *linear scale of frequency* provided the ratiometer is so designed that the *deflection is directly proportional to the ratio of two currents.*

— Since the supply voltage V does not appear in the expression for deflection, the instrument *may be used for a fairly wide range of voltage below the maximum specified.* However, the voltage *should not be too low* otherwise distortions are introduced which make the meter read wrongly.

- This meter may be employed *upto a frequency of 5 kHz.*

3.11.2.7. Saturable core frequency meter

Fig. 3.124 shows the construction and connection diagram of a saturable core frequency meter.

Construction :

— This instrument has a saturable core transformer T as its primary detector.

— The core assembly consists of the following parts :

(i) An outer core made of magnetic material;

(ii) Leg A made of non-magnetic material;

(iii) Leg B made of magnetic material, it *saturates at low values of m.m.f.*

The cross-section of leg A being same as that of B.

— The primary winding P is wound around legs A and B. The secondary winding S consists of two coils, one around leg A and the other around B as shown in the figure. The two coils are connected in *series. The e.m.fs. induced in these coils oppose each other.*

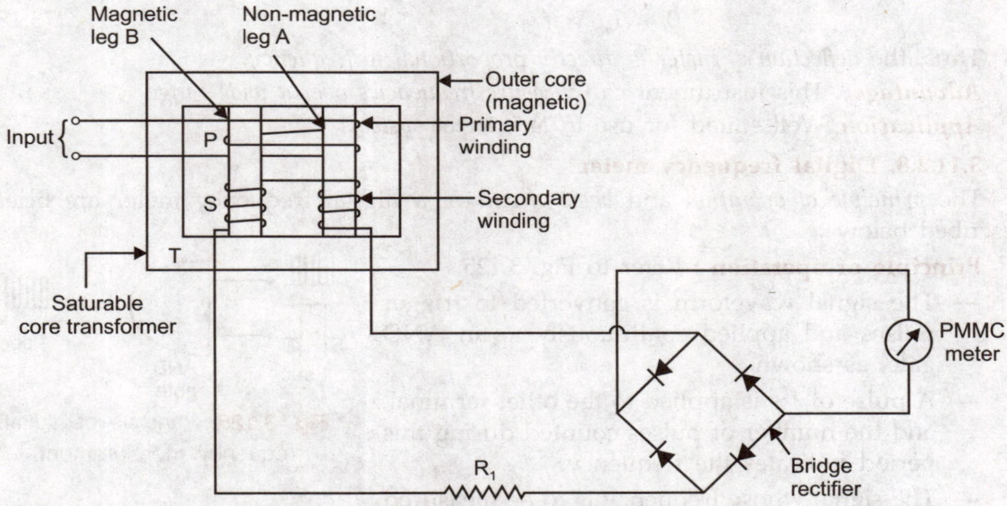

Fig. 3.124. Saturable core frequency meter.

Theory of operation :

When there is *saturation,* the rate of increase of induced voltage in the secondary coil over leg A will equal the rate of increase of induced voltage in secondary coil wound over the magnetic leg B. Thus the rate of increase in these two coils will cancel, and *secondary output voltage will not be a function of the primary voltage but will be a function of frequency only.*

If ϕ is the difference between the flux of leg A and leg B, then the induced e.m.f. in the secondary winding is given by :

$$e = N \cdot \frac{d\phi}{dt}$$

Now, the average secondary voltage per half cycle,

$$E_{av.} = \frac{\omega}{\pi} \int_{0}^{\pi/\omega} e.dt$$

$$= \frac{\omega}{\pi} N \int_{-\phi_m}^{+\phi_m} d\phi$$

or, $$E_{av.} = 4Nf\phi_{max}$$

$\therefore$ $$f = \frac{E_{av.}}{4N\phi_{max}} = kE_{av.} \quad \text{...(3.187)}$$

where k is a constant after the magnetic leg B becomes saturated.

The output of this circuit is rectified and the current in the D.C. meter is :

$$I_{av.} = \frac{E_{av.}}{R}$$

where R includes the resistance of the bridge rectifier, the meter and the resistance R_1.

$\therefore$ $$f = kRI_{av.} \quad \text{...(3.188)}$$

The deflection θ is directly proportional to current $I_{av.}$ as current is detected by PMMC meter, or

$$\theta \propto I_{av.} \propto f \quad \text{...(3.189)}$$

Thus, the *deflection of meter is directly proportional to frequency.*

Advantage. This instrument can *measure frequencies over a wide range.*

Application. Well suited for use in *tachometer systems.*

3.11.2.8. Digital frequency meter

The *principle of operation* and *basic circuit* of a digital frequency meter are briefly described below :

Principle of operation : Refer to Fig. 3.125

— The signal waveform is converted to trigger pulses and applied continuously to an AND gate, as shown.

— A pulse of 1 s is applied to the other terminal, and the number of pulses counted during this period indicates the frequency.

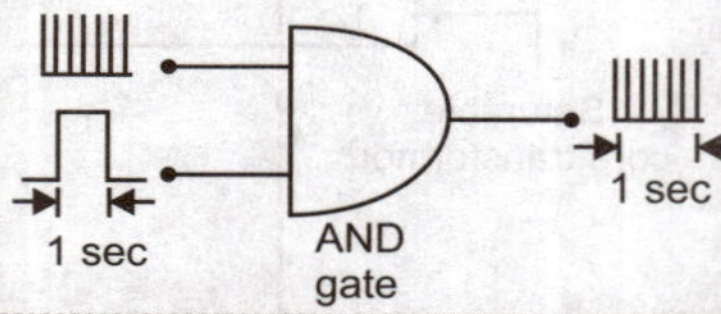

Fig. 3.125. Principle of digital frequency measurement.

— The signal whose frequency is to be measured is *converted into a train of pulses one pulse for each cycle of the signal.* The *number of pulses* occuring in a definite interval of time is then *counted* by an electronic counter.

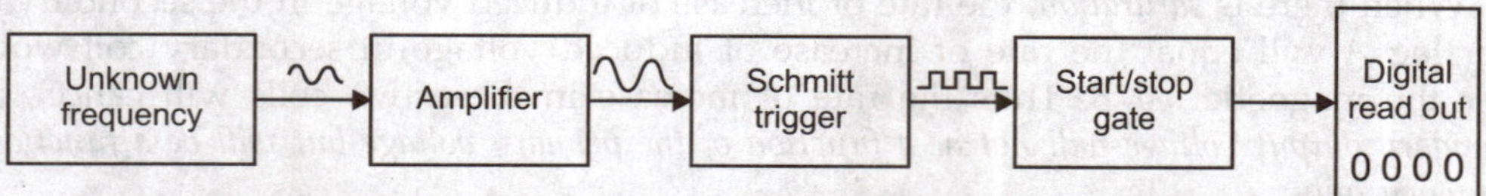

Fig. 3.126. Basic circuit of a digital frequency meter.

Since each pulse represents the cycle of the unknown signal the number of counts is a direct indication of the frequency of the unknown signal.

Since electronic counters have a high speed of operation, high frequency signals can be measured.

Basic circuit :

Fig. 3.126 shows the block diagram of a basic circuit of a digital frequency meter :

— The signal may be amplified before being applied to Schmitt trigger.

— The Schmitt trigger converts the input signal into square wave with fast rise and fall times, which is then differentiated and clipped. As a result, the output from a Schmitt trigger is a train of pulses, one pulse for each cycle of the signal.

— The output pulses from the Schmitt trigger are fed to 'Start'/'Stop' gate. When this gate is *enabled,* the input pulses pass through this gate and are fed directly to the electronic counter, which counts the number of pulses. When this gate is *disabled,* the counter stops counting the incoming pulses. The counter displays the number of pulses that have passed through it in the time interval between start and stop. If this interval is known, the unknown frequency can be measured.

3.11.2.9. Measurement of Audio and Radio Frequencies

The audio and radio frequencies (AF and RF) can be measured by using the following techniques/methods :

1. Bridge technique.
2. Heterodyne technique.
3. Wavemeters.
4. Interpolation technique.

Some of these methods/techniques are described below :

1. Bridge technique :

An unknown frequency can be determined by applying it to bridge in which the balance dependent upon frequency and then calculating the frequency from the circuit constants is required to give balance. Such arrangements find considerable use for measuring AF.

Any type of bridge or null network in which balance depends upon frequency can be used. Typical examples are : *Wien, Hay* and *resonance bridges* and the *null network.*

- Fig. 3.127 shows "*null network*" for measuring AF; the frequency is calculated using the relation :

$$\omega = \frac{1}{R_2}\sqrt{\frac{2}{C_1 C_2}} \qquad ...(3.190)$$

- The "*Wien bridge*" is particularly satisfactory for measuring AF because it can be brought into balance merely by varying resistive elements containing no inductance, and can be conveniently proportioned to cover a wide frequency range (for details refer to Art. 6.3.11).

An accuracy in the order of 0.5 to 1.0 % can be readily obtained in frequency bridges. In the frequency range 300 to 500 Hz, balance can be mostly easily made by the use of telephone receivers. At frequencies outside this range some indicating arrangements, such as an amplifier and VTVM combination, is normally required.

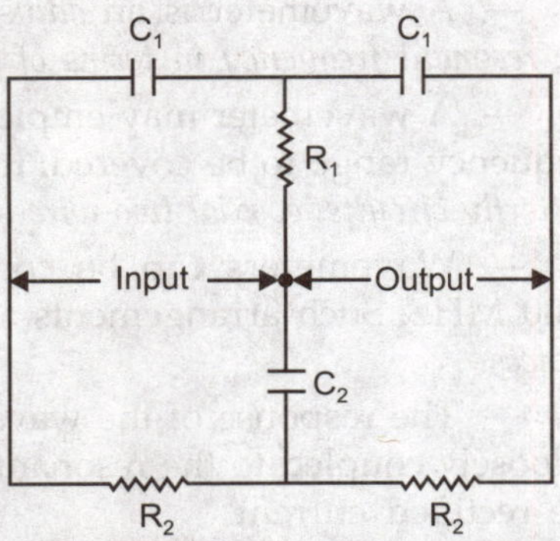

Fig. 3.127. Null network for measuring AF.

2. Heterodyne technique :

- The basis of heterodyne frequency meter is a *stable and tunable oscillator*.
- Heterodyne frequency meters are commonly provided with a *detector, audio amplifier,* and *headphones* as shown in Fig. 3.128 :

— Energy from the *oscillator* is supplied to the *detector* by internal coupling, and in addition provision is made whereby energy from an external source may be superimposed upon the detector input. The frequency of the heterodyne meter can then be made the same as the unknown frequency by adjusting so that the difference (or beat) frequency of the two oscillators as observed on the headphones is reduced to zero.

— These frequency meters are sometimes provided with a *crystal oscillator calibrator*. This is simple but well-designed crystal oscillator utilizng a *low temperature coefficient crystal to ensure good frequency stability*. The crystal oscillator produces a known frequency which can be used to check one or more points on the frequency calibration of the heterodyne meter with the aid of *detector audio amplifier*.

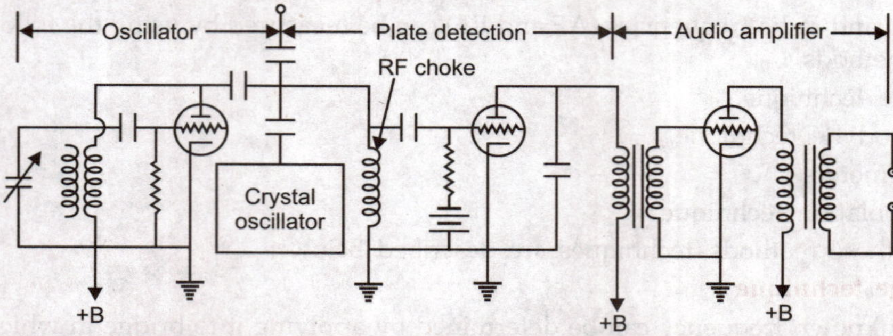

Fig. 3.128. Heterodyne frequency meter.

Uses : Heterodyne frequency meters are employed for the following *purposes* :

1. To produce known frequencies.
2. To measure an unknown frequency by comparison with the controllable and known frequency of heterodyne frequency meter.
3. To interpolate between fixed frequencies.

3. Wavemeters :

Wavemeters are used to measure frequency when high *accuracy is not required* and where *simplicity and portability are important*.

— A wavemeter is an *adjustable resonant circuit provided with a calibration that gives the resonant frequency in terms of setting of the tuning adjustments*.

— A wavemeter may employ any type of resonant circuit that is convenient for the frequency range to be covered; thus resonant circuits based on *lumped constants including butterfly circuits, coaxial two-wire lines, and cavities,* all find use in wavemeter applications.

— Wavemeters can be constructed with lumped circuit elements up to about 1200 MHz. Such arrangements are normally employed either as *absorption or as reaction devices*.

— The response of the wavemeter is commonly indicated by a crystal detector that is loosely coupled to the resonant circuit and associated with a micrometer that indicates the rectified current.

— Arrangements employing a fixed inductance and variable capacitance are suitable for use from the lowest radio frequencies up to about 100 MHz; this entire range can be covered by single variable condenser provided with a suitable set of plug-in coils as shown in Fig. 3.129. In this the coupling between the crystal and resonant circuit is made small by making capacitance C_1 small and C_2 large.

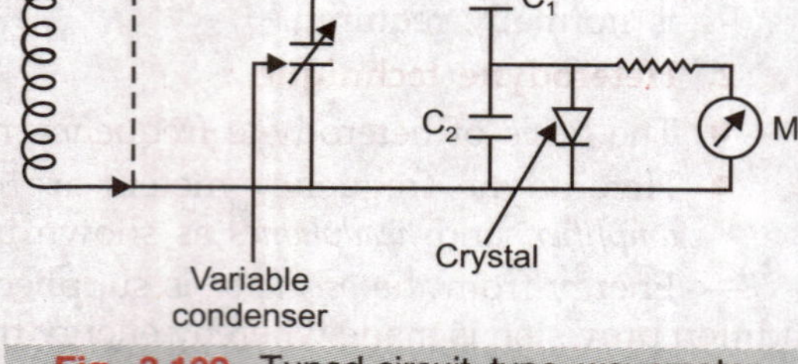

Fig. 3.129. Tuned-circuit type wavemeter.

— A variation of the inductance-capacitance wavemeter that is useful at *very high frequencies consists in varying both the coil and the condenser simultaneously*. This increases the frequency range that can be covered with a single coil and condenser combination, and is particularly useful at frequencies of the order of 50 to 400 MHz.

- The *accuracy* obtainable with such type of wavemeters depends upon the circuit Q and upon details of construction. In general, accuracies of the order of 1% or, better are readily obtained but it is quite difficult to make the accuracy as great as 0.1 %.

Cavity wavemeters :

- The *resonant cavities are extensively used as wavemeter at* ***microwave frequencies****. They have high accuracy, mechanical simplicity, and a large physical size in proportion to the wavelength being measured.*
- Practical cavity wavemeters are always in the form of cylinders which are tunable by means of a piston that varies the cavity length, as shown in Fig. 3.130.

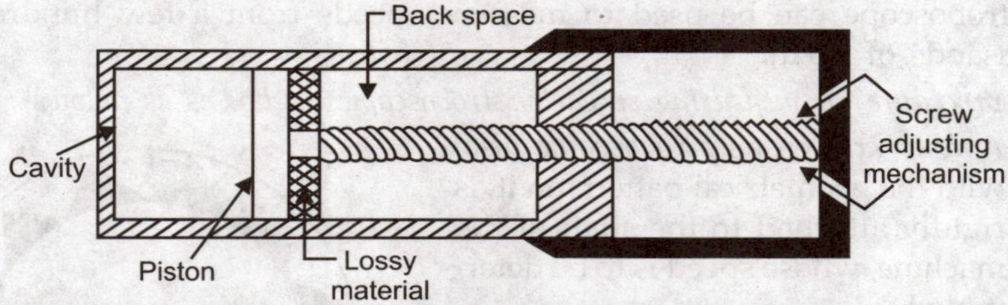

Fig. 3.130. Simple cylindrical cavity.

- Coupling to a wavemeter is accomplished by a loop or an orifice. A loop is used when co-axial systems are used to excite a cavity or carry power from it, while orifice is normally used in a waveguides system.
- The accuracy with which frequency can be determined using a well made cavity wavemeter is *quite high* (Precision of the order of 1 part in 1000 is easy to obtain and 1 part in 100,000 can be obtained by careful mechanical design combined with temperature compensation and correction).

Frequency standards :

The fundamental standard of frequencies is the period of rotation of earth. It can be measured with great accuracy by astronomical methods and is taken as a standard frequency of 1 cycle/day. The error in the measurement is approximately 164 μs. All standards of frequency must ultimately be referred to this fundamental source for calibration purpose.

Practical standards can be *classified* as follows :

(i) Primary standards. A primary standard of frequency is an oscillator which generates a frequency that is constant over a long period and is checked directly with the earth's rotation.

(ii) Secondary standards. These standards of frequency are *very stable oscillators* which have their frequencies checked periodically against the primary standards.

- Recent developments in microwave spectroscopy have introduced the *possibility of using spectrum lines as frequency standard.*

3.11.3. Rotational Speed Measurements

3.11.3.1. Introduction

The rotational speed of a machine can be measured by using electrical methods or methods involving electrical-operated measuring equipment. If this machine is an alternator the speed is a measure of the frequency of the A.C. voltage generated, the number of the poles of the alternator being known.

One simple and direct method is to use a small generator with a permanent-magnet field, and to *measure, with a voltmeter, the generated voltage. This is proportional to the speed,* and *the moving-coil voltmeter, with a uniform scale, can be calibrated directly in rotational speed (r.p.m. or r.p.s.).*

3.11.3.2. Stroboscopic method for measurement of speed

***Stroboscope** is a device which emits periodic flashes of light of very short duration.*

— The stroboscopes are generally *calibrated to read the speed directly in r.p.m.* The frequencies of the flashes is adjusted until the body appears stationary and the speed is read directly from the dial which is employed for adjusting the frequency. When the adjustment is calibrated to read the speed directly, the stroboscope is called *"Tachometer"*.

— A stroboscope can be used to measure speeds from a few hundred to many thousands of r.p.m.

- *The procedure for measuring speed by stroboscopic method is as follows :*

— A disc known as *stroboscopic disc*, with the geometrical pattern on it, is rigidly attached to the shaft of the machine, whose speed is to be determined.

— The disc is illuminated by a series by rapid flashes, each being of very short direction in comparison to time interval between successive flashes. The series of flashes for illuminating the stroboscopic disc are obtained with the help of a stroboscope.

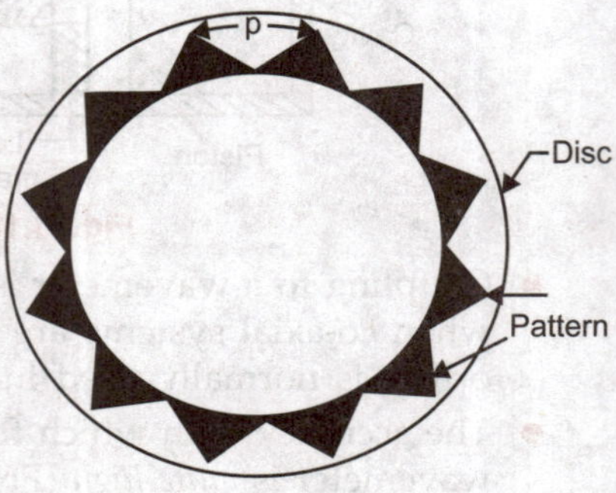

Fig. 3.131. Disc with a geometrical pattern.

— The frequency of the flashes is adjusted such that disc seems to be stationary. This will happens when the point adjacent to that viewed in the preceeding flash, has taken up the same position and since all points are exactly identical, the eye cannot mark the change. In other words, the disc will *seem to be stationary when the disc moves one point pitch during the time interval between the two successive flashes.* Same effect will be caused if disc moves through $2p, 3p, ... np$ between the successive flashes.

Let, n = Number of points of the disc,

F = Number of flashes per minute,

N = The speed (primary) of the machine in r.p.m., and

d = Diameter of the circular pattern.

The pitch of the points, $p = \dfrac{\pi d}{n}$ ...(*i*)

The distance moved by the disc during the successive flashes $= \dfrac{\pi d N}{F} = p.$ ...(*ii*)

From (*i*) and (*ii*), we get

$$\frac{\pi d}{n} = \frac{\pi d N}{F}$$

or, $$N = \frac{F}{n}$$ ...(3.191)

Thus, if frequency of flashes is known, the rotating speed (N) of the shaft can be determined.

- The fact that the pattern will be stationary even if the actual speed is a whole multiple of N, causes no confusion since the approximate speed can be measured by another but less accurate method. So, the difference between N and its multiple can be easily known *e.g.* if the speed determined by stroboscopic method is 300 and with a less accurate method 890, the accurate speed of the machine is 900 r.p.m., multiplying factor being 3.

When the actual speed is a whole number of N then N is known as the primary speed of the machine.

- A neon lamp with a tripping circuit is generally used with stroboscope to provide periodic flashes of light of very short duration.

Disadvantages :

(*i*) This method is not as accurate as the methods using digital meters.

(*ii*) The stroboscope cannot be used for measuring speed in surroundings having ambient light above a certain level.

Applications :

(*i*) This method is especially suitable for measuring the speed of *small-power rating machines.*

(*ii*) Also suitable for determining the speed of machine parts which are *not readily accessible or where it is not practible to use mechanical methods or where the speed is variable.*

3.11.4. Synchroscopes

3.11.4.1. Introduction

It may be logically assumed that one alternator is placed in parallel with one or more other alternators only when an additional load requires it. Those alternators already carrying load are known as the *running machines,* while that which is to be placed in the system is known as the *incoming machine.*

At the time of synchronising the following *conditions* must be met :

1. The effective voltage of the incoming alternator *must be exactly equal to that of the others, or of the busbars connecting them.*
2. The *phase rotation,* or sequence of the running and incoming alternators, *must be the same.*
3. The individual phase voltages which are to be connected to each other *must be in exact phase opposition.* This is the same as saying that D.C. generators must be connected + to + and – to –.
4. The *frequencies should be the same,* although it is more desirable that the frequencies at the instant of paralleling be almost, but not quite, identical.

- *A* **synchroscope** *is used to determine the correct instant for closing the switch which connects an alternator with another alternator* (*or to the power station busbars*).
    — The voltages can be checked with the help of a voltmeter.
    — For a 3-phase machine the phase sequence of the two should be the same. This condition is verified by a *phase sequence indicator.*

— *The function of a synchroscope is to indicate the difference in phase and frequency of voltages of the busbar and the incoming machine.*

Types of synchroscopes :

There are two types synchroscopes :

1. *Moving iron type.*
2. *Electrodynamic (Weston) type.*

Both are special forms of respective power factor meters.

3.11.4.2. Moving iron type synchroscope

Fig. 3.132 shows the construction and connection diagram of a moving iron type synchroscope, which is due to Lipman :

Construction :

— It consists of a *fixed coil A* divided into two parts. This coil is designed for a small value of current and is connected in series with a resistance across two phases of the busbar.

— There are two iron cylinders C_1 and C_2 mounted on the spindles. Each iron cylinder is provided with two iron vanes whose axes are 180° out with each other.

— The iron cylinders are excited by two pressure coils P_1 and P_2 which are connected to two phases of the incoming machine one of the coils has series resistance R and the other has a series inductance L_1. This is done *to create an artificial phase difference of 90° between the currents of two pressure coils.*

— There are *no control springs*.

— The pointer (provided in the instrument) moves over a dial marked *"Fast"* and *"Slow"*

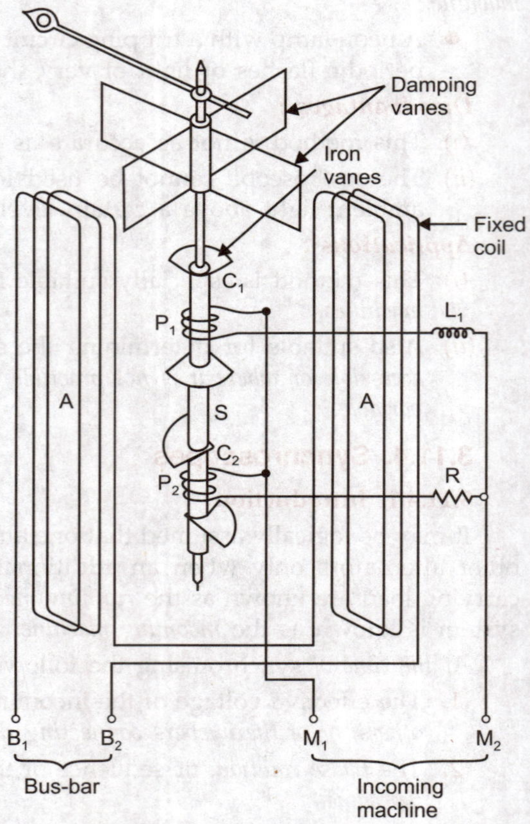

A = Fixed coil; C_1, C_2 = Iron cylinders; P_1, P_2 = Pressure coils

Fig. 3.132. Moving-iron alternating-field synchroscope.

Working :

— When the frequency of the incoming machine is equal to that of busbar, the synchroscope works as a single-phase power factor meter, and so indicates the phase difference between the two e.m.fs.

— When the frequency of the incoming machine is different from the frequency of busbar, the spindle rotates continuously at a speed in r.p.s. equal to difference between the frequencies of the two systems. The direction of rotation depends upon whether the incoming machine is running too fast or too slow.

— The instrument has 360° scale. The displacement from the zero position indicates the phase difference between the e.m.fs. When pointer of the instrument is stationary at the *centre or zero position, the parallel switch should be closed.*

- The moving iron type synchroscopes are commonly used because of the following *reasons* :
  - (*i*) They are cheap and robust.
  - (*ii*) They are inherently suitable for continuous operation.
  - (*iii*) No slip rings are necessary since there are no electrical connections to the moving system.

3.11.4.3. Electro-dynamometer (Weston) type synchroscope

Weston synchroscope is an instrument of dynamometer type and is *spring-controlled*. Fig. 3.133 shows the construction and connection diagram of this instrument.

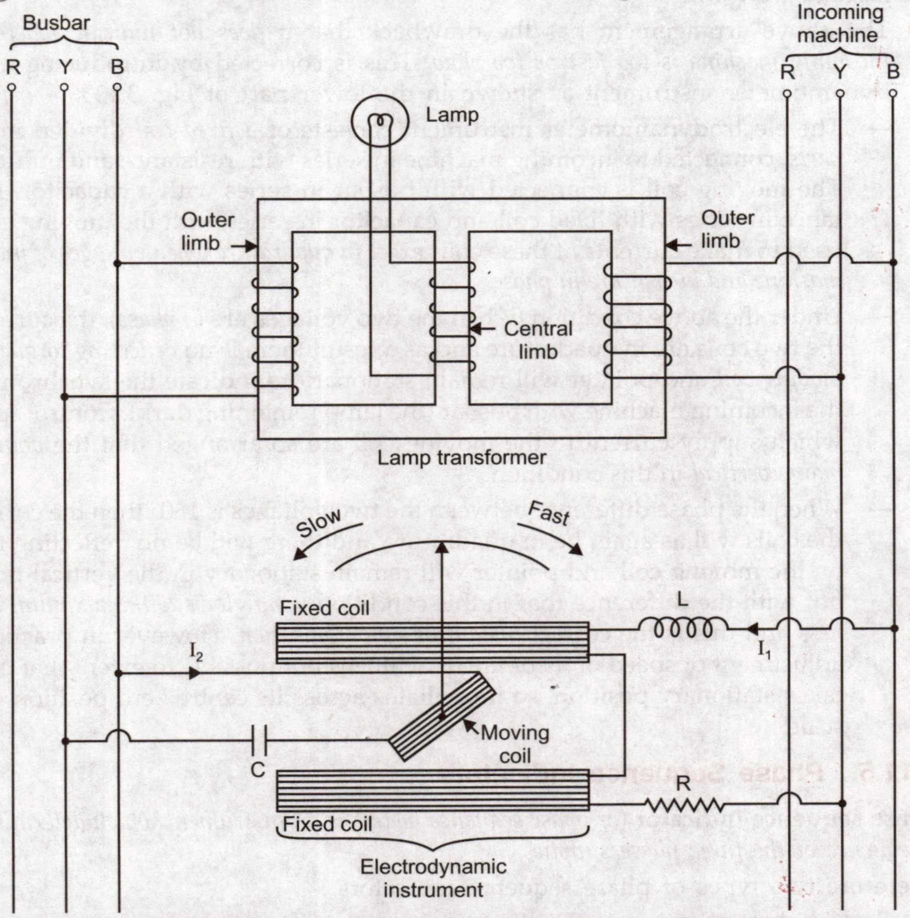

Fig. 3.133. Electro-dynamometer (Weston) type synchroscope.

- It consists of a three limbed transformer (also called lamp transformer). The winding on one of the outer limbs is excited from bus-bars and that on the other limb by incoming machine. The winding on the central limb is connected to a lamp.
  - — The two fluxes produced by the outer limbs are forced through the central limb. *The resultant flux through the central limb is equal to the phasor sum of these fluxes.* This resultant flux induces an e.m.f. in the windings of the central limb.

— The windings on the two outer limbs are so arranged that when the bus-bar and the incoming machine voltages are in phase, the two fluxes through the central limb are *additive* and thus e.m.f. induced in the central limb winding is *maximum.* Hence under these conditions the *lamp glows with maximum brightness.*

— When the two voltages are *180° out of phase* with each other the resultant flux is *zero* and hence *no e.m.f. is induced* in the central limb *winding*, consequently the lamp *does not glow at all and is dark.*

— When the frequency of incoming machine is different from that of busbar, the lamp will flicker (*i.e.* will be bright and dark alternately). *The frequency of flickering is equal to the difference in frequencies of the busbar and the incoming machine.*

The correct instant of synchronising is when the lamp is flickering at a very slow rate and is at its maximum brightness.

- The above arrangement has the drawback that it *does not indicate whether the incoming machine is too fast or too slow.* This is corrected by introducing electrodynamometer instrument as shown in the lower part of Fig. 3.133.

  — The electrodynamometer instrument consists of a *fixed coil* divided into *two parts,* connected to incoming machine in series with resistance and inductance. The moving coil is connected with busbar in series with a capacitor. Inductance in series with fixed coil and capacitor in series with the moving coil are used to make currents of these coils *exact in quadrature when voltages of incoming machine and busbar are in phase.*

  — Under the above condition (when the two voltages are *in phase*), the currents in the two coils are in quadrature and as a result there is no *deflecting torque on the moving coil* and pointer will remain stationary to indicate the synchronism of the incoming machine with busbar (the lamp remaining dark). Control springs, which supply current to the moving coil are so arranged that the *pointer remains vertical* in this condition.

  — When the phase difference between the two voltages is 180° then the current in the coils will as again be in quadrature and there will be no deflecting torque on the moving coil and pointer will remain stationary in the vertical position but with the difference that in this condition *lamp glows with maximum brightness* and this is the correct *instant of synchronisation.* However in practice, the adjustment of speed of incoming machine is not possible to extent that pointer takes stationary position, so it oscillates across its centre zero position on the scale.

3.11.5. Phase Sequence Indicators

Phase sequence indicator (*or phase sequence meter*) *is an instrument which determines the phase sequence of the three phase supplies.*

There are two types of phase sequence indicators :

1. *Rotating type phase sequence indicator.*
2. *Static type sequence indicator.*

3.11.5.1. Rotating type phase sequence indicator

This indicator works on the *principle similar to that of three-phase induction motors.*

— It consists of three coils mounted 120° apart (*i.e.* in star connection) in space. The three ends of the coils are brought out and connected to three terminals marked R Y B as shown in Fig. 3.134.

— These star-connected coils are excited by the supply whose frequency is to be determined. On the top of the coils is mounted an aluminium disc.

— The coils produce a rotating magnetic field and eddy current e.m.fs. are induced in the disc. These e.m.fs. cause eddy currents to flow in the aluminium disc. *A torque is produced due to the interaction of the eddy currents with the field.* Due to this torque the disc revolves and its *direction of rotation depends upon the phase sequence of the supply.*

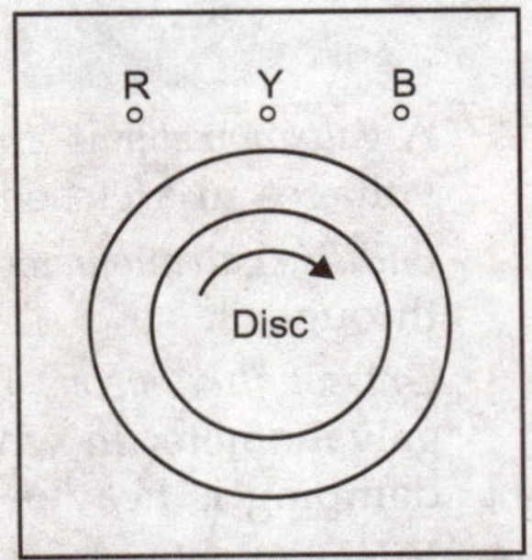

Fig. 3.134. Phase sequence indicator.

— The direction of the rotation of the disc is indicated by an arrow. *When the direction of the rotation is the same as that indicated by the arrow head, the phase sequence of the supply is the same as marked on the terminals of the instrument.* However, if the disc revolves *opposite* to the direction indicated by the arrow head, the sequence of suply is *opposite* to that marked on the terminals.

3.11.5.2. Static type sequence indicator

A static type sequence indicator may have the following *two* types of arrangement :

- One arrangement consists of two lamps and an inductor as shown in Fig. 3.135.
  — When the phase sequence is RYB, lamp-1 will be dim and lamp-2 will glow brightly.
  — In case, the phase sequence is RBY, lamp-1 will glow brightly and lamp-2 will be dim.
- In another arrangement, similar phase sequence indicator consists of two neon lamps and a capacitor basically as shown in Fig. 3.135(*a*). In practice, a resistor must be added in series with each lamp *to limit the current* as shown in Fig. 3.135(*b*).

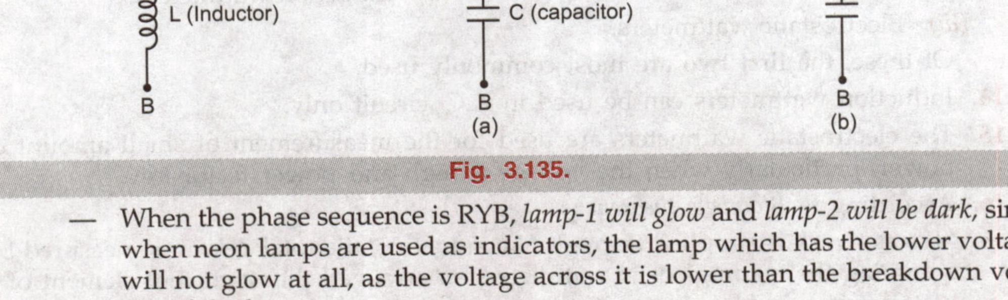

Fig. 3.135.

— When the phase sequence is RYB, *lamp-1 will glow* and *lamp-2 will be dark*, since when neon lamps are used as indicators, the lamp which has the lower voltage will not glow at all, as the voltage across it is lower than the breakdown voltage of the lamp.

HIGHLIGHTS

1. A *galvanometer* is an instrument used for detecting presence of small currents or voltages in a closed circuit or for measuring their magnitudes.
2. *Ballistic galvanometer* is used to measure the quantity of electricity (charge) passed through it.
3. *Grassot flux meter*, used for measurement of flux, is a special type of ballistic galvanometer in which the controlling torque is very small and electromagnetic damping is heavy.
4. *Vibration galvanometers* are of D'Arsonval type having a moving coil suspended between the pieces of a permanent magnet. When an alternating current is passed through the moving coil, an alternating deflecting torque is produced which makes the coil vibrate with a frequency equal to the frequency of the current passing.

   These galvanometers are the most widely used "tuned detectors" particularly as null point detectors in the A.C. bridges.
5. *Duddell's oscillograph* is essentially a vibration galvanometer having a low period of vibrations.
6. *Moving iron instruments* are commonly used in laboratories and switch boards at commercial frequencies because they are very cheap and can be manufactured with required accuracy.
7. The permanent moving-coil instrument is the most accurate type for D.C. measurements.
8. *Meggers* (megohm meters) are instruments which measure the insulation resistance of electric circuits relative to earth and one another.
9. *Hot-wire instruments* utilise the expansion of wire heated by current being measured.
10. *Thermocouple instruments* are suitable for very high frequencies upto about 50 Hz.
11. A *rectifier type instrument* is more sensitive than any other type of A.C. meter.
12. *Electrostatic instruments* are always used as *voltmeters* and that too more as a laboratory rather than as industrial instruments.
13. The wattmeters are of the following three types :

    (*i*) Dynamometer wattmeters; (*ii*) Induction wattmeters;

    (*iii*) Electrostatic wattmeters.

    Of these, the first two are most commonly used.
14. Induction wattmeters can be used in A.C. circuit only.
15. The electrostatic wattmeters are used for the measurement of small amount of power, particularly when the voltage is high and power factor low.
16. According to *Blondel's theorem* :

    "If a network is supplied through N conductors, the total power is measured by summing the readings of N wattmeters so arranged that a current element of a wattmeter is in each line and the corresponding voltage element is connected between that line and a common point. If the common point is located on one of the lines, then the power may be measured by N-1 wattmeters".
17. A *polyphase meter* is used to measure power in a polyphase (3-phase) system.
18. *Energy meters* are used to measure the quantity of electric energy supplied to a circuit in a given time.

19. The various types of energy meters are :
   (i) *Electrolytic meters* ...... Their operation depends on electrolytic action.
   (ii) *Motor meters* ...... There are really small electric motors
      — Mercury motor meters ...... Normally used on D.C. circuits.
      — Commutator motor motors ...... Can be used both for D.C. as well as A.C. work
      — Induction motor/energy meters ...... used only on A.C. circuits.
   (iii) *Clock meters* ...... They function as clock mechanisms.

OBJECTIVE TYPE QUESTIONS

Choose the Correct Answer :

Galvanometers

1. A d'Arsonval galvanometer uses a light and scale arrangement. The light source is placed 1 m away from the moving system of galvanometer. The arrangement uses a circular scale calibrated in mm. The deflection indicated by the scale is
   (a) 1000 mm (b) 2000 mm
   (c) 500 mm (d) none of the above.
   if a current of 1 mA is passed through the coil. The spring stiffness is 2×10^{-6} Nm/rad and the displacement constant is 2 Nm/A.
2. The time period of free oscillations in a galvanometer having a relative damping of 0.6 is 2 s. The frequency of damped oscillations is :
   (a) 0.5 rad/s (b) 0.3 rad/s
   (c) 0.4 rad/s (d) none of the above.
3. The resistance required for critical damping in a circuit is 1000 Ω. The galvanometer circuit has a resistance of 800 Ω. Is the galvanometer circuit
   (a) underdamped ? (b) undamped ?
   (c) overdamped ? (d) none of the above.
4. A galvanometer has a ratio of 0.9 for damped frequency oscillations to undamped frequency oscillations. Suppose moment of inertia, stiffness constant and damping constant are made twice their original value, what would be the new ratio of damped frequency oscillations to undamped frequency oscillations,
   (a) 0.9 (b) 1.11
   (c) 4 (d) 2.
5. In d'Arsonval galvanometer, an iron core is usually used between the permanent magnet pole faces. This is used so that
   (a) flux density in the air gap becomes high thereby a large deflecting torque is produced.
   (b) the effect of stray magnetic fields is reduced.
   (c) moment of inertia of moving parts becomes smaller.
   (d) none of the above.
6. Sometimes, the d'Arsonval galvanometers, do not use ferromagnetic cores between poles of the permanent magnet. In this case
   (a) the flux density becomes smaller resulting in low deflecting torque.
   (b) the dimensions of the moving coil can be made smaller thereby reducing the moment of inertia.
   (c) the magnetic field may not be radial resulting in a non-uniform scale even if spring control is used.
   (d) all of the above.

7. A current of 2 μA is passed though the moving coil of an undamped d'Arsonval galvanometer which has a displacement constant of 2 Nm/A and a control constant of 10×10^{-4} Nm/rad. The moving oscillates with an amplitude of :
   (*a*) 0.2 rad.
   (*b*) 0.4 rad.
   (*c*) 0.8 rad.
   (*d*) none of the above.
8. If the damping in a d'Arsonval galvanometer is only due to electromagnetic effects, the resistance required for critical damping is :
   (*a*) $G^2/\sqrt{CJ}$
   (*b*) $G/\sqrt{CJ}$
   (*c*) $G/2\sqrt{CJ}$
   (*d*) $G^2/2\sqrt{CJ}$
9. The relative damping in a galvanometer is 0.8. Its logarithmic decrement is approximately :
   (*a*) 0.48
   (*b*) 1.25
   (*c*) 4.19
   (*d*) −4.19.
10. In a critically damped galvanometer the deflection at a time 0.9 times the time of free oscillations after a current is passed through the moving coil is approximately
    (*a*) 0.986 times the final deflection.
    (*b*) 0.901 times the final deflection.
    (*c*) 0.866 times the final deflection.
    (*d*) none of the above.
11. A d'Arsonval galvanometer uses a lamp and scale arrangement. Its current sensitivity is 250 mm/μA. If the resistance of the coil is 100 Ω and the external resistance required for critical damping is 900 Ω, (which is connected in the circuit), the voltage sensitivity is
    (*a*) 0.25 mm/μV
    (*b*) 2.5 mm/V
    (*c*) 25 mm/μV
    (*d*) 250 mm/μV.
12. Ayrton shunt is used in d'Arsonval galvanometers so as to limit the current in the galvanometer coil to its maximum permissible value. The relative value of current through the galvanometer coil and the shunt
    (*a*) depends upon the value of resistance of galvanometer coil only.
    (*b*) depends upon the values of resistance of galvanometer coil and the shunt.
    (*c*) does not depend upon the value of resistance of galvanometer coil.
    (*d*) none of the above.
13. A Ballistic galvanometer should be designed with
    (*a*) a large period of natural oscillations and a negligble damping contant.
    (*b*) a small period of natural oscillations and a high damping constant.
    (*c*) a large period of natural oscillations and a high damping factor.
    (*d*) a small period of natural oscillations and a low damping factor.
14. A Duddell's oscillograph will give no amplitude disortion and phase displacement if its
    (*a*) moment of inertia and stiffnes constant are zero.
    (*b*) stiffness constant and damping factor are zero.
    (*c*) moment of inertia and damping factor are zero.
    (*d*) all of the above.
15. In a Duddell's oscilloscope the phase displacements of fundamental and 13th harmonic are calculated to be 3° and 36° respectively. The oscillograph will show the 13th harmonic to be
    (*a*) 39° ahead of its true position.
    (*b*) 38° ahead of its true position.
    (*c*) 3° ahead of its true position.
    (*d*) 3° behind its true position.

16. A vibration galvanometer is to be tuned to a frequency of 50 Hz. The ratio of its control constant to inertia constant should be

(*a*) 98696 (*b*) 2500

(*c*) 10.132×10^{-6} (*d*) none of the above.

17. A vibration galvanometer is tuned

(*a*) by changing the length and tension of vibrating coil.

(*b*) by attaching weight to the vibrating coil.

(*c*) by changing its damping constant.

(*d*) all of the above.

18. A Duddell's oscillograph can be used for frequencies

(*a*) upto 50 Hz (*b*) upto 500 Hz

(*c*) above 500 Hz (*d*) upto 10 kHz.

19. In a fluxmeter

(*a*) the controlling torque is produced by weights attached to moving coil.

(*b*) the controlling torque is produced by springs.

(*c*) there is no controlling torque.

(*d*) none of the above.

20. In an unshunted flux meter, the sensitivity is dependent upon

(*a*) resistance of moving coil.

(*b*) resistance of search coil.

(*c*) resistance of both moving coil and search coil.

(*d*) none of the above.

Ammeters, Voltmeters, Wattmeters and Energy meters

21. The heater wire of thermocouple instruments is made very thin in order to have

(*a*) a high value of resistance.

(*b*) to reduce skin effects at high frequencies.

(*c*) to reduce the weight of the instrument.

(*d*) to decrease the over-ranging capacity of the instrument.

22. Electrostatic type instruments are primarily used as

(*a*) ammeters (*b*) wattmeters

(*c*) voltmeters (*d*) ohmmeters.

23. Creeping in a single-phase induction type energy meter may be due to :

(*a*) overcompensation for friction

(*b*) overvoltage

(*c*) vibrations

(*d*) all of the above.

24. A Merz Price maximum demand Indicator Indicates :

(*a*) maximum demand

(*b*) average maximum demand over a specified period of time

(*c*) maximum energy consumption

(*d*) all of the above.

25. The range of an electrostic voltmeter can be extended by using

(*a*) a capacitor in series with the voltmeter whose capacitance is greater than the capacitance of the voltmeter.

(*b*) a capacitor in series with the voltmeter whose capacitance is smaller than the capacitance of the voltmeter.

(*c*) a resistor in series with the voltmeter.

(*d*) an inductor in series with the voltmeter.

26. The sensitivity of a PMMC instrument is 10 kΩ/V. If this instrument is used in a rectifier type voltmeter with half-wave rectification, what would be the sensitivity ?

(*a*) 10 kΩ/V (*b*) 9 kΩ/V

(*c*) 4.5 kΩ/V (*d*) 22.2 kΩ/V.

27. The high torque to weight ratio in an analog indicating instrument indicates

(*a*) high friction loss (*b*) low friction loss

(*c*) nothing as regards friction loss (*d*) none of the above.

28. A PMMC instrument produces a deflection of 90° when a current 2 mA passes through its moving coil. The instrument is spring controlled. Supposing the instrument is redesigned with its spring constant as before, and its permanent magnet is replaced with a new permanent magnet which produces twice the flux density as before, what would be the deflection if a current of 1 mA is passed through the coil ? The other features of the PMMC instrument remain the same.

(*a*) 180° (*b*) 90°

(*c*) 45° (*d*) 135°.

29. A PMMC meter rated at 50 μA is used in a rectifier type of instrument which uses full wave rectification. What is the sensitivity on sinusoidal A.C. ?

(*a*) 20 kΩ/V (*b*) 9 kΩ/V

(*c*) 22.2 kΩ/V (*d*) 18 kΩ/V.

30. A reverse biased diode in addition to a forward biased diode is used in rectifier type voltmeters using half-wave rectification. The use of the reverse biased diode is :

(*a*) it does not allow any current to flow through the meter during negative half cycle

(*b*) it short circuits the meter during the negative half cycle

(*c*) it does not allow the reverse leakage current to flow through the meter during the negative half cycle

(*d*) all of the above.

31. VAh metering can be done by using

(*a*) a ball and disc friction gearing.

(*b*) Trivector meter.

(*c*) Bridge connected rectifiers.

(*d*) all of the above.

32. Light load adjustments for induction type energy meters are usually done at

(*a*) 10% of full load current.

(*b*) 5% of full load current.

(*c*) 50% of full load current.

(*d*) 1% of full load current.

33. In order to reduce errors on account of temperature, instruments are provided with shunts and series connected swamping resistance. In order that errors on account of temperature changes be low, the swamping resistance

(*a*) should be made of a material having a high resistance temperature co-efficient.

(*b*) should be made of a material having a low resistance temperature co-efficient with the value of swamping resistance equal to meter resistance.

(*c*) should be made of a material having low resistance temperature co-efficient and should have a value of about 20 to 30 times that of meter resistance.

(*d*) should have an infinite value.

34. A 4-range milliammeter having ranges of 0—10 mA, 0—50 mA, 0—100 mA and 0—500 mA is used for measurement of current whose magnitude is not known. In order to prevent damage to the instrument, the selector switch of the ammeter should be first placed at

(*a*) 0—50 mA range (*b*) 0—100 mA range
(*c*) 0—10 mA range (*d*) 0—500 mA range.

35. The braking torque provided by a permanent magnet in a single-phase energy meter is proportional to the
(*a*) square of the flux of the permanent magnet.
(*b*) speed of the meter.
(*c*) distance of the permanent magnet from the centre of the revolving disc.
(*d*) all of the above.

36. The braking torque provided by a permanent magnet in a single-phase energy meter can be changed by
(*a*) providing a magnetic shunt and changing its position.
(*b*) changing the distance of the permanent magnet from the centre of the revolving disc.
(*c*) both (*a*) and (*b*).
(*d*) none of the above.

37. Which meter has the highest accuracy in the prescribed limit of frequency range :
(*a*) PMMC (*b*) Moving iron
(*c*) Electrodynamometer (*d*) Rectifier.

38. Which instrument is the cheapest disregarding the accuracy ?
(*a*) PMMC (*b*) Moving Iron
(*c*) Electrodynamometer (*d*) Rectifier.

39. Which instrument has the highest frequency range with accuracy within reasonable limits ?
(*a*) Moving Iron (*b*) Electrodynamometer
(*c*) Thermocouple (*d*) Rectifier.

40. In the single-phase induction meter, in order to obtain true value of energy, the shunt magnet flux should lag behind the applied voltage by
(*a*) 90° (*b*) 0°
(*c*) 45° (*d*) none of the above.

41. In an induction type of meter, maximum torque is produced when the phase angle, between the two fluxes is
(*a*) 0° (*b*) 45°
(*c*) 60° (*d*) 90°.

42. In an induction type meter, maximum torque is obtained when the parameters of rotating disc are :
(*a*) $R = 0$ (*b*) $X = 0$
(*c*) $R = X$ (*d*) none of the above.
where R and X are respectively the resistance and reactance of eddy current paths in the disc.

43. A 1 mA ammeter has a resistance of 100 Ω. It is to be converted to a 1 A ammeter. The value of shunt resistance is :
(*a*) 0.001 Ω (*b*) 0.1001 Ω
(*c*) 100000 Ω (*d*) 100 Ω.

44. Swamping resistance is connected
(*a*) in series with the shunt to reduce temperature error in shunted ammeters.
(*b*) in series with the ammeters to reduce errors on account of friction.
(*c*) in series with meter and have a high resistance temperature co-efficient in order to reduce temperature errors in ammeters.
(*d*) in series with the meter and have a negligible resistance co-efficient in order to reduce temprature errors in shunted ammeters.

45. A quadrant type electrostatic instrument uses two types of connections (*i*) Heterostatic and (*ii*) Idiostatic. An external battery is used

(*a*) for Idiostatic connection

(*b*) for Heterostatic connection

(*c*) for both Idiostatic and Heterostatic connections

(*d*) none of the above.

46. The size of electrostatic voltmeters is large as compared with that of corresponding electromagnetic voltmeters because

(*a*) they are used for measurement of high voltages.

(*b*) energy density in an electostatic field is consideraly smaller than that in an electromagnetic field.

(*c*) their frequency range is smaller.

(*d*) they have a high dielectric loss.

47. In a single-phase induction type energy meter, the lag adjustment is done

(*a*) to make the current coil flux to lag 90° behind the applied voltage.

(*b*) to make the pressure coil flux to lag 90° behind the applied voltage.

(*c*) to bring the pressure coil flux in phase with the applied voltage.

(*d*) none of the above.

48. In a circuit of a single-phase induction energy meter, the pressure coil lags the voltage by 88°, the errors while measuring power in two circuits having power factors of unity and 0.5 lagging are respectively are :

(*a*) −0.061%, + 6.1% (*b*) + 0.061%, − 6.1%

(*c*) −0.061%, −6.1% (*d*) −6.1%, − 6.1%.

49. Horizontally mounted moving iron instruments use

(*a*) eddy current damping (*b*) electromagnetic damping

(*c*) fluid friction damping (*d*) air friction damping.

50. The reason why eddy current damping cannot be used in moving iron instruments is :

(*a*) they have a strong operating magnetic field.

(*b*) they are not normally used in vertical position.

(*c*) they need a large damping force which can only be provided by air friction.

(*d*) they have a very weak operating magnetic field and introduction of a permanent magnet required for eddy current damping would distort the operating magnetic field.

51. Phantom loading for testing of energy meters is used

(*a*) to isolate the current and potential circuits.

(*b*) to reduce power loss during testing.

(*c*) for meters having low current ratings.

(*d*) to test meters having a large current rating for which loads may not be available in the laboratory. This also reduces power losses during testing.

52. In an induction type energy meter, compensation for static friction is provided by

(*a*) shading bands which are actuated to provide a constant torque irrespective of load.

(*b*) lag circuits.

(*c*) drilling holes in the disc.

(*d*) none of the above.

53. An hypothetical electrostatic voltmeter with spring control is designed with one fixed and another movable plate. Both plates are in the form of a circle with one plate over the other so as to produce a rotatry motion. If a voltage is applied between them, no rotatory motion is produced because

(*a*) there is no change in the voltage.

(*b*) there is no change in the capacitance irrespective of the relative position of the two plates.

(*c*) the spring force is very large.

(*d*) the applied voltage is small to produce any noticeable deflection.

54. An electrostatic voltmeter draws a small value of current on d.c.

(*a*) under steady state condition respective of the applied voltage.

(*b*) when switched on irrespective of the applied voltage.

(*c*) when measuring low voltages.

(*d*) when measuring high voltages.

55. In a household single-phase single phase induction type wattmeter, the meter can be reversed by

(*a*) reversing the supply terminals.

(*b*) reversing the load terminals.

(*c*) opening the meter connections and reversing either the potential coil terminals or current coil terminals.

(*d*) opening the meter and reversing connections of both current and potential coil circuits.

56. If an induction type energy meter runs fast, it can be slowed down by

(*a*) lag adjustment.

(*b*) light load adjustment.

(*c*) by adjusting the position of braking magnet and making it come closer to the centre of the disc.

(*d*) by adjusting the position of braking magnet and making it move away from the centre of the disc.

57. The PMMC meters used in rectifier type voltmeters are provided with shunts even at the expense of lowering sensitivity. Why ?

(*a*) The shunting of the meter causes the circuit current to be high thereby making the rectifier work in the linear portion of its $v - i$ characteristics.

(*b*) The shunt prevents the meter against sudden overloads.

(*c*) The provision of shunt increases the sensitivity of the circuit.

(*d*) All of the above.

58. Why are multimeters provided with separate scale for low a.c. voltages ?

(*a*) To improve the readability of the scale.

(*b*) To have high accuracy.

(*c*) To take into account the high value of resistance of rectifier at low voltages (and currents) and also the fact that at low voltages (and currents) the value of rectifier resistance is not constant but varies considerably even for small changes in voltages (or current).

(*d*) None of the above.

59. A voltmeter has a resistance of 2000 Ω. When it is connected across a d.c. circuit its power consumption is 2 mW. Suppose this voltmeter is replaced by a voltmeter of 4000 Ω resistance, the power consumption will be :

(*a*) 4 mW (*b*) 1 mW

(*c*) 2 mW (*d*) none of the above.

The circuit conditions remain undisturbed.

60. A milliammeter of resistance 100 Ω is connected in series with a circuit. Its power consumption is 0.1 mW. Supposing it is replaced with a milliammeter of 200 Ω resistance the power consumption will be

(*a*) 0.2 mW (*b*) 0.05 mW

(*c*) 0.1 mW (*d*) none of the above.

61. A 1 mA d'Arsonval movement has a resistance of 100 Ω. It is to be converted to a 10 V voltmeter. The value of multiplier resistance is

(a) 999 Ω (b) 9999 Ω
(c) 9900 Ω (d) 990 Ω.

62. A d'Arsonval movement is rated at 50 μA. Its sensitivity is
(a) 20000 Ω/V (b) 200000 Ω/V
(c) 200 Ω/V (d) cannot be determined.

63. An Astatic movement in an electrodynamometer type of instrument is used for
(a) eliminating errors on account of external magnetic fields.
(b) increasing the operating torque of the instruments.
(c) providing eddy current damping.
(d) none of the above.

64. In order that an electrodynamometer type of instrument exhibits a pure square law response, the meter range should be limited to
(a) –45° to + 45° about the position for zero mutual inductance between fixed and moving coils.
(b) –45° to + 45° about the position for maximum mutual inductance between fixed and moving coils.
(c) –22.5° to 22.5° about the position for zero inductance between fixed and moving coils.
(d) –90° to 90° about the position for maximum mutual inductance between fixed and moving coils.

65. A d'Arsonval movement has a sensitivity of 40,000 Ω/V and its lateral resistance is 4000 Ω. The resistance of multiplier to convert it to 1 V voltmeter is :
(a) 44000 Ω (b) 36000 Ω
(c) 3600 Ω (d) none of the above.

66. A PMMC instrument is spring controlled. The control spring stiffness decreases by about 0.04 percent per °C rise in temperature and the strength of magnets goes down by 0.02 percent per °C rise in temperature. The rise in temperature is 10°C. The deflection
(a) decreases by about 0.2 percent (b) decreases by about 0.4 percent
(c) increases by about 0.4 percent (d) increases by about 0.2 percent.

67. Moving iron type of instrument can be used as
(a) standard instruments for calibration of other instruments.
(b) transfer type instruments.
(c) indicator type instruments as on panels.
(d) all of the above.

68. Moving iron instruments when measuring voltages or currents
(a) indicate the same values of the measurement for both ascending and descending values.
(b) indicate higher value of measurand for ascending values.
(c) indicate higher value of measurand for descending values.
(d) none of the above.

69. An electrodynamometer type of instruments finds its major use as
(a) standard instrument only.
(b) transfer instrument only.
(c) both as standard and transfer instrument.
(d) an indicator type of instrument.

70. The operating magnetic field in an electrodynamometer type of instrument has a flux density typically about
(a) 1 Wb/m^2 (b) 0.5 Wb/m^2
(c) 0.05 Wb/m^2 (d) 0.005 Wb/m^2.

71. The power consumption of PMMC instruments is typically about

(a) 0.25 W to 2 W
(b) 0.25 mW to 2 mW
(c) 25 μW to 200 μW
(d) none of the above.

72. In a series type ohmmeter

(a) 0 marking is on the left hand side of scale while ∞ marking is on the right hand side.
(b) 0 marking is on the right hand side of scale and ∞ infinity marking on the left hand side.
(c) any of the above two markings can be on left or right side of the scale.
(d) 0 marking is in the middle of scale.

73. The frequency range of moving iron instruments is

(a) audio frequency band 20 Hz to 20 kHz.
(b) very low frequency band 10 Hz to 30 kHz.
(c) low frequency band 30 Hz to 300 kHz.
(d) power frequencies 0 to 125 Hz.

74. Spring controlled moving iron instruments exhibit a square law response resulting in a non-linear scale. The shape of the scale can be made almost linear by

(a) keeping rate of change of inductance, L, with deflection, θ, as constant.
(b) keeping $\frac{1}{\theta} \cdot \frac{dL}{d\theta}$ as constant.
(c) keeping $\frac{1}{K\theta}$ as constant where K is the spring constant.
(d) keeping $\theta \cdot \frac{dL}{d\theta}$ as constant.

75. A megger is used for measurement of

(a) low valued resistances.
(b) medium valued resistances.
(c) high valued resistances, particularly insulation resistance.
(d) all of the above.

76. The controlling torque in a megger is provided by

(a) springs
(b) weights attached to the moving system
(c) it does not need any controlling torque
(d) none of the above.

77. A voltage of 200 V at 5 Hz is applied to an electrodynamometer type of instrument which is spring controlled. The indication on the instruments is

(a) 200 V
(b) 0 V
(c) the instrument follows the variations in voltage and does not give a steady response
(d) upto 1 MHz.

78. Thermocouple instruments can be used for a frequency range

(a) upto 100 Hz
(b) upto 5000 Hz
(c) upto 1 MHz
(d) 50 MHz and above.

79. A series type ohmmeter is designed with a d'Arsonval movement having full scale current of 0.5 mA and an internal resistance. It uses a battery of 3V and has a half scale resistance of 3000 Ω. The battery voltage goes down to 2.7 V. What is the value of shunt resistance required so that full scale current is obtained if the meter terminals are short circuited.

(a) 62.5 Ω
(b) 2700 Ω
(c) 55 Ω
(d) 130 Ω.

80. A make break switch is provided to disconnect the battery when the meter is not in use in

(*a*) both series and shunt type ohmmeters. (*b*) only in series type ohmmeters.

(*c*) only in shunt type ohmmeters. (*d*) none of the above.

81. A moving iron instrument can be used for current and voltage measurements

(*a*) in a.c. circuits only (*b*) in d.c. circuits only

(*c*) in both a.c. and d.c. circuits for any value of frequency (in case of a.c. circuits)

(*d*) in both a.c. and d.c. circuits for frequencies upto about 125 Hz (in case of a.c. circuits).

82. In spring controlled moving iron instruments, the scale is

(*a*) uniform

(*b*) cramped at the lower end and expanded at the upper end

(*c*) expanded at the lower end and cramped at the upper end

(*d*) cramped both at the lower and the upper ends.

83. The moving iron voltmeters indicate

(*a*) the same value for d.c. and a.c. voltages.

(*b*) lower values for a.c. voltages than for corresponding d.c. voltages.

(*c*) higher values for a.c. voltages than for corresponding d.c. voltages.

(*d*) none of the above.

84. A moving iron voltmeter reads low for a.c. voltages than for corresponding values of d.c. voltages. The meter can be made to read equally for both a.c. and d.c. voltages

(*a*) if the resistance of the multiplier is made very high.

(*b*) if the inductance of the coil is made small.

(*c*) if the resistance of the coil is made very large.

(*d*) if the multiplier resistance is shunted by a capacitor of appropriate value to make the circuit non-inductive.

85. The power in a 3-phase 4-wire circuit can be measured by using

(*a*) 2 wattmeters (*b*) 4 wattmeters

(*c*) 3 wattmeters (*d*) 1 wattmeters.

86. The power in a 3-phase circuit is measured with the help of 2 wattmeters. The readings of one of wattmeters is positive and that of the other is negative. The magnitude of readings is different. It can be concluded that the power factor of the circuit is

(*a*) unity (*b*) zero (lagging)

(*c*) 0.5 (lagging) (*d*) less than 0.5 (lagging)

87. A capacitor is connected across a portion of resistance of the multiplier in order to make the pressure coil circuit non-inductive. The value of this resistance is r while the total resistance and inductance of pressure circuit are respectively R_p and L. The value of capacitance C is :

(*a*) $0.41\ L/r^2$ (*b*) $0.41\ L/R_p^2$

(*c*) L/R_p^2 (*d*) L/r^2.

88. When measuring power with an electrodynamometer wattmeter in a circuit where the load currents is small

(*a*) the current coil should be connected on the load side

(*b*) the pressure coil should be connected on the load side

(*c*) it is immaterial whether current coil or pressure coil is connected on the load side.

89. The power in a d.c. circuit is measured with the help of an ammeter and a voltmeter. The voltmeter is connected on the load side. The power indicated by the product of readings of two instruments (VI) is :

(a) the power consumed in the load.
(b) the sum of power consumed by load and the voltmeter.
(c) the sum of power consumed by load and the ammeter.
(d) none of the above.

90. When measuring power with an electrodynamometer wattmeter in a circuit where the load current is large
(a) the current coil should be connected on the load side.
(b) the pressure coil should be connected on the load side.
(c) the pressure coil should be connected on the supply side.
(d) it is immaterial whether the pressure coil or the current coil is on the load side.

91. When measuring power with an electrodynamometer wattmeter in a circuit having a low power factor
(a) the current coil should be connected on the load side.
(b) the current coil should be connected on the supply side.
(c) the pressure coil should be connected on the load side.
(d) a compensated wattmeter with pressure coil connected on the load side should be used.

92. In electrodynamometer type wattmeters, current coils designated for carring heavy currents use standard wire or laminated conductors
(a) to reduce iron loses. (b) to reduce hysteresis losses.
(c) to reduce eddy current losses in conductors.
(d) all of the above.

93. In electrodynamometer type of wattmeters, the inductance of presure coil circuit produces error
(a) which is constant irrespective of the power factor of the load.
(b) which is higher at low power factors.
(c) which is lower at low power factors.
(d) none of the above.

94. In an electrodynamometer type of wattmeters
(a) the current coil is made fixed. (b) the pressre coil is made fixed.
(c) any of the two coils i.e. current coil or pressure coil can be made fixed.
(d) both the coils should be movable.

ANSWERS

1. (b)	**2.** (c)	**3.** (c)	**4.** (a)	**5.** (a)	**6.** (d)	**7.** (c)
8. (d)	**9.** (c)	**10.** (a)	**11.** (a)	**12.** (c)	**13.** (a)	**14.** (c)
15. (c)	**16.** (a)	**17.** (a)	**18.** (b)	**19.** (c)	**20.** (d)	**21.** (b)
22. (c)	**23.** (d)	**24.** (b)	**25.** (b)	**27.** (c)	**27.** (b)	**28.** (b)
29. (d)	**30.** (d)	**31.** (d)	**32.** (b)	**33.** (c)	**34.** (d)	**35.** (d)
37. (c)	**37.** (c)	**38.** (b)	**39.** (c)	**40.** (a)	**41.** (d)	**42.** (b)
43. (b)	**44.** (d)	**45.** (b)	**46.** (b)	**47.** (b)	**48.** (c)	**49.** (d)
50. (d)	**51.** (d)	**52.** (a)	**53.** (b)	**54.** (a)	**55.** (c)	**56.** (d)
57. (a)	**58.** (c)	**59.** (b)	**50.** (a)	**61.** (c)	**62.** (a)	**63.** (a)
64. (a)	**65.** (b)	**66.** (d)	**67.** (c)	**68.** (c)	**69.** (c)	**70.** (d)
71. (c)	**72.** (b)	**73.** (d)	**74.** (d)	**75.** (c)	**76.** (c)	**77.** (c)
78. (d)	**79.** (a)	**80.** (c)	**81.** (d)	**82.** (b)	**83.** (b)	**84.** (d)
85. (c)	**86.** (d)	**87.** (d)	**88.** (a)	**89.** (b)	**90.** (b)	**91.** (d)
92. (c)	**93.** (b)	**94.** (a).				

THEORETICAL QUESTIONS

Galvanometers

1. What is the purpose of an instrument ?
2. How are instruments classified ?
3. Give the comparison between mechanical and electrical/electronic instruments.
4. What factors should be taken into considerations while selecting the instruments ?
5. Explain briefly the functions of instruments.
6. Give the comparison between analog and digital instruments.
7. How are electrical instruments classified ?
8. Discuss briefly essential features of indicating instruments.
9. Explain briefly the following damping methods :
   (*i*) Air damping; (*ii*) Eddy current damping;
   (*iii*) Fluid friction damping;
10. What is a galvanometer ?
11. List the various types of galvanometers.
12. Describe the constuctional details and principle of operation of a D'Arsonval galvanometer. Derive the expression for steady state deflection.
13. How are the constants of a D'Arsonval galvanometer determined ? Why the knowledge of these constants is necessary ?
14. Explain with the help of a neat sketch the construction, and give the theory of operation of a D'Arsonval galvanometer.
15. Name the four intrinsic constants of a galvanometer.
16. Explain briefly the effect of galvanometer resistance on damping.
17. Discuss briefly the following sensitivities of galvanometer :
    (*i*) Current sensitivity;
    (*ii*) Voltage sensitivity;
    (*iii*) Megohm sensitivity.
18. Explain briefly the following three general types of galvanometers :
    (*i*) Portable point type.
    (*ii*) Laboratory reflecting type.
    (*iii*) Box type.
19. Explain the function of Ayrton Universal shunt used for D'Arsonval galvanometers. Prove that the current through the galvanometer does not depend upon the resistance of the galvanometer in case an Ayrton shunt is used.
20. Define the terms "current sensitivity", "voltage sensitivity" and "megohm sensitivity" as applied to D'Arsonval galvanometers. Explain how current sensitivity of a galvanometer can be increased.
21. Dscribe the construction and working of a ballistic galvanometer. Explain the different in constructional details of a ballistic galvanometer and a D'Asonval galvanometer.
22. Prove that is a ballistic galvanometer, the charge is proportional to first swing of the moving coil.
23. Describe the different methods used for calibration of a ballistic galvanometer.
24. Describe the construction and working of an unshunted fluxmeter. Prove that the flux is proportional to the deflection of the fluxmeter.
25. Describe the function of a shunted fluxmeter. Prove that the deflection of a shunted fluxmeter is independent of the resistance of its coil.

26. Explain the construction and working of a vibration galvanometer. Explain the term 'tuning' and describe low a vibration galvanometer is tuned.
27. Derive the expression for steady state deflection of a vibration galvanometer when a sinusoidal voltage is applied to its coil.
28. Give the comparison of ballistic galvanometer with vibration galvanometer.
29. Explain the construction and working of Duddell's oscillograph. List is applications.

Ammeters, Voltmeters and Ohmmeters

30. Explain the working principle of a moving iron indicating instrument. Show that this type of instrument can be used both of d.c. and a.c. measurements. Indicate also the errors involed.
31. Describe the constructional details and working of a moving iron attraction type meter. Derive its torque equation.
32. Derive the expression for torque equation for a moving iron instrument and comment upon the nature of scale.
33. With the help of neat diagrams explain the working principle of a repulsion type moving iron instrument.
34. Explain the construction and operating principle of moving iron instruments. List the errors and methods of compensation.
35. What is essential difference between a moving coil and a moving iron instrument ?
36. Explain the construction of a PMMC meter with the help of a neat sketch. How are different forces produced ? Derive the torque equation. Enlist the advantages and disadvantages of PMMC metres.
37. Describe the construction and working of a PMMC type of instrument. Derive the expression for deflection for a PMMC ammeter if it is (*i*) spring controlled (*ii*) gavity controlled. Comment upon the shape of scale in both cases.
38. Describe the principle of operation, construction, main features and applications of a moving coil instrument.
39. Give a sketch showing the construction of a moving coil voltmeter. Derive an expression for the torque of moving coil instruments.
40. State the causes of change of accuracy in moving coil instruments with change of temperature. Explain how compensation is made in ammeters for change of electrical resistance of the moving coil with change of temperature.
41. Explain with the help of a neat sketch the construction of an electrodynamic type moving coil ammeter. Give the equation for torque of the ammeter and the principal advantages and disadvantages of these instruments. State the causes of change of accuracy with change of temperature and frequency.
42. Describe in details the construction and principle of operation of a transfer instrument generally used in the laboratory for measurement of current.
43. How are thermal instruments classified ? Explain their principles of operation. Explain the operation of a hot wire ammeter with neat diagrams.
44. Draw an illustrative diagram and explain the working of a hot wire instrument. Why is it not normally used as a standard meter ?
45. Describe briefly the constructional features of a thermocouple instrument and the principle of operation. Point out the advantages and disadvantages of this type over the other types.
46. Draw the circuit diagram and explain the principle of operation of a rectifier type voltmeter employing a bridge rectifier.
47. Explain with the help of a neat diagram the principle of operation of a rectifier ammeter. Also briefly discuss the factors which influence the performance of a rectifier ammeter. Specify the suitable current and frequency range of a rectifier ammeter.

48. Describe the construction and working principle of the electrostatic voltmeters.
49. Give the theory of quadrant electrometer. Explain how the instrument can be used for measurement of voltage.
50. Describe the construction of a quadrant type of electrostatic voltmeter. Derive the expression for deflection when idiostatic type of connections are used. The meter is spring controlled.
51. Derive an expression for the force of attraction between the plates in a parallel plate electrostatic voltmeter.

Wattmeters

52. Draw the possible methods of connecting the pressure coil of a wattmeter and compare the errors. Explain the meaning of "compensation winding" in a wattmeter and show how it helps to reduce the error.
53. Discuss the erors introduced in the dynamometer type wattmeter due to the resitance of current and potential coils. How is the error due to resistance of presure coil compensated for by using a compensating coil ?
54. Discuss the various types of errors and their methods of compensation in the dynamometer type wattmeters.
55. Describe the three-ammeter method for measurement of power and power factor in a single phase circuit. Derive the expressions for power and power factor.
56. With neat connection and vector diagrams explain how power can be measured with three ammeters in a single-phase a.c. circuit.
57. Give connection and vector diagrams for the 3-voltmeter method of power measurement.
58. Show with the help of neat diagram how would you extend range of a wattmeter using C.T. and P.T.
59. Draw the circuit diagram of a measurement set-up using a curent transformer, a potential transformer and a wattmeter.
60. Explain how you measure the total power in a 3-phase circuit with the help of two-wattmetes only.
61. Draw the necessary circuit diagram for measurement of three phase power by two wattmeter method. Make necessary derivations. In case of balanced load, discuss the effects of the following load power factors on the two wattmeter readings : Zero, 0.5 and unity.
62. Deduce an expression for the power factor of a balanced three phase load with the help of two wattmeter readings. What will be the power factors :
    (*i*) When the readings of the two ammeters are equal.
    (*ii*) When the reading of one of the wattmeter is zero.
    (*iii*) When the readings of two wattmeters are equal and opposite.
63. Show that in two-wattmeter method of 3-phase power measurement the sum of the readings of the two wattmeters gives the total power consumed in a 3-phase circuit. Hence prove,

$$\tan \Phi = \sqrt{3}\,\frac{W_1 - W_2}{W_1 + W_2}$$

where Φ is the phase angle of the load and W_1 and W_2 are the readings of the wattmeters.
64. What is Blondel's theorem ? Explain with vector diagram how the total power and power factor of a balanced 3-phase 3-wire circuit, can be measured by means of two wattmetes, proving the formula employed.
65. With the help of circuit and phasor diagrams and using usual notations show that the total power in a 3-pase, 3-wire, star-connected balanced load can be measured with the help of only one wattmeter.

66. Show that the power in a 3-phase system is measured by the use of (*i*) only one wattmeter and (*ii*) two wattmeters. Indicate how the power factor is determined. Comment on the accuracy of the measuremens when the load is unbalanced and the supply is a four-wire system.
67. Explain the construction and principle of operation of a dynamometer type wattmeter. How it can be made to read d.c. as well as a.c. ?
68. Explain the construction and working of an electro-dynamometer type of wattmeter.
69. Describe the constructional details and working of a single phase electro-dynamometer type of wattmeter. Derive the expression for deflection for a.c. operation if the instrument is spring controlled.
70. What is special feature of a wattmeter suitable for working on low power factor circuits? Explain.
71. Describe the working principle and construction of an induction type wattmeter. How does this instrument compare with the dynamometer type ?
72. What are the errors in induction type wattmeter, and how are they compensated ?
73. Describe the construction of an induction wattmeter with the help of a neat diagram. What is the function of copper shading band ?

    Explain how torque is produced in an induction type wattmeter. Prove that the torque is proportional to the power in the circuit.
74. Explain the theory of electrostatic wattmeter and discuss its merits over other wattmeters.
75. Draw a circuit diagram of thermal wattmeter and explain its principle of operation.
76. Explain with neat diagram as to how wattmeters are tested.
77. Describe, with connection diagram, the method of calibrating a dynamometer type wattmeter by means of an electrostatic meter.

Energy meters and Measurement of energy

78. Give the essential components of motor meters. Describe the errors common to all meters.
79. Describe, with a neat sketch, the construction and working of a mercury motor meter. How it can be used as a watt-hour meter ?
80. Describe with a neat sketch the construction and working of Elihu Thomson commutator meter. How is braking torque provided in this instrument ? Give its merits also.
81. Describe with the help of diagrams the construction and working of a domestic single phase energy meter.
82. Describe with neat sketch the working of single-phase energy meter. Discuss the friction compensation and creep prevention in an energy meter.
83. Explain working principle of a single-phase induction type energy meter with the help of connection diagam. Give the various sources of errors and their compensation.
84. Describe the construction and working of a single-phase induction type energy meter. Show that the total number of revolutions made by its disc during a particular time is proportional to the energy consumed.
85. Draw a sketch showing the essential features of construction of a single-phase energy meter. Give the theory of operation of the instrument explaining clearly the function of the quadrature loops of the shunt magnet. Describe the mechanism of friction compensation.
86. Describe the expresion for torque for a single-phase induction type of energy meter. Explain that this meter can measure the true energy consumed only if the phase angle between the applied voltage and the pressure coil flux is 90°. Describe a method for making this adjustment.
87. Draw the connection diagram of a 3-phase energy meter and explain its working. How do you correct it, if it is found to be moving fast?
88. Draw the circuit diagram for the measurement of the energy in a 3-phase, 3-wire system using a 2-element energy meter. Is the above arrangement suitable for the measurement of energy in a 3-phase, 4-wire system. Explain.

89. What are the errors in energy meter and how they are compensaed in multi-element induction type energy meter ?

How these energy meter errors will compare with that of watt-meter ?

90. Classify different types of integrating electricity meters. What type of meter is used for measuring reactive kVAh ? How does it differ from a meter used for measuring kWh on the same circuit ?

91. Explain the advantages of using "phantom load" in meter testing. Describe with the aid of diagram how a single-phase energy meter is tested and calibrated with the help of rotating sub-standard watt-hour meter.

92. Explain briefly how would you carry out single phase meter testing in the laboratory.

93. How will you test the single-phase energy meter ? Explain the procedure of adjustments of different compensation systems.

94. Explain, with the help of a neat diagram the principle and working of a thermal type maximum demand indicator.

95. Explain the working of Merz Price demand indicator.

96. Explain how the following adjustments are made in a single-phase induction type energy meter :

(*i*) Lag adjustment

(*ii*) Adjustment for friction compensation

(*iii*) Creep

(*iv*) Overload compensation

(*v*) Temperature compensation.

97. Explain the sources of errors in single-phase induction type energy meters.

98. Describe the construction and working of a two element induction energy meter.

99. Explain the different types of tariffs used. Explain the underlying principle for using different types of tariffs for different types of consumers.

100. Describe the construction and working of a Merz Price maximum demand indicator.

101. Explain the different types of methods used for VAh metering.

102. What is a Trivector meter ?

103. What is Phantom loading ? Explain with an example how is it more advantageous than testing with direct loading.

104. Describe a circuit for testing of a single-phase induction type energy meter at different loads and power factors.

105. Enumerate the different types of tests that are carried out on single-phase Induction type watt-hour meters. Describe the Long period dial test.

106. Describe the crossed phase method for measurement of VARh in 3-phase 3-wire circuits.

Miscellaneous Instruments

107. Explain the principle of operation of the moving iron power factor meter.

108. How the power factor of a single phase circuit is measured ? Describe one type of meter for this purpose.

109. Explain with a neat sketch the construction and principle of a dynamometer type single phase power factor meter. How would the accuracy of such an instrument be affected by frequency and waveform variation ?

110. Discuss the construction, principle and theory of one type of frequency meter.

111. What methods can be used for the measurement of frequency ? Explain any one frequency meter with its advantages and disadvantages.

112. Explain the working of a frequency meter, which depends for its action on the phenomenon of electrical resonance.

113. Describe the working of a frequency meter, which depends on mechanical resonance for its action.
114. Describe the construction and give the theory of operation of one type of frequency meter. Explain clearly whether the instrument performance will be affected or not by normal changes in the supply voltage.
115. Draw a neat single sketch of a Weston synchroscope and explain its working.
116. Explain clearly and briefly the construction and working of a synchroscope.
117. Draw synchroscope internal circuit diagram and explain the working and uses. In a synchroscope it is observed that the pointer is revolving once in every second. What is frequency of the incoming machine ?
118. What are the uses of synchroscope ?
119. Describe the construction and working of rotating and static type phase sequence indicators.

UNSOLVED EXAMPLES

Galvanometers

1. The scale of a galvanometer is placed at a distance of 600 mm. A deflection of 42 mm is observed. What is the angle through which the coil has turned ? [**Ans.** 2°]
2. Determine the displacement constant of a suspended coil in a galvanometer of moving coil vibrating type having 28 turns, 4 mm × 2 mm mean area and situated in a magnetic field of 0.8 tesla. [**Ans.** 179.2×10^{-6} Nm/A]
3. Fig. 3.10 shows a galvanometer system in which a wire of 0.1 mm in diameter is used as a torsion member. The wire is made of phosphor bronze for which modulus of rigidity is 40 GN/m^2. The polar moment of inertia for a round wire is $\frac{\pi}{32} d^4$, where d is the diameter of the wire.

   A collimated ray of light is reflected by the galvanometer mirror shown. The radius from the mirror to the scale is 200 mm.

   Calculate : (*i*) Torsional constant; (*ii*) Maximum allowable torque if the shear stress is limited to 100 MN/m^2; (*iii*) Sensntivity of the instrument in m/Nm; (*iv*) Maximum possible scale deflection.

   [**Ans.** (*i*) 52.6×10^{-6} Nm/rad; (*ii*) 39.3×10^{-6} Nm; (*iii*) 7600 m/Nm; (*iv*) 298.7 mm]
4. The coil of a moving coil galvanometer is wound on a non-magnetic former whose height and width are both 20 mm. It moves in a constan field of 0.12 Wb/m^2. The moment of inertia of its moving parts is 0.25×10^{-6} kg-m^2, and the control spring constant is 30×10^{-6} Nm/rad. Calculate : (*i*) The number of turns that must be wound on the coil to produce a deflection of 150° with a current of 10 mA, and (*ii*) The resistance of the coil to produce critical damping, all damping being assumed as electromagnetic.

   [**Ans.** (*i*) 164; (*ii*) 11.28 Ω]
5. In a moving coil galvanometer with free time period of 3 seconds, a current of 2.5 μA gives a deflection of 300 mm on a scale at a distance of 1.5 m. The moving system has a moment of inertia of 4×10^{-6} kg-m^2. Determine the total resistance in the galvanometer circuit for critical damping. [**Ans.** 29,400 Ω]
6. A moving coil galvanometer deflects 200 mm on a scale at a distance of 1 m from the mirror when a current of 2 μA is passed through it. The free time period of the galvanometer is 4 seconds. The galvanoeter is dead beat when the total resistance in galvanometer circuit is 20,000 Ω.

   Determine the moment of inertia of the galvanometer moving system.

   [**Ans.** 4.13×10^{-6} kg-m^2]

7. The following data relate to a moving coil galvanometer which has a former of a non-conducting material :

The current sensitivity of the instrument = 0.001 μA/mm at 1 m; The period of undamped oscillation = 6 seconds; The displacement constant of the instrument = 5×10^{-3} Nm/A.

Calculate : (*i*) Control constant;

(*ii*) Inertia constant;

(*iii*) Total circuit resistance for instrument to be dead beat.

[**Ans.** (*i*) 10×10^{-9} Nm/rad; (*ii*) 9.12×10^{-9} kg-m^2; (*iii*) 1309 Ω]

8. A d'Arsonval galvanometer has a rectangular coil wound on an aluminium former of resistivity 27×10^{-9} Ωm. The sides of the former, each of effective length 30 mm, move in a radial field of uniform flux density 0.15 Wb/m^2 against a control torque of 18×10^{-6} Nm/rad. The width of the former is 20 mm, the ends being effectively outside the magnetic field, and the moment of inertia of the moving system is 80×10^{-9} kg-m^2. Determine the cross-sectional area of the aluminium former if it is to provide critical damping, other sources of damping being neglected. [**Ans.** 0.8 mm^2]

9. The following data relate to a moving coil galvanometer :

Number of turns of the coil = 250; Flux density in the uniform magnetic field = 0.05 Wb/m^2; Torsion constant = 1.5×10^{-7} Nm per radian; Dimensions of the coil = 15 mm (width) × 20 mm (height); Moment of inertia = 1.5×10^{-7} kg-m^2; Galvanometer resistance = 200 Ω; Total damping = 4×10^{-8} Nm/radian/second.

Determine :

(*i*) The deflection of galvanometer in radian and in mm when a current of 2.5 μA is flowing through the galvanometer, the scale is at a distance of 1.5 m from the mirror.

(*ii*) The curent sensitivity.

(*iii*) The voltage sensitivity.

(*iv*) The megohm sensitivity.

(*v*) Frequency of oscillations.

(*vi*) Relative damping.

(*vii*) Logarithmic decrement.

[**Ans.** (*i*) 0.0625 radian; 187.5 mm; (*ii*) 75 mm/μA; (*iii*) 0.375 mm/μV; (*iv*) 75 MΩ/mm; (*v*) 0.158 Hz; (*vi*) 0.189; (*vii*) 0.6]

10. The following data relate to a moving coil galvanometer :

Number of turns of the coil = 300; Flux density in the magneic field = 0.1 Wb/m^2; Control constant = 0.2×0^{-6} Nm/rad; Dimensions of the coil = 20 mm (wide) × 25 mm (high); Moment of inertia = 0.15×10^{-6} kg-m^2; Galvanometer resistance = 300 Ω; Resistance connected in series with the galvanometer = 1000 Ω.

Determine :

(*i*) The resistance to be connected to galvanometer for critical damping.

(*ii*) Relative damping.

(*iii*) Logarithmic decrement.

(*iv*) Frequency of free oscillations.

(*v*) First maximum deflection.

(*vi*) Period of undamped and damped oscillations.

[**Ans.** (*i*) 350 Ω; (*ii*) 0.5; (*iii*) 1.814; (*iv*) 0.184 Hz; (*v*) 1.163 θ_f ; (*vi*) 5.44s ; 6.3 s]

11. The Ayston universal shunt has a total resistance of 6,000 Ω and galvanometer has a resistance of 2000 Ω. Determine the multiplying power of shunt for 1000 Ω, 2000 Ω and 3000 Ω tapping. [**Ans.** 8, 4, 2.667]

12. A ballistic galvanometer has a resistance of 150 Ω and an undamped period of 7.5 seconds. A steady e.m.f. of 3.5 mV produces a deflection of 210 mm. Determine the quantitiy of electricity discharged from a capacitor if the deflection produced is 750 mm. The relative damping is 0.8. **[Ans.** 307.5 μC]

13. A solenoid 1m long and wound with 800 turns has a search coil of 50 turns and cross-sectional area 300 mm^2 at its centre. Reversing a current of 4A in the solenoid causes a deflection of 5 divisions in a ballistic galvanometer connected to the search coil. Calculate : (*i*) The galvanometer constant in flux linkages per division; (*ii*) The flux linkage sensitivity.
[Ans. (*i*) 24.1 × 10^{-6} weber turns per divisions ;
(*ii*) 41.5 × 10^{-3} divisions per weber turn]

14. The discharge of a capacitor through a ballistic galvanometer produces a damped frequency of 0.125 Hz and successive swings of 120, 96 and 76.8 mm.

    Calculate : (*i*) The damping ratio; (*ii*) The logarithmic decrement; (*iii*) The period of undamped and damped oscillations. **[Ans.** (*i*) 0.0708; (*ii*) 0.223; (*iii*) 7.98s; 8s]

15. A moving coil ballistic galvanometer of 150 Ω resistance gives a throw of 75 dicisions when the flux through a search coil to which it is connected is reversed. Find the flux density in which the reversal of the coil takes place, given that the galvanometer constant = 110 μC/division, and the search coil has 1400 turns, a mean area of 5000 mm^2 and a resistance of 20 Ω.

    Calculate also the ballistic and flux linkages sensitivities of the galvanometer.
    [Ans. 0.1 Wb/m^2; 9100 div./C; 60.67 weber turns/div.]

16. A galvanometer with an undamped period of 9 seconds and a current sensitivity of 1 μA/div. is connected to a search coil in a circuit of 4000 Ω resistance. Determine the flux linkage change in the search coil to produce a first swing of 100 divisions if the logarithmic decrement is 0.2. **[Ans.** 632 × 10^{-3} weber turn]

17. A ballistic galvanometer gives a first maximum deflection of 60° for a discharge of 1000 μC. Find the quantity of electricity which when discharged through this galvanometer gives rise to a spot deflection of 10 divisions on a millimeter circular scale 1 metre away.
[Ans. 4.77 μC]

18. A certain fluxmeter has the following constants :

    Air gap flux density = 0.05 Wb/m^2; Turns on moving coil = 40, Area of moving coil = 750 mm^2.

    If the flux linking a 10 turn search coil of 200 mm^2 area connected to the fluxmeter is reversed in a uniform field of 0.5 Wb/m^2, calculate the deflection of the fluxmeter.
    [Ans. 76.5°]

19. A fluxmeter when used with single turn search coil gives full scale deflection if a given uniform field linking with the search coil is reversed. It is proposed to extend the range of fluxmeter to measure flux densities four times the above value using the same search coil and method of reversals. If the resistance of search coil is 1.1 Ω, find the resistance of shunt to be connected in parallel with the fluxmeter. **[Ans.** 0.367 Ω.]

20. A magnetic circuit excited by an m.m.f. of 800 AT has a reluctance of 1,20,000 AT/Wb. A fluxmeter is used to measure the flux. If the fluxmeter has 120 divisions and flux linkages required for deflection of one division are 0.15 × 10^{-3} weber turn, calculate the resistance of the shunt required for use with the search coil. Number of tuns of search coil is one and its resistance is 0.025 Ω. The measurement is made by switching off the excitation. **[Ans.** 0.00926 Ω]

21. The following data relate to a vibration galvanometer tuned to a frequency of 50 Hz :

Inertia constant = 0.01 kg-m^2; Damping constant = 0.025 Nm/rad.-s^{-1}

Calculate the ratio of sensitivity for fundamental to that for third harmonic for the galvanometer. **[Ans.** 1000]

Ammeters, Voltmeter and Ohmmeters

22. The coil of instrument has 42½ turns. The mean width of the coil is 25 mm and the axial length of the coil is 20 mm. If the flux density is 0.1 Wb/m^2, calculate the torque on the moving coil for a current of 15 mA through the coil. **[Ans.** 3.187 × 10^{-5} Nm]

23. A moving coil voltmeter with a resistance of 20 Ω gives a full scale deflection of 120° when a potential difference of 100 mV is applied across it. The moving coil has dimensions of 30 mm × 25 mm and is wound with 100 turns. The control spring constant is 0.378 × 10^{-6} Nm/degree. Find the flux density in the gap. Find also the diameter of copper wire of coil winding if 30 percent of instrument resistance is due to coil winding. The specific resistance for copper = 1.7 × 10^{-8} Ωm. **[Ans.** 0.12 Wb/m^2; 0.2 mm]

24. The inductance of a moving iron ammeter in μH is given by the expresion :

$$L = 20 + 10\,\theta - 3\theta^2$$

where, θ is the deflection in radians from zero position.

Determine the deflection of the ammeter for a current of 10 A if the spring constant is 8 × 10^{-6} Nm/rad. **[Ans.** 93°]

25. For a certain dynamometer ammeter the mutual inductance M varies with deflection θ (expressed in degrees) as M = – 6 cos (θ + 30°) mH. Find the deflecting torque produced by a direct current of 50 mA corresponding to a deflection of 60°. **[Ans.** 15 μNm]

26. A PMMC meter gives reading of 40 mA when connected across two opposite corners of a bridge rectifier, the other two corners of which are connected in series with a capacitor to 100 kV, 50 Hz supply. Determine the capacitance. **[Ans.** 1413 pF]

27. A PMMC meter is connected across appropriate points of bridge having resistance of each arm of 35 Ω and connected across a supply of voltage, v = 5 sin θ + 0.2 sin 3θ. Determine the reading of the ammeter if its resistance is 30 Ω. **[Ans.** 32.25 mA]

28. An absolute electrometer uses a movable circular plate 70 mm in diameter. During a voltage measurement, the distance between the plates is 3.2 mm and the force of attraction is 0.003 N. The medium is air. ϵ_0 = 8.85 × 10^{-12} F/m. Find V. **[Ans.** 1343.1 V]

29. A moving-coil instrument gives full-scale deflection with 15 mA and has a resistance of 5 Ω. Calculate the resistance to be connected :

(*i*) in parallel to enable the instrument to read upto 1 A.

(*ii*) in series to enable it to read upto 100 V. **[Ans.** (*i*) 0.076 Ω, (*ii*) 6661.7 Ω]

30. A moving-coil instrument has a resistance of 10 ohms and gives a full-scale deflection when carrying 50 mA. Show how it can be adopted to measure voltage upto 750 volts and current upto 100 amperes. **[Ans.** R = 14990 Ω, R_s = 0.005 Ω]

31. A 20 V moving-iron voltmeter reads correctly when put on D.C. and the instrument has a resistance of 800 ohms and inductance of 0.15 H. Find out the reading on 20 V A.C. mains (*i*) at 250 Hz (*ii*) 50 Hz. **[Ans.** (*i*) 18.6 V, (*ii*) 19.87 V]

32. A 15 V moving-iron voltmeter has a resistance of 500 Ω and inductance of 0.12 H. Assuming that this instrument reads correctly on D.C. what will be its reading on A.C. at 15 V when the frequency is (*i*) 25 Hz and (*ii*) 100 Hz ?

[Ans. (*i*) 14.99 V, (*ii*) 14.83 V]

33. The total resistance of a moving-iron voltmeter is 1000 Ω and coil has an inductance of 0.765 H. The instrument is calibrated with a full-scale deflection of 50 V D.C. Calculate the percentage error when the instrument is used on (*i*) 25 Hz supply, (*ii*) 50 Hz supply, the applied voltage being 50 V in each case. [**Ans.** (*i*) 0.72%, (*ii*) 36%]

34. In a moving-coil instrument, the moving coil has 40 turns and is of square shape with a mean length of 40 mm along each side. The coil hangs in a uniform radial field of 0.08 Wb/m^2. Find the turning moment of the coil when it is carrying a current of 10 mA. [**Ans.** 512×10^{-7} Nm]

35. A moving-coil instrument has a resistance of 5 Ω between terminals and full scale deflection is obtained with a current of 0.015 A. The instrument is to be used with a manganin shunt to measure 100 A full scale. Calculate the error caused by a 10°C rise in temperature. (*i*) when the internal resistance of 5 Ω is due to copper only, (*ii*) when a 4 Ω manganin swamping resistor is used in series with a copper resistor of 1 Ω.

Take : α_{copper} = 0.004°C and $\alpha_{manganin}$ = 0.00015/°C. [**Ans.** (*i*) 3.7%, (*ii*) 0.8%]

Wattmeters

36. The volt circuit of an electrodynamic wattmeter has an inductance of 8 mH and a resistance of 2000 Ω. What is the percentage error of the instrument when measuring an inductive load having phase angle of 89° and supply frequency is 50 Hz. Neglect the impedance of the current coil and assume that the volt circuit current is negligible as compared with the load current. [**Ans.** 7.2%]

37. A dynamometer type wattmeter with its voltage coil connected across the load side of the instrument reads 250 watts. If the load voltage be 200 volts, what power is being taken by load ? The voltage coil branch has a resistance of 2000 Ω. [**Ans.** 230 W]

38. An electro-dynamic wattmeter has a voltage circuit of resistance of 8,000 Ω and inductance of 63.6 mH which is connected directly across a load carrying 8A at a 50 Hz voltage of 240 V and pf of 0.1 lagging. Estimate the percentage error in the wattmeter reading caused by the loading and inductance of the voltage circuit. [**Ans.** 6.46% (high)]

39. The reading on the ammeters connected for the 3-ammeter method of power measurement are 2.5 A, 4A and 5.6 A in the non-inductive resistor, the load and the main respectively. The terminal voltage is 300 volts. Calculate the value of (*i*) non-inductive resistor (*ii*) the load impedance (*iii*) the power absorbed by the load and (iv) power factor of the load. [**Ans.** (*i*) 120 Ω (*ii*) 75 Ω (*iii*) 546.6 W (*iv*) 0.455]

40. You are supplied with a wattmeter, the current coil of which can take a maximum current of 5A and whose voltage coil can take a maximum voltage of 150 V. You are required to measure the total power of 10 kW in a 250 V, single phase supply at 0.8 pf. What additional equipment you need ? Connect them in proper circuit and find out the multiplying factor which has to be used.

[**Ans.** CT of ratio 50/5A, PT of ratio 250/150V; 16.667]

41. The power in a single phase high voltage circuit is measured by using instrument transformers with voltmeter, ammeter and wattmeter. Observed readings of the instruments (assuming no error) are 115V, 4.5 A, 200W. Characteristics of the transformers are PT : nominal ratio 11500/115V, ratio correction factor 0.995, phase angle–25 minutes; CT : nominal ratio 25/5A, ratio correction factor 0.997, phase angle +15 minutes. Neglecting the voltage coil phase angle in the wattmeter, calculate the true power.[**Ans.** 99.86 kW]

42. Two wattmeters connected to read total power in a three phase system supplying a balanced load gives readings of 8.5 kW and –3.5 kW respectively. Calculate total power and power factor. [**Ans.** 5 kW; 0.2339 (agging)]

43. Two wattmetes used to measure the power input in a 3-Φ circuit indicate 1000W and 5000W respectively. Find the power factor of the circuit (*i*) when both wattmeter readings are positive, (*ii*) when the latter is obtained by reversing the current coil connections.

[**Ans.** 0.866 (lag); 0.189 (lag)]

44. Two wattmeters connected to measure the input in a balanced 3-phase circuit indicate 2000 watts and 500 watts respectively. Find power factor

(*i*) when both readings are obtained in same direction of connecting switch.

(*ii*) when second reading is obtained after reversing the connection of current coil of the instrument. [**Ans.** (*i*) 0.694 (lag); (*ii*) 0.325 (lag)]

45. The power input to a 2,000V, 50 Hz, 3-phase motor running on full load at an efficiency of 90% is measured by two wattmeters which indicate 300 kW and 100 kW respectively. Calculate (*i*) input (*ii*) power factor (*iii*) line current and (*iv*) hp output.

[**Ans.** (*i*) 400 kW (*ii*) 0.756(lag) (*iii*) 152A (*iv*) 490(metric)]

46. A 30phase motor draws a line current of 50A from 220V source while starting. The pf is 0.4. Find the readings of the two wattmeters connected to measure power.

[**Ans.** 8.85 kW; (–)1.23 kW]

47. A balanced load is supplied from a 3-Φ, 400V, 3-wire system whose power is measured by two wattmeters. If the total power supplied is 26 kW at 0.75 power factor lagging, find the readings of the each of the two wattmeters. [**Ans.** 19.62 kW; 6.38 kW]

48. The power flowing in a 3-Φ, 3-wire balanced load system is measured by the two-wattmeter method. The reading of wattmeter A is 5000 watts and on wattmeter B is –1,000 watts. (*a*) what is the power factor of the system ? (*b*) If the voltage of the circuit is 440 volts, what is the value of capacitance which must be introduced into each phase to cause the whole of the power measured to appear on wattmeter A? [**Ans.** (*a*) 0.359 (*b*) 5.4 Ω]

49. In a particular measurement, the wattmeter readings were 5 kW and 1 kW. Calculate the power and power factor, if (*a*) both meters read direct, (*b*) one of the meters has been reversed. [**Ans.** (*a*) 6 kW; 0.656 (*b*) 4 kW, 0.359]

50. In balanced three phase system power is measured by two-wattmeter method and the ratio of two wattmeter readings is 2 : 1. Determine the power factor of the system and deduce the relation used. [**Ans.** 0.866]

51. In a dynamometer wattmeter the moving coil has 500 turns of means diameter 3 cm. Estimate the torque if the axis of the field and moving coils are at (*a*) 60° (*b*) 90° when the density in the field coils is 15 mWb/m^2, the current in the moving coil is 0.5 A and power being measured has a power factor of 0.866.

[**Ans.** (*a*) 199×10^{-6} Nm (*b*) 230×10^{-6} Nm]

Energy meters

52. In a test of 30 minute duration with a constant current of 5 A such a meter registered 0.50 kWh. If the meter is subsequently used on a 200 V supply, determine its error, and state whether the meter is running fast or slow. [**Ans.** No error]

53. A d.c. ampere-hour meter is rated at 10 A, 250 V. The declared constant is 5 A-s/revolution. Express this constant in revolutions per kWh. Also determine the full load speed of the meter. (**Ans.** 2,880 revolution per kWh; 2 revolutions per second)

54. A meter, whose constant is 600 revolutions per kWh, makes 5 revolutions in 20 seconds. Calculate the load in kW. [**Ans.** 1.5 kW]

55. An energy meter is designed to make 100 revolutions of the disc for one unit of energy. Calculate the number of revolutions made by it when connected to a load carrying 20 A at 230 volts at 0.8 pf for an hour. If it actually makes 360 revolutions; find the percentage error. [**Ans.** 368 revolutions, 2.175% slow]

56. The constant of a 3-phase 2-element energy meter is 0.24 revolution of disc per hour. If the meter is normally used with a PT of ratio 22 kV/220 V and a CT of ratio 500/5 A, determine the error expressed as a percentage of the correct reading, from the following test results :

Line voltage 220 V, current 5.25 A, power factor 1.0, time to complete 80 revolutions 61 seconds. [**Ans.** 1.65% slow]

57. The declared constant of a 5 A, 220 V d.c. watthour meter is 3275 revolutions per kWh. Calculate the speed of the disc at full load.

In a test run at half load, the meter takes 59.5 s to complete 30 revolutions. Calculate the error of the meter. [**Ans.** 1 r.p.s., 0.84% fast]

58. A 50 A, 230 V meter on full load test makes 61 revolutions in 37 s. If the normal disc speed is 520 revolutions per kWh, find the percentage error. [**Ans.** 0.75% low]

59. A wattour meter is calibrated to measure energy on a 250 V supply. On test a steady current of 15 A is passed through it for 5 hours at unity power factor. If the meter readings before and after the test are 8234.21 kWh and 8253.13 kWh respectively, calculate the percentage error.

If the spindle turns through 290 revolutions during 5 minutes when a current of 20 A is passing through the meter at 250 V and 0.87 power factor, calculate the meter constant.

[**Ans.** 0.9% high ; 800 revolutions/kWh]

CHAPTER

4 Instrument Transformers

4.1 INTRODUCTION

For the measurement of *large currents in D.C. circuits,* it is usual to use low-range ammeters with suitable *shunts*. For measuring *high voltages,* low range voltmeters are used with *high resistance connected in series with them*. However, this method is neither convenient nor practical for alternating current and voltage instruments. For this purpose, *specially constructed accurate-ratio* ***"instrument transformers"*** *are employed in conjunction with standard low-range A.C. instruments. Their purpose is to reduce the line current or supply voltage to a value small enough to be easily measured with meters of moderate size and capacity.* In other words, they are used for *extending the range of A.C. ammeters and voltmeters.*

The following are the *two basic advantages* inherent in the method using instrument transformers for measurement purposes :

(*i*) Standard rated instruments may be used.

(*ii*) Operating personnel coming in contact with the instruments are not subjected to high voltage and current of the lines, and so there is less danger to them. Even, with a low-voltage system, instrument transformers are used for measuring large currents, so that heavy loads to the instrument panel and to the ammeter and other current terminals are avoided.

The principle of the instrument transformer is fundamentally the same as that of the power transformer.

The instrument transformers are *classified* as follows :

1. *Current transformers (CT)* : These are used for the measurement of "*current*".
2. *Potential transformers (PT)* : These are used for the measurement of "*voltage*".

These transformers are extensively used for *precise measurements.*

- The instrument transformers are widely *used* in protection circuits of power systems *for the operation of over-current, under-voltage, earth-fault* and *various other relays.*

Fig. 4.1 shows the wiring diagrams for current and potential instrument transformers.

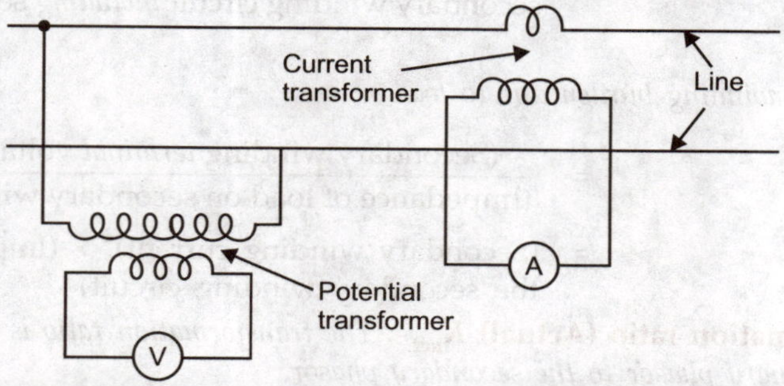

Fig. 4.1. Wiring diagram for current and potential instrument transformers.

Advantages and disadvantages of instrument transformers :

The advantages and disadvantages of instrument transformers as *compared with ammeter shunts and voltmeter multipliers* are as follows :

***Advantages* :**

1. The instrument transformer enables the single range instrument to cover a *large* current or voltage range (normal ranges : 0-5A; 0-110V).
2. It is possible to locate the measuring instruments far away from the high voltage circuit by using long leads connecting the instrument transformer to the instrument.
3. When instruments are employed in conjunction with instrument transformers, their readings *do not depend* upon their constants *viz.*, R, L and C as in the case of shunts and multipliers.
4. Many instruments can be operated from a single instrument transformer.
5. Small power loss.
4. By employing current transformer with a suitable split and hinged core, upon which the secondary winding is wound, the current in a heavy current busbar can be measured without breaking the current circuit.

***Disadvantages* :** The instrument transformers *can not* be used for D.C. measurements.

- In using instrument transformers *for current (or voltage) measurements*, we must know the *ratio of primary current (or voltage) to the secondary current (or voltage)*. These ratios give the *multiplying factor* for finding the primary values from the instrument readings on the secondary side.

However, for *energy or power measurements*, it is essential to know not only the transformation ratio *but also the phase angle between primary and secondary currents (or voltages) because it necessitates further correction to the meter reading.*

4.2 TERMS RELATING TO INSTRUMENT TRANSFORMERS

The various terms relating to instrument transformers are discussed below :

1. **Burden of an instrument transformer :** The load across the secondary winding can be conveniently expressed as the output in volt-ampere at the rated secondary voltage. *The rated burden is the volt-ampere loading which is permissible without errors exceeding the limits for the particular class of accuracy.*

$$\textit{Total secondary burden} = \frac{(\text{Secondary winding } \textit{induced} \text{ voltage})^2}{\text{(Impedance of the secondary winding circuit } \textit{including} \text{ impedance of secondary winding)}}$$

= (Secondary winding current)^2 × (Impedance of secondary winding circuit *including* secondary winding) ...(4.1)

Secondary winding burden due to load

$$= \frac{(\text{Secondary winding } terminal \text{ voltage})^2}{(\text{Impedance of load on secondary winding})}$$

= (Secondary winding current)^2 × (Impedance of *load* in the secondary winding circuit) ...(4.2)

2. **Transformation ratio (Actual) $K_{act.}$:** *The transformation ratio is defined as the ratio of the primary phasor to the secondary phasor.*

   Transformer ratio (Actual) $K_{act.}$

$$= \frac{|\text{Primary phasor}|}{|\text{Secondary phasor}|} \quad ...(4.3)$$

$$= \frac{\text{Primary winding current (Actual)}}{\text{Secondary winding current (Actual)}} \text{ for a } CT \quad ...(4.4)$$

$$= \frac{\text{Primary winding voltage (Actual)}}{\text{Secondary winding voltage (Actual)}} \text{ for a } PT \quad ...(4.5)$$

3. **Nominal transformation ratio ($K_{nom.}$) :** *It is the ratio of rated primary winding current (or voltage) to the rated secondary winding current (or voltage).*

i.e.,
$$K_{nom.} = \frac{\text{Rated primary winding current}}{\text{Rated secondary winding current}} \text{ ... for a } CT \quad ...(4.6)$$

$$= \frac{\text{Rated primary winding voltage}}{\text{Rated secondary winding voltage}} \text{ ... for a } PT \quad ...(4.7)$$

4. **Turns ratio (K_t) :** This is defined as under :

$$\text{Turns ratio, } K_t = \frac{\text{Number of turns of secondary winding}}{\text{Number of turns of primary winding}} \text{ ...for a } CT \quad ...(4.8)$$

$$= \frac{\text{Number of turns of primary winding}}{\text{Number of turns of secondary winding}} \text{ ...for a } PT \quad ...(4.9)$$

- *Normally,* $K_{nom.} = K_t$

5. **Ratio correction factor (*RCF*) :** The ratio correction factor is *the transformation (actual) ratio divided by nominal ratio.*

   In other words,

$$RCF = \frac{K_{act.}}{K_{nom.}}$$

or,
$$K_{act.} = RCF \times K_{nom.} \quad ...(4.10)$$

- The ratio marked on transformers is their *nominal ratio.*

4.3 CURRENT TRANSFORMER (CT)

4.3.1. General Aspects

- Just as a shunt extends the range of a D.C. ammeter, so does the current transformer perform the same function in A.C. circuits. Thus a *high magnitude alternating current can be easily measured by a combination of a current transformer and a low range ammeter.*

- The primary of a current transformer (CT) consists of a few turns of thick cross-section connected in *series* with the high current line. Very often the primary is just one turn formed by taking the line conductor through the secondary winding (Fig. 4.2). The secondary winding consists of a large number of turns of fine wire designed for either 5A or 1A rating. Thus a current transformer is *step-up* transformer. The current transformer has the secondary effectively short-circuited through the low impedance of the ammeter. Fig. 4.3 shows the *current transformer connections.*

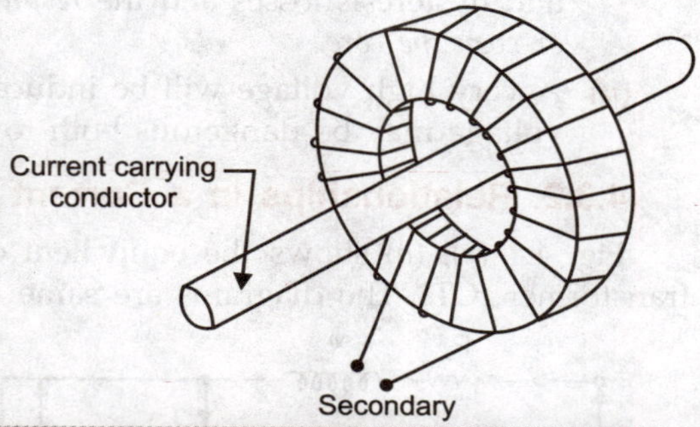

Fig. 4.2. Line conductor acting as primary.

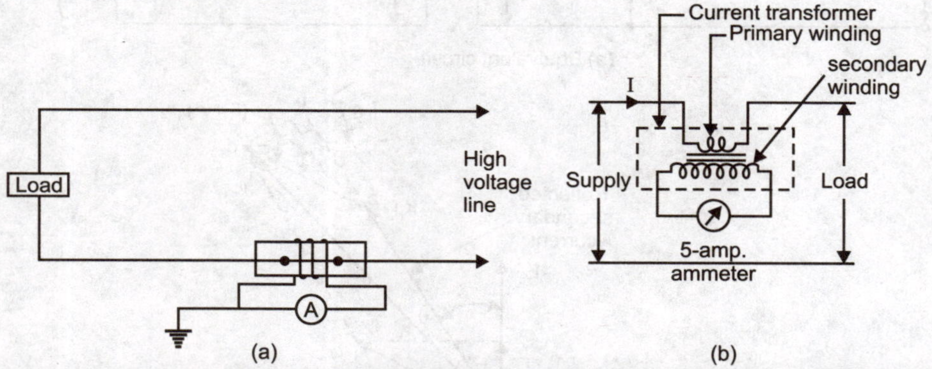

Fig. 4.3. Current transformer connections.

— *Core material* for CT is *usually low-loss silicon steel, mumetal.*

— Core should have *high permeability* and *low loss.*

— To keep losses low, flux density is kept low (≈ 0.1 tesla).

— Core lamination can be *rectangular or ring shaped*. Ring type core is jointless but difficult to wind.

— Windings must be rigid to withstand extremely large mechanical forces produced by short circuit currents.

— For low and medium voltages, tape and varnish insulations are used while for high voltage working CT may be compound filled or oil immersed.

- The *current transformer ratio is not equal to the ratio of secondary to primary turns, mainly because of the effect of the magnetising current*. The primary current can be thought of as the sum of two currents, the first to balance secondary current so that primary and secondary m.m.fs may balance and the second being the no-load current I_0. The component I_0 besides being responsible for a slight error in the current ratio, is also responsible for a phase angle error. The transformer *must be carefully designed to minimise the ratio and phase angle errors.*

☞ • It may be noted that *current transformer* ***must never be operated on open-circuit*** *for the* following two *reasons* :

(*i*) There will be no secondary m.m.f. and since the primary current (and m.m.f.) is fixed, the core flux will increase enormously. This will cause large eddy current and hysteresis losses and the *resulting high temperature may damage the insulation or even the core.*

(*ii*) A very high voltage will be induced in the multi-turn secondary and this high voltage may be dangerous both to life and to the insulation.

4.3.2. Relationships in a Current Transformer

Fig. 4.4 (*a*), (*b*) shows the equivalent circuit and the phasor diagram of a current transformer (CT). The diagrams are same as for any other transformer. Here,

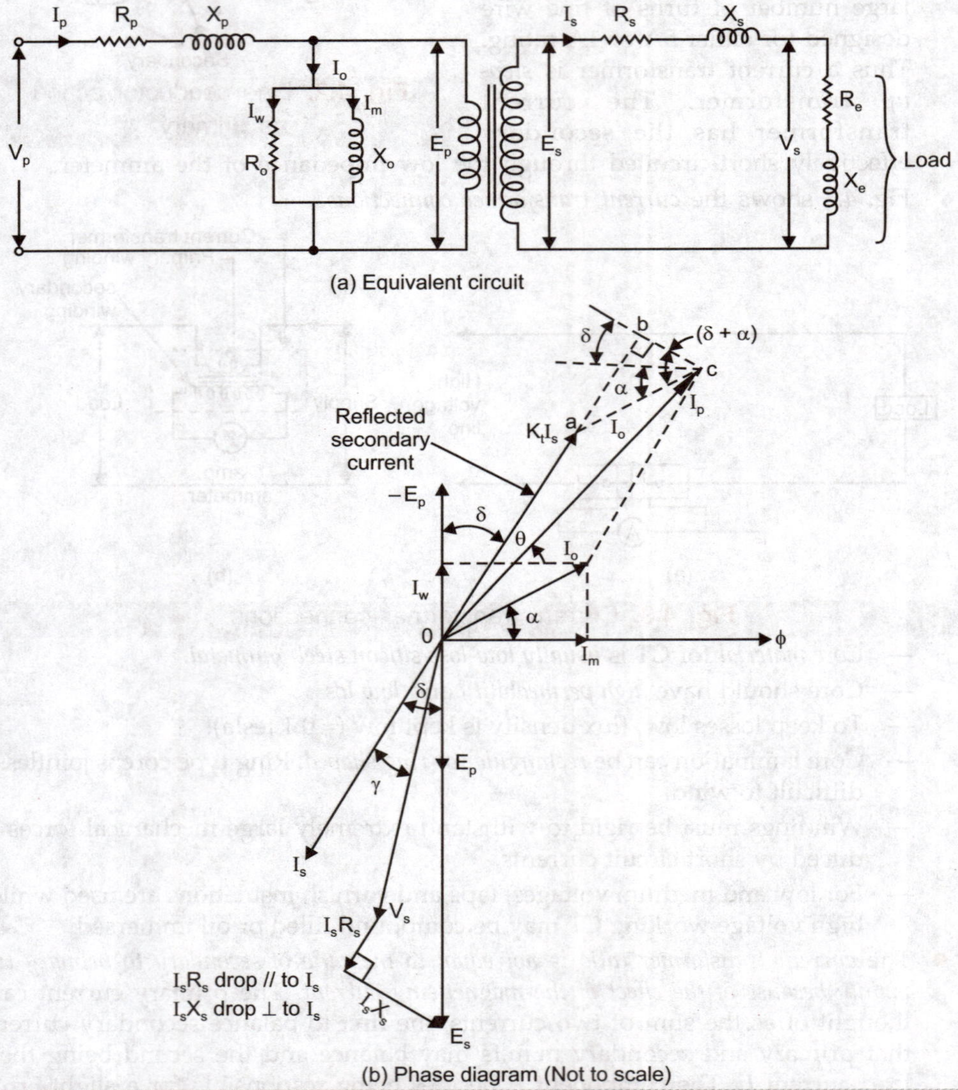

Fig. 4.4. Current transformer (CT).

ϕ = Working flux,

$$K_t = \text{Turns ratio} = \frac{\text{Number of secondary winding turns}}{\text{Number of primary winding turns}},$$

R_p = Resistance of primary winding,
X_p = Reactance of primary winding,
R_s = Resistance of secondary winding,
X_s = Reactance of secondary winding,
R_e = Resistance of external burden *i.e.*, load on secondary,
X_e = Reactance of external burden *i.e.*, load on secondary,
E_p = Primary induced voltage,
E_s = Secondary induced voltage,
V_p = Primary supply voltage,
V_s = Secondary terminal voltage,
I_0 = Exciting current,
I_p = Primary winding current,
I_s = Secondary winding current,
I_m = Magnetising component of I_0
I_w = Loss component of I_0.
α = Angle between exciting current I_0 and working flux ϕ,
θ = Phase angle of the transformer,
δ = Angle between secondary winding induced voltage (E_s) and secondary winding current (I_s)
= Phase angle of total impedance of secondary burden.

$$= \tan^{-1}\left(\frac{X_s + X_e}{R_s + R_e}\right), \text{ and}$$

γ = Phase angle of load or burden *i.e.*, $R_e + jX_e$

$$= \tan^{-1}\frac{X_e}{R_e}$$

Actual transformer ratio (current) ($K_{act.}$) :

Consider the phasor diagram of CT with a lagging power factor, as shown in Fig. 4.4 (*b*).

Consider the right-angled triangle *obc* :

$$(oc)^2 = (ob)^2 + (bc)^2$$
$$= (oa + ab)^2 + (bc)^2$$

or,
$$I_p^2 = \left[K_t I_s + I_o \sin(\delta+\alpha)\right]^2 + \left[I_o \cos(\delta+\alpha)\right]^2$$
$$= K_t^2 I_s^2 + 2K_t I_s I_o \sin(\delta+\alpha) + I_o^2 \sin^2(\delta+\alpha) + I_o^2 \cos^2(\delta+\alpha)$$
$$= K_t^2 I_s^2 + 2K_t I_s I_o \sin(\delta+\alpha) + I_o^2 [\sin^2(\delta+\alpha) + \cos^2(\delta+\alpha)]$$
$$= K_t^2 I_s^2 + 2K_t I_s I_o \sin(\delta+\alpha) + I_o^2$$

or,
$$I_p = \sqrt{K_t^2 I_s^2 + 2K_t I_s I_o \sin(\delta+\alpha) + I_o^2} \quad ...(4.11)$$

$\therefore$ Actual transformation (current) ratio,

$$K_{act.} = \frac{I_p}{I_s} = \frac{\sqrt{K_t^2 I_s^2 + 2K_t I_s I_o \sin(\delta+\alpha) + I_o^2}}{I_s} \quad ...(4.12)$$

In a well designed transformer $I_o << K_t I_s$. Usually I_o is less than 1% of I_p and I_p is, therefore, very nearly equal to $K_t I_s$.

∴ Eqn. (4.12) can be written as :

$$K_{act.} = \frac{\sqrt{K_t^2 I_s^2 + 2K_t I_s I_o \sin(\delta+\alpha) + I_o^2 \sin^2(\delta+\alpha)}}{I_s}$$

... Adjusting I_o^2 as $I_o^2 \sin^2(\delta+\alpha)$

or, $$K_{act.} = \frac{K_t I_s + I_o \sin(\delta+\alpha)}{I_s} = K_t + \frac{I_o}{I_s}\sin(\delta+\alpha) \quad ...(4.13)$$

Although (4.13) is approximate, yet it is *sufficiently accurate for all practical purposes.* The above *theory is applicable when the burden is "inductive" which is normally the case.* Eqn. (4.13) can be further expanded as,

$$K_{act.} = K_t + \frac{I_o}{I_s}(\sin\delta\cos\alpha + \cos\delta\sin\alpha)$$

$$= K_t + \frac{I_o \cos\alpha \,.\, \sin\delta + I_o \sin\alpha \,.\cos\delta}{I_s}$$

or, $$K_{act.} = K_t + \frac{I_m \sin\delta + I_w \cos\delta}{I_s} \quad ...(4.14)$$

$(\because I_o\cos\alpha = I_m; I_o \sin\alpha = I_w)$

Note that δ is *positive* for *lagging* p.f. load while *negative* for *leading* p.f. load.

Phase angle (θ) :

The angle by which the secondary current phasor, when reversed, differs in phase from the primary current, is known as the ***phase angle*** *of the transformer.*

This angle is taken to be +ve if the secondary current reversed leads the primary current. The angle is taken as –ve if the secondary current reversed lags behind the primary current.

The angle between I_s reversed and I_p is θ. Therefore, the phase angle is θ.

From the phasor diagram [Fig. 4.4 (*b*)], we have :

$$\tan\theta = \frac{bc}{ob} = \frac{bc}{oa+ab} = \frac{I_o\cos(\delta+\alpha)}{K_t I_s + I_o \sin(\delta+\alpha)}$$

As θ is very small, we can write,

$$\theta = \frac{I_o\cos(\delta+\alpha)}{K_t I_s + I_o \sin(\delta+\alpha)} \text{ rad.} \quad ...(4.15)$$

Since $I_o << K_t I_s$, therefore, we can neglect the term $I_o \sin(\delta+\alpha)$

$$\theta \simeq \frac{I_o\cos(\delta+\alpha)}{K_t I_s} \text{ rad.} \quad ...(4.16)$$

$$\simeq \frac{I_o\cos\delta\cos\alpha - I_o\sin\delta\sin\alpha}{K_t I_s}$$

$$\simeq \frac{I_m\cos\delta - I_w\sin\delta}{K_t I_s} \text{ rad.} \quad ...(4.17)$$

$$\simeq \frac{180}{\pi}\left(\frac{I_m\cos\delta - I_w\sin\delta}{K_t I_s}\right) \text{ degrees} \quad ...(4.18)$$

4.3.3. Errors in Current Transformer

It is evident from the eqn. 4.14 that the value of transformation ratio (actual ratio) is *not* equal to the turns ratio (K_t). Also the value is not constant and depends upon magnetising and loss components of the exciting current, the secondary winding load current and its power factor. This means the secondary winding current is not a constant fraction of the primary winding current but depends upon the above mentioned factors. This introduces *considerable errors* into current measurements.

In power measurements, due to the use of CT transformers, the following two types of errors are introduced :

1. *Ratio error* : Due to actual transformation ratio being *different* from the turns ratio (K_t).

2. *Phase angle error* : Due to secondary winding current *not* being exactly 180° out of phase with the primary winding current (difference being angle θ).

1. **Ratio error.** It is defined as :

$$\text{Percentage ratio error} = \frac{\text{Nominal ratio} - \text{Actual ratio}}{\text{Actual ratio}} \times 100$$

$$= \frac{K_{nom.} - K_{act.}}{K_{act.}} \times 100 \qquad ...(4.19)$$

2. **Phase angle error.** This error is denoted by angle θ, by which the phase difference between primary and secondary is different from 180°. It is given by :

$$\theta = \frac{180}{\pi}\left[\frac{I_m \cos\delta - I_w \sin\delta}{K_t I_s}\right] \text{ degrees}$$

- ***Approximate formulae for errors :***

Usually, the *instrument burden* is largely resistive with some inductance and, therefore, δ *is +ve and is generally small.*

Hence, $\sin\delta \simeq 0$ and $\cos\delta \simeq 1$

Thus, from eqns. (4.14) and (4.18), we have :

$$K_{act.} = K_t + \frac{I_w}{I_s} \qquad ...(4.20)$$

and,

$$\theta = \left(\frac{180}{\pi}\right)\left(\frac{I_m}{K_t I_s}\right) \qquad ...(4.21)$$

But $I_p \simeq K_t I_s$, and, therefore, eqns. (4.20) and (4.21) can be rewritten as :

$$K_{act.} \simeq K_t + \frac{K_t I_w}{I_p} \simeq K_t\left(1 + \frac{I_w}{I_p}\right) \qquad ...(4.22)$$

and,

$$\theta = \left(\frac{180}{\pi}\right)\left(\frac{I_m}{I_p}\right) \qquad ...(4.23)$$

4.3.4. Characteristics of Current Transformers

The performance of current transformers under different operating conditions may be estimated by considering the phasor diagram. Brief statements of the performance variation with load conditions are as follows :

1. **Effect of change in power factor of secondary burden :**

The power factor of the secondary circuit depends on the power factor of the burden on secondary, this directly affects the two errors of the transformer as under :

- *Ratio error* : As δ is positive for all *inductive* loads, therefore, sin (δ + ∝) is positive and hence actual ratio $K_{act.}$ is always *greater* than the turns ratio K_t [refer to eqn. (4.13)]. δ is negative for *capacitive* burdens and, consequently $K_{act.}$ is *less* than K_t.
- *Phase angle error* : When load is *inductive* and δ is small positive then θ is positive. But as δ approaches 90° as load becomes highly inductive then θ becomes negative. In case of *capacitive* load, δ is negative and θ is always positive.

2. **Effect of change in burden on secondary :**

— An increase in secondary winding circuit burden impedance means an increase in volt ampere rating. This necessitates an increase in the secondary winding induced voltage which can be generated by an increased flux and flux density. Therefore, both magnetising component I_m and loss component I_w are increased; subsequently *errors also increase.*

— A great burden impedance, in general, not only increases the transformation ratio, but also *shifts* the phase angle between primary winding current and secondary winding current reversed to *more positive values.*

3. **Effect of change in primary winding current (I_p) :**

— When the primary current I_p changes, the secondary current I_s changes proportionately.

— At *low values* of I_p (or I_s), the exciting current component I_m and I_w are a greater proportion of I_p and, therefore, the errors are *greater*.

— When I_p increases, I_s also increases and there is a *decrease in ratio error and phase angle* [refer to eqns. (4.22) and (4.23)].

4. **Effect of change in frequency :** The effect of increase in frequency is similar to that produced by a decrease in secondary burden, as increase in frequency will produce a corresponding *decrease in flux for the given e.m.f.* Consequently, there is reduction in I_m and I_w and hence *errors also get reduced.*

4.3.5. Clip-on Type Current Transformer

The main feature of this type of transformer is that *it can be used instantaneously for measurement of current in a current carrying conductor* (cable, insulated bus-bar or fuse link) *without interrupting the circuit for connection of ammeter in series with the circuit.*

Fig. 4.5, shows the schematic diagram of a clip-on type current transformer.

— It consists of a laminated core which is so arranged that it can be opened out at a hinged section by merely pressing a trigger-like projection. When the core is thus opened, it permits the admission of very heavy current-carrying busbars or feeders where upon the trigger is released and the core is tightly closed by a spring.

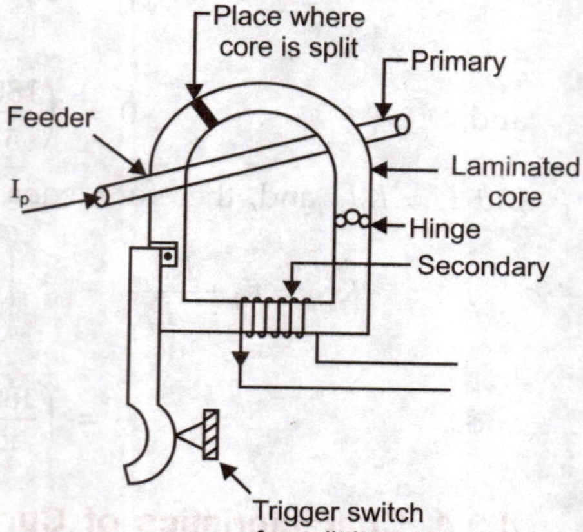

Fig. 4.5. Schematic diagram of a clip-on type current transformer.

— The current-carrying conductor of feeder acts as a single-turn primary whereas the secondary is connected across the standard ammeter conveniently mounted in the handle itself; the ammeter gives the value of current flowing through the conductor.

- The instrument can be used in A.C. circuits only.
- If can be operated by one hand.
- The accuracy claimed by the manufacturers is within 3% of full-scale with the conductor in a central position in the core.
- There transformers are designed for a voltage of 5 kV between the primary and secondary windings.

Example 4.1. *The following data relate to a current transformer :*

Number of primary turns = *single turn*
Number of secondary turns = 240
Secondary winding current = 5 *A*
External burden = 1.2 Ω *(non-inductive)*
Magnetomotive force (mmf) = 96 *AT*
for setting up flux in the core
Cross-section of the core = 1200 mm^2
Supply frequency = 50 *Hz*

Neglecting the effects of magnetic leakage, iron losses and I^2R *losses, calculate :*

(i) Actual transformation ratio.
(ii) Phase angle.
(iii) Maximum flux density in the core.

Solution. *Given* : $N_p = 1$; $N_s = 240$; $I_s = 5$ A; $R_e = 1.2\ \Omega$; mmf = 96 AT; $A_c = 1200\ \text{mm}^2$; $f = 50$ Hz.

(i) Actual transformation ratio ($K_{act.}$) :

$$\text{Turns ratio, } K_t = \frac{N_s}{N_p} = \frac{240}{1} = 240$$

Voltage induced in the secondary winding,

E_s = Current in the secondary winding (I_s) × impedance of secondary winding (R_e)
= 5 × 1.2 = 6.0 V

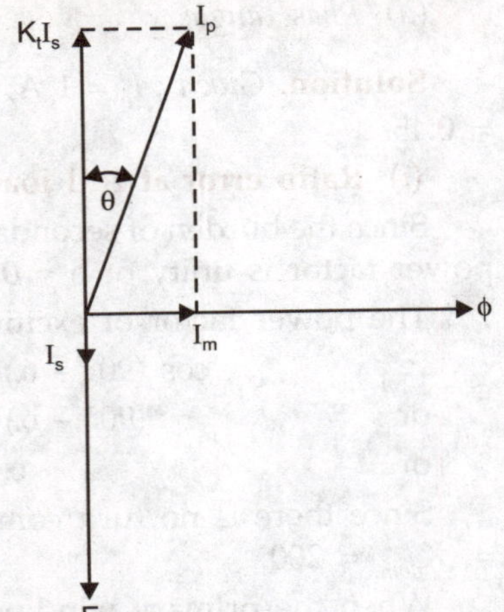

Fig. 4.6. Phasor diagram.

As the secondary burden is purely resistive, the I_s is in phase with E_s, and the secondary power factor is unity or δ = 0 (Refer to Figs. 4.6 and 4.4).

The loss component (I_w) of no-load current (I_o) is to be neglected and, therefore, $I_w = 0$

$$\therefore \quad I_o = \sqrt{I_m^2 + I_w^2} = I_m = \text{magnetising component}$$

and, $\alpha = 0$

Also, $$I_m = \frac{mmf}{N_p} = \frac{96}{1} = 96\,A$$

Now, reflected secondary winding current = K_tI_s = 240 × 5 = 1200 A

Referring to phasor diagram (Fig. 4.6), we have :

$$\text{Primary current, } I_p = \sqrt{(K_t I_s)^2 + (I_m)^2}$$

$$= \sqrt{(1200)^2 + (96)^2} = 1203.8\,A$$

$\therefore$ *Actual transformation ratio,* $K_{act.} = \frac{I_p}{I_s} = \frac{1203.8}{5} = \mathbf{240.76}$ **(Ans.)**

(*ii*) **Phase angle, θ :**

From Fig. 4.6, we get :

$$\tan\theta = \frac{I_m}{K_t I_s} = \frac{96}{1200} = 0.08$$

$$\therefore \quad \theta = \tan^{-1}(0.08) = \mathbf{4° \; 34'} \quad \textbf{(Ans.)}$$

(*iii*) **Maximum flux density in the core, B_{max} :**

We know that : $E_s = 4.44\, f\, \phi_{max}\, N_s$

$$\therefore \quad \phi_{max} = \frac{E_s}{4.44\, fN_s} = \frac{6.0}{4.44 \times 50 \times 240} = 0.1126 \times 10^{-3}\, Wb$$

Hence, $B_{max} = \frac{\phi_{max}}{A_c} = \frac{0.1126 \times 10^{-3}}{1200 \times 10^{-6}} = \mathbf{0.0938\ Wb/m^2}$ **(Ans.)**

Example 4.2. *A ring core current transformer of ratio 1000/5 A is operating at full primary current with a secondary burden of non-inductive resistance of 1.1 Ω. Its exciting current is 1 A at a power factor of 0.45. Calculate :*

(*i*) *The ratio error at full load, assuming that there has been no compensation.*

(*ii*) *Phase angle.*

Solution. *Given* : $I_o = 1$ A; Nominal ratio, $K_{nom.} = \frac{1000}{5} = 200; R_e = 1.1\Omega$, Power factor = 0.45.

(*i*) **Ratio error at full load :**

Since the burden of secondary winding is purely resistive therefore, secondary winding power factor is unity or δ = 0.

The power factor of exciting current is 0.45,

$$\therefore \quad \cos(90° - \alpha) = 0.45$$

$$\text{or} \quad (90° - \alpha) = \cos^{-1}(0.45) = 63.26°$$

$$\text{or} \quad \alpha = 90° - 63.26° = 24.74°$$

Since there is no turn compensation, the turns ratio is equal to nominal ratio or $K_t = K_{nom.} = 200$.

When the primary winding carries rated current of 1000 A, the secondary winding carries a current of 5 A.

Rated secondary winding current, $I_s = 5$ A

$$\therefore \quad K_t I_s = 200 \times 5 = 1000 \text{ A.}$$

Actual transformation ratio,

$$K_{act.} = K_t + \frac{I_o}{I_s}\sin(\delta + \alpha) \quad \text{...[Eqn. (4.13)]}$$

$$= 200 + \frac{1}{5}\sin(0 + 26.74°) = 200.09$$

$$\therefore \quad \text{Ratio error} = \frac{\text{Nominal ratio} - \text{Actual ratio}}{\text{Actual ratio}} \times 100 = \frac{K_n - K_T}{R} \times 100$$

$$= \frac{200 - 200.09}{200.09} \times 100 = -\mathbf{0.045\%} \quad \textbf{(Ans.)}$$

(ii) **Phase angle, θ :** Refer to Fig. 4.4.

$$\theta = \frac{180}{\pi}\left[\frac{I_o \cos(\delta + \propto)}{K_t I_s}\right] \text{degrees} \qquad \text{...[Eqn. (4.16)]}$$

$$= \frac{180}{\pi}\left[\frac{1 \times \cos(0 + 26.74°)}{200 \times 5}\right] = \mathbf{3'4''} \quad \textbf{(Ans.)}$$

Example 4.3. *The primary winding of a 1000/5A, 50 Hz current transformer has a single turn. Its secondary burden consists of a non-inductive impedance of 1.4 Ω. If the iron loss in the core is 1.4 W at full load and magnetising mmf is 80 AT, calculate :*

(i) Flux in the core.

(ii) Ratio error at full load.

Neglect leakage reactance.

Solution. *Given* : Nominal ratio, $K_{nom.} = \frac{1000}{5} = 200$; $N_p = 1$; Secondary impedance = 1.4Ω; mmf = 80 AT; Iron loss in the core = 1.4 W.

(i) **Flux in the core, ϕ_{max} :**

Let us *assume* turns ratio to be equal to nominal, in the absence of any other data.

Then, Turns ratio, $K_t = K_{nom.} = 200$.

$$\therefore \quad N_s = K_t N_p = 200 \times 1 = 200 \qquad (\because N_p = 1)$$

Now, secondary induced voltage, $E_s = 5 \times 1.4 = 7$ V

Also, $E_s = 4.44 f \phi_{max} N_s$

$$\therefore \quad \phi_{max} = \frac{E_s}{4.44 \times f \times N_s} = \frac{7}{4.44 \times 50 \times 200} = \mathbf{0.1576 \times 10^{-3}\ Wb}\ \textbf{(Ans.)}$$

(ii) **Ratio error at full load :**

$$\text{Primary induced voltage, } E_p = \frac{E_s}{K_t} = \frac{7}{200}\text{ V}$$

$$\text{Loss component of exciting current, } I_w = \frac{\text{Iron loss}}{E_p} = \frac{1.4}{(7/200)} = 40\text{ A}$$

$$\text{Magnetising component, } I_m = \frac{mmf}{N_p} = \frac{80}{1} = 80\text{ A}$$

As the burden is purely resistive, therefore, the secondary load angle is zero or δ = 0.

We know that,

$$\text{Actual ratio, } K_{act.} = K_t + \frac{I_m \sin\delta + I_w \cos\delta}{I_s} \qquad \text{...[Eqn. (4.14)]}$$

$$= 200 + \frac{40}{5} = 208\text{ V}$$

$$\therefore \quad \text{Ratio error} = \frac{K_{nom.} - K_{act.}}{K_{act.}} \times 100$$

$$= \frac{200 - 208}{208} \times 100 = -\mathbf{3.846\%} \quad \textbf{(Ans.)}$$

Example 4.4. *A current transformer with a bar primary has 250 turns in its secondary winding. The resistance and reactance of the secondary circuit are 1.4 Ω and 1.1 Ω respectively including the transformer winding. When 5A currents flows in the secondary winding, the magnetising mmf is 80 AT and the iron loss is 1.1 W. Determine the following :*

(i) Ratio error.

(ii) Phase angle error.

Solution. *Given* : $N_p = 1$; $N_s = 250$; $R_p = 1.4\ \Omega$; $X_s = 1.1\ \Omega$; $I_s = 5\text{A}$; Magnetising mmf = 80 AT; Iron loss = 1.1 W.

(i) Ratio error :

$$\text{Turns ratio, } K_t = \frac{N_s}{N_p} = \frac{250}{1} = 250$$

$$\text{Secondary circuit impedance} = \sqrt{(1.4)^2 + (1.1)^2} = 1.78\,\Omega$$

For secondary winding circuit :

$$\cos\delta = \frac{1.4}{1.78} = 0.7865; \; \sin\delta = \frac{1.1}{1.78} = 0.618$$

Secondary induced voltage, $E_s = 5 \times 1.78 = 8.9$ V

$$\text{Primary induced voltage, } E_p = \frac{E_s}{250} = \frac{8.9}{250} = 0.0356\,\text{V}$$

Loss of component referred to primary winding,

$$I_w = \frac{\text{Iron loss}}{E_p} = \frac{1.1}{0.0356} = 30.9\,\text{A}$$

$$\text{Magnetising current, } I_m = \frac{mmf}{N_p} = \frac{80}{1} = 80\text{A}$$

$$\text{Actual transformation ratio, } K_{act.} = K_t + \frac{I_m \sin\delta + I_w \cos\delta}{I_s} \quad \text{...[Eqn. (4.14)]}$$

$$= 250 + \frac{80 \times 0.618 + 30.9 \times 0.7865}{5} = 264.75$$

Assuming, $K_{nom.} = K_t = 250$, since no other contrary information is given, we have :

$$\text{Ratio error} = \frac{K_{nom.} - K_{act.}}{K_{act.}} \times 100$$

$$= \frac{250 - 264.75}{264.75} \times 100 = -\mathbf{5.57\%} \quad \textbf{(Ans.)}$$

(ii) Phase angle error, θ :

We know that,

$$\theta = \frac{180}{\pi}\left[\frac{I_m \cos\delta - I_w \sin\delta}{K_t I_s}\right] \text{degrees} \quad \text{...[Eqn. (4.18)]}$$

$$= \frac{180}{\pi}\left[\frac{80 \times 0.7865 - 30.9 \times 0.618}{250 \times 5}\right] = \mathbf{2.01°} \quad \textbf{(Ans.)}$$

Example 4.5. *The secondary winding burden of current transformer, having a bar primary and 300 secondary turns, is an ammeter of resistance 1.0 Ω and reactance 0.55 Ω. The resistance and reactance of secondary winding are 0.3 Ω and 0.25 Ω respectively. The core requires the equivalent of an mmf of 90 AT for magnetisation and 45 AT for core losses. Determine the following :*

(i) The primary winding current and actual transformation ratio when the ammeter in the secondary winding circuit indicates 5 A.

(ii) The number of turns which could be reduced in the secondary winding so that the ratio error be zero for this condition.

Solution. *Given* : $N_p = 1$; $N_s = 300$; $R_e = 1.0\ \Omega$; $X_e = 0.55\ \Omega$; $R_s = 0.3\ \Omega$; $X_s = 0.25\ \Omega$; mmf for magnetisation = 90 AT; mmf for core loss = 45 AT; $I_s = 5$ A.

(i) I_p ; $K_{act.}$:

Total secondary circuit resistance = $R_s + R_e = 0.3 + 1.0 = 1.3\ \Omega$

Total secondary circuit reactance = $X_s + X_e = 0.25 + 0.55 = 0.8\ \Omega$

$$\text{Secondary circuit phase angle, } \delta = \tan^{-1}\left(\frac{0.8}{1.3}\right) = 31.6°$$

$$\cos\delta = \cos 31.6° = 0.8517, \text{ and}$$

or,

$$\sin\delta = \sin 31.6° = 0.524$$

$$\text{Turn ratio, } K_t = \frac{N_s}{N_p} = \frac{300}{1} = 300$$

$$\text{Magnetising current, } I_m = \frac{\text{Magnetising } mmf}{N_p} = \frac{90}{1} = 90\text{ A}$$

$$\text{Loss component, } I_w = \frac{mmf \text{ equivalent to iron loss}}{N_p} = \frac{45}{1} = 45\text{A}$$

$\therefore$

$$\textit{Actual ratio, } K_{act} = K_t + \frac{I_m \sin\delta + I_w \cos\delta}{I_s} \quad \text{...[Eqn. (4.14)]}$$

$$= 300 + \frac{90 \times 0.524 + 45 \times 0.8517}{5} = \mathbf{317.1} \quad \textbf{(Ans.)}$$

$$\text{Primary current, } I_p = K_{act.} \times I_s = 317.1 \times 5 = \mathbf{1585.5\ A} \quad \textbf{(Ans.)}$$

(ii) Number of turns to be reduced :

For eliminating the ratio error, we must reduce the secondary winding turns or in other words we must *reduce the turn ratio.*

For zero ratio errors :

$$K_{nom.} = K_{act.} \quad (\text{where, } K_{nom.} = K_t = 300)$$

or,

$$300 = K_t' + \frac{I_m \sin\delta + I_w \cos\delta}{I_s}$$

or,

$$300 = K_t' + \frac{90 \times 0.524 + 45 \times 0.8517}{5} = K_t' + 17.1$$

or

$$K_t' = 282.9$$

Hence secondary winding turns, $N_s = K_t' \times N_p = 282.9 \times 1 = 282.9.$

$\therefore$ *Reduction in secondary winding turns* = 300 − 282.9 ≃ **17** **(Ans.)**

Example 4.6. *A current transformer has a bar primary, 100 secondary winding turns, and 100/1 nominal ratio. It is operating on an external burden of 1.45 Ω non-inductive, the secondary winding resistance being 0.25 Ω. The primary winding exciting current is 1.8 A, lagging 38.4° to the secondary voltage reversed. With 1A current flowing in the secondary winding, calculate :*

(i) The actual ratio of primary winding current to secondary winding current;

(ii) The phase angle between these currents.

Solution. *Given* : $N_p = 1$; $N_s = 100$; $K_{nom.} = 100/1$; $R_e = 1.45\ \Omega$; $R_s = 0.25\ \Omega$; $I_o = 1.8$A; Lagging angle, with secondary voltage reversed = 38.4°; $I_s = 1$A.

(*i*) **The actual ratio, $K_{act.}$:**

$$\text{Turn ratio, } K_t = \frac{N_s}{N_p} = \frac{100}{1} = 100$$

$$\text{Total secondary circuit resistance} = 0.25 + 1.45 = 1.7\ \Omega$$

$$\text{Secondary circuit reactance} = 0$$

$\therefore$ The secondary p.f. is unity and its phase angle, $\delta = 0$

Now,
$$\alpha = \text{Angle between } I_o \text{ and } \phi$$
$$= 90° - (\text{angle between } I_o \text{ and } E_s \text{ reversed})$$
$$= 90° - 38.4° = 51.6°$$

Also,
$$K_{act.} = K_t + \frac{I_o}{I_s}\sin(\delta+\alpha) \quad \text{...[Eqn. (4.13)]}$$

$$= 100 + \frac{1.8}{1.0}\sin(0 + 51.6°) = \mathbf{101.41} \quad \textbf{(Ans.)}$$

(*ii*) **Phase angle, θ :**

$$\theta = \frac{180}{\pi} \times \frac{I_o \cos(\delta+\alpha)}{K_t I_s} \text{ degrees} \quad \text{...[Eqn. (4.16)]}$$

$$= \frac{180}{\pi} \times \frac{1.8\cos(0+51.6°)}{100 \times 1} = \mathbf{0.641°} \quad \textbf{(Ans.)}$$

Example 4.7. *A current transformer rated as 1000/5A, 25 VA has a turn ratio of 1 : 200. The core loss and magnetising components of the primary current are 5A and 8A under rated conditions. Determine the ratio and phase angle errors for the rated burden and rated secondary current at 0.8 p.f. lagging and 0.8 p.f. leading.*

Neglect the resistance and leakage reactance of secondary winding.

Solution. *Given* : $N_p = 1$, $N_s = 200$ *i.e.*, $K_t = \frac{200}{1} = 200$; Magnetising current, $I_m = 8$A; Loss component, $I_w = 5$A; $\cos\phi = 0.8$ lagging and leading.

Case-1 : *0.8 p.f. lagging :*

For lagging p.f. the secondary phase angle δ is + ve.

$$\cos\delta = \text{secondary p.f.} = 0.8, \quad \sin\delta = \sqrt{1^2 - (0.8)^2} = 0.6$$

$$\text{Actual ratio, } K_{act.} = K_t + \frac{I_m \sin\delta + I_w \cos\delta}{I_s} \quad \text{...[Eqn. (4.14)]}$$

$$= 200 + \frac{8\times0.6+5\times0.8}{5} = 201.76$$

$$\therefore \quad \textit{Ratio error} = \frac{K_{nom.} - K_{act.}}{K_{act.}} \times 100$$

$$= \frac{200-201.76}{201.76} = -\mathbf{0.872\%} \quad \textbf{(Ans.)} \quad (\because K_{nom.} = K_t = 200)$$

$$\textit{Phase angle, } \theta = \frac{180}{\pi}\left[\frac{I_m \cos\delta - I_w \sin\delta}{K_t I_s}\right] \text{degrees} \qquad \text{...[Eqn. (4.18)]}$$

$$= \frac{180}{\pi}\left[\frac{8\times0.8-5\times0.6}{200\times5}\right] = \mathbf{0.1948°} \quad \textbf{(Ans.)}$$

Case-2 : *0.8 p.f. leading* :

For leading p.f. δ is –ve. $\cos\delta = 0.8$ and $\sin\delta = -0.4$.

Now, $\quad$ Actual ratio, $K_{act.} = K_t + \dfrac{I_m \sin\delta + I_w \cos\delta}{I_s}$

$$= 200 + \frac{8\times(-0.6)+5\times0.8}{5} = 199.84$$

$$\therefore \quad \textit{Ratio error} = \frac{K_{nom.} - K_{act.}}{K_{act.}} \times 100$$

$$= \frac{200-199.84}{199.84} \times 100 = +\mathbf{0.08\%} \quad \textbf{(Ans.)}$$

$$\textit{Phase angle, } \theta = \frac{180}{\pi}\left[\frac{I_m \cos\delta - I_w \sin\delta}{K_t I_s}\right] \text{degrees}$$

$$= \frac{180}{\pi}\left[\frac{8\times0.8-5(-0.6)}{200\times5}\right] = \mathbf{0.538°} \quad \textbf{(Ans.)}$$

Example 4.8. *A 500/5A, 50 Hz current transformer has a one turn primary and an output of 20 VA. The number of secondary turns is 99, and the resistance and leakage reactance of the secondary winding are 0.4 Ω and 0.35 Ω respectively. At rated load with non-inductance burden, the inphase and quadrature components (referred to the flux) of the exciting mmf are 6AT and 8AT respectively.*

Calculate the current and phase angle errors.

Solution. *Given* : $N_p = 1$; $N_s = 99$; $\therefore$ $K_t = 99$; $K_{nom.} = \dfrac{500}{5} = 100$; $R_s = 0.4\ \Omega$; $X_s = 0.35\ \Omega$; Magnetising mmf = 6 AT; Loss mmf = 8 AT.

(*i*) **Ratio (current) error :**

$$\text{Magnetising current, } I_m = \frac{\text{Magnetising mmf}}{\text{Primary winding turns}} = \frac{6}{1} = 6 \text{ A}$$

$$\text{Loss component, } I_w = \frac{\text{Loss mmf}}{\text{Primary winding turns}} = \frac{8}{1} = 8 \text{ A}$$

Resistance of external burden (purely resistive – given), R_e

$$= \frac{VA}{I_s^2} = \frac{20}{5^2} = 0.8\ \Omega$$

Reactance of external burden, $X_e = 0\ \Omega$

Resistance of total secondary circuit burden

$$= R_s + R_e = 0.4 + 0.8 = 1.2\ \Omega$$

Reactance of total secondary circuit burden

$$= X_s + X_e = 0.35 + 0 = 0.35\ \Omega$$

$$\text{Secondary phase angle, } \delta = \tan^{-1}\frac{0.35}{1.2} = 16.26°$$

$\therefore$ $\cos\delta = \cos 16.26° = 0.96$ and $\sin\delta = \sin 16.26° = 0.28$

$$\text{Actual ratio, } K_{act.} = K_t + \frac{I_m \sin\delta + I_w \cos\delta}{I_s}$$

$$= 99 + \frac{6 \times 0.28 + 8 \times 0.96}{5} = 100.87$$

$$\textit{Ratio (current) error} = \frac{K_{nom.} - K_{act.}}{K_{act.}} \times 100$$

$$= \frac{100 - 100.87}{100.87} \times 100 = \mathbf{-0.86\%} \quad \textbf{(Ans.)}$$

***(ii)* Phase angle error, θ :**

$$\theta = \frac{180}{\pi}\left[\frac{I_m \cos\delta - I_w \sin\delta}{K_t I_s}\right] \text{degrees}$$

$$= \frac{180}{\pi}\left[\frac{6 \times 0.96 - 8 \times 0.28}{99 \times 5}\right] = \mathbf{0.407°} \quad \textbf{(Ans.)}$$

Example 4.9. *A 100/5A current transformer, at its rated load of 20VA, has an iron loss of 0.18 W and a magnetising current of 1.4 A. It is supplying rated output to a meter having a ratio of resistance to reactance of 4. Calculate :*

(i) Ratio error.

(ii) Phase angle.

Solution. *Given :* $K_{nom.} = \frac{100}{5} = 20$; Burden of secondary circuit = 20 VA; Iron loss = 0.18 W; $I_m = 1.4$ A; $\frac{X}{R} = 4$.

***(i)* Ratio error :**

Assuming $K_{nom.} = K_t = 20$ ... Any other data being absent.

Neglecting the secondary winding burden,

the total burden of the secondary circuit = Burden of the meter.

$\therefore$ Burden of secondary circuit = 20 VA

$$\text{Voltage across primary winding, } E_p = \frac{VA}{I_p} = \frac{20}{100} = 0.2\,V$$

$$\text{Loss component of exciting current, } I_w = \frac{\text{Iron loss}}{E_p} = \frac{0.18}{0.2} = 0.9\,\text{A}$$

$$\text{Secondary circuit phase angle, } \delta = \tan^{-1}\left(\frac{X}{R}\right)$$

or,
$$\delta = \tan^{-1}\left(\frac{1}{4}\right) = 14.04°$$

$\therefore \quad \cos\delta = 0.97$ and $\sin\delta = 0.2426$

$$\text{Actual ratio, } K_{act.} = K_t + \frac{I_m \sin\delta + I_w \cos\delta}{I_s}$$

$$= 20 + \frac{1.4 \times 0.2426 + 0.9 \times 0.97}{5} = 20.24$$

$$\therefore \quad \textit{Ratio error} = \frac{K_{nom.} - K_{act.}}{K_{act.}} \times 100$$

$$= \frac{20 - 20.24}{20.24} \times 100 = -\mathbf{1.186\%} \quad \textbf{(Ans.)}$$

(*ii*) Phase angle, θ :

$$\theta = \frac{180}{\pi}\left[\frac{I_m \cos\delta - I_w \sin\delta}{K_t I_s}\right] \text{ degree}$$

or,
$$\theta = \frac{180}{\pi}\left[\frac{1.4 \times 0.97 - 0.9 \times 0.2426}{20 \times 5}\right] = \mathbf{0.653°} \quad \textbf{(Ans.)}$$

4.4 POTENTIAL TRANSFORMER (PT)

4.4.1. General Aspects

A **potential transformer** *is a step down transformer used along with a low range voltmeter for measuring a high voltage.*

- The primary is connected across the high voltage supply and the secondary to the voltmeter or potential coil of the wattmeter.
- Since the voltmeter (or potential coil) impedance is very high, the secondary current is very small and the potential transformer behaves as an ordinary two winding transformer operating on no-load.
- Fig. 4.7. shows a potential transformer used to measure the voltage of a circuit. It may be noted that the secondary is *grounded. This is done so that if the insulation breaks down, the high voltage does not endanger personnel who may be reading the meters.*
- These transformers are *made with high quality iron core operating at very low flux densities so that the magnetising current may be very small.*
- Careful design ensures minimum variation of voltage ratio with load and minimum phase shift between input and output voltages.

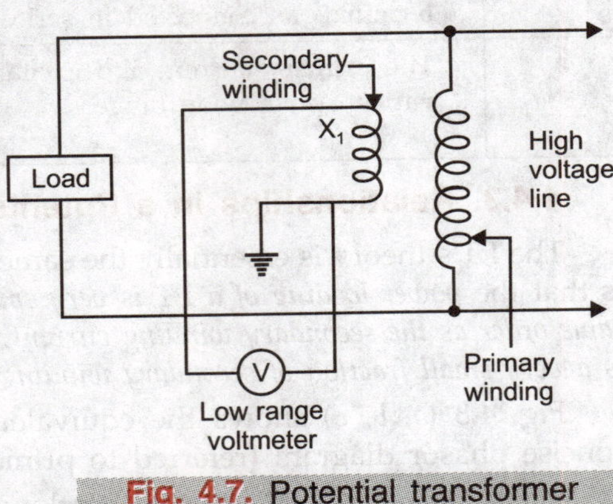

Fig. 4.7. Potential transformer connections.

- Potential transformer secondary is commonly designed for an output of **110 V.**
- PTs are usually of the following types :
  — Upto voltage of 5000 V ..... Usually of *dry type.*
  — Between 5000 V and 1,38000 V .... *Either dry type or oil immersed.*
  — Voltages exceeding 1,38000 V .... *Always immersed type.*
- As a rule, the PTs are of *shell type* because such a construction develops a *high degree of accuracy.*
- Potential transformer (PT) is different from power transformer. In power transformer current is 5% of full-load current, while in *case of PT, secondary current is of the same order as magnetising current.* Thus *PT has a small volt-ampere ratings (say about 15 VA).*
- The primary and secondary winding of the PT are *coaxial* in order to reduce the *leakage reactance to a minimum.* The low voltage winding (secondary winding) is placed next to the core.
- Oil filled bushings are used for oil filled PTs to minimise the overall size of the transformer. When neither side of the line is at ground potential, two bushings are used.

4.4.2. Comparison of Current and Potential Transformers

The comparison between current and potential transformers is given in the table below :

S.No.	Current transformer (CT)	Potential transformer (PT)
1.	Secondary must *always be shorted.*	Secondary is *nearly under open circuit conditions.*
2.	The winding carries *full-line current.*	The winding is impressed with *full-line voltage.*
3.	The primary current is *independent* of the secondary circuit conditions.	The primary current depends on the secondary circuit conditions.
4.	It can be treated as series transformer under short circuit conditions.	It can be treated as parallel transformer under open circuit secondary.
5.	A small voltage exists across its terminals as connected in series.	Full line voltage appears across its terminals.
4.	The primary current and excitation varies over a wide range.	The line voltage is almost constant hence exciting current and flux density varies over a limited range.

4.4.3. Relationships in a Potential Transformer (PT)

The PT's theory is essentially the same as that of a power transformer. The difference is that the *power loading of a PT is very small and consequently the exciting current is of the same order as the secondary winding current, while in a power transformer the exciting current is a very small fraction of secondary winding load current.*

Fig. 4.8 (*a, b, c*) shows the equivalent circuit, phasor diagram and enlarged and concise phasor diagram (referred to primary) of a potential transformer. Here,

ϕ = Working flux,

I_m = Magnetising current,

I_w = Iron loss component of no load (exciting) current,
I_o = No load (exciting) current,
E_s = Secondary winding induced voltage,

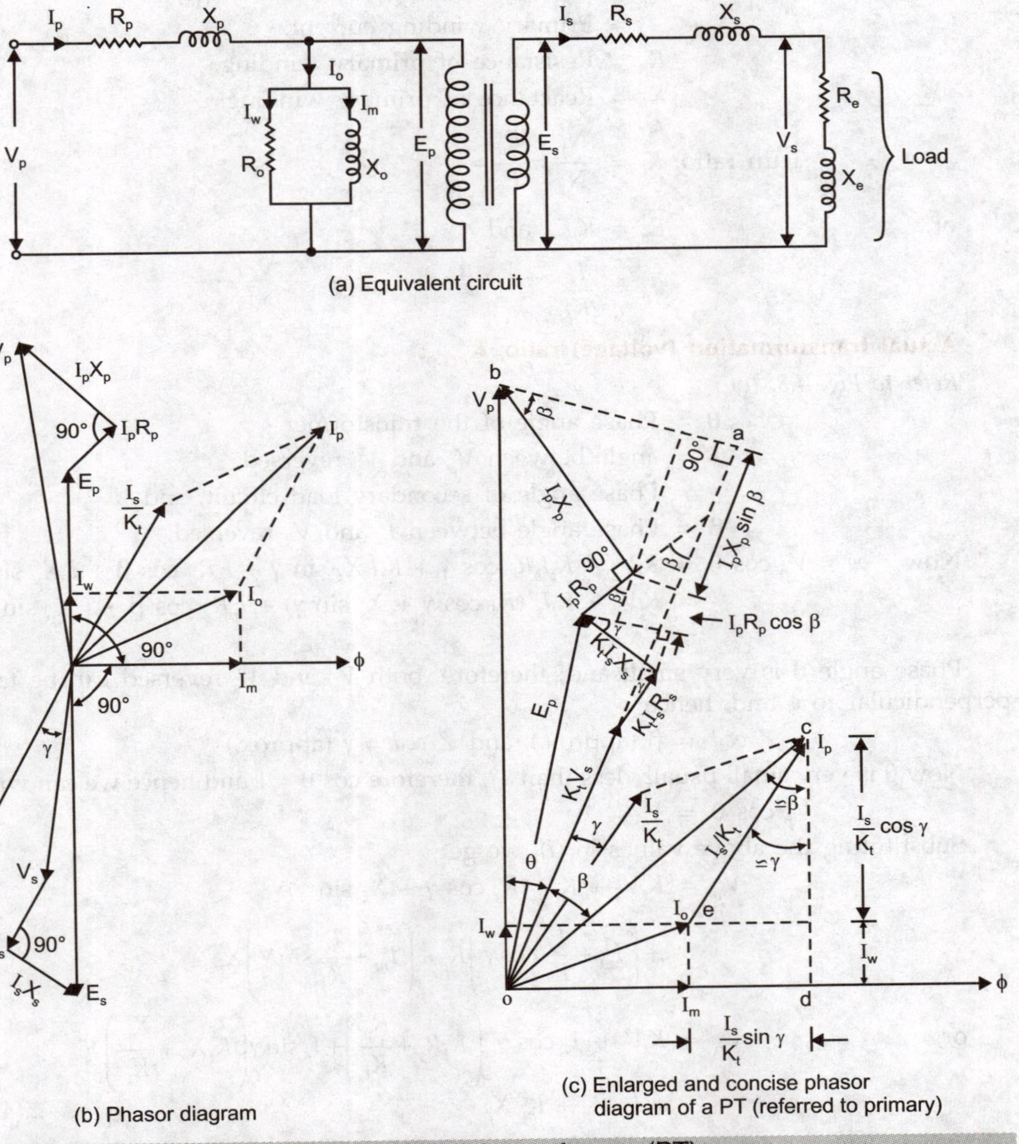

Fig. 4.8. Potential transformer (PT).

V_s = Secondary winding terminal voltage,
I_s = Secondary winding current,
R_s = Resistance of secondary winding,
X_s = Reactance of secondary winding,
R_e = Resistance of secondary load circuit,
X_e = Reactance of secondary load circuit,
γ = Phasor angle of secondary load circuit,

$$= \tan^{-1}\left(\frac{X_e}{R_e}\right)$$

E_p = Primary winding induced voltage,

I_p = Primary winding current,

R_p = Resistance of primary winding,

X_p = Reactance of primary winding.

$$\text{Turn ratio, } K_t = \frac{N_p}{N_s} = \frac{E_p}{E_s} = \frac{I_s}{I_p}$$

or,
$$E_p = K_t E_s, \text{ and}$$

$$I_p = \frac{I_s}{K_t}$$

Actual transformation (voltage) ratio, $K_{act.}$:

Refer to Fig. 4.8. (c) :

θ = Phase angle of the transformer
= angle between V_p and V_s reversed,

γ = Phase angle of secondary load circuit, and

β = Phase angle between I_p and V_s reversed,

Now,
$$oa = V_p \cos\theta = K_t V_s + K_t I_s R_s \cos\gamma + K_t I_s X_s \sin\gamma + I_p R_p \cos\beta + I_p X_p \sin\beta$$
$$= K_t V_s + K_t I_s (R_s \cos\gamma + X_s \sin\gamma) + I_p R_p \cos\beta + I_p X_p \sin\beta \quad ...(i)$$

Phase angle θ is very small and, therefore, both V_p and V_s reversed can be taken perpendicular to ϕ and, hence

$$\angle ocd = \beta \text{ (approx.), and } \angle ecd = \gamma \text{ (approx.)}$$

Now θ is very small usually less than 1°, therefore $\cos\theta = 1$ and hence we can write :

$$V_p \cos\theta = V_p$$

Substituting the above values in (*i*), we get :

$$V_p = K_t V_s + K_t I_s (R_s \cos\gamma + X_s \sin\gamma) + \left(I_w + \frac{I_s}{K_t}\cos\gamma\right)R_p + \left(I_m + \frac{I_s}{K_t}\sin\gamma\right)X_p$$

or,
$$V_p = K_t V_s + I_s \cos\gamma\left(K_t R_s + \frac{R_p}{K_t}\right) + I_s \sin\gamma\left(K_t X_s + \frac{X_p}{K_t}\right) + I_w R_p + I_m X_p \quad ...(4.24)$$

$$= K_t V_s + \frac{I_s}{K_t}\cos\gamma\left(K_t^2 R_s + R_p\right) + \frac{I_s}{K_t}\sin\gamma\left(K_t^2 X_s + X_p\right) + I_w R_p + I_m X_p$$

$$= K_t V_s + \frac{I_s}{K_t}\cos\gamma\,(R_p)_{eq} + \frac{I_s}{K_t}\sin\gamma\,(X_p)_{eq} + I_w R_p + I_m X_p$$

[where, $K_t^2 R_s + R_p = (R_p)_{eq}$ = equivalent resistance of transformer referred to 'primary'
and, $K_t^2 X_s + X_p = (X_p)_{eq}$ = equivalent reactance of transformer referred to 'primary']

$$\therefore \quad V_p = K_t V_s + \frac{I_s}{K_t}\left[(R_p)_{eq}\cos\gamma + (X_p)_{eq}\sin\gamma\right] + I_w R_p + I_m X_p \quad ...(4.25)$$

∴ *Actual transformation (voltage) ratio,*

$$K_{act.} = \frac{V_p}{V_s} = K_t + \frac{\frac{I_s}{K_t}\left[(R_p)_{eq}\cos\gamma + (X_p)_{eq}\sin\gamma\right] + I_w R_p + I_m X_p}{V_s} \quad ...(4.26)$$

Eqn. (4.24) can be rewritten as :

$$V_p = K_t V_s + K_t I_s \cos\gamma\left(R_s + \frac{R_p}{K_t^2}\right) + K_t I_s \sin\gamma\left(X_s + \frac{X_p}{K_t^2}\right) + I_w R_p + I_m X_p$$

$$= K_t V_s + K_t I_s \cos\gamma\,(R_s)_{eq} + K_t I_s \sin\gamma\,(X_s)_{eq} + I_w R_p + I_m X_p$$

[where, $R_s + \frac{R_p}{K_t^2} = (R_s)_{eq}$ = equivalent resistance of transformer *referred to 'secondary'*

and, $X_s + \frac{X_p}{K_t^2} = (X_s)_{eq}$ = equivalent reactance of transformer *referred to 'secondary'*]

$$\therefore \quad V_p = K_t V_s + K_t I_s\,[(R_s)_{eq}\cos\gamma + (X_s)_{eq}\sin\gamma) + I_w R_p + I_m X_p \quad ...(4.27)$$

∴ *Actual transformation (voltage) ratio,*

$$\mathbf{K}_{act.} = \frac{V_p}{V_s} = K_t + \frac{K_t I_s\left[(R_s)_{eq}\cos\gamma + (X_s)_{eq}\sin\gamma\right] + I_w R_p + I_m X_p}{V_s} \quad ...(4.28)$$

Using eqns. 4.26 and 4.28, the difference between $K_{act.}$ and K_t is :

$$K_{act.} - K_t = \frac{\frac{I_s}{K_t}\left[(R_p)_{eq}\cos\gamma + (X_p)_{eq}\sin\gamma\right] + I_w R_p + I_m X_p}{V_s} \quad ...(4.29)$$

$$= \frac{K_t I_s\left[(R_s)_{eq}\cos\gamma + (X_s)_{eq}\sin\gamma\right] + I_w R_p + I_m X_p}{V_s} \quad ...(4.30)$$

Phase angle, θ :

From the phasor diagram [Fig. 4.8 (*c*)], we have :

$$\tan\theta = \frac{ab}{oa} = \frac{I_p X_p \cos\beta - I_p R_p \sin\beta + K_t I_s X_s \cos\gamma - K_t I_s R_s \sin\gamma}{K_t V_s + K_t I_s R_s \cos\gamma + K_t I_s X_s \sin\gamma + I_p R_p \cos\beta + I_p X_p \sin\beta}$$

The terms in denominator involving I_p and I_s are small and, therefore, they can be neglected as compared with $K_t V_s$

$$\therefore \quad \tan\theta = \frac{I_p X_p \cos\beta - I_p R_p \sin\beta + K_t I_s X_s \cos\gamma - K_t I_s R_s \sin\gamma}{K_t V_s}$$

$$= \frac{X_p\left(I_w + \frac{I_s}{K_t}\cos\gamma\right) - R_p\left(I_m + \frac{I_s}{K_t}\sin\gamma\right) + K_t I_s X_s \cos\gamma - K_t I_s R_s \sin\gamma}{K_t V_s}$$

$$= \frac{I_s \cos\gamma\left(\frac{X_p}{K_t} + K_t X_s\right) - I_s \sin\gamma\left(\frac{R_p}{K_t} + K_t R_s\right) + I_w X_p - I_m R_p}{K_t V_s}$$

$$= \frac{\frac{I_s \cos\gamma}{K_t}(X_p + K_t^2 X_s) - \frac{I_s \sin\gamma}{K_t}(R_p + K_t^2 R_s) + I_w X_p - I_m R_p}{K_t V_s}$$

$$= \frac{\frac{I_s \cos\gamma}{K_t}(X_p)_{eq} - \frac{I_s \sin\gamma}{K_t}(R_p)_{eq} + I_w X_p - I_m R_p}{K_t V_s}$$

$$= \frac{\frac{I_s}{K_t}\left[(X_p)_{eq}\cos\gamma - (R_p)_{eq}\sin\gamma\right] + I_w X_p - I_m R_p}{K_t V_s}$$

Since θ is small, therefore, θ = tan θ

$$\therefore \quad \theta = \frac{\frac{I_s}{K_t}\left[(X_p)_{eq}\cos\gamma - (R_p)_{eq}\sin\gamma\right] + I_w X_p - I_m R_p}{K_t V_s} \text{ rad.} \qquad \text{...(4.31)}$$

In terms of quantities *referred to secondary,*

$$(R_s)_{eq} = \frac{R_p}{K_t^2} \quad \text{and} \quad (X_s)_{eq} = \frac{X_p}{K_t^2}$$

$$\therefore \quad \theta = \frac{K_t I_s\left[(X_s)_{eq}\cos\gamma - (R_s)_{eq}\sin\gamma\right] + I_w X_p - I_m R_p}{K_t V_s}$$

or,
$$\theta = \frac{I_s}{V_s}\left[(X_s)_{eq}\cos\gamma - (R_s)_{eq}\sin\gamma\right] + \frac{I_w X_p - I_m R_p}{K_t V_s} \text{ rad.} \qquad \text{...(4.32)}$$

Ratio (voltage) error. The actual ratio of transformation varies with operating conditions and the error in secondary voltage may be defined as :

$$\textit{Percentage ratio error} = \frac{K_{nom.} - K_{act.}}{K_{act.}} \times 100 \qquad \text{...(4.33)}$$

Phase angle error. In an ideal transformer there should not be any phase difference between primary winding voltage and the secondary winding voltage *reversed*. However, in an actual transformer there exists a phase difference between V_p and V_s reversed.

Phase angle error = θ, is given by : same as given in eqns. 4.31 and 4.32. The phase angle is taken as *+ve* when secondary winding voltage reversed *leads* the primary winding voltage. The angle is *–ve* when secondary winding voltage reversed *lags* the primary winding voltage.

- It may be remembered that *while measuring 'voltage', ratio error is only important while for 'power measurements' both ratio and phase angle errors are involved.*

4.4.4. Characteristics of Potential Transformers

1. **Effect of change in Secondary Burden :**
   - With the increase in secondary burden, the secondary current is increased and consequently the primary current is increased. Both the primary and secondary voltage drops increase and thus for a given value of V_p, there is a decrease in the value of V_s and hence the *actual ratio increases as the burden increases*. This variation of error is *almost linear* with change in burden.
   - Further, owing to increased voltage drop, V_p advances while V_s retards in phase. As a result, the *phase angle* between V_p and V_s reversed *increases* and becomes *more negative as the current increases*.
2. **Effect of power factor of secondary burden :**
   - When the power factor of a secondary circuit burden is reduced, angle γ increases. As a consequence, current I_p shifts towards I_o [see Fig. 4.8 (*b*)]. The voltages V_p and V_s come more nearly into phase with E_p and E_s respectively since the voltage drops are almost constant. The results in an increase in V_p relative to E_p. But, since V_p is constant, therefore, E_p reduces relative to V_p; V_s also reduces relative to E_s. Thus the *transformation ratio increases as the p.f. of secondary burden reduces.*
   - While V_s advances in phase and V_p retards hence *negative phase angle reduces as lagging p.f. of the secondary decreases.*
3. **Effect of change in frequency :**
   - The flux is inversely proportional to frequency, for a constant voltage. Thus, with increase in frequency, flux is reduced and, therefore I_m and I_w are decreased leading to *decrease in voltage ratio*. The decrease is not so much, as with increase in frequency the leakage reactance increases and, therefore, leakage reactance drops are increased giving an *increase in the ratio*. Thus changes in voltage ratio (due to change in frequency) are dependent upon relative values of I_o and leakage reactances (as they produce opposite effects).
   - With the *increase in frequency*, the *phase angle is increased* since both the effects advance V_s and the increase in secondary reactance retards V_s.
4. **Effect of primary voltage.** The variations of errors due to change in primary voltage are of no importance, as supply-system voltages do not vary widely.

4.4.5. Reduction of Errors in Instrument Transformers

The errors (ratio and phase angle) in instrument transformer may be reduced by using the following methods :

1. ***Reduction of magnetising and loss components (of exciting current I_o).*** From the equations (derived earlier) of ratio and phase angle errors it is evident that these errors depend on I_m and I_w (components of I_o). These components can be *reduced* by :

   (*i*) Using materials of high permeability; (*ii*) Keeping the flux density in the core to low value; (*iii*) Choosing low reactance core; (*iv*) Using a core of large cross-section; (*v*) Providing smaller magnetic paths to the flux; (*vi*) Designing the assembly and interleaving the core with proper precautions.
2. ***Reduction of resistance and leakage reactance.*** The errors depend on the voltage drops which in turn depend on the values of resistance and leakage reactance.
   - The *resistance* can be reduced by using thick conductors (*i.e.*, of large cross-section) and using the smallest length of mean turn.

- The *leakage reactance* of windings depends upon the leakage fluxes of primary and secondary windings and, therefore, windings *must be kept as close as possible.*
  — Further, a high flux density means a high flux in the core and so the windings have lesser number of turns, which obviously results in smaller leakage reactance of the windings.

3. ***Turns compensation*** **:**

- At no load (when $I_s = 0$) the actual ratio ($K_{act.}$), in case of PT, *exceeds* the turns ratio (K_t) by an amount $\frac{(I_w R_p + I_m X_p)}{V_s}$ (Refer to eqn. 4.26). The ratio further increases with an inductive or resistive load because of voltage drops in resistance and leakage reactance of these windings. If $K_t = K_{nom.}$, the $K_{act.}$ differs from the nominal ratio and *errors creep* in.

  The remedy lies in making the K_t less $K_{nom.}$ This can be accomplished by either *reducing* the number of turns of primary winding or *increasing* the number of secondary winding turns. By doing so, it becomes possible to make the actual ratio $K_{act.}$ equal to $K_{nom.}$ for one particular value of burden. At the same time, there is a general reduction in the ratio error over the entire working range of the burden. This is called *"turns compensation"*.
- *The phase angle θ, practically, remains unaffected due to a small change in* K_t.

Example 4.10. *The following data relate to a 1000/100 volts potential transformer :*

Primary resistance = 84.4 Ω

Primary reactance = 62.5 Ω

Secondary resistance = 0.78 Ω

Total equivalent reactance = 102 Ω

No-load current = 0.03 A *at* 0.42 *power factor*

Calculate :

(i) Phase angle error at no load.

(ii) Burden in VA at unity power factor at which the phase angle will be zero.

Solution. *Given :* $K_t = \frac{1000}{100} = 10$; $R_p = 84.4\ \Omega$; $X_p = 62.5\ \Omega$; $R_s = 0.78\ \Omega$; $X_e = 102\ \Omega$; $I_o = 0.03$ A at 0.42 p.f.

(i) Phase angle error at no load, θ :

No load p.f. $\cos\phi_o = 0.42$;

$$\therefore \quad \sin\phi_o = \sqrt{1-(0.42)^2} = 0.9075$$

Now,
$$I_w = I_o \cos\phi_o = 0.03 \times 0.42 = 0.0126 \text{ A}$$
$$I_m = I_o \sin\phi_o = 0.03 \times 0.9075 = 0.0272 \text{ A}$$

$$\text{Phase angle, } \theta = \frac{\frac{I_s}{K_t}\left[(X_p)_{eq}\cos\gamma - (R_p)_{eq}\sin\gamma\right] + I_w X_p - I_m R_p}{K_t V_s} \text{ rad} \quad \text{...[Eqn. (4.31)]}$$

At no load, $I_s = 0$

$$\therefore \qquad \theta = \frac{I_w X_p - I_m R_p}{K_t V_s} \text{rad}$$

$$\text{or,} \qquad \theta = \frac{0.0126 \times 62.5 - 0.0272 \times 86.4}{10 \times 100} = \mathbf{-0.00156 \text{ rad}} \quad \textbf{(Ans.)}$$

(*ii*) Burden in VA :

$$\theta = 0 \qquad \text{...(Given)}$$

At unity p.f., $\cos\gamma = 1$, $\sin\gamma = 0$

$$\theta = \frac{\frac{I_s}{K_t}\left[(X_p)_{eq}\cos\gamma - (R_p)_{eq}\sin\gamma\right] + I_w X_p - I_m R_p}{K_t V_s} \text{rad.}$$

Substituting the values, we have :

$$\text{or,} \qquad \theta = \frac{\frac{I_s}{10}(102 \times 1 - 0) + 0.0126 \times 62.5 - 0.0272 \times 86.4}{10 \times 100}$$

$$\text{or,} \qquad \theta = 10.2\, I_s - 1.563$$

$$\therefore \qquad I_s = \frac{1.563}{10.2} = 0.1532\, A$$

$\therefore$ Burden load in $VA = V_s I_s = 100 \times 0.1532 =$ **15.32 VA** **(Ans.)**

Example 4.11. *A single-phase potential transformer has a turns ratio of 3810/63. The nominal secondary voltage is 63 V and the total equivalent resistance and leakage reactance referred to secondary are 2 Ω and 1 Ω respectively. Calculate the ratio and phase angle errors when the transformer is supplying a burden of 100 + j 200 Ω.* **(PTU – Dec. 2005)**

Solution. *Given* : $K_t (= K_{nom.}) = \frac{3810}{63} = 60.476$; $V_s = 63$ V; $(R_s)_{eq} = 2\Omega$; $(X_s)_{eq} = 1\ \Omega$; Burden = 100 + *j* 200 Ω.

Ratio error :

Neglecting no load components of exciting current,

Transformation (actual) ratio,

$$K_{act.} = K_t + \frac{K_t I_s\left[(R_s)_{eq}\cos\gamma + (X_s)_{eq}\sin\gamma\right]}{V_s}$$

$$\text{or,} \qquad K_{act.} = K_t + \frac{K_t\left[(R_s)_{eq}\cos\gamma + (X_s)_{eq}\sin\gamma\right]}{(V_s / I_s)} \qquad \text{...[Eqn. (4.28)]}$$

$$\text{Burden} = R_e + jX_e = 100 + j\,200\ \Omega \qquad \text{...}(i)$$

$$\therefore \qquad \gamma = \tan^{-1}\left(\frac{X_e}{R_e}\right) = \tan^{-1}\left(\frac{200}{100}\right) = 63.43°$$

$$\text{Also,} \qquad \frac{V_s}{I_s} = Z_s = \sqrt{100^2 + 200^2} = 223.6\,\Omega$$

Substituting the various values in eqn. (*i*), we get :

$$K_{act.} = 60.476 + \frac{60.476[2 \times \cos 63.43° + \sin 63.43°]}{223.6} = 60.9598$$

$$\therefore \quad \% \text{ Ratio error} = \frac{K_{nom.} - K_{act.}}{K_{act.}} \times 100$$

$$= \frac{60.476 - 60.9598}{60.9598} \times 100 = -\mathbf{0.7936\%} \quad \textbf{(Ans.)}$$

Phase angle error, θ :

$$\theta = \frac{I_s}{V_s}\left[(X_s)_{eq}\cos\gamma - (R_s)_{eq}\sin\gamma\right] \text{ rad.} \qquad \text{...[Eqn. (4.32)]}$$

$$= \frac{1}{223.6}[1 \times \cos 63.43° - 2\sin 63.43°] \text{ rad.}$$

$$= -5.999 \times 10^{-3} \text{rad.} = -5.999 \times 10^{-3} \times \frac{180}{\pi} \text{degrees}$$

$$= -\ \mathbf{0.3437°} \quad \textbf{(Ans.)}$$

4.5 CAPACITIVE POTENTIAL TRANSFORMERS

The conventional electro-magnetic type of potential transformer becomes *expensive* at voltages above about 100 kV (phase), due to insulation requirements. A capacitive voltage transformer is a less expensive alternative, this transformer is illustrated in Fig. 4.9.

— It consists of a capacitance potential divider used in conjunction with a conventional auxiliary transformer.

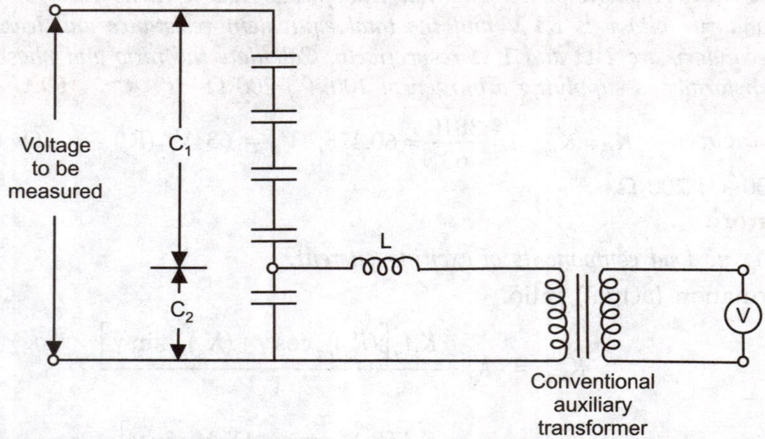

Fig. 4.9. Capacitive potential transformer.

— The capacitance potential divider *steps down the voltage to be measured*, say to about 10 kV. The output voltage of the capacitance divider is *further stepped down by the auxiliary transformer* to the desired secondary voltage, say 110 V.

— The auxiliary transformer consists of an inductance *L*, which may consist wholly or partly of leakage inductance of the windings of auxiliary transformer. The value of this *L may be adjusted to* $1/[\omega^2(C_1 + C_2)]$ *so that the voltage drop due to the current drained from the divider is largely compensated. Thus the overall voltage transformation ratio is independent of burden.*

The overall ratio is the product of the divider and transformer ratio.

— *The auxiliary transformer must be of very small ratio error and phase angle to give satisfactory performance of the complete unit.*

4.6 TESTING OF INSTRUMENT TRANSFORMERS

The methods of testing of instrument transformers *i.e.*, finding their ratio and phase angle errors may be broadly classified into the following two groups :

1. **Absolute methods :**
   - In these methods, the errors of transformer under test are determined in terms of constants – resistance, inductance and capacitance of the testing circuit.
2. **Comparison method :**
   - In this case, the transformer (under test) errors are compared with those of a standard current transformer whose errors are known.

The above two methods, *according to their measurement techniques*, can be *classified* as follows :

1. ***Deflection methods :***
   - These methods *employ the deflections of suitable instruments* like electrodynometer wattmeters for measuring quantities related to the phasors under consideration or to their deflection. The required ratio and phase angles are then found out from the magnitudes of deflections.
   - The deflection methods, in some cases, may be made direct reading.
2. ***Null methods :***
   - These methods employ a network in which *appropriate phasor quantities are balanced against one another. The ratio and phase angle errors are then found from the magnitudes of deflection.*
   - The null methods may be made direct reading in terms of calibrated scales on the adjustable elements in the network.

4.6.1. Testing of Current Transformers (CTs)

The following two methods of testing of CTs will be discussed :

1. Mutual inductance method ... *Absolute method*
2. Arnold method ... *Comparison method*

4.6.1.1. Mutual inductance method (*Absolute*)

This absolute method of testing CTs uses *null technique.*

Fig. 4.10 shows the circuit and phasor diagrams for the mutual inductance method :

— R_p and R_s are low resistances, non-inductive shunts. R_p is fixed while R_s is variable and has a slide wire for fine adjustment of the resistance.

— The voltage drop across R_p is matched against voltage drop across R_s. The balance conditions are indicated by a vibration galvanometer (V.G.) placed in the circuit.

— If there is no phase difference between I_p and I_s, the virbration galvanometer will indicate zero deflection when;

$$I_pR_p = I_sR_s, \quad \text{or,} \quad \frac{I_p}{I_s} = \frac{R_s}{R_p} \qquad ...(4.34)$$

Thus, R_p and R_s should be so selected that the ratio $\frac{R_s}{R_p}$ is nearly equal to nominal ratio ($K_{nom.}$) of the current transformer.

— The adjustment of resistance R_s is so made that two voltage drops are *equal*. In order *to obtain zero deflection, the magnitude as well as phase of the voltage drops should be same*. Consequently, a mutual inductance M is put *to compensate for phase* *difference between I_p and I_s as without any phase compensating device it is impossible to obtain the balance with resistance alone.*

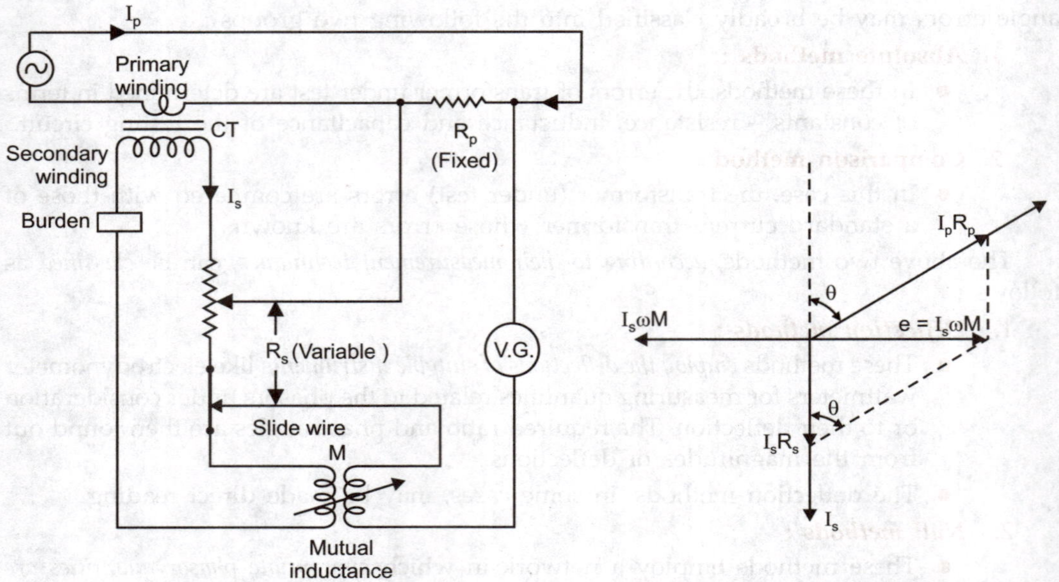

(a) Circuit diagram (b) Phasor diagram (under balance conditions)

Fig. 4.10. Mutual inductance method.

Referring to Fig. 4.10(*b*) representing phasor diagram under balance conditions, we get :

$$\tan \theta = \frac{I_s \omega M}{I_s R_s} = \frac{\omega M}{R_s} \qquad ...(4.35)$$

and,

$$\cos \theta = \frac{I_s R_s}{I_p R_p} \qquad ...(4.36)$$

From the above relations, we have :

$$\textit{Phase angle, } \theta \simeq \frac{\omega M}{R_s} \text{ rad.} \qquad ...(4.37)$$

$$\textit{Actual ratio, } K_{act.} = \frac{I_p}{I_s} = \frac{R_s}{R_p \cos \theta} \simeq \frac{R_s}{R_p} \qquad ...(4.38)$$

...θ being very small

- While stating the burden at which errors have been measured, the secondary load circuit, which includes R_s, impedance of primary winding of mutual inductance and impedance marked burden, must be taken into account.

4.6.1.2. Arnold method (*Comparison*)

This comparison method of CT testing involves *null techniques.*

- This method provides *very accurate results.*

Fig. 4.11 shows the circuit and phasor diagrams for Arnold method of CT testing:

— In this method, CT under test (*X*) is *compared with a known standard CT (S) of the same nominal ratio* ($K_{nom.}$)

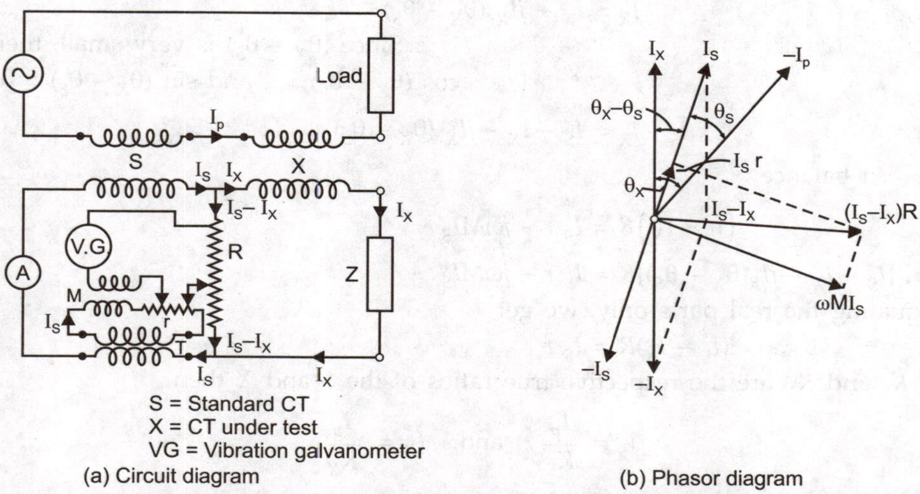

Fig. 4.11. Arnold method of CT testing.

— S and X are connected with primaries in series and, secondaries also in series. The secondary circuit is completed by an ammeter the primary of a 5/5 isolating CT, T and a suitable burden Z for the CT under test.

— R is a non-inductive resistance; its value is decided according to the required sensitivity and range.

— M is a variable mutual inductor with a range of ± 2.4 μH and r is a variable resistance ranging ± 500 μΩ.

— Since the inductive inteference in measuring circuit causes error during measurement, therefore, it is essential to eliminate it. The *elimination procedure* involves the following *steps*:

(i) The leads of resistance R are disconnected and short-circuited, r and M are set at zero and full rated primary current is allowed to flow.

(ii) The connecting leads and apparatus of the measuring circuit are so arranged that vibration galvanometer shows no deflection.

But doing so, the circuit becomes free from inductive interference.

— The vector difference of two secondary currents $(I_S - I_X)$ flows through the non-inductive resistance R. The potential difference across R is balanced against e.m.f. in the secondary of mutual inductance M and the voltage drop across the portion r of the low resistance centre tapped slide wire.

— The current in the primary mutual inductance M and slide wire is due to current transformer T. As T is of unity ratio so the current flowing in the primary of mutual inductance M and slide wire is equal to secondary current of S (standard CT). Balance is indicated by the vibration galvanometer and is obtained by adjustment of M and r.

- If θ_x and θ_s are the respective phase angles of the test and standard current transformers, then the angles between the *reversed* vector of I_p and the vectors I_S and I_X will be as shown in Fig. 4.11(*b*).

Taking I_S as the referene vector, we have :

$$\bar{I}_S = I_S + j0 \quad \text{...(4.39)}$$

$$\bar{I}_X = I_X \cos(\theta_X - \theta_S) + jI_X \sin(\theta_X - \theta_S) \quad \text{...(4.40)}$$

or, $$\bar{I}_X \simeq I_X + jI_X(\theta_X - \theta_S) \qquad ...(4.41)$$

...Since $(\theta_X - \theta_S)$ is very small, therefore, $\cos(\theta_X - \theta_S) \simeq 1$ and $\sin(\theta_X - \theta_S) \simeq \theta_X - \theta_S$

or, $$\bar{I}_S - \bar{I}_X = I_S - I_X - jI_X(\theta_X - \theta_S) \qquad ...(4.42)$$

∴ At balance :

$$\left(\bar{I}_S - \bar{I}_X\right)R = I_S\, r - j\omega M I_S \qquad ...(4.43)$$

or, $[I_S - I_X - jI_X(\theta_X - \theta_S)]R = I_S\, r - j\omega M I_S$ ...(4.44)

Equating the real parts only, we get

$$(I_S - I_X)R = I_S\, r \qquad ...(4.45)$$

If K_S and K_X are the respective true ratios of the S and X then,

$$I_S = \frac{I_p}{K_S}, \quad \text{and,} \quad I_X = \frac{I_p}{K_X}$$

Inserting these values in eqn. (4.45) we get :

$$\left(\frac{I_p}{K_S} - \frac{I_p}{K_X}\right)R = \frac{I_p}{K_S}\, r$$

or, $$1 - \frac{K_S}{K_X} = \frac{r}{R}$$

or, $$\frac{K_S}{K_X} = 1 - \frac{r}{R}$$

or, $$\frac{K_S}{K_X} = \frac{1}{1-(r/R)} = 1 + \frac{r}{R} \qquad ...(4.46)$$

...since $\frac{r}{R}$ is very small.

Hence, K_X can be determined from the known values of K_S, r and R. Now, equating imaginary parts of eqn (4.44) we get :

$$-jI_X R(\theta_X - \theta_S) = -j\omega M I_S \qquad ...(4.47)$$

or, $$(\theta_X - \theta_S) = \frac{\omega M I_S}{R I_X} \simeq \frac{\omega M}{R} \qquad ...(4.48)$$

...since $\frac{I_S}{I_X}$ is very *close to unity*.

4.6.2. Testing of Potential Transformers (PTs)

The following tests are conducted on the potential transformers, for the determination of the ratio and phase angle errors :

4.6.2.1. Absolute null methods

The principle of most absolute methods of testing potential transformers is to balance the secondary voltage against a fraction of the primary voltage obtained from a potential divider and the errors are measured in terms of circuit constants.

Fig. 4.12 shows the circuit and phasor diagrams for such a method :

— The burden (Z) with which the transformer is to be tested is connected across the secondary winding, and the normal primary voltage at normal frequency, is applied to the primary winding. One end of the secondary winding is connected to one end of the primary winding.

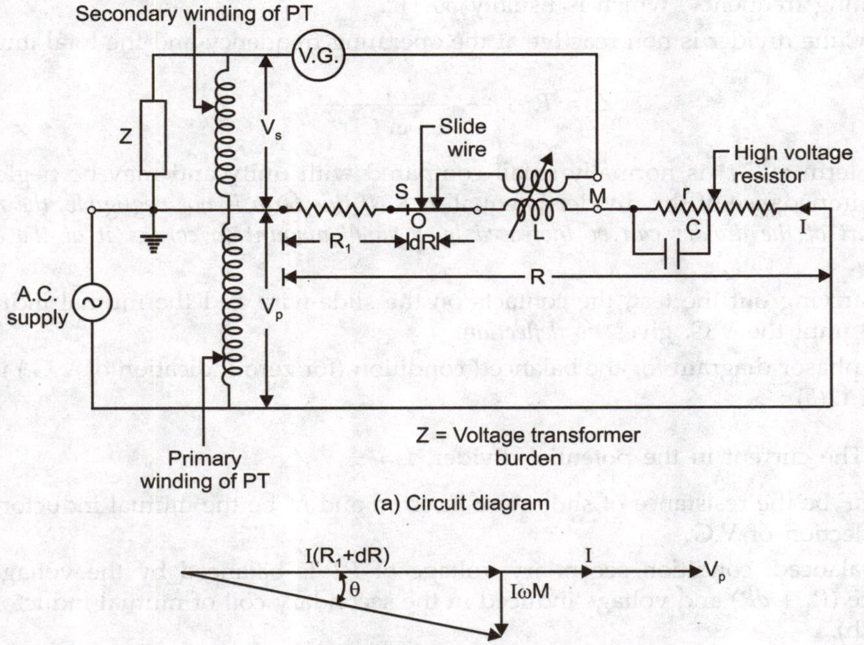

(b) Phasor diagram

Fig. 4.12. Absolute method of testing potential transformers.

— A non-inductive and non-capacitive potential divider is connected across the primary winding. This divider consists of fixed resistance R_1 in series with slide wire S, which is connected to *h.v.* resistance r in series with a primary coil of a variable mutual inductor M. A variable capacitor C shunts a small portion of the resistance of the potential divider.

— V.G. is a vibration galvanometer. The resistances are so chosen that nominal ratio $(K_{nom.})$ of the test transformer is equal to $\frac{R}{R_1}$.

L is the inductance of the primary winding of the mutual inductor and will cause a phase error in the divider ; the capacitor C compensates for this error in the following way :

The impedance of the whole divider,

$$\bar{Z} = R - r + j\omega L + \frac{r}{1 + j\omega Cr} \quad \text{...(4.49)}$$

or,

$$\bar{Z} = R - r + j\omega L + \frac{r}{1 + j\omega Cr} \times \frac{1 - j\omega Cr}{1 - j\omega Cr}$$

$$= R - r + j\omega L + \frac{r(1 - j\omega Cr)}{1 + \omega^2 C^2 r^2}$$

$$= R - r + \frac{r}{1 + \omega^2 C^2 r^2} + j\omega \left(L - \frac{Cr^2}{1 + \omega^2 C^2 r^2} \right) \quad \text{...(4.50)}$$

Now, in this divider, the values of C anf r are so chosen that the term $\frac{Cr^2}{1+\omega^2C^2r^2} = L$ at operating frequency, which is usually 50 Hz.

Now, the divider is non-reactive at the operating frequency and the total impedance,

$$Z = R - r + \frac{r}{1+\omega^2C^2r^2} \quad \text{...(4.51)}$$

The term $\omega^2C^2r^2$ is normally small compared with unity and may be neglected, so that the impedance of the divider is equal to R. *If this term is not negligible, the resistance of the part of the divider can be increased by a small amount to correct it at the operating frequency.*

In carrying out the test, the contacts on the slide-wire and the mutual inductor are adjusted until the V.G. gives *no deflection.*

The phasor diagram for the balanced condition (for zero indication of V.G.) is shown in Fig. 4.12(*b*) :

The current in the potential divider, $I = \frac{V_s}{R}$

Let *dF* be the resistance of slide wire inserted and *M* be the mutual inductor to have zero deflection of V.G.

In balanced, condition secondary voltage of *PT* is balanced by the voltage across resistance $(R_1 + dR)$ and voltage induced in the secondary coil of mutual inductor *M* (see Fig. 4.12b).

$$\therefore \quad \bar{V}_s = I\,(R_1 + dR) - jI\omega M \quad \text{...(4.52)}$$

The phase angle θ of the *PT* under test is usually very small, therefore, it can be neglected and,

$$V_s = I\,(R_1 + dR)$$

$$\text{or,} \quad V_s = \frac{V_p}{R}(R_1 + dR) \quad \text{...(4.53)}$$

Thus, the transformer ratio is, $\frac{V_p}{V_s} = \frac{R}{R_1 + dR}$

- The *ratio error* (of PT) $= \frac{K_{nom.} - K_{act.}}{K_{act.}}$

$$= \frac{\frac{R}{R_1} - \frac{R}{R_1 + dR}}{\frac{R}{R_1 + dR}} = \frac{R_1 + dR}{R_1} - 1 = \frac{dR}{R_1} \quad \text{...(4.54)}$$

R_1 is a fixed quantity in the circuit so that ratio error of PT can be obtained directly from the slide wire calibrated in values of ratio error.

- *Phase angle*, θ. From the phasor diagram (Fig. 4.12b),

$$\tan\theta = \frac{I\omega M}{I(R_1 + dR)} = \frac{\omega M}{R_1 + dR} \quad \text{...(4.55)}$$

Since *dR* is very small in comparison to R_1, therefore, it can be neglected and θ being very small, tan θ can be taken equal to θ.

$\therefore$ *Phase angle error* (of PT), $\theta = \dfrac{\omega M}{R_1}$ ...(4.56)

Thus *mutual inductor M can be calibrated directly in terms of the phase angle error.*

4.6.2.2. Comparison method

In this method, the PT under test (X) with burden across the secondary is *compared* with standard PT (*S*) whose errors are small and known.

Fig. 4.13 shows the circuit diagram and phasor diagram respectively for comparison method of testing potential transformers :

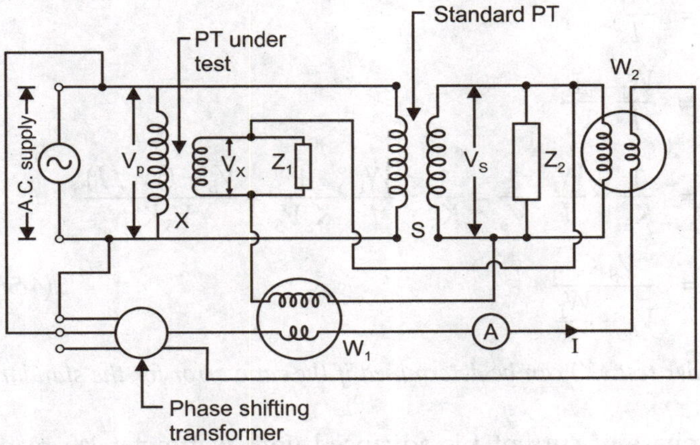

(a) Circuit diagram

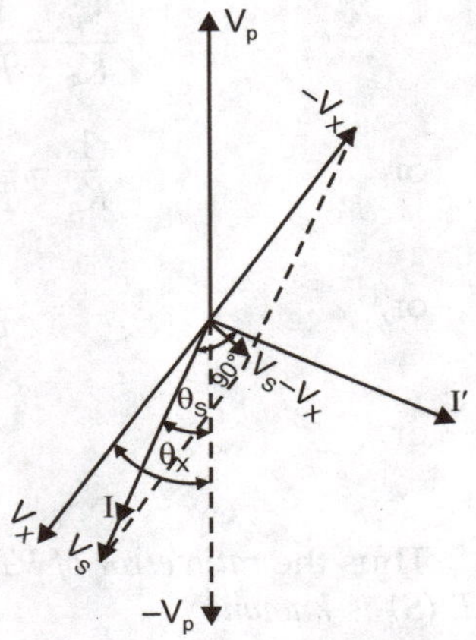

(b) Phasor diagram

Fig. 4.13. Comparison method of testing potential transformers.

— The primary windings of both the PTs (*i.e.*, *S* and *X*) are connected in parallel; the secondary windings are so connected that the voltage acting across the potential coil of wattmeter W_1 is the vector difference of the two secondary voltages V_S and V_X. The potential coil of wattmeter W_2 is directly connected across the secondary of the standard PT (*S*).

— The current coils of both the wattmeters are connected in series so as to carry the same current and are *supplied from a phase shifting transformer.*

— Burden impedances Z_1, and Z_2 are adjusted to the required values.

— The wattmeter W_2 acts as a phase indicating device and also indicates the voltages at which transformers are tested.

The phasor diagram is shown in Fig. 4.13(b).

• During experimentation, since the phase shifting transformer is first adjusted to give the maximum value, therefore, the current *I* will be in phase with secondary voltage of standard transformer (*S*), V_S as shown in phasor diagram (Fig. 4.13b) and reading of the wattmeter $W_2 = V_S I$. The *reading of wattmeter W_1 for this setting* is also noted which will be equal to the *product of current I and the component of the voltage which is in phase with this current*. Since in this position V_S and V_X are nearly in phase opposition across the potential coil of wattmeter W_1, so the voltage component of voltage circuit in phase with currents *I* is $(V_S - V_X)\cos(\theta_X - \theta_S)$, and

$$W_1 = I\,[(V_S - V_X)\cos(\theta_X - \theta_S)] \quad \text{...(4.57)}$$

$$\simeq I\,(V_S - V_X)$$

...since $\sin(\theta_X - \theta_S)$ is very small, therefore, $\cos(\theta_X - \theta_S) \simeq 1$.

or, $$V_S - V_X = \frac{W_1}{I}$$

If R_S and R_X are true voltage transformation ratios of S and X potential transformers, then

$$\frac{V_p}{V_S} = K_S, \text{ and, } \frac{V_p}{V_X} = K_X$$

Inserting these values in eqn. (4.57), we get

$$\frac{V_p}{K_S} - \frac{V_p}{K_X} = \frac{W_1}{I}$$

or, $$\frac{1}{K_S} - \frac{1}{K_X} = \frac{W_1}{I} \cdot \frac{I}{V_p}$$

or, $$\frac{1}{K_X} = \frac{1}{K_S} - \frac{W_1}{I} \cdot \frac{1}{V_p} = \frac{1}{K_S} - \frac{W_1}{I} \cdot \frac{1}{K_S V_S} = \frac{V_S - (W_1/I)}{V_S K_S}$$

or, $$K_X = \frac{V_S K_S}{V_S - \frac{W_1}{I}} \quad ..(4.58)$$

Thus the *ratio error of PT under test (X) can be determined if the ratio error for the standard PT (S) is known.*

- In second position the phase of current I is advanced until wattmeter W_2 reads zero and reading of wattmeter W_1 is noted, say W'_1. In this position current I is in quadrature with V_S and new reading of wattmeter W_1 is given as :

$$W'_1 = IV_X \cos [90° - (\theta_X - \theta_S)] \quad ...(4.59)$$
$$= IV_x \sin (\theta_X - \theta_S)$$

or, $$\sin (\theta_X - \theta_S) = \frac{W'_1}{IV_x}$$

Since $V_X = V_S$ and $(\theta_X - \theta_S)$ is very small, therefore, $\sin (\theta_X - \theta_S) = \theta_X - \theta_S$

$$\therefore \quad \theta_X - \theta_S = \frac{W'_1}{IV_s} \quad ...(4.60)$$

Thus, *the phase angle of PT under test (X) can be determined if the phase angle for standard PT (S) is known.*

HIGHLIGHTS

1. The principle of the instrument transformer is fundamentally the same as that of the power transformer.
2. Current transformers (CTs) are used for the measurement of *'current'*.
3. Potential transformers (PTs) are used for the measurement of *'voltage'*.
4. CT and PT are extensively used for precise measurements.
5. Ratio Correction Factor (RCF) is the ratio of actual transformation ratio to nominal ratio (*i.e.*, RCF = $\frac{K_{act.}}{K_{nom.}}$).

4. The *CT must never be operated on open circuit* because : (*i*) There will be no secondary mmf and since the primary current (and mmf) is fixed, the core flux will increase enormously. This will cause large eddy current and hysteresis losses and the resulting high temperature may damage the insulation or even the core; (*ii*) A very high voltage will be induced in the multi-turn secondary and this high voltage may be dangerous both to life and to the insulation.
7. A potential transformer (PT) is a *step down transformer* used along with a low range voltmeter for measuring a high voltage.
8. PT has a small volt-ampere ratings (say 15 VA).

OBJECTIVE TYPE QUESTIONS

Choose the Correct Answer :

1. The ratio of transformation in the case of potential transformers
   (*a*) increases with increase in power factor of secondary burden.
   (*b*) remains constant irrespective of the power factor of secondary burden.
   (*c*) decreases with increase in power factor of secondary burden.
   (*d*) none of the above.
2. In case of potential transformers
   (*a*) the phase angle error is always positive.
   (*b*) the phase angle error is always negative.
   (*c*) the phase angle error is usually zero.
   (*d*) the phase angle error is positive when the secondary winding voltage reversed leads the primary winding voltage and is negative when the secondary winding voltage reversed lags behind the primary winding voltage.
3. The disadvantages of using multipliers with voltmeters at high voltages are :
   (*a*) The power consumption of multipliers becomes large at large voltages.
   (*b*) The multipliers at high voltages have to be shielded in order to prevent capacitive currents.
   (*c*) The metering circuit is not electrically isolated from the power circuit.
   (*d*) all of the above.
4. The nominal ratio of a current transformer is
   (*a*) primary winding current/secondary winding current
   (*b*) rated primary winding current/rated secondary winding current
   (*c*) number of secondary winding turns/number of primary winding turns
   (*d*) all of the above.
5. The errors in current transformers can be reduced by designing them with
   (*a*) high permeability and low loss core materials, avoiding any joints in the core and also keeping the flux density to a low value.
   (*b*) using primary and secondary windings as close to each other as possible.
   (*c*) using large cross-sections for both primary and secondary winding conductors.
   (*d*) all of the above.
6. Capacitive potential transformers are used
   (*a*) for primary winding phase voltages above 100 kV.
   (*b*) for keeping the value of transformation ratio constant irrespective of the burden by making certain adjustment.

(c) because they are cheaper than the electromagnetic transformers above a certain voltage range.
(d) all of the above.

7. The size of potential transformers
(a) is the same as that of power transformers of the same VA rating.
(b) is much greater than that of power transformers of the same VA rating because they are designed for low ratio and phase angle errors which require large sized cores and winding conductors.
(c) is smaller than that of power transformers of the same VA rating.
(d) none of the above.

8. The transformation ratio in the case of a potential transformer is defined as ratio of
(a) primary winding voltage/secondary winding voltage
(b) rated primary winding voltage/rated secondary winding voltage
(c) number of turns of primary winding/number of turns of secondary winding
(d) all of the above.

9. The burden of current transformers is expressed in terms of
(a) secondary winding current
(b) VA rating of transformer
(c) voltage, current and power factor of secondary winding circuit
(d) none of the above.

10. The current in the primary winding of a current transformer depends upon
(a) burden of the secondary winding of transformer.
(b) load connected to the system in which the C.T. is installed.
(c) both burdens of the transformer secondary winding and load connected to the system.
(d) none of the above.

11. Turns compensation is used in current transformers primarily for reduction of
(a) phase angle error.
(b) both ratio and phase angle errors.
(c) ratio error, reduction in phase angle error is incidental.
(d) none of the above.

12. The advantages of instrument transformers are :
(a) The readings of instruments used in conjunction with them do not depend upon their resistance, inductance etc.
(b) The ratings of instrument transformers have been standardized and the ratings of instruments used in conjunction with them also get standardized. Therefore, there is reduction of cost and ease in replacement.
(c) The metering circuit is electrically isolated from the power circuit thereby providing safety to operating personnel.
(d) all of the above.

13. The disadvantages of shunts for use at high currents are :
(a) It is difficult to achieve good accuracy with shunts.
(b) Power consumption of the shunts is large.
(c) The metering circuit is not electrically isolated from the power circuit.
(d) All of the above.

14. When the secondary winding of a current transformer is open-circuited with the primary winding energized
(a) the whole of the primary current produces large value of flux in the core (limited to

only saturation) thereby producing a large voltage in the secondary winding.

(*b*) the large voltage may act as safety hazard for the operators and may even rupture the insulation.

(*c*) when the large magnetizing force is taken off it leaves a large value of residual magnetism.

(*d*) all of the above.

15. Clamp on meters are used because

(*a*) with their use it is possible to measure current flowing in a line without breaking the circuit.

(*b*) their accuracy is high.

(*c*) it is not possible to measure voltage without them.

(*d*) all of the above.

16. A short-circuiting link is provided on the secondary side of a current transformer because :

(*a*) when the secondary winding of the CT is short-circuited by the link with the primary winding energized a very high current flows on the primary side.

(*b*) when the secondary winding of the CT is short circuited by the link with the primary winding energized, it is possible to make any adjustments in the secondary winding circuit like replacing a faulty ammeter.

(*c*) when the short-circuiting link is opened with the secondary winding open-circuited, the current on the primary side falls to almost zero.

(*d*) all of the above.

17. When the secondary winding of a potential transformer is suddenly open circuited with primary winding excited

(*a*) large voltages are produced in the secondary winding which may be a safety hazard to operating personnel.

(*b*) the large voltages so produced may rupture the insulation.

(*c*) the primary winding draws only the no load current.

(*d*) none of the above.

18. The ratio and phase angle errors in potential transformers may be reduced by

(*a*) increasing the exciting current.

(*b*) increasing the resistance and leakage reactance in the transformer.

(*c*) not employing turns compensation.

(*d*) none of the above.

ANSWERS

1. (*c*)	**2.** (*d*)	**3.** (*d*)	**4.** (*b*)	**5.** (*a*)	**4.** (*d*)	**7.** (*b*)
8. (*a*)	**9.** (*b*)	**10.** (*b*)	**11.** (*c*)	**12.** (*d*)	**13.** (*d*)	**14.** (*d*)
15. (*a*)	**14.** (*b*)	**17.** (*c*)	**18.** (*d*).			

THEORETICAL QUESTIONS

1. Explain the use of instrument transformers.
2. What are the basic advantages inherent in the method of using instrument transformers for measurement purposes ?
3. List the advantages and disadvantages of instrument transformers.
4. Draw the wiring diagrams for current and potential transformers.
5. What are the advantages of instrument transformers over ammeter shunts and voltmeter multipliers ?

4. Explain the meaning of the term "burden" in the instrument transformer.
5. What is meant by "ratio error" and "phase angle error" in CTs ?
6. What is meant by turns compensation and why is it done ?
7. Derive expressions for the ratio and phase angle errors of a current transformer.
8. Draw the phasor diagram of a CT and discuss the effect of variation of power factor of the secondary burden upon the performance.
9. What is phase angle error and on which factors does it depend ?
10. Define the following terms as applied to instrument transformers :
    (*i*) Burden of an instrument transformer.
    (*ii*) Actual transformation ratio.
    (*iii*) Nominal transformation ratio.
    (*iv*) Turn ratio.
11. What is "Ratio correction factor" ? Explain briefly.
12. State the reasons why current transformer must never be operated on open circuit.
13. Write a short note on 'clip-on type current transformer'.
14. How does a PT differ from a power transformer ?
15. Explain the use of PT with the help of a connection diagram.
16. Give the comparison between CT and PT.
17. Derive the expressions for 'actual transformation (voltage), ratio and phase angle' in case of a potential transformer.
18. Explain the following in relation to a PT :
    (*i*) Effect of change in secondary burden.
    (*ii*) Effect of power factor of secondary burden.
    (*iii*) Effect of change in frequency.
19. How errors can be reduced in instrument transformers ?
20. Describe the working of a capacitive type potential transformer. Describe how the transformation ratio can be made independent of the burden of the transformers.
21. Describe the following methods for testing of a current transformer:
    (*i*) Mutual inductance method;
    (*ii*) Arnold comparison method.
22. Explain the 'Absolute null method' for testing of a potential transformer.

UNSOLVED EXAMPLES

Current transformer (CT)

1. The turns of primary and secondary windings of a current transformer are 1 and 200 respectively. The secondary winding supplies a current of 5A to a non-inductive burden of 1 Ω resistance. The requisite flux is set up in the core by an mmf of 80 AT. The frequency is 50 Hz and net cross-section of the core is 1000 mm^2. Neglecting the effects of magnetic leakage, iron losses and I^2R losses, calculate :
    (*i*) The actual transformation ratio.
    (*ii*) Phase angle.
    (*iii*) Maximum flux density in the core.
    [**Ans.** (*i*) 200.64; (*ii*) 4°34′; (*iii*) 0.1126 Wb/m^2]
2. A ring core current transformer of ratio 1000/5A is operating at full primary current, with a secondary burden of non-inductive resistance of 1 Ω. Its exciting current is 1 A at a p.f. of 0.4.

    Calculate : (*i*) The ratio error at full load, assuming there has been no compensation; (*ii*) Phase angle.
    [**Ans.** (*i*) – 0.04%; (*ii*) 3′9″.]
3. The primary winding of a 1000/5A, 50 Hz current transformer has a single turn. Its secondary burden consists of a non-inductive impedance of 1.6 Ω. If the iron loss in the core is 1.5 W at full load and magnetising mmf is 100 AT, calculate : (*i*) Flux in the core; (*ii*) Ratio error at full load. Neglect leakage reactance.
    [**Ans.** (*i*) 0.18×10^{-3} Wb; (*ii*) – 3.61%]

4. A current transformer with a bar primary has 300 turns in its secondary winding. The resistance and reactance of the secondary circuit are 1.5 Ω and 1.0 Ω respectively including the transformer winding. With 5A flowing in the secondary winding, the magnetising mmf is 100 AT and the iron loss is 1.2 W. Calculate :

(*i*) Ratio error; (*ii*) Phase angle error. [**Ans.** (*i*) – 5.54%; (*ii*) 2.34°]

5. The secondary winding burden of a current transformer, having a bar primary and 200 secondary winding turns, is an ammeter of resistance 1.2 Ω and reactance 0.5 Ω. The resistance and reactance of secondary winding are 0.2 Ω and 0.3 Ω respectively. The core requires the equivalent of an mmf of 100 AT for magnetisation and 50 AT for core losses. Determine the following :

(*i*) The primary winding current and actual transformation ratio when the ammeter in the secondary winding indicates 5 A.

(*ii*) The number of turns which could be reduced in the secondary winding so that the ratio error be zero for this condition.

6. A current transformer has a bar primary, 100 secondary winding turns and 100/1 nominal ratio. It is operating on an external burden of 1.6 Ω non-inductive, the secondary winding resistance being 0.2 Ω. The primary winding current is 1.9 A, lagging 40.6° to the secondary voltage reversed. With 1 A current flowing in the secondary winding, calculate :

(*i*) The actual ratio of primary winding current to secondary winding current;

(*ii*) The phase angle between these currents. [**Ans.** (*i*) 101.44; (*ii*) 0.708°]

7. The current transformer rated as 1000/5A, 25 VA has a turn ratio of 1 : 200. The core loss and magnetising components of the primary current are 4A and 7A under rated conditions. Determine the ratio and phase angle errors for the rated burden and the rated secondary current at 0.8 p.f. lagging and 0.8 p.f. leading.

Neglect the resistance and leakage reactance of secondary winding.

[**Ans.** – 0.24%, + 0.185°; + 0.603%, 0.46°]

8. A 500/5A, 50 Hz current transformer has a one turn primary and an output of 15 VA. The number of secondary turns is 98 and the resistance and leakage reactance of the secondary winding are 0.35 Ω and 0.3 Ω respectively. At rated load with non-inductive burden, the in-phase and quadrature components (referred to the flux) of the exciting mmf are 13 AT and 10 AT respectively. Calculate the current and phase angle errors.

[**Ans.** – 0.38%; 0.537°]

9. A 100/5A current transformer, at its rated load of 25 VA, has an iron loss of 0.2 W and a magnetising current of 1.5 A. It is supplying rated output to a meter having a ratio of resistance to reactance of 5. Calculate it ratio error and phase angle.

[**Ans.** – 1.075%; 0.75°]

Potential transformer (PT)

10. The following data relate to a 1000/100 volts potential transformer :

Primary resistance = 94.5 Ω; Primary reactance = 66.2 Ω; Secondary resistance = 0.86 Ω; Total equivalent reactance = 110 Ω. Calculate :

(*i*) Phase angle error at no load;

(*ii*) Burden in VA at unity p.f. at which the phase angle will be zero. [**Ans.** –4.1′; 10.9 VA]

CHAPTER

5 Potentiometers

5.1. Introduction; 5.2. D.C. potentiometers – Simple D.C. potentiometer – Direct reading potentiometer – Crompton potentiometer – Practical applications of D.C. potentiometers – Measurement of resistance – Measurement of current – Measurement of high e.m.fs (say 250 V) – Measurement of power – Calibration of voltmeter – Calibration of ammeter – Calibration of wattmeter – Protection of standard cell and galvanometer – Self balancing potentiometers. **5.3. A.C. potentiometers** – Standardisation of A.C. potentiometers – Classification of A.C. potentiometers – Drysdale polar potentiometer – Gall co-ordinate potentiometer – Advantages and disadvantages of A.C. potentiometers – Application of A.C. potentiometers – Measurement of self inductance – Calibration of voltmeter – Calibration of ammeter – Calibration of wattmeter – *Highlights – Objective Type Questions – Theoretical Questions – Unsolved Examples.*

5.1 INTRODUCTION

A **potentiometer** *is an instrument used for measuring and comparing the e.m.fs of different cells and for calibrating and standardising voltmeters, ammeters, etc.*

Since a potentiometer measures voltage, it *can also be used to determine current* simply by measuring the voltage drop produced by the unknown current passing through a known standard resistance.

- Some *significant features/advantages* of potentiometers are :

(*i*) Measurements using comparison methods are capable of *high degree of accuracy* because the result obtained does not depend on the actual deflection of a pointer, as is the case in deflection methods, but only upon the accuracy with which the voltage of the reference source is known.

(*ii*) A potentiometer makes use of a *balance or null condition, no current flows and hence no power is consumed* in the circuit containing the unknown e.m.f. when the instrument is balanced. Thus the determination of voltage by a potentiometer is quite *independent of the source resistance.*

Uses of potentiometer. The following are the *applications/uses of potentiometers* :

1. Measurement of small e.m.fs. (upto 2 volts).
2. Comparison of e.m.fs. of two cells.
3. Measurement of high e.m.fs. (say 250 volts).
4. Measurement of resistance.
5. Measurement of current.
6. Calibration of ammeter.
7. Calibration of voltmeter.

5.2 D.C. POTENTIOMETERS

5.2.1. Simple D.C. Potentiometer

- For the *accurate measurement of potential difference, current and resistance the potentiometer is one of the most useful instruments.*
- *Its principle of action is that an unknown e.m.f. or p.d. is measured by balancing it, wholly or in part, against a known difference of potential.*
- Fig. 5.1, shows a *simple D.C. potentiometer.*

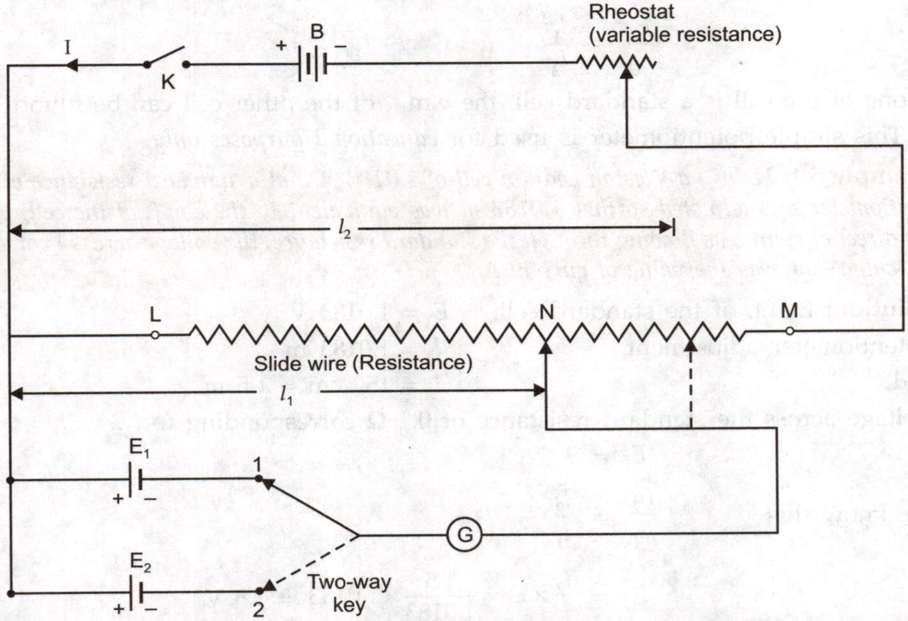

Fig. 5.1. Simple D.C. potentiometer.

Construction :

— In its simplest form, it consists of a German silver or manganin wire usually one metre long and stretched between two terminals as shown.

— This wire is connected in series with a suitable rheostat and battery B which sends a steady current through the resistance wire LM. As the wire is of uniform cross-section throughout, the fall in potential across it is uniform and the drop between any two points is *proportional to the distance between them.*

— As seen, the battery voltage is spread over the rheostat and the resistance wire LM. As we go along LM, there is a progressive fall of potential. If r is the resistance per cm of this wire, l is the length, then for a current I amperes, the fall of potential over the whole length of wire is Irl volts.

Working :

— The two cells whose e.m.fs. are to be compared are joined as shown in Fig. 5.1, always remembering the *positive terminals of the cells and battery must be joined together.*

— The cells can be joined with the galvanometer in turn through a two-way key. The other end of the galvanometer is connected to a movable contact on LM. By this movable contact, a point like M is found when there is no current in and hence no deflection of galvanometer G. Then it means that the e.m.f. of the cell just balances the potential fall LN due to the battery current passing through it.

Suppose that the balance or null point for first cell of e.m.f. E_1 occurs at a length l_1 as measured from point L.

Then, $$E_1 = Ir\, l_1 \quad ...(i)$$

Similarly, if the balance point is at l_2 for the other cell,

then, $$E_2 = Ir\, l_2 \quad ...(ii)$$

Dividing (*i*) by (*ii*), we get :

$$\frac{E_1}{E_2} = \frac{l_1}{l_2} \quad ...(5.1)$$

If one of the cell is a standard cell, the e.m.f. of the other cell can be found.

- This simple potentiometer is used for *educational purposes only.*

Example 5.1. *Using a Weston cadium cell of 1.0183 V and a standard resistance of 0.1 Ω a potentiometer was adjusted so that 1.0183 m was equivalent to the e.m.f. of the cell; when a certain direct current was flowing through the standard resistance, the voltage across I correspond to 150 cm. What was the value of current ?*

Solution: E.m.f. of the standard cell, $E_1 = 1.0183$ V

Potentiometer adjustment, $l_1 = 1.0183$ m

and, $l_2 = 150$ cm $= 1.5$ m

Voltage across the standard resistance of 0.1 Ω corresponding to l_2,

$$E_2 = ?$$

We know that $$\frac{E_2}{E_1} = \frac{l_2}{l_1}$$

$$\therefore \quad E_2 = \frac{l_2}{l_1} \times E_1 = \frac{1.5}{1.0183} \times 1.0183 = 1.5 \text{ V}$$

Current flowing through the standard resistance,

$$I_2 = \frac{E_2}{R} = \frac{1.5}{0.1} = \mathbf{15\ A.}\ \textbf{(Ans.)}$$

5.2.2. Direct-reading Potentiometer

The simple potentiometer (described above) in its commercial form is so calibrated that the readings of the potentiometer give the voltage ***directly***, thereby eliminating tedious arithmetical calculations and so saving appreciable time.

Fig. 5.2, shows a *direct-reading potentiometer* :

— The resistance *R* consists of 14 equal resistances joined in series, the resistance of each unit being equal to that of the whole slide wire *S*, which is divided into 100 equal parts.

— The current of the battery *B* is controlled by the slide wire resistance W.

Standardisation of the potentiometer : Refer to Fig. 5.2. The procedure for standardising the potentiometer is as follows :

— A standard cell *i.e.,* Weston cadium cell of e.m.f. 1.0183 V is connected to sliding contacts *L* and *M* through a sensitive galvanometer *G*.

— First, *L* is put on stud 10 and *M* on 18.3 divisions on *S* and then *W* is adjusted for zero deflection in *G*. In that case, potential difference between *L* and *M* is

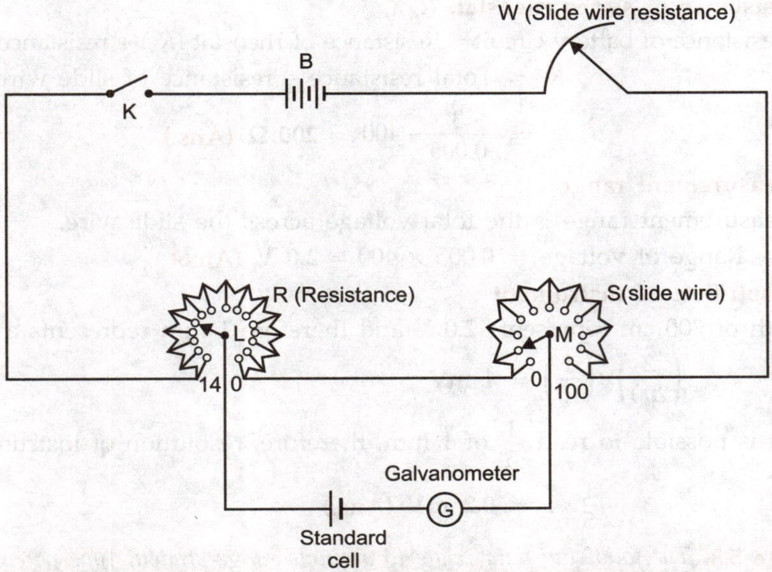

Fig. 5.2. Direct-reading potentiometer.

equal to cell voltage, *i.e.*, 1.0183 V so that potential drop on each resistance of R is $\frac{1}{10} = 0.1$ V and every division of S represents $\frac{0.1}{100} = 0.001$ V. After standardising this way, the *position of W is not to be changed in any case otherwise the whole adjustment would go wrong*. After this the instrument becomes *direct reading*.

Suppose in a subsequent experiment, for balance, L is moved to stud no. 6 and M to 75 divisions, then voltage would be = (6 × 0.1) + (75 × 0.001) = 0.675 V.

- It may be noted that since most potentiometers have 14 steps on R, it is usually not possible to measure p.ds. exceeding 1.5 V. For measuring *higher voltages* it is necessary to use a ***volt-ratio box.***

Example 5.2. *A basic slide wire potentiometer has a working battery voltage of 3.0 V with negligible resistance. The resistance of the slide wire is 400 Ω and its length is 200 cm. A 200 cm scale is placed along the slide wire. The slide wire has 1 mm scale divisions and it is possible to read upto 1/5 of a division. The instrument is standardised with 1.018 V standard cell with sliding contact at the 101.8 cm mark on scale. Calculate :*

(*i*) *Working current.*

(*ii*) *Resistance of series rheostat.*

(*iii*) *Measurement range.*

(*iv*) *Resolution of instrument.* **(Nagpur University)**

Solution: (*i*) Working current, I_m :

Because the instrument is standardised with an e.m.f. of 1.018 V with sliding contact at 101.8 cm, it is obvious that a length 101.8 cm represents a voltage of 1.018 V.

Resistance of 101.8 cm length of wire $= \frac{101.8}{200} \times 400 = 203.6\ \Omega$.

$\therefore$ Working current, $I_m = \frac{101.8}{203.6} = 0.005$ A or **5 mA (Ans.)**

(ii) Resistance of series rheostat, R_h :

Total resistance of battery circuit = Resistance of rheostat (R_h) + resistance of slide wire

$$\therefore \quad R_h = \text{Total resistance} - \text{resistance of slide wire}$$

$$= \frac{3}{0.005} - 400 = \mathbf{200\ \Omega} \text{ (Ans.)}$$

(iii) Measurement range :

The measurement range is the total voltage across the slide wire.

$$\therefore \quad \text{Range of voltage} = 0.005 \times 400 = \mathbf{2.0\ V} \text{ (Ans.)}$$

(iv) Resolution of instrument :

A length of 200 cm represents 2.0 V and therefore 1 mm represents a voltage of

$$\left(\frac{2}{200}\right) \times \left(\frac{1}{10}\right) = 1 \text{ mV}$$

Since it is possible to read $\frac{1}{5}$ of 1 mm, therefore, resolution of instrument is

$$\frac{1}{5} \times 1 = \mathbf{0.2\ mV} \text{ (Ans.)}$$

Example 5.3. *The following data relate to a single range student type potentiometer :*

18 step dial switch–each step represents 0.1 V;

The dial resistors are 10 Ω;

The slide wire of the potentiometer is circular; and has 11 turns and a resistance of 11 Ω each;

The slide wire has 100 divisions and interpolation can be done to one fourth of a division.

The working battery has a voltage of 6.0 V and negligible internal resistance.

Calculate :

(i) Measuring range of instrument.

(ii) Resolution of instrument.

(iii) Working current.

(iv) Setting of rheostat. **(Roorkee University)**

Solution: (i) Measuring range of instrument :

Total resistance of measuring circuit, R_m

$$= \text{Resistance of dial} + \text{resistance of slide wire}$$

or, $$R_m = 18 \times 10 + 11 = 191\ \Omega$$

Since each step of 10 Ω represents a voltage drop of 0.1 V, range of instrument

$$= 191\left(\frac{0.1}{10}\right) = 1.91 \text{ V}.$$

(ii) Resolution of instrument :

The slide wire has a resistance of 11 Ω and therefore voltage drop across slide wire

$$= 11 \times \left(\frac{0.1}{10}\right) = 0.11 \text{ V}$$

The slide wire has 11 turns and therefore voltage drop across each turn

$$= \frac{0.11}{11} = 0.01 \text{ V}$$

Each turn is divided into 100 divisions and therefore each divisions represents a voltage drop of $\frac{0.01}{100} = 0.0001$ V.

Since each turn can be interpolated to $\frac{1}{4}$ of a division,

$\therefore$ *Resolution* of instrument $= \frac{1}{4} \times 0.0001\ \text{V} = 0.000025 = \mathbf{25\ \mu V}$ **(Ans.)**

(iii) **Working current, I_m :**

The voltage drop across working circuit of potentiometer is 1.91 V and the resistance is 191 Ω.

$\therefore$ Working current, $I_m = \frac{1.91}{191} = 0.01\ \text{A} = \mathbf{10\ mA}$ **(Ans.)**

(iv) **Setting of rheostat :**

Total resistance across battery circuit $= \frac{6}{0.01} = 600\ \Omega$.

The resistance of potentiometer circuit is 191 Ω.

$\therefore$ Resistance of series rheostat, $R_h = 600 - 191 = \mathbf{409\ \Omega}$ **(Ans.)**

5.2.3. Crompton Potentiometer

Fig. 5.3 shows the general arrangement of a crompton potentiometer :

— It consists of a graduated slide-wire AC connected in series with fourteen (or more) coils each of which has a resistance equal to that of the slide-wire (of the order of 10 ohms).

— There are two moving contacts P_1 and P_2 sliding over the slide-wire; and the studs of the resistance coils respectively.

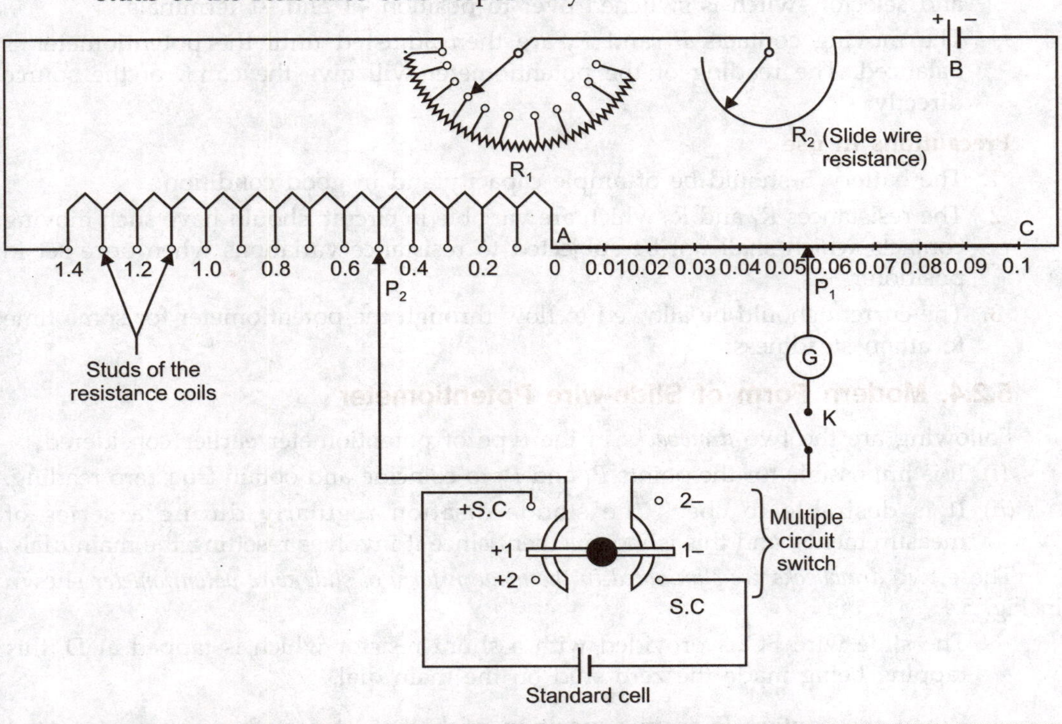

Fig. 5.3. Crompton potentiometer.

— B is the supply battery (2 V), and R_1 and R_2 are two variable resistors, the former consisting of a number of coils for *coarse adjustment* of the potentiometer current, and the latter taking the form of a slide-wire for fine adjustment.

— The galvanometer G is connected in series with a key K, and multiple circuit switch, by means of which either the standard cell or e.m.f. to be measured can be connected in the galvanometer circuit.

— The terminals to which the apparatus under test is connected are marked *+ve* and *–ve* to avoid the possibility of damage to the potentiometer due to the wrong polarity being used. The standard cell and supply battery terminals are marked similarly.

Standardisation. Before putting the potentiometer in use, it is standardised, *i.e.*, made direct reading by adjustment of the current from the supply battery as follows :

— A Weston type standard cell S.C., whose e.m.f. is 1.0183 V is connected to the terminals of selector switch marked as S.C.

— The moving contact P_2 is set at 1.00 stud of the coils and moving contact P_1 is set at 0.0183 on the slide wire.

— Then the resistances R_1 and R_2 are so adjusted that there is no deflection in the galvanometer when they key K is pressed. The potentiometer is now said to be *standardised.* After standardisation the positions of R_1 and R_2 *should not be disturbed in any case* otherwise the whole adjustment would go wrong.

Measurement of unknown e.m.fs. The unknown e.m.fs. can be measured as follows:

— The source whose e.m.f. is to be measured is connected to terminals +1 and –1 and selector switch is switched over to position +1 and –1 terminals.

— The moving contacts P_1 and P_2 are then adjusted until the potentiometer is balanced. The reading of the potentiometer will give the e.m.f. of the source directly.

Precautions in use :

1. The battery B should be of ample capacity and in good condition.
2. The resistances R_1 and R_2 which are variable in circuit, should have such moving contacts which shall not be subjected to resistance variations when once set in position.
3. The current should be allowed to flow through the potentiometer for some time to attain steadiness.

5.2.4. Modern Form of Slide-wire Potentiometer

Following are the two *drawbacks* in the type of potentiometer earlier considered :

(*i*) It is impossible for the points P_1 and P_2 to coincide and obtain true zero reading.

(*ii*) It is desirable to check the standardisation regularly during a series of measurements, and this is inconvenient since it involves resetting the main dials.

These two *drawbacks are eliminated in the modern form of slide wire potentiometer* shown in Fig. 5.4.

— The slide wire BC is provided with a shunt resistor which is tapped at D, this tapping being made the zero stud on the main dial.

— When the contact P_1 is in a position such that $\frac{r_1}{r_3} = \frac{r_2}{r_4}$ there is no potential difference between the zero stud and P_1; hence the slide-wire has a *true zero.* The

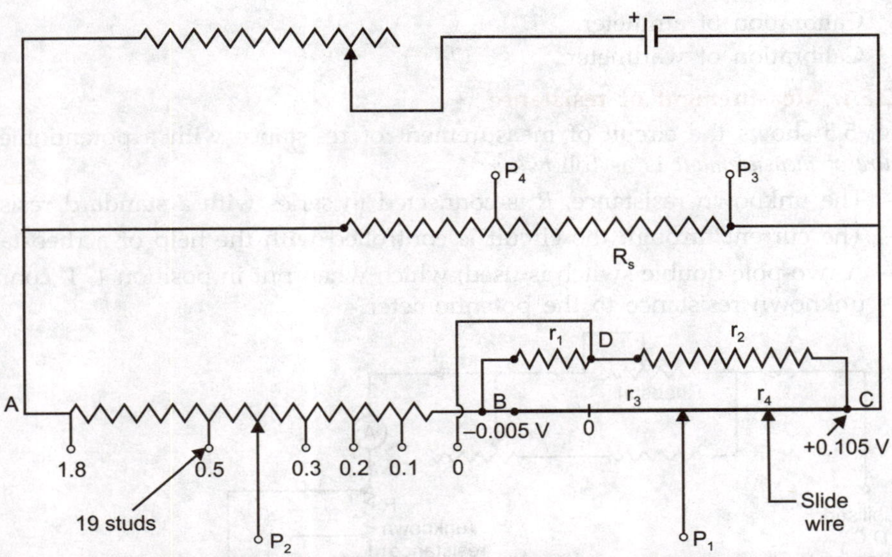

Fig. 5.4. Circuit diagram for modern form of slide-wire potentiometer.

slider can travel a short distance to the left of the zero position, giving a small negative reading, and movement to the right gives positive readings. The range of the slide wire is usually from –0.005 V to 0.105 V, and the resistance of the slide-wire and its shunt is slightly greater than that of one step on the main dial.

Standardisation :

— Standardisation in this potentiometer is carried out with the aid of the shunt resistor R_s between A and C. This resistor is tapped at points P_3 and P_4, which are so chosen that a potential difference equal to the e.m.f. of a standard cell exists between them when the potentiometer is correctly standardised. Thus the potentiometer may be standardised independently of the main dial by connecting a galvanometer and standard cell between P_3 and P_4 and adjusting the potentiometer current until balance is achieved. This circuit is known as an *independent standardising circuit.*

— Sometimes for quick standardisation, instead of shunt resistor R_s, a *standard coil* in series with the slide wire is provided. The resistance of this coil is so chosen that the current flowing in the slide wire in standardised position gives the voltage drop across the standard coil *equal to the e.m.f. of the standard cell.* Now the potentiometer can be standardised independently of the main dial by connecting a galvanometer and standard cell across this coil terminals and then adjusting the potentiometer current with rheostat until balance is obtained.

5.2.5. Practical Applications of D.C. Potentiometers

Some of the important applications of D.C. potentiometers are enumerated and described below :

1. Measurement of resistance.
2. Measurement of current.
3. Measurement of high e.m.fs. (say 250 V).
4. Measurement of power.
5. Calibration of voltmeter.

6. Calibration of ammeter.
7. Calibration of wattmeter.

5.2.5.1. Measurement of resistance

Fig. 5.5 shows the circuit of measurement of resistance with a potentiometer. The *procedure of measurement* is as follows :

— The unknown resistance, R is connected in series with a standard resistor S.

— The current through the circuit is controlled with the help of a rheostat.

— A two-pole double switch is used; which when put in position 1, 1′ connects the unknown resistance to the potentiometer.

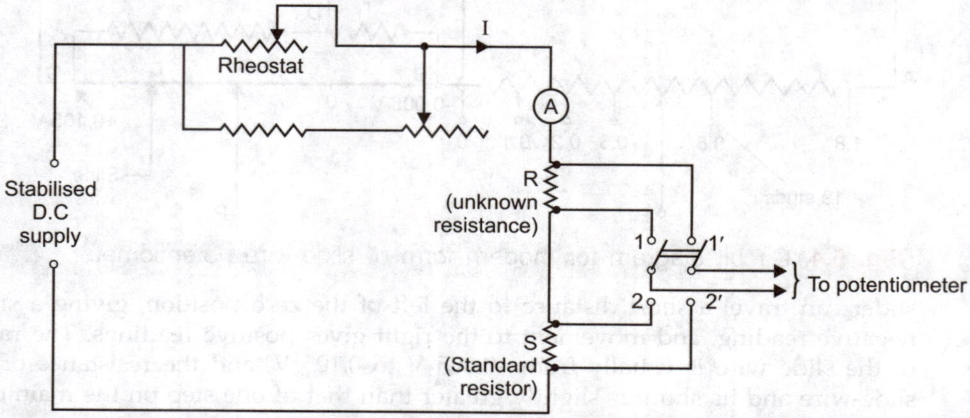

Fig. 5.5. Measurement of resistance with potentiometer.

If the reading of the potentiometer is V_R, then,

$$V_R = IR \quad ...(i)$$

Now the switch is thrown to position 2, 2′, this connects the standard resistor S to the potentiometer. If the reading of potentiometer is V_S, then,

$$V_S = IS \quad ...(ii)$$

Dividing (*i*) by (*ii*), we get

$$\frac{V_R}{V_S} = \frac{IR}{IS}$$

or,
$$R = \frac{V_R}{V_S} \cdot S \quad ...(5.2)$$

The value of R can be accurately known since the value of standard resistance S is accurately known.

Following points are *worth noting* :

(*i*) It is desirable that the current flowing through the circuit be so adjusted that the value of voltage drop across each resistor is of the order of 1 V.

— The ammeter inserted in the circuit is merely for indicating that whether the current flowing through the circuit is within the capacity of the resistors or not otherwise the exact value of current flowing need not not be known.

(*ii*) The resistance of the *standard resistor S*, which must be accurately known, should be of the *same order as the resistance, R* under the measurement.

- The potentiometer method of measurement of resistance is *suitable for measurement of "low resistances"*.

5.2.5.2. Measurement of current

Fig. 5.6 shows the circuit of measurement of current with a potentiometer. The *procedure* is as follows :

— The unknown current I, whose value is to be determined, is passed through a standard resistor S as shown. The standard resistance should be of such a value that *voltage drop* across it caused by flow of current to be measured, may *not exceed* the range of the potentiometer.

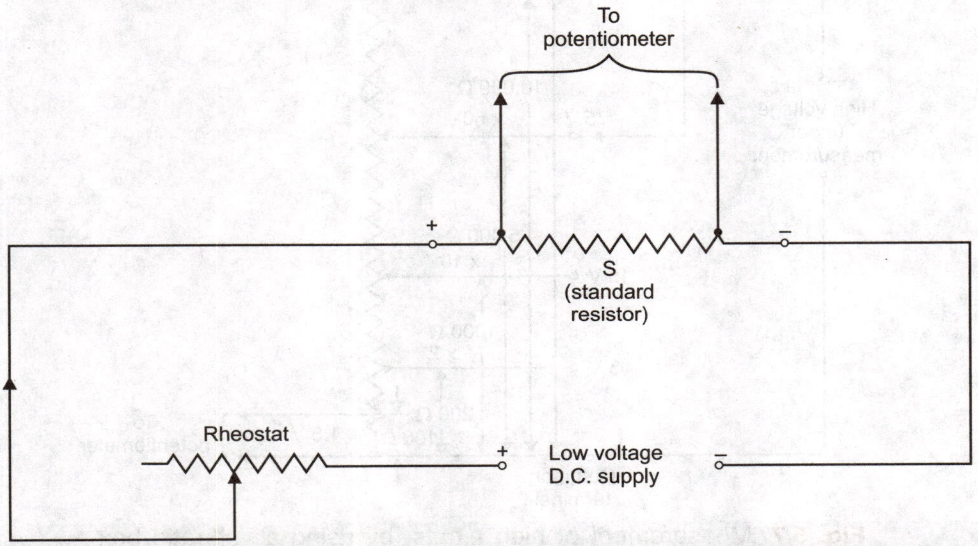

Fig. 5.6. Measurement of current with potentiometer.

— Voltage drop across S measured in volts divided by the value of S in ohms gives the value of unknown current in amperes.

i.e., Unknown current (in amps.), $I = \dfrac{\text{Voltage drop across } S \text{ (measured in volts)}}{\text{Value of } S, \text{ in ohms}}$ ...(5.3)

5.2.5.3. Measurement of high e.m.fs. (say 250 V)

A potentiometer *cannot* be used to measure high e.m.fs. of the order of 250 V *directly*, as it is quite beyond its range.

The voltage above the direct range of potentiometer (normally 1.5 V) can be measured by using a *volt-ratio box* in conjunction with the potentiometer. The volt-ratio box consists of a simple resistance potential divider with various tappings on the input side, as shown in Fig. 5.7. Each input terminal is marked with the maximum voltage which can be applied and with the corresponding multiplying factor for the potentiometer scale.

The *procedure* of measurement is as follows :

— High e.m.f. to be measured is applied across the suitable input terminals of volt-ratio box and leads to the potentiometer are taken from *two tapping points* meant for this purpose.

— The p.d. across these two points is measured on potentiometer as earlier described.

Let, v = Potential difference (p.d.) measured by potentiometer, and

k = Multiplying factor,

Then, Voltage to be measured, $V = kv$ volts ...(5.4)

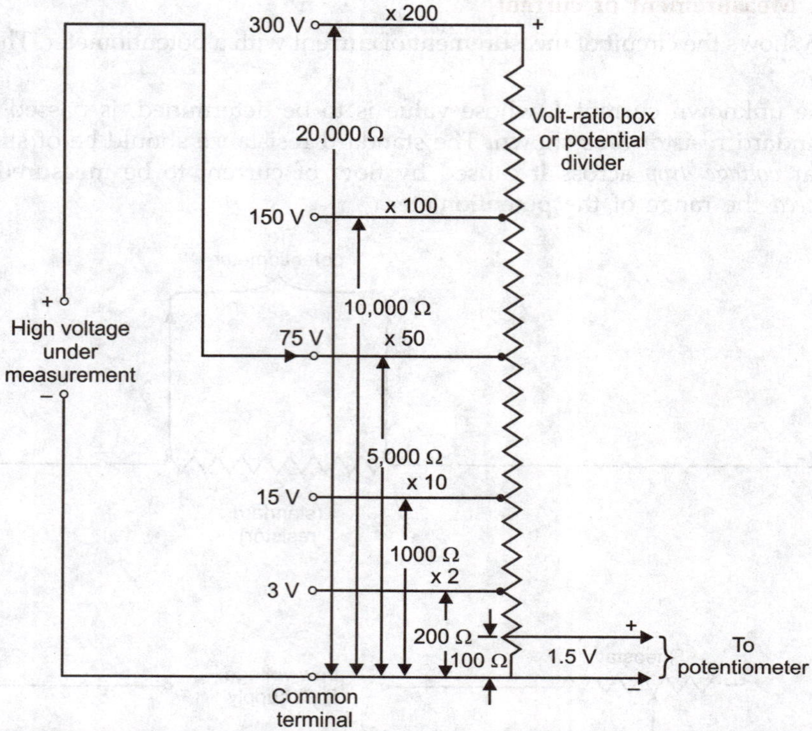

Fig. 5.7. Measurement of high e.m.fs. by using a volt-ratio box in conjunction with the potentiometer.

5.2.5.4. Measurement of power

Fig. 5.8, shows the circuit for measurement of power with a potentiometer. The *procedure* of measurement is as follows :

— The measurements are made, one across the standard resistance *S*, connected in series with the load and other across at output terminals of volt-ratio box.

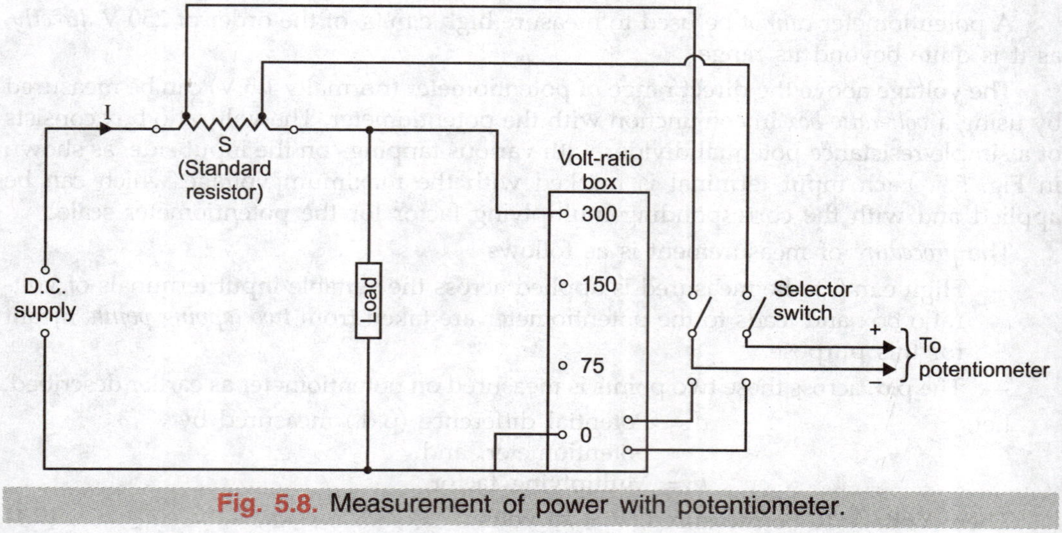

Fig. 5.8. Measurement of power with potentiometer.

— The current is computed from the voltage drop across the standard resistance and the voltage across the load is computed from the potentiometer reading across the output terminals of volt-ratio box.

If V_S is the reading of potentiometer when connected across S and V_R is the reading of potentiometer when connected across volt-ratio box, then :

$$\text{Current through load, } I = \frac{V_S}{S}, \text{ and}$$

$$\text{Voltage across load, } V = k\,V_R, \quad \text{where } k \text{ is the multiplying factor of volt-ratio box.}$$

$$\therefore \quad \text{Power consumed, } P = VI = kV_R \times \frac{V_S}{S} \quad \text{...(5.5)}$$

5.2.5.5. Calibration of voltmeter

Fig. 5.9 shows the circuit for calibration of a voltmeter. The *procedure* of calibration is as following :

— The primary requirement in this calibration process is that a *suitable stable D.C. voltage supply* is available otherwise any changes in the supply voltage will cause a corresponding change in the calibration of voltmeter.

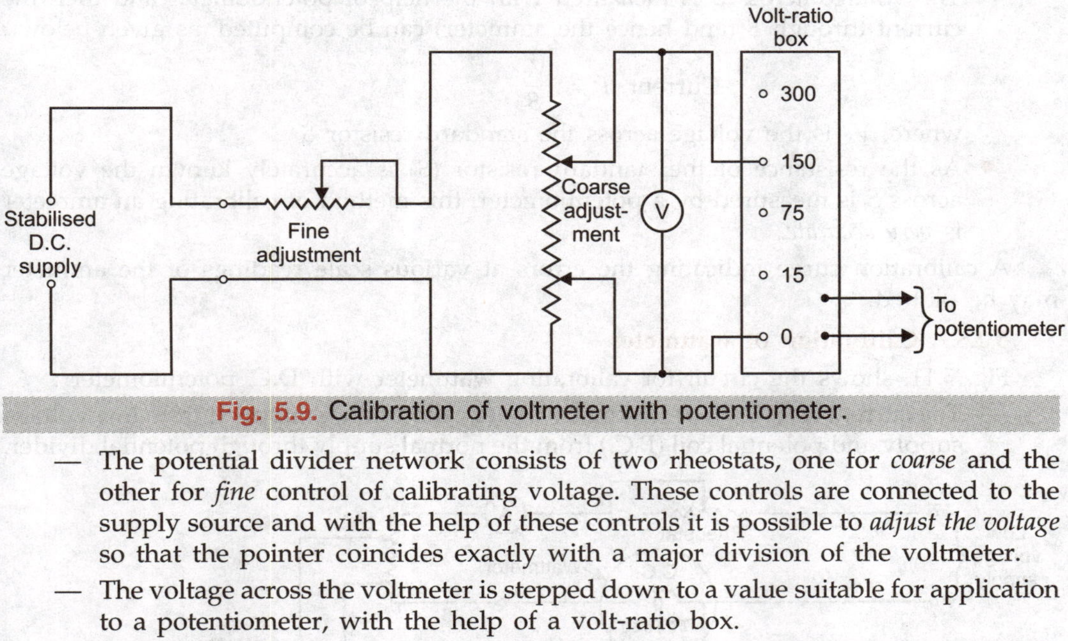

Fig. 5.9. Calibration of voltmeter with potentiometer.

— The potential divider network consists of two rheostats, one for *coarse* and the other for *fine* control of calibrating voltage. These controls are connected to the supply source and with the help of these controls it is possible to *adjust the voltage* so that the pointer coincides exactly with a major division of the voltmeter.

— The voltage across the voltmeter is stepped down to a value suitable for application to a potentiometer, with the help of a volt-ratio box.

In order to get accurate measurements, it is necessary to *measure voltages near the maximum range of the potentiometer*, as far as possible.

— The potentiometer measures the *true value of voltage.* If the reading of the potentiometer does not agree with the voltmeter reading, a *–ve* or *+ve* error is indicated. A calibration curve may be drawn with the help of the readings of voltmeter and potentiometer.

5.2.5.6. Calibration of ammeter

Fig. 5.10. shows the circuit for calibrating an ammeter. The *procedure* of calibration is as follows :

— A standard resistor, S, of suitable value and sufficient current carrying capacity is placed in *series* with the ammeter under calibration.

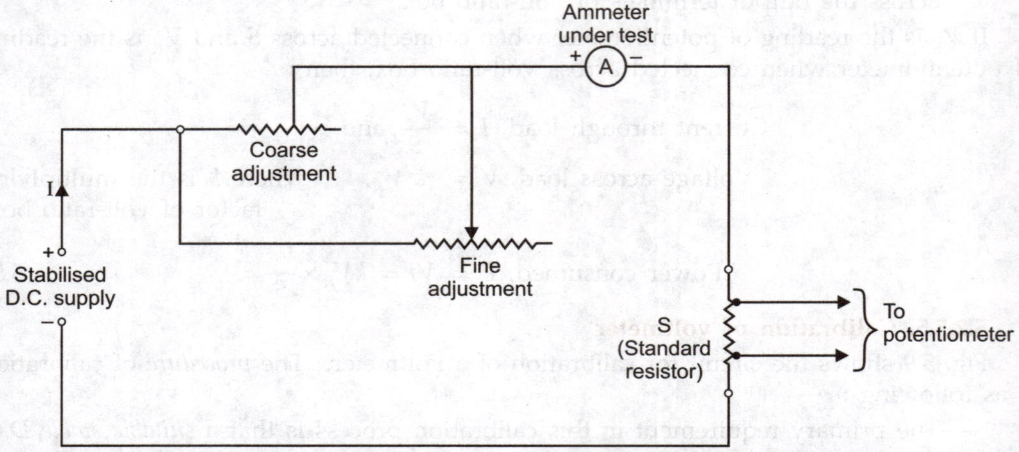

Fig. 5.10. Calibration of an ammeter with potentiometer.

— The voltage across S is measured with the help of potentiometer and then the current through S (and hence the ammeter) can be computed, as given below :

$$\text{Current, } I = \frac{V_S}{S}$$

where, V_S is the voltage across the standard resistor S.

- As the resistance of the standard resistor (S) is accurately known the voltage across S is measured by a potentiometer, this method of calibrating an ammeter is *very accurate.*

A calibration curve indicating the errors at various scale readings of the ammeter may be plotted.

5.2.5.7. Calibration of wattmeter

Fig. 5.11. shows the circuit for calibrating wattmeter with D.C. potentiometer :

— The current coil (C.C.) of wattmeter under calibration is supplied from low voltage supply and potential coil (P.C.) from the normal supply through potential divider.

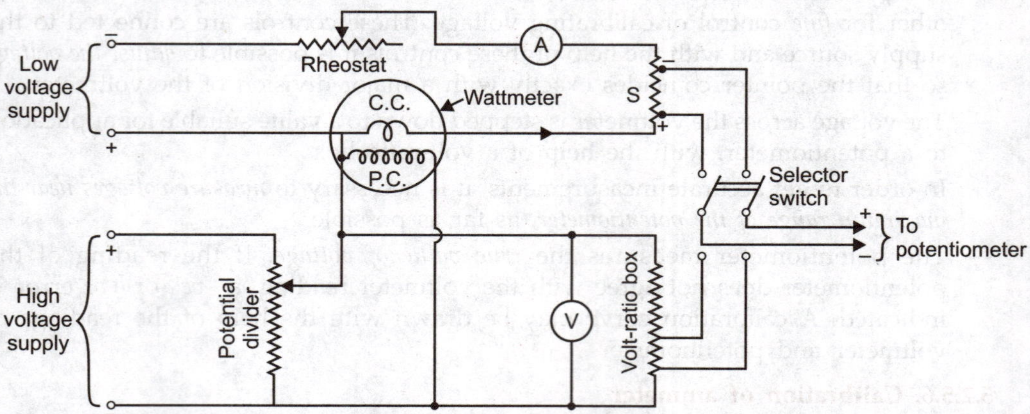

Fig. 5.11. Calibration of wattmeter with D.C. potentiometer.

— The voltage V across the potential coil and the current I through current coil are measured in turn by the potentiometer employing a volt-ratio box and standard resistor, S, as shown.

The true power is then VI and the wattmeter reading may be compared with this value.

This method of calibration of wattmeter is quite *precise*.

Example 5.4. *A simple slide wire is used for the measurement of current in a circuit. The voltage across a standard resistance of 1.0 Ω is balanced at 75 cm. Find the magnitude of the current if the standard cell having an e.m.f. of 1.45 volts is balanced at 50 cm.* **(AMIE)**

Solution: *Given* : $S = 1.0\ \Omega$; $l_1 = 75$ cm; $E = 1.45$ volts; $l_2 = 50$ cm.

Magnitude of current, *I* :

Voltage per cm length of potentiometer wire,

$$v = \frac{E}{l_2} = \frac{1.45}{50} = 0.029 \text{ V}$$

The p.d. which balances at 75 cm,

$$V = v \times l_1 = 0.029 \times 75 = 2.175 \text{ V}$$

$$\therefore \quad \text{Magnitude of current, } I = \frac{2.175}{1.0} = \mathbf{2.175\ A} \text{ (Ans.)}$$

Example 5.5. *The e.m.f. of standard cell used for standardisation is 1.0183 volts. If the balance is obtained at a length of 55 cm, determine :*

(i) The e.m.f. of the cell which balances at 70 cm.

(ii) The current flowing through a standard resistance of 2 Ω if the p.d. across it balances at 65 cm.

(iii) The voltage of a supply main which is reduced by a volt-ratio box to one hundredth and balance is obtained at 85 cm.

(iv) The percentage error in ammeter reading 0.35 ampere when balance is obtained at 45 cm with the p.d. across a 2.5 Ω resistor in the ammeter circuit.

(v) The percentage error in a voltmeter reading 1.42 volts when balance is obtained at 75 cm.

Solution: *Given* : E.m.f. of the standard cell = 1.0183 V

The voltage drop per cm length of potentiometer wire,

$$v = \frac{1.0183}{55} = 0.0185145$$

(i) The e.m.f. of a cell which balances at 70 cm

$$= v\ .\ l = 0.0185145 \times 70 = \mathbf{1.29601\ volts} \text{ (Ans.)}$$

(ii) The potential difference which balance at 65 cm

$$= v\ .\ l = 0.0185145 \times 65 = 1.20344 \text{ V}$$

Magnitude of the standard resistor, $S = 2\ \Omega$

$$\therefore \quad \textit{Current flowing through 2 Ω resistance} = \frac{V}{S} = \frac{1.20344}{2} \text{ volts} = \mathbf{0.60172\ A} \text{ (Ans.)}$$

(iii) The p.d. which balances at 85 cm,

$$V = v\ .\ l = 0.0185145 \times 85 = 1.57373 \text{ volts (Ans.)}$$

$$\therefore \quad \textit{Voltage of supply main} = V \times \text{Ratio of volt-ratio box}$$

$$= 1.57373 \times 100 = \mathbf{157.373\ volts} \text{ (Ans.)}$$

(*iv*) The p.d. which balances at 45 cm,

$$V = v \cdot l = 0.0185145 \times 45 = 0.83315 \text{ volts}$$

Current through 2.5 Ω resistance,

$$I = \frac{V}{S} = \frac{0.83315}{2.5} = \mathbf{0.33326\ A}$$ **(Ans.)**

∴ *Percentage error in ammeter reading*

$$= \frac{0.35 - 0.33326}{0.33326} \times 100 = \mathbf{5\%\ high}$$ **(Ans.)**

(*v*) The p.d. which balances at 75 cm

$$= 0.0185145 \times 75 = 1.38859 \text{ volts}$$

Voltmeter reading = 1.42 volts

∴ Percentage error in voltmeter reading

$$= \frac{1.42 - 1.38859}{1.38859} \times 100 = \mathbf{2.26\%\ high}$$ **(Ans.)**

Example 5.6. *A Crompton's potentiometer consists of a resistance dial having 15 steps of 10 Ω each and a series connected slide wire of 10 Ω which is divided into 100 divisions. If the working current of the potentiometer is 10 mA and each division of slide wire can be read accurately upto $\frac{1}{5}$ of its span, calculate the resolution of the potentiometer in volt.*

(Punjab University)

Solution. Resolution of the potentiometer :

Total resistance of the potentiometer,

$$R = \text{Resistance of the dial + resistance of wire}$$
$$= 15 \times 10 + 10 = 160\ \Omega$$

Working current, I = 10 mA = 0.01 A

Voltage range of the potentiometer = Working current × total resistance of potentiometer

$$= 0.01 \times 160 = 1.6 \text{ V}$$

Voltage drop across slide-wire = Working current × slide-wire resistance

$$= 0.01 \times 10 = 0.1 \text{ V}$$

Since slide wire has 100 divisions, therefore, each division represents $\frac{0.1}{100}$ or 0.001 V

As each division of slide wire can be read accurately upto $\frac{1}{5}$ of its span, therefore,

Resolution of potentiometer $= \frac{0.01}{5} = \mathbf{0.0002\ V}$ **(Ans.)**

Example 5.7. *The following readings were obtained during the measurement of a low resistance using a potentiometer :*

Voltage drop across a 0.1 Ω standard resistance = 1.0439 V

Voltage drop across the low resistance under test = 0.4305 V

Calculate the value of unknown resistance, current and power lost in it.

Solution. *Given* : $S = 0.1\ \Omega$; $V_S = 1.0439$ V; $V_R = 0.4305$ V

Resistance of unknown resistor, $R = \frac{V_R}{V_S} \times S$ [Eqn. (5.2)]

$$= \frac{0.4305}{1.0439} \times 0.1 = \mathbf{0.041239\ \Omega} \quad \textbf{(Ans.)}$$

Current through the resistor, $\quad I = \frac{V_S}{S} = \frac{1.0439}{0.1} = 10.439\text{ A}$

Power loss in unknown resistance, $\quad P = I^2R = (10.439)^2 \times 0.041239 = \mathbf{4.494\ W}$ **(Ans.)**

Example 5.8. *The following readings were obtained during the measurement of low resistance by means of a potentiometer :*

Voltage drop across a 0.1 Ω standard resistance connected in series with the unknown resistance = 1.03606 V

Voltage drop across low resistance under test = 0.85621 V

The resistance of standard resistor at the temperature of test is 0.10016 Ω. Upon setting the potentiometer dial to zero and breaking the current passing through the unknown resistance, the thermal e.m.f. of latter produced a galvanometer deflection equivalent to 24 µV, the direction of deflection being the same as that produced by an increase of the potentiometer reading during the voltage measurement.

Calculate the unknown resistance. **(Bengalore University)**

Solution. Unknown resistance, *R* :

Indicated voltage drop across the unknown resistance = 0.85621 V

Thermal e.m.f. with unknown resistance = 24 µV = 0.000024 V

∴ Actual voltage drop across unknown resistance,

$$V_R = 0.85621 - 0.000024 = 0.856186\text{ V}$$

∴ Unknown resistance, $R = \frac{V_R}{V_S} \times S = \frac{0.856186}{1.03606} \times 0.10016 = \mathbf{0.08277\ \Omega}$ **(Ans.)**

Example 5.9. *The current is measured through a standard resistance of 0.1 ± 0.1 Ω% (standard deviation) by a potentiometer that is accurate to ± 0.0001 V (standard deviation). The voltage across the resistance is measured to be 0.25643 V.*

What is the current and to what accuracy it has been determined ?

Solution. *Current,* $\quad I = \frac{V_S}{S} = \frac{0.25643}{0.1} = \mathbf{2.5643\ A}$ **(Ans.)**

The fractional standard deviation of current,

$$\frac{\delta I}{I} = \sqrt{\left(\frac{\delta V}{V}\right)^2 + \left(\frac{\delta S}{S}\right)^2}$$

$$= \sqrt{\left(\frac{0.0001}{0.25643}\right)^2 + (0.001)^2} = 0.00107$$

or, *percentage accuracy of measurement of current* = ± 0.00107 × 100 = **± 0.107%** **(Ans.)**

5.2.6. Protection of Standard Cell and Galvanometer

— If the potentiometer is unbalanced, current flows through the galvanometer circuit, which may be excessively high if the working point is far away from balance conditions.

In a standard cell, the current must be limited to 0.1 mA otherwise it will be damaged.

— Normally, a *galvanometer is protected against overcurrents by using a shunt of low resistance.* However, if this practice is followed in a potentiometer, it will *not only fail to protect the standard cell but also increase the current in the standard cell during the standardising process* (since the circuit resistance decreases considerably on account of low resistance put in parallel with the galvanometer). Thus a *series resistance is used to limit the current in both the standard cell and galvanometer. This resistance should be sufficiently large to protect the circuit if the polarity of the standard cell should accidently be reversed.*

Usually, a *series resistance of at least 20,000 Ω be inserted in the standard cell to protect it against overcurrents.*

5.2.7. Self-balancing Potentiometers

The self-balancing potentiometers find wide applications in *"industry"* due to their following *special features* :

1. Their action is automatic and thus the constant attention of an operator is eliminated.
2. They draw a curve of quantity being measured with the help of recording mechanism.
3. They can be mounted on a switchboard or panel and thus acts as monitoring devices for the quantity under measurements.
   - These potentiometers are very useful for *instrumentation purposes.*
   - They are also used in *multiplying and dividing networks in analogue computers and in special function generators.*

5.3 A.C. POTENTIOMETERS

The potentiometer method is an exceedingly useful one for the accurate measurement of alternating currents and voltages, since such measurements are not easily carried out by other methods.

- The principle of A.C. potentiometer is the *same* as that of D.C. potentiometer, the *most important difference in operation being that :*

"Whereas in the D.C. potentiometer *only the magnitudes* of the "unknown" e.m.f. and slide-wire voltage drop must be made equal to obtain balance, in the A.C. potentiometer the *phases of the two voltages, as well as their magnitudes, must be equal for balance to be obtained."*

This condition obviously necessitates modification of the potentiometer as constructed for D.C. work, and means that the operation is somewhat more complicated.

The following *factors* must be considered for the *operation of an A.C. potentiometer :*

(*i*) In all A.C. potentiometers, the potentiometer circuit must be supplied from the *same source as the voltage or current being measured.*

(*ii*) The sources of A.C. supply should be *free from harmonics* because in the presence of harmonic it may not be possible to achieve a balance. The supply source should be as *sinusoidal as possible.*

(*iii*) There being no reference source (the reference source in D.C. being a standard cell or a Zener source) the absolute accuracy with which an A.C. voltage can be measured in an A.C. potentiometer cannot be comparable with corresponding type of D.C. measurement.

(*iv*) Extraneous or stray e.m.fs. picked up from stray fields or couplings between portions of the potentiometer circuit, which seriously affect the result, must be eliminated, compensated for or measured (since they may add vectorially to the e.m.f. being measured).

- In order to avoid error in readings, the slide wire and resistance coils should be *non-reactive.*

5.3.1. Standardisation of A.C. Potentiometers

— The A.C. potentiometers are made direct reading type, *i.e.*, the readings are read off directly from the dial settings. In order to do this, the A.C. potentiometer must be standardised as is done in case of D.C. instruments.

— The standardisation is done with the help of a standard D.C. source, *i.e.*, a standard cell or a Zener source and a "*transfer instrument*". This transfer instrument is usually an electrodynamometer milliammeter, so constructed that its response to alternating currents is the same as its D.C. response. Such an instrument *can be calibrated on D.C. and then brought to the same setting on A.C.* Alternatively a *thermocouple type of instrument may be used as a transfer instrument.*

5.3.2. Classification of A.C. Potentiometers

A.C. potentiometers are usually *classified* according to the manner in which the instrument dials (or scales) present the value of unknown voltage. There are two general types of A.C. potentiometers :

1. Polar potentiometers :

— In these potentiometers, the unknown e.m.f. is measured in "*polar form*" *i.e.*, in *terms of its magnitude and relative phase.* The magnitude is given by one scale and its phase angle with respect to some reference vector is given by the second scale.

— Provision is made for reading phase angles upto 360°.

— The voltage is read in the form $V\angle\theta$ (See Fig. 5.12).

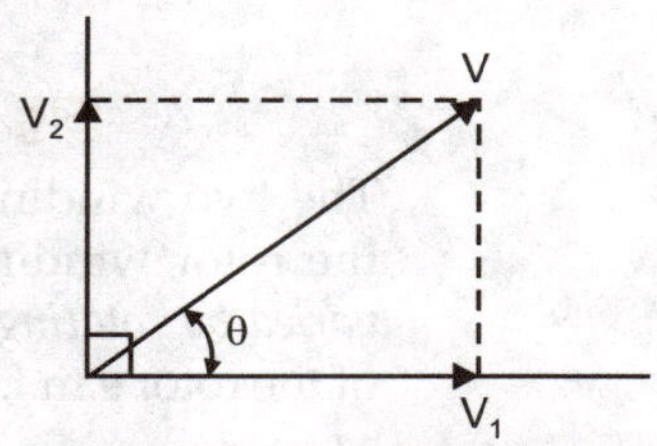

Fig. 5.12.

- *Example* : Drysdale instruments.

2. Co-ordinate potentiometers :

— In these instruments, the unknown e.m.f. is measured in *terms of its components along and perpendicular to a standard axis.* The two components of the unknown e.m.f. are given directly by two different scales known as "*in-phase*" (V_1) and "*quadrature*" (V_2) scales.

— Provision is made in these instruments to read both +ve and –ve values of voltages so that all angles upto 360° are covered.

— Voltage, $V = \sqrt{V_1^2 + V_2^2}$; $\theta = \tan^{-1}\left(\frac{V_2}{V_1}\right)$

- *Examples* : Gall-Tinsley and Campbell-Larsen type instruments.

5.3.3. Drysdale Polar Potentiometer

Fig. 5.13 shows a Drysdale polar potentiometer :

— For A.C. measurements, the slide wire MN is supplied from a phase shifting circuit so arranged that the *magnitude of the voltage supplied by it remains constant while its phase can be varied through 360°*. Consequently, slide-wire current *I* can be maintained constant in magnitude but varied in phase.

— The *"phase shifting circuit"* consists of two stator coils supplied from the same source in parallel, their currents are made to differ by 90° by using well-known *phase-splitting technique*.

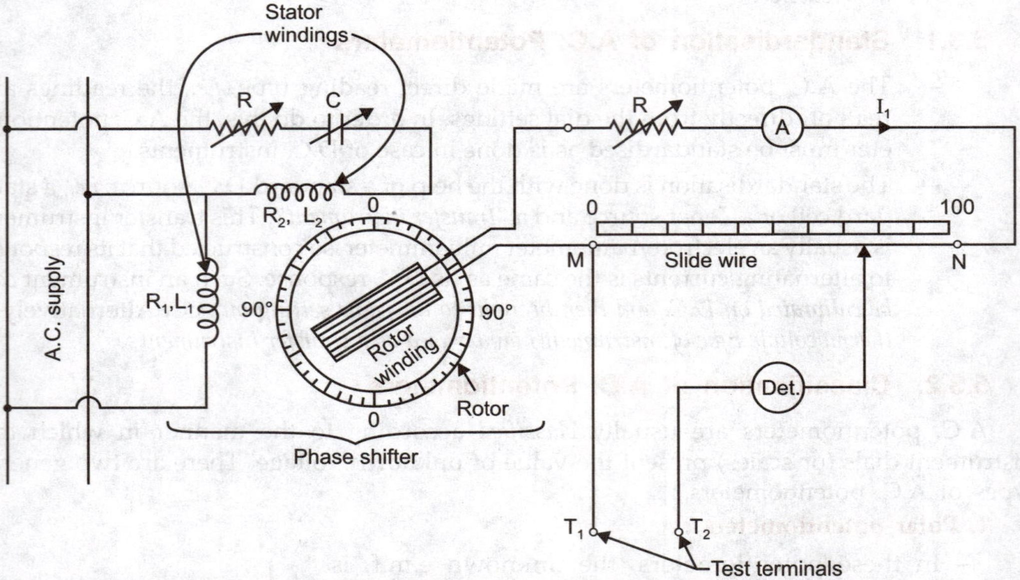

Fig. 5.13. Drysdale polar potentiometer.

The two windings produce rotating flux which induces a secondary e.m.f. in the rotor winding which is of *constant magnitude but the phase of which can be varied by rotating the rotor in any position either manually or otherwise*. The phase of the rotor e.m.f. is read from the circular graduated dial provided for the purpose.

— The ammeter A in the slide-wire circuit is of electrodynamic or thermal type.

— *Before using it for A.C. measurements, the potentiometer is first calibrated by using D.C. supply for slide wire and standard cell for test terminals T_1 and T_2.*

— The unknown alternating voltage to be measured is applied across test terminals T_1 and T_2 and the balance is effected by the *alternate adjustment of the slide-wire contact and the position of phase-shifting rotor. The slide-wire reading represents the magnitude of the test voltage and the voltage phase-shifter reading gives the phase with reference to an arbitrary reference vector.*

5.3.4. Gall Co-ordinate Potentiometer

Fig. 5.14 shows a Gall co-ordinate potentiometer :

— This potentiometer uses two slide-wires *XY* and *MN* with their currents I_1 and I_2 having a *mutual phase difference of 90°*.

— The two currents are obtained from single-phase supply through isolating transformers, the *circuit for "quadrature" slide wire MN incorporating a phase shift arrangement*.

— Before using the instrument, the current I_1 is first standardised as described for Drysdale potentiometer (Art. 5.3.3). Next current I_2 is standardised with the help

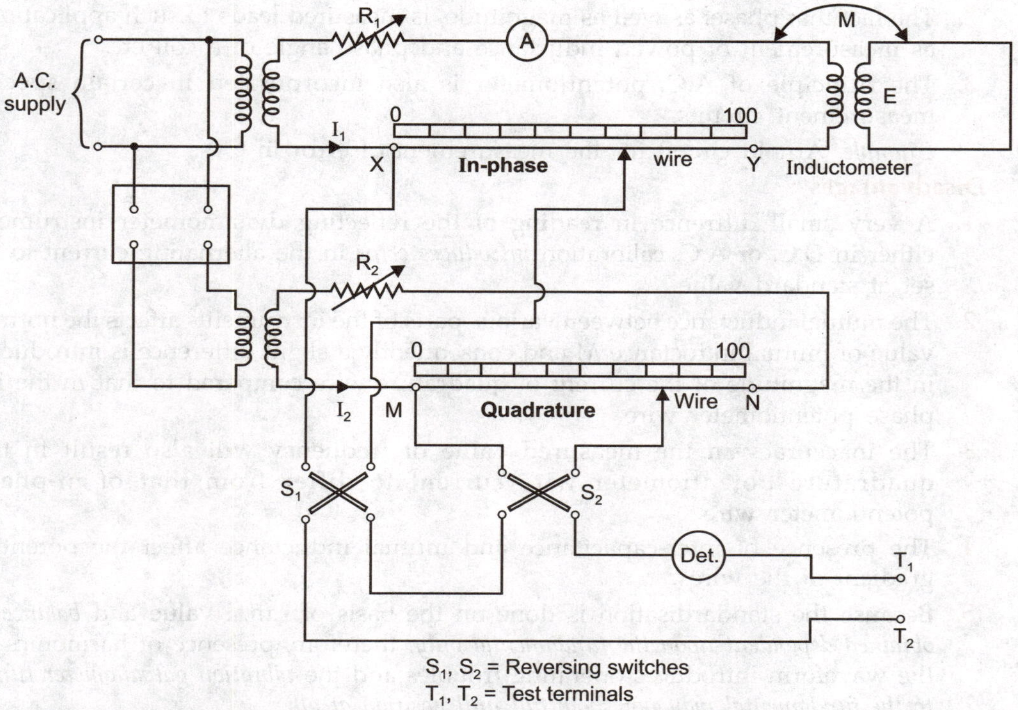

Fig. 5.14. Gall co-ordinator potentiometer.

of the mutually induced e.m.f. in inductometer secondary. This e.m.f. $E = \omega MI_1$, and is in quadrature phase with I_1.

— Now, E is balanced against the voltage drop on the slide-wire MN. This balance will be obtained only when I_2 is of current magnitude and is in exact quadrature with I_1. *Balance is obtained with the help of phase-shifter and rheostat* R_2.

— The *unknown voltage* is applied across the test terminals T_1 and T_2.

— 'Slide-wire MN measures that component of unknown voltage which *is in phase with* I_2. Similarly, slide-wire XY measures that component of the unknown voltage which is *in phase with* I_1. Since I_1 and I_2 are in quadrature, the two measured values are quadrature components of the unknown voltage. If V_1 and V_2 are these values, then :

$$\text{Unknown voltage, } V = \sqrt{V_1^2 + V_2^2}, \text{ and, } \theta = \tan^{-1}\left(\frac{V_2}{V_1}\right) \text{ w.r.t. } I_1$$

— Reversing switches S_1 and S_2 are used for measuring both +ve and –ve in-phase and quadrature components of the unknown voltage.

5.3.5. Advantages and Disadvantages of A.C. Potentiometers

A.C. potentiometers entail the following *advantages* and *disadvantages* :

Advantages :

1. *Very versatile instrument.* By employing shunts and volt-ratio boxes its use may be extended to cover current, voltage and resistance measurements over a wide range of values.

2. The fact that phase, as well as magnitude, is measured leads to such applications as measurement of power, inductance and phase angle of a coil etc.
3. The principle of A.C. potentiometer is also incorporated in certain special measurement circuits.

   Example. Arnold circuit for the measurement of error in CTs.

Disadvantages :

1. A very small diffrence in reading of the reflecting dynamometer instrument either in D.C. or A.C. calibration *introduces error* in the alternating current to be set at standard value.
2. The mutual inductance between various parts of the instruments affects the normal value of mutual inductance M and consequently a slight difference is introduced in the magnitude of the current of quadrature wire compared to that in the in-phase potentiometer wire.
3. The inaccuracy in the measured value of frequency will also result in the quadrature potentiometer wire current to differ from that of in-phase potentiometer wire.
4. The presence of inter-capacitance and mutual inductance affect the potential gradient of the wires.
5. Because the standardisation is done on the basis of r.m.s. value and *balance is obtained dependent upon the fundamental only,* therefore, presence of harmonics in the waveform introduces operating troubles and the *vibration galvanometer tuned to the fundamental may not show full null position at all.*

   — The indicating instruments of high quality may give very nearly as high accuracy and be much simpler to use for measurement of current, voltage and power.

   — For the measurement of electrical properties of components, A.C. bridges can be arranged to give higher accuracy and freedom from frequency errors.

5.3.6. Applications of A.C. Potentiometers

Out of the several applications, only the following important applications of A.C. potentiometers will be described here :

1. Measurement of self inductance.
2. Calibration of voltmeter.
3. Calibration of ammeter
4. Calibration of wattmeter.

5.3.6.1. Measurement of self inductance

Fig. 5.15 (*a, b*) shows the circuit diagram and vector diagram for the measurement of inductance and resistance of a coil under test by using an A.C. potentiometer.

— A standard non-inductive resistor is connected in series with the coil under test, and two potential differences V_1 and V_2 are measured in magnitude and phase by the potentiometer.

Refer to Fig. 5.15(*b*) :

Voltage drop across standard resistor R_s, $V_2 = IR_s$

where, I = Current flowing through the coil and R_s, and

R_s = Resistance of the standard non-inductive resistor.

or, $$I = \frac{V_2}{R_s} \quad ...(5.6)$$

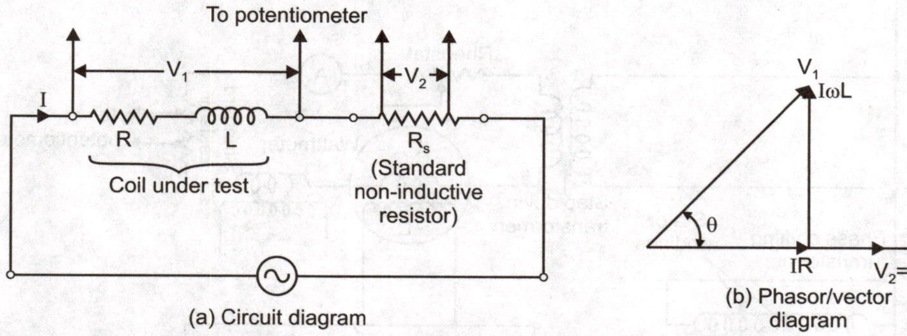

Fig. 5.15. Measurement of self inductance.

Voltage drop across inductive coil = V_1 (By measurement)

Phase angle between current and voltage drop across the coil = θ.

Voltage drop due to resistance of coil, $IR = V_1 \cos\theta$

$$\therefore \quad \text{Resistance of the coil, } R = \frac{V_1 \cos\theta}{I} = \frac{V_1 \cos\theta}{(V_2/R_s)} = \frac{V_1 R_s \cos\theta}{V_2} \quad ...(5.7)$$

Voltage drop due to inductance of coil, $I\omega L = V_1 \sin\theta$

$$\therefore \quad \textit{Inductance of coil,} \quad L = \frac{V_1 \sin\theta}{I\omega} = \frac{V_1 \sin\theta}{\omega(V_2/R_s)} = \frac{V_1 R_s \sin\theta}{V_2 \omega} \quad ...(5.8)$$

- The *accuracy obtained* by this method largely depends on the *accuracy of the potentiometer*.

5.3.6.2. Calibration of voltmeter

The method of calibration of an A.C. voltmeter by using A.C. potentiometer is similar to that adopted for calibration of D.C. voltmeter by using D.C. potentiometer (refer to Art. 5.2.5.5).

5.3.6.3. Calibration of ammeter

The method of calibration of an A.C. ammeter is similar to D.C. potentiometer method for D.C. ammeter *i.e.* A.C. ammeter under calibration is connected in series with a control resistance (non-inductive variable resistor), for varying the current, and a non-inductive standard resistor and voltage drop across standard resistor is measured on A.C. potentiometer.

However, the standardising A.C. potentiometer *involves the use of a suitable transfer instrument*.

5.3.6.4. Calibration of wattmeter

The method of calibration of wattmeter by A.C. potentiometer is similar to D.C. potentiometer method (refer to Art. 5.3.5.7).

Fig. 5.16 shows the circuit for calibration of wattmeter by A.C. potentiometer :

— The current coil (C.C.) of the wattmeter is supplied through a stepdown transformer and the potential coil (P.C.) from the secondary of a variable transformer whose primary is supplied from the rotor of a phase shifting transformer.

— The voltage V across the P.C. of wattmeter and current I through the C.C. of wattmeter are measured by the potentiometer, employing a volt-ratio box and a standard resistor as shown.

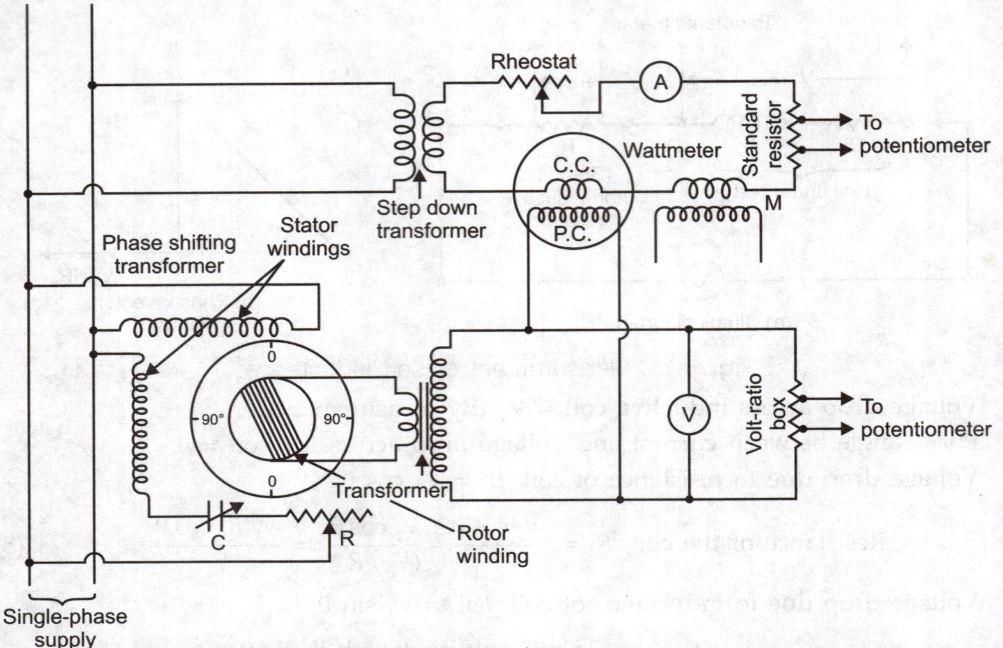

Fig. 5.16. Calibration of wattmeter by A.C. potentiometer.

— The *power factor* cos ϕ *is varied by rotating the phase-shift rotor*, the phase angle between voltage and current, ϕ being given by the reading on the dial of the phase-shifter. The *power is then VI cos* ϕ and the wattmeter reading may be compared with this value.

— A *small mutual inductance M is included to ensure accuracy of measurement of zero p.f.*

Example 5.10. *The following readings were obtained during the measuement of power by a polar potentiometer :*

Voltage across 0.2 Ω standard resistance in series with the load = 1.52 ∠35°
Voltage across 200 : 1 potential divider across the line = 1.43 ∠53°
Determine the current, voltage, power and power factor of the load.

Solution. (*i*) Current through the load, $\bar{I} = \dfrac{1.52\angle 35^\circ}{0.2} = 7.6\ \angle 35^\circ$ A

Magnitude of the current, I = **7.6 A** **(Ans.)**

(*ii*) Voltage across the load, $\bar{V} = 200\ (1.43\ \angle 53^\circ) = 286\ \angle 53^\circ$ V

Magnitude of voltage, V = **286 V** **(Ans.)**

(*iii*) Phase angle of load = 53° – 35° = 18°

∴ *Power factor of the load,* **cos ϕ** = cos 18° = **0.951 lagging** **(Ans.)**

(*iv*) *Power consumed by the load,*

$P = VI \cos \phi = 286 \times 7.6 \times 0.951 =$ **2067 W** **(Ans.)**

Example 5.11. *The following results were obtained for determination of impedance of a coil by using a co-ordinate type potentiometer :*

Voltage across 1.0 Ω resistor in series with coil

= + 0.2404 V in-phase dial and 0.0935 V on quadrature dial

Voltage across 10 : 1 potential divider used with the coil
= + 0.3409 V on in-phase dial and + 0.2343 V on quadrature dial
Calculate the resistance and reactance of the coil.

Solution. Current through the coil, $\bar{I} = \dfrac{(+0.2404 - j\,0.0935)}{1.0} = (0.2404 - j\,0.0935)$ A

Voltage across the coil, $\bar{V} = 10\,(0.3409 + j\,0.2343) = (3.409 + j\,2.343)$ V

$\therefore$ Impedance of the coil, $\bar{Z} = \dfrac{\bar{V}}{\bar{I}} = \dfrac{3.409 + j\,2.343}{0.2404 - j\,0.0935} = \dfrac{4.136\,\angle 34.5°}{0.258\,\angle -21.2°}$

$= 16.03\,\angle 55.7° = 9.03 + j13.24)\ \Omega$

$\therefore$ Resistance of coil, R = **9.03 Ω** **(Ans.)**

Reactance of coil, X = **13.24 Ω** **(Ans.)**

Example 5.12. *The following readings were obtained during measurement of inductance of a coil on an A.C. potentiometer :*

Voltage drop across 0.1 Ω standard resistor conneced in series with the coil
= 0.613 ∠12° 6′ V
Voltage across the test coil through a 100 : 1 volt-ratio box
= 0.781 ∠50° 48′
Frequency = 50 Hz
Determine the value of inductance of the coil. **(U.P. Tech. University)**

Solution. Current through the coil, $\bar{I} = \dfrac{0.613\,\angle 12°6'}{0.1} = 6.13\,\angle 12°\,6'$ A

Voltage across the coil, $\bar{V} = 100 \times 0.781\,\angle 50°\,48' = 78.1\,\angle 50°\,48'$ V

$\therefore$ Impedance of the coil, $\bar{Z} = \dfrac{\bar{V}}{\bar{I}} = \dfrac{78.1\angle 50°48'}{6.13\angle 12°6'} = 12.74\angle 38°42'\ \Omega$

Resistance of the coil, $R = 12.74 \cos 38°42' = 9.94\ \Omega$

Reactance of the coil, $X = 12.74 \sin 38°42' = 7.96\ \Omega$

$\therefore$ Inductance of coil, $L = \dfrac{X}{2\pi f} = \dfrac{7.96}{2\times\pi\times 50} = 0.0253$ H or **25.3 mH** **(Ans.)**

Example 5.13. *The current taken by a small iron core choke coil is measured by a rectangular co-ordinate A.C. potentiometer. A 1.0 Ω non-inductive resistance is connected in series with the choke coil. The voltages measured across the resistance and the coil are (0.8 – j 0.75) V and (1.2 + j 0.3) V respectively. Determine the iron loss in the coil.*

Assume the voltage and current to be sinusoidal. **(AMIE)**

Solution. Standard resistance, $R_s = 1.0\ \Omega$

Voltage drop across the resistance, $\bar{V}_s = (0.8 - j\,0.75)$ V

Current through the coil, $\bar{I} = \dfrac{0.8 - j\,0.75}{1.0} = (0.8 - j\,0.75)$ A

$= 1.0966\ \angle -43.15°$

Voltage across the coil, $\bar{V}_{coil} = (1.2 + j\,0.3)\text{ V} = 1.237\,\angle 14°$ V

Phase angle between voltage across the coil and the current in the coil

$= 14° - (-43.15°) = 57.15°$

$$\text{Power loss in the coil} = VI \cos \phi$$
$$= 1.237 \times 1.0966 \times \cos 57.15° = 0.7358 \text{ W}$$

Assuming copper loss in the coil to be negligible and hence taking the total losses occuring in the coil as iron loss, we have :

Iron loss in the coil = **0.7358 W** **(Ans.)**

Example 5.14. *In calibrating a wattmeter with the help of an A.C. rectangular co-ordinate potentiometer (as shown in Fig. 5.17) using a 300 : 1 volt-ratio box and a 0.1 Ω standard resistor the voltage across the resistor was found to be $V_1 = (0.35 - j\,1.0)$ V and low voltage tap of volt-ratio box gave a reading of $V_2 = (0.80 + j\,0.15)$ V.*

Determine :

(i) True reading of the wattmeter.

(ii) Load power factor. **(AMIE)**

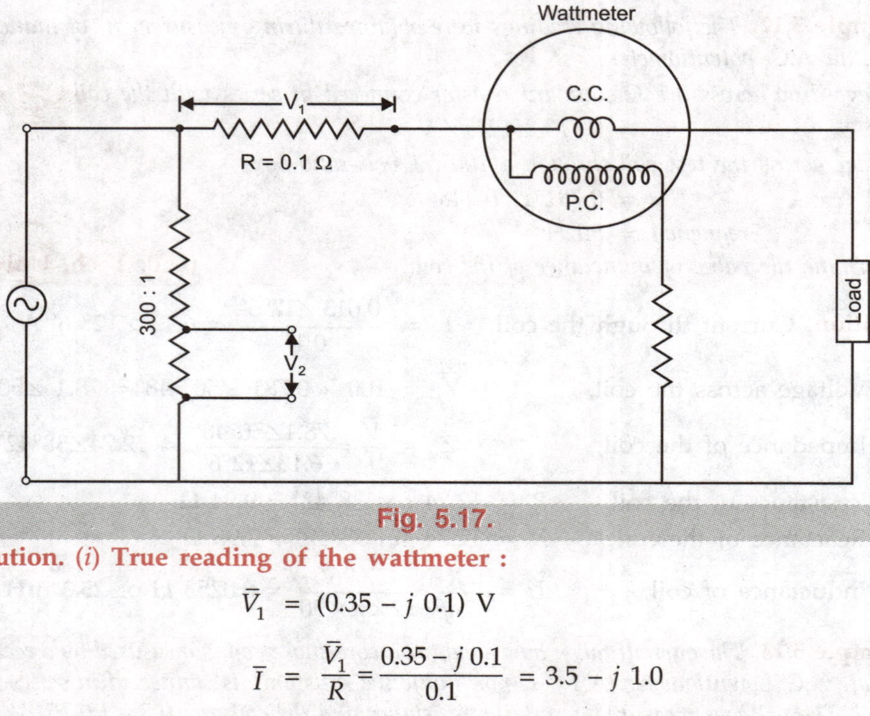

Fig. 5.17.

Solution. (*i*) True reading of the wattmeter :

$$\bar{V}_1 = (0.35 - j\,0.1) \text{ V}$$

$$\bar{I} = \frac{\bar{V}_1}{R} = \frac{0.35 - j\,0.1}{0.1} = 3.5 - j\,1.0$$

$$\bar{V}_2 = (0.80 + j\,0.15) \text{ V}$$

Supply voltage, $\bar{V} = 300\,\bar{V}_2 = 300(0.80 + j\,0.15) = (240 + j\,45)$

Now, product of $\bar{V}$ and $\bar{I}$ (Apparent power) $= (\bar{V}) \times (\bar{I}) = (240 + j\,45) \times (3.5 - j\,1.0)$

$$= (244.2 \angle 10.62°)\,(3.64 \angle -15.94)$$

$$888.89 \angle 26.56° = \underbrace{795.1}_{\text{True or real power}} + \underbrace{j\,397.4}_{\text{Reactive power}}$$

∴ True reading of the wattmeter (true or real power) = **795.1 W** **(Ans.)**

(*ii*) Power factor of the load circuit,

$$\cos \phi = \cos 26.56° = \mathbf{0.894} \quad \textbf{(Ans.)}$$

HIGHLIGHTS

1. A *potentiometer* is an instrument used for measuring and comparing the e.m.fs. of different cells and for calibrating and standardising voltmeters, ammeters, etc.
2. Simple potentiometer is used for educational purposes only.
3. Some of the important *applications* of D.C. potentiometers are :
   (*i*) Measurement of resistance;
   (*ii*) Measurement of current;
   (*iii*) Measure of high e.m.fs. (say 250 V);
   (*iv*) Measurement of power;
   (*v*) Calibration of voltmeter;
   (*vi*) Calibration of ammeter;
   (*vii*) Calibration of wattmeter.
4. A.C. potentiometers are *classified* as follows :
   (*i*) *Polar potentiometers.* In these potentiometers, the unknown e.m.f. is measured in *polar form i.e.* in terms of its magnitude and relative phase. The magnitude is given by one scale and its phase angle with respect to some reference vector is given by the second scale. Provision is made for relative phase angle upto 360°.

   Example. Drysdale instrument.

   (*ii*) *Co-ordinate potentiometers.* In these potentiometers, the unknown e.m.f. is measured in terms of its components along and perpendicular to a standard axis. The two components of the known e.m.f. are given directly by two different scales known as 'in-phase' and 'quadrature' scales.

   Examples. Gall-Tinsley and Campbell-Larsen type instruments.

OBJECTIVE TYPE QUESTIONS

Choose the Correct Answer :

1. Standardization of potentiometers is done in order that, they become
   (*a*) accurate (*b*) precise
   (*b*) accurate and direct reading (*d*) accurate and precise.
2. A Crompton's potentiometer is provided with a dial having 15 steps of 10 Ω each and a slide wire of 10 Ω resistance. The slide wire is divided into 100 divisions and one fifth of a division can be read with certainity. The working current of potentiometer is 10 mA. The range and resolution of the potentiometer are respectively :
   (*a*) 1.6 V, 0.2 mV (*b*) 1.6 V, 0.5 mV
   (*c*) 1.5 V, 0.1 mV (*d*) 1.6 V, 0.1 mV.
3. The readings of a polar type A.C. potentiometer when measuring the reactance of a coil are $I = 12\angle 13.8°$, $V = 27.8\ \angle 29.7°$. The reactance of the coil is
   (*a*) 2.317 Ω (*b*) 0.634 Ω
   (*c*) 2.22 Ω (*d*) – 2.22 Ω.
4. A co-ordinate type potentiometer is used for determination of impedance of a coil and the results obtained are : voltage across 1 Ω resistor in series with coil, 0.24 V on in-phase dial and – 0.09 V on quadrature dial,

   Voltage across 10 : 1 potential divider used with coil + 0.37 V on in-phase dial and + 0.32 on quadrature dial. The resistance of coil is :

(a) 9.14 Ω (b) 2.34 Ω
(c) 10.76 Ω (d) none of the above.

5. A potentiometer is basically a
(a) deflectional type instrument (b) null type instrument
(c) deflection as well as null type instrument (d) a digital instrument.

6. When a potentiometer is used for measurement of voltage of an unknown source, the power consumed in the circuit of the unknown source under null conditions
(a) is very high (b) is high
(c) is small (d) is ideally zero.

7. A volt-ratio box is designed with a resistance 100 Ω/V. It has input terminals of 300 V, 150 V and 75 V, and an output terminal of 1.5 V which is connected to the potentiometer. Supposing a voltage of 270 V is connected to the 300 V terminal, what is voltage indicated by the potentiometer and what is the power consumption under null conditions ?
(a) 1.5 V, 3 W (b) 1.35 V, 3 W
(c) 1.5 V, 2.43 W (d) 1.35 V, 2.43 W.

8. The standardisation of A.C. potentiometers is done by
(a) directly using A.C. standard voltage source.
(b) using D.C. standard sources and transfer instruments.
(c) using D.C. standard sources and d'Arsonval galvanometer.
(d) using A.C. standard sources and transfer instruments.

ANSWERS

1. (c) 2. (a) 3. (b) 4. (a) 5. (b) 6. (d) 7. (d)
8. (b).

THEORETICAL QUESTIONS

1. What is a potentiometer ?
2. Describe the construction and working of a simple D.C. potentiometer.
3. Describe with the help of a diagram the principle and working of one form of D.C. potentiometer. Explain how this potentiometer is standardised.
4. Give some significant features/advantages of potentiometers.
5. List the applications/uses of potentiometers.
6. Draw a connection diagram of Crompton potentiometer and bring out its salient features. How is it standardised ?
7. Draw a neat sketch of a modern D.C. potentiometer and discuss how the potentiometer is standardised.
8. Draw the circuit of a D.C. potentiometer. Explain how you can calibrate the same against a standard cell. Discuss the effect of room temperature on this calibration.
9. Explain how the potentiometer can be used for measurement of resistance.
10. Name the different types of potentiometer and describe one out of them.
11. Briefly explain how you will calibrate a given D.C. ammeter in the laboratory using a direct reading D.C. potentiometer.
12. Explain how the potentiometer may be used for precise measurement of voltage (240 V D.C.).
13. What is a volt-ratio box ? Explain briefly.

14. Describe with the help of suitable diagrams, how a D.C. potentiometer can be used for :
    (*i*) Calibration of a voltmeter;
    (*ii*) Calibration of an ammeter;
    (*iii*) Calibration of wattmeter;
    (*iv*) Determination of an unknown reistance.
15. What are the applications of self-balancing potentiometers ?
16. How does an A.C. potentiometer differ from a D.C. potentiometer ?
17. What factors must be considered for the operation of an A.C. potentiometer ?
18. What are the problems associated with an A.C. potentiometer ? Describe the working of any one A.C. potentiometer.
19. Name the different forms of A.C. potentiometers and bring out the difference between them.
20. Describe the construction and working of a polar type potentiometer. Explain the method for standardising it.
21. Describe with the help of diagram, the principle and construction of polar potentiometer.
22. Describe the construction and working of a co-ordinate type of potentiometer. Explain the procedure for standardisation of the potentiometer. Describe how this potentiometer can be used for :
    (*i*) Measurement of self inductance of a coil.
    (*ii*) Calibration of a voltmeter and A.C. energy meters.
23. Explain the principle of a rectangular type co-ordinate A.C. potentiometer. Draw a diagram of scheme of connections and describe how the potentiometer is standardised.
24. What are the advantages and disadvantages/limitations of A.C. potentiometers ?
25. Describe briefly the applications of A.C. potentiometer.

UNSOLVED EXAMPLES

1. The e.m.f. of a standard cell used for standardisation is 1.0186 volts. If the balance is obtained at 60 cm determine :
   (*i*) The e.m.f. of the cell which balances at 75 cm.
   (*ii*) The current flowing through a standard resistance of 2 Ω if the p.d. across it balances at 66 cm.
   (*iii*) The voltage of the supply main which is reduced by a volt-ratio box to one hundredth and balance is obtained at 84 cm.
   (*iv*) The percentage error in ammeter reading 0.28 ampere when balance is obtained at 40 cm with p.d. across a 2.5 Ω resistance in the ammeter circuit.
   (*v*) The percentage error in a voltmeter reading 1.25 volts when balance is obtained at 72 cm. [**Ans.** (*i*) 1.27325 V; (*ii*) 0.56023 A; (*iii*) 142.604 V; (*iv*) 3% high; (*v*) 2.25% high]
2. The following readings were obtained during the measurement of a low resistance using a potentiometer :
   Voltage drop across a 0.1 Ω standard resistance = 1.0235 V
   Voltage drop across the low resistance under test = 0.4221 V
   Calculate the value of unknown resistance, current and power lost in it.
   [**Ans.** 0.041208 Ω; 10.235 A; 4.316 W]
3. The following readings were obtained during measurement of low resistance by means of a potentiometer :
   Voltage drop across a 0.1 Ω standard resistance connected in series with the unknown resistance = 1.01575 V.

Voltage drop across low resistance under test = 0.83942 V

The resistance of standard resistor at the temperature of test is 0.10014 Ω. Upon setting the potentiometer dials to zero and breaking the current passing through the unknown resistance, the thermal e.m.f. of latter produced a galvanometer deflection equivalent to 23 μV, the direction of the deflection being the same as that produced by an increase of the potentiometer reading during the voltage measurement.

Calculate the unknown resistance. [**Ans.** 0.08275 Ω]

4. The current is measured through a standard resistance of 0.1 ± 0.1 Ω% (standard deviation) by a potentiometer that is accurate to ± 0.0001 V (standard deviation). The voltage across the resistance is measured to be 0.2514 V. To what accuracy current has been determined ? [**Ans.** ± 0.108%]

5. The following readings were obtained during the measurement of power by a polar potentiometer :

   Voltage across a 0.2 Ω standard resistance in series with the load = $1.46\angle 32°$

   Voltage across a 200 : 1 potential divider across the line = $1.37\angle 56°$

   Determine the current, voltage, power and power factor of the load.

   [**Ans.** 7.3 A; 274 V; 1827 W; 0.9135]

6. A co-ordinate type potentiometer is used for determination of impedance of a coil and the results obtained are :

   Voltage across 1.0 Ω resistor in series with coil = 0.238 V on in-phase dial and
   – 0.085 V on quadrature dial

   Voltage across a 10 : 1 potential divider used with coil
   = + 0.3375 V on in-phase dial and
   + 0.232 V on quadrature dial

   Calculate the resistance and reactance of the coil. [**Ans.** R = 9.49 Ω; X = 13.13 Ω]

CHAPTER

6 Measurement of Resistance, Inductance and Capacitance

6.1 Introduction; 6.2. Measurement of resistance – Classification of resistances – **Measurement of *low resistance*** – Ammeter – voltmeter method – potentiometer method – Kelvin double bridge method – Measurement of ***medium resistance*** – Ammeter-voltmeter method – Substitution method – Wheatstone bridge – Carey-foster slide-wire bridge method – Measurement of ***high resistances*** – Direct deflection method – Loss of charge method – Megohm bridge – Megger – Measurement of *"insulation resistance"* when power is *on* – two-wire system – Three-wire system (Russell's method) – Measurement of earth resistance – *"Location of cable faults"* – Murray loop test – Varley loop test – Location of faults in high-volage cables; **6.3. Measurement of Inductance – A.C. bridges –** A.C. bridge-general form – Maxwell bridge – Hay bridge – Anderson bridge – Owen bridge – De sauty bridge – Schering bridge – Heaviside mutual inductance bridge – Carey Foster (Heydweiller) bridge – Campbell bridge – Wien bridge – Resonance bridge – Wagner earthing device – Types of detectors – Sources of errors in bridge circuits and the elimination of errors – *Highlights – Objective Type Questions – Theoretical Questions – Unsolved Examples.*

6.1 INTRODUCTION

For several years the precision measurements of component values have been made by using various forms of *bridges*.

- **Bridge circuits** *are extensively used for measuring component values such as* **R** *(Resistance)*, ***L*** *(Inductance)* and ***C*** *(Capacitance)*.
- A "bridge circuit" in its simplest form consists of a network of four resistance arms forming a closed circuit, with a D.C. source of current applied to two opposite junctions and a current detector connected to the other two junctions.
- Since the bridge circuit merely compares the value of an unknown component with that of an accurately known component (a standard), its measurement accuracy can be very high. This is because the readout of this comparison is based on the null indication at bridge balance, and is essentially independent of the characteristics of the null detector. The *measurement accuracy is, therefore, directly related to the accuracy of the bridge component and not to that of the null detector used.*

Advantages of the bridge circuits are :

(*i*) High measurement accuracy (as the measurement is done by comparing the unknown value with the standard value).

(*ii*) The accuracy is independent of null detector's characteristics (it depends on the values of the components).

(*iii*) The balance equation is independent of the magnitude of the input voltage or its source impedance, the sensitivity and impedance of the null detector or any impedance shunting the detector.

(*iv*) The interchange of the source and detector does not affect the balance condition.

(*v*) The bridge circuit can be used in control circuits.

Types of bridges :

Bridges are of the following two types :

1. *D.C. bridges :*

- These bridges are used to measure the "*resistances.*"
- They use the D.C. *voltage* as the excitation voltage.

Examples : (*i*) Wheatstone bridges.

(*ii*) Kelvin bridges.

2. *A.C. bridges :*

- These bridges are used to measure the "*impedances*" consisting of inductances and capacitances.
- They use the *alternating voltage* as the excitation voltage.

Examples : (*i*) Maxwell bridge;

(*ii*) Hay bridge;

(*iii*) Schering bridge;

(*iv*) Anderson bridge etc.

6.2 MEASUREMENT OF RESISTANCE

6.2.1. Classification of Resistance

From the point of view of measurement, resistances can be *classified* as follows :

1. *Low resistances.* All resistances of the *order of 1 Ω and under* may be classified as low resistances.
   — Such resistances, in practice, may be met with in the *armatures and series winding of large machines, in ammeter shunts, cable lengths, contacts etc.*
2. *Medium resistances.* This class includes resistances from about *1 Ω upwards to about 100 kΩ.*
   — In practice, the majority of the pieces of electrical apparatus used have resistances which lie between these limits.
3. *High resistances.* Resistances of the order of *100 kΩ and upwards* are classified as high resistances.

Measurements of high resistances are required for determination of :

(*i*) Insulation resistance of components and built-up electrical equipment of all types.

(*ii*) Resistance of high resistance circuit elements.

(*iii*) Volume resistivity and surface resistivity of a material.

6.2.2. Measurement of Low Resistances

Following methods are used for the measurement of *low resistances* :

1. Ammeter-voltmeter method.
2. Potentiometer method.
3. Kelvin double bridge method.

6.2.2.1. Ammeter-voltmeter method

- This method, which is the *simplest* of all, is in *very common use* for the measurement of *low resistances* when an *accuracy of the order of 1% is sufficient*.
  — However, it must be realised that it is essentially a comparatively *rough* method, the accuracy being limited by those of the ammeter and voltmeter used, even if corrections are made for the "shunting" effect of the voltmeter.

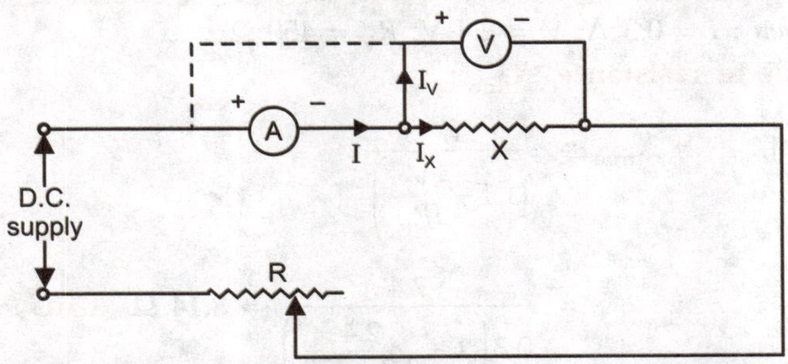

Fig 6.1. Ammeter-voltmeter method of measuring low resistances.

- In this method, current through the resistor (X) under test and the potential drop across it are simultaneously measured. The readings are obtained by ammeter and voltmeter respectively. The attainable accuracy depends primarily on the accuracy and ranges of the instruments employed for measurement of current and voltage.

From Fig. 6.1, it may be observed that there are *two ways* in which the ammeter and voltmeter may be connected for measurement, and that, in either case the resistance of one of the instruments will affect the measurement to some extent.

Case 1 : When the voltmeter is connected directly across the resistor, the ammeter measures current flowing through the unknown resistance X and the voltmeter. In this case :

Current through ammeter = Current through unknown resistance (X) + Current through voltmeter

or, $$I = I_X + I_V \quad ...(6.1)$$

or, $$I_X = I - I_V$$

∴ *True value* of unknown resistance,

$$X_{true} = \frac{V}{I_X} = \frac{V}{I - I_V} = \frac{V}{I - \frac{V}{R_V}} = \frac{V}{I\left(1 - \frac{V}{IR_V}\right)} \quad ...(6.2)$$

where, V = The voltmeter reading,

R_V = The resistance of the voltmeter, and

I = The current indicated by the ammeter.

Case 2 : When the ammeter is connected so that it indicates only the current flowing through the unknown resistance, the voltmeter measures voltage drop across the ammeter and unknown resistance X. In this case :

$$V = IR_A + IX = I(R_A + X)$$

or, $$X_{true} = \frac{V}{I} - R_A \quad ...(6.3)$$

where, R_A = The resistance of the ammeter.

Example 6.1. *The ammeter-voltmeter method is used to measure a resistance. With the voltmeter connected across the resistance the readings on the ammeter and voltmeter are 0.3 A and 2.4 V respectively. The resistance of the voltmeter is 450 Ω.*

Calculate :

(i) True value of resistance.

(ii) Percentage error in the value of resistance, if the voltmeter current is ignored.

Solution. *Given* : $I = 0.3$ A; $V = 2.4$ V; $R_V = 450\ \Omega$

(i) **True value of resistance, X_{true} :**

We know that,
$$X_{true} = \frac{V}{I\left(1 - \frac{V}{IR_V}\right)} \quad \text{[Eqn. (6.2)]}$$

$$= \frac{2.4}{0.3\left(1 - \frac{2.4}{0.3 \times 450}\right)} = \mathbf{8.14\ \Omega} \quad \textbf{(Ans.)}$$

(ii) **Percentage error in the value of resistance :**

Measured value of unknown resistance,

$$X_m = \frac{V}{I} \quad \text{....ignoring } I_V$$

$$= \frac{2.4}{0.3} = 8\ \Omega$$

$$\therefore \quad \%\text{ age error} = \frac{X_m - X_{true}}{X_{true}} \times 100 = \frac{8 - 8.14}{8.14} \times 100$$

$$= -1.72\%\ \textit{i.e.}\ \mathbf{1.72\%\ (low)} \quad \textbf{(Ans.)}$$

6.2.2.2. Potentiometer method

In this method the unknown resistance is *compared with a standard resistance of the same order of magnitude.*

Fig. 6.2 shows the circuit diagram for this method :

— The unknown resistance X, an ammeter A, a rheostat R to limit the current and a standard resistance S are connected, all in series with a *low voltage, high current supply source.* The value of S should be accurately known and of the same order of resistance and must be of the same or higher current rating than the one under test (X).

— The current flowing through the circuit is adjusted so that potential difference, if possible, across each of the resistor is about 1 V.

— The voltage drop across both the unknown resistor X and standard resistor S are measured by a potentiometer. The ratio of the two potentiometer readings gives the ratio of X to S.

Mathematically,
$$\frac{X}{S} = \frac{\text{Potentiometer reading across } X}{\text{Potentiometer reading across } S} = \frac{V_X}{V_S} \quad \text{...(6.4)}$$

As the accuracy of this method depends on there being no change in current between the two readings, therefore, it is necessary that source of supply of current through the circuit be *extremely stable.* When the necessary precautions are taken, and a good potentiometer and sensitive galvanometer are used, very high accuracy can be obtained.

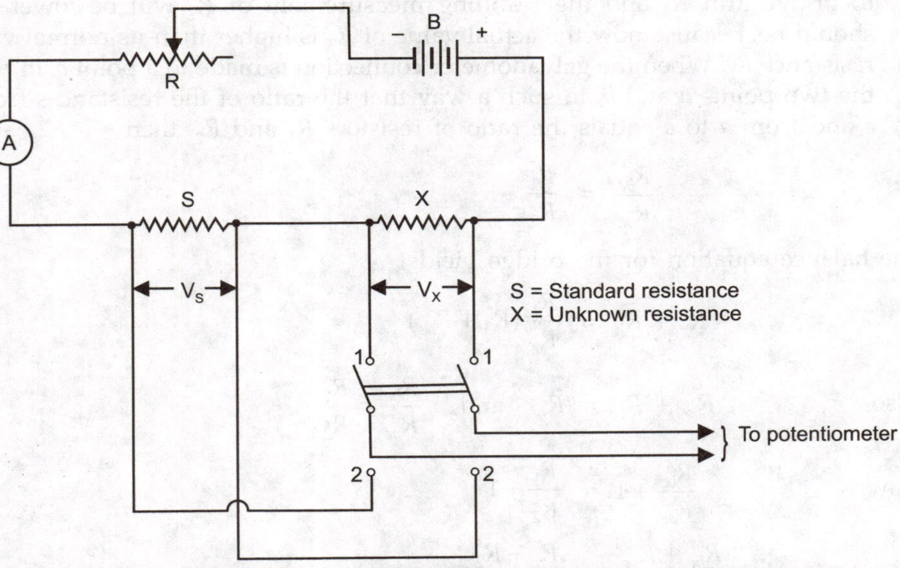

Fig. 6.2. Potentiometer method of measuring low resistances.

Example 6.2. *In a measurement of resistance by potentiometer, the voltage drops across a resistor under test and across 0.024 Ω standard resistor were found to be 0.885 V and 1.1 V respectively. Determine the value of resistor under test.*

Solution. *Given* : $S = 0.024\ \Omega$; $V_X = 0.885$ V; $V_S = 1.1$ V

Resistance of resistor under test, X :

We know that,
$$\frac{X}{S} = \frac{V_X}{V_S} \qquad \text{[Eqn. (6.4)]}$$

or,
$$X = \frac{SV_X}{V_S} = \frac{0.024 \times 0.885}{1.1} = \mathbf{0.0193\ \Omega} \quad \textbf{(Ans.)}$$

6.2.2.3. Kelvin double bridge method

Firstly we shall discuss Kelvin bridge and thereafter Kelvin double bridge method will be described in detail.

Kelvin bridge :

This bridge is a modified version of Wheatstone bridge and provides greatly increased accuracy in the measurement of *low-value resistances*, generally below 1 Ω.

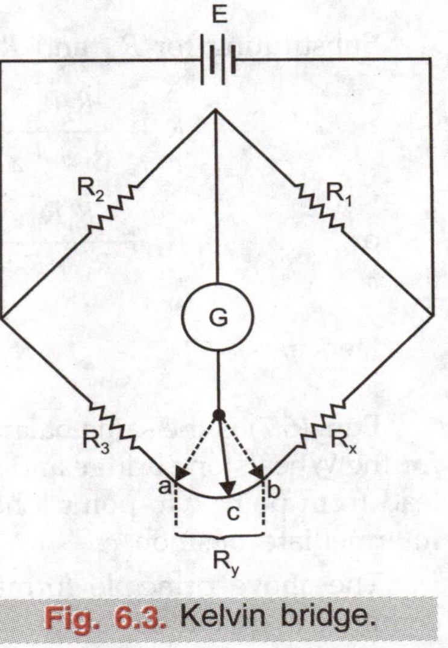

Fig. 6.3. Kelvin bridge.

Fig. 6.3 shows the schematic of Kelvin bridge :

— R_y represents the *resistance of connecting lead* from R_3 to R_x

— Two galvanometer connections are available, to point *a* or to point *b*. When the galvanometer is connected to point *a*, the resistance R_y (of connecting lead) is added to the unknown resistor R_x, resulting in too high an indication for R_x. When connection is made to *b*, R_y is added to bridge arm R_3 and the resulting measurement of R_x will be lower than it should be, because now the actual value of R_3 is higher than its normal value by resistance R_y. When the galvanometer connection is made to a point c, in between the two points a and b, in such a way that the ratio of the resistances from b to c and from a to c equals the ratio of resistors R_1 and R_2, then

$$\frac{R_{bc}}{R_{ac}} = \frac{R_1}{R_2} \qquad \text{...(6.5)}$$

The balance equation for the bridge yields,

$$R_x + R_{bc} = \frac{R_1}{R_2}(R_3 + R_{ac}) \qquad \text{...(6.6)}$$

Also, $$R_{ac} + R_{bc} = R_y, \text{ and, } \frac{R_{bc}}{R_{ac}} = \frac{R_1}{R_2}$$

Now, $$\frac{R_{bc}}{R_{ac}} + 1 = \frac{R_1}{R_2} + 1$$

or, $$\frac{R_{bc} + R_{ac}}{R_{ac}} = \frac{R_1 + R_2}{R_2}$$

or, $$\frac{R_y}{R_{ac}} = \frac{R_1 + R_2}{R_2}$$

or, $$R_{ac} = \frac{R_2\, R_y}{R_1 + R_2} \qquad \text{...}(i)$$

and, $$R_{bc} = R_y - R_{ac}$$

$$= R_y - \frac{R_2\, R_y}{R_1 + R_2}$$

$\therefore$ $$R_{bc} = \frac{R_1\, R_y + R_2 R_y - R_2 R_y}{(R_1 + R_2)} = \frac{R_1 R_y}{R_1 + R_2} \qquad \text{....}(ii)$$

Substituting for R_{ac} and R_{bc} in eqn. (6.6), we get

$$R_x + \frac{R_1 R_y}{R_1 + R_2} = \frac{R_1}{R_2}\left(R_3 + \frac{R_2\, R_y}{R_1 + R_2}\right)$$

or, $$R_x + \frac{R_1 R_y}{R_1 + R_2} = \frac{R_1 R_3}{R_2} + \frac{R_1 R_y}{(R_1 + R_2)}$$

Hence, $$R_x = \frac{R_1 R_3}{R_2} \qquad \text{...(6.7)}$$

Eqn. (6.7) is the same balance equation [Eqn. (6.24), suffices to be noted] as developed for the Wheatstone bridge and it indicates that the effect of the resistance of the connecting lead from point a to point b has been eliminated by connecting the galvanometer to the intermediate position c.

The above principle forms the basis for construction of the Kelvin double bridge, commonly known as *Kelvin bridge*.

Kelvin double bridge :

This bridge is so called (*i.e.* double bridge) since the circuit contains a *second set of ratio arms* (l and m) as shown in Fig. 6.4. This second set of ratio arms connects the galvanometer to a point f at the appropriate potential between c and d and it eliminates the effect of the yoke resistance R_y. The ratio of the resistances of arms l and m is the same as the ratio of R_1 and R_2.

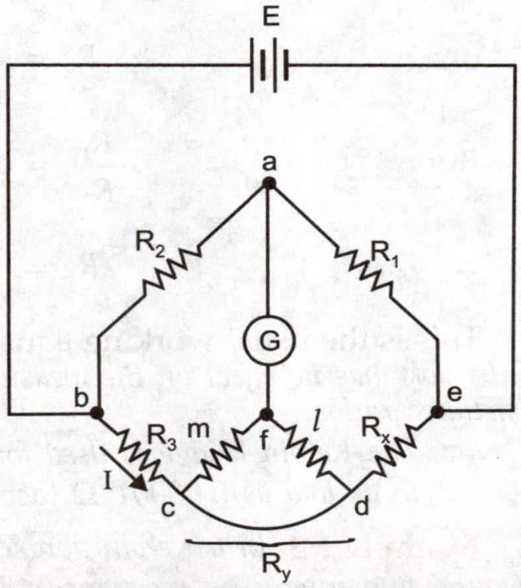

Fig. 6.4. Kelvin double bridge.

The galvanometer will read zero when the potential at a equals the potential at f.

i.e., $$E_{ab} = E_{bcf}$$

But, $$E_{ab} = \frac{R_2}{R_1 + R_2} \times E \quad ...(6.8)$$

and, $$E = I\left[R_3 + R_x + \frac{(l+m)\,R_y}{(l+m)+R_y}\right] \quad ...(6.9)$$

Substituting for E in eqn. (6.8), we get

$$E_{ab} = \frac{R_2}{R_1 + R_2} \times I\left[R_3 + R_x + \frac{(l+m)\,R_y}{(l+m)+R_y}\right] \quad ...(6.10)$$

Similarly, $$E_{bcf} = I\left[R_3 + \frac{m}{l+m}\left\{\frac{(l+m)\,R_y}{(l+m)+R_y}\right\}\right] \quad ...(6.11)$$

But, $$E_{ab} = E_{bcf}$$

$$\therefore \quad \frac{IR_2}{R_1 + R_2}\left[R_3 + R_x + \frac{(l+m)\,R_y}{(l+m)+R_y}\right] = I\left[R_3 + \frac{m}{l+m}\left\{\frac{(l+m)\,R_y}{(l+m)+R_y}\right\}\right]$$

or, $$R_3 + R_x + \frac{(l+m)\,R_y}{(l+m)+R_y} = \frac{R_1 + R_2}{R_2}\left[R_3 + \frac{m\,R_y}{(l+m)+R_y}\right]$$

or, $$R_3 + R_x + \frac{(l+m)\,R_y}{(l+m)+R_y} = \left(\frac{R_1}{R_2} + 1\right)\left(R_3 + \frac{mR_y}{l+m+R_y}\right)$$

or, $$R_x + \frac{(l+m)\,R_y}{l+m+R_y} + R_3 = \frac{R_1R_3}{R_2} + R_3 + \frac{m\,R_1R_y}{R_2\,(l+m+R_y)} + \frac{m\,R_y}{l+m+R_y}$$

or, $$R_x = \frac{R_1R_3}{R_2} + \frac{m\,R_1R_y}{R_2\,(l+m+R_y)} + \frac{m\,R_y}{l+m+R_y} - \frac{(l+m)\,R_y}{l+m+R_y}$$

or, $$R_x = \frac{R_1R_3}{R_2} + \frac{m\,R_1R_y}{R_2\,(l+m+R_y)} + \frac{m\,R_y - l\,R_y - m\,R_y}{l+m+R_y}$$

or, $$R_x = \frac{R_1R_3}{R_2} + \frac{m\,R_1R_y}{R_2\,(a+b+R_y)} - \frac{l\,R_y}{l+m+R_y}$$

or, $$R_x = \frac{R_1 R_3}{R_2} + \frac{m R_y}{l + m + R_y}\left(\frac{R_1}{R_2} - \frac{l}{m}\right) \quad ...(6.12)$$

But, $$\frac{R_1}{R_2} = \frac{l}{m}$$

$\therefore$ $$R_x = \frac{R_1 R_3}{R_2}$$

This is the usual working equation for the Kelvin bridge. *It indicates that the resistance of the yoke has no effect on the measurement, provided that the two sets of ratio arms have equal resistance ratios.*

- *The Kelvin bridge is used for measuring very low resistances, from approximately 1 Ω to as low as 0.00001 Ω* (accuracy varying from ± 0.05 % to ± 0.2%).

Example 6.3. *In a Kelvin double bridge, there is error due to mismatch between the ratios of outer and inner arm resistances. The following data relate to this bridge :*

Standard resistance = 100.03μ Ω; Inner ratio arms = 100.31Ω and 200Ω; Outer ratio arms = 100.24Ω and 200Ω; The resistance of connecting leads from standard to unknown resistor is 680μ Ω.

Calculate the unknown resistance.

Solution. Refer to Fig. 6.4. *Given* : R_3 = 100.03 μ Ω; l = 100.31 Ω; m = 200 Ω; R_1 = 100.24 Ω; R_2 = 200 Ω; R_y = 680μ Ω

Unknown resistance, R_x :

We know that, $$R_x = \frac{R_1 R_3}{R_2} + \frac{m R_y}{l + m + R_y}\left(\frac{R_1}{R_2} - \frac{l}{m}\right) \quad \text{[Eqn. (6.12)]}$$

Inserting the various values, we get

$$R_x = \frac{100.24 \times 100.03 \times 10^{-6}}{200} + \frac{200 \times 680 \times 10^{-6}}{100.31 + 200 + 680 \times 10^{-6}}\left(\frac{100.24}{200} - \frac{100.31}{200}\right)$$

$$= 50.135 \times 10^{-6} + 4.528 \times 10^{-4} (-3.5 \times 10^{-4})$$

$$= 49.97 \times 10^{-6}\ \Omega = \mathbf{49.97\ \mu\Omega} \quad \textbf{(Ans.)}$$

Kelvin bridge ohmmeter :

- This is a modified form of Kelvin bridge and is *intended for the rapid measurement of the winding resistances of machines and transformers, and for the measurement of contact and earth conductor resistances.*
- The accuracy is of the order of ± 0.2 per cent.
- This instrument is *direct reading* and balance is obtained by rotating a single dial.
- Six standard resistors are used and these range from 0.01 Ω to 1000 Ω. The ratio dial is calibrated from 0.5 Ω to 10.5 Ω using 100 Ω standard resistor. The overall range of the instrument is from 0.00005 Ω to 105 Ω.

6.2.3. Measurement of Medium Resistances

The methods used for measurement of *medium resistances* are :

1. Ammeter-voltmeter method.
2. Substitution method.
3. Wheatstone bridge.
4. Carey-Foster slide-wire bridge method.

6.2.3.1. Ammeter-voltmeter method

This method has already been explained in Art. 6.2.2.1.

6.2.3.2. Substitution method

The connections for measurement of resistance by substitution are shown in Fig. 6.5. Let R be a variable resistance which can be changed in small steps, say of 0.1 Ω.

(*i*) Refer to Fig. 6.5 (*a*). First resistance X is put in the circuit and the value of current noted.

Then resistance X is *removed* and it is *substituted* by a known variable resistance R which is varied so that the value of the current is same in both the cases. *This value of R is equal to the unknown resistance.*

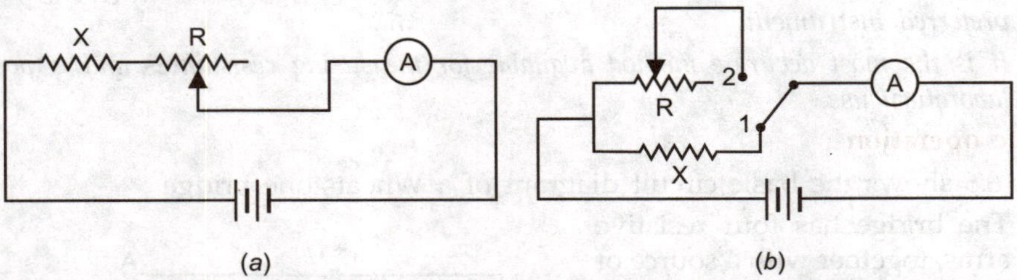

Fig. 6.5. Measurement of resistance by substitution.

(*ii*) If R is of fixed value, note the *readings* of the ammeter for the following cases :

(*a*) For resistances X and R in series.

(*b*) When resistance X is removed.

Let the readings of the ammeter for these cases be I_1 and I_2

$$\therefore \quad I_1 = \frac{V}{R+X} \qquad ...(6.13)$$

$$\text{and,} \quad I_2 = \frac{V}{R} \qquad ...(6.14)$$

$$\frac{I_2}{I_1} = \frac{R+X}{R} = 1 + \frac{X}{R} \qquad ...(6.15)$$

$$\text{or,} \quad \frac{X}{R} = \frac{I_2}{I_1} - 1$$

$$\text{or,} \quad X = R\left(\frac{I_2 - I_1}{I_1}\right) \qquad ...(6.16)$$

(*iii*) Refer to Fig. 6.5(*b*). The two-way switch first makes contact with 1 and then with 2 and let these readings be I_1 and I_2.

$$I_1 = \frac{V}{X} \text{ (where } X \text{ is in the circuit)} \qquad ...(6.17)$$

$$I_2 = \frac{V}{R} \text{ (when } R \text{ is in the circuit)} \qquad ...(6.18)$$

$$\therefore \quad \frac{I_2}{I_1} = \frac{X}{R}$$

$$\text{or,} \quad X = R \times \frac{I_2}{I_1} \qquad ...(6.19)$$

6.2.3.3. Wheatstone bridge

The *simplest form of bridge is for the purpose of measuring resistance and is called the Wheatstone bridge.*

- The Wheatstone bridge is widely used for precision measurement of resistance from *approximately 1Ω to the low megaohm range.*
  - The *"upper limit"* is set by the reduction in sensitivity to unbalance, caused by high resistance values, thus reducing the galvanometer current.
  - The *"lower limit"* is set by the resistance of the connecting leads and the contact resistance at the binding posts. The resistance of the leads could be calculated or measured, and the final result modified, but *contact resistance is very hard to compute or measure. For low resistance measurements, therefore, the Kelvin bridge is generally the preferred instrument.*
- *It is the most accurate method available for measuring resistances and is popular for laboratory use.*

Basic operation :

Fig. 6.6 shows the basic circuit diagram of a Wheatstone bridge :

— The bridge has four resistive arms, together with a source of e.m.f. (a battery) and null detector, usually a galvanometer (G) or other sensitive current meter.

— The current through the galvanometer depends on the potential difference between the points C and D. The bridge is said to be *balanced* when the potential difference across the galvanometer is *zero volt* so that there is no current through the galvanometer. Hence the bridge is balanced when potential difference between points *C* and *D* is equal,

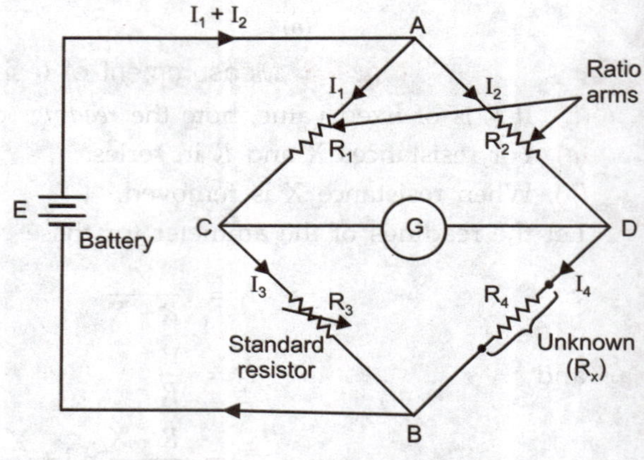

Fig. 6.6. Wheatstone bridge.

i.e.
$$I_1R_1 = I_2R_2 \quad \text{...(6.20)}$$

When the current through the galvanometer is zero, the following conditions should be satisfied :

$$I_1 = I_3 = \frac{E}{R_1 + R_3} \quad \text{...(6.21)}$$

and,
$$I_2 = I_4 = \frac{E}{R_2 + R_4} \quad \text{...(6.22)}$$

Substituting the values of I_1 and I_2 in Eqn. (6.20), we get

$$\frac{E}{R_1 + R_3} \times R_1 = \frac{E}{R_2 + R_4} \times R_2$$

or,
$$\frac{R_1}{R_1 + R_3} = \frac{R_2}{R_2 + R_4}$$

$$\text{or,} \quad R_1(R_2 + R_4) = R_2(R_1 + R_3)$$

$$\text{or,} \quad R_1R_2 + R_1R_4 = R_1R_2 + R_2R_3$$

$$\text{or} \quad R_1R_4 = R_2R_3 \quad \text{...(6.23)}$$

Eqn. (6.23) is the well known expression for balance of the Wheatstone bridge. When the three of the resistors are known, the fourth may be determined. Thus if R_4 is the unknown resistor, its resistance R_x can be found as follows :

$$R_x = \frac{R_2}{R_1} . R_3 \quad \text{...(6.24)}$$

The resistors R_2 and R_1 are called *ratio arms* while the resistor R_3 is called the *standard arm* of the bridge.

When the null detector has sufficient sensitivity to indicate the balance position of the bridge with the required degree of position, the measurement of the unknown resistance R_x is *independent* of the characteristics or calibration of the null-detecting galvanometer.

Measurement errors :

The various sources of measurement errors are :

(*i*) The limiting errors of three known resistances–main measurement error.

(*ii*) Insufficient sensitivity of the null detector.

(*iii*) Changes in resistance of the bridge arms due to heating effect of the current through the resistors.

— In order to check this error the power dissipation in the bridge arms must be computed in advance, particularly when low-resistance values are to be measured, and the *current must be limited to a safe value.*

(*iv*) Thermal e.m.fs. in the bridge circuit or the galvanometer circuit.

— To prevent thermal effects, the more sensitive galvanometers sometimes have copper coils and copper suspension systems to avoid having dissimilar metals in contact with one another and generating thermal e.m.fs.

(*v*) Errors due to the resistance of leads and contacts exterior to the actual bridge circuit. These errors can be *reduced* by using a Kelvin bridge.

Applications of Wheatstone bridge :

Following are the *applications* of Wheatstone bridge :

1. To measure the D.C. resistance of various types of wire either for the purpose of quality control of the wire itself, or of some assembly in which it is used.

Example : Measurement of the resistance of motor windings, transformers, solenoids, and relay coils.

2. This bridge is also used extensively by telephone companies and others to *locate cable faults.*

Limitations :

The Wheatstone bridge entails the following *limitations* :

1. In case of measurement of low resistance, the resistance of the leads and contacts becomes significant, resulting in introduction of an error. This error can be *eliminated* by using Kelvin's double bridge.

2. While measuring high resistances, the resistance of the bridge becomes so large that the galvanometer becomes *insensitive to imbalance*. Consequently, a power supply has to replace the battery and a D.C. VTVM replaces the galvanometer. The Wheatstone bridge *cannot* be used for measuring high resistances in megaohms.

3. The *change in resistance* of the bridge arms due to the heating effect of current through the resistance–the temperature rise causes a change in the value of the resistance, and *excessive current may cause a permanent change in value.*

Example 6.4. *Fig. 6.6 consists of the following parameters :*
$R_1 = 20\ k\ \Omega$; $R_2 = 30\ k\ \Omega$; $R_3 = 80\ k\ \Omega$.
Determine the unknown resistance R_x.

Solution. Using the bridge balance equation, we have :

$$R_1 R_x = R_2 R_3$$

or,
$$R_x = \frac{R_2 R_3}{R_1} = \frac{30 \times 80}{20}$$
$$= \mathbf{120\ k\ \Omega}\ \textbf{(Ans.)}$$

Example 6.5. *Each of the ratio arms of a laboratory type Wheatstone bridge has guaranteed accuracy of ± 0.05%, while the standard arm has a guaranteed accuracy of ± 0.1%. The ratio arms are both set at 1000 Ω and bridge is balanced with standard arm adjusted to 3154 Ω.*

Determine the upper and the lower limits of the unknown resistance, based upon the guaranteed accuracies of the known bridge arms. **(Punjab University)**

Solution. Refer to Fig. 6.6. *Given* : $R_1 = R_2 = 1000\ \Omega$, $R_3 = 3154\ \Omega$

We know that :
$$\frac{R_1}{R_2} = \frac{R_3}{R_x}$$

or, Value of unknown resistance,
$$R_x = \frac{R_2 \times R_3}{R_1}$$
$$= \frac{1000 \times 3154}{1000} = 3154\ \Omega \qquad \text{[Eqn. (6.24)]}$$

∴ Percentage error in determination of R_x,

$$\frac{\delta R_x}{R_x} = \pm \frac{\delta R_1}{R_1} \pm \frac{\delta R_2}{R_2} \pm \frac{\delta R_3}{R_3}$$
$$= \pm 0.05 \pm 0.05 \pm 0.1 = \pm 0.2\%$$

∴ *Limiting value of* R_x = 3154 ± 0.2% = **3147** *to* **3160 Ω** **(Ans.)**

Wheatstone bridge with digital readout :

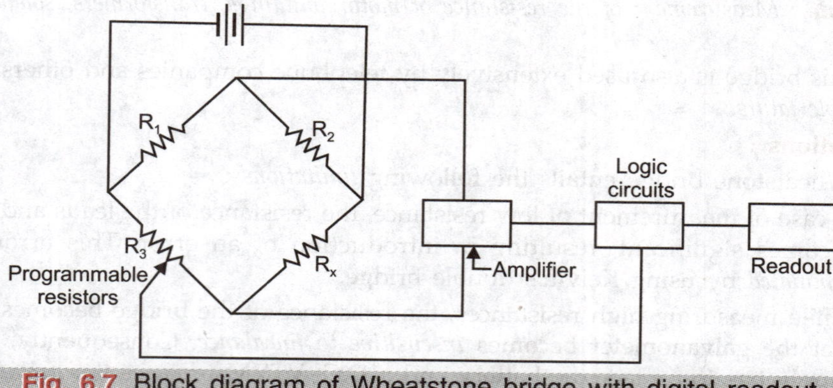

Fig. 6.7. Block diagram of Wheatstone bridge with digital readout.

Fig. 6.7 shows the block diagram of Wheatstone bridge with digital readout. Here a logic circuit is used to provide a signal to R_3, reuse the null, and provide a digital readout representing the value of R_x.

6.2.3.4. Carey-Foster slide-wire bridge method

This bridge circuit is an elaboration of the Wheatstone bridge and is *specially suited for comparing two nearly equal resistances.*

Fig. 6.8, shows the connection for Carey-Foster bridge :

— R_1 and R_2 are nominal ratio arms, R_x the resistance under test, and R_3 the standard resistance.

— A slide-wire of length L and uniform cross-section is included between R_x and R_3 as shown.

— Resistances R_1 and R_2 are first adjusted so that the ratio $\frac{R_1}{R_2}$ is approximately equal to the ratio $\frac{R_x}{R_3}$. Exact balance is obtained by adjustment of the sliding contact on the slide-wire. Let l_1 be the distance of the sliding contact from the left-hand end of the slide-wire. The resistances R_x *and* R_3 *are then interchanged* and balance again obtained. Let the distance now be l_2.

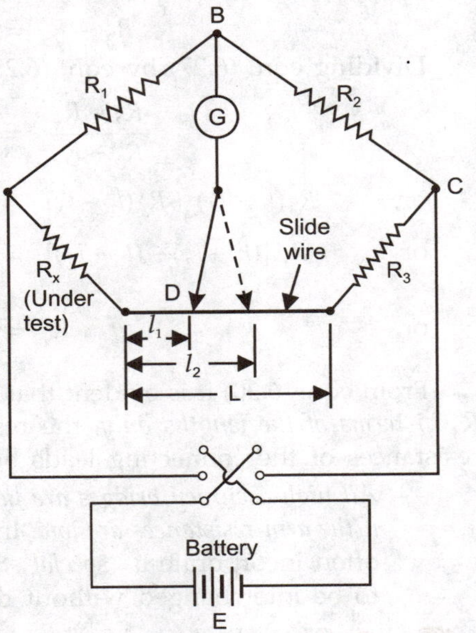

Fig. 6.8. Carey-Foster slide-wire bridge.

Then, for the *first balance,* $$\frac{R_1}{R_2} = \frac{R_x + l_1 r}{R_3 + (L - l_1)r} \quad ...(6.25)$$

where, r is the resistance per unit length of the slide-wire

For the *second balance,* $$\frac{R_1}{R_2} = \frac{R_3 + l_2 r}{R_x + (L - l_2)r} \quad ...(6.26)$$

Comparing eqns. (6.25) and (6.26), we get :

$$\frac{R_x + l_1 r}{R_3 + (L - l_1)r} = \frac{R_3 + l_2 r}{R_x + (L - l_2)r}$$

or, $$\frac{R_x + l_1 r}{R_3 + (L - l_1)r} + 1 = \frac{R_3 + l_2 r}{R_x + (L - l_2)r} + 1$$

or, $$\frac{R_x + l_1 r + R_3 + Lr - l_1 r}{R_3 + (L - l_1)r} = \frac{R_3 + l_2 r + R_x + Lr - l_2 r}{R_x + (L - l_2)r}$$

or, $$\frac{R_x + R_3 + Lr}{R_3 + (L - l_1)r} = \frac{R_3 + R_x + Lr}{R_x + (L - l_2)r}$$

or, $$R_3 + (L - l_1)r = R_x + (L - l_2)r$$

or, $$R_3 + Lr - l_1 r = R_x + Lr - l_2 r$$

or, $$R_3 - R_x = r(l_1 - l_2) \quad ...(6.27)$$

Thus the difference between R_3 and R_x is obtained from the resistance of the slide-wire between the two balance points.

The slide wire is calibrated *i.e.* r is obtained by shunting R_3 by a known high resistance, thus reducing the effective value to R_3'. Now balance points l_1' and l_2' are then obtained by repeating the whole procedure previously described, so that :

$$R_3' - R_x = r(l_1' - l_2') \quad ...(6.28)$$

Dividing eqn. (6.27) by eqn. (6.28), we get

$$\frac{R_3 - R_x}{R_3' - R_x} = \frac{l_1 - l_2}{l_1' - l_2'} \quad ...(6.29)$$

or, $$R_3(l_1' - l_2') - R_x(l_1' - l_2') = R_3'(l_1 - l_2) - R_x(l_1 - l_2)$$

or, $$R_x\,[(l_1' - l_2') - (l_1 - l_2)] = R_3\,(l_1' - l_2') - R_3'(l_1 - l_2)$$

or, $$R_x = \frac{R_3\,(l_1' - l_2') - R_3'(l_1 - l_2)}{(l_1' - l_2' - l_1 + l_2)} \quad ...(6.30)$$

From eqn. (6.30) it is evident that this method gives a *direct comparison between R_x and R_3 in terms of the lengths only,* the resistances of R_1 and R_2, contact resistances and the resistances of the connecting leads being eliminated.

— *All high-accuracy bridges are liable to error owing to thermoelectric e.m.fs. particularly if the arm resistances are low.* In order to *minimise this trouble,* Carey-Foster bridges often incorporate a "*special switch,*" which enables the connections to R_x and R_3 to be interchanged without the necessity for handling the coils.

☞ • *The methods used for measurement of medium resistances are unsuitable for measurement of low resistances i.e. resistances having a value under 1 Ω. The reason being that the resistance of leads and contacts, though small, are appreciable in comparison in the case of low resistances; hence special techniques have to be used for the measurement of low resistances.*

6.2.4. Measurement of High Resistances

The various methods employed for the measurement of *high resistances* are :

1. Direct deflection method.
2. Loss of charge method.
3. Megohm bridge.
4. Megger (Refer to Art. 3.8.7.3)

6.2.4.1. Direct deflection method

- In this method a very sensitive and high resistance (1 kΩ or more) moving coil galvanometer is connected in series with the resistance to be measured, and to a battery supply. The deflection of the galvanometer gives a measure of the insulation resistance. This is *very simple but rough method of measuring insulation resistance.*
- Fig. 6.9 illustrates the direct deflection method for measuring insulation resistance of a cable :
  — The galvanometer G measures the current I_R between conductor core and metal sheath; leakage currents I_L over the surface of the insulating material are carried by the guard wire wound on the insulation and does not flow through the galvanometer [Fig. 6.9 (*a*)].

— The cables without metal sheath can be tested in a similar way if the cable, except the end or ends on which connections are made, is immersed in water in a tank [Fig. 6.9 (*b*)] The water and tank then form the return path for the current. The cable is immersed in saline water for about 24 hours and temperature in kept constant (at about 20°C) and then the measurement is taken as in Fig. 6.9(*b*). The resistance of the cable, $R = \frac{V}{I_R}$.

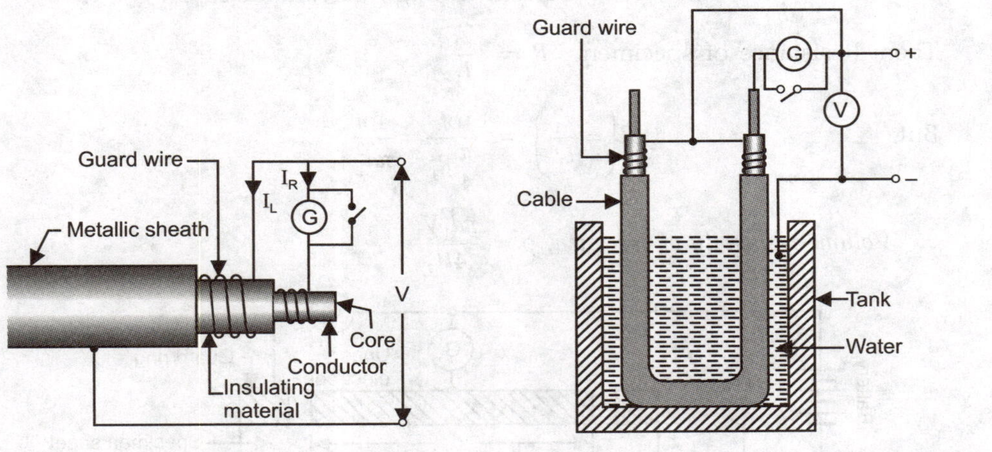

(a) Cable having sheath. (b) Cable having no conducting sheath.

Fig. 6.9. Measurement of insulation resistance of cable.

The deflection of the galvanometer, in some cases, is observed and its scale is afterwards calibrated by replacing the insulation by a standard high resistance (usually 1 MΩ), the galvanometer shunt being varied, as required to give a deflection of the same order as before.

— While conducting tests on cables, the galvanometer should be short-circuited before applying the voltage. *The short-circuiting connection is removed only after sufficient time has elapsed so that charging and absorption currents cease to flow.*

The galvanometer (of D'Arsonval type) employed should be *very sensitive* (at least 1000 mm/microampere at a scale distance of 1 metre), should be *of high resistance,* and also have *uniform scale.* The galvanometer and its circuit, together with the keys, *must be well insulated to prevent leakage currents.*

— The battery should be of about 500 V and its e.m.f. should remain constant throughout the test.

Measurement of volume and surface resistivities :

The insulation resistance of insulating material samples available in *sheet form* is often measured by the *direct deflection method.* In such cases, *volume resistivity* and *surface resistivity* of the material are measured.

Fig. 6.10 shows the schematic diagram for measurement of volume and surface resistivities of a specimen of insulating material. The specimen is provided with tin foil or colloidal graphite electrodes; the upper electrode having a guard ring.

Volume resistivity :

In order to measure *"volume resistivity"*, which is *in fact the specific resistance,* readings of the voltage applied and the current through the galvanometer are taken. Leakage currents over the specimen's edge will flow between the guard ring and the lower electrode and hence will *not* introduce error into the measurement.

The *volume resistivity* is calculated as under :

Let, d_1 = Diameter of upper electrode,

t = Thickness of the specimen sheet,

V_1 = Voltmeter reading, and

I_1 = Current through galvanometer G_1.

Then, Resistance of specimen, $R = \dfrac{V_1}{I_1}$

But, $$R\left(=\frac{V_1}{I_1}\right) = \frac{\rho t}{\frac{\pi}{4}d_1^2} = \frac{4\rho t}{\pi d_1^2}$$

$\therefore$ *Volume resistivity of specimen,* $$\rho = \frac{\pi d_1^2 V_1}{4tI_1} \quad ...(6.31)$$

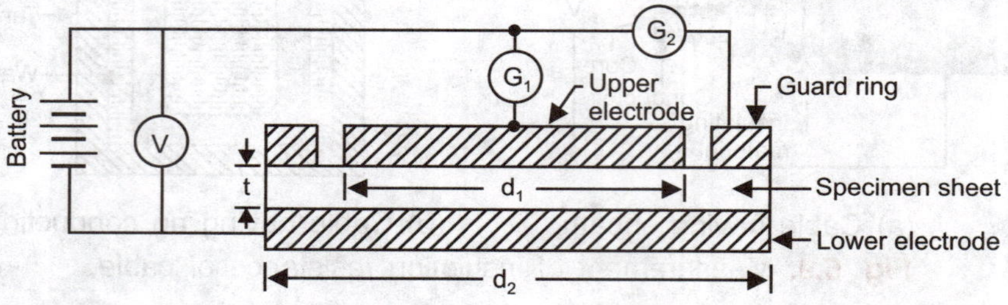

Fig. 6.10. Measurement of volume and surface resistivities of insulating material specimen.

— The resistivity of a thin layer of dielectric materials is different from volume resistivity, not only because of an *adherent humidity layer but also of contamination, chemical alterations, absorption of gases, or structural modification.*

The resistance R_t between two electrodes embedded in or attached to a dielectric medium is given by :

$$\frac{1}{R_t} = \frac{1}{R_v} + \frac{1}{R_s} \quad ...(6.32)$$

where, R_v = Volume resistance, and

R_s = Surface resistance.

— R_v can be measured separately from R_s by means of guard rings.

Surface resistivity :

The *"surface resistivity" is defined as the resistance between opposite edges of a square area of the surface of the material.* This quantity depends upon the general condition of the surface and upon humidity and is not a constant.

— In order to measure surface resistivity, the galvanometer is placed in position G_2 (See Fig. 6.10). In this position, the *galvanometer measures the leakage current and current flowing between upper and lower electrodes will be eliminated from measurement.*

Let, d_2 = Diameter of lower electrode disc,

V_2 = Voltmeter reading, and

I_2 = Current through the galvanometer G_2.

$\therefore$ Surface resistance, $R_s = \dfrac{V_2}{I_2}$

The leakage current flows along a path of length t and width πd_2 and, therefore, surface resistivity,

$$\rho_s = \frac{R_s \times \pi d_2}{t} = \frac{\pi d_2}{t} \cdot \frac{V_2}{I_2} \quad \text{...(6.33)}$$

- Other forms of *specimen and electrodes* (*e.g.* the electrodes and guard ring may be mercury, either placed in specially machined recesses, in moulded insulating materials, or retained by metal rings on the surface of sheet materials) are also used.

6.2.4.2. Loss of charge method

Fig. 6.11 shows the connection diagram for loss of charge method for measurement of high resistance :

C is a capacitor of known capacitance, V is electrostatic voltmeter, R_{leak} is the total leakage resistance of the capacitor and voltmeter and R_x is the resistance to be measured.

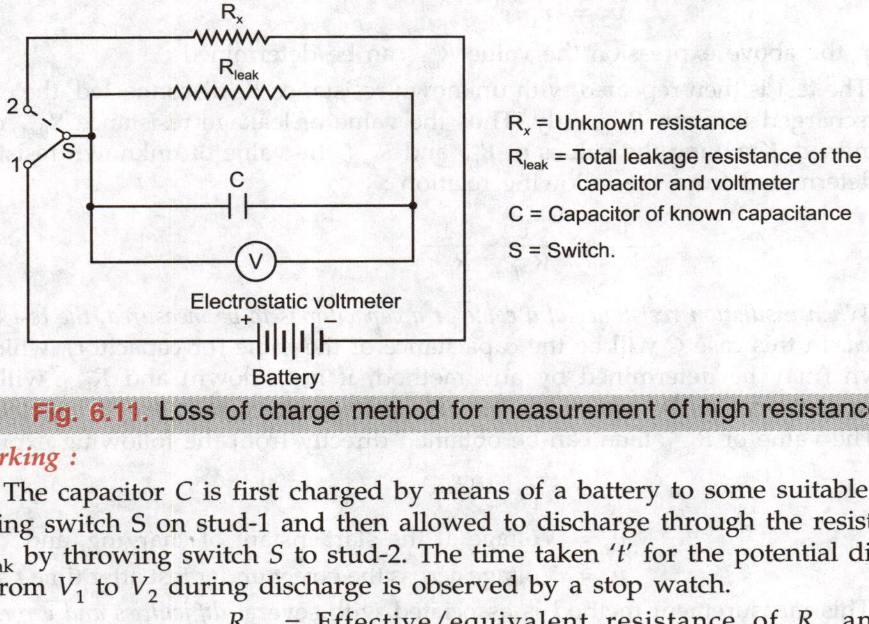

Fig. 6.11. Loss of charge method for measurement of high resistance.

Working :

— The capacitor C is first charged by means of a battery to some suitable voltage by putting switch S on stud-1 and then allowed to discharge through the resistance R_x and R_{leak} by throwing switch S to stud-2. The time taken 't' for the potential difference to fall from V_1 to V_2 during discharge is observed by a stop watch.

Let, $R_{eff.}$ = Effective/equivalent resistance of R_x and R_{leak} connected in parallel.

— If at any instant, say time 't', the voltage across the discharging capacitor is v volts, the charge on the discharging capacitor is q couloumbs and the capacity of the capacitor is C farads then current (i) at this instant,

$$i = -\frac{dq}{dt} = -C \cdot \frac{dv}{dt} \quad \text{...(6.34)}$$

Also,
$$i = \frac{\text{Potential drop across } R_{eff.}}{R_{eff.}} = \frac{v}{R_{eff.}} \quad \text{...(6.35)}$$

Comparing eqns. (6.34) and (6.35), we get :

$$\frac{v}{R_{eff.}} = -C \cdot \frac{dv}{dt}$$

or, $$\frac{dv}{v} = -\frac{dt}{CR_{eff.}} \quad ...(6.36)$$

Integrating both sides, we have :

$$\int_{v_1}^{v_2} \frac{dv}{v} = \int_0^t -\frac{dt}{CR_{eff.}}$$

or, $$[\ln v]_{v_1}^{v_2} = -\frac{t}{CR_{eff.}}$$

or, $$l_n\left(\frac{v_2}{v_1}\right) = -\frac{t}{CR_{eff.}}$$

or, $$\frac{v_2}{v_1} = e^{(-t/CR_{eff.})}$$

or, $$v_2 = v_1 . e^{(-t/CR_{eff.})} \quad ...(6.37)$$

From the above expression the value $R_{eff.}$ can be determined :

— The test is then repeated with unknown resistance R_x disconnected, the capacitor being discharged through R_{leak} only. Thus the value of leakage resistance R_{leak} can also be determined. Knowing the values of $R_{eff.}$ and R_{leak} the value of unknown resistance R_x can be determined from the following relation :

$$\frac{1}{R_x} = \frac{1}{R_{eff.}} - \frac{1}{R_{leak}} \quad ...(6.38)$$

— *When insulation resistance of a cable or a capacitor is to be measured, the test need not be repeated.* In this case C will be the capacitance of the cable (or capacitor), which must be known (may be determined by any method, if not known) and R_{leak} will be the insulation resistance to be measured. No external resistance need to be inserted in the circuit. The value of R_{leak} then can be obtained directly from the following expression :

$$v_2 = v_1 e^{(-t/CR_{leak})} \quad ...(6.39)$$

where, v_1 = Voltage at the start instant of charging, and

v_2 = Voltage across the object under test after time t seconds.

- This measurement method is associated with several *difficulties and errors.* Some important among them are :

(i) Effect of temperature upon insulation resistance :

— Usually, the resistance of insulating materials falls rapidly with increasing temperature, so temperature change and operating temperature play very important roles in the determination of the resistance.

— The value of insulation resistance should always be stated together with the temperature at which the test was carried out.

(ii) Leakage and absorption effects :

— Since the leakage currents are of comparable magnitude to the current under measurement, therefore, to eliminate leakage, the capacitor and voltmeter having exceedingly high resistance must be employed.

— The current which flows through the resistance is due to resistance as well as absorption. In order to eliminate absorption effect the observations should be made over a long period (several hours).

(iii) Effect of time of electrification :

— All insulating materials have absorption effects. Although the absorption current falls fairly steeply in the beginning, yet its fall becomes gradual thereafter.

— Thus the determination of insulation resistance by the ratio of applied voltage to the current flowing through insulation depends upon the time of application of the voltage.

Example 6.6. *The value of a high resistance is measured by loss of charge method. A capacitor having a capacitance of 2.5 μF is charged to a potential of 500 V D.C. and is discharged through the high resistance. An electrostatic voltmeter, kept across the high resistance, reads the voltage as 300 V at the end of 60 seconds.*

Calculate the value of high resistance. **(AMIE)**

Solution. *Given :* C = 2.5 μF = 2.5 × 10^{-6} F; v_1 = 500 V; v_2 = 300 V; t = 60 s

Value of unknown resistance, R_x :

We know that, $$v_2 = v_1\, e^{(-t/CR_x)} \quad \text{[Eqn. (6.37)]}$$

or, $$\frac{v_2}{v_1} = e^{(-t/CR_x)} = \frac{1}{e^{(t/CR_x)}}$$

or, $$e^{(t/CR_x)} = \frac{v_1}{v_2}$$

Taking log on both sides, we get :

$$t/CR_x = \ln\left(\frac{v_1}{v_2}\right)$$

or, $$R_x = \frac{t}{C\ln\left(\frac{v_1}{v_2}\right)}$$

$$= \frac{60}{2.5\times10^{-6}\times\ln\left(\frac{500}{300}\right)} = \mathbf{46.98\ M\Omega}\quad \textbf{(Ans.)}$$

Example 6.7. *A 2.5 μF capacitor is charged to a potential difference of 450 volts. The capacitor is disconnected from the supply and the reading on an electrostatic voltmeter in parallel with the capacitor is observed to fall to 280 volts in 15.2 minutes. The test is repeated with a resistance R in parallel with the capacitor and the voltmeter. The voltmeter's reading is now found to fall from 450 V to 280 V in 10.8 minutes.*

Calculate the value of R.

Solution. *Given :* C = 2.5 μF = 2.5 × 10^{-6} F; v_1 = 450 V; v_2 = 280 V; t_1 = 15.2 min. = 15.2 × 60 = 912 s; t_2 = 10.8 × 60 = 646 s

Value of *R* :

***Case-1* :** Total leakage resistance of capacitor and voltmeter,

$$R_{leak} = \frac{t}{C\ln\left(\frac{v_1}{v_2}\right)} = \frac{912}{2.5\times10^{-6}\ln(450/280)} \simeq 768.9\ M\Omega$$

Case-2 : Equivalent resistance of R_{leak} and insulation resistance R to be measured,

$$R_{eff.} = \frac{t}{C\ln\left(\frac{v_1}{v_2}\right)} = \frac{648}{2.5\times10^{-6}\ \ln\left(\frac{450}{280}\right)} = 546.3\ \text{M}\Omega$$

Unknown resistance R can be determined from the following relation :

$$\frac{1}{R} = \frac{1}{R_{eff.}} - \frac{1}{R_{leak}} \qquad \text{[Eqn. (6.38)]}$$

or,
$$R = \frac{R_{eff.}\times R_{leak}}{R_{leak} - R_{eff.}} = \frac{546.3\times768.9}{(768.9-546.3)} = \mathbf{1887\ M\Omega} \quad \textbf{(Ans.)}$$

Example 6.8. *The following results were obtained by loss of charge method of testing a cable :*
Discharged immediately after charging, the deflection = 200 divisions
Discharged 30 seconds after charging, the deflection = 125 divisions
Discharged 30 seconds after charging,
when in parallel with a resistance of 10 MΩ, the deflection = 100 divisions
Calculate insulation resistance of the cable. **(Roorkee University)**

Solution.
$$v_2 = v_1 . e^{(-t/CR)} \qquad \text{[Eqn. (6.37)]}$$

or,
$$CR = \frac{t}{\ln\left(\frac{v_1}{v_2}\right)} \qquad ...(i)$$

where,
v_1 = Voltage at $t = 0$,
v_2 = Voltage at time t seconds,
C = Capacitance of the cable, and
R_x = Resistance of the cable to be calculated.

Assuming 1 scale division deflection of galvanometer = 1V, then,

Case-1 :
v_1 = 200 V, when $t = 0$
v_2 = 125 V, when $t = 30$ s

Substituting these values in eqn. (*i*) we have :

or
$$CR_x = \frac{t}{\ln\left(\frac{v_1}{v_2}\right)} = \frac{30}{\ln\left(\frac{200}{125}\right)} = 63.83 \qquad ...(i)$$

Case-2 : Effective resistance in MΩ,

$$R_{eff.} = \frac{1}{\frac{1}{R_x}+\frac{1}{10}} \qquad \text{[From Eqn. (6.38)]}$$

$$= \frac{10R_x}{10+R_x}$$

and,
$$CR_{eff.} = \frac{30}{\ln\left(\frac{v_1}{v_2}\right)} = \frac{30}{\ln\left(\frac{200}{100}\right)} = 43.28 \qquad ...(ii)$$

Dividing (*ii*) by (*i*), we get :

$$\frac{R_{eff.}}{R_x} = \frac{43.28}{63.83} = 0.678$$

or, $$\frac{10R_x}{(10 + R_x)\,R_x} = 0.678$$

or, $$10 = (10 + R_x) \times 0.678$$

or, $$R_x = \frac{10}{0.678} - 10 = \mathbf{4.75\ M\Omega}\ \textbf{(Ans.)}$$

6.2.4.3. Megohm bridge

- Fig. 6.12 shows the circuit of a completely self-contained Megohm bridge; it includes the following :
  - — Power supplies;
  - — Bridge members;
  - — Amplifiers;
  - — Indicating instrument.
- Sensitivity for balancing against high resistance is obtained by use of adjustable high voltage supplies of 500 V or 1000 V and the use of a sensitive null indicating arrangement such as a high gain amplifier with an electronic voltmeter or a CRO.

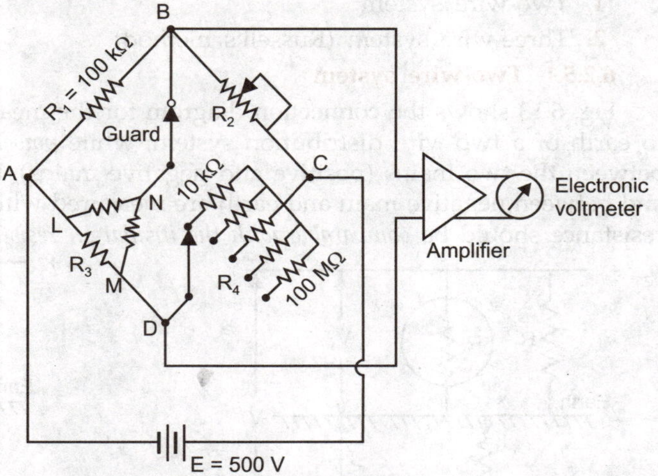

Fig. 6.12. Megohm bridge.

— The dial on R_2 is calibrated 1--10–100–1000 MΩ with main decade 1–10 occupying greater part of the dial space.

— Since unknown resistance $R_3 = \frac{R_1 R_4}{R_2}$, the arm R_2 is made tapered, so that the dial calibration is approximately logarithmic in main decade 1–10.

— Arm R_4 gives five multipliers 0.1, 1, 10, 100 and 1000.

— The junction of ratio arms R_1 and R_2 is brought on the main panel and is designated as 'Guard' terminal.

Example 6.9. *A test voltage is applied for several minutes between the conductor of a 100 metres long cable and earth. A reflecting galvanometer connected in series reads 280 scale divisions, the universal shunt being 2.5. With a standard resistance of 1 MΩ in the circuit the scale reading is 350, the value of the universal shunt being 1000.*

Calculate the insulation resistance of the cable. **(Bangalore University)**

Solution. Let, I_R = Current through the circuit with unknown resistance *R* connected, and

I_S = Curent through the circuit with standard resistance *S* connected.

Then, $I_R \propto$ galvanometer deflection × universal shunt multiplier

or, $I_R \propto 280 \times 2.5 \propto 700$

and, $I_S \propto 350 \times 1000 \propto 3{,}50{,}000$

Since current is inversely proportional to the resistance of the circuit, the circuit voltage remaining the same,

$$R = \frac{I_S}{I_R} \times S = \frac{3,50,000}{700} \times 1\ \text{M}\Omega = \mathbf{500\ M\Omega} \quad \textbf{(Ans.)}$$

6.2.5. Measurement of Insulation Resistance When Power is On

Following two cases will be discussed :

1. Two-wire system.
2. Three-wire system (Russell's method).

6.2.5.1. Two-wire system

Fig. 6.13 shows the connection diagram for the measurement of insulation resistance to earth of a two-wire distribution system, while *power is on.* In this case, the voltages between the two mains (positive and negative mains), between positive main and earth and between negative main and earth are measured with a *high resistance voltmeter* whose resistance should be *comparable with the insulation resistance under test.*

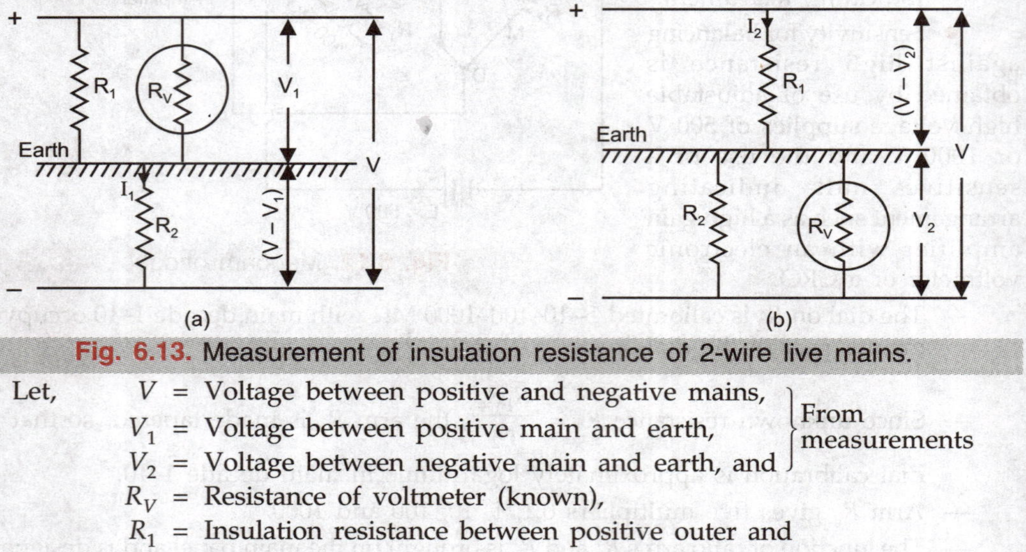

Fig. 6.13. Measurement of insulation resistance of 2-wire live mains.

Let, V = Voltage between positive and negative mains,
V_1 = Voltage between positive main and earth,
V_2 = Voltage between negative main and earth, and } From measurements
R_V = Resistance of voltmeter (known),
R_1 = Insulation resistance between positive outer and earth, and
R_2 = Insulation resistance between negative outer and earth.

Case-1 : Let I_1 be the current flowing from positive main to negative main in the first case. Then,

$$V_1 = I_1 \times \left(\frac{1}{\frac{1}{R_1} + \frac{1}{R_V}} \right) \quad \text{... Since } R_1 \text{ and } R_V \text{ are in parallel across positive main and earth}$$

or,
$$V_1 = I_1 \times \left(\frac{R_1 R_V}{R_1 + R_V} \right) \quad ...(i)$$

and,
$$V - V_1 = I_1 R_2 \quad ...(ii)$$

Dividing (*ii*) by (*i*), we get :

$$\frac{V - V_1}{V_1} = \frac{R_2 (R_1 + R_V)}{R_1 R_V}$$

or,

$$\frac{V}{V_1} - 1 = \frac{R_2 (R_1 + R_V)}{R_1 R_V}$$

or,

$$\frac{V}{V_1} = \frac{R_2 (R_1 + R_V)}{R_1 R_V} + 1$$

or,

$$\frac{V}{V_1} = \frac{R_1 R_V + R_2 (R_1 + R_V)}{R_1 R_V} = \frac{R_1 R_V + R_1 R_2 + R_2 R_V}{R_1 R_V}$$

or,

$$\frac{V}{V_1} = \frac{R_1 R_2 + R_V (R_1 + R_2)}{R_1 R_V} \quad ...(iii)$$

Case-2 : Let I_2 be the current flowing from positive main to negative main. Then,

$$V_2 = I_2 \times \left(\frac{R_2 R_V}{R_2 + R_V} \right) \quad ...(iv)$$

and

$$V - V_2 = I_2 R_1 \quad ...(v)$$

Dividing (*v*) by (*iv*), we have :

$$\frac{V - V_2}{V_2} = \frac{R_1 (R_2 + R_V)}{R_2 R_V}$$

or,

$$\frac{V}{V_2} - 1 = \frac{R_1 (R_2 + R_V)}{R_2 R_V}$$

or,

$$\frac{V}{V_2} = \frac{R_2 R_V + R_1 (R_2 + R_V)}{R_2 R_V}$$

or,

$$\frac{V}{V_2} = \frac{R_1 R_2 + R_V (R_1 + R_2)}{R_2 R_V} \quad ...(vi)$$

Dividing (*iii*) by (*vi*), we get :

$$\frac{V_2}{V_1} = \frac{R_2}{R_1}$$

or,

$$R_2 = R_1 \cdot \frac{V_2}{V_1}$$

Substituting this value of R_2 in (*iii*), we get :

$$\frac{V}{V_1} = \frac{R_1 \times R_1 \cdot \frac{V_2}{V_1} + R_V \left(R_1 + R_1 \cdot \frac{V_2}{V_1} \right)}{R_1 R_V}$$

or,

$$\frac{V}{V_1} = \frac{R_1 \cdot \frac{V_2}{V_1} (R_1 + R_V) + R_1 R_V}{R_1 R_V}$$

or, $$\frac{V}{V_1} = \left(\frac{R_1 + R_V}{R_V}\right)\frac{V_2}{V_1} + 1$$

or, $$R_V V = R_1 V_2 + R_V V_2 + R_V V_1$$

or, $$R_1 V_2 = R_V (V - V_1 - V_2)$$

or $$\boldsymbol{R_1} = R_V \left(\frac{V - V_1 - V_2}{V_2}\right) \quad ...(6.39)$$

Similarly, $$\boldsymbol{R_2} = R_V \left(\frac{V - V_1 - V_2}{V_1}\right) \quad ...(6.40)$$

- The *combined insulation resistance* of the system can be determined by assuming that the two mains are bunched together so that the resultant vlaue is the combined resistance of R_1 and R_2 in *parallel*. Then,

$$\textit{Insulation resistance of the system} = \frac{R_1 R_2}{R_1 + R_2}$$

6.2.5.2. Three-wire system (Russell's method)

Fig. 6.14 shows the connection diagrams for measurement of insulation resistance of three-wire live mains.

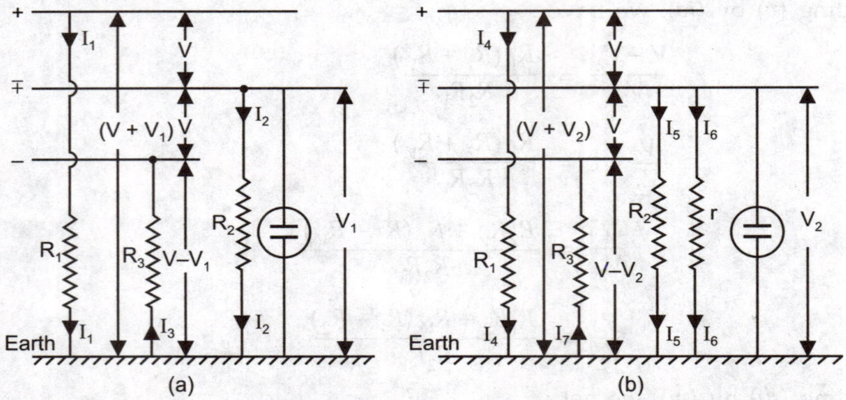

Fig. 6.14. Measurement of insulation resistance of 3-wire line mains.

— In this case an *electrostatic voltmeter* is used and connection of middle wire to earth is temporarily removed.

— Firstly, the voltmeter is connected between the middle wire and earth [See Fig. 6.14(*a*)] and reading of voltmeter is noted; let it be V_1. Then, a known resistance r of a fairly high value (about 2 kΩ) is connected between the middle wire and earth and the new reading of the voltmeter is noted; let it be V_2. Since the resistance of an electrostatic voltmeter can be taken as infinity, therefore, no current will flow through it.

Using Kirchhoff's laws, we get :

In case of *first test* :

$$I_1 + I_2 = I_3$$

or, $$\frac{V + V_1}{R_1} + \frac{V_1}{R_2} = \frac{V - V_1}{R_3} \quad ...(i)$$

In case of *second test* : Refer to Fig. 6.14(*b*).

$$I_4 + I_5 + I_6 = I_7$$

or, $$\frac{V+V_2}{R_1}+\frac{V_2}{R_2}+\frac{V_2}{r} = \frac{V-V_2}{R_3} \quad ...(ii)$$

Subtracting (*ii*) from (*i*), we get :

$$\left(\frac{V+V_1}{R_1}-\frac{V+V_2}{R_1}\right)+\left(\frac{V_1}{R_2}-\frac{V_2}{R_2}\right)-\frac{V_2}{r} = \left(\frac{V-V_1}{R_3}-\frac{V-V_2}{R_3}\right)$$

$$\frac{V_1-V_2}{R_1}+\frac{V_1-V_2}{R_2}-\frac{V_2}{r} = \frac{V_2-V_1}{R_3}$$

or, $$(V_1-V_2)\left(\frac{1}{R_1}+\frac{1}{R_2}+\frac{1}{R_3}\right) = \frac{V_2}{r} \quad ...(iii)$$

The combined insulation resistance of the system is given by the resultant of the resistances R_1, R_2 and R_3 connected in parallel.

i.e. $$R = \frac{1}{\frac{1}{R_1}+\frac{1}{R_2}+\frac{1}{R_3}}$$

Hence, the expression (*ii*) reduces to,

$$\frac{V_1-V_2}{R} = \frac{V_2}{r}$$

or, $$(V_1-V_2)r = V_2R$$

or, $$\frac{V_1-V_2}{V_2} = \frac{R}{r}$$

or, $$R = r\left(\frac{V_1}{V_2}-1\right) \quad ...(6.41)$$

6.2.6. Measurement of Earth Resistance

In any electrical system, the provision of an earth electrode is made due to the *reasons* mentioned below :

1. It is of significant importance that all the parts of electrical equipment (such as switches, circuit breakers, machines, tanks of transformers etc.) must be at earth potential and therefore they *must be connected to an earthed electrode.* When all these parts are connected to an earth electrode, a continuous low resistance path is available for leakage currents to flow to earth; these currents operate the protective devices and thus the faulty circuit is isolated in case a fault occurs.

Thus, *the purpose of earthing is to protect the various parts of the installation, as well as persons working against damage in case the insulation of a system fails at any point.*

2. The purpose of an earth electrode is to ensure that in the event of overvoltage on the system due to lightning discharges or other system faults, those parts of equipment which are normally *dead* as far as voltages are concerned do not attain dangerously high potentials.

3. In case of 3-phase circuit the neutral of the system is earthed so that the potential of the circuit with respect to earth is stabilised.

- The resistance of any earthing system depends on the following *main factors* :
  (*i*) Shape and material of earth electrode or electrodes used.
  (*ii*) Depth in the soil at which the electrodes are burried.
  (*iii*) Specific resistance of soil surrounding and in the neighbourhood of electrodes.

Megger earth tester :

The earth resistance can be measured with the help of *Earth tester* or a *Megger earth tester*.

The *Megger earth tester* (Fig. 6.15) is essentially a *direct reading ohmmeter and a hand driven generator which supplies the testing current.*

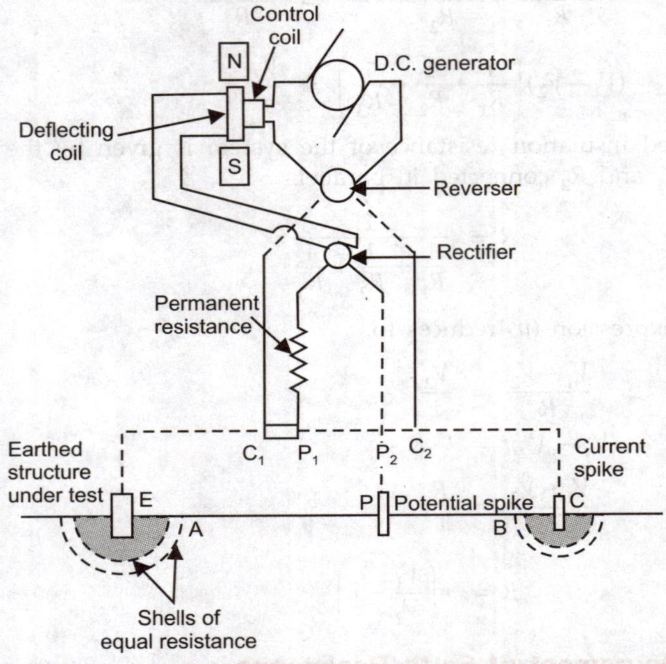

Fig. 6.15. Megger tester circuit containing direct current shown by full lines; circuit containing A.C. current shown by dotted lines.

Construction :

— The ohmmeter consists of two coils (current coil and potential coil) mounted at a fixed angle to each other on a common axle. The current coil carries current proportional to the current flowing in the test circuit, while the potential coil carries current proportional to the potential across the resistance under test. Thus the potential coil acts as a voltmeter while current coil acts as an ammeter. Since the *deflection* of the needle is proportional to the ratio of the current in the two coils, it gives *resistance directly*.

— The hand-operated generator produces the direct current, but to eliminate the effect of electrolytic e.m.f., it is necessary to pass alternating current through the soil, so to change the D.C. into an alternating supply a rotary current reverser is mounted on the same shaft of the generator. The alternating current in the soil will produce an alternating drop in the soil but the potential to be applied across the moving coil must be direct because the ohmmeter in a moving coil instrument works on D.C. alone, so for changing the alternating drop into the direct drop a synchronous rotary rectifier is also attached as shown in Fig. 6.15. Although the instrument is immune to stray currents, yet while taking readings sometimes it will be observed that the needle vibrates; this happens only at the instant when the stray alternating current has the same frequency as the frequency generated. For such occasions either *increase the speed of the generator a little or lower its speed.*

Working :

- For *measurement of earth resistance* two spikes acting as current and potential electrodes are driven into the ground at a suitable distance (C_2 at a distance of about 25 m from earth electrode under test). P_1 and C_1 terminals of the meter are shorted and are connected to the earth electrode under test.
- First spike P_2 is fixed between the spike C_2 and the earth electrode and the resistance is determined; then this spike taken 3 to 4 m near to the earth electrode and then for the same distance near to electrode C_2 and the two readings are again noted. The average of these three readings is the required earth resistance.

Specifications required for earthing as per ISI :

Following are *recommended specifications as per I.S.I.* for providing good earthing :

1. The earthing electrode should be situated at a place at least $1\frac{1}{2}$ metres away from the building (outside) whose installation system is being earthed.
2. The earth wire should be of same material as that of earth electrode used.
3. The minimum sectional area of earth lead wire should not be less than 0.02 sq. inch (No. 8 S.W.G.) and not more than 0.1 sq. inch.
4. The size of earth conductor as a general rule should not be less than half of the section of the line conductor.
5. The earth wire should be taken through G.I. pipe of 12 mm dia. for at least 32 cm length above and below ground surface to the earth electrode to safeguard against mechanical wear and tear.
6. Loose earth and coal salt mixture should be filled around the earth electrode for effective earthing.
7. The earth wire connected to the earth electrode *shall not be* necessarily run along the whole wiring system. All the earth wire run along the sub-circuits should be terminated and hooked firmly at the main board and from where the main earth wires should be run to the earth electrode. The loop earth wires should be of 14 S.W.G. copper wire.
8. All the joints in the earth wire should be firmly done with nuts and bolts of the same material as of earth wire.

6.2.7. Location of Cable Faults

Out of several methods available for locating the faults in underground cables, the *two commonly used methods are* :

1. Murray loop test.
2. Varley loop test.
   - Both these tests use the principle of Wheatstone bridge for locating the faults.
   - These tests can be employed to locate the *earth fault* or *short circuit fault in* underground cables *provided that a sound cable runs along the faulty cable.*

6.2.7.1. Murray loop test

This is the *most common and accurate method* of locating earth or short-circuit faults in the underground cables.

(*i*) **Earth fault.** Fig. 6.16 shows the circuit diagram for locating the earth fault by Murray loop test. *AB* is the sound cable and *CD* is the faulty cable. *F* is the point at which earth fault occurs. The far end *D* of the faulty cable is joined to the far end *B* of the sound cable through a low resistance link. Two *variable* resistances *P* and *Q* are joined to ends *A* and *C* respectively and serve as the ratio arms of the Wheatstone bridge.

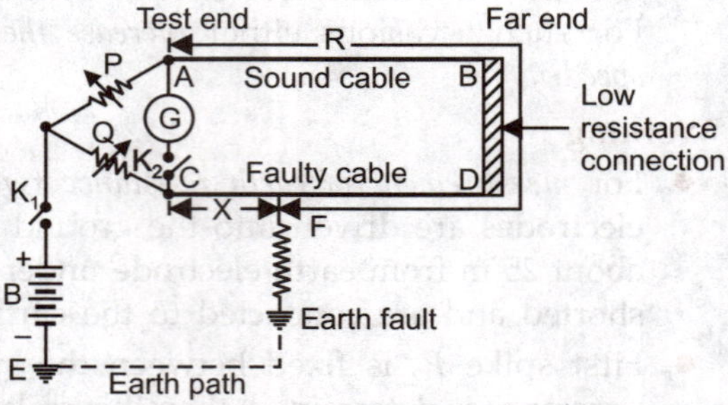

Fig. 6.16. Murray loop test for locating earth

Let, R = Resistance of the conductor loop up to the fault from the test end, and

X = Resistance of the other length of loop.

P, *Q*, *R* and *X* constitute the four arms of the Wheatstone bridge. The resistances P and Q are varied till the galvanometer (*G*) *indicates zero deflection.*

When the bridge is balanced, we have

$$\frac{P}{Q} = \frac{R}{X}, \quad \text{or,} \quad \frac{P}{Q} + 1 = \frac{R}{X} + 1,$$

or, $$\frac{P+Q}{Q} = \frac{R+X}{X}$$

If *r* is the resistance of each cable, then

$$R + X = 2r$$

$$\frac{P+Q}{Q} = \frac{2r}{X}, \quad \text{or,} \quad X = \frac{Q}{P+Q} \times 2r \qquad ...(6.42)$$

If *l* is the length of each cable in metres, then resistance per metre length of the cable $= \frac{r}{l}$

∴ Distance of fault point (*F*) from the test end,

$$d = \frac{X}{r/l} = \frac{Q}{P+Q} \times 2r \times \frac{l}{r} = \frac{Q}{P+Q} \times 2l$$

or, $$d = \frac{Q}{P+Q} \times \text{(loop length) metres} \qquad ...(6.43)$$

It may be noted that resistance of the fault is in the battery circuit and not in bridge circuit. Therefore, *fault resistance does not affect the balancing of the bridge. However, in case the fault resistance is high, the sensitivity of the bridge is reduced.*

(*ii*) **Short-circuit fault.** Fig. 6.17 shows the circuit diagram for locating the short-circuit fault by Murray loop test. Again, *P*, *Q*, *R* and *X* constitute the arms of the bridge. The fault resistance is in battery circuit and not in the bridge circuit. The bridge is balanced by varying the resistances *P* and *Q*. When the bridge is balanced, we have :

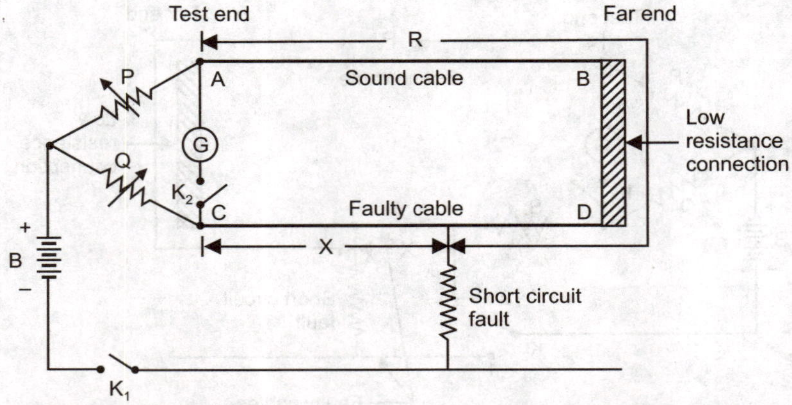

Fig. 6.17. Murray loop test for locating short-circuit fault.

$$\frac{P}{Q} = \frac{R}{X}, \quad \text{or,} \quad \frac{P+Q}{Q} = \frac{R+X}{X} = \frac{2r}{X}$$

$$X = \frac{Q}{P+Q} \times 2r$$

or, $$X = \frac{Q}{P+Q} \times \text{(loop length) metres} \quad \text{...(6.44)}$$

Thus the position of the fault is located.

6.2.7.2. Varley loop test

This test which is also used to locate earth fault or short-circuit fault in the underground cables uses Wheatstone bridge principle.

It *differs* from Murray loop test in that here the *ratio arms P and Q have fixed resistances.* Balance is obtained by adjusting the *variable resistance* S connected to the test end of the faulty cable. The circuit diagrams for locating earth fault and short-circuit fault by Varley loop test are shown in Figs. 6.18 and 6.19 respectively.

- The key K_2 is first thrown to position 1. The variable resistance S is varied till the bridge is balanced for resistance value of S_1. Then,

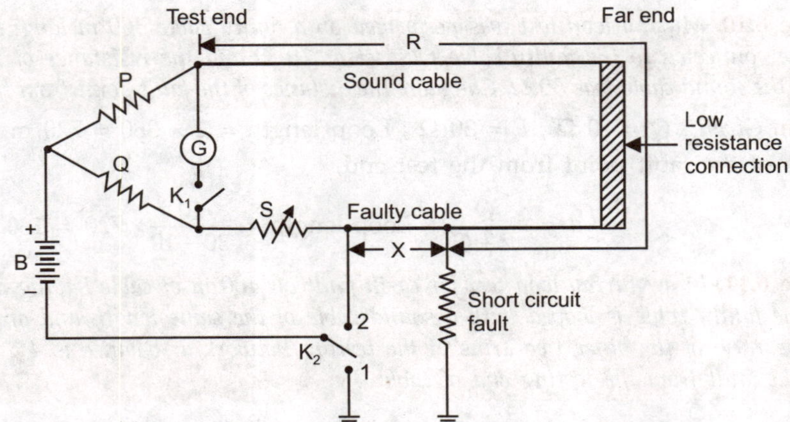

Fig. 6.18. Varley loop test for locating earth fault.

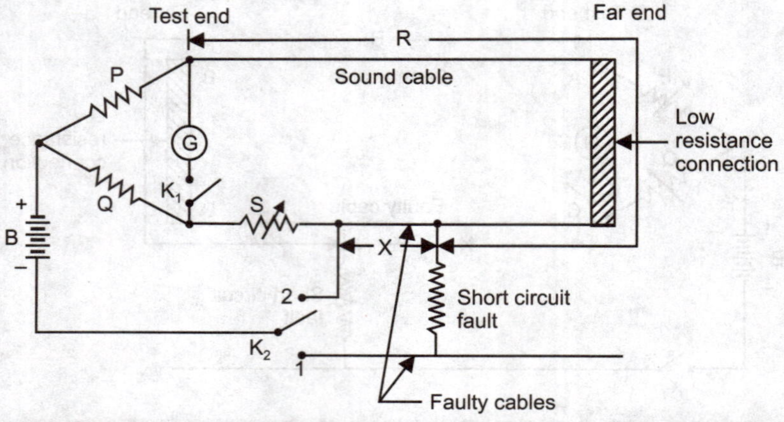

Fig. 6.19. Varley loop test for locating short-circuit fault.

$$\frac{P}{Q} = \frac{R}{X + S_1}, \quad \text{or,} \quad \frac{P+Q}{Q} = \frac{R + X + S_1}{X + S_1}$$

or,
$$X = \frac{Q(R+X) - PS_1}{P+Q} \qquad \text{...(6.45)}$$

- The key is now thrown to position 2 and bridge is balanced with the new value of resistance S_2. Then,

$$\frac{P}{Q} = \frac{R+X}{S_2} \quad \text{or,} \quad Q\,(R + X) = PS_2$$

Since the values of P, Q, S_1 and S_2 are known, the value of X can be determined.

Loop resistance $= R + X = \frac{P}{Q} S_2$

If r is the resistance of the cable per metre length, then,
Distance of fault from the test end,

$$d = \frac{X}{r} \text{ metres.} \qquad \text{...(6.46)}$$

Example 6.10. *Murray loop test was performed on a faulty cable 360 m long. At balance, the resistance connected to the faulty cable was set at 10 Ω and the resistance of the resistor connected to the sound cable was 30 Ω. Calculate the distance of the fault point from the test end.*

Solution. *Given* : $Q = 10\ \Omega$; $P = 30\ \Omega$; Loop length = 2 × 360 = 720 m

Distance of the fault point from the test end,

$$d = \frac{Q}{P+Q} \times \text{Loop length} = \frac{10}{30+10} \times 720 = \mathbf{180\ m} \quad \textbf{(Ans.)}$$

Example 6.11. *In a Murray loop test for earth fault on 400 m of cable having a resistance 1.5 Ω/km, the faulty cable is looped with a sound cable of the same length and area of cross-section. If the ratio of the other two arms of the testing network at balance is 4 : 1, find the distance of the fault from the testing end of cable.*

Solution. *Given* : Loop length = 2 × 400 = 800 m; $\frac{P}{Q} = 4$

$$\frac{P}{Q} = \frac{4}{1} \quad \text{or} \quad \frac{P+Q}{Q} = \frac{4+1}{1} = 5$$

Distance of fault from test end,

$$d = \frac{Q}{P+Q} \times \text{loop length} \qquad \text{[Eqn. (6.43)]}$$

$$= \frac{1}{5} \times 800 = \mathbf{160\ m.} \quad \textbf{(Ans.)}$$

Example 6.12. *In Murray loop test for a fault to earth on a 400 m length of cable having a resistance of 1.5 Ω/km, the faulty cable is looped with a sound cable of the same length but having a resistance of 3.5 Ω/km. The resistances of the other two arms of the testing network at balance are in the ratio 3 : 1. Calculate the distance of the fault from the testing end of the cable.*

Solution.

$$\frac{P}{Q} = \frac{3}{1}, \quad \text{or,} \quad \frac{P+Q}{Q} = \frac{3+1}{1} = 4$$

$$\text{Loop resistance} = \frac{1.5}{1000} \times 400 + \frac{3.5}{1000} \times 400 = 2\ \Omega$$

Resistance of the faulty cable from the test end up to fault point,

$$X = \frac{Q}{P+Q} \times \text{loop resistance} = \frac{1}{4} \times 2 = 0.5\ \Omega$$

Distance of fault point from the testing end,

$$d = \frac{X}{15/1000} = \frac{0.5}{15/1000} = \mathbf{333.33\ m} \quad \textbf{(Ans.)}$$

Example 6.13. *Varley loop test is conducted to locate an earth fault on a 25 km long cable. The resistance per km of single conductor is 15 Ω. The loop is completed with a similar sound conductor. At balance, the variable resistance connected to the faulty conductor is 150 Ω. The fixed resistors have equal values. Calculate the distance of the fault from the test end.*

Solution. Resistance of faulty cable from test end to fault point,

$$X = \frac{Q(R+X) - PS}{P+Q} \qquad \text{[Eqn. (6.45))]}$$

Here, $P = Q\ ;\ S = 150\ \Omega\ ;\ R + X = 25\ (15 + 15) = 750\ \Omega$

$$\therefore \quad X = \frac{Q \times 750 - Q \times 150}{Q+Q} = 300\ \Omega$$

The resistance per km = 15 Ω

∴ Distance of fault from the test end,

$$d = \frac{X}{15} = \frac{300}{15} = \mathbf{20\ km.} \quad \textbf{(Ans.)}$$

6.2.7.3. Location of faults in high-voltage cables

As the resistance presented by high-voltage cable under fault condition is very high, therefore, such faults cannot be located with accuracy by methods already described earlier.

In high-voltage cables breakdown occurs at the time of application of the high voltage especially when the oil is in the cable and cable joints breaks down and therefore presents a low resistance. However, when the applied voltage is withdrawn the oil *recovers its insulating property and the cable again presents high insulation resistance. The fault resistance may be* ***reduced*** *by a continued application of the high voltage so as to carbonise or burn out the insulation at the point of the fault. After the insulation resistance is so reduced, ordinary loop-tests may then be applied to locate the fault position.*

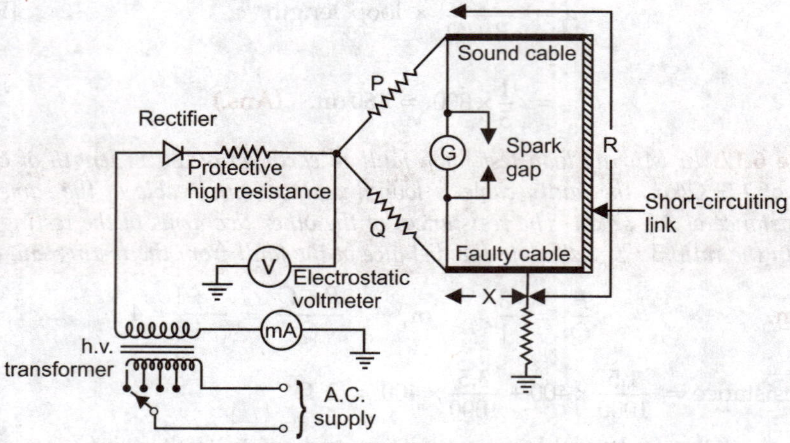

Fig. 6.20. Location of faults in high voltage cables.

- Fig. 6.20 shows a more satisfactory method for location of faults in high voltages. This method involves performance of Murray loop test at high voltage obtained from either half-wave or full-wave rectifier :
  - — The rectified supply at about 60 kV is obtained from a high-vacuum diode; a higher voltage, if desired, may be obtained with two valves.
  - — The rectifier is fed from a step-up h.v. transformer.
  - — A current limiting high resistance and a milliammeter are included in the supply circuit for the limitation and measurement of the fault current respectively.
  - — An electrostatic voltmeter is employed to measure the voltage at the junction point of the ratio arms. A loop is made connecting together the faulty and sound cables with the help of a short-circuiting link. In order to vary the ratio arms (P, Q) a highly insulated slide-wire, having a sliding contact which can be operated by a long insulating handle is used.
  - — The spark gap is used to protect the galvanometer.

Test procedure : The test is conducted as follows :

— The circuit is energised and the tap at the primary of the h.v. transformer is adjusted to give a low fault current of the order of 5 mA at the time of cable breakdown.

— The bridge is balanced by varying P/Q ratio.

— The tap of primary of h.v. transformer is then changed to increase the fault current to about 50 mA.

— The bridge is *rebalanced*.

It may be mentioned here that at the time of breakdown, a very large voltage occurs across the protective high resistance and therefore the voltage across the cable is low. However, if the fault clears, the voltage across the cable would rise suddenly and it is on this account that highly insulated slide-wire is used for operator's safety.

— The position of the fault can be obtained using equations given for Murray loop test.

6.3. MEASUREMENT OF INDUCTANCE AND CAPACITANCE

A.C. BRIDGES

6.3.1. A.C. Bridge—General form

An A.C. bridge in its basic form consists of the following elements/components :

1. *Four bridge arms.*

2. *A source of excitation.* The power source supplies an A.C. voltage to the bridge at the desired frequency.

- For power measurements at low frequencies, the power line may serve as the source of excitation.
- At higher frequencies, the electronic oscillators are universally used as the source of supply due to the following *reasons* :

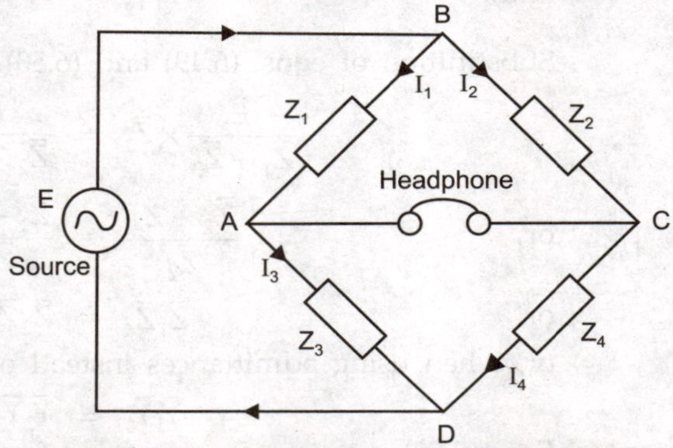

Fig. 6.21. General form of the A.C. bridge

(*i*) The output waveform is very close to sine wave.

(*ii*) The output frequency is very stable.

(*iii*) The output frequency can be determined with accuracy and is also easily adjustable.

(*iv*) The output power is adequate to drive the logic circuits.

— A typical oscillator has a frequency range of 40 Hz to 125 kHz with power output of 7W.

3. *Null detector.* It must respond to A.C. unbalance currents. The null/balance detectors commonly used for A.C. bridges are :

(*i*) *Headphones.* These are used as detectors at the frequencies of 250 Hz to 3 to 4 kHz.

(*ii*) *Tunable amplifier circuits* :

— A tuned detector is the most sensitive detector while working with single frequency.

— Tunable amplifier detectors are used for frequency range of 10 Hz to 100 Hz.

(*iii*) *Vibration galvanometers.* These are useful for low audio frequency range from 5 Hz to 1000 kHz but are commonly used below 200 Hz.

General equation for bridge balance :

Fig. 6.21 shows the general form of an A.C. bridge.

- The four bridge arms $\bar{Z}_1, \bar{Z}_2, \bar{Z}_3$ and $\bar{Z}_4$ (Bar *i.e.* – indicates complex notation) are indicated as unspecified impedances and detector is represented by headphones (*cheapest and effective form*).
- The balance condition is reached when the detector response is zero, or indicates a null. Balance adjustment to obtain a null response is made by varying one or more of the bridge arms.
- The *condition for bridge balance* requires that potential difference from *A* to *C* in Fig. 6.21 be zero. This will be the case when the voltage drop from *B* to *A* equals the voltage drop from *B* to *C* in both *magnitude* and *phase*. In complex notation we can write :

$$\bar{E}_{BA} = \bar{E}_{BC} \quad ...(6.47)$$

or,
$$\bar{I}_1\bar{Z}_1 = \bar{I}_2\,\bar{Z}_2 \quad ...(6.48)$$

Also at balance,
$$\bar{I}_1 = \bar{I}_3 = \frac{\bar{E}}{\bar{Z}_1 + \bar{Z}_3} \quad ...(6.49)$$

and,
$$\bar{I}_2 = \bar{I}_4 = \frac{\bar{E}}{\bar{Z}_2 + \bar{Z}_4} \quad ...(6.50)$$

Substitution of eqns. (6.49) and (6.50) into eqn. (6.48) yields :

$$\frac{\bar{E}}{\bar{Z}_1 + \bar{Z}_3} \times \bar{Z}_1 = \frac{\bar{E}}{\bar{Z}_2 + \bar{Z}_4} \times \bar{Z}_2$$

or,
$$\frac{\bar{Z}_2 + \bar{Z}_4}{\bar{Z}_2} = \frac{\bar{Z}_1 + \bar{Z}_3}{\bar{Z}_1} \quad \text{or} \quad 1 + \frac{\bar{Z}_4}{\bar{Z}_2} = 1 + \frac{\bar{Z}_3}{\bar{Z}_1}$$

or,
$$\bar{Z}_1\bar{Z}_4 = \bar{Z}_2\bar{Z}_3 \quad ...(6.51)$$

or, when using admittances instead of impedance, we have

$$\bar{Y}_1\bar{Y}_4 = \bar{Y}_2\bar{Y}_3 \quad ...(6.52)$$

Eqn. (6.51) is the most convenient form in most cases and is the *general equation for balance* of the A.C. bridge. Eqn. (6.52) is useful when *dealing wih parallel components in bridge arms.*

— Eqn. (6.51) states the product of the impedances of one pair of opposite arms must be equal to the product of impedances of the other pair of opposite arms, with the impedances expressed in complex notation. If the impedance is written in the form $\bar{Z} = Z\angle\theta$, where Z represents the magnitude and θ the phase angle of the complex impedance, Eqn. (6.52) can be written in the form

$$(Z_1\angle\theta_1)(Z_4\angle\theta_4) = (Z_2\angle\theta_2)(Z_3\angle\theta_3) \quad ...(6.53)$$

or,
$$Z_1Z_4\angle(\theta_1 + \theta_4) = Z_2Z_3\angle(\theta_2 + \theta_3) \quad ...(6.54)$$

— Eqn. (6.54) shows that following *two conditions* must be met *simultaneously* when balancing an A.C. bridge :

(*i*) The magnitudes of the impedances should satisfy the relationship :

$$Z_1Z_4 = Z_2Z_3 \quad ...(6.55)$$

(*ii*) The phase angles of the impedances should satisfy the relationship :

$$\angle\theta_1 + \angle\theta_4 = \angle\theta_2 + \angle\theta_3 \quad ...(6.56)$$

Example 6.14. *The following data relate to the basic A.C. bridge of Fig. 6.21 :*

$\bar{Z}_1 = 50\angle 80°$ *(inductive impedance);*

$\bar{Z}_2 = 125\Omega$ *(pure resistance) ;*

$\bar{Z}_3 = 200\angle 30°$ *(inductive impedance);*

$\bar{Z}_4 =$ *Unknown*

Determine the constants of the unknown arm :

Solution. *Given* : $\bar{Z}_1 = 50\angle 80°$; $\bar{Z}_2 = 125\ \Omega$, $\bar{Z}_3 = 200\angle 30°$

$\bar{Z}_4$:

We know that :
$$Z_1Z_4 = Z_2Z_3 \quad ...(6.55)$$

or,
$$Z_4 = \frac{Z_2Z_3}{Z_1} = \frac{125 \times 200}{50} = 500\Omega$$

Also, $\angle\theta_1 + \angle\theta_4 = \angle\theta_2 + \angle\theta_3$ ...(6.56)

or, $80° + \theta_4 = 0° + 30°$

or, $\theta_4 = 0° + 30° - 80° = -50°$

Hence, $\bar{Z}_4 = \mathbf{500\angle{-50°}\ \Omega}$,

indicating that we are dealing with a *capacitive* element, possibly consisting of a series combination of a resistor and a capacitor. **(Ans.)**

Example 6.15. *The following data relate to the balanced A.C. bridge shown in Fig. 6.22.*

Arm AB, $R_1 = 225\ \Omega$; Arm BC, $R_2 = 150\ \Omega$ in series with $C_2 = 0.53\ \mu F$; arm CD, unknown; arm DA, $R_3 = 100\ \Omega$ in series with $L = 7.95$ mH. The oscillator frequency = 1 kHz.

Calculate the constants of arm CD.

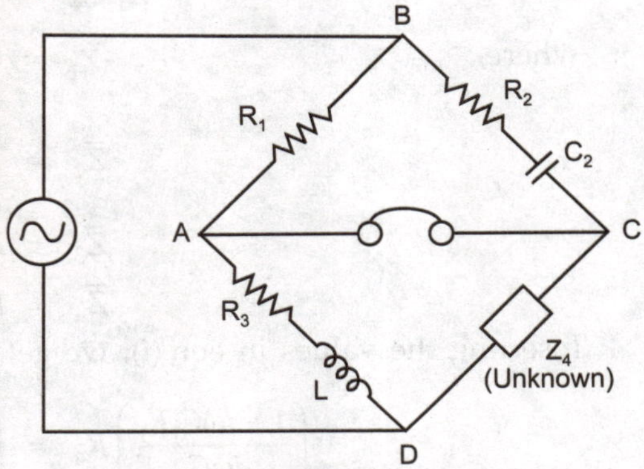

Fig. 6.22.

Solution. *Given* : $f = 1$ kHz, $R_1 = 225\ \Omega$; $R_2 = 150\ \Omega$, $C = 0.53\ \mu F$ or $\omega C_2 = 2\pi f \times 0.53 \times 10^{-6} = 2\pi \times 1000 \times 0.53 \times 10^{-6}\ \Omega = 3.33 \times 10^{-3}\ \Omega$; $R_3 = 100\ \Omega$, $L = 7.95$ mH or $\omega L = 2\pi f L = 2\pi \times 1000 \times 7.95 \times 10^{-3} \simeq 50\Omega$.

Arm *CD*, $\bar{Z}_4$:

Now, $\bar{Z}_1 = R_1 = 225\ \Omega$

$$\bar{Z}_2 = R_2 - \frac{j}{\omega C_2} = 150 - \frac{j}{3.33 \times 10^{-3}} = (150 - j\,300)\ \Omega$$

$$\bar{Z}_3 = R_3 + j\omega L = (100 + j\,50)\ \Omega$$

$\bar{Z}_4$ = Unknown

The general equation for bridge balance states that :

$$\bar{Z}_1\,\bar{Z}_4 = \bar{Z}_2\,\bar{Z}_3$$

or,

$$\bar{Z}_4 = \frac{\bar{Z}_2\,\bar{Z}_3}{\bar{Z}_1} = \frac{(150 - j300)(100 + j\,50)}{225}$$

$$= \frac{(15000 + j\,7500 - j\,30000 + 15000)}{225}$$

$$= \frac{30000 - j\,22500}{225} = \mathbf{133.33 - j\,100}$$

i.e. $R_4 = 133.33\ \Omega$ and $X_{C_4} = \dfrac{1}{\omega C_4} = \dfrac{1}{2\pi f C_4} = 100$

or, $$C_4 = \frac{1}{2\pi \times 1000 \times 100} = 1.59 \times 10^{-6}\ F \text{ or } 1.59\ \mu F$$

Hence, $\mathbf{R_4 = 133.33\ \Omega}$, $\mathbf{C_4 = 1.59\ \mu F}$ **(Ans.)**

Example 6.16. *In a four-arm bridge network (Fig. 6.23), the arm AB consists of an imperfect condenser, BC and CD are non-reactive resistances of 1000 Ω each and DA is a standard capacitor of 0.0115 μF capacitance in series with a resistance of 140 Ω. If the bridge is balanced* *for frequency $\omega = 7500$ radians/sec., find the shunt-less resistance and capacitance of the imperfect condenser.* **(UPTU, 2004-05)**

Solution. Refer to Fig. 6.23 : *Given* : $\omega = 7500$ rad/s; $R_2 = 140\ \Omega$; $R_3 = R_4 = 1000\ \Omega$; $C_2 = 0.0115\ \mu$ F.

We know that for balance,

$$\bar{Z}_1 \bar{Z}_4 = \bar{Z}_2 \bar{Z}_3 \quad ...(i)$$

where,

$$\bar{Z}_1 = R_1 + \frac{1}{j\omega C_1} = \frac{1 + j\omega C_1 R_1}{j\omega C_1}$$

$$\bar{Z}_2 = R_2 + \frac{1}{j\omega C_2} = \frac{1 + j\omega C_2 R_2}{j\omega C_2}$$

$$\bar{Z}_3 = R_3$$

$$\bar{Z}_4 = R_4$$

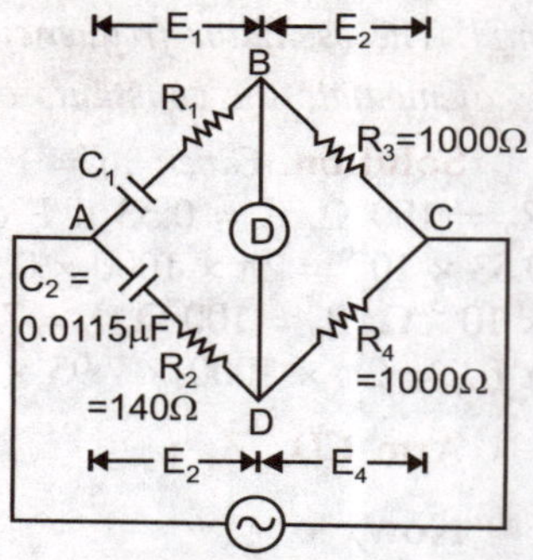

Fig. 6.23.

Inserting the values in eqn.(*i*), we get

$$\left(\frac{1 + j\omega C_1 R_1}{j\omega C_1}\right) R_4 = \left(\frac{1 + j\omega C_2 R_2}{j\omega C_2}\right) R_3$$

or,

$$j\omega C_2 R_4 = \omega^2 R_1 C_1 C_2 R_4$$
$$= j\omega C_1 R_3 - \omega^2 C_1 C_2 R_2 R_3$$

Comparing the *real part*, we have :

$$\omega^2 R_1 C_1 C_2 R_4 = \omega^2 C_1 C_2 R_2 R_3$$

$$R_1 = \frac{R_2 R_3}{R_4} \quad ...(ii)$$

or,

$$\mathbf{R_1} = \frac{140 \times 1000}{1000} = \mathbf{140\ \Omega}\ \textbf{(Ans.)}$$

Comparing the *imaginary part*, we get :

$$\omega C_2 R_4 = \omega C_1 R_3$$

or,

$$C_2 R_4 = C_1 R_3 \quad ...(iii)$$

or,

$$\mathbf{C_1} = \frac{C_2 R_4}{R_3} = \frac{(0.0115 \times 10^{-6}) \times 1000}{1000}$$

$$= 0.0115 \times 10^{-6}\ \text{F} \quad \text{or} \quad \mathbf{0.0115\ \mu\ F}\ \textbf{(Ans.)}$$

MEASUREMENT OF INDUCTANCE

6.3.2. Maxwell Bridge (*for measurement of* "inductance")

Maxwell bridge, shown in Fig. 6.24 *measures an unknown inductance in terms of a known capacitor* :

— The use of standard arm offers the advantage of compactness and easy shielding.

— The capacitor is almost a loss-less component.

— One arm has a resistance R_1 in parallel with C_1, and hence it is easier to write the balance equation using the admittance of arm 1 instead of the impedance.

The general equation for bridge balance is,

$$\bar{Z}_1 \bar{Z}_x = \bar{Z}_2 \bar{Z}_3$$

or,
$$\bar{Z}_x = \frac{\bar{Z}_2 \bar{Z}_3}{\bar{Z}_1} = \bar{Z}_2 \bar{Z}_3 \bar{Y}_1 \quad \text{...(6.57)}$$

where, $\bar{Z}_1 = R_1$

in parallel with C_1 *i.e.* $\bar{Y}_1 = \dfrac{1}{\bar{Z}_1}$

$$\bar{Y}_1 = \frac{1}{R_1} + j\omega C_1$$

$$\bar{Z}_2 = R_2$$

$$\bar{Z}_3 = R_3$$

$\bar{Z}_x = R_x$ in series with L_x $(= R_x + j\omega L_x)$

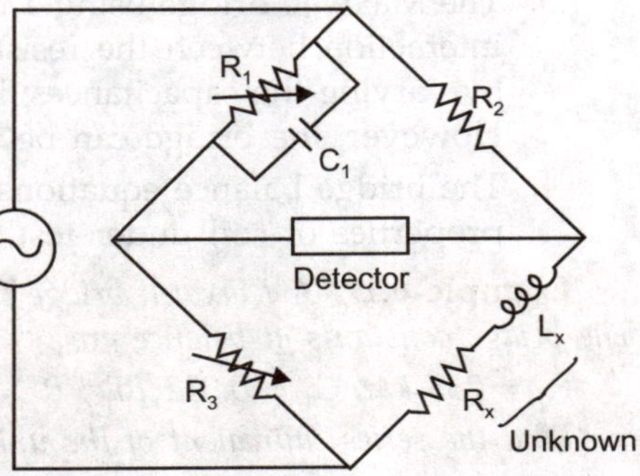

Fig. 6.24. Maxwell bridge for inductance measurements.

Substituting these values in eqn. (6.57), we get

$$R_x + j\omega L_x = R_2 R_3 \left(\frac{1}{R_1} + j\omega C_1\right)$$

or,
$$R_x + j\omega L_x = \frac{R_2 R_3}{R_1} + j\omega C_1 R_2 R_3$$

Equating *real terms* and *imaginary terms*, we have

$$R_x = \frac{R_2 R_3}{R_1}; \quad \text{...(6.58)}$$

$$L_x = C_1 R_2 R_3$$

To obtain the bridge balance, first R_3 is adjusted for inductive balance and R_1 is adjusted for resistive balance.

The *"quality factor"* of the coil is given by,

$$Q = \frac{\omega L_x}{R_x} = \frac{\omega . C_1 R_2 R_3}{(R_2 R_3 / R_1)} = \frac{\omega . C . R_2 R_3 R_1}{R_2 R_3}$$

i.e.
$$Q = \omega C_1 R_1 \quad \text{...(6.59)}$$

- Maxwell bridge is limited to the measurement of *low Q values (1-10).*
- The bridge is *particularly suited for inductance measurements,* since comparison with a capacitor is more ideal than with another inductance. Commercial bridges measure from *1-1000 H, with ± 2 % error* (If Q is very large, R_1 becomes excessively large and it is impractical to obtain a satisfactory variable standard resistance in the range of values required).

Advantages :

1. The balance equation is independent of losses associated with inductance.
2. The measurement is independent of the excitation *frequency.*
3. The scale of the resistance can be calibrated to read inductance directly.
4. A wide range of inductance at power and audio frequencies can be measured.

Disadvantages :

1. High Q values cannot be measured.
2. It cannot be used for measurement of very low Q values, because of balance coverage problem.
3. The Maxwell bridge using a fixed capacitor has the disadvantage that there is an interaction between the resistance and reactance balances. This can be avoided by varying the capacitances, instead of R_2 and R_3, to obtain a reactance balance. However, the bridge can be made to read directly in Q.
4. The bridge balance equations are independent of frequency. But practically, the properties of coil under test vary with frequency which can cause error.

Example 6.17. *A Maxwell bridge (Fig. 6.24) is used to measure an inductive impedance. The bridge constants at balance are :*

$R_1 = 235\ k\Omega,\ C_1 = 0.012\ \mu F\ ;\ R_2 = 2.5\ k\Omega;\ R_3 = 50\ k\Omega$

Find the series equivalent of the unknown impedance.

Solution. Refer to Fig. 6.26.

R_x : ; L_x :

We know that :
$$R_x = \frac{R_2 R_3}{R_1} = \frac{2.5 \times 50}{235}$$
$$= \mathbf{0.53\ k\ \Omega\ (Ans.)}$$
$$L_x = C_1 R_2 R_3 = 0.012 \times 10^{-6} \times (2.5 \times 10^3) \times (50 \times 10^3)$$
$$= \mathbf{1.5\ H\ (Ans.)}$$

The *equivalent series circuit* is shown in Fig. 6.25.

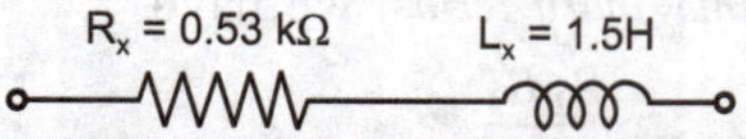

Fig. 6.25.

6.3.3. Hay Bridge (*For measurement of "inductance"*)

The Hay bridge shown in Fig. 6.26 differs from the Maxwell bridge by having resistor R_1 in series with standard capacitor C_1 instead of in parallel. It is apparent that for large phase angles, R_1 should have a very low value. This bridge is, thus, *more convenient for measuring high – Q coils.*

The general equation for bridge balance is,

$$\bar{Z}_1 \bar{Z}_x = \bar{Z}_2 \bar{Z}_3 \quad ...(i)$$

where,
$$\bar{Z}_1 = R_1 - \frac{j}{\omega C_1};$$
$$\bar{Z}_2 = R_2\ ;$$
$$\bar{Z}_3 = R_3\ ;$$
$$\bar{Z}_x = R_x + j\omega L_x$$

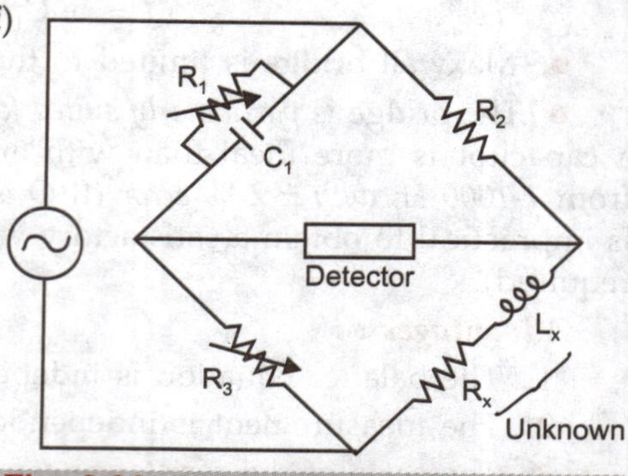

Fig. 6.26. Hay bridge for inductance measurements.

Substituting these values in eqn. (*i*), we get :

$$\left(R_1 - \frac{j}{\omega C_1}\right)(R_x + j\omega L_x) = R_2 R_3$$

$$R_1R_x + j\omega L_x R_1 - \frac{jR_x}{\omega C_1} + \frac{L_x}{C_1} = R_2 R_3$$

Equating the *real* and *imaginary* terms, we get :

$$R_1 R_x + \frac{L_x}{C_1} = R_2 R_3 \quad ...(6.60)$$

and,

$$\frac{R_x}{\omega C_1} = \omega L_x R_1 \quad ...(6.61)$$

or,

$$R_x = \omega^2 L_x C_1 R_1 \quad ...[6.61\ (a)]$$

Substituting for R_x in eqn. (6.60), we have

$$R_1(\omega^2 L_x C_1 R_1) + \frac{L_x}{C_1} = R_2R_3$$

or,

$$\omega^2 R_1^2 C_1 L_x + \frac{L_x}{C_1} = R_2R_3$$

Multiplying both sides by C_1, we get

$$\omega^2 R_1^2\, C_1^2\, L_x + L_x = R_2R_3C_1$$

or,

$$L_x\,(1+ \omega^2 R_1^2\, C_1^2\, L_x) = R_2R_3C_1$$

or,

$$L_x = \frac{R_2R_3C_1}{1+\omega^2 R_1^2\, C_1^2} \quad ...(6.62)$$

Substituting for L_x in eqn. (6.61*a*), we have

$$R_x = \frac{\omega^2 C_1^2\, R_1R_2R_3}{1+\omega^2 R_1^2\, C_1^2} \quad ...(6.63)$$

The term ω appears in the expressions for both L_x and R_x. This indicates that the bridge is *frequency sensitive*.

— This bridge is also used in the *measurement of incremental inductance*. The inductance balance equation depends on the losses of the inductor (or Q) and also on the operating frequency.

- A *commercial bridge measures from 1 μH – 100H with ± 2 % error.*
- An inconvenient feature of this bridge is that equation giving the balance condition for inductance contains the multiplier $\frac{1}{(1+1/Q^2)}$ (where $Q = \frac{1}{\omega R_1 C_1}$). The inductance balance thus depends on its Q and frequency.

$$\therefore \quad L_x = \frac{R_2R_3C_1}{1+(1/Q^2)} \quad ...(6.64)$$

For a value of Q greater than 10, the term $\frac{1}{Q^2}$ will be smaller than $\frac{1}{100}$ and can therefore be *neglected*. Then,

$L_x = R_2R_3C_1$, which is the same as Maxwell's equation. But for inductors with a Q less than 10, the $\frac{1}{Q^2}$ term cannot be neglected. Hence *this bridge is not suited for measurements of coils having $Q < 10$.*

Example 6.18. *Find the series equivalent inductance and resistance of the network that causes an opposite angle (Hay bridge) to null the following bridge arms (Fig. 6.27)*

$\omega = 3000$ *rad/s* ; $R_2 = 9\ k\Omega$; $R_1 = 1.8\ k\Omega$; $C_1 = 0.9\ \mu F$, $R_3 = 0.9\ k\Omega$.

Solution. We know that,

$$R_x = \frac{\omega^2 C_1^2 R_1 R_2 R_3}{1 + \omega^2 R_1^2 C_1^2} \quad ...(6.63)$$

$$= \frac{(3000)^2 (0.9 \times 10^{-6})^2 \times (1.8 \times 1000)(9 \times 1000)(0.9 \times 1000)}{1 + (3000)^2 (1.8 \times 1000)^2 (0.9 \times 10^{-6})^2}$$

$$= \frac{106.27 \times 10^3}{1 + 14.58}$$

$$= 6822\ \Omega \quad \text{or} \quad 6.822\ k\Omega$$

Also, $$L_x = \frac{R_2 R_3 C_1}{1 + \omega^2 R_1^2 C_1^2} \quad ...(6.62)$$

$$= \frac{(9 \times 1000)(0.9 \times 1000) \times (0.9 \times 10^{-6})}{1 + (3000)^2 (1.8 \times 1000)^2 (0.9 \times 10^{-6})^2}$$

$$= \frac{7.29}{1 + 14.58} = 0.468 \text{H}$$

i.e. R_x = **6.822 kΩ**; L_x = **0.468 H (Ans.)**

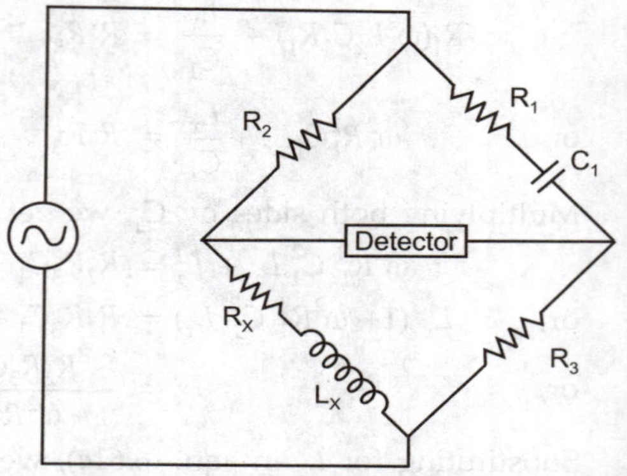

Fig. 6.27.

6.3.4. Anderson Bridge (For *measurement of* "Inductance")

This method is one of the commonest and best bridge methods for precise measurement of inductance over a wide range.

— Here, the self-inductance is *measured in terms of a standard capacitor.*

Fig. 6.28 shows the circuit and the phasor diagrams of the bridge for *balanced conditions.*

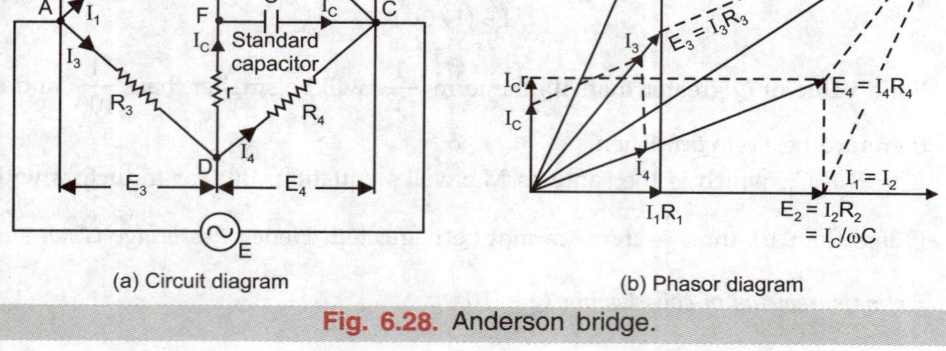

Fig. 6.28. Anderson bridge.

Let,

$$L_1 = \text{Self-inductance to be measured,}$$
$$R_1 = \text{Resistance of the coil under test,}$$
$$R_2, R_3, R_4, r = \text{Known non-inductive resistances, and}$$
$$C = \text{Standard known capacitor.}$$

At balance :

$$\bar{I}_1 = \bar{I}_2 \quad \text{and,} \quad \bar{I}_3 = \bar{I}_C + \bar{I}_4$$

Now,

$$\bar{I}_1 R_2 = I_C \times \frac{1}{j\omega C}$$

or,

$$I_C = \bar{I}_1 j\omega C R_2 \qquad ...(i)$$

Other balance equations are :

$$\bar{I}_1 (R_1 + j\omega L_1) = \bar{I}_2 R_3 + I_C \,.\, r \qquad ...(ii)$$

and,

$$\bar{I}_C \,.\, r + \bar{I}_C \,.\, \frac{1}{j\omega C} = (\bar{I}_3 - \bar{I}_C)\, R_4$$

or,

$$I_C \left(r + \frac{1}{j\omega C} \right) = (\bar{I}_3 - \bar{I}_C)\, R_4 \qquad ...(iii)$$

Substituting the value of $\bar{I}_C$ from (*i*) in eqns. (*ii*) and (*iii*), we get :

$$\bar{I}_1 (R_1 + j\omega L_1) = \bar{I}_3 R_3 + \bar{I}_1 j\omega C R_2 \,.\, r$$

or,

$$\bar{I}_1 (R_1 + j\omega L_1 - j\omega C R_2 \,.\, r) = \bar{I}_3 R_3 \qquad ...(iv)$$

and,

$$\bar{I}_1 j\omega C R_2 \left(1 + \frac{1}{j\omega C} \right) = (\bar{I}_3 - \bar{I}_1 \,.\, j\omega C R_2)\, R_4$$

or,

$$\bar{I}_1 j\omega C R_2 r + \bar{I}_1 R_2 = \bar{I}_3 R_4 - \bar{I}_1 j\omega C R_2 R_4$$

or,

$$\bar{I}_1 (j\omega C R_2 r + j\omega C R_2 R_4 + R_2) = \bar{I}_3 R_4 \qquad ...(v)$$

Substituting the value of R_4 from (*v*) in (*iv*), we get

$$\bar{I}_1 (R_1 + j\omega L_1 - j\omega C R_2 \,.\, r) = \bar{I}_1 \left(\frac{j\omega C R_2 R_3 r}{R_4} + j\omega C R_2 R_3 + \frac{R_2 R_3}{R_4} \right)$$

Equating the *real* and *imaginary terms,* we get :

$$\mathbf{R_1} = \frac{R_2 R_3}{R_4} \qquad ...(6.65)$$

$$L_1 - C R_2 r = \frac{C R_2 R_3 r}{R_4} + C R_2 R_3$$

or,

$$L_1 = C R_2 r + \frac{C R_2 R_3 r}{R_4} + C R_2 R_3$$

or,

$$L_1 = C R_2 \left[r \left(1 + \frac{R_3}{R_4} \right) + R_3 \right]$$

or, $$L_1 = \frac{CR_2}{R_4}\left[r(R_3 + R_4) + R_3R_4\right] \quad ...(6.66)$$

Advantages :

1. A fixed capacitor can be used instead of a variable capacitor as in the case of a Maxwell bridge.
2. It is much easier to obtain balance in case of Anderson bridge than a Maxwell bridge for low Q-coils.
3. This bridge may be used for accurate determination of capacitance in terms of inductance.

Disadvantages/Limitations :

1. It is more complex than its prototype Maxwell bridge.
2. The shielding of the bridge becomes more difficult due to an additional junction point.

- In view of the above complications of the Anderson bridge, in all the cases where a variable capacitor is permissible the more simple Maxwell bridge is used (instead of Anderson bridge).

Example 6.19. *Fig. 6.29 shows the connection of an Anderson bridge for measuring the inductance L and resistance R of the coil. Find R and L, if balance is obtained when Q = S = 1,000 Ω, P = 500 Ω, k = 100 Ω and C = 0.5 μF.* **(AMIE)**

Solution. *Given :* Q = S = 1,000 Ω, P = 500 Ω, r = 100 Ω, C = 0.5 μF.

R, L :

In the balance condition of bridge :

$$R = \frac{P \times Q}{S} \quad \text{[Eqn. (6.65)]}$$

$$= \frac{500 \times 1000}{1000} = 500\ \Omega \quad \textbf{(Ans.)}$$

$$L = \frac{CP}{S}[r(Q+S) + QS] \quad \text{[Eqn. (6.67)]}$$

$$= \frac{0.5 \times 10^{-5} \times 500}{1000}[100(1000+1000) + 1000 \times 1000]$$

$$= \textbf{0.3 H} \quad \textbf{(Ans.)}$$

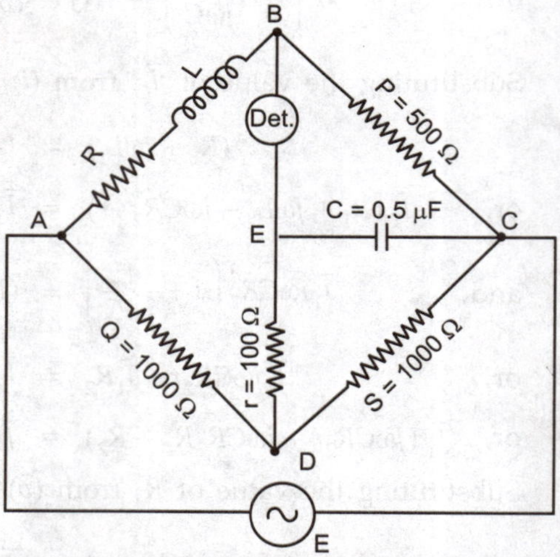

Fig. 6.29.

Example 6.20. *The following data relate to Anderson bridge shown in Fig. 6.30.*

The arms BC, CD and DA consist of resistances having values 1000 Ω, 1000 Ω and 500 Ω respectively.

A resistance of 100 Ω and a capacitance of 3 μF are connected respectively to the arms DF and CF.

An A.C. supply of 100 Hz is applied between the terminals A and C and a detector is connected between the terminals B and F.

The detector indicates null under the above conditions. Determine the values of R and L connected to the arm AB. **(UPTU)**

Solution. *Given :* $R_2 = 1000\ \Omega$; $R_3 = 500\ \Omega$; $R_4 = 1000\ \Omega$; $r = 100\ \Omega$; $C = 3\ \mu F$.

***R, L* :**

At balance condition of the bridge :

We know that :

$$R = \frac{R_2 R_3}{R_4} \quad \text{[Eqn. (6.65)]}$$

$$= \frac{1000 \times 500}{1000} = \mathbf{500\ \Omega} \quad \textbf{(Ans.)}$$

$$L = \frac{CR_2}{R_4}[r(R_3 + R_4) + R_3R_4] \quad \text{[Eqn. (6.66)]}$$

$$= \frac{3 \times 10^{-6} \times 1000}{1000}[100(500 + 1000) + 500 \times 1000]$$

$$= \mathbf{1.95\ H} \quad \textbf{(Ans.)}$$

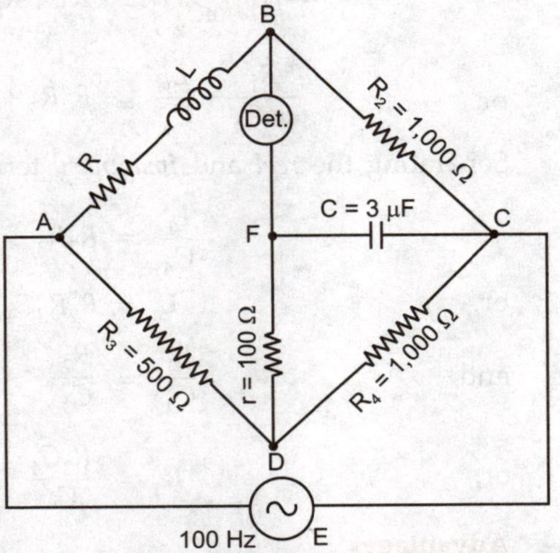

Fig. 6.30.

6.3.5. Owen Bridge (For *measurement of* "Inductance")

This bridge may be used *for measurement of an inductance in terms of capacitance.*

Fig. 6.31 shows the circuit and phasor diagrams, for this bridge, under balance condition.

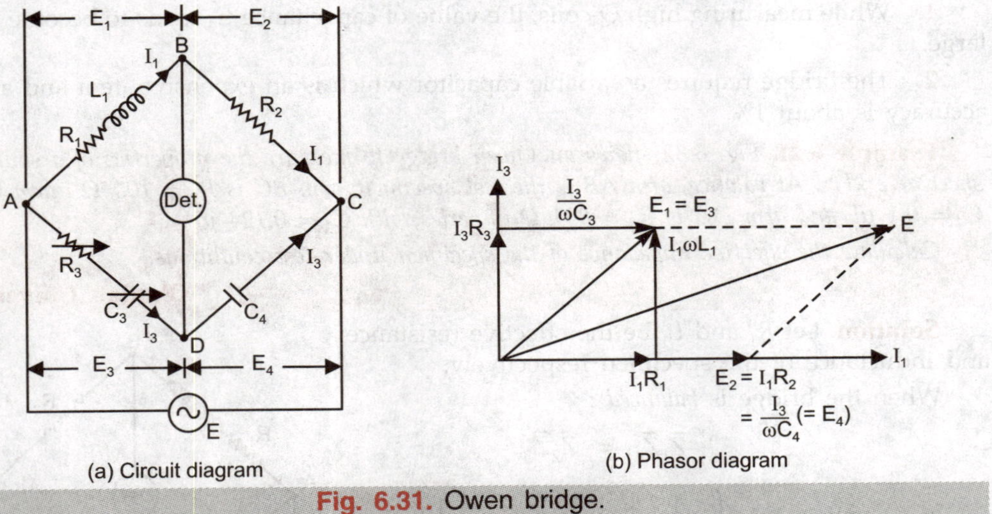

Fig. 6.31. Owen bridge.

Let,
L_1 = Unknown self-inductance of resistance R_1,
R_2 = Fixed non-inductive resistance,
R_3 = Variable non-inductive resistance,
C_3 = Variable standard capacitor, and
C_4 = Fixed standard capacitor.

When the bridge is balanced :

$$\bar{Z}_1 \cdot \bar{Z}_4 = \bar{Z}_2 \cdot \bar{Z}_3$$

$$(R_1 + j\omega L_1)\left(\frac{1}{j\omega C_4}\right) = R_2 \cdot \left(R_3 + \frac{1}{j\omega C_3}\right)$$

or $$\frac{R_1}{j\omega C_4} + \frac{L_1}{C_4} = R_2 R_3 + \frac{R_2}{j\omega C_3}$$

Separating the *real* and *imaginary* terms, we have :

$$\frac{L_1}{C_4} = R_2 R_3$$

or, $$\boldsymbol{L_1 = R_2 R_3 C_4} \quad \text{...(6.67)}$$

and, $$\frac{R_1}{C_4} = \frac{R_2}{C_3}$$

or, $$\boldsymbol{R_1 = \frac{R_2 C_4}{C_3}} \quad \text{...(6.68)}$$

Advantages :

1. The balance equations are quite simple and do not contain any frequency component.
2. Since R_3 and C_3, the variable elements, are in the same arm, convergence to balance condition is much easier.
3. The bridge can be used over a wide range of measurement of inductances.

Disadvantages :

1. While measuring high Q-coils, the value of capacitance C_3 tends to become rather large.
2. The bridge requires a variable capacitor which is an expensive item and also its accuracy is about 1%.

Example 6.21. *Fig. 6.32 shows an Owen bridge to measure the properties of a sample of sheet at 2 kHz. At balance, arm AB is the test specimen; arm BC is $R_2 = 100\ \Omega$; arm CD is $C_4 = 0.1\ \mu F$ and arm DA is $R_3 = 834\ \Omega$ in series with $C_3 = 0.124\ \mu F$.*

Calculate the effective impedance of the specimen under test conditions.

(Punjab University)

Solution. Let R_1 and L_1 be the effective resistance and inductance of the specimen respectively.

When the bridge is *balanced* :

$$\bar{Z}_1 \bar{Z}_4 = \bar{Z}_2 \bar{Z}_3$$

or, $$(R_1 + j\omega L_1)\left(\frac{1}{j\omega C_4}\right) = R_2 \left(R_3 + \frac{1}{j\omega C_3}\right)$$

or, $$R_1 \cdot \frac{1}{j\omega C_4} + \frac{L_1}{C_4} = R_2 R_3 + R_2 \cdot \frac{1}{j\omega C_3}$$

Equating *real* and *imaginary* terms, we get :

$$\frac{L_1}{C_4} = R_2 R_3$$

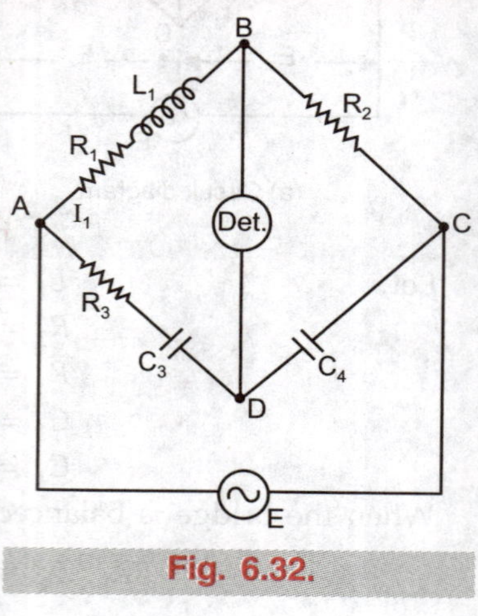

Fig. 6.32.

or, $L_1 = R_2R_3C_4 = 100 \times 834 \times 0.1 \times 10^{-6}$

$= 8.34 \times 10^{-3}$ H or **8.34 mH** **(Ans.)**

and, $\dfrac{R_1}{C_4} = \dfrac{R_2}{C_3}$

or, $R_1 = R_2 \times \dfrac{C_4}{C_3} = 100 \times \dfrac{0.1}{0.124} = \mathbf{80.6\ \Omega}$ **(Ans.)**

Reactance of specimen at 2 kHz,

$$X_1 = 2\pi f L_1 = 2\pi \times (2 \times 1000) \times 8.34 \times 10^{-3} = 104.8\ \Omega$$

∴ Impedance of the specimen, $Z_1 = \sqrt{R_1^2 + X_1^2}$

$= \sqrt{(80.6)^2 + (104.8)^2} = \mathbf{132.2\ \Omega}$ **(Ans.)**

MEASUREMENT OF CAPACITANCE

6.3.6. De Sauty Bridge (For *measurement of* "Capacitance")

This bridge is the simplest method of *comparing two capacitances.*

Fig. 6.33(*a* & *b*) shows the circuit and phasor diagrams of the bridge.

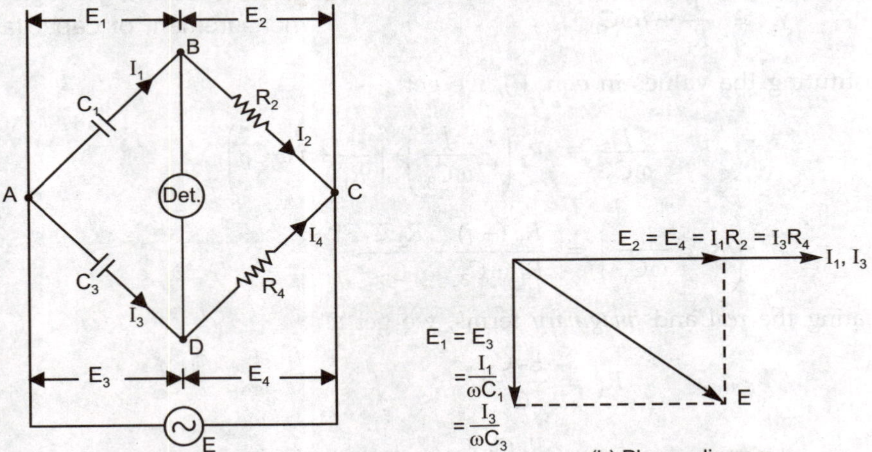

(a) Circuit diagram (b) Phasor diagram

Fig. 6.33. De Sauty bridge.

Let, C_1 = Capacitor whose capacitance is to be measured,

C_3 = Standard capacitor, and

R_3, R_4 = Non-inductive resistors.

At balance : $I_1 = I_2$; $I_3 = I_4$

and, $\left(\dfrac{1}{j\omega C_1}\right) R_4 = \left(\dfrac{1}{j\omega C_3}\right) R_2$

or, $\dfrac{R_4}{C_1} = \dfrac{R_2}{C_3}$

or, $$C_1 = \frac{C_3 R_4}{R_2} \qquad ...(6.69)$$

6.3.7. Schering Bridge (*for Measurement of* "Capacitance")

It is one of the most important A.C. bridges, widely used for the *measurement of unknown capacitors, dielectric loss and power factor.*

Fig. 6.34 shows the basic circuit arrangement of Schering bridge. The capacitor C_3 is a high quality mica capacitor (low-loss) for general measurements, or an air capacitor (having a very stable value and a very small electric field) for insulation measurement.

The general equation for balance is,

$$\bar{Z}_1 \bar{Z}_x = \bar{Z}_2 \bar{Z}_3$$

or, $$\bar{Z}_x = \frac{\bar{Z}_2 \bar{Z}_3}{\bar{Z}_1} = \bar{Z}_2 \bar{Z}_3 \bar{Y}_1 \quad ...(i)$$

where, $$\bar{Z}_x = R_x - \frac{j}{\omega C_x}$$

$$\bar{Z}_2 = R_2$$

$$\bar{Z}_3 = -\frac{j}{\omega C_3}$$

$$\bar{Y}_1 = \frac{1}{R_1} + j\omega C_1$$

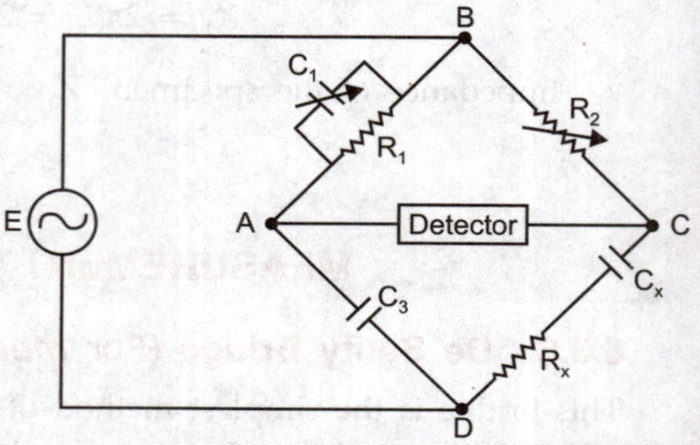

Fig. 6.34. Schering bridge for the measurement of capacitance.

Substituting the values in eqn. (*i*), we get

$$R_x - \frac{j}{\omega C_x} = R_2\left(-\frac{j}{\omega C_3}\right) \times \left(\frac{1}{R_1} + j\,\omega\,C_1\right)$$

or $$\left(R_x - \frac{j}{\omega C_x}\right) = \frac{R_2\,(-j)}{R_1\,\omega\,C_3} + \frac{R_2\,C_1}{C_3}$$

Equating the *real* and *imaginary* terms, we get :

$$R_x = \frac{R_2 C_1}{C_3} \quad ...(6.70)$$

and, $$C_x = \frac{R_1 C_3}{R_2} \quad ...(6.71)$$

The dial of capacitor C_1 can be calibrated directly to give the dissipation factor at a particular frequency.

- The lower junction of the bridge is grounded. At the frequency normally used on this bridge, the reactances of capacitors C_3 and C_x are much higher than the resistances of R_1 and R_2. Hence if the junction of R_1 and R_2 is grounded, the detector is effectively at ground potential. This reduces any stray-capacitance effect, and makes the bridge more stable.

Power factor. The *power factor* (p.f.) of a series R.C. combination is defined as *the cosine of the phase angle of the circuit.*

∴ The power factor of the unknown, p.f. $= \frac{R_x}{Z_x}$

For phase angles very close to 90°, the reactance is almost equal to the impedance and we can approximate the power factor to,

$$\text{p.f.} \simeq \frac{R_x}{X_x} = \frac{R_x}{\frac{1}{\omega C_x}} = \omega R_x C_x \quad ...(6.72)$$

Dissipation factor (D). The *dissipation factor* of a series *R-C* circuit is defined *as the cotangent of the phase angle.*

i.e.,
$$D = \frac{R_x}{X_x} = \omega C_x R_x \quad ...(6.73)$$

Also, since the quality of a coil is defined by $Q = \frac{X_L}{R_L}$, we find that the dissipation factor, *D*, is the *reciprocal of the quality factor, Q,*

i.e.
$$D = \frac{1}{Q} \quad ...(6.74)$$

The dissipation factor tells us something about the quality of the capacitor; *i.e.*, how close the phase angle of the capacitor is to the ideal value of 90°.

Substituting the value of C_x in eqn. (6.71) and of R_x in eqn. (6.70) in the eqn. (6.74), we get

$$D = \omega \times \frac{R_1 C_3}{R_2} \times \frac{R_2 C_1}{C_3} = \omega R_1 C_1 \quad ...(6.75)$$

— When resistor R_1 in the Schering bridge of Fig. 6.34 has a fixed value, the dial of capacitor C_1 may be calibrated directly in dissipation factor *D*. This is the usual practice in a Schering bridge. The appearance of the term ω in the expression for the dissipation factor [eqn. (6.75)] means, of course, that the calibration of the C_1 dial holds for only one particular frequency at which dial is calibrated. A different frequency can be used, provided that a correction is made by multiplying the C_1 dial reading by the ratio of the two frequencies.

- Commercial units *measure from 100 pF–1μF, with ± 2% accuracy.*
- The Schering bridge is *widely used for testing small capacitors at low voltages with very high precision.*

Example 6.22. *The Schering bridge shown in Fig. 6.34 has the following constants :*
Arm AB ... Capacitor of 0.4 μF in parallel with 1.5 kΩ resistance;
Arm BC ... Resistance of 3 kΩ;
Arm CD ... Unknown capacitor C_x and R_x in series
Arm DA ... Capacitor of 0.4 μF;
Frequency ... 1 kHz.
Determine the following :
(i) R_x and C_x.
(ii) Dissipation factor.

Solution. *Given:* R_1 = 1.5 kΩ; C_1 = 0.4 μF ; R_2 = 3 kΩ; C_3 = 0.4 μF.

(i) R_x; C_x :

We know that,
$$R_x = \frac{R_2 C_1}{C_3} = \frac{3k\Omega \times 0.4\,\mu F}{0.4\mu F} \quad \text{[Eqn. (6.70)]}$$
$$= \mathbf{3\ k\Omega}\ \textbf{(Ans.)}$$

and, $C_x = \frac{R_1 C_3}{R_2} = \frac{1.5\,k\Omega \times 0.4\,\mu F}{3\,k\Omega}$

$= \mathbf{0.2\ \mu F}$ **(Ans.)**

(ii) Dissipation factor, *D* :

$$D = \omega\, C_x\, R_x \qquad \text{[Eqn. (6.73)]}$$
$$= 2\,\pi\, f.\, C_x.\, R_x$$
$$= 2\,\pi \times 1000 \times 0.2 \times 10^{-6} \times (3 \times 1000) = 3.77$$

i.e. $D = \mathbf{3.77}$ **(Ans.)**

Example 6.23. *A bakelite sheet of 5 mm thickness is tested at 50 Hz between the electrodes 12 cm in diameter. The Schering bridge (Fig. 6.35) used has an air capacitor C_2 of 106 pF, a non-reactive resistance of R_4 of $\left(\frac{1000}{\pi}\right)$ Ω in parallel with a variable capacitor C_4 and a non-reactive variable resistance R_3. Balance is obtained with C_4 = 0.55μF and R_3 = 270Ω.*

Determine the following :

(i) Capacitance ;

(ii) Power factor, and

(iii) Relative permittivity of the sheet.

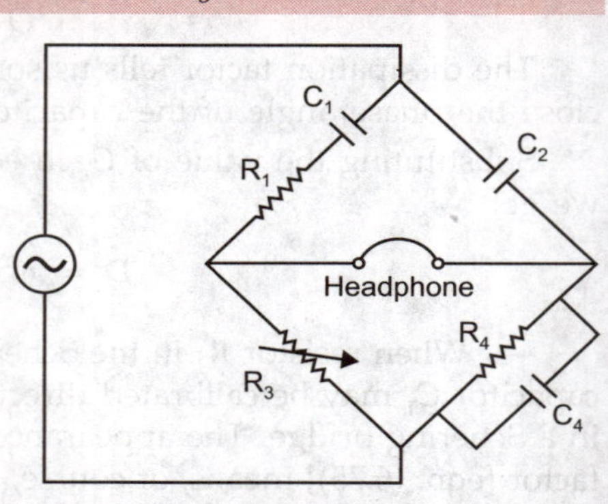

Fig. 6.35.

Solution. Refer to Fig. 6.35.

Given :

$$C_2 = 106\text{pF} = 106 \times 10^{-12}\ \text{F};$$
$$R_4 = \frac{1000}{\pi}\ \Omega;$$
$$C_4 = 0.55\mu\ \text{F} = 0.55 \times 10^{-6}\ \text{F};$$
$$R_3 = 270\ \Omega$$

(i) Capacitance *C* :

When the bridge is balanced, we have

$$R_1 = \frac{R_3\, C_4}{C_2} \qquad \text{...[Eqn. (6.70)]}$$

$$= \frac{270 \times 0.55 \times 10^{-6}}{106 \times 10^{-12}} = 1.4\ \text{M}\Omega$$

and, $$C_1 = \frac{R_4\, C_2}{R_3} \qquad \text{...[Eqn. (6.71)]}$$

$$= \frac{(1000/\pi) \times 106 \times 10^{-12}}{270}$$

$= \mathbf{124.9\ pF}$ **(Ans.)**

(ii) Power factor :

$$\text{Power factor} = \omega R_1 C_1 \qquad \text{...[(Eqn. (6.72)]}$$
$$= 2\pi \times 50 \times (1.4 \times 10^6) \times (124.9 \times 10^{-12})$$

$= \mathbf{0.0549}$ **(Ans.)**

(iii) Relative permittivity ϵ_r :

We know that $$C = \frac{\epsilon_0 \epsilon_r A}{t} \quad ...(i)$$

where, $C (= C_1)$ = Capacitance,

$\epsilon_o = 8.854 \times 10^{-12}$

A = Area of the electrodes

$$= \frac{\pi}{4} \times \left(\frac{12}{100}\right)^2 = 0.0113\, m^2, \text{ and}$$

t = Thickness of bakelite sheet

= 5 mm or 0.005 m.

Substituting the values in eqn. (*i*), we get

$$124.9 \times 10^{-12} = \frac{8.854 \times 10^{-12} \times \epsilon_2 \times 0.0113}{0.005}$$

or $$\epsilon_r = \frac{124.9 \times 10^{-12} \times 0.005}{8.854 \times 10^{-12} \times 0.0113}$$

= **6.24 (Ans.)**

Example 6.24. *The four arms of an A.C. bridge network are as follows :*

Arm AB : an unknown impedance; arm BC: a standard capacitor C_3 of 1000 pF; arm CD : a non-inductive resistor R_4 of 100 Ω in parallel with a capacitor C_4 of 0.01 μF; arm DA : a non-inductive resistor R_2 of 1000 Ω.

The A.C. supply is connected across terminals B, D and the supply frequency is 50 Hz. If the bridge is balanced with the above values, determine the components of the unknown impedance, while deriving the balance conditions. **(AMIE)**

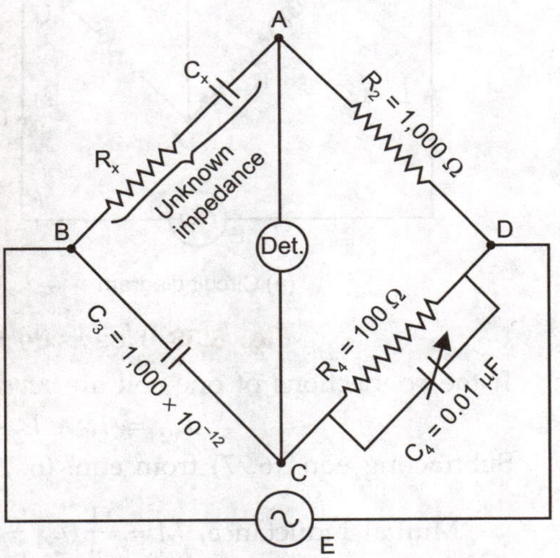

Fig. 6.36.

Solution. The configuration of the bridge is that of *Schering bridge* (Fig. 6.36).

Given : $R_2 = 1000\ \Omega$; $C_3 = 1000 \times 10^{-12}$ pF;

$R_4 = 100\ \Omega$; $C_4 = 0.01\ \mu F$.

We know that, $$R_x = \frac{R_2 C_4}{C_3}$$

$$= \frac{1000 \times 0.01 \times 10^{-6}}{1000 \times 10^{-12}}$$

= **0.01 MΩ (Ans.)**

and, $$C_x = \frac{R_4 C_3}{R_2} = \frac{100 \times 1000 \times 10^{-12}}{1000}$$

= **100 pF (Ans.)**

Measurement of Mutual Inductance

6.3.8. Heaviside Mutual Inductance Bridge

Measurement of mutual inductance. If the terminals of two coils whose mutual inductance is to be measured are available for series connections, measurement may be made by one of the self-inductance bridges discussed earlier. If the connections are made so that the magnetic fields of two coils are additive as shown in Fig. 6.37(a), the effective inductance of the two coils in series is,

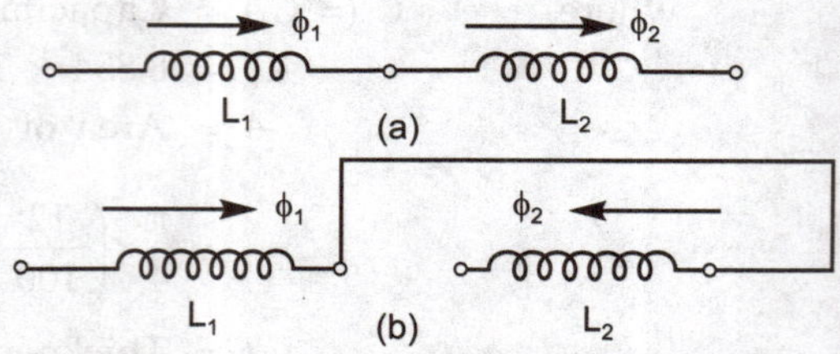

Fig. 6.37. Positive and negative couplings for mutual inductors.

$$L_{e1} = L_1 + L_2 + 2M \quad ...(6.76)$$

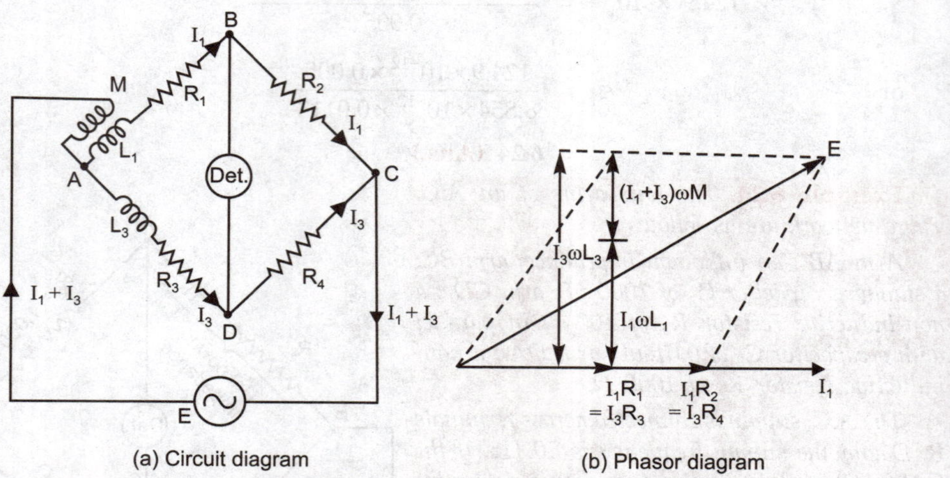

Fig. 6.38. Heaviside mutual inductance bridge.

If the connections of one coil are *reversed* as shown in Fig. 6.37(b), then

$$L_{e2} = L_1 + L_2 - 2M \quad (6.77)$$

Subtracting eqn. (6.77) from eqn. (6.76), we get

$$\text{Mutual inductance, } M = \frac{1}{4}(L_{e1} - L_{e2}) \quad ...(6.78)$$

Hence mutual inductance is obtained as one-fourth of difference of self-inductance measured with *series additive* and *series subtractive* connections.

This method is of advantage only when fairly high coupling between the two coils is obtained otherwise it results in poor accuracy due to nearly equal terms of eqn. (6.78).

- Fig. 6.38 shows the circuit and phasor diagrams of Heaviside mutual inductance bridge, which *measures mutual inductance in terms of a known self-inductance.*

Let,
M = Unknown mutual inductance,
L_1 = Self-inductance of secondary of mutual inductance,
L_3 = Known self-inductance, and
R_2, R_3, R_3, R_4 = Non-inductive resistors.

At balance voltage drop between B and C must equal the voltage drop between D and C. Also voltage drop across A–B–C equals the voltage drop across A–D–C. Thus we have the following equations at balance :

$$\bar{I}_1 R_2 = \bar{I}_2 R_4 \qquad ...(i)$$

and, $$(\bar{I}_1 + \bar{I}_2)(j\omega M) + \bar{I}_1 (R_1 + R_2 + j\omega L_1) = \bar{I}_2 (R_3 + R_4 + j\omega L_3) \qquad ...(ii)$$

Inserting the value of $\bar{I}_1 = \dfrac{\bar{I}_1 R_4}{R_2}$ from (*i*) in (*ii*), we get

$$\left(\frac{\bar{I}_2 R_4}{R_2} + \bar{I}_2\right)(j\omega M) + \frac{\bar{I}_2 R_4}{R_2}(R_1 + R_2 + j\omega L_1) = \bar{I}_2 (R_3 + R_4 + j\omega L_3)$$

or, $$I_2\left(\frac{R_4}{R_2} + 1\right)(j\omega M) + \frac{\bar{I}_2 R_4}{R_2}(R_1 + R_2 + j\omega L_1) = \bar{I}_2 (R_3 + R_4 + j\omega L_3)$$

or, $$j\omega M\left(\frac{R_4}{R_2} + 1\right) + \frac{R_4 R_1}{R_2} + R_4 + j\omega L_1 \frac{R_4}{R_2} = R_3 + R_4 + j\omega L_3$$

Separating the *real* and *imaginary* terms, we have :

$$\frac{R_4 R_1}{R_2} + R_4 = R_3 + R_4$$

or, $$\frac{R_4 R_1}{R_2} = R_3$$

or, $$R_1 = \frac{R_2 R_3}{R_4} \qquad ...(6.79)$$

and, $$\left(\frac{R_4}{R_2} + 1\right) M + L_1 \frac{R_4}{R_2} = L_3$$

$$\left(\frac{R_4}{R_2} + 1\right) M = L_3 - \frac{L_1 R_4}{R_2}$$

or, $$\left(\frac{R_4 + R_2}{R_2}\right) M = \frac{R_2 L_3 - R_4 L_1}{R_2}$$

or, $$M = \frac{R_2 L_3 - R_4 L_1}{R_2 + R_4} \qquad ...(6.80)$$

It is clear from eqn. (6.80), that L_1, the self-inductance of the secondary of the mutual inductor must be known in order that M be measured by this method.

In case, $R_2 = R_4$, we get, $$M = \frac{L_2 - L_1}{2} \qquad ...(6.81)$$

and $$R_1 = R_3 \qquad ...(6.82)$$

- This method *can be used for measurement of self-inductance*. Let L_3 be the self-inductance to *be determined*.

Then, from eqns. (6.79) and (6.80), we get :

$$L_3 = \frac{M(R_2 + R_4) + R_4 L_1}{R_2}$$

$$= M\left(1 + \frac{R_4}{R_2}\right) + \frac{R_4}{R_2} L_1 \quad ...(6.83)$$

and, $$R_3 = \frac{R_1 R_4}{R_2} \quad ...(6.84)$$

In case $R_4 = R_2$, we have

$$L_3 = 2M + L_1$$

and $$R_3 = R_1$$

Campbell's Modification of Heaviside bridge :

This bridge is used to meaure self-inductance in terms of a mutual inductance.

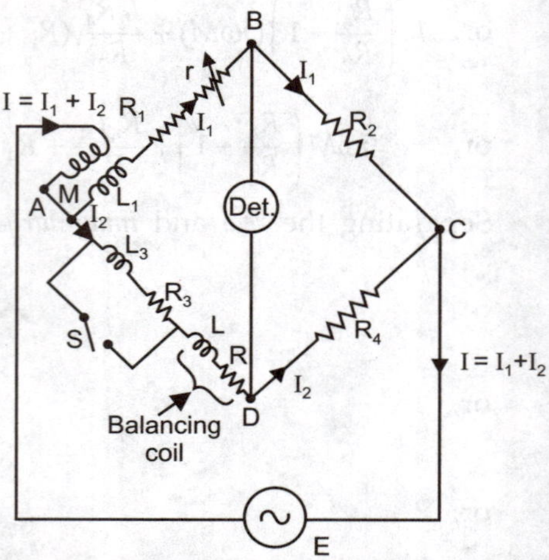

Fig. 6.39. Campbell's modification of Heaviside bridge.

Fig. 6.39 shows a modified Heaviside bridge; this modification is due to Campbell :

— In this case an additional balancing coil L, R is included in arm AD in series with the inductor under test.

— An additional resistance r is put in arm AB. Balance is obtained by varying M and r.

— A short-circuiting switch S is placed across the coil R_3, L_3 *under measurement*.

— Two sets of readings are taken, one with switch being open and the other with switch being closed. Let values of M and r be M_1 and r_1 with switch open and M_2, r_2 with switch closed.

∴ With switch *open*, from eqn. (6.83), we have :

$$L_3 + L = \frac{M_1(R_2 + R_4) + R_4 L_1}{R_2} \quad ...(i)$$

and, with switch *closed*,

$$L = \frac{M_2(R_2 + R_4) + R_4 L_1}{R_2} \quad ...(ii)$$

Subtracting (*ii*) from (*i*), we get :

$$L_3 = (M_1 - M_2)\left(1 + \frac{R_4}{R_2}\right) \quad ...(6.85)$$

With switch open, from eqn. (6.84), we have :

$$R_3 + R = (R_1 + r_1) \times \frac{R_4}{R_2} \quad ...(iii)$$

and, with switch closed, $$R = (R_1 + r_2)\frac{R_4}{R_2} \quad ...(iv)$$

Subtracting (*iv*) from (*iii*), we get :

$$R_3 = (r_1 - r_2) \times \frac{R_4}{R_2} \quad \text{...(6.86)}$$

- When we have *"equal ratio arms"*, $R_2 = R_4$

and, therefore from eqns. (6.85) and (6.86), we get

$$\boldsymbol{L_3} = 2(M_1 - M_2) \quad \text{...(6.87)}$$

and $$\boldsymbol{R_2} = r_1 - r_2 \quad \text{...(6.88)}$$

6.3.9. Carey Foster (Heydweiller) Bridge

Basically, Carey Foster used this bridge, but it was subsequently modified by Heydweiller for use on A.C.

This bridges is *used* for the following purposes :

(*i*) *To measure the capacitance in terms of a standard mutual inductance.* The bridge in this case is known as *Carey Foster bridge.*

(*ii*) *To measure the mutual inductance in terms of a standard capacitance.* The bridge in this case is known as *Heydweiller bridge.*

Fig. 6.40 shows the circuit and phasor diagrams for the bridge under balance conditions.

The unusual feature of this bridge is that one of its arms, *AD*, is *short-circuited* and therefore the potential drop across this arm is *zero.* For obtaining the balance, the potential drop across arm *AB should also be equal to zero* and for this reason *negative coupling is needed for the mutual inductance.*

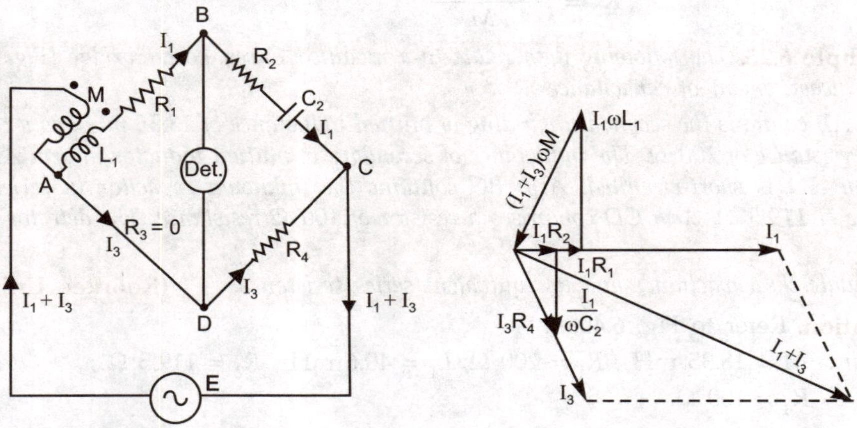

(a) Circuit diagram (b) Phasor diagram

Fig. 6.40. Carey Foster (Heydweiller) bridge.

When the bridge is balanced :

$$\bar{I}_1 (R_1 + j\omega L_1) - (\bar{I}_1 + \bar{I}_3)\, j\omega M = 0$$

or, $$\bar{I}_1 (R_1 + j\omega L_1 - j\omega M) = \bar{I}_3 j\omega M \quad \text{...}(i)$$

and, $$\bar{I}_1 \left(R_2 + \frac{1}{j\omega C_2} \right) = \bar{I}_3 R_4 \quad \text{...}(ii)$$

Substituting the value of $\bar{I}_3$ from (*ii*) in (*i*), we get :

$$\bar{I}_1 (R_1 + j\omega L_1 - j\omega M) = \frac{\bar{I}_1}{R_4}\left(R_2 + \frac{1}{j\omega C_2}\right) j\omega M$$

or, $$R_1 + j\omega L_1 - j\omega M = \frac{R_2}{R_4} j\omega M + \frac{M}{R_4 C_2}$$

Equating the *real* and *imaginary* terms, we get :

$$R_1 = \frac{M}{R_4 C_2}$$

or, $$\mathbf{M} = R_1 R_4 C_2 \quad ...(6.89)$$

and, $$L_1 = M\left(1 + \frac{R_2}{R_4}\right)$$

or, $$L_1 = \frac{M(R_2 + R_4)}{R_4} = R_1 C_2 (R_2 + R_4) \quad ...(6.90)$$

- If the bridge is used for *measurement of capacitance*, eqns. (6.89) and (6.90) may be written as :

$$\mathbf{C_2} = \frac{M}{R_1 R_4} \quad ...(6.91)$$

and, $$\mathbf{R_2} = \frac{R_4 (L - M)}{M} \quad ...(6.92)$$

Example 6.25. *The following data relate to a modified Carey Foster bridge [Fig. 6.40(a)], used for measurement of capacitance :*

Arm AB contains the secondary winding of mutual inductance of 18.35 mH and a total non-reactive resistance of 200 Ω. The inductance of secondary of mutual inductor in arm AB is 40.6 mH. Arm AD is short-circuited. Arm BC contains the unknown capacitor in series with a resistance of 119.5 Ω. Arm CD comprises a resistor of 100 Ω resistance. The detector is across BD.

Calculate the capacitance and its equivalent series resistance. **(Roorkee University)**

Solution. Refer to Fig. 6.40(*a*).

Given : $M = 18.35$ mH ; $R_1 = 200\ \Omega$; $L_1 = 40.6$ mH ; $R_2 = 119.5\ \Omega$; $R_4 = 100\ \Omega$.

We know that : $$\mathbf{C_2} = \frac{M}{R_1 R_4} \quad \text{[Eqn. (6.91)]}$$

$$= \frac{18.35 \times 10^{-3}}{200 \times 100} F = \mathbf{0.9175\ \mu F} \quad \textbf{(Ans.)}$$

and, $$\mathbf{R_2} = \frac{R_4 (L - M)}{M}$$

$$= \frac{100\,(40.6 - 18.35)}{18.35} = 121.25\ \Omega \quad \text{[Eqn. (6.92)]}$$

∴ *Series resistance of capacitor* = 121.25 − 119.5 = **1.75 Ω** **(Ans.)**

6.3.10. Campbell Bridge (*For measurement of "Mutual Inductance"*)

This bridge *measures an unknown mutual inductance in terms of a standard mutual inductance.*

Fig. 6.41 shows the circuit diagram of Campbell bridge.

Let, M_1 = *Unknown* mutual inductance,

L_1 = Self-inductance of secondary of mutual inductance M_1,

M_2 = Variable standard mutual inductance,

L_2 = Self-inductance of secondary of mutual inductance M_2, and

R_1, R_2, R_3, R_4 = Non-inductive resistance.

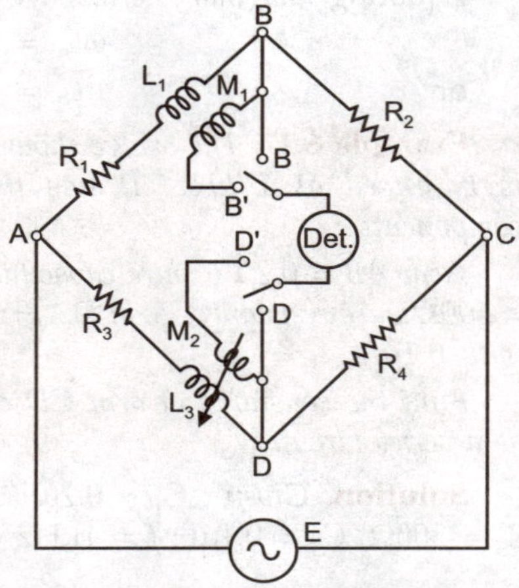

Fig. 6.41. Campbell's bridge.

For the balancing process, the following *two steps* are required :

Step 1. Detector is connected between B and D. The circuit now becomes a simple self-inductance comparison bridge. Then, condition for balance is :

$$\frac{L_1}{L_2} = \frac{R_1}{R_3} = \frac{R_2}{R_4} \qquad ...(6.93)$$

The bridge may be balanced by adjustment of R_2 (or R_4) and R_1 (or R_3).

Step 2. Detector is connected between B' and D'. Keeping the adjustments as in step 1 above, the variable inductance M_2 is varied to get balance. Then,

$$\frac{M_1}{M_2} = \frac{R_2}{R_4}$$

or,

$$M_1 = M_2 \cdot \frac{R_2}{R_4} \qquad ...(6.94)$$

Additional Worked Examples on Bridges

Example 6.26. *In a bridge shown in Fig. 6.42 the balance was obtained with L_1 = 52.6 mH and R_2 = 1.68Ω. Find the resistance and inductance of the coil.*

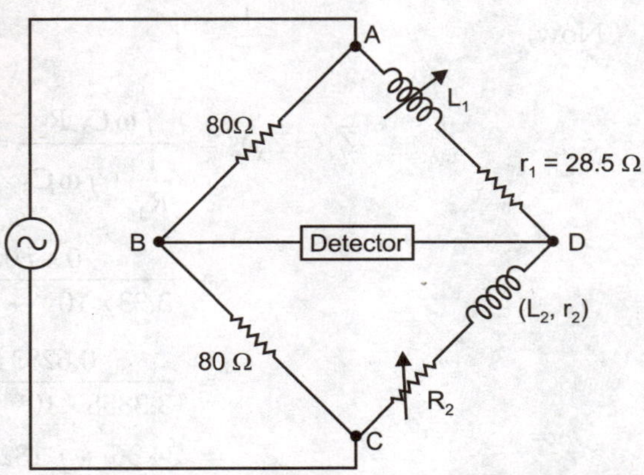

Fig. 6.42.

Solution : *Given* : L_1 = 52.6 mH; R_2 = 1.68Ω ; r_1 = 28.5Ω.

Refer to Fig. 6.42.

r_2; L_2;

Balance of the bridge was obtained when L_1 = 52.6 mH and R_2 = 1.68Ω

At balance :

$80\,(r_1 + j\,\omega L_1) = 80\,[(R_2 + r_2) + j\omega L_2]$

or, $r_1 + j\omega L_1 = (R_2 + r_2) + j\omega L_2$

Equating the *real terms*, we get,

$$r_1 = R_2 + r_2$$

$$\therefore \quad r_2 = r_1 - R_2 = 28.5 - 1.68 = \mathbf{26.82\ \Omega}\ \textbf{(Ans.)}$$

Equating *imaginary terms*, we have :

$$\omega L_1 = \omega L_2$$

$$\text{or} \quad L_1 = L_2 = \mathbf{52.6\ mH}\ \textbf{(Ans.)}$$

Example 6.27. *The bridge shown in Fig. 6.43 is balanced at 1 kHz. It has the following components :*

Arm AB = 0.2μ F pure capacitance; Arm BC = 500Ω pure resistance; Arm DA = 300Ω parallel with 0.1μF.

Find the constants of arm CD considering it as a series circuit.

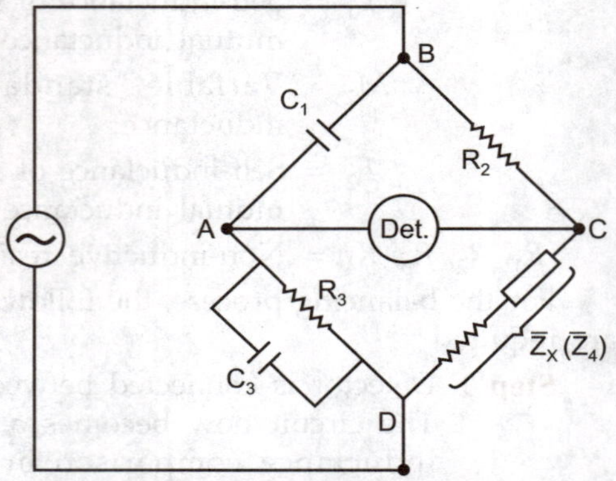

Fig. 6.43.

Solution. *Given* : $C_1 = 0.2\mu F$: $R_2 = 500\Omega$; $R_3 = 300\Omega$, $C_3 = 0.1\mu F$; f = 1kHz or 1000 Hz.

$$\text{Now,} \quad \bar{Z}_1 = 0 - j\left(\frac{1}{\omega C_1}\right)$$

$$\bar{Z}_2 = R_2$$

$$\bar{Z}_3 = R_3 \parallel X_{C3}$$

$$\text{or,} \quad \bar{Y}_3 = \frac{1}{R_3} + j\left(\frac{1}{X_{C3}}\right) = \frac{1}{R_3} + j(\omega C_3)$$

$$\bar{Z}_4 = \bar{Z}_x$$

The basic balance equation is given by :

$$\bar{Z}_1 \bar{Z}_4 = \bar{Z}_2 \bar{Z}_3$$

$$\therefore \quad \bar{Z}_4 = \frac{\bar{Z}_2 \bar{Z}_3}{\bar{Z}_1} = \frac{\bar{Z}_2}{\bar{Z}_1 \bar{Y}_3} = \frac{R_2}{-\dfrac{j}{\omega C_1}\left[\dfrac{1}{R_3} + j(\omega C_3)\right]}$$

$$\text{Now,} \quad -\frac{1}{j} = j$$

$$\therefore \quad \bar{Z}_4 (= \bar{Z}_x) = \frac{j\,\omega C_1 R_2}{\dfrac{1}{R_3} + j\,\omega C_3} = \frac{0 + j(2\pi \times 1000) \times 0.2 \times 10^{-6} \times 500}{\dfrac{1}{300} + j(2\pi \times 1000) \times 0.1 \times 10^{-6}}$$

$$= \frac{0 + j\,0.6283}{3.33 \times 10^{-3} + 6.283 \times 10^{-4}}$$

$$= \frac{0.6283\angle 90^\circ}{3.3888 \times 10^{-3}\ \angle +10.68^\circ} = 185.4\angle 79.32^\circ$$

$$= 34.36 + j\ 182.19\ \Omega$$

$$\therefore \quad \bar{Z}_x = (34.36 + j\ 182.19)\ \Omega$$

or, $R_x = \mathbf{34.36\ \Omega}$ **(Ans.)**

and, $X_{L_x} = 2\pi f L_x = 182.19$

$$\therefore \quad L_x = \frac{182.19}{2\pi \times 1000} \simeq 0.029 \text{ H or } \mathbf{29\ mH} \textbf{ (Ans.)}$$

Example 6.28. *The following data relate to the bridge shown in Fig. 6.44.*

$R_1 = 200\ \Omega$; $R_2 = 200\ \Omega$; $C_2 = 5\mu F$; $R_3 = 500\Omega$; $C_3 = 0.2\mu F$.

Find the constants of Z_x considering it as a series circuit.

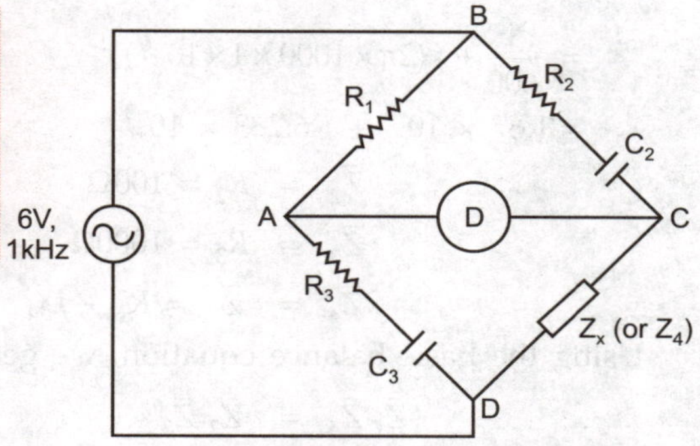

Fig. 6.44.

Solution. Refer to Fig. 6.44.

$$\bar{Z}_1 = R_1 = 200\Omega$$

$$\bar{Z}_2 = R_2 - j\left(\frac{1}{\omega C_2}\right)$$

$$= 200 - j\left[\frac{1}{2\pi \times 1 \times 10^3 \times 5 \times 10^{-6}}\right] = 200 - j\ 31.83\ \Omega$$

$$\bar{Z}_3 = R_3 - j\left(\frac{1}{\omega C_3}\right)$$

$$= 500 - j\left[\frac{1}{2\pi \times 1 \times 10^3 \times 0.2 \times 10^{-6}}\right] = 500 - j\ 795.77\ \Omega$$

When the bridge is balanced,

$$\bar{Z}_1 \bar{Z}_4 = \bar{Z}_2 \bar{Z}_3$$

$$\therefore \quad \bar{Z}_4 \text{ (or } \bar{Z}_x) = \frac{\bar{Z}_2 \bar{Z}_3}{\bar{Z}_1} = \frac{(200 - j\,31.83)(500 - j\,795.77)}{200}$$

$$= \frac{(202.517 \angle -9.04°)(939.81 \angle -57.86°)}{200}$$

or, $\bar{Z}_x = 951.64 \angle -66.9° = (373.36 - j\ 875.3)\ \Omega$

$\therefore$ $R_x = \mathbf{373.36\ \Omega}$ **(Ans.)**

and, $X_C = \dfrac{1}{2\pi f C_x} = 875.3$

or, $C_x = \dfrac{1}{2\pi \times 1 \times 10^3 \times 875.3} = \mathbf{0.1818\mu F}$ **(Ans.)**

Example 6.29. *The following data relate to the bridge shown in Fig. 6.45.*

$R_1 = 600\Omega$; $C_1 = 1\mu F$; $R_2 = 100\Omega$; $R_3 = 1000\Omega$

Find the values of R_x and L_x at balance.

Solution. Refer to Fig. 6.45.

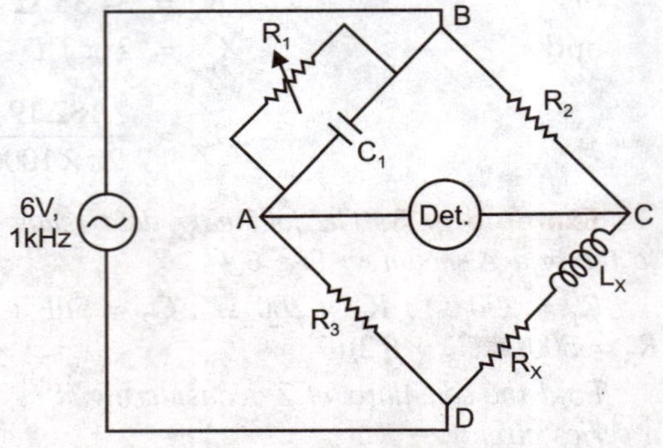

Fig. 6.45.

$$\bar{Z}_1 = R_1 \parallel X_{C_1}$$

or, $$\bar{Y}_1 = \frac{1}{R_1} + j\omega C_1$$

$$= \frac{1}{600} + j(2\pi \times 1000 \times 1 \times 10^{-6})$$

$$= 1.67 \times 10^{-3} + j\ 6.283 \times 10^{-3}$$

$$\bar{Z}_2 = R_2 = 100\Omega$$

$$\bar{Z}_3 = R_3 = 1000\Omega$$

$$\bar{Z}_4 = \bar{Z}_x = R_x + jX_L$$

Using the basic balance equation, we get :

$$\bar{Z}_1 \bar{Z}_4 = \bar{Z}_2 \bar{Z}_3$$

or, $$\bar{Z}_4 (= \bar{Z}_x) = \frac{\bar{Z}_2 \bar{Z}_3}{\bar{Z}_1} = \bar{Z}_2 \bar{Z}_3 \bar{Y}_1$$

$$= 100 \times 1000 \times [1.67 \times 10^{-3} + j\ 6.283 \times 10^{-3}]$$

$$= (167 + j\ 628.3)\ \Omega = (R_x + j\ X_L)\ \Omega$$

$\therefore$ $$R_x = \mathbf{167\ \Omega}\ \text{(Ans.)}$$

$$X_{L_x} = 628.3\ \Omega = 2\pi f L_x$$

$\therefore$ $$L_x = \frac{628.3}{2\pi \times 1000} \simeq \mathbf{0.1\ H}\ \text{(Ans.)}$$

6.3.11. Wien Bridge (*For measurement of* "Frequency")

The Wien bridge is *primarily known as a "frequency" determining bridge*. It can also be used for the *measurement of an unknown capacitor with great accuracy*.

Fig. 6.46 shows a Wien bridge. It has a series *R-C* combination in one arm and a parallel combination in the adjoining arm.

The impedance of one arm is,

$$\bar{Z}_1 = R_1 - \frac{j}{\omega C_1}$$

The admittance of parallel arm is

$$\bar{Y}_3 = \frac{1}{R_3} + j\omega C_3$$

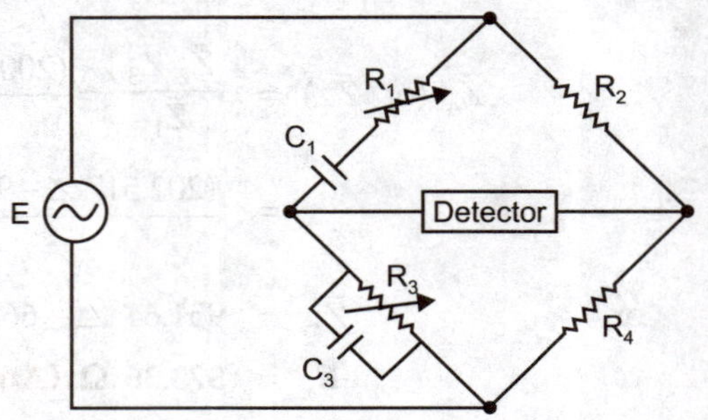

Fig. 6.46. Wien bridge for measuring frequency.

Using the bridge balance equation, we have

$$\bar{Z}_1 \bar{Z}_4 = \bar{Z}_2 \bar{Z}_3$$

or, $$\bar{Z}_1 \bar{Z}_4 = \bar{Z}_2 \times \frac{1}{\bar{Y}_3}$$

or, $$\bar{Z}_2 = \bar{Z}_1 \bar{Z}_4 \bar{Y}_3$$

or, $$R_2 = \left(R_1 - \frac{j}{\omega C_1}\right) R_4 \left(\frac{1}{R_3} + j\,\omega\, C_3\right) \qquad \text{...(6.95)}$$

Expanding this expression, we get

$$R_2 = \frac{R_1\, R_4}{R_3} + j\,\omega\, R_1\, R_4\, C_3 - \frac{jR_4}{\omega\, C_1\, R_3} + \frac{R_4\, C_3}{C_1}$$

or, $$R_2 = \left(\frac{R_1\, R_4}{R_3} + \frac{R_4\, C_3}{C_1}\right) - j\left(\frac{R_4}{\omega\, C_1\, R_3} - \omega\, C_3\, R_1\, R_4\right) \qquad \text{...(6.96)}$$

Equating the *real terms*, we obtain

$$R_2 = \frac{R_1\, R_4}{R_3} + \frac{R_4\, C_3}{C_1}$$

or, $$\frac{R_2}{R_4} = \frac{R_1}{R_3} + \frac{C_3}{C_1} \qquad \text{...(6.97)}$$

Equating the *imaginary terms*, we have

$$\omega\, C_3 R_1 R_4 = \frac{R_4}{\omega\, C_1\, R_3} \qquad \text{...(6.98)}$$

or, $$\omega^2 = \frac{1}{C_1\, C_3\, R_1\, R_3} \qquad \text{...[6.98 (a)]}$$

or, $$\omega = \frac{1}{\sqrt{C_1 C_3 R_1 R_3}}$$

Since, $\omega = 2\pi f$, therefore,

$$2\pi f = \frac{1}{\sqrt{C_1\, C_3\, R_1\, R_3}}$$

or, $$f = \frac{1}{2\pi\sqrt{C_1\, C_3\, R_1\, R_3}} \qquad \text{...(6.99)}$$

We observe that the two conditions for bridge balance, (eqns. 6.97, 6.99), result in an expression determining the required resistance ratio $\frac{R_2}{R_4}$, and another expression determining the frequency of the applied voltage. In other words, if we satisfy eqn. (6.97), and also excite the bridge with a frequency described by eqn. (6.99), the bridge will be in balance.

In most Wien bridge circuits, the components are chosen such that $R_1 = R_3$ and $C_1 = C_3$; this reduces eqn. (6.97) to $\frac{R_2}{R_4} = 2$ and eqn. (6.99) to

$$f = \frac{1}{2\pi RC} \qquad \text{...(6.100)}$$

which is *general expression* for the frequency of the Wien bridge.

— In a practical bridge, capacitors C_1 and C_3 are *fixed* capacitors, and resistors R_1 and R_3 are variable resistors *controlled by a common shaft*. Provided now that $R_2 = 2R_4$, the bridge may be used as a frequency-determining device balanced by a single *control. This control may be calibrated directly in terms of frequency.*

- By the use of this bridge an *accuracy of 0.5% - 1% can be easily obtained.*

- Because it is frequency sensitive, it is difficult to balance unless the waveform of the applied voltage is purely sinusoidal.

Applications :

1. The bridge is employed for *measuring frequency in the audio range*.

— The audio range is normally divided into 20Hz - 200Hz - 2kHz - 20kHz ranges. In this case, the resistances can be used for range changing and capacitors C_1 and C_3 for fine frequency control within the range.

2. The bridge can also be used for *measuring capacitances*; in this case operational frequency must be known.
3. The bridge is also used in a harmonic distortion analyser, as a notch filler, and in audio frequency and radio frequency oscillators as a frequency determining element.

Example 6.30. *Determine the equivalent parallel resistance and capacitance that causes a Wien bridge (Fig. 6.46) to null with the following component values :*

$R_1 = 2.8k\Omega$; $C_1 = 4.8\mu F$; $R_2 = 20k\Omega$; $R_4 = 80k\Omega$; $f = 2$ kHz;

Solution. *Given* : $f = 2\text{kHz}$ or $\omega = 2\pi f = 2\pi \times (2 \times 1000) = 12.57 \times 10^3$ rad/s.

$$\mathbf{R_3 = ?,\ C_3 = ?}$$

We know that :

$$\frac{R_2}{R_4} = \frac{R_1}{R_3} + \frac{C_3}{C_1} \qquad \text{...(Eqn. 6.97)}$$

and

$$\omega^2 = \frac{1}{C_1 C_3 R_1 R_3} \qquad \text{...(Eqn. 6.98 }(a))$$

or,

$$C_3 = \frac{1}{w^2 C_1 R_1 R_3}$$

Substituting for C_3 in eqn. (6.97), we get

$$\frac{R_2}{R_4} = \frac{R_1}{R_3} + \frac{1}{\omega^2 C_1^2 R_1 R_3}$$

$$\frac{R_2}{R_4} = \frac{1}{R_3}\left(R_1 + \frac{1}{\omega^2 C_1^2 R_1}\right)$$

or,

$$R_3 = \frac{R_4}{R_2}\left(R_1 + \frac{1}{\omega^2 C_1^2 R_1}\right) \qquad \text{...}(i)$$

$$= \frac{80\times10^3}{20\times10^3}\left[2.8\times10^3 + \frac{1}{(12.57\times10^3)^2\,(4.8\times10^{-6})^2\,(2.8\times10^3)}\right]$$

$$= 4\ (2.8 \times 10^3 + 0.098) = \mathbf{11.2\ k\Omega}\ \textbf{(Ans.)}$$

Now,

$$C_3 = \frac{1}{\omega^2 C_1 R_1 R_3} \qquad \text{Rearranging (Eqn. 6.98}a)$$

$$= \frac{1}{(12.57\times10^3)^2\times(4.8\times10^{-6})\times(2.8\times10^3)\times11.2\times10^3}$$

$$= \mathbf{42.04\ pF}\ \textbf{(Ans.)}$$

6.3.12. Resonance Bridge

Fig. 6.47 shows a resonance bridge, consisting of a series resonance circuit :

— The series resonance circuit is formed by R_4, C_4 and L_4 in series.

— All the other arms consist of resistors only.

Using the bridge balance equation, we have

$$Z_1 Z_4 = Z_2 Z_3$$

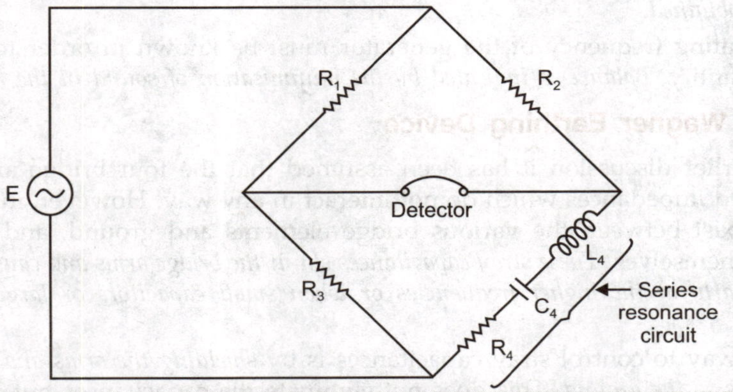

Fig. 6.47. Resonance bridge

Here, $\bar{Z}_1 = R_1$; $\bar{Z}_2 = R_2$; $\bar{Z}_3 = R_3$; $\bar{Z}_4 = R_4 + j\omega L_4 - \dfrac{j}{wC_4}$

Substituting the values in the balance equation, we get

$$R_1\left(R_4 + j\omega L_4 - \frac{j}{\omega C_4}\right) = R_2 R_3$$

or, $$R_1 R_4 + j\,\omega\, L_4 R_1 - \frac{jR_1}{\omega C_4} = R_2 R_3$$

Equating the *real* and *imaginary terms*, we have

$$R_1 R_4 = R_2 R_3$$

$$\therefore \qquad R_4 = \frac{R_2 R_3}{R_1}$$

and, $$\omega L_4 - \frac{1}{\omega C_4} = 0$$

or, $$\omega^2 = \frac{1}{L_4 C_4}$$

or, $$(2\pi f)^2 = \frac{1}{L_4 C_4} \quad \text{or} \quad 2\pi f = \frac{1}{\sqrt{L_4 C_4}}$$

or *Resonance frequency*, $$f = \frac{1}{2\pi\sqrt{L_4 C_4}} \qquad ...(6.101)$$

- This bridge can be *used to measure unknown inductances or capacitances*. The resistor R_4 can be measured by keeping a fixed ratio $\frac{R_3}{R_1}$ and using a standard variable resistance to obtain balance.
  — If an "*inductance*" is *being measured, a standard capacitor is varied unitl balance is obtained.*
  — If a "*capacitance*" is *being measured, a standard inductor is varied until balance is obtained.*

The operating frequency of the generator must be known in order to calculate the unknown quantity. *Balance is indicated by the minimisation of sound in the headphones.*

6.3.13. Wagner Earthing Device

In the earlier discussion it has been assumed that the four bridge arms consist of simple lumped impedances which do not interact in any way. However, in practice, *stray capacitances* exist between the various bridge elements and ground, and also between bridge arms themselves. *These stray capacitances shunt the bridge arms and cause measurement error, particularly at the higher frequencies or when small capacitors or large inductors are measured.*

- One way to control stray capacitances is by *shielding the arms and connecting the shields to the ground*. This does not eliminate the capacitances but at least makes them constant in value, and they can therefore be *compensated*.
- One of the most widely used methods for eliminating some of the effects of stray capacitance in a bridge circuit is the *Wagner earth (ground) connection*. This circuit *eliminates the troublesome capacitance which exists between the detector terminals and the ground.*

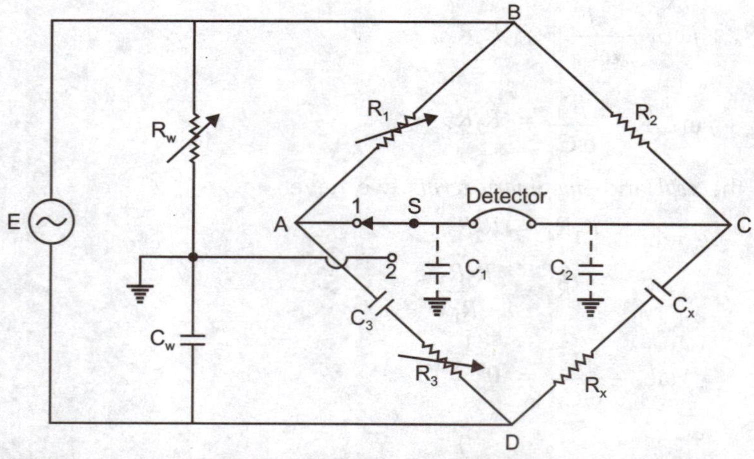

Fig. 6.48. Wagner earthing device.

Fig. 6.48 shows the circuit of a capacitance bridge. C_1 and C_2 represent the stray capacitances. The oscillator is removed from its usual ground connection and bridged by a series combination of resistor R_w and capacitor C_w. The junction of R_w and C_w is grounded and is called *Wagner ground connection*.

The initial adjustment of the bridge involves the following *procedure* :

— The detector is connected to point 1, and R_1 is adjusted for null or minimum sound in the headphones.

— The switch is then thrown to position 2, which connects the detector to Wagner ground point. Resistor R_w is now adjusted for minimum sound.

— When the switch is thrown to position 1 again, some unbalance will probably exist. Resistors R_1 and R_3 are then adjusted for minimum detector response, and the switch is again thrown to position 2.

This procedure is repeated till a null is reached on both switch positions; under this condition points 1 and 2 are at the same potential and this is *ground potential*. Stray capacitances C_1 and C_2 are then effectively shorted out and have no effect on normal bridge balance.

The capacitances from point *B* to *D* to ground are also eliminated by the addition of Wagner ground connection, since current through these capacitances will enter the Wagner ground connection.

The capacitances across the bridge arms are *not* eliminated by this earthing device and they will still *affect* the accuracy of the measurement.

- Because the addition of Wagner earthing device does not affect the balance conditions, the *measurement procedure* remains *unchanged*.

6.3.14. Types of Detectors

Following types of detectors are used in various applications :

1. *For low frequency* ... Vibration galvanometer – A moving magnet vibration galvanometer has a range of 1.5 k Hz
2. *For ordinary laboratory work at frequencies up to a few 100 Hz...* Moving coil type of instrument; it has a high sensitivity.
3. *For higher frequencies (> 800 Hz)* ... Telephone (headphone)–*best detector* (Vibration galvanometers and headphones have no phase selectivity, *i.e.*, they do not indicate whether it is resistance or reactance adjustments that are required).
4. *For frequencies above about 3 kHZ, and particularly at high AF or RF...* Heterodyne or beat tone detector.

☞ Following points are worth noting :

- The A.C. galvanometer and separately excited dynamometer are phase selective, and are best suited at low frequencies.
- An electrodynamometer can also be used as an A.C. detector.
- Electrometers are used as detectors because small capacitances possess a very large impedance when used with A.C. circuits at low frequency, and when measured in the bridge they form a high impedance branch. Hence, this detector is used to *increase sensitivity*.
- In several cases, especially in bridges for routine use, a pointer instrument is used. It is advantageous if it can be made phase selective. These pointer instruments are generally moving coil milliammeters operated with some arrangement of copper oxide rectifiers (frequency range 40 Hz - 1 kHz)
- *Modern bridge techniques employ the amplifier as a regular feature.*
- With all detectors, the impedance should be selected to suit that of the bridge. A higher sensitivity can be obtained by using an interbridge transformer.

— When *headphones* are used as detectors, precautions should be taken *to eliminate capacitance effects* between the observer and the headphones.

6.3.15. Sources of Errors in Bridge Circuits and the Elimination of Errors

Factors causing errors :

The following *factors cause errors in A.C. bridge circuit* :

1. Stray-conductance effects, due to imperfect insulation.
2. Mutual-inductance effects, due to magnetic coupling between various components of the bridge.
3. Stray-capacitance effects, due to electrostatic fields between conductors at different potentials.
4. 'Residues' in components – *e.g.*, the existence of small amount of series inductance or shunt capacitance in nominally non-reactive resistors.

Methods of reducing bridge errors :

The important *methods of reducing bridge errors are* :

1. *Use of high-quality components*.
2. *Proper bridge lay-out*.

The bridge lay-out should take care of the following *requirements* :

— All the leads to a single corner should be brought as nearly as possible to the same point.

— A pair of leads to a component should not form a large loop; if possible the leads should be co-axial.

— If there is more than one inductor in the bridge, leads may have to be quite long.

By doing so, the stray mutual inductance coupling and lead self-inductances are reduced to minimum and the stray capacitances are definitely located across individual arms of bridge; they can be measured and compensated for relatively easily.

3. *Operation of bridge at sufficient sensitivity*. In order to obtain a high degree of accuracy *e.g.* 1 part in 10^4 or better, it is necessary that a bridge must be operated at sufficient sensitivity which implies that the change in voltage at the detector terminals from the balance conditions should be large when the detector is considered open.

4. *Use of insulating stands*. When the insulation between various components/elements of a bridge is not good, trouble may arise because of leakage currents from one arm to another. This is specially true in case of high impedances bridges. It can be avoided by mounting the various bridge components and other pieces of apparatus on insulating stands.

5. In order to *avoid 'eddy current errors'* the presence of large conducting masses near the bridge work should be avoided.

6. *Residual errors*. Residues mean small inherent inductance or capacitance of the resistors. In precise work it becomes necessary to evaluate them in order to eliminate them or compensate for errors due to these. The self-inductance is important only when the coils used are multi-turn coils and supply used is of high frequency.

7. *Frequency and waveform errors*. These errors are overcome by employing *wave filters* which eliminate the unwanted harmonics from the source or by employing *tuned detectors* in place of headphones such as vibration galvanometers which do not respond to harmonics and respond readily only to the fundamental for which they are tuned.

HIGHLIGHTS

1. Resistances are classified as follows :
   (i) Low resistances — of the order of 1 Ω and under
   (ii) Medium resistances — From about 1 Ω upwards to about 100 kΩ
   (iii) High resistances — Of the order of 100 kΩ and upwards
2. When balancing an A.C. bridge, the following two conditions must be met simultaneously
   (i) The magnitudes of the impedances should satisfy the relationship :
   $$Z_1Z_4 = Z_2Z_3$$
   (ii) The phase angle of impedances should satisfy the relationship :
   $$\angle\theta_1 + \angle\theta_4 = \angle\theta_2 + \angle\theta_3$$

OBJECTIVE TYPE QUESTIONS

Choose the Correct Answer :

1. The value of resistance of an earthing electrode depends upon
   (a) shape and material of electrode.
   (b) depth to which electrode is driven into earth.
   (c) specific resistance of soil.
   (d) all of the above.
2. From the point of view of safety, the resistance of earthing electrode should be :
   (a) low
   (b) high
   (c) medium
   (d) the value of resistance of earth electrodes does not affect the safety.
3. A Wheatstone bridge cannot be used for precision measurements because errors are introduced into it on account of
   (a) resistance of connecting leads (b) thermo-electric emfs
   (c) contact resistances (d) all of the above.
4. Low resistances are provided with four terminals
   (a) to facilitate the connection of current and potential circuits.
   (b) in order that the resistance value becomes definite irrespective of the nature of contacts at the current terminals.
   (c) to eliminate the effect of thermo-electric emfs.
   (d) to eliminate the effect of leads.
5. In a Kelvin's double bridge two sets of readings are taken when measuring a low resistance, one with the current in one direction and the other with direction of current reversed. This is done to
   (a) eliminate the effect of contact resistance.
   (b) eliminate the effect of resistance of leads.
   (c) correct for changes in battery voltage.
   (d) eliminate the effect of thermo-electric emfs.
6. A resistance of value 10 Ω approximately is to be measured by ammeter-voltmeter method with resistance of ammeter as 0.02 Ω and that of voltmeter as 5000 Ω. The resistance should be measured

(a) by connecting the ammeter on the side of unknown resistance as this connection gives better accuracy.
(b) by connecting the voltmeter on the side of unknown resistance as this connection gives better accuracy.
(c) by any of the two connections, as both of them give equal accuracy.
(d) none of the above.

7. Equal resistances of 100 Ω each are connected in each arm of a Wheatstone bridge which is supplied by a 2 V battery source. The galvanometer of negligible resistance connected to the bridge can sense as low current as 1 μA. The smallest value of resistance that can be measured is
(a) 20 mΩ (b) 2 μΩ
(c) 20 μΩ (d) none of the above.

8. A Wheatstone bridge has ratio arms of 1000 Ω and 100 Ω resistances, the standard resistance arms consist 4 decade resistance boxes of 1000, 100, 10, 1 Ω steps. The maximum and minimum values of unknown resistance which can be determined with this setup are
(a) 111100 Ω, 1 Ω (b) 11110 Ω, 10 Ω
(c) 111100 Ω, 10 Ω (d) none of the above.

9. High resistances are provided with a guard terminal. This guard terminal is used to
(a) bypass the leakage current.
(b) guard the resistance against stray electrostatic fields.
(c) guard the resistance against overloads.
(d) none of the above.

10. When measuring insulation resistance of cables using d.c. source, the galvanometer used should be initially short-circuited because
(a) cables have a low value of initial resistance.
(b) cables have a high value of capacitance which draws a high value of charging current.
(c) cables have a low value of capacitance which draws a high value of charging current.
(d) none of the above.

11. The advantage of Varley loop tests over Murray loop tests is
(a) they can be used for localizing of short circuit faults.
(b) they can be used for localizing of earth fault.
(c) the loop resistance can be experimentally determined.
(d) their accuracy is higher.

12. When localizing ground fault with the help of loop tests, the resistance of the fault
(a) affects the balance conditions.
(b) affects the value of cable resistance.
(c) affects the sensitivity of the bridge.
(d) all of the above.

13. A bridge circuit works at a frequency of 2 kHz. The following can be used as detectors for detection of null conditions in the bridge :
(a) Vibration galvanometers and headphones.
(b) Headphones and tunable amplifiers.
(c) Vibration galvanometers and tunable amplifiers.
(d) Vibration galvanometers, headphones and tunable amplifiers.

14. Wagner's Earth Devices are used in A.C. bridge circuits for
(a) eliminating the effect of earth capacitances.
(b) eliminating the effect of inter-component capacitances.

(c) eliminating the effect of stray electrostatic fields.
(d) shielding the bridge elements.

15. Maxwell's inductance-capacitance bridge is used for measurement of inductance of
(a) low Q coils (b) medium Q coils
(c) high Q coils (d) low and medium Q coils.

16. The advantage of Hay's bridge over Maxwell's inductance-capacitance bridge is because
(a) its equations for balance do not contain any frequency term.
(b) it can be used for measurement of inductance of high Q coils.
(c) it can be used for measurement of inductance of low Q coils.
(d) none of the above.

17. The equations under balance conditions for a bridge are
$R_1 = R_2R_3/R_4$ and $L_1 = R_2R_3C_4$
where R_1 and L_1 are respectively unknown resistance and inductance. In order to achieve converging balance
(a) R_2 and R_3 should be chosen as variables (b) R_2 and C_4 should be chosen as variables
(c) R_4 and C_4 should be chosen as variables (d) R_3 and C_4 should be chosen as variables.

18. Frequency can be measured by using
(a) Maxwell bridge (b) Schering bridge
(c) Heaviside Campbell bridge (d) Wien bridge.

19. The resonant frequency of an RLC series circuit is 1.5 MHz with the tuning capacitor set at 150 pF. The bandwidth is 10 kHz. The effective resistance of the circuit is :
(a) 29.5 Ω (b) 14.75 Ω
(c) 9.4 Ω (d) 4.7 Ω

ANSWERS

1. (d)	**2.** (a)	**3.** (d)	**4.** (b)	**5.** (d)	**6.** (c)	**7.** (a)
8. (c)	**9.** (a)	**10.** (b)	**11.** (c)	**12.** (c)	**13.** (b)	**14.** (a)
15. (b)	**16.** (b)	**17.** (c)	**18.** (d)	**19.** (d).		

THEORETICAL QUESTIONS

Measurement of Resistances

1. How are resistances classified ?
2. Explain what do you mean by low, medium and high resistances. Suggest various suitable methods for measuring them giving justification.
3. State the various methods of measurement of low resistance. Why is the voltmeter-ammeter method unsuitable for the precise measurement of low resistance ?
4. What do you understand by low, medium and high resistance ? Describe Kelvin's double bridge method for measurement of small resistances.
5. What are the problems associated with measurement of low resistances ? How are they overcome through use of Kelvin's double bridge ? Derive the expression for the unknown resistance in the case of Kelvin's double bridge.
6. Explain the procedure of measuring a low resistance with the help of Kelvin's double bridge. Derive the relation for finding unknown resistance.
7. Explain the theory and working principle of Kelvin's double bridge method for measurement of low resistance.
8. Explain with the help of a connection diagram how you would determine the value of

low resistance shunt by Kelvin's double arm bridge method. What are the important precautions that are to be taken in such an experiment ?

9. Describe Kelvin's double bridge for the comparison of two low resistances. Give the theory of the bridge and arrangement necessary in order that the greatest precision possible may be obtained.
10. Describe Carey Foster's slide wire bridge for the measurement of medium resistances.
11. Describe a method for measurement of insulation resistance of a cable.
12. What are the special problems associated with the measurement of high resistances of the order of mega ohms ?
13. Describe with the help of diagram the loss of charge method for determining the insulation resistance of a length of cable.
14. Describe price-guard wire method of measuring high resistance with the help of diagram.
15. Describe the loss of charge method for the determination of high resistance.
16. Describe the 'loss of charge' method of measuring high resistance and prove the formula used. Mention the possible errors and suggest methods to minimise these.
17. Explain the loss of charge method for the measurement of insulation resistance of a cable.
18. Describe and explain with the help of neat sketches the construction and working of megger.
19. Explain how will you measure insulation resistance (i) 2-wire D.C. live main and (ii) 3-wire D.C. live main.
20. What is a *"bridge circuit"* ?
21. What are the applications of bridge circuits?
22. What are the advantages of bridge circuits ?
23. What are the types of bridges ?
24. Explain briefly how a 'Wheatstone bridge' is used for measurement of resistance.
25. What are the sources of measurement errors in a Wheatstone bridge?
26. State the limitations of "Wheatstone bridge".
27. Explain briefly the following bridges :
    (*a*) Kelvin bridge; (*b*) Kevin double bridge.
28. Draw the block diagram of Wheatstone bridge with digital readout and explain it briefly.
29. Draw the general form of an A.C. bridge and derive the general equation for its balance.
30. Explain briefly how *"inductance"* can be measured by using a 'Maxwell bridge'.
31. What are the advantages and disadvantages of a Maxwell bridge ?
32. Discuss briefly how "Hay bridge" can be used for the measurement of inductance.
33. With the help of circuit diagram explain how capacitance can be measured by the use of a 'Schering bridge'.
34. Describe briefly how Wien bridge can be used for the measurement of frequency.
35. State the applications of a Wien bridge.
36. Explain briefly *"Resonance bridge"* with the help of a neat circuit diagram. Name the parameters which can be measured by this bridge.
37. Explain with a neat circuit diagram the working of a "Wagner earthing device".
38. How are detectors classified ?
39. What precautions should be taken while using a bridge ?
40. Draw a neat diagram of the Anderson bridge. Deduce the equations when the bridge is under balance condition. Explain clearly how you may measure self-inductance by using this bridge. Draw the phasor diagram of the voltages and currents of the bridge arms at balance.

41. Prove that in Wien bridge the value of the resistance of the capacitor at balance is given by :

$$R_1 = \frac{R_3 (1 + \omega^2 R_2^2 C_2^2)}{\omega^2 R_2 R_4 C_2^2}$$

where ω, C_2, R_2, R_3, R_4 are known values with usual notation.

What are the uses of the above bridge in electrical engineering ?

42. Describe with connection diagram the Heaviside Campbell bridge for precise measurement of self-inductance of a coil. Deduce the formula used.

43. Discuss the common sources of error in A.C. bridges. How are they eliminated ?

44. Write short notes on the following :

(*i*) De Sauty bridge

(*ii*) Carey Foster bridge

(*iii*) Wagner earthing.

UNSOLVED EXAMPLES

1. A resistance is measured by the ammeter-voltmeter method. With the voltmeter connected across the resistance the readings on the ammeter and voltmeter are 0.4 A and 3.2 V respectively. The resistance of the voltmeter is 500 Ω.

Calculate :.

(*i*) True value of resistance.

(*ii*) Percentage error in the value of the resistance if the voltmeter current is ignored.

[**Ans.** (*i*) 8.13 Ω ; (*ii*) –1.6%]

2. In a measurement of resistance by potentiometer, the voltage drops across a resistor under test and across a 0.02 Ω standard resistor were found to be 0.735 V and 0.98 V respectively. Determine the value of resistor under test. [**Ans.** 0.015 Ω]

3. Determine the insulation resistance of a short length of cable in which voltage falls from 125 to 100 volts in 25 seconds. The capacity of the condenser is 600×10^{-12} F.

[**Ans.** 1,86,726 MΩ]

4. The insulation resistance of 2 metre cable was measured by the loss of charge method. The voltage across the standard capacitor of 0.003 μF falls from 222 V to 155 V in one minute. Calculate the insulation resistance of the cable. Derive the formula used.

What will be the insulation resistance if the length of the cable is doubled ?

[**Ans.** 55,670 MΩ]

5. Calculate the insulation resistance of a short length of cable which was tested for insulation resistance by 'loss of charge' method. An electrostatic voltmeter of infinite resistance is connected between the cable and earth forming a joint capacitance of 0.0006 μF. It is observed that after charging voltage falls from 300 volts to 100 volts in one minute.

[**Ans.** 91,024 MΩ]

6. Calculate the insulation resistance of a length of cable in which voltage falls from 100 to 80 volts in 20 seconds, the capacity being 0.0003 μF. [**Ans.** 299,000 MΩ]

7. A cable is tested by loss of charge method using a ballistic galvanometer with the following results : discharged immediately after electrification, deflection 250 divisions; discharged 25 seconds after electrification (*i*) deflection 150 divisions (*ii*) When in parallel with a resistance of 7 MΩ deflection 100 divisions. Calculate the insulation resistance of the cable. [**Ans.** 5.56 MΩ]

8. Fig. 6.6 consists of the following parameters :

R_1 = 10 kΩ; R_2 = 15 kΩ ; R_3 = 40 kΩ

Determine the unknown resistance R_x. [**Ans.** 60 kΩ]

9. In a Kelvin double bridge, there is error due to mismatch between the ratios of outer and inner arm resistances. The following data relate to this bridge :

Standard resistance = 100.03 μΩ; Inner ratio arms = 100.31 Ω and 200 Ω; Outer ratio arms = 100.24 Ω and 200 Ω; The resistance of the connecting leads from standard to unknown resistance is 700 μΩ.

Calculate the unknown resistance. **[Ans.** 49.97 μΩ]

10. The impedances of the basic A.C. bridge of Fig. 6.21 are given as follows :

$\bar{Z}_1 = 100\ \Omega\angle 80°$ (inductive impedance); $\bar{Z}_2 = 250\ \Omega$ (pure resistance);

$\bar{Z}_3 = 400\ \Omega\ \angle 30°$ (inductive impedance); $\bar{Z}_4$ = Unknown

Determine the constants of the unknown arm. **[Ans.** 1000 $\angle -50°$]

11. A Maxwell bridge (Fig. 6.24) is used to measure an inductive impedance. The bridge constants at balance are :

$R_1 = 470$ kΩ; $C_1 = 0.01$ μF; $R_2 = 5.1$ kΩ ; $R_3 = 100$ kΩ

Find the series equivalent of the unknown impedance. **[Ans.** $R_x = 1.09$ kΩ; $L_x = 5.1$ H]

12. The Schering bridge shown in Fig. 6.34 has the following constants :

Arm AB ... Capacitor of 0.5 μF in parallel with 1 kΩ resistance;

Arm BC ... Resistance of 3 kΩ;

Arm CD ... Unknown capacitor C_x and R_x in series

Arm DA ... Capacitor of 0.5 μF

Frequency ... 1 kHz

Determine the following :

(*a*) Unknown resistance and capacitance.

(*b*) Dissipation factor. **[Ans.** (*a*) 2Ω, 0.25 μF; (*b*) 3.1416]

13. An A.C. bridge is connected as shown in Fig. 6.49. Determine the resistance and inductance of AB. **[Ans.** 400 Ω ; 1.12 H]

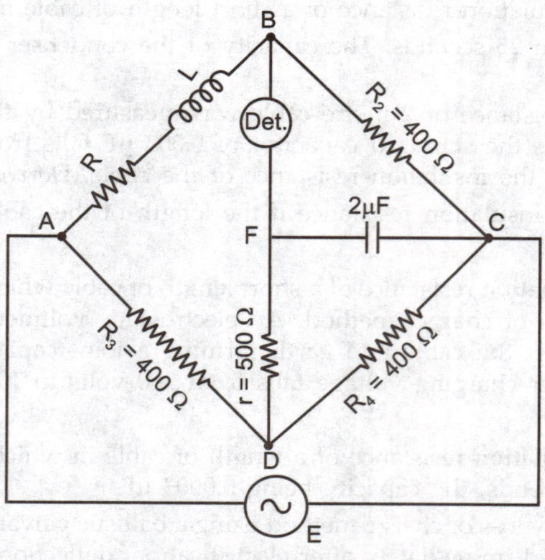

Fig. 6.49.

14. A dielectric sheet of thickness 1 mm is tested at 50 Hz between two electrodes of 10 cm diameter. The Schering bridge employed has a standard compressed air capacitor C_3 of 100 pF; a non-inductive resistor R_4 of 350 Ω in parallel with a variable capacitor C_4 and a non-inductive variable resistor R_2. At balance $C_4 = 0.4$ μF, $R_2 = 250$ Ω.

Calculate power factor and permittivity of the sheet. **[Ans.** 0.04398 ; 2]

15. A capacitor is tested by a Schering bridge. It forms one arm *AB* of the bridge. The other arms are : *AD* – a non-reactive resistance of 100 Ω ; *DC* – a non-reactive resistance of 300 Ω shunted by a capacitor of 0.5 μF ; *BC* – a standard loss-free capacitor of 100 pF, the supply frequency is 50 Hz. The bridge is at balance with the above components.

Deduce the balance contions of the bridge and hence find out the capacitance and power factor of the capacitor under test. [**Ans.** 298.7×10^{-12} F ; 0.0471]

16. A sample of bakelite was tested by the Schering bridge method at 25 kV, 50 Hz. Balance was obtained with a standard condenser of 109 pF capacitance, a condenser of capacitance 0.5 μF in parallel with a non-reactive resistor of 309 Ω and a non-reactive resistor of 100 Ω. Determine the equivalent series resistance and the power factor of the specimen. [**Ans.** 0.458 MΩ ; 0.0485]

17. A balanced 1 kHz bridge has the following configuration :

Arm *AB* : R_1 = 1000 Ω in parallel with C_1 = 0.053 μF ; *BC* : R_2 = 1500 Ω in series with C_2 = 0.53 μF ; *CD* : the unknown, *DA* : pure capacitance C_4 = 0.265 μF.

Determine *R* and *L* or *C* constants of the unknown. Draw the phasor diagram of the bridge at above frequencey. [**Ans.** *R* = 119.46 Ω ; 0.1656 μF]

18. Determine the equivalent parallel resistance and capacitance that causes a Wien bridge (Fig. 6.46) to null with the following component values :

R_1 = 3.1 kΩ; C_1= 5.2 μF ; R_2 = 25 kΩ; R_4 = 100 kΩ; *f* = 2.5 kHz

[**Ans.** 12.4 kΩ ; 20.3 pF]

19. The following data were obtained during the measurement of the distributed or self-capacitance of a coil :

First measurement : f_1 = 2 MHz and C_1 = 460 pF;

Second measurement : f_2 = 4 MHz and C_2 = 100 pF.

Find the distributed capacitance, C_d. [**Ans.** 20 pF]

20. Calculate the value of distributed capacitance of a coil when the following measurements are made :

At frequency f_1 = 2MHz, the tuning capacitor is set at 450 pF;

At frequency f_2 = 5MHz, the tuning capacitor is tuned at 60 pF. [**Ans.** 14.28 pF]

CHAPTER

7 Magnetic Measurements

7.1. Introduction to magnetic materials – Terms connected with magnetic materials – Magnetic parameters – Classification of magnetic materials – Principal ferromagnetic elements – Ferromagnetism – Magnetic domains. **7.2. Classification of magnetic measurements; 7.3. Ballistic (D.C.) tests** – Measurement of flux/flux density – Measurement of value of magnetising force (H) – Determination of B-H curve – Determination of hysteresis loop – Determination of leakage factor in dynamo-electric machinery – Magnetic potentiometer – Testing bar specimens of magnetic materials – Bar specimens versus ring specimens. **7.4. Permeameters** – Hopkinson permeameter (Bar and Yoke method) – Ewing double bar permeameter – Illiovici permeameter – Burrows permeameter – Fahy's simplex permeameter – The National Physical Laboratory (England) form of permeameter. **7.5. A.C. magnetic testing; 7.6. Methods of measurement of Iron losses** – Wattmeter method – A.C. bridge methods – A.C. potentiometer method – Oscillographic method. **7.7. Methods of measurement of air gap flux and field strength** – Measurement of air gap flux – Measurement of field strength by withdrawal of search coil – By nuclear magnetic resonance – Use of bismuth spiral – Use of mumetal wires – Use of Hall effect devices. **7.8. Testing of permanent magnets; 7.9. Testing of Feebly magnetic materials; 7.10. Precautions in Magnetic Testing** – *Highlights – Objective Type Questions – Theoretical Questions – Unsolved Examples.*

7.1 INTRODUCTION TO MAGNETIC MATERIALS

- ***Magnetic materials** are those materials in which a state of magnetisation can be induced.* Such materials when magnetised create a magnetic field in the surrounding space.
- In a material all the molecules contain electrons which orbit around the nucleus; these orbits are therefore, equivalent to circulating currents and so develop an m.m.f. Each m.m.f. due to an individual orbit, in most molecules, is neutralised by an opposite one. But in magnetic materials such as iron and steel *there are a number of unneutralised orbits,* such that a resultant axis of m.m.f. exists which *produces a magnetic dipole.*
- *In unmagnetised specimens,* because of mutual attraction and repulsion among the dipoles, the *molecular m.m.f. axes lie along continuous closed paths, and no external magnetic effect can be detected.*

In *magnetised specimens,* the dipoles line up parallel with the exciting m.m.f. and when the exciting m.m.f. is removed, a *number of dipoles may remain aligned in the direction of the external field* and thus exhibit *permanent magnetism.*

- The *readiness of a material to accept magnetism is expressed by its permeability.* For all materials except a few magnetic ones, the permeability is that of free space and is constant. It is denoted by μ_0 (= $4\pi \times 10^{-7}$).

For magnetic materials the permeability equals μ_0 times the relative permeability which is denoted by μ_r(*i.e.*, $\mu = \mu_0\mu_r$).

The *relative permeability varies with the degree of magnetisation of the material and may have a value* as high as 2500.

7.1.1. Terms Connected with Magnetic Materials

1. **Magnetic force.** It is the force exerted by one magnet on another either to attract it or repel it.
2. **Unit pole strength.** It is defined as the strength of that pole which when placed in vacuum at a distance of one metre from a similar and equal pole, repels it with a force of one newton.
3. **Magnetic flux density.** It is usually represented by letter B and is defined as the flux or lines of force passing per unit area through any substance through a plane at right angles to the direction of magnetic flux; it is measured in Wb/m^2 (or T, *i.e.*, Tesla).
4. **Magnetic field strength.** It may be defined in the following *two ways* :
   (*i*) Field strength at any point within a magnetic field is the number of lines of force passing through a unit area round the point considered and held perpendicular to the lines.
   (*ii*) Field strength at any point within a magnetic field is the force exerted by a unit north pole at that point.
5. **Relative permeability.** It is the ratio of the flux density (B) produced in that material to the flux density produced in vacuum by the same magnetising force (H). It is denoted by μ_r.
6. **Absolute permeability.** Absolute permeability of the material is the ratio of flux density in that material to the magnetising force producing that flux density and is denoted by μ : $\mu = \mu_0\mu_r$ where μ_0 is the permeability of free space having a value of $4\pi \times 10^{-7}$ H/m.
7. **Magnetic potential.** The magnetic potential at any point within a magnetic field is measured by the work done in carrying a unit north pole from infinity to that point against the force of magnetic field.
8. **Intensity of magnetisation.** It is defined as the pole strength per unit area of the bar or magnetic moment per unit volume of the bar. It is denoted by letter I.
9. **Susceptibility.** It is denoted by letter χ and is defined as the ratio of intensity of magnetisation (I) to magnetising force (H). In order words $\chi = \dfrac{I}{H}$.
10. **Magnetomotive force.** It is that force which drives or tends to drive the flux through a magnetic circuit. In short it is written as m.m.f. It is the product of number of conductors (N) and current (I) passing through them, *i.e.*, m.m.f. = NI.
11. **Magnetic reluctance.** It is that property of the material which opposes the production of magnetic flux in it.
12. **Coercive force.** It may be defined as the demagnetising force which is necessary to neutralise completely the magnetism in an electromagnetic after the value of magnetising force becomes zero.
13. **Remanence.** It is defined as the magnetic flux density which still persists in magnetic material even when the magnetising force is completely removed. It is expressed in Wb/m^2.

14. **Retentivity.** It is that property of magnetic material which is measured by its maximum value of the residual induction.
15. **Core loss, incremental.** The core loss in a magnetic material when subjected simultaneously to a D.C. biasing magnetising force and an alternating magnetising force.
16. **Core loss, specific.** The power expended per unit mass of magnetic material in which there is symmetrical, sinusoidally varying induction of a specified maximum value, B, at a specified frequency, f.
17. **Core loss, total.** The power expended in a magnetic specimen in which there is cyclically alternating induction.

Note. *Measurements of core loss are normally made with sinusoidally alternating induction, or the results are corrected for deviations from the sinusoidal condition.*

18. **Curie temperature.** The temperature above which a ferromagnetic material becomes paramagnetic.
19. **Cyclically magnetised condition.** A magnetic material is in a cyclically magnetised condition when, after having been subjected to a sufficient number of identical cycles of demagnetising force. It follows identical hysteresis loops on successive cycles which are not necessarily symmetrical with respect to the origin of the axes.
20. **Hysteresis loop, normal.** A closed curve obtained with a ferromagnetic material by plotting (usually to rectangular co-ordinates) corresponding D.C. value of magnetic induction B for ordinates and magnetising force H for abscissa when the material is passing through a complete cycle between equal definite limits of either magnetising force $\pm H_m$ or magnetic induction $\pm B_m$. In general the normal hysteresis loop has inversion symmetry with respect to the origin of the B and H axes, but this may not be true for special materials.
21. **Hysteresis loss, incremental.** The hysteresis loss occurring in a ferromagnetic material when subjected simultaneously to constant biasing magnetic force and cyclic excitation.
22. **Hysteresis loss, normal.** (*i*) The power expended in a ferromagnetic material as a result of hysteresis, when the material is subjected to an SCM (symmetrically cyclically magnetised) excitation. (*ii*) The energy loss per cycle in a magnetic material as a result of magnetic hysteresis when the induction is cyclic (but not necessarily periodic).
23. **Magnetic hysteresis.** The property of a ferromagnetic magnetic material revealed by the lack of correspondence between the changes in induction resulting from increasing magnetising force and from decreasing magnetising force.
24. **Induction, intrinsic.** The vector difference between the magnetic induction in a magnetic material and the magnetic induction that would exist in a vacuum under the influence of the same magnetising force.
25. **Induction, normal.** The maximum induction in a magnetic material that is in symmetrically cyclically magnetised condition.

Note. *Normal induction is a magnetostatic parameter usually measured by ballistic methods.*

26. **Induction curve, normal.** A curve of a previously demagnetised specimen depicting the relation between normal induction and corresponding ascending values of magnetising force. This curve starts at the origin of B and H axes.

27. **Magnetic field of induction.** A state of a region such that a conductor carrying a current in the region would be subjected to mechanical force and an e.m.f. would be induced in an elementary loop rotated with respect to the field in such a manner as to change the flux linkage.
28. **Ageing coefficient.** The percentage change in specific magnetic property resulting from a specified ageing treatment.
29. **Magnetic ageing.** The change in magnetic properties of a material resulting from metallurgical change. This term applies whether the change results from a continued normal or specified accelerated ageing condition.

7.1.2. Magnetic Parameters

The various *magnetic parameters* are described below :

1. Magnetic moment :

It has been observed that a coil carrying a current I, when located in a region of uniform magnetic flux density, will experience, a *torque* whose magnitude depends on the *area of the coil, the current* and the *component of flux density in the plane of the coil.* The magnetic dipole moment is defined by the following relation :

$$p_m = I \times A \quad ...(7.1)$$

where, p_m = magnetic dipole moment–Ampere metre^2 (A-m^2) and is a vector quantity,

I = Current, and

A = Area of the coil.

A *permanent bar magnet* also experiences a torque when placed in a uniform field, which tends to align it with the field. If the magnet is supposed to have free poles of opposite kind, we can define the dipoles moment, *as the product of pole strength and distance between them.*

2. Magnetisation :

It has been found that all materials are affected by the presence of a magnetic field, in that they are found experimentally to *acquire magnetic moments.* The *magnitude of this moment per unit volume, is called the magnetisation of the medium* and is described by the vector M.

The magnetisation M of a material may be expressed in terms of its elementary magnetic dipole moments p_m by

$$M = Np_m \quad ...[7.1(a)]$$

where, N is the number of magnetic dipoles per unit volume.

To prove for a magnetic material : $B = \mu_0(M + H)$

Consider a solenoid.

Let, l = Length of the solenoid,

A = Area of cross-section of the solenoid,

N = Number of turns, and

I = Current through the solenoid.

If the solenoid is placed in vacuum the flux density will be :

$$B_0 = \mu_0 H \quad ...(7.2)$$

$$\left(\text{where } H = \frac{NI}{l}, H \text{ being magnetising force}\right)$$

If the vacuum is replaced by a homogeneous magnetic medium the flux density will increase to

$$B = \mu_0 \mu_r H \quad \text{...(7.3)}$$

Thus, there will be an increase in flux density of

$$B - B_0 = \mu_0(\mu_r - 1) H = \mu_0 H' \quad \text{...(7.4)}$$

Now this increase in flux density can be thought of due to an increase in H by H' rather than due to the change in the medium. If I' is the corresponding increase in current then we have

$$H' = \frac{NI'}{l} = \frac{NI' A}{lA} \quad \text{...(7.5)}$$

The right hand side of eqn. (7.5) represents the *total magnetic dipole moment per unit volume and is the magnetisation of the medium* denoted by the letter M. Hence

$$H' = (\mu_r - 1) H = M$$

or,
$$\frac{M}{H} = \chi = \mu_r - 1 \quad \text{...(7.6)}$$

where χ is a *dimensionless quantity* defined as the *magnetic susceptibility* of the medium. Under this condition

$$B = B_0 + \mu_0 H = \mu_0(1 + \chi)H \quad \text{...(7.7)}$$

$$= \mu_0\left(1 + \frac{M}{H}\right)H$$

i.e.,
$$B = \mu_0(M + H) \quad \text{...(7.8)}$$

3. Magnetic susceptibility :

The eqn. (7.7) can be rewritten as

$$B = \mu_0(1 + \chi)H = \mu_0 \mu_r H \quad \text{...(7.9)}$$

where $\mu_r = (1 + \chi)$ is defined as the *relative permeability* of the medium.

- The *magnetic susceptibility* depends on the *nature of the magnetic material and on its state, i.e.,* temperature etc. The susceptibility of a sample may change on *cold working.* If copper is cold worked, the susceptibility may change from a –ve to a +ve value and on annealing after cold work, it may again become –ve.
- The susceptibility may be determined by *measuring the force exerted on a magnetic material* when it is placed in a magnetic field. The susceptibility of a ferromagnetic substance is very strongly dependent on the field strength. Therefore, the magnetic properties of ferromagnetic materials are best described by giving the magnetisation or the flux density as a function of the field strength.
- The M–H and B–H curves of the ferromagnetic materials are very similar in nature. The slope of the former when divided by the magnetic space constant, μ_0 gives the *magnetic susceptibility* and the slope of the latter gives the *magnetic permeability,* $\mu (= \mu_0 \mu_r)$.
- For ferromagnetic materials the *susceptibility* (or, the permeability) is *not constant.* It has a low value at weak fields. As the field intensity is increased, the susceptibility increases, reaches a maximum value and then begins to drop, *ultimately reaching a constant value on saturation.*

7.1.3. Classification of Magnetic Materials

All materials possess magnetic properties to a greater or lesser degree and these are determined by the facts that (*i*) a magnetic field exerts forces and torques on the bodies, (*ii*) a body placed in a magnetic field distorts the field. The magnetic properties of the materials are characterised by their relative permeabilities. In accordance with the value of relative permeability the materials may be classified in the following *three* ways :

1. Ferromagnetic materials :

The relative permeabilities of these materials are *much greater than unity* and are dependent on the field strengths. They attract the lines of force strongly [Fig. 7.1(*c*)]. The principal ferromagnetic elements are *iron, cobalt, nickel.* Gadolinium however, also comes under this classification. These have high susceptibility.

2. Paramagnetic materials :

These have relative permeability *slightly greater than unity* and are *magnetised slightly.* They attract the lines of forces weakly [Fig. 4.1(*b*)]. Aluminium, platinum and oxygen belong to this category.

3. Diamagnetic materials :

The relative permeability of these materials is *slightly less than unity.* They *repel the lines of force slightly* [Fig. 4.1(*a*)]. The examples are bismuth, silver, copper and hydrogen.

Fig. 4.2. shows the arrangement of atomic magnetic moment.

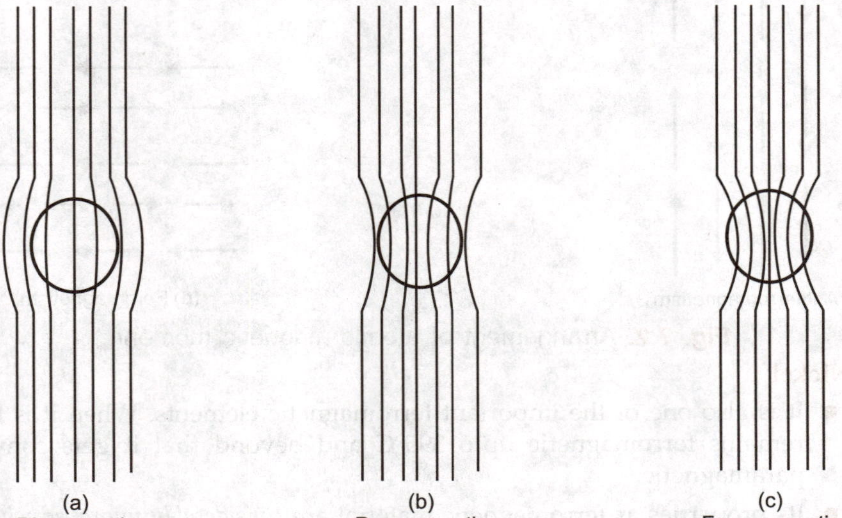

(a) Diamagnetic (b) Paramagnetic (c) Ferromagnetic

Fig. 7.1

7.1.4. Principal Ferromagnetic Elements

The principal ferromagnetic elements are discussed below :

1. Iron :

- It has the highest susceptibility comparatively. This property added with low cost makes iron most suitable for commercial purposes.
- Its magnetic properties are considerably affected by the presence of traces of carbon, oxygen and nitrogen, therefore, these undesirable elements should be reduced to the possible attainable limits. High purity means high permeability and reduced hysteresis loss. Its *permeability is to the tune of 2000.*

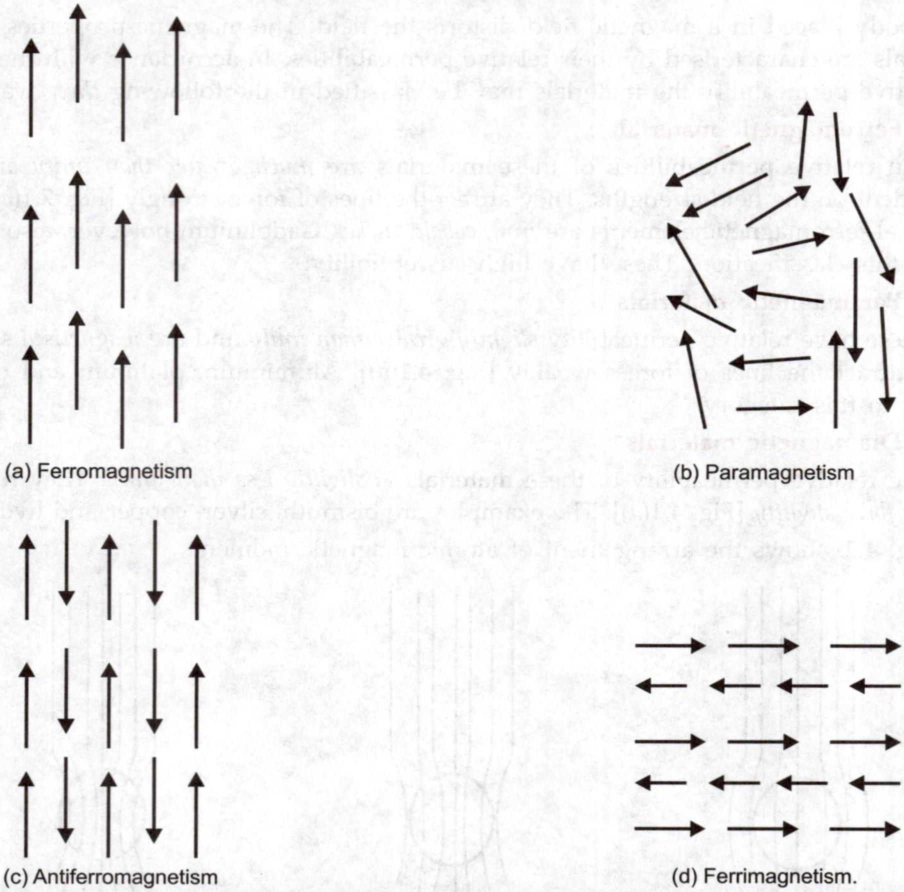

Fig. 7.2. Arrangement of atomic magnetic moment.

2. Nickel :

- It is also one of the important ferromagnetic elements. When it is heated, it remains ferromagnetic upto 395°C and beyond that it gets converted to paramagnetic.
- Its properties as ferromagnetic material are *considerably improved when alloyed with iron and cobalt. Its permeability is 300.*

3. Cobalt :

- Its permeability is 250 and increases with the temperature upto 300°C. It loses its magnetic properties at about 1130°C.
- When in pure state it does not prove to be an important ferromagnetic material.
- Its cost is adequately high.

The only elements which are strongly ferromagnetic at normal temperature are iron, cobalt and nickel. In Fig. 7.3 are shown typical *normal-induction curves* of annealed samples of iron, nickel and cobalt of comparatively high purity. These curves are given only for the purpose of general comparison and should not be considered at representing critical values. Small variations in the degree of purity or in the annealing procedure lead to substantial differences in normal induction. It may be noted that iron has greatest permeability, and this factor coupled with its low cost makes it the only element of commercial importance by itself.

7.1.5. Ferromagnetism

The outstanding characteristic of a ferromagnetic material is that it is very strongly attracted by the magnet. Importance of iron as a magnetic material has led to the name ferromagnetism. *Magnetic properties even exist in the absence of applied magnetic field.*

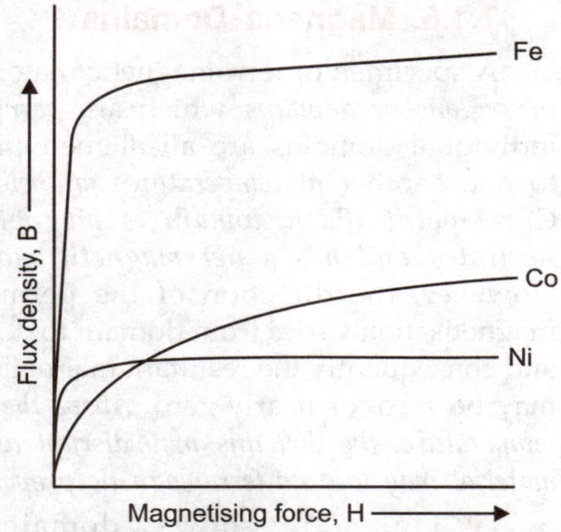

Fig. 7.3. Typical normal-induction curves of annealed samples of iron, nickel and cobalt.

Ferromagnetism arises out of the electronic structure within the atoms. According to Pauli's exclusion principle, not more than two electrons can occupy each energy level of an isolated atom and the same holds true for atoms in a crystal structure. The two electrons having the same energy levels have "spins" in opposite direction, and since each spinning electron can be considered equivalent to a moving charge, each electron acts as an extremely small magnet having north and south poles. A material having even number of electrons in general has as many electrons spinning in one direction as in the other and the sum total effect is magnetically neutral in structure. In case of a material with unfilled subvalence shell, *more electrons spin in one direction than in other resulting in a net magnetic moment* [Fig. 7.4]. In materials like α-iron, cobalt and nickel, these moments are quite strong and atoms are sufficiently close together so that there is *spontaneous magnetic alignment of adjacent atoms.* These conditions cause *ferromagnetism.*

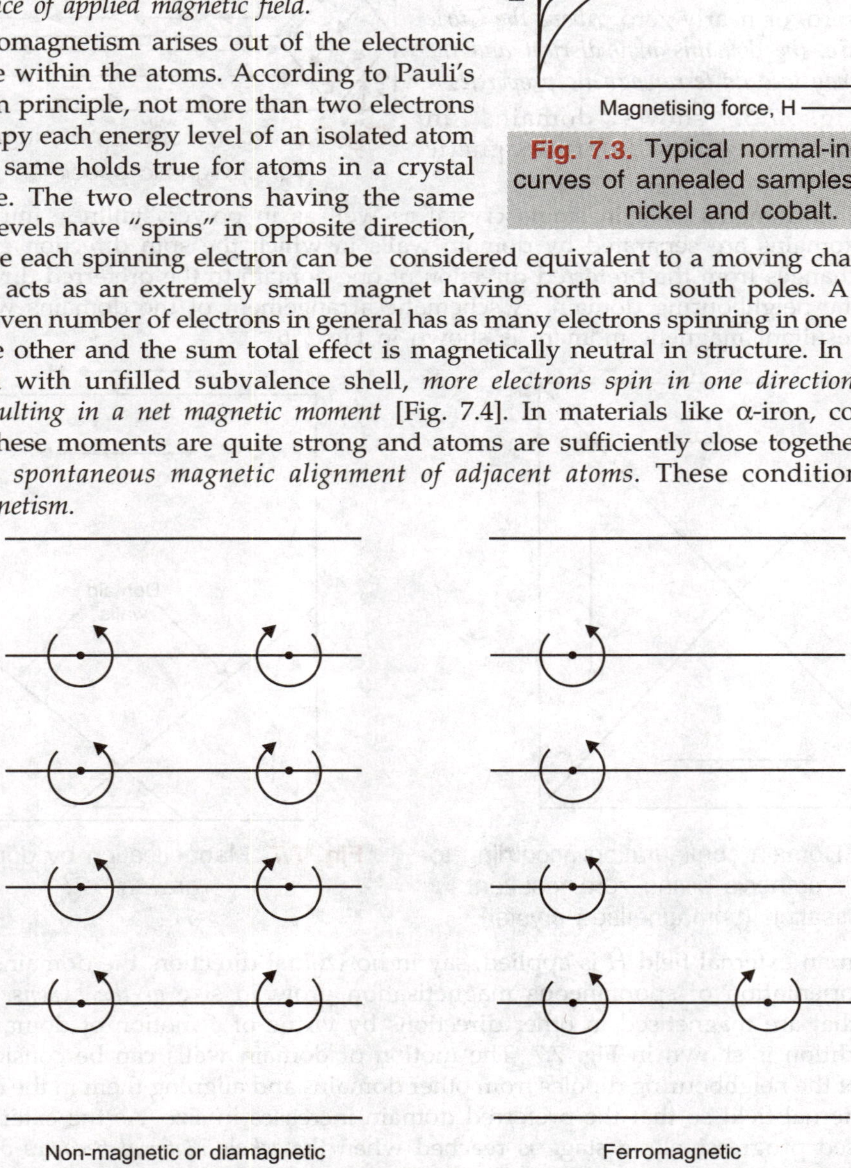

Fig. 7.4

7.1.6. Magnetic Domains

A specimen of ferromagnetic material, according to Weiss, consists of a large number or *regions or domains* which are *permanently magnetised*. The atomic moments in the individual domains are all aligned parallel to one another *at temperatures far below the Curie point. Each domain is magnetically saturated and has a net magnetic moment.* However, the direction of the permanent magnetisation varies from domain to domain, and consequently the resultant magnetisation may be zero, or nearly zero. *Above the Curie temperature, the domains may disrupt and the material may lose its ferromagnetic properties.*

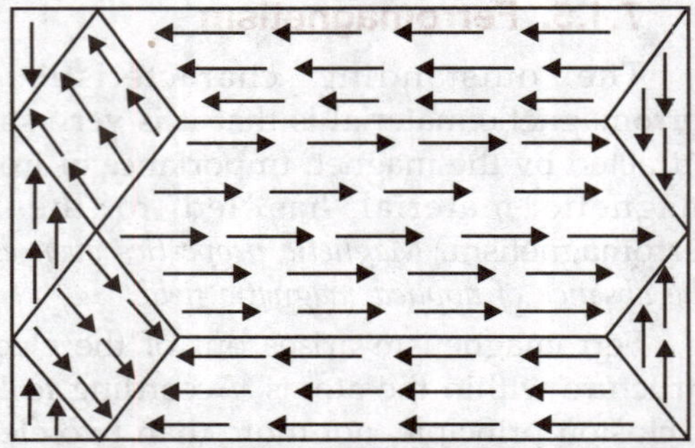

Fig. 7.5. Domains in an unmagnetised ferromagnetic specimen.

- Fig. 7.5. shows domains in an unmagnetised ferromagnetic substance.
- The domains exist in single crystal as well as in polycrystalline samples. The domains are separated by domain walls in which the spin direction gradually changes from the preferred direction of one domain to the preferred direction of the neighbouring domain. A schematic arrangement of the domains with zero resultant magnetic moment is shown in Fig. 7.6.

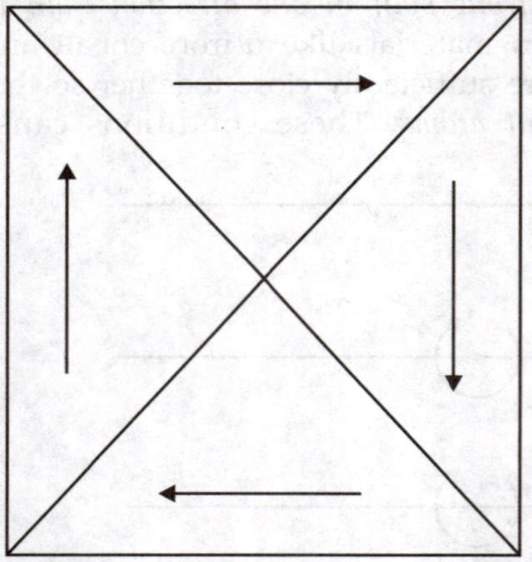

Fig. 7.6. Domain configuration according to Weiss hypothesis giving zero resultant magnetisation (unmagnetised crystal).

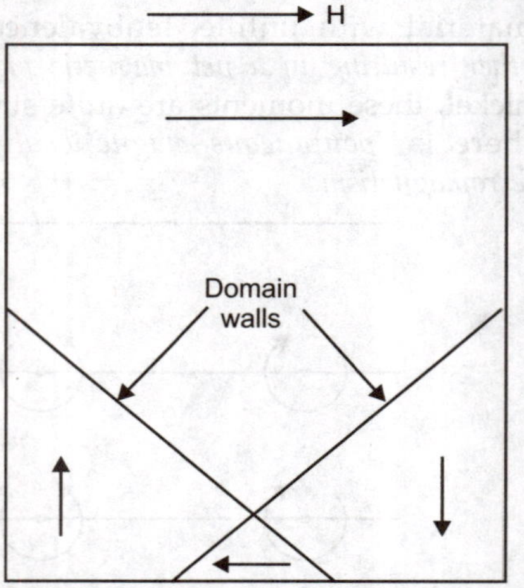

Fig. 7.7. Magnetisation by domain growth.

When an external field H is applied, say in horizontal direction, the domains having parallel orientation of spontaneous magnetisation grow in size *at the expense of other domains* that are magnetised in other directions by virtue of a motion of domain walls. This condition is shown in Fig. 7.7. The motion of domain walls can be considered as stealing of the neighbouring dipoles from other domains and aligning them in the direction of the external field so that the preferred domain increases in size. As the external field is increased progressively, a stage is reached when the *whole material becomes one single domain giving rise to saturation of the magnetisation.* Thus the *hysteresis results due to motion of domain walls and also due to domain rotation to some extent.*

- The *spontaneous magnetisation takes place in a material in certain preferred directions.* When an external field is applied in an arbitrary direction, some domains which are easy to rotate, orient themselves in the direction of the field and become nearly immobile.
- The coercive force needed to neutralise the spontaneous magnetisation in a ferromagnetic material, can be interpreted in terms of the *mobility of* the domain walls. The mobility of domain walls depends on the impurities in the material lattice imperfections etc. *Thus it is possible to design a material by adding or subtracting impurities to obtain a desired value of coercive force.*

The domain structure of a material can be *observed* under a powerful microscope *by spreading a colloidal suspension of iron particles on nicely prepared surface of the material. These particles are attracted towards the domain boundaries where the magnetisation is quite large.* This technique of observing the domain structure is known as the *magnetic powder technique.*

7.2 CLASSIFICATION OF MAGNETIC MEASUREMENTS

Measurements of various magnetic phenomena are called **magnetic measurements.**

The magnetic measurements and a thorough knowledge of characteristics of magnetic materials are of significant importance in designing and manufacturing electrical equipment.

The magnetic measurements are *more inaccurate* than other types of measurements in electrical engineering due to the following ***reasons :***

(*i*) The magnetic flux *cannot* be measured directly, by the mere insertion of instrument in a circuit, as an electric current, voltage and power. *The flux unlike current does not confine itself to a definite path.*

(*ii*) The magnetic materials are *not* homogeneous. There is *no uniformity* between different batches of test specimens even if such batches are of the same composition.

In *magnetic measurements, the* ***principal requirements*** are :

(*i*) Measurement of magnetic field strength in air.

(*ii*) Determination of B-H curve and hysteresis loop for soft ferromagnetic materials.

(*iii*) Determination of eddy current and hysteresis losses of soft ferromagnetic materials subjected to alternating magnetic fields.

(*iv*) Testing of permanent magnets.

- The two most important quantities to be measured are : The flux density, **B** in a specimen of ferromagnetic material and magnetising force **H** producing the flux density.
  - — The flux density B is measured by a *ballistic galvanometer or a fluxmeter* which is a special type of ballistic galvanometer. Ballistic galvanometer or fluxmeter does not measure B in a magnetic material directly, instead measures the changes in the flux.
  - — Magnetising force H is measured by *permeameter.*

Types of tests :

The study of magnetic measurements has been divided into three categories :

1. ***D.C. tests.*** These tests are often termed as *"ballistic tests".*
   - — These are used for *determination of B.H. curves and hysteresis loops* of ferromagnetic materials.
   - — These tests provide an adjustable m.m.f. on the magnetic circuit and a ballistic galvanometer or flux meter for measurement of flux density.

2. *A.C. tests.* These tests may be carried at power, audio or radio frequencies.
   — A.C. tests are used to determine iron losses or core losses in strip (sheet) material when it is subjected to alternating field.
   — These losses can be separated into hysteresis losses and eddy current losses by a separate method involving variation of frequency and form factor.
   — The strip material is assembled as a close magnetic circuit in the form of square. There are two types of such squares : (*i*) Epstein square; and (*ii*) Lloyd Fisher square.
3. *Steady state tests.*
   — These are used to measure the *steady state value of flux in the air gap of magnetic circuit.*

- The above tests are normally carried out on *ring* specimens of the ferromagnetic materials although *bar* specimens are much easier to construct. However, some difficulties and accuracies arise in performing the test on bar specimens; *"permeaters" overcome these difficulties and inaccuracies in testing bar specimens of magnetic materials.*

7.3 BALLISTIC (OR D.C.) TESTS

These tests are used for :

1. Determination of flux/flux density in a specimen.
2. Determination of B.H. curves.
3. Plotting of hysteresis loop.

7.3.1. Measurement of Flux/Flux Density

Fig. 7.8 shows the connection diagram for the measurement of flux/flux density in a *ring specimen.*

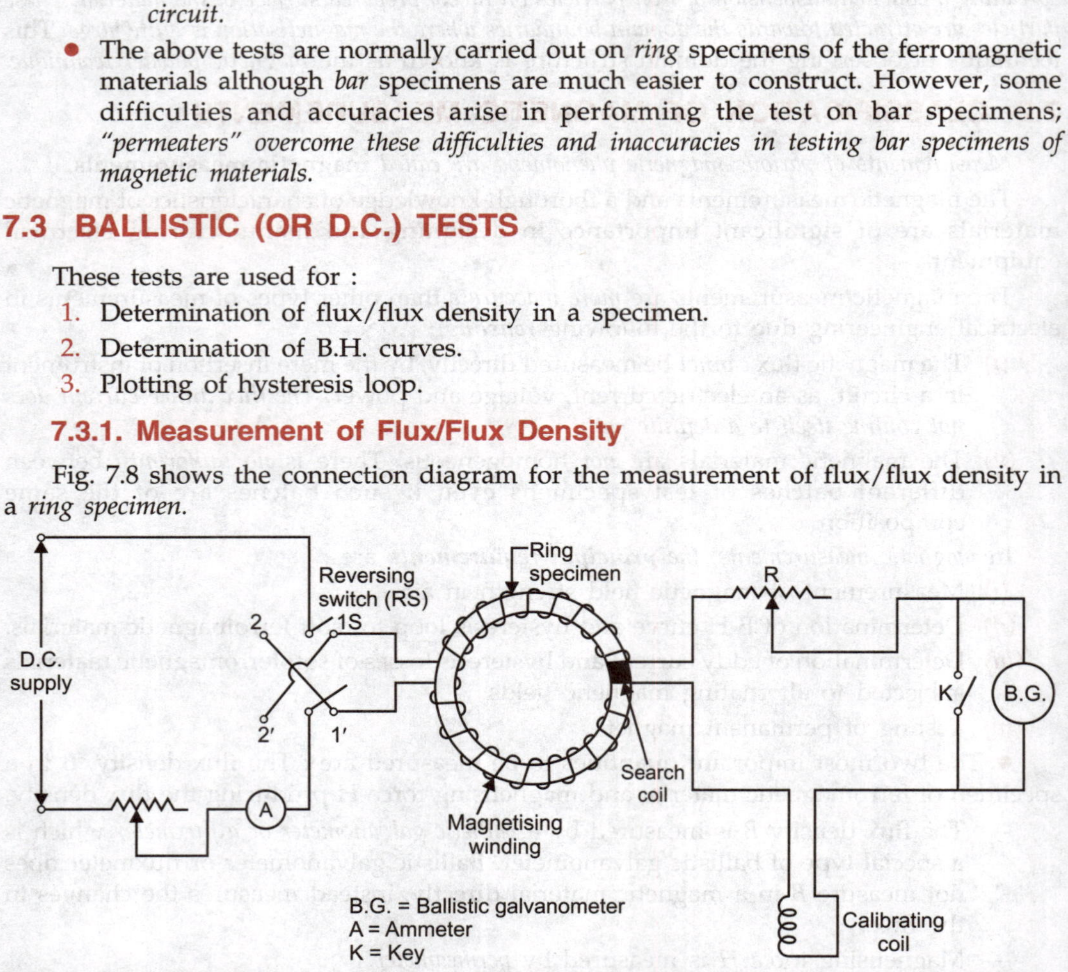

Fig. 7.8. Measurement of flux/flux density in ring specimens.

— The ring specimen is wound with a magnetising winding which carries a current *I*.

— A search coil (known as **"B coil"**) of convenient number of turns is wound on the specimen and connected through a resistance and calibrating coil, to a ballistic galvanometer as shown.

— The current through the magnetising coil is reversed and therefore the flux linkages of the search coil change inducing an e.m.f. e in it. This e.m.f. sends a current through the ballistic galvanometer causing it to deflect.

Let, ϕ = Flux linking with search coil,
R = Resistance of the ballistic galvanometer circuit,
N = Number of turns in the search coil, and
t = Time taken to reverse the flux.

∴ Average e.m.f. induced in the search coil,

$$e = N \cdot \frac{d\phi}{dt} = \frac{2N\phi}{t}$$

Average current through the ballistic galvanometer is,

$$i = \frac{2N\phi}{Rt}$$

Charge passing is : $Q = it = \frac{2N\phi}{Rt} \times t = \frac{2N\phi}{R}$

Let θ_1 be the throw of the galvanometer and G_q be the constant of galvanometer expressed in coulomb per unit deflection.

Charge indicated by the galvanometer = $G_q\ \theta_1$.

$$\therefore \quad \frac{2N\phi}{R} = G_q \cdot \theta_1$$

or, Flux, $\boldsymbol{\phi} = \frac{R \cdot G_q \cdot \theta_1}{2N}$ ...(7.10)

In a uniform field with search coil turns at right angles to the flux density vector, we have,

Flux density, $\mathbf{B} = \frac{\text{Flux}}{\text{Area}} = \frac{\phi}{A_{sp.}}$ ...(7.11)

where, $A_{sp.}$ is the cross-sectional area of the specimen.

Correction for air flux :

In the above calculations we have assumed that the flux is *uniform* throughout the specimen, and that the effective area of cross-section of the search coil is *equal* to the cross-sectional area of the specimen. However, usually the area of the search coil is larger than that of the specimen and thus the flux linking with the search coil is the sum of the flux existing in the specimen and the flux which is present in the air space between the specimen and the search coil.

∴ The value of flux observed = True value of flux in the specimen + flux in the air space between the specimen and search coil.

or, $B'\ A_{sp.} = B\ A_{sp.} + \mu_0 H\ (A_{sc} - A_{sp})$

∴ *True value of flux density,* $B = B' - \mu_0 H\left(\frac{A_{sc}}{A_{sp}} - 1\right)$ ...(7.12)

where, B' = Observed or apparent value of flux density, Wb/m^2,
B = True value of flux density in specimen, Wb/m^2,
A_{sp} = Cross-sectional area of speciment, m^2, and
A_{sc} = Cross-sectional area of search coil, m^2.

7.3.2. Measurement of Value of Magnetising Force (H)

The value of H *inside a specimen* can be calculated from the formula, given by :

$$H = \frac{NI}{l} \text{ AT/m} \quad ...(7.13)$$

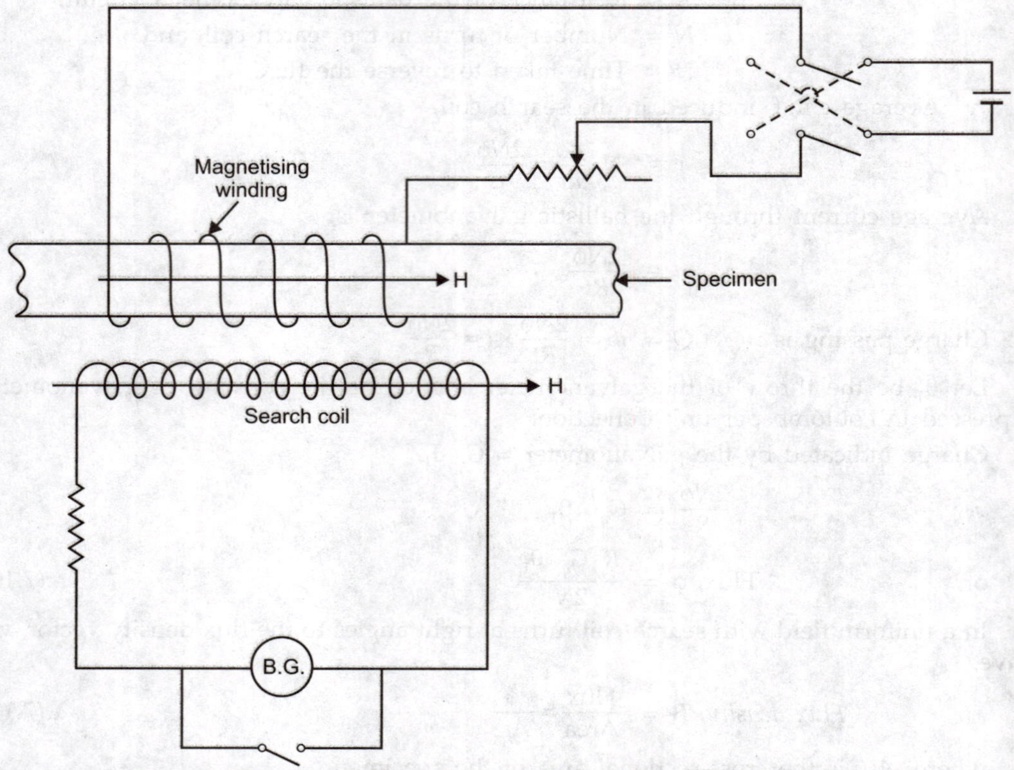

Fig. 7.9. Determination of magnetising force H.

where, N = Number of turns in the specimen,
I = Current flowing through the specimen, and
l = Mean circumference of the ring, *m*.

H can also be determined from measurements made *outside the specimen* from the equation,

$$H = \frac{B_0}{\mu_0} \quad ...(7.14)$$

where, $\mu_0 = 4\pi \times 10^{-7}$ henry/metre and the value of B_0 can be determined by the method already described.

That is *H cannot be measured directly.*

Thus the magnetising force of a constant magnetic field can be measured *indirectly by a ballistic galvanometer and a search coil* as shown in Fig. 7.9.

— If magnetising force is to be determined in the air gap, the search coil (called **H-coil**) is placed in the air gap itself ; while magnetising force *H* within the speciment can be determined by measuring the *H* on its surface, since the tangential components of the field are of equal magnitude for both sides on the interface.

— The search coil as positioned in Fig. 7.9 measures the value of flux density B_0 in air in a similar manner described earlier. After finding out the value of B_0, H can be readily determined from eqn. 7.14.

7.3.3. Determination of B-H Curve

The following two methods are available for the determination of B-H curve of a specimen :

1. *Method of reversals.*
2. *Step-by-step method.*

1. Method of reversals :

— For the determination of B-H curve, a ring shaped specimen whose dimensions are known is used.

— A layer of thin tape is put on the ring and a search coil insulated by paraffined wax is wound over the tape. Another layer of tape is put over the search coil and the magnetising winding is uniformly wound over this tape.

The test circuit is shown in Fig. 7.8.

Procedure :

— First of all the specimen is demagnetised and then the magnetising current I is set to its lowest value.

— The ballistic galvanometer key K is closed and the reversing switch S is operated about twenty times backward and forward. This is done to bring the specimen into a *reproducible cyclic magnetic state.*

— Key K is now opened and the value of flux corresponding to this value of H is measured by reversing the switch S and noting the throw of galvanometer.

The value of *flux density* corresponding to this H can be calculated by dividing the flux by the area of the speciment.

— The above procedure is repeated for various values of H upto the maximum testing point.

The B-H curve may be plotted from the measured values of B corresponding to the various values of H.

2. Step-by-step method :

The special feature of this method of determination of B-H curve is that there is *no reversal of magnetising current.*

The connection diagram of this method remains the same as for the previous method (see Fig. 7.8) except that the direct current is now supplied to the magnetising coil through a *potential divider having a number of tapping* as shown in Fig. 7.9. The tappings are so arranged that the value of magnetising current I and so magnetising force H is increased upto the desired maximum value, in a number of suitable steps.

Procedure :

— The tapping switch S_2 is set on tapping 1 and switch S_1 is closed. The throw of galvanometer corresponding to this increase in flux density in the specimen, from zero to some value B_1 is observed. The value of B_1 can be calculated from the throw of the galvanometer. The value of the corresponding magnetising force H_1 may be calculated from the value of current flowing in the magnetising winding at tapping 1.

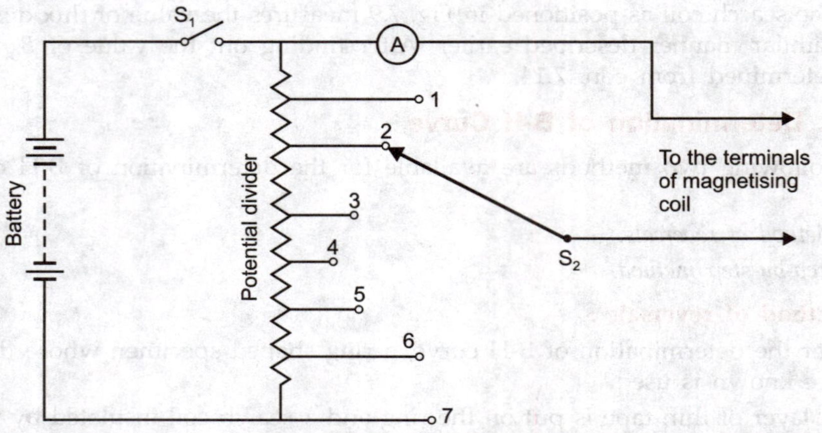

Fig. 7.10

— The magnetising force is then increased suddenly to H_2 by suddenly changing the position of switch S_2 from tapping 1 to tapping 2 and the corresponding increase in flux density is determined from the galvanometer throw observed. The flux density B_2 corresponding to magnetising force H_2 will be equal to B_1 *plus* increase in flux density (δB) determined from the galvanometer throw.

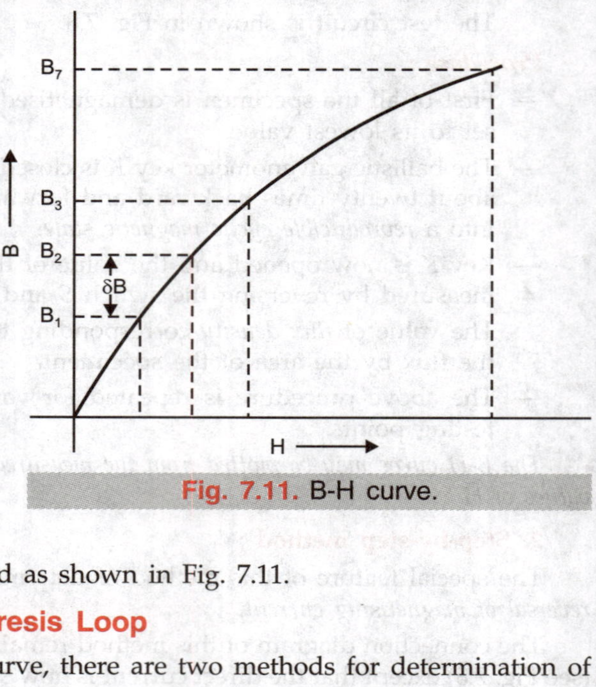

Fig. 7.11. B-H curve.

— The process is repeated for other values of H to the maximum point and complete B-H curve is obtained as shown in Fig. 7.11.

7.3.4. Determination of Hysteresis Loop

Similar to determination of B-H curve, there are two methods for determination of hysteresis loop of a magnetic material specimen :

1. *Method of reversals.*
2. *Step-by-step method.*

1. Method of reversals. Fig. 7.12 shows the connection diagram for determination of hysteresis loop by method of reversals.

R_1, R_2 and R_3 are the *variable resistances* for adjusting the resistances in the ballistic galvanometer and magnetising coil circuits, R_4 is a variable *shunting resistance,* which can be connected across the magnetising coil by means of switch S_2 thus reducing the magnetising current from its maximum value down to any desired value depending upon the value of R_4.

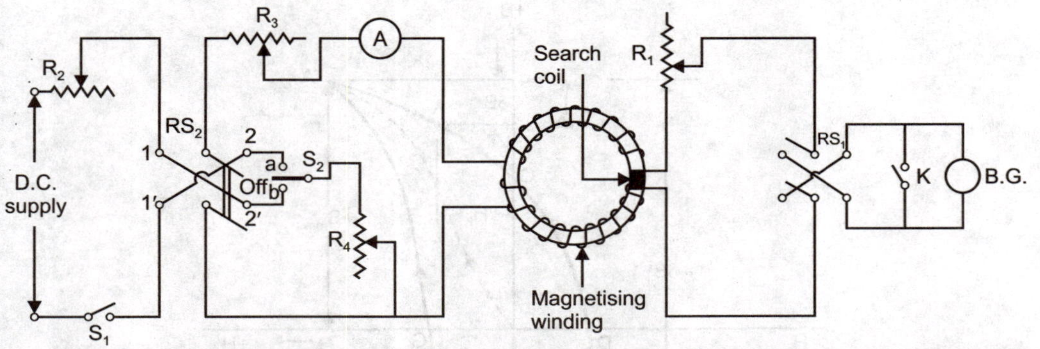

Fig. 7.12. Connection diagram for determination of hysteresis loop by method of reversals.

Procedure : Refer to Figs. 7.12 and 7.13.

— The value of magnetising force H_{max} required to produce flux density B_{max} to be used during the test is obtained from the B-H curve of the specimen.

— The resistances R_2 and R_3 are adjusted to give such a current in the magnetising coil that magnetising force H_{max} (determined from B-H curve) is produced with S_2 in 'off' position. The resistance in the galvanometer circuit R_1 is adjusted to obtain suitable deflection in ballistic galvanometer on reversing the maximum magnetising current. The shunting resistance R_4 is adjusted to give required reduction in magnetising current when connected across magnetising winding.

— The reversing switch RS_2 is placed on contacts 1, 1′ and ballistic galvanometer is connected to the circuit by opening short-circuiting key K. The value of B_{max} is determined corresponding to H_{max} from the deflection of galvanometer observed on reversing switch RS_2 and point A on the hysteresis loop is obtained.

— The switch S_2 is then thrown from off position to contact b in order to connect resistance R_4 across the magnetising winding and reduce the magnetising force to H_K (say). The corresponding reduction in flux density δB is obtained from the galvanometer throw and thus point K is obtained on the loop.

— The galvanometer is then short-circuited by closing key K and reversing switch RS_2 is reversed to contacts 2, 2′. Switch S_2 is moved to the "off" position and reversing switch S_2 is moved back to contacts 1, 1′. This procedure passes the specimen through the cycle of magnetisation and back to the point A.

The specimen is now ready for the next step in the test. The part AC of the loop is obtained by adjusting the shunting resistance R_4 to give different reduced values of H and determining corresponding reduction in B.

— In order to obtain part CDE of the loop, the galvanometer is short-circuited, switch S_2 is put in the "off" position, and reversing switch RS_2 moved back again to contacts 1, 1′. Now place switch S_2 on contact a, open the key K and rapidly reverse RS_2 to contacts 2, 2′. This causes the magnetising force to change from $+H_{max}$ to $-H_L$ (say). From the throw of the galvanometer change in flux density δB′ can be obtained. Thus point L on the hysteresis loop can be located. The magnetisation of the specimen is brought back to point A by reversing switch RS_2 on to contacts, 1, 1′ with key K closed. By continuing this procedure, other parts on part CDE of the hysteresis loop are obtained. Thus part $ACDE$ of the loop can be traced.

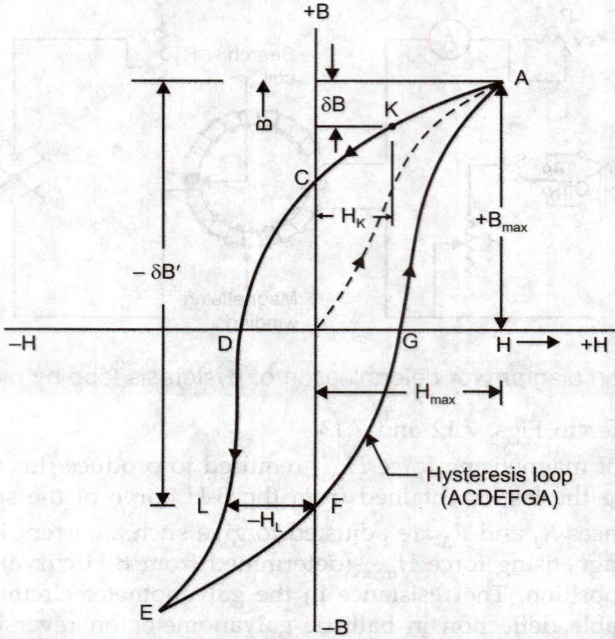

Fig. 7.13. Hysteresis loop of a magnetic material.

— The part *EFGA* of the loop may be obtained by drawing in the reverse of the part *ACDE* as the two halves are identical.

7.3.5. Determination of Leakage Factor in Dynamo-Electric Machinery

In the dynamo-electric machine, the *leakage factor* is defined as :

$$\text{Leakage factor} = \frac{\text{Total flux}}{\text{Useful flux}}$$

where *"useful flux" is the flux in the armature* (*i.e., flux crossing the air gap*) and *"total flux"* in the pole bodies which in turn is equal to the useful flux *plus* leakage flux existing on the pole body at its root.

Thus in order to measure the leakage factor, we have to measure flux in the pole bodies and flux in the armature. *For measurement of leakage factor the 'flux meter' is used, as ballistic galvanometer is unsuitable due to high inductance of the field system of the machine.*

— The yoke carries *half* of the total flux and therefore it is possible to measure the value of *"total flux per pole" by using two search coils on the yoke and connecting them in series across the fluxmeter.*

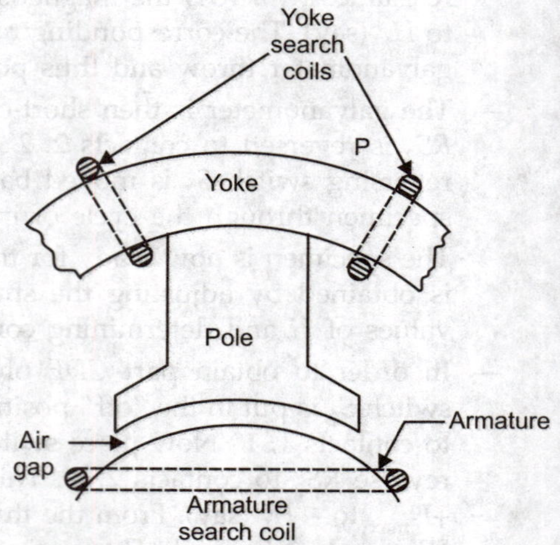

Fig. 7.14. Measurement of leakage factor.

The armature is kept stationary and another search coil is put on it. The coil is so positioned that it embraces the useful flux per pole. *This search coil is then connected to the fluxmeter and this way the "useful flux" is measured.*

Leakage factor is then calculated from the above two readings of the fluxmeter.

Search coils of usually *one turn* are used so that fluxmeter reading gives the flux directly.

In large electrical machines, the flux is quite large and hence it becomes necessary to use a *shunt* along with fluxmeter in order that its *range may be increased.*

7.3.6. Magnetic Potentiometer

The **magnetic potentiometer** *measures the magnetic potential drop along any part of a magnetic circuit and hence distribution of m.m.f. round the circuit.*

It can also be used *to determine magnetic potential drop across a joint* in the magnetic circuit or between two points between which *magnetic leakage is suspected.*

Construction :

— It consists of a one metre long flat and uniform coil made of two or four layers of thin wire wound unidirectionally on a strip of flexible non-magnetic material.

— The coils ends are brought out at the middle of the strip as shown in Fig. 7.15 and connected to a ballistic galvanometer.

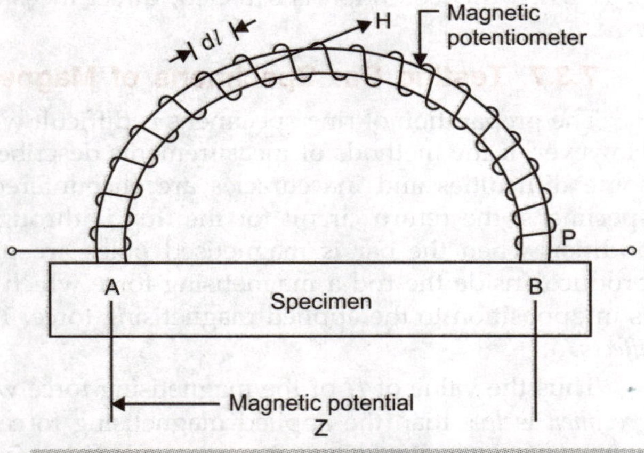

Fig. 7.15. Magnetic potentiometer.

Working :

— The magnetic potential difference between points *A* and *B* of the filed is measured by placing the ends of the strip at these points and observing the throw of the ballistic galvanometer when the flux through the specimen is changed, assuming the flux is changed by reversing the field.

Let, A = Area of *strip*, m^2,

n = Number of turns per unit length of the strip,

H' = Tangential component of magnetising force, AT/m, and

R = Resistance of the ballistic galvanometer.

$\therefore$ *'Flux linkages' of a small infinitesimal portion of strip* of length *dl* are given as :

$$\text{Flux turns} = (\mu_0 H'A)n{\cdot}dl = \mu_0 H'An{\cdot}dl$$

$$\text{Total flux linkages of the strip} = \int \mu_0 H'Andl = \mu_0 An \int H'dl$$

The *change in flux linkages* when the current in the magnetising winding is *reversed*

$$= 2\,\mu_0\, An\, \int H'dl$$

But, $\int H'\cdot dl = Z$ = Magnetic potential difference between *A* and *B*.

$$\therefore \text{Change in flux linkages} = 2\,\mu_0\, An\, Z$$

Now, Change $$Q = it = \frac{e}{R}t = \frac{2\mu_0\, An\, Z}{R \times t}\cdot t = \frac{2\mu_0 An}{R} Z \text{ coulomb} \quad ...(i)$$

Also, Change indicated by galvanometer,

$$Q = G_q \theta_1 \text{ coulomb} \quad \text{...(ii)}$$

Equating eqns. (*i*), and (*ii*) we have :

$$\frac{2\mu_0 AnZ}{R} = G_q \cdot \theta_1$$

or, Magnetic potential, $Z = \dfrac{RG_q \theta_1}{2\mu_0 An}$ ...(7.15)

The value of G_q (galvanometer constant) can be found with the help of a calibrating circuit.

- The above analysis is correct only if the portion *AB* of the specimen does not carry any part of the magnetising winding. Since it is practically impossible to achieve uniformity of *H* when this condition is satisfied, direct measurement of *H* by this method is *rarely used.*

7.3.7. Testing Bar Specimens of Magnetic Materials

The preparation of ring specimens is difficult while *bar specimens are easier to construct.* However, if the methods of measurements described above are applied to bar specimens some difficulties and inaccuracies are encountered. This is because when we use bar specimens, the return circuit for the flux is through air whose *resistance is very high.* In addition when the bar is magnetised poles are produced at its ends, and these poles produce, inside the rod a magnetising force which acts from North-pole to South which is in opposition to the applied magnetising force. This is called *self-demagnetisation or end effects.*

Thus the value of *H* of the magnetising force which *is effective in producing the flux in specimen* is *less* than the applied magnetising force as calculated from the m.m.f. of the magnetising winding. Thus the *true magnetising force H* is given by :

$$H = H_a - H_d$$

where, H_a = Applied magnetising force

$$= \frac{NI}{l} \quad \text{...for a } \textit{long solenoid}$$

$$= \frac{NI}{\sqrt{l^2 + d^2}} \quad \text{...for a } \textit{sort solenoid} \text{ of length } l \text{ and diameter } d.$$

H_d = The *magnetising force due to self-demagnetisation or end effects.*

- For obtaining accuracy in measurements with bar specimens, we must apply *corrections for the demagnetising force or reduce the effect to negligible proportions.* The self-demagnetising force H_d, however, is somewhat an uncertain quantity.

— Since the effect of H_d is small when the ratio $\frac{l}{d}$ of the bar is large, the dimensions of the specimen should be so chosen as to make $\frac{l}{d}$ *as large as possible.* It has been observed that the demagnetising has negligible effect on the value of magnetising force when the ratio $\frac{l}{d}$ of bar specimens is 25 or more.

7.3.8. Bar Specimens Versus Ring Specimens

- The ballistic methods which use specimen in the form of a ring give the *highest accuracy* under proper conditions.

- The *main advantage of "ring specimens" is that the end effects are absent in them and they are free from errors resulting from magnetic leakage.*
- The magnetising winding in ring specimen, however, has to be wound *uniformly* on the ring otherwise the m.m.f. distribution will not be uniform which may lead to magnetic leakage resulting in errors in the measurements.

— The magnetising force is calculated in terms of mean diameter of the ring and the m.m.f. in the magnetising winding. But in the inner circumference of the ring is smaller than the outer, hence the magnetising force and the flux density are not uniform over the cross-section of the ring. *If the ratio of mean diameter to the radial width is more than eight, both magnetising force and flux density are reasonably uniform over the cross-section of the specimen and consequently the error involved is not appreciable.*

Types of ring specimens. There are three types of ring specimen used, depending upon the material to be tested :

1. *Laminated type :*
   — This is the normal form of "isotropic sheet materials."
   — The ring is assembled with the directions of rolling of individual laminations distributed radially so that the test should yield result corresponding to *mean properties of the material.*
2. *Clock-spring type :*
   — A long strip of material is wound in the form of a spiral to form a ring.
   — This type is used for testing *"anisotropic materials" i.e.,* grain oriented cold rolled steel.
3. *Solid type :*
   — Solid rings are prepared either by casting or forging.

Uses of ring specimens. The use of ring specimen in magnetic testing is limited owing to following *reasons* :

(*i*) The labour involved in preparing them, because the magnetising winding has to be put with hand for each individual specimen.

(*ii*) The value of the magnetising force is limited by heating.

The ring specimens are therefore used for the following purposes :

1. Testing of small quantities of materials of very high permeability.
2. Testing powder or dust core materials.

7.4 PERMEAMETERS

— Normally it is not a practical arrangement to a have $\frac{l}{d}$ as large as possible, for a bar specimen to be tested. ***Permeameters*** have been developed *to overcome the difficulties and inaccuracies encountered in performing measurements on bar specimens of magnetic materials.* Thus pemeameters test bar specimens under controlled field conditions.

— The permeameter generally consists of a fixed steel frame to which bar specimens can be connected.

— There are various types of permeameters differing widely in the arrangements of magnetising and search coils and in the means of guarding against leakage fluxes and m.m.f. drop at joints.

— Most of the permeameters incorporate modifications of the *bar and yoke* arrangements invented by Hopkinson.

The following permeameters will be discussed :

1. Hopkinson permeameter (Bar and yoke method).
2. Ewing double bar permeameter.
3. Illiovici permeameter.
4. Burrow's permeameter.
5. Fahy's simplex permeameter.
6. The National Physical Laboratory (England) form of permeameter.

7.4.1. Hopkinson Permeameter (Bar and Yoke Method)

The bar and yoke method is generally employed for the testing of bar specimens in which the demagnetising effect of free poles is largely eliminated by the use of heavy section yokes. Bars give the advantage of simple arrangement for the application of the magnetising force.

Fig. 7.16 shows the schematic arrangement of bar and yoke method :

— A test coil is wound upon the central part of a bar specimen.

— The bar is clamped between two halves of a massive iron yoke whose reluctance is *low* compared to that of the bar specimen. This yoke provides a return path for the flux.

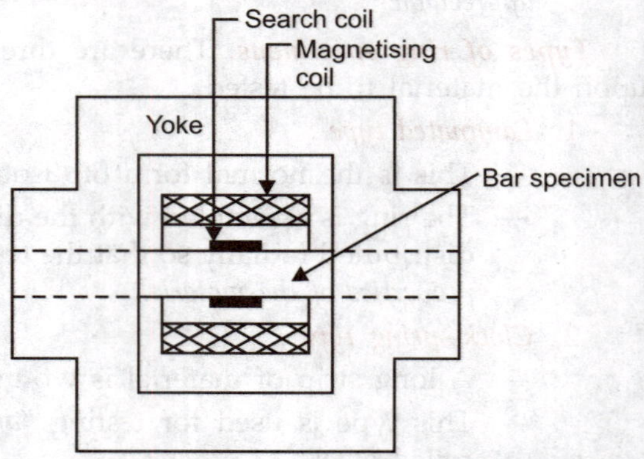

Fig. 7.16. Bar and yoke method (Hopkinson permeameter).

— The specimen is wound with a magnetising winding as shown.

Let, N = Number of turns on the magnetising winding,

I = Current in the magnetising winding,

l = Length of the bar specimen between two halves of the yoke,

A_{sp} = Cross-sectional area of specimen,

μ_{sp} = Permeability of the specimen when the magnetising current is I,

S_y = Reluctance of the yoke,

S_j = Reluctance of the joints between the bar specimen and the yoke, and

ϕ = Flux in the magnetic circuit.

Then, Reluctance of specimen,

$$S_{sp} = \frac{l}{\mu_{sp} A_{sp}}$$

$$\therefore \quad \text{Flux, } \phi = \frac{\text{m.m.f.}}{\text{Reluctance of magnetic circuit}}$$

$$= \frac{NI}{S_y + S_j + (l/\mu_{sp} A_{sp})}$$

Flux density in the specimen,

$$B = \frac{\phi}{A_{sp}} = \frac{NI}{A_{sp}\left[S_y + S_j + (l/\mu_{sp} A_{sp})\right]}$$

$$\text{Magnetising force, } H = \frac{B}{\mu_{sp}} = \frac{NI}{\mu_{sp} A_{sp}\left[S_y + S_j + (l/\mu_{sp} A_{sp})\right]}$$

Let,

$$m = \frac{\text{Reluctance of yoke and joints}}{\text{Reluctance of specimen}}$$

$$= \frac{S_y + S_j}{(l/\mu_{sp} A_{sp})} = \frac{\mu_{sp} A_{sp}}{l}(S_y + S_j) \quad ...(7.16)$$

$$\therefore \quad H = \frac{NI}{l(1+m)} \quad ...(7.17)$$

m is made small by reducing S_y and S_j *i.e.*, by making the yokes of large section and by fitting the specimen to the yoke so that the *air gaps at the joints are as small as possible.* If m is small, the eqn. (7.17) will become :

$$H = \frac{NI}{l}(1+m)^{-1} = \frac{NI}{l}(1-m+....) = \frac{NI}{l}(1-m) \quad ...(7.18)$$

i.e., the actual value of the magnetising force is *less* than that computed from ampere-turns per unit length $\left(i.e., \frac{NI}{l}\right)$, by the amount $\frac{m\,NI}{l}$ which *depends on the reluctance of the return circuit.*

The flux density B may be measured by ballistic galvanometer as explained earlier.

7.4.2. Ewing Double Bar Permeameter

Fig. 7.17 shows an Ewing double bar permeameter :

— It consists of two exactly similar bar specimens of the material under test and two pairs of magnetising coils, one pair being exactly of half the length of the other pair. The number of turns per metre axial length for both pairs of coil is the *same.*

— The bar specimens fit tightly into the holes of two yokes. The yokes are made of annealed soft iron. The position these yokes on the bars can be adjusted. The

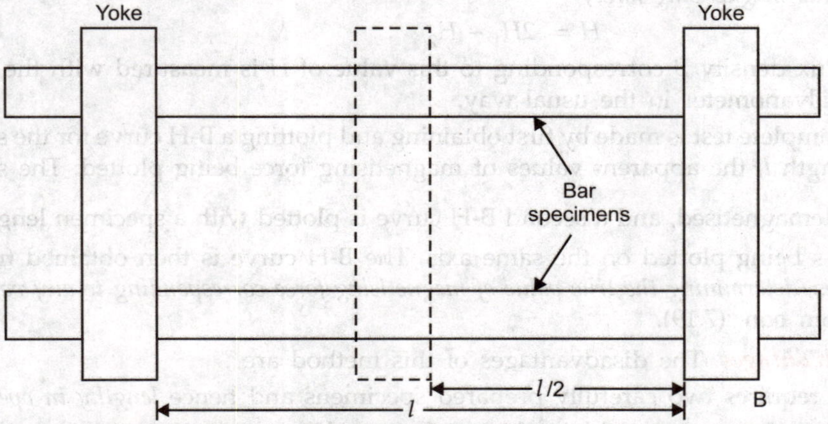

Fig. 7.17. Ewing double bar permeameter.

purpose of the arrangement of bar specimens and yokes (shown in Fig. 7.17) is the elimination of the reluctance of yoke and joints.

— The tests are made, one with length of specimen l and the other with length $\frac{l}{2}$, assuming that for both positions, the reluctance of yokes and the joints is the *same* for a given flux density.

Let, n = Number of turns per unit length of magnetising coils,

I_1 = Current in the coils when the specimen length is l,

I_2 = Current in the coils when the specimen length is $\frac{l}{2}$,

H_1 = Apparent magnetising force when length is l,

H_2 = Apparent magnetising force when length is $\frac{l}{2}$,

a = m.m.f. required for the yokes and joints in each case, and

B = Flux density in the specimen (same in both the cases).

Then,
$$H_1 = \frac{nlI_1}{l} = nI_1,$$

and,
$$H_2 = \frac{n(l/2)\cdot I_2}{(l/2)} = nI_2$$

If H is the true magnetising force in the iron for a flux density B, then, we have :

$$H_1 \cdot l = H \cdot l + a \quad ...(i)$$

or,
$$H_2 \cdot l/2 = H \cdot l/2 + a \quad ...(ii)$$

($\because$ Total applied m.m.f. = m.m.f. required for iron + m.m.f. required for yoke and joints).

Subtracting expressions (*i*) from (*ii*), we get

$$a = (H_2 - H_1)l$$

Inserting this value of a in (*i*), we have :

$$H_1 l = Hl + (H_2 - H_1)l$$

or,
$$H = \frac{H_1 l - (H_2 - H_1)l}{l} = \frac{H_1 l - H_2 l + H_1 l}{l} = 2H_1 - H_2$$

i.e., True magnetising force,

$$H = 2H_1 - H_2 \quad ...(7.19)$$

The flux density B corresponding to this value of H is measured with the help of a ballistic galvanometer in the usual way.

The complete test is made by first obtaining and plotting a B-H curve for the specimens with a length l, the apparent values of magnetising force being plotted. The specimens are then demagnetised, and a second B-H curve is plotted with a specimen length $\frac{l}{2}$, the two curves being plotted on the same axis. The B-H curve is then obtained from these two curves, *determining the true value of magnetising force corresponding to any value of flux density* from eqn. (7.19).

Disadvantages. The disadvantages of this method are :

1. It requires two carefully prepared specimens and hence *lengthy in operation.*
2. The reluctance of yokes and joints for two different positions of the yokes are *not* quite the same unless the specimens are very precisely machined.

7.4.3. Illiovici Permeameter

Fig. 7.18 shows the construction of Illiovici permeameter :

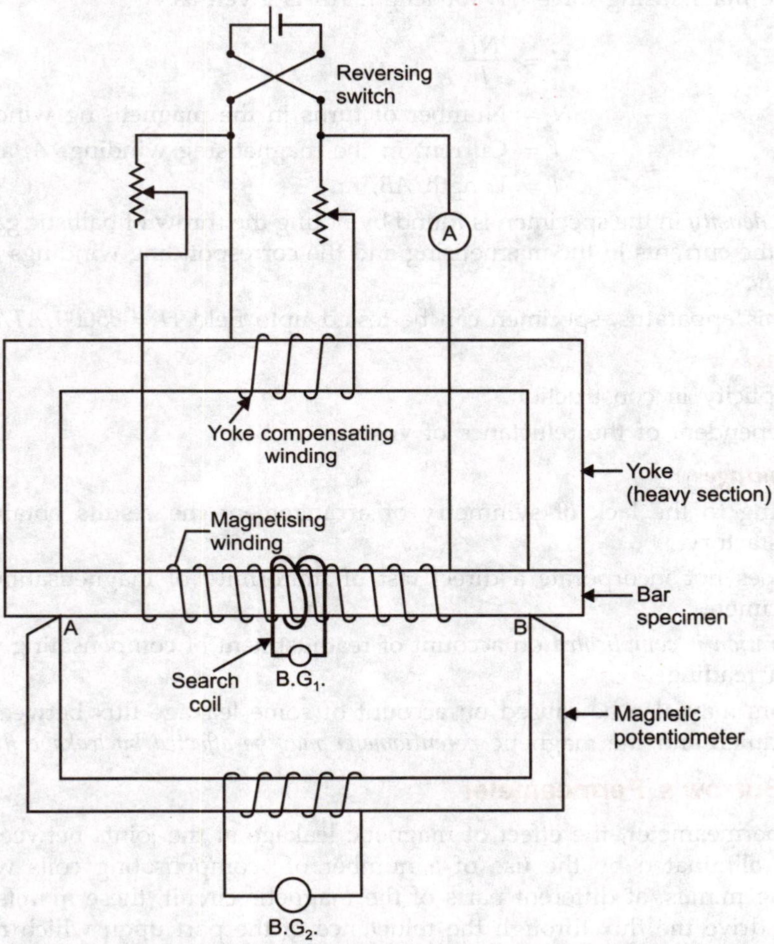

Fig. 7.18. Illiovici permeameter.

— It consists of a bar specimen clamped against a yoke of heavy section.

— A magnetising winding and a search coil (connected across a ballistic galvanometer $B.G_1$.) are wound on the specimen.

— A magnetic potentiometer is connected to a section *AB* of the specimen.

— A compensating winding is wound on the yoke; this winding is connected in parallel with the magnetising winding across a reversing switch.

— When the magnetic potentiometer indicates *no* magnetic potential drop, the m.m.f. for part *AB* is provided by the *magnetising winding*, while for the remaining part of the specimen, the yoke and the joints, is provided by the *compensating winding*. This condition is obtained by adjusting the current in the main winding to the required test value and then adjusting the current in the compensating winding so that ballistic galvanometer $B.G_2$. shows no throw when currents are reversed.

Under this condition there is *no flux* through the *magnetic potentiometer and therefore *no magnetic potential difference* exists between A and B.

Thus the magnetising force (H) for length AB is given as :

$$H = \frac{NI}{l}$$

where, N = Number of turns in the magnetising winding,

I = Current in the magnetising winding, A, and

l = Length AB, m.

The *flux density* in the specimen is found by noting the throw of ballistic galvanometer $B.G_1$. when the currents in the magnetising and the corresponding windings are *reversed-simultaneously.*

- In this apparatus, specimen can be tested upto field $H = 35000\ AT/m$.

Advantages :

1. Simplicity in construction.
2. Independent of the reluctance of yoke.

Disadvantages :

1. Owing to the lack of symmetry of arrangement the results obtained are *not* satisfactory.
2. It does not incorporate a direct test of uniformity of magnetisation along the specimen.
3. *Operation is complicated* on account of readjustment of compensating winding for each reading.
4. Errors may also be caused on account of some leakage flux between yoke and specimen and the magnetic *potentiometer may be affected by leakage flux.*

7.4.4. Burrow's Permeameter

In this permeameter, the effect of magnetic leakage at the joints between yoke and specimen is eliminated by the use of a number of compensating coils which apply compensating m.m.fs. at different parts of the magnetic circuit, these m.m.fs. being just sufficient to drive the flux through the reluctance of the part upon which the coils are placed.

Fig. 7.19 shows the arrangement of the magnetic circuit and coils.

— S_1 is the bar specimen to be tested, S_2 being a bar of similar dimensions to S_1.

— These bars are surrounded by magnetising windings M_1 and M_2, which are uniformly wound along the lengths of the bars.

— A_1, A_2, B_1 and B_2 are *compensating coils* for the elimination of leakage effects at the joints between the two bars and the massive yoke, in which the bars fit.

— c and c' are two exactly *similar search coils* wound at the *centres* of the two bars, while d_1 and d_2 are two *similar search coils* wound in the *position* shown, on the test bar, and each having exactly *half* the number of turns of search coil c. Coils d_1 and d_2 are *connected permanently in series,* as are also the four coils A_1, A_2, B_1, B_2.

---

* In general, the cores of magnetic potentiometer should be of non-magnetic materials but in Illiovici permeameter *ferromagnetic cores are allowable since it is used as a null detector and not for actual measurement.*

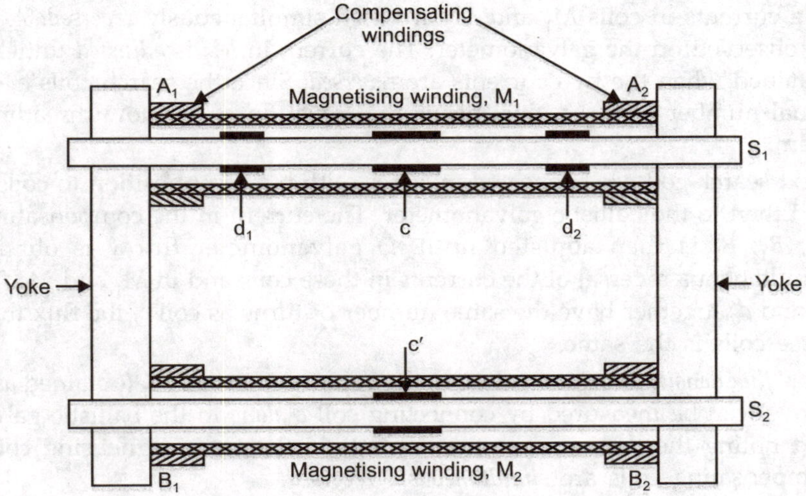

Fig. 7.19. Burrow's permeameter.

— The dimensions of the coils M_1, M_2 are such that H in the specimen is approximately 10^4 times the current in the windings, the maximum value of H for which the apparatus is used being about *40,000 AT/m.*

— The dimensions of the bar specimens are approximately 30 cm long and 1 cm in diameter.

— The coils A_1, A_2, B_1, B_2 are supplied from a separate battery supply, coil M_1 from another separate supply, and M_2 from another.

— In carrying out the test it is necessary, first of all, to ensure that, for a given value of the current in M_1–*i.e.*, of H in the test bar—the flux threading all four search coils c, c', d_1 and d_2, is the *same, the current in the compensating coils, and in* M_2, *being adjusted until this condition is fulfilled.* If the flux threading coils d_1 and d_2 is the *same* as that threading c and c', there can be *no* appreciable leakage of flux through the air in the neighbourhood of the joints. This *means that the m.m.f. for the joints is being supplied by the compensating coils, and that the m.m.f. in coil* M_1 *is used up merely in driving the flux through the bar specimen* S_1 *inside it.*

Thus, H in the specimen is given by :

$$H = \frac{NI}{l}$$

where,
N = Number of turns on coil M_1,
I = Current in M_1, amperes, and
l = Length of specimen in metres.

In some cases, it may be necessary to make corrections for the fact that the magnetised solenoids are not infinitely long, but such corrections are usually negligible.

- Following is the *"procedure" for obtaining equal flux threading all four search coils :*

— After demagnetising the specimen, the current in the magnetising coil M_1 is set at the required testing value. Search coils c and c' are then connected in series, but in opposition, to a ballistic galvanometer.

— The currents in coils M_1 and M_2 are then simultaneously *reversed.* A throw will be observed on the galvanometer. The current in M_2 is *adjusted* until *no throw* is obtained when the two currents are *reversed.* Since the search coils *c* and *c'* have equal number of turns, this means that *equal fluxes* are now threading through them.

— Next search coil *c* is connected in series with but in opposition to coils d_1 and d_2, and then to the ballistic galvanometer. The current in the compensating coils A_1, A_2, B_1, B_2 is then adjusted until *no* galvanometer throw is obtained upon simultaneous reversal of the currents in these coils and in M_1 and M_2. Then, since d_1 and d_2 together have the same number of turns as coil *c*, the flux threading all three coils is the same.

— The *"flux density"* corresponding to the value of *H* in coil M_1 (obtained as described above) can be measured by connecting coil *c alone* to the ballistic galvanometer, and noting the throw when the currents in the two magnetising coils and the compensating coils are *simultaneously reversed.*

Advantages :

1. This method results in *uniform conditions along the specimen* provided that the material itself is uniform.
2. This method can be used *for magnetic testing of many kinds and also for calibration of specimens that serve as reference standards in some of the simpler methods.*

7.4.5. Fahy's Simplex Permeameter

Fig. 7.20 shows the outline of construction of Fahy's simplex permeameter :

— Two iron clamping posts clamp the specimen against a laminated steel yoke. The yoke carries the magnetising winding, and the specimen is surrounded by a search coil for the measurement of *flux density* in it.

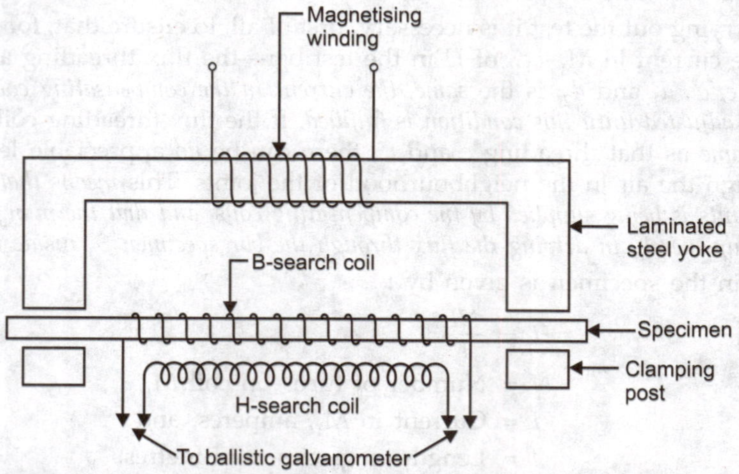

Fig. 7.20. Fahy's simplex permeameter.

— The magnetising force *H* acting on the specimen is measured, like the flux density, by a ballistic galvanometer, utilising an air-used search coil of several thousand turns located between the two clamping posts as shown. The values of *H* so measured are corrected by the calibration of *H*-search coil, utilising a specimen of known magnetic characteristics in place of the test specimen.

- An hysteresis loop may be determined equally well by application of the method for the ring specimen described earlier. Some Fahy meters have been equipped with a flip coil for H for use in hysteresis loop determination.

Advantages :

1. *Simpler* in both construction and operation as compared to Burrows permeameter (while its accuracy is of the same order as Burrows apparatus).
2. *Less sensitive* to the effects of magnetic inhomogeneities in the specimen (since the B coil is wound uniformly over the entire length of the specimen and, therefore, it tends to measure the average flux density in the specimen; eventually the effects of local magnetic irregularities are diluted).
3. *Only test specimen is required.*

7.4.6. The National Physical Laboratory (England) Form of Permeameter

This is one of the most accurate forms of permeameter and is comparatively easy to use in practice.

The arrangement of the permeameter is shown schematically in Fig. 7.21 :

— The specimen is clamped between two similar rectangular yokes Y_1 and Y_2. To keep the flux path of low reluctance, cross-sectional area of these yokes is taken *large.*

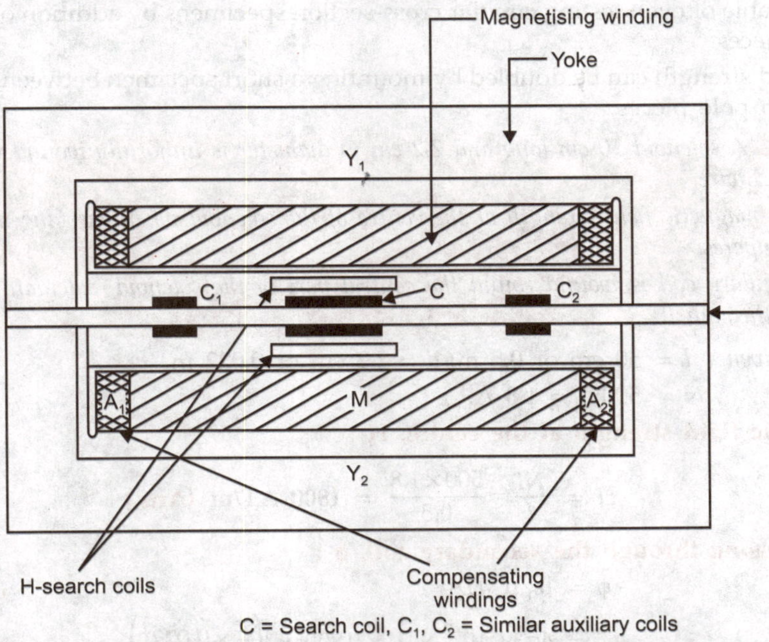

Fig. 7.21. National physical laboratory permeameter.

— For the measurement of *flux density* in the specimen, a search coil C is bound in middle part of the specimen.

— Two similar auxiliary coils C_1 and C_2, each having *half* the number of turns of that in search coil C, are bound at either side of the specimen. Coils C_1 and C_2 are used to check the uniformity in flux in the specimen (in a similar way as described earlier in Burrows permeameter).

— The main magnetising winding M is bound such that it surrounds the specimen. Both ends of M are provided with compensating windings A_1 and A_2, which supply necessary m.m.fs. to reluctance of the joints.

— Now series combination of windings C_1 and C_2 is connected in series opposition with search coil C and then connected with ballistic galvanometer. Current in the main magnetising winding is set for the required value of H. Current in the compensating windings is so adjusted that there is *no throw* in the ballistic galvanometer, during simultaneous reversal of current in these windings A_1 and A_2 and main winding M. This condition shows the *uniform distribution of flux density in the specimen.* Now this flux density is measured with the help of ballistic galvanometer by connecting it to search coil C.

— **H-coils** are provided for the measurement of value of H in the specimen. Measurement of H is done by connecting H *coils* to the ballistic galvanometer.

Significant features. This permeameter may be taken as typical of modern precision permeameters because of its following *features* :

1. It gives accurate results over a wide range of magnetising force. This range extends from *10 AT/m to 300,000 AT/m.*
2. It uses an *H-coil* for determination of value of magnetising force H instead of using a calculated value of H and these values are normally erroneous.
3. It is capable of even testing circular cross-section specimens by addition of shaped poles pieces.
4. The field strength can be doubled by mounting a short specimen between massive soft iron pole pieces.

Example 7.1. *A solenoid 50 cm long and 2.2 cm in diameter is uniformly wound with 500 turns of insulated wire.*

(i) Find the magnetic field strength at the centre of the solenoid when carrying a current of 1.8 amperes.

(ii) If a secondary coil is wound round the central part of the solenoid, calculate the flux passing through it.

Solution. *Given :* l = 50 cm or 0.5 m; d = 2.2 cm or 0.022 m,
N = 500; I = 1.8 A.

(*i*) **Magnetic field strength at the centre, H :**

$$H = \frac{NI}{l} = \frac{500 \times 1.8}{0.5} = \mathbf{1800\ AT/m}\ \textbf{(Ans.)}$$

(*ii*) **Flux passing through the secondary coil, ϕ :**

$$\phi = \mu_0\, \mu_r\, HA$$
$$= 4\pi \times 10^{-7} \times 1 \times 1800 \times \left(\frac{\pi}{4} \times 0.022^2\right)$$
$$= \mathbf{85.98 \times 10^{-8}\ Wb}\ \textbf{(Ans.)}$$

(assuming air-cored solenoid *i.e.*, μ_r = 1)

Example 7.2. *A ballistic galvanometer having a circuit resistance of 5000 Ω and a constant of 0.1 μC per scale division is connected in turn (i) to a coil of 4 turns wound round the field coil of a D.C. machine and (ii) to a 6 turns coil placed on the armature surface embracing the total flux per pole entering armature. When normal field current is broken the galvanometer readings are 226 and 272 divisions respectively.*

Calculate :

(i) Flux per pole, and

(ii) Leakage coefficient. **(Punjab University)**

Solution. *Given* : $R = 5000\ \Omega$; $G_q = 0.1\ \mu C$ per scale division.

***(i)* Total flux per pole, ϕ :**

Electricity passed through the galvanometer,

$$Q = G_q \theta \quad ...(i)$$

Electricity passed through the galvanometer, when current is broken also,

$$Q = \frac{\phi N}{R} \quad ...(ii)$$

where, ϕ = Flux per pole,

N = Number of turns on search coil, and

R = Circuit resistance of the galvanometer.

Comparing expression (*i*) and (*ii*), we get

$$G_q \theta = \frac{\phi N}{R}$$

or,
$$\phi = \frac{R G_q \theta}{N}$$

$$= \frac{5000 \times 0.1 \times 10^{-6} \times 226}{4} = \mathbf{0.02825\ Wb} \text{ (Ans.)}$$

***(ii)* Leakage coefficient :**

Useful flux = Flux measured by galvanometer when connected to a search coil wound on the armature surface

$$= \frac{5000 \times 0.1 \times 10^{-6} \times 272}{6} = 0.02266\ \text{Wb}$$

$$\therefore \quad \textit{Leakage coefficient} = \frac{\text{Total flux}}{\text{Useful flux}} = \frac{0.02825}{0.02266} = \mathbf{1.247} \text{ (Ans.)}$$

Example 7.3. *A moving coil ballistic galvanometer of 150 ohms resistance gives a throw of 75 divisions when the flux through a search coil to which it is connected is reversed. Find the flux density given that the galvanometer constant is 110 μC per scale division and the search coil has 1400 turns, a mean area of 50 cm² and a resistance of 20 Ω.* **(Madras University)**

Solution. *Given* : $R = 150 + 20 = 170\ \Omega$; $\theta = 75$ divisions; $G_q = 110\ \mu C$ per scale division; $N_{sc} = 1400$; $A = 50\ \text{cm}^2 = 50 \times 10^{-4}\ \text{m}^2$.

Flux density, B :

We know that,
$$B = \frac{R \cdot G_q \cdot \theta_1}{2NA} \quad ...[\text{Eqns, (7.10, 7.11)}]$$

or,
$$B = \frac{170 \times 110 \times 10^{-6} \times 75}{2 \times 1400 \times 50 \times 10^{-4}} = \mathbf{0.1\ Wb/m^2} \text{ (Ans.)}$$

Example 7.4. *An iron ring of 400 mm² cross-sectional area with a mean length of 1.2 m is wound with a magnetising winding of 120 turns. A secondary coil with 240 turns of wire is connected to a ballistic galvanometer having a constant of 1.1 μC per scale division, the total* *resistance of the secondary circuit being 2200 Ω. On reversing a current of 12 A in the magnetising winding, the galvanometer shows a deflection of 100 scale divisions. Calculate :*

(i) The flux density in the specimen;

(ii) The value of permeability, at this flux density.

Solution. *Given* : Area, $A = 400\ \text{mm}^2 = 400 \times 10^{-6}\ \text{m}^2$; $l = 1.2$ m; $N_1 = 120$; $N_2 = 240$; $G_q = 1.1\ \mu\text{C}$ per scale division; $R = 2200\ \Omega$; $I = 12$ A; $\theta_1 = 100$ scale divisions.

(i) The flux density in the specimen, B :

$$\text{Total m.m.f. of coil} = N_1 I_1 = 120 \times 12 \text{ AT}$$

$$\text{Magnetising force, } H = \frac{N_1 I_1}{l} = \frac{120 \times 12}{1.2} = 1200 \text{ AT/m}$$

Charge through the galvanometer, $Q = G_q\ \theta_1 = 1.1 \times 100 = 110\ \mu\text{C}$ ...(*i*)

Let ϕ is the flux passing through the ring,

Then, Flux linkages of search coil,

$$\psi = N_2 \phi = 240\ \phi$$

Change in flux linkages of search coil due to reversal of current,

$$d\psi = 2\psi = 480\ \phi$$

Average e.m.f. induced in the search coil,

$$e = \frac{d\psi}{dt} = \frac{480\ \phi}{dt}, \text{ where, } dt = \text{time of reversal.}$$

∴ Current through the search coil,

$$Q = i.dt = \frac{e}{R} \cdot dt$$

$$= \frac{480\ \phi}{dt\ R} \cdot dt = \frac{480\ \phi}{R} = \frac{480\ \phi}{2200} \quad \text{...(ii)}$$

From (*i*) and (*ii*), we have :

$$\frac{480\ \phi}{2200} = 110 \times 10^{-6}$$

or,

$$\phi = \frac{110 \times 10^{-6} \times 2200}{480} = 504 \times 10^{-6} \text{ Wb}$$

∴ Flux density in the specimen,

$$B = \frac{\phi}{A} = \frac{504 \times 10^{-6}}{400 \times 10^{-6}} = \mathbf{1.26\ Wb/m^2}\ \textbf{(Ans.)}$$

(ii) The value of permeability, μ_r :

We know that, $B = \mu_0\ \mu_r\ H$

$$\therefore \quad \mu_r = \frac{B}{\mu_0 H} = \frac{1.26}{4\pi \times 10^{-7} \times 1200} = \mathbf{836}\ \textbf{(Ans.)}$$

Example 7.5. *An iron ring with a mean diameter of 0.12 m and a cross-section of 35.5 mm² is wound with a magnetising winding of 350 turns and a secondary winding of 250 turns. On reversing a current of 12 A in the magnetising winding, a ballistic galvanometer gives a throw of 280 scale divisions, while a Hibbert magnetic standard with 12 turns and a flux of 0.28 × 10⁻³ Wb gives a reading of 105 scale divisions, other conditions remaining the same.*

Find the relative permeability of the specimen. **(Rorkee University)**

Solution. *Given* : Mean diameters of iron ring, $D = 0.12$ m; $A = 35.5\ mm^2$; $N_1 = 350$; $N_2 = 250$, $I = 12\ A$; $\theta = 280$ scale divisions; Turns on Hibbert magnetic standard, $N_{st} = 12$; $\phi_{st} = 0.28 \times 10^{-3}$ Wb; Throw of galvanometer with Hibbert's standard, $\theta_{st} = 105$ scale divisions.

Relative permeability of the specimen, μ_r :

In a Hibbert's magnetic standard, the coil slides through the gap of a permanent magnet.

∴ Change in flux linkages of the coil when it goes down the Hibbert's standard

$$= \text{Flux } (\phi_{st}) \times \text{turns of the Hibbert's standard } (N_{st})$$

$$= (0.28 \times 10^{-3}) \times 12 = 3.36 \times 10^{-3} \text{ weber turns}$$

Constant of galvanometer,

$$G_\phi = \frac{3.36 \times 10^{-3}}{105} = 32 \times 10^{-6} \text{ Weber turns/division.}$$

The galvanometer shows a deflection of 280 scale divisions when connected to search coil.

∴ Change in flux linkages of search coil,

$$d\psi = 280 \times 32 \times 10^{-6} = 8960 \times 10^{-6} \text{ Weber turn}$$

This change is caused by the reversal of flux.

$$\therefore \text{ Flux linkages of coil, } \psi = \frac{d\psi}{2} = \frac{8960 \times 10^{-6}}{2} = 4480 \times 10^{-6} \text{ weber turn}$$

$$\text{Flux linking with search coil, } \phi = \frac{\text{Flux linkage of coil}}{\text{Turns of coil}}$$

$$= \frac{4480 \times 10^{-6}}{250} = 17.92 \times 10^{-6} \text{ Wb}$$

$$\text{Flux density in the ring specimen, } B = \frac{17.92 \times 10^{-6}}{35.5 \times 10^{-6}} = 0.505 \text{ Wb/m}^2$$

$$\text{Length of flux path in ring, } l = \pi D = \pi \times 0.12 = 0.377 \text{ m}$$

$$\text{Total m.m.f. of magnetism winding} = N_1 I = 350 \times 12 = 4200$$

$$\text{Magnetising force, } H = \frac{N_1 I}{l} = \frac{4200}{0.377} = 11141 \text{ AT/m}$$

$$\therefore \text{ Relative permeability of specimen, } \mu_r = \frac{B}{\mu_0 H} = \frac{0.505}{4\pi \times 10^{-7} \times 11141} = \mathbf{36.07}$$ **(Ans.)**

Example 7.6. *A permeameter, used to test a ferromagnetic material, is provided with a coil wound over the specimen, this coil being known as "B-coil" and another coil wound near the surface of the coil called "H-coil". These two coils are connected to two separate ballistic galvanometers having a constant of 28×10^{-6} weber turn per scale division. The testing is done by reversing the current of the magnetising winding wound over the specimen. The reading of the ballistic galvanometer connected to 'B'-coil is 100 divisions and that connected to 'H'- coil is 10 divisions.*

The data of the two coils is :

B-coil :	*Turns = 100;*	*Area = 60 mm^2*
H-coil :	*Turns = 10,000;*	*Area = 6 mm^2*

Calculate :

(i) Flux density in specimen.

(ii) Magnetising force.

(iii) Relative permeability of the specimen.

Solution. *(i)* **Flux density in speciment, B_{sp} :**

Let ϕ and N represent the flux and the number of turns respectively.

Then, Flux linkage, $\psi = N\phi$

Change in flux linkages due to reversal of current,

$$d\psi = 2\psi = 2\,N\phi$$

Flux linkages indicated by galvanometer $= G_\phi\,\theta_1$

$$\therefore \quad 2\,N\phi = G_\phi\,\theta_1$$

or, $$\text{Flux } \phi = \frac{G_\phi\,\theta_1}{2N}, \text{ and}$$

Flux density $$B = \frac{\phi}{A} = \frac{G_\phi\,\theta_1}{2NA}$$

Since *'B-coil'* is wound over the specimen, therefore it gives flux density in the *specimen*,

$\therefore$ Flux density in specimen,

$$B_{sp} = \frac{G_\phi\,\theta_1}{2NA} = \frac{28\times10^{-6}\times100}{2\times100\times60\times10^{-6}} = \mathbf{0.233\ Wb/m^2}\ \textbf{(Ans.)}$$

(ii) **Magnetising force, H :**

Since *'H'*-coil is put at the surface of the coil, therefore, it indicates the flux density in the *air.*

$$\text{Flux density in air, } B_{air} = \frac{28\times10^{-6}\times10}{2\times10000\times6\times10^{-6}} = 0.00233\ \text{Wb/m}^2$$

$$\text{Magnetising force, } H = \frac{B_{air}}{\mu_0} = \frac{0.00233}{4\pi\times10^{-7}} = \mathbf{1854\ AT/m}\ \textbf{(Ans.)}$$

(iii) **Relative permeability of the specimen, μ_r :**

$$\mu_r = \frac{B_{sp}}{\mu_0\,H} = \frac{0.233}{4\pi\times10^{-7}\times1854} = \mathbf{100}\ \textbf{(Ans.)}$$

Example 7.7. *The constant of a given potentiometer is obtained by aid of a coil of 320 turns in which a current of 0.7A is reversed. The resulting throw of galvanometer is 160 scale divisions. It is then used to measure the magnetic potential difference between two points and throw is 310 divisions.*

Calculate the magnetic potential difference.

Solution. *Given* : $N = 320$; $I = 0.7$ A; $\theta_1 = 160$ scale divisions; $\theta_2 = 310$ scale divisions.

Magnetic potential difference :

M.m.f. of coil $= NI = 320 \times 0.7 = 224$ AT

Change in m.m.f. due to reversal of current in the coil $= 2 \times 224 = 448$ AT

A deflection of 160 scale divisions is caused by an m.m.f. of 448 AT,

$\therefore$ m.m.f. causing a deflection of 310 divisions $= \frac{310}{160}\times448 = 868$ AT

Hence, *magnetic potential difference* = **868 AT (Ans.)**

7.5 A.C. MAGNETIC TESTING

The A.C. magnetic testing is carried out for the following *purposes* :

(*i*) To determine the iron losses in magnetic materials at different values of flux density and frequency;

(*ii*) To separate two components of iron losses, *i.e.*, eddy current losses and hysteresis losses.

Iron losses :

When ferromagnetic materials are subjected to an alternating field, power loss due to *hysteresis effects* and *eddy currents* occurs.

(*i*) ***Hysteresis loss*** **(P_h).** This loss depends upon the *frequency*, and *maximum flux density* of the magnetic field to which the specimen is subjected. It is expressed as :

$$P_h = K_h f B_{max}^k \text{ watts per m}^3 \text{ or per kg} \qquad ...(7.20)$$

where, K_h = Hysteresis coefficient for the given material,

f = Frequency of alternation, Hz,

B_{max} = Maximum flux density of the alternating magnetic field, Wb/m^2,

k = Steinmetz coefficient; its value varies from 1.6 to 2.

(*ii*) ***Eddy current loss*** **(P_e).** This loss depends upon the *wave shape of the variation of flux with time, frequency* and *maximum flux density* of the magnetic field to which the specimen is subjected.

This loss is given by the expression :

$$P_e = K_e K_f^2 f^2 B_{max}^2 \text{ watts per m}^3 \text{ or per kg} \qquad ...(7.21)$$

where, K_e = Eddy current loss constant for the given material, and

K_f = Form factor of alternating value of flux.

— The "*hysteresis loss*" for a given specimen can be easily determined by plotting hysteresis loop with direct current supply. But when the ferromagnic materials are subjected to alternating currents, the actual hysteresis loss differs from that determined by plotting hysteresis loop.

— The "*eddy current loss*" occurs only when the magnetic materials are subjected to alternating currents. Hence the ferromagnetic materials used in manufacture of all A.C. electrical machinery and other equipment are tested with alternating currents only.

• "*Total iron loss*" is the sum of hysteresis and eddy current losses and is given by the relation :

$$P_i = P_h + P_e$$

or, $$P_i = K_h f B_{max}^k + K_e K_f^2 f^2 B_{max}^2 \text{ watts per m}^3 \text{ or per kg} \qquad ...(7.22)$$

The total iron loss can be calculated with the aid of above expression but it *gives good results only in the case of static machinery and not in case of dynamo electric machinery. The reason being that this expression is applicable when the field is simply alternating, but when the reversals of flux are due to rotational magnetic field, the losses do not follow the same laws.* Therefore, in practice, ***iron loss curves*** *are used which represent the loss in watt per kg as a function of flux density.*

Fig. 7.22 shows loss curves for thickness t = 0.5 mm and t = 0.3 mm for a particular frequency say 50 Hz.

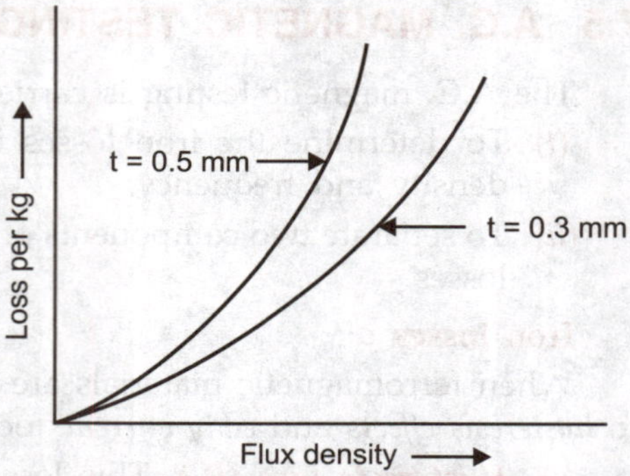

Fig. 7.22. Loss curves.

Separation of iron losses :

As the hysteresis and eddy current losses have different laws of variation with both frequency and form factor, K_f, it is *possible to separate the losses by variation of frequency or form factor if B_{max} can be maintained constant.*

1. *Variation of frequency.* In this test the form factor K_f and maximum flux density B_{max} remain constant. When the *frequency is varied* the total iron loss may be expressed as :

$$P_i = Af + Bf^2 \quad \text{...(7.23)}$$

where, $$A = K_h\, B_{max}^k \quad \text{...(7.24)}$$

and, $$B = Ke\, K_f^2\, B_{max}^2 \quad \text{...(7.25)}$$

Both A and B being constants for the test.

Now, we have : $$\frac{P_i}{f} = A + Bf \quad \text{...(7.26)}$$

These constants may be determined by plotting $\frac{P_i}{f}$ against f, as shown in Fig. 7.23. The intercept on the vertical axis gives value of constant A and slope of the line gives value of constant B. Thus knowing constants A and B the hysteresis and eddy current loss components for any given frequency can be determined.

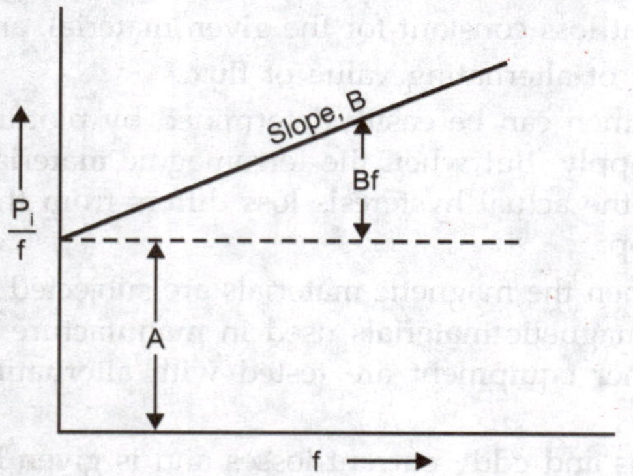

Fig. 7.23. Variation of iron loss with frequency.

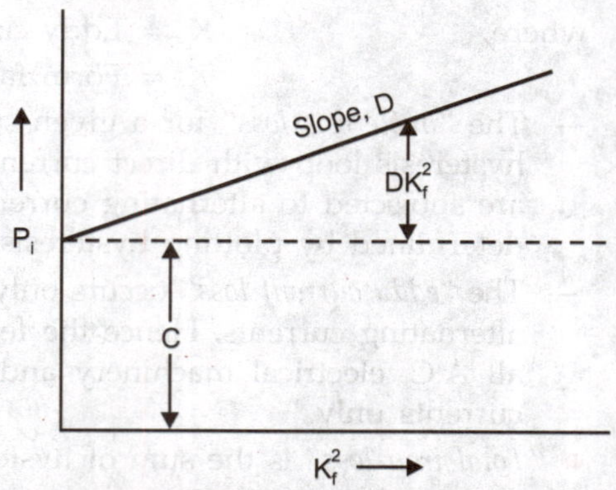

Fig. 7.24. Variation of iron loss with K_f^2.

2. *Variation of form factor.* The value of maximum flux density B_{max} and frequency f are kept constant in this test. When form factor K_f is varied, the total iron loss may be written as :

$$P_i = C + D\,K_f^2 \quad \text{...(7.27)}$$

where, $$C = K_h\, f\, B_{max}^k \quad \text{...(7.28)}$$

and, $$D = K_e\, f^2\, B_{max}^2 \quad \text{...(7.29)}$$

Both C and D being constant for the test.

The constants C and D can be determined by plotting P_i against K_f^2, as shown in Fig. 7.24. The intercept upon the vertical axis gives the value of constant C and the slope of the line drawn gives the value of constant D. Knowing the constants C and D the components of iron loss can be determined.

Factors affecting permeability and hysteresis loss :

- Generally *if the initial permeability is high, the hysteresis loss is low and vice versa.* The permeability and the hysteresis loss depend upon the following conditions :
  (*i*) Physical condition of the sample.
  (*ii*) Chemical purity of the sample.
- When the crystals of a ferromagnetic material are *cold worked*, they experience *deformation* as a result of which the material has *very poor magnetic properties.* Due to internal strains on the domains, greater magnetic field is required to give a definite magnetisation, as a consequence the *permeability decreases and the hysteresis loss is increased.* A material which has *suffered magnetic damage due to cold work may be treated to a sufficiently high temperature when the magnetic properties will be restored.*
- The impurity content of the material exercises a limit on the highest magnetic permeability and the lowest hysteresis loss that can be obtained. *Impurities bring harm to the magnetic properties by affecting the regular geometric pattern of the crystals.*
- The main impurities in the magnetic materials used for transformer cores and electrical machinery are *carbon, sulphur, oxygen* and *nitrogen. Carbon is the most detrimental* and the amount of carbon is commercial materials is kept *as low as 0.01 per cent.*

7.6 METHODS OF MEASUREMENT OF IRON LOSSES

The following methods are used to measure iron losses in ferromagnetic materials :

1. Wattmeter method.
2. A.C. bridge methods.
3. A.C. potentiometer method.
4. Oscillographic method.

7.6.1. Wattmeter Method

This is *perhaps the commonest method of measuring the total loss in sheet steel with alternating current.*

— The sheet material to be tested is arranged in the form of a ***magnetic square*** [of which there are several forms, Epstein being the originator of the arrangement] with fixed magnetising and search coils into which the strips of sheet material to be tested are slipped and clamped.

— The clamping arrangement should be such that *air gaps* in the circuits are reduced to a *minimum possible.*

— Each winding is divided into *four parts,* each part being on one side of the square.

— The windings extend as far as possible along the whole length of the side and a reasonable approximation to *uniform flux density is obtained.*

1. Epstein square. Fig. 7.25 shows the arrangement of an Epstein square :

— It consists of four stacks of strips. These stacks are bound and then taped.

— The individual strips are insulated from each other and each strip is in the plane of the square.

— The stacks are slipped into four magnetising coils with the strips projecting beyond the coils.

— The ends of the four strips are interleaved (as in the construction of transformer core) and clamped at corners.

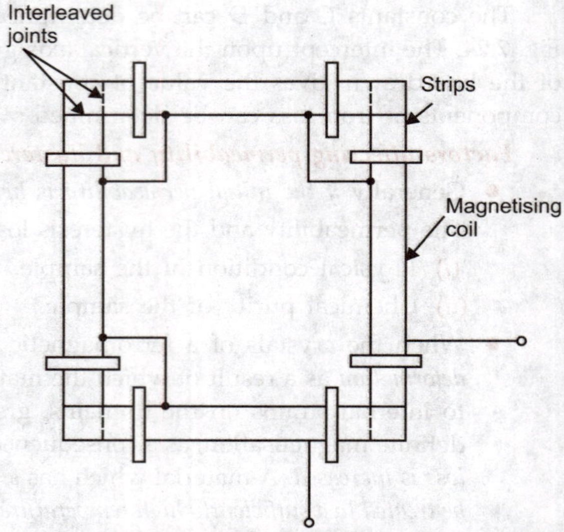

Fig. 7.25. Epstein square.

2. Lloyd-Fisher square. Fig. 7.26 shows the construction of Lloyd-Fisher square :

— Each strip (usually 0.25 m long and 50 to 60 mm wide) is perpendicular to the plane of the square. The strips are built up into four stacks which are placed inside four similar magnetising coils of large cross-sectional area. These four coils are connected in series to form the *primary winding*.

— Each magnetising coil has two similar single layer coils underneath it; these are called *secondary coils*. These secondary coils are connected in series in group of four, one from each core, to form *two separate secondary windings*.

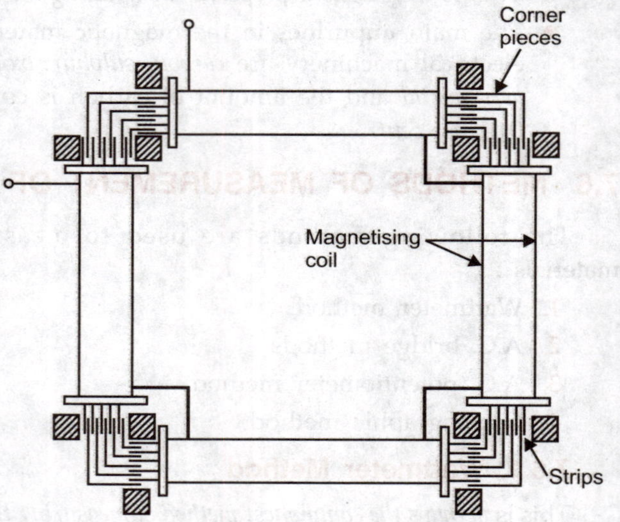

Fig. 7.26. Lloyd-Fisher square.

— The magnetic circuit is completed by bringing the four stacks together in the form of a square and joining them at the corners. The corner joints are made by a set of standard right angled corner pieces.

— There is an overlapping of corner pieces and strips at the corners due to which cross-section of iron is doubled at the corners; therefore, a correction must be applied for this. Also the measured losses has to be *corrected* for loss in the corner pieces.

Advantages :

(*i*) This square gives *rather more reliable* results than Epstein square, in case allowance for corner pieces is known with adequate accuracy.

(*ii*) The use of corner pieces in this type of square makes it *superior for testing anisotropic materials.* (The Epstein square is inferior for tests on anisotropic materials as the direction of flux at the corners is partially perpendicular to the path of flux in other portions of the strip).

Set up for the test :

Fig. 7.27 shows the connection diagram for finding the total iron loss by *wattmeter method* :

— The test specimen is weighed before assembly and its cross-sectional area is determined.

— The primary winding, which contains the current coil (C.C.) of the wattmeter, is connected to an A.C. supply.

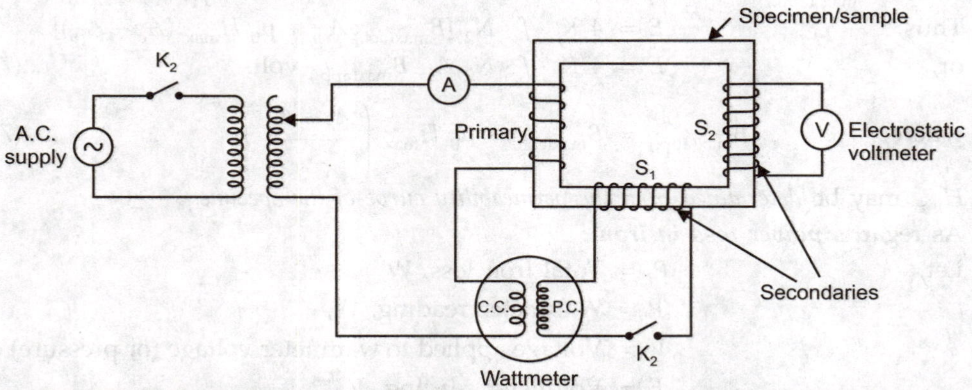

Fig. 7.27. Connections for "wattmeter method" of iron loss measurement.

— The test specimen has two secondary windings S_1 and S_2, S_1 is connected to the pressure coil of the wattmeter through switch K_2. S_2 is connected to an electrostatic voltmeter or an electrodynamic voltmeter of very high impedance. The supply frequency is adjusted to the correct value.

— The voltage applied to the primary winding is adjusted, till the magnetising current is adjusted to give the required value of B_{max}. The readings of the wattmeter and voltmeter are observed.

Theory of the method :

The voltage induced in secondary winding S_2, whose r.m.s. value is measured by the voltmeter, is given by the expression :

$$E = 4\ K_f\ \phi_{max} \cdot f \cdot N_2 \text{ volts}$$

or,

$$E = 4\ K_f B_{max} \cdot A \cdot f N_2 \text{ volts} \quad ...(7.30)$$

where,
K_f = Form factor (1.11 for sinusoidal waves),
ϕ_{max} = Maximum value of flux, Wb,
B_{max} = Maximum flux density Wb/m^2,
A = Cross-section of the specimen, m^2,
f = Frequency, Hz, and
N_2 = Number of turns on the secondary winding.

$$\therefore \quad B_{max} = \frac{E}{4\,K_f \cdot A \cdot f \cdot N_2} \qquad \text{...[7.30(a)]}$$

The expression may need to be corrected, especially at high values of B_{max}, for the fact that the coil S_2 encloses some air flux as well as the flux in the specimen/sample, since the cross-sectional area of the coil must be greater than that of specimen itself.

Let, A_c = Cross-sectional area of the coil,

A_{sp} = Cross-sectional area of the specimen,

H_{max} = The maximum magnetising force (equal to that in the air space within the coil)

$B_{max\ (act.)}$ = Actual maximum flux density in the specimen.

Then, the *total flux within the coil is :*

$$B_{max(act.)} \cdot A_{sp} + \mu_0 H_{max} (A_c - A_{sp}) = B_{max(app.)} \cdot A_{sp}$$

where, $B_{max\ (app.)}$ is the *apparent maximum flux* density in the specimen.

Thus, $E = 4\,K_f \cdot f \cdot N_2\,[B_{max(act.)} \cdot A_{sp} + \mu_0 H_{max} (A_c - A_{sp})]$

or, $$E = 4\,K_f \cdot f \cdot N_2\,A_{sp}\,B_{max(app.)} \text{ volts} \qquad \text{...(7.31)}$$

where, $$B_{max(app.)} = B_{max(act.)} + \mu_0 H_{max} \left(\frac{A_c - A_{sp}}{A_{sp}} \right) \qquad \text{...(7.32)}$$

H_{max} may be *determined from the permeability curve of the specimen.*

As regards ***power loss in iron :***

Let, P_i = Total iron loss, W,

P = Wattmeter reading, W,

V = Voltage applied to wattmeter voltage (or pressure) coil,

E = Voltmeter reading, V

= Voltage induced in coil S_1, since S_1 and S_2 have equal number of turns.

r_p = Resistance of wattmeter voltage (or pressure) coil, Ω,

r_c = Resistance of coil S_1, Ω, and

I_p = Current in the voltage (or pressure) coil circuit, A.

Then, $$E = I_p(r_p + r_c) \qquad \text{...}(i)$$

$$V = I_p r_p \qquad \text{...}(ii)$$

∴ *Total iron loss in specimen + total copper loss in the secondary winding circuit*

$$= P \cdot \frac{E}{V} = P\left[\frac{I_p(r_p + r_c)}{I_p\,r_p}\right] = P\left(\frac{r_p + r_c}{r_p}\right)$$

$$= P\left(1 + \frac{r_c}{r_p}\right) \qquad \text{...}(iii)$$

Again, the *copper losses* in r_p and r_c are :

$$I_p^2(r_p + r_c) = \left(\frac{E}{r_p + r_c}\right)^2 (r_p + r_c) = \frac{E^2}{r_p + r_c} \qquad \text{...}(iv)$$

$$\therefore \qquad \textit{Total iron loss, } \mathbf{P}_i = P\left(1+\frac{r_c}{r_p}\right)-\frac{E^2}{r_p+r_c} \text{ watts} \qquad ...(7.33)$$

...[Subtracting (*iv*) from (*iii*)]

- *Specific iron loss i.e.,* iron loss per kg can be calculated by dividing the total iron loss by the weight of the specimen.
- *The hysteresis and eddy current components of the loss can be graphically determined from the results of power measurements* such as the above, *at different frequencies.*

Measurement of iron loss at different from factors :

Fig. 7.28, shows the connections for the measurement of power loss at different form factors.

— The form factor is obtained from the readings of two voltmeters V_1 and V_2. The voltmeter V_1 (may be an electrostatic or electrodynamic type) is connected across S_1 and it reads the *r.m.s. value* of the voltage. Voltmeter V_2 (a moving coil instrument) is connected across S_2 through a bridge rectifier and, therefore, it reads the *average value of the voltage.*

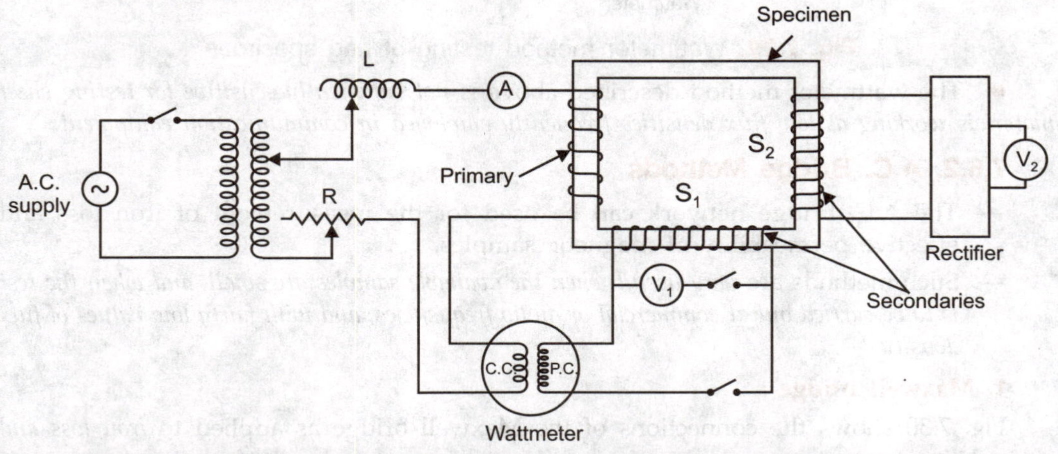

Fig. 7.28. Measurement of power loss at different form factors—Wattmeter method.

$$\text{Form factor, } K_f = \frac{\text{R.m.s. value}}{\text{Average value}} = \frac{\text{Reading of voltmeter } V_1}{\text{Reading of voltmeter } V_2}$$

(since S_1 and S_2 have *equal* number of turns)

The variation of form factor is obtained by adjusting the values of series resistance (R) and inductance (L).

Wattmeter method for testing ring specimens :

Since sheet steel in the form of rings is used for making current transformers and also spiral cores are used for construction of current transformers and variacs, therefore it becomes necessary to measure iron loss in ring specimens. The methods employed and experimental set up is the same as that for strips.

Fig. 7.29 shows the connection diagram for measurement :

— S_1, S_2 are the secondary windings;

— Voltmeter V_1 reads the *r.m.s. value;*

— Voltmeter V_2 reads the *average value* of voltage.

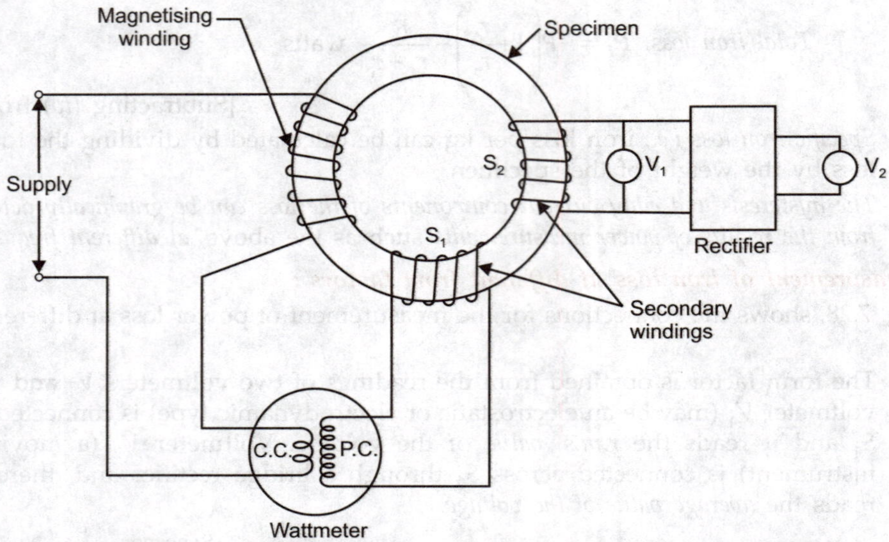

Fig. 7.29. Wattmeter method testing of ring specimen.

- The wattmeter method described above is *not sufficiently sensitive for testing sheet materials working at low flux densities frequently employed in communication equipment.*

7.6.2. A.C. Bridge Methods

— The A.C. bridge network can be used for the measurement of iron loss and effective permeability of magnetic samples.

— Such methods are *very useful when the available samples are small, and when the test is to be carried out at commercial or audio frequencies, and with fairly low values of flux density.*

1. Maxwell bridge :

Fig. 7.30 shows the connections of the Maxwell bridge as applied to *iron loss and permeability measurements :*

— R_1, R_2 and R_3 (variable) are resistors. In series with R_3, in the same arm, is a variable inductor L of resistance r. It may be necessary to connect R_3 in series with the winding on the specimen if the resistance of the latter is small.

— The specimen, in ring form, is wound with a winding whose inductance is L_{sp} and effective resistance R_{sp}, this effective resistance containing an iron-loss component. R_w is the actual resistance of the winding on the ring.

— G is the vibration galvanometer or telephone, and A an ammeter.

— The supply is from an alternator having a pure sine waveform.

— The bridge is balanced by adjustment of L and R_3.

Theory :

When the bridge is balanced :

$$I_1 R_1 = I_2 R_2$$

and, $$I_1(R_{sp} + j\omega L_{sp}) = I_2[(r + R_3) + j\omega L]$$

where, $$\omega = 2\pi f$$

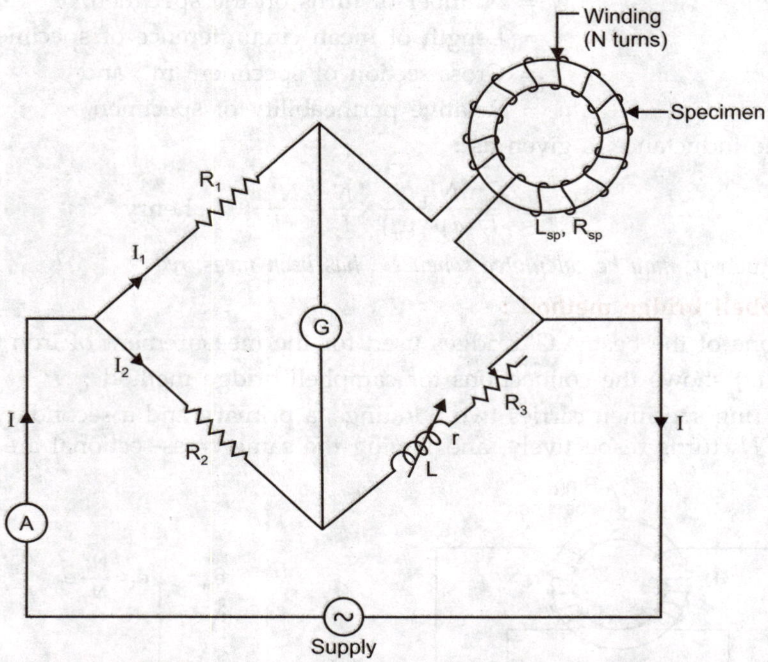

Fig. 7.30. Maxwell bridge for iron loss measurements.

Then, $$\frac{I_1}{I_2} = \frac{R_2}{R_1} \quad ...(i)$$

$$I_1 R_{sp} = I_2(r + R_3) \quad ...(ii)$$

and, $$I_1 L_{sp} = I_2 L \quad ...(iii)$$

From (*i*) and (*ii*), $$\frac{R_2}{R_1} = \frac{r + R_3}{R_{sp}}$$

or, $$R_{sp} = (r + R_3)\frac{R_1}{R_2} \quad ...(iv)$$

From (*i*) and (*iii*), $$L_{sp} = L \cdot \frac{R_1}{R_2} \quad ...(v)$$

- The *iron loss* (P_i) *in the sample* is given by :

$$\text{Iron loss, } P_i = I_1^2 R_{sp} - I_1^2 R_w \text{ or } I_1^2(R_{sp} - R_w) \quad ...(vi)$$

Now, current $I = I_1 + I_2$ (I_1 and I_2 being in phase)

Thus, $$I = I_1 + \frac{R_1}{R_2} I_1 = I_1\left(\frac{R_1 + R_2}{R_2}\right)$$

or, $$I_1 = I \cdot \left(\frac{R_2}{R_1 + R_2}\right) \quad ...(vii)$$

Hence, the iron loss is given by :

$$P_i = I^2\left(\frac{R_2}{R_1 + R_2}\right)^2 (R_{sp} - R_w) \quad ...(7.34)$$

[Using eqn. (*vi*) and (*vii*)]

Let, N = Number of turns on the specimen,

l = Length of mean circumference of specimen, m,

a = Cross-section of specimen, m^2, and

μ_r = Relative permeability of specimen.

Then, the inductance is given as :

$$L_{sp} = \frac{NI_1}{l/(a\mu_0\mu_r)} \times \frac{N}{I_1} = \frac{N^2 a}{l}\mu_0\mu_r \text{ henry} \quad ...(7.35)$$

From which μ_r *may be calculated when* L_{sp} *has been measured.*

2. Campbell bridge method :

This is one of the best A.C. bridges used for the measurement of iron loss.

Fig. 7.31(*a*) shows the connections for campbell bridge method :

— The ring specimen carries two windings, a primary and a secondary, having N_1 and N_2 turns respectively, and having the same cross-sectional area.

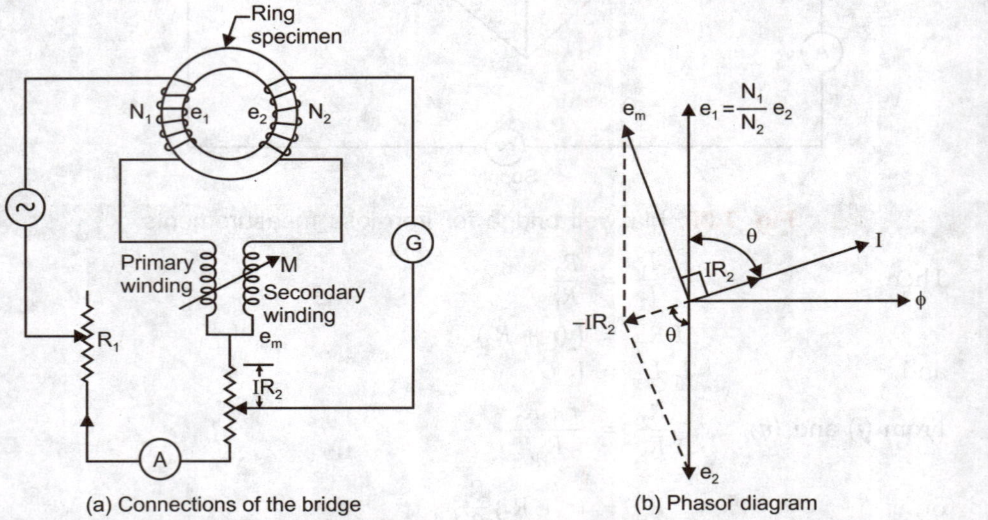

Fig. 7.31. Campbell bridge method.

— *M* is a variable mutual inductor connected as shown. R_1 and R_2 are variable resistors, and *A* is an ammeter. *G* is a vibration galvanometer or telephone according to the supply frequency. The waveform of the supply voltage should be sinusoidal, and R_1 should be made *sufficiently large* to ensure that current waveform is also *sinusoidal*.

— *M* and R_2 are adjusted until the galvanometer shows no deflection or until the telephone gives minimum sound.

- *Under balance conditions* the phasor diagram is as shown in Fig. 7.31(*b*) :

— ϕ is the flux in the specimen, and *I* the current in the primary winding, leading the flux by a small angle on account of iron loss.

— e_1, e_2, and e_m are the *induced voltages* in the primary winding, the secondary winding, and the secondary of the mutual inductor *M* respectively. Balance is obtained when the vector sum e_m and e_2 is equal, and opposite, to the vector IR_2 representing the voltage drop in resistor R_2.

Let M' = Mutual inductance between the primary and secondary windings of the specimen,

$$e_2 = \omega M'I, \text{ and}$$

$$e_m = \omega MI.$$

Since IR_2 is very small, we can write,

$$e_2 \simeq e_m$$

or, $$\omega M'I = \omega MI \quad \text{or} \quad M' \simeq M$$

Flux in the specimen, $$\phi = \frac{N_1 I}{l_{sp}/(\mu_{sp} A_{sp})} = \frac{N_1 I \mu_{sp} A_{sp}}{l_{sp}}$$

Mutual inductance between primary and secondary windings of specimen is :

$$M = \frac{N_2 \phi}{I} = \frac{N_1 N_2 \mu_{sp} A_{sp}}{l_{sp}}$$

or, A.C. permeability $$\mu_{sp} = \frac{M l_{sp}}{N_1 N_2 A_{sp}} \qquad \text{...(7.36)}$$

Iron loss, $$P_i = e_1 I \cos\theta = e_1 I \cdot \frac{IR_2}{e_2} = \frac{I^2 R_2 e_1}{e_2}$$

But, $$\frac{e_1}{e_2} = \frac{N_1}{N_2}$$

$\therefore$ $$P_i = \frac{I^2 R_2 N_1}{N_2} \qquad \text{...(7.37)}$$

Current I is measured with the help of ammeter A and thus the iron loss can be computed.

7.6.3. A.C. Potentiometer Method

The iron loss and A.C. permeability of ferromagnetic materials working at low flux densities can be determined by using an A.C. potentiometer.

- Fig. 7.32 shows the connections for iron loss measurement with an A.C. potentiometer.
  - — The ring specimen carries two windings. The primary windings has N_1 turns and the secondary N_2 turns, the supply to the former being through a regulating transformer, from an alternator (having sinusoidal voltage waveform) which also supplies current to the two potentiometer slide-wire circuits.
  - — A variable resistor R and a standard resistor are connected in series with the primary winding, the latter being for the purpose of measuring the current in the primary winding by measuring the magnitude and phase of the voltage drop across it.
  - — G is a vibration galvanometer.
- The flux density B_{max} in the specimen is obtained from the expression :

$$B_{max} = \frac{E_2}{4K_f \cdot A \cdot f \cdot N_2}$$

where, E_2 = E.m.f. (r.m.s. value) induced in the secondary winding,

K_f = Form factor = 1.11 for sinusoidal supply,

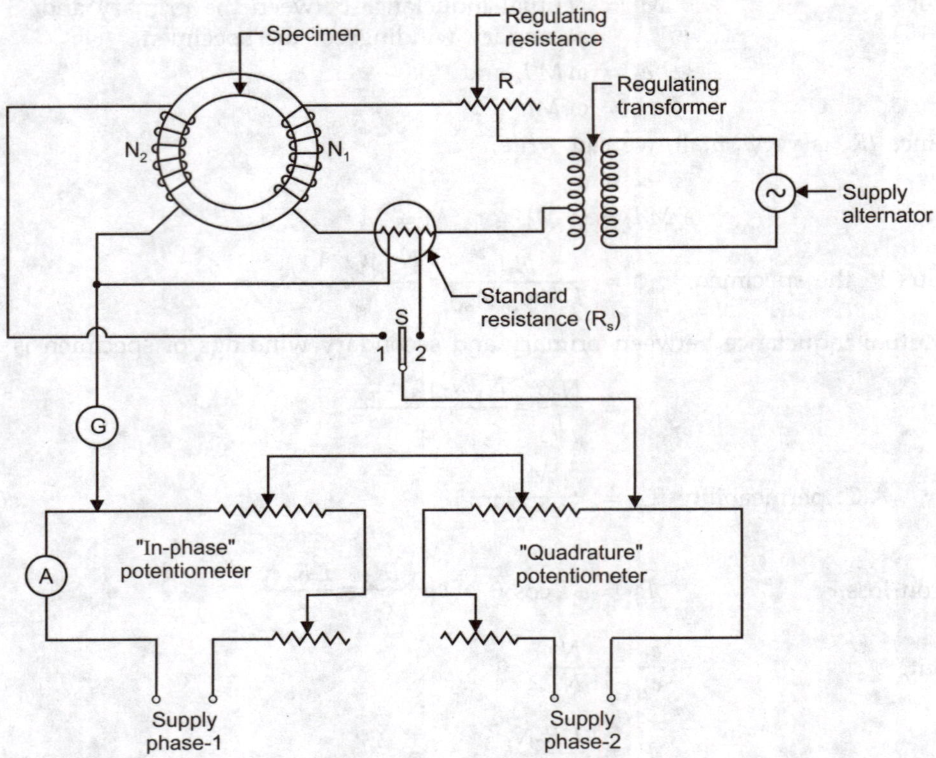

Fig. 7.32. Iron-loss measurement by A.C. potentiometer

A = Cross-sectional area of the specimen,

f = Supply frequency, and

N_2 = Number of turns in the secondary winding.

— The potentiometer is standardised first, and then the supply is given to the primary winding of the specimen.

— The voltage E_2 is measured by moving the switch S on to contact 1, setting the *"quadrature"* potentiometer at *zero*, and adjusting the *"in-phase"* potentiometer until the vibration galvanometer shows no deflection. The setting of the "in-phase" potentiometer for balance then gives the value of voltage E_2 directly.

— The switch S is then thrown over to contact 2, and both the "in-phase' and "quadrature" potentiometers are adjusted to give zero galvanometer deflection. The reading of the *"in-phase"* potentiometer then gives the value $I_e R_s$, where I_e is the loss component of the current in the primary winding and R_s is the value of the standard resistance (I_e = reading of in-phase potentiometer/R_s). The reading of the *"quadrature"* potentiometer gives the value $I_m R_s$ where I_m is the magnetising component of the primary current (I_m = reading of quadrature potentiometer/R_s).

$$\text{Iron loss} = I_e \cdot E_1 = I_e \cdot E_2 \cdot \frac{N_1}{N_2}$$

- This method is *very suitable for A.C. magnetic testing at low flux densities since magnetising and iron loss components of exciting current are measured separately.*

7.6.4. Oscillographic Method

- Fig. 7.33 shows the connection diagram for measurement of iron losses by cathode ray oscilloscope (CRO) :
  - — Test specimen is taken in ring form and two windings, *magnetising winding* having N_1 turns and *search coil* having N_2 turns are wound on it.
  - — Magnetising winding is supplied by an autotransformer with an ammeter and non-inductive resistance R_p in series. Auto-transformer is supplied by an alternator.

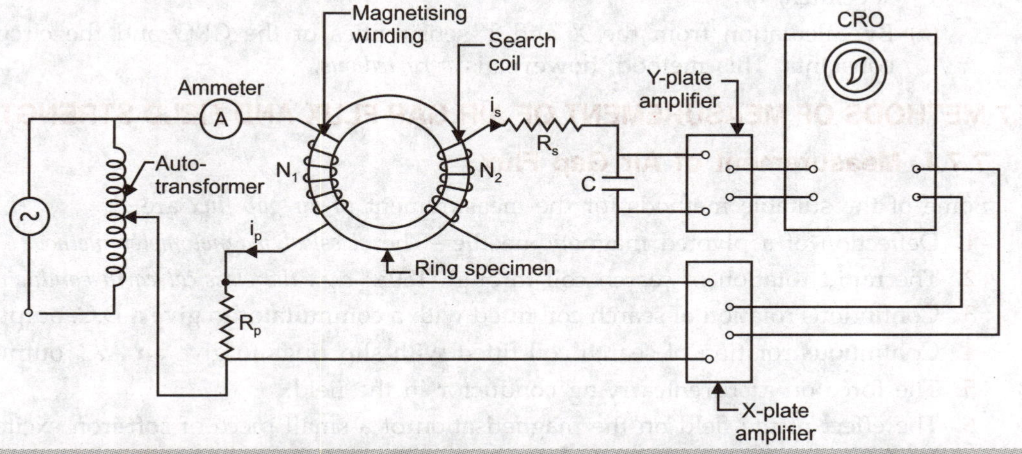

Fig. 7.33. Magnetic material testing with cathode ray oscillograph (CRO).

- Voltage $V_p (= i_p\ R_p)$ across resistance R_p is directly proportional to magnetising current i_p and, consequently, to the strength of magnetic field in the test specimen. This voltage V_p is applied to the *X-plates* of the CRO through an amplifier.

Output of search coil is supplied to a capacitor C with a high resistance R_s in series. The e.m.f. induced in the search coil,

$$e_{sc} = N_2 \cdot \frac{d\phi}{dt} = N_2 \cdot A_{sc} \frac{dB}{dt}$$

when resistance R_s is higher than that of the remaining section of the circuit, the current in the measuring circuit is :

$$i_s = \frac{e_{sc}}{R_s} = N_2 \cdot \left(\frac{A_{sc}}{R_s}\right)\left(\frac{dB}{dt}\right)$$

The voltage across capacitor C is :

$$V_c = \frac{1}{C}\int i_s\, dt = \frac{1}{C}\frac{N_2 A_{sp}}{R_s}\int dB = \frac{1}{C}\frac{N_2 A_{sp}}{R_s} B \quad ...(7.38)$$

Therefore voltage V_c across the capacitor C is proportional to flux density in the test specimen. This voltage V_c is applied to *Y-plates* of the CRO through an amplifier.

The *X-deflection* is proportional to the voltage drop across R_p carrying the magnetising current, the deflection is therefore proportional to the *magnetising force H*.

The *Y-deflection* is proportional to the voltage V_c and therefore to the *flux density B* in the specimen.

(*i*) ***Plotting of hysteresis loop.*** According to the arrangement shown in the figure, the X-deflection is proportional to H and the Y-deflection is proportional to B. Thus the figure traced on the screen will be the *hysteresis loop* of the material.

(ii) ***Plotting B-H curve.*** When the peak flux density in the specimen is slowly and continuously varied (*e.g.*, by varying the excition of the alternator supplying the magnetising current) the locus of the tips of the hysteresis loops gives the *B-H curve* for the material.

(iii) ***Determination of hysteresis loop.*** The hysteresis loss for the material may be estimated :

(*a*) From the area of the loop either by comparison with a loop for a standard specimen, or,

(*b*) By calculation from the X and Y sensitivities of the CRO and the circuit constants. This method, however is *very tedious.*

7.7 METHODS OF MEASUREMENT OF AIR GAP FLUX AND FIELD STRENGTH

7.7.1. Measurement of Air Gap Flux

Some of the suitable methods for the measurement of *air gap flux* are :

1. Deflection of a pivoted magnetic needle—The *classical magnetometer method.*
2. The rapid rotation of search coil through 180°—*e.g., the classical earth conductor.*
3. Continuous rotation of search coil fitted with a commutator to give a D.C. output.
4. Continuous rotation of search coil fitted with slip rings to give an A.C. output.
5. The force on a currentcarrying conductor in the field.
6. The effect of the field on the magnetisation of a small piece of soft-iron excited by A.C. in a magnetising winding.

7.7.2. Measurement of Field Strength

The following methods of measurement of field strength will be described :

1. By withdrawal of search coil.
2. By nuclear magnetic resonance.
3. Use of bismuth spiral.
4. Use of mumetal wires.
5. Use of Hall effect devices.

7.7.2.1. By withdrawal of search coil

— In this method a search coil is placed in a magnetic field. The terminals of the search coil are connected to a ballistic galvanometer.

— This search coil is suddenly withdrawn from the magnetic field, as a consequence the flux linking with the search coil changes. This induces an e.m.f. in the search coil which causes the galvanometer to deflect. The flux density in the field may be known from the throw of the galvanometer.

7.7.2.2. By nuclear magnetic resonance

— When a high frequency alternating magnetic field is superimposed on atomic nuclei located in a constant magnetic field, resonance absorption of energy from the high-frequency field is liable to occur at a specific ratio of the constant field strength H to frequency f of the alternating field.

— According to laws of quantum mechanics the magnetic moment of, say, the hydrogen nucleus (proton) may be oriented either in the same direction as that of external magnetic field or in *opposition to it.*

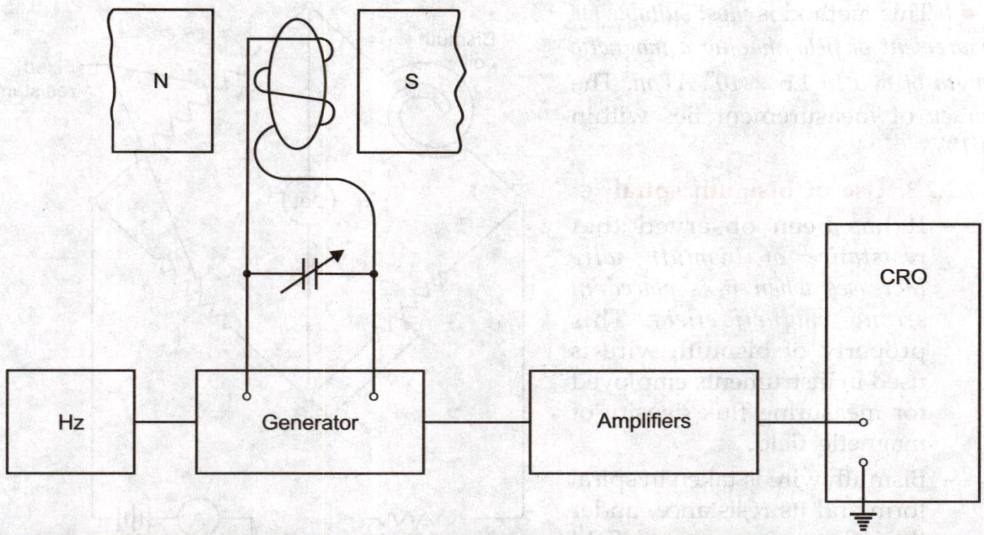

Fig. 7.34. Measurement of field strength by nuclear magnetic resonance.

— The orientation of the proton magnetic moment can be changed by application of an energy quantum *hf*, where *h* is Planck's constant and *f* is the frequency. The difference between these two energy states of the proton amounts to $2m_pH$ where m_p is the proton magnetic moment.

Thus, $$hf = 2m_pH$$

or, $$f = \frac{2m_pH}{h}$$

Now, $$\omega = 2\pi f = 2\pi \times \left(\frac{2m_p H}{h}\right) = \frac{4\pi m_p H}{h} = \gamma_p H$$

or, $$H = \frac{2\pi f}{\gamma_p} \quad ...(7.39)$$

where, γ_p = *"Gyromagnetic ratio" of the nucleus i.e., the ratio of the magnetic moment to the mechanical moment.*

The value of γ_p of a number of substances is known to the nearest thousandth of a percent.

According to this method the magnetic strength of a constant field is determined by measuring the frequency at which nuclear magnetic resonance occurs.

- **Fig. 7.34** shows the block diagram of the test installation for the measurement of field strength by nuclear magnetic resonance :

— A proton sample, say, an aqueous solution of $Fe(NO_3)_3$ contained in a glass ampule is surrounded by a small coil and is inserted in the magnetic field to be measured, the axis of the coil being at right angles to the field.

— The coil and capacitor constitute an oscillatory circuit, which is connected to a high frequency oscillator.

— *Resonance is detected by a sharp change in the amplitude of the resonance curve traced on the screen of CRO at a change in the oscillator frequency.*

The resonance frequency is measured by means of a *frequency meter or wavemeter.*

- This method is *most suitable for measurement of fields having a magnetic strength of 800 to 1.6 × 10^6 AT/m.* The *accuracy* of measurement lies within *±0.01%*.

7.7.2.3. Use of bismuth spiral

— It has been observed that *resistance of bismuth wire increases when it is placed in strong magnetic field.* This property of bismuth wire is used in instruments employed for measuring flux density of magnetic field.

— Bismuth wire is taken in spiral form and its resistance, under the influence of magnetic field is measured with a Wheatstone bridge, as shown in Fig. 7.35.

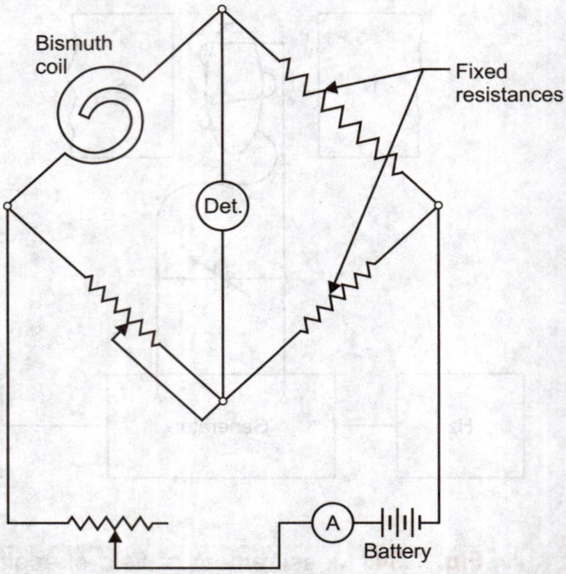

Fig. 7.35. Measurement of flux density with bismuth spiral.

— The change in the value of resistance is converted into the flux density *by calibration.*

Advantages :

(*i*) It is simple to use.

(*ii*) It takes a small area, being in spiral form, for flux measurement.

(*iii*) With this method, even flux density in air gap between poles and armature of D.C. machine can be measured.

Disadvantage. The change in resistance under influence of magnetic field is very small and a large change takes place with even small change in temperature.

7.7.2.4. Use of mumetal wires

— This method is based on the principle that *impedance of mumetal wire changes when it is kept in magnetic field.*

— The change in impedance is measured with A.C. bridges and *with proper calibration, change in impedance is converted into the value of magnetic field strength.*

— This method is free from disadvantages associated with bismuth spiral method.

- *Because of high sensitivity it is used in modern instrumentation.*

7.7.2.5. Use of Hall effect devices

- When a current carrying conductor is placed in a magnetic field a *transverse effect* (discovered by Hall in 1879) is noted. This called the *Hall effect.*

— Hall found that : *"When a magnetic field is applied at right angles to the direction of the electric current, an electric field is set up which is perpendicular to both the direction of electric current and the applied magnetic field."*

In other words : *"When any specimen carrying a current I is placed in the transverse magnetic field B, then an electric field E is induced in the specimen in the direction perpendicular to both I and B (the phenomenon is known as the Hall effect)."*

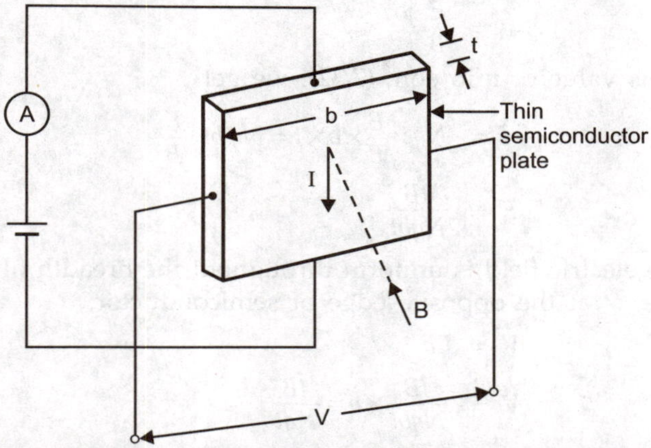

Fig. 7.36. Field strength measurement using Hall effect.

This e.m.f. E is very small in conductors and is difficult to measure, but in some semiconductors such as germanium, the e.m.f. is sufficiently large to be measured with a sensitive moving-coil instrument.

— This phenomenon now forms the *basis* of some simple and elegant field measuring instruments.

- Consider the thin plate of semiconductor illustrated in Fig. 7.36.

The current is made up from electrons and positive holes travelling in opposite directions, and since current is rate of flow of charge, it can be expressed as :

$$I = N.q.v.b.t \quad \text{...(7.40)}$$

where, N = Number of charge carriers per unit volume,

q = Charge on each charge carrier,

v = Average velocity of the charge carriers, and

b, t = Breadth and thickness of the semiconductor plate respectively.

Now, if there is a magnetic field, of flux density B, crossing the thin plate, there will be a force on each charge, given by the expression :

$$F_B = q.v.B \quad \text{..(7.41)}$$

where, B is flux density of the magnetic field.

This force F_B will deflect the carriers to one side in accordance with Fleming's left hand rule. Thus, there is tendency for charge to *accumulate* on one side of the semiconductor, but *existence of this charge establishes on electric field which, in turn, exercises an opposing force on the carriers.* Finally a state of equilibrium is reached when the force due to the magnetic field is equal to that due to electric field and there is then no accumulation. The polarity of the electric field depends upon whether conduction is due to negative electrons or positive holes of the charge.

If the electric-field strength is E, then the force on the charge carriers is given by :

$$F_E = q \cdot E \quad \text{...(7.42)}$$

Then, *at equilibrium,*

$$F_E = F_B$$

or, $$q{\cdot}E = q{\cdot}v{\cdot}B$$

or, $$E = v{\cdot}B$$

or, $$v = \frac{E}{B}$$

Substituting this value of v in eqn. (7.32), we get

$$I = Nq \times \frac{E}{B} \times b \times t = Nq\,bt \cdot \frac{E}{B}$$

or, $$E = \frac{IB}{Nqbt} \qquad ...(7.43)$$

Now, since the electric field is uniform throughout the breadth of semiconductor, so potential difference, V at the opposite edge of semiconductor,

$$V = Eb$$

or, $$V = \frac{IB}{Nqbt} \times b = \frac{IB}{Nqt} \qquad ...(7.44)$$

The term $\frac{1}{Nq}$ is called the *Hall coefficient* of material. It can been seen from eqn. (7.44) that for a *given current* the *voltage* appearing at the opposite edges of semiconductor is *directly proportional to the flux density* of given magnetic field.

7.8 TESTING OF PERMANENT MAGNETS

The method of testing permanent magnet depends upon the shape of the magnet which in turn varies greatly according to the purpose for which the magnet is to be used.

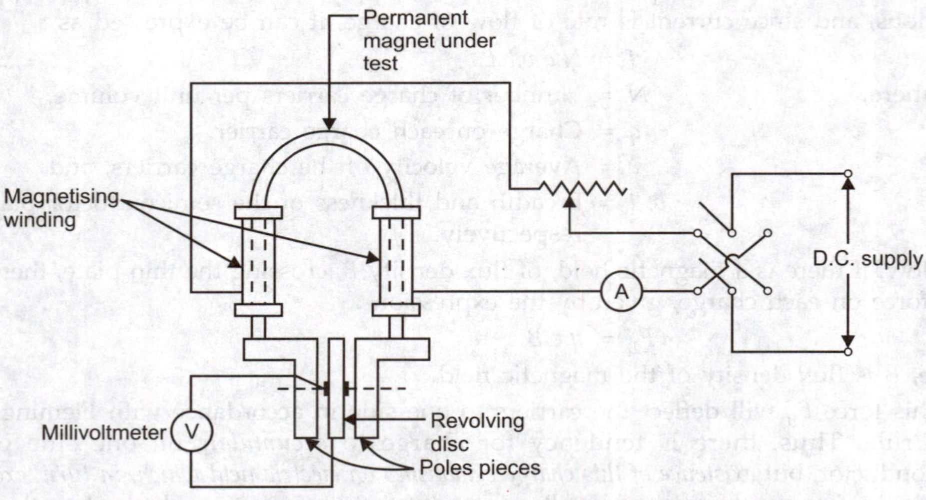

Fig. 7.37. Testing permanent magnet by batteridge apparatus.

Because of the self-demagnetising force in permanent magnets, it is necessary, in testing such magnets, *to measure H by means of search coils, of small cross-section, laid flat against the surface of the magnet*. These search coils give the value of the air flux density at the magnet surface at different points, and thus measure H for these points. The flux density B is measured by search coils in the ordinary way.

Fig. 7.37 shows ***Batteridge apparatus*** for magnet testing :

— In this method of magnet testing, the magnet to be tested is placed with its straight ends inside two magnetising coils and so that its ends press against two poles pieces, in the air gap between which a thin iron disc, plated with copper is mounted.

— The disc is mounted in ball bearings and clearances between it and the pole pieces are small. It is driven round at constant speed by a small motor, and has two small brushes making contact with its spindle and its rim.

Working. When a current flows in the magnetising windings, flux crosses the gap between the pole pieces, and the flux density in the gap will be proportional to that in the magnet. An e.m.f. will be induced in the revolving disc, this e.m.f. being proportional to the gap flux density and being measured by the millivoltmeter V. Thus the voltmeter reading gives a measure of flux density in the specimen, while H is obtained by the usual formula from the constants of magnetising coils and from the current in these coils.

Some leakage exists between the two arms of the magnet, but with proper design of apparatus its effect upon the results is small.

7.9 TESTING OF FEEBLY MAGNETIC MATERIALS

Fig. 7.38 shows the apparatus employed for measurements upon material which is distinctly magnetic, but whose permeability is low (of the order of *unity*) :

— The magnetising winding has a search coil at its centre, this search coil being fixed so that the specimen can be slipped inside it or withdrawn without disturbing the search coil itself.

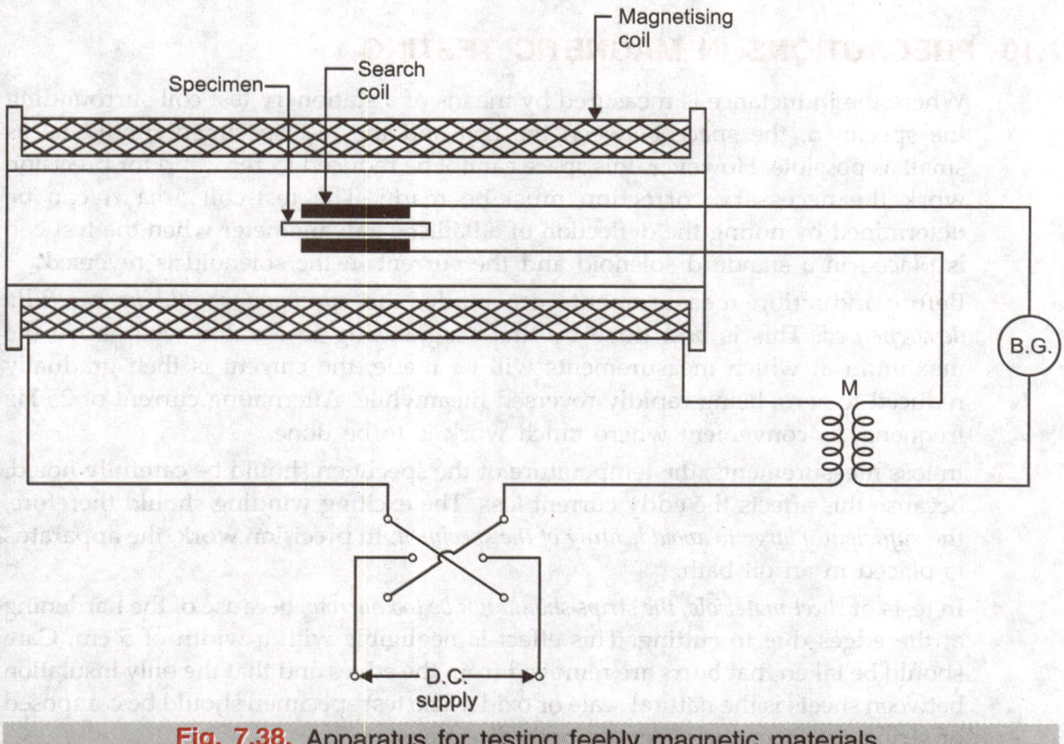

Fig. 7.38. Apparatus for testing feebly magnetic materials.

— The search coil is connected through a ballistic galvanometer (B.G.), to the secondary of a mutual inductor, the primary of which is connected in the magnetising circuit.

— As the material under test is only feebly magnetic, it is assumed that there is no demagnetisation due to end effect.

Testing procedure :

— Current is passed through the magnetising coil to give the required field strength H, which is given by $H = \frac{NI}{l}$ (NI being the number of ampere-turns on the magnetising coil and l its axial length), since the demagnetising effect of the ends of the specimen is negligible.

— Before placing the specimen inside the search coil the mutual inductance M is adjusted to give no galvanometer throw when the magnetising current is reversed. The specimen is then inserted, and the galvanometer throw upon reversal of the current observed. The flux density (B_f) obtained from this throw will be that due to the ferric induction in the specimen.

When the field strength is H, the total flux density,

$$B = B_f + \mu_0 \cdot H,$$

and, for the relative permeability μ_r, we have

$$\mu_0\, \mu_r\, H = B_f + \mu_0 H \qquad \text{for the relative permeability } \mu_r$$

or, $$\text{Relative permeability, } \mu_r = \frac{B_f + \mu_0 H}{\mu_0 H} = \left(\frac{B_f}{\mu_0 H} + 1\right) \quad ...(7.45)$$

7.10 PRECAUTIONS IN MAGNETIC TESTING

1. Where the inductance is measured by means of a stationary test coil surrounding the specimen, the space between the coil and the test specimen should be as small as possible. However, this space cannot be reduced to zero, and for precision work the necessary correction must be made. The test-coil area A can be determined by noting the deflection of a ballistic galvanometer when the test coil is placed in a standard solenoid and the current in the solenoid is reversed.
2. Before induction measurements are made, the *specimen should be carefully demagnetised.* This is best done by first magnetising to a value well above the maximum at which measurements will be made; the current is then gradually reduced to zero, being rapidly reversed meanwhile. Alternating current of 25 Hz frequency is convenient where much work is to be done.
3. In loss measurements, the temperature of the specimen should be carefully noted, because this affects the eddy current loss. The exciting winding should therefore, the *sufficiently large to avoid heating of the specimen.* In precision work, the apparatus is placed in an oil bath.
4. In tests of *sheet materials, the strips should not be too narrow,* because of the hardening at the edges due to cutting. This effect is negligible with a width of 5 cm. Care should be taken that burrs are removed from the edges and that the only insulation between sheets is the natural scale or oxide. The test specimen should be composed of strips cut from the sheets in both directions.
5. The readings of a ballistic galvanometer, should be kept at about the same magnitude by varying the resistance in series with it. The observational error is thus kept constant.

Example 7.8. *(a) What do you mean by eddy current losses ? Explain briefly.*

(b) The iron loss in the sample is 300 watts at 50 Hz with the eddy current component 5 times as big as the hysteresis loss component. At what frequency will the iron loss be doubled if the flux density is kept same ? **(Punjab University)**

Solution. *(a)* **Eddy current losses :**

- The term "eddy currents" is applied to those electric currents which circulate within a mass of conducting material when the latter is situated in a varying magnetic field. The conducting material may be considered as consisting of large number of closed conducting paths, each of which behaves like a short circuited winding. The varying magnetic field induces eddy e.m.fs. in these closed elemental paths giving rise to eddy currents. These eddy currents produce loss in power resulting in heating of materials. This loss is of considerable importance as it affects the efficiency and heating of electrical machines.
- *The eddy currents produce a magnetic field of their own which opposes the main magnetic field. As the effect of eddy currents is not uniform over the cross-section of the material, this results in a flux distribution which is not uniform, the flux density in the outer portions being greater than that at the centre. Thus there is a reduction in effective cross-section. The effect of eddy currents upon flux distribution is chiefly of importance in transformers where the material otherwise would be worked at a uniform flux density.*
- The magnetic materials used for varying magnetic fields are laminated (made up of thin sheets insulated from each other) so as to reduce eddy currents and associated losses, as by laminating, the *area of path of eddy currents is reduced giving rise to a large value of resistance.*

(b) Given : $P_i = 300$ W; $f = 50$ Hz; $P_e = 5\ P_h$;

Frequency, f':

Total iron loss, $P_i = P_h + P_e = 300$

or, $P_h + 5P_h = 300$

or, $P_h = 50$ W

and $P_e = 5\ P_h = 5 \times 50 = 250$ W

Let iron loss be *doubled* at supply frequency f'.

Since hysteresis loss is *proportional to the frequency,*

$\therefore$ New hysteresis loss, $P_h' = \dfrac{f'}{f} \times P_h = \dfrac{f'}{50} \times 50 = f'$ watts

Also, as eddy current loss various as *square of supply frequency,*

$\therefore$ New eddy current loss,

$$P_e' = \left(\frac{f'}{f}\right)^2 \times P_e$$

$$= \left(\frac{f'}{50}\right)^2 \times 250 = 0.1\ f'^2$$

Total iron loss, at new frequency f',

$$P_i' = P_h' + P_e' = 2P \qquad \text{...(Given)}$$

or, $f' + 0.1f'^2 = 2 \times 300 = 600$

or, $f'^2 + 10f' - 6000 = 0$

or, $f' = \dfrac{-10 \pm \sqrt{10^2 + 4 \times 6000}}{2}$ = **72.6 Hz (Ans.)**

Example 7.9. *(a) Derive an expression for eddy current loss in thin sheets.*

(b) Calculate the loss per kg in a specimen of alloy steel for a maximum flux density of 1.1 Wb/m^2 and a frequency of 50 Hz (sinusoidal), using 0.5 mm thick sheets. The resistivity of alloy steel is 30 × 10^{-8} Ωm. The density is 7800 kg/m^3. Hysteresis loss in each cycle is 380 W-s/m^3.

Solution. (*a*) **Eddy current loss in thin sheets :**

Fig. 7.39 shows a thin plate of thickness t and width b, thickness being *considerably smaller than the width.* Let us suppose that this sheet carries a flux, $B = B_{max} \sin \omega t$, the field running parallel to the axis of the sheet. Eddy currents would flow in the sheet in the elemental paths as shown in Fig. 7.39.

Let us consider 1 metre axial length and also 1 metre width of sheet. Take an elemental path of thickness dx at a distance x from the axis. Flux enclosed by 1 metre length of this path.

$$\phi = B(\text{flux density}) \times 2x \times 1 \text{ (area)} = 2xB$$
$$= 2xB \sin \omega t$$

Instantaneous values of eddy e.m.f. induced in the elemental path,

$$e_x = \frac{d\phi}{dt} = \frac{d}{dt}(2x\, B_{max} \sin \omega t)$$
$$= 2xB_{max} \cos \omega t = 2x\omega B_{max} \cos \omega t.$$

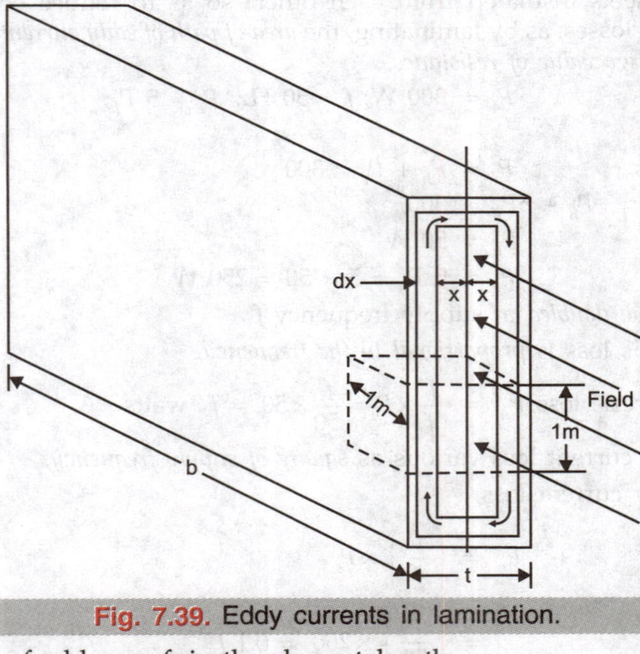

Fig. 7.39. Eddy currents in lamination.

R.M.S. value of eddy e.m.f. in the elemental path,

$$E_{r.m.s.} = \frac{1}{\sqrt{2}} \times E_{max}$$
$$= \frac{1}{\sqrt{2}} \times 2x\omega B_{max} \qquad [\therefore E_{max} = 2x\omega B_{max} \text{ i.e., when } \cos \omega t = 1]$$
$$= \sqrt{2}\, x\omega B_{max} = \sqrt{2}x \times 2\pi f\, B_{max} = 2\sqrt{2}x\pi f\, B_{max}$$

If ρ is the resistivity of material, the resistance of each elemental path,

$$R_{xe} = \frac{2\times\rho}{1\times dx} = \frac{2\rho}{dx}$$

(Neglecting portions of path parallel to edges of plates).

∴ Eddy current in element path,

$$I_{xe} = \frac{E_{xe}}{R_{xe}} = \frac{2\sqrt{2}x\pi f B_{max}}{2\rho/dx} = \frac{\sqrt{2}\pi f B_{max} x dx}{\rho}$$

Eddy current loss in this path with 1 metre length and 1 metre width is

$$P_{xe} = I_{xe}^2 R_{xe} = \frac{2\pi^2 f^2 B_{max}^2 (xdx)^2}{\rho^2} \times \frac{2\rho}{dx}$$

$$= \frac{4\pi^2 f^2 B_{max}^2}{\rho} \cdot x^2 dx$$

∴ Total eddy current loss in a plate 1 metre wide, 1 metre long and t metre thick is

$$= \int_{x=0}^{x=t/2} \frac{4\pi^2 f^2 B_{max}^2}{\rho} \cdot x^2\, dx = \frac{\pi^2 f^2 B_{max}^2}{6\rho} t^3 \text{ watt} \quad \text{...(i)}$$

Eddy current loss *per unit volume*

$$= \frac{\pi^2 f^2 B_{max}^2 t^2}{6\rho} \quad \text{...(ii)}$$

(the volume being t m^3)

$$= K_e f^2 B_{max}^2 \text{ watt} \quad \text{...(iii)}$$

where, $$K_e = \frac{\pi^2 t^2}{6\rho} \quad \text{...(iv)}$$

- The thickness t of laminated material is of considerable importance as regards to eddy current loss. Eqn. (*ii*) shows *that eddy current losses are reduced by using thinner plates and a material of higher resistivity.*
- Eqn. (*ii*) is derived by *assuming that magnetic field is uniform throughout and parallel to the axis and also the sheets are very thin.* Therefore, as such, it *should* ***not*** *be applied to cases where frequency is* ***abnormal****, the sheets thick or the variation of flux density is* ***not sinusoidal*** *with time. The eddy current loss can be reduced in the following ways :*

1. It has been found that this loss can be minimised by building up the required cross-section for the flux path by stacking thin pieces known as *"laminations"* as shown in Fig. 7.40. Since the e.m.fs. set up in the material by the varying flux are usually of small magnitude, the natural oxide on the surface of the sheet iron or steel from which the laminations are punched will effectively insulate the laminations from one another, and thus limit each eddy current path to a single lamination.

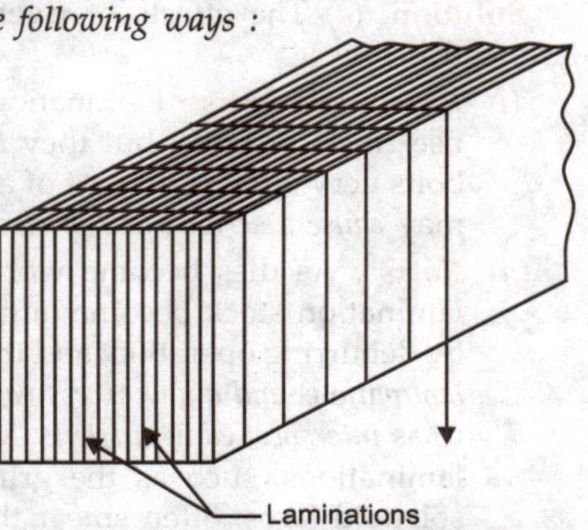

Fig. 7.40

2. This loss may also be reduced *by grinding the ferromagnetic material to a powder and mixing it with a binder that effectively insulates the particles from one another.* This mixture is then formed under pressure into the desired shape and heat treated. Magnetic cores for use of *communication equipment* are frequently made by this process.

- It may be noted that *iron losses* (hysteresis and eddy current losses) *if allowed to take place unchecked, not only reduce the efficiency of electrical equipment but also raise the temperature of the core.* Hence these losses should be kept *as small as is economically possible.*

(*b*) *Given :* B_{max} = 1.1 Wb/m^2; f = 50 Hz; t = 0.5 mm; $\rho = 30 \times 10^{-8}$ Ωm
Density = 7800 kg/m^3; P_h/cycle = 380 W-s/m^3.

Total iron loss :

Eddy current loss is given by

$$P_e = \frac{\pi^2 f^2 B_{max}^2 t^2}{6\rho} \text{ watt per m}^3$$

$$= \frac{\pi^2 f^2 B_{max}^2 t^2}{6\rho \times 7800} \text{ watt/kg}$$

$$= \frac{\pi^2 \times 50^2 \times 1.1^2 \times (0.5/1000)^2}{6 \times 30 \times 10^{-8} \times 7800} = 0.5316 \text{ W/kg}$$

$$\textit{Hysteresis loss, } P_h = \frac{380 \times 50}{7800} = 2.4359 \text{ W/kg at 50 Hz}$$

$$\textit{Total iron loss} = P_e + P_h = 0.5316 + 2.4539$$

$$= \mathbf{2.9855 \text{ W/kg}} \text{ (Ans.)}$$

Example 7.10. *(a) Discuss briefly the effect of the following factors on core loss : (i) Rivets and bolts; (ii) Burrs; (iii) Pressure.*

(b) Specimen of cold rolled grain oriented 0.3 mm thick stampings has a resistivity of 4.8 × 10^{-7} Ωm. The published hysteresis loop is essentially rectangular in form, with a coercive force of 12 AT/m for all values of peak flux densities upto 1.6 Wb/m^2. The manufacturer quotes the iron loss in material as 1.2 W/kg, with a sinusoidal flux density of 1.0 Wb/m^2 (peak) at 100 Hz. Calculate the loss in the material from its properties and compare it with the quoted value. The density is 7650 kg/m^3.

Solution. (*a*) The effects of rivets and bolts, burrs and pressure on core loss are described below :

(*i*) ***Rivets and bolts :*** Laminations must be held together. A common method is to use rivets or bolts, but they *tend to short circuit lamination insulation.* Rivets and bolts very often form part of a closed magnetic loop which, under A.C. operation, may *cause a serious loss of electrical energy.*

(*ii*) ***Burrs :*** As dies become worn, the burrs that they form on punched edges of lamination stock become larger and larger. Thus burrs are not always removed by deburring operation, and they *cut through sheet insulation forming contacts between laminations and allow excessive eddy currents to flow.* These are commonly referred to as *intersheet eddy currents.* Machining operations that are performed on stacked laminations, such as the grinding of induction motor stators or broaching of solenoid faces, often smear the laminations together, forming good connections between sheets. Many schemes, such as treating the faces with acid or alkali, have been tried in an attempt to eat away the metal particles after laminations have been assembled into apparatus, but this attempted cure often causes other troubles, such as rusting. Burning of the burrs with a flame has also been attempted, with varying success.

(*iii*) ***Pressure :*** Mechanical strains set up in magnetic materials either from *cold working during rolling or punching or distortion during final assembly operations,* affect magnetic properties in several ways. The hysteresis and eddy current losses as well as the magnetising current are increased. *The increased hysteresis loss due to punching depends upon the material, its thickness, and the configuration of the grains.* The loss may be as high as 300 percent of the loss in the unpunched sheet. The strains produced by punching operations increase the loss primarily in the narrow section next to the punched edge. The percent increase in hysteresis loss is much lower for high flux densities than for low ones.

The *effect of tension on silicon steel is to increase the permeability until high flux densities are reached. The effect of compression, however, is much more marked than that of tension and reduces the permeability considerably.*

(*b*) *Given :* $t = 0.3$ mm; $\rho = 4.8 \times 10^{-7}$ Ωm; Density = 7650 kg/m^3

Total iron loss :

The material is subjected to a peak flux density of 1 Wb/m^2 and, therefore, the value of coercive force is 12 AT/m for all densities. The hysteresis loop is thus a rectangle with sides 2 × 12 = 24 AT/m and 2 × 1 = 2 Wb/m^2.

$$\text{Loss per cycle} = \text{Area of hysteresis loop} = 24 \times 2 = 48 \text{ W-s/m}^3$$

$$\text{Hysteresis loss/kg at 100 Hz} = \frac{48 \times 100}{7650} = 0.627 \text{ W/kg}$$

Eddy current loss,

$$P_e = \frac{\pi^2 f^2 B_{max}^2 t^2}{6\rho} \text{ W/m}^3 = \frac{\pi^2 f^2 B_{max}^2 t^2}{6\rho \times 7650} \text{ W/kg}$$

$$= \frac{\pi^2 \times (100)^2 \times (1.0)^2 \times (0.3/1000)^2}{6 \times 4.8 \times 10^{-7} \times 7650} = 0.403 \text{ W/kg}$$

Total iron loss,

$$P_i = P_h + P_e = 0.627 + 0.403$$

$$= \mathbf{1.03 \text{ W/kg}} \text{ (Ans.)}$$

The calculated loss is *smaller* than the quoted. **(Ans.)**

Example 7.11. *The measured values of iron loss of a magnetic specimen of weight 13 kg are 17.2 W and 28.9 W at 40 Hz and 60 Hz respectively at a constant peak flux density. Determine the values of hysteresis and eddy current losses in W/kg at 50 Hz for the same value of flux density.* **(AMIE)**

Solution. *Given :* W = 13 kg; P_i = 17.2 W at 40 Hz; P_i = 28.9 W at 60 Hz.

P_h, P_e at f = 50 Hz

At a given flux density, $P_h \propto f$ and $P_e \propto f^2$

Let, $P_h = Af$, and $P_e = Bf^2$

where, A and B are constants.

Total iron loss, $P_i = P_h + P_e = Af + Bf^2$ per kg

Substituting $P_i = 17.2$ W when $f = 40$ Hz, and

$P_i = 28.9$ W when $f = 60$ Hz, we get:

$$\frac{17.2}{13} = 40A + (40)^2 B \quad ...(i)$$

$$\frac{28.9}{13} = 60A + (60)^2 B \quad ...(ii)$$

Solving expressions (*i*) and (*ii*), we get

$$A = 0.0254;\ B = 0.000194$$

∴ When $f = 50$ Hz,

$$P_h = Af = 0.0254 \times 50 = \mathbf{1.27\ W\ per\ kg}\ \textbf{(Ans.)}$$

and, $$P_e = Bf^2 = 0.000194 \times (50)^2 = \mathbf{0.485\ W\ per\ kg}\ \textbf{(Ans.)}$$

Example 7.12. *A laminated iron cylinder is rotated in a magnetic field. The iron loss is 500 W at 480 r.p.m. and 700 W at 600 r.p.m. Find the loss if the laminations were twice as thick, the induction density increased by 25% and the speed were 700 r.p.m. Take steinmetz coefficient as 1.6. Assume the waveform to be sinusoidal.*

Solution. *Given* : Iron loss at 480 r.p.m. = 500 W; Iron loss at 600 r.p.m. = 700 W; Steinmetz coefficient = 1.6

The frequency of pulsations is proportional to rotational speed.

Let N be the speed in r.p.m. and V_i be the volume of iron.

$$\text{Eddy current loss, } P_e = \frac{\pi^2 f^2 B_{max}^2 t^2}{6\rho} V_i$$

$$= \frac{\pi^2 Z^2 N^2 B_{max}^2 t^2}{6\rho} V_i = K_e' N^2 \quad ...(i)$$

(where $f = ZN$ and Z is constant)

$$\text{Hysteresis loss, } P_h = K_h f B_{max}^{1.6} V_i = K_h ZN B_{max}^{1.6} V_i = K_h' N \quad ...(ii)$$

where K_e' and K_h' are constants if B_{max} and t remain constant.

$$\therefore \quad \text{Total iron loss, } P_i = K_e' N^2 + K_h' N \quad ...(iii)$$

Substituting the value of P_i and N in the expression (*iii*), we get

$$500 = K_e' (480)^2 + K_h' (480)$$

or, $$1.04 = 480 K_e' + K_h' \quad ...(1)$$

$$700 = K_e'(600)^2 + K_h' (600)$$

or, $$1.166 = 600 K_e' + K_h' \quad ...(2)$$

Subtracting (1) from (2), we get

$$0.126 = 120 K_e'$$

∴ $$K_e' = 0.00105$$

and, $$K_h' = 0.536$$

Total loss with B_{max} increased by 25%, thickness increased to twice the original value and N = 700 r.p.m.,

$$P_i = 0.00105 \times (700)^2 \times (1.25)^2 \times 2^2 + 0.536 \times 700 \times (1.25)^{1.6}$$

$$= 3215.6 + 536.2 = \mathbf{3751.8\ W}\ \textbf{(Ans.)}$$

Example 7.13. *In a power loss test on a 15 kg specimen of sheet steel laminations, the maximum flux density and wave form factor are maintained constant and the following results are obtained :*

Frequency (Hz) :	25	40	50	60	80
Total loss (watts) :	18.5	36	50	66	104

Calculate the eddy current and hysteresis loss per kg at 50 Hz. **(AMIE)**

Solution. At a given flux density and waveform factor, total iron losses (P_i) are given as :

$$P_i = Af + Bf^2$$

or, $$\frac{P_i}{f} = A + Bf$$

The value of $\frac{P_i}{f}$ for various value of f are :

f :	25	40	50	60	80
$\frac{P_i}{f}$:	0.74	0.9	1.0	1.1	1.3

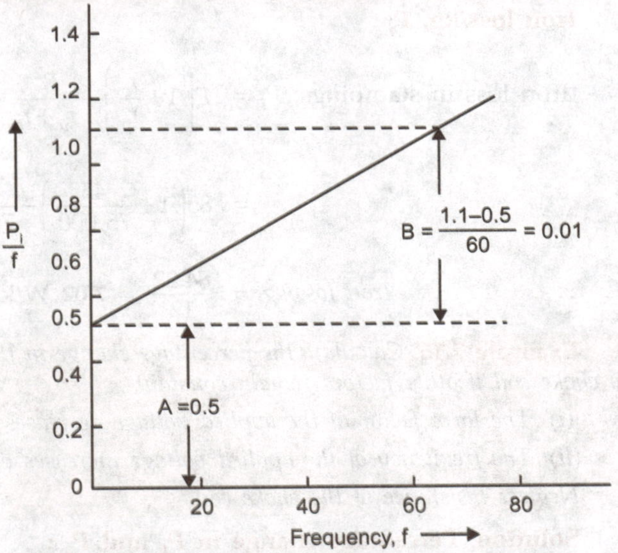

Fig. 7.41

A graph is plotted between $\frac{P_i}{f}$ and f is shown in Fig. 7.41.

From the graph $A = 0.5$, and, $B = 0.01$.

Eddy current loss at 50 Hz $= Bf^2 = 0.01 \times (50)^2 = 25$ W

$\therefore$ *Eddy current loss per kg* $= \frac{25}{15} =$ **1.67 W/kg (Ans.)**

Hysteresis loss at 50 Hz $= Af = 0.5 \times 50 = 25$ W

$\therefore$ Hysteresis loss per kg $= \frac{25}{15} =$ **1.67 W/kg (Ans.)**

Example 7.14. *The following data relate to a test conducted on a specimen of iron stampings weighing 12 kg by an Epstein square shown in Fig. 7.25 :*

Cross-sectional area of the specimen = 1700 m^2

Number of turns on each of the two windings S_1 and S_2 = 520

Resistance of each of the windings S_1 and S_2 = 45 Ω

Resistance of pressure coil of the wattmeter = 75,000 Ω

Frequency of sinusoidal voltage supply = 50 Hz

Current indicated by an ammeter in the primary circuit = 0.4 A

Voltage indicated by the voltmeter across a secondary winding = 250 V

Calculate the value of flux density in the specimen and iron loss/kg at this flux density if the wattmeter connected on the primary side indicates 85 W.

Solution. *Given* : Weight of specimen = 12 kg; A_{sp} = 1700 mm^2; N_2 =520; r_c = 45 Ω; r_p = 75,000 Ω; E = 250 V; P = 85 W.

Flux density in the specimen, B_{max} :

$$E = 4.44\, f\, B_{max}\, A_{sp} \times N_2$$

or,
$$B_{max} = \frac{E}{4.44\, f\, A_{sp}\, N_2} = \frac{250}{4.44 \times 50 \times 1700 \times 10^{-6} \times 520}$$
$$= \mathbf{1.27\ Wb/m^2}\ \textbf{(Ans.)}$$

Iron loss/kg, P_i :

Iron loss in stampings, $$P_i = P\left(1+\frac{r_c}{r_p}\right) - \frac{E^2}{r_p + r_c}$$ ...[Eqn. (7.33)]

$$= 85\left(1+\frac{45}{75{,}000}\right) - \frac{(250)^2}{75{,}000+45} = 84.22\ \text{W}$$

∴ $$\textit{Iron loss/kg} = \frac{84.22}{12} = \mathbf{7.02\ W/kg}\ \textbf{(Ans.)}$$

Example 7.15. *Calculate the percentage change in the hysteresis and eddy current losses of a choke coil if other factors remain constant :*

(i) The form factor of the applied voltage decreases by 15%;

(ii) The frequency of the applied voltage increases by 15%.

Neglect resistance of the choke coil.

Solution. Percentage change in P_h and P_e :

(i) When the form factor of the applied voltage decreases by 75% :

Since P_h is independent of form factor of supply voltage, so remains the same *i.e.,* percentage change in P_h *is zero* **(Ans.)**

P_e varies as square of the form factor of supply voltage,

∴ $$\frac{P_{e2}}{P_{e1}} = \left(\frac{K_{f2}}{K_{f1}}\right)^2 = (0.85)^2 = 0.7225$$

or, $$P_{e2} = 0.7225\ P_{e1}$$ (∵ Form factor decreases by 15%)

Hence, *percentage change in* P_e

$$= \frac{P_{e2} - P_{e1}}{P_{e1}} \times 100 = \frac{0.7225 P_{e2} - P_{e1}}{P_{e1}} \times 100$$

$$= \mathbf{27.75\%}\ \textbf{(\textit{decrease})}\ \textbf{(Ans.)}$$

(ii) When the frequency of the applied voltage increases by 15% :

Now supply frequency, $f_2 = 1.15\ f_1$

Since, $$P_h \propto f$$

∴ $$\frac{P_{h2}}{P_{h1}} = \frac{f_2}{f_1} = 1.15$$

or, $$P_{h2} = 1.15\ P_{h1}$$

$$\therefore \quad \text{Percentage change in } P_h = \frac{P_{h2} - P_{h1}}{P_{h1}} \times 100$$

$$= \frac{1.15\, P_{h1} - P_{h1}}{P_{h1}} \times 100 = \textbf{15\% (increase)} \textbf{ (Ans.)}$$

Since, $P_e \propto f^2$

$$\therefore \quad \frac{P_{e2}}{P_{e1}} = \left(\frac{f_2}{f_1}\right)^2 = (1.15)^2 = 1.3225$$

or, $P_{e2} = 1.3225\, P_{e1}$

$$\therefore \quad \text{Percentage change in } P_e = \frac{P_{e2} - P_{e1}}{P_{e1}} \times 100$$

$$= \frac{1.3225\, P_{e1} - P_{e1}}{P_{e1}} \times 100 = \textbf{32.25\% (increase)} \textbf{ (Ans.)}$$

Example 7.16. *While testing a sample of sheet steel stampings at a frequency of 50 Hz the following results obtained :*

Volts	*4.9*	*69.3*	*91.8*	*100.5*	*110.5*	*118.0*
Amperes	*0.2*	*0.3*	*0.46*	*0.52*	*0.64*	*0.77*
Watts	*9.5*	*16.8*	*27.5*	*32.5*	*39.0*	*44.8*

Mean width of plates = 3 cm; Mean thickness = 0.0489 cm; Number of plates = 51; Total weight = 10 kg; Number of magnetising turns on coil = 600.

Allowing 4 watts for copper loss in the magnetising winding calculate the iron loss in watts per kg at a maximum flux density of 1 Wb/m² at a frequency of 50 Hz.

(Punjab University)

Solution. *Given :* b = 3 cm or 0.03 m; t = 0.0489 cm or 0.000489 m; N_p = 51; Total weight = 10 kg; N = 600; Copper loss = 4 watts; $B_{max} = 1$ Wb/m²; f = 50 Hz.

Iron loss/kg :

Mean area of plates,

$$A_p = N_p \times b \times t = 51 \times 0.03 \times 0.000489 = 7.48 \times 10^{-4}\ \text{m}^2$$

$$\text{Induced voltage, } E = 4.44\, f\, \phi_{max}\, N$$

$$= 4.44 \times 50 \times (1 \times 7.48 \times 10^{-4}) \times 600 \simeq 100\ \text{V}$$

From the graph plotted (Fig. 7.42), corresponding to a voltage of 100 V, power supplied is 30.5 W.

Now, Total losses = Iron loss + copper loss

$\therefore$ Iron loss = Total loss – copper loss = 30.5 – 4 = 26.5 W

Hence, $\textit{Iron loss/kg} = \frac{26.5}{10}$ = **2.65 W** **(Ans.)**

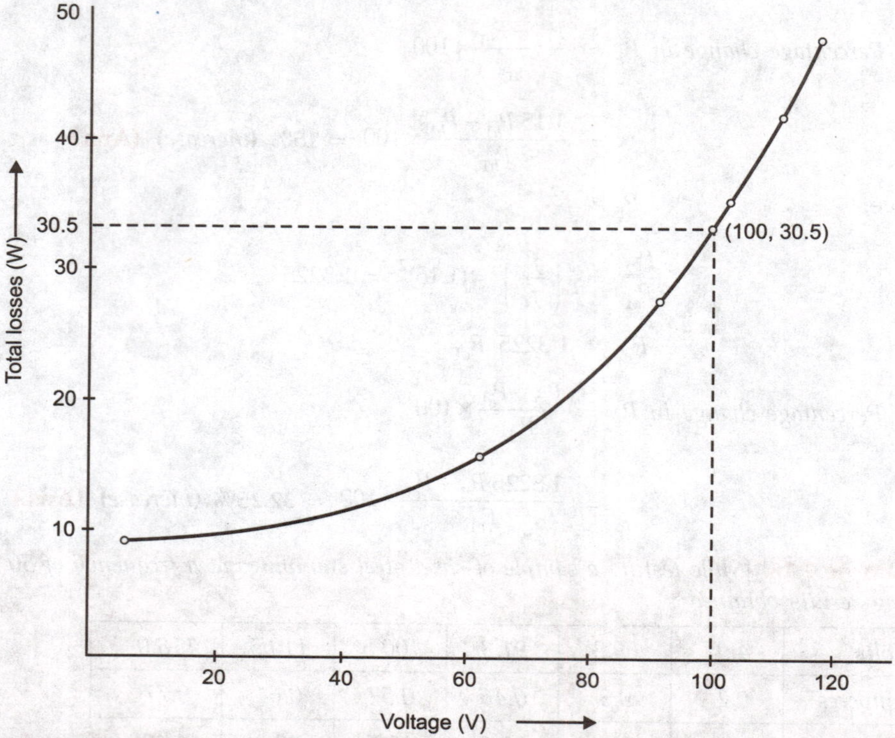

Fig. 7.42

Example 7.17. *An iron cored coil and a standard 2.5 Ω resistor are connected in series and the potential difference is measured across each on an A.C. potentiometer of co-ordinate type. The readings are 3.5 ±j 1.1 V for the coil and 4.6–j1.8 V for the resistor.*

Calculate the magnetising component and loss component of current through the coil.

Solution. I_e, I_m :

Current through the coil, $I = \dfrac{4.6 - j1.8}{2.5} = 1.84 - j0.72 = 1.976 \angle -21.4°$

Voltage across the coil, $V = 3.5 + j1.1 = 3.67\angle 17.45°$

Angle between V and I, $\phi = 17.45° - (-21.4°) = 38.85°$

Loss component of current, $I_e = I \cos \phi = 1.976 \cos 38.85°$

$= \mathbf{1.54\ A}$ **(Ans.)**

Magnetising component of current,

$I_m = I \sin \phi = 1.976 \sin 38.85°$

$= \mathbf{1.24\ A}$ **(Ans.)**

HIGHLIGHTS

1. Measurements of various magnetic phenomena are called *magnetic measurements.*
2. *Types of tests* : (*i*) D.C. tests; (*ii*) A.C. tests; (*iii*) Steady state tests.
3. Ballistic (or D.C.) tests are used for :

   (*i*) Determination of flux/flux density; (*ii*) Determination of B.H curves; (*iii*) Plotting of hysteresis loop.

4. *Magnetic potentiometer* measures the magnetic potential along any part of a magnetic circuit and hence distribution of m.m.f. round the circuit. It can also be used to determine magnetic potential drop across a joint.
5. *Permeameters* overcome the difficulties and inaccuracies encountered in performing measurements on bar specimens of magnetic materials. Commonly used permeameters are :
   (*i*) Hopkinson permeameter (Bar and yoke method)
   (*ii*) Ewing double bar permeameter
   (*iii*) Illiovici permeameter
   (*iv*) Burrows permeameter
   (*v*) Fahy's simplex permeameter
   (*vi*) The National Physical Laboratory form of permeameter.
6. Total iron losses are given by :

$$P_i = P_h + P_e$$
$$= K_h f\, B_{max}^k + K_e f^2 B_{max}^2 \text{ watts/m}^3 \text{ or per kg.}$$

7. *Methods of measurement of iron losses :*
   (*i*) Wattmeter method.
   (*ii*) A.C. bridge methods.
   (*iii*) A.C. potentiometer method.
   (*iv*) Oscillographic method.
8. *Methods of measurement of field strength :*
   (*i*) By withdrawal of search coil.
   (*ii*) By nuclear magnetic resonance.
   (*iii*) Use of bismuth spiral.
   (*iv*) Use of mumetal wires.
   (*v*) Use of Hall effect devices.

OBJECTIVE TYPE QUESTIONS

Choose the Correct Answer :

1. Iron losses in a specimen of iron are determined by using a co-ordinate type of potentiometer. The loss component of no load current is read by :
   (*a*) In-phase potentiometer
   (*b*) Quadrature potentiometer
   (*c*) Either by in-phase or quadrature potentiometer
   (*d*) None of the above.
2. Magnetizing force is measured by Nuclear Magnetic Resonance method. The expression for magnetising force is
   (*a*) $H = f/\nu_p$ (*b*) $H = 2\pi f/h$
   (*c*) $H = fh/4\pi$ (*d*) $H = 2\pi f/\nu_p$
   f = frequency in Hz, h = Planck's constant and ν_p = gyrometric ratio.
3. The reason for using ring type specimens for ballistic tests is :
   (*a*) they are free from and effects.
   (*b*) they require less labour to prepare as compared with bar type specimens.
   (*c*) they can be used for large quantities of materials of very small permeability.
   (*d*) all of the above.

4. Ballistic tests are used in magnetic measurements for
   (*a*) determination of *B-H* curve of the specimen only.
   (*b*) determination of hysteresis loop of the specimen only.
   (*c*) determination of flux density, magnetizing force and *B-H* curve and hysteresis loop of the specimen.
   (*d*) finding out iron losses in the specimen.
5. In a Lloyd-Fisher square used for determination of iron loss in a specimen of iron laminations, the current coil of the wattmeter is connected in the primary circuit, while the pressure coil is connected in the secondary winding circuit. This is done so that
   (*a*) the I^2R losses in the current coil are not included in wattmeter reading.
   (*b*) the I^2R losses in the pressure coil circuit are not included in the wattmeter reading.
   (*c*) the I^2R losses both in current coil and pressure coil circuit are not included in the wattmeter reading.
   (*d*) the wattmeter reads only the iron losses in the specimen of laminations.
6. The true value of flux density in a ring specimen is :
   (*a*) $B = B' + \mu_0 H\left(\frac{A_c}{A_s} - 1\right)$ (*b*) $B = B' - \mu_0 H\left(\frac{A_c}{A_s} - 1\right)$
   (*c*) $B = B' + \mu_0 H\left(\frac{A_s}{A_c} - 1\right)$ (*d*) $B = B' - \mu_0 H\left(\frac{A_s}{A_c} - 1\right)$

   where B' = measured value of flux density, H = magnetizing force, A_s = area of cross-section of specimen, A_c = area of cross-section of coil producing magnetizing force.
7. Burrow's permeameter is preferred to other permeameters as it
   (*a*) incorporates arrangement for uniform distribution of mmf over the entire length of the specimen.
   (*b*) incorporates arrangements for elimination of leakage flux at the joints.
   (*c*) is very accurate and can serve as reference standard.
   (*d*) all of the above.
8. In an Epstein square, used for finding the iron loss in a stack of laminations, the laminations are so arranged that
   (*a*) plane of each lamination is perpendicular to the plane of the square.
   (*b*) plane of each lamination is parallel to the plane of the square.
   (*c*) the plane of laminations may be either parallel or perpendicular to the plane of the square.
   (*d*) the plane of laminations may be at any angle to the plane of square.
9. Permanent magnets are tested by
   (*a*) ballistic methods.
   (*b*) using an electric circuit having a mutual inductance.
   (*c*) potentiometric methods.
   (*d*) Betteridge apparatus..
10. The iron losses when measured in a specimen of laminations at a constant maximum flux density of 1.3 Wb/m^2 are : 20 W at 40 Hz and 40 W at 60 Hz. What are the hysteresis and eddy current losses at 50 Hz and a flux density of 1.3 Wb/m^2.
   (*a*) 10 W, 20 W (*b*) 7.38 W, 212.8 W
   (*c*) 8.38 W, 20.8 W (*d*) 5 W, 25 W

ANSWERS

1. (*a*)	2. (*d*)	3. (*d*)	4. (*c*)	5. (*a*)	6. (*b*)	7. (*d*)
8. (*b*)	9. (*c*)	10. (*d*).				

THEORETICAL QUESTIONS

1. Define the following terms relating to magnetic materials :
   (*i*) Magnetic field strength; (*ii*) Magnetic potential; (*iii*) Coercive force; (*iv*) Remanence; (*v*) Curie temperature.
2. How are magnetic materials classified ?
3. Give the classification of magnetic measurements.
4. Discuss the procedure for the determination of flux density in a ring specimen.
5. Explain the method of determination of magnetising force *H*.
6. Describe with the help of a neat diagram, a method for the determination of B-H curve of a magnetic sample. Point out various sources of errors and methods of minimising them.
7. Describe a method for the measurement of B-H curve of a magnetic substance of a bar form.
8. Give the theory underlying the use of magnetic potentiometer.
9. How is the leakage factor in a dynamo-electric machinery determined ?
10. Write a short note on "Testing bar specimens of magnetic materials".
11. Give the comparison between bar specimens and ring specimens in relation to magnetic measurements.
12. Give the advantages and disadvantages of ring and bar specimens used in magnetic testing of materials.
13. What are permeameters ?
14. Describe the principle and construction of a parameter for the determination of B-H curve of a sample.
15. Describe Fahy's simplex permeameter.
16. Describe the 'Lloyd Fisher square' method of measuring iron losses in ferromagnetic material.
17. Describe the method for the measurement of total loss in sheet steel with alternating current. Give the theory of the method.
18. Explain with a neat circuit diagram the Epstein or Lloyd Fisher square method of testing core loss of laminated sheet steel. Make necessary derivations and explain how eddy current and hysteresis losses may be evaluated separately. Can you explain why a separate sine wave generator may be required for this test.
19. Explain how transformer steel available in strips of 25 cm length and 5 cm width is tested for iron losses by the wattmeter method. Indicate the method of stacking the strips.
20. Explain the "method of reversals" for experimental determination of a magnetic specimen.
21. Describe the method for determination of B-H curve of a magnetic materials using :
   (*i*) Method of reversals, and (*ii*) Step-by-step method.
22. Describe the construction and working of the following types of permeameters :
   (*i*) Illiovici permeameter; (*ii*) Burrow's permeameter.
23. Explain the methods of separation of iron losses into their two components : eddy current and hysteresis losses if the maximum value of flux density is maintained constant and (*i*) frequency is varied keeping the form factor constant; (*ii*) form factor is varied keeping the frequency constant.
24. Explain how the value of A.C. permeability of magnetic materials is determined through the use of :
   (*i*) Maxwell's bridge, and (*ii*) Campbell's bridge.
25. Describe how magnetising and loss components of no load current of a transformer be determined by using an A.C. potentiometer.

26. What is 'Nuclear magnetic resonance"? Explain how the strength of a magnetic field in air can be determined using this principle.
27. Explain the oscillographic method of determination of hysteresis loss of a magnetic material.
28. Explain a method for testing of permanent magnets.
29. Describe a method for testing of feebly magnetic materials.
30. What precautions should be taken in magnetic testing ?

UNSOLVED EXAMPLES

1. A solenoid 60 cm long and 2.5 cm is diameter is uniformly wound with 600 turns of insulated wire.
   (*i*) Find the magnetic field strength at the centre of the solenoid when carrying a current of 1 ampere;
   (*ii*) If a secondary coil is wound round the central part of the solenoid, calculate the flux passing through it. **[Ans.** 123.2×10^{-8} Wb]
2. An iron ring of 350 mm^2 cross-sectional area with a mean length of 1 m is wound with a magnetising winding of 100 turns. A secondary coil with 200 turns of wire is connected to a ballistic galvanometer having a constant of 1 μC per scale division, the total resistance of the secondary circuit being 2000 Ω. On reversing a current of 10 A in the magnetising winding, the galvanometers shows a deflection of 100 scale divisions.
   Calculate : (*i*) The flux density in the specimen; (*ii*) The value of permeability at this flux density. **[Ans.** (*i*) 1.429 Wb/m^2; (*ii*) 1137]
3. An iron ring has a mean diameter of 0.1 m and a cross-section of 33.5 mm^2. It is wound with a magnetising winding of 320 turns and secondary winding of 220 turns. On reversing a current of 10 A in the magnetising winding, a ballistic galvanometer gives a throw of 272 scale divisions, while a Hibbert magnetic standard with 10 turns and a flux of 0.25 $\times 10^{-3}$ Wb gives a reading of 102 scale divisions, other conditions remaining the same. Find the relative permeability of the specimen. **[Ans.** 35.3]
4. A ferromagnetic material is tested with the help of a permeameter provided with a coil wound over the specimen, this coil being known as "B-coil" and another coil wound near the surface of the coil called "H-coil". These two coils are connected to two separate ballistic galvanometers having a constant of 25×10^{-6} Weber turn per scale division. The testing is done by reversing the current of the magnetising winding wound over the specimen. The reading of the ballistic galvanometer connected to 'B'-coil is 100 divisions and that connected to 'H'-coil is 10 divisions. The data of the two coils is :
   'B'-coil : Turns = 100; Area = 50 mm^2
   'H'-coil : Turns = 10000; Area = 5 mm^2
   Find : (*i*) Flux density in the specimen; (*ii*) Magnetising force; (*iii*) Relative permeability of the specimen. **[Ans.** (*i*) 0.25 Wb/m^2; (*ii*) 2000 AT/m; (*iii*) 100]
5. The constant of a given magnetic potentiometer is obtained by aid of a coil of 300 turns in which a current of 0.6 A is reversed. The resulting throw of the galvanometer is 157 scale divisions. It is then used to measure the magnetic potential difference between two points and throw is 304 divisions. Calculate the magnetic potential difference. **[Ans.** 697 AT]
6. Calculate the loss of energy caused by hysteresis in one hour in 50 kg of iron when subjected to cyclic magnetic changes. The frequency is 25 Hz, the area of the hysteresis loop represents 240 joules/m^3 and the density of iron is 7800 kg/m^3. **[Ans.** 138240J]
7. Calculate the loss per kg in a specimen of alloy steel for a maximum flux density of 1.2 Wb/m^2 and a frequency of 50 Hz, using 0.5 mm thick sheets. The resistivity of alloy steel is 30×10^{-8} ohm metre. The density is 7800 kg/m^3. Hysteresis loss in each cycle is 400 W-s/m^3. **[Ans.** 3.2 W/kg]

8. A specimen of cold rolled grain oriented 0.3 mm thick stamping has a resistivity of 5 × 10^{-7} Ωm. The published hysteresis loop is essentially rectangular in form, with a coercive force of 12 AT/m for all values of peak flux densities upto 1.6 Wb/m^2. The manufacturer quotes the iron loss in the material as 1.2 W/kg, with a sinusoidal flux density of 1.0 Wb/m^2 (peak) at 100 Hz. Calculate the loss in the material from its properties and compare it with the quoted value. The mass density is 7650 kg/m^3. [**Ans.** 1.015 W/kg]

9. A laminated iron cylinder is rotated in a magnetic field. The iron loss is 500 W at 480 r.p.m. and 700 W at 600 r.p.m. Find the loss if the laminations were twice as thick, the induction density increased by 20% and speed were 720 r.p.m. Take Steinmetz coefficient as 1.6. [**Ans.** 3622 W]

10. A specimen of iron stampings weighing 10 kg and having a cross-sectional area of 1680 mm^2 is tested by an Epstein square shown in Fig. 7.25. Each of the two windings S_1 and S_2 have 515 turns. A sinusoidal voltage supply at 50 Hz is given to the primary winding of Epstein square and an ammeter connected in the primary circuit indicates a current of 0.35 A and the electrostatic voltmeter connected across a secondary winding indicates a voltage of 250 V. The resistance of each of the windings S_1 and S_2 is 40 Ω. The resistance of the pressure coil of the wattmeter is 80,000 Ω.

    Calculate the value of flux density in the specimen and the iron loss/kg at this flux density if the wattmeter connected on the primary side indicates 80 W.

    [**Ans.** 1.3 Wb/m^2; 7.926 W/kg]

11. Calculate the percentage change in the hysteresis and eddy current losses of choke coil if other factors remain constant :

    (*i*) The form factor of the applied voltage decreases by 20%.

    (*ii*) The frequency of the applied voltage increases by 20%.

    [**Ans.** (*i*) Zero, 36% decrease; (*ii*) 20% increase; 44% increase)

12. An iron-cored coil and a standard 2 Ω resistor are connected in series and the potential difference is measured across each of an A.C. potentiometer of co-ordinate type. The readings are 3.6 + *j*1.2 V for the coil and 4.8 – *j*2 V for the resistor. Calculate the magnetising component and loss component of current through the coil.

    [**Ans.** I_e = 1.96 A; I_m = 1.72 A]

CHAPTER

8 High Voltage Measurements

8.1 Introduction; 8.2. High voltage tests – Sustained low-frequency tests – Constant direct-current tests – High-frequency tests – Surge or impulse tests; **8.3. Testing apparatus** – High voltage transformers – Apparatus for voltage regulation – Control gear and protective devices – Apparatus for voltage measurement – Low-frequency high-voltage tests – High voltage D.C. testing of cables – High-frequency high-voltage tests – Surge or impulse testing; **8.4. High voltage testing of insulating materials, porcelain insulators and insulating oil.** *Highlights – Objective Type Questions – Theoretical Questions.*

8.1 INTRODUCTION

- The rapid advance in the use of high voltages (to the tune of 220 kV and 440 kV) for transmission purposes has necessarily been accompanied by extensive research upon high-voltage phenomena. The subject of high voltage testing is, therefore, a very important one at the present time, and includes a large variety of testing methods.
- A high voltage encountered in practice involve a range of frequencies from zero, in the case of D.C. to the very high frequencies of lightning impulses.
- In measurement of high voltages, some methods measure its *peak value* while other methods measure *r.m.s. value.* For convenience, a *mean value is sometimes measured instead of an r.m.s. value.*
  - — ***Peak values** govern the maximum voltage stress and have an important bearing on ultimate breakdown or the corona formation.*
  - — ***R.m.s. values** have a more direct bearing on dielectric loss;* but through internal heating effect that also has a close connection with dielectric strength.

High voltage sources. These are of two types : *A.C.* and *D.C.* high voltage sources.

- ***A.C. sources*** can be categorised into *high* frequencies and *low* frequencies sources.
  - — *"High frequency high voltage A.C. sources"* are *lightning* and *high voltage transmission line sparks.*
  - — *"Low frequency high voltage A.C. sources"* are set up by *transformers.*
- ***High voltage D.C. sources*** can be further divided into *low current* and *high current sources.*
  - — *"High voltage high current D.C. sources"* are *A.C. step-up transformer plus rectifier.*
  - — High voltage low current D.C. sources are *high frequency oscillator plus rectifier.* This type of voltage is required for X-ray and CROs.

Crest factor. The crest factor of a waveform is defined as the ratio of *peak voltage to r.m.s. voltage.*

i.e. $$\text{Crest factor} = \frac{\text{Peak voltage}}{\text{R.m.s. voltage}} \quad \text{...(8.1)}$$

The value of crest factor for sinusoidal waveform is 1.414.

When the crest factor of a waveform is known, its r.m.s. voltage can be determined if its peak voltage is given with the help of eqn. (8.1). Similarly peak voltage of a waveform can be evaluated if its r.m.s. voltage is known with the knowledge of crest factor of a waveform. Thus, *the knowledge of crest factor is quie helpful in high voltage measurements.*

8.2 HIGH VOLTAGE TESTS

Although there are a large number of methods of high voltage testing, but due to limitation of space we shall be considering a few important methods only.

The high voltage tests may be *classified* as follows :

1. Sustained low-frequency tests.
2. Constant direct-current tests.
3. High-frequency tests.
4. Surge or impulse tests.

8.2.1. Sustained Low-frequency Tests

- These tests are *commonest* of all high-voltage tests. They are conducted at the power frequency *i.e.*, 50 Hz.
- In these tests, test voltage should be of *sine waveform*.
- Voltages of 2 kV or 3 kV are used for routine voltage tests upon motors, switchgear, and other apparatus after manufacture, and in some cases after installation, in accordance with specifications.

  These tests are also carried out on :

  — Specimens of insulation for the determination of dielectric constant and dielectric loss.

  — Supply mains for routine testing.

  — High voltage transformers, porcelain insulators and high voltage cables, etc. for works testing. The test voltage in such cases (and research work) may be of the order of 2000 kV.

8.2.2. Constant Direct-current Tests

- In order to transmit large amounts of power efficiently, high transmission voltages are essential.
- The cables and other equipment employed for transmission of electric power at extra high voltage, according to standard specifications, are required to withstand a test-voltage of at least 10 kV above the maximum voltage to which it is intended to be subjected in use continuously, for at least 30 minutes for insulation stability. Such tests may be performed with D.C. voltage or A.C. voltage. However *D.C. testing is preferred to A.C. testing* because of the following *reasons* :

  — With an A.C. voltage supply, the high voltage transformer required for the purpose would have to be of a *very large capacity* owing to the heavy charging (capacitive) currents drawn by the lines or the cables. Consequently, a *large expenditure* has to be incurred for procuring the high voltage testing transformer.

  — Furthermore, *great difficulties* will be experienced in *transporting* the testing transformer to the testing site due to its large size.

8.2.3. High-frequency Tests

— The high-frequency tests, at frequencies (varying from several kHz to MHz or more) are important in work upon insulators for radio purposes where such frequencies are usual.

— It has been observed that in the case of porcelain insulators in service on transmission lines, breakdown, or flashover occurs in most cases as a result of *high-frequency disturbances in the line,* these being either due to *switching operations* or *external causes.* Thus, *insulators which may be satisfactory under high voltage of low frequency are not necessarily satisfactory under the different conditions existing when the frequency is of much* higher order.

Therefore, to ensure reliable operation of insulators under disturbances caused by switching and arcing earth must be subjected to high frequency tests.

8.2.4. Surge or Impulse Tests

— The surge or impulse tests are carried out in order to investigate the influence of *"surges"* in a transmission line upon breakdown of the line insulators and of the end turns of transformers connected to the line.

— Such surges are usually produced by lightning in the neighbourhood of the transmission line, and are the result of the sudden discharge of the electricity on the suspended particles of a thunder cloud, which takes the form of a lightning stroke.

An indirect stroke (direct stroke being rare) occurs when a charged cloud above and near to the line, electrostatically induces charges of opposite sign to that of its own charge, in the line. These charges are bound as long as the cloud remains near without discharging its electricity by lightning stroke. But if the cloud is suddenly discharged, as it is when the lightning stroke occurs, the induced charges in the line are no longer bound, but travel with the velocity of light along the line to equalize the potential at all points on the line. This means that a steep-fronted voltage wave travels along the line, its forms being as shown in Fig. 8.1. The wave is characterised by a steep rise of voltage upto its peak value followed by a slower drop to zero voltage (The probable maximum voltage appearing on a line due to direct stroke is 10 to 15 MV and that by indirect stroke is 500 kV). The increase in voltage occurs in a very short time of the order of *one millionth of a second.* Thus, very severe stresses are imposed upon the line insulators and upon the transformer windings, in case the wave is allowed to reach them. Such surges invariably cause rupture of insulators and other equipment.

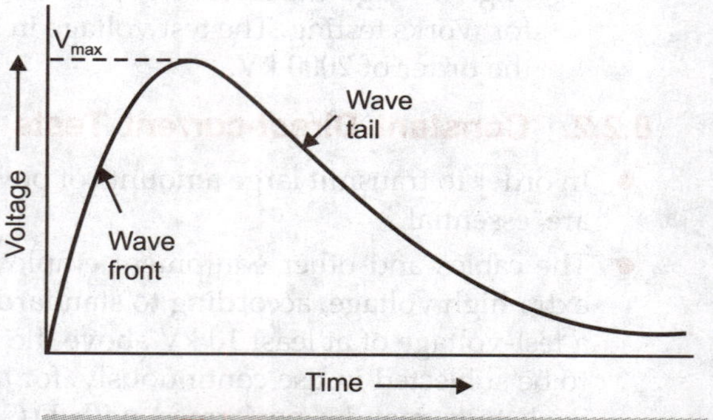

Fig. 8.1. Surge waveform.

— Impulse test is necessary to test the insulation of the power system so that voltage surges, due to switching or lightning strokes etc. are not permitted to change the insulation of power system equipment.

- *The testing plant for this purpose of testing must be capable of producing a high unidirectional voltage, suddenly applied to the apparatus under test.*

8.3. TESTING APPARATUS

For all types of high-voltage tests, certain apparatus is common. Special apparatus is required, in addition, for direct current, high frequency, and impulse tests.

The *following apparatus is usually essential for all tests* :

1. High-voltage transformers.
2. Apparatus for voltage regulation.
3. Control gear and high-voltage connections, including safety and protective devices.
4. Apparatus for voltage measurement.

8.3.1. High Voltage Transformers

— These transformers need to be carefully designed since they are subjected to transient voltages and surges during their normal operation when the insulation under test breaks down. *To withstand the internal streses set up by such disturbances the transformer insulation must be carefully proportioned.*

— The transformers are usually *single-phase core type,* and are *oil immersed,* Bakelite cylinders being used for insulation between high- and low-voltage windings.

— Transformers for cable testing, and certain other purposes, may also be required to give a considerable current, and the question of regulation and cooling must therefore be carefully considered.

— *In insulation testing the current taken from the transformer, when the test specimen breaks down, is limited by the insertion of water resistances in the test circuit, so that for such purposes the transformer need not have a large kVA capacity.*

— Equipments giving a maximum voltage upto *500 kV have usually a single high-voltage transformer. For the higher voltages, 1000 kV and upwards, two or more transformers are generally used,* connected in *"cascade". This arrangement is found to be more convenient than the single transformer for very high voltages, owing to the very large size and high cost of a single transformer for the purpose.*

The general arrangement of connecting transformers in cascade is shown in Fig. 8.2. :

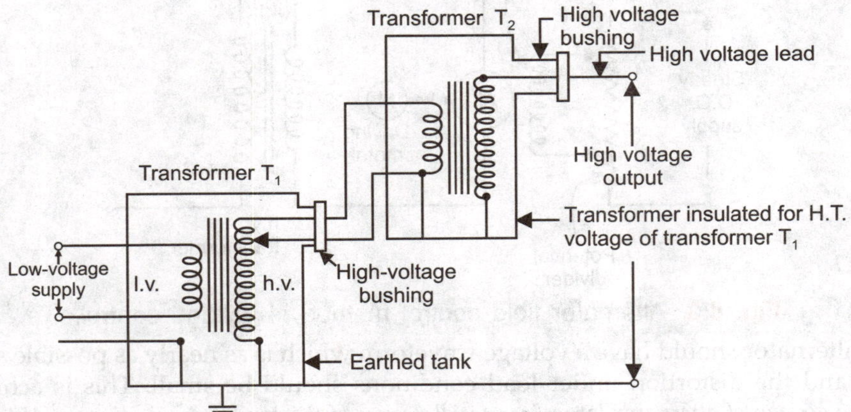

Fig. 8.2. Cascade connected transformers.

— The low-voltage supply is connected to the primary winding of transformer T_1, the tank of which is earthed. One end of the *h.v.* secondary winding of this transformer is conneted to the earthed tank. From the other end of the secondary winding, a lead passes through a high voltage bushing, which provides insulation for the full secondary voltage between this lead and the tank. Through this bushing also, runs a second lead from a tapping point on the secondary winding, the voltage between this point and high voltage end of the secondary winding being that required for the primary winding of transformer T_2.

— One end of the secondary winding of transformer T_2 is connected to its tank, which is insulated from earth for the full secondary voltage of transformer T_1. The other end of the secondary winding provides the high-voltage terminal of the equipment, *the total voltage-to-earth obtainable being approximately the sum of the secondary voltages of the two transformers.*

8.3.2. Apparatus for Voltage Regulation

The *requirements of voltage-regulating apparatus are* :

(*i*) The regulation of voltage should be *smooth* – to avoid surges on *h.v.* side of the transformer, and also for accuracy of voltage measurement.

(*ii*) The voltage-regulating apparatus *shall not cause distortion of the waveform.*

The secondary voltage of the transformer is regulated by variation of the voltage applied to its primary winding. This may be done as follows :

1. By variation of the alternator field current.
2. By resistance, or inductance control.
3. By induction regulators.
4. By a tapped transformer.

8.3.2.1. Variation of the alternator field current

This method is applicable when a separate alternator is used to supply the testing transformer. Except in the case of comparatively small plants for routine testing, a separate alternator is generally used.

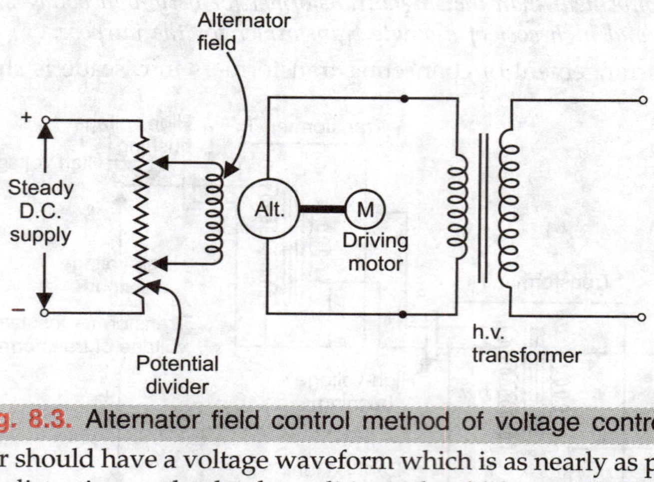

Fig. 8.3. Alternator field control method of voltage control.

The alternator should have a voltage waveform which is as nearly as possible *sinusoidal* on load, and the distortion under load conditions should be small. This is achieved by *making the air gap of alternator large, by special design of the armature windings; and by suitable regulation of the number, and shape, of the slots.*

In the case of *fairly small machines,* the field current of the alternator is varied *directly,* but when a large alternator is used the field current of the exciter may be varied, thus varying the alternator field current indirectly. Fig. 8.3. shows the connections for *direct variation of the alternator field current :*

— A potential divider, connected across a steady D.C. supply is used, and the connections are arranged so that a small field current in reverse direction to the normal one may be obtained. By this arrangement, zero voltage may be obtained by neutralisation of the residual magnetism in the field.

— Smooth and gradual voltage regulation is obtained by a special design of the potential divider, which should have a *large number of turns and some provision for very gradual and steady movement of the sliding contact.*

This method of voltage control has following *"advantages"* :

1. Convenient and simple.
2. Smooth regulation of voltage from zero to full voltage.
3. Output waveform is not distorted due to absence of impedance used in series of circuit for regulation.
4. Freedom from any disturbance caused by frequent short circuiting of secondary winding of h.v. transformer when insulation of power system under test breaks down.

8.3.2.2. Resistance or inductance control

In the case of small equipments, when a separate alternator is not used, resistance or inductance must be inserted in the A.C. supply circuit for voltage regulation.

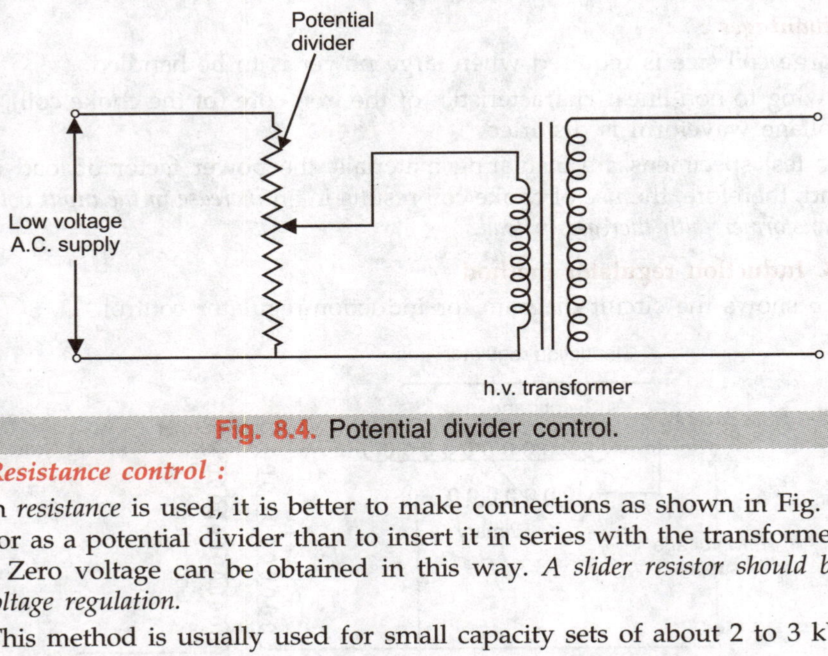

Fig. 8.4. Potential divider control.

1. ***Resistance control :***

When *resistance* is used, it is better to make connections as shown in Fig. 8.4, using the resistor as a potential divider than to insert it in series with the transformer primary winding. Zero voltage can be obtained in this way. *A slider resistor should be used for smooth voltage regulation.*

- This method is usually used for small capacity sets of about 2 to 3 kV rating.

Advantages :

(*i*) Cheap and convenient.
(*ii*) Smooth voltage variation is obtained.
(*iii*) Small distortion of voltage waveform.

Disadvantages :

(*i*) Large power loss in potential divider.
(*ii*) Because of very large size and heavy cost of the resistance potential divider, the arrangement cannot be employed for large capacity plants.

2. ***Inductance control :***

Voltage regulation may be obtained by connecting a choke coil in series with the primary winding of the h.v. testing transformer. The arrangement is shown in Fig. 8.5 wherein a *choke of variable inductance is used. Voltage regulation is obtained by the withdrawal or insertion of the iron core of the coil.*

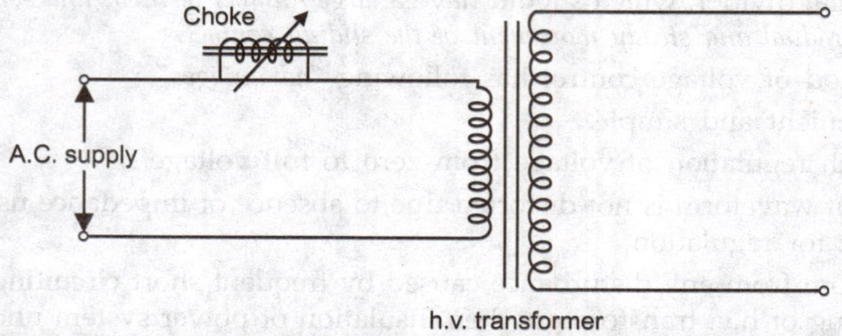

Fig. 8.5. Voltage variation by using series reactance.

Advantages :

This method is *more efficient* than resistance method, on account of its low power factor (and hence considerably smaller power loss).

Disadvantages :

(*i*) Large coil size is required when large power is to be handled.

(*ii*) Owing to non-linear characteristics of the iron core (of the choke coil), the input voltage waveform is distorted.

(*iii*) As test specimens are insulating materials, the power factor of load is leading and, therefore, the use of choke coil results in an *increase in the input voltage to the transformer with increase in load.*

8.3.2.3. Induction regulator method

Fig. 8.6 shows the circuit diagram for induction-regulator control.

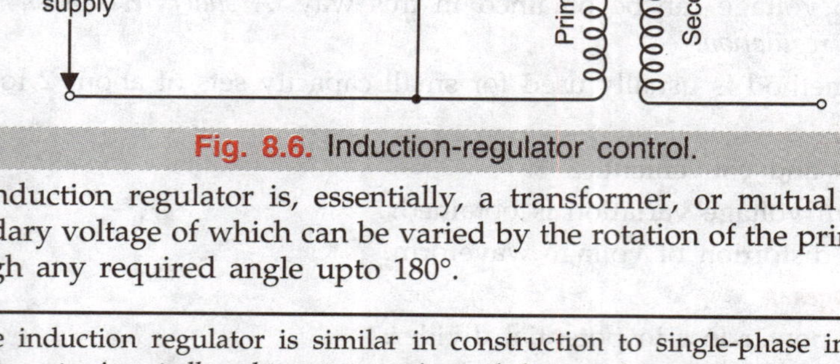

Fig. 8.6. Induction-regulator control.

— An *induction regulator is, essentially, a transformer, or mutual inductor, the secondary voltage of which can be varied by the rotation of the primary winding through any required angle upto 180°.

* Single phase induction regulator is similar in construction to single-phase induction motor except that the rotor is *not allowed to rotate continuously* but can be turned through one pole pitch by hand wheel or small rotor.

— If the primary and secondary windings have equal number of turns, by rotation of primary winding, the voltage induced in the secondary winding may be varied from $-V$ to $+V$ (where V is the low voltage supply). Thus the voltage applied to *h.v.* transformer primary way be *varied from zero to 2V*.

Advantages :

(*i*) Voltage regulation is gradual and smooth.

(*ii*) Can be used for all loads and power factors.

(*iii*) It is possible to avoid distortion of the voltage waveform by using a carefully designed regulator.

Uses. This method is often *used in cable-testing equipments,* since its gradual voltage variation at loads of any magnitude is advantageous for such work.

8.3.2.4. Voltage variation by means of a tapped transformer

Fig. 8.7 shows the connections for voltage regulation by a *tapped transformer* :

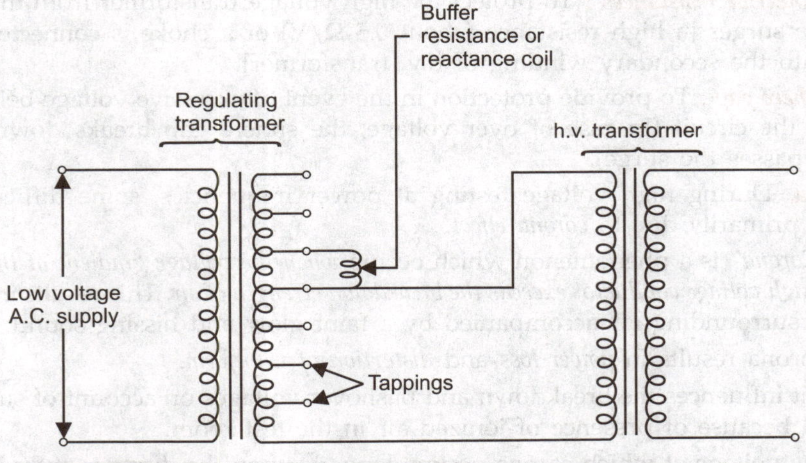

Fig. 8.7. Voltage regulation by a tapped transformer.

— The primary of the intermediate regulating transformer is fed from a low voltage A.C. supply. The secondary winding of the transformer has a large number of *tappings* whereby the voltage applied to the primary of the *h.v.* transformer may be *varied*.

— To *avoid 'surges'* due to the opening of the secondary circuit of the regulating transformer when the tapping switch is moved, *two contact brushes are used,* making contact with adjacent studs, and with a *"buffer"* resistance or reactance coil between them to *prevent short-circuit of a section of transformer winding.*

An auto-transformer may be used instead of the double-wound transformer shown.

— *In order to obtain smooth voltage regulation a large number of tappings should be used.*

Advantages :

(*i*) High efficiency.

(*ii*) Very small distortion of voltage waveform.

Disadvantages :

(*i*) Unless a large number of tappings is used, the voltage regulation is not smooth.

(*ii*) When the power to be handled is large, the equipment becomes expensive.

8.3.3. Control Gear and Protective Devices

On the ***low-voltage side*** (*i.e.* in the primary circuit) of the high voltage transformer the following protective devices should be provided :

(i) *Main switch :* To isolate the testing apparatus from the supply.

(ii) *Fuses.*

(iii) *A circuit breaker :* To protect the testing apparatus from damage in case of supply failure.

(iv) *Over-voltage relay :* To protect the apparatus from the effects of over voltages.

(v) *Interlocks :* To guard against the supply to the transformer being switched on except at a low value.

(vi) *Earth connections :* To provide safety and to protect the apparatus.

On the ***high-voltage side*** of the transformer, the following protective devices should be provided :

(i) *Protective resistances :* To protect the high voltage transformer from the effects of the surges [a high resistance (about 0.5 Ω/V) or a choke is connected in series with the secondary winding of h.v. transformer].

(ii) *Sphere gap :* To provide protection in the event of excessive voltage being applied to the circuit (In case of over voltage, the sphere gap breaks down and thus bypasses the surge).

Corona. During high voltage testing at power frequencies, some difficulties arise which are primarily due to *corona effect*.

— *"Corona"* is a phenomenon which occurs *when the voltage gradient at the surface of a high voltage conductor exceeds the breakdown strength of air.* This results in ionisation of surrounding air accompanied by a faint glow and hissing sound.

— Corona results in *power loss* and *distortion of waveform.*

Thus, it influences the breakdown and flashover voltages on account of surges in the circuit and because of presence of ionized air in the test room.

— The voltage at which corona occurs depends upon the *diameter of the high-voltage conductor,* the voltage increasing with increasing diameter. Thus to avoid such effects, the conductors in the high-voltage circuit should be made *much larger* than is necessitated by current-carrying considerations.

Fig. 8.8 shows a simplified diagram of connections for a *high-voltage testing equipment having a separate supply alternator.*

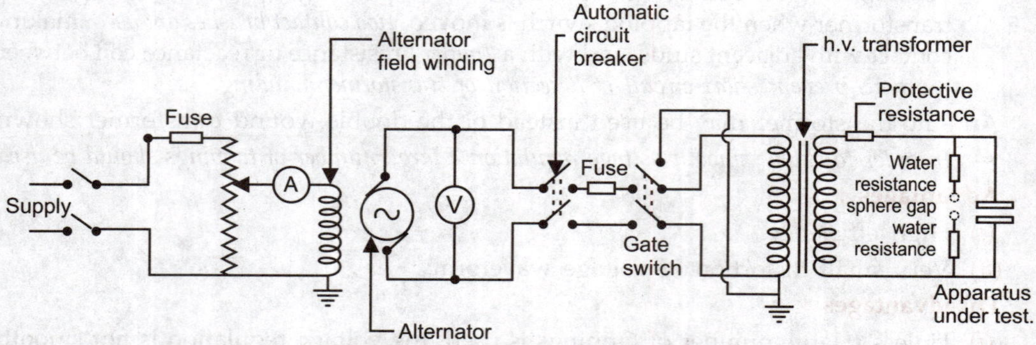

Fig. 8.8. Circuit for h.v. testing at power frequencies.

8.3.4. Apparatus for Voltage Measurement

Because of the difficulty of designing electrostatic voltmeters, for the measurement of extra high voltages, which will be free from errors due to corona effects within the instrument and to external electrostatic fields, a number of special methods of measurement have been devised for the purpose. Some methods must be available, also, for calibrating such voltmeters when constructed.

The methods of measurement of voltage are divided into *two categories* :

1. Methods used for measurement of *r.m.s. values of voltages.*
2. Methods of measurement of *peak values of the voltages.*

However, if voltage under test is of sinusoidal shape then the peak value can be determined by multiplying r.m.s. value of voltage with peak factor or *vice versa.*

8.3.4.1. Measurement of r.m.s. values of voltages

The r.m.s. value of voltage can be measured by the following *methods/instruments* :

1. Electrostatic voltmeters.
2. Transformation ratio method.
3. Potential divider methods.
4. Standard impedance methods.
5. Potential transformer method.
6. Sphere gaps.

1. Electrostatic voltmeter :

The electrostatic voltmeters have been discussed in chapter 3. However, the design of this type of meter has to be modified for high voltage use.

The following instruments may be used for measuring r.m.s. values of high voltages :

(i) Attracted disc voltmeters.
(ii) Attracted sphere voltmeter.
(iii) Oscillating ellipsoid voltmeter.

(i) Attracted disc voltmeters :

Fig. 8.9 shows one form of these instruments :

— This voltmeter consists of two hollow metal mushroom shaped discs. One of the discs has a portion of its centre removed and a small movable disc enclosed. The movable disc is connected to the pointer of the instrument through a gear system. The other disc, which is fixed in position, is connected to the high voltage supply.

— The distance between the two discs remains *fixed for any measurement and can be changed for changing the range of the instrument.*

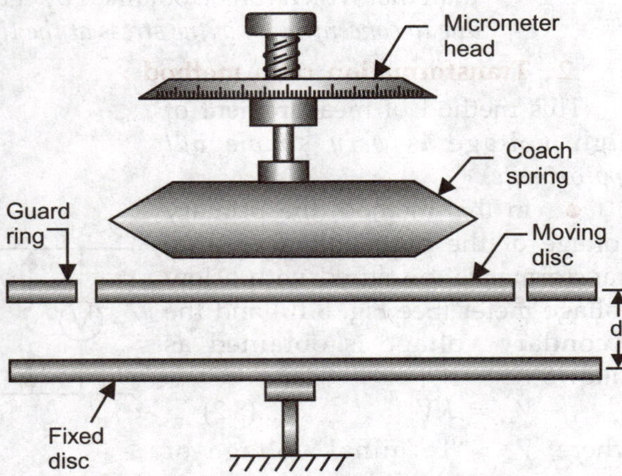

Fig. 8.9. Attracted disc type electrostatic voltmeter.

— Normally, a scale is provided on the movable disc stand which gives the proper range of the multiplying factor with which the deflection is to be multiplied to obtain the value of high voltage. *The disc's large size provides adequate protection for the working parts of the instrument against the electrostatic disturbances.*

- These meters are designed to measure *upto 200 kV directly* by connecting them across the source and *can be used for measuring higher voltage in conjunction with potential dividers.*

(ii) Attracted sphere voltmeter :

— In this voltmeter, the voltage is measured based on the *measurement of electrostatic force between two spheres.*

— Forces of 4.2 to 7.5 N can be produced with direct voltage which will flashover gap of 100 mm and 500 mm between spheres of 1 m diameter. The attractive forces between the spheres can be measured with a stiff elastic support. *The errors involved are less than 1 percent for voltage as high as 1 MV.*

(iii) Oscillating ellipsoid voltmeter :

— In this type of voltmeter, the voltage to be measured is applied to a pair of discharge free parallel plate pair of electrodes lying in vertical planes and a small conducting electrode of accurate prolate spheroidal shape is suspended at its mid-point by a long unspun silk thread so that it hangs in the uniform field between the central parts of the plates with its long horizontal axis.

— On the application of the voltage, an electrostatic field is produced due to which charges of opposite polarity and of calculable magnitude accumulate on the two ends of the ellipsoid. The ellipsoid, therefore, tends to align its long axis along the direction of the field. The *ellipsoid oscillates about its final position on account of the moment of inertia.*

— The controlling torque produced by the silk thread is *very small.* Therefore, *frequency of oscillations is approximately proportional to the applied voltage.* The constant of proportionality can be calculated from the mass and dimensions of the ellipsoid and the geometry of the plate electrodes.

- This voltmeter has the *drawback* that the average *electrostatic stress is much smaller* than that which can be obtained by the two voltmeters described above. This is *due to concentration of the stress at the tips of the ellipsoid.*

2. Transformation ratio method :

This method of measurement of high voltage is *very simple but approximate.*

- In this method, the primary voltage of the high voltage testing transformer is measured with a low voltage meter (See Fig. 8.10) and the secondary voltage is obtained as follows :

$$V_s = KV_p \qquad ...(8.2)$$

where, V_s = Terminal voltage of secondary winding,

K = Turns ratio, and

V_p = Terminal voltage of primary winding.

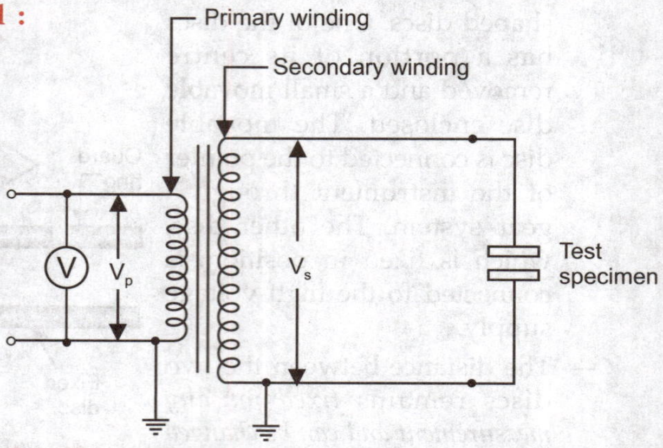

Fig. 8.10. Measurement of r.m.s. voltage by transformation method.

It is assumed, in this method, that *K remains constant* (hence the method though simple, gives wrong results).

- In some high voltage transformers, a separate voltage coil is provided whose voltage is measured with the help of a low voltage voltmeter. The secondary voltage (V_s) in this case is given by :

$$V_s = \left(\frac{N_s}{N_c}\right).V_c \qquad \text{...(8.3)}$$

where, N_s = Number of turns of secondary winding,

N_c = Number of turns of the coil, and

V_c = Voltage across coil.

— This method cannot be used for cascade arrangement of transformers. It gives an accuracy of 1 to 2 per cent.

— The advantage of this system lies in its convenience as a means of taking continuous reading after being calibrated by comparison with one of direct systems of measurements.

3. Potential divider methods :

In such methods, either a *high value resistor* or an *air capacitor* must be used, *connected across the high-voltage winding, a definite fraction of the total secondary voltage being measured by means of a low-voltage electrostatic voltmeter.*

(i) Resistance potential divider :

Fig. 8.11 shows an arrangement for measurement of high voltage by resistance potential divider which uses a high voltage transformer.

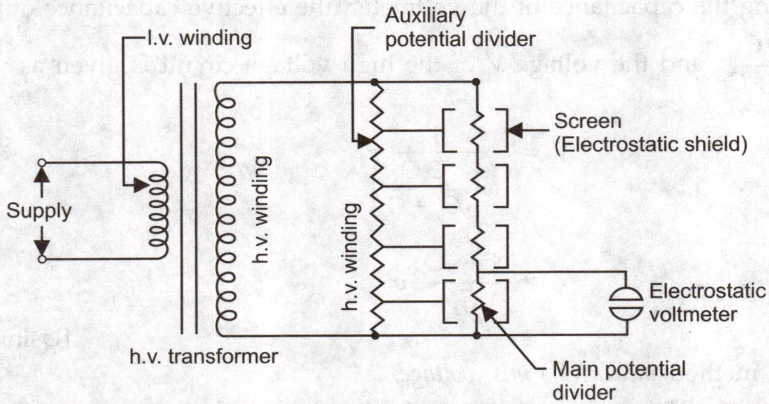

Fig. 8.11. Resistance potential divider.

Electrostatic voltmeter can be used successfully upto about 10 kV, with an unshielded resistance potential divider with resistance units wound on flat insulating sheets. However, for voltages *above 10 kV*, resistance potential divider *requires careful electrostatic shielding to minimise stray capacitance effects.* An arrangement where the shield potentials are correctly maintained by an auxiliary potential divider is shown in Fig. 8.10.

(ii) Capacitance potential divider :

Fig. 8.12 shows an arrangement for *capacitance potential divider* :

— In this method, an electrostatic voltmeter, whose capacitance is large compared with that of the high-voltage standard capacitor, is connected in series with the latter, and is shunted by a capacitance of much greater value than its own.

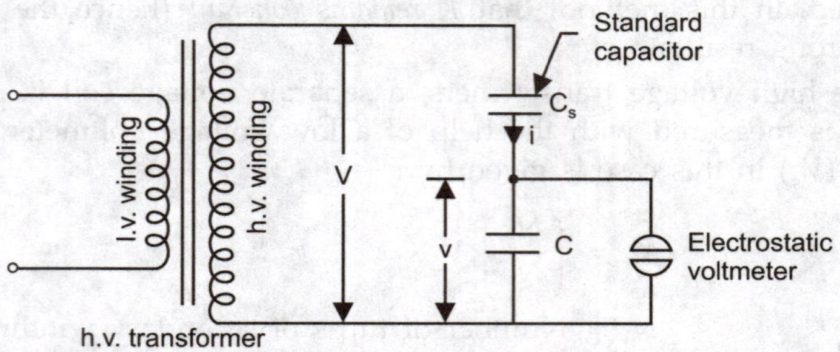

Fig. 8.12. Capacitance potential divider.

— The *capacitance of the standard capacitor must be accurately known,* and it is important that it shall be *free from dielectric loss.* For the latter reason, *air capacitors* (usually from 50 to 100 μμF) are always used for the purpose.

Let, C_s = Capacitance of the standard capacitor,

C = Capacitance of the capacitor which shunts the electrostatic voltmeter,

v = Voltmeter reading,

V = Voltage of the high voltage circuit, and

i = Current passing through the capacitors.

Then, $$v = \frac{i}{\omega C} \quad ...(8.4)$$

(where $\omega = 2\pi f$, f being the supply frequency)

Neglecting the capacitance of the voltmeter, the effective capacitance of C_s and C, in series is $\frac{C_s C}{C_s + C}$, and the voltage V of the high voltage circuit is given as :

$$V = \frac{i}{\omega\left(\frac{C_s C}{C_s + C}\right)}$$

or, $$V = \frac{C_s + C}{C_s} . v \quad ...(8.5)$$

[Using eqn. (8.4)]

- This method *measures r.m.s. voltage.*

Advantages : The use of capacitance potential divider as compared to resistance potential divider for use at high voltage has the following *advantages* :

(*i*) Simple in construction.
(*ii*) Free from heating effects.
(*iii*) Comparatively free from effects of stray magnetic fields and can be easily shunted.

4. Standard impedance methods :

Fig. 8.13 shows the connections for the measurement of voltage by a standard impedance method :

— In this method, a very high standard impedance is connected across the h.v. terminals of the transformer. The impedance may be of a known value of resistance of high value or an air capacitor.

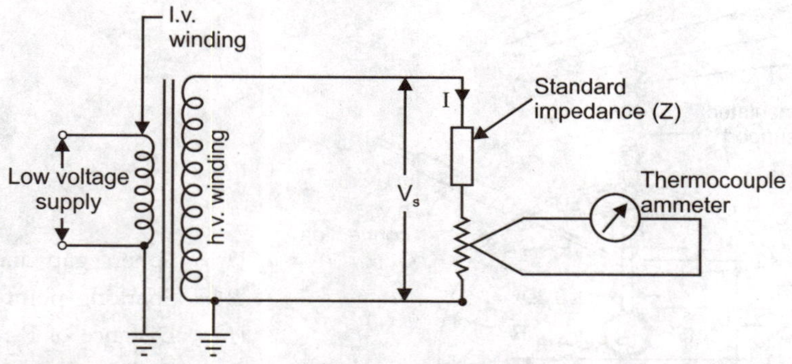

Fig. 8.13. Measurement of voltage by standard impedance method.

— The current flowing through the standard impedance may be measured by means of a thermocouple ammeter by measuring voltage across a known resistance; using an electrostatic voltmeter or by using a valve voltmeter.

Secondary voltage, $V_s = I \times Z$

where, Z = Impedance connected across the *h.v.* winding terminals

— A small heater of fine wire is connected in series with standard impedance, through which current flows. This current heats up the wire along with the thermo-junction, which is made up of two dissimilar metals. This induces e.m.f. in the thermo-junction circuit and the magnitude of the induced e.m.f. depends of the temperature of the heater. So r.m.s. value of A.C. current flowing through the standard impedance can be conveniently measured by measuring the induced e.m.f. in the thermocouple. After measuring the current through the standard impedance, the r.m.s. value of high voltage is obtained by multiplying it by the value of standard impedance.

5. Potential transformer method :

Modern potential transformers claim the following *advantages* over the potential dividers :

(*i*) The indicating instrument is isolated from the high voltage circuit.

(*ii*) The instrument impedance has negligible effect on the transformation ratio, within limits.

— Two state reduction by means of a combination of a potential divider and potential transformer proves very useful for the measurement of very high voltages.

— Potential transformers may be used with electrodynamometer, moving iron, or electrostatic instruments for measurement of r.m.s. values of A.C. voltages upto 200 kV.

The range may be extended by capacitor potential transformers.

6. Sphere gaps :

The *breakdown voltage of a spark gap between two metallic spheres may be used for the measurement of voltages upto the highest encountered in high voltage testing.*

— In this method of measurement of extra high voltage, two equal-sized metal spheres made of brass, bronze, steel, aluminium, aluminium alloy or copper (usually of copper), separated by an air gap, are mounted vertically one above the other, with the lower sphere earthed (See Fig. 8.14).

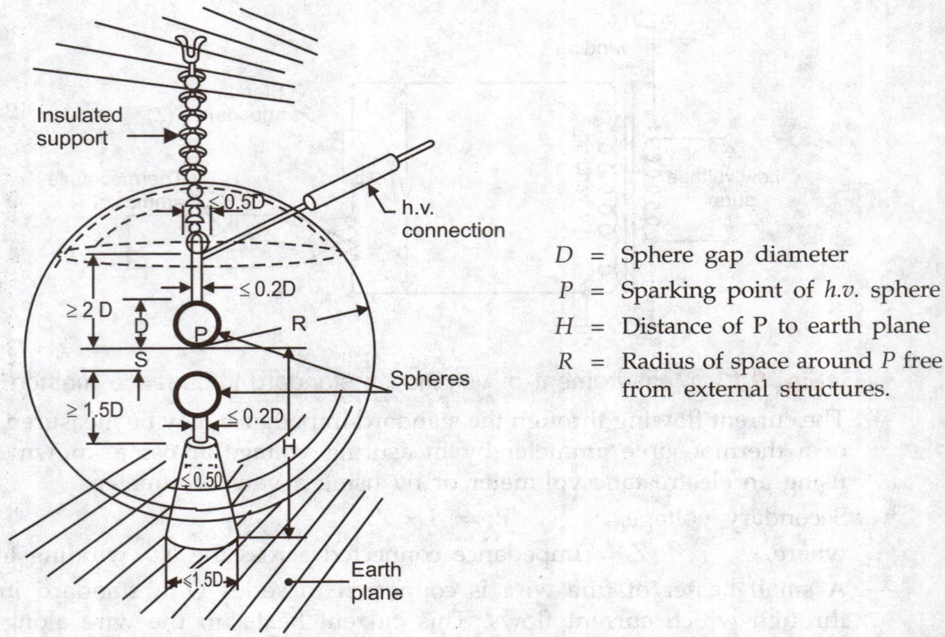

Fig. 8.14. Measurement of voltage with sphere gaps.

— The voltage between the spheres is revised till a spark passes between the two spheres. The value of voltage required to *spark over* (breakdown) *depends upon the dielectric strength of air, the size of spheres the distance between the spheres, humidity and temperature.*

— Calibration tables have been compiled on the basis of experimental work with different sizes of spheres, different gap lengths at specified conditions of pressure and temperature. The breakdown voltages corresponding to actual working conditions may be directly read off from the calibration tables. However, *corrections must be applied for temperature and pressure conditions.*

- *The spark-over voltages given in the calibration tables are in terms of r.m.s. voltage for sine wave. To obtain the peak value of the voltage under measurement, the r.m.s. values given are multiplied by the peak factor,* $\sqrt{2}$.

Advantages :

(*i*) The method is cheap, simple and reliable.

(*ii*) Peak voltages of values ranging from 2 kV to 2500 kV can be measured by this method and calibration of high voltage voltmeters and voltage measuring devices can be easily done.

(*iii*) Sphere gaps may also be used for measurement of voltage in surge (impulse) tests.

Disadvantages/Limitations :

(*i*) This method does not provide a continuous record of voltage.

(*ii*) The voltage measuremens carried out by sphere gaps can be relied upon to be accurate within 2 to 3 per cent, provided suitable precautions are taken. This is not a high accuracy in comparison with other electrical measurements. However, even this accuracy is acceptable owing to difficult conditions of working in *h.v.* testing.

Precautions :

(*i*) The spheres should not be highly polished.

(*ii*) A current limiting resistance of 1 ohm per volt may be inserted in the sphere gap circuit to avoid pitting of spheres.

(*iii*) Tolerance on diameter of sphere should be 1% for the diameter upto 1.0 m and 2% for spheres having diameter more than 1.0 m.

(*iv*) Ionization of air gap lowers the value of spark-over voltage for any sphere gap so ultraviolet or any other type of radiation should be avoided.

(*v*) The time interval between consecutive flash overs should be large enough to avoid appreciable heating of spheres.

(*vi*) No conductor or body having a conducting surface (except the supporting shanks) shall be near the sparking point of the h.v. sphere than the distance given by the expression $\left(0.25 + \frac{V}{300}\right)$ metres, where V is the voltage in kV (peak) which is to be measured.

(*vii*) After testing, the spheres should be properly cleaned to remove any extraneous matter.

8.3.4.2. Measurement of peak values of voltages

— Due to the fact that the *maximum peak* value of applied voltage is that which produces the actual breakdown stress in the material under test, it is important that this value should be known in most cases. If methods of voltage measurement are known which give r.m.s. voltage values to be used, the *peak voltage may be obtained from the peak factor of the voltage wave.*

— It is often more satisfactory, however, to use some method of voltage measurement *which gives the peak value of the voltage directly.* From the fact that the breakdown of air is involved in the *"sphere-gap method"*, this is obviously one method which *depends upon the peak values,* and the calibration curves for sphere-gaps, already referred to, give such peak values.

Other methods of measurement of peak voltage are :

1. Rectified capacitor-charging current method.
2. Ryall crest voltmeter.
3. Ionic wind voltmeter.

1. Rectified capacitor-charging current method :

This method is based upon the fact that the *average charging current of a standard air capacitor is proportional to peak voltage.* The charging current is therefore rectified and its average value is then measured by a D.C. moving-coil permanent-magnet milliammeter or by a D'Arsonval galvanometer.

Fig. 8.15(*a*) shows the connections of an arrangement for peak voltage measurement using a full-wave rectification :

— C_s is a high voltage compressed gas capacitor across which practically whole of the voltage is dropped.

— The rectifier bridge (specially designed) is used for the rectification of alternating currents.

— The average charging current is measured by the milliammeter.

Assuming the impedance of the rectifier to be negligible, as compared to that of capacitor, the impedance offered to the path of current is thus on account of capacitor only and thus the charging current leads the applied voltage by 90°.

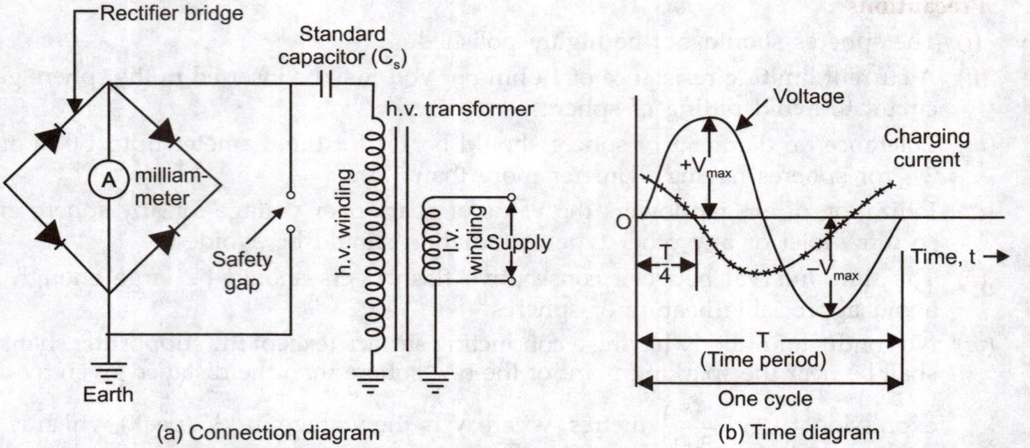

Fig. 8.15. Rectified capacitor-charging current method.

If v is the voltage at any instant, then,

$$v = V_{max} \sin \omega t,$$

where, V_{max} is the peak voltage.

Charging current at any instant,

$$i = \frac{V_{max}}{(1/\omega C)} \sin\left(\omega t + \frac{\pi}{2}\right) = V_{max}\, \omega C \cos \omega t \quad \text{...(8.6)}$$

The average charging current [See Fig. 8.15(*b*)],

$$I_{av} = \frac{1}{(T/4)} \int_0^{(T/4)} V_{max}\, \omega C \cos \omega t \, dt = 4V_{max}\, f\, C$$

or,

$$V_{max} = \frac{I_{av.}}{4Cf} \quad \text{...(8.7)}$$

$$\left(\because T = \frac{1}{f}\right)$$

Thus, if the values of capacitance of the capacitor and frequency f (which must be constant throughout the test) are known, the peak value of an alternating voltage can be determined.

The *accuracy of voltage measurement* by this method depends upon the following *factors/requirements* :

(*i*) Impedance of the rectifier should be very small compared to the impedance of the capacitor.

(*ii*) Voltage to be measured should be sinusoidal.

(*iii*) Milliammeter should indicate the correct reading of the average current.

(*iv*) Rectifier should suppress fully the one-half wave of the alternating wave.

- This method gives satisfactory result for all types of wave shapes only when positive and negative halves of the wave shapes are *equal* and wave shape has *only one peak in each half cycle.*

2. Ryall crest voltmeter :

Fig. 8.16 shows an arrangement for measuring peak voltage by Ryall crest voltmeter.

Construction :

— Two capacitors, one a variable one (C_1) of much greater capacitance than the other (C_2), are connected in series across the voltage whose peak value is to be measured.

— A neon lamp is connected across the variable capacitor as shown. Neon lamps have the characteristics that they only commence to glow when the voltage applied to them reaches a certain definite value, called the *"striking voltage"*.

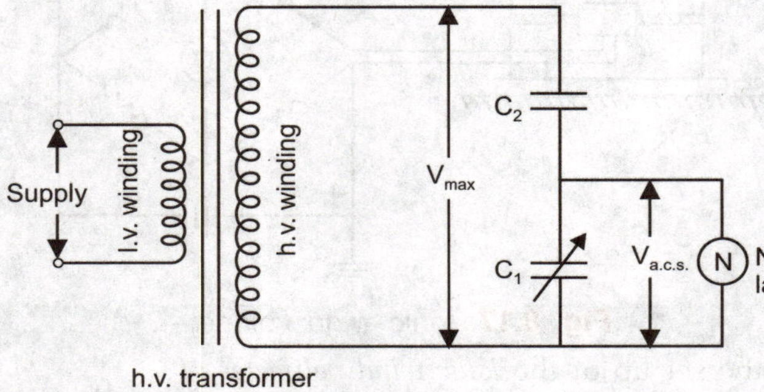

Fig. 8.16. Ryall crest voltmeter.

Working :

— The capacitance of capacitor C_1 is gradually reduced so that there is an increased voltage across it. This voltage is increased until the lamp strikes. Then peak voltage is given by :

$$V_{max} = V_{a.c.s.}\left(1+\frac{C_1}{C_2}\right) \quad ...(8.8)$$

where, $V_{a.c.s.}$ = A.C. striking voltage of the lamp (which is a constant quantity if the type of lamp is suitably chosen)

— The accuracy of the method depends upon the accuracy with which the values of C_1 and C_2 and $V_{a.c.s.}$ for neon lamp are determined.

- *This method can be used for measurement of peak value of voltage upto 150 kV (maximum).*

3. Ionic wind voltmeter :

This voltmeter which is used for direct measurement of high voltage, can measure both r.m.s. and peak values of A.C. voltages as well as D.C. voltages.

Its *principle of working* is as follows :

"When a highly-charged point is situated in air or other gas, a movement of the air surrounding the point is observed. This is referred to as the *"electric wind"*, and is brought about by the repulsion of ions from the surface of the point by the intense electrostatic field. These ions, colliding with uncharged molecules of air in the neighbourhood of the point, carry the latter with them and setup the *"electric wind"*. If an earthed electrode is placed *near the highly-charged one, an intense electric field exists of course, between the two, and a similar wind is observed also at the earth electrode."*

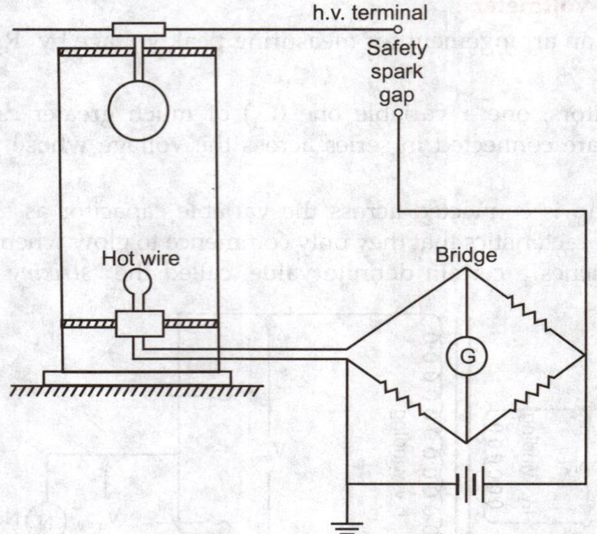

Fig. 8.17. Ionic wind voltmeter.

Fig. 8.17 shows the set up of the *ionic wind voltmeter* :

— A hot wire, of platinum-gold alloy, included in one arm of a wheatstone bridge network, is used as the earthed electrode of a high-voltage gap.

— Before the high voltage is applied to the gap the bridge network is balanced *i.e.*, galvanometer G is set to zero.

— When the voltage across the gap exceeds a certain value, called the "*threshold voltage", the *electric wind cools the hot wire and hence reduces its resistance.* A reduction of 25% in resistance is obtainable under suitable conditions, and this is *sufficient to cause an appreciable out-of-balance voltage* in the bridge network, as indicated by the galvanometer deflection.

— The *cooling and hence the galvanometer deflection, depends upon the potential gradient (and hence upon the applied voltage), the temperature of the electrode, the nature and pressure of gas in which the electrostatic field exists, and upon the frequency.*

— When the voltage is alternating, only that portion of the voltage wave which lies above the threshold value is effective from the viewpoint of cooling of the hot wire. The area of the effective portion of the voltage wave and hence cooling, varies for different wave shapes, even though the waves have the same r.m.s. value. Therefore, voltage waveform influences the instrument readings.

Calibration of the instrument is on a sine waveform.

Advantages :

(i) Robust in construction.

(ii) The high voltage may be measured by an observer at some distance from the charged conductors.

(iii) Freedom from disturbance by weather and temperature conditions.

- These voltmeters are used upto *300 kV*.

* The "*Threshold voltage*" is the voltage which is required before the potential gradient is sufficient for ionization to commence.

8.3.4.3. Measurement of D.C. voltages

The methods described in previous articles are suitable for measurement of high voltage D.C. and A.C. systems but for a few *exceptions* :

— The following methods *cannot be used for D.C. measurements* :

(*i*) Capacitance potential dividers;

(*ii*) Potential transformers;

(*iii*) Capacitor charging current method.

- Resistance potential dividers can be used for high voltage D.C. measurements. The series resistance method is also satisfactory when a high resistance is used in series with a current operated D.C. instrument.

8.3.4.4. Measurement of instantaneous voltage

The instantaneous voltage can be measured by using high voltage cathode ray oscilloscope (CRO) with suitable potential divider, where necessary. CRO is discussed in detail in chapter 11.

8.3.5. Low-frequency High-voltage Tests

Fig. 8.18 shows the arrangement for low frequencey high voltage testing (on insulating materials).

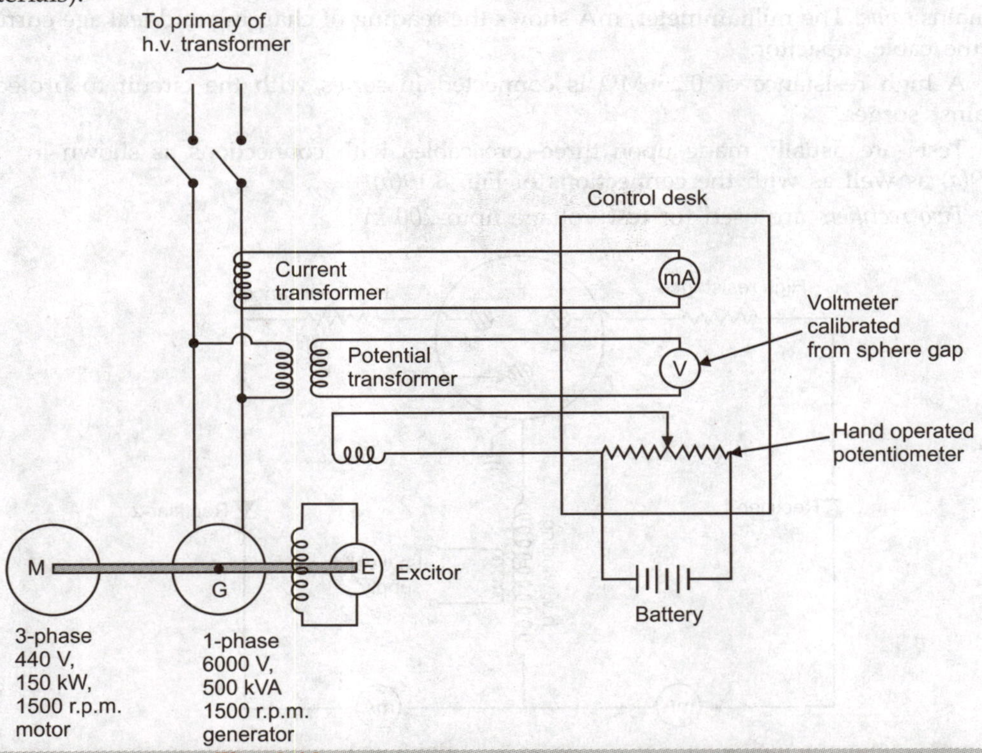

Fig. 8.18. Low-frequency high-voltage testing.

8.3.6. High Voltage D.C. Testing of Cables

Fig. 8.19 shows the connections for D.C. high-voltage cable tests.

In this test, current passes through the rectifier only for half wave of the voltage cycle. During this period, capacitor formed by the sheath and core is *charged to a voltage equal to that of maximum value,* V_{max} of high voltage transformer secondary voltage.

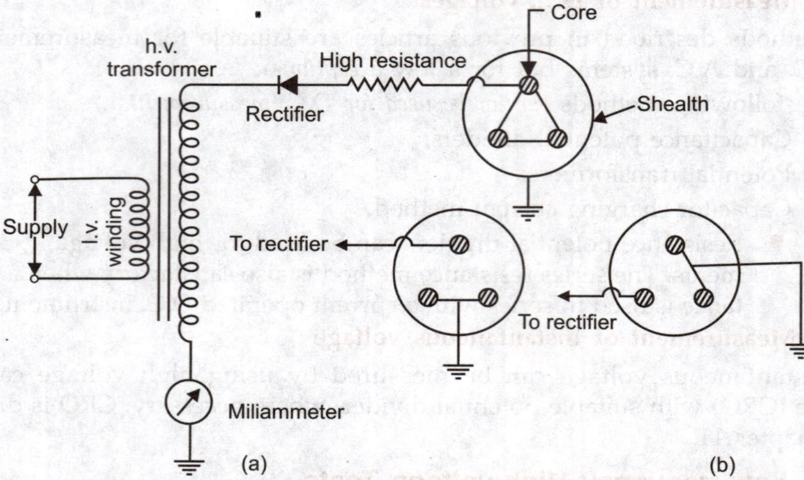

Fig. 8.19. Connections for high-voltage cables (3-core) tests.

During next half cycle, rectifier *stops conduction* and voltage of the cable capacitor remains *same.* The milliammeter, mA shows the reading of charging and leakage currents of the cable capacitor.

A high resistance of 0.25 MΩ is connected in series with the circuit to protect it against surges.

Tests are usually made upon three-core cables with connections as shown in Fig. 8.19(*a*) as well as with the connections of Fig. 8.19(*b*).

Two rectifiers are used for test voltage upto *200 kV.*

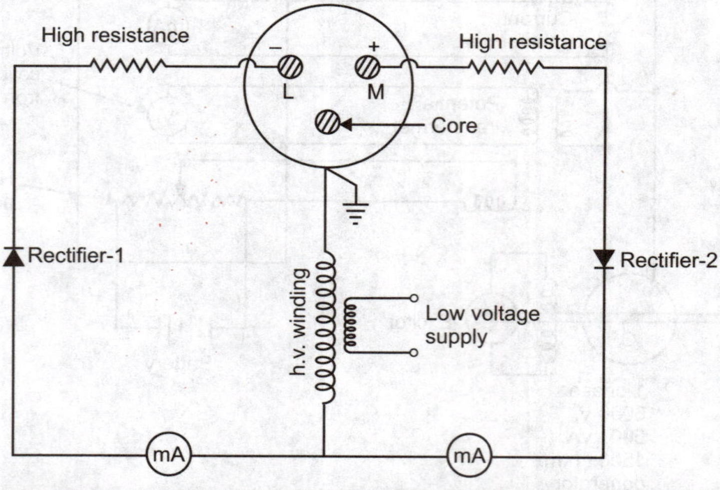

Fig. 8.20. Connections for high-voltage D.C. test upto 200 kV.

— Fig. 8.20 shows the connections for testing a 3-core cable with two rectifiers. In this case, test is being performed between cores *L* and *M*. During first half of the voltage cycle, rectifier-1 conducts and core *L* of the cable is charged to a potential V_{max} below earth and rectifier-2 does not permit any current flow through it. During next half wave of the voltage cycle rectifier-2 conducts and core *M* of the cable is charged to potential V_{max} above earth. Thus D.C. voltage between the cores is $2V_{max}$ and from each core to earth is V_{max}.

— In the case of single-core cable, if the same voltage ($2V_{max}$) is to be applied between core and sheath as between cores in the 3-core cable, *two auxiliary capacitors must be used,* connected as shown in Fig. 8.21.

- In some portable high-voltage D.C. testing sets a single-phase low-voltage alternator, driven by a petrol engine, is used for the supply.
- For avoiding *trouble from surges,* which may cause breakdown of the apparatus, it is advisable to increase the testing voltage to its full value *gradually,* and also to *discharge the cable through a high resistance.* Due to dielectric absorption, it is necessary, also, to maintain the discharging connection for a considerable length of time, in order to avoid a subsequent dangerous rise of voltage.

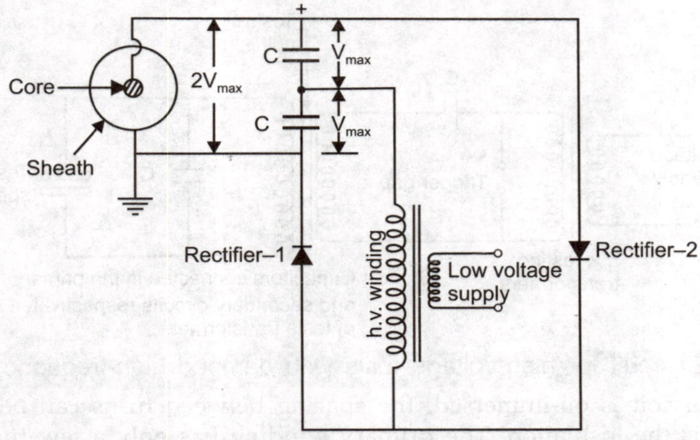

Fig. 8.21. Connections for high voltage D.C. testing for 1-core cables.

8.3.7. High-frequency High-voltage Tests

It has been observed that the behaviour of insulating materials at high frequencies is very different from that at ordinary commercial frequencies. This is mainly due to the greater dielectric power loss, within the material, at high frequency. *The heat produced by this power loss tends to produce breakdown of the insulation at voltages which are smaller than those at which breakdown occurs.*

Such tests are also *useful in the detection of lack of homogeneity in compound-filled porcelain insulators.*

Two kinds of high-frequency high-voltage tests are carried out, these are :

1. Tests with undamped high-frequency oscillations.
2. Tests with damped high-frquency oscillations.

1. Tests with undamped high-frequency oscillations :

— Undamped insulations do not occur in power systems, but are useful for insulation testing purposes, especially for insulation to be used in radio work.

— High frequency alternators have been used for frequency upto 30 kHz, and high-frequency arc generators have also been used, but these have the *disadvantage* that it is *difficult to obtain smooth voltage variation.*

— For the production of high-frequency high-voltages, usually electronic circuits are used; the voltage obtainable is 150 kV, and the frequency 100 kHz.

2. Tests with damped high-frequency oscillations :

Fig. 8.22. shows a circuit for high-voltages, usually electronic circuits are used; the voltage obtainable is 150 kV, and the frequency 100 kHz.

Circuit details :

— Damped high-frequency oscillations are obtained by the use of a Tesla coil, together with a circuit containing a quenched spark-gap as shown.

— The Tesla coil constitutes the h.v. tansformer. It consists of two air-cored coils which are placed concentrically. The high voltage secondary coil has a large number of turns, and is wound on a frame of insulating material, the insulation between turns being air, or in some cases, oil.

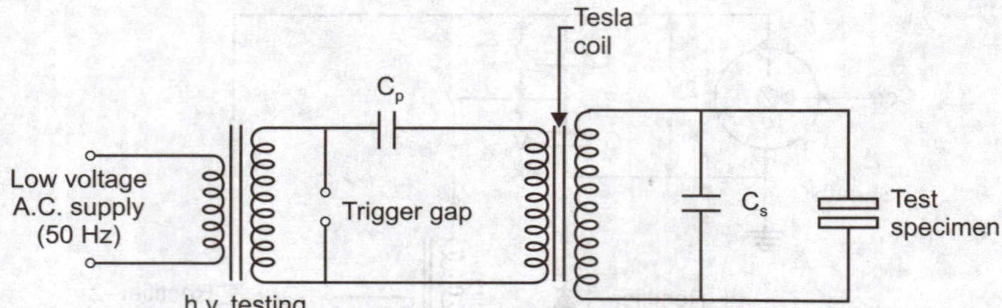

Fig. 8.22. Circuit for high-voltage tests with damped high-frequency oscillations.

If the Testla coil is oil-immersed, the spacing between turns can be made smaller than when air is the insulation. The primary winding has only a few turns, wound on an insulating frame.

— The supply is usually 50 Hz A.C. to the primary a *h.v.* testing transformer.

— Two capacitors C_p (air-cored) and C_s (made up of sphere gap) are connected in the primary and secondary circuits respectively, of the Tesla transformer. A trigger gap for sparking is also induced in the primary circuit of Tesla coil as shown.

Working :

— When the primary of Tesla coil is connected to A.C. supply, the capacitor C_p is charged to some maximum voltage depending upon the voltage on the secondary side of the supply transformer and upon the setting of the "trigger gap". The trigger gap across the secondary breaks at the pre-set value of voltage, the capacitor C_p discharges and thus high frequency damped oscillations are set up in the circuit containing C_p, the spark gap and primary of the Tesla coil. The spark gap will be conducting due to formation of arc across it during the time interval of dying out of oscillations.

— The time taken for the train of oscillations to die away is very small fraction of a second. The charge and discharge of primary capacitor C_p occurs twice per cycle of voltage (*i.e.* if supply frequency is 50 Hz then 100 trains of damped oscillations per second will be set up).

The frequency of the oscillations will be very high, usually of the order of *100 kHz and is given by :

* The highest frequency from which any trouble can be expected in a transmission system is about 80 kHz and hence 100 kHz is the maximum frequency required for testing.

$$f = \frac{1}{2\pi\sqrt{L_p C_p}} \text{ Hz} \qquad ...(8.8)$$

where L_p and C_p are the inductance and capacitance respectively of the oscillatory circuit.

As the primary winding of the Tesla coil carries an oscillatory current, oscillations are induced in the secondary winding. If the primary and secondary windings are *tuned i.e.* $L_pC_p = L_sC_s$ the frequency of oscillations in the secondary winding will be the *same* as in the primary. These damped high-frequency oscillations are impressed upon the test specimen as shown in Fig. 8.22.

The voltage is measured by a sphere gap, while a CRO may be used to record the waveform of secondary voltages of the Tesla coil.

8.3.8. Surge or Impulse Testing

8.3.8.1. Terms relating to surge testing :

The following terms relate to surge testing :

1. ***Impulse voltage :*** Refer to Fig. 8.23 :

- An impulse voltage is a unidirectional voltage which rises rapidly to its maximum value and falls more or less rapidly to zero.

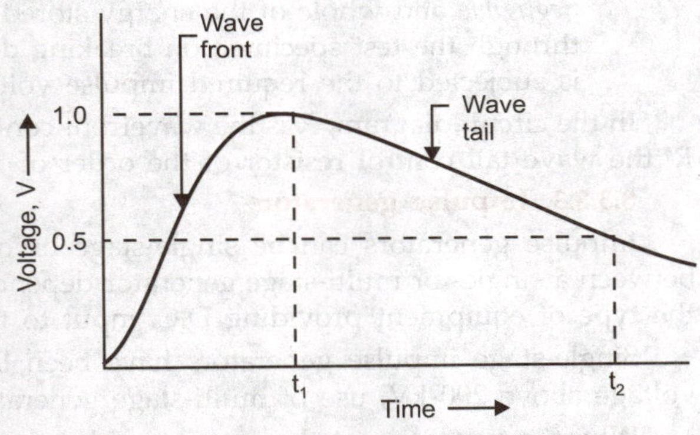

Fig. 8.23. Full surge wave.

If an impulse voltage develops without causing a flashover or puncture, it called a *"full impulse voltage"*, however, if a flashover or puncture occurs, it is termed as a *"chopped impulse voltage"* (See Fig. 8.24).

- The impulse voltage is conveniently expresed as t_1/t_2 wave :

  Time interval t_1 is the time taken by the impulse voltage to rise to its peak value and t_2 is time taken to fall to 50% of its peak value (See Fig. 8.23).

- Normally *1/50 μs impulse voltages are used for testing*. This voltage indicates that it *rises to its peak value in 1 μs and falls to 50% of its peak value in 50 μs.*

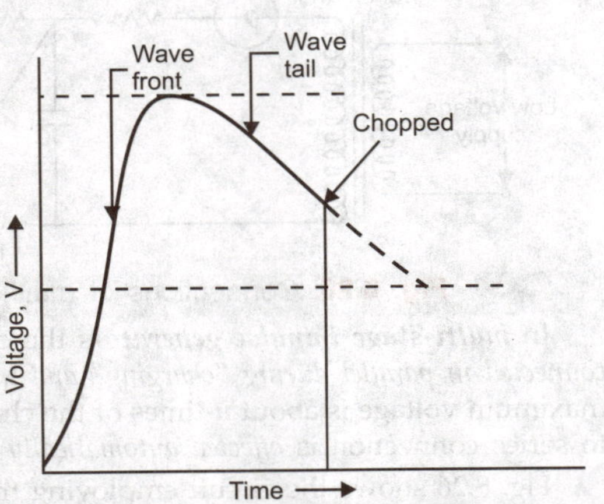

Fig. 8.24. Chopped surge curve.

2. ***Impulse flashover voltage :***

- 50% impulse flashover voltage – it is the peak value of that impulse voltage which causes flashover of the object under test for about half the number of applied impulses.

3. ***Impulse puncture voltage :***

- It is the peak value of that impulse voltage which causes the puncture of the specimen under test *when puncture occurs on wave tail and is the value of the voltage at the instant of puncture when puncture occurs on the wavefront.*

4. ***Impulse ratio for flashover :*** It is the *ratio of impulse flashover voltage to the peak value of power frequency flashover voltage.*

8.3.8.2. Basic impulse test circuit

Fig. 8.25 shows a basic impulse test circuit from which test impulse surge voltage can be obtained :

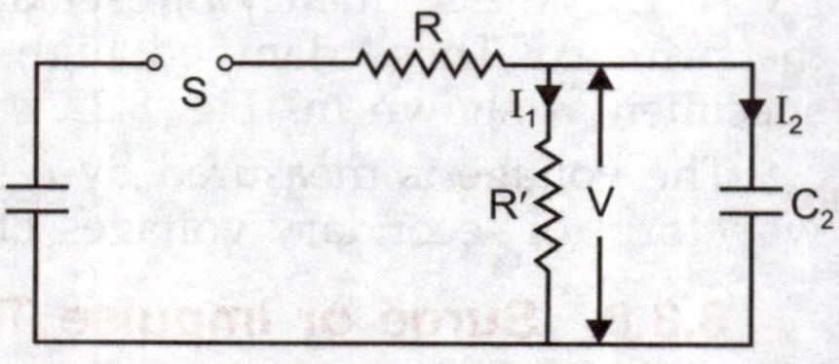

Fig. 8.25. Basic impulse test

— Capacitor C_1 is first charged from a D.C. source till the breakdown of spark-gap occurs. Now the voltage to which C_1 is charged is available.

— The discharge takes place in such a *short time* that *energy loss in the resistors is negligible* and whole of the energy stored in capacitor C_1 is suddenly discharged through the test specimen on breaking down of trigger gap. Thus the specimen is subjected to the required impulse voltage.

In the circuit diagram, *R* is the wavefront control resistor of a few hundred ohms and *R′* the wave tail control resistor of the order of 100 kΩ.

8.3.8.3. Impulse generators

Impulse generators can be single-stage or of multi-stage construction. The choice between a single- or multi-stage generator depends upon the output voltage desired and the type of equipment providing D.C. input to the generator.

Single-stage impulse generators have been built for voltages upto 200 kV but for voltage above 200 kV, use of multi-stage generators is usually made.

Where expense is not the main consideration, ***single-stage generators*** may be used upto a voltage output of 200 kV (or even higher).

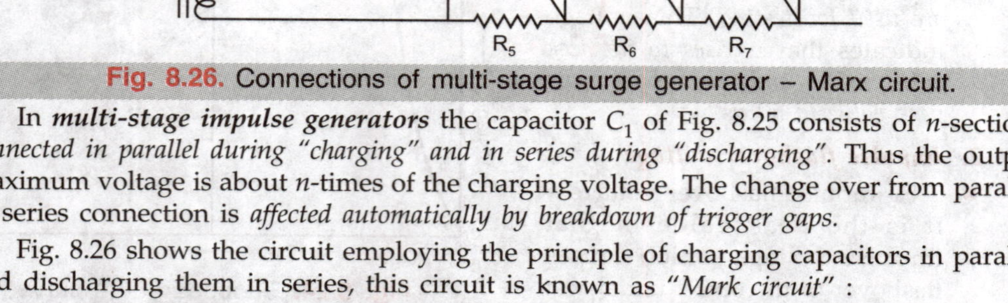

Fig. 8.26. Connections of multi-stage surge generator – Marx circuit.

In ***multi-stage impulse generators*** the capacitor C_1 of Fig. 8.25 consists of *n*-sections *connected in parallel during "charging" and in series during "discharging"*. Thus the output maximum voltage is about *n*-times of the charging voltage. The change over from parallel to series connection is *affected automatically by breakdown of trigger gaps.*

Fig. 8.26 shows the circuit employing the principle of charging capacitors in parallel and discharging them in series, this circuit is known as *"Mark circuit"* :

— In this method of obtaining impulse/surge voltage number of capacitors (C_1, C_2, C_3, C_4 etc.) connected in *parallel* are charged from a high voltage transformer through a rectifying device, as shown in the Fig. 8.26.

— At a certain voltage, which depends upon the setting of the trigger gaps, the trigger gaps breakdown and connect the capcitors in *series* simultaneously. Thus the voltage equal to the sum of the voltages, to which each of the capacitors is charged, is applied to the apparatus under test.

— The discharge takes place so rapidly that energy loss in resistors is negligible and whole of the energy stored in the capacitors is discharged through the apparatus under test, on breaking down of trigger gaps. The trigger gaps are set in such an order that S_1 breaks down at voltage slightly lower than S_2 and S_2 at a voltage slightly lower than S_3 and so on. Thus the trigger gap S_1 breaks down first, and S_2, S_3 and S_4 follow instantly.

— In order to obtain a suitable impulse voltage, any number of trigger gaps (usually upto 50) can be used.

- Fig. 8.27 shows a simplified circuit for 3-stage impulse generator, used now-a-days.

— The capacitors C_1, C_2 and C_3 are charged in parallel through high resistance resistors R_1, R_2, R_3, R_4 and R_5.

— On breakdown of spark gaps S_1, S_2 and S_3, the points L and P, M and Q, N and T etc. are connected together and therefore all the capacitors are connected in series and discharge through the circuit or apparatus under test.

— The resistors R_1, R_2, R_3, R_4 and R_5 are taken of such a high value that there is no significant discharge through them. When suitably chosen, these resistors *can also be used as wave tail control resistors. The control of wave front can also be obtained by including low value resistance in series with each trigger gap.*

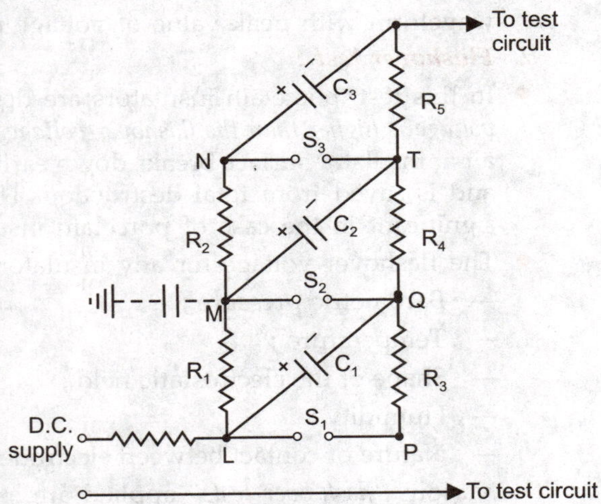

Fig. 8.27. Circuit diagram for 3-stage impulse generator.

8.4. HIGH VOLTAGE TESTING OF INSULATING MATERIALS, PORCELAIN INSULATORS AND INSULATING OIL

8.4.1. Testing of Insulating Materials

- The *"dielectric strength" of an insulating material is the minimum voltage gradient at which electrical failure or breakdown of the material occurs under prescribed conditions.* Although breakdown of insulation may be caused by leakage current, it is normally caused by voltage stress exceeding the dielectric strength of the material. The dielectric strength is expressed in terms of *voltage per unit thickness.*

- Dielectric strength varies with structure homogeneity and quality of sample, thickness of sample, shape of electrodes, waveform of test voltage and duration and rate of application of voltage, moisture content of specimen and heat capacity of electrodes. Dielectric strength is *not* proportional to thickness but is relatively greater for thin sheets. It is also a *function of time of application,* being greater, lesser is the time.

- Measurement of capacitance, permittivity, dielectric loss and loss angle of insulating materials is very important. These measurements are *usually made at high voltages.* All these quantities vary with voltage used for measurements and for some applications, values measured at low voltages are of no use at all. *Normally, "***Schering bridge***" is used for determination of these quantities.*

8.4.2. High Voltage Tests on Porcelain Insulators

The following tests are applied to *porcelain insulators* are :

1. Power frequency tests.
2. Flashover tests.
3. Sample tests.
4. Routine tests.

1. ***Power frequency tests :***

- In this test, frequency used is between 25 to 100 Hz of approximately sine waveform with peak value of voltage not to exceed 1.45 times the r.m.s. value.

2. ***Flashover test :***

- In this test, porcelain insulators are designed in such a way that their *puncture voltage is higher than the flashover voltage.* Thus in case of sudden over-voltage, the air at insulator surface breaks down earlier and thus the insulator is not punctured and is saved from total destruction. Therefore, flashover voltage tests are very significant in the case of porcelain insulators.
- The flashover voltage for any insulator depends upon :
  — Barometric pressure ;
  — Temperature ;
  — Shape of the electrostatic field ;
  — Humidity ;
  — Nature of contact between electrodes and insulators.

The various *"flash-over tests"* applied are :

(*i*) *50 percent dry impulse flashover tests with 1/50 μs impulse voltage.*

(*ii*) *Dry flashover and dry one minute test.* In this test the voltage is gradually raised to a specified value in approximately 10 seconds and is maintained for one minute. The voltage is then gradually raised until flashover occurs.

(*iii*) *Wet flashover and one minute rain test.* During this test, the insulator is sprayed constantly with water and the same procedure as for dry flashover and dry one minute test is followed.

3. ***Sample tests :***

The sample tests include the following :

(*i*) *Mechanical strength test :*

— This test is performed on pin-type insulators to determine their ultimate mechanical strength.

— *During performance of this test the pin-type insulator is mounted on a steel pin and load* is applied to the tie-wire groove by means of a steel bridle. This applied load is then gradually increased till the failure takes place.

— Suspension type insulators are tested in pure tension.

(ii) Temperature cycle test :

— To perform this test, the insulator is subjected to several temperature cycles by immersing it alternately in hot and cold water.

— The insulator or its glaze must not be affected and after drying it out it must withstand the usual flashover test.

(iii) Puncture test :

— This test is performed to determine puncture voltage.

— In this test the insulator is totally immersed in oil and is subjected to the voltage being gradually increased to a certain specified figure or till puncture takes place.

4. ***Routine tests :***

— These tests are applied to all the insulators.

— The insulators are initially subjected to a low voltage which is rapidly increased till a flash-over occurs every few seconds. The voltage is maintained at this value for a minimum of 5 minutes or if failure occurs, for five minutes after the last punctured piece has been removed. The voltage is gradually reduced to $\frac{1}{3}$rd of its value before supply is switched off.

— *The porcelain insulators used for communication purposes are tested with high frequency undamped oscillations.*

8.4.3. Testing of Insulating Oil (Liquid insulant)

The dielectric strength of liquid insulant is tested in a special equipment comprising a cell made up of glass or plastic. The *cell material is transparent and non-absorbent.* Two polished electrodes made up of copper, brass, bronze, or stainless steel are used. These electrodes are either spherical (12.5 to 13 mm dia.) or of the shape shown in the Fig. 8.28 IS 6792—1972. The electrodes are arranged horizontally and the gap between them is accurately adjusted to 2.5 ± 0.1 mm. A variable high voltage supply through a step-up transformer is connected to the electrodes. A voltmeter will indicate the breakdown voltage. The meter will give r.m.s. voltage in kV.

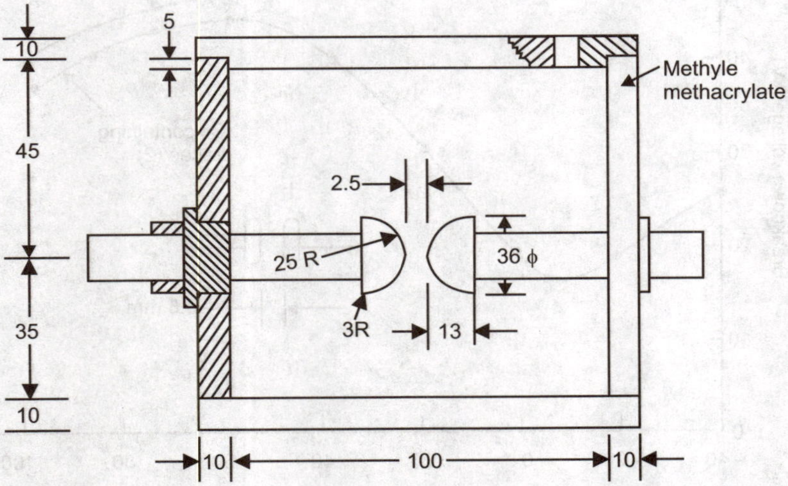

Fig. 8.28. Testing Electric strength of liquid insulant.

The test consists of increasing the test voltage at a uniform rate of 2 kV/sec. from zero upto the value producing breakdown. *The breakdown voltage is the voltage reached during the test at the time the first spark occurs between the electrodes.* The test is carried out six times on the same cell filling (oil) and the *arithmetic mean of six breakdown voltages is taken to be the breakdown voltage of the oil.*

In order to obtain accurate and reliable results attention should be paid to the following facts :

- The distance between the end faces of the two electrodes should be accurately set.
- The electrodes should be wiped, clean and rinsed out with clean dry oil or to be dried with thermostat.
- Right before the test the jar containing the oil should slowly turned upside down several times to mix the oil (it should not be shaken to avoid possible air bubbles).
- After each test the oil should be gently stirred with a clean dry glass stick.

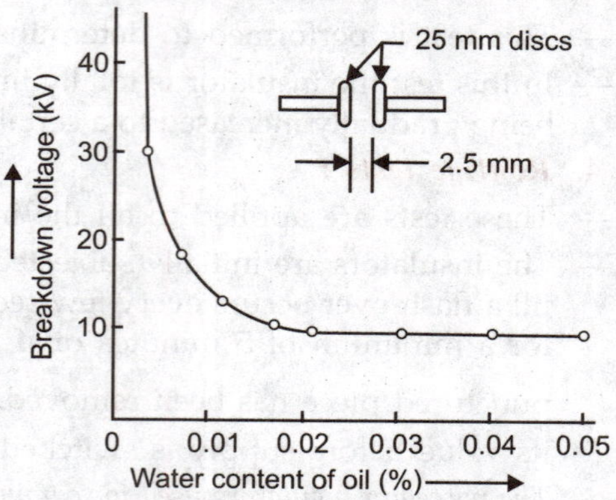

Fig. 8.29. Breakdown voltage of transformer oil as a function of the water content.

- Fig. 8.29 shows the breakdown voltage of transformer oil as a function of the water content.
- Fig. 8.30 shows the temperatures dependence of breakdown voltage of transformer oil, dry oil (1), oil containing traces of water (2).

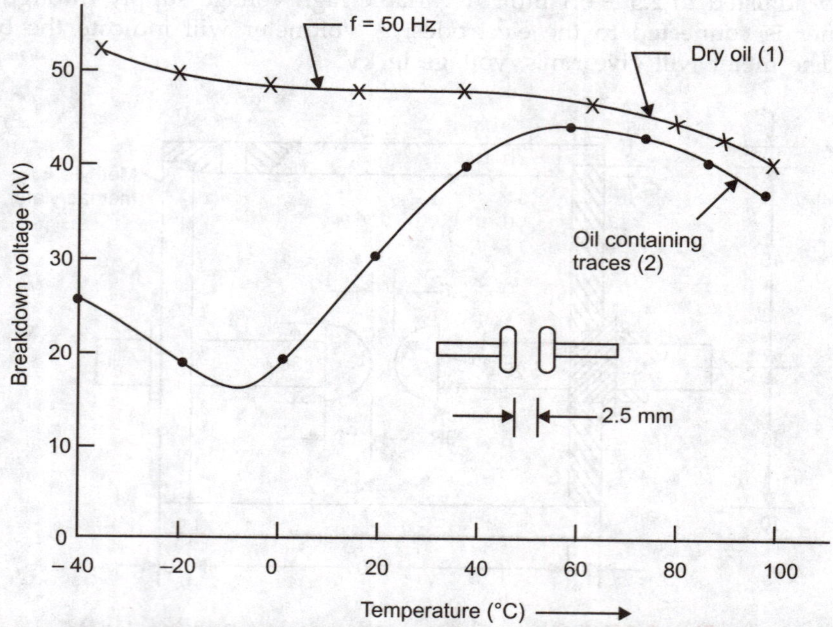

Fig. 8.30. Temperature dependence of breakdown voltage of transformer oil, dry oil (1), oil containing traces of water (2).

HIGHLIGHTS

1. High voltage tests may be classified as follows :
   (*i*) Sustained low-frequency tests;
   (*ii*) Constant direct-current tests;
   (*iii*) High-frequency tests;
   (*iv*) Surge or impulse tests.
2. *Testing Apparatus* :
   (*i*) High voltage transformers;
   (*ii*) Apparatus for voltage regulation;
   (*iii*) Control gear and high-voltage connections, including safety and protective devices;
   (*iv*) Apparatus for voltage measurement.

OBJECTIVE TYPE QUESTIONS

Choose the Correct Answer :

1. The voltage control circuits do not use resistance potential dividers because
   (*a*) they involve a large power loss.
   (*b*) they cause distortion of waveform.
   (*c*) they do not give a smooth variation of voltage.
   (*d*) they have non-linear characteristics.
2. Sphere gaps are used for measurement of
   (*a*) instantaneous values of voltage.
   (*b*) r.m.s. values of voltage.
   (*c*) peak values of voltage.
   (*d*) average values of voltage.
3. A.C. transmission lines and cables are tested by using d.c.
   (*a*) they may be used for d.c. if required.
   (*b*) they draw high values of charging current and hence would require a large sized h.v. transformer if tested with a.c.
   (*c*) h.v. testing transformers cannot be used for long times that are required for testing.
   (*d*) all of the above.
4. Cascade connection of transformers is normally used when a single transformer cannot be used because the minimum test voltage
   (*a*) is about 100 kV (*b*) is about 500 kV
   (*c*) is about 1000 kV (*d*) all of the above.
5. A Cockcroft Walton circuit for generation of high voltage d.c. has n stages. If the peak voltage of the secondary winding of the high voltage transformer is V_m, the output voltage of the circuit is :
   (*a*) $n\,V_m$ (*b*) $2\,(n-1)\,V_m$
   (*c*) $2\,(n+1)\,V_m$ (*d*) $2\,n\,V_m$
6. The impulse voltage wave shape is defined as t_1/t_2 where
   (*a*) t_1 = time taken by the impulse voltage to rise to its peak and
   t_2 = time taken to fall to zero

(*b*) t_1 = time taken by the impulse voltage to rise to its peak and
t_2 = time taken to fall to 50% of its peak

(*c*) t_1 = time taken by the impulse voltage to rise to 50% of its peak and
t_2 = time taken to fall to zero

(*d*) none of the above.

7. The electric strength of new insulating oil should have a minimum strength (r.m.s.) of

(*a*) 30 kV/mm (*b*) 3 kV/mm
(*c*) 50 kV/mm (*d*) 5 kV/mm

8. High-frequency high-voltage tests are carried out on porcelain insulators because

(*a*) they can withstand an a.c. voltage at low frequency but fail to withstand that voltage when applied at high frequency.

(*b*) the insulators, though used on power frequency circuits, may be subjected to high frequency voltages on account of disturbances caused by switching which produce high dielectric loss and heating that may cause failure of insulator at a lower voltage.

(*c*) the insulators may be used for communication purposes wherein the frequencies may be of the order of MHz.

(*d*) all of the above.

9. If the sizes of power transformers and high voltage testing transformers of equal kVA rating are compared :

(*a*) the size of a h.v. testing transformer is greater than that of a power transformer

(*b*) the size of a h.v. testing transformer is smaller than that of a power transformer

(*c*) the size of a h.v. transformer is the same as that of a power tansformer.

10. In Marx's circuit,

(*a*) the capacitors are charged in series and discharged in series.

(*b*) the capacitors are charged in parallel and discharged in parallel.

(*c*) are charged in series and discharged in parallel.

(*d*) are charged in parallel and dicharged in series.

ANSWERS

1. (*a*) **2.** (*c*) **3.** (*b*) **4.** (*b*) **5.** (*d*) **6.** (*b*) **7.** (*b*)
8. (*d*) **9.** (*a*) **10.** (*d*).

THEORETICAL QUESTIONS

1. Write a brief account of different types of high voltage tests usually performed in a h.t. laboratory.

2. How are high voltages classified ?

3. Explain briefly any two of the following high voltage tests :

(*i*) Sustained low-frequency tests.

(*ii*) Constant direct-current tests.

(*iii*) High-frequency tests.

(*iv*) Surge or impulse tests.

4. What are the apparatus essential for all high voltage tests ? What are the methods of voltage regulation in such tests ? Describe these methods in detail.

5. With a neat diagram of connections explain the working of the cascade connected transformers for h.t. testing. Why is it considered advantageous to use this method rather than a single transformer for obtaining high voltages ?

6. Why is high voltage testing important for electrical power apparatus and insulating materials ? Describe briefly the high voltage tests done on electrical apparatus.
7. Explain the reasons for connecting h.v. transformers in cascade. Explain with the help of a diagram, the cascade connections of two transformers.
8. Describe the different methods of voltage control used for h.v. test circuits. Explain their applications, advantages and disadvantages.
9. Explain the following potential divider methods for measurement of r.m.s. value of high voltages.
   (*i*) Resistance potential divider.
   (*ii*) Capacitive potential divider.
   (*iii*) Compensated capacitive potential divider.
10. How will you measure the peak value of an A.C. high voltage ?
11. Explain briefly the use of sphere gaps for high voltage measurement bringing out their advantages and disadvantages in such measurement.
12. Explain, how can you measure high voltage by sphere gap method. Discuss the effect of humidity, if any, on your measurement. Do you measure the r.m.s. value of the voltage or the peak value ? Can you suggest any limitations of this method ?
13. Give the construction, principle and theory of electrostatic voltmeters for the measurement of high voltage.
14. Show that by measuring the rectified charging current of a standard condenser, the peak value of power frequency voltage can be measured.
15. Describe one method of measurement of high voltage of the order of 100 kV. Give the theory of the method.
16. Describe the working, advantages and disadvantages of the following types of voltmeters used for measurement of peak values of high voltages.
   (*i*) Ryall crest voltmeter;
   (*ii*) Ionic wind voltmeter.
17. Explain why transmission lines and cables, which are used for power transmission and distribution of A.C. voltages are tested with high D.C. voltage. Describe the circuits used for high voltage D.C. testing of cables. Describe why is it necessary to establish an equivalence between D.C. and A.C. test voltages ?
18. Describe the circuit of a single-stage impulse generator.
19. Describe a high voltage impulse generator.
20. Explain, with the help of neat diagrams, the working of a surge generator.
21. How is a standard impulse voltage wave shape specified ? Explain clearly with suitable connection diagram the functioning of a multi-stage surge generator. How is the shape of the same controlled ?
22. Explain briefly how the testing of electrical strength of insulating oil is carried out.
23. Explain the terms 'standard impulse voltage' and 'impulse ratio'. Describe in detail the method of generating impulse voltage using Marx circuit for surge or impulse testing.
24. Explain the nature of surge waveform. Describe a high voltage multistage impulse generator.
25. Give a brief account of tests to which porcelain insulators are subjected before commissioning.
26. What is meant by routine testing of porcelain insulators ?
27. Write short notes on :
   (*i*) High voltage testing.
   (*ii*) Crest voltmeter.
   (*iii*) Sphere gap voltmeter.

CHAPTER

9 Illumination

9.1 Introduction; 9.2. Definitions; 9.3. Laws of illumination or luminance; 9.4. Polar curves; 9.5. Photometry – Photometer heads – Photocells (for photometry measurements) – Distribution photometry – Measurement of M.S.C.P. by integrating sphere – Measurement of brightness or luminance – Measurement of illumination – Illumination photometers. *Highlights – Objective Type Questions – Theoretical Questions – Unsolved Examples.*

9.1 INTRODUCTION

Light is *a form of electromagnetic energy radiated from a body which is capable of being perceived by the human eye.*

- The sensation of light results from a flow of energy into the eye and the light will appear to vary if the rate of this flow of energy varies. Light radiations form only a very small part of the complete range of electromagnetic radiations. Light can be of different colours, which depend on the wavelength of the radiation causing it.

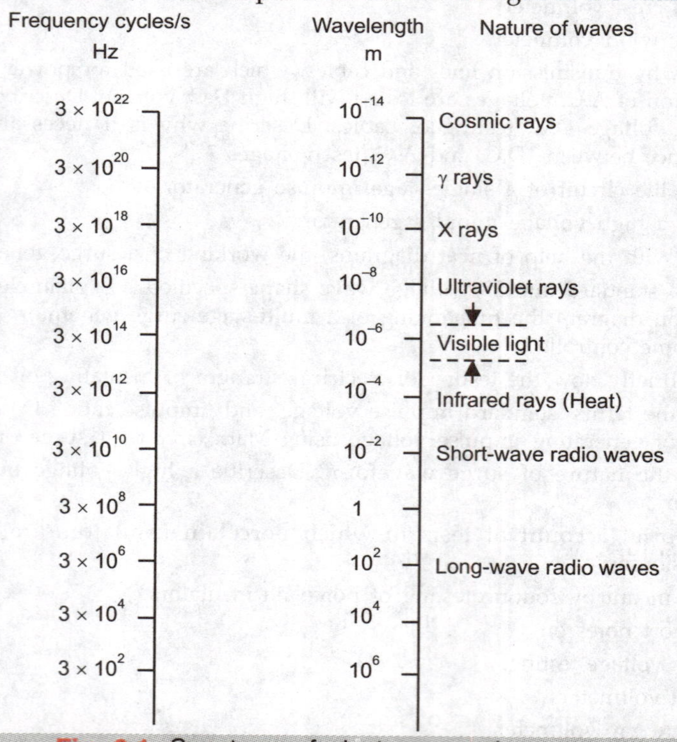

Fig. 9.1. Spectrum of electromagnetic waves.

- Light can be described as a *vibratory motion*, which is transmitted in the form of waves through space. Visible light travels in the form of transverse waves of electromagnetic oscillations. The speed of all electromagnetic waves is 3×10^8 m/s in free space. The wavelength and frequency are different for different waves. The velocity with which these waves travel is related to the wavelength and frequency, by the relation : $v = \lambda f$.

The complete range of waves along with their frequency and wavelength is illustrated in Fig. 9.1.

Fig. 9.2 shows the light emitted together with its colour for the wavelengths within the visible spectrum.

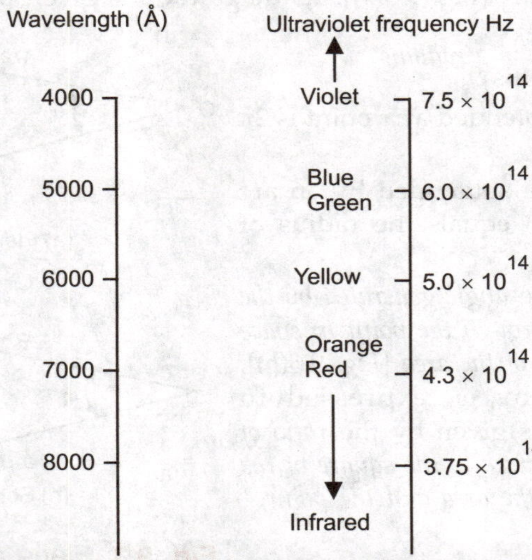

Fig. 9.2. Wavelength and colour of light.

Since the wavelengths of light are very short, smaller unit of length called Angstrom Unit (AU) named after a Sweedish scientist is used.

Angstrom Unit (Å) = 10^{-10} m

Another smaller unit is *micron* (1 micron = 10^{-6} m).

A list of colours with their wavelengths is given below :

Colour	*Wavelengths*
Violet	4100 Å
Blue	4700 Å
Green	5500 Å
Yellow	5800 Å
Red	6000 Å
Orange	6100 Å

- All the studies and research in lighting engineering try to achieve a good lighting scheme to make the occupants feel happy in the case of interior design and factory lighting and make the pedestarians and motorists and other road users comfortable in the case of highway lighting. The following are the *essentials of any good lighting system* :
  (*i*) Adequate illumination of suitable colour on the working surfaces.
  (*ii*) Good maintenance.

(*iii*) Avoidance of hard shadows.

(*iv*) Avoidance of glare.

- The *aim of artificial lighting* is to supplement the daylight or to replace it in modern offices, homes, industries, workplaces etc. ***Good illumination*** *ensures increased production, effectivity of work and reduced accidents.*

9.2. DEFINITIONS

1. Solid angle :

- *A* plane angle is subtended at a point and is enclosed by two straight lines lying in the same plane. [Fig. 9.3(*a*)]. Its magnitude is given by,

$$\theta = \frac{\text{Arc}}{\text{Radius}} \text{ 'radians'}$$

The largest angle subtended at a point is 2π radians.

A *radian* is the angle subtended by an arc of a circle whose length equals the radius of the circle.

- ***Solid angle*** is the *angle generated by the surface passing through the point in space and the periphery of the area [Fig. 9.3(b)].* It is denoted by ω, expressed in '*steradians*' and is given by the *ratio of the area of the surface to the square of the distance between the area and the point,*

i.e.,
$$\omega = \frac{\text{Area}}{(\text{Radius})^2} = \frac{A}{r^2}$$

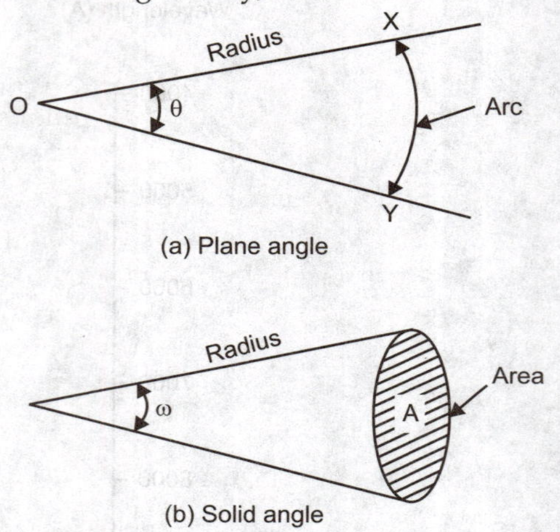

Fig. 9.3. Plane angle and solid angle.

The largest solid angle subtended at a point is due to a sphere at its centre, and is equal to $\frac{4\pi r^2 \text{ (Area of sphere)}}{r^2} = 4\pi$ steradians.

Relationship between ω and θ is obtained as follows :

Consider a curved surface of a spherical segment *ABC* of height *h* and radius *r* (Fig. 9.4)

Surface area of segment $ABC = 2\pi rh$

Here, $h\ (BD) = OB - OD = r - r\cos\frac{\theta}{2} = r\left(1 - \cos\frac{\theta}{2}\right)$

$\therefore$ Surface area of segment $ABC = 2\pi r^2\left(1 - \cos\frac{\theta}{2}\right)$

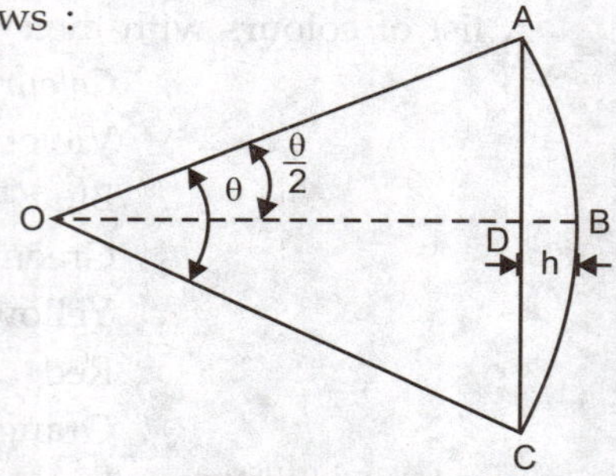

Fig. 9.4. Relation between solid and plane angles.

$$\text{Solid angle, } \omega = \frac{\text{Surface area}}{(\text{Radius})^2} = \frac{2\pi r^2\left(1 - \cos\frac{\theta}{2}\right)}{r^2}$$

$$= 2\pi\left(1 - \cos\frac{\theta}{2}\right) \quad ...(9.1)$$

2. **Light.** *The radiant energy from a hot body which produces the visual sensation upon the human eye is called* ***light.***

It is denoted by the symbol Q, expressed in lumen-hours (analogous to watt-hours)

3. **Luminous flux.** *The total quantity of light energy emitted per second from a luminous body is called* ***luminous flux.***

It is represented by the symbol F and measured in *lumens.* The concept of luminous flux assists us to specify the output and efficiency of a given light source.

4. **Luminous intensity.** Luminous intensity in a given direction is the *luminous flux emitted by the source per unit solid angle.*

It is denoted by the symbol I and is measured in *'candela'* (cd) or *lumens/stearadian i.e.,*

$I = \frac{F}{\omega}$ lumens/steradian or candela, where ω in the solid angle.

5. **Lumen.** It is the unit of luminous flux and is defined as the *amount of luminous flux given out in a space represented by one unit solid angle by a source having an intensity of one candle power in all directions.*

i.e., Lumens = Candle power (C.P.) × Solid angle (ω)

Total lumens given out by source of one candela is 4π lumens.

6. **Candle power.** It is defined as the *number of lumens emitted by a source in a unit solid angle in a given direction.* It is denoted by symbol C.P.

i.e.,
$$C.P. = \frac{\text{Lumens}}{\omega}$$

7. **Illumination.** It is the *luminous flux received by a surface per unit area.* It is denoted by symbol *E* and is measured in *'lumens per square metre' or lux or 'metre-candle'* (*i.e.,* $E = \frac{F}{A}$ lumens/m^2 or lux, where *A* is the area of the surface).

Illumination differs from light very much, though generally these terms are used more or less synonymously. Strictly speaking *light is the cause and illumination is the result of the light* on the surfaces it falls. Thus illumination makes surfaces more or less bright with a certain colour and it is this brightness and colour which the eye sees and interprets as something useful or pleasant or otherwise.

8. **Brightness (or luminance).** Brightness of a surface is defined as the *luminous intensity per unit projected area of the surface in the given direction.* It is denoted by the symbol *L.*

When a surface of area A has an effective luminous intensity of *I* candelas in a direction θ to the normal (Fig. 9.5), then luminance or brightness of that surface,

$$L = \frac{I}{A\cos\theta} \text{ candela/square metre } (Cd/m^2)$$

A *"uniform diffuse source"* is one in which the *intensity per unit projected area is the same from all directions of view.*

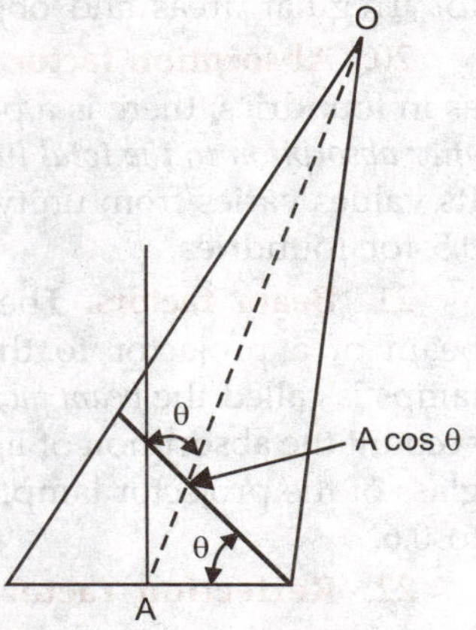

Fig. 9.5.

- ***Relation beween I, L and E :***

Consider a uniform diffuse spherical source with radius *r* metres and luminous intensity *I* candela.

Then,
$$L = \frac{I}{\pi r^2}$$

and,
$$E = \frac{I}{4\pi r^2} \times 4\pi = \frac{I}{r^2}$$

$$\therefore \quad E = \pi L \quad \text{...(9.2)}$$

9. Mean horizontal candle power (M.H.C.P.). It is defined as the *mean of candle power in all directions in the horizontal plane containing the source of light.*

10. Mean spherical candle power (M.S.C.P.). It is defined as the *mean of candle powers in all directions and in all planes from the source of light.*

11. Mean hemi-spherical candle power. It is defined as the *mean of all candle powers in all directions above or below the horizontal plane passing through the source of light.*

12. Reduction factor. Reduction factor of a source of light is the *ratio of its mean spherical candle power to its mean horizontal candle power, i.e.,*

$$\text{Reduction factor} = \frac{\text{M.S.C.P.}}{\text{M.S.H.P.}}.$$

13. Lamp efficiency. It is defined as the *ratio of the luminous flux to the power input.* It is expressed in *lumens per watt.*

14. Specific consumption. It is defined as the *ratio of power input to the average candle power.* It is expressed in *watts per candle.*

15. Space-height ratio. It is defined as the *ratio of horizontal distance between adjacent lamps and height of their mountings.*

16. Utilisation factor (UF). The *ratio of total lumens reaching the working plane to total lumens given out by the lamp* is called *utilisation factor (or co-efficient of utilisation).*

17. Maintenance factor (MF). It is the *ratio of illumination under normal working conditions to the illumination when the things are perfectly clean.*

18. Depreciation factor. This is merely the reverse of the maintenance factor and is defined as the *ratio of initial metre-candles to the ultimate maintained metre-candles on the working plane.* It value is *more than unity.*

19. Waste light factor. Whenever a surface is illuminated by a number of sources of light, there is always a certain amount of wastage of light on account of overlapping and falling of light outside at the edges of the surface. The effect is taken into account by multiplying the theoretical value of lumens required by 1.2 for rectangular areas and 1.5 for irregular areas and objects such as statues, monuments etc.

20. Absorption factor. In the places where atmosphere is full of smoke fumes, such as in foundries, there is a possibility of absorption of light. *The ratio of total lumens available after absorption to the total lumens emitted by the source of light is called the* ***absorption factor.*** Its values varies from unity for clean atmosphere to 0.5 for foundries.

21. Beam factors. The ratio of lumens in the beam of a projector to the lumens given out by lamps is called the *beam factor.* This factor takes into account the absorption of light by reflector and front glass of the projector lamp. Its value varies from 0.3 to 0.6.

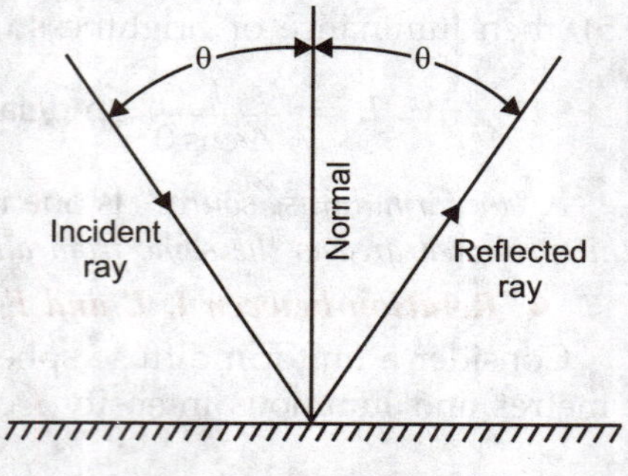

Fig. 9.6.

22. Reflection factor. When a ray of light impinges on a surface, it is reflected from the surface at an angle of incidence, as shown in Fig. 9.6. A certain portion of incident light is absorbed by the surface. The ratio of reflected light to the incident light is called the *'reflection factor'*. It is *always less than unity.*

23. **Glare.** The opening of the pupil in the human eye is controlled by the iris. If a bright object comes into the view of the eye, large amount of light produces an intense image on the retina and the iris automatically protects the eye by contracting the pupil, thus reducing the intensity of the image. When the eye is towards another object which is less bright as compared to the bright object already in the field of view, the iris will contract reducing the amount of light received on the retina from every object in the field of view and making it difficult to see the object desired. At the same time, the portion of the retina which received image of the bright object remains fatigued. This phenomenon is called ***"Glare"***, and is familiar in connection with motor-car head lights.

In other words,"**glare**" *may be defined as the brightness within the field of vision of such a character that causes annoyance, discomfort, interference with vision or eye-fatigue.*

9.3. LAWS OF ILLUMINATION OR LUMINANCE

The illumination (E) of a surface depends upon the following *factors* (The source is assumed to be a point source or otherwise sufficiently away from the surface to be regarded as such).

1. E is *directly proportional to the luminous intensity* (I) of the source. In other words, $E \propto I$.

2. **Inverse square law.** The *illumination of a surface is inversely proportional to the square of the distance of the surface from the source.*

In other words, $E \propto \frac{1}{r^2}$.

Proof : Consider surface areas A_1 and A_2 at distances r_1 and r_2 respectively from the point source S of luminous intensity I and *normal* to the rays as shown in Fig. 9.7. Let the solid angle subtended be ω.

Total luminous flux radiated $= I\omega$ lumens

Illumination of the surface of area A_1,

$$E_1 = \frac{I\omega}{A_1} = \frac{I\omega}{\omega r_1^2} = \frac{I}{r_1^2} \text{ lumens per unit area}$$

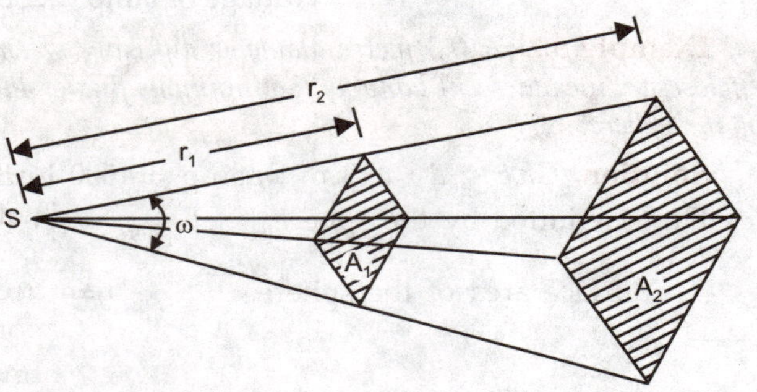

Fig. 9.7. Inverse square law.

Similarly, illumination on the surface of area A_2,

$$E_2 = \frac{I\omega}{A_2} = \frac{I\omega}{\omega r_2^2} = \frac{I}{r_2^2} \text{ lumens per unit area.}$$

Hence, the *illumination of a surface is inversely proportional to the square of the distance between the surface and the light source provided that the distance between the surface and the source is sufficiently large so that source can be regarded as a point source.*

3. **Lambert's cosine law.** According to this law, E *is directly proportional to the cosine of the angle made by the normal to the illuminated surface with the direction of the incident flux.*

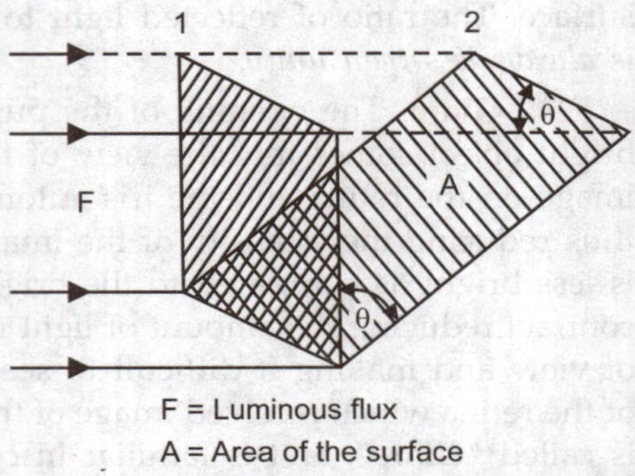

Fig. 9.8. Lambert's cosine law.

As shown in Fig. 9.8, let F be the flux incident on the surface of area A when in position 1. When this surface is turned back through an angle θ, then flux incident on it is $F \cos \theta$. Hence, illumination of the surface when in position 1 is $E_1 = \frac{F}{A}$. But when in position 2, $E_2 = \frac{F \cos \theta}{A}$.

$$\therefore \quad E_2 = E_1 \cos \theta$$

Combining all these factors together, we get

$$E = \frac{I \cos \theta}{r^2}.$$

The unit is lumens per unit area.

Example 9.1. *A 250 V lamp has a total flux of 1500 lumens and takes a current of 0.4 A. Calculate :*

(i) Lumens per watt. *(ii) M.S.C.P. per watt.*

Solution. *Given* : V = 250 volts ; F = 1500 lumens ; I = 0.4 A

Mean spherical candle power of lamp,

$$\text{M.S.C.P.} = \frac{F}{4\pi} = \frac{1500}{4\pi} = 119.4$$

(*i*) **Lumens per watt** $= \frac{\text{Output of lamp in lumens}}{\text{Wattage of lamp in watts}} = \frac{1500}{250 \times 0.4} = \mathbf{15.}$ **(Ans.)**

(*ii*) **M.S.C.P. per watt** $= \frac{\text{M.S.C.P. of lamp}}{\text{Wattage of lamp}} = \frac{119.4}{250 \times 0.4} = \mathbf{1.194.}$ **(Ans.)**

Example 9.2. *A 0.3 metre diameter diffusing sphere of opal glass having 15% absorption encloses an incandescent lamp with a luminous flux of 4500 lumens. Calculate average luminance of the sphere.*

Solution. *Given* : d = 0.3 m ; Lamp = 4500 lumens

Flux 1 emitted by the sphere, $F_{sphere} = (1 - 0.15) \times 4500 = 3825$ lumens

Surface area of the sphere $= 40\pi r^2 = 4\pi \times \left(\frac{0.3}{2}\right)^2 = 0.2827 \text{ m}^2$

$\therefore$ Average luminance of sphere $= \frac{F_{sphere}}{\text{Surface area of sphere}}$

$= \frac{3825}{0.2827} = \mathbf{13530 \text{ lumens/m}^2}.$ **(Ans.)**

Example 9.3. *The candle power of a lamp is 120. A plane surface is placed at a distance of 2.5 metres from this lamp. Calculate the illumination on the surface when it is (i) normal, (ii) inclined to 45°, and (iii) Parallel to rays.*

Solution. (*i*) $E = \frac{C.P.}{d^2} = \frac{120}{(2.5)^2} = \mathbf{19.2 \text{ lux.}}$ **(Ans.)**

(*ii*) $E = \frac{C.P.}{d^2} \times \cos 45° = \frac{120}{(2.5)^2} \times \cos 45° = \mathbf{13.58 \text{ lux.}}$ **(Ans.)**

(*iii*) E = 0, since the rays of light are parallel to the surface, they cannot illuminate it. **(Ans.)**

Example 9.4. *Derive the relation to find the illumination at any point on the plane surface due to light source suspended at height h from the plane surface.*

Solution. Refer to Fig. 9.9. Consider a point *P* on the plane surface *AB* where illumination due to light source *S* of candle power C.P. at a height h from the surface *AB* is to be determined.

Let *d* be the distance between source *S* and point *P*.

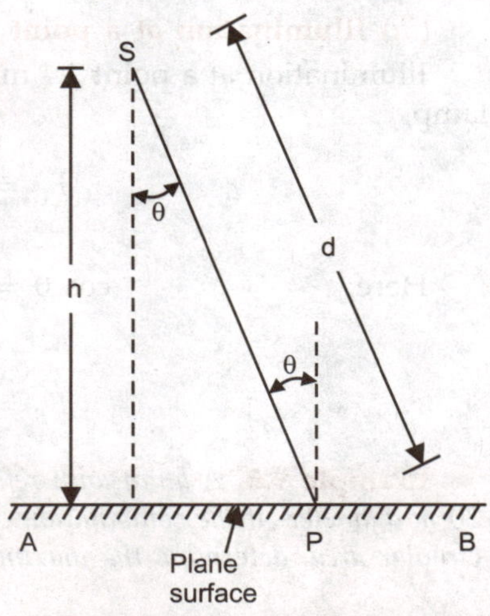

Fig. 9.9.

Then, $\cos\theta = \frac{h}{d}$, or, $d = \frac{h}{\cos\theta}$

Illumination at point *P*, by laws of illumination

$$= \frac{\text{C.P.}}{d^2}\cos\theta = \frac{\text{C.P.}}{(h/\cos\theta)^2}\cos\theta$$

$$= \frac{\text{C.P.}}{h^2}\cos^3\theta,$$

and illumination at any point *O*, *vertically below the source of light* $= \frac{\text{C.P.}}{h^2}$.

Hence, *illumination at any point on a plane is* $\cos^3\theta$*-times of illumination at a point just vertically below the light source, where θ is the angle between the normal to the surface at the point and rays of light*.

Example 9.5. *A 500 W lamp having M.S.C.P. of 800 is suspended 3 m above the working plane :*

(*i*) *Illumination directly below the lamp at the working plane.*

(*ii*) *Lamp efficiency.*

(*iii*) *Illumination at a point 2.4 m away on the horizontal plane from vertically below the lamp.*

Solution. Wattage of the lamp = 500 W

M.S.C.P. of the lamp, I = 800

Height of the lamp, h = 3 m

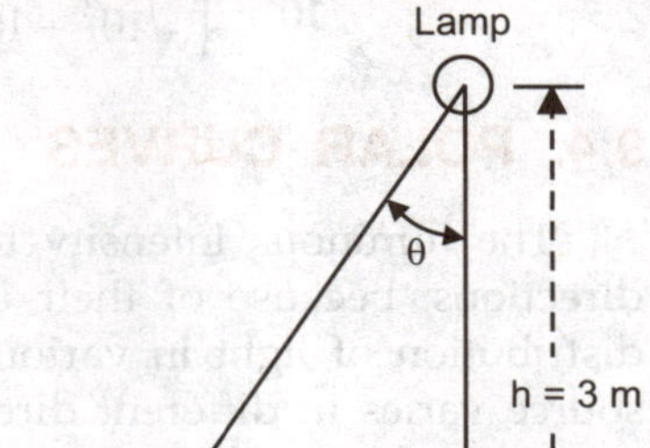

Fig. 9.10.

(*i*) Illumination directly below the lamp at the working plane :

Illumination directly below the lamp,

$$E_A = \frac{I}{h^2} = \frac{800}{3^2} = \textbf{88.9 lux.} \quad \textbf{(Ans.)}$$

(*ii*) Lamp efficiency :

$$\text{Lamp efficiency} = \frac{\text{Luminous flux}}{\text{Power input}}$$

$$= \frac{4\pi \times \text{M.S.C.P.}}{500}$$

$$= \frac{4\pi \times 800}{500}$$

$$= \textbf{20.1 lumens/watt.} \quad \textbf{(Ans.)}$$

***(iii)* Illumination at a point 2.4 m away :**

Illumination at a point 2.4 m away on the horizontal plane from vertically below the lamp,

$$E_B = \frac{I}{h^2}\cos^3\theta$$

Here,
$$\cos\theta = \frac{3}{\sqrt{3^2 + 2.4^2}} = 0.7808$$

$$\therefore \quad E_B = \frac{800}{3^2}\times(0.7808)^3 = \mathbf{42.3\ lux.}\quad \textbf{(Ans.)}$$

Example 9.6. *A lamp with reflector is mounted 10 m above the centre of a circular area of 20 m diameter. If the combination of the lamp and reflector gives a uniform C.P. of 800 over the circular area, determine the maximum and minimum illumination produced on the area.*

(Punjab University)

Solution. Candle power of the lamp, C.P. = 800

Height of the lamp, h = 10 m

Diameter of the circular area = 20 m

The *maximum illumination* will occur *directly below the lamp i.e.,* at point *A* and

$$= \frac{\text{C.P.}}{h^2} = \frac{800}{10^2} = \mathbf{8\ lux.}\quad \textbf{(Ans.)}$$

The minimum illumination will occur at the periphery of the circular area *i.e.,* at point *B* and

$$= \frac{\text{C.P.}}{h^2}\cos^3\theta$$

$$= \frac{800}{10^2}\times\left(\frac{10}{\sqrt{10^2+10^2}}\right)^3 = \mathbf{2.83\ lux.}\quad \textbf{(Ans.)}$$

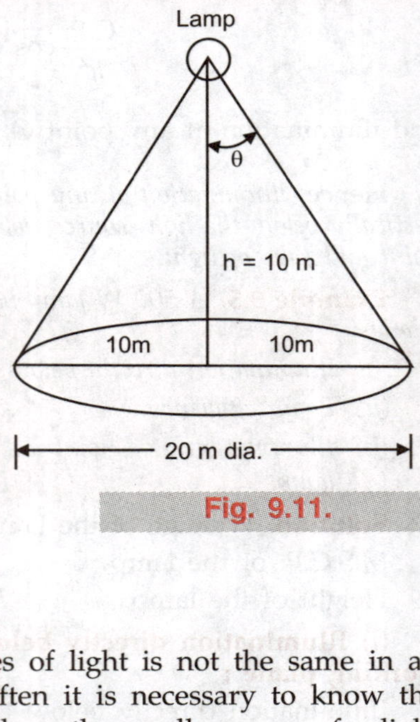

Fig. 9.11.

9.4. POLAR CURVES

The luminous intensity for most lamps or sources of light is not the same in all directions, because of their un-symmetrical shape. Often it is necessary to know the distribution of light in various directions to ascertain how the candle power of a light source varies in different directions. The luminous intensity in all the directions can be represented by ***polar curves.***

— If the luminous intensity *i.e.,* candle power is measured in a *horizontal plane* about a vertical axis and a curve is plotted between the candle power and the angular position, a *"horizontal polar curve"* or diagram is obtained.

— If the candle power is measured at angular position in a vertical plane, a polar curve in the vertical plane, called *"vertical polar curve"*, is obtained.

Fig. 9.12 shows typical polar curves for an ordinary lamp.

The *polar curves are used to determine the following :*

(i) The mean horizontal candle power (M.H.C.P.) and mean spherical candle power (M.S.C.P.).

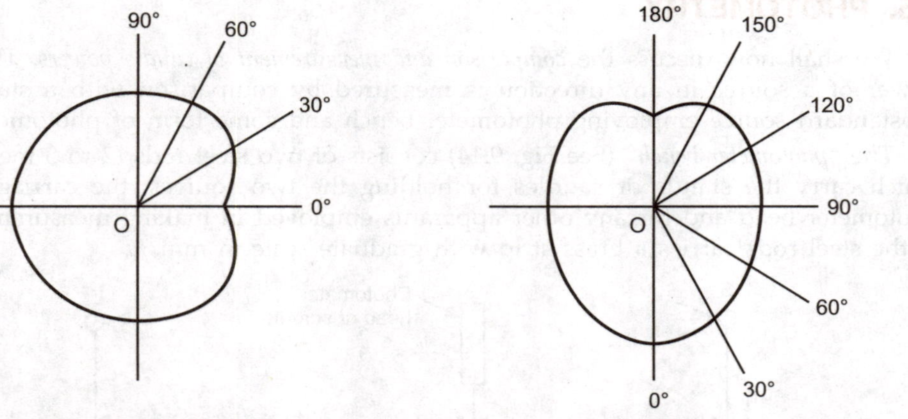

(a) For horizontal plane (b) For vertical plane

Fig. 9.12. Polar curves.

(*ii*) The actual illumination of a surface by employing the candle power in that particular direction as read from the vertical polar curve in illumination calculations.

- The M.H.C.P. of a lamp can be determined from the horizontal polar curve by taking the mean value of the candle power in a horizontal direction.
- The M.S.C.P. can be determined from the vertical polar curve by *Roussean's construction*.

Roussean's construction. Suppose the vertical polar curve is in the form of two lobes symmetrical about *YOY′* axis. The Roussean's construction for this polar curve is illustrated in Fig. 9.13.

- Draw a circle with any convenient radius with centre *O* as the centre.
- Draw *PQ* parallel to *YOY′* and equal to the vertical diameter of the circle.
- Draw any line *OEA* meeting the polar curve in *E* and the circle in *A*. Let the projection be *S*.
- At S erect an ordinate ST = OE.
- By similar construction draw other ordinates. The curve PSQUTVP obtained by joining these ordinates is known as *Roussean's curve*. The *mean ordinate of this curve gives the mean spherical candle power* (M.S.C.P.).

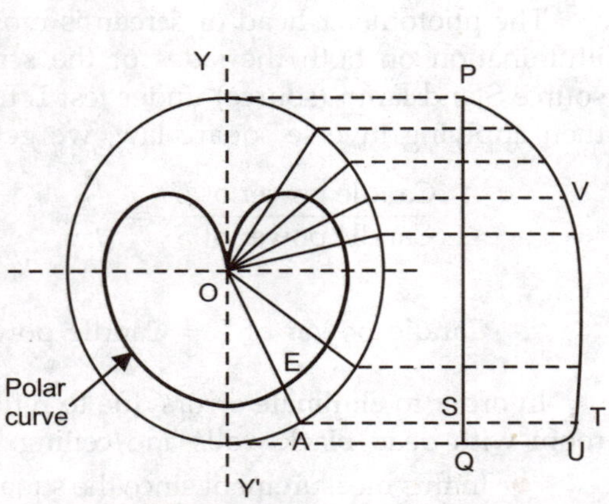

Fig. 9.13. Roussean's construction.

$$\text{The mean ordinate of the curve} = \frac{\text{Area PSQUTVP}}{\text{Length of } PQ}$$

The area under the curve can either be determined on a graph paper or by using Simpson's rule.

9.5. PHOTOMETRY

We shall now discuss the *comparison and measurement of candle powers.* The candle power of a source in any direction is measured by comparison with a standard or substandard source employing photometer bench and some form of photometer.

The *"photometer bench"* (See Fig. 9.14) consists of two steel rods (2 to 3 metres long) which carry the stands or saddles for holding the two sources, the carriage for the photometer head and for any other apparatus employed in making measurement. One of the steel rods carries a brass strip with graduate scale in mm.

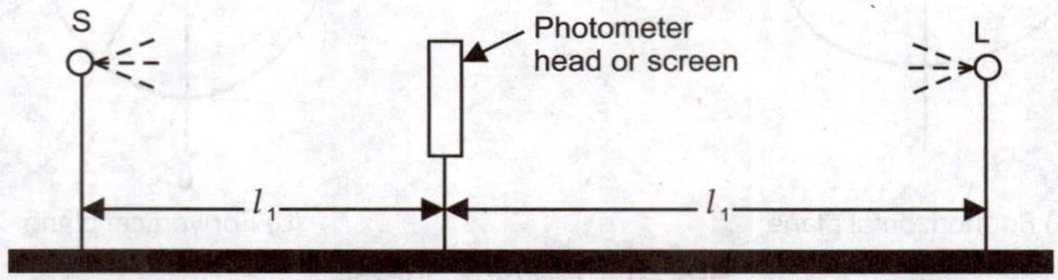

S = Standard source (known C.P.)
L = Lamp under test.

Fig. 9.14. Photometer bench for measurement of candle power.

The bench should be rigid so that source being compared may be free from vibrations and the carriage holding the photometer head should be capable of moving smoothly and with very little effort. The photometer head acts as a *screen* for comparison of illumination of the standard source and the source under test.

The photometer head or screen is moved in between the two fixed sources until the illumination on both the sides of the screen is same. If the distances of the standard source S and lamp (source) under test L from photometer head are l_1 and l_2 respectively, then applying inverse square law, we get

$$\frac{\text{Candle power of } L}{\text{Candle power of } S} = \frac{l_2^2}{l_1^2}$$

$$\therefore \quad \text{Candle power of } L = \text{Candle power of } S \times \frac{l_2^2}{l_1^2} \qquad ...(9.3)$$

In order to eliminate errors due to reflected light the experiment is performed in dark room with dead black walls and ceiling.

- In this measurement since the square of distances are involved, therefore, distances should be accurately measured. For obtaining distance exactly, two points are determined at which there is a perceptible difference in illumination from the two sides and the point half way between them is taken as the position of equal illumination.

9.5.1. Photometer Heads

Although, quite a few types of photometer heads are in use, yet most common in use are *Bunsen* and *Lummer-Brodhun type.* These heads give quite accurate results when the lamps of the same or very similar colour are to be compared. However, when two lamps with different colours are to be compared, a *"Flicker photometer"* gives better results.

Bunsen head. It is also called the *"Grease spot photometer"*. It consists of a piece of thin opaque paper at the centre of which is a transluscent "spot", made by treating the paper with oil or wax.

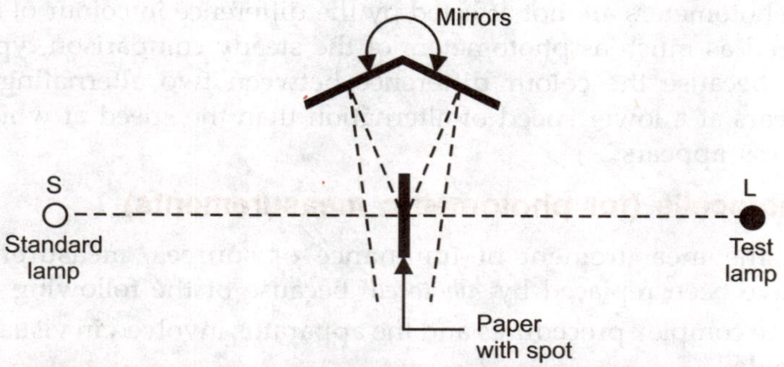

Fig. 9.15. Grease spot photometer head.

- Light falls on the paper from both the lamps. The opaque part of the paper will be illuminated on either side by the lamp on that side only. But the transparent "spot" will be illuminated from both the lamps. The photometer head is adjusted at such positions that the *transparent spot is not perceptible.* The paper is viewed from one side first and then from the other. We get two distances l_1 and l_1' of the photometer head from the test lamp (L) and l_2 and l_2' the distances of the photometer head from the standard lamp (S). The C.P. (candle power) of the test lamp is calculated from the following relation :

$$\frac{\text{C.P. of test lamp } (L)}{\text{C.P. of the standard lamp } (S)} = \frac{l_1 l_1'}{l_2 l_2'}$$

- In another method, one side of the paper is illuminated from a fixed lamp. On the other side of the photometer first the test lamp (L) is used and then the standard lamp (S) is used to make the spot to vanish. If l_1 and l_2 are these distances, then we have,

$$\frac{\text{C.P. of test lamp } (L)}{\text{C.P. of standard lamp } (S)} = \frac{l_1^2}{l_2^2}$$

The *use of two mirrors with the photometer head is perhaps the most accurate method* (Fig. 9.15). The two sides of the spot can be viewed simultaneously and the position of the head for equal "contrast" in illumination between the opaque part and the transparent spot of the paper on the two sides is located. The C.P. of the test lamp (L) can then be calculated by the expression given earlier.

Lummer-Brodhun photometer head. There are two types of Lummer-Brodhun photometer heads :

(*i*) Equality-of-brightness type.

(*ii*) Contrast type.

Contrast type is more accurate and, therefore *more used in photometric measurements.*

Flicker photometers :

- These photometers are used *when the two sources giving light of different colours are to be compared.*
- They operate on the fact that if two illuminated surfaces are presented to the eye with rapid alternations, the *flicker disappears when the surfaces are of equal brightness.* The speed of alternation should be kept as low as possible at which the disappearance of the flicker can be obtained for the small variation in brightness.

- These photometers are not affected by the difference in colour of two lights to be compared as much as photometers of the steady comparison types are affected. This is because the colour difference between two alternating fields of light disappears at a lower speed of alternation than the speed at which difference in brightness appears.

9.5.2. Photocells (for photometric measurements)

Except for the measurement of luminance of sources, measurement by visual comparisons have been replaced by *'photocell'* because of the following facts :

(*i*) The more complex procedures and the apparatus involved in visual measurements are avoided.

(*ii*) These cells give *more accurate and faster measurements.*

(*iii*) The measurements are consistent.

The two types of photocells used in photometric measurements are :

1. Photovoltaic cells
2. Photoemissive cells.

1. **Photovoltaic cell.** Fig. 9.16 shows the construction of a photovoltaic cell (also known as the barrier layer or rectified cell). It consists of a base plate made of either steel or aluminium and carries a layer of metallic selenium which is light sensitive. An electrically conducting layer of cadmium oxide is applied by sputtering over the selenium layer. The layer is *sufficiently thin* to allow light to reach the selenium and is electrically continuous as it acts as a negative pole. The negative contact is formed of a strip of wood's metal sprayed on to the edge of the top surface. The base plate forms the positive contact. The front surface of the cell is protected by a transparent varnish.

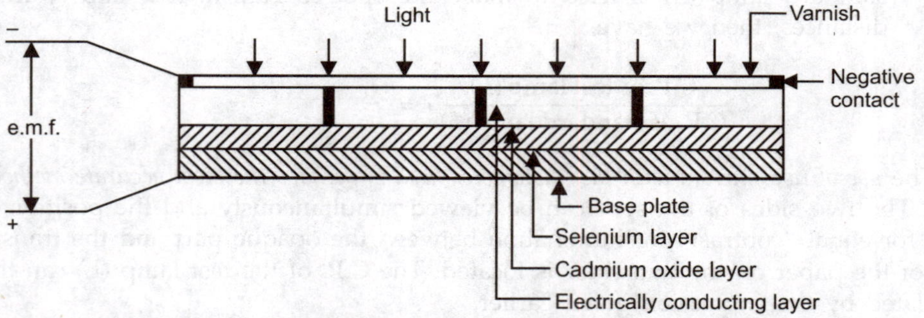

Fig. 9.16. Photovoltaic cell.

As the light falls on the upper surface of selenium, electrons are released from the surface which maintain a flow of current through the external circuit connected between the +ve and –ve contacts.

- *In photometry, the current output of a photocell should be proportional to the illumination which is achieved by keeping the external resistance at a low value.* For achieving greater accuracy, the illumination should not be allowed to exceed 275 lumens/m^2.
- The smaller the size of the cell (of course compatible with obtaining sufficient current to be properly measured) a better linear relation is obtained between current and illumination. This is because, for such a cell, the resistance of the electrically conducting film is at a minimum. Also, the current being small, the voltage drop due to the circuit resistance will be kept low.

2. **Photoemissive cell.** In order to get greater precision in terms of linearity and stability it is better to use photoemissive cell rather than photovoltaic cell. However, the circuit involved is more complex and requires the use of some sort of valve amplifiers.

9.5.3. Distribution Photometry

- ***Distribution photometry*** *assists in finding out the variations of luminous intensity around a lighting fitting so that polar curves and other diagrams for the fitting can be obtained and also the performance of the fitting is studied.*

Following points are important for proper distribution photometry :

(*i*) The *temperature of the test area should be kept stable* (since the output flux of some sources like fluorescent tubes depends upon the temperature of the surroundings).

(*ii*) *Only that light which is to be measured should be allowed to reach the photocell.*

(*iii*) *Optical path should be made sufficiently long* so that the inverse square law may be applied with a reasonable degree of accuracy.

- In a very simple ***distribution photometer,*** the photocell fitting moves around the fitting on a semicircular track and its position is controlled by means of a flexible cable. A complete series of curves at different angles of azimuth can be obtained by rotating the fitting about a vertical axis. The mounting of the fitting on the apparatus is facilitated by mounting a gear system on the gentry which turns the fittings so that it can be brought down to easy access.

9.5.4. Measurement of M.S.C.P. by Integrating Sphere

The ***integrating sphere*** consists of a hollow sphere whose diameter (one or more than one metre) is large compared to the lamp to be tested, having a smooth inner surface with a uniform coating of white paint. If the lamp is hung inside the sphere, the light is so diffused that a uniform illumination is produced over the whole surface. A small window of translucent glass provided at one side of the sphere is illuminated by reflected light from the inner surface of the sphere. A small *screen* is inserted in between the lamp and window to prevent from the lamp reaching the window *directly*.

The mean spherical candle power (M.S.C.P.) is measured as follows :

- The lamp whose M.S.C.P. is to be determined, is placed at the centre of the sphere and brightness of window is measured with the help of some form of illuminometer.

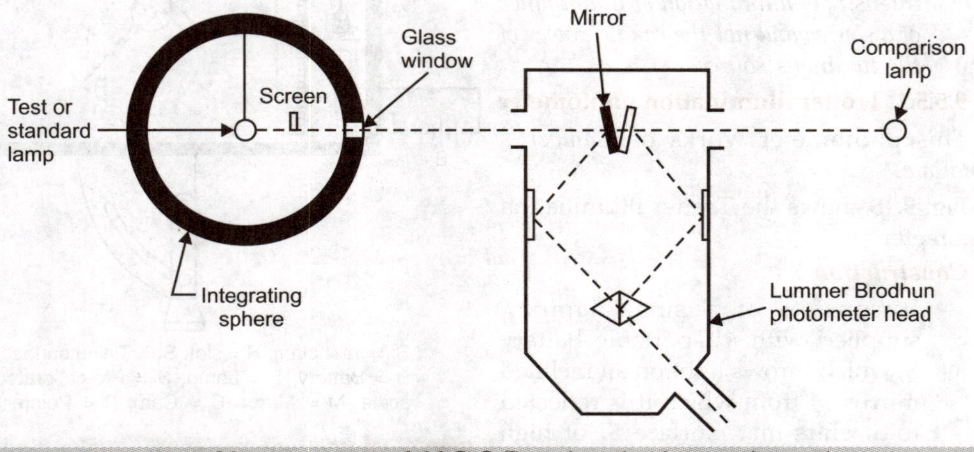

Fig. 9.17. Measurement of M.S.C.P. using the integrating sphere.

- The test lamp is replaced by a standard lamp whose M.S.C.P. is known and brightness of the window is again measured.

Since the M.S.C.P. of the two lamps are proportional to the respective brightness of the window, therefore, the M.S.C.P. of the lamp under test can be determined by knowing the M.S.C.P. of standard lamp and brightness of window in the two cases.

Fig. 9.17 shows an arrangement in which a Lummer-Brodhun photometer head is used in which a mirror is placed as indicated :

— The light from the glass window is reflected by the mirror as shown.

— Another lamp is used for comparison and for obtaining balance by shifting the position of the comparison lamp from the photometer head.

— We get two distances of comparison lamp from the photometer head : *First distance* when the glass window is illuminated by the test lamp and the *second distance* when the glass window is illuminated by the standard or sub-standard lamp.

— The squares of the corresponding distances of the comparison lamp from the photometer head are *inversely proportional to the M.S.C.Ps. of the lamps.*

9.5.5. Measurement of Illumination – Illumination Photometers

The simplest and best method of measuring illumination is by a *portable type illumination photometer calibrated to read directly in "lux (lumens/m^2)".*

— No shadow should be allowed to fall on the meter during measurements and meter should be held perfectly horizontal and stationary for accurate results.

— The meter should be *colour-corrected type* so that it can be used for all kinds of light sources.

- A number of such instruments is use are :

Trotter Illumination Photometer, Macbeth Illuminometer, Holophane lumeter etc.

— These are portable photometers which are built *compact and light in weight* and are used for measurements of illumination in *homes, offices, commercial houses* and *industries.*

— These photometers *measure the intensity of illumination or illuminance of a source and not the candle power of the luminous sources.*

9.5.5.1. Trotter illumination photometer

This photometer works on *Lambert's cosine law.*

Fig. 9.18 shows the Trotter illumination photometer :

Construction :

— It consists of a small lamp L supplied with $4V$ portable battery B, which throws light on an inclined mirror M from which it is reflected to a white matt surface S_1 of high diffusion quality which can be *tilted.*

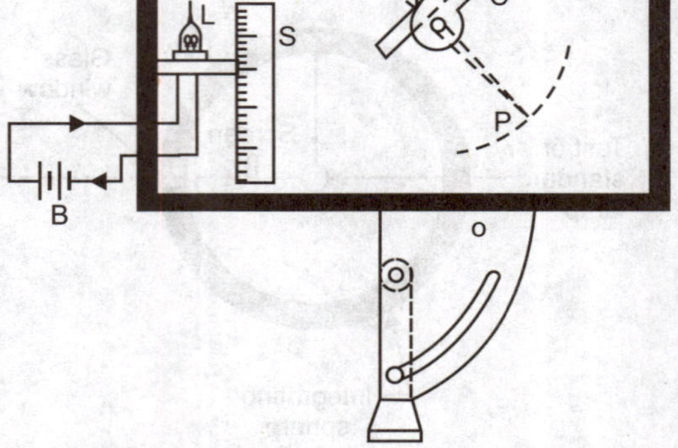

S_2 = Test plate; H = slot; S_1 = Tilting surface; B = Battery; L = Lamp; S = Pre-calibrated scale; M = Mirror; C = Cam; P = Pointer

Fig. 9.18. Trotter illumination photometer.

— The top of the box containing the lamp and the tilting surface S_1 carries a test plate S_2 which receives the illumination to be measured. The test plate S_2 is exactly the same with diffusing material as S_1. Like S_1, it is also of high diffusing quality so that *intensity variations in different directions obey "cosine law"*. It is necessary that *test plate should be made out of a material which does not introduce errors due to selective absorption.*

— The test plate has a slot H cut in it, through which the surface S_1 can be viewed.

Working :

— For measurement, S_2 is placed where illumination is required.

— Then the surface S_1 is tilted by means of a cam C, through a roller attached to the tilting surface, until its illumination as seen through slot H is the same as that of the test plate S_2.

— A pointer P moving over a pre-calibrated scale S then gives the illumination of S_2 in lumens/m^2.

— The scale S is then calibrated by producing known intensities of illumination on S_2 from a standard lamp and marking the position of pointer for photometric balance between S_1 and S_2. Thus, it is not necessary to know the candle power of the lamp although for constancy of candle power it must be aged before use in the photometer.

— The purpose of cam C is to provide a *fairly graduated scale.*

- The range of this photometer is normally from *0.1 to 42 lumens/m²*.

9.5.5.2. Macbeth illuminometer

It is a sensitive photometric device and is one of the most accurate instruments of this type. It is very simple and easy to operate.

Its operation is based upon the *inverse square law.*

Fig. 9.19 shows the Macbeth illuminometer.

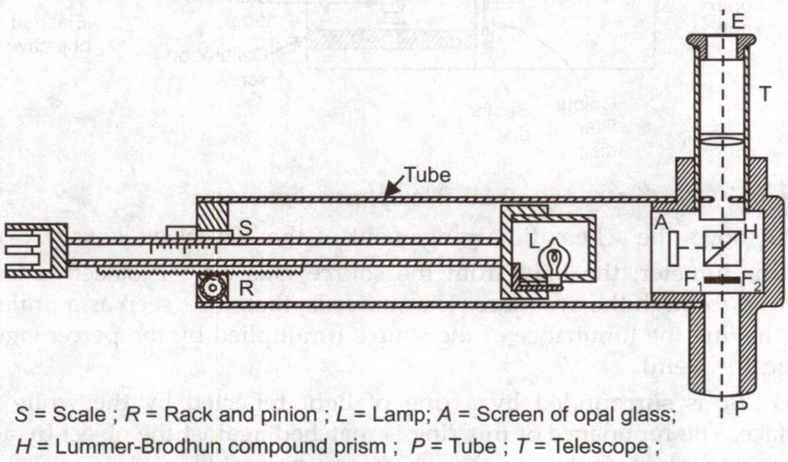

S = Scale ; *R* = Rack and pinion ; *L* = Lamp; *A* = Screen of opal glass; *H* = Lummer-Brodhun compound prism ; *P* = Tube ; *T* = Telescope ; *E* = Eye piece ; F_1, F_2 = neutral filters

Fig. 9.19. Macbeth illuminometer.

Construction :

— A Lummer-Brodhun compound prism H is mounted in a box and viewed through a telescope T.

— On the opposite side of the prism from the telescope is a tube *P*, which is pointed towards a test plate of opal glass, situated at the point at which the illumination is to be measured.

— *A* is a screen of opal glass, which is illuminated by a lamp *L* in an enclosure having an aperture in the side nearest the screen *A*.

Working :

— For measurement, the lamp *L* is moved along a tube which is at right angles to the telescope, by a rack and pinion *R*, until balance is obtained at the photometer head. A scale *S* then gives the illumination of the test plate *directly*.

— The lamp *L* is supplied from a dry battery or accumulator, and a milliammeter and rheostat are provided with instrument for adjustment of the lamp current to its correct value (*i.e.* that which makes the scale *S* direct reading).

- The normal range of this photometer is from *5 to 250 lumens/m²*. Its range can be extended either way by inserting neutral filters F_1 and F_2.

9.5.5.3. Holophane lumeter

The average brightness or luminance of a source in a given direction is given by the *ratio of its intensity to the projected area of the source in that direction.* For achieving greater sensitivity it is necessary to use some kind of optical system to collect more light flux.

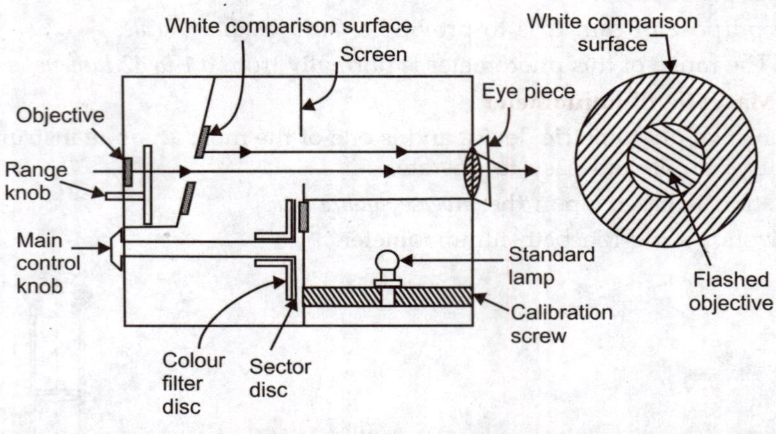

Fig. 9.20. Holophane lumeter.

Fig. 9.20 shows the schematic arrangement of the *"holophane lumeter"*:

— In this lumeter, the light from the source falls on the objective lens which is viewed through the eye piece. The source is, therefore, seen as a uniformly bright *disc* having the luminance of the source (multiplied by the percentage loss in the optical system).

— This *disc* is surrounded by a ring of light reflected by the white comparison surface. The luminance of this ring is matched against the object by adjusting the range and main control knobs which are graduated.

— A standard lamp placed in the whitened compartment, illuminates the white screen. This lamp is moved with the help of a calibration screw only when the instrument requires recalibration.

— Light reaches the white comparison surface after passing through the opal window and amount of this light is adjusted by adjusting the sector disc by means of main control knob.

HIGHLIGHTS

1. *Light* is defined as the radiant energy from a hot body which produces the visual sensation upon the human eye.
2. *Law of inverse square.* The illumination of a surface is inversely proportional to the square of the distance between the surface and the light source provided that the distance between the surface and the source is sufficiently large so that source can be regarded as a point source.
3. *Lambert's cosine law.* According to this law, the illumination of any point on a surface is proportional to the cosine of the angle between the normal at that point and the direction of luminous flux.
4. The luminous intensity in all the directions can be represented by *polar curves.*
5. Most commonly used photometer heads are :
   Bunsen head ; Lummer-Brodhun photometer head.
6. Flicker photometers are used when the two sources giving light of different colours are to be compared.
7. The two types of photocells used in photometric measurements are :
   (*i*) Photovoltaic cells; (*ii*) Photoemissive cell
8. *Distribution photometry* assists in finding out the variations of luminous intensity around a lighting fitting so that polar curves and other diagrams for the fitting can be obtained and also the performance of the fitting is studied.

OBJECTIVE TYPE QUESTIONS

Choose the Correct Answer :

1. Which of the following statements is *correct* ?
   (*a*) Light is a form of heat energy.
   (*b*) Light is a form of electrical energy.
   (*c*) Light consists of shooting particles.
   (*d*) Light consists of electromagnetic waves.
2. Luminous efficiency of a fluorescent tube is
   (*a*) 10 lumens/watt (*b*) 20 lumens/watt
   (*c*) 40 lumens/watt (*d*) 60 lumens/watt.
3. Candela is the unit of which of the following ?
   (*a*) Wavelength (*b*) Luminous intensity
   (*c*) Luminous flux (*d*) Frequency.
4. Colour of light depends upon
   (*a*) frequency (*b*) wave length
   (*c*) both (*a*) and (*b*) (*d*) speed of light.
5. Illumination of one lumen per sq. metre is called ......
   (*a*) lumen metre (*b*) lux
   (*c*) foot candle (*d*) candela.
6. A solid angle is expressed in terms of ......
   (*a*) radians/metre (*b*) radians
   (*c*) steradians (*d*) degrees.

7. The unit of luminous flux is ......
   (*a*) watt/m^2 (*b*) lumen
   (*c*) lumen/m^2 (*d*) watt.
8. In houses, the illumination is in the range of
   (*a*) 2–5 lumens/watt (*b*) 10–20 lumens/watt
   (*c*) 35–45 lumens/watt (*d*) 60–65 lumens/watt.
9. "The illumination is directly proportional to the cosine of the angle made by the normal to the illuminated surface with the direction of the incident flux."
   Above statement is associated with
   (*a*) Lambert's cosine law (*b*) Planck's law
   (*c*) Bunsen's law of the illumination (*d*) Macbeth's law of illumination.
10. Desired illumination of level of the working plane depends upon
    (*a*) age group of observers.
    (*b*) whether the object is stationary or moving.
    (*c*) size of the object to be seen and its distance from the observer.
    (*d*) whether the object is to be seen for longer duration or shorter duration of time.
    (*e*) all above factors.
11. Which of the following instruments is used for the comparison of candle powers of different sources ?
    (*a*) Radiometer (*b*) Bunsen meter
    (*c*) Photometer (*d*) Candle meter.
12. ...... photometer is used for comparing the lights of different colours ?
    (*a*) Grease spot (*b*) Bunsen
    (*c*) Lummer brodhun (*d*) Guilds flicker.
13. Luminous flux is
    (*a*) the light energy radiated by sun.
    (*b*) the part of light energy, radiated by sun, which is received on the earth.
    (*c*) the rate of energy radiation in the form of light waves.
    (*d*) none of the above.
14. Stroboscopic effect due to use of discharge lamps in workshops results in moving machinery appearing
    (*a*) stationary
    (*b*) stationary running slow
    (*c*) stationary running in reverse direction
    (*d*) all of the above.
15. When a electric bulb is broken it produces bang, this is due to
    (*a*) vacuum inside the bulb
    (*b*) pressure of air in the bulb
    (*c*) pressure inside is equal to that outside
    (*d*) none of the above.
16. Due to which of the following reasons the light of a tube appears cooler than that of a bulb ?
    (*a*) Tungsten is not used in the tube.
    (*b*) Tube is painted with milky colour.
    (*c*) Tube consumes less power.
    (*d*) Surface area of the tube is more than that of bulb.
    (*e*) none of the above.

17. Soft shadows are produced by
(*a*) using surface source of light instead of point source of light.
(*b*) increasing the number of lamps.
(*c*) both (*a*) and (*b*).
(*d*) none of the above.

18. The level of illumination on a surface *least* depends on
(*a*) ambient temperature (*b*) candle power of the source
(*c*) distance of the source (*d*) type of reflector used.

19. The errors involved in photometric measurements are of the order of
(*a*) 1% (*b*) 1 – 5%
(c) 10 – 20% (*d*) 40 – 50%.

20. The desirable properties of a laser beam are :
(*a*) it has a high radiant energy
(*b*) it can be polarised
(*c*) it is monochromatic and coherent
(*d*) all of the above.

21. A photoconductive cell may be called as :
(*a*) photoresistive cell
(*b*) photo resistor
(*c*) light dependent resistor
(*d*) all of the above.

22. The resistance of a photoresistive cell
(*a*) increase with increase in light intensity
(*b*) remains constant irrespective of the light intensity incident upon it
(*c*) decreases with increase in light intensity in a linear manner
(*d*) decreases with increase in light intensity in an exponential manner.

23. The response of human eye under normal light conditions is called
(*a*) stopopic vision and it peaks at a wavelength of 555 nm
(*b*) stopopic vision and it peaks at a wavelength of 507 nm
(*c*) photopic vision and it peaks at a wavelength of 507 nm
(*d*) photopic vision and it peaks at a wavelength of 555 nm.

24. A light source may be considered as a point source if its diameter is
(*a*) less than 10% of the distance between it and the detector
(*b*) less than 1% of the distance between it and the detector
(*c*) more than 10% of the distance between it and the detector
(*d*) more than 1% of the distance between it and the detector.

25. In case of measurement of intensity of a test source, a standard source is used. The photometric head used when comparing two lights of different colours, is
(*a*) Bunsen photometer head.
(*b*) Constast type Lummer Brodhun photometer head.
(*c*) Comparison type Lummer Brodhun photometer head.
(*d*) Flicker photometer head.

26. The current through a photoemissive tube depends upon :
(*a*) intensity of light incident upon it.
(*b*) colour and wavelength of incident light.
(*c*) voltage applied between cathode and anode of tube.
(*d*) all of the above.

27. The luminous sensitivity of gas filled photoelectric cells varies between
(*a*) 10 to 100 μA/lm (*b*) 40 to 150 μA/lm
(*c*) 0.002 to 0.1 μA/μW (*d*) 0.01 to 0.15 μA/μW.

28. Photodiodes
(*a*) do not have any response in darkness
(*b*) have a response even in darkness
(*c*) have a leakage current even in darkness which is called dark current that doubles for every 10°C increase in the ambient temperature which decreases the signal to noise ratio.
(*d*) have a leakage current even in darkness which is called dark current that doubles for every increase of 5°C in temperature and which increases the signal to noise ratio.

29. The sensitivity of a photovoltaic cell
(*a*) is almost constant when it is short circuited.
(*b*) decreases with increase of load resistance.
(*c*) is almost negligible when the load resistance is about 10000 Ω.
(*d*) all of the above
(*e*) none of the above.

ANSWERS

1. (*d*)	**2.** (*d*)	**3.** (*b*)	**4.** (*c*)	**5.** (*b*)	**6.** (*c*)	**7.** (*b*)
8. (*d*)	**9.** (*a*)	**10.** (*e*)	**11.** (*c*)	**12.** (*d*)	**13.** (*c*)	**14.** (*d*)
15. (*a*)	**16.** (*a*)	**17.** (*c*)	**18.** (*a*)	**19.** (*c*)	**20.** (*d*)	**21.** (*d*)
22. (*d*)	**23.** (*d*)	**24.** (*a*)	**25.** (*d*)	**26.** (*d*)	**27.** (*b*)	**28.** (*c*)
29. (*d*).						

THEORETICAL QUESTIONS

1. What is the basic nature of light ? Explain.
2. Define the following terms :
Luminous flux, Lumen, Illumination, Lamp efficiency.
3. Explain briefly the following :
(*i*) Space-height ratio (*ii*) Utilisation factor
(*iii*) Depreciation factor.
4. What is a solid angle ?
5. State the laws of illumination.
6. What is a polar curve ? How is it useful to an illumination engineer ?
7. What do you understand by polar curves ? Explain Rousseau's construction for calculating M.S.C.P. of a lamp.
8. Explain briefly the laws of illumination or luminance.
9. Discuss the following laws of illumination :
(*i*) Inverse square law.
(*ii*) Lambert's cosine law.
10. What do you mean by the term 'photometry' ?
11. Explain briefly with a neat diagram "Photometer bench" for measurement of candle power.
12. Describe the working of the "Grease spot photometer".
13. Write a short note on "Flicker photometers".
14. Explain with a neat sketch the working of a "photovoltaic cell" used in photometry measurements.

15. Write short notes on the following :
   (*i*) Photoemissive cell.
   (*ii*) Distribution photometry.
16. Describe with a neat diagram the measurement of M.S.C.P. (mean spherical candle power) by integrating sphere.
17. Explain briefly the working of a "*holophane lumeter*" with a neat sketch.
18. Write a short note on "*Measurement of illumination*".
19. Explain briefly any two of the following illumination photometers :
   (*i*) Trotter illumination photometer.
   (*ii*) Macbeth illuminometer.
   (*iii*) Holophane lumeter.

UNSOLVED EXAMPLES

1. A 250 V lamp has a total flux of 3000 lumens and takes a current of 0.8 A. Calculate (*i*) lumens/watt, and (*ii*) M.S.C.P./watt. [**Ans.** (*i*) 15 ; (*ii*) 1.2]
2. A 0.4 metre diameter diffusing sphere of opal glass, having 20% absorption, encloses an incandescent lamp with a luminous flux of 4850 lumens. Calculate the average luminance of the sphere. [**Ans.** 7720 lumens/m^2]
3. A filament lamp of 500 W is suspended at a height of 5 metres above working plane and gives uniform illumination over an area of 8 m diameter. Assuming efficiency of reflector as 60% and efficiency of lamp as 0.9 watt per candle power, determine the illumination on the working plane. [**Ans.** 83.33 lumens/m^2]
4. The candle power of a lamp is 100. A plane surface is placed at a distance of 2 metres from this lamp. Calculate the illumination on the surface when it is (*i*) normal, (*ii*) inclined to 45°, and (*iii*) parallel to the rays. [**Ans.** (*i*) 25 lux, (*ii*) 17.67 lux (*iii*) zero]
5. A 500 W lamp having M.S.C.P. of 100 is suspended 2.7 metres above the working plane. Calculate :
   (*i*) Illumination directly below the lamp at the working plane.
   (*ii*) Lamp efficiency.
   (*iii*) Illumination at a point 2.5 metres away on the horizontal plane from vertically below the lamp.
   [**Ans.** (*i*) 137.17 lux ; (*ii*) 25.13 homens/watt ; (*iii*) 54.2 lux]
6. A lamp with reflector is mounted 12 m above the centre of a circular area of 24 m diameter. If the combination of the lamp and reflector gives a uniform C.P. of 1000 over the circular area, determine the maximum and minimum illumination produced on the area. [**Ans.** 6.94 lux ; 2.46 lux]
7. A lamp having a uniform C.P. of 200 in all directions is provided with a reflector which directs 60% of the total light uniformly on to a circular area of 10 m diameter. The lamp is hung 6 m above the area. Calculate :
   (*i*) The illumination at the centre and edge of the surface with and without the reflector.
   (*ii*) The average illumination over the area without the reflector.
   [**Ans.** (*i*) 5.56 lux, 2.52 lux (without reflector), 19.2 lux (without reflector), (*ii*) 3.7 lux]

CHAPTER

10 Digital Measurements of Electrical Quantities

10.1 GENERAL ASPECTS

• *A **digital instrument** can be considered as a counter which counts the pulses in a predetermined time.* Digital transducers whose output is in the form of pulses are used to monitor the desired parameter.

"Accuracy" of a digital instrument is dependent on the number of pulses generated by the transducer because the fraction of pulse cannot be generated and in counting there can be ambiguity of only one pulse at start/stop. Hence *more are pulses* corresponding to a measure less the possibility of error corresponding to one pulse and *more the accuracy.*

— The digital instruments use *logic circuits and techniques* for carrying out measurements of quantities.

— The information in the electronic digital read-out (display) devices is *presented as a series of digits on the tubes, screen, or printed on a piece of paper.* The relevant characters (letters of alphabet from A to Z, numerals from 0 to 9, punctuation mark and other symbols in common use) can be generated by :

(*i*) Semiconductor light emitting diodes (LED).

(*ii*) Liquid crystal displays (LCD).

(*iii*) Numerical indicators tubes (NIT).

(*iv*) Hot filament or bar tubes.

— Digital instruments provide better resolution than analog instruments. A typical value of *resolution* of a digital instrument may be one part in 10^6 *i.e.,* 1 μV can be read on the 1-V input range.

— *The **digital meters** work on the principle of "quantization".*

The analog quantity to be measured is first subdivided or quantized into a number of small intervals upto many decimal places. The objective of the digital instrument is then to determine in which portion of the subdivision the measurand can thus be identified as an integral multiple of the smallest unit called the *quantum,* chosen for subdivision.

The measuring procedure thus *reduces to one of counting the number of quanta present in the measurand.*

— Fig. 10.1. shows the building block of a digital instrument. Any digital instrument invariably consists of a an analog-to-digital converter (ADC) in its input stage. The display block may be analog or digital in nature. If an analog read out is desired, it becomes necessary to include a stage involving digital-to-analog (D/A) conversion.

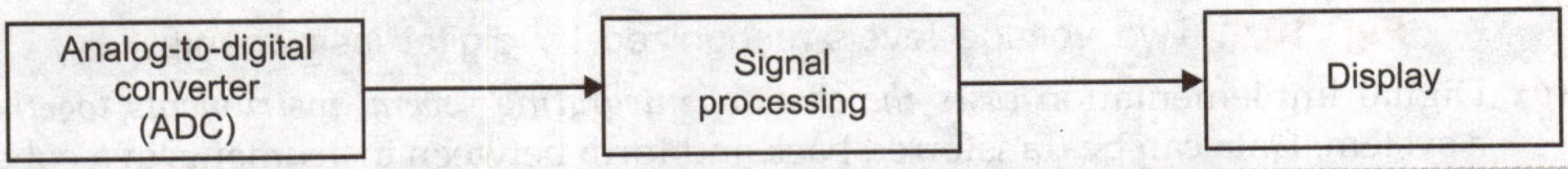

Fig. 10.1. Building block of a digital instrument.

10.2 ANALOG AND DIGITAL SYSTEMS

An ***analog system*** comprises devices that manipulate the physical quantities represented in *analog form.* Examples of analog systems are : Automobile speedometer, Magnetic tape recording and playback equipment etc.

A ***digital system*** is a combination of devices designed for manipulating physical quantities or information represented in *digital form.* Examples of digital systems are : Digital computers, typewriters, digital watches etc.

Advantages of digital systems/techniques :

1. Easier to design.
2. Easy storage of information.
3. Greater accuracy and precision.
4. Operation is programmable.
5. Digital circuits are less affected by the noise.

- Digital system may consist of the following *components* :

(*i*) Resistors; (*ii*) Capacitors;
(*iii*) Transistors; (*iv*) Linear ICs;
(*v*) Digital ICs; (*vi*) Digital devices;
(*vii*) Analog-to-digital converters (ADC) (*viii*) Digital-to-analog converters (DAC).

— *Data in digital form facilitates various operations that are normally required in signal processing.*

— An increase in the availability and type of computer facilities and a decrease in the cost of various methods required for digital systems is accelerating the development of digital instrumentation for measurement and signal processing.

10.3 BASIC CONCEPTS OF DIGITAL INSTRUMENTS

10.3.1. Introduction

A truly digital instrument uses *binary* (*i.e.,* base 2) *logic elements* such as *gates, flip-flops,* and so on, to perform the measurement function. These circuits operate with only two voltage levels (Fig. 10.2), and these represent the binary digits 0 and 1.

Digital implementation entails the following *"advantages"* :

(*i*) The instrument is more *easily modified,* especially where the complexity is in embedded software.

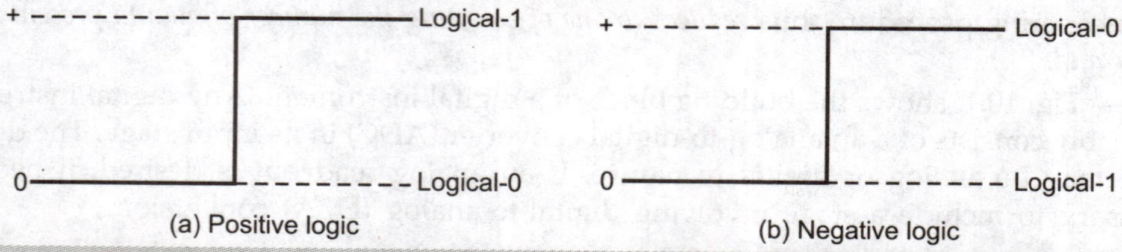

Fig. 10.2. Two voltage levels-recognized by digital instruments.

(ii) Digital implementation *eases the chore of integrating several instruments together* in a system. Data can be transferred back and forth between instruments in a coherent manner much more easily than analog signals, especially over long distances.

— For larger instrumentation systems, the use of general purpose interface bus (IEEE-488 GPIB) makes the chore of integrating instruments easy.

(iii) Even though software digital solutions are also very complex as in the case of analog solutions, yet the *hardware implications are minimal.* One can add RAM or ROM or use a hard drive to store a larger, more complex program that would be impossibly difficult to implement in analog hardware.

10.3.2. Binary Counters

The binary counter forms the heart of many "discrete logic" digital implementations.

- *A **binary counter** consists of a chain of locked flip-flops, each state is designed to count by 2.*
- *A **decoder** is used to indicate the state of the count* by examining the unique Q and not-Q states at the output of each stage.
- A 4-bit binary counter will count from 0000_2 to 1111_2 (*i.e.,* 0_{10} to 15_{10}) while a *"binary coded decimal"* **(BCD)** counter counts only from 0000_2 to 1001_2 (0_{10} to 9_{10}). Table 10.1 shows both binary and BCD coding, the latter of which is a subset of 4-bit binary coding.

Table 10.1. Binary and BCD Codes

Decimal	Binary	BCD		
0	0000		0000	
1	0001		0001	
2	0010		0010	
3	0011		0011	
4	0100		0100	
5	0101		0101	
6	0110		0110	
7	0111		0111	
8	1000		1000	
9	1001		1001	
10	1010	0001	0000	
11	1011	0001	0001	
12	1100	0001	0010	
13	1101	0001	0011	Two 4-bit
14	1110	0001	0100	counters
15	1111	0001	0101	required

10.3.3. Decimal Counting Unit (DCUs)

Decimal counting unit *is the circuit of any electronic counter.* It consists of the following components :

(*i*) Decode counter;

(*ii*) Decoder/display-driver;

(*iii*) Data latch.

Fig. 10.3. shows the block diagram of an example of a decimal counting unit (DCU) :

— This circuit is constructed of TTL integrated circuit logic elements. The decode counter uses a chain of flip-flops, but it also contains some internal gates that sense the tenth state (*i.e.*, "9") and reset the counter to 0000 if another input pulse is received. In fact, now, there are several MSI and LSI integrated circuits available that contain all circuitry needed to form a complete electronic counter of several decades.

— The four output lines from the *"decade counter"* are coded in 8421 BCD format and they change state with every pulse.

— *"Input"* and *"reset"* are other two lines, in which the input line accepts the pulses being counted, while the reset line will reset the counter to 0000 when brought to ***High.*** For normal counting the line is kept ***Low***, and to clear the counter it is brought momentarily ***High.***

— The *"decoder"* is used to convert the 4-bit BCD code to the code needed to correctly drive the read-out device.

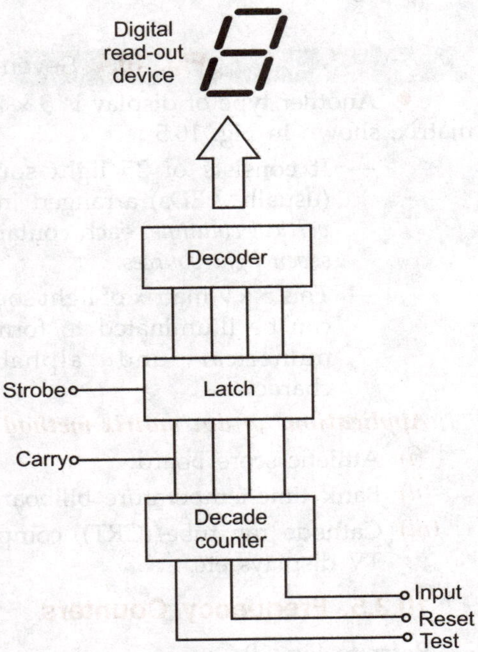

Fig. 10.3. Decimal decode counter.

— A quad *"latch"* circuit is used to eliminate the rolling effect (this effect is created because the display changes with the count).

— The *"strobe"* line, in practice, is held low while the count is being accumulated, and at the end of the count period the strobe line is brought high momentarily, just long enough to transfer the count data from the input lines to the output lines.

- The individual DCU can count only from 0 to 9, but if two DCUs are used in cascade, the assembly can count from 0 to 99, and three DCUs can count from 0 to 999 etc.; the assembly is known as ***decimal counting assembly (DCA).***

10.3.4. Display Devices

- In modern equipment the most common display is the *seven-segment* readout shown in Fig. 10.4(*a*). This type of display consists of seven illuminated bars arranged in a *figure-eight pattern.* It is possible to *form all ten decimal digits by lighting appropriate bars* [Fig. 10.4(*b*)]. A BCD-to-seven-segment decoder will ground those terminals *a* through *g* required to form the decimal digit represented in BCD form at the decoder input. Probably the most popular types as of this writing are the light-emitting diode (LED) and liquid crystal diode (LCD) displays.

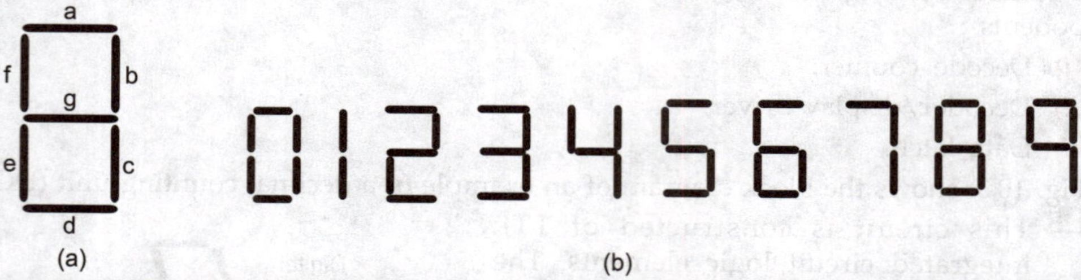

Fig. 10.4. Seven-segment decimal readout.

- Another type of display is 5 × 7 dot matrix, shown in Fig. 10.5 :
  - — It consists of 35 light sources (usually LEDs) arranged in *five vertical columns,* each containing *seven light sources.*
  - — This 5 × 7 matrix of light sources can be illuminated to form all numerical and alphabetic characters.

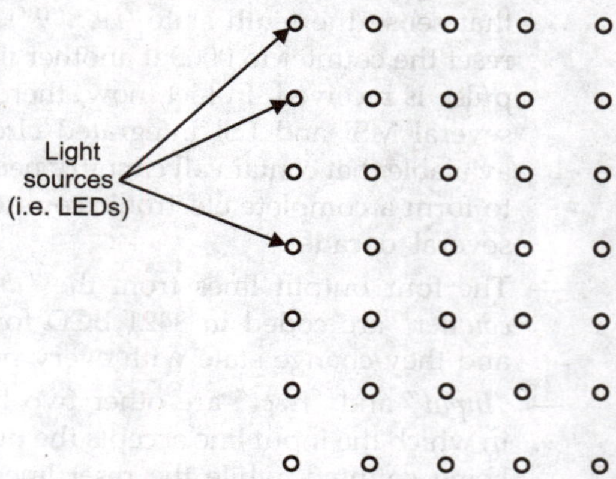

Fig. 10.5. 5 × 7 dot matrix.

Applications of dot matrix method :

(*i*) Athletic score board;

(*ii*) Bank time-temperature billboards;

(*iii*) Cathode ray tube (CRT), computer TV displays etc.

10.3.5. Frequency Counters

Refer to Fig. 10.6 :

- **Fig. 10.6(*a*)** shows the basic block diagram of a **"frequency counter"**. It consists of the following *sections* :

(*i*) Decode counter assembly (DCA); (*ii*) Main gate;

(*iii*) Trigger; (*iv*) Input amplifier;

(*v*) Main gate flip-flop; (*vi*) Time base;

(*vii*) Display clock.

- — The frequency counter measures *"events per unit of time (EPUT)"* (*i.e.,* cycles per second), so the DCA must be tuned on only for a given period of time (*e.g.,* 0.1, 1, or 10 s).
- — The main gate, main gate flip-flop, and time base sections are used *to allow input pulses into the DCA* for the designed period of time.
- — The time base section consists of a crystal *oscillator* that produces pulses at precise rate such as 100 kHz, 1 MHz, 4 MHz, 10 MHz, and so on. A chain of *decode dividers* is used to reduce the crystal oscillator frequency *to a lower frequency.* The time base output frequency will be 10 Hz for 0.1 s, 1 Hz for 1 s and 0.1 Hz for 10 s measuring periods.

- **Fig. 10.6(*b*)** shows the timing diagram for one complete interval of EPUT counter.

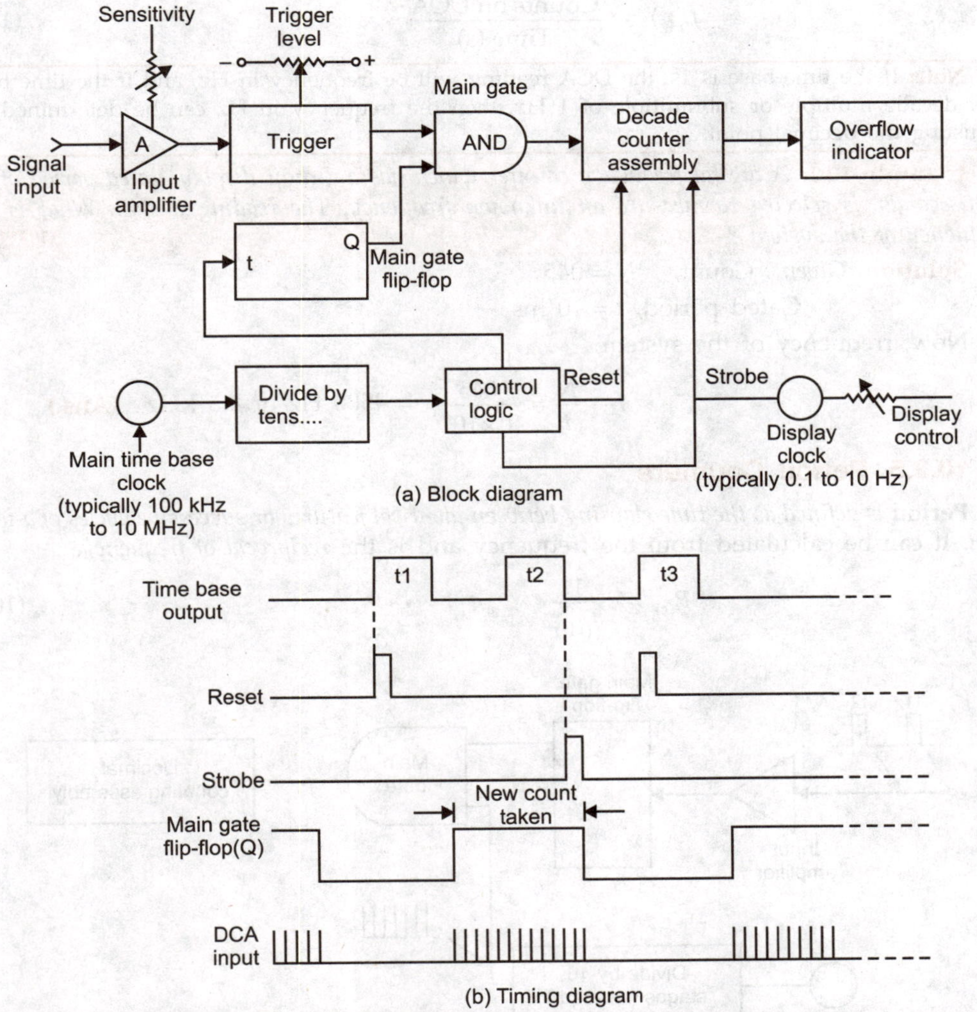

Fig. 10.6. Frequency counter.

— Pulses *t1*, *t2* and *t3* are output from the time base section. When pulse *t1* goes high, the control logic section generates a short pulse to reset the DCA to zero. When *t1* goes low again, the *Q* output of the J-K main gate flip-flop will go high. The main (AND) gate has one input tied to the *Q* output of the flip-flop and the other input is tied to the signal being counted. As a result, the main gate passes input pulses to the DCA *only* when the *Q* terminal of the flip-flop is high.

— Until the negative transition of *t2* occurs, the flip-flop remains set. At that time the *Q* output of the flip-flop drops low, turning off the flow of pulses into the DCA and causing the control logic section to generate a strobe pulse. This pulse tells the DCU latches to transfer data from the counter to the decoders. The display, then, shows *only completed count cycles and will hold the previous count until the end of the next interval.*

The frequency of the input signal is the number of counts accommodated on the DCA divided by the time base period in seconds.

i. e., $$f_{(\text{Hz})} = \frac{\text{Counts on DCA}}{\text{Time (s)}} \qquad ...(10.1)$$

Note: If the time base is 1s, the DCA reading will be frequency in Hz, and if the time base is a decade multiple or submultiple of 1 Hz, then the frequency in Hz can be determined by adjusting the decimal point.

Example 10.1. *A digital frequency counter which has a 3-digit display, gated period of 10 milliseconds, is selected to measure an unknown frequency. The reading is 045. What is the frequency of the system ?* **(UPTU)**

Solution. *Given :* Count, $N = 045$

Gated period, $t = 10$ ms

Now, frequency of the system,

$$f = \frac{N}{t} = \frac{045}{10 \times 10^{-3}} = 4500 \text{ Hz or } \mathbf{4.5\ kHz} \quad \textbf{(Ans.)}$$

10.3.6. Period Counters

Period *is defined as the time elapsing between identical features or successive cycles of a wave form.* It can be calculated from the frequency and is the *reciprocal of frequency.*

$$P_{(s)} = \frac{1}{f_{(\text{Hz})}} \qquad ...(10.2)$$

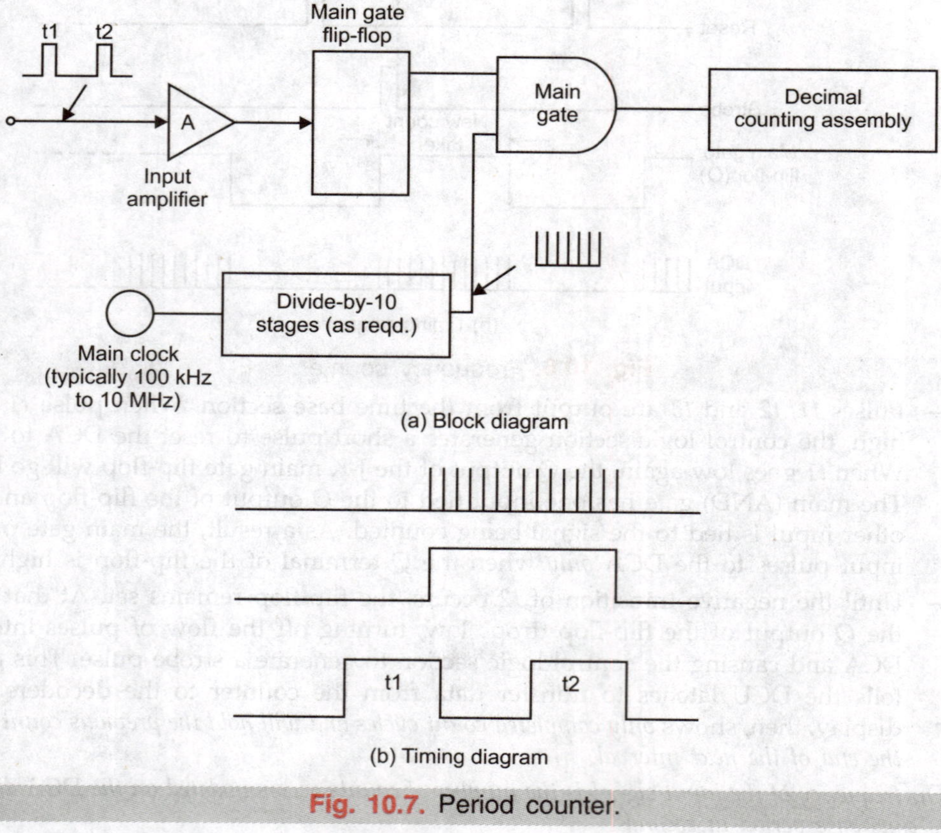

Fig. 10.7. Period counter.

Refer to Fig. 10.7.

- A period counter can be made by *reversing* the roles of the input stage and time base; the input amplifier is connected to the main gate flip-flop, and the time base is connected to the main gate as shown in **Fig. 10.7(*a*).**
- **Fig. 10.7(*b*)** shows the timing diagram for a period counter. The flip-flop is set (*i.e.*, *Q* goes high) on the negative-going transitions of *t1*, and this allows the gate to pass pulses from the time base to DCA.

The time base frequency determines the time interval represented by each pulse. The period is given by the count accumulated by the DCA and the time base frequency :

$$P_{(s)} = \frac{\text{DCA count}}{f_{(\text{Hz})}} \quad ...(10.3)$$

Note: A time base frequency of 1000 Hz yields a period counter resolution of $\frac{1}{1000}$ s or 1 ms; similarly, one 10 kHz time base yields 0.1 ms resolution.

10.3.7. Characteristics of Digital Meters

Following are the few specifications which characterise digital meters :

1. Resolution; 2. Sensitivity; 3. Accuracy.

1. Resolution. *It is defined as the number of digit positions or simply the number of digits used in a meter.*

If number of full digits is *n*, then resolution,

$$R = \frac{1}{10^n} \quad ...(10.4)$$

For $n = 4$, $$R = \frac{1}{10^4} = 0.0001 \text{ or } 0.01\%.$$

- A three-digit display on the digital meter for 0-1 V range will be able to indicate from 000 to 999 mV, with *smallest increment (resolution)* of 1 mV.

2. Sensitivity. *It is the smallest change in input which a digital meter is able to detect.* Thus, it is the full-scale value of the lowest voltage range multiplied by the resolution of the meter. In other words,

$$\text{Sensitivity, } S = (fs)_{min} \times R \quad ...(10.5)$$

where, $(fs)_{min}$ = Lowest full-scale value of digital meter, and

R = Resolution is decimal.

3. Accuracy. In the accuracy specifications of digital meters, the following two quantities are included :

(*i*) A percentage of range.

(*ii*) A percentage of reading.

Following solved examples, illustrate these characteristics :

Example 10.2. *What is the resolution of $3\frac{1}{2}$ digit display on 1 V and 10 V ranges? A $3\frac{1}{2}$ digital voltmeter has an accuracy specification of ± 0.5% of reading ± 1 digit. What is possible error in volt when the instrument is reading 2.00 V on 5-V scale.*

Solution. • Number of full digits on $3\frac{1}{2}$ digit display, $n = 3$

$\therefore$ Resolution, $R = \dfrac{1}{10^n} = \dfrac{1}{10^3} = 0.001$

It means, the meter cannot distinguish between values that differ from each other by less than 0.001 of full scale (*fs*).

For full-scale range of 1 V:

The *resolution* = (*fs*) × *R* = 1 × 0.001 = **0.001 V** **(Ans.)**

For full-scale range of 10 V :

The *resolution* = 10 × 0.001 = **0.01 V** **(Ans.)**

- The display for 2 V reading on 10 V scale of $3\frac{1}{2}$ digital meter would be 0.0200.

The digit in the least significant digit (LSD) has a value of :

$$(fs) \times R = 5 \times \frac{1}{10^3} = 0.005 \text{ V}$$

$$\textit{Total possible error} = \frac{0.5}{100} \times \text{reading} + \text{value of digit in LSD}$$

$$= \frac{0.5}{100} \times 2 + 0.005 = 0.01 + 0.005 = \mathbf{0.015\ V} \quad \textbf{(Ans.)}$$

Example 10.3. *What is the resolution of a $4\frac{1}{2}$ digit display? How would 16.95 V be displayed on a 10 V range and 0.6564 V on a 1 V and 10 V ranges ?*

Solution. • Number of full digits on $4\frac{1}{2}$ digit display, $n = 4$

$\therefore$ *Resolution,* $R = \dfrac{1}{10^n} = \dfrac{1}{10^4} = \mathbf{0.0001}$ **(Ans.)**

- There are 5 digit faces in $4\frac{1}{2}$ digit display, so *16.95 would be displayed as* **16.950** **(Ans.)**

Resolution on 1 V range is 0.000 × 1 = 0.0001 V.

Any reading upto the *4th decimal can be displayed.*

Hence, *0.6564 will be displayed as* **0.6564** **(Ans.)**

Resolution on 10 V range is 0.0001 × 10 = 0.001 V.

Hence, decimals upto *third decimal place can be displayed.*

Hence, *0.6564 will be displayed as* **0.656** **(Ans.)**

Example 10.4. *Explain briefly the $3\frac{1}{2}$-digit display in a digital meter.*

Solution. The number of digit positions in a digital meter determines the resolution. Hence a 3-digit display on digital voltmeter (DVM) for 0–1 V range will indicate values from 0–999 mV with a smallest increment of 1 mV.

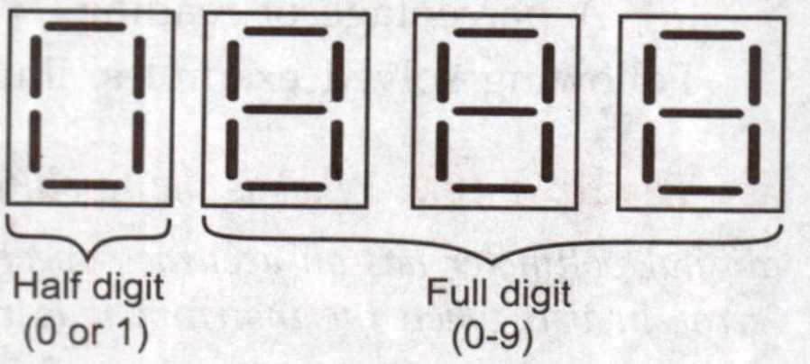

Fig. 10.8. 3½-digit display.

Normally, a *fourth* digit capable of indicating 0 or 1 (hence called a *half digit*) is placed to the left. This permits the digital meter to read values above

999 upto 1999, to give overlap between ranges for convenience, a process called *"over-ranging"*. This type of display is called a $3\frac{1}{2}$ *digit display*, shown in Fig. 10.8.

10.3.8. Analog-to-Digital (A/D) and Digital-to-Analog Conversion

Refer to Article 9.4

10.4 DIGITAL VOLTMETERS (DVMs)

10.4.1. Introduction

A **digital voltmeter** *(DVM) displays measurements of D.C. or A.C. voltage as discrete numerals* instead of pointer deflections on a continuous scale as in analog instruments.

- It is a *versatile and accurate instrument* that is employed in many laboratory measurement applications.
- Owing to the development and perfection of IC modules, the size, power requirements and cost of the digital voltmeter have been drastically reduced and, therefore, DVMs can actively compete with conventional analog instruments, both in price and portability.
- The various varieties of DVM differ in the following ways :
  (*i*) Number of digits; (*ii*) Number of measurements;
  (*iii*) Accuracy; (*iv*) Speed of reading;
  (*v*) Digital output of several types.

10.4.2. Advantages of DVMs

Following are the *advantages of DVM over other voltmeters* :

1. Higher accuracy and resolution.
2. Greater speed.
3. No parallax.
4. Reduced human error.
5. Compatibility with other digital equipment for further processing and recording.

10.4.3. Characteristic Features of DVMs

The characteristic features of digital voltmeters (DVMs) are :

1. *Input range* : From ± 1.000 V to ± 1000 V with automatic range selection and overload indication.
2. *Absolute accuracy* : As high as ± 0.005% of the reading.
3. *Resolution* : 1 part in million (1 μV reading can be read or measured on 1 V range)
4. *Stability* : Short-term, 0.002% of the reading for a 24-hr period; long term, 0.008% of the reading for a 6-month period.
5. *Input resistance* : Typically 10 MΩ;
   Input capacitance : 40 pF
6. *Calibration* : Internally from stabilised reference sources, independent of measuring circuit.
7. *Output signals* : In BCD (Binary-coded-decimal) form, for print output and further digital processing.

- *Optional features* may include additional circuitry to measure current, ohms and voltage ratio.

10.4.4. Application of DVMs

DVMs are often used in *"data processing systems"* or *"data logging systems"*.

— In such systems, a number of analog input signals are scanned sequentially by an electronic system and then each signal is *converted to an equivalent digital value by the A/D (analog-to-digital) converter in the DVM.*

— The digital value is then transmitted to a printer along with the information about the input line from which the signal has been derived.

— The whole data is then printed out.

In this way, *a large number of input signals can be automatically scanned or processed and their values either printed or logged.*

10.4.10. Classification of Digital Voltmeters (DVMs)

The digital voltmeters can broadly be *classified* as follows :

1. Ramp-type DVM.
2. Dual-slope integrating type DVM.
3. Integrating type DVM (Voltage-to-frequency conversion).
4. Successive–approximation DVM.

10.4.6. Ramp-type DVM

The *principle of operation* of the ramp-type DVM is based *on the measurement of the time it takes for a linear ramp voltage to rise from 0 V to the level of input voltage, or decrease from the level of the input voltage to zero.* This interval of time is measured with an *electronic time-interval counter*, and the count is displayed as a *number of digits* on electronic indicating tubes.

- *Fig. 10.9* shows a ramp-type DVM.

Fig. 10.9(*a*) shows the *"voltage-to-time conversion"* using gated clock pulses.

— At the start of the measuring cycle, a ramp voltage is initiated; this voltage can be positive-going or negative going. The negative-going ramp, shown in Fig. 10.9(*a*) is continuously compared with the unknown input-voltage.

— At the instant that the ramp voltage equals the unknown voltage, a coincidence circuit, or comparator, generates a pulse which opens a gate **[See Fig. 10.9(*b*)]**. The ramp voltage continues to decrease with time until it finally reaches 0 V (or ground potential) and a second comparator generates an output pulse which closes the gate.

— An oscillator generates clock pulses which are allowed to pass through the gate to a number of decade counting units (DCUs) which totalise the number of pulses passed through the gate.

— The *decimal number,* displayed by the indicator tubes associated with the DCUs, *is a measure of the magnitude of the input voltage.*

— The sample-rate multivibrator (MV) determines the rate at which the measurement cycles are initiated. The sample-rate circuit provides an initiating pulse for the ramp generator to start its next ramp voltage. At the same time, a reset pulse is generated which returns all the DCUs to their *zero state,* removing the display momentarily from the indicator tubes.

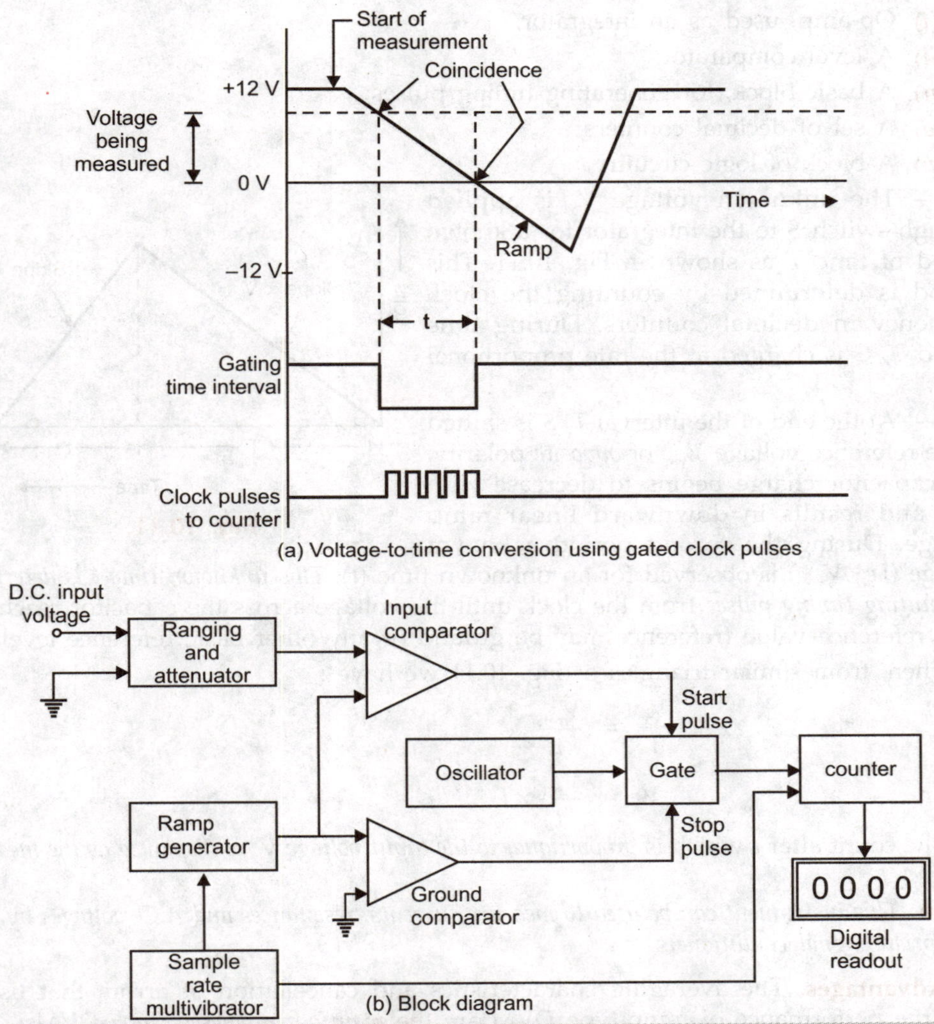

Fig. 10.9. Ramp-type DVM.

10.4.7. Dual-slope Integrating type (DVM)

Fig. 10.10 shows the block diagram of a dual-slope integrating type DVM. It consists of the following *five* basic blocks:

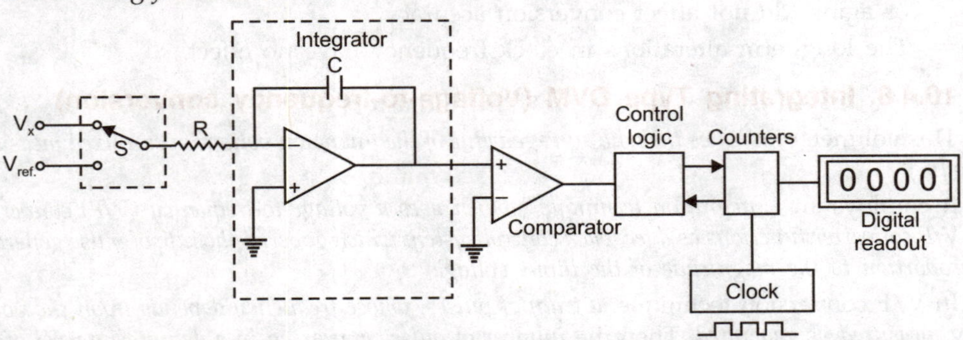

Fig. 10.10. Block diagram of a dual-slope integrating type DVM.

(i) Op-amp, used as an integrator;

(ii) A level comparator;

(iii) A basic block, for generating timing pulses;

(iv) A set of decimal counters;

(v) A block of logic circuitry.

— The unknown voltage V_x is applied through switch S to the integrator for a known period of time T as shown in Fig. 10.11. This period is determined by counting the clock frequency in decimal counters. During time period T, C is charged at the rate proportional to V_x.

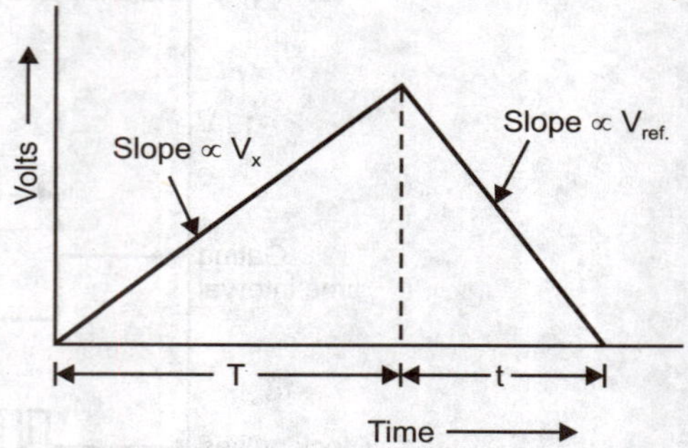

Fig. 10.11.

— At the end of the interval T, S is shifted to the reference voltage $V_{ref.}$ of *opposite* polarity. The capacitor charge begins to decrease with time and results in downward linear ramp voltage. During the second period a known voltage (*i.e.*, $V_{ref.}$) is observed for an unknown time (t). *This unknown time t is determined by counting timing pulses* from the clock until the voltage across the capacitor reaches its basic reference value (reference may be ground or any other basic reference level).

Then, from similar triangles of Fig. 10.11 we have :

$$\frac{V_x}{T} = \frac{V_{ref.}}{t}$$

or,
$$V_x = \frac{V_{ref.}}{t} \times T \qquad ...(10.5)$$

The count after t which is *proportional to the input voltage V_x is displayed as the measured voltage.*

- *This instrument can be used to measure currents, resistances and A.C. voltages by using appropriate signal conditioners.*

Advantages. The averaging characteristics and cancellation of errors that usually limit the performance of ramp-type DVM are the *main advantages of such DVMs:*

— The integration characteristics provide the average value of the input signal during the period of first integration. Consequently, disturbance, such as spurious noise pulses, are *minimised.*

— Long-term drifts in the time constant, as may result from temperature variations or aging, do not affect conversion accuracy.

— The long-term alterations in clock frequency have no effect.

10.4.8. Integrating Type DVM (Voltage-to-frequency conversion)

This voltmeter *indicates the true average value of the unknown voltage over a fixed measuring period.*

It employs an *"integration technique" which uses a voltage-to-frequency (V/F) conversion. The V/F converter functions as a feedback control system which governs the rate of pulse generation in proportion to the magnitude of the input voltage.*

In V/F conversion technique, *a train of pulses, whose frequency depends upon the voltage being measured, is generated.* Then the *number of pulses appearing in a definite interval of time is measured.*

The basic requirement of this technique is *"operational integrator"*, the circuit diagram and symbolic representation of which is shown in [Fig. 10.12(*a*, *b*)], it is a modification of operational amplifier. The operational integrator consists of an amplifier (having high gain) with a capacitor C connected between output and input. If V_i is the input voltage, then V_o, of the operational integrator is given by :

$$V_o = \frac{1}{RC}\int V_i \cdot dt \qquad \text{...(10.6)}$$

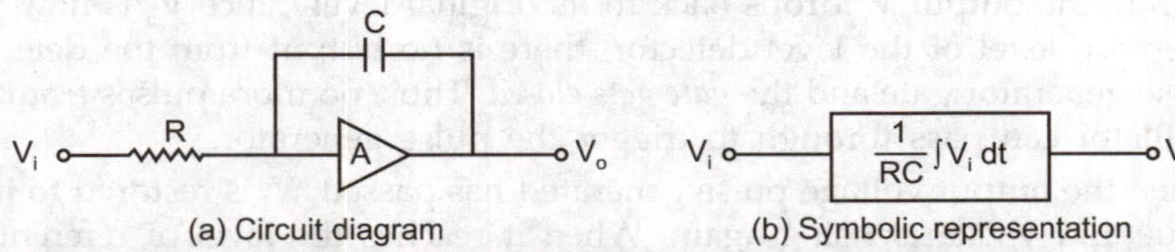

Fig. 10.12. Operational integrator.

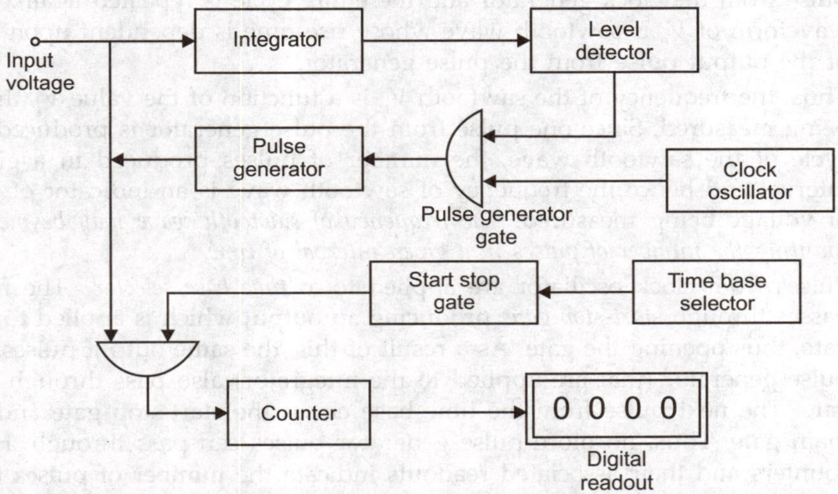

Fig. 10.13. Block diagram of an integrating type DVM (Voltage-to-frequency conversion).

If the input voltage is constant, we have :

$$V_o = -\frac{V_i}{RC} \cdot t \qquad \text{...(10.7)}$$

Thus, if a constant input voltage V_i is applied, an output voltage V_o is produced which rises at a uniform rate and has a polarity *opposite* to that of input voltage. That is, for a constant input voltage the integrator produces a ramp output voltage of opposite polarity such that the greater the input voltage, the sharper is the rate of rise, or step of the output voltage.

Fig. 10.13. shows the *block diagram* of this voltmeter :

— The unknown voltage, V_i is applied to the input of the integrator and the output voltage V_o starts to rise. *The slope of V_o is determined by the value of input voltage V_i.* This voltage is fed to a level detector, and, when V_o reaches a certain reference level, the *detector* sends a pulse to *pulse generator gate*. The level detector is a device similar to a voltage comparator.

— The output voltage from integrator, V_o is compared with the fixed voltage of an internal reference source and when V_o reaches that level, the level detector produces an *output pulse.* This output pulse opens the pulse level gate, permitting pulse from a fixed frequency clock oscillator to pass through the *pulse generator*. This generator is similar to schmitt trigger that *produces an output pulse of fixed amplitude and width for every pulse it receives.*

— The output pulse whose polarity is *opposite* to that of V_i and has greater amplitude is fed back to the input of the integrator. Hence, the net output to the integrator is now of reversed polarity (opposite to that of V_i). As a result of this reversed output, the output V_o drops back to its original level. Since V_o is now below the reference level of the level detector, there is no output from the detector to the pulse generator gate and the *gate gets closed.* Thus, no more pulses from the clock oscillator can pass through to trigger the pulse generator.

— When the output voltage pulse generated has passed, V_i is restored to its original value and V_o starts rising again. When it reaches the level of reference voltage again, the pulse generator gate is opened. The pulse generator is triggered by a pulse from the clock generator and the entire cycle is repeated again. Thus, the waveform of V_o is sawtooth wave whose rise time is dependent upon the value of the output pulse from the pulse generator.

— Thus, the frequency of the sawtooth V_n is a function of the value V_i, the voltage being measured. Since one pulse from the pulse generator is produced for each cycle of the sawtooth wave, the number of pulses produced in a given time interval and hence the frequency of sawtooth wave is an indicator of the value of voltage being measured. *The frequency of sawtooth wave may be measured by counting the number of pulses in a given interval of time.*

— Pulses from clock oscillator are applied to a *'time base selector'*. The first pulse passes through *start-stop gate,* producing an output which is applied to the main gate, thus opening the gate. As a result of this, the same output pulses from the pulse generator (that are applied to the integrator) also pass through the main gate. The next pulse from the time base closes the start-stop gate and also the main gate. Thus, no more pulse generator pulses can pass through. Hence the counters and their associated readouts indicate the number of pulses that have passed during a known interval of time. This count is an indication of the voltage being measured. *In order to make the counter read directly in terms of voltage the amplitude and width of pulse generator can be suitably adjusted.*

- The drawback of this DVM is that it *needs excellent characteristics in the linearity of the ramp.* The A.C. noise and supply noise are averaged out.

10.4.9. Successive-approximation DVM

Fig. 10.14, shows the block diagram of a successive-approximation DVM, in which the electronic implementation of the successive-approximation technique is illustrated :

— A *"digital-to-analog (D/A) converter"* is used to provide the estimates. The "equal to or greater than" or "less than" decision is made by the *"comparator"*. The D/A converter provides the estimate and is compared to the input signal.

— A special shift register called a *"successive-approximation register (SAR)"* is used to control the D/A converter and consequently the estimates. At the *beginning* of the conversion all the outputs from the SAR are at *logic zero.* If the estimate is *greater than the input*, the comparator output is *high* and the first SAR output reverses state and the second output changes to a logic "one". If the comparator output is *low,* indicating that the estimate is *lower than the input signal* the first-output remains in the logic one state and the second output assumes the logic one state. *This continues to all the states until the conversion is complete.*

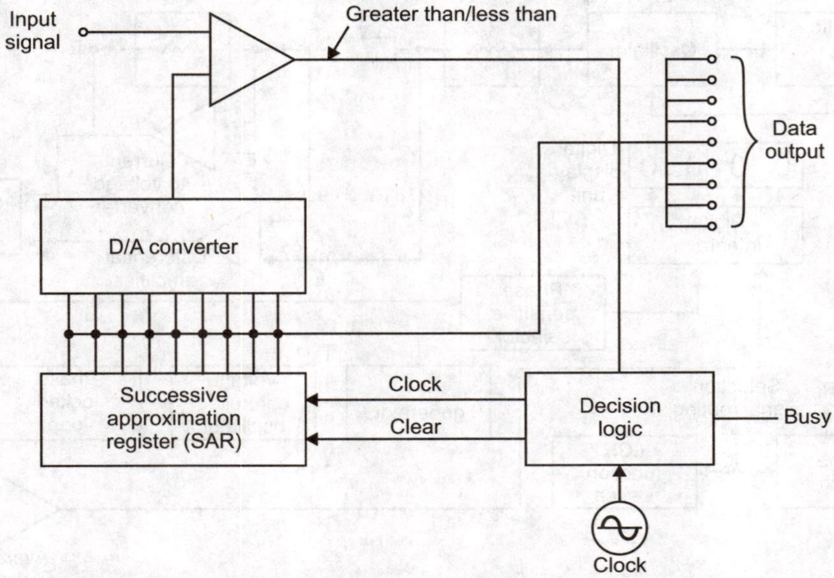

Fig. 10.14. Block diagram of a successive-approximation DVM.

— The above sequence of events are performed electronically. *For an N-bit conversion after N clocks, the actual value of the input is known.* The least significant bit is the state of the comparator. In some systems an additional clock is employed to store the last bit in the SAR and thus N + 1 clocks are required for a conversion.

10.5 DIGITAL LCR METER

This type of meter is used *to measure the resistance, inductance, capacitance and dissipation factor*. The desired function can be selected by using a *rotary switch*. The various ranges available are :

(*i*) 200 μH/pF/Ω; (*ii*) 2000 μH/pF/Ω; (*iii*) 20 mH/nF/kΩ; (*iv*) 200 mH/nF/kΩ; (*v*) 2H/μF/MΩ.

With the help of this meter, the following ranges of various measurements can be made :

- *Resistance :* From 200 Ω to 20 MΩ;
- *Inductance :* From 2000 μH to 2H;
- *Capacitance :* From 2000 pF to 2 μF.

Fig. 10.110. shows the block diagram of the digital LCR meter :

In this meter, the *basic principle used is to measure the voltage across the component and the current passing through the component under test*, when the test signal is *fed to the component*. These processed voltage and current signals (from the component under test) are then fed to the *digital integrator unit* which finally enables a display unit to provide directly the value of the test component.

— The oscillator (in built) applies the measuring test-signal to the test component, through a source resistor R_s (selectable). The typical test frequency is 1 kHz (for the above mentioned five ranges).

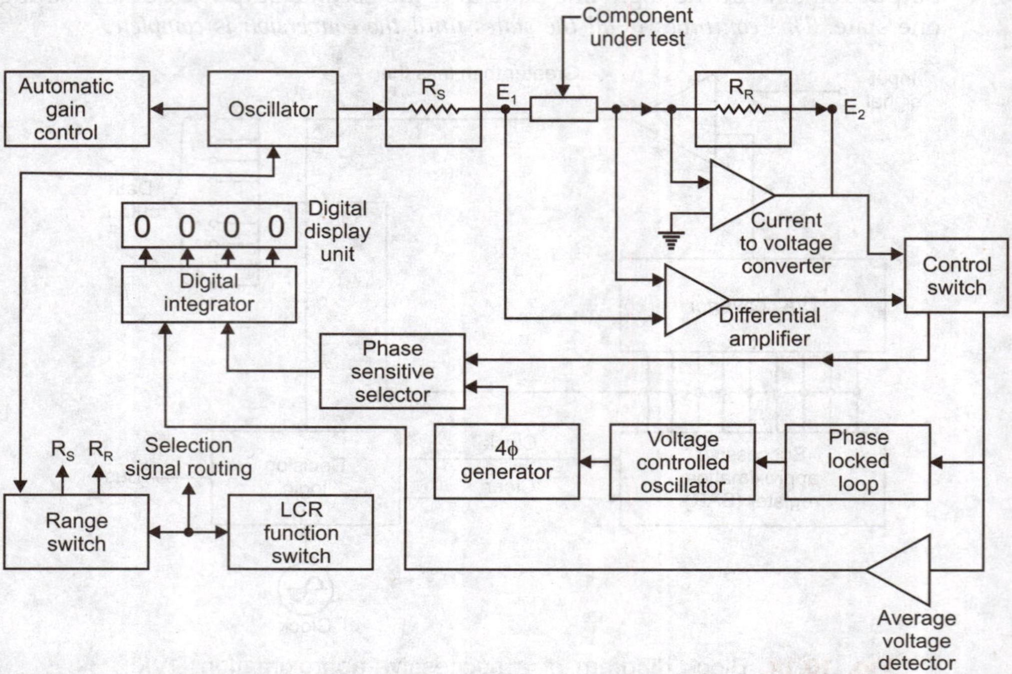

Fig. 10.15. Block diagram of a digital LCR meter.

— The signal current then flows to current-to-voltage converter, which is essentially an op-amp with range resistor R_R, connected in the feedback path.

— The op-amp drives the junction of component and R_R to a virtual ground and hence R_R does not change the current through the component. Thus, *signal current develops a voltage E_2 proportional to the current through the component.*

The voltage E_1 and E_2 are the vector quantities and as such they define the characteristics of the component at a particular test frequency and signal level.

Mathematically : $E_1 \propto V$ and $E_2 \propto I$

Thus, *Capacitance,* $C \propto \frac{I}{V} \propto \frac{E_2}{E_1}$; *Inductance,* $L \propto \frac{V}{I} \propto \frac{E_1}{E_2}$.

The above ratios are adopted in measurement modes and are obtained by dual-slope integration method. For *inductance* measurements, a *series* equivalent circuit of an inductor is assumed while for capacitance measurements, a parallel equivalent circuit of capacitor is assumed for five ranges.

— The values of R_S and R_R are selected, based on the impedance of the unknown component. For the measurement of *inductance,* the component impedance is usually low and hence R_S is chosen much higher. This achieves a *constant current drive* to the component. The R_S decides the value of current. For *capacitance* measurement, the component impedance is high and R_S is chosen to be much low. Thus, *constant voltage drive* is provided to the test component.

— The voltage E_1 is applied as one input to *'differential amplifier'*; E_1 is then fed to a *control switch,* along with E_2. The greater of the two is fed to *'average voltage detector (AVD)'* and lesser to the *'phase sensitive detector (PSD)'*.

— The signal given to AVD is also given to *'phase locked loop (PLL)'* and voltage controlled oscillator (VCD). These two produce the clock signals, which are locked in phase with the reference signal. The clock signal is then divided into phase shifted 90° and 270° signals. These are given to PSD, which detects the phase angle.

— The D.C. voltage outputs, from AVD and PSD are then given to *'digital integrator unit'* and finally to *'digital display unit'* where the component value is displayed.

- For wide range of values, the basic accuracy of this meter is ± 0.15 percent of the reading.

  — The range selection is fully automatic.

  — The display used is normally 4 digit LEDs with automatic decimal point.

  — The inductors and capacitors are automatically differentiated.

Commercial digital LCR meter :

Important features available in commercial digital LCR meter are :

(*i*) Auto-ranging and auto-computing facility.

(*ii*) High basic accuracy, upto 0.2 percent.

(*iii*) High frequency range, from 0.1 to 1 kHz.

(*iv*) Hold function.

(*v*) Multipoint measurement techniques.

(*vi*) Series or parallel modes of operation.

- Most of these meters are *"microprocessor controlled"* and multipoint measurement technique ensures better accuracies in measurement. Apart from the basic measurement of LCR and *Q* they can also be used for series and parallel measurement *i.e.,* they can measure series and parallel components of *L* and *C*.

— The microprocessor configuration provides operation flexibility *i.e.,* comparison with previous recorded values, storage of these values for future references (memory function).

— The user software provides user prompt facility by means of flashing LED for respective functions and to select appropriate, test frequencies to achieve optimum results.

— The *'Hold function'* is useful for selecting the components within a specified range.

10.6 DIGITAL MULTIMETER (DMM)

Digital multimeter offers *increased versatility due to its additional capability to measure A.C. voltage and current, D.C. current, and resistance.*

Fig. 10.16, shows the block diagram of a digital multimeter (DMM) :

- In the *"A.C. voltage mode"*, the applied input is fed through a calibrated, compensated attenuator, to a precision full-wave rectifier circuit followed by a ripple reduction filter. The resulting D.C. is fed to ADC and the subsequent display system.
- For *current measurements,* the drop across an internal calibrated shunt is measured, directly by the ADC in the *"D.C. current mode"*, and after A.C. to D.C. conversion in the *"A.C. current mode"*. This drop is often in the range of 200 mV (corresponding to full scale).

  — Due to the lack of precision in the A.C.–D.C. conversions, the accuracy in the A.C. range is generally of the order of 0.2 to 0.5%. In addition, the measurement range is often limited to about 50 Hz at the lower frequency end due to the ripple in the rectified signal becoming a non-negligible percentage of the display and hence results in fluctuation of the displayed number. At the higher frequency end, deterioration of the performance of the AC/DC converter limits the accuracy.

— The A.C. measurement, is *often average reading, r.m.s. calibrated.*

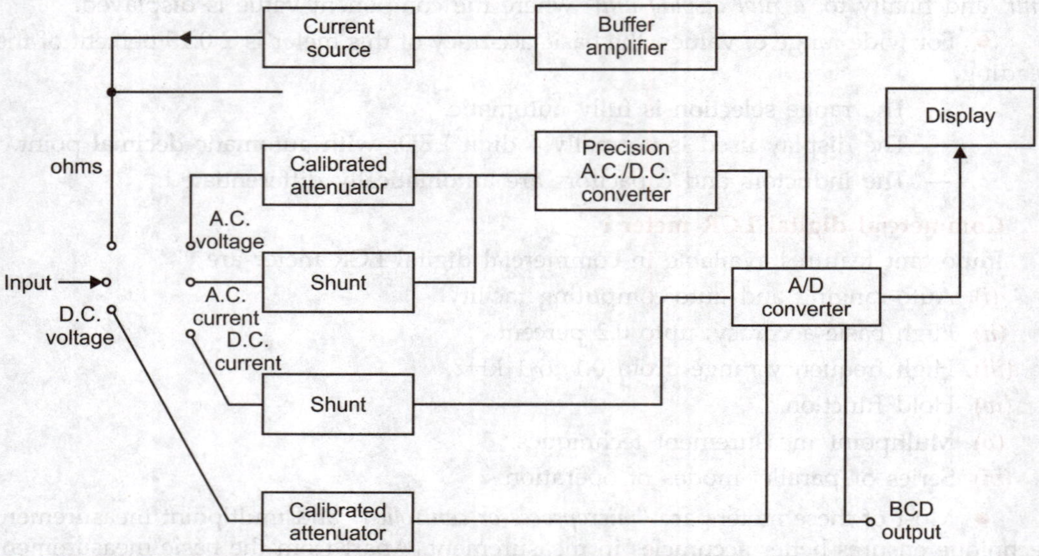

Fig. 10.16. Block diagram of a digital multimeter.

- In the *"resistance range"* the digital multimeter operates by measuring the voltage across the externally connected resistance, resulting from a current forced through it from a calibrated internal current source.

— The accuracy of resistance measurement is of the order of 0.1 to 0.5% depending on the accuracy and stability of the internal current sources. The accuracy may be *proper* in the *highest range* which is often about 10 to 20 MΩ. In the *lowest range*, the full scale may be $\simeq$ 200 Ω with a resolution of about 0.01 Ω for a $4\frac{1}{2}$ digit digital multimeter. In this range of resistance measurement, the effect of the lead resistance will have to be carefully considered.

Comparison between "Analog" and "Digital" Multimeters

S. No.	Aspects	Analog multimeters	Digital multimeters
1.	*Power supply*	Not required	Required
2.	*Visual indication of changes in the reading*	Better	Not that much better
3.	*Effect of electric noise*	Less	More
4.	*Isolation problems*	Less	More
5.	*Accuracy*	Less	High
6.	*Interface of the output with external equipment*	Not possible	Possible

7.	*Construction*	Simple	Complicated
8.	*Size*	Bigger	Smaller
9.	*Cost*	Less	More
10.	*Nature of output*	Ambiguous–many times	Unambiguous reading

10.7 MICROPROCESSOR-BASED INSTRUMENTS-INTELLIGENT INSTRUMENTS

"Intelligent instruments", a new class of instruments, involve the *use of microprocessors (μP) as an integral part of the measuring instruments.*

Fig. 10.17. shows a block diagram of *microprocessor based impedance measuring instrument* :

— The operation makes interface with the instrument *via* the IEEE 488 bus to allow control by, or to make the measurement available to a large external computer system.

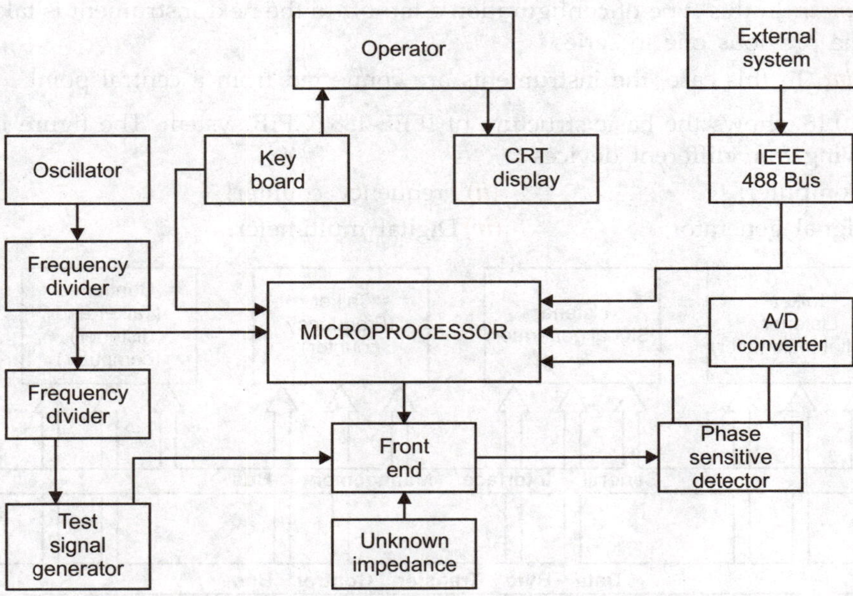

Fig. 10.17. Block diagram of microprocessor based impedance measuring instrument.

— The *frequency division of the oscillator signal* provides the timing clock signal and the A.C. test signal.

— The *front end circuit* applies the test signal to the unknown impedance and a standard impedance provides an output signal, proportional to the voltage across each, to the *phase sensitive detector* (PSD).

— PSD, which is also controlled by the microprocessor, converts the A.C. inputs of the impedance in vector form to a D.C. output.

— The A/D converter provides the digital data, which is used by the microprocessor to compute the value of the unknown impedance. This is then displayed on the CRT or sent as output to the IEEE 488 bus.

10.8 IEEE-488 GENERAL PURPOSE INTERFACE BUS (GPIB) INSTRUMENTS

Now-a-days automatic test equipment is one of the leading methods for testing electronic equipment in factory production and troubleshooting situations. *The basic method is to use a programmable digital computer to control a bank of test instruments.* The bank of instruments can be configured for a special purpose or for general use.

The IEEE-488 bus or General Purpose Interface Bus (GDIB) is a tool that is based on the IEEE (Institution of Electrical and Electronics Engineers) specifications.

The digital signals on IEEE-488 bus are generally similar to TTL (transistor-transistor logic), *i.e.*, a logic LOW is less than 0.8 V and a logic HIGH is greater than 2.0 V.

The digital signals can be connected to the instruments through a multiconductor cable upto 20 metres is length provided that an instrument load is placed every 2 metres.

Most IEEE-488 bus/GPIB systems operate unrestricted to 250 bilobytes per second or faster with some restrictions.

There are *two basic configurations* for the IEEE-488/GPIB system :

1. *Linear.* In this type of configuration a tap-off to the next instrument is taken from the previous one in series.
2. *Star.* In this case, the instruments are connected from a central point.

Fig. 10.18. shows the basic structure of IEEE-488/GPIB system. The figure indicates the following four different devices :

(i) Computer; *(ii)* Frequency counter;
(iii) Signal generator; *(iv)* Digital multimeter.

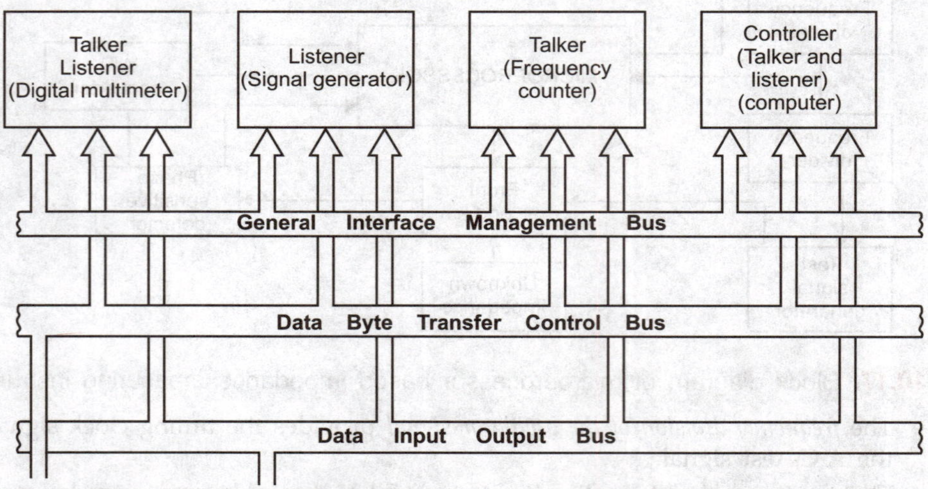

Fig. 10.18. Basic structure of IEEE-488/GPIB system.

The IEEE-488/GPIB system itself consists of *three major buses* :

(i) General Interface Management (GIM) bus. This bus co-ordinates the whole system and ensures an orderly flow of data over the data input/output (DIO) bus.

(ii) Data I/O (DIO) bus. This bus is a bi-directional 8-bit data bus that carries data, interface messages and device-dependent messages between the *controllers, talkers* and *listeners*. This bus sends asynchronously in byte-serial format.

(iii) *Data Byte Transfer (DBT) bus.* This bus controls the sending of data along the DIO bus.

- The signals defined for the above three buses are implemented as conductors in a system interface cable.
- Each IEEE-488/GPIB system compatible instrument will have a female 36-pin Amphenol-style connector on the rear panel.
- The devices connected to IEEE-488 GPIB system (*i.e.*, computer, frequency counter, signal generator and digital multimeter) are categorised as : *Controller; Listener and/or Talker.*

(i) ***Controller :***

— Its function is to communicate device addresses and other inface buses to instrument in the system.

(ii) ***Listener :***

— Its function is to receive commands from other instruments (usually the controller) when the correct address is placed on the base.

— The listener acts on the message received but does not send back any data to the controller.

(iii) ***Talker :***

— Its function is to respond to the message sent to it by the controller.

- ***V × I bus*** is another widely growing platform for instrumentation systems. It uses a mainframe chasis with a maximum of 13 slots to hold *modular instruments on plug-in boards.*

— The *V × I* backplane includes the 32-bit VME data bus, high performance instrumentation, buses for precision timing and synchronisation between instrument components, standardised initialisation and resource management to ease configuration.

HIGHLIGHTS

1. A *digital instrument* can be considered as a counter which counts the pulses in a predetermined time.
2. The digital meters work on the principle of *"quantization"*,
3. Digital voltmeters (DVM) are broadly classified as :
   (*i*) Ramp-type DVM;
   (*ii*) Dual-slope integrating type DVM;
   (*iii*) Integrating type DVM (Voltage-to-frequency);
   (*iv*) Successive-approximation DVM.
4. Intelligent instruments involve the use of microprocessor as an integral part of the measuring instruments.

OBJECTIVE TYPE QUESTIONS

Choose the Correct Answer :

1. The D' Arsonval meter movement can be converted into an audio-frequency A.C. ammeter by adding a .... to it.
   (*a*) thermocouple (*b*) rectifier
   (*c*) chopper (*d*) transducer.

2. In a linear meter, half-scale deflection occurs when there is .... per cent of the rated current through its coil

(*a*) 100 (*b*) 25
(*c*) 50 (*d*) 75.

3. It is desired to convert a 0–1000 μA meter movement, with an internal resistance of 100 Ω into a 0–100 mA meter. The required value of shunt resistance is about

(*a*) 1 Ω (*b*) 10 Ω
(*c*) 99 Ω (*d*) 100 Ω.

4. Loading effect is principally caused by .... instruments.

(*a*) high resistance (*b*) low-sensitivity
(*c*) high-sensitivity (*d*) high-range.

5. A multimeter is used to measure

(*a*) resistance (*b*) current
(*c*) voltage (*d*) all of the above.

6. Digital instruments are those which

(*a*) have numerical readout (*b*) use LED or LCD displays
(*c*) have a circuitry of digital design (*d*) use deflection type meter movement.

7. The main difference between the electronic and electrical instruments is that an electronic instrument contains.

(*a*) an electronic device (*b*) a transducer
(*c*) a digital readout (*d*) electrons.

8. A sinusoidal voltage of rms value 10 V is applied to a D'Arsonval movement connected in series with a half-wave rectifier. It will show a reading of .... volt.

(*a*) 9 (*b*) 4.5
(*c*) 10 (*d*) 7.7

9. A VTVM produces negligible loading effect on a circuit under test primarily because

(*a*) it virtually draws no current from the circuit
(*b*) of its very high internal resistance
(*c*) it uses high vacuum tubes
(*d*) it is a null deflection instrument.

10. In a 3½ digit voltmeter, the largest number that can be read is

(*a*) 0999 (*b*) 1999
(*c*) 4999 (*d*) 9999.

11. A 3½ digit voltmeter having a resolution of 100 mV can be used to measure maximum voltage of

(*a*) 100 V (*b*) 200 V
(*c*) 1000 V (*d*) 5000 V.

12. The essential elements of an electronic instrument are

(*a*) transducer (*b*) signal conditioner
(*c*) indicating device (*d*) all of the above.

13. The current sensitivity of a meter is expressed in

(*a*) ampere (*b*) ohm/ampere
(*c*) ohm/volt (*d*) ampere/division.

14. The basic meter movement can be converted into an ohmmeter by connecting a .... with it.

(*a*) high resistance in series (*b*) low resistance in parallel
(*c*) battery in series (*d*) battery and a variable resistance in series.

15. A 0–1 mA meter has a sensitivity of

(*a*) 1 kΩ/V (*b*) 1 mA

(*c*) 1 kΩ (*d*) 1000 A.

16. A moving coil instrument has a resistance of 10 Ω and takes 40 mA to produce full-scale deflection. The shunt resistance required to convert this instrument for use as an ammeter of range 0 to 2A is

(*a*) 0.1021 Ω (*b*) 0.2041 Ω

(*c*) 0.2561 Ω (*d*) 0.4210 Ω.

17. A moving coil ammeter has a fixed shunt of 0.02 ohm resistance. If the coil resistance of the meter is 1000 Ω, a potential difference of 500 mV is required across it for full-scale deflection. Under this condition, the current in the shunt would be

(*a*) 2.5 A (*b*) 2.5 A

(*c*) 0.25 A (*d*) 0.25 A.

ANSWERS

1. (*b*)	**2.** (*c*)	**3.** (*a*)	**4.** (*b*)	**10.** (*d*)	**6.** (*c*)	**7.** (*a*)
8. (*b*)	**9.** (*b*)	**10.** (*b*)	**11.** (*c*)	**12.** (*d*)	**13.** (*c*)	**14.** (*d*)
15. (*a*)	**16.** (*b*)	**17.** (*b*)				

THEORETICAL QUESTIONS

1. What is a digital instrument? Explain.
2. Explain briefly the principle on which the digital meters work.
3. What are the building blocks of a digital instrument? Explain briefly.
4. How does a digital system differ from an analog system? Explain.
5. What are the advantages of digital systems/techniques?
6. Enumerate the components of a digital system.
7. What are the advantages of digital implementation?
8. Explain briefly the following :
   (*i*) Binary counters;
   (*ii*) Decimal counting units (DCUs);
   (*iii*) Display devices.
9. Explain with a block diagram the working of a 'frequency counter'.
10. What do you mean by the term 'Period' ? How is it measured by a 'Period counter'?
11. Explain briefly the following characteristics of digital meters :
    (*i*) Resolution;
    (*ii*) Sensitivity;
    (*iii*) Accuracy.
12. What is the function of a digital voltmeter?
13. State the advantages of digital voltmeters.
14. State the characteristic features of DVMs.
15. List the applications of DVMs.
16. Give the classification of digital voltmeters.
17. Explain briefly with neat diagrams the working of any two of the following instruments :
    (*i*) Ramp-type DVM;
    (*ii*) Dual-slope integrating type DVM;
    (*iii*) Integrating type DVM (Voltage-to-frequency conversion).

18. With the help of a block diagram, describe briefly the working of a successive-approximation digital voltmeter.
19. What is the function of a digital LCR meter? Explain its working with the help of a neat block diagram.
20. State the important features available in a commercial digital LCR meter.
21. Describe briefly a "digital multimeter" with the help of a block diagram.
22. Give the comparison between analog and digital multimeters.
23. Give the description of a microprocessor based impedance measuring instruments.
24. Discuss briefly the working of IEEE-488 bus system.

CHAPTER

11 Cathode Ray Oscilloscope (CRO)

11.1 INTRODUCTION

Cathode ray oscilloscope *(CRO) is a device that allows the amplitude of electrical signals, whether they be voltage, current, power, etc., to be displayed primarily as a function of time.*

It is a very useful and versatile laboratory instrument used for display, measurement and analysis of waveforms and other phenomena in electrical and electronic circuits.

— It is basically an *electron beam voltmeter*.

- The oscilloscope depends on the movement of an *electron beam* to impinge on a phosphor surface, which produces a *visible spot*. If the electron beam is deflected in either of two orthogonal axes, such as the familiar X and Y axes used in conventional graph construction, the luminous spot can be used to create *two-dimensional displays*. Typically, the X-axis of the oscilloscope is deflected at a constant rate, relative to time, and the vertical or Y-axis is deflected in response to an input stimulus such as a voltage. This produces the time-dependence variation of the input voltage, which is very important to the design and development of electronic circuits.
- The oscilloscope is *capable of displaying events that take place over periods of microseconds and nanoseconds*.
- CROs *operate on voltages*. However it is possible to convert current, strain, acceleration, pressure and other physical quantities into voltages with the help of transducers and thus to present visual representations of a wide variety of dynamic phenomena on CROs.

These oscilloscopes have been evolved continuously, and they are now available which can measure frequencies upto 1GHz and observe events as small as 20 Hz in duration.

- With some oscilloscopes several additional features, such as digital multimeters and counters, are available. Several oscilloscopes are *microprocessor controlled*. They have the ability to calculate several features, such as rise time or pulse width of the measured waveform, and to display these values along with the display of waveforms. They are *easier to use and internal routines often act as a guide for the user, and display a warning if there is any error in setting*.

Although, most oscilloscopes are monochromatic, *colour oscilloscopes* are finding increasing applications in computers and in television.

- Most present day oscilloscopes are *capable of accepting two or more inputs displaying them simultaneously*. This may be achieved through using a *split beam* or *by using a multiple beam tube*.

11.2 APPLICATIONS OF CRO

There are variety of applications in which CRO is used. Some of these applications are:

1. Tracing of an actual waveform of current or voltage.
2. Determination of amplitude of a variable quantity.
3. Comparison of phase and frequency.
4. Measurement of capacitance and inductance.
5. In televisions.
6. In radar.
7. For finding B-H curves for hysteresis loop.
8. For engine pressure analysis.
9. For studying the heart beats, nervous reactions etc.
10. For tracing transistor curves.
11. For determining the modulation characteristics and to detect the standing waves in transmission lines.
12. It can be used to check the diodes and the faulty components in the various circuits.

11.3 BLOCK DIAGRAM OF AN OSCILLOSCOPE

Fig. 11.1. shows the block diagram of general-purpose oscilloscope. Its consists of the following major systems :

1. ***Cathode ray tube :***

— It generates the electron beam, accelerates the beam to a high velocity, deflects the beam to create the image, and contains a phosphor screen where the electron beam eventually becomes visible.

— It displays the quantity being measured.

2. ***Vertical amplifier.*** It amplifies the signal waveform to be viewed.

3. ***Horizontal amplifier.*** It is fed with a sawtooth voltage which is then applied to horizontal deflection plates.

4. ***Sweep or time base generator.*** It produces sawtooth voltage waveform used for horizontal deflection of the electron beam.

5. ***Trigger circuit.*** It produces trigger pulses to start horizontal sweep.

6. ***High and low-voltage supply :***

— The power supply block provides the voltages required by the cathode ray tube to generate and accelerate the beam, as well as to supply the required operating voltages for the other circuits of the oscilloscope.

— Relatively high voltages are required by cathode ray tubes, on the order of a few thousand volts, for acceleration, as well as a low voltage for the heater of the electron gun, which emits the electrons. Supply voltages for other circuits are various values, usually not more than a few hundred volts.

7. ***Delay line.*** High frequency oscilloscopes always include delay line in the vertical amplifiers. *The purpose of such lines is to delay the vertical signal enough to keep it from reaching the CRT deflection plates before the horizontal sweep circuits are running.*

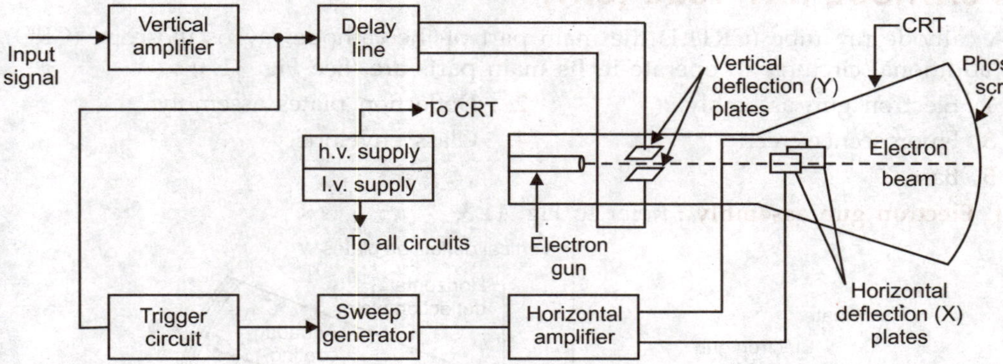

Fig. 11.1. Block diagram of a general purpose CRO.

Working of a CRO :

- When signal is to be displayed or viewed on the screen it is applied across the Y-plates of a cathode ray tube (CRT). But to see its waveform or pattern, it is essential to spread it horizontally from left to right. This is achieved by applying a sawtooth voltage wave to X-plates.

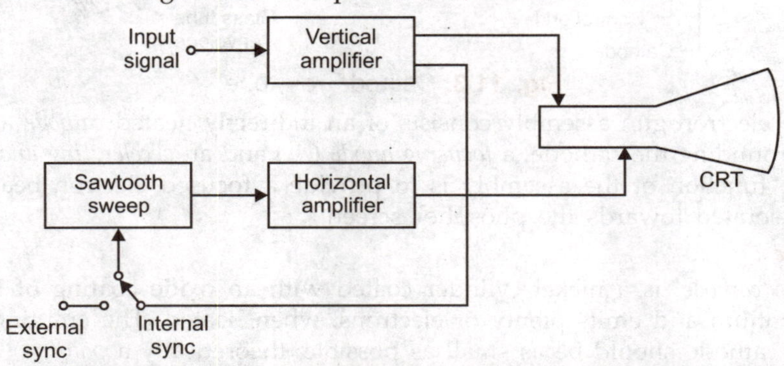

Fig. 11.2

Under these conditions the electron beam would move uniformly from left to right thereby graphing vertical vibrations of the input signal versus time. Due to repetitive tracing of the viewed waveform, we get a continuous display because of persistence of vision.

- However, to get a stable stationary display on the screen it is essential to *synchronize the horizontal sweeping of the beam* (sync) with the input signal across the Y-plates. The signal will be properly synced only when *its frequency equals the sweep generator frequency*.

The usual method of synchronizing the input signal is to *use a portion of the input signal to trigger the sweep generator so that the frequency of the sweep signal is locked or synchronized to the input signal*. It is called internal sync because the synchronization is obtained by internal wiring connection as shown in Fig. 11.2.

— In the *external synchronization*, an external trigger is used to trigger the signal being measured.

— In the *line synchronization*, the trigger is obtained from the power supply to CRO which is normally 230 V, 50Hz.

11.4 CATHODE RAY TUBE (CRT)

A cathode ray tube (CRT) is the main part of the cathode ray oscilloscope (CRO), with additional circuitry to operate it. Its main parts are (See Fig. 11.3) :

1. Electron gun assembly;
2. Deflection plates assembly;
3. Fluorescent screen;
4. Glass envelope;
5. Base.

1. Electron gun assembly : Refer to Fig. 11.3.

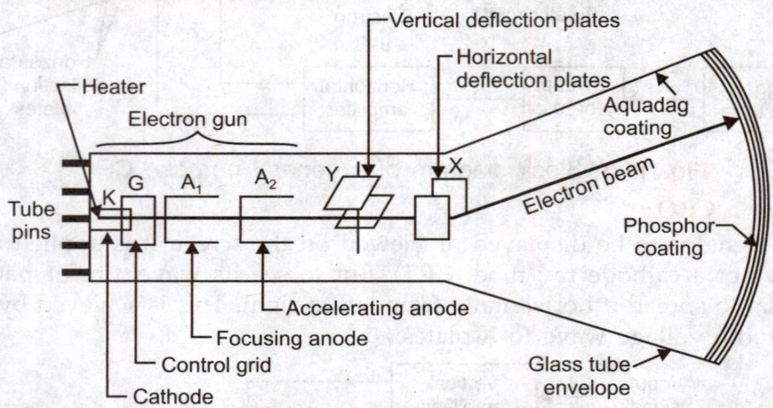

Fig. 11.3. Cathode ray tube.

- The electron gun assembly consists of an indirectly heated *cathode* a *control grid* surrounding the cathode, a *focusing anode* (A_1) and an *accelerating anode* (A_2). The sole function of the assembly is to provide a focused electron beam which is accelerated towards the phosphor screen.

Cathode :

- The '*cathode*' is a nickel cylinder coated with an oxide coating of barium and strontium and emits plenty of electrons, when heated. The emitting surface of the cathode should be as small as possible, theoretically a *point*.
- Rate of emission of electrons (or say the intensity of electron beam) depends on the magnitude of cathode current, which can be controlled by the control grid in a manner similar to a conventional vacuum tube.
- The typical values of current and voltage required by an indirectly heated cathode are 600 mA at 6.3 V (A.C. or D.C.). The special low power designs use 140 mA at 1.5 V.

Control grid :

- The '*control grid*' is usually a metal cylinder covered at one end but with a small hole in the cover. This is usually a metal cap of low permeability steel, about

15 mm in diameter and 15 mm long. An aperture of approximate 0.25 mm is drilled in the grid cap for the passage of electrons through it.

- The grid is kept at *negative potential* (variable) w.r.t. cathode and its function is to vary the electron emission and so the brilliancy of the spot on the phosphor screen. The hole in the grid is provided to allow passage for electrons through it and concentrate the beam of electrons along the axis of tube.

Focusing and accelerating anodes :

- Electron beam comes out from the control grid through a small hole in it and enters a *pre-accelerating anode,* which is a hollow cylinder in shape and is at a potential of few hundred volts more positive than the cathode so as to accelerate the electron beam in the electric field.
- The electron beam is then focused on the screen by an electrostatic lens consisting of two more cylindrical anodes called the *'focusing anode'* and *'accelerating anode'* apart from the pre-accelerating anode. The function of these anodes is to concentrate and focus the beam on the screen and also to accelerate the speed of electrons.
- The pre-accelerating and accelerating anodes are connected to a common positive high voltage of about 1500V; the focusing anode is connected to a lower adjustment voltage of 500 V.

2. Deflection plates assembly :

- The electron beam, after leaving the electron gun, passes through two pairs of deflection plates. One pair of plates is mounted horizontally and produces an electric field in the vertical plane. This pair produces a vertical deflection and is thus called *Vertical deflection plates* or *Y-plates*. The other pair of plates is mounted vertically and produces a horizontal deflection. This pair of plates is called *Horizontal deflection plates or X-plates.*

  The plates are flared so as to allow the beam to pass through them without striking the plates.
- When a sinusoidal voltage is applied to Y-plates, the beam will be moved up and down according to the variation of plate potential. If one frequency of variation is more than 16 Hz the deflection will be a vertical line in the centre of the screen. In case the sinusoidal voltage is applied to X-plates and frequency of variation is more than 16 Hz the deflection will be a horizontal line. If potentials are applied to both sets of plates simultaneously, the deflection will be an oblique line. The amount of deflection is in proportion to the voltage applied to the pair of plates.

It must be remembered that :

— The signal to be displayed on the screen is always applied across the Y-plates.

— The voltage applied across X-plates is a ramp voltage *i.e.* a voltage which increases linearly with time. It has a *'sawtooth waveform'* as shown in Fig. 11.4. It is also called *horizontal time-base or sweep voltage.* It has a sweep time of T_{sw}.

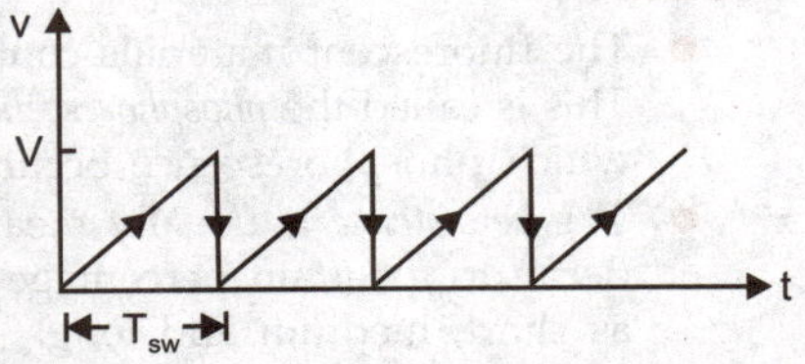

Fig. 11.4

3. Fluorescent screen :

- The end wall of the CRT, called *'screen'*, is coated with *phosphor*. When struck by high-energy electrons, it glows. In other words, it absorbs the kinetic energy of the electrons and converts it into light–the process being known as *fluorescence.*

That is why the screen is called *fluorescent screen*. The colour of the emitted light depends on the type of phosphor used.

The intensity of light emitted from the CRT screen, called *luminance*, depends upon the following *factors* :

(*i*) The number of bombarding electrons striking the screen per second.

(*ii*) The energy with which the bombarding electrons strike the screen which is determined by the accelerating potential (more the accelerating potential more the luminance).

(*iii*) The time the beam strikes a given area of the phosphor *i.e.*, sweep speed.

(*iv*) The physical characteristics of the phosphor itself.

- Among the fluorescent materials used are :
  (*i*) *Zinc or orthosilicate* ... for giving a green trace very suitable for visual observations;
  (*ii*) *Calcium tungstate* ... For giving blue and ultra-violet radiations very suitable for photography;
  (*iii*) *Zinc sulphide* (with other materials) ... For giving a white light suitable for T.V.;
  (*iv*) *Zinc phosphate* ... It gives a pronounced after glow and is useful when studying transient phenomena because the trace persists for short while after the transient has disappeared. .
- The phosphor of the oscilloscope is designated as follows :

P_1 = Green medium; P_2 = Blue green medium;

P_5 = Blue very short; P_{11} = Blue short ect.

— The material P_{31} is used commonly for general purpose oscilloscopes due to the following *characteristics* :

(*i*) High illumination level. (*ii*) High writing speed.

(*iii*) High *burn resistance (to avoid accidental damage)

(*iv*) It gives colour to which human eye response is maximum.

(*v*) It gives short persistence required to avoid multiple image display.

* When an electron beam strikes the screen its K.E. is absorbed by the phosphor material. This creates heating. The heat produced gives rise to *"phosphor burn"* which is damaging and may cause phosphor destruction. So the phosphor material should have a *high burn resistance*.

Note :

- The fluorescent materials continue to emit light even after radiation exposure is cut off. This is called the *phosphorescence* characteristics of the materials. The length of time during which phosphorescence occurs is called the *persistence* of the phosphor.
- The *persistence* is usually measured in terms of the time required for the CRT image to decay to a certain percentage (usually 10%) of the original light output. It is expressed as short, medium and long.
  (*i*) *"Short persistence"*. It lasts for *microseconds*. It is required for extremely high speed phenomena, to avoid smearing and interference caused when one image persists and overlaps with next one.
  (*ii*) *"Medium persistence"*. It lasts for *milliseconds*. These traces are mostly used for general purpose applications.
  (*iii*) *"Long persistence"*. It lasts for seconds to several minutes. These traces are used for transients as they keep the fast transient on the screen for observation after the transient has disappeared.

4. Glass envelope :

- The glass envelope is funnel-shaped having a phosphor-coated screen at its flared end. It is *highly-evacuated in order to permit the electron beam to traverse the tube easily.*
- The inside part of the flared part of the tube is coated with a conducting graphite layer called *"aquadag"* which is maintained at the same potential as the second anode. This layer performs the two functions : (*i*) It accelerates the electron beam after it passes between the deflecting plates, and (*ii*) It collects the electrons produced by secondary emission when electron beam strikes the screen. Hence it prevents the formation of negative charge on the screen.

— In some tubes, particularly CRTs with magnetic focusing (such as TV picture tubes), the accelerating anode is dispersed with entirely and the conductive coating is used as the final accelerating anode.

- *Horizontal and vertical marks* an marked on the screen of the CRT to provide user a correct measurement. These marks, *usually in rectangular form*, are called *graticule.*

— The graticule may be *external, internal* and *projected.*

— The graticules are generally laid out in an 8 × 10 pattern as shown in Fig. 11.5. with 8 major divisions along the vertical axis and 10 major divisions along the horizontal axis. The tick marks on the centre graticule represent minor divisions or subdivisions.

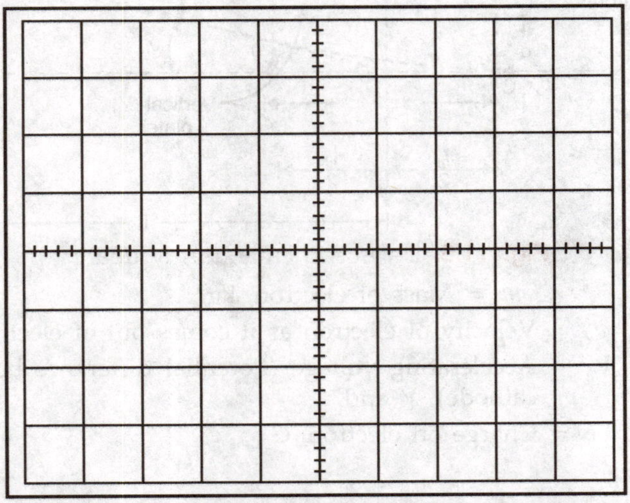

Fig. 11.5. The CRT graticule.

Size : Size refers to *screen diameter*. CRTs for Oscilloscopes are available in sizes 1, 2, 3, 5 and 7 inches. 3 inch tube is most common for portable instruments.

Example : *A CRT having a number $5GP_1$ is 5 inch tube with a medium persistence green trace. The first number indicates that it is a 5 inch Tube and P_1 indicates a medium persistence green trace.*

— Both round and rectangular CRTs are found is oscilloscopes. Vertical viewing size is 8 cms and horizontal is 10 cms (See Fig. 11.5).

- **Operating voltages :**

  The CRT requires a heating voltage of 6.3 V A.C. or D.C. at 600 mA. Several D.C. voltages are listed below; the voltages vary with the types of tube used :

(*i*) Negative grid (control) voltage : –14 V to –200 V

(*ii*) Positive anode A_1 (Focusing anode) : –100 V to –1100 V

(*iii*) Positive anode A_2 (Accelerating anode) : 600 V to 6000 V

A_3 : 200 V to 20000 V in some cases

5. Base. The base is provided to the CRT through which the connections are made to the various parts.

11.5 ELECTROSTATIC DEFLECTION

Fig. 11.6. shows the upward deflection of an electron beam when it passes between the vertical or *Y*-plates of a CRT. The beam deflects upwards because the upper *Y*-plate has been made positive w.r.t. the lower plate. *Reversing* the polarity of the *"applied voltage"* would, obviously, cause the beam to deflect downwards.

The velocity of an electron coming out of the electron gun can be found by equating the kinetic and potential energies.

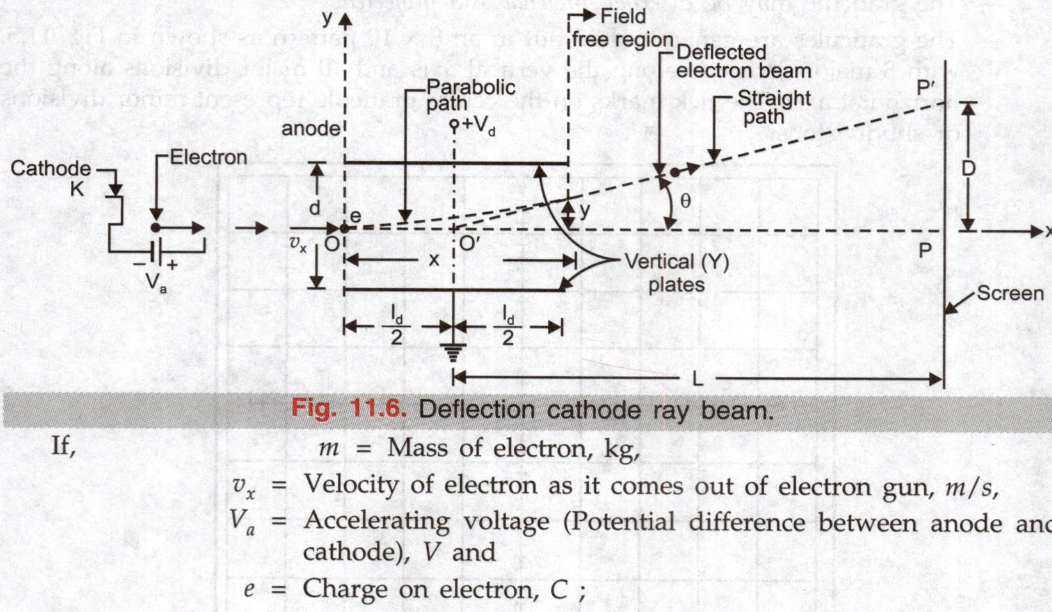

Fig. 11.6. Deflection cathode ray beam.

If, m = Mass of electron, kg,

v_x = Velocity of electron as it comes out of electron gun, m/s,

V_a = Accelerating voltage (Potential difference between anode and cathode), *V* and

e = Charge on electron, *C* ;

Then,

$$\frac{1}{2} m v_x^2 = eV_a$$

or,

$$v_x = \sqrt{\frac{2eV_a}{m}} \qquad ...(11.1)$$

This velocity v_x *remains constant* because the fields due to *X*-and *Y*-plates act at right angles to each other.

Fig. 11.6. shows the *electrostatic deflection* due to the *Y*-plates.

Let, d = Distance between *Y*-plates,

l_d = Length of each *Y*-plate,

L = Distance from O' to screen,

V_d = Deflecting voltage (*i.e.*, potential difference between *Y*-plates), and

D = Deflection on the screen.

(The straight line of travel of the electron is tangent to the parabola at $x = l_d$)

Now, electric field between Y-plates

$$= \frac{V_d}{d} \text{ volts/m} \qquad ...(11.2)$$

$$\therefore \quad \text{Force on electron} = \frac{eV_d}{d} \text{ newtons} \qquad ...(11.3)$$

Also, force on electron = $m \times$ acceleration in y-direction (*i.e.* a_y)

$$\therefore \quad a_y = \frac{eV_d}{m\,d} m/s^2 \qquad ...(11.4)$$

The time taken by the electron to traverse the length l is $\frac{l_d}{v_x}$. Since the initial velocity in the y-direction is zero, v_y, *i.e.* velocity in y-direction when electron leaves the field between plates, is

$$v_y = 0 + a_y.t$$

$$= \frac{eV_d}{md}.\frac{l_d}{v_x}$$

$$i.e. \quad v_y = \frac{eV_d}{md}.\frac{l_d}{v_x} \qquad ...(11.5)$$

After the electron leaves the field of the Y-plates, it moves in a *straight line* because it is in field-free space/region. *During the motion between the deflecting plates, a uniform acceleration is acting on the electron and therefore, its path is parabolic.* If the straight line is extended backward, it can be shown that it passes through O′, midway between the plates.

The deflection of the electron when it leaves the field is,

$$y = (\text{Average velocity in } y\text{-direction}) \times \frac{l_d}{v_x}$$

$$= \left(\frac{1}{2}.\frac{eV_d}{md}.\frac{l_d}{v_x}\right) \times \frac{l_d}{v_x}$$

$$\text{or,} \quad y = \frac{1}{2}\left(\frac{eV_d}{md}\right)\left(\frac{l_d}{v_x}\right)^2 \qquad ...(11.6)$$

$$\text{From Fig. 11.6;} \quad \frac{y}{D} = \frac{(l_d/2)}{L} \quad or \quad \frac{l_d}{2L}$$

$$\text{or,} \quad D = \frac{2L}{l_d}.y = \frac{2L}{l_d} \times \frac{1}{2} \times \frac{eV_d}{md} \times \frac{l_d^2}{v_x^2}$$

$$i.e. \quad D = \frac{L\,e\,V_d\,l_d}{m\,d\,v_x^2} \qquad ...(11.7)$$

Substituting the values of v_x from eqn. (11.1), we get

$$D = \frac{L\,eV_d l_d}{m\,d} \times \frac{m}{2eV_a} = \frac{LV_d l_d}{2dV_a}$$

i.e. $$D = \frac{LV_d\, l_d}{2\, dV_a} \qquad \text{...(11.8)}$$

It is seen from eqn. (11.8) that D (vertical deflection on screen) is proportional to V_d (deflection voltage). This direct proportionality indicates that CRT may be used as a *linear voltage-indicating device*. The deflection voltage, although *assumed fixed* in the above discussion, usually varies and the image on the screen follows the variations of the deflection voltage in a linear manner.

- **Electrostatic deflection sensitivity (S)** *is defined as the deflection per volt of the deflection voltage.*

  i.e., $$S = \frac{D}{V_d} = \frac{L\, l_d}{2\, d\, V_a} \qquad \text{...(11.9)}$$

  Its units are m/V.
- **Deflection factor (G)** of a cathode ray tube is given by the *reciprocal of the sensitivity*.

  i.e. $$G = \frac{1}{S} = \frac{2dV_a}{L\, l_d} \qquad \text{...(11.10)}$$

  Its units are V/m.

— It is seen that S is independent of the deflection voltage V_d and the ratio $\frac{e}{m}$. Moreover, S varies inversely with V_a (acceleration voltage); thus by decreasing V_a, sensitivity can be increased.

On the other hand high V_a value, produces a bright spot. But for high V_a, high V_d is required and such a beam which is highly accelerated is *difficult to deflect*. Such a beam is called *"hard beam"*.

— The typical values of deflection sensitivities are 1.0 mm/V to 0.1 mm/V corresponding to the deflection factors of 10 V/cm to 100 V/cm respectively.

11.6 MAGNETIC DEFLECTION

When a conductor of length l carrying a current I is placed in a magnetic field of flux density B, a force acts on the conductor. This force F is given by (assuming the directions of I and B to be perpendicular),

$$F = BIl \text{ newtons} \qquad \text{...(11.11)}$$

The direction F is perpendicular to the directions of I and B. If I and B are not perpendicular, only the component of I perpendicular to B contributes to the force.

Let the length l contains n electrons; if these electrons take t seconds to travel length l then the charge per second *i.e.* current is given as :

$$I = \frac{ne}{t} \qquad \text{...(11.12)}$$

where, e is the charge on an electron, C.

∴ Force on the n electrons is,

$$F = \frac{Bnel}{t} \qquad \text{...(11.13)}$$

$\frac{l}{t}$ is the drift speed (*i.e.* v) of the electrons.

$\therefore$ Force *per electron* $= B\,e\,v$ ...(11.14)

Equ. (11.14) indicates that if v is zero, F is zero. This means that if electron is at rest, it does not experience any force and it remains at rest. If the velocity is along the direction of magnetic flux, there is no force on the electron and it continues to move at the same velocity.

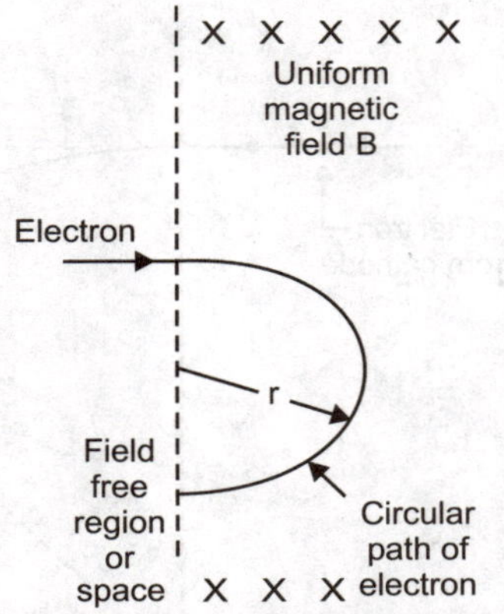

Fig. 11.7. Electron moving in a magnetic field.

Refer to **Fig. 11.7.** When an electron enters a magnetic field, having flux density directed into the page, with a speed v the force F is perpendicular to v and therefore the speed v remains constant. This forces causes the electron to move in a circular motion with constant speed v. If r is the radius of the circular path, the acceleration towards the centre is $\frac{v^2}{r}$.

Then, $$F = \frac{mv^2}{r} = Bev$$

or, $$r = \frac{mv}{eB} \qquad ...(11.15)$$

Also, angular velocity,

$$\omega = \frac{v}{r} = \frac{v}{(mv/eB)} = \frac{eB}{m} \text{ rad./sec.} \qquad ...(11.16)$$

and, Time period T (*i.e.*, to complete one revolution) is given as :

$$T = \frac{2\pi}{\omega} = \frac{2\pi}{(eB/m)} = \frac{2\pi m}{eB} \text{ rad./sec.} \qquad ...(11.17)$$

From above discussion, we find that the radius of the circular path is directly proportional to v. ω and T do not depend on r or v. Thus, the fast moving particles travel in a bigger circle while the slow moving particles travel in smaller circle.

The *time period for both the slow and the fast particles is the same*. This result forms the basis of operation of *cyclotron, magnetic focusing apparatus and many other apparatus*.

- A magnetic field can be used for deflection of electron beam in a cathode ray tube.

Fig. 11.8 shows a transverse magnetic field of flux density B for detecting an electron beam. The field is pointing into page. The electron moves in a straight line from cathode to boundary A of the magnetic field. In the magnetic field region, the electron experiences a force F (eqn. 11.14) and moves in a circular path AB. The radius of the circular path is r (eqn. 11.15) and the centre is at point C. The angle of deflection θ is very small and is given by,

$$\theta \simeq \frac{l}{r} \qquad ...(11.18)$$

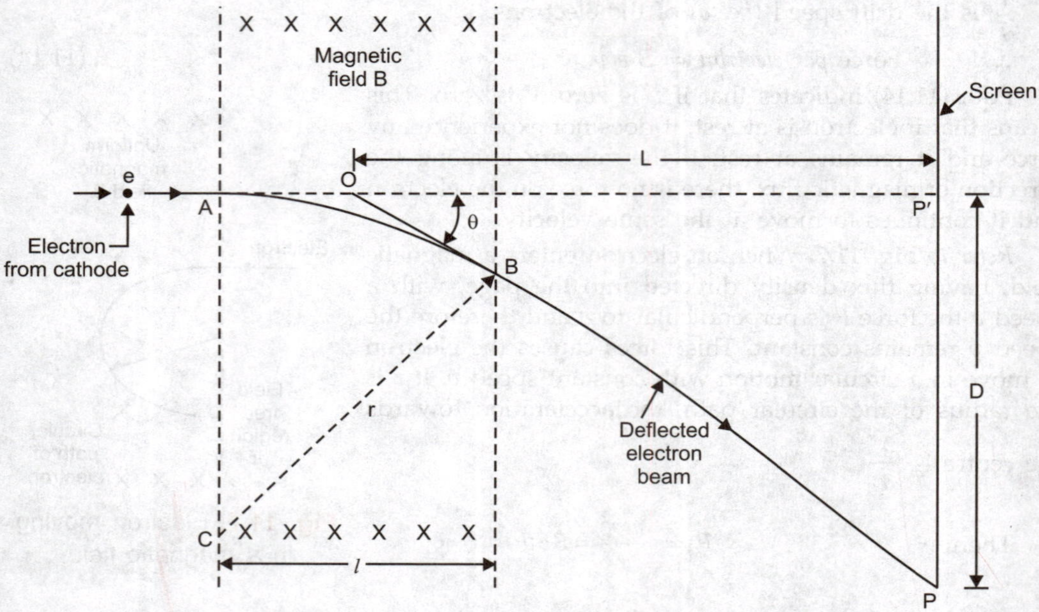

Fig. 11.8. Deflection of electron beam by a magnetic field.

As distance L is very large as conspaired to l, therefore, the straight line PB when projected backward passes through point O at the centre of the magnetic field. The deflection D is given as :

$$D = L\tan\theta \simeq L\theta = \frac{Ll}{r} = \frac{LleB}{mv} \qquad ...(11.19)$$

The velocity v is v_x as given by equ. (11.1). Substituting this value in eqn. (11.19), we get :

$$D = \frac{LleB}{\sqrt{\frac{2eV_a}{m}}} = BLl\left(\sqrt{\frac{e}{2mV_a}}\right) \qquad ...(11.20)$$

Magnetic deflection sensitivity (S) *is defined as the deflection on the screen in meters per tesla of magnetic flux density.*

i.e. $$S = \frac{D}{B} = Ll\sqrt{\frac{e}{2mV_a}} \qquad ...(11.21)$$

- It is seen that magnetic deflection and magnetic deflection sensitivity are inversely proportional to $\sqrt{V_a}$. Therefore, magnetic deflection is preferable in tubes with high accelerating voltage like *TV tube* and *radar* (while *electrostatic deflection is preferred for general purpose oscilloscope*).

Comparison between deflection methods :

The comparison between 'electrostatic deflection' and 'magnetic deflection' method is given as follows :

S.N.	Aspects	Electrostatic deflection	Magnetic deflection
1.	*Method used to obtain deflection.*	By the application of voltage to the plates.	By controlling the magnetic field by varying the current through the coils.
2.	*Relationship between the deflection and V_a.*	Inversely proportional to V_a.	Invesely proportional to $\sqrt{V_a}$.
3.	*Size of the beam for given V_a.*	Small	Large (wide).
4.	*Sensitivity (S) formula.*	$S = \frac{Ll_d}{2dV_a}$	$S = Li\sqrt{\frac{e}{2mV_a}}$
5.	*Dependance of deflection on the ratio e/m.*	No	Yes
6.	*Size of tubes necessary for given display area.*	Longer tubes.	Shorter tubes (can be built for the given display area).
7.	*Effect on the spot brightness with reduction in V_a (for greater sensitivity)*	Reduced brightness.	Increased brightness.
8.	*Applications.*	General propose oscilloscopes.	T.V. and radar.

- ***Differences between CRT and television picture tube.***

1. The CRT used in CRO uses electrostatic method of focussing as compared to a TV picture tube which employs electromagnetic envelope.

2. In CRT used in CRO, connection to accelerating anode is brought out through pin in the base along with other electrode connections as shown in Fig. 11.3. However in TV picture tube the lead in from accelerating anode is brought out through the sides of glass envelope.

3. In CRT used in CRO, EHT, (*i.e.*, extra high tension transformer) to accelerating anode is given from mains supply whereas in TV picture tube EHT supplied to accelerating anode is generated in the line output stage during flyback. During flyback the anode current of line output stage is suddenly cut off and anode voltage generates a spike of 2 to 3 kV amplitude. By increasing the number of turns on the primary winding to form an auto-transformer of say 1 : 5 turns pulses of 15 kV or more is obtained. After rectification and smoothing, EHT is obtained and applied to accelerating anode.

This method of applying EHT has an advantage because if input to line and valve fails, no EHT is produced and thus the tube is saved.

Example 11.1 : *A CRT has an anode voltage of 2 kV and parallel deflecting plates 2 cm long and 5 mm apart. The screen is 30 cm from the centre of the plates. Find the input voltage required to deflect the beam through 3 cm. The input voltage is applied to the deflecting plates through amplifiers having an overall gain of 100.* **(Nagpur University S - 2005)**

Solution. *Given :* V_a = 2 kV or 2000 V; l_d = 2 cm or 0.02 m; d = 5 mm or 0.005 m; L = 30 cm or 0.3 m; D = 3 cm or 0.03 m; Overall gain = 100.

We know that, $$D = \frac{Ll_d V_d}{2dV_a} \quad \text{...[Eqn. (11.1)]}$$

or, $$V_d = \frac{2dV_a D}{Ll_d}$$

$$= \frac{2\times 0.005\times 2000\times 0.03}{0.3\times 0.02} = 100 \text{ V}$$

$\therefore$ Input voltage required for deflection of 3 cm

$$= \frac{V_d}{gain} = \frac{100}{100} = \mathbf{1\ V}\ \textbf{(Ans.)}$$

Example 11.2. *An electrostatically deflected cathode ray tube has plane parallel deflecting plates which are 2.5 cm long and 0.5 cm apart, and the distance from their centre to the screen is 20 cm. The electron beam is accelerated by a potential difference of 2500 V and is projected centrally between the plates.*

Calculate the deflection voltage required to cause the beam to deflect through 3 cm and find the corresponding deflection sensitivity of the screen. **(PTU Dec. - 2005)**

Solution. *Given :* l_d = 2.5 cm = 0.025 m; d = 0.5 cm = 0.005 m; L = 20 cm = 0.2 m; V_a = 2500 V ; D = 3 cm = 0.03 m

$\mathbf{V_d\ ;\ S:}$

We know that, $$D = \frac{Ll_d V_d}{2dV_a} \quad \text{...[Eqn. (11.8)]}$$

or, *Deflection voltage,* $$V_d = \frac{D\times 2dV_a}{Ll_d} = \frac{0.03\times 2\times 0.005\times 2500}{0.2\times 0.025} = \mathbf{150\ V}\ \textbf{(Ans.)}$$

$\therefore$ *Deflection sensitivity,* $$S = \frac{D}{V_d} = \frac{0.03}{150} \text{ m/V} = 2\times 10^{-4} \text{ m/V or } \mathbf{0.2\ mm/V}\ \textbf{(Ans.)}$$

Example 11.3. *The horizontal deflection plates in a CRT are 20 mm long and 5 mm apart. The centre of the plates is 20 cm from the screen. Accelerating voltage is 2500 V. Determine the deflection sensitivity.* **(UPTU - 2004)**

Solution. *Given:* l_d = 20 mm = 0.02 m; d = 5 mm = 0.005 m; L = 20 cm = 0.2 m; V_a = 2500 V.

We know that, deflection sensitivity,

$$S = \frac{Ll_d}{2dV_a} \quad \text{...[Eqn. (11.9)]}$$

or, $$S = \frac{0.2\times 0.02}{2\times 0.005\times 2500} \text{ m/V or } \mathbf{0.16\ mm/V}\ \textbf{(Ans.)}$$

Example 11.4. *In a CRT, the anode to cathode voltage is 2500 V. The parallel deflection plates are 1.5 cm long and spaced 5 mm. The screen is 50 cm from the centre of the deflecting plates. If mass of electron is* 9.109×10^{-31} *kg and charge on electron* 1.602×10^{-19}*C, find :*

(*i*) *The beam speed.*

(*ii*) *The deflection sensitivity of the tube.*

Solution. *Given :* V_a = 2500 V; l_d = 1.5 cm = 0.015 m; d = 5 mm = 0.005 m; L = 50 cm = 0.5 m; $m = 9.109\times 10^{-31}$ kg; $e = 1.602\times 10^{-19}$ C.

(*i*) **The beam speed, v :**

We know that, $$v = \sqrt{\frac{2eV_a}{m}} \quad \text{...(Eqn. (11.1)]}$$

$$= \sqrt{\frac{2 \times 1.602 \times 10^{-19} \times 2500}{9.109 \times 10^{-31}}}$$

$$= \mathbf{29.65 \times 10^6\ m/s} \quad \text{(Ans.)}$$

(*ii*) **Deflection sensitivity, S :**

$$S = \frac{Ll_d}{2dV_a} \quad \text{...[Eqn. (11.2)]}$$

$$= \frac{0.5 \times 0.015}{2 \times 0.005 \times 2500} \text{ m/V}$$

or $$= \mathbf{0.3\ mm/V} \quad \text{(Ans.)}$$

Example 11.5. *A cathode-ray tube, with magnetic deflection, has its screen 22 cm from the centre of the magnetic field. The width of the uniform magnetic field is 3.3 cm and the final anode potential is 6 kV. Calculate the density of the magnetic field to produce a deflection of 4.4 cm on the screen.*

Take : Mass of electron = 9.107×10^{-31} kg; charge on electron = 1.6×10^{-19} C.

Solution. *Given :* $L = 22$ cm $= 0.22$ m; $l = 3.3$ cm $= 0.033$ m; $V_a = 6$ kV;
$D = 4.4$ cm $= 0.044$ m; $m = 9.107 \times 10^{-31}$ kg; $e = 1.6 \times 10^{-19}$ C.

Density of magnetic field, B :

We know that, $$D = BLl\left(\sqrt{\frac{e}{2mV_a}}\right) \quad \text{...[Eqn. (11.20)]}$$

Substituting the values we get,

$$0.044 = B \times 0.22 \times 0.033\left(\sqrt{\frac{1.6 \times 10^{-19}}{2 \times 9.107 \times 10^{-31} \times (6 \times 10^3)}}\right)$$

or, $$0.044 = B \times 27.779$$

or, $$B = 1.584 \times 10^{-3}\ \text{Wb/m}^2 \quad \text{or} \quad \mathbf{1.584\ m\ W\ b/m^2} \text{ (Ans.)}$$

Example 11.6. *In a cathode-ray tube, the electron beam is displaced vertically by a magnetic field of flux density 1.8×10^{-4} Wb/m². The length of the magnetic field along the tube axis is same as that of the electrostatic deflection plates. The final anode voltage is 800 V.*

Calculate the voltage which should be applied to the Y-deflection plates 10 mm apart, to return the spot back to the centre of the screen.

Take : Mass of electron = 9.107×10^{-31} kg; charge on electron = 1.6×10^{-19} C.

Solution. *Given :* $B = 1.8 \times 10^{-4}$ Wb/m²; $V_a = 800$ V; $d = 10$ mm $= 0.01$ m.

Deflection voltage, V_d :

Case I. *Magnetic deflection :*

In this case, deflection, $$D = BLl\left(\sqrt{\frac{e}{2mV_a}}\right) \quad \text{...[Eqn. (11.20)]}$$

$$= 1.8 \times 10^{-4} L l \left(\sqrt{\frac{1.6 \times 10^{-19}}{2 \times 9.107 \times 10^{-31} \times 800}} \right)$$

or, $$D = 1.886\ lL \text{ metres} \qquad ...(1)$$

Case II. *Electrostatic deflection* :

In this case, $$D = \frac{LV_d l}{2dV_a} \qquad ...[\text{Eqn. } (11.8)]$$

or $$D = \frac{lLV_d}{2 \times 0.01 \times 800} = 0.0625\ l\ M\ Vd \text{ metres} \qquad ...(2)$$

Equating (1) and (2), we get

$$1.886\ lL = 0.0625\ lLV_d$$

or, $$V_d = \frac{1.886}{0.0625}$$

$$= \mathbf{30.176\ V\ (Ans.)}$$

11.7 VERTICAL AMPLIFIER

- The input signal is fed to the *Y*-deflecting plates through a vertical amplifier.
- The function of vertical amplifier is to *amplify the input signal.*
- It may consist of several stages in cascade and generally, operates with fixed gain. The output stage of this amplifier is invariably a push-pull amplifier. It is necessary to adjust the magnitude of the input signal so that a large signal may not cause the display to go out of the screen. Therefore, the input signal is fed to the vertical amplifier through an *attenuator*.
- The *"attenuator"* is a network of resistances and capacitances and it *reduces the input signal* so that the display of large input signals *may not go out of the screen*.

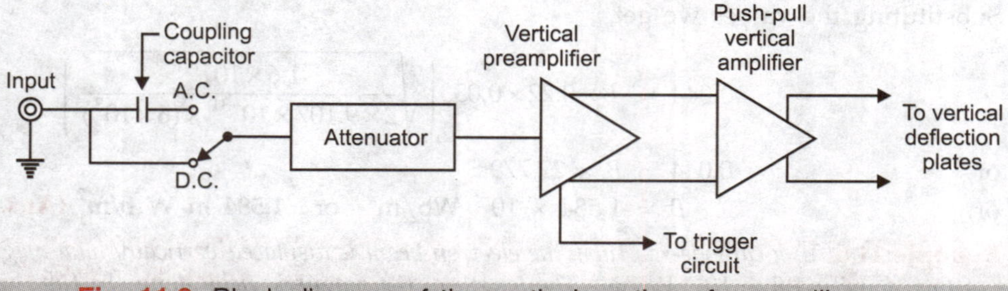

Fig. 11.9. Block diagram of the vertical section of an oscilloscope.

Fig. 11.9 shows the block diagram of a complete *vertical deflection system*.

— The input convector feeds an input attenuator, after which follows the vertical amplifier. The input impedance of an oscilloscope is rather high, being on the order of 1 MΩ, which is desirable for measuring voltages in high impedance circuits.

— The attenuator sets the sensitivity of the oscilloscope in the common 1-2-5 sequence. As an example, the input attenuator could provide for 10, 20, 50, 100, 200 mV, etc. per centimetre.

— Practically, all oscilloscopes provide a switchable input coupling capacitor. This is provided so that measurements of A.C. signals may be viewed in the presence of high D.C. voltages by including the coupling capacitor. When D.C. measurements are to be made, the capacitor may be removed. The value of the capacitor is chosen so that the frequency response of the oscilloscope is preserved down to a few Hz.

- *The vertical amplifier is the principal factor in determining the sensitivity and bandwidth of an oscilloscope.*

— The gain of the vertical amplifier determines the smallest signal that the oscilloscope can satisfactorily reproduce on the CRT screen. The sensitivity of an oscilloscope is directly proportional to gain of the vertical amplifier that is, as gain increases, sensitivity increases, which allows us to observe smaller-amplitude signal.

— The vertical sensitivity of an oscilloscope is the smallest deflection factor that can be selected with the rotary switch.

— The *bandwidth* of an oscilloscope determines the range of frequencies that can be observed with the instrument. The bandwidth of an oscilloscope is the range of frequencies over which the gain of the vertical amplifier is within 3 dB of the midband frequency gain.

11.8 TIME BASE (SWEEP OR RAMP) GENERATOR

If the horizontal deflecting plates, X-plates are not energised, the electron beam would draw a vertical line on the screen.

- The *function of the time base generator* is to drive the beam at a steady speed across the screen and, when it reaches the right-hand side of the screen, the beam is made to fly back to the starting position on the left-hand side of the screen. The time base generator produces a *saw-tooth wave* (Fig. 11.10) of *the same frequency as the* input signal to the *Y*-plates.

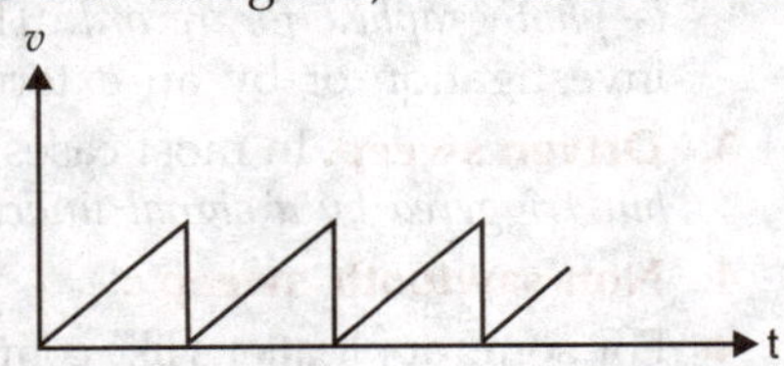

Fig. 11.10. Saw-tooth voltage wave.

- The horizontal deflection of the beam is proportional to the instantaneous voltage of the saw-tooth wave, therefore, the beam sweeps at a uniform rate across the screen in the horizontal direction, running to the starting position almost instantaneously and sweeping again.
- The time base generator gives an output sweep signal of adjustable frequency so that the *Y* input signals of a wide range of frequencies may be displayed on the screen.

There are many applications of a CRO in which a time base generator is *not* needed and it is necessary to apply an external input signal to the *X*-plates (through a switch).

11.9 TYPES OF SWEEPS

There are *four* types of sweeps :

1. Free running sweep :

- In very basic oscilloscopes, the sweep generator is continuously charging and discharging a capacitor. One ramp voltage is followed *immediately* by another, hence the sawtooth pattern appears. A sweep generator operating in this manner is said to be *"free running"*.
- In order to present a stationary display on the screen, the sweep generator

signal must be forced to run in *synchronization* with a vertical input signal. In basic oscilloscopes, this is accomplished by carefully adjusting the sweep frequency to a value very close to the exact frequency of the vertical input signal, or submultiple of this frequency. With signals at the same frequency, an *internal synchronization pulse will lock the sweep generator into the vertical input signal*. This method of synchronization has some serious limitations when an attempt is made to observe *low-amplitude* signals. However, the most serious limitation is probably the inability of the instrument to maintain synchronization when the amplitude or frequency of the vertical signal is not constant, such as voice or music signals.

2. Triggered sweep :

- A waveform to be observed on the CRO may not be periodic but may perhaps occur at irregular intervals. In this case, it is desirable that the sweep circuit remain inoperative and the sweep be initiated by the waveform under examination. In some cases, the waveform may be periodic, but it may be that the interesting part of the waveform is of very short duration compared to the period of the waveform. Under such cases, a *triggered switch is used*.
- In triggered sweep or single sweep, the spot is swept once across the screen in response to a trigger signal.
- The *triggered sweep is used for examination of transient or one time signals and waveform is photographed for record*. The trigger can be obtained from the signal under investigation or by an external source.

3. Driven sweep. In most cases, a *'driven sweep'* is used where the sweep is *recurrent but triggered by a signal under test*.

4. Non-sawtooth sweep :

- For some applications like comparison of two frequencies or finding phase shift between two voltages non-sawtooth sweep voltages are utilised for the sweep circuit.
- Sweep frequencies vary with the type of the oscilloscope.

A laboratory oscilloscope may have sweep frequencies upto several MHz; a simple oscilloscope for audio work has an upper limit 100 kHz. Most T.V. services require a sweep voltage frequency upto 1 MHz.

11.10 HORIZONTAL AMPLIFIER

- The function of the horizontal amplifier is to *amplify the signal applied to the X-plates*.
- This amplifier consists of several stages in the cascade.
- For measurement of *voltages and currents*, the input to the horizontal amplifier is a *sawtooth wave*. For *frequency and phase* measurements, the input is an *external signal*. The horizontal amplifier should be capable of amplifying both these inputs without distortion.

The input impedance of most general purpose oscilloscopes is about 1 MΩ shunted by a capacitance of about 20 pF.

11.11 DELAY LINE

The *purpose of a "delay line" is to delay the vertical signal enough to keep it from reaching the CRT deflection plates before the horizontal sweep circuits are running*.

- In an oscilloscope, all electronic circuitry (attenuators, amplifiers, pulse shapers, etc.) causes a certain amount of line delay in the transmission of signal voltages to the deflection plates. Almost all of this delay is created in circuits that *switch*, *shape* or *generate*.

Referring to Fig. 11.11, we observe that the horizontal signal (time base, or sweep voltage) is initiated or *triggered*, by a portion of the output signal applied to the vertical plates of the CRT. In horizontal channel, the signal processing consists of generating and shaping a trigger pulse (trigger pick off) that starts the sweep generator, whose output is fed to the horizontal amplifier and then to the horizontal deflection plates. This whole process consumes time on of the order of 80 ns or so. Thus, to allow the operator to observe the leading edge of the signal waveform, the signal drive for the vertical CRT plates must therefore be *delayed* by at least the same amount of time. This is the function of the vertical delay line. In the Fig. 11.11 is shown a 200 ns delay line which has been added to the vertical channel. It may be remembered that a trigger pick off must *precede* the delay line, although the delay line can appear almost anywhere along the vertical signal path.

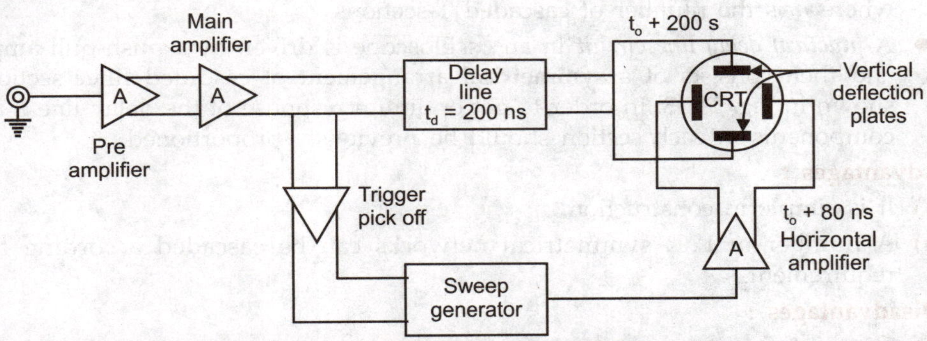

Fig. 11.11. Delay line circuit.

In CRO, the following two types of delay lines are used :

1. Lumped-parameter delay line.
2. Distributed-parameter delay line.

1. Lumped-parameter delay line :

- This type of delay line consists of a cascaded symmetrical LC networks, such as the so called T-section of Fig. 11.12.

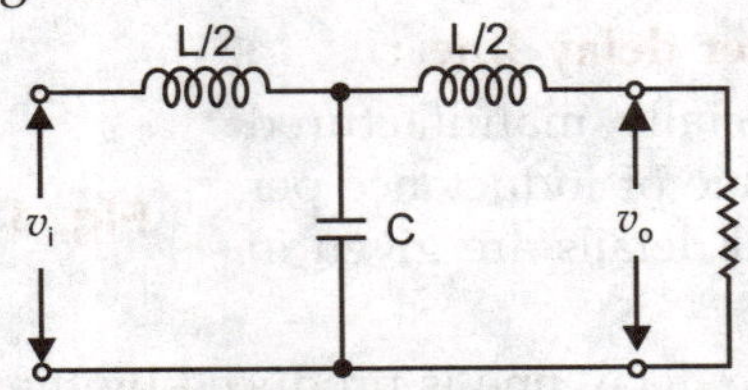

Fig. 11.12. T-filter section.

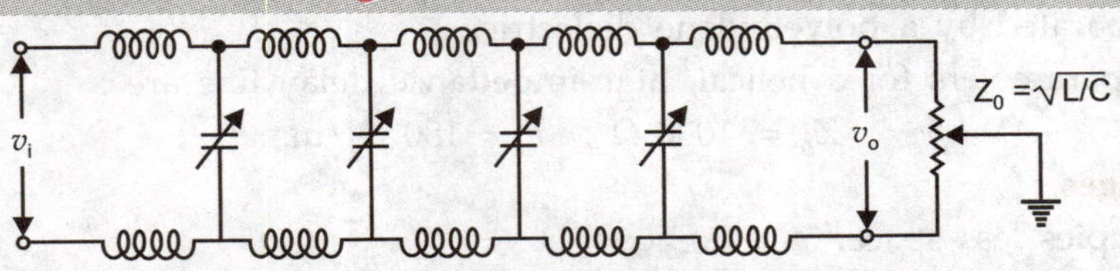

Fig. 11.13. Practical delay line.

- Each section is capable of delaying the signal by 3 to 6 ns.

- The T-section is terminated into its characteristic impedance Z_o. So, it acts as a low-pass filter, its cut-off frequency is given as :

$$f_c = \frac{1}{\pi\sqrt{LC}} \quad \text{...(11.22)}$$

If the spectrum of input signal v_i consists of frequencies much less than the cut-off frequency, output signal v_o will be a faithful reproduction of v_i but delayed by a time,

$$t_s \simeq \frac{1}{\pi f_c} = \sqrt{LC} \quad \text{...(11.23)}$$

where, t_s is the delay time for a single T-section. However, when a number of T-sections are cascaded into a so-called *lumped-parameter* delay line, the total delay time,

$$t_d = n\, t_s \quad \text{...(11.24)}$$

where n is the number of cascaded T-sections.

- A *practical delay line circuit* in an oscilloscope is driven by a push-pull amplifier and then consists of a symmetrical arrangement of cascaded filter sections as shown in Fig. 11.13. In order to get optimum response of the delay line L and C components in each section should be previously proportioned.

Advantages :

(*i*) It is simple in construction.

(*ii*) A number of L.C. symmetrical networks can be cascaded according to the requirement.

Disadvantages :

(*i*) Since it has a sharp cut-off frequency therefore, as the input frequency increases the amplitude and phase *distortion* takes place.

(*ii*) In order to get better transient response the delay line should be matched to the characteristic impedance at the input and output side. This makes the termination circuiry *complex*.

2. Distributed-parameter delay line :

- It consists of a specially manufactured coaxial cable with a high value of inductance per unit length. Its constructional details are given in Fig. 11.14.

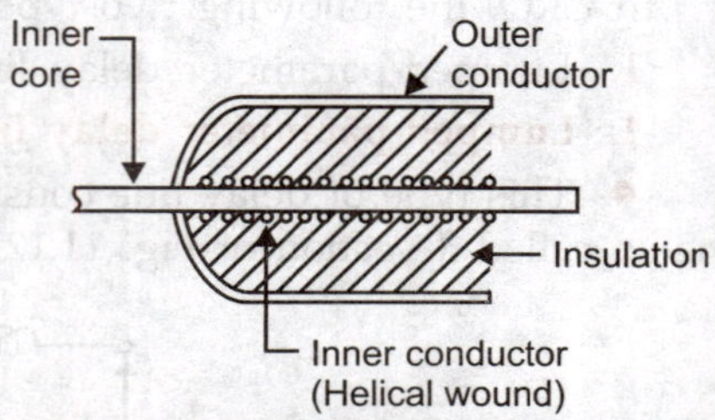

Fig. 11.14. Distributed parameter delay line.

- The inductance of the delay line is produced by the inner coil, and it equals that of solenoid with n turns per meter. The capacitance of the delay line is that of two coaxial cylinders separated by a polyethylene dielectric.

Typical parameters for a helical, high-impedance delay line are :

$$Z_o = 1000\ \Omega\ ; \quad t_d = 180 \text{ ns/m}.$$

Advantages :

(*i*) Occupies less space.

(*ii*) It does not require careful adjustment of a lumped-parameter delay line.

11.12 BASIC CONTROLS OF CRO

Fig. 11.15. shows the detailed block diagram of a general purpose oscilloscope, exhibiting a number of controls required for the proper functioning of CRO.

1. Vertical (Y) deflection system :

● The signals to be examined are usually applied to the vertical or Y-deflection plates through an attenuator and a number of amplifier stages. Vertical amplifier is required because the signals are not strong enough to produce measurable deflection on CRT screen.

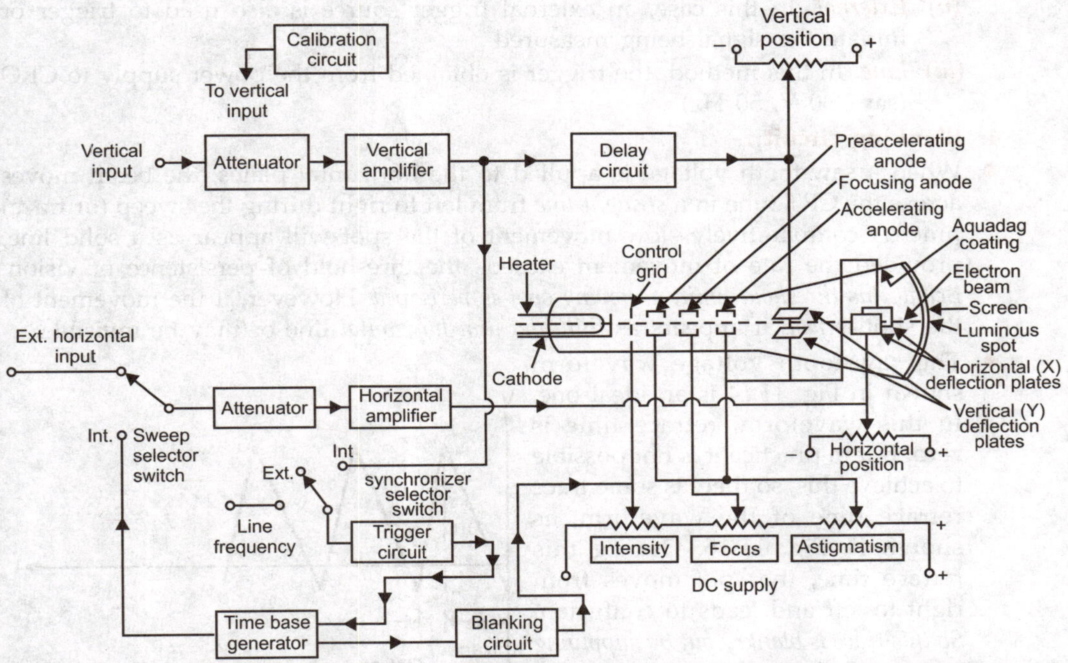

Fig. 11.15. Detailed block diagram of a general purpose oscilloscope.

— The *amplifier response must be wide enough to pass faithfully the entire band of frequencies to be measured.*

● The high voltage signals (to be examined) must be attenuated to bring them as within the range of vertical amplifiers. The output of the vertical amplifier is also applied to the synchronizing amplifier through the *synchronizer selector switch in internal position (Int.).* This permits the horizontal sweep circuit to be triggered by the signal being investigated.

2. Horizontal (X) deflection system :

● The horizontal or X-deflection plates are fed by a sweep voltage that provides a *time base*. The horizontal plates are supplied through an amplifier, but they can be fed directly when voltages are of sufficient magnitudes.

● When external signals are to be applied to the horizontal deflection system, they can also be fed through the horizontal amplifier, *via* the sweep selector switch which is in the external position (Ext.). When the sweep selector switch is in the internal position, the horizontal amplifier receives an input from the saw-tooth sweep generator which is triggered by the synchronizing amplifier.

3. Synchronization :

- In order to obtain a *stationary pattern* synchronization [irrespective of the type of sweep (*e.g.* free running, triggered sweep, driven sweep, non saw-tooth sweep)] has to be done. This requires that the time base be operated at a submultiple frequency of the signal under measurement (applied to *Y*-plates).

The following three usual sources for synchronization can be selected by synchronizing selector :

(*i*) *Internal.* Here the trigger is obtained from the signal being measured through the vertical amplifier.

(*ii*) *External.* In this case, an external trigger source is also used to trigger or initiate the signal being measured.

(*iii*) *Line.* In this method, the trigger is obtained from the power supply to CRO (say 230 V, 50 Hz)

4. Blanking circuit :

- When a saw-tooth voltage is applied to the horizontal plates, the beam moves across the CRT tube in a *straight line* from left to right during the sweep (or trace) time. A comparatively slow movement of the spot will appear as a solid line, provided the rate of movement exceeds the threshold of persistence of vision. *Below this threshold limit, a moving spot is perceived*. However, if the movement of the spot is *fast*, it appears as *thin and dim horizontal* line or may be invisible.
- The saw-tooth voltage waveform shown in Fig. 11.14 is an ideal one. In this waveform, retrace time is zero. But in practice it is not possible to achieve this, so there is some trace retrace time of the waveform, as shown in Fig. 11.16. During this retrace time, the spot moves from right to left and leads to confusion. *So this trace is blanked out by supplying a high -ve voltage to the grid during the retrace time.* The blanking voltage is usually developed/triggered by time-base generator.

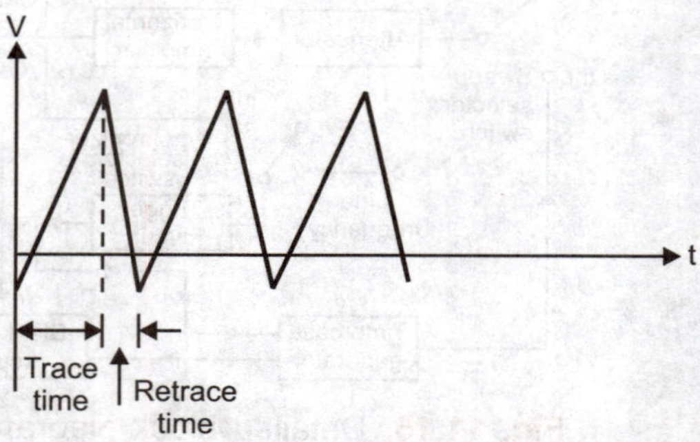

Fig 11.16. Sawtooth voltage waveform.

5. Positioning controls. These controls involve two knobs–one for controlling the horizontal position and another for controlling the vertical position.

— The spot can be moved to *left or right i.e.*, horizontally with the help of a knob, which regulates the D.C. potential applied to the *X*-deflection plates in addition to the usual saw-tooth wave. Similarly, the spot can be moved *up and down i.e.* vertically with the help of another knob, which regulates the D.C. potential applied to the *Y*-deflection plates in addition to the signal.

6. Intensity (Z-axis) modulation :

- The intensity modulation is done by inserting a signal between the ground and the cathode/control grid.
- It is applied during normally visible portion of the trace.
- It can be used for *brightening the display*.

— The periodically brightened spots may be used as markers for time calibration of the main waveform.

7 Intensity control :

- The electron-beam intensity is varied by the *intensity control potentiometer* which changes the grid potential w.r.t. cathode. The grid potential determines the amount of electrons leaving the cathode and thus controls the intensity of the beam.
- It must be ensured that the electron beam does not form burning spots on the screen. A stationary spot should be kept on very low intensity. If the intensity is kept high, the spot must be kept moving. If a *"halo"* appears around the spot, the intensity is too high; the intensity should be turned before turning the oscilloscope on.

8. Focus control :

- The middle anode, in the electron gun of a CRT, is kept at a lower potential w.r.t. other two anodes; it acts like an electrostatic lens and focal length of this lens can be varied by varying the potential of the middle anode w.r.t. other two anodes. So the focusing of an electron beam is done by varying the potential of the middle anode with the help of a potentiometer, as shown in Fig. 11.15.
- *By increasing the +ve potential applied to the focusing anode the electron beam can be narrowed and the spot on the screen can be made a pin point.*

9. Astigmatism:

- *Astigmatism* is an additional focusing control (in most modern oscilloscopes) and is analogous to as astigmatism in optical lenses.
- In order to focus the spot correctly, it is necessary to stop it near the centre of the screen by switching off the time base and adjusting the *X* and *Y* positioning controls. The spot is then made as *sharp as possible by successive adjustment of focus and astigmatism controls.*

10. Calibration circuit :

- Laboratory oscilloscopes normally have an internally generated and stablised voltage of known amplitude which is used for calibration purposes.
- The calibrating voltage usually has a square waveform.

11.13 DUAL TRACE OSCILLOSCOPE

The comparison of two or more voltages is invariably necessary in the analysis and study of many electronic circuits and systems. This is possible by using more than one oscilloscope but in such a case it is difficult to trigger the sweep of each oscilloscope precisely at the same time. A common and less costly method to solve this problem is to use dual trace or multitrace oscilloscopes. In this method, the same electron beam is *used to generate "two traces" which can be deflected from two independent* vertical sources.

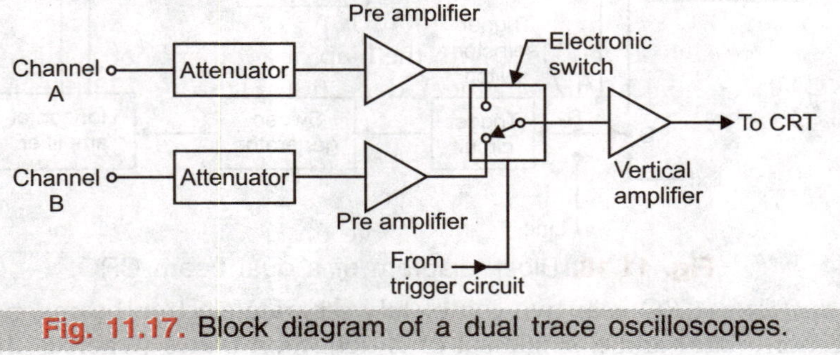

Fig. 11.17. Block diagram of a dual trace oscilloscopes.

Dual trace oscilloscopes are used extensively by *industrial firms and research laboratories.*

These oscilloscopes produce a dual-trace display by means of electronic switching of two separate input signals.

Fig. 11.17 shows the block diagram of a dual trace oscilloscope :

- There are two vertical input circuits marked channels *A* and *B* with identical preamplifiers.
- The outputs of these preamplifiers are fed to an electronic switch which alternately connects them to the main vertical amplifier of the oscilloscope.
- In this type of oscilloscope, there is *only one electron beam.*
- There are two common operating modes for the electronic switch, called *alternate* and *chop;* these are selected from the instrument's front panel.
  - In ***alternate mode,*** the electronic switch alternates between channels A and B, letting each through for one cycle of the horizontal sweep. The display is blanked during the flyback and hold-off periods, as in conventional oscilloscopes provided the sweep speed is much greater than the decay time of the CRT phosphor, the screen will show a steady display of both the waveforms at channels *A* and *B*. *This mode cannot be used for displaying very low frequency signals.*
  - In the ***chop mode,*** the electronic switch free runs at a high frequency of the order of 100 kHz to 300 kHz. The result is that small segments from channels *A* and *B* are connected alternately to the vertical amplifier, and displayed on the screen. Provided the chopping rate is much faster than the horizontal sweep rate, the display will show a continuous line for each channel. If the sweep rate approaches the chopping rate then the individual segments will be visible and the alternate mode should now be used.

11.14 DUAL BEAM OSCILLOSCOPE

Fig. 11.18 illustrates a dual beam oscilloscope.

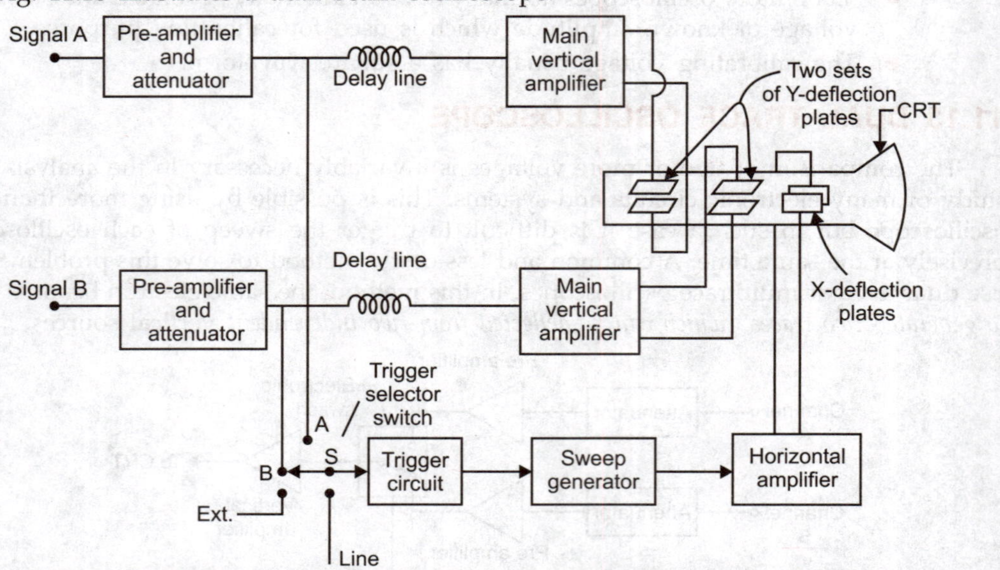

Fig. 11.18. Block diagram of a dual beam CRO.

- The *dual beam CRO* uses two completely separate electron beams, two sets of *Y*-deflection plates and a single set of *X*-deflection plates (whereas the dual trace oscilloscope has one cathode ray gun, and an electronic switch which switches two signals to a single vertical amplifier).

- Only one beam can be synchronized at one time, since the sweep is the same for both signals, *i.e.*, a common time base is used for both beams. Therefore, the signals must have the same frequency or must be related harmonically, in order to obtain both beams locked on the CRT screen, *e.g.*, the input signal of an amplifier can be used as signal *A* and its output signal as signal *B*.
- ***Multiple beam oscilloscope*** has a *single tube but several beam producing systems inside. Each system has separate vertical deflecting pair of plates and generally a common time base system*. The triggering can be done internally using either of the multiple inputs or externally by an external signal or line voltages.

Comparison of dual trace and dual beam oscilloscopes :

S. No.	Aspects	Dual trace CRO	Dual beam CRO
1.	*No. of electron beams used*	One (to generate two traces)	Two
2.	*No. of vertical amplifiers used*	One	Two
3.	*Display of two signals*	Not simultaneous (the two signals are not displayed simultaneously in real time but appears to be displayed simultaneously.	Simultaneous
4.	*Switching between traces*	Difficult (since beam is shared between the two signals)	Easy
5.	*Frequencies of signals*	Different	Same frequency or their frequencies must be integer multiples of each other
6.	*Size and weight*	Less	More
7.	*Modes of operation*	Alternate and chop	The two different types are using double gun tube or split beam using single electron gun

11.15 STORAGE OSCILLOSCOPE

In the conventional CRT, the persistence of the phosphor ranges from a *few milliseconds to several seconds*, so that an event that occurs only once will disappear from the screen after a relatively short period of time.

- A "*storage CRT can retain the display much longer, upto several hours after the image was first written on the phosphor*. This *retention* feature can also be useful when displaying the waveform of a very low-frequency signal. In the conventional (non-storage) oscilloscope, the start of such a display would fade before the end is written.
- A storage oscilloscope uses the phenomenon of "*secondary electron emission*" *to build up and store electrostatic charges on the surface of an insulated target*.

— Storage CRT tubes can be classified as : *Bistable tubes* and *halftone tubes*. The "*bistable tube*" will either store or not store an event and produces only one level of image brightness. The "*halftone tube*" can retain an image for varying lengths of time (variable persistence) and at different levels of image brightness.

- The storage oscilloscopes are specially useful in the following cases :
  (*i*) For real-time observation of events that occur only once.
  (*ii*) For displaying the waveform of a very low-frequency signal.

11.16 SAMPLING OSCILLOSCOPE

Above the range of 50-300 MHz, *sampling techniques* have to be employed to obtain suitable display and CRO employing such sampling methods are called "*sampling oscilloscopes*". The display is made up from as many as 1000 dots of luminescence. The vertical deflection for each dot is obtained from progressively later points in each successive cycle of input waveform as shown in Fig. 11.19. The horizontal deflection of the electron beam is obtained by applications of a staircase waveform to X-deflection plates. Fig. 11.20 shows the display of dots.

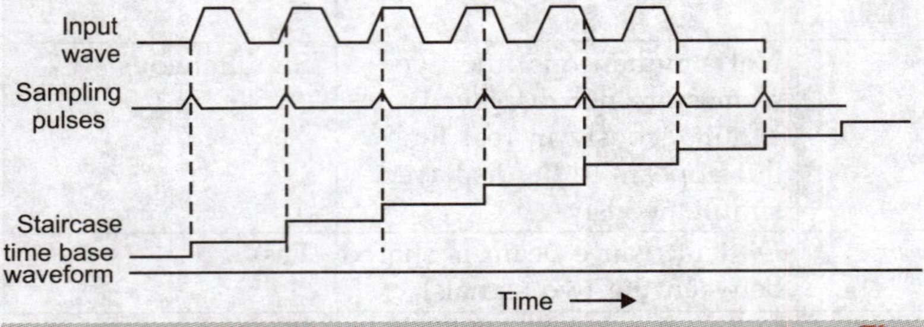

Fig. 11.19. Sampling principle.

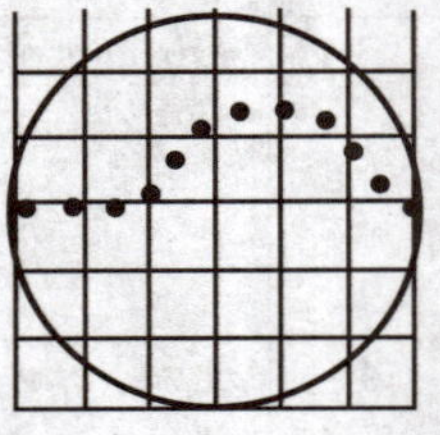

Fig. 11.20. Display of dots.

Fig. 11.21 shows the block diagram of a sampling oscilloscope :

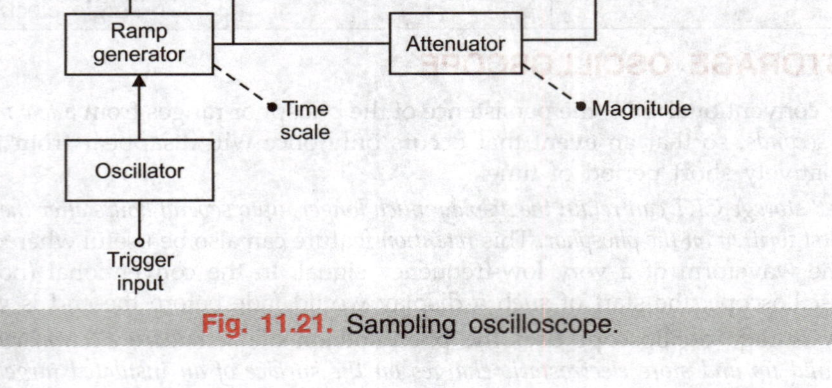

Fig. 11.21. Sampling oscilloscope.

— The input is applied to the *diode sampling gate*.

— At the start of each sampling cycle, a trigger input pulse is generated which activates the blocking *oscillator*. The oscillator output is given to the *ramp generator* which generates the linear ramp signal.

— Since the sampling must be synchronized with the input signal frequency, the signal is delayed in the *vertical amplifier*.

— The *staircase generator* produces a staircase waveform which is applied to an *attenuator*. The attenuator controls the magnitude of the staircase signal and then it is applied to a *voltage comparator*. Another input to the voltage comparator is the output of the ramp generator. The voltage comparator compares the two signals and produces the output pulse when the two voltages are equal. This is nothing but a *sampling pulse* which is applied to sampling gate through the gate control circuitry. This pulse opens the diode gate and the sample is taken in. This sampled signal is then applied to the vertical amplifier and the Y-deflecting plates.

— The output of the staircase generator is also applied to the X-deflecting plates. During each step of staircase, the spot moves on the screen. *The comparator output advances the staircase output through one step*. After the certain number of pulses about 100 or so, the staircase generator resets. The smaller the size of the steps of the staircase generator, larger is the number of samples and higher is the resolution of image.

— The sampling oscilloscope responds and stores rapid bits of information and displays them continuously. It is this quantity of sampling oscilloscopes which has limited sensitivity, bandwidth and small display size. The *sampling method transforms the input signal into lower frequency range, where low frequency circuitry is then capable of producing an effective display*.

Advantages :

1. The display produced is clear.
2. High speed electrical signals can be produced.
3. Very high frequency performance can be achieved.
4. By controlling the size of the steps of the staircase generator, the number of samples and hence the resolution can be controlled.
5. The sampling technique allows the design of the oscilloscope with wide bandwidth, high sensitivity even for low duty cycle pulses.

Limitation. This oscilloscope *cannot be used to display* the *transient waveforms*.

11.17 DIGITAL READOUT OSCILLOSCOPE

This oscilloscope provides readout of the signal information such as voltage, time etc., in addition to the conventional CRT display.

It consists of a *high-speed laboratory CRO and an electronic counter, both in one cabinet*. Fig. 11.22. shows the block diagram of a digital CRO for voltage measurement :

— The input signal is first sampled using sampling circuit.

— The sampled signal is given as one output to each start and step comparators. When input is sampled, the sampling circuit advances the sampling position by a fixed amount. This process is known as *strobing*. The selected sweep time per cm control and number of samples taken per cm decide the equivalent time between two samples.

— The CRT trace is used to identify 0% and 100% zone positions. The positions can be shifted anywhere on the display. The potential divider is used which taps the voltage between 0% and 100% levels. The 0% level is used to produce a pulse for opening of gate while the 100% is used to produce a pulse of closing of gate.

— The coincidence of any of the input waveforms with the selected percentage point is sensed by the voltage comparator. When 0% level gives start pulse, clock gate opens and counter starts counting the pulses. When 100% level gives stop pulse, clock gate closes and counter stops counting.

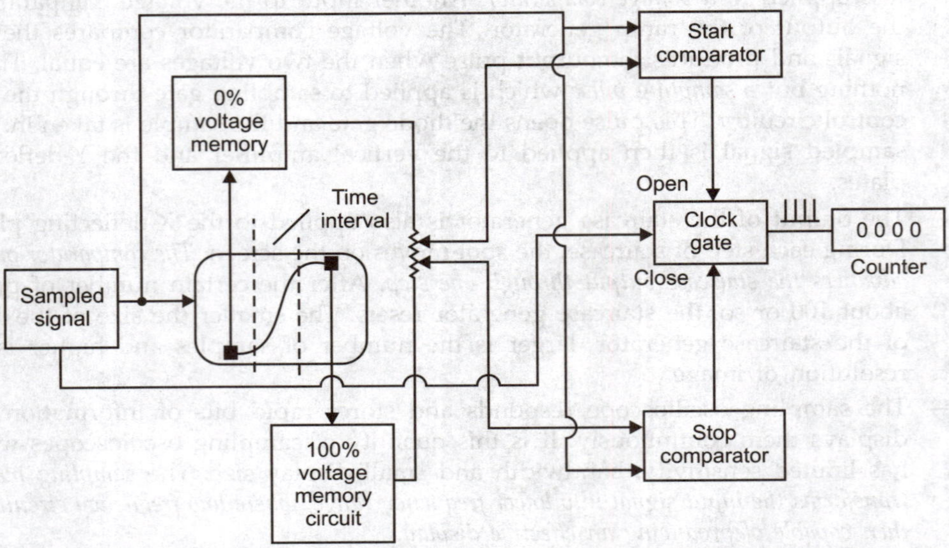

Fig. 11.22. Block diagram of a digital CRO for voltage measurement.

— *The number of pulses counted by a counter are proportional to actual sample taken*. The digital readout is obtained by using Nixie display tube.

11.18 DIGITAL STORAGE OSCILLOSCOPE

Following are the distinct *disadvantages of the storage cathode ray tube* :

(*i*) The storage tube can preserve a stored waveform for a finite amount of time; eventually, waveform is lost. The power to the storage tube must be present as long as the image is to be stored.

(*ii*) The trace of a storage tube is, generally, not as fine as a normal cathode ray tube.

(*iii*) The writing rate of the storage tube is less than a conventional cathode ray tube, which limits the speed of the storage oscilloscope.

(*iv*) The storage cathode ray tube is considerably more expensive than a conventional tube and requires additional power supplies.

(*v*) Only one image can be stored. If two traces are to be compared, they must be superimposed on the same screen and displayed together.

The *"digital storage oscilloscope"* is a superior method of trace storage. In this technique, the waveform to be stored is *digitized, stored in a digital memory, and retrieved for display on the storage oscilloscope*. The stored display can be displayed indefinitely as long as power is applied to the memory, which can be supplied with a small battery.

Fig. 11.23 shows the block diagram of a digial storage oscilloscope :

— The input is amplified and attenuated with input amplifiers (as in any oscilloscope).

— The digital storage oscilloscope uses the same types of input circuitry and oscilloscope probes as a conventional oscilloscope, and many digital storage oscilloscopes can operate in a conventional mode, bypassing the digitizing and storing features.

— The output of the input signal amplifiers feeds an A/D (analog-to-digital) converter. The main requirement of the A/D converter is its speed, while accuracy and resolution are of secondary impedance.

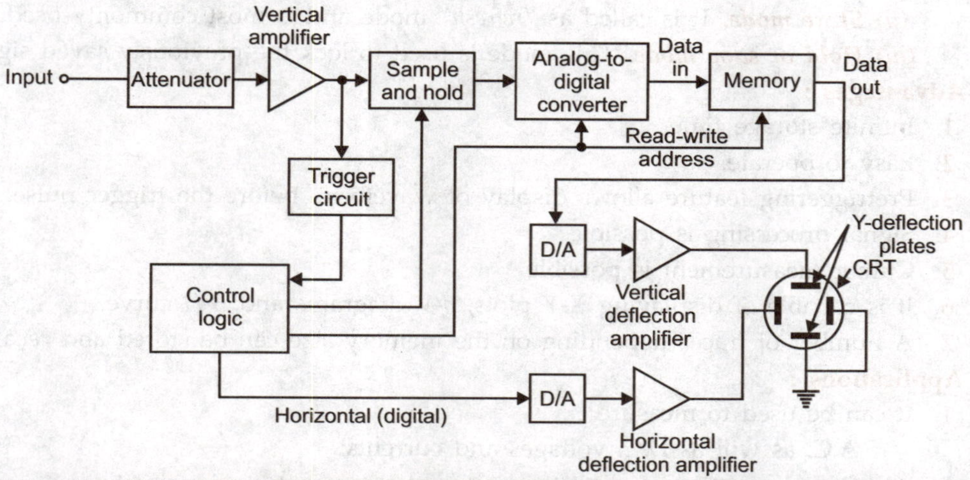

Fig. 11.23. Digital storage oscilloscope.

— The digitized output is in the binary form and not BCD form. Generally, successive approximation type of ADC is used in digital storage oscilloscope. If a 12 bit converter is used then accuracy of about 0.1% and resolution of about 0.025% is obtained, while if a 10 bit ADC is used then the resolution of about 0.1% is obtained.

— Digitizing the analog signal means to take samples of the input signal at periodic intervals of time. The sampling rate should be greater than twice the highest frequency present in the input signal, so that the sampling theory is satisfied. This ensures that aliasing effect will not occur and there is no loss of information. The sampling rate may be as high as 100,000 samples per second. For such high sampling rates, a fast conversion A/D convertor is required. *'Flash' A/D converters* are used at the cost of decreased resolution with increase in sampling rate.

— The size of the memory is related to the amount of horizontal segments of the trace that can be divided into one sweep of the time base, *e.g.*, if memory size is 1 K then the trace will be divided into 1000 horizontal segments. If a 10 bit ADC is used then the frequency response of 25 kHz is obtained; the total memory storage capacity is 4096 for a single channel, 2048 for two channels, 1024 for four channels each.

— The selection of the sampling rate and memory size depends on the type of waveform being recorded. As the analog signal is sampled at suitable rates, the data points are readable on to the memory. Once the signal is stored, the required manipulations can be done.

— A *continuous storage oscilloscope* consists of a feature called ***pretrigger view***. This indicates that the events that occurred before the trigger input signal was applied, are displayed. This selection is a percentage selection.

This mode is useful when failure occurs.

— *Single shot events* are sometimes transient in nature and hence they are quickly lost. Unless the wavefrom is stored, the observer cannot see such events. *Such events are stored in the memory of the digital storage oscilloscope and memory can be rapidly read and on the screen a continuous waveform can be displayed.*

- The digital storage oscilloscope has the following threee modes of operation:
  - *(i)* ***Roll mode.*** This is used is observe the fast varying signal.
  - *(ii)* ***Store mode.*** It is called as "*refresh*" mode and is most commonly used.
  - *(iii)* ***Hold or save mode.*** This mode is used to lock the previously saved signal.

Advantages :

1. Infinite storage time.
2. Easy to operate.
3. Pretriggering feature allows display of waveform, before the trigger pulse.
4. Signal processing is possible.
5. Cursor measurement is possible.
6. It is capable of displaying *X-Y* plots, *p-V* diagrams and *B-H* curve.
7. A number of traces depending on the memory size can be stored and recalled.

Applications :

1. It can be used to measure :
   — A.C. as will as D.C. voltages and currents;
   — Frequency, time period, time interval between two signals etc.,
   — Inductance, capacitance.
2. It is used to give the visual representation for a target of radar such as aeroplane, ship etc.
3. In medical fields, it is used to display cardiograms that are useful for diagnosis of heart disease in the patient.
4. It is used to observe the radiation pattern generated by the transmitting antenna.
5. It can be used to check the faulty components in various circuits. It can also be used to analyse TV waveforms.
6. It can be used to determine the modulation characteristics and detect the standing waves in transmission lines.

11.19 HIGH FREQUENCY OSCILLOSCOPE

A general purpose oscilloscope has an upper frequency limits of about *300 kHz.* However, there are many applications which involve measurement or examination of signals whose frequencies are much higher than this upper limit. For such high-frequency (HF) applications, conventional HF oscilloscopes are built. These HF oscilloscopes can display signals with frequencies upto *500 MHz.* They differ in their internal construction from low-frequency (LF) oscilloscopes by their special HF CRTs and high-frequency amplifiers.

Fig. 11.24 shows a high frequency CRT.

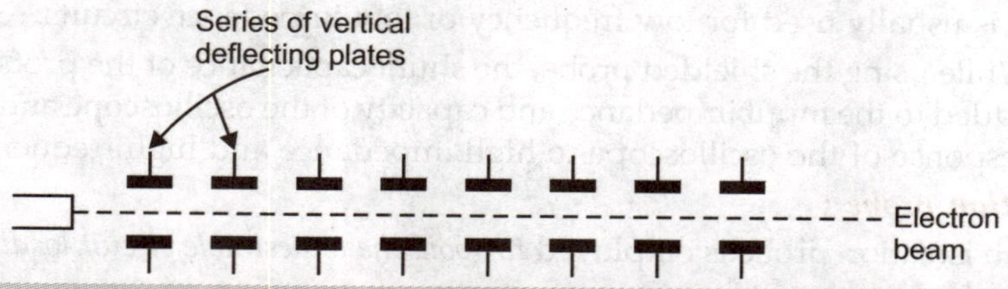

Fig. 11.24. High-frequency CRT.

— In an ordinary cathode ray oscilloscope, there is only one pair of vertical deflection plates. When the HF signal is to be displayed, the electron beam does not get sufficient time to pick up the instantaneous level of the signal. Also, at high frequencies the number of electrons striking the screen in a given time and the intensity of the beam is reduced. Hence, instead of one set of vertical deflection plates, a *series of vertical deflection plates* are employed.

— The vertical deflection plates are so *shaped* and *spaced that an electron moving along CRT receives from each set of plates an additional deflection force in proper time sequence.* This synchronization is obtained by making the signal travel from one plate to the next plate at the *same speed as the transit time of electrons.*

— The signal is applied to each pair of plates, and as the electron beam moves the signal also moves through the delay times. The time delays are so arranged that the same electrons are deflected by the input signal. Thus the electron beam picks up the level of the input signal. *The time delays between the plates correspond exactly to the electron transit times.* In addition, new fluorescent materials have now been developed to increase the brightness at high frequencies.

11.20 ACCESSORIES OF CATHODE RAY OSCILLOSCOPES

The CRO is one of the most useful instruments in the electronic industry. Its usefulness is further extended by the provision of accessories or auxiliary equipment. Source of the accessories are enumerated and described below :

1. Probes
2. Electronic switch
3. Calibrators
4. Cameras.

11.20.1. Probes

The *function of a probe is to connect the test circuit to the oscilloscope without altering, loading, or otherwise disturbing the test circuit.*

Probes are of the following types :

1. Direct probe;
2. Isolation probe;
3. High impedance or 10:1 probe;
4. Active probes;
5. Current probe.

1. *Direct probe :*

- It is the simplest of all the probes and uses a shielded co-axial cable.
- It avoids stray pick-ups which may create problems when low level signals are being measured.

- It is usually used for low frequency or low impedance circuits.
- While using the shielded probe, the shunt capacitance of the probe and cable is added to the input impedance and capacity of the oscilloscope and acts to lower response of the oscilloscope to high impedance and high frequency circuits.

2. *Isolation probe* :

- An isolation probe is employed *to avoid the undesirable circuit loading effect of the shielded probe.*
- It is made by placing a carbon resistor in series with the test lead, as shown in Fig. 11.25.

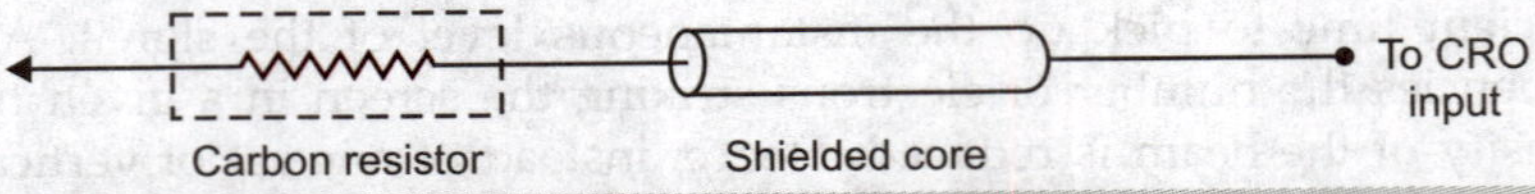

Fig. 11.25. Isolation probe.

- This probe causes a slight change in the amplitude of the waveform and a slight change in the wave shape. To avoid this possibility, a high impedance compensated probe, called a *low capacitance* or a 10 : 1 probe is employed.

3. *High Impedance or 10:1 Probe* :

These probes are used *to increase the input resistance and reduce the effective input capacitance of an oscilloscope.*

Fig. 11.26 (*a*, *b*) shows the layout of a high impedance or 10:1 probe and its equivalent circuit.

The capacitor is adjusted so that the elements of the bridge are balanced. Under conditions of balance we have :

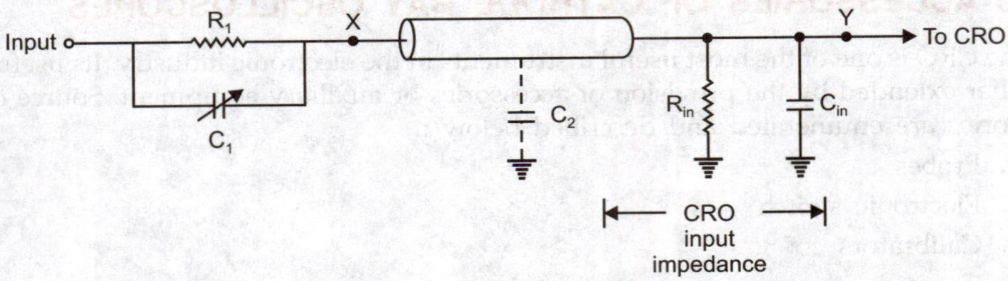

(a) Lay out of a 10:1 probe.

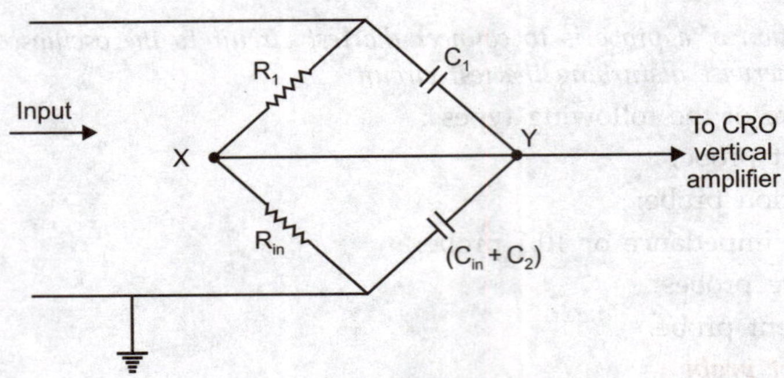

(b) Equivalent circuit of 10:1 probe.

Fig. 11.26. High impedance or 10:1 probe.

$$R_1 X_{(C_{in}+C_2)} = R_{in} X_{(C_1)}$$

or, $$\frac{R_1}{\omega (C_{in}+C_2)} = \frac{R_{in}}{\omega C_1}$$

or, $$R_1 C_1 = R_{in} (C_{in} + C_2) \quad \text{...(11.25)}$$

Therefore, X and Y are equipotential and the effect of the probe is equivalent to placing a potential divider consisting of R_1 and R_{in} across, the input circuit.

- The attenuation of the signal is 10:1, *i.e.*, $\frac{R_1 + R_{in}}{R_{in}}$ = 10:1 over a wide frequency range. Hence, it is called a compensated 10:1 probe.
- Capacitor C_1 is adjusted by connecting the probe tip to a square-wave of 1 kHz and observing the CRT display. When the CRT display has *optimum response* the value of C_1 is deemed to be appropriate.

Then, $$V_{out} = 0.1\, V_{in} = \frac{V_{in} \times R_{in}}{R_1 + R_{in}} \quad \text{...(11.26)}$$

4. *Active probes :*

- These probes are designed *to provide an efficient method of coupling high frequency, fast rise time signals to the CRO input.*
- Usually active probes have *very high input impedance with less attenuation than passive probes.*
- Active devices may be diodes, FETs, BJTs, etc.
- These probes are *more expensive and bulky than passive probes, but they are useful for small signal measurements, because their attenuation is less.*

Fig. 11.27 shows a basic circuit of an active probe using a FET and BJT.

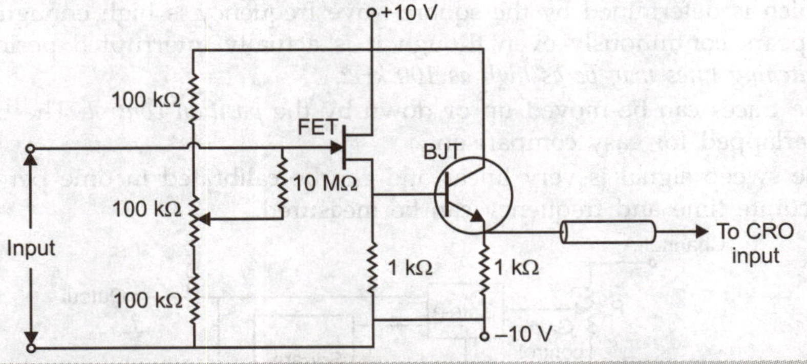

Fig. 11.27. FET probe.

- The FET is used as the active element to amplify the input signal. Although voltage gain of FET follower circuit shown is unity, the follower circuit provides a power gain so that input impedance can be increased.
- To be effective, the FET must be mounted directly in the voltage probe tip, so that the capacitance of the interconnecting cable can be eliminated. This requires that the power for the FET be supplied from the oscilloscope to the FET in the probe tip. The FET voltage follower drives a coaxial cable but instead of the cable connecting directly to the high input impedance of the oscilloscope it is terminated in its characteristic impedance.

- Because there is no signal attenuation between FET amplifier and probe tip, the range of the signals that can be handled by the FET probe is limited to the dynamic range of the FET amplifier and is typically *less than a few volts.* This is a *significant drawback of the FET probe.* Therefore, for *handling a large dynamic range, external attenuators are added at the probe tip.*
- *Active probes have limited use because the FET probe effectively becomes an attenuator probe. So the oscilloscopes are typically used with 10:1 attenuator probe.*

5. ***Current probe :***

- It is a very valuable probe and can be clamped around a wire carrying current *without any physical contact to the probe,* allowing the oscilloscope to be employed to measure the magnitude of the current with a frequency from 0 to 50 Hz (D.C.).
- The current sensor consists of two parts : A *conventional transformer* for transforming alternating current to voltage, and a *Hall effect device* for converting direct current to voltage.

11.20.2. Electronic switch

The electronic switch is an extremely useful accessory for displaying the two signals simultaneously on single trace oscilloscopes. Fig. 11.28 shows the block diagram of an electronic switch :

— Basically, the electronic switch is a *square-wave generator.*

— Two signals which are to be displayed are each applied to the separate gain control and gate stage. The gate stages are alternately biased to cut-off by square-wave signals from the square-wave generator. Thus only one gate stage is in a condition to pass its signal at any given time.

— The output of both stages are applied directly to the oscilloscope input. Amplitudes of the two signals are adjusted by their gain controls. When the switching rate, which is determined by the square-wave frequency is high enough each signal appears continuously even though it is actually interrupted periodically. *The switching rates may be as high as 100 kHz.*

— The traces can be moved up or down by the *position control.* The traces can be overlapped for easy comparison.

— The sweep signal is very linear and can be calibrated in time per mm so that accurate time and frequency can be measured.

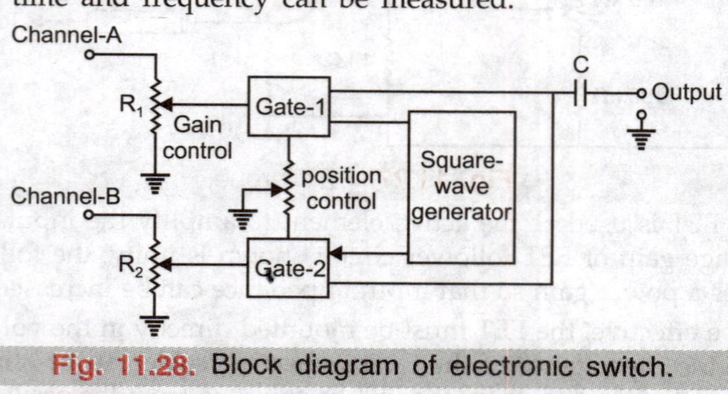

Fig. 11.28. Block diagram of electronic switch.

11.20.3. Calibrators

Several oscilloscopes have a *built-in reference source of voltage* which has usually a square waveform with a frequency of 1 kHz. The voltage may have a single magnitude or may be of several selectable magnitudes. *Provision of this facility enables the oscilloscope amplifiers and time base to be calibrated and checked for accuracy.*

11.20.4. Cameras

The use of photography is the best method of permanent records of oscilloscope traces for analysis.

- The special cameras used for keeping the records are of the following two types : (*i*) 35 mm film; (*ii*) Polaroid film.
- *35 mm unit* is usually prefocused and bolted over the tube face. Some cameras have a fixed lens aperture, the exposure is simply controlled by hand operation of flapper shutter.
- *Polaroid film* photographic technique approaches *instant photography*. A permanent record is obtained within 10–20 s. However, this process is comparatively *expensive*.

11.21 CRO MEASUREMENTS

CRO can be used to measure various characteristics of an input signal and the properties of the output signal. The various parameters which can be measured by using CRO are :

(*i*) Voltage; (*ii*) Current;
(*iii*) Period; (*iv*) Frequency;
(*v*) Phase; (*vi*) Amplitude;
(*vii*) Peak-to peak value (*viii*) Duty cycle etc.

11.21.1. Voltage measurement

The procedure for voltage measurement is as follows :

— The voltage to be measured is applied to Y-deflection plates through the vertical amplifier. Generally, a coaxial cable is used for connection.

— The X-deflection plates are excited by the time-base generator.

— To aid the measurement, a set of squares is marked on the transparent screen cover. Using these markings and the gain of the vertical amplifier, the peak-to-peak (p-p) value (and hence r.m.s. value) of the voltage wave can be measured.

— After noting down the selection in *volts/division* from the front panel (selected for measurement) the peak-to-peak value, amplitude and r.m.s value of sinusoidal signal can be obtained as given below :

(*i*) Peak-to-peak value, $V_{p-p} = \left(\frac{\text{Volts}}{\text{div.}}\right) \times \text{(number of divisions)}$

(The volts/div. is nothing but *deflection sensitivity* of CRO)

(*ii*) Amplitude, $V_{max} = \frac{V_{p-p}}{2}$

(*iii*) R.M.S. value, $V_{r.m.s.} = \frac{V_{p-p}}{2\sqrt{2}}$ (for sinusoidal signals only)

- It must be remembered that an oscilloscope is very complex device and that the accuracy of measurements is not better than ± 3%. It is basically a *display device so that we can observe the waveform in detail and find out the extent of harmonics in the wave.*

Example 11.7. *For the waveform shown in Fig. 11.29, if vertical attenuation is 3 mV/div., determine :*

(*i*) *Peak-to-peak value,*

(*ii*) *Amplitude, and*

(*iii*) *R.M.S. value of the signal.*

Solution. Vertical attenuation = 3 mV/div.

From the figure, it is evident that the screen is divided such that one part is subdivided into 5 units.

$$\therefore \quad 1 \text{ subdivision} = \frac{1}{5} = 0.2 \text{ units (see Fig. 11.29)}$$

(*i*) **Peak-to-peak value-V_{p-p} :**

It can be observed that positive peak of the signal corresponds to two full divisions and three subdivisions. Hence positive peak is 2 + 3 × 0.2 = 2.6 units. While the negative peak also corresponds to 2.6 units.

$$\therefore \quad V_{p-p} = 2.6 + 2.6 = 5.2 \text{ divisions}$$

$$\text{Also,} \quad V_{p-p} = \frac{\text{Volts}}{\text{division}} \times \text{No. of divisions} = 3 \times 5.2 = \mathbf{15.6\ mV} \text{ (Ans.)}$$

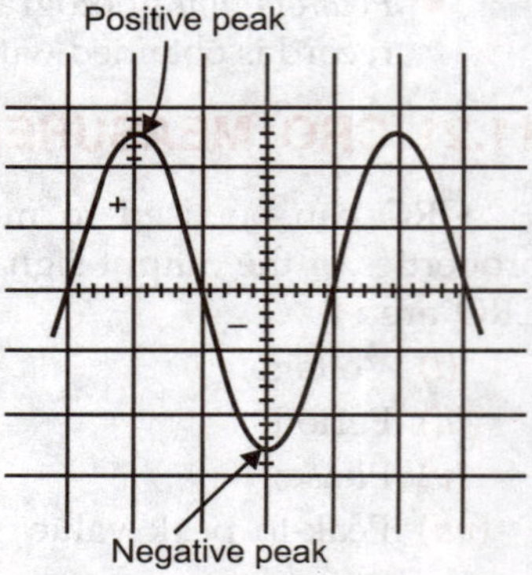

Fig. 11.29.

(*ii*) **Amplitude, V_{max} :**

$$V_{max} = \frac{V_{p-p}}{2} = \frac{15.6}{2} = \mathbf{7.8\ mV} \text{ (Ans.)}$$

(*iii*) **R.m.s. value, $V_{r.m.s.}$:**

$$V_{r.m.s.} = \frac{V_{max}}{\sqrt{2}} = \frac{7.8}{\sqrt{2}} = \mathbf{5.515\ mV} \text{ (Ans.)}$$

11.21.2. Current measurement

A CRO has a very high input impedance and cannot be used for direct measurement of current. However, the current can be measured in terms of voltage drop across a standard resistance as shown in Fig. 11.30; *i.e.*, $I = \frac{V \text{ (measured on CRO)}}{R}$.

- For measurement of *high frequency currents, the shunt should be non-inductive.*

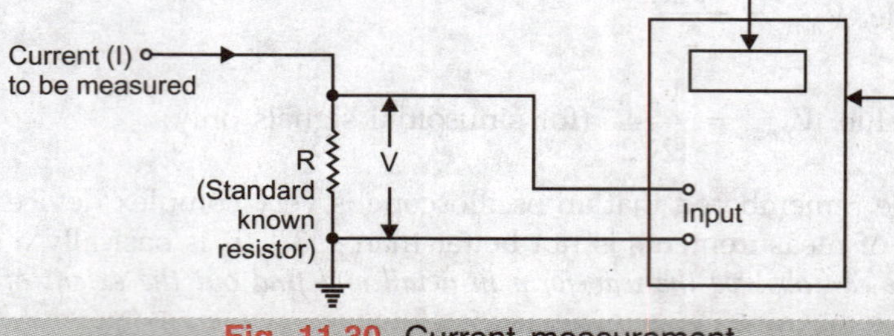

Fig. 11.30. Current measurement.

11.21.3. Time period measurement

For the measurement of time period (*T*) of the waveform, it is displayed on the screen such that *one complete cycle is visible on the screen*. The accuracy increases if a single cycle occupies as much as the horizontal distance on the screen.

After noting the time/division selected on the front panel, the time period of the waveform can be obtained as :

Time period, $T = \dfrac{\text{Time}}{\text{Division}} \times$ number of divisions occupied by *one cycle*.

The frequency (*f*) is then given as :

$$f = \frac{1}{T}$$ (This is the method of frequency measurement without Lissajous pattern)

11.21.4. Lissajous figures/patterns

Lissajous patterns (named in the honour of French scientist who first obtained them geometrically and optically) *are formed when two sine waves are applied simultaneously to the vertical and horizontal deflecting plates of CRO*. The two sine waves may be obtained from two audio oscillators. Obviously, in this case, a *sine wave sweeps a sine-wave input signal.*

- *The shape of the Lissajous pattern depends on the frequency and phase relationship of the two sine waves.*
- Two sine waves of the *same* frequency and amplitude may produce a line, an ellipse or a circle depending on their *phase difference* as shown in Fig. 11.31.

  In general, the shape of Lissajous figures depend on :

(*i*) Amplitude ;

(*ii*) Phase difference ;

(*iii*) Ratio of frequency of the two waves.

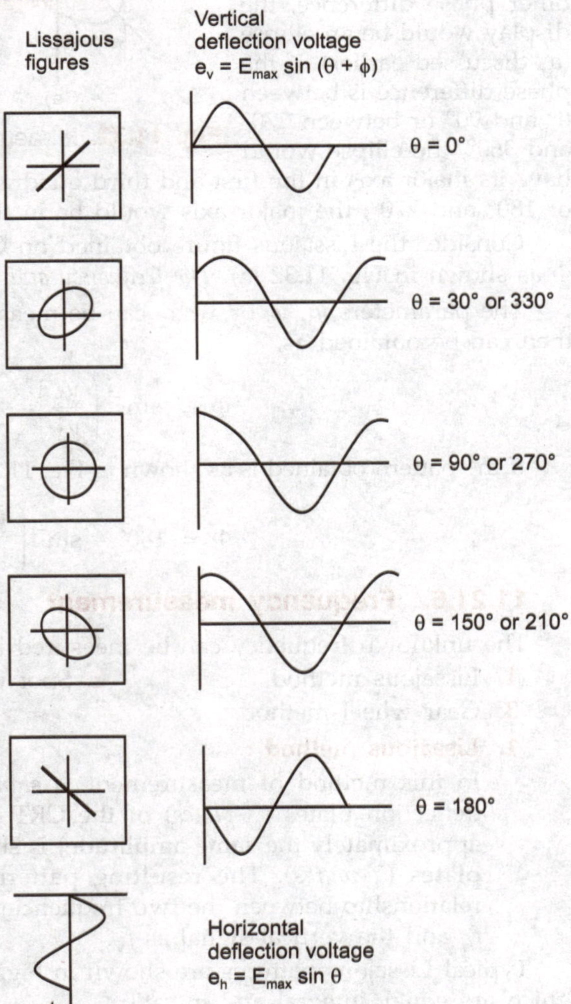

Fig. 11.31. Lissajous patterns for same frequency, different phase shift.

Uses. The Lissajous figures are used for the following purposes :

1. To determine an unknown frequency by comparing it with a known frequency.
2. To check audio oscillator with a known-frequency signal.

3. To check audio amplifiers and feedback networks for phase shift.

11.21.5. Phase-angle measurement

An oscilloscope can be used to find the phase angle between the two sinusoidal quantities of the same frequency. One of the signals is fed to the *Y*-plates. The time-base generator is switched out and the second signal is fed to the *X*-plates. It is necessary that *X* and *Y* are of equal magnitudes. This fact can be checked by measuring the amplitudes separately (as in voltage measurement).

If the two signals are in phase, the display would be a straight line at 45° to the horizontal. If the phase angle is 90°, the display would be a circle. For any other phase difference, the display would be an ellipse (as discussed earlier). If the phase difference is between 0° and 90° or between 270° and 360°, the ellipse would have its major axis in the first and third quadrants. For difference between 90° and 180° or 180° and 270°, the major axis would be in the second and fourth quadrants.

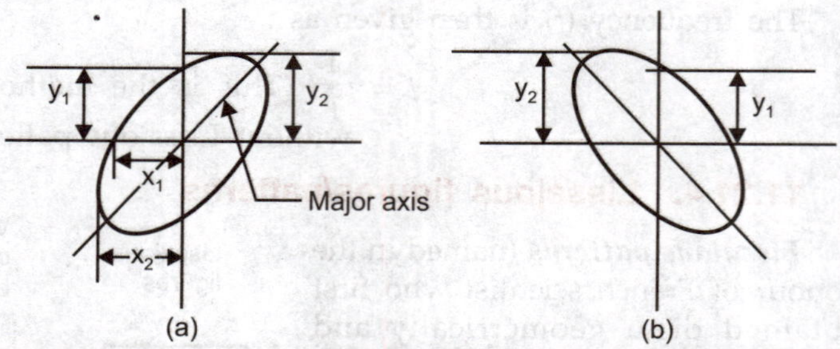

Fig. 11.32. Measurement of phase angles by using CRO.

Consider the Lissajous figure obtained on CRO with an unknown phase difference Φ as shown in Fig. 11.32 (*a*). *The frequency and amplitude of the two waves are same.*

The parameters y_1, y_2 or x_1, x_2 can be measured (See Fig. 11.32(*a*)); the phase angle then can be obtained as,

$$\Phi = \sin^{-1}\left(\frac{y_1}{y_2}\right) = \sin^{-1}\left(\frac{x_1}{x_2}\right)$$

If the pattern obtained is as shown in Fig. 11.32 (*b*), then the phase angle Φ is given by:

$$\Phi = 180° - \sin^{-1}\left(\frac{y_1}{y_2}\right)$$

11.21.6. Frequency measurement

The unknown frequency can be measured by the following methods :

1. Lissajous method.
2. Spot wheel method.
3. Gear wheel method.

1. Lissajous method :

In this method of measurement, a *standard frequency* is applied to one set of deflection plates (*X-plates*) of the CRT tube while the unknown frequency (of approximately the same amplitude) is simultaneously applied to the other set of plates (*Y-plates*). The resulting patterns depend on the integral and phase relationship between the two frequencies. The horizontal signal is designated as f_H and the vertical signal as f_V.

Typical Lissajous patterns are shown in Fig. 11.31 (*a* & *b*) for sinusoidal frequencies which are equal, integral and in ratio.

Example 11.8. *Voltage E_1 is applied to the horizontal input and voltage E_2 is applied to the vertical input of a CRO. E_1 and E_2 have the same frequency. The trace is an ellipse. The slope* *of the major axis is positive. The maximum vertical value is 2.5 divisions and the point where the ellipse crosses the vertical axis is 1.25 divisions. The ellipse is symmetrical about the horizontal and vertical axis. Determine the possible phase angles of E_2 with rasped to E_1.*

(Nagpur University, W-2005)

Solution. Referring to Fig. 11.32 (*a*),

$$\sin \Phi = \frac{y_1}{y_2} = \frac{1.25}{2.5} = 0.5$$

$$\therefore \quad \Phi = 30°$$

Thus possible phases are **30° or 330° (Ans.)**

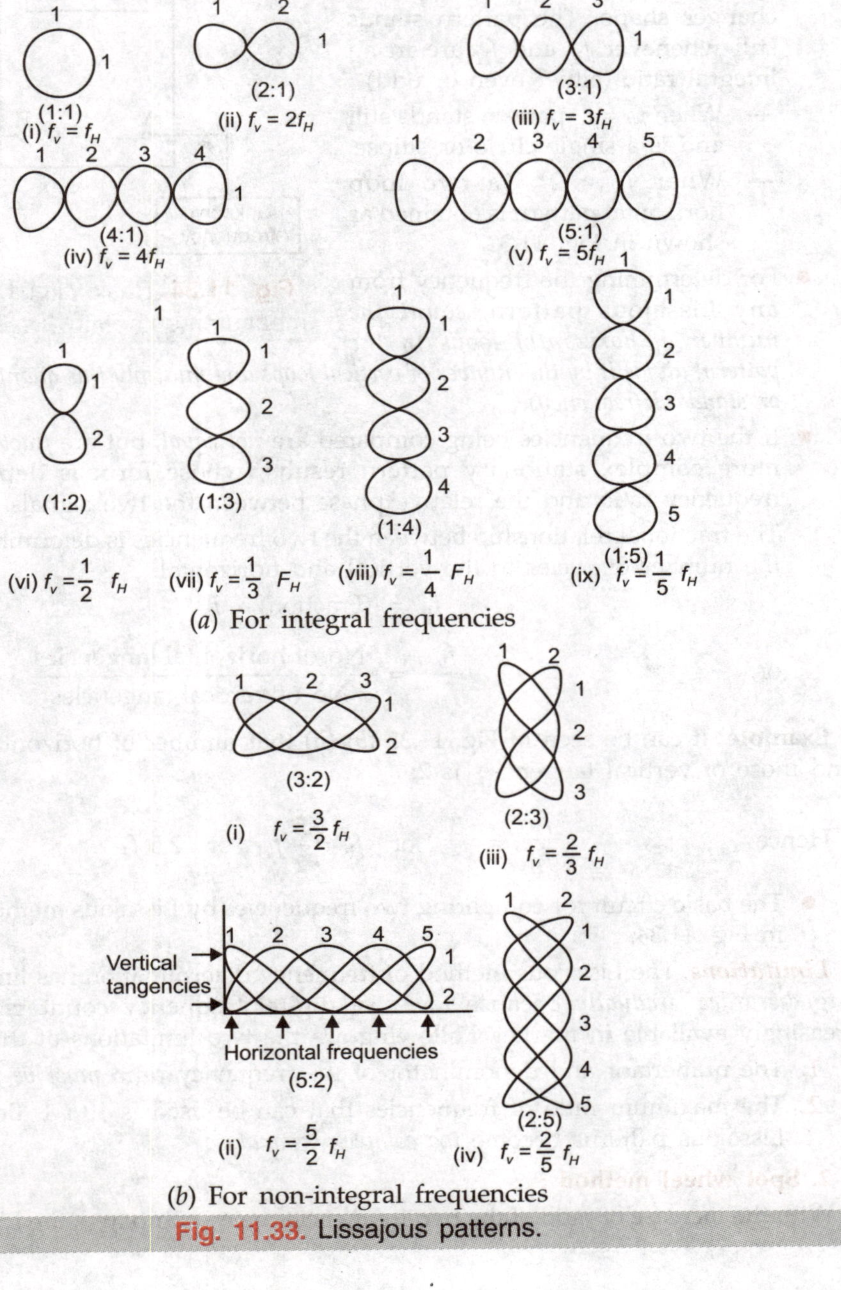

Fig. 11.33. Lissajous patterns.

Measurement procedure :

- Set up the oscilloscope and switch off the internal sweep (change to Ext.)
- Switch off sync. control.
- Connect the signal source as shown in Fig. 11.34.
- Set the horizontal and vertical gain control for the desired width and height of the pattern.
- Keep frequency f_V constant and vary frequency f_H, noting that the pattern spins in alternate directions and changes shape. The pattern stands still whenever f_V and f_H are in an integral ratio (either even or odd)
  — When $f_V = f_H$ pattern stands still and is a single circle or ellipse.
  — When $f_V = 2f_H$, a two loop horizontal pattern is obtained as shown in Fig. 11.33.

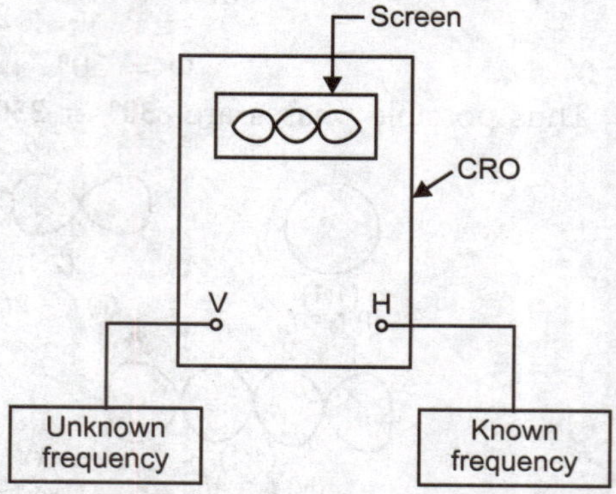

Fig. 11.34. Basic circuit for frequency measurements with Lissajous patterns.

- For determining the frequency from any Lissajous pattern, *count the number of horizontal loops in the pattern, divide it by the number of vertical loops and multiply this quantity by f_H (known or standard frequency).*
- If the two frequencies being compared are *not equal*, but are *fractionally related*, a more complex stationary pattern results, whose form is dependent on the frequency ratio and the relative phase between the two signals.

  The fractional relationship between the two frequencies is determined by counting the number of cycles in the vertical and horizontal.

$$f_V = (\text{Fraction}) \times f_H$$

or,

$$\frac{f_V}{f_H} = \frac{\text{No. of horizontal tangencies}}{\text{No. of vertical tangencies}}$$

Example. It can be seen in Fig. 11.33 (*b*) (*ii*) that number of horizontal tangencies is 5 and those of vertical tangencies is 2;

Hence,

$$\frac{f_V}{f_H} = \frac{5}{2}, \text{ or } f_V = \frac{5}{2} f_H \text{ or } 2.5\, f_H.$$

- The basic circuit for comparing two frequencies by Lissajous method is illustrated in Fig. 11.34.

Limitations. The Lissajous method of frequency determination has limitations and is being *discarded gradually* because low-cost digital frequency counters are becoming increasingly available in market. Following are the two limitations of this method :

1. The numerator and denominator of the frequency ratio *must be whole number*.
2. The maximum ratio of frequencies that can be used is 10 : 1. Beyond that, the Lissajous patterns become *too complex* to *analyse*.

2. Spot wheel method :

With the increase of ratio of the two frequencies being compared, the Lissajous pattern becomes more complicated hence more difficult to interpret. In such cases, it is advantageous to use spot wheel method of display.

Fig. 11.35 illustrates the *spot wheel method of frequency determination* :

— *Fig. 11.35 (a)* shows the *basic circuit* for frequency measurement, whereby an intensity modulated circular or elliptical pattern is produced on the face of the CRT.

— The *lesser* of the two frequencies being compared is applied to the deflecting plates of the CRT through resistance-capacitance phase shifter as illustrated, thereby producing the circular or elliptical sweep.

— The higher frequency is applied to the control grid of the CRT (biased slightly below cut off), as shown in *Fig. 11.35 (b)*, thus modulating the intensity of beam. As a consequence, the pattern appears as a series of alternate bright and dark spots, as shown in *Fig. 11.35 (c)*, where the ratio of high to low frequencies is given by the *number of bright spots*.

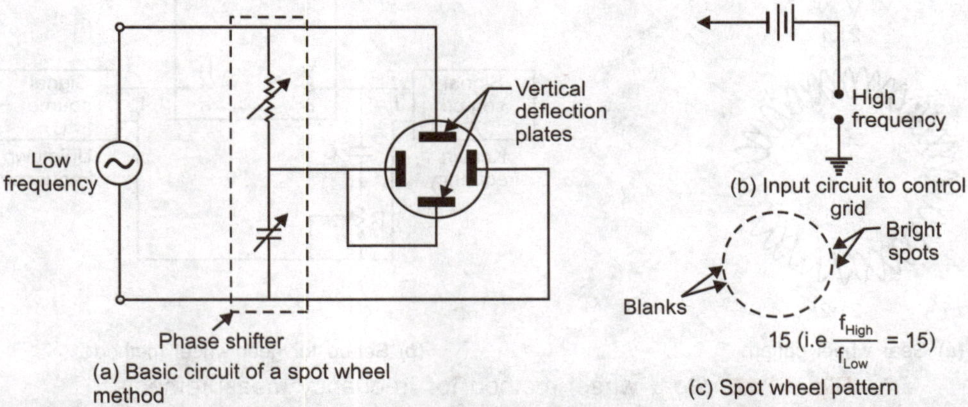

Fig. 11.35. Spot wheel method for frequency measurements.

— When the voltages across *R* and *C* are equal, the pattern is circle; if they are unequal, the pattern is elliptical.

— The variation of the grid-cathode potential changes the density of the electron stream within the CRT and *determines the intensity*. When the grid bias of the CRT reaches cut-off potential, there is no electron stream and fluorescence at the screen cannot occur. *This property is used for the comparison of two frequencies, provided their ratio is an integer. The high frequency signal is applied to the grid.* Sufficient grid must be present so that the negative peaks of the sine wave cut off the electron beam.

— The high frequency signal need not be a pure sine wave, it can be a square wave.

Fig. 11.35(*c*) shows that the frequency ratio is 15 : 1 $\left(i.e. \frac{f_{High}}{f_{Low}} = 15 \right)$ and there are 15 blanks in the pattern :

3. Gear wheel method :

When the number of loops in a Lissajous figure is large, the accurate counting becomes difficult.

- A test method that uses a *modulated ring pattern* in place of looped figure and permits a higher count is shown in Fig. 11.36 (*a*). This pattern is also called a 'gear wheel' or 'toothed wheel', because of its shape. The unknown frequency is determined by multiplying the known frequency by the number of teeth in the pattern.

- Fig. 11.36 (*b*) shows the set-up for gear wheel method for frequency measurements :

— In this method, a phase shift network (RC) introduces a 90° phase shift between the horizontal and vertical channels of the oscilloscope, which is needed to produce a ring or circle pattern with a known frequency f_V. This ring is *modulated* by the voltage from the unknown frequency.

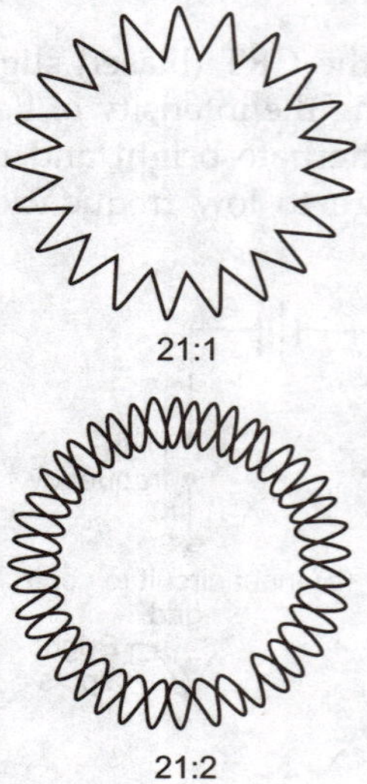

(*a*) Gear wheel pattern

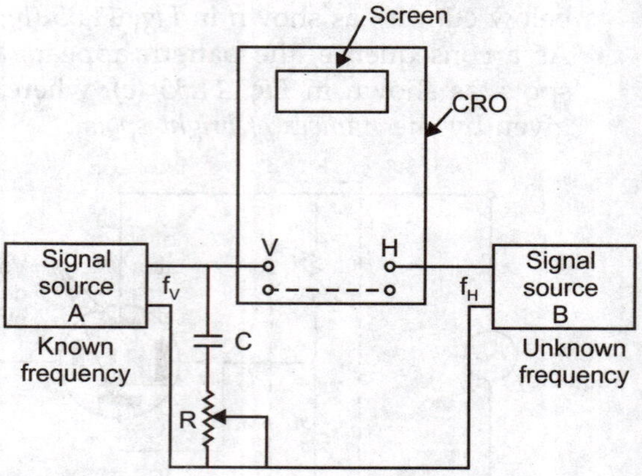

(*b*) Set-up for gear wheel method

Fig. 11.36. Gear wheel method for frequency measurements.

— When the *voltages across* R and C are unequal, the pattern is elliptical. If they are equal, the pattern is circle.

— *The unknown frequency must be higher than the known frequency and the amplitude of the unknown must be below that of the known to prevent distortion.*

Measurement procedure :

- After setting up the oscilloscope, switch off and Sync–control and connect the circuit as shown in Fig. 11.36 (*b*). Switch off the signal *B* temporarily.
- *Switch on the signal source A* and adjust *R* to obtain a ring pattern on the screen. Adjust the horizontal and vertical gain controls to spread the ring over the screen as much as possible.
- *Switch on the signal source B*, noting that the ring becomes *toothed by the signal*. Adjust the known frequency f_V to stop the ring from rotating. Adjust the amplitude of f_H to obtain a distinct teeth pattern. The unknown frequency is then given by :

$$f_H = nf_V$$

where f_H is an integral multiple (odd or even) of f_V.

— *The wheel rotates or spins if f_H is not exact multiple of f_V, therefore adjust the known frequency to stop the wheel from rotating.*

11.21.7. Dielectric loss measurement

Fig. 11.37 (*a*) shows the circuit diagram or measurement of dielectric loss by cathode ray oscilloscope :

- C is capacitor formed by dielectric material under test and C_s is the *standard capacitor of negligible loss*; these capacitors C and C_s are connected in series across A.C. supply source.
- The voltage across the capacitor C or a known fraction of it applied to the X-deflection plates and the voltage across the standard capacitor C_s is applied to the Y-deflection plates of CRO. There will be phase difference between the voltages across capacitors C and C_s, because voltage across standard capacitor C_s will lag 90° behind the screen flowing through it but the voltage across capacitor C will not *lag exactly 90°* behind the current flowing through it *owing to its losses*.

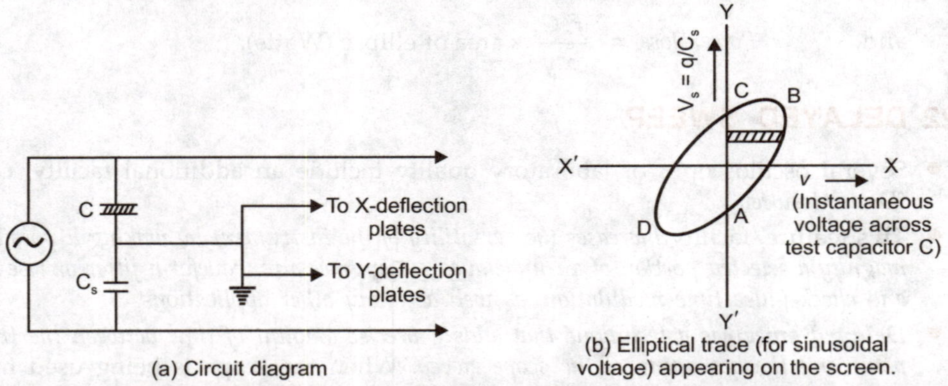

Fig. 11.37. Dielectric loss measurement by CRO.

- When the voltage is sinusoidal then the trace appearing on the screen of CRT will be an ellipse, as shown in Fig. 11.37 (*b*), and the *area of this trace will represent the power loss per cycle of the capacitor C under test*. More the power loss in the test capacitor C, more will be the phase difference of the two voltages and so more will be the area of the trace.

The work done on the test capacitor C over any period of time is given as :

$$dw = \int vidt = v\int \frac{dq}{dt}.dt = \int v.\,dq$$

where, v = Instantaneous voltage across the test capacitor C and

q = Instantaneous charge on the standard capacitor C_s.

If the deflection sensitivity CRO is x and y mm/V in X-and Y-axes respectively, then the area of the hatched section shown in the figure will be :

$$x.\ v.\ y.\frac{dq}{C_s}\ mm^2$$

— When the *spot moves from point A to point B*, the area between curve AB and the Y-axis is

$$\int x.v.y.\frac{dq}{C_s} = \frac{xy}{C_s}\int v.dq = \frac{xy}{C_s}\times \text{work done on capacitor } C$$

— When the *spot moves from point B to point C*, both the voltages across the capacitor C and charge on the capacitor C_s are being reduced so that *energy is being restored* to the circuit and this amount of energy is represented by the curve BC and the Y-axis. But the area enclosed by curve ABC and the Y-axis represents the work done on the dielectric and *not returned to the circuit*. Similarly the area enclosed by *CDA* represents work done during the remaining half cycle, so that the *area of the ellipse represents the work done on the dielectric per cycle*.

$$\text{Since area of ellipse} = \frac{xy}{C_x} \times \text{work done on capacitor C, therefore:}$$

Work done or dilectric loss per cycle

$$= \frac{C_s}{x.y} \times \text{area of ellipse}$$

and, $$\textit{Power loss} = \frac{f.C_s}{x.y} \times \text{area of ellipse (Watts).}$$

11.22 DELAYED SWEEP

- Several oscilloscopes of laboratory quality include an additional facility, called *'Delayed sweep'*.
- This feature/facility *increases the versatility of the instrument by making it possible to magnify a selected portion of an undelayed sweep, measure waveform jitter or rise time, and check pulse time modulation, as well as many other applications.*
- ***Delayed sweep*** *is a technique that adds a precise amount of time between the trigger point and the beginning of the scope sweep.* When the scope is being used in the sweep mode, the start of the horizontal sweep can be delayed, typically from a few μs to perhaps 10 seconds or more. Delayed sweep operation allows the users to view a small segment of the waveform, *e.g.* an oscillation or ringing that occurs during a small portion of a low frequency waveform.

The following are the most common approaches used by the oscilloscope manufacturers for delayed sweep operations :

(*i*) Intensified sweep, where the delayed sweep acts as a positional magnifier.

(*ii*) Normal triggering sweep after the desired time delay, which is set from the panel controls.

(*iii*) A Delay Plus Trigger mode, where a visual indication, such as light, indicates that the delay time has elapsed and sweep is ready to be triggered.

HIGHLIGHTS

1. The *cathode ray oscilloscope* is a device that allows the amplitudes of electrical signals, whether they be voltage, current, power, etc., to be displayed primarily as a function of time.
2. Cathode ray tube (CRT) consists of the following parts :

   (*i*) Electron gun assembly;

   (*ii*) Deflection plates assembly;

   (*iii*) Fluorescent screen;

   (*iv*) Glass envelope

   (*v*) Base

3. *Electrostatic deflection* :

$$v_x = \sqrt{\frac{2eV_a}{m}} \quad ...(i)$$

$$D = \frac{LV_d l_d}{2dV_a} \quad ...(ii)$$

where, v_x = Velocity of electron as it comes out of electron gun, m/s,
e = Charge on electron, C,
m = Mass of electron, kg,
V_a = Accelerating voltage,
D = Deflection on the screen,
V_d = Deflection voltage,
l_d = Length of each Y-plate,
d = Distance between Y-plates, and
L = Distance of middle of Y-plate to the screen.

Electrostatic deflection sensitivity, $S = \frac{D}{V_d}$; its units are m/V.

4. Magnetic deflection :

$$D = BLl\left(\sqrt{\frac{e}{2mV_a}}\right)$$

where, B = Flux density.

5. The function of *vertical amplifier* is to amplify the input signal.
6. The function of *time base generator* is to drive the beam at a steady speed across the screen and, when it reaches the right-hand side of the screen the beam is made to fly back to the starting position on the left-hand side of the screen. This generator produces a *saw-tooth wave of the same frequency as the input signal* to the Y-plates.
7. The purpose of "*delay line*" is to delay the vertical signal enough to keep it from reaching the CRT deflection plates before the horizontal sweep circuits are running.
8. Basic controls of CRO are :

   (*i*) Vertical (Y) deflection system; (*ii*) Horizontal (X) deflection system; (*iii*) Synchronization; (*iv*) Blanking circuit; (*v*) Positioning controls; (*vi*) Intensity (Z-axis) modulation; (*vii*) Intensity control; (*viii*) Focus control; (*ix*) Astigmatism; (*x*) Calibration circuit.
9. A "*storage oscilloscope*" uses the phenomenon of "*secondary electron emission*" to build up and store electrostatic charges on the surface of an insulated target.
10. "*Digital readout oscilloscope*" consists of a *high-speed laboratory CRO and electronic counter, both in one cabinet.*
11. The function of a *probe* is to connect the test circuit to the oscilloscope without altering, loading, or otherwise disturbing the test circuit.
12. The "*electronic switch*" is an extremely useful accessory for *displaying the two signals simultaneously on single trace oscilloscopes.*
13. The various parameters which can be measured by using CRO are :

    (*i*) Voltage; (*ii*) Current; (*iii*) Period; (*iv*) Frequency; (*v*) Phase : (*vi*) Amplitude; (*vii*) Peak-to-peak value ; (*viii*) Duty cycle etc.

14. *Lissajous figures/patterns* are formed when two sine waves are applied simultaneously to the vertical and horizontal deflecting plates of CRO.
15. The unknown frequency can be measured by the following methods :
(*i*) Lissajous method; (*ii*) Spot wheel method : (*iii*) Gear wheel method.
16. *"Delayed sweep"* is a technique that adds a precise amount of time between the trigger point and the beginning of the scope sweep.

OBJECTIVE TYPE QUESTIONS

Choose the Correct Answers :

1. A thin aluminium film is usually deposited on the non-viewing side of the phosphor because
(*a*) it acts as a heat-sink and prevents phosphor burn
(*b*) the light scatter from the phosphor is reduced
(*c*) it does not allow the screen to be negatively charged
(*d*) all of the above.
2. During the retrace time, the electrons forming the horizontal beam
(*a*) move from left to right on the screen
(*b*) move from right to the left on the screen
(*c*) move from bottom to top of screen
(*d*) move from top to bottom of screen.
3. The deflection of an electron beam on a CRT screen is 10 mm. Suppose the pre-accelerating anode voltage is halved and the potential between deflecting plates is doubled. The deflection of the electron beam will be :
(*a*) 80 mm (*b*) 40 mm
(*c*) 20 mm (*d*) 10 mm.
4. If the distance of screen from a CRT to centre of deflection plates is 15 cm, the length of deflection plates is 2 cm, the distance between plates is 1 cm and the accelerating voltage is 500 V, the deflection sensitivity is :
(*a*) 33.2 V/cm (*b*) 0.03 cm/V
(*c*) 66.4 V/cm (*d*) 0.015 cm/V.
5. Post-acceleration is needed in a CRO if the frequency of the signal is
(*a*) less than 1 MHz (*b*) more than 1 MHz
(*c*) more than 10 MHz (*d*) more than 10 Hz.
6. In a digital storage oscilloscope, the input signals are
(*a*) directly applied to the oscilloscope
(*b*) multiplexed, converted to digitial form and stored and applied to oscilloscope
(*c*) multiplexed, converted to digital form and stored, converted to analog form and applied to oscilloscope
(*d*) applied to amplifier, stored as analog signals, multiplexed, converted to digital form, stored in digital form, converted to analog form, and applied to CRO through an amplifier.
7. The source of emission of electrons in a CRT is
(*a*) PN junction diode
(*b*) a barium and strontium oxide coated cathode
(*c*) accelerating anodes
(*d*) post-accelerating anodes.

8. In a CRT, the focusing anode is located
   (*a*) between pre-accelerating and accelerating anodes
   (*b*) after accelerating anode
   (*c*) before pre-accelerating anode (*d*) none of the above.
9. The horizontal amplifier should be designed for
   (*a*) high frequency signals with a fast rise time
   (*b*) high amplitude signals with a slow rise time
   (*c*) high amplitude signals with a fast rise time
   (*d*) low amplitude signals with a fast rise time.
10. A vertical amplifier for a CRO can be designed for
    (*a*) only a high gain (*b*) Only a broad bandwidth
    (*c*) a constant gain times bandwidth product
    (*d*) all of the above.
11. In an oscilloscope, when the unknown signal applied to the vertical plates is being synchronized with the sweep signal applied to horizontal plates, the pattern seen on the CRO screen moves towards the right. It means that
    (*a*) the frequency of the signal is lower than that of the sweep signal
    (*b*) the frequency of the signal is greater than that of the sweep signal
    (*c*) the frequency of the signal is equal to the frequency of that sweep signal
    (*d*) none of the above.
12. P_1 phosphor material is used for display in CRTs for
    (*a*) photographic applications (*b*) general purpose applications
    (*c*) television applications (*d*) all of the above.
13. An aquadag is used in a CRO to collect
    (*a*) primary electrons (*b*) secondary emission electrons
    (*c*) both primary and secondary emission electrons
    (*d*) none of the above.

ANSWERS

1. (*d*)	2. (*b*)	3. (*b*)	4. (*b*)	5. (*c*)	6. (*d*)
7. (*b*)	8. (*a*)	9. (*b*)	10. (*c*)	11. (*a*)	12. (*b*)
13. (*b*)					

THEORETICAL QUESTIONS

1. What is a 'CRO'? Explain briefly.
2. What are the applications of 'CRO'?
3. Draw the block diagram of an oscilloscope and explain briefly its major systems.
4. Explain briefly the working of a CRO.
5. What is a 'CRT' ? Explain briefly.
6. With the help of a neat diagram, explain briefly the main parts of a cathode ray tube (CRT).
7. Derive an expression for vertical deflection of an electron beam in a 'CRT'.
8. Describe the different types of phosphor materials used in a CRO and list their applications.
9. Explain the functioning of a time-base generator in a CRO.

10. What are the different types of amplifiers used for CROs? Describe the basis on which they are classified.
11. Describe in details the vertical amplifier used in a CRO.
12. Describe the different types of sweeps used in a CRO. Explain their fields of application.
13. Describe the function of attenuators in CROs.
14. Describe the following :
    (*a*) Blanking circuit (*b*) Astigmatism control
    (*c*) Sources of synchronization (*d*) Z-axis modulation.
15. What do you understand by *"Delay line"*? Explain briefly.
16. Enumerate the basic controls of CRO and explain them briefly.
17. Describe briefly the following :
    (*a*) Dual trace oscilloscope (*b*) Dual beam oscilloscope.
18. Give the comparison between *'dual trace'* and *'dual beam'* oscilloscopes.
19. Discuss briefly any two of the oscilloscopes :
    (*a*) Sampling oscilloscope (*b*) Digital readout oscilloscope
    (*c*) High frequency oscilloscope
20. What is a *'probe'* ?
21. Explain briefly high impedance (or 10 : 1) probe.
22. What is an *"Electronic switch"*? Explain its working with the help of a block diagram.
23. Discuss the advantages and disadvantages of analog and digital type of oscilloscopes.
24. Discuss in detail the *'delayed sweep'*.
25. Describe briefly how the following measurements can be made with the use of CRO.
    (*a*) Frequency; (*b*) Phase angle;
    (*c*) Voltage.
26. Explain briefly any two of the following methods of frequency measurement :
    (*a*) Lissajous figures (*b*) Spot wheel method
    (*c*) Gear wheel method.

UNSOLVED EXAMPLES

1. In a CRT, the anode to cathode voltage is 2000 V. The parallel deflector plates are 1.5 cm long and spaced 5 mm. The screen is 50 cm from the centre of the deflection plates. Find the following :
   (*a*) The beam speed.
   (*b*) The deflection sensitivity of the tube.
   Take : Mass of electron = 9.107×10^{-31} kg : charge on electron = 1.602×10^{-19} C.
   [**Ans.** (*i*) 26.5×10^6 m/s; (*ii*) 0.325 mm/V]
2. A cathode ray tube, with magnetic deflection, has its screen 20 cm from the centre of the magnetic field. The width of the uniform magnetic field is 3 cm and the final anode potential is 6 kV. Calculate the density of magnetic field to produce a deflection of 4 cm on the screen.
   Take : Mass of electron = 9.107×10^{-31} kg : charge on electron = 1.6×10^{-19} C.
   [**Ans.** 1.7423 Wb/m^2]
3. In a cathode-ray tube, the electron beam is displaced vertically by a magnetic field of flux density 2×10^{-4} wb/m^2. The length of the magnetic field along the tube axis is the same as that of the electrostatic deflection plates. The final anode voltage is 800 V.
   Calculate the voltage which should be applied to the *Y*-deflection plates 1 cm apart, to return the spot back to the centre of the screen. [**Ans.** 33.53 V]

CHAPTER

12 Universities' Questions (Latest) –With Solutions

SECTION A : Short Answer Questions

Q.1. *What is an instrument? What are the primary requirements of any measuring instrument?*

Ans. • ***Instrument*** may be defined as a device for measuring the value or magnitude of a quantity or variable.

• The primary requirements for any measuring instrument are :

(*i*) Power consumption should be as small as possible.

(*ii*) No alteration in the circuit conditions with the induction of instrument in the circuit.

Q.2. *List the types of instruments that are mainly used as ammeters and voltmeters.*

Ans. The following types of instruments are mainly used as ammeters and voltmeters :

(*i*) PMMC; (*ii*) Moving iron; (*iii*) Electrodynamometer; (*iv*) Hot wire and thermocouple; (*v*) Electrostatic; (*vi*) Induction type; (*vii*) Rectifier.

Q.3. *What are the main advantages of 'Rectifier instruments'?*

Ans. Following are the main advantages of rectifier instruments :

(*i*) They can operate well in the RF (radio-frequency) range with proper use of silicon or germanium rectifiers

(*ii*) Measurements of current and voltage at several hundred megahertz are possible.

(*iii*) This type of meter is more sensitive than any other type of A.C. meter.

Q.4. *What is meant by the term "Measurand"?*

Ans. The quantity to be measured is known as the ***measurand***.

Q.5. *Why absolute instruments are rarely used in practice?*

Ans. Absolute instruments are rarely used in practice because working with these instruments is time-consuming.

Q.6. *What is difference between analog and digital signals?*

Ans. *"Analog signals"* vary in the continuous fashion and assimilate an infinite number of values in any given range, while the 'digital signals' vary in discrete steps and comprise only finite different values in a given range.

Q.7. *What is meant by resolution of an instrument?*

Ans. The ***resolution*** of any instrument is the smallest change in the input signal (quantity under measurement) which can be detected by the instrument. It may be expressed as an actual value or as a fraction or percentage of the full-value scale.

Q.8. *What are systematic errors?*

Ans. 'Systematic errors' remain constant or change according to a definite law on repeated measurement of given quantity. These errors can be evaluated and their influence on the results of measurement can be eliminated by the introduction of proper corrections.

Q.9. *What is the difference between fundamental and derived units?*

Ans. The size of the 'fundamental units' is quite arbitrary and can be selected to fit a certain set of circumstances i.e., mass, length, time, etc. whereas the 'derived units' can be expressed in terms of fundamental units with the help of physical equations *i.e.*, area, volume, force etc.

Q.10. *What is the difference between primary and secondary standards?*

Ans. *"Primary standards"* are absolute standards of such high accuracy that they can be used as the ultimate reference standards while the *"secondary standards"* are the basic reference standards used in industrial measurement laboratories. Secondary standards are calibrated and compared against the primary standards in National standard laboratories on periodic basis.

Q.11. *What is Josephon effect?*

Ans. A flow of current between two pieces of a superconducting material separated by an insulator, in known as ***Josephon effect***.

Q.12. *Why is mica used as dielectric in precision capacitors?*

Ans. Mica capacitors are used as precision capacitors because of their 'small tolerances' and 'high stability' under varying temperature and electric stress. Mica is easily available in the form of very thin sheets and have a high resistivity and 'dielectric strength' both of which characteristics are necessary for the purpose for which these are used.

Q.13. *How is gravity control inferior to spring control?*

Ans. The scales of gravity controlled instruments are not uniform but are crowded in the beginning. Secondly, these are to be used in vertical position in order that the control may operate.

Q.14. *On what principle does D'Arsonval galvanometer operate?*

Ans. The following principle is employed in galvanometers.

'When a current carrying conductor is placed in a magnetic field, a mechanical force is experienced on a conductor which tends to move the conductor in a direction given by Flemming's left hand rule'.

Q.15. *What is ballistic sensitivity?*

Ans. ***Ballistic sensitivity*** is defined as the first maximum deflection in mm caused on a scale at a distance of one metre from mirror by a unit charge suddenly passed in the galvanometer.

Q.16. *What is the difference between an ammeter and a voltmeter?*

Ans. An *"ammeter"* is a low resistance indicating instrument while a *"voltmeter"* is high resistance one.

Q.17. *List the advantages of Electrodynamic instruments.*

Ans. *Advantages*:

1. Can be used on both D.C. as well as A.C. systems.
2. Free from hysteresis and eddy current errors.
3. It is possible to construct ammeters upto 10A and voltmeters upto 600 V with precision grade accuracy.

Q.18. *Write different methods of measurement of medium resistance.*

Ans. The methods of measuring medium resistances are :

(*i*) Voltmeter ammeter method; (*ii*) Ohm-meter method;
(*iii*) Bridge method; (*iv*) Substitution method.

Q.19. *What is meant by drift?*

Ans. By "*drift*" we mean variation in the value of quantity under measurement with time. When it is repeatedly measured, if there is no change in the value, it means zero drift.

Q.20. *What are the essentials of indicating instruments?*

Ans. Indicating instruments consist essentially of a pointer moving over a calibrated scale and attached to the moving system pivoted on jewelled bearings. For satisfactory working of indicating instruments the torques required are :

(*i*) Deflecting torque; (*ii*) Controlling torque; (*iii*) Damping torque.

Q.21. *What are prepayment meters?*

Ans. "*Prepayment meter*" is special type of house supply meter, which itself is essentially a standard pattern energy element with the addition of a coin collector box and the prepayment mechanism geared to the recorder train.

Q.22. *Why are the rectifier instruments indispensable for measurements in communication circuits?*

Ans: The "*rectifier instruments*" are indispensable for measurements in communication circuits' since they have very high sensitivity and as such the power required for their operation is small enough to be within the capacities of the circuits.

Q.23. *What materials are used for ammeter shunts?*

Ans. 'Manganin' for D.C. circuits and 'constantan' for A.C. circuits.

Q.24. *How is range of an electrostatic voltmeter increased?*

Ans. The range of an electrostatic voltmeter is increased by employing a 'multiplier' in the form of either a 'resistance' or 'capacitor potential divider'.

Q.25. *How does a current transformer differ from a conventional transformer?*

Ans. The current transformer (CT) is usually a step-up transformer in terms of primary-secondary turn-ratio. It differs slightly in operation from that conventional transformer. In case of a CT, as the load impedance or 'burden' on the secondary is very small so the CT operates on short-circuit conditions. Also the current in secondary is not governed by load impedance on the secondary but depends upon the current flowing in the primary winding (i.e. main circuit current).

Q.26. *What are instrument transformers? How do they differ from power transformers?*

Ans.

- Instrument transformers are devices used with instruments (Ammeter, voltmeter etc.) for A.C. measurement. They *increase* the range of the measuring instruments.
- The power transformers work at 50 Hz and used for stepping up and down the voltage or current. These are connected between generating station and the receiving station.

Q.27. *Define transformation ratio of a current transformer.*

Ans. ***Transformer ratio*** in case of instrument transformers, in general, is defined as the ratio of the primary phasor to the secondary phasor.

Transformation ratio (*Actual*) for a CT,

$$K_{act} = \frac{\text{Primary winding current (Actual)}}{\text{Secondary winding current (Actual)}}$$

Transformation ratio (*Nominal*) for a CT,

$$K_{\text{nom.}} = \frac{\text{Rated primary winding current}}{\text{Rated secondary winding current}}$$

Q.28. *What is ratio correction factor (RCF)?*

Ans. The ***ratio correction factor*** is the transformation (actual) ratio divided by nominal ratio.

In other words, $$\text{RCF} = \frac{K_{\text{act.}}}{K_{\text{nom.}}}$$

- The ratio marked on transformers is their nominal ratio.

Q.29. *What do you understand by the term 'Burden of CT'?*

Ans. The product of voltage and current on the secondary side when the CT is supplying the instrument with its maximum rated value of current, is known as ***rated burden*** and is expressed in volt-ampere.

Q.30. *Why the secondary of current transformer is never kept open-circuited?*

Ans. The secondary of current transformer is never kept open-circuited because of the following two *reasons* :

(*i*) There will be no secondary m.m.f. and since the primary current (and m.m.f) is fixed, the core flux will increase enormously. This will cause large eddy current and hysteresis losses and the resulting high temperature may damage the insulation or even the core.

(*ii*) A very high voltage will be induced in the multi-turn secondary and this high voltage may be dangerous both to life and to the insulation.

Q.31. *How can average reading VTVM be converted to a RMS reading voltmeter?*

Ans. The average reading of VTVM can be converted into a RMS reading voltmeter by calibrating its meter scale in terms of the r.m.s. value of a sinusoidal wave.

Q.32. *What is meant by the deflection factor of a CRO?*

Ans. The deflection factor of a CRO is the reciprocal of the deflection sensitivity.

Q.33. *What is a Lissajous pattern?*

Ans. Lissajous is a pattern that results from applying periodic signals to the deflection plates of a CRO.

Q.34. *List the factors that may lead to inaccuracies in measurement by A.C. bridges.*

Ans. The factors that may lead to inaccuracies in measurement by A.C. bridges are :

(*i*) Stray conductance effect due to imperfect insulation;

(*ii*) Mutual inductance effect due to magnetic coupling between various components of the bridge.

Q.35. *Name different types of frequency meters.*

Ans. The frequency meters are of following types :

(*i*) Mechanical resonance type; (*ii*) Electrical resonance type; (*iii*) Weston type; (*iv*) Ratiometer type; (*v*) Saturable core type; (*vi*) Digital frequency meters.

Q.36. *Why is sensitivity of the A.C. section of multimeter less than the D.C. section?*

Ans. Due to complex circuitry, the sensitivity of A.C. section of a multimeter is less than the D.C. section.

Q.37. *For what VTVM stands for?*

Ans. The VTVM stands for vacuum tube volt meter.

Q.38. *What is Cathode Ray Oscilloscope (CRO)?*

Ans. *CRO* is a device that allows, the amplitude of electrical signals, whether they be voltage, current, power, etc., to be displayed primarily as a function of time. It is basically an electron beam voltmeter.

It is very useful and versatile laboratory instrument used for display, measurement and analysis of waveforms, and other phenomena in electrical and electronic circuits.

Q.39. *Name main components of a cathode ray tube (CRT).*

Ans. The main components of a CRT are :

(*i*) Electron gun assembly; (*ii*) Deflection plates assembly;
(*iii*) Fluorescent screen; (*iv*) Glass envelope;
(*v*) Base.

Q.40. *Define a transducer and distinguish between a sensor and a transducer.*

Ans.
- A ***transducer*** is a device which converts the energy from one form to another.
- A ***sensor*** senses the condition, state or value of the process variable and produces an output which reflects this condition, state or value. A *transducer* transforms the energy of the process variable to an output of some other type of energy which is able to operate some control device. Sometimes a *secondary transducer* may be employed to transform the output of *primary sensor* to still another type of energy.

Q.41. *What is 'LVDT'?*

Ans. ***LVDT*** (Linear-variable-differential transformer) is a passive inductive transducer and is commonly employed to measure force (or weight, pressure and acceleration, etc. which depend on force) in terms of the amount and direction of displacement of an object.

Q.42. *What is frequency counter?*

Ans. A ***frequency counter*** is a digital instrument that can measure and display the frequency of any periodic waveform.

Q.43. *What is a load cell and what for is it used?*

Ans. A ***load cell*** is an electromechanical device that converts weight or force into an electrical signal and is widely used for measurement of static and dynamic forces.

Q.44. *What is IEEE 488 bus system?*

Ans. For a set of instruments to be interfaced with a single computer, a general purpose bus called the "IEEE 488 bus system", has been developed to which a number of instruments can be attached so that communications between the computer and the instruments are possible over a common cable.

Q.45. *What is meant by loading effect?*

Ans. The incapability of the system to faithfully measure, record, or control the input signal (measurand) in undistorted form is known as ***loading effect***.

Q.46. *What are the special features of vibration galvanometers?*

Ans. The special features of vibration galvanometers are large sensitivity, small damping, and large restoring constant.

Q.47. *What are the shortcomings of multimeter?*

Ans. It may be noted that A.C. sensitivity of a half-wave A.C. meter is only 45% of the D.C. sensitivity, therefore, the loading effect is greater while measuring A.C.

voltages than that while measuring corresponding D.C. voltages.

The rectifying action depends upon the frequency to some extent due to capacitance effect and rectifier instruments show lower readings. Instrument indications may be in error by as much as 0.5% decrease for every 1 kHz rise in frequency.

Q.48. *What is the usual current-rating of CT's secondary?*

Ans. The usual current rating CT's secondary is 5*A*.

Q.49. *How is electrostatic wattmeter superior to other types of wattmeters?*

Ans. An electrostatic wattmeter is free from errors on account of waveforms, frequency, and eddy currents.

Q.50. *What is the difference between an energy meter and a wattmeter?*

Ans. The essential difference between an 'energy meter' and a 'wattmeter' is that the former is fitted with some type of registration mechanism whereby all the instantaneous readings of power are summed over a definite period of time, whereas the latter indicates the value of power at particular instant when it is read.

Q.51. *What is meant by meter constant of an energy meter?*

Ans. The number of revolutions made by the aluminium disc for 1 kWh of energy consumption is called 'meter constant' of an energy meter.

Q.52. *What is trivector meter?*

Ans. *"Trivector meter"* is an instrument which measures kVAh and also kVR of maximum demand. It consists of a kWh meter and a reactive kivolt ampere-hour meter in the common case with a special summation between them.

Q.53. *On what principle does Weston frequency meter operate?*

Ans. *"Weston frequency meter"* is a moving iron instrument which operates on the principle of variation of impedance in an inductive circuit with the variation in supply frequency.

Q.54. *Distinguish between, D.C. and A.C. potentiometers.*

Ans. In 'D.C. potentiometers' only magnitude of unknown e.m.f. is measured whereas in 'A.C. potentiometers' magnitude and phase of the unknown e.m.f. is measured. Accuracy of measurement of D.C. potentiometer is higher as compaerd to a A.C. potentiometer.

Q.55. *Why is screening of bridge elements required?*

Ans. If each component in a bridge has a definite screen connected to one end, a very high accuracy in measurement can be achieved by using 'Wagner's earthing device' in the bridge circuit. This device removes all the earth capacitances from the bridge network.

Q.56. *Why are Ballastic (or D.C) tests conducted?*

Ans. (*i*) To determine flux/flux density in a specimen; (*ii*) To determine B.H. curves; (*iii*) To plot hysteresis loop.

Q.57. *Name the two methods available for the determination of B.H. curve of a specimen.*

Ans. (*i*) Method of reversals; (*ii*) Step-by-step method.

Q.58. *What is the function of a magnetic potentiometer?*

Ans. The 'magnetic potentiometer' measures the magnetic potential drop along any part of a magnetic circuit and hence distribution of m.m.f. round the circuit. It can also be used to determine magnetic potential drop across a joint in the magnetic circuit or between two points between which magnetic leakage is suspected.

Q.59. *What are permeameters?*

Ans. Normally it is not a practical arrangement to have a $\frac{l}{d}$ as large as possible for a bar specimen to be tested. *"Permeameters"* have been developed to overcome the difficulties and inaccuracies encountered in performing measurements on bar specimens of magnetic materials. Thus permeameters test bar specimens under controlled field conditions.

Q.60. *Enumerate the methods which are used for measurement of iron losses.*

Ans. (*i*) Wattmeter method; (*ii*) A.C. bridge methods; (*iii*) A.C. potentiometer method; (*iv*) Osillographic method.

Q.61. *Why magnetic measurements are not as accurate as other types of measurements in electrical engineering?*

Ans. This is mainly due to the fact that magnetic flux cannot be measured directly by the mere insertion of instrument in a circuit as an electric current, voltage, and power. The other difficulties arise from the fact that flux unlike current does not confine itself to a definite path.

Q.62. *Why underground cables meant for operation on A.C. systems are tested with direct current systems?*

Ans. If alternating current is employed for the test, heavy charging current will be drawn by the cable and so the transformer required for the purpose would be of inconveniently large capacity and cause transportation difficulties of moving the testing transformer to the site of installation. In case, direct current is used for the tests, the plant required would only be of small capacity and can be constructed in portable form.

Q.63. *What are the main components of a 'Universal bridge'?*

Ans. A 'Universal bridge' consists of signal sources (D.C. and 1 kHz) bridges (Wheatstone bridge, Maxwell's and Hay's bridges and De Sauty bridge), imbalance amplifier and metered detector in addition to power supply.

Q.64. *Why is a triggering circuit provided in a CRO?*

Ans. In a CRO, a 'triggering circuit' is provided for synchronising two types of deflections so that horizontal deflection starts at the same point of input vertical signal each time it sweeps.

Q.65. *What do you mean by the deflection factor of a CRO?*

Ans. The deflection factor of a CRO is the reciprocal of the deflection sensitivity.

Q.66. *What is an oscilloscope probe?*

Ans. Probe is a means that connects the test circuit to the oscilloscope without altering loading, or disturbing the test circuit.

Q.67. *What is a digital instrument?*

Ans. A 'digital instrument' indicates the measured value in the form of decimal digits.

Q.68. *What are the essential components of a DVM?*

Ans. A DVM comprises a clock generator, a circuit that converts the voltage to be measured into pulses, and a digital frequency counter.

Q.69. *Enumerate the various types of high voltage tests.*

Ans. (*i*) Sustained low-frequency tests; (*ii*) Constant direct-current tests; (*iii*) High-frequency tests; (*iv*) Surge or impulse tests.

Q.70. *Name the apparatus which are usually essential for all hight voltage tests.*

Ans. (*i*) High-voltage transformers; (*ii*) Apparatus for voltage regulation; (*iii*) Control gear and high-voltage connections, including safety and protective devices; (*iv*) Apparatus for voltage measurement.

SECTION B : Conventional Questions with Solutions

Q.1. (*a*) *Explain 'Josephen-Effect' Standard.*

(*b*) *A standard cell has a voltage rating of 1.018500 V; Internal resistance = 480 Ω; Insulation resistance between its terminals is 5.4 MΩ. Calculate :*

(i) The current drain due to insulance resistance; (ii) The terminal voltage.

Sol. (*a*) Refer to Article 1.10.4

Given: Voltage rating of the cell = 1.018500 V; Internal resistance = 480 Ω; Insulation resistance between the terminals = 5.4 MΩ.

(*i*) **The current drain:**

The current flowing through the insulation resistance

$$= \frac{1.018500}{480+5.4\times10^6} = 0.1886\ \mu A$$

Thus, there is *constant drain of current of* **0.1886 μA (Ans.)**

(*ii*) **The terminal voltage:**

Voltage drop in internal resistance = $0.1886 \times 10^{-6} \times 480 = 90.53\ \mu V$

∴ Terminal voltage = Voltage rating of the cell (e.m.f.) – Voltage drop in internal resistance

$= 1.018500 - 90.53 \times 10^{-6} =$ **1.0184095 V (Ans.)**

Q.2. (*a*) *What are the applications of the instruments and measurement systems?*

(*b*) *A pressure indicator showed a reading as 38 bar on a scale range of 0-50 bar. If the true value was 37.5 bar, determine :*

(i) Static error; (ii) Static correction; (iii) Relative static error

Sol. (*a*) Refer to Article 2.7.

(*b*) *Given:* Measured value, V_m = 38 bar, True value, V_t = 37.5 bar, scale range = 0.50 bar.

(*i*) **Static error E_s :**

$E_s = V_m - V_t = 38 - 37.5 =$ **+ 0.5 bar (Ans.)**

(*ii*) **Static correction, C_s :**

$C_s = -E_s = -(+0.5) =$ **– 0.5 bar (Ans.)**

(*iii*) **Relative error, E_r :**

$E_r = \frac{E_s}{V_t} = \frac{0.5}{37.5} = 0.0133$ or **1.33% (Ans.)**

Q.3. (*a*) *What are limiting and known errors?*

(*b*) *During a test the current was measured as 24.4 A flowing in resistor of 0.13 Ω. Later it was discovered that the ammeter reading was low by 0.95% and marked resistance was high by 0.22%. Determine the true power as percentage of the power that was originally calculated.*

Sol. **(a)** *Limiting errors :*

In most instruments the accuracy is generated to be within a certain percentage of full scale reading. The manufacturer has to specify the deviations from the nominal value of a particular quantity. The *limits of these deviations* from the specified values are defined as "limiting error or guarantee errors".

Known errors :

When the error of a quantity of an instrument is known the effect of this error when combined with errors can be computed in a manner similar to the combinations of the limiting errors. However, the difference is that in case of known errors the *signs* of relative errors are given and must be presented in the calculations.

(b) True value of current, $I = 24.4\left(1+\frac{0.95}{100}\right) = 24.632$ A

True value of resistance, $R = 0.13\left(1-\frac{0.22}{100}\right) = 0.1297\ \Omega$

True value of power $= I^2R = (24.632)^2 \times 0.1297 = 78.694$ W

Measured value of power $= (24.4)^2 \times 0.13 = 77.397$ W

$$\therefore \frac{\text{True power}}{\text{Measured power}} \times 100 = \frac{78.694}{77.397} \times 100 = \mathbf{101.67\%}$$ **(Ans.)**

Q.4. **(a)** *Name the various types of D.C. galvanometers.*

(b) *A coil of a moving coil galvanometer is wound on a non-magnetic former whose height and width are 25 mm and 20 mm respectively. It moves in a field of 0.11 Wb/m^2. The moment of inertia of its moving parts is 0.28 × 10^{-6} kg-m^2 and the control spring constant is 31 × 10^{-6} Nm/rad.*

Determine : (i) The number of turns that must be wound on the former to produce a deflection of 135° with a current of 11 mA; (ii) The resistance of coil to produce critical damping, all damping being assumed as electromagnetic.

Sol. **(a)** D'Arsonval type galvanometers, which are D.C. galvanometers, are made in three general types :

(*i*) Portable point type; (*ii*) Laboratory reflecting type; (*iii*) Box type.

(b) *Given:* Height, l = 25 mm, 0.025 m; Width, b = 20 mm or 0.02 m; B = 0.11 Wb/m^2; Moment of inertia, $J = 0.28 \times 10^{-6}$ kg-m^2;

Control spring constant, $C = 31 \times 10^{-6}$ Nm/rad; $\theta = 135° = 135 \times \frac{\pi}{180} = 2.356$ rad; current, I = 11 mA or 0.011 A

(*i*) **Number of turns, N :**

For steady deflection, $GI = C\theta$ ...[Eqn. (3.11)]

$$\therefore \text{ Displacement constant, } G = \frac{C\theta}{I} = \frac{31\times10^{-6}\times2.356}{0.011} = 6.64 \times 10^{-3} \text{ Nm/A}$$

Also, $G = NBlb$

$$\therefore \quad N = \frac{G}{Blb} = \frac{6.64\times10^{-3}}{0.11\times0.025\times0.02} = 120.7 \text{ say } \mathbf{121}$$ **(Ans.)**

(ii) Resistance of coil for critical damping, R_c :

We know that, $$R_c = \frac{G^2}{2\sqrt{CJ}} \quad \text{...[Eqn. (3.31)]}$$

$$= \frac{(6.64\times10^{-3})^2}{2\sqrt{31\times10^{-6}\times0.28\times10^{-6}}} = \frac{4.409\times10^{-5}}{5.892\times10^{-6}}$$

$$= \mathbf{7.48\ \Omega}\ \textbf{(Ans.)}$$

Q.5. (a) *What is Blondel's theorem? What are its common applications?*

(b) *Two wattmeters are used to measure input to a 1.5 kV, 50 Hz, 3-phase motor running on full load at an efficiency of 82 per cent. Their readings are 240 kW and 75 kW respectively. Calculate : (i) Input; (ii) Power factor; (iii) Line current, and (iv) Output.*

Sol. (a) **Blondel's theorem** is used in the measurement of power in a poly-phase system. According to this theorem: "If a network is supplied through N conductors, the total power is measured by summing the readings of N wattmeters so arranged that a current element of a wattmeter is in each line and the corresponding voltage element is connected between that line and a common point. If the common point is located on one of the lines, then the power may be measured by N – 1 wattmeters".

The most common applications of this theorem include the 'two-wattmeter method' of measurement of power in 3-phase, 3 wire load circuits and 'three-wattmeter method' of measurement of power in 3-phase, 4-wire circuits.

(b) Since the motor is running at full-load, its power factor must be *greater than 0.5*. Hence W_2 reading is *positive*.

$\therefore$ W_1 = 240 kW, and W_2 = +75 kW

(i) **Input** = $W_1 + W_2$ = 240 + 75 = **315 kW** **(Ans.)**

(ii) **Power factor; cos θ :**

$$\tan\theta = \frac{\sqrt{3}(W_1 - W_2)}{W_1 + W_2} \quad \text{...[Eqn. (3.157)]}$$

$$= \frac{\sqrt{3}(240-75)}{(240+75)} = 0.907$$

or, $$\phi = \tan^{-1}(0.907) = 42.2°$$

$\therefore$ $$\cos\phi = \mathbf{0.74}\ \textbf{(Ans.)}$$

(iii) **Line current; I_2 :**

$$P = \sqrt{3}\ E_L I_L \cos\phi$$

$\therefore$ $$I_2 = \frac{P}{\sqrt{3}\,E_L\cos\phi} = \frac{315\times1000}{\sqrt{3}\times(1.5\times1000)\times0.74} = \mathbf{163.8A}\ \textbf{(Ans.)}$$

(iv) **Output:**

$$\text{Output} = \text{Input} \times \text{efficiency}$$

$$= 315 \times 0.82 = \mathbf{258.3\ kW}\ \textbf{(Ans.)}$$

Q.6. (*a*) *Discuss working of a Q-meter.*

(*b*) *A coil with a resistance of 10 Ω is connected in the direct measurement mode. Resonance occurs when oscillating frequency is 1×10^6 Hz and resonating capacitor is set at 60 pF. Calculate the % error introduced in the calculated value of Q by 0.02 Ω insertion resistance.*

Sol. (*a*) Refer to Art. 2.6.14

(*b*) *Given:* $R = 10\ \Omega$, $f = 1 \times 10^6$ Hz ; $C = 60$ pF; Resistance used across the oscillatory circuit = 0.02 Ω.

Percentage error :

The true or effective Q of the coil,

$$Q_{\text{true}} = \frac{1}{\omega CR} \quad \text{... [Eqn. (2.70)]}$$

$$= \frac{1}{(2\pi \times 1 \times 10^6) \times 60^{-12} \times 10} = 265.3$$

The indicated or calculated Q of the coil,

$$Q_{\text{indicated}} = \frac{1}{\omega C(R + 0.02)}$$

$$= \frac{1}{(2\pi \times 1 \times 10^6) \times 60 \times 10^{-12}\ (10 + 0.02)} = 264.7$$

$$\therefore \quad \text{Percentage error} = \frac{Q_{\text{true}} - Q_{\text{indicated}}}{Q_{\text{true}}}$$

$$= \frac{265.3 - 264.7}{265.3} \times 100 = \mathbf{0.226\%} \text{ (Ans.)}$$

Q.7. *A potential transformer of ratio 1000 / 100 has primary resistance 94.5 Ω, secondary resistance 0.86 Ω, primary reactance 66.2 Ω, total equivalent reactance 110 Ω, and no-load current 0.02 A at 0.4 power factor.*

Calculate the phase angle error at no-load.

Sol. *Given:* Turn ratio, $K_t = \frac{1000}{100} = 10$; $R_p = 94.5\ \Omega$; $X_p = 66.2\ \Omega$; $R_s = 0.86\ \Omega$; $(X_p)_{eq} = 110\ \Omega$; $I_0 = 0.02$ A at 0.4 p.f.

Phase angle error at no-load, θ :

No-load p.f. $\cos \phi = 0.4$;

$$\therefore \quad \sin \theta = \sqrt{1 - \cos^2 \phi_0} = \sqrt{1 - (0.4)^2} = 0.9165$$

Now,

$$I_w = I_o \cos \phi_0 = 0.02 \times 0.4 = 0.008 \text{ A}$$

$$I_m = I_o \sin \phi_0 = 0.02 \times 0.9165 = 0.0183 \text{ A}$$

$$\text{Phase angle } \theta = \frac{\frac{I_s}{K_t}\left[(X_p)_{eq} \cos \gamma - (R_p)_{eq} \sin \gamma\right] + I_w X_p - I_m R_p}{K_t V_s} \text{ rad}$$

...[Eqn. (6.31)]

At no load, $I_s = 0$

$$\therefore \qquad \theta = \frac{I_w X_p - I_m R_p}{K_t V_s} \text{ rad}$$

$$\text{or} \qquad \theta = \frac{0.008 \times 66.2 - 0.0183 \times 94.5}{10 \times 100} = -0.0012 \text{ rad}$$

$$= -0.0012 \times \frac{180}{\pi} \times 60 = -\mathbf{4.12'} \quad \textbf{(Ans.)}$$

Q.8. **(a)** *Explain the principle of working of a current transformer (CT) and give expressions for the ratio and phase angle errors.*

(b) *A 1000/5 A, 50 Hz bar type current transformer has secondary burden of 1.5 Ω (non-inductive). Calculate the flux in the core and the ratio error at rated condition of the CT. Assume iron loss in the core to be 1.5 watts. Neglect leakage flux and the magnetising current.*

Sol. **(a)** Refer to Arts. 4.3.1 and 4.3.3.

(b) *Given:* Nominal ratio, $K_{nom} = \frac{1000}{5} = 200$; $N_p = 1$;

Secondary impedance = 1.5 Ω (non-inductive); Iron loss in the core = 1.5 Ω.

Flux in the core, ϕ_{max} :

Let us *assume* turns ratio to be equal to nominal, in the absence of any other data.

Then, turn ratio, $K_t = K_{nom.} = 200$

$\therefore \qquad N_s = K_t N_p = 200 \times 1 = 200$

Now, secondary induced voltage, $E_s = 5 \times 1.5 = 7.5$ V

$$\therefore \qquad \phi_{max} = \frac{E_s}{4.44 \times f \times N_s} = \frac{7.5}{4.44 \times 50 \times 200} = \mathbf{0.1689 \times 10^{-3}\ Wb} \quad \textbf{(Ans.)}$$

Ratio error :

$$\text{Primary induced voltage, } E_p = \frac{E_s}{K_t} = \frac{7.5}{200} \text{ V}$$

$$\text{Loss component of exciting current, } I_w = \frac{\text{Iron loss}}{E_p} = \frac{1.5}{(7.5/200)} = 40 \text{ W}$$

As the burden is purely resistive, therefore, the secondary load angle is zero or $\delta = 0$

$$\text{We know that: Actual ratio, } K_{act} = K_t + \frac{I_w \sin\delta + I_w \cos\delta}{I_s}$$

$$= K_t + \frac{I_w}{I_s} = 200 + \frac{40}{5} = 208$$

$$\text{Ratio error} = \frac{K_{nom.} - K_{act.}}{K_{act.}} \times 100$$

$$= \frac{200 - 208}{208} = \mathbf{-3.846\%} \quad \textbf{(Ans.)}$$

Q.9. **(a)** *What are the advantages of bridge circuits?*

(b) *A circuit consisting of an unknown coil, a resistance and a variable capacitor connected in series is tuned to resonance using a Q-meter. If the frequency is 400 kHz, the resonating capacitor is set at 210 pF, the resistance is 0.9 Ω and Q-meter indicates 105, determine the effective inductance and resistance of the unknown coil.*

Ans. **(a)** The *advantages of bridge circuits* are :

(*i*) High measurement accuracy (as the measurement is done by comparing the unknown value with the standard value).

(*ii*) The accuracy is independent of null detector's characteristics (it depends on the values of the components).

(*iii*) The balance equation is independent of the magnitude of the input voltage or its source impedance, the sensitivity and impedance of the null detector, or any impedance shunting the detector.

(*iv*) The interchange of the source detector does not affect the balance condition.

(*v*) The bridge circuit can be used in control circuits.

(b) *Given*: f_1 = 400 kHz; C = 210 pF; R_{sh} = 0.9 Ω; Q = 105.

Inductance (L); resistance (R) :

Under *resonant conditions,* we have :

$$L_{coil} = \frac{1}{\omega^2 C} = \frac{1}{(2\pi \times 400 \times 1000)^2 \times 210 \times 10^{-12}}$$

$$= \mathbf{753.9\ \mu H} \ \textbf{(Ans.)}$$

$$R_{coil} = \frac{\omega L}{Q} - R_{sh} = \frac{2\pi \times 400 \times 1000 \times 753.9 \times 10^{-6}}{105} - 0.9$$

$$= \mathbf{17.14\ \Omega} \ \textbf{(Ans.)}$$

Q.10. **(a)** *List the various types of errors which occur in moving iron instruments.*

(b) *The inductance of a moving-iron ammeter is given by the expression :*

$$L = (10 + 5\theta - 2\theta^2)\ \mu H$$

Find the deflection for a current of 5 A, if control spring constant is 7.5 × 10^{-6} Nm/rad.

Sol. **(a)** The various types of errors which occur in moving iron instruments are :

1. *Errors with both A.C. and D.C. :*

(*i*) Errors due to hysteresis.

(*ii*) Errors due to stray fields.

(*iii*) Errors due to temperature changes.

2. *Errors with A.C. only :*

(*i*) Errors due to change in frequency.

(*ii*) Waveform error.

(b) *Given*: $L = (10 + 5\theta - 2\theta^2)$ μH; I = 5 A; C = 7.5 × 10^{-6} Nm/rad

Deflection, θ :

Now, $$\frac{dL}{d\theta} = (5 - 4\theta)\ \mu H/rad \text{ or } (5 - 4\theta) \times 10^{-6}\ H/rad$$

We know that, $$\theta = \frac{1}{2}\frac{I^2}{C}\frac{dL}{d\theta} \quad \text{...[Eqn. (3.78)]}$$

or, $$\theta = \frac{1}{2} \times \frac{(5)^2}{7.5 \times 10^{-6}} \times (5 - 4\theta) \times 10^{-6}$$

$$= 1.67 \times 10^6 (5 - 4\theta) \times 10^{-6} = 1.67 (5 - 4\theta)$$

or, $$\theta = 8.35 - 6.68\theta$$

or $$\theta = \frac{8.35}{7.68} = 1.087 \text{ rad} = \mathbf{62.3°} \textbf{ (Ans.)}$$

Q.11. *A ballistic galvanometer having a circuit resistance of 4800 Ω and a costant of 0.11 μC per scale division is connected in turn (i) to a coil of 4 turns wound round the field coil of a D.C. machine, and (ii) to a 6 turns coil placed on the armature surface embracing the total flux per pole entering the armature. When normal field current is broken the galvanometer readings are 222 and 270 divisions respectively.*

Calculate : (i) Flux per pole, and (ii) Leakage coefficient.

Sol. *Given*: R = 4800 Ω; Galvanometer's constant = 0.11 μC per scale division

(*i*) **Total flux per pole, ϕ :**

Electricity passed through the galvanometer,

$$Q = G_q \theta \quad \text{...(}i\text{)}$$

Where, θ is the throw of the galvanometer.

Electricity passed through the galvanometer, when current is *broken* also,

$$Q = \frac{\phi N}{R} \quad \text{...(ii)}$$

where, ϕ = flux per pole,

N = Number of turns on the search coil, and

R = Circuit resistance of the galvanometer.

Comparing (*i*) and (*ii*), we get :

$$G_q \theta = \frac{\phi N}{R}$$

or, $$\phi = \frac{R G_q \theta}{N}$$

$$= \frac{4800 \times 0.11 \times 10^{-6} \times 222}{4} = \mathbf{0.02930 \text{ Wb}}$$

(*ii*) **Leakage coefficient :**

Useful flux = Flux measured by galvanometer when connected to a search coil wound on the armature surface

$$= \frac{4800 \times 0.11 \times 10^{-6} \times 270}{6} = 0.02376 \text{ Wb}$$

$$\therefore \quad \text{Leakage coefficient} = \frac{\text{Total flux}}{\text{Useful flux}} = \frac{0.02930}{0.02376} = \mathbf{1.233} \textbf{ (Ans.)}$$

Q.12. **(*a*)** *What are the applications of a vector voltmeter?*

(*b*) *In a CRT, the anode to cathode voltage is 2400 V. The parallel deflection plates are 15 mm long and spaced 5 mm. The screen is 450 mm from the centre of the deflection plates. If mass of electron is 9.109 × 10^{-31} kg and charge on electron 1.602 × 10^{-19} C, find :*

(i) The beam speed; (ii) The deflection sensitivity of the tube.

Sol. (a) A *'vector voltmeter'* measures the magnitude of a signal at two points in a circuit and simultaneously measures the phase difference between the voltage waveforms at these two points. It is useful in VHF applications and can be used successfully in measurements, such as :

(i) Complex insertion loss; *(ii)* Amplifier gain and phase shift; *(iii)* Two-post network parameters; *(iv)* Filter transfer function; *(v)* Radio frequency distortion.

(b) *Given:* Acceleration voltage, V_a = 2400 V; Length of each Y-plate, l_d = 15 mm = 0.015 m; Distance between Y-plates, d = 5 mm = 0.005 m; Distance of screen from the centre of the deflecting plates, L = 450 mm = 0.45 m; m = 9.109×10^{-31} kg; e = 1.602×10^{-9} C.

(i) The beam speed, v :

We know that: $$v = \sqrt{\frac{2eV_a}{m}} \quad \text{...[Eqn. (2.1)]}$$

$$= \sqrt{\frac{2 \times 1.602 \times 10^{-19} \times 2400}{9.109 \times 10^{-31}}} = \mathbf{29.05 \times 10^6\ m/s}\ \textbf{(Ans.)}$$

(ii) Deflection sensitivity, S :

$$S = \frac{Ll_d}{2dV_a} \quad \text{...[Eqn. (2.2)]}$$

$$= \frac{0.45 \times 0.015}{2 \times 0.005 \times 2400} \text{ m/V} = 2.81 \times 10^{-4} \text{ m/V}$$

$$= \mathbf{0.281\ mm/V}\ \textbf{(Ans.)}$$

Q.13. (a) *How are "known errors" treated?*

(b) *In a wheatstone bridge, the unknown resistance is measured by using three resistances such that :* $R_4 = \frac{R_2 R_3}{R_1}$, *where,* R_1 = 60 ± 0.5% Ω; R_2 = 600 ± 0.5% Ω; R_3 = 460 ± 0.5% Ω

Determine the following :

(i) The magnitude of unknown resistance (R_4), and

(ii) The limiting error in per cent and in ohms (for the unknown resistance).

Sol. (a) Known error: When the error of quantity of an instrument is known, the effect of this error, when combined with errors, can be computed in a manner similar to the combinations of the limiting errors. However, the difference is that in case of known errors the *signs* of relative errors are given and must be presented in the calculations.

(b) *Given:* R_1 = 60 ± 0.5% Ω; R_2 = 600 ± 0.5% Ω; R_3 = 460 ± 0.5% Ω.

(i) Magnitude of unknown resistance, R_4:

$$R_4 = \frac{R_2 R_3}{R_1} = \frac{600 \times 460}{60} = \mathbf{4600\ \Omega}\ \textbf{(Ans.)}$$

(ii) The limiting error:

The relative limiting error of the unknown resistance,

$$\frac{\delta R_4}{R_4} = \pm\left(\frac{\delta R_2}{R_1} + \frac{\delta R_3}{R_3} + \frac{\delta R_1}{R_1}\right)$$

$$= \pm (0.5 + 0.5 + 0.5) = \pm 1.5\% \quad \textbf{(Ans.)}$$

$$\text{Limiting error in ohms} = \pm \frac{1.5}{100} \times 4600 = \pm 69\ \Omega \quad \textbf{(Ans.)}$$

Q.14. (*a*) *What do you mean by voltage sensitivity of a galvanometer?*

(*b*) *A ballistic galvanometer has a resistance of 120 Ω and an undamped period of 8.2 seconds. A steady e.m.f. of 3.7 mV produces a deflection of 205 mm. If the deflection produced is 690 mm and relative damping is 0.76, determine the quantity of electricity discharged from the capacitor.*

Sol. (*a*) 'Voltage sensitivity' is defined in terms of deflection in scale divisions caused by unit voltage applied across the galvanometer.

In case of a galvanometer attached with a mirror, the voltage sensitivity is defined as the deflection produced in mm on a scale at a distance of 1 m from the mirror by a unit voltage impressed on the galvanometer.

i.e. Voltage sensitivity, $S_V = \dfrac{d}{V \times 10^6}$ mm/μV

Where, d is the deflection on scale in mm.

(*b*) *Given*: R = 120 Ω; Undamped period, T_0 = 8.2 s; steady e.m.f. V = 3.7 mV; Steady deflection, θ_s = 205 mm; θ_1 = 690 mm;

Relative damping, γ = 0.76.

Quantity of electricity discharged, Q :

$$\text{Steady state current, } I_s = \frac{3.7 \times 10^{-3}}{120} = 30.83 \times 10^{-6}\ \text{A}$$

Under steady state conditions: $C\theta_s = GI_s$ ...[Eqn. (3.45)]

$$\therefore \quad \frac{C}{G} = \frac{I_s}{\theta_s} = \frac{30.83 \times 10^{-6}}{205} = 0.15 \times 10^{-6}\ \text{A/mm}$$

$$\text{Logarithmic decrement,} \quad \lambda = \frac{\pi\gamma}{\sqrt{1-\gamma^2}} = \frac{\pi \times 0.76}{\sqrt{1-(0.76)^2}} = 3.67$$

$$\text{Hence,} \quad Q = \frac{T_o}{2\pi} \cdot \frac{I_s}{\theta_s}\left(1 + \frac{\lambda}{2}\right)\theta_1 \quad \text{...[Eqn. (3.48)]}$$

$$= \frac{8.2}{2\pi} \times (0.15 \times 10^{-6})\left(1 + \frac{3.67}{2}\right) \times 690$$

$$= 382.9 \times 10^{-6}\ \text{C} = \mathbf{382.9\ \mu C} \quad \textbf{(Ans.)}$$

Q.15. (*a*) *What is loading effect?*

(*b*) *The coil of a 250 V moving iron voltmeter has a resistance of 450 Ω and inductance of 0.95 H. The current taken by the instrument when placed on 250 V, D.C. supply is 0.04 A. Determine the percentage error when the instrument is placed on 250 V, A.C. supply at 100 Hz.*

Sol. (*a*) *"Loading effect"*:

The incapability of the system to faithfuly measure the input signal in undistorted form is called the *loading effect*. This results in loading errors. Loading effects occur due to extraction of energy from the measurement.

(b) Total ohmic resistance, $R = \dfrac{250}{0.04} = 6250\ \Omega$

(Original calibration of the instrument is with direct current).

Reactance of the coil, $X_L = 2\pi fL = 2\pi \times 100 \times 0.95 = 596.9\ \Omega$

Coil impedance, $Z = \sqrt{R^2 + X_L^2} = \sqrt{(6250)^2 + (596.9)^2} = 6278.4\ \Omega$

Voltage reaching on A.C. $= \dfrac{250 \times 6250}{6278.4} = 248.87$ V

$\therefore$ Error $= 248.87 - 250 = -1.13$ V

Percentage error $= \dfrac{1.13}{250} \times 100 =$ **0.452%** **(Ans.)**

Q.16. (a) *What is burden of an instrument transformer?*

(b) *A current transformer rated at 500/5A, 50 Hz with an output of 15 VA has 1 turn primary. At rated load with non-inductive burden, the in-phase and quadrature components (referred to the flux) of the exciting mmf are 8 AT and 10 AT respectively. The number of turns in the secondary are 98, and the resistance and leakage reactance of the secondary winding are 0.35 Ω and 0.3 Ω respectively.*

Determine the following:

(i) *Current error.*

(ii) *Phase angle error.*

Sol. (a) ***Burden of an instrument transformer :***

The load across the secondary winding can be conveniently expressed as the output in volt-ampere at the rated secondary voltage. The *'rated burden'* is the volt-ampere loading which is permissible without errors exceeding the limits for particular class of accuracy.

$$\text{Total secondary burden} = \frac{(\text{Secondary winding } \textit{induced} \text{ voltage})^2}{(\text{Impedance of the secondary winding circuit } \textit{including} \text{ impedance of secondary winding})}$$

$= (\text{Secondary winding current})^2 \times$ (Impedance of secondary winding circuit *including* secondary winding)

Secondary winding burden due to load

$$= \frac{(\text{Secondary winding } \textit{terminal} \text{ voltage})^2}{(\text{Impedance of load on secondary winding})}$$

$= (\text{Secondary winding current})^2 \times$ (Impedance of *load* in the secondary winding circuit)

(b) *Given:* $N_p = 1$; $N_s = 98$; Output = 15 VA; Magnetising *m.m.f.* = 8 AT; Loss mmf = 10 AT; $R_s = 0.35\ \Omega$; $X_s = 0.3\ \Omega$

(i) Ratio (current) error:

$$K_t = \frac{N_s}{N_p} = \frac{98}{1} = 98; \quad K_{nom.} = \frac{500}{5} = 100$$

$$\text{Magnetising current,} \quad I_m = \frac{\text{Magnetising mmf}}{\text{Primary winding turns}} = \frac{8}{1} = 8\text{A}$$

Loss component, $I_w = \dfrac{\text{Loss mmf}}{\text{Primary winding turns}} = \dfrac{10}{1} = 10 \text{ A}$

Resistance of external burden (purely resistive-given), R_e

$$= \frac{VA}{I_s^2} = \frac{15}{5^2} = 0.6\ \Omega$$

Reactance of external burden, $X_e = 0\ \Omega$

Resistance of total secondary circuit burden

$$= R_s + R_e = 0.35 + 0.6 = 0.95\ \Omega$$

Reactance of total secondary circuit burden

$$= X_s + X_e = 0.3 + 0 = 0.3\ \Omega$$

Secondary phase angle, $\delta = \tan^{-1}\left(\dfrac{0.3}{0.95}\right) = 17.53°$

$\therefore$ $\cos\delta = \cos 17.53° = 0.954$ and $\sin\delta = 0.301$

Actual ratio, $K_{act.} = K_t + \dfrac{I_m \sin\delta + I_w \cos\delta}{I_s}$

$$= 98 + \frac{8\times0.301+10\times0.954}{5} = 100.39$$

$\therefore$ *Ratio (current) error* $= \dfrac{K_{nom.} - K_{act.}}{K_{act.}}\times100$

$$= \frac{100-100.39}{100.39}\times100 = \mathbf{-0.388\%} \quad \textbf{(Ans.)}$$

(*ii*) **Phase angle error, θ :**

We know that, $\theta = \dfrac{180}{\pi}\left[\dfrac{I_m\cos\delta - I_w\sin\delta}{K_t\, I_s}\right]$ degress ... [Eqn. 4.18]

$$= \frac{180}{\pi}\left[\frac{8\times0.954 - 10\times0.301}{98\times5}\right] = \mathbf{0.54°} \quad \textbf{(Ans.)}$$

Q.17. (*a*) *What are the applications of Schering bridge?*

(*b*) *The Schering bridge shown in the Fig.1 has the following constants :*

Arm AB : Capacitor of 0.35 μF in parallel with 1.4 kΩ resistance;

Arm BC : Resistance of 2.8 kΩ; Arm CD : Unknown capacitor C_x and R_x in series;

Arm DA : Capacitor of 0.42 μF; Frequency 1 kHz.

Determine : (i) R_x and C_x; (ii) Dissipation factor.

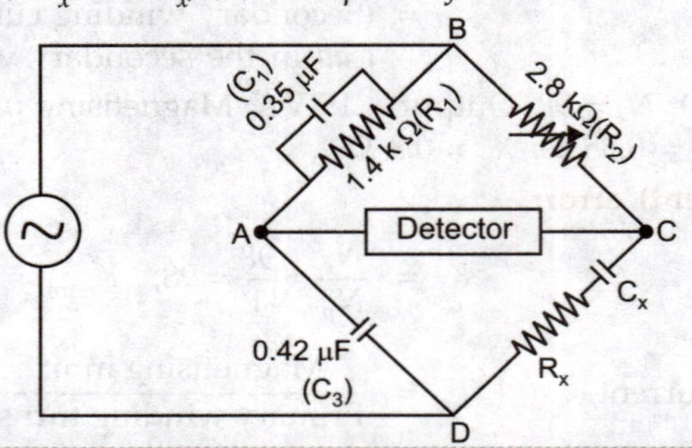

Fig. 1

Sol. (a) Schering bridge (one of the most important A.C. bridges) is used for the measurement of *unknown capacitors, dielectric loss* and *power factor*. It is widely used for testing *small capacitors at low voltages with very high precision*.

(b) *Given*: R_1 = 1.4 kΩ; C_1 = 0.35 μF; R_2 = 2.8 kΩ; C_3 = 0.42 μF

(i) R_x, C_x :

We know that: $$R_x = \frac{R_2C_1}{C_3} \quad \text{...[Eqn. (2.58)]}$$

$$= \frac{2.8\,\text{k}\Omega \times 0.35\,\mu\text{F}}{0.42\,\mu\text{F}} = \mathbf{2.33\ k\Omega} \quad \textbf{(Ans.)}$$

and, $$C_x = \frac{R_1C_3}{R_2} \quad \text{...[Eqn. (2.59)]}$$

$$= \frac{1.4 \times 0.42}{2.8} = \mathbf{0.21\ \mu F} \quad \textbf{(Ans.)}$$

(ii) Dissipation factor, *D*:

$$D = \omega C_x R_x \quad \text{...[Eqn. (2.61)]}$$

$$= 2\pi f.C_x.R_x$$

$$= 2\pi \times (1 \times 1000) \times (0.21 \times 10^{-6}) \times (2.33 \times 10^3)$$

$$= \mathbf{3.07} \quad \textbf{(Ans.)}$$

Q.18. (a) *What is a 'Grassot Fluxmeter'?*

(b) *A magnetic circuit excited by an m.m.f. of 8200 AT has a reluctance of 1,18,000 AT/Wb. A fluxmeter is used to measure the flux. If the fluxmeter has 125 divisions and flux linkages required for a deflection of one division is 0.15 × 10^{-3} weber turn, calculate the resistance of the shunt required for use with the search coil. Number of turns of each coil is 1 and its resistance is 0.028 Ω. The measurement is made by switching off the excitation.*

Sol. (a) A ***Grassot fluxmeter*** is a special type of ballistic galvanometer in which the *controlling torque is very small* and *electromagnetic damping is heavy*. It is used for the measurement of flux.

(b) *Given*: *m.m.f.* = 8200 AT; Reluctance = 1,18,000 AT/Wb; Number of turns of each coil, N_{sc} = 1; R_{sc} = 0.028 Ω; Number of flux divisions on fluxmeter = 125; Flux linkages required for deflection of one division = 0.15 × 10^{-3} weber turn

Resistance of shunt, R_{sh}:

$$\text{Flux to be measured} = \frac{\text{m.m.f.}}{\text{relactance}} = \frac{8200}{1{,}18{,}000} = 0.0695\ \text{Wb}$$

Value of flux when the fluxmeter shows full scale deflection

$$= \frac{0.15 \times 10^{-3} \times 125}{1} = 18.75 \times 10^{-3}\ \text{Wb}$$

$$\text{Ratio } \frac{\text{flux to be measured}}{\text{flux for full scale division}} = \frac{0.0695}{18.75 \times 10^{-3}} = 3.7$$

$$\text{Thus, we have: } \frac{R_{sc} + R_{sh}}{R_{sh}} = \frac{0.028 + R_{sh}}{R_{sh}} = 3.7$$

$$\therefore \quad R_{sh} = \mathbf{0.01037\ \Omega} \quad \textbf{(Ans.)}$$

Q.19. A sinusoidal alternating voltage of 240 V peak is applied to a circuit containing a rectifying device which entirely presents current flowing in one direction, and offers a non-reactive resistance of 150 Ω to flow of current in other direction. Determine the readings on : (*i*) Hot-wire ammeter; (*ii*) Moving coil in the circuit.

Sol. *Given:* V_{max} = 240 V; Resistance of rectifying device, R = 150 Ω.

Maximum value of current flowing through the instruments,

$$I_{max} = \frac{V_{max}}{R} = \frac{240}{150} = 1.6\text{A}$$

Reading of hot-wire ammeter, $I_{r.m.s.}$:

The hot-wire ammeter *reads r.m.s. value*, thus,

$$I_{r.m.s.} = \sqrt{\frac{1}{2\pi}\int_0^{\pi} I_{max}^2 \sin^2\theta \, d\theta}$$

$$= \sqrt{\frac{I_{max}^2}{2\pi}\int_0^{\pi} \frac{(1-\cos 2\theta)}{2} d\theta}$$

$$= \sqrt{\frac{I_{max}^2}{2\pi} \times \frac{\pi}{2}} = \frac{I_{max}}{2} = \frac{1.6}{2} = \mathbf{0.8\ A}$$

Reading of moving coil ammeter, I_{av}:

The moving coil ammeter *reads average value*, thus,

$$I_{av} = \frac{1}{2\pi}\int_0^{\pi} I_{max} \sin\theta \, d\theta$$

$$= \frac{I_{max}}{2\pi}\left| -\cos\theta \right|_0^{\pi} = \frac{I_{max}}{\pi} = \frac{1.6}{\pi} = \mathbf{0.509\ A\ (Ans.)}$$

Q.20. (*a*) *What is 'Wagner earthing device'?*

(*b*) *A circuit having a capacitance of 150 pF is tuned to a frequency of 1.1 MHz. In this circuit the current falls to 70.7% of its resonant value when the frequency of an e.m.f. of constant magnitude injected in series with the circuit deviates from the resonant frequency by 5.5 kHz. Calculate the Q factor and effective resistance of circuit.*

Sol (*a*) ***Wagner earthing connection*** is one of the most widely used methods for eliminating some of the effects of stray capacitance in a bridge circuit. This circuit *eliminates* the troublesome capacitance which exists between the detector terminals and the ground.

(*b*) **Given:** C = 150 pF; f_0 = 1.1 MHz

Values of *Q* and *R* (effective):

We know that, $$Q = \frac{\omega_0}{\omega_1 - \omega_2} = \frac{f_0}{f_1 - f_2}$$ ...[Eqn. (2.105)

The two frequencies at which current is 70.2% or $\frac{1}{\sqrt{2}}$ time the current at resonant frequency are:

$$f_1 = 1.1 \times 10^6 + 5.5 \times 10^3$$

and $$f_2 = 1.1 \times 10^6 - 5.5 \times 10^3$$

∴ $$f_1 - f_2 = 11 \times 10^3 \text{ Hz}$$

Hence, $$Q = \frac{1.1 \times 10^6}{11 \times 10^3} = \mathbf{100} \quad \textbf{(Ans.)}$$

Now, $$R = \frac{\omega_1 - \omega_2}{\omega_0^2 C} \quad \text{...[Eqn. (204)]}$$

$$= \frac{f_1 - f_2}{2\pi f_0^2 C} = \frac{11 \times 10^3}{2\pi \times (1.1 \times 10^6)^2 \times 150 \times 10^{-12}}$$

$$= \mathbf{9.64\ \Omega} \quad \textbf{(Ans.)}$$

Q.21. *(a) What is the working principle of electrostatic instruments?*

(b) Following data relate to an electrostatic voltmeter:

Full scale deflection = 90°; Control constant of the spring = 7.5 × 10^{-6} Nm/rad; Capacitance with zero reading = 11 pF.

Determine the capacitance when the instrument reads 1400 V and the deflection is 90°.

Sol. (a) The underlying principle of operation of electrostatic instruments is the force of attraction between electric charges on neighbouring plates between which potential difference is maintained. This force gives rise to a deflecting torque. Unless the p.d. is adequately large, the force is small. Hence such instruments are used for measurement of *very high voltages*.

The instruments are almost always used as *voltmeters* and that too more as a laboratory than as industrial instruments.

(b) *Given:* $\theta = 90° = \frac{\pi}{2}$ rad; Spring constant, $K = 7.5 \times 10^{-6}$ Nm/rad;

$C = 11$ pF $= 11 \times 10^{-12}$ F, when reading is zero; $V = 1400$ volts.

New value of capacitance, C_{new} :

At steady deflection state:

$$T_d = T_c$$

or, $$\frac{1}{2} V^2 \frac{dC}{d\theta} = K\theta \quad \text{...[Eqn. (3.111)]}$$

or, $$\frac{dC}{d\theta} = \frac{2K\theta}{V^2} = \frac{2 \times 7.5 \times 10^{-6} (\pi/2)}{(1400)^2}$$

$$= 0.1202 \times 10^{-10} \text{ F/rad}$$

Change in capacitance, $dC = 0.1202 \times 10^{-10} \times \frac{\pi}{2} = 0.1888 \times 10^{-10}$ F

$$\therefore \quad C_{new} = C + dC = 11 \times 10^{-12} + 0.1888 \times 10^{-10}$$

$$= 29.88 \times 10^{-12} \text{ F or } \mathbf{29.88\ pF} \ \textbf{(Ans.)}$$

Q.22. *A bakelite sheet of 4.5 mm thickness is tested at 50 Hz between the electrodes 110 mm diameter. The schering bridge shown in Fig. 2 has an air capacitor C_2 of 104 pF, a non-reactive resistance of R_4 of $(1000/\pi)$ Ω in parallel with a variable capacitor C_4 and non-reactive variable resistance R_3. Balance is obtained with C_4 = 0.52 μF and R_3 = 250 Ω.*

Determine the following:

(i) Capacitance; (ii) Power factor; (iii) Relative permittivity of sheet.

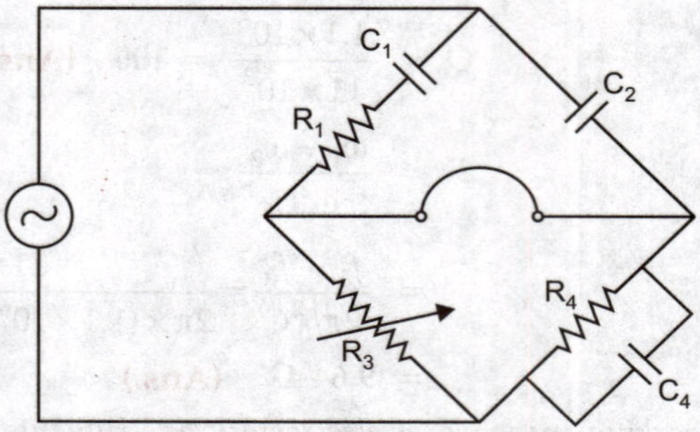

Fig. 2

Sol. *Given:* C_2 = 104 pF = 104 × 10^{-12} F; R_4 = (1000/π) Ω; C_4 = 0.52 μF = 0.52 × 10^{-6} F; R_3 = 250 Ω.

(i) Capacitance *C*:

When the bridge is balanced, we have:

$$R_1 = \frac{R_3 C_4}{C_2} \quad \text{...[Eqn. (2.58)]}$$

$$= \frac{250 \times 0.52 \times 10^{-6}}{104 \times 10^{-12}} = 1.25 \text{ M}\Omega$$

and,

$$C_1 = \frac{R_4\, C_2}{R_3} \quad \text{...[Eqn. (2.59)]}$$

$$= \frac{(1000/\pi_1) \times 104 \times 10^{-12}}{250} = \mathbf{132.4\ pF} \quad \textbf{(Ans.)}$$

(ii) Power factor :

$$\text{Power factor} = \omega R_1 C_1 \quad \text{...[Eqn. (2.61)]}$$

$$= (2\pi \times 50) \times (1.25 \times 10^6) \times (132.4 \times 10^{-12})$$

$$= \mathbf{0.052} \ \textbf{(Ans.)}$$

(iii) Relative permittivity, ϵ_r:

We know that:

$$C = \frac{\epsilon_0 \epsilon_r A}{t}$$

where,

$C (= C_1)$ = Capacitance,

$\epsilon_0 = 8.854 \times 10^{-12}$

A = Area of the electrodes $= \frac{\pi}{4} \times \left(\frac{110}{1000}\right)^2$

$= 9.5 \times 10^{-3}$ m^2

t = Thickness of bakelite sheet = 4.5 mm = 0.0045 m

Substituting the values, in the above equation, we get:

$$132.4 \times 10^{-12} = \frac{8.854 \times 10^{-12} \times t_r \times (9.5 \times 10^{-3})}{0.0045}$$

or,

$$\epsilon_r = \frac{132.4 \times 10^{-12} \times 0.0045}{8.854 \times 10^{-12} \times (9.5 \times 10^{-3})} = \mathbf{7.08} \ \textbf{(Ans.)}$$

Q.23. *(a) What are the various types of errors in dynamometer wattmeters?*

(b) An electrodynamic wattmeter is used to measure power in a single-phase circuit. The load voltage is 210 V and load current is 4.8 A at a logging power factor of 0.12. The resistance and inductance of the wattmeter potential coil are 13000 Ω and inductance 110 mH respectively.

Calculate the % age error in the wattmeter reading.

Sol. *(a)* The various types of *"errors"* in dynamometer wattmeters are :

(*i*) Error due to pressure-coil inductance; (*ii*) Error due to pressure-coil capacitance; (*iii*) Error due to eddy currents; (*iv*) Error due to power loss in pressure coil or current coil; (*v*) Error due to friction; (*vi*) Error due to temperature; (*vii*) Error due to stray fields.

(b) Given: V = 210 volts; I = 4.8 A; cos ϕ = 0.12 (lagging); R_p = 13000 Ω; L = 110 mH = 0.11 H.

%age error:

$$\text{Load power} = VI\cos\phi = 210 \times 4.8 \times 0.12 = 121 \text{ W}$$

$$\phi = \cos^{-1}(0.12) = 83.11°$$

$$\text{Power lost in the coil} = \frac{V^2}{R_P} = \frac{(210)^2}{13000} = 3.39 \text{ W}$$

Neglecting the inductance of the voltage coil the reading of wattmeter would be 121 + 3.39 = 124.39 W

Reactance of pressure coil circuit,

$$X_p = 2\pi fL = 2\pi \times 50 \times 0.11 = 34.56\ \Omega$$

$$\theta = \tan^{-1}\left(\frac{X_p}{R_p}\right) = \tan^{-1}\left(\frac{34.56}{13000}\right) = 0.15°$$

$$\text{Wattmeter reading} = \text{True reading} \times \frac{\cos\theta \times \cos(\phi - \theta)}{\cos\phi}$$

...[Eqn. 3.133(*a*)]

$$= 124.39 \times \frac{\cos(0.15°) \times \cos(83.11° - 0.15°)}{\cos(83.11°)}$$

$$= 127.1 \text{ W}$$

$$\therefore \quad \%\text{ age error} = \frac{127.1 - 124.39}{100} \times 100 = \mathbf{2.71\%\ (high)} \quad \textbf{(Ans.)}$$

Q.24. *(a) Describe briefly 'Bolometer method' of power measurement.*

(b) An A.C. energy meter is tested for half hour run at a supply voltage of 230 V and a load current of 9.5 A at 0.82 lag power factor. The dial reading at the start was 56.85 and at the end of the test was 57.70. The meter constant is 1200 revolutions / kWh. The meter made 875 revolutions in 30 minutes run.

Determine : (i) Error in registration, and (ii) Error in r.p.m. of the meter.

Sol. *(a)* Refer to Article 3.9.10.

(b) Given: Supply voltage, V = 230 volts; I = 9.5 A; cos ϕ = 8.2 (lagging);

Energy consumption recorded by the meter = 57.70 – 56.85 = 0.85 kWh

(*i*) Error in registration:

Actual energy consumed in half an hour

$$= \frac{VI\cos\phi.t}{1000} = \frac{230 \times 9.5 \times 0.82 \times (30/60)}{1000}$$

$$= 0.896 \text{ kWh}$$

∴ Error in registration = Energy recorded – energy actually consumed

$= 0.85 - 0.896 = -0.046$ kW

or, **0.046 kW (less) (Ans.)**

(ii) Error in r.p.m. of the meter:

Actual revolutions required to be made by the meter for an energy consumption of 0.85 kWh

$= 1200 \times 0.85 = 1020$ rev.

Revolution made by the meter = 875

$$\therefore \quad \text{Error in r.p.m.} = \frac{875 - 1020}{30} = -\mathbf{4.83\ r.p.m.}\ \textbf{(Ans.)}$$

Q.25. **(a)** *What are the practical applications of D.C. potentiometers?*

(b) *The current is measured through a standard resistance of 0.1 ± 0.1 Ω% (standard deviation) by a patentiometer that is accurate to ± 0.0001 V (standard deviation). The voltage across the resistance is measured to be 0.25598 V.*

Calculate the current and the accuracy with which it has been determined.

Sol. **(a)** Following are the important *applications* of D.C. potentiometers :

(*i*) Measurement of resistance; (*ii*) Measurement of current, (*iii*) Measurement of high e.m.f. (say 250 V); (*iv*) Measurement of power; (*v*) Calibration of voltmeter; (*vi*) Calibration of ammeter; and (*vii*) Calibration of wattmeter.

(b) *Current*, $I = \frac{V_S}{S} = \frac{0.25598}{0.1} = \mathbf{2.5598\ A}$ **(Ans.)**

The fractional standard deviation of current,

$$\frac{\delta I}{I} = \sqrt{\left(\frac{\delta V}{V}\right)^2 + \left(\frac{\delta S}{S}\right)^2}$$

$$= \sqrt{\left(\frac{0.0001}{0.25598}\right)^2 + (0.001)^2} = 0.001074$$

or, % *accuracy* of measurement of current

$= \pm 0.001074 \times 100 = \mathbf{\pm 0.1074\%}$ **(Ans.)**

Q.26. **(a)** *Draw the circuit for calibrating wattmeter with D.C. potentometer and explain it briefly.*

(b) *A rectangular co-ordinate A.C. potentiometer is used to measure the current taken by a small iron core choke coil. A 1.0 Ω non-inductive resistance is connected in series with the choke coil. The voltages measured across the resistance and the coil are (0.75 – j 0.68) V and (1.15 + j 0.26) V respectively.*

Determine the iron loss in the coil, assuming the voltage and current to be sinusoidal.

Sol. **(a)** Refer to Article 5.2.5.7

(b) *Given:* Standard resistance, R_s = 1.0 Ω; Voltage across the resistance, $\bar{V}_s$ = (0.75 – j 0.68) V; Voltage across the coil, $\bar{V}_{coil}$ (1.15 + j 0.26) V.

Iron loss in the coil:

$$\text{Current through the coil} = \frac{V_S}{R_S} = \frac{(0.75 - j\,0.68)}{1.0}$$

$$= (0.75 - j\,0.68)\,A = 1.0124 \angle -42.2°$$

$$\bar{V}_{coil} = (1.15 + j\,0.26)\text{ V} = 1.179 \angle 12.7°$$

Phase angle between voltage across the coil and current in the coil

$$= 12.7° - (-42.2°) = 54.9°$$

$$\text{Power loss in the coil} = \text{VI}\cos\phi$$

$$= 1.179 \times 1.0124 \times \cos 54.9° = 0.686 \text{ W}$$

Assuming copper loss in the coil to be negligible and taking the total losses occurring in the coil as iron loss, we have :

Iron loss in the coil = **0.686 W** **(Ans.)**

Q.27. *The value of a high resistance is measured by loss of charge method. A capacitor havin a capacitance of 2.8 μF is charged to a potential of 500 V D.C. and is discharged through the high resistance. An electrostatic voltmeter, kept across the high resistance, reads the voltage as 320 V at the end of one minute. Determine the value of high resistance.*

Sol. *Given*: C = 2.8 μF = 2.8 × 10^{-6} F ; v_1 = 500 V ; v_2 = 320 V; t = 1 min. = 60 s.

Unknown resitance, R_x:

We know that, $$v_2 = v_1 e^{(-t/CR_x)} \quad \text{...[Eqn. (6.37)]}$$

or, $$\frac{v_2}{v_1} = e^{(-t/CR_x)} = \frac{1}{e^{(t/CR_x)}}$$

or, $$e^{(t/CRx)} = \frac{v_1}{v_2}$$

Taking log on both sides, we get:

$$\frac{t}{CR_x} = \ln\left(\frac{v_1}{v_2}\right)$$

or, $$R_x = \frac{t}{C\ln\left(\frac{v_1}{v_2}\right)}$$

$$= \frac{60}{2.8\times10^{-6}\times\ln\left(\frac{500}{320}\right)} = 48 \text{ M}\Omega \quad \textbf{(Ans.)}$$

Q.28. **(a)** *Explain briefly how leakage factor is measured in dynamo-electric machinery.*

(b) *The constant of a given potentiometer is obtained by a coil of 310 turns in which a current of 0.6 A is reversed. The resulting throw of galvanometer is 145 scale divisions. It is then used to measure the magnetic potential difference between two points and throw is 290 divisions. Calculate the magnetic potential difference.*

Sol. (a) Refer to Article 7.3.5.

(b) *Given: N* = 310 ; *I* = 0.6 A ; θ_1 = 145 scale divisions; θ_2 = 290 scale divisions.

Magnetic potential difference:

M.m.f. of coil = *N* = 310 × 0.6 = 186 AT

Change in m.m.f. due to reversal of current in the coil

$$= 2 \times 186 = 372 \text{ AT}$$

A deflection of 145 scale divisions is caused by an m.m.f. of 372 AT,

∴ M.m.f. causing a deflection of 290 divisions

$$= \frac{290}{145} \times 372 = 744 \text{ AT}$$

Hence, magnetic potential difference = **744 AT** **(Ans.)**

Q.29. **(a)** *List the various methods of measuring magnetic field strength.*

(b) *The measured values of iron loss of a magnetic specimen of weight 10 kg are 13.5 W and 23.4 W at 40 Hz and 60 Hz respectively at a constant peak flux density.*

Determine the values of hysteresis and eddy current losses in W/kg at 50 Hz for the same value of flux density.

Sol. **(a)** The various methods of measurement of magnetic field strength are:

(*i*) By withdraw of search coil ; (*ii*) By nuclear magnetic resonance; (*iii*) Use of bismuth sprial; (*iv*) Use of mumetal wires; (*v*) Use of Hall effect devices.

(b) *Given*: W = 10 kg; Iron loss, P_i = 13.5 W at 40 Hz; P_i = 23.4 W at 60 Hz.

P_h, P_e at f = 50 Hz :

At a given flux density, $P_h \propto f$, and $P_e = f^2$

(where, P_h, P_e and f are hysteresis loss, eddy current loss and frequency respectively)

Let, $P_h = A_f$ and $P_e = Bf^2$

where, A and B are constants.

Total iron loss, $P_i = P_h + P_e = Af + Bf^2$ per kg

Substituting the values, we get:

$$\frac{13.5}{10} = 40\,A + (40)^2 B$$

$$0.03375 = A + 40\,B \quad ...(i)$$

and, $$\frac{23.4}{10} = 60\,A + (60)^2 B$$

or, $$0.039 = A + 60\,B \quad ...(ii)$$

From (*i*) and (*ii*), we get:

$$A = 0.02325, \quad B = 0.0002625$$

∴ When f = 50 Hz

$$P_h = Af = 0.02325 \times 50 = \mathbf{1.1625\ W\ per\ kg}$$ **(Ans.)**

$$P_i = Bf^2 = 0.0002625 \times (50)^2 = \mathbf{0.6562\ W\ per\ kg}$$

(Ans.)

Q.30. **(a)** Explain briefly the following:

(*i*) Space-height ratio; (*ii*) Utilisation factor; (*iii*) Defreciation factor.

(b) Describe the working of "Trotter illumination photometer".

Sol. **(a)** Refer to Article 9.2.

(b) Refer to Article 9.5.5.

INDEX

A

B

C

D

E

O

P

R

S

T

W

Notes

Notes

Notes